BOOKS | À LA CARTE EDITION

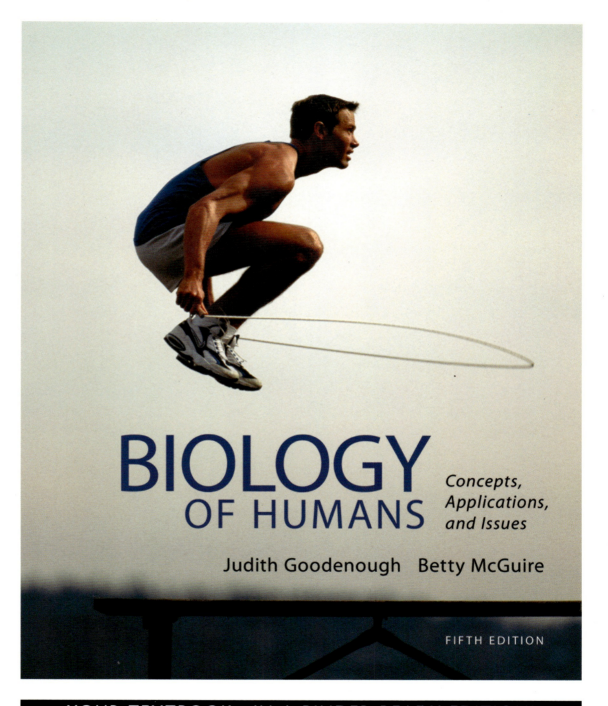

BIOLOGY OF HUMANS
Concepts, Applications, and Issues

Judith Goodenough   Betty McGuire

FIFTH EDITION

YOUR TEXTBOOK—IN A BINDER-READY EDITION!

This unbound, three-hole punched version of your textbook lets you take only what you need to class and incorporate your own notes—all at an affordable price!

ISBN-13: 978-0-321-88669-9
ISBN-10: 0-321-88669-0

Brief Contents

PART I — The Organization of the Body

1. Humans in the World of Biology 1
1a. **SPECIAL TOPIC:** Becoming a Patient: A Major Decision 13
2. Chemistry Comes to Life 20
3. The Cell 42
4. Body Organization and Homeostasis 64

PART II — Control and Coordination of the Body

5. The Skeletal System 85
6. The Muscular System 100
7. Neurons: The Matter of the Mind 113
8. The Nervous System 126
8a. **SPECIAL TOPIC:** Drugs and the Mind 142
9. Sensory Systems 150
10. The Endocrine System 171
10a. **SPECIAL TOPIC:** Diabetes Mellitus 190

PART III — Maintenance of the Body

11. Blood 198
12. The Cardiovascular and Lymphatic Systems 213
12a. **SPECIAL TOPIC:** Cardiovascular Disease 232
13. Body Defense Mechanisms 239
13a. **SPECIAL TOPIC:** Infectious Disease 259
14. The Respiratory System 268
15. The Digestive System and Nutrition 286
15a. **SPECIAL TOPIC:** Food Safety and Defense 314
16. The Urinary System 322

PART IV — Reproduction and Development

17. Reproductive Systems 342
17a. **SPECIAL TOPIC:** Sexually Transmitted Diseases and AIDS 362
18. Development throughout Life 372
18a. **SPECIAL TOPIC:** Autism Spectrum Disorder 393

PART V — Genes and DNA

19. Chromosomes and Cell Division 400
19a. **SPECIAL TOPIC:** Stem Cells—A Repair Kit for the Body 418
20. Genetics and Human Inheritance 425
21. DNA and Biotechnology 442
21a. **SPECIAL TOPIC:** Cancer 463

PART VI — Evolution and Ecology

22. Evolution and Our Heritage 476
23. Ecology, the Environment, and Us 497
24. Human Population, Limited Resources, and Pollution 514

Jump into learning about the *Biology of Humans*

Known for its unique "Special Topics" chapters and emphasis on everyday health concerns, the **Fifth Edition** of *Biology of Humans: Concepts, Applications, and Issues* continues to personalize the study of human biology with a conversational writing style, stunning art, abundant applications, and tools to help students develop critical-thinking skills. The authors give students a practical and friendly introduction for understanding how their bodies work and for preparing them to navigate today's world of rapidly expanding—and shifting—health information.

Abundant Applications

Engaging Media

More Visual Appeal

Apply Your Learning to everyday life

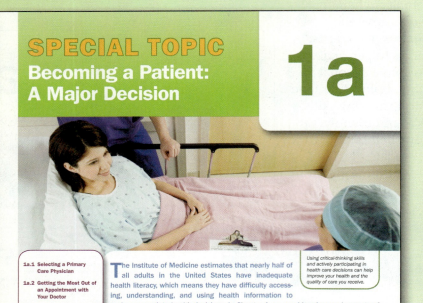

◀ *Ten Special Topic* chapters explore high-interest health topics more thoroughly than can be accomplished in a brief essay.

NEW! Special Topic Chapter 1a

The Fifth Edition features a new "Special Topic" chapter (1a) titled "Becoming a Patient: A Major Decision," which discusses how to select a doctor or a hospital, how to research health conditions, and more.

A new design distinguishes the Special Topic chapters within the text.

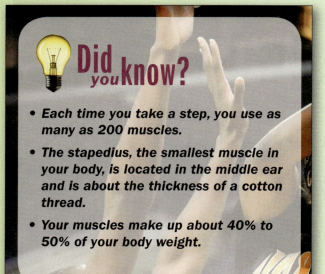

◀ **NEW!** *Did You Know* feature
Each chapter now opens with new "Did You Know" questions that pique students' interest with intriguing and little-known facts about the topic that follows, and the expanded online resources within MasteringBiology are now referenced at the end of each chapter.

Consider the issues before forming an opinion

Health, Ethical, and Environmental Issues essays **explore** contemporary topics that directly relate to students' lives and motivate them to make connections to the world around them.

ENVIRONMENTAL ISSUE

Medicinal Plants and the Shrinking Rain Forest

The healing powers of many plants have been known for centuries. Historically, such knowledge was gained by trial and error and passed along by word of mouth. For example, many cultures have long known that tea made from willow bark relieves pain and reduces fever. Scientists learned that willow bark contains salicylic acid. They isolated the compound and developed it into the drug we know today as aspirin. Similarly, digitalis, a heart medication, was discovered after a patient with an untreatable heart condition was seen to benefit from an herbal drink provided as a folk remedy. The potion contained purple foxglove, which, like willow bark, is frequently mentioned in ancient texts as a healing herb. Broccoli, a more familiar plant, con-

Figure 1.A *The rosy periwinkle (Catharanthus roseus) is a source of two anticancer drugs.*

more healing chemicals first discovered in plants

Most plants that have proved to be medically useful are found in the tropics, regions where the human population is growing rapidly. Unfortunately, the forests in these regions are being cut to create living space and foster economic development. For example, Madagascar is the home of the rosy periwinkle (Figure 1.A), which is the source of two anticancer drugs. Humans have destroyed 90% of the vegetation in that nation. Considering that 155,000 of the known 250,000 plant species are from tropical rain forests, and that fewer than 2% of the known plant species have been tested for medicinal value, we have no way of knowing what potential new medicines are being destroyed.

Questions to Consider
- Should indigenous people be compensated for plants found in their locality if extracts of the plants become drugs?
- What steps might be taken to preserve biodiversity within the rain forest?

HEALTH ISSUE

Acne: The Miseries and Myths

Acne and adolescence go hand in hand. In fact, about four out of five teenagers have acne, a skin condition that will probably annoy, if not distress, them well into their twenties and possibly beyond.

Simple acne is a condition that affects hair follicles associated with oil glands. During the teenage years, oil glands increase in size and produce larger amounts of oily sebum. These changes are prompted, in both males and females, by increasing levels of "male" hormones called *androgens* in the blood; the androgens are secreted by the testes, ovaries, and adrenal glands. The changes thus induced in the activity and structure of oil glands set the stage for acne.

It should come as no surprise, then, that acne occurs most often on areas of the body where oil glands are largest and most numerous: the face, chest, upper back, and shoulders.

Acne is the inflammation that results when sebum and dead cells clog the duct where the oil gland opens onto the hair follicle (Figure 4.B). A follicle obstructed by sebum and cells is called a whitehead. Sometimes the sebum in plugged follicles mixes with the skin pigment melanin, forming a blackhead. Thus, melanin, not dirt or bacteria, lends the dark color to these blemishes. The next stage of acne is pimple formation, beginning with the formation of a red, raised bump, often with a white dot of pus at the center. The bump occurs when obstructed follicles rupture and spew their contents into the surrounding epidermis. Such ruptures may occur naturally by the general buildup of sebum and cells or may be induced by squeezing the area. The sebum, dead cells, and bacteria that thrive on them then cause a small infection—a pimple or pustule—that will usually heal within a week or two without leaving a scar.

There are many misconceptions about the causes of acne. Eating nuts, chocolate, pizza, potato chips, or any of the other "staples" of the teenage diet does not cause acne. Also, acne is not caused by poor hygiene. Follicles plug from below, so dirt or oil on the skin surface is not responsible. (Most doctors do, however, recommend washing the face two or three times a day with hot water to help open plugged follicles.)

Questions to Consider
- Why do you think that there are so many misconceptions about the causes of acne?
- If a new medication for acne were marketed, how would you decide whether to use it?

(d) **A blackhead**
Sebum in the clogged follicle oxidizes and mixes with melanin.

(e) **An inflamed pimple**
The follicle wall ruptures, releasing the contents of a whitehead or blackhead into the surrounding epidermis.

ETHICAL ISSUE

Conducting Research on Our Relatives

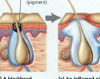

Chimpanzees look and behave somewhat differently from the way we do. Still, it is hard to look into their eyes and not see something of ourselves. Chimpanzees are our closest living relatives, sharing a remarkably high percentage of our DNA sequence. Nevertheless, we use them and other nonhuman primates in invasive scientific research that might benefit us. Is this ethical?

Using nonhuman primates in research is costly and controversial. Even so, they often are preferred as subjects because they are so similar to humans. For example, human and nonhuman primates possess brains with similar organization, develop comparable plaques in their arteries, and experience many of the same changes in anatomy, physiology, and behavior with age. In some cases, Nobel Prize–winning research has resulted from the contributions of research with nonhuman primates, including development of vaccines for yellow fever (1951) and polio (1954) and insight into how visual information is processed in the brain (1981). Research with nonhuman primates has also led to significant advances in our understanding of Alzheimer's disease, AIDS, and severe acute respiratory syndrome (SARS).

The care and use of nonhuman primates (and other vertebrate animals) in research is regulated by federal agencies such as the Public Health Service and the U.S. Department of Agriculture. Animal research also is regulated at the local level. Each college, university, or research center has an Institutional Animal Care and Use Committee whose members include veterinarians, researchers, and members of the public. In addition to federal and local oversight, scientists and animal care personnel are striving to improve housing for captive nonhuman primates and to consider their psychological well-being. Even with such efforts, controversy and questions remain.

Questions to Consider
- Should we ban the use of nonhuman primates in medical research? If we do, will such a ban slow the progress in fighting diseases such as AIDS and Alzheimer's? If you or a loved one had one of these illnesses, would you feel differently about research using nonhuman primates?
- Nonhuman primates represent a fraction of the animals used in research. More than 90% of research animals are rodents, such as rats, mice, and guinea pigs. Where would you draw the line when deciding which (if any) animals are acceptable for use in research that might benefit us?

Questions to Consider conclude ▶ every *Environmental Issue*, *Ethical Issue*, and *Health Issue* box to encourage readers to carefully consider their own position on various issues.

Questions to Consider
- Should we ban the use of nonhuman primates in medical research? If we do, will such a ban slow the progress in fighting diseases such as AIDS and Alzheimer's? If you or a loved one had one of these illnesses, would you feel differently about research using nonhuman primates?
- Nonhuman primates represent a fraction of the animals used in research. More than 90% of research animals are rodents, such as rats, mice, and guinea pigs. Where would you draw the line when deciding which (if any) animals are acceptable for use in research that might benefit us?

Take a *Visual Cue* in understanding how your body works

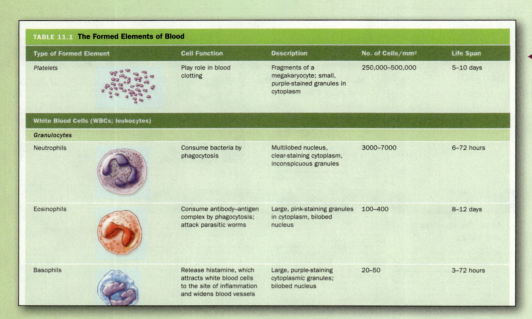

Visual Summary Tables include visual cues when appropriate, making the tables more appealing and easier to read.

Vibrant, three-dimensional figures enhance explanations throughout the text.

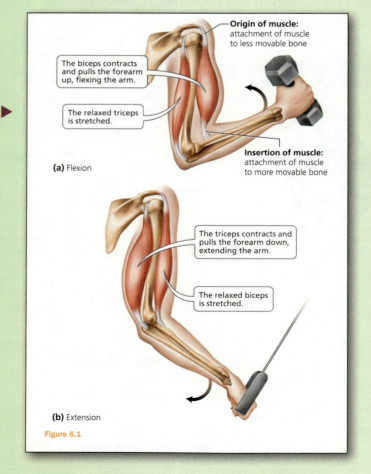

Figure 6.1

Stop. Think. Apply.

Active Learning features engage students in learning the chapter material, including:

stop and think
Frostbite is damage to tissues exposed to cold temperatures. Given what you know about the body's response to cold temperature, why are fingers and toes particularly susceptible to frostbite?

◀ *Stop and Think* **features** engage students in the learning process and promote active learning. Students are invited to pause, think about the concept previously explained, and apply that information to a new situation.

"What Would You Do?" **questions** raise ethical questions about issues that society faces today. Students are able to see the relevance of information learned in a biology classroom to current real-life dilemmas or decisions people must make. ▶

what would you do?
An increasing number of advertisers use neuromarketing to measure consumers' interest in their product. Neuromarketers use a brain scanner consisting of a cap containing electrodes to measure brain activity in all parts of a consumer's brain when he is presented with a product or advertisement. By noting the pattern of brain activity, a neuromarketer can determine a person's interest in, emotional response to, and perhaps even memories associated with a product. This information can then be used to manipulate desire for the product. Do you think neuromarketing is an invasion of privacy? Should it be legal?

Figure legend questions added to select illustrations make the art program as instructive as it is beautiful and encourage students to think critically while reviewing the art.

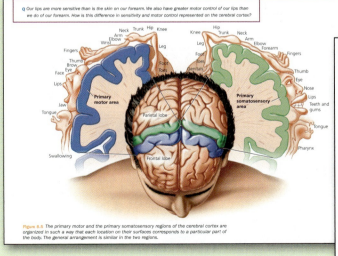

Q Our lips are more sensitive than is the skin on our forearm. We also have greater motor control of our lips than we do of our forearm. How is this difference in sensitivity and motor control represented on the cerebral cortex?

Figure 8.5 The primary motor and the primary somatosensory regions of the cerebral cortex are organized in such a way that each location on their surfaces corresponds to a particular part of the body. The general arrangement is similar in the two regions.

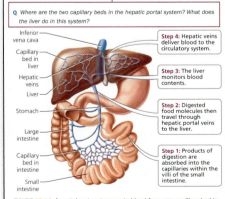

Q Where are the two capillary beds in the hepatic portal system? What does the liver do in this system?

FIGURE 15.11 *A portal system transports blood from one capillary bed to another. In the hepatic portal system, the hepatic portal vein carries blood from the capillary network of the villi of the small intestine to the capillary beds of the liver. The liver monitors blood content and processes nutrients before they are delivered to the bloodstream.*

A The capillary beds are in the small intestine and the liver. The liver monitors and adjusts blood content.

Expanded MasteringBiology resources engage students

MasteringBiology is an online assessment and tutorial system designed to help instructors teach more efficiently and is pedagogically proven to help students learn. It helps instructors maximize class time with customizable, easy-to-assign, and automatically graded assessments that motivate students to learn outside of class and arrive prepared for lecture. The powerful gradebook provides unique insight into student and class performance even before the first test. As a result, instructors can spend class time where students need it most. The Mastering system empowers students to take charge of their learning through activities aimed at different learning styles and engages them in learning science through practice and step-by-step guidance—at their convenience, 24/7.

NEW! Assignable tutorials from Interactive Physiology for Human Biology provide coaching through helpful wrong-answer feedback and hints.

Dozens of additional animation activities engage students in learning basic human anatomy and physiology topics and are accompanied by assignable assessment questions.

NEW! Assignable vocabulary review exercises give students an opportunity to practice using new human biology terms in context.

The complete Interactive Physiology for Human Biology CD-Rom resources can now be accessed through the MasteringBiology study area for students to review the most challenging topics in the course.

ABC News Video clips discuss recent news stories such as "Henrietta Lacks Cells" and "Modern Humans and Neanderthal DNA". Each video includes assessment questions that can be assigned to students as homework.

Regularly-updated "Current Events" activities contain links to articles from the *New York Times* and related assessment questions.

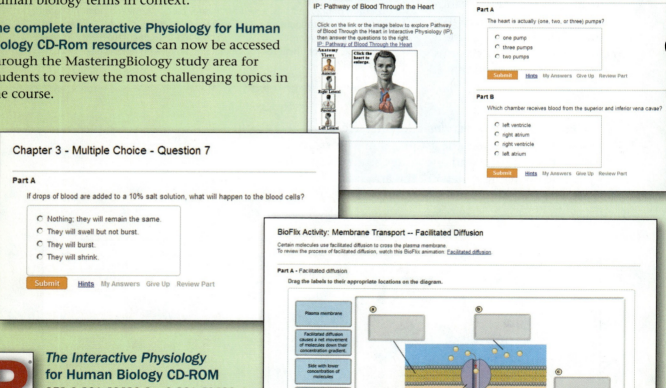

The Interactive Physiology for Human Biology CD-ROM
978-0-321-59539-3 • 0-321-59539-4

Provides outstanding animations, activities, and additional resources.

Can be packaged with the textbook at no additional charge.

Fifth Edition

Biology of Humans

Concepts, Applications, and Issues

Fifth Edition

Biology of Humans

Concepts, Applications, and Issues

Judith Goodenough
University of Massachusetts, Amherst

Betty McGuire
Cornell University

PEARSON

San Francisco   Boston   New York
Cape Town   Hong Kong   London   Madrid   Mexico City   Montreal
Munich   Paris   Singapore   Sydney   Tokyo   Toronto

Vice President, Editor-in-Chief: Beth Wilbur
Executive Director of Development: Deborah Gale
Senior Acquisitions Editor: Star MacKenzie
Project Editor: Nicole George-O'Brien
Assistant Editor: Frances Sink
Text Permissions Project Manager: Joanna Green
Text Permissions Specialist: Electronic Publishing Services
Managing Editor: Michael Early
Production Project Manager: Lori Newman
Production Management: Element LLC
Copyeditor: Sally Peyrefitte
Compositor: Element LLC
Design Manager: Marilyn Perry
Interior Designer: Cenveo Publisher Services
Cover Designer: Hespenheide Design
Illustrators: Scientific Illustrators
Photo Permissions Management: Bill Smith Group
Photo Researcher: Bill Smith Group
Photo Editor: Travis Amos
Manufacturing Buyer: Michael Penne
Executive Marketing Manager: Lauren Harp
Cover Photo Credit: Bob Peterson/UpperCut Images/Getty Images

Credits and acknowledgments for materials borrowed from other sources and reproduced, with permission, in this textbook appear on the appropriate page within the text [or on p. C-1].

Copyright © 2014, 2012, 2010 Pearson Education, Inc. All rights reserved. Manufactured in the United States of America. This publication is protected by Copyright, and permission should be obtained from the publisher prior to any prohibited reproduction, storage in a retrieval system, or transmission in any form or by any means, electronic, mechanical, photocopying, recording, or likewise. To obtain permission(s) to use material from this work, please submit a written request to Pearson Education, Inc., Permissions Department, 1900 E. Lake Ave., Glenview, IL 60025. For information regarding permissions, call (847) 486-2635.

Many of the designations used by manufacturers and sellers to distinguish their products are claimed as trademarks. Where those designations appear in this book, and the publisher was aware of a trademark claim, the designations have been printed in initial caps or all caps.

Benjamin Cummings is a trademark, in the U.S. and/or other countries, of Pearson Education, Inc. or its affiliates.

Library of Congress Cataloging-in-Publication Data is available upon request.

1 2 3 4 5 6 7 8 9 10—CRK—16 15 14 13 12

ISBN 10: 0-321-82171-8; ISBN 13: 978-0-321-82171-3 (Student Edition)
ISBN 10: 0-321-88637-2; ISBN 13: 978-0-321-88637-8 (Instructor's Review Copy)
ISBN 10: 0-321-88669-0; ISBN 13: 978-0-321-88669-9 (Books a la Carte Edition)

Brief Contents

PART I — The Organization of the Body

1. Humans in the World of Biology 1
1a. SPECIAL TOPIC: Becoming a Patient: A Major Decision 13
2. Chemistry Comes to Life 20
3. The Cell 42
4. Body Organization and Homeostasis 64

PART II — Control and Coordination of the Body

5. The Skeletal System 85
6. The Muscular System 100
7. Neurons: The Matter of the Mind 113
8. The Nervous System 126
8a. SPECIAL TOPIC: Drugs and the Mind 142
9. Sensory Systems 150
10. The Endocrine System 171
10a. SPECIAL TOPIC: Diabetes Mellitus 190

PART III — Maintenance of the Body

11. Blood 198
12. The Cardiovascular and Lymphatic Systems 213
12a. SPECIAL TOPIC: Cardiovascular Disease 232
13. Body Defense Mechanisms 239
13a. SPECIAL TOPIC: Infectious Disease 259
14. The Respiratory System 268
15. The Digestive System and Nutrition 286
15a. SPECIAL TOPIC: Food Safety and Defense 314
16. The Urinary System 322

PART IV — Reproduction and Development

17. Reproductive Systems 342
17a. SPECIAL TOPIC: Sexually Transmitted Diseases and AIDS 362
18. Development throughout Life 372
18a. SPECIAL TOPIC: Autism Spectrum Disorder 393

PART V — Genes and DNA

19. Chromosomes and Cell Division 400
19a. SPECIAL TOPIC: Stem Cells—A Repair Kit for the Body 418
20. Genetics and Human Inheritance 425
21. DNA and Biotechnology 442
21a. SPECIAL TOPIC: Cancer 463

PART VI — Evolution and Ecology

22. Evolution and Our Heritage 476
23. Ecology, the Environment, and Us 497
24. Human Population, Limited Resources, and Pollution 514

Special Interest Essays

ENVIRONMENTAL ISSUES

Medicinal Plants and the Shrinking Rain Forest   2
Radon Gas: A Killer That Can Be Stopped   27
Toilet to Tap   35
The Deadly Interaction between Asbestos and Lysosomes   55
Noise Pollution   164
Lead Poisoning   205
Environment and Epigenetics   451
Air Pollution and Human Health   524

ETHICAL ISSUES

Anabolic Steroid Abuse   110
Hormone Therapy   178
Kidney Donation and Trafficking   336
Making Babies   386
Trisomy 21   414
Gene Testing   439
Forensic Science, DNA, and Personal Privacy   459
Conducting Research on Our Relatives   485
Maintaining Our Remaining Biodiversity   510

HEALTH ISSUES

Mitochondrial Diseases   60
Fun in the Sun?   77
Acne: The Miseries and Myths   78
Osteoporosis: Fragility and Aging   90
Neurotransmitters and Disease   121
Brain Injury: A Silent Epidemic   132
Benefits of Cardiovascular Exercise   226
Rejection of Organ Transplants   253
Surviving a Common Cold   281
Smoking and Lung Disease   283
Peptic Ulcers   297
Urinalysis   329
Breast Cancer   354
Disparities in Health and Health Care at All Life Stages   388
Genetically Modified Food   456

Detailed Contents

About the Authors xvii
Preface xviii

PART I THE ORGANIZATION OF THE BODY

1 Humans in the World of Biology 1

1.1 Basic Characteristics of All Living Things 1
1.2 Evolution: A Unifying Theme in Biology 4
1.3 Levels of Biological Organization 5
1.4 Scientific Method 5
Inductive and Deductive Reasoning 8
Clinical Trials 8
Epidemiological Studies 9
1.5 Critical Thinking to Evaluate Scientific Claims 9

ENVIRONMENTAL ISSUE
Medicinal Plants and the Shrinking Rain Forest 2

1a Becoming a Patient: A Major Decision 13

1a.1 Selecting a Primary Care Physician 13
1a.2 Getting the Most Out of an Appointment with Your Doctor 15
1a.3 Finding a Specialist and Getting a Second Opinion 18
1a.4 Appointing a Health Care Agent 18
1a.5 Selecting a Hospital and Staying Safe 18
1a.6 Researching Health Conditions on Your Own 19

2 Chemistry Comes to Life 20

2.1 The Nature of Atoms 20
Elements 21
Isotopes and Radioisotopes 21
2.2 Compounds and Chemical Bonds 23
Covalent Bonds 23
Ionic Bonds 24
2.3 The Role of Water in Life 25
Properties of Water 27
Acids and Bases 28
The pH Scale 28
Buffers 28
2.4 Major Molecules of Life 29
Carbohydrates 31
Lipids 31
Proteins 35
Nucleic Acids and Nucleotides 37

ENVIRONMENTAL ISSUE
Radon Gas: A Killer That Can Be Stopped 27
Toilet to Tap 35

3 The Cell 42

3.1 Eukaryotic Cells Compared with Prokaryotic Cells 42
3.2 Cell Size and Microscopy 44
3.3 Cell Structure and Function 44
3.4 Plasma Membrane 45
Plasma Membrane Structure 45
Plasma Membrane Functions 46
Movement Across the Plasma Membrane 47
3.5 Organelles 50
Nucleus 50
Endoplasmic Reticulum 51
Golgi Complex 52
Lysosomes 52
Mitochondria 54
3.6 Cytoskeleton 55
3.7 Cellular Respiration and Fermentation in the Generation of ATP 56
Cellular Respiration 57
Fermentation 59

ENVIRONMENTAL ISSUE
The Deadly Interaction between Asbestos and Lysosomes 55

HEALTH ISSUE
Mitochondrial Diseases 60

4 Body Organization and Homeostasis 64

4.1 From Cells to Organ Systems 64
Tissues 64
Cell Junctions 70
Organs and Organ Systems 70
Body Cavities Lined with Membranes 70

4.2 Skin: An Organ System 72
Skin Functions 72
Skin Layers 72
Skin Color 76
Hair, Nails, and Glands 76

4.3 Homeostasis 78
Negative Feedback Mechanisms 79
Hypothalamus and Body Temperature 79

HEALTH ISSUE
Fun in the Sun? 77
Acne: The Miseries and Myths 78

PART II CONTROL AND COORDINATION OF THE BODY

---

5 The Skeletal System 85

5.1 Bone Functions 85
5.2 Bone Structure 86
5.3 Bone as a Living Tissue 87
Cartilage Model 87
Hormones and Bone Growth 88

5.4 The Role of Fibroblasts and Osteoblasts in Repairing Bone Fractures 88
5.5 Bone Remodeling 89
5.6 Axial Skeleton 91
Skull 92
Vertebral Column 93
Rib Cage 94

5.7 Appendicular Skeleton 94
Pectoral Girdle 94
Pelvic Girdle 95

5.8 Joints 95
Synovial Joints 95
Damage to Joints 96
Arthritis 97

HEALTH ISSUE
Osteoporosis: Fragility and Aging 90

---

6 The Muscular System 100

6.1 Function and Characteristics of Muscles 100
6.2 Skeletal Muscles Working in Pairs 101
6.3 Contraction of Muscles 101
Sliding Filament Model 103
Calcium Ions and Regulatory Proteins 104
Role of Nerves 104
Muscular Dystrophy 105

6.4 Voluntary Movement 106
Motor Units and Recruitment 106
Muscle Twitches, Summation, and Tetanus 107

6.5 Energy for Muscle Contraction 107
6.6 Slow-Twitch and Fast-Twitch Muscle Cells 108
6.7 Building Muscle 109

ETHICAL ISSUE
Anabolic Steroid Abuse 110

---

7 Neurons: The Matter of the Mind 113

7.1 Cells of the Nervous System 113
Neuroglial Cells 113
Neurons 114

7.2 Structure of Neurons 114
Axons and Dendrites 114
Myelin Sheath 114

7.3 Nerve Impulses 116
Plasma Membrane of a Neuron 116
Resting Potential 116
Action Potential 116

7.4 Synaptic Transmission 118
Release of the Neurotransmitter and the Opening of Ion Channels 119
Summation of Input from Excitatory and Inhibitory Synapses 120
Removal of Neurotransmitter 120
Roles of Different Neurotransmitters 122

HEALTH ISSUE
Neurotransmitters and Disease 121

---

8 The Nervous System 126

8.1 Organization of the Nervous System 126
8.2 The Central Nervous System 127
Protection of the Central Nervous System 127
Brain: Command Center 128
Spinal Cord: Message Transmission and Reflex Center 134

8.3 The Peripheral Nervous System 135
Somatic Nervous System 135
Autonomic Nervous System 135

8.4 Disorders of the Nervous System 138
Headaches 138

Strokes 138
Coma 138
Spinal Cord Injury 138

HEALTH ISSUE

Brain Injury: A Silent Epidemic 132

8a Drugs and the Mind 142

8a.1 Psychoactive Drugs and Communication between Neurons 142
8a.2 Drug Dependence 143
8a.3 Alcohol 143
Absorption and Distribution 144
Rate of Elimination 144
Health-Related Effects 145
8a.4 Marijuana 146
THC Receptors in the Brain 146
Health-Related Effects of Long-Term Use 146
Medical Marijuana 147
8a.5 Stimulants 147
Cocaine 147
Amphetamines 148
Nicotine 148
8a.6 Hallucinogens 148
8a.7 Opiates 149

9 Sensory Systems 150

9.1 Sensory Receptors 150
9.2 Classes of Receptors 151
9.3 The General Senses 151
Touch, Pressure, and Vibration 152
Temperature Change 152
Body and Limb Position 152
Pain 152
9.4 Vision 153
Wall of the Eyeball 153
Fluid-Filled Chambers 154
Focusing and Sharp Vision 154
Light and Pigment Molecules 157
Rods: Vision in Dim Light 157
Cones: Color Vision 158
9.5 Hearing 159
Form and Function of the Ear 159
Loudness and Pitch of Sound 162
Hearing Loss 162
Ear Infections 164
9.6 Balance and the Vestibular Apparatus of the Inner Ear 165
9.7 Smell and Taste 165

ENVIRONMENTAL ISSUE

Noise Pollution 164

10 The Endocrine System 171

10.1 Functions and Mechanisms of Hormones 171
Hormones as Chemical Messengers 172
Feedback Mechanisms and Secretion of Hormones 172
Interactions between Hormones 174
10.2 Hypothalamus and Pituitary Gland 175
Anterior Lobe 176
Posterior Lobe 179
10.3 Thyroid Gland 180
10.4 Parathyroid Glands 181
10.5 Adrenal Glands 182
Adrenal Cortex 182
Adrenal Medulla 184
10.6 Pancreas 184
10.7 Thymus Gland 185
10.8 Pineal Gland 186
10.9 Locally Acting Chemical Messengers 186

ETHICAL ISSUE

Hormone Therapy 178

10a Diabetes Mellitus 190

10a.1 General Characterization and Overall Prevalence 190
10a.2 Type 1 and Type 2 Diabetes 191
Characterization and Risk Factors 191
Symptoms and Complications 192
Diagnosis 193
Treatments 194
Lifestyle Changes and Key Recommendations after Diagnosis 194
Prognoses 195
10a.3 Gestational Diabetes 196
10a.4 Other Specific Types of Diabetes 197

PART III — MAINTENANCE OF THE BODY

11 Blood 198

- 11.1 Functions of Blood 198
- 11.2 Composition of Blood 199
  - Plasma 199
  - Formed Elements 199
  - Platelets 200
  - White Blood Cells and Defense against Disease 201
  - Red Blood Cells and Transport of Oxygen 202
- 11.3 Blood Cell Disorders 204
  - Disorders of Red Blood Cells 204
  - Disorders of White Blood Cells 204
- 11.4 Blood Types 206
  - ABO Blood Types 206
  - Rh Factor 206
  - Blood Donation 208
- 11.5 Blood Clotting 208

ENVIRONMENTAL ISSUE
Lead Poisoning 205

12 The Cardiovascular and Lymphatic Systems 213

- 12.1 Cardiovascular System 213
- 12.2 Blood Vessels 215
  - Arteries 215
  - Capillaries 216
  - Veins 218
- 12.3 Heart 219
  - Two Circuits of Blood Flow 221
  - Coronary Circulation 222
  - Cardiac Cycle 222
  - Internal Conduction System 223
  - Electrocardiogram 224
- 12.4 Blood Pressure 224
- 12.5 Lymphatic System 225

HEALTH ISSUE
Benefits of Cardiovascular Exercise 226

12a Cardiovascular Disease 232

- 12a.1 The Prevalence of Cardiovascular Disease 232
- 12a.2 Blood Clots 232
- 12a.3 Problems with Blood Vessels 233
  - High Blood Pressure 233
  - Aneurysm 233
  - Atherosclerosis 234
- 12a.4 Heart Attack and Heart Failure 236
- 12a.5 Cardiovascular Disease and Cigarette Smoking 237
- 12a.6 Preventing Cardiovascular Disease 238

13 Body Defense Mechanisms 239

- 13.1 The Body's Defense System 239
- 13.2 Three Lines of Defense 240
  - First Line of Innate Defense: Physical and Chemical Barriers 240
  - Second Line of Innate Defense: Defensive Cells and Proteins, Inflammation, and Fever 240
  - Third Line of Defense: Adaptive Immune Response 244
- 13.3 Distinguishing Self from Nonself 245
- 13.4 Antibody-Mediated Responses and Cell-Mediated Responses 246
- 13.5 Steps of the Adaptive Immune Response 247
- 13.6 Active and Passive Immunity 251
- 13.7 Monoclonal Antibodies 252
- 13.8 Problems of the Immune System 252
  - Autoimmune Disorders 253
  - Allergies 254

HEALTH ISSUE
Rejection of Organ Transplants 253

13a Infectious Disease 259

- 13a.1 Pathogens 259
  - Bacteria 259
  - Viruses 261
  - Protozoans 263
  - Fungi 264
  - Parasitic Worms 264
  - Prions 264
- 13a.2 Spread of a Disease 265
- 13a.3 Infectious Diseases as a Continued Threat 266
  - Emerging Diseases and Reemerging Diseases 266
  - Global Trends in Emerging Infectious Diseases 267
  - Epidemiology 267

14 The Respiratory System 268

14.1 Structures of the Respiratory System 268
Nose 269
Sinuses 271
Pharynx 272
Larynx 272
Trachea 272
Bronchial Tree 273
Alveoli 274
Lungs 274

14.2 Mechanism of Breathing 274
Inhalation 275
Exhalation 275
The Volume of Air Moved Into or Out of the Lungs during Breathing 275

14.3 Transport of Gases between the Lungs and the Cells 276
Oxygen Transport and Hemoglobin 276
Carbon Dioxide Transport and Bicarbonate Ions 277

14.4 Respiratory Centers in the Brain 278
Basic Breathing Pattern 278
Chemoreceptors 279

14.5 Respiratory Disorders 279
Common Cold 279
Flu 280
Pneumonia 280
Strep Throat 280
Tuberculosis 280
Cystic Fibrosis 281
Bronchitis 281
Emphysema 281
Lung Cancer 282

HEALTH ISSUE
Surviving a Common Cold 281
Smoking and Lung Disease 283

15 The Digestive System and Nutrition 286

15.1 The Gastrointestinal Tract 286

15.2 Specialized Compartments for Food Processing 287
Mouth 287
Pharynx 290
Esophagus 290
Stomach 290
Small Intestine 292
Accessory Organs: Pancreas, Liver, and Gallbladder 294
Large Intestine 296

15.3 Nerves and Hormones in Digestion 298

15.4 Planning a Healthy Diet 299

15.5 Nutrients 299
Lipids 301
Carbohydrates 302
Proteins 303
Vitamins 303
Minerals 304
Water 304

15.6 Food Labels 304

15.7 Energy Balance 307

15.8 Obesity 307

15.9 Weight-Loss Programs 308

15.10 Eating Disorders 309

HEALTH ISSUE
Peptic Ulcers 297

15a Food Safety and Defense 314

15a.1 Foodborne Illnesses 314
General Symptoms, Diagnosis, and Treatment 314
Common Foodborne Infections 316
How Does Food Become Contaminated? 316
Methods of Combating Food Contamination 317

15a.2 Keeping Food Safe at International and National Levels 318
International Oversight 318
National Oversight 318

15a.3 Food Defense and Bioterrorism 318

15a.4 Personal Food Safety 319
Food Selection 319
Food Handling and Storage 320

16 The Urinary System 322

16.1 Eliminating Waste 322

16.2 Components of the Urinary System 323

16.3 Kidneys and Homeostasis 324
Structure of the Kidneys 324
Nephrons 324
Acid–Base Balance 329
Water Conservation 329
Hormones and Kidney Function 331
Red Blood Cells and Vitamin D 333

16.4 Dialysis and Transplant Surgery 333
Dialysis 334
Kidney Transplant Surgery 334

16.5 Urination 336

16.6 Urinary Tract Infections 338

HEALTH ISSUE
Urinalysis 329

ETHICAL ISSUE
Kidney Donation and Trafficking 336

PART IV — REPRODUCTION AND DEVELOPMENT

17 Reproductive Systems 342

- **17.1 Gonads** 342
- **17.2 Male and Female Reproductive Roles** 342
- **17.3 Form and Function of the Male Reproductive System** 343
  - Testes 343
  - Duct System 343
  - Accessory Glands 343
  - Penis 345
  - Sperm Development 346
  - Hormones 347
- **17.4 Form and Function of the Female Reproductive System** 348
  - Ovaries 348
  - Oviducts 348
  - Uterus 348
  - External Genitalia 348
  - Breasts 348
  - Ovarian Cycle 349
  - Coordination of the Ovarian and Uterine Cycles 350
  - Menopause 353
- **17.5 Disorders of the Female Reproductive System** 354
- **17.6 Stages of the Human Sexual Response** 355
- **17.7 Birth Control** 356
  - Abstinence 356
  - Sterilization 356
  - Hormonal Contraception 356
  - Intrauterine Devices 357
  - Barrier Methods 358
  - Spermicidal Preparations 358
  - Fertility Awareness 358
  - Emergency Contraception 358

HEALTH ISSUE
Breast Cancer 354

17a Sexually Transmitted Diseases and AIDS 362

- **17a.1 Long-Lasting Effects of STDs and STIs** 362
- **17a.2 STDs Caused by Bacteria** 363
  - Chlamydia and Gonorrhea 363
  - Syphilis 364
- **17a.3 STDs Caused by Viruses** 366
  - Genital Herpes 366
  - HPV and Genital Warts 366
- **17a.4 HIV/AIDS** 367
  - Global Pandemic 368
  - Form of HIV 368
  - Replication of HIV 368
  - Transmission of HIV 368
  - Sites of HIV Infection 369
  - Stages of HIV Infection 369
  - Treatments 371

18 Development throughout Life 372

- **18.1 Periods of Development in Human Life** 372
- **18.2 Prenatal Period** 372
  - Pre-embryonic Period 373
  - Embryonic Period 378
  - Fetal Period 381
- **18.3 Birth** 383
- **18.4 Birth Defects** 384
- **18.5 Milk Production by Mammary Glands** 385
- **18.6 Postnatal Period** 386
  - Possible Causes of Aging 388
  - High-Quality Old Age 389

ETHICAL ISSUE
Making Babies 386

HEALTH ISSUE
Disparities in Health and Health Care at All Life Stages 388

18a Autism Spectrum Disorder 393

- **18a.1 Characterization and Prevalence** 393
- **18a.2 Diagnosis** 395
- **18a.3 Possible Causes** 396
- **18a.4 Treatment and Therapy** 397
- **18a.5 Fear That Vaccines Cause Autism Spectrum Disorder** 398

PART V — GENES AND DNA

19 Chromosomes and Cell Division 400

- **19.1 Two Types of Cell Division** 400
- **19.2 Form of Chromosomes** 401
- **19.3 The Cell Cycle** 401
  - Interphase 402
  - Division of the Nucleus and the Cytoplasm 402
- **19.4 Mitosis: Creation of Genetically Identical Diploid Body Cells** 402
- **19.5 Cytokinesis** 405
- **19.6 Karyotypes** 405
- **19.7 Meiosis: Creation of Haploid Gametes** 405
  - Functions of Meiosis 406
  - Two Meiotic Cell Divisions: Preparation for Sexual Reproduction 407

Genetic Variability: Crossing Over and Independent Assortment 409
Extra or Missing Chromosomes 411

ETHICAL ISSUE
Trisomy 21 414

19a Stem Cells—A Repair Kit for the Body 418

19a.1 Stem Cells: Unspecialized Cells 418
19a.2 Sources of Human Stem Cells 419
Adult Stem Cells: Unipotent and Multipotent 419
Umbilical Cord and Placental Stem Cells: Multipotent 419
Embryonic Stem Cells: Pluripotent 420
Induced Pluripotent Stem Cells 423

19a.3 Potential Uses for Stem Cells 423
Replacement for Damaged Cells 423
Growing New Organs 423
Testing New Drugs 424

20 Genetics and Human Inheritance 425

20.1 Principles of Inheritance 425
Gamete Formation 426
Mendelian Genetics 426
Pedigrees 429
Dominant and Recessive Alleles 431
Codominant Alleles 432
Incomplete Dominance 432
Pleiotropy 433
Multiple Alleles 433
Polygenic Inheritance 433
Genes on the Same Chromosome 435
Sex-Linked Genes 435
Sex-Influenced Genes 436

20.2 Breaks in Chromosomes 436
20.3 Detecting Genetic Disorders 437
Prenatal Genetic Testing 437
Newborn Genetic Testing 438
Adult Genetic Testing 439

ETHICAL ISSUE
Gene Testing 439

21 DNA and Biotechnology 442

21.1 Form of DNA 442
21.2 Replication of DNA 443
21.3 Gene Expression 444
RNA Synthesis 444
Protein Synthesis 446

21.4 Mutations 449
21.5 Regulating Gene Activity 450
Gene Activity at the Chromosome Level 450
Regulating the Transcription of Genes 450

21.6 Genetic Engineering 450
Recombinant DNA 450
Applications of Genetic Engineering 453
Gene Therapy 454

21.7 Genomics 457
Human Genome Project 457
Microarray Analysis 458
Comparison of Genomes of Different Species 459

ENVIRONMENTAL ISSUE
Environment and Epigenetics 451

HEALTH ISSUE
Genetically Modified Food 456

ETHICAL ISSUE
Forensic Science, DNA, and Personal Privacy 459

21a Cancer 463

21a.1 Uncontrolled Cell Division 463
Benign or Malignant Tumors 463
Stages of Cancer Development 464

21a.2 Development of Cancer 465
Lack of Restraint on Cell Division 465
DNA Damage and Cell Destruction 467
Unlimited Cell Division 467
Blood Supply to Cancer Cells 468
Adherence to Neighboring Cells 468
Body Defense Cells 468

21a.3 Multiple Mutations 469
21a.4 Cancer Stem Cell Hypothesis 469
21a.5 Known Causes of Cancer 470
Viruses 470
Chemicals 471
Radiation 471

21a.6 Reducing the Risk of Cancer 471
21a.7 Diagnosing Cancer 472
21a.8 Treating Cancer 474
Surgery 474
Radiation 474
Chemotherapy 474
Immunotherapy 474
Inhibition of Blood Vessel Formation 475
Gene Therapy 475

xvi Detailed Contents

PART VI EVOLUTION AND ECOLOGY

22 Evolution and Our Heritage 476

22.1 Evolution of Life on Earth 476
Small Organic Molecules 477
Macromolecules 477
Early Cells 478

22.2 Scale of Evolutionary Change 478
Microevolution 478
Macroevolution 480

22.3 Evidence of Evolution 482
Fossil Record 482
Geographic Distributions 483
Comparative Molecular Biology 485
Comparative Anatomy and Embryology 486

22.4 Human Evolution 487
Primate Characteristics 487
Misconceptions 489
Trends in Hominin Evolution 490

ETHICAL ISSUE
Conducting Research on Our Relatives 485

23 Ecology, the Environment, and Us 497

23.1 Earth as an Ecosystem 497
23.2 Biosphere 498
23.3 Ecological Succession 498
23.4 Energy Flow 500
Food Chains and Food Webs 500
Energy Transfer through Trophic Levels 502
Ecological Pyramids 502
Health and Environmental Consequences of Ecological Pyramids 502

23.5 Chemical Cycles 504
The Water Cycle 504
The Carbon Cycle 506
The Nitrogen Cycle 507
The Phosphorus Cycle 508

23.6 Biodiversity 508

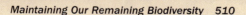

ETHICAL ISSUE
Maintaining Our Remaining Biodiversity 510

24 Human Population, Limited Resources, and Pollution 514

24.1 Population Changes 514
Population Growth Rate 514
Age Structure 516
Immigration and Emigration 516

24.2 Patterns of Population Growth 516
24.3 Environmental Factors and Population Size 518
24.4 Earth's Carrying Capacity 518
24.5 Human Impacts on Earth's Carrying Capacity 518
Agricultural Advances 518
Depletion of Resources 519
Pollution 521

24.6 Global Climate Change 522
Global Warming and Greenhouse Gases 522
Carbon Footprint 524

24.7 Looking to the Future 526

ENVIRONMENTAL ISSUE
Air Pollution and Human Health 524

About the Authors

Judith Goodenough

Judith Goodenough

Judith received her B.S. in biology from Wagner College (Staten Island, NY) and her doctorate in biology from New York University. She has more than 35 years of teaching experience at the University of Massachusetts, Amherst, until recently specializing in introductory-level courses. In 2009, she was selected as a College of Natural Sciences Fellow for Blended Learning and developed a hybrid course in introductory physiology. Her insights into student concerns and problems—gained from more than 30 years of teaching human biology and 20 years of team-teaching the biology of social issues—have helped shape this book. In 1986, Judith was honored with a Distinguished Teaching Award from the University of Massachusetts. In addition to teaching, she has written articles in peer-reviewed journals, contributed chapters to several introductory biology texts, and authored numerous laboratory manuals. With the team of McGuire and Jakob, she is also the coauthor of *Perspectives on Animal Behavior,* Third Edition.

Betty McGuire

Betty McGuire

Betty received her B.S. in biology from Pennsylvania State University, where she also played varsity basketball. She went on to receive an M.S. and Ph.D. in zoology from the University of Massachusetts, Amherst, and then spent two happy years as a postdoctoral researcher at the University of Illinois, Champaign-Urbana. Her field and laboratory research emphasizes the social behavior and reproduction of small mammals. She has published more than 50 research papers, coauthored the text *Perspectives on Animal Behavior* as well as several introductory biology study guides and instructor manuals, and served as an associate editor for *Mammalian Species,* a publication of the American Society of Mammalogists. At Smith College, Betty taught human biology, introductory biology, vertebrate biology, and animal behavior. Now at Cornell University, she teaches mammalogy and vertebrates: structure, function, and evolution.

Preface

Humans are curious by nature. This book was written to stimulate that curiosity, inspiring appreciation for the intricacy of human biology and the place of humans in the ecosphere. To satisfy that curiosity with solid and current information, we provide students with a conceptual framework for understanding how their bodies work and for dealing with issues relevant to human health in the modern world. We sustain the student's interest by continually illustrating the connections between biological concepts and issues of current social, ethical, and environmental concern. Our central belief is that the application of biological concepts to familiar experiences is the key to helping students see the excitement of science and understand its importance in their lives.

This edition builds on the fourth edition's strengths of clarity, liveliness, consistency, currency, and relevance. The writing is engaging, the explanations straightforward, and the pedagogical framework meticulously constructed. All features are designed to help students identify important facts and ideas, understand them, and appreciate why they matter.

Application of the material to students' interests brings concepts to life and illustrates the ethical and social relevance of human biology. This strategy is especially apparent in the "Special Topic" chapters and the dozens of Special Interest Essays distributed throughout the other chapters.

Practical Goals and Special Features

The principal goals of this textbook are (1) to give *a clear presentation of the fundamental concepts* of human anatomy, physiology, development, genetics, evolution, and ecology; (2) to *apply these concepts* in ways that will both interest and benefit students; (3) to help students *develop reasoning skills* so they can make use of their newly acquired knowledge in the situations they face in daily life; (4) to help students *evaluate the many sources of information* available to them and to select those that are reliable and accurate; and (5) to give students an understanding of how the choices they make can *affect society and the planet*, as well as their own quality of life.

Much of the material covered in human biology has a bearing on ethical, social, and environmental issues that are important to us all. Connections between human biology topics and ethical, social, and environmental issues help students develop a global perspective on their impact on the biosphere and will prepare them to be responsible citizens of their country and the world. Society is currently immersed in many pressing biological debates, and students need the tools to understand these issues and make informed decisions.

New to This Edition

The Fifth Edition includes new features and information that increase application of the material to students' everyday lives and make it more accessible.

- A new **Special Topic** chapter, titled **Becoming a Patient: A Major Decision** has been added. Focusing on how to be a well-informed patient, this new chapter covers topics such as (1) how to select a doctor or hospital; (2) how to get the most out of an appointment with your doctor; (3) the importance of second opinions; and (4) how to research health conditions on your own.
- New **Did You Know?** questions open each chapter and pique students' interest with interesting and little-known facts about the topic that follows.
- New and expanded multiple choice questions are included in the chapter review sections, providing students the opportunity to more thoroughly test their understanding of the material covered.
- Several new **What Would You Do?** boxes place greater focus on the ethical issues associated with each chapter by posing questions that ask the student to take an informed stand on complicated and controversial topics.
- An open, welcoming new design makes the Fifth Edition more student-friendly, easier to read, and more appropriate for this course level.
- The primary section heads in each chapter are now numbered to help students navigate the material and organize concepts as they read through the chapter.
- The MasteringBiology online homework, tutorial, and assessment system replaces the Human Biology Place companion website of previous editions. Each end-of-chapter review includes a reminder that directs students to MasteringBiology to access related quizzes, activities, and more.

Special Topic Chapters

The text contains 10 **Special Topic chapters:** Chapter 1a, Becoming a Patient: A Major Decision; Chapter 8a, Drugs and the Mind; Chapter 10a, Diabetes Mellitus; Chapter 12a, Cardiovascular Disease; Chapter 13a, Infectious Disease; Chapter 15a, Food Safety and Defense; Chapter 17a, Sexually Transmitted Diseases and AIDS; Chapter 18a, Autism Spectrum Disorder; Chapter 19a, Stem Cells—A Repair Kit for the Body; and Chapter 21a, Cancer. Created to further motivate students to learn, each of these short chapters builds on the "pure biology" presented in the immediately preceding chapter to cover issues likely to be of personal interest. The discussions these chapters contain are more thoroughly developed than would be possible in a boxed essay. Even if instructors do not include these special topics in their reading assignments, we believe the issues are so pertinent to students that they will read the special topic chapters of their own volition, or at least refer to them occasionally as guides to a healthier lifestyle.

Much of the information offered in the text is practical: What can be done to prevent the spread of sexually transmitted diseases? How should food be selected, prepared, and stored to reduce the chances of foodborne illness? The body each of us is born with is a most intricate machine, but it does not come equipped with an owner's manual. In a sense, this book can be the students' owner's manual. Studying and applying the lessons to their individual lifestyles and health issues can help students live longer, happier, and more productive lives.

Special Interest Essays

Three categories of **Special Interest Essays** use the basic scientific content of the chapters to explore issues having broader impact on individual health, society, and the environment.

Environmental Issue essays deal with ways in which human activities alter the environment or, conversely (sometimes simultaneously), ways in which the environment influences human health. Among the topics discussed in Environmental Issue essays are asbestos, genetically modified foods, and noise pollution.

Ethical Issue essays explore ethical and social issues related to the topics in a chapter. They explore questions concerning such subjects as anabolic steroid use, gene testing, and the use of primates in research.

Finally, **Health Issue** essays deal primarily with personal health topics. They provide current information on certain health problems that students, their families, or their friends might encounter. Topics discussed in Health Issue essays include acne, osteoporosis, treatments for the common cold, and disparities in health and health care.

All of these essays include **Questions to Consider** that ask students to think about the ethical implications of certain behaviors (such as taking anabolic steroids) or medical procedures (such as generating extra embryos as part of infertility treatments).

Stop and Think Questions

The **Stop and Think** questions scattered throughout each chapter are intended to promote active learning. They invite the students to pause in their reading to think about the information that was just presented and apply it to a new and interesting situation. These periodic checks allow the students to determine whether they have followed and understood the basic chapter content. In the Fifth Edition we have increased the number of **Stop and Think** questions in each of the main chapters and added these questions to the Special Topic chapters.

What Would You Do? Questions

The **What Would You Do?** questions, which are also placed throughout each chapter, challenge the student to form an opinion or to take a stand on a particular issue that society faces today, as well as to identify the criteria used in reaching that opinion or decision. These questions help students see the relevance of biology to real-life problems and foster the practice of thinking through such complicated issues as the use of sperm-sorting technology by parents to select the gender of their offspring and strategies for slowing the growth of human populations. When the subject of one of these questions is controversial, the text presents examples of arguments from both sides, as well as evidence in support of competing arguments.

Did You Know? Boxes

The new **Did You Know?** boxes at the start of each chapter pique students' interest with little-known facts about the topic that follows. These features provide an engaging introduction to the material, most often by providing information that students can relate to their everyday lives.

Enticing Illustration and Design Program

Users of previous editions—instructors and students alike—were unreservedly enthusiastic in praising the illustrations for their appeal and helpfulness. The visual program consists of simple but elegantly rendered illustrations that have been carefully designed for effective pedagogy. Their very beauty stimulates learning. This is particularly true of the many vibrant, three-dimensional anatomical figures, whose realistic style and appropriate depth and detail make them easy for students to interpret and use for review. Micrographs often appear side by side with illustrations to aid interpretation and understanding and to give the actual view of a structure or process being studied.

Within each category of illustrations—from molecular models to depictions of human tissues and organs—the figures are consistent in plan and style throughout the text. Numerous key figures pull concepts together to present the "big picture." Reference figures help students locate particular structures within the body. Flowcharts walk students through a process one step at a time so they can visually follow the progress of a discussion after they have read an explanation in the body of the text. Similarly, step-by-step figures break complex concepts down into simpler components. Finally, color is used in the visual program as an effective means of organizing information and maintaining consistency throughout the text.

Figure Questions

A question accompanies at least one figure in each chapter. This feature asks a question prompting students to pause and critically examine the information in the figure. Answers are provided below the figure legend.

Engaging Design

This Fifth Edition of *Biology of Humans* presents an engaging design that was created to complement the vibrancy of the illustrations, clarify the organizational structure of the chapters, and increase overall readability.

Organization and Pedagogy

After an introductory chapter on the science of biology, the text presents a discussion of the chemistry of life; proceeds through cells, tissues, organs, and organ systems; and ends with discussions of genetics, evolution, and ecology. As teachers ourselves, we understand the difficulty of covering all the topics in a human biology text in one semester. Instructors are inevitably forced to make difficult decisions concerning what to include and what to leave out. We also know there are many equally valid ways of organizing the material. For this reason, the chapters in this text are written so as not to depend heavily on material covered in earlier chapters. The independence of each chapter allows the instructor to tailor the use of this text to his or her particular course. At the same time, we provide cross-references where they may be helpful to direct students to relevant discussions in other chapters.

The pedagogical features that provide a consistent framework for every chapter have been designed not only to help students understand the information presented in their human biology course, but also to help them study more effectively. Some of the most important of these elements are described next.

Chapter Outlines and Introductions

Each chapter begins with a list of the chapter's main topics constructed from the major headings. Because it identifies the chapter's important concepts and the relationships between them, this feature provides a conceptual framework on which students can mentally organize new information as they read. Special interest essay boxes are also included in this outline.

Key Terms and Glossary

Because this text is intended for students who are not science majors, we have held the use of technical language to a minimum. Important terms are set in bold type where they are formally introduced, and they are listed as key terms at the end of each chapter. Other terms of lesser importance are set in italics. The **Recognizing Key Terms** list also provides chapter page numbers indicating where each term is defined. The **Glossary** at the end of the book contains definitions for all the key terms and many of the terms set in italics.

Looking Ahead (and Back)

It's widely known that students often compartmentalize chapters and have trouble seeing how one chapter relates to the next. To address this issue, we conclude each chapter with a **Looking Ahead** box to show the students how the following chapter will build on the one they have just finished reading. In addition, we begin each chapter with an introductory paragraph that clearly explains how the material from the previous chapter relates to what they're about to read in the present chapter. This **Looking Ahead** (and back) approach draws explicit ties between chapters.

End-of-Chapter Questions

The questions provided at the end of each chapter are designed in several formats to encourage students to review and understand the relevant material instead of simply memorizing a few salient facts. Some, specifically **Reviewing the Concepts**, are intended simply as content review. Others—particularly those under the heading **Applying the Concepts**—require critical thinking and challenge the students to apply what they have learned to new situations. The third type of end-of-chapter question, **Becoming Information Literate**, prompts students to explore and evaluate resources beyond the text and can be used as a starting point for developing research papers or reports. In response to reviewer feedback, this Fifth Edition features many new multiple choice questions in each chapter. Review questions that require a written answer are followed by the page number(s) containing the relevant discussion. Answers to all Reviewing the Concepts questions are provided in an appendix, as are hints for answering the Applying the Concepts questions. These hints, which help students identify the information needed to answer each question, are intended to guide students in their thinking process instead of simply providing a quick answer.

Chapter Updates

All of the material in the book has been carefully reviewed, revised, and updated. The latest statistical information and medical advances have been incorporated throughout. The following is a list of some of the more significant changes in each chapter.

Chapter 1 The discussion of classification has been modified to emphasize evolution as a unifying theme in biology. The example of the scientific method has been modified to illustrate the proper use of a histogram and line graph. The art program has been modified to support the new example. The accompanying figure has been revised to include the step of forming a question about the observation that prompts the experiment. The discussion of epidemiological experiments has been updated using the 2011 study on cell phone use and cancer. New discussions of information literacy and information technology literacy emphasize these important skills. A new table provides hints to identify reliable websites.

Chapter 2 A new figure depicting why water is an excellent solvent has been added. An updated version of the periodic table is now included, and new text has been added to the figure showing levels of protein structure.

Chapter 3 The essay on mitochondrial diseases has been streamlined, and some questions at the end of the chapter have been revised.

Chapter 4 New art and micrographs showing epithelial tissue, connective tissue, and muscle tissue have all been chosen to more clearly demonstrate the different structures. A discussion of Lipodissolve, a new nonsurgical treatment for destroying unwanted pockets of fat, has been added. A new **What Would You Do?** prompts students to consider their willingness to be organ donors.

Chapter 5 A discussion of bone fractures has been added. There is a new figure illustrating the growth of a long bone. A new **What Would You Do?** asks students to think about the rights of children who are suitable bone marrow donors for their siblings.

Chapter 6 The **Stop and Think** question about curare was further developed to help students think about the role of acetylcholine in muscle contraction. There is a short discussion on the new phenomenon "text neck," which is caused by the increased frequency of texting.

Chapter 7 A new **What Would You Do?** asks about the ethics of performing controlled experiments to develop drugs to treat schizophrenia and whether participants in such studies are capable of giving informed consent.

Chapter 8 A new **What Would You Do?** prompts students to consider the ethics of neuromarketing—monitoring brain activity to evaluate consumer interest in products.

Chapter 8a Short discussions on medical marijuana and on oxycodone abuse have been added.

Chapter 9 Changes in the art program enhance understanding. A figure of muscle spindles and tendon receptors has been added. New photographs illustrate differences in normal vision, farsightedness, and nearsightedness help the student understand these differences in vision. A photograph of a cataract has been added.

Chapter 10 Two new figures depict hormonal regulation of calcium and blood glucose. These new figures are stylistically consistent with those in the chapter on body organization and homeostasis (Chapter 4). Minor changes have been made to the essay on hormone therapy and the **What Would You Do?** on melatonin.

Chapter 10a A new figure shows how insulin regulates blood glucose levels and where in the process problems occur for type 1 and type 2 diabetics. All statistics regarding prevalence of diabetes have been updated in the text and the accompanying figure. There is a new **What Would You Do?** that asks students to consider whether health insurance companies should charge type 1 diabetics and overweight people at risk for developing type 2 diabetes more for coverage, given their very high medical expenses. The section on other types of diabetes has been updated.

Chapter 11 The discussions of plasma and bone marrow have been streamlined. Transport of carbon dioxide as a bicarbonate ion has been emphasized.

Chapter 12 The discussions of capillary exchange and of lymphoid organs have been enhanced. Several figures have been modified to increase clarity. A new **What Would You Do?** prompts students to consider legal and ethical aspects of heart transplants.

Chapter 12a This chapter has been reorganized to increase clarity. New figures of an aneurysm and of coronary angioplasty enhance the related discussions. The discussion of coronary bypass surgery now includes surgery using computer-guided robotic arms.

Chapter 13 The discussions of interferons, clonal selection, and adaptive immune responses have been modified to improve clarity. Specific references to chapters containing additional discussions of the safety of vaccinations and of type 1 diabetes as an autoimmune disease enhance the cohesiveness of the text.

Chapter 13a Updates and additions to the discussion of antibiotic resistance increase interest and currency of the chapter. A new **What Would You Do?** asks students to balance the costs and benefits of reducing the prevalence of drug-resistant tuberculosis.

Chapter 14 Cystic fibrosis is now discussed in this chapter. All text and figures were carefully reviewed for currency.

Chapter 15 This chapter now contains information on both the digestive system and nutrition. The chapter also has been updated to reflect *Dietary Guidelines for Americans, 2010* and ChooseMyPlate, which have replaced the older guidelines and MyPyramid. The art program has been revised accordingly and also to reflect more current statistics on obesity. A new **What Would You Do?** prompts students to consider who should bear responsibility for obesity.

Chapter 15a In response to reviewer comments, this chapter on food safety and defense now follows the digestive system chapter. Examples of foodborne illnesses have been updated, and new suggestions have been added to the section on personal food safety. A new **What Would You Do?** addresses the debate regarding purchasing eggs from battery versus free-range chickens, given no consistent evidence that the eggs differ in likelihood of carrying *Salmonella*.

Chapter 16 The introductory paragraphs have been revised to emphasize the role of the kidneys in homeostasis and to stress that kidneys do more than produce urine. The section on eliminating wastes has been reorganized; it first describes all the different types of wastes and then covers the organs that remove each type. In response to reviewer comments, the figure showing the structure of the nephron has been simplified and reorganized for better flow with the text, and the figure depicting regulation of blood volume and pressure by antidiuretic hormone has been completely revised. The **Health Issue** essay on urinalysis now includes information on urobilin and drug testing. A new **What Would You Do?** focuses on whether the mental abilities of a person needing a kidney transplant should be considered when prioritizing who should receive a donated kidney.

Chapter 17 The breast cancer essay is enhanced. Nexplanon is added to the birth control discussion.

Chapter 17a The difference between sexually transmitted infections and sexually transmitted diseases is now emphasized. The text now includes the updated recommendations that young boys receive HPV vaccinations. The discussion of HIV treatment is updated and enhanced.

Chapter 18 The section on birth defects now includes discussion of the use of folic acid to help prevent spina bifida, anencephaly, and other neural tube defects.

Chapter 18a This chapter has been completely revised in line with the fifth edition of the Diagnostic and Statistical Manual of Mental Disorders (DSM-5), which has an expected publication date of May 2013. The DSM-5 will contain significant changes from the previous edition of the DSM (DSM-IV-TR). In particular, four previously distinct neurodevelopmental disorders—autistic disorder, Asperger's disorder, pervasive developmental disorder not otherwise specified (PDD-NOS), and childhood disintegrative disorder—now are subsumed under a single category, autism spectrum disorder (ASD). This major change is reflected throughout the chapter. Also, a new table highlights the new diagnostic criteria for ASD that have been drafted by the American Psychiatric Association and posted on their DSM-5 Development website for public review and comment. Because the diagnostic criteria for ASD now include overreaction or underreaction to sensory stimuli, we have added a new photograph depicting sensory integration therapy. A new **What Would You Do?** focuses on the erroneous belief that vaccines cause autism spectrum disorder.

Chapter 19 New micrographs of mitosis improve the art program. All text and figures were carefully reviewed for currency.

Chapter 19a The recent success of somatic cell nuclear transfer using an egg cell and a skin cell is described. **Questions to Consider** have been added to the **Ethical Issue** essay on gene testing. The chapter also describes improvements in techniques for rebuilding organs.

Chapter 20 All material was reviewed to confirm inclusion of the most recent information available.

Chapter 21 The chapter includes a discussion of the relationship between the decline in monarch butterfly populations and the use of genetically modified crops. The chapter also covers new developments in gene therapy for HIV.

Chapter 21a New advances in cancer treatment are included. All text and figures were carefully reviewed for currency.

Chapter 22 This chapter has been updated to include several exciting discoveries and advances. The section on biogeography and dispersal, and the associated figure, now reflect new fossil evidence suggesting that marsupials originated in China, not North America. The figure summarizing primate relationships has been modified to show that the split between gorillas and (humans + chimps) occurred about 10 million years ago, as indicated by the gorilla genome published in March 2012. The section on comparative molecular biology also has been updated to reflect this new information. The section covering the evidence of evolution now includes a new **Stop and Think** on which organisms are likely to be represented in the fossil record.

Chapter 23 Biological magnification of radiation in tuna following the 2011 earthquake and tsunami in Japan and Texas's efforts to desalinate water from aquifers are now included.

Chapter 24 Population statistics have been updated throughout the chapter.

Teaching and Learning Solutions for Instructors and Students

Biology of Humans, Fifth Edition, is supported by a full complement of carefully designed materials for both students and instructors.

For Instructors

Instructor Resource DVD
9780321886606 | 0321886607

The Instructor Resource DVD provides a range of ready-to-use media supplements to help instructors teach the course, engage students, and accommodate different learning styles. Instructors can augment their lectures, show students the relevance of the subject matter, and increase student comprehension using the following tools:

- An image library of all the art, tables, and photographs from the book
- A selection of images with customizable labels and stepped-out art
- Editable PowerPoint lecture presentation slides with embedded links to *ABC News* videos and Human Biology Animations
- Clicker questions
- Human Biology Animations
- BioFlix and BioFlix PowerPoint slides
- 28 *ABC News* video clips
- PowerPoint slides for *Scientific American: Concepts and Current Issues in Biology,* Volumes 1–5
- *Interactive Physiology* for human biology slides, worksheets, and answer sheets
- Interactive Quiz Show games
- Microsoft Word files for the Instructor's Guide and Test Bank
- Computerized test bank
- An all-electronic set of ready-to-print Transparency Acetate Masters of selected illustrations from the text

Instructor's Guide
9780321886545 | 0321886542

The Instructor's Guide provides tips for making the material relevant, interesting, and interactive, especially for nonmajors. Each chapter includes the following:

- Learning Objectives that identify goals for students and instructors
- Lecture activity suggestions
- Suggestions for class demonstrations and student activities
- Resource listings of relevant websites

Test Bank

The Test Bank includes over 1000 multiple-choice, fill-in-the-blank, short answer, and essay test questions—and answers—originally created and reviewed by a panel of educators. All questions are correlated to Bloom's Taxonomy of learning. Microsoft Word and TestGen versions of the files are available on the Instructor Resource DVD and can be downloaded from MasteringBiology.

Human Biology Place Companion Website
www.humanbiology.com

This Human Biology Place companion website is an open-access website offering animated web activities that provide a highly interactive way for students to get more involved in their studies. Accompanying quizzes, book-specific self-tests and essay questions, crossword puzzles, and flashcards are featured. *ScienceDaily* RSS feeds of summaries and links to human biology–related news stories are updated six times a day. Accompanying **Becoming Information Literate** essay questions give students the opportunity to investigate and comment on current biology news.

The password-protected instructor portion of the site contains an image library, lecture presentations, clicker questions, Instructor's Guide, Test Bank, and other assets from the Instructor Resource DVD.

A password-protected eText is also available on the website 24/7 that instructors can annotate, highlight passages to direct students to important content, and hide sections that aren't relevant to their course.

BlackBoard Course Cartridge
9780321886323 | 0321886321

BlackBoard Course Cartridge is a course management system that contains a range of preloaded content, such as testing and assessment question pools, chapter-level overviews and objectives, interactive web-based activities, animations, RSS feeds, flashcards, and exercises in **Becoming Information Literate**—all designed to help students master core course objectives. This course management system also includes access to chapter guides, a Test Bank, and animations.

For Students

MasteringBiology

Now integrated with the Fifth Edition, the **MasteringBiology**® online homework, tutorial, and assessment system helps instructors teach more efficiently and is pedagogically proven to help students learn. It helps instructors maximize class time with customizable, easy-to-assign, and automatically graded assessments that motivate students to learn outside class and arrive prepared for lecture. The powerful gradebook provides unique insight into student and class performance even before the first test. As a result, instructors can spend class time where students need it most. The Mastering system empowers students to take charge of their learning through activities aimed at different learning styles and engages them in learning science through practice and step-by-step guidance—at their convenience, 24/7.

Interactive Physiology for Human Biology
0-321-59539-4

The *Interactive Physiology* for Human Biology (IP for HB) CD-ROM reinforces readings and lectures with a wealth of outstanding animations, engaging activities, helpful self-testing, and much more. Students can now access the complete IP for HB resources through the MasteringBiology Study Area, and, using the MasteringBiology item library, instructors may assign related tutorials that provide coaching through helpful wrong-answer feedback and hints.

Blackboard Student Access Kit
9780321886385 | 0321886380

Acknowledgments

It takes more than authors to get a book to the readers, and many dedicated people have helped get this text to your hands.

The project was enthusiastically launched by Star McKenzie, senior acquisitions editor, who helped us plan this edition and assembled a team of professionals who brought our vision to reality.

We are especially thankful for the opportunity to work with Nicole George-O'Brien, who kept our vision for this edition in mind through the entire project and was involved at every level. Sally Peyrefitte copyedited the book with a detailed, accurate, and consistent touch. The art program is essential not only to the appearance of the book but also to its usefulness as an instrument for learning and teaching. Changes to the art program were ably carried out by Brian Morris at Scientific Illustrators. Our photo researcher, Sarah Bonner, was diligent in her pursuit of striking and pedagogically important photographs.

The team at Element was skillfully led by our production editor Heidi Allgair, and all aspects of the book's production were expertly overseen by our production project manager at Pearson, Lori Newman.

Many thanks go out to all of the instructors who reviewed the book and provided us with the useful feedback that helped shape this new edition:

Andrea Abbas, *Washtenaw Community College*
Gary Arnet, *California State University, Chico*
Samantha Butler, *University of Southern California*
Edward Gabriel, *Lycoming College*
Noah Henley, *Rowan-Cabarrus Community College*
Cynthia Littlejohn, *University of Southern Mississippi*
Dennis McCracken, *University of North Carolina at Pembroke*
Jean Shingle, *Immaculata University*
Renato Tameta, *Schenectady County Community College*
Jessica Thomas, *Eastern Connecticut State University*
Greg Thurman, *Central Methodist University*
Rebecca Vance, *University of Alabama at Birmingham*
Naomi Waissman, *El Paso Community College*
Muhammed A. Wattoo, *Cornell Medical College*

From Judith Goodenough

I thank my family and friends, who supported and encouraged me at every stage of this project. My husband, Steve, was a cheerleader and convinced me that I would complete this project. Without his witty quips, I would have lost my sanity. He reassures me that I'll always be his first wife. My daughters, Aimee and Heather, inspire me and continually remind me that the people you love should always come first. Aimee, a rehabilitation counselor for clients with traumatic brain injury, contributed the Chapter 8 essay, "Brain Injury: A Silent Epidemic." The willingness of my mother, Betty Levrat, to help in any way allowed me to focus on writing.

Margaret Ludlam cheered me up and helped me cope when things got tense. Lee Estrin, one of "the group" of dear friends who have always provided moral support and advice, was always willing to visit for an "I need a break" weekend, even when I couldn't actually stop working completely.

From Betty McGuire

I thank my husband, Willy Bemis, for support and encouragement throughout this edition, and my daughter Kate and son Owen, for (usually) waiting patiently for me to complete just one more sentence or paragraph. Dora, Kevin, and Cathy McGuire were endlessly encouraging and understanding, as they have been throughout my life. Dora and Willy reviewed the new Special Topic chapter on becoming a patient. I also thank Muhammad A. Wattoo, MD, Internal Medicine, and Clinical Assistant Professor of Medicine, Weill Cornell Medical College NYC, for reviewing the new Special Topic chapter and meeting with me to discuss it. Finally, Lowell Getz, my longtime friend and research colleague, waited with good humor as one after another of our papers took a backseat to a book chapter.

To Stephen, my husband,
best friend, personal hero,
and the funniest person I know.
To Aimee and Heather, my daughters,
who fill me with love, wonder, and amazement.
To Betty Levrat, my mother,
an excellent role model
and endless source of support and encouragement.
To "The Group," friends for more than forty years,
who help me hold it all together.
—J. G.

In loving memory of James Patrick McGuire.
To Willy, Kate, and Owen Bemis,
and to Dora, Kevin, and Cathy McGuire.
—B. M.

Humans in the World of Biology

1

- Scientists have discovered two planets about the size of Earth orbiting a distant star, Keplar-20e and Keplar-20f. These planets may have environmental conditions that could sustain life.
- Estimates of the total number of species on Earth range from 5 to 30 million, and only 1.7 to 2 million species have been formally identified (Millennium Ecosystem Assessment).

1.1 Basic Characteristics of All Living Things

1.2 Evolution: A Unifying Theme in Biology

1.3 Levels of Biological Organization

1.4 Scientific Method

1.5 Critical Thinking to Evaluate Scientific Claims

ENVIRONMENTAL ISSUES

Medicinal Plants and the Shrinking Rain Forest

In this chapter, we see that life has many levels of organization: individual, population, community, ecosystem, and biosphere. Throughout most of this book, we focus on the human individual—how the individual human body functions and the biological principles that govern those functions. However, we also examine many of the larger health, social, and environmental issues that we must be aware of, because they can affect all of us.

Humans are a small, but important, part of the diverse life on Earth.

1.1 Basic Characteristics of All Living Things

We will begin by exploring the Amazon rain forest—a place that is teeming with life. Given the biodiversity of the rain forest, it not surprising that scientists are exploring it in search of any secrets it may reveal, including any plants that may have healing qualities (see the Environmental Issue essay, *Medicinal Plants and the Shrinking Rain Forest*).

We say that life is abundant in the rain forest, but how do you determine whether something unfamiliar to you is alive? In most cases, the question is easy to answer. Although the leaves around you have different shapes and sizes, a brief examination assures you that they are leaves and that the tree whose trunk you are exploring is clearly a tree—thus telling you that the tree specimen you are examining is indeed alive. But what about the gray material adhering to the trunk? Is it also alive?

Defining *life* might seem to be easy, but it is not. In fact, no single definition satisfies all life scientists. For example, if we say that something is alive if it reproduces, someone is likely to note that a page with a wet ink spot can fall on top of another page and reproduce itself almost exactly. If we say something is alive if it grows, what should we conclude about crystals? They grow, but they are not alive. And so it goes.

ENVIRONMENTAL ISSUE

Medicinal Plants and the Shrinking Rain Forest

The healing powers of many plants have been known for centuries. Historically, such knowledge was gained by trial and error and passed along by word of mouth. For example, many cultures have long known that tea made from willow bark relieves pain and reduces fever. Scientists learned that willow bark contains salicylic acid. They isolated the compound and developed it into the drug we know today as aspirin. Similarly, digitalis, a heart medication, was discovered after a patient with an untreatable heart condition was seen to benefit from an herbal drink provided as a folk remedy. The potion contained purple foxglove, which, like willow bark, is frequently mentioned in ancient texts as a healing herb. Broccoli, a more familiar plant, contains the anticancer chemical sulforaphane.

More than 25% of the prescription medicines sold in the United States today contain chemicals that came from plants, and 70% of the newly developed drugs are from natural sources. Many more healing chemicals first discovered in plants used medicinally by native people are now routinely synthesized in laboratories. Unfortunately, however, there are many plants whose medicinal compounds scientists have **not** been able to synthesize.

FIGURE 1.A *The rosy periwinkle* (Catharanthus roseus) *is a source of two anticancer drugs.*

Most plants that have proved to be medically useful are found in the tropics, regions where the human population is growing rapidly. Unfortunately, the forests in these regions are being cut to create living space and foster economic development. For example, Madagascar is the home of the rosy periwinkle (Figure 1.A), which is the source of two anticancer drugs. Humans have destroyed 90% of the vegetation in that nation. Considering that 155,000 of the known 250,000 plant species are from tropical rain forests, and that fewer than 2% of the known plant species have been tested for medicinal value, we have no way of knowing what potential new medicines are being destroyed.

Questions to Consider

- Should indigenous people be compensated for plants found in their locality if extracts of the plants become drugs?
- What steps might be taken to preserve biodiversity within the rain forest?

No single definition applies to all forms of life, so we find that instead of defining life, we can only characterize it. That is, we can only list the traits associated with life. Most biologists agree that, in general, the following statements characterize life.

1. **Living things contain nucleic acids, proteins, carbohydrates, and lipids.** The same set of slightly more than 100 elements is present in various combinations in everything on Earth—living or nonliving. However, living things can combine certain of these elements to create molecules that are found in all living organisms. These molecules include nucleic acids, proteins, carbohydrates, and lipids. The nucleic acid DNA (deoxyribonucleic acid) is especially important because DNA molecules can make copies of themselves, an ability that enables organisms to reproduce (Figure 1.1). The molecules of life are discussed in Chapter 2.

2. **Living things are composed of cells.** Cells are the smallest units of life. Some organisms have only a single cell (*unicellular organisms*); others, such as humans, are composed of trillions of cells (*multicellular organisms*). All cells come from preexisting cells. The ability of cells to divide to form new cells makes reproduction, growth, and repair possible. Cells are discussed in Chapter 3 and cell division in Chapter 19.

3. **Living things grow and reproduce.** Living things grow and ultimately generate new individuals that carry some of the genetic material of the parents. Some organisms, such as bacteria, reproduce simply by making new and virtually exact copies of themselves. Other organisms, including humans, reproduce by combining genetic material with another individual. Many organisms have stages of life. Humans progress from embryo to fetus, child, adolescent, and then adult. Reproduction and development are discussed in Chapters 17 and 18, respectively.

4. **Living things use energy and raw materials.** The term **metabolism** refers to all chemical reactions that occur within the cells of living things. Through metabolic activities, organisms extract energy from various nutrients and transform it to do many kinds of work. Metabolism maintains life and allows organisms to grow. Chemical reactions involved in the transformation of energy are discussed in Chapter 2.

5. **Living things respond to their environment.** A boxer weaves and ducks to avoid the blow of an opponent. A chameleon takes aim at and captures its prey. For a living thing to respond, it must first detect a stimulus and then have a way to react. As later chapters explain, your sensory organs detect stimuli. Your nervous system processes sensory input, and your skeletal and muscular systems enable you to respond. The skeletal and muscular systems are discussed in Chapters 5 and 6, respectively. The nervous system is discussed in Chapter 8, and sensory organs in Chapter 9.

6. **Living things maintain homeostasis. Homeostasis** is the relatively constant and self-correcting internal environment of a living organism. We generally find that life can exist only within certain limits and that living things tend to behave in ways that will keep their body systems functioning within those limits. For example, if you become too cold, you shiver (a metabolic response). Shivering produces heat that warms your body. Alternatively, if you become too hot, your sweat glands will be activated to cool you down. In addition to these and other physiological

1.1 Basic Characteristics of All Living Things

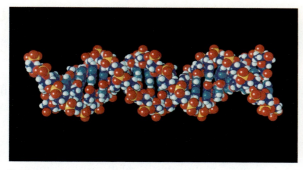

1. Living things contain nucleic acids, proteins, carbohydrates, and lipids.
This is a computer-generated model of the nucleic acid DNA, which carries genetic information.

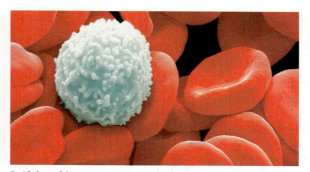

2. Living things are composed of cells.
These are red blood cells (disks) and a white blood cell (sphere).

3. Living things grow and reproduce.
All organisms reproduce their own kind.

4. Living things use energy and raw materials.
This father is feeding his child.

5. Living things respond to their environment.
This chameleon sees and catches its prey.

6. Living things maintain homeostasis.
This person's body temperature will remain about 37°C (98.6°F) in spite of extreme environmental temperature.

7. Populations of living things evolve and have adaptive traits.
The orchid is adapted to live perched on branches of trees. It uses other plants for support so that it can receive enough sunlight to produce its own food by photosynthesis.

FIGURE 1.1 *Characteristics of life*

responses, the sensation of being hot or cold may motivate you to behave in ways that cool you down or warm you up. We discuss homeostasis in Chapter 4, where we make an initial survey of body systems.

7. **Populations of living things evolve and have adaptive traits.** Members of a population of reproducing organisms possessing beneficial genetic traits will survive and reproduce better than members of the population that lack these traits. As a result of this process, called *natural selection*, each of the amazing organisms you see around you has **adaptive traits**—that is, traits that help it survive and reproduce in its natural environment. For example, most plants in the rain forest have shallow root systems, because the topsoil in the Amazon is thin and nutrients are near the surface. As a result, tall trees have developed, through evolution, supports like cathedral buttresses to hold them up, while vines climb over both roots and trees to reach the light. Many plants do not grow in the ground at all but live high above it in the canopy for greater exposure to sunlight, which provides energy to produce sugar. These plants, called *epiphytes*, are rooted on the surfaces of other plants. Rain forest animals also have adaptations—the ability to fly or climb, for example—that enable them to reach the plants for food. Adaptive traits and evolution are discussed in Chapter 22.

stop and think

Scientists have discovered water and methane on Mars. Water is necessary for life. Solar radiation would quickly destroy methane, so "something" must be producing the methane we detect. If samples of water or soil from Mars were brought back to scientists on Earth, what characteristics of life could they look for to determine whether the samples contain anything that is or was once alive?

1.2 Evolution: A Unifying Theme in Biology

Evolution is a common theme in biology because it explains the unity and diversity of the at least 10 million species of organisms that live on Earth. Organisms are unified because all species descended from the first cells. However, as organisms adapted to different environments through evolution, diversity among species arose.

Scientists organize, or classify, living organisms in a way that shows evolutionary relationships among them. This means, for the most part, that organisms with the greatest similarity are grouped together.

Several classification systems have been proposed over the years. One system recently favored by many biologists recognizes three domains. Two of the domains, Bacteria and Archaea, consist of the various kinds of prokaryotes—all of which are very small, single-celled organisms that lack a nucleus or other internal compartments. All other organisms, including humans, belong to the third domain, Eukarya. Organisms in domain Eukarya have eukaryotic cells, which contain a nucleus and complex internal compartments called *organelles*. Domain Eukarya is subdivided into four kingdoms—protists, fungi, plants, and animals—as shown in Figure 1.2. Within each kingdom, organisms are further categorized into groups whose members share characteristics that distinguish them from members of other groups in the kingdom. These groups in turn are subdivided into smaller groups to show successively closer relationships.

As humans, we belong to a subdivision of the animal kingdom called *vertebrates* (animals with a nerve cord protected by a backbone) and, more specifically, to the group called *mammals*. Two characteristics that make us mammals are that we have hair and that we feed our young milk produced by mammary glands. However, we are further defined as belonging to the *primates*, along with lemurs, monkeys, and apes, because

Domain Bacteria	Domain Archaea	Domain Eukarya			
Unicellular prokaryotic organisms	Unicellular prokaryotic organisms; most live in extreme environments	Eukaryotic cells that contain a membrane-bound nucleus and internal compartments			
		Kingdoms			
		Protists	Fungi	Plants	Animals
		Protozoans, algae, diatoms	Molds, mushrooms	Mosses, ferns, seed plants	Invertebrates and vertebrates

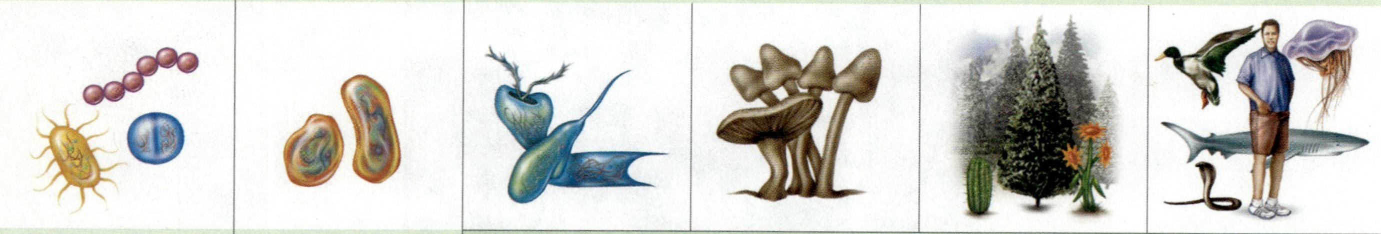

FIGURE 1.2 *One classification scheme showing three domains and four kingdoms of life*

we share a suite of features that includes forward-looking eyes and a particularly well-developed brain. Humans, monkeys, and apes also have opposable thumbs (a thumb that can touch the tips of the other four fingers). Smaller details, such as tooth structure and skeletal characteristics, serve to divide the primates into smaller subgroupings.

stop and think
If a new organism were discovered in the rain forest, what characteristic would you look for to decide whether the animal was a mammal?

1.3 Levels of Biological Organization

As we study human biology, we learn that life can be organized on many levels (Figure 1.3). Cells, the smallest unit of life, are themselves composed of molecules. A multicellular organism may consist of different tissues, groups of similar cells that perform specific functions. Organs also may consist of different types of tissue that work together for a specific function. Two or more organs working together to perform specific functions form an organ system. Humans have 11 organ systems, as we see in Chapter 4.

Life can also be organized at levels beyond the individual organism. A **population** is individuals of the same species (individuals that can interbreed) living in a distinct geographic area. Examples of a population include yellow-bellied marmots living in an alpine meadow or four-eyed butterfly fish living in a coral reef. A **community** is all living species that can potentially interact in a particular geographic area. Examples of a community include *all the species* that live and interact in an alpine meadow or *all the species* living in a coral reef.

An **ecosystem** includes all the living organisms in a community along with their physical environment. The size of the locality that defines the ecosystem varies with the interest of the person studying it. In other words, an ecosystem can be defined as the whole Earth, a particular forest, or even a single rotting log within a forest. Whatever its size, an ecosystem is viewed as being relatively self-contained.

The **biosphere** is that part of Earth where life is found. It encompasses all of Earth's living organisms and their habitats. In essence, the biosphere is the narrow zone in which the interplay of light, minerals, water, and gases produces environments where life can exist on Earth. The biosphere extends only about 11 km (7 mi) above sea level and the same distance below, to the deepest trenches of the sea. If Earth were the size of a basketball, the biosphere would have the depth of about one coat of paint. In this thin layer covering one small planet, we find all of the life we currently know of in the entire universe.

1.4 Scientific Method

Humans are an irrepressibly curious species, constantly asking questions about the things they observe. **Science** is a systematic approach to answering those questions, a way of acquiring knowledge through carefully documented investigation and experimentation—the scientific method.

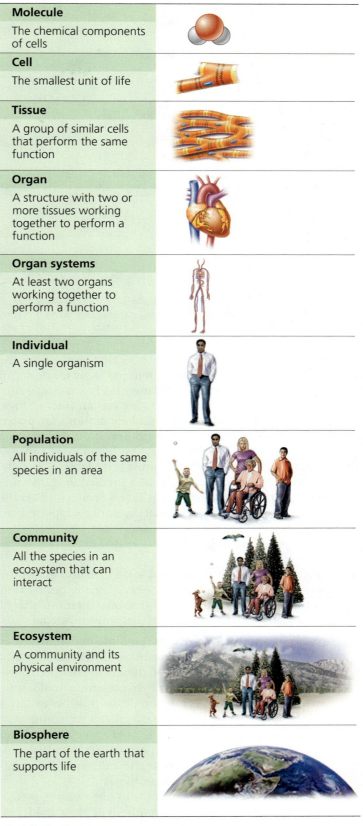

FIGURE 1.3 *Levels of organization of life*

Q How would you modify this diagram to indicate that the scientific process also includes testing alternative hypotheses?

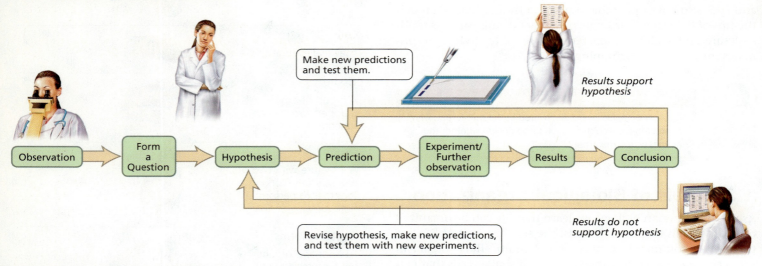

FIGURE 1.4 The scientific process consists of making observations, formulating questions, creating testable hypotheses, conducting experiments, drawing conclusions, revising hypotheses, and designing new experiments.

A Add another arrow from Form a Question indicating Hypothesis 2. All the steps that are currently shown leading from "Hypothesis" would be repeated for Hypothesis 2.

There is no such thing as *the* scientific method in the sense of a single, formalized set of steps to follow for doing an experiment. Instead, the **scientific method** is a way of learning about the natural world by applying certain rules of logic to the way information is gathered and conclusions are drawn (Figure 1.4). It often begins with an observation that raises a question. Next, a possible explanation is formulated to answer the question, but that explanation must be testable. Generally, the tentative explanation will lead to a prediction. If the prediction holds true when it is tested, the test results support the explanation. If the original explanation is not supported, an alternative explanation is generated and tested.

Let's review these steps in a bit more detail.

1. **Make careful observations, and ask a question about the observations.** The process of science usually begins with an observation that prompts a question. Questions should be reasonable and consistent with existing knowledge.

2. **Develop a testable hypothesis (possible explanation) as a possible answer to your question.** The next step is to make an educated guess about the answer to that question, called a **hypothesis**. The hypothesis should be a statement, not a question. It should be possible to test a hypothesis and to prove it false. Keep in mind that although a hypothesis can be shown to be false, it can never be proved to be true. You can collect data that support a hypothesis, but you must also rule out other possible explanations (alternative hypotheses).

   Generally, the hypothesis leads to one or more predictions that will support the hypothesis if they hold true when tested. Depending on the hypothesis, the test may involve further observations or experimental manipulation.

   Different hypotheses can sometimes lead to identical predictions. In such a case, a test can support or refute both hypotheses, depending on the outcome. In this event, it is necessary to make other predictions that will allow us to reject one of the hypotheses. When we find that the results of various tests are more consistent with one hypothesis than with others, we must still be cautious. New evidence may come to light that will disprove the hypothesis, or a new hypothesis may be proposed that is also consistent with the observations.

3. **Make a prediction based on your hypothesis, and test it with a controlled experiment.** Now you make a prediction regarding what should occur if the hypothesis is correct. This prediction will determine the experiment or observations that are necessary to test the hypothesis.

   Ideally, your experiment will be designed in such a way that there can be only one explanation for the results. In such an experiment, called a **controlled experiment**, the research subjects are randomly divided into two groups. One group is designated as the **control group**, and the other one is designated as the **experimental group**. Both groups are treated in the same way except for the *one* factor, called the independent **variable**, whose effect the experiment is designed to reveal.

   In a scientific study, additional variables that have not been controlled for, and may have affected the outcome, are called *confounding variables*. When there are confounding variables, we cannot say for sure which variable or variables caused the effect.

   Let's see how the scientific method works. An advertisement on television proclaims that eating a daily bowl of oatmeal lowers blood cholesterol levels. Lowering levels of blood cholesterol is desirable because elevated cholesterol is related to atherosclerosis, a condition in which fatty deposits clog blood vessels. In turn, atherosclerosis increases one's risk of having a heart attack or stroke.

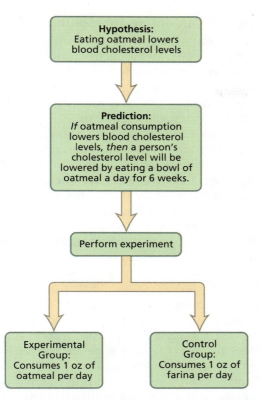

FIGURE 1.5 *The design of an experiment to test the prediction that oatmeal lowers blood cholesterol levels*

What observation(s) led to this claim? Oatmeal contains the soluble fiber ß-glucan. We begin to gather information about soluble fiber and learn the following:

- Soluble fiber binds to bile in the intestines, preventing bile from being reabsorbed into the body.
- Bile is high in cholesterol.
- Bile bound to soluble fiber is removed from the body in feces.
- The liver then removes cholesterol from the blood to synthesize new bile.

What experiment would support the claim that oatmeal lowers blood cholesterol? Scientists would first formulate a hypothesis. In this case, the hypothesis might be that ß-glucan in oatmeal lowers blood cholesterol levels. They would then make a prediction that will hold true if the hypothesis is correct: *If* oatmeal consumption lowers blood cholesterol levels, *then* a person will lower his or her cholesterol level by eating a bowl of oatmeal a day for 6 weeks (Figure 1.5). The component of total cholesterol measured was the LDL (low-density lipoprotein) cholesterol-carrier, because this "bad" form of cholesterol promotes atherosclerosis.

To test this hypothesis, we could gather adults whose LDL cholesterol levels are similar—between 4.5 mmol/L and 5.5 mmol/L—and divide them randomly into two groups. The experimental group of volunteers consumes a 1 oz packet of oatmeal per day for 6 weeks. The control group eats a 1 oz packet of farina, a wheat cereal lacking ß-glucan. At the end of 6 weeks, the blood level of LDL cholesterol of each volunteer is measured again.

4. **Draw a conclusion based on the results of the experiment.** Next, you arrive at a **conclusion**, which is an interpretation of the data. The results of a scientific

stop and think

Why is randomly dividing the volunteers into groups a better experimental design than allowing the volunteers to choose their group?

experiment are often summarized in a graph, such as the one shown in Figure 1.6, which presents the results of the experiment we just described. When you read a graph, first look at the axes. The horizontal line, or x-axis, shows the independent variable—the variable altered by the researcher. In this case, the independent variable is the amount of soluble fiber consumed. The vertical line is the y-axis; it presents the dependent variable, that is, the variable that was changed by the independent variable. In this experiment, the dependent variable was the blood level of LDL cholesterol. Always read the labels on the axes to see what the graph pertains to, and notice the scale so that you can appreciate the extent of variation. In this case, we use a bar graph to present the results, because each treatment is a discrete category. Notice in Figure 1.6 that blood levels of LDL cholesterol declined more in the experimental group than in the control group. Thus, we might conclude that eating oatmeal lowers blood LDL cholesterol.

Could these results be due to chance alone? Scientists base conclusions on the **statistical significance** of the data, which is a measure of the possibility that the results were due to chance. A probability of less than 5% (written as $p < 0.05$) that the results are due to chance is generally acceptable. The lower the number, the more confidence we have in the accuracy of the results. In this experiment, the differences in blood LDL cholesterol at the end of the treatment phase were statistically significant from the starting values.

5. **Make new predictions, and test them.** The experiment supports our hypothesis: ß-glucan in oatmeal lowers blood LDL cholesterol levels. We might, therefore, make two additional predictions: (1) If one consumes greater amounts

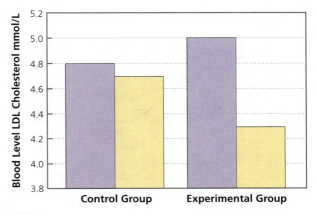

FIGURE 1.6 *Cholesterol level in blood decreases with increased consumption of oatmeal.*

of oatmeal, then the cholesterol-lowering effect should be greater; and (2) if one stops eating oatmeal, then blood LDL cholesterol levels should rise again.

We could test these predictions by gathering another, larger group of volunteers with blood LDL cholesterol levels between 4.5 mmol/L and 5.5 mmol/L and dividing them randomly into four groups. In the three experimental groups, volunteers consume one, two, or three 1 oz packets of oatmeal per day. Volunteers in the control group eat a 1 oz packet of farina. Levels of LDL cholesterol in the blood are measured 6 weeks after the volunteers begin eating their designated type and amount of cereal, and again 6 weeks after the volunteers stop eating cereal.

This time we use a line graph to indicate the changes in LDL cholesterol over time. In this experiment, the dependent variable, indicated on the vertical y-axis, is blood level of LDL cholesterol. The independent variable is the amount of soluble fiber consumed over time. Notice in Figure 1.7 that the cholesterol-lowering effect of oatmeal increases with the amount of oatmeal consumed and that effect slowly reverses once cereal consumption stops.

Another requirement of scientific inquiry is that experiments must be repeated and yield similar results. Other scientists following the same procedure should obtain a similar outcome. Note, however, that it can be very difficult to identify all the factors that might affect the outcome of an experiment.

The testing and refinement of a hypothesis represents one level of the scientific process. As time passes, related hypotheses that have been confirmed repeatedly can be fit together to form a **theory**—a well-supported and wide-ranging explanation of some aspect of the physical universe. Because of its breadth, a theory cannot be tested by a single experiment but instead emerges from many observations, hypotheses, and experiments. Nevertheless, a theory, like a hypothesis, leads to additional predictions and continued experimentation. Among the few explanations that have been tested thoroughly enough to be considered theories are the cell theory of life (which says all cells come from preexisting cells) and the theory of evolution by natural selection (which you learn about in Chapter 22).

Inductive and Deductive Reasoning

Scientific investigation usually involves two types of reasoning: inductive reasoning and deductive reasoning.

In **inductive reasoning**, facts are accumulated through observation until the sheer weight of the evidence allows some logical general statement to be made. You use inductive reasoning to develop a testable hypothesis.

Deductive reasoning begins with a general statement that leads logically to one or more deductions, or conclusions. The process can usually be described as an "if-then" series of associations. We used deductive reasoning when we predicted, "*If* oatmeal consumption lowers blood cholesterol levels, *then* a person will lower his or her cholesterol level by eating a bowl of oatmeal a day for 6 weeks." This prediction helped us decide whether the results of the experiment supported or refuted the hypothesis.

Clinical Trials

Before testing a new drug or treatment on humans, scientists must take steps to ensure that it will not do more harm than good (Table 1.1). Usually a drug is tested first on animals, such as laboratory rodents. Rats and mice are mammals, so some aspects of their physiology are similar to, and can be generalized to, human physiology. Using rodents to test drugs offers a number of advantages: they are relatively inexpensive to use, have short life spans, and reproduce quickly. Research on animals also helps determine how the body handles the drug, which helps determine dosage. Most medical advances, including vaccinations, chemotherapy, new surgical techniques, and organ transplants, began with animal studies. Strict rules safeguard the care and use of animals in research and testing.

If no ill effects are discovered in animals receiving the drug, then studies on humans, called *clinical trials,* may begin. In all phases of clinical testing, the studies are done on people who volunteer. In phase I of a clinical trial, the drug is screened for safety on fewer than 100 healthy people. At this stage, researchers hope to learn whether they can safely give the drug to humans, determine the effective dosage range, and identify side effects.

If the drug is found to be safe, it is tested further. In phase II of a clinical trial, a few hundred people with the target disease are given the drug to see whether it works for its intended purpose. If it does, the new drug will be compared with alternative treatments in phase III trials. Thousands of participants are

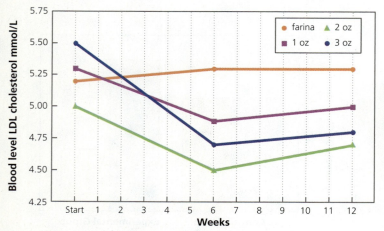

FIGURE 1.7 *Oatmeal's cholesterol-lowering effect increases with the amount of oatmeal consumed.*

TABLE 1.1 Tests Performed on a New Drug before It Is Approved by the Food and Drug Administration (FDA)	
Tests on laboratory animals	
Is the drug safe for use on animals?	
Clinical trials	
Phase I	Is the drug safe for humans?
Phase II	Does the drug work for its intended purpose?
Phase III	How does the new drug compare with other available treatments?

involved in phase III of a clinical trial. The U.S. Food and Drug Administration (FDA) approves only those drugs or treatments that have passed all three phases of human-subjects testing.

what would you do?

The job of the FDA is to ensure the safety and effectiveness of new drugs and treatments. It must balance the patients' desires for access to new treatments against the government's desire to protect patients from treatments that may be unsafe or ineffective. The drug approval process is painstakingly slow, usually taking more than 8 years. Do you think that the FDA should bypass certain steps of the approval process to make new drugs available to critically ill patients who may not be able to wait? If you do, what criteria should be used to decide the degree of illness that would warrant treatment with a drug that was not yet approved? Who should be held responsible if early access to a drug of unknown safety causes a patient to suffer serious side effects or premature death?

FIGURE 1.8 There are more than 4.6 billion people in the world who use cell phones.

Recall that a well-designed experiment has both an experimental group and a control group. Clinical studies are no different. In a drug trial, the experimental group receives the drug under consideration. The control group receives a **placebo**, an innocuous, nondrug substance made to look like the drug being tested. Study participants are randomly assigned to either the control group or the experimental group and do not know whether they are receiving the treatment or a placebo. When neither the researchers nor the study participants know which people are receiving treatment and which are receiving the placebo, the study is described as being *double-blind*. It is important that participants not know whether they are receiving the placebo or the drug because their expectations about the drug could affect the way they respond. Similarly, researchers should not know which people are in the experimental or control groups because their expectations or desire for a particular result could affect their interpretation of the data.

Finally, it is extremely important, and legally required, that study participants give their **informed consent** before the trial begins. An informed consent document lists all the possible harmful effects of the drug or treatment and must be signed before a person can take part in the study. To give informed consent, study participants must be mentally capable of understanding the treatment and risks, so they cannot be mentally impaired as a result of mental retardation, mental or other illness, or substance abuse.

Epidemiological Studies

Human health can also be studied without clinical trials. In an epidemiological study, researchers look at patterns that occur within large populations. For example, an epidemiological study to investigate the effects of air pollution on asthma (a condition in which airway constriction causes breathing problems) would look for a correlation of some kind between the variable of interest (air pollution) and its suspected effects (worsening of asthma). If the researchers' hypothesis is that air pollution aggravates asthma, they might predict and then look for evidence that the number of people admitted to hospitals for asthma-related problems increases with increased levels of air pollution.

Recent epidemiological studies have asked the question, "Does using a cell phone increase your risk of developing brain cancer?" Cell phones emit radiofrequency waves and are usually held to one's ear (Figure 1.8). Based on a 2011 review of epidemiologic studies of cell phone use and brain cancer, the World Health Organization (WHO) stated that cell phone use is "possibly carcinogenic to humans." The WHO statement is a cautious recommendation based largely on a study in which researchers tracked nearly 13,000 cell phone users from 13 countries over 10 years. A comparison of brain cancer rates of *all* people who used cell phones and *all* people who never used a cell phone did not show a difference in brain tumor rates. However, when the heaviest cell phone users were compared to all others (cell phone users and non-users), researchers found a slight increase of brain cancer among heavy cell phone users. Thus, this study does not conclusively demonstrate a link between cell phone use and cancer, but it doesn't rule out the possibility.

Soon after the WHO report, a new epidemiological study was released—the largest to date. In the new study, researchers followed more than 350,000 cell phone users for over 10 years. They did not find a dose-response relationship between cell phone use and the development of cancer. In other words, the incidence of brain tumors did not increase with the length of cell phone use and was not greater in the regions of the brain most heavily exposed to cell phone radiation. Nearly all studies to date have failed to show a link between cell phone use and brain cancer. Although these results are reassuring, additional research must be done. Cell phone use is a relatively new practice, and many cancers take years to develop. You can find current information on cell phones and cancer on the website of the American Cancer Society.

1.5 Critical Thinking to Evaluate Scientific Claims

Few of us perform controlled experiments in our everyday lives, but all of us must evaluate the likely validity of scientific claims. We encounter them in many forms—as advertisements, news stories, and anecdotes told by friends. We often must make decisions based on these claims, but how can we decide whether they are valid? Critical-thinking skills can help us analyze the information and make prudent decisions.

The key to becoming a critical thinker is to ask questions. The following list is not exhaustive, but it may help guide your thinking process.

1. **Is the information consistent with information from other sources?** The best way to answer this question is to gather as much information as possible from a variety of sources. Do not passively accept a report as true. Do some research.

2. **How reliable is the source of the information?** Investigate the source of the information to determine whether that person or group has the necessary scientific expertise. (Table 1.2). Is there any reason to think the claim may be biased? Who stands to gain if you accept it as true? For example, the FDA is probably a more reliable source of information on the effectiveness of a drug than is the drug company marketing the drug. If a claim is controversial, listen to both sides of the debate, and be aware of who is arguing on each side.

3. **Was the information obtained through proper scientific procedures?** Information gathered through controlled experiments is more reliable than anecdotal evidence, which cannot be verified. For example, your friend might tell you that his muscles have gotten larger since he started using some special exercise equipment. But you cannot be sure unless measurements were taken before and after he began to use it. Even if your friend can prove his muscles are bigger with such measurements, there is no guarantee that exercising with this equipment will build *your* muscles.

4. **Were experimental results interpreted correctly?** Consider, for example, a headline advertising capsules containing fish oils: "Fish Oils Increase Longevity." It may be tempting to conclude from this headline that you will live longer if you take fish oil supplements, but in fact the headline is referring to an experiment in which *dietary* fish oil increased longevity in *rats*. Rats fed a diet high in fish oils lived longer than did rats on a diet low in fish oil. The claim that taking fish oil supplements will increase longevity is not a valid conclusion based on the experiment. First, the study was done on rats, not on humans. Not all aspects of rat physiology generalize to humans. For example, rats are more resistant to heart disease than are humans. Second, the amount of dietary fish oil, not the amount of fish oil from capsules, was the variable in the study the headline refers to. Supplements of fish oil may not have the same effect as dietary fish oils. It could be that taking fish oil supplements would boost the amount of fish oil in your body to unhealthy levels.

5. **Are there other possible explanations for the results?** Suppose you learn that the fish oil headline is based on a study showing that people who eat fish at least three times a week live longer than people who eat fish less frequently. In this case, the data indicate that there is a correlation between fish in the diet and length of life. However, a correlation between two factors does not prove that one *caused* the other. Instead, the two factors may *both* be caused by a third factor. In this case, the difference in longevity may be due to other differences in the lifestyles of the two groups. For instance, people who eat fish may exercise more frequently or have less stressful jobs or live in areas with less pollution.

As your critical thinking skills develop, so should your information literacy skills. **Information literacy** involves the ability to recognize what you need to know, locate relevant information, evaluate it, apply it to the problem at hand, and communicate it effectively. These skills are important in our personal lives, as well as in education, the workforce, and society.

We are presented with a wealth (or some might say a glut) of information, much of which comes to us through technology. When in need of an answer, many of us find "an app for that." Thus, **information technology literacy** is also essential.

Throughout your life you will be asked to make many decisions about scientific issues. Some will affect your community and even beyond. For example, should we eliminate genetically modified food? Should stem cell research be permitted? Should companies polluting the atmosphere be taxed at a higher rate? We will raise these and similar questions throughout this textbook. You will encounter questions like these in your studies, and you will find others every day in the local and national news media.

Although you may never be one of the lawmakers deciding these issues, you are a voter who can help choose the lawmakers and voice your opinions to the lawmakers who will decide. Scientists can provide facts that may be useful as we all struggle to answer the complex questions facing society, but they cannot provide simple answers. As scientific knowledge grows and our choices become increasingly complex, each of us must stay informed and review the issues critically.

TABLE 1.2　How to Identify a Reliable Website

Who is the author?	Does the website provide information about the author? If not, try "Googling" the author's name to learn more about his or her affiliations.
Who is the publisher of the site?	Notice the suffix on the domain name (.edu = educational; .gov = government, .org = nonprofit organization; .com = commercial). A commercial site is likely to be more biased.
What is the purpose of the website?	Is it a report on a study by a reliable organization? Is its purpose to promote scholarship? Is its purpose to market a product? Is its purpose to present personal opinion?
When was the information posted?	Are there older or more recent websites that support or contradict the information presented in this website?

looking ahead

In this chapter, we considered the characteristics of life and the nature of scientific thinking. In the next chapter, we will explore ways to use critical thinking skills to make wise decisions as health care consumers.

HIGHLIGHTING THE CONCEPTS

1.1 Basic Characteristics of All Living Things (pp. 1–4)
- Life cannot be defined, only characterized.
- Living things contain nucleic acids, proteins, carbohydrates, and lipids; are made of cells; grow and reproduce; metabolize; detect and respond to stimuli; and maintain homeostasis. Populations of living things evolve over long periods of time.

1.2 Evolution: A Unifying Theme in Biology (pp. 4–5)
- Classifications of living organisms reflect the evolutionary relationships among them. One currently popular classification system recognizes three domains: Bacteria, Archaea, and Eukarya. Domain Eukarya consists of four kingdoms: protists, fungi, plants, and animals.
- Humans are classified as animals, vertebrates, mammals, and primates.

1.3 Levels of Biological Organization (p. 5)
- Human biology can be studied at different levels. Within an individual, the levels of increasing complexity are molecules, cells, tissues, organs, and organ systems.
- Beyond the level of the individual are populations, communities, ecosystems, and the biosphere.

1.4 Scientific Method (pp. 5–9)
- The scientific method consists of making observations, formulating a question and a good hypothesis (a testable, possible explanation), conducting experiments (performed with controls), and drawing a conclusion, which may lead to further experimentation.
- As evidence mounts in support of related hypotheses, the hypotheses may be organized into a theory, which is a well-supported explanation of nature.
- Inductive reasoning uses a large number of specific observations to arrive at a general conclusion. Deductive reasoning, in contrast, uses "if-then" logic to progress from the general to the specific.
- There are strict rules concerning experiments on humans and other animals. Drugs are usually tested on animals before they are tested on humans. If no ill effects in animals are observed, the drug is then tested on humans. Phase I trials determine whether the drug is safe for humans, phase II determines whether it works for its intended purpose, and phase III determines whether it is more effective than existing treatments.
- The design of a human experiment often includes an experimental group that receives the treatment and a control group that receives a placebo. In what is known as a double-blind experiment, neither the study participants nor the researchers know who is receiving the real treatment.
- Study participants must sign an informed consent document indicating that they were made aware of the possible harmful consequences of the treatment.
- Epidemiological studies examine patterns within populations to find a correlation between a variable and its suspected effects.

1.5 Critical Thinking to Evaluate Scientific Claims (pp. 9–10)
- Critical thinking consists of asking questions, gathering information, and evaluating evidence and its source carefully before drawing conclusions.
- Information literacy involves the ability to identify what you need to know, locate relevant information, evaluate it, apply it to the problem at hand, and communicate it effectively.
- Information technology literacy involves knowing how to effectively use technology to answer questions and solve problems.

RECOGNIZING KEY TERMS

cell *p. 2*
metabolism *p. 2*
homeostasis *p. 2*
adaptive trait *p. 4*
population *p. 5*
community *p. 5*
ecosystem *p. 5*

biosphere *p. 5*
science *p. 5*
scientific method *p. 6*
hypothesis *p. 6*
controlled experiment *p. 6*
control group *p. 6*
experimental group *p. 6*

variable *p. 6*
conclusion *p. 7*
statistical significance *p. 7*
theory *p. 8*
inductive reasoning *p. 8*
deductive reasoning *p. 8*
placebo *p. 9*

informed consent *p. 9*
information literacy *p. 10*
information technology literacy *p. 10*

REVIEWING THE CONCEPTS

1. List seven traits that characterize life. *pp. 2–4*
2. Name two characteristics that classify humans as mammals. Name two that classify them as primates. *pp. 4–5*
3. Define *controlled experiment*. *p. 6*
4. Differentiate inductive reasoning from deductive reasoning. *p. 8*
5. What is a placebo? *p. 9*
6. What is meant by a *double-blind experiment*? *p. 9*
7. What is the highest level in the classification of life?
    a. genus
    b. kingdom
    c. family
    d. domain
8. Forward-looking eyes are characteristic of
    a. eukaryotes.
    b. mammals.
    c. primates.
    d. animals.
9. Which of the following is an example of a population?
    a. all individuals of the same species in an area
    b. all the species that can interact in an ecosystem
    c. a community and its interaction with the physical environment
    d. the part of Earth that supports life

10. Which of the following is an example of an epidemiological experiment?
    a. To test the effectiveness of a drug, researchers administer it to an experimental group of subjects but not to a control group of subjects.
    b. Researchers look at the pattern of cancer development across populations in New England.
    c. Researchers conduct a clinical trial to determine whether a treatment for multiple sclerosis is effective.
    d. Researchers conduct a double-blind study to determine whether fish oil supplements reduce the frequency of heart attacks.
11. A theory is
    a. a testable explanation for an observation.
    b. a conclusion based on the results of an experiment.
    c. a wide-ranging explanation for natural events that has been extensively tested over time.
    d. the factor that is altered in a controlled experiment.
12. Which of the following is *not* an example of a good scientific question?
    a. Does obesity reduce one's life expectancy?
    b. Does dietary fat increase a woman's chance of developing breast cancer?
    c. Does a full moon cause people to behave irrationally?
    d. Does temperature alter the rate of an enzymatic reaction?
13. A trait that increases the chance that an organism will survive and reproduce in its natural environment is described as being
    a. adaptive.
    b. inductive.
    c. hypothetical.
    d. deductive.
14. A hypothesis is
    a. a conclusion that has been tested repeatedly.
    b. a significant result from an experiment.
    c. a testable explanation for an observation.
    d. a relationship between two factors.

APPLYING THE CONCEPTS

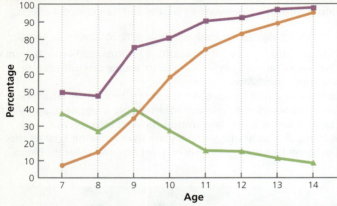

- Do you have access to a mobile phone?
- Do you use a mobile phone that you share in the family?
- Do you have a mobile phone of your own?

The percentage of Swedish children questioned who answered yes to three questions about cell phone use.
"Ownership and Use of Wireless Telephones: A Population-Based Study of Swedish Children Aged 7–14 Years" by Fredrik Soderqvist et al. in BMC PUBLIC HEALTH 7 (2007): 105–13, fig. 1, p. 107.

1. Interpret the accompanying graph to answer the questions.
   a. What percentage of these 7-year-old children have access to a cell phone?
   b. At a young age, these children share a cell phone with their family. At what age does sharing begin to decline?
   c. What percentage of 10-year-old children own their own cell phone?
2. A native you met in the rain forest told you that one of the plants you collected brings relief to people who are having difficulty breathing. You suspect that it might be a good treatment for asthma, a condition in which constriction of airways causes breathing problems. You are able to isolate a component of this plant as a drug. Tests on animals show that it is effective. Phase I and phase II of clinical trials on humans show that the drug can be given safely to humans. Design an experiment to test the hypothesis that this drug eases breathing during an asthma attack.
3. Find an article or advertisement that makes a scientific claim. Use your critical-thinking skills to evaluate the claim.

BECOMING INFORMATION LITERATE

Humans have an impact on the world at every level of the organization of life: molecule, cell, tissue, organ system, individual, population, community, ecosystem, and biosphere. Use reliable sources of information (books, newspapers, magazines, journals, or reliable websites) to identify at least one current issue or concern to humans at each level of life's organization. Provide a citation for each source you use.

Begin by planning a strategy for your search. Keep a log of each source you use, and evaluate it for reliability and helpfulness.

MasteringBiology®

Go to MasteringBiology for practice quizzes, activities, eText, videos, current events, and more.

SPECIAL TOPIC
Becoming a Patient: A Major Decision

1a

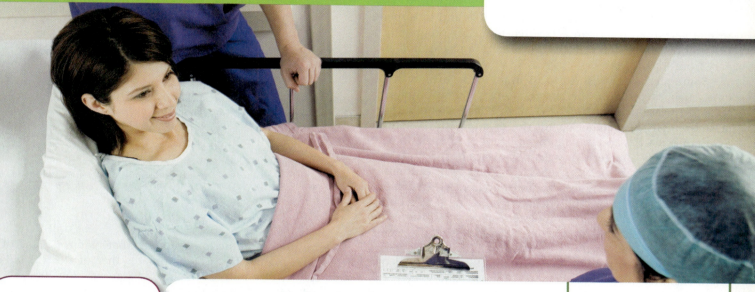

Using critical-thinking skills and actively participating in health care decisions can help improve your health and the quality of care you receive.

- 1a.1 Selecting a Primary Care Physician
- 1a.2 Getting the Most Out of an Appointment with Your Doctor
- 1a.3 Finding a Specialist and Getting a Second Opinion
- 1a.4 Appointing a Health Care Agent
- 1a.5 Selecting a Hospital and Staying Safe
- 1a.6 Researching Health Conditions on Your Own

The Institute of Medicine estimates that nearly half of all adults in the United States have inadequate health literacy, which means they have difficulty accessing, understanding, and using health information to make appropriate health decisions. In Chapter 1 we considered several aspects of critical thinking, including the importance of asking questions, gathering information, and evaluating sources of information to make an informed decision. In this chapter, we show how you can apply your critical-thinking skills when you become a patient. We emphasize the importance of asking questions about your personal health and health care, and we describe how and where to find reliable health and medical information. Our goal is to help you be an active and informed participant in decisions regarding your health and health care.

1a.1 Selecting a Primary Care Physician

Do you have a primary care provider? If your answer is "no" or "I don't know," then here are some steps that you can follow to become an active participant in your own health care. We hear a great deal about health care in the media every day, especially about the need for taking personal responsibility for our health. Finding a primary care provider you can work with to maintain and improve your health is a critical step in this process. Here we focus on primary care physicians. However, some patients have a physician assistant or a nurse practitioner, rather than a doctor, as their primary care provider. These providers consult with physicians. Therefore, the general steps for selecting a physician assistant or a nurse practitioner are similar to those we describe for doctors.

There are two types of physicians: doctors of medicine (MDs) and doctors of osteopathic medicine (DOs). The education and training requirements for an MD degree and a DO degree are nearly identical, so we will describe them generally and note any differences. Typically, a physician's education begins with four years of premedical education at a college or university. This is followed by four years of medical school to earn either an MD or a DO degree. The major difference in training occurs during medical school, when DOs receive training in manual therapy for the treatment of musculoskeletal pain and disability. After medical school, MDs and DOs spend several years training in an accredited residency program at a hospital or clinic. States license physicians, and requirements vary from state to state.

Some physicians wish to focus in a specific area of medicine, such as pediatrics or anesthesiology. These physicians seek certification from independent specialty boards. There are more than 20 member boards (for example, the American Board of Pediatrics and the American Board of Anesthesiology) that are part of a larger oversight organization, the American Board of Medical Specialties (ABMS). To attain certification in a specialty or subspecialty, a doctor must be licensed to practice medicine, complete specific education and training requirements, and pass an examination designed and administered by the specialty board. Doctors who meet all of these requirements are board certified. Doctors who have obtained the required education and training but have not completed the specialty board examination are board eligible. To maintain board certification, a doctor must stay up-to-date with advances in his or her specialty and demonstrate ongoing expertise in areas that include medical knowledge, patient care, and communication. The Maintenance of Certification program, also overseen by ABMS, is an assessment program that promotes continuous learning by board-certified specialists.

A primary care physician is a doctor with training (and possibly board certification) in family medicine, general internal medicine, or general pediatrics. A primary care physician treats common medical conditions and advises patients on broad aspects of health care, including preventive care. He or she provides care in non-emergency situations, usually in an office or clinic. Patients requiring urgent medical attention typically visit an urgent care facility or an emergency room, commonly known as the ER. Visits to the ER are appropriate for severe illnesses or injuries but not for routine examinations.

A primary care physician is often your gateway to more specialized services, should you need referral to a medical specialist. If you are admitted to the hospital, your primary care physician may oversee or participate in your care. Clearly, your doctor will be an important person in your life, so it is important to think carefully about what you want in a primary care physician. One measure of the importance of primary care physicians comes from the Centers for Disease Control and Prevention (CDC), which periodically publishes statistics on ambulatory care medical visits. Ambulatory care visits include personal health care consultations, procedures, and treatments delivered on an outpatient basis (no overnight stay) in hospitals, clinics, urgent care centers, or physician offices (for comparison, the term *inpatient* refers to a patient who is admitted to a hospital or clinic for treatment that requires a stay of at least one night). The CDC estimates that of the 1.2 billion ambulatory care visits in the United States during 2007, about one-half (48.1%) were made to primary care physicians in office-based practices (Figure 1a.1).

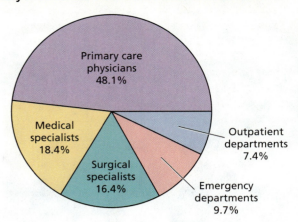

Ambulatory Medical Care Utilization Estimates for 2007

FIGURE 1a.1 *Distribution of ambulatory care visits by setting in the United States. Visits to primary care physicians, medical specialists, and surgical specialists occurred in office-based practices. Visits to emergency departments and outpatient departments occurred in nonfederal general and short-stay hospitals. These 2007 data were published in 2011 and represent the most recent information available.*

Data from S. M. Schappert and E. A. Rechtsteiner, "Ambulatory medical care utilization estimates for 2007," National Center for Health Statistics, *Vital Health Statistics* 13 (2011), Number 169.

There are approximately 353,000 primary care physicians in the United States, ranging from individuals in their own private practice to teams of primary care physicians working in large health centers and health maintenance organizations. Large health centers offer the convenience of "one-stop shopping" because they provide most basic medical services, including blood drawing, x-ray exams, and pharmacy, with the separate departments housed under one roof. The downside of such an arrangement is that you may not always see the same primary care physician. At the other extreme, if your primary care physician is in private practice, then you may have to visit several physical locations, cope with different billing arrangements, and take greater personal responsibility for following through with recommended tests. The big advantage is that you will always see the same physician, and this may lead to a more rewarding and long-term relationship with your doctor that promotes continuity of care. For example, it may be convenient for you and other members of your family to have one primary care physician. Although this used to be the norm in the United States, this pattern is changing for many reasons, not least of which is the need for age-specific specializations, mobility of individual family members, and insurance requirements that may limit access to a physician because of insurance network restrictions. One thing is certain: at some point, you will need to select a primary care physician.

The keys to finding a primary care physician in this wealth of choices include obtaining a list of providers from your insurance organization, knowing the absolute minimum requirements, and then considering features related to style of care. Asking for a list of providers early in the process helps to ensure that your insurance company will not contest future medical bills. At a minimum, your primary care physician should have excellent record keeping, a patient-oriented approach, and strength of character, for doctors sometimes have to deliver bad news. Having an accessible location is also important. Most physicians do not communicate with their patients via e-mail or other electronic media. Nor do they generally answer personal health questions over the telephone. You will need to visit the office, so it should be conveniently located. Finally, make sure that you can get an appointment with your doctor in a reasonable amount of time after you request one.

Next, you should consider style of care, which varies among doctors. For example, one physician may emphasize friendly, informal conversation as a way to engage you in your care, inviting you to be involved in decision making. Another may sit formally with you in the examination room or office, in a more authoritarian style. Excellent medical outcomes may result from either of these strategies, and of course there is a continuum between them. The important point is to be comfortable with your physician and his or her style. If you are to engage their training and knowledge in helping you to be well, you need to be completely honest with your doctor about all aspects of your health and lifestyle. Above all, a primary care physician must be someone you trust to put you and your health concerns first.

With all the choices out there, where do you start? Here are some ideas. If you are in good health and need only general medical care, then perhaps the easiest and best strategy is to ask another physician whom you already know and respect for recommendations. For example, if you are leaving a pediatric practice, your current doctor may be an excellent reference for a primary care physician. It is also important to ask around. Find out which doctors your friends see and what they have to say about them. Go to the website of a prospective practice to learn about office hours, after-hours care, and what happens when the doctor is unavailable. You also can access public information on physicians. For example, the database of the Federation of State Medical Boards (www.fsmb.org) provides information on the professional qualifications of doctors (e.g., education, board certifications, and states in which they are licensed to practice) and any disciplinary actions by state medical boards against them. Visit the office, talk with the staff, and assess their courtesy, professionalism, and organization. Ask whether the practice has fully functional systems for maintaining electronic medical records (data pertaining to the services provided a patient at a particular care delivery organization, which owns this legal record) and electronic health records (a subset of data from various care delivery organizations that have treated a particular patient; the patient owns this record and can access it and append information).

Once you have narrowed your search for a primary care physician, how do you get in to visit a prospective doctor? The first step is to call his or her office and ask whether they are accepting new patients. If the answer is yes, then ask whether it would be possible to come in to meet the doctor for an initial consultation. In some cases, you may have to pay for this consultation. Another strategy is to wait until you have a minor health problem, call the doctor's office, and ask whether you can see the doctor about it. Either way, you need to proceed fairly deliberately in gaining access to the doctor of your choice. If you are not satisfied with a physician, then continue your search until you find one to suit your needs.

1a.2 Getting the Most Out of an Appointment with Your Doctor

A visit to your doctor is going to be a short yet important interaction. Know ahead of time the length of your appointment, especially the duration of face-to-face time with your doctor. Keep in mind that most appointments last less than 20 minutes (this will vary depending on whether it is an initial visit or a follow-up) and might occur only once a year, as in an annual physical examination. By preparing for your visit in advance, you can make the most of the opportunity to learn from your physician and stay healthy.

Before your visit, take time to prepare a list of medical questions and concerns; simple checklists are available to help you prepare for your doctor's appointment (Figure 1a.2). If some aspect of your physical or mental health is bothering

FIGURE 1a.2 Completing a checklist, such as the one shown here, before your doctor's appointment can help you to get the most out of your appointment.

From www.partnersforhealthtn.gov.

you, then you should include it on the list. Prioritize the list to make sure that your most serious two or three concerns are addressed. Along with your list of concerns, bring a list (or the actual containers) of any medications (prescribed and over-the-counter) that you are currently taking and the doses; include vitamins and supplements because they can interact with over-the counter and prescription medications. Think about your personal medical history, including current medical conditions as well as past illnesses, medical procedures and surgeries, and your family's medical history. Make notes on any health challenges that either you or your immediate family have experienced. Be sure to bring relevant documents, such as test results, records from other physicians, and insurance information. Be conservative when bringing documents from your personal research into your health concerns. This is not meant to discourage you from researching health issues on your own, but to warn you against bringing stacks of printouts to an appointment. Doctors need the freedom to perform their own independent assessment without being sidetracked by other information. Finally, you may want to ask a family member or friend to go with you to your appointment (Figure 1a.3). Although you may not want your companion in the examining room, having someone with you can be very helpful during the consultation phase of the appointment. Companions can provide support and take notes, both of which can result in more information being exchanged during the appointment and remembered afterwards. Consistency is important, so bring the same family member or friend to your various appointments. Ideally, your companion should be the health care agent you have chosen to act on your behalf should you be unable to speak for yourself (see section 1a.4).

Typical preliminaries before a medical examination include recording weight, height, and blood pressure. These preliminaries take only a few minutes and are usually performed by staff. If the staff member doesn't tell you, be sure to ask for these results as they are recorded so that you will be prepared to discuss them with the doctor.

FIGURE 1a.3 *Many people find it helpful to have a friend or family member present when consulting with a physician.*

Although every appointment is unique, most involve both consultation and examination. A typical pattern is to have a few minutes of discussion with your physician during which you explain the reason for your visit, followed by any examination that may be necessary, followed by a review of findings and the treatment plan, if needed. During discussions with your doctor, it is important to speak about your medical concerns as clearly, concisely, and directly as possible. For example, be specific when describing your symptoms and their time course: what symptoms are you experiencing, and when did they start? Have the symptoms improved or worsened? Aim for as open a discussion as possible. Many medical conditions can be difficult to discuss, but for maximum benefit you need to overcome shyness. Remember that your doctor is a professional who has probably heard similar stories. If you do not feel comfortable talking with your doctor, then you might have the wrong doctor. The point of visiting the doctor is to learn about your health, but it is also important that you get your points across and questions answered.

During the consultation phase of your appointment, your primary care physician may recommend a particular laboratory test or screening procedure. Ask your doctor how the test or procedure will be performed, what information will be gained, what are the risks, and how accurate are the results of the test. Other important things to ask the doctor (or the doctor's staff, if time runs short) include how and when will you receive the results and what to do if you don't receive them. It is not a good strategy to assume that "no news is good news," so take personal responsibility and follow up if test results do not arrive within a reasonable amount of time.

During your appointment, the doctor may write a prescription for a new medication. Be sure to tell your doctor about any reactions you have had to medications in the past. Also ask questions so that you can learn as much as possible about your new prescription. For example, ask the name of the new medication, and ask the doctor to write it down; be sure that you can read the medication name written on the prescription that you will bring to a pharmacist. Ask what the new medication is supposed to do and whether it might interact with any supplements or other medications (prescription or over-the-counter) that you are taking. Inquire about the mechanics of taking the medication, including how much to take, when to take it, and how long to take it. Also ask about side effects and what to do if they occur. Some medications require that you avoid certain foods, drinks, or activities, and you should ask about this, too, as well as what to do if you miss a dose or accidentally take too large a dose. Ask what will happen if you don't take the medication. To save money, ask whether a generic form of the medication is available.

When you pick up your medication from the pharmacy, check to make sure it is the medication prescribed by your doctor. If you have remaining questions about the new medication, then ask the pharmacist. Finally, take personal responsibility for reading and understanding drug labels, whether the medication is prescribed by your doctor or sold over the counter (Figure 1a.4). Read labels at least every time you open a new bottle of medication.

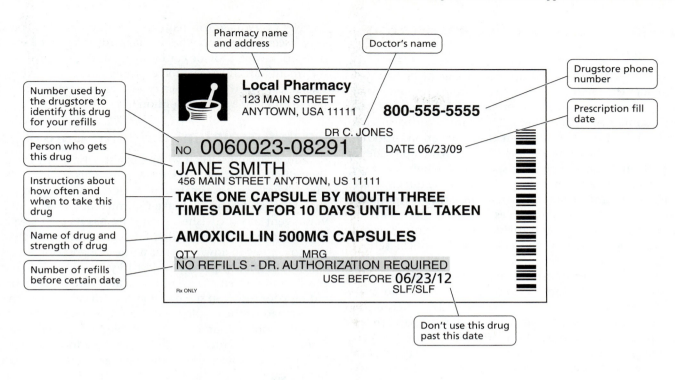

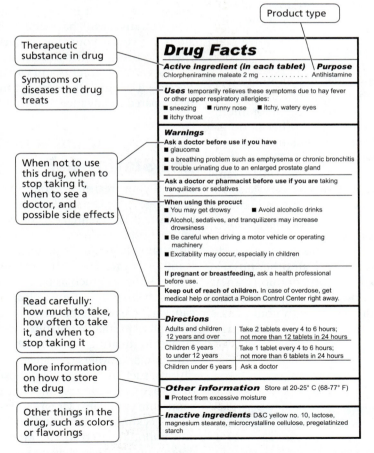

FIGURE 1a.4 *Tips for reading drug labels. (a) Prescription drug label. (b) Over-the-counter drug label.*

From the United States Department of Health and Human Services, "How to read drug labels."

1a.3 Finding a Specialist and Getting a Second Opinion

Should you need a specialist, your primary care physician typically provides the referral. There also are online resources that you can personally search for information. For example, the database maintained by the ABMS lists board-certified specialists and includes information about their education. Another option is to check the for-profit WebMD site, which provides a directory of physicians, including specialists, when you type in your city and state or zip code. Included in the directory are physician names and contact information, areas of specialty, hospital affiliations, and the names of other physicians in their practice. Some listings provide personal information, such as age and gender of the doctor and medical school attended.

There are some key questions to ask when meeting with a specialist. If you have not personally researched the physician's background, then you may wish to ask whether he or she is board certified. If you need the doctor to perform a certain procedure, then ask how often he or she has performed the procedure in the past year—you want a doctor who performs the procedure frequently. You might also ask how many of these procedures he or she has performed overall and what the success rate has been. Be sure to ask about other treatment options and the benefits and risks of the proposed treatment.

Once a physician provides you with a diagnosis and treatment plan, you may wish to get a second opinion before starting treatment. This opinion, provided by another specialist, may confirm the first doctor's diagnosis and treatment plan or suggest modifications. Seeking a second opinion provides reassurance that you have researched other options and offers additional opportunity for any remaining questions to be answered and concerns addressed. The doctor who provided your initial diagnosis and treatment plan can suggest a specialist for a second opinion, or you can personally search for one. To get a truly independent assessment, it is best to obtain a second opinion from a physician who is not connected with your doctor; for example, find a doctor in a different practice. Additionally, the second doctor should examine you and then look at records from your other physician. Most health insurance plans cover second opinions, and many plans require second opinions before they will pay for a major medical procedure. To be sure, check with your plan. Even if you have to pay for it, getting a second opinion may still be very worthwhile.

Not every diagnosis requires a second opinion. You should seek a second opinion when a diagnosis is based on a test that does not always provide conclusive results and when suggested treatments are invasive (for example, surgery) or long term. The severity of the disease may influence whether there is sufficient time for a second opinion.

1a.4 Appointing a Health Care Agent

All adults should select a willing health care agent with whom they discuss their health care wishes. This requires that you understand various treatment options so you can evaluate their costs and benefits, make informed decisions, and inform your agent of your wishes. Each state has its own laws regarding advance directives, which are legal documents that allow you to convey in advance your wishes for end-of-life care. Typically, though, any person at least 18 years old, other than your physician or an employee of the health care organization providing your care, can serve as your health care agent; an exception is made if the employee is a family member. You formalize the selection of your agent by completing a health care proxy form, copies of which should be provided to your agent, family members, loved ones, and primary care physician. The health care proxy form is a legal document, and health care providers are required to follow its instructions. You can update the form if necessary. Your agent, acting as your legal representative, will have the right to make health care decisions for you if your doctor decides you are unable to make them yourself. Such decisions include whether to have medical tests and treatments and when to start and stop treatments. Because it is difficult to predict medical situations, it is best to be prepared in advance.

1a.5 Selecting a Hospital and Staying Safe

Should your treatment require hospitalization, a major step in your care is selecting the hospital. Choosing a hospital is a decision that you make with your doctor, so start by asking which hospital your doctor thinks is best for you and at which hospital(s) he or she is permitted to admit patients. Because it is important for you to be an active participant in the decision and to know as much as possible about the hospital where you will be treated, take time to learn about hospitals because they can vary greatly in patient populations and levels of experience with particular medical conditions. Some hospitals specialize in treating patients of a certain age group, such as children, and others specialize in treating certain medical conditions, such as cancer. Many hospitals, however, treat patients of all ages with diverse medical conditions. Here we provide some things to consider when researching hospitals.

Checking the accreditation status of a hospital is an early and important step in the process of selecting a hospital. To determine whether they meet national standards of quality, most hospitals invite the Joint Commission to evaluate their staff, facilities and equipment, information management, medication management, infection control, and success in treating patients. Reviews are conducted at least once every three years

what would you do?

Your primary care physician refers you to a specialist, who wants to repeat an array of expensive medical tests just completed under the direction of your primary care physician. Your insurance company is willing to pay for this second round of tests. What would you do? Had your insurance company refused in advance to pay for these tests, what would you do?

stop and think

When some people first hear that they have a serious medical condition requiring inpatient treatment, they want the best hospital, no matter how far it is from home. What challenges might distant inpatient care pose for a person?

to track improvements or declines in quality. Reports prepared by the Joint Commission include accreditation status (a six-level rating system, ranging from "Not accredited" to "Accredited with commendation"), areas evaluated and those needing improvement, and comparison to national results. By going to www.qualitycheck.org, you can check the accreditation status of hospitals. Federal and state governments, nonprofit organizations, and private companies and organizations also offer report cards and other tools you can use to compare hospitals. For example, the United States Department of Health and Human Services offers Hospital Compare, an online resource that locates hospitals in the vicinity of the city, state, or zip code that you type in and then compares up to three hospitals of your choice on measures such as outcomes of care and surveys of patient experiences. You can refine your comparison by specifying particular medical conditions and surgical procedures.

You should also base your choice of hospital on the hospital's extent of experience and level of success in treating your medical condition. Because hospitals that perform many of the same types of procedures typically have higher success rates, it is important to ask your doctor or the hospital about the number of patients who have the procedure there. Also ask for information on patient outcomes and satisfaction surveys. You should select a hospital where the procedure you need is performed frequently on large numbers of patients who do well after the procedure and rate highly their experience at the hospital. Finally, check to make sure that the hospital is covered by your health plan.

A 1999 report from the Institute of Medicine (IOM) titled "To Err is Human: Building a Safer Health System" raised public awareness of medical errors. The IOM estimated that between 44,000 and 98,000 people die each year in the United States as a result of medical errors in hospitals. Even at the lower end, the number of deaths per year was higher than those from either car accidents (about 43,500) or breast cancer (about 42,300). A 2011 report issued by HealthGrades, a company that focuses on health care quality and ratings, found that medical errors in hospitals are still common. Such errors include incorrect or incomplete diagnosis or treatment and may involve lab tests, equipment, surgery, or medication (for example, a patient is given the wrong medication or given the correct medication but the wrong dose). Medical errors can happen anyplace that you receive health care or medications. We focus on those occurring in hospitals, but errors also can occur at doctors' offices and pharmacies and in your own home.

The best way to prevent errors is to be vigilant and informed about your health care. For example, when someone brings you a medication in the hospital, be sure to check that the medication is indeed meant for you and not someone else and that it is the correct medication and the correct dose (ask the nurse and personally check the label). Also make sure that you are receiving the medication by the correct route (for example, orally versus intravenously) and at the right time interval (for example, once every 12 hours). Almost 20% of medication doses in hospitals are given incorrectly; dispensing medications at the wrong time and omitting a dose are the most common mistakes.

Information about your diagnosis and treatment should be communicated clearly throughout your hospital stay among you, your hospital health care team, and your primary care physician. It is especially critical as your time to leave the hospital nears. Be sure that you understand all aspects of your discharge plan. Ideally, this written plan should include a list of follow-up appointments and tests, a medication plan, a timetable for resuming normal activities, and steps to take if a problem arises. It is important to ask questions. Hospitals can be confusing and stressful places, so having a friend or family member with you at the time your discharge plan is discussed can be very helpful. Once discharged, take personal responsibility, and comply with medication and other discharge instructions.

1a.6 Researching Health Conditions on Your Own

In addition to learning about your health from your primary care physician and health care team (should you have a medical condition and be undergoing treatment), find other reliable sources of information. The Internet can provide information around-the-clock, but the information is not always accurate or unbiased. When using the Internet, be sure to scrutinize websites and avoid those that are selling products or making claims inconsistent with information from other sources. Check that the person or group making the claims has the necessary scientific or medical training and expertise. When investigating specific medical conditions, search for articles in the primary medical literature. Although reading articles in medical journals can appear daunting at first, give it a try because you may be able to understand enough of the information to ask your doctor about it.

looking ahead

In this chapter, we discussed how to apply critical-thinking skills to keeping yourself healthy and getting involved in your health care. In Chapter 2, we focus on how basic chemical principles help us to understand how the human body is put together and how it functions. Like the material in this chapter, the information provided in Chapter 2 lays a foundation for discussions throughout the text.

2 Chemistry Comes to Life

Did you Know?

- Almost 99% of the mass of the human body is made up of just six elements: oxygen (65%), carbon (18%), hydrogen (10%), nitrogen (3%), calcium (1.5%), and phosphorus (1%).
- By weight, the human brain is about 78% water, and human bones are about 25% water.

2.1 The Nature of Atoms

2.2 Compounds and Chemical Bonds

2.3 The Role of Water in Life

2.4 Major Molecules of Life

ENVIRONMENTAL ISSUES

Radon Gas: A Killer That Can Be Stopped
Toilet to Tap

Understanding basic chemical principles can help us understand how the human body is put together and how it works.

Chemistry is the branch of science concerned with the composition and properties of substances, including the stuff our bodies are built from. In Chapter 1a, you learned that becoming an informed patient means knowing your basic medical history and understanding any treatments recommended and medications prescribed. In this chapter, we consider basic chemical concepts, such as the way atoms bond and the way molecules join to form the major molecules of life. By understanding fundamental concepts of chemistry, we gain insights into how our bodies work and how foods and medications relate to health. This chapter lays a foundation for discussions throughout the rest of the book.

2.1 The Nature of Atoms

The world around you contains an amazing variety of physical substances: the grass or concrete you walk on, the water you drink, the air you breathe, and even this book you are reading. All of these substances that make up our world are called matter. Fundamentally, **matter** is anything that takes up space and has mass. The three traditional states of matter are solids, liquids, or gases. All forms of matter are made up of atoms.

Atoms are units of matter that cannot be broken down into simpler substances by ordinary chemical means. Each atom is composed of even smaller, subatomic particles, such as protons, neutrons, and electrons. These subatomic particles are characterized by their location within the atom, their electrical charge, and their mass.

Each atom has a nucleus at its center and a surrounding spherical "cloud" of electrons. As you can see in Figure 2.1, the nucleus contains protons and neutrons. Electrons, in contrast, move around the nucleus and occur at certain energy levels called *shells*. Note that shells are

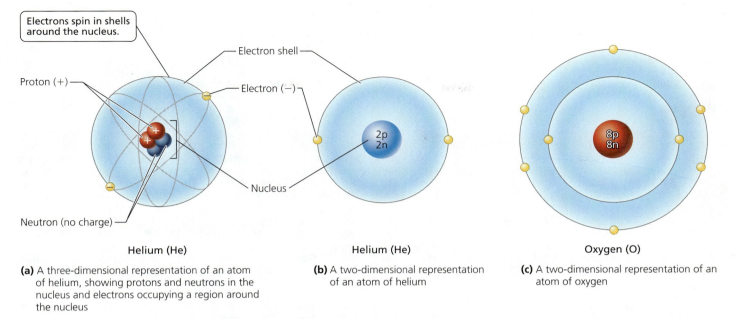

FIGURE 2.1 Atoms can be represented in different ways.

three-dimensional spaces, although they are often depicted in textbook figures as two-dimensional circles for convenience. The shell closest to the nucleus can hold up to 2 electrons. The next shell out can hold up to 8 electrons. Atoms with more than 10 electrons have additional shells. Electrons have different amounts of energy; those with the most energy are found farthest from the nucleus. The number of electrons in the outermost shell determines an atom's chemical properties. As we will see, those atoms whose outermost shells are not full tend to interact with other atoms.

Neutrons, as their name implies, are neutral: they have no electrical charge. In contrast, protons have a positive charge, and electrons have a negative charge. The negatively charged electrons stay near the nucleus because they are attracted to the positively charged protons in it. Most atoms have the same number of positively charged protons and negatively charged electrons. As a result, they are "neutral," having no net charge. Table 2.1 summarizes the basic characteristics of protons, neutrons, and electrons.

Elements

An **element** is a pure form of matter containing only one kind of atom. You are probably familiar with many elements, such as gold, silver, iron, and oxygen. Earth and everything on it or in its atmosphere is made up of a little more than 100 elements. Only about 20 elements are found in the human body,

which consists mostly of oxygen, carbon, hydrogen, and nitrogen. Each element consists of atoms containing a certain number of protons in the nucleus. For example, all carbon atoms have six protons. The number of protons in the atom's nucleus is called the *atomic number*.

The periodic table lists the elements and describes many of their characteristics. Figure 2.2 depicts the periodic table. Note that each element has a name and a one- or two-letter symbol. The symbol for the element carbon is C, and that for chlorine is Cl. (The abbreviations are not always as intuitive; for example, the abbreviation for gold is Au, based on the Latin name for the metal.) Besides an atomic number, each atom also has an atomic weight. Each proton and neutron has an approximate mass of 1 atomic mass unit, or amu. The mass of an electron is so small that it is usually considered zero. Because electrons have negligible mass, and protons and neutrons each have an atomic mass of 1, the atomic weight for any atom equals the number of protons plus the number of neutrons. Oxygen has an atomic weight of 16, indicating that it has eight protons (we know this from its atomic number) and eight neutrons in its nucleus.

Isotopes and Radioisotopes

All the atoms of a particular element contain the same number of protons; they can, however, have different numbers of neutrons. Such differences result in atoms of the same element having slightly different atomic weights. Atoms that have the same number of protons but differ in the number of neutrons are called **isotopes**. More than 300 isotopes occur naturally on Earth. The element carbon, for example, has three isotopes. All carbon atoms have six protons in the nucleus. Most carbon atoms also have six neutrons, but some have seven or eight. The isotopes of carbon thus have atomic weights of 12, 13, and 14, respectively, depending on the number of neutrons in the nucleus. These isotopes are written ^{12}C (the most common form in nature), ^{13}C, and ^{14}C.

TABLE 2.1	Review of Subatomic Particles		
Particle	Location	Charge	Mass
Proton	Nucleus	1 positive unit	1 atomic mass unit
Neutron	Nucleus	None	1 atomic mass unit
Electron	Outside the nucleus	1 negative unit	Negligible

FIGURE 2.2 *The periodic table*

Radiation is energy moving through space. Examples include radio waves, light, heat, and the excess energy or particles given off by unstable isotopes as they break down. Some elements have both stable and unstable isotopes. Unstable, radiation-emitting isotopes are called **radioisotopes**. About 60 occur naturally, and many more have been made in laboratories.

Depending on the context, radiation can be dangerous (Figure 2.3) or useful (Figure 2.4). Absorption of harmful radiation may lead to damage to organs, such as the skin, and development of some cancers. In other cases, radiation may not produce any noticeable injury to the person who was exposed, but it may alter the hereditary material in the cells of the reproductive system, possibly causing defects in the individual's offspring. For an example of the harmful effects of radiation and how to protect yourself from radiation that occurs naturally in the environment, see the Environmental Issue essay, *Radon Gas: A Killer That Can Be Stopped*.

In stark contrast to the harmful effects of radiation are its medical uses. Medical professionals use radiation for diagnosis and therapy. Perhaps the most familiar diagnostic use of radiation is x-ray imaging. A less common diagnostic procedure is the use of small doses of radiation to generate visual images of internal body parts. Radioactive iodine, for example, is often used to identify disorders of the thyroid gland. This gland, located in the neck, normally accumulates the element iodine, which it uses to regulate growth and metabolism. Small doses of iodine-131 (^{131}I), a radioactive isotope, can be given to a patient suspected of having metabolic problems. The radioactive iodine is taken up by the patient's thyroid gland and detected by medical instruments, as shown in Figure 2.4. The small amount of radioactive iodine used in imaging does not damage the thyroid gland or surrounding structures. However, larger doses can be used to kill thyroid cells when the gland is enlarged and overactive.

Radiation can also be used to kill cancer cells. Cancer cells divide more rapidly and have higher rates of metabolic activity than do most normal cells. For these reasons, cancer cells are more susceptible to the destructive effects of radiation. Still, when medical professionals aim an outside beam of radiation at a tumor to kill the cancer cells inside it, they must also take steps to shield the surrounding healthy tissue. Sometimes a radiation source is placed within the body to treat a cancer. For example, one treatment for prostate cancer (the prostate gland is

FIGURE 2.3 *Sunburn, perhaps the most common burn from radiation, is caused by overexposure to ultraviolet radiation from the sun's rays.*

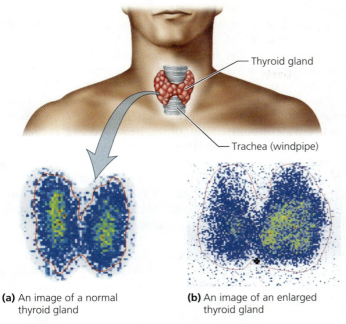

(a) An image of a normal thyroid gland

(b) An image of an enlarged thyroid gland

FIGURE 2.4 *Radioactive iodine can be used to generate images of the thyroid gland for diagnosing metabolic disorders.*

an accessory reproductive gland in males) involves placing radioactive seeds (pellets) directly in the prostate gland (Figure 2.5). Once in place, the seeds emit radiation that damages or kills nearby cancer cells. In most cases, the seeds are left in place, even though they stop emitting radiation within 1 year.

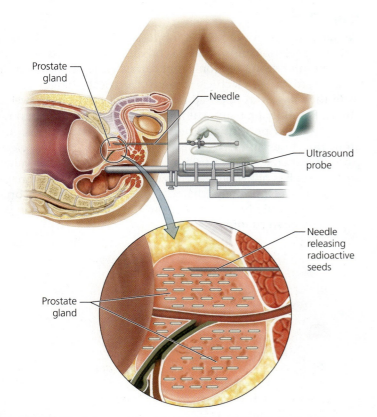

FIGURE 2.5 *Prostate cancer can be treated by implanting radioactive seeds in the prostate gland. A physician injects the seeds through needles, with guidance from an ultrasound probe placed in the rectum.*

what would you do?

Irradiation is the process in which an item is exposed to radiation. Many foods today are intentionally irradiated to delay spoilage, increase shelf life, and remove harmful microorganisms, insect pests, and parasites. The food does not become radioactive as a result. Supporters of the practice note that test animals fed on irradiated food show no adverse effects. Opponents, however, point to the environmental risks of building and operating food irradiation plants and the lack of carefully controlled, long-term experiments verifying that irradiated food is safe for people of all ages and nutritional states. Several foods, including white potatoes, wheat flour, fresh meat and poultry, and fresh spinach and iceberg lettuce, can be irradiated in the United States. If the entire product is irradiated, then a distinctive logo (Figure 2.6) must appear on its packaging. If an irradiated food is an ingredient in another product, then it must be listed as irradiated in the ingredients statement, but the logo is not required. Do you think irradiating food is a safe practice? Would you eat irradiated food?

FIGURE 2.6 *Logo for irradiated foods. This logo and words such as "Treated with radiation" must appear on food that has been irradiated in its entirety.*

2.2 Compounds and Chemical Bonds

Two or more elements may combine to form a new chemical substance called a **compound**. A compound's characteristics are usually different from those of its elements. Consider what happens when the element sodium (Na) combines with the element chlorine (Cl). Sodium is a silvery metal that explodes when it comes into contact with water. Chlorine is a deadly yellow gas. In combination, however, they form a crystalline solid called sodium chloride (NaCl)—plain table salt (Figure 2.7).

The atoms (or, as we will soon see, ions) in a compound are held together by chemical bonds. There are two types of chemical bonds: covalent and ionic. Recall that atoms have outer shells, which are the regions surrounding the nucleus where the electrons are most likely to be found. Figure 2.8 depicts the first two shells as concentric circles around the nucleus. A full innermost shell contains 2 electrons. A full second shell contains 8 electrons. Atoms with a total of more than 10 electrons have additional shells. When atoms form bonds, they lose, gain, or share the electrons in their outermost shell.

Covalent Bonds

A **covalent bond** forms when two or more atoms *share* electrons in their outer shells. Consider the compound methane (CH_4). Methane is formed by the sharing of electrons between one atom of carbon and four atoms of hydrogen. Notice in Figure 2.9a that the outer shell of an isolated carbon atom contains only four electrons, even though it can hold as many as eight. Also note that hydrogen atoms have only one electron, although the first shell can hold up to two electrons. A carbon atom can fill its outer shell by joining with four atoms of hydrogen. At the same time, by forming a covalent bond with

(a) The element sodium is a solid metal.

(b) Elemental sodium reacts explosively with water.

(c) The element chlorine is a yellow gas.

(d) When the elements sodium and chlorine join, they form table salt, a compound quite different from its elements.

FIGURE 2.7 *The characteristics of compounds are usually different from those of their elements.*

the carbon atom, each hydrogen atom fills its first shell. We see, then, that the covalent bonds between the carbon atom and hydrogen atoms of methane result in filled outer shells for all five atoms involved.

A **molecule** is a chemical structure held together by covalent bonds. Compounds are formed by two or more elements, so molecules that contain only one kind of atom are not considered compounds. For example, oxygen gas, which is formed by the joining of two oxygen atoms, is *not* a compound, but it *is* a molecule. Molecules are described by a formula that contains the symbols for all of the elements included in that molecule. If more than one atom of a given element is present in the molecule, a subscript is used to show the precise number of that kind of atom. For example, the molecular formula of sucrose (table sugar) is $C_{12}H_{22}O_{11}$, showing that one molecule of sucrose contains 12 atoms of carbon, 22 atoms of hydrogen, and 11 atoms of oxygen. Numbers placed in front of the molecular formula indicate that more than one molecule is present. For example, three molecules of sucrose are described by the formula $3C_{12}H_{22}O_{11}$.

As shown in the methane molecule in Figure 2.9a, the bond between each of the four hydrogen atoms and the carbon atom consists of a single pair of electrons. A bond in which a single pair of electrons is shared is called a *single bond*; in this example, the methane molecule contains four single covalent bonds. Sometimes, however, two atoms share two or three pairs of electrons. These bonds are called *double* and *triple covalent bonds*, respectively. For example, carbon dioxide, CO_2, produced by chemical reactions inside our cells, has double covalent bonds between the carbon atom and each of two oxygen atoms (Figure 2.9b). And the nitrogen atoms in nitrogen gas, N_2, are joined by triple covalent bonds (Figure 2.9c).

Covalent bonds in molecules are sometimes depicted by a structural formula. Notice in the box at the right of Figure 2.9a that one straight line is drawn between the carbon atom and each hydrogen atom in the structural formula for the methane molecule. The single line indicates a single covalent bond resulting from a pair of shared electrons. In the box to the right of Figure 2.9b, the double lines between the carbon and oxygen atoms in the carbon dioxide molecule indicate a double covalent bond, or two pairs of shared electrons. In Figure 2.9c, three lines drawn between the two nitrogen atoms in gaseous nitrogen depict a triple covalent bond, or three pairs of shared electrons.

Ionic Bonds

We have all heard the phrase "opposites attract" in reference to human relationships—and it is no different for ions. An **ion** is an atom or group of atoms that carries either a positive (+) or a negative (−) electrical charge. Electrical charges result from the *transfer* (as opposed to sharing) of electrons between atoms. Recall that a neutral atom has the same number of positively charged protons and negatively charged electrons. An atom that loses an electron has one more proton than electrons and therefore has a positive charge. An atom that gains an electron has one more electron than protons and has a negative charge. Oppositely charged ions are attracted to one another. An **ionic bond** results from the mutual attraction of oppositely charged ions.

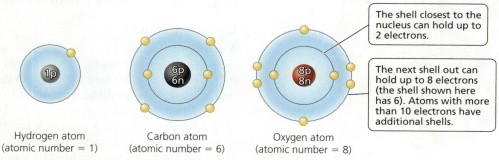

FIGURE 2.8 *Atoms of hydrogen, carbon, and oxygen. Each of the concentric circles around the nucleus represents a shell occupied by electrons.*

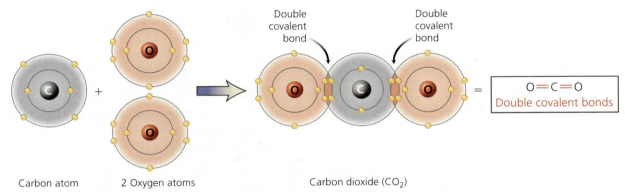

(a) The molecule methane (CH_4) is formed by the sharing of electrons between one carbon atom and four hydrogen atoms. Because in each case one pair of electrons is shared, the bonds formed are single covalent bonds.

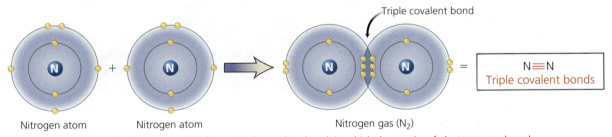

(b) The oxygen atoms in a molecule of carbon dioxide (CO_2) form double covalent bonds with the carbon atom. In double bonds, two pairs of electrons are shared.

(c) The nitrogen atoms in nitrogen gas (N_2) form a triple covalent bond, in which three pairs of electrons are shared.

FIGURE 2.9 Covalent bonds form when atoms share electrons. Shown here are examples of single, double, and triple covalent bonds. For each example, the structural formula is shown on the far right.

Ions form because of the tendency of atoms to attain a complete outermost shell. Consider, again, the atoms of sodium and chlorine that join to form sodium chloride. As shown in Figure 2.10, an atom of sodium has one electron in its outer shell. An atom of chlorine has seven electrons in its outer shell. Sodium chloride is formed when the sodium atom transfers the single electron in its outer shell to the chlorine atom. The sodium atom now has a full outer shell. This comes about because the sodium atom loses its third shell, making the second shell its outermost shell. The sodium atom, having lost an electron, has one more proton than electrons and therefore now has a positive charge (Na^+). The chlorine atom, having gained an electron to fill its outer shell, has one more electron than protons and now has a negative charge (Cl^-). These oppositely charged ions are attracted to one another, and an ionic bond forms. Because they do not contain shared electrons, ionic bonds are weaker than covalent bonds.

2.3 The Role of Water in Life

Water is such a familiar part of our everyday lives that we often overlook its unusual qualities. Unique properties of water include its virtuosity as a dissolving agent, its high heat capacity, and its high heat of vaporization. Water's unusual qualities can be traced to its *polarity* (tendency of its molecules to have

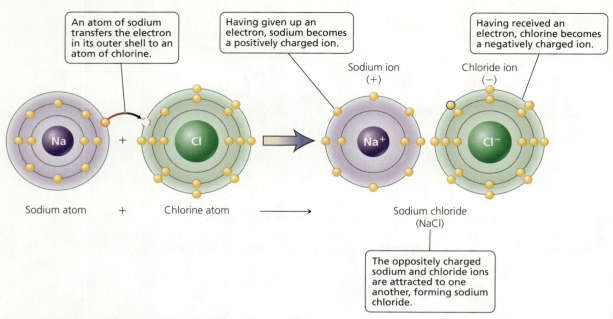

FIGURE 2.10 An ionic bond involves the transfer of electrons between atoms. Such a transfer creates oppositely charged ions that are attracted to one another.

positive and negative regions) and the hydrogen bonds between its molecules.

Polarity and Hydrogen Bonds In covalently bonded molecules, electrons may be shared equally or unequally between atoms. When the sharing of electrons is unequal, different ends of the same molecule can have slight opposite charges. Unequal covalent bonds are called polar, and molecules with unevenly distributed charges are called polar molecules. In water (H_2O), for example, the electrons shared by oxygen and hydrogen spend more time near the oxygen atom than near the hydrogen atom. As a result, the oxygen atom has a slight negative charge; each hydrogen atom has a slight positive charge; and water molecules are polar (Figure 2.11a). The hydrogen atoms of one water molecule are attracted to the oxygen atoms of other water molecules. The attraction between a slightly positively charged hydrogen atom and a slightly negatively charged atom nearby is called a **hydrogen bond**. In the case of water, the hydrogen bonding occurs between hydrogen and oxygen. However, sometimes hydrogen bonds form between hydrogen and atoms of other elements.

Hydrogen bonds are weaker than either ionic or covalent bonds. For this reason, we illustrate them by dotted lines rather than solid lines, as shown in Figure 2.11b. Even though individual hydrogen bonds are very weak, collectively they can be significant. Hydrogen bonds maintain the shape of proteins and our hereditary material, DNA, and they account for some of the unique physical properties of water.

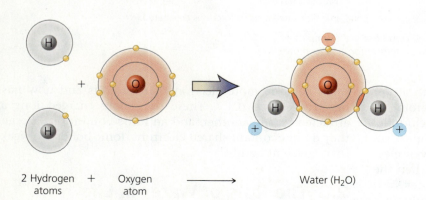

(a) Water is formed when an oxygen atom covalently bonds (shares electrons) with two hydrogen atoms. Because of the unequal sharing of electrons, oxygen carries a slight negative charge, and the hydrogen atoms carry a slight positive charge.

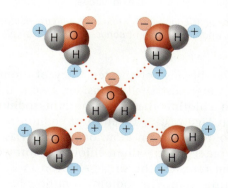

(b) The hydrogen atoms from one water molecule are attracted to the oxygen atoms of other water molecules. This relatively weak attraction (shown by dotted lines) is called a hydrogen bond.

FIGURE 2.11 The hydrogen bonds of water

ENVIRONMENTAL ISSUE

Radon Gas: A Killer That Can Be Stopped

Some lethal substances are obvious because you can see, smell, or taste them. Others are not obvious, even when surrounding you in your own home. Such is the case for radon, a radioactive gas produced when uranium breaks down in rock, soil, or water. This invisible, odorless gas moves up through the soil, eventually seeping through cracks in the foundations of buildings. Once inside a building, radon may accumulate and reach harmful levels. The same techniques that help to make a building energy efficient also tend to limit the exchange of air from inside a building to outside, and, as a result, modern, energy-efficient buildings can accumulate radon.

Although radon can enter any type of building, the greatest exposure threat usually occurs in homes because people typically spend more time in their home than in their school or workplace. Most radon enters buildings as gas from the soil, but radon can also enter a home through well water. The greatest risk from radon in water occurs when it is released into the air during showers or other household uses of water. The U.S. Environmental Protection Agency (EPA) estimates that 1 in 15 homes in the United States has an unacceptable level of radon, defined as 4 or more picocuries of radon per liter of air. (A picocurie is a measure of radioactivity that equals about 2.2 disintegrations of radioactive particles per minute.)

Each year in the United States, radon contributes to the death of about 21,000 people; this number exceeds deaths from drunk driving. Radon kills by causing lung cancer. After cigarette smoking, radon is the next leading cause of lung cancer. Smoking combined with exposure to radon is particularly risky.

The EPA recommends that all homes be tested for radon (Figure 2.A). Residents can test for radon themselves using a do-it-yourself kit available at most hardware stores. Alternatively, residents can hire a certified radon professional to do the testing. If testing reveals high levels of radon in a home, then a contractor trained to fix radon problems can install a venting system that pulls radon from beneath the house and releases it to the outside. New homes can be built with features to reduce radon levels; even so, testing should be performed to make sure that the construction techniques are effective.

Questions to Consider

- If you live in a campus dormitory, would you be willing to pay more in fees to fix a newly discovered radon contamination problem in your building? If not, whose responsibility should this be?

- Beginning in 2012, landlords in Maine must test for radon in their buildings every 10 years. Do you think that other states should follow Maine's example, or should radon testing be an individual responsibility?

FIGURE 2.A Home radon-testing kits are available from the National Safety Council and are sold in many hardware and retail stores.

TABLE 2.2	Review of Chemical Bonds		
Type	**Basis for Attraction**	**Strength**	**Example**
Covalent	Sharing of electrons between atoms; the sharing between atoms may be equal or unequal	Strongest	CH_4 (methane)
Ionic	Transfer of electrons between atoms creates oppositely charged ions that are attracted to one another	Strong	NaCl (table salt)
Hydrogen	Attraction between a hydrogen atom with a slight positive charge and another atom (often oxygen) with a slight negative charge	Weak	Between a hydrogen atom on one water molecule and an oxygen atom on another water molecule

Covalent bonds, ionic bonds, and hydrogen bonds are summarized in Table 2.2.

Properties of Water

Life depends on the properties of water. Many of these properties result from water's polarity and hydrogen bonding. Let's consider some of the properties that make water so vital to life.

Because of the polarity of its molecules, water interacts with many substances. This interactivity makes it an excellent solvent, easily dissolving both polar and charged substances. Ionic compounds, such as NaCl, dissolve into independent ions in water. The sodium ions and chloride ions separate from one another in water because the sodium ions are attracted to the negative regions of water molecules and the chloride ions are attracted to the positive regions (Figure 2.12). Because of its excellence as a solvent, water serves as the body's main transport medium. As the liquid component of blood, it carries dissolved nutrients, gases, and wastes through the circulatory

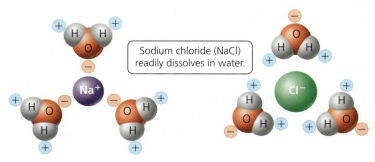

FIGURE 2.12 Water is an excellent solvent because of the polarity of its molecules. When a compound such as table salt (NaCl) is placed in water, it readily dissolves into independent ions.

system. Metabolic wastes are excreted from the body in urine, another watery medium.

Because of its hydrogen bonds, water helps prevent dramatic changes in body temperature. About 67% of the human body is water (thus, if a person weighs 68 kg [150 lb], water makes up about 45 kg [100 lb] of the body weight). Because humans, as well as many other organisms, are made up largely of water, they are well suited to resist changes in body temperature and to keep a relatively stable internal environment. This ability comes from water's *high heat capacity,* which simply means that a great deal of heat is required to raise water's temperature. Hydrogen bonds hold multiple water molecules together, so a large amount of heat is required to break these bonds (a higher temperature corresponds to an increase in the movement of the molecules). Water in blood also helps redistribute heat within our bodies. Our fingers don't usually freeze on a frigid day because heat is carried to them by blood from muscles, where the heat is generated.

Another property of water that helps prevent the body from overheating is its *high heat of vaporization,* which means that a great deal of heat is required to make water evaporate (that is, change from a liquid to a gas). Water's high heat of vaporization is also due to its hydrogen bonds, which must be broken before water molecules can leave the liquid and enter the air. (These bonds, by the way, remain broken as long as water is in the gaseous phase known as *water vapor.*) Water molecules that evaporate from a surface carry away a lot of heat, cooling the surface. We rely on the evaporation of water in sweat to cool the body surface and prevent overheating. By the same token, water vapor in the air can inhibit the evaporation of sweat—which is why we tend to feel hotter on a humid day.

It is clear that water is essential for human life and for the lives of many other organisms we share the planet with. Despite our reliance on water, we continue to pollute both seawater and freshwater. Equally alarming is the global shortage of freshwater, caused largely by the burgeoning human population. For one community's creative response to the water shortage, see the Environmental Issue essay, *Toilet to Tap.*

stop and think
Sharp increases in body temperature can cause heat stroke, a condition that may damage the brain. Explain why heat stroke is more likely to occur on a hot, humid day than on an equally hot, dry day.

Acids and Bases

Sometimes a water molecule dissociates, or breaks up, forming a positively charged hydrogen ion (H+) and a negatively charged hydroxide ion (OH−):

$$\text{H—O—H} \longleftrightarrow \text{H}^+ + \text{OH}^-$$
Water   Hydrogen ion   Hydroxide ion

(Note that in equations describing chemical reactions, an arrow should be read as "yields.")

In any sample of water, the fraction of water molecules that are dissociated is extremely small, so water molecules are much more common in the human body than are H+ and OH−. In fact, the amount of H+ in the body must be precisely regulated. Substances called acids and bases influence the concentration of H+ in solutions.

Acids and bases are defined by what happens when they are added to water. An **acid** is anything that releases hydrogen ions (H+) when placed in water. A **base** is anything that releases hydroxide ions (OH−) when placed in water. Hydrochloric acid (HCl), for example, dissociates in water to produce hydrogen ions (H+) and chloride ions (Cl−). Because HCl increases the concentration of (H+) in solution, it is classified as an acid. Sodium hydroxide (NaOH), by contrast, dissociates in water to produce sodium ions (Na+) and hydroxide ions (OH−). Because NaOH increases the concentration of OH− in solution, it is classified as a base. The OH− produced when NaOH dissociates reacts with H+ to form water molecules and thus reduces the concentration of H+ in solution. Therefore, acids increase the concentration of H+ in solution, and bases decrease the concentration of H+ in solution.

The pH Scale

We often want to know more than simply whether a substance is an acid or a base. For example, how strong an acid is battery acid? How strong a base is household ammonia? Questions like these can be answered by knowing the pH of these solutions and understanding the pH scale (Figure **2.13**). The **pH** of a solution is a measure of hydrogen ion concentration. The **pH scale** ranges from 0 to 14; a pH of 7 is neutral (the substance does not increase H+ or OH−), a pH less than 7 is acidic, and a pH greater than 7, basic.

Usually, the amount of H+ in a solution is very small. For example, the concentration of H+ in a solution with a pH of 6 is [1×10^{-6}] (or 0.000001) moles per liter (a mole, here, is not a small furry animal with a star-shaped nose but a unit of measurement that indicates a specific number of atoms, molecules, or ions). Similarly, the concentration of H+ in a solution with a pH of 5 is [1×10^{-5}] (or 0.00001) moles per liter. Technically, pH is the negative logarithm of the concentration of H+ in a solution. According to the pH scale, the lower the pH, the greater the acidity—or concentration of H+—in a solution. Each reduction of pH by one unit represents a tenfold increase in the amount of H+. So a solution with a pH of 5 is 10 times more acidic than a solution with a pH of 6. And a solution with a pH of 4 is 100 times more acidic than one with a pH of 6. Some characteristics of acids and bases, including their values on the pH scale, are summarized in Table **2.3**.

Buffers

Most biological systems must keep their fluids within a narrow range of pH values. Substances called **buffers** keep pH values from changing dramatically. Buffers remove excess H+ from solution when concentrations of H+ increase. Buffers add H+ when concentrations of H+ decrease. For example, an important buffering system that keeps the pH of blood at about 7.4 is the carbonic acid–bicarbonate system. When

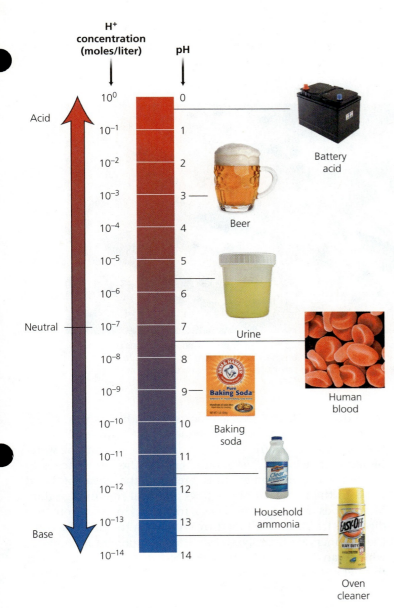

FIGURE 2.13 The pH scale and the pH of some body fluids and other familiar substances

TABLE 2.3	Review of the Characteristics of Acids and Bases	
Characteristic	**Acid**	**Base**
Behavior in water	Releases H⁺	Releases OH⁻
Ph	Less than 7	Greater than 7
Example	HCl (hydrochloric acid)	NaOH (sodium hydroxide)

carbon dioxide is added to water it forms carbonic acid (H_2CO_3), which dissociates into hydrogen ions and bicarbonate ions (HCO_3^-):

$$CO_2 + H_2O \longleftrightarrow H_2CO_3 \longleftrightarrow H^+ + HCO_3^-$$
Carbon dioxide • water • Carbonic acid • Hydrogen ion • Bicarbonate ion

Carbonic acid and bicarbonate have a buffering action because when levels of H⁺ decrease in the blood, carbonic acid dissociates, adding H⁺ to solution. When levels of H⁺ increase in the blood, the H⁺ combines with bicarbonate and is removed from solution. Such action is essential because even slight changes in the pH of blood—say, a drop from 7.4 to 7.0 or an increase to 7.8—can cause death in a few minutes.

In the human body, almost all biochemical reactions occur around pH 7 and are maintained at that level by powerful buffering systems. An important exception occurs in the stomach, where hydrochloric acid (HCl) produces pH values from about 1 to 3. In the stomach, HCl kills bacteria swallowed with food or drink and promotes the initial breakdown of proteins. These activities require an acid stomach, and the stomach has several ways of protecting itself from the acid (discussed in Chapter 15). However, sometimes stomach acid backs up into the esophagus, and "heartburn" is the uncomfortable result. Taking an antacid can ease the discomfort of heartburn. Antacids consist of weak bases that temporarily relieve the pain of stomach acid in the esophagus by neutralizing some of the hydrochloric acid.

The critical link between pH and life is illustrated by the impact of acid rain on our environment and health. *Acid rain* is usually defined as rain with a pH lower than 5.6, the pH of natural precipitation. Burning of fossil fuels in cars, factories, and power plants is a major cause of acid rain. The sulfur dioxide and nitrogen oxides produced by these activities react with water in the atmosphere to form sulfuric acid (H_2SO_4) and nitric acid (HNO_3), which fall to Earth as rain, snow, or fog.

The effects of acid rain on the environment have been devastating. On land, acid rain has been linked to the decline of forests. Trees become stressed and more susceptible to disease when their nutrient uptake is disrupted by increased acidity in the soil (Figure 2.14a). In aquatic environments, acid rain has been linked to declines in populations of fish and amphibians. Embryos of spotted salamanders, for example, develop abnormally under somewhat acidic conditions and die when pH values are less than about 5 (Figure 2.14b). Acid rain is harmful to human health, as well. The pollutants that cause acid rain form fine particles of sulfate and nitrate that are easily inhaled. Once inside us, they cause irritation and respiratory illnesses such as asthma and bronchitis.

Because most acid rain is caused by human activity, it is within our power to reduce, if not eliminate, the problem. Power plants and automobile manufacturers have taken steps to reduce emissions of sulfur dioxide and nitrogen oxides, leading to significant progress in reducing acid rain.

2.4 Major Molecules of Life

Most of the molecules we have discussed so far have been small and simple. Many of the molecules of life, however, are enormous by comparison and have complex architecture. Some proteins, for example, are made up of thousands of atoms linked together in a chain that repeatedly coils and folds upon itself. Exceptionally large molecules, including many important biological molecules, are known as **macromolecules**.

(a) Acid rain has destroyed parts of our forests.

(b) In some areas, acid rain has reduced or eliminated populations of aquatic organisms. Acidic conditions have been shown, for example, to disrupt the development of spotted salamanders or to kill the embryos outright.

FIGURE 2.14 *Effects of acid rain*

Macromolecules that consist of many small, repeating molecular subunits linked in a chain are called **polymers**. **Monomers** are the small molecular subunits that form the building blocks of the polymer. We might think of a polymer as a pearl necklace, with each monomer representing a pearl. As we shall see, a protein is a polymer, or chain, of amino acid monomers linked together. And glycogen, the storage form of carbohydrates in animals, is a polymer of glucose monomers.

Polymers form through **dehydration synthesis** (sometimes called the *condensation reaction*). In this process, the reaction that bonds one monomer covalently to another releases a water molecule: one of the monomers donates OH, and the other donates H. The reverse process, called **hydrolysis**, which the body uses to break many polymers apart, requires the addition of water across the covalent bonds. The H from the water molecule attaches to one monomer, and the OH attaches to the adjoining monomer, thus breaking the covalent bond between the two. Hydrolysis plays a critical role in digestion. Most foods consist of polymers too large to pass from our digestive tract into the bloodstream and on to our cells. Thus, the polymers are hydrolyzed into their component monomers, which can be absorbed into the bloodstream for transport throughout the body. Dehydration synthesis and hydrolysis are summarized in Figure 2.15.

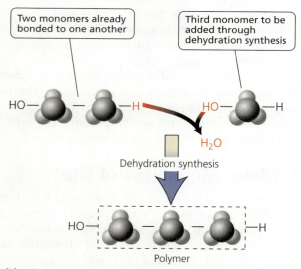
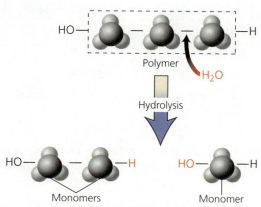

(a) Polymers are formed by dehydration synthesis, in which a water molecule is removed and two monomers are joined.

(b) Polymers are broken down by hydrolysis, in which the addition of a water molecule disrupts the bonds between two monomers.

FIGURE 2.15 *Formation and breaking apart of polymers*

Carbohydrates

The **carbohydrates**, known commonly as sugars and starches, provide fuel (energy) for the human body. Carbohydrates are compounds made entirely of carbon, hydrogen, and oxygen, with each molecule having twice as many hydrogen atoms as oxygen atoms. Sugars and starches can be classified by size into monosaccharides, oligosaccharides, and polysaccharides.

Monosaccharides **Monosaccharides**, also called simple sugars, are the smallest molecular units of carbohydrates. They contain from three to seven carbon atoms and, in fact, can be classified by the number of carbon atoms they contain. A sugar that contains five carbons is *pent*ose; one with six carbons is *hex*ose; and so on. Glucose, fructose, and galactose are examples of six-carbon sugars. Monosaccharides can be depicted in several ways (Figure 2.16).

Oligosaccharides **Oligosaccharides** (*oligo* means "few") are chains of a few monosaccharides joined together by dehydration synthesis. A **disaccharide**, one type of oligosaccharide, is a double sugar that forms when two monosaccharides covalently bond to each other. The disaccharide sucrose (table sugar) consists of the monosaccharides glucose and fructose (Figure 2.17). Two glucose molecules form the disaccharide maltose, an important ingredient of beer. Another disaccharide is lactose, the principal carbohydrate of milk and milk products. The joining of glucose and galactose forms lactose.

Polysaccharides A **polysaccharide** (*poly* means "many") is a complex carbohydrate that forms when monosaccharides (most commonly glucose) join together in long chains. Most polysaccharides store energy or provide structure. In plants, the storage polysaccharide is **starch**; in animals, it is **glycogen**, a short-term energy source that can be broken down to release energy-laden glucose molecules. Humans store glycogen mainly in muscle and liver cells (Figure 2.18a).

Cellulose is a structural polysaccharide found in the cell walls of plants (Figure 2.18b). Humans lack the enzymes necessary to digest cellulose, and, as a result, it passes unchanged through our digestive tract. (Enzymes are discussed later in this chapter.) Nevertheless, cellulose is an important form of dietary fiber (roughage) that helps fecal matter move through the large intestines. Including fiber in our diet may reduce the incidence of colon cancer.

Lipids

Lipids, such as fats, are compounds that do not dissolve in water. Lipids are nonpolar (having no electrical charges),

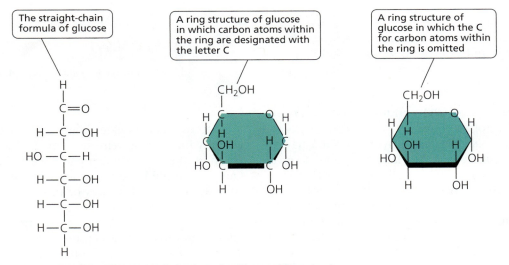

FIGURE 2.16 Monosaccharides are simple sugars, generally having a backbone of three to six carbon atoms. Many of these carbon atoms are also bonded to hydrogen (H) and a hydroxyl group (OH). In the fluid within our cells, the carbon backbone usually forms a ringlike structure. Here, three representations of the monosaccharide glucose ($C_6H_{12}O_6$) are shown.

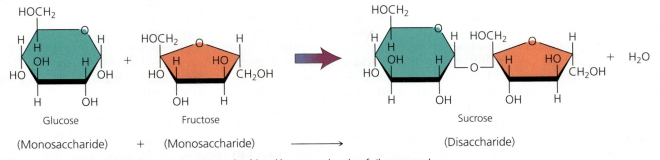

FIGURE 2.17 Disaccharides are built from two monosaccharides. Here, a molecule of glucose and one of fructose combine to form sucrose.

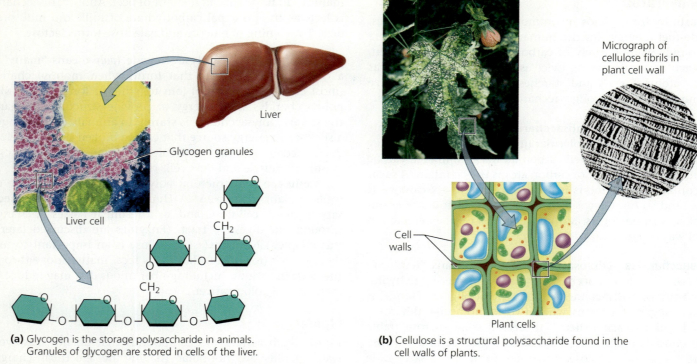

(a) Glycogen is the storage polysaccharide in animals. Granules of glycogen are stored in cells of the liver.

(b) Cellulose is a structural polysaccharide found in the cell walls of plants.

FIGURE 2.18 *Polysaccharides may function in storage (for example, glycogen) or provide structure (for example, cellulose).*

whereas water is polar. Because of this difference, water shows no attraction for lipids and vice versa, so water and lipids do not mix. Three types of lipids that are important to human health are triglycerides, phospholipids, and steroids.

Triglycerides Fats and oils are **triglycerides**, compounds made of one molecule of glycerol and three fatty acids. **Fatty acids** are chains of carbon atoms also bonded to hydrogens and having the acidic group COOH at one end. The fatty acids bond to glycerol through dehydration synthesis (Figure 2.19a). Triglycerides are classified as saturated or unsaturated, depending on the presence or absence of double bonds between the carbon atoms in their fatty acids (Figure 2.19b). *Saturated fatty acids* have only single covalent bonds linking the carbon atoms. They are described as "saturated with hydrogen" because their carbon atoms are bonded to as many hydrogen atoms as possible. Saturated fats are made from saturated fatty acids. Butter, a saturated fat, is solid at room temperature because its fatty acids can pack closely together. Fatty acids with one or more double bonds between carbon atoms are described as *unsaturated fatty acids*—that is, not saturated with hydrogen—because they could bond to more hydrogen atoms if the double bonds between their carbon atoms were broken. The double bonds cause "kinks" in the fatty acids and prevent molecules of unsaturated fat from packing tightly into a solid. Thus, unsaturated fats, such as olive oil, are liquid at room temperature. Sometimes hydrogens are added to unsaturated fats to stabilize them, with the goal of lengthening their shelf life, or to solidify them. For example, hydrogens are added to vegetable oil to make margarine. These partially hydrogenated fats are called *trans* fats and are found in many packaged snacks, such as cookies and potato chips.

stop and think
What would the hydrolysis of a fat yield?

Fats and oils provide about twice the energy per gram that carbohydrates or proteins do. This high energy density makes fat an ideal way for the body to store energy for the long term. Our bulk would be much greater if most of our energy storage consisted of carbohydrates or proteins, given their relatively low energy yield compared with fat.

In preparation for long-term energy storage, excess triglycerides, carbohydrates, and proteins from the foods we consume are converted into small globules of fat that are deposited in the cells of adipose tissue. There, the fat remains until our bodies need extra energy; at that time, our cells break down the fat to release the energy needed to keep vital processes going. Besides long-term energy storage, fat serves a protective function in the body. Thin layers of fat surround major organs such as the kidneys, cushioning the organs against physical shock from falls or blows. Fat also serves as insulation and as a means of absorbing lipid-soluble vitamins from the intestines and transporting them in the bloodstream to cells. Despite the importance of fats and oils to human health, in excess they can be dangerous, particularly to our circulatory system (discussed in Chapter 12a).

Phospholipids A **phospholipid** consists of a molecule of glycerol bonded to two fatty acids and a negatively charged phosphate group. Another small molecule of some kind—usually

Q Would the triglyceride shown in part (b) be a solid or a liquid at room temperature? Why?

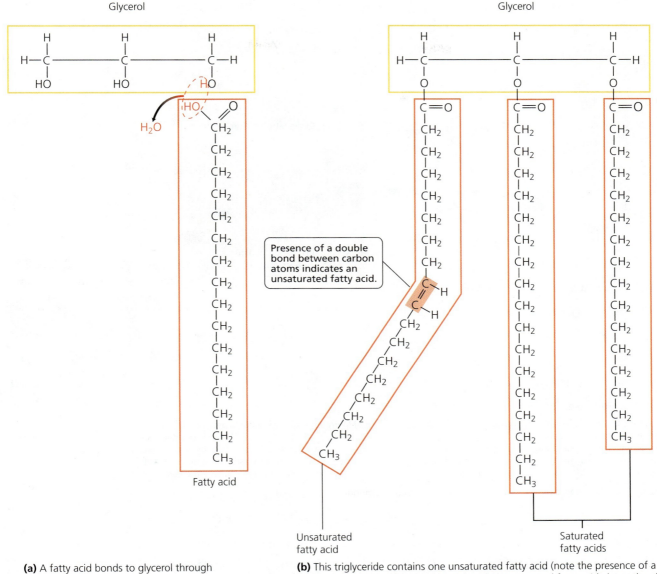

(a) A fatty acid bonds to glycerol through dehydration synthesis.

(b) This triglyceride contains one unsaturated fatty acid (note the presence of a double bond between the carbon atoms) and two saturated fatty acids (note the absence of any double bonds between the carbon atoms).

FIGURE 2.19 Triglycerides are composed of a molecule of glycerol joined to three fatty acids.

▲ The triglyceride would be a liquid at room temperature because the kink in the unsaturated fatty acid in part (b) would prevent close packing of adjacent molecules.

polar, called the variable group—is linked to the phosphate group. This general structure provides phospholipids with two regions having very different characteristics. As you will notice in Figure 2.20a, the region made up of fatty acids is nonpolar; it is described as a **hydrophobic**, or "water-fearing," tail. The other region is polar, and it makes up the **hydrophilic**, or "water-loving," head. The tails, being hydrophobic, do not mix with water. The heads, being hydrophilic, interact readily with water. The hydrophilic heads and hydrophobic tails of phospholipids are responsible for the structure of plasma (cell) membranes. In cell membranes, phospholipids are arranged in a double layer, called a *bilayer* (Figure 2.20b), with the hydrophilic heads of each layer facing away from each other. That way, each surface of the membrane consists of hydrophilic heads in contact with the watery solutions inside and outside the cell. The hydrophobic tails of the two layers point toward each other and help hold the membrane together.

Steroids A **steroid** is a type of lipid made up of four carbon rings attached to molecules that vary from one steroid to the next. Cholesterol, one of the most familiar steroids, is a component of the plasma membrane and is also the foundation from

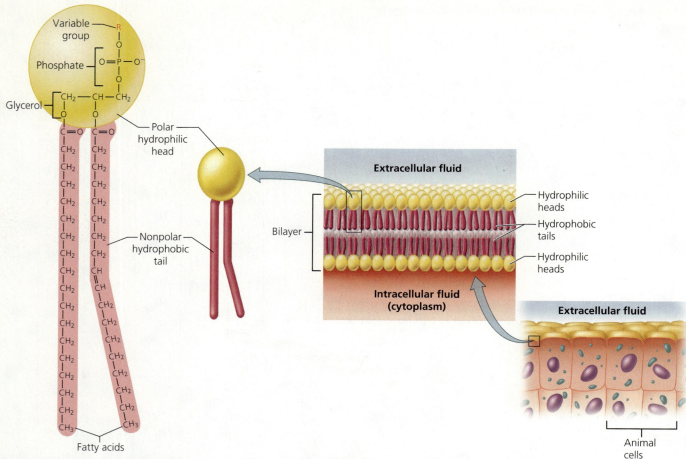

(a) A phospholipid consists of a variable group designated by the letter R, a phosphate, a glycerol, and two fatty acids. Because the variable group is often polar and the fatty acids nonpolar, phospholipids have a polar hydrophilic ("water-loving") head and a nonpolar hydrophobic ("water-fearing") tail.

(b) Within the phospholipid bilayer of the plasma membrane, the hydrophobic tails point inward and help hold the membrane together. The outward-pointing hydrophilic heads mix with the watery environments inside and outside the cell.

FIGURE 2.20 *Structure of a phospholipid. Phospholipids are the main components of plasma membranes.*

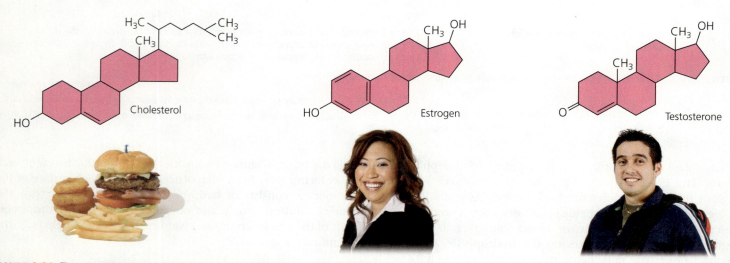

FIGURE 2.21 *The steroid cholesterol is a component of cell membranes, and it is the substance used to make steroid hormones, such as estrogen and testosterone. All steroids have a structure consisting of four carbon rings. Steroids differ in the groups attached to these rings.*

ENVIRONMENTAL ISSUE

Toilet to Tap

We depend on a steady supply of freshwater. Yet, experts tell us that our water supply is dwindling. Most of Earth's water (97.4%) is in the oceans; this water is too salty for us to use without first treating it. About 2% is locked away in glaciers and polar ice. This leaves 0.6% of Earth's water, in the form of freshwater, for humans and other organisms to use. The available water is classified as surface water (rivers, lakes, reservoirs) and groundwater (water in porous rock layers below Earth's surface). Over the last century, we have damaged our supplies of freshwater by redirecting the flow of rivers, constructing dams, draining wetlands, and extracting groundwater at rates that exceed its replacement (as part of the water cycle, groundwater is naturally replaced when water that falls as precipitation seeps into the ground). We will have an ever-greater impact on our water supply as the human population continues to grow.

Some communities in California have taken a unique approach to the problem. Like many western states, California has been experiencing prolonged drought. At the same time, it has a steadily growing population, and it faces the threat of seawater moving into its groundwater. Orange County, in southern California, has responded to the impending water crisis by building the Groundwater Replenishment System. This new system, which began operations in 2008, accomplishes two very important things. First, it recharges the groundwater supply with treated wastewater rather than sending the wastewater out to sea. Second, it builds up the county's seawater intrusion barrier—a series of wells into which water is pumped to create an "underground water dam" that blocks seawater from entering the groundwater basin.

Here is how the system works. Industrial and household sewage is treated at the Orange County Sanitation District, as it has been in the past. The sewage, or primary effluent, undergoes several treatments designed to break down organic material and to remove particulate matter. After these treatments, the water, called secondary effluent, would normally be discharged to the ocean. But under the new system, at this point the treated wastewater enters the Advanced Water Purification Facility (AWPF). At the AWPF, the water undergoes a several-step purification process. About half of the water exiting the AWPF is injected into Orange County's seawater intrusion barrier. The other half is piped to a giant percolation pond, where it moves through gravel, sand, and clay into the groundwater supply, filtering naturally just as rainwater finds its way into groundwater. Eventually, this water will enter drinking well intakes. Hence, the water has been described as going from "toilet to tap" (Figure 2.B).

The end result of Orange County's Groundwater Replenishment System is water that meets or exceeds all existing standards for drinking water. Although some people may cringe at the thought of drinking water that has been recycled from sewage, it is a reality that we may have to accept as global and regional water supplies continue to tighten.

FIGURE 2.B *The toilet-to-tap program replenishes groundwater supplies with reclaimed wastewater.*

Questions to Consider

- Would you be hesitant to drink water that has been recycled from sewage, even if it met current drinking water standards? If yes, what is the reason for your hesitation?
- Do you think that human ingenuity will always be able to produce the technology needed to avoid major environmental crises, such as a global shortage of freshwater?

---

which steroid hormones, such as estrogen and testosterone, are made (Figure 2.21). Cholesterol in our blood comes from two sources, our liver and our diet. A high level of cholesterol in the blood is considered a risk factor for heart disease (as described in Chapter 12a).

Proteins

A **protein** is a polymer made of one or more chains of amino acids. In many proteins, the chains are twisted, turned, and folded to produce complicated structures. Thousands upon thousands of different proteins are found in the human body, contributing to structural support, transport, movement, and regulation of chemical reactions. Despite their great diversity in structure and function, all proteins are made from a set of only 20 different amino acids.

Amino acids Amino acids are the building blocks of proteins. They consist of a central carbon atom bound to a hydrogen atom (H), an amino group (NH_2), an acidic carboxyl group (COOH), and a side chain, often designated by the letter R (Figure 2.22). Amino acids differ in their side chains. *Nonessential* amino acids are those that our bodies can synthesize. *Essential* amino acids are those that cannot be synthesized by our bodies and must be obtained from the foods we eat.

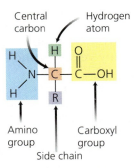

FIGURE 2.22 *Structure of an amino acid. Amino acids differ from one another in the type of R group (side chain) they contain.*

stop and think

Which three elements are found in carbohydrates, fats, and proteins? Which additional element do proteins always contain?

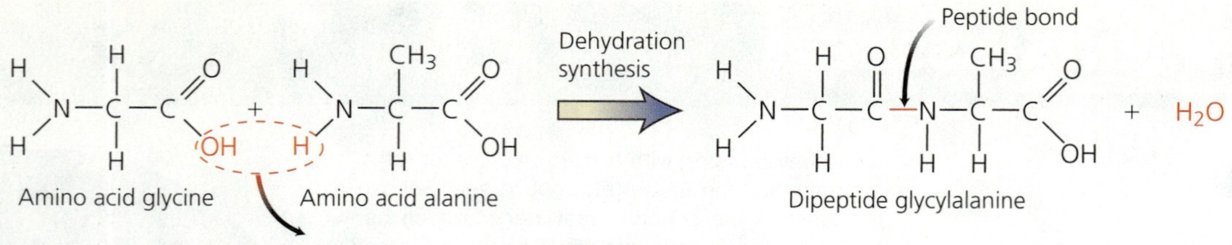

FIGURE 2.23 Formation of a peptide bond between two amino acids through dehydration synthesis. The carboxyl group (COOH) of one amino acid bonds to the amino group (NH₂) of the adjacent amino acid, releasing water.

The amino acids that form proteins are linked by peptide bonds, which are formed through dehydration synthesis. A peptide bond links the carboxyl group (COOH) of one amino acid to the amino group (NH₂) of the adjacent amino acid, as shown in Figure 2.23. Chains containing only a few amino acids are called **peptides**. *Di*peptides contain two amino acids, *tri*peptides contain three amino acids, and so on. Chains containing 10 or more amino acids are called **polypeptides**. The term *protein* is used for polypeptides with at least 50 amino acids.

Protein structure Proteins have four levels of structure: primary, secondary, tertiary, and quaternary (Figure 2.24). The **primary structure** of a protein is the particular sequence of amino acids. This sequence, determined by the genes, dictates a protein's structure and function. Even slight changes in primary structure can alter a protein's shape and ability to function. The inherited blood disorder sickle-cell anemia provides an example. This disease results from the substitution of one amino acid for another during synthesis of the protein hemoglobin, which carries oxygen in our red blood cells. This single substitution in a molecule that contains hundreds of amino acids creates a misshapen protein, which alters the shape of red blood cells. Death can result when the oddly shaped cells clog the tiny vessels of the brain and heart (see Chapter 11).

The **secondary structure** of proteins consists of patterns known as pleated sheets and helices, which are formed by certain kinds of bends and coils in the chain, as a result of hydrogen bonding. Alterations in the secondary structure of a protein normally found on the surface of nerve cells can transform the protein into an infectious agent known as a *prion*. Prions have been implicated in several diseases, including Creutzfeldt-Jakob disease in humans and mad cow disease in cattle (see Chapter 13a).

The **tertiary structure** is the overall three-dimensional shape of the protein. Hydrogen, ionic, and covalent bonds between different side chains may all contribute to tertiary structure. Changes in the environment of a protein, such as increased heat or changes in pH, can cause the molecule to unravel and lose its three-dimensional shape. This process is called **denaturation**. Even a minor change in the shape of a protein can result in loss of function.

Finally, some proteins consist of two or more polypeptide chains. Each chain, in this case, is called a *subunit*. **Quaternary structure** is the structure that results from the assembled subunits. The forces that hold the subunits in place are largely the attractions between oppositely charged side chains.

Enzymes Life is possible because of enzymes. Without them, most chemical reactions within our cells would occur far too slowly to sustain life. **Enzymes** are substances—almost always proteins—that speed up chemical reactions without being consumed in the process. Typically, reactions with enzymes proceed 10,000 to 1 million times faster than the same reactions without enzymes.

The basic process by which an enzyme speeds up a chemical reaction can be summarized by the following equation:

$$E + S \longrightarrow ES \longrightarrow E + P$$

Enzyme + Substrate → Enzyme–substrate complex → Enzyme + Product

During an enzymatic reaction, the substance at the start of the process is called the *substrate*, and the substance at the end is called the *product*. For example, the enzyme maltase speeds up the reaction in which maltose is broken down into glucose. In this reaction, maltose is the substrate, and glucose is the product. Similarly, the enzyme sucrase speeds up the breakdown of sucrose (the substrate) into molecules of glucose and fructose (the products). From these examples you can see that an enzyme's name may resemble the name of its substrate. These particular examples are decomposition reactions, in which a substance is broken down into its component parts. Enzymes also increase the speed of many synthesis reactions.

During reactions promoted by enzymes, the substrate binds to a specific location, called the **active site**, on the enzyme, to form an **enzyme–substrate complex** (Figure 2.25). This binding orients the substrate molecules so they can react. The substrate is converted to one or more products, which leave the active site, allowing the enzyme to bind to another substrate molecule. The entire process occurs very rapidly. One estimate suggests that a typical enzyme can convert about 1000 molecules of substrate into product every second.

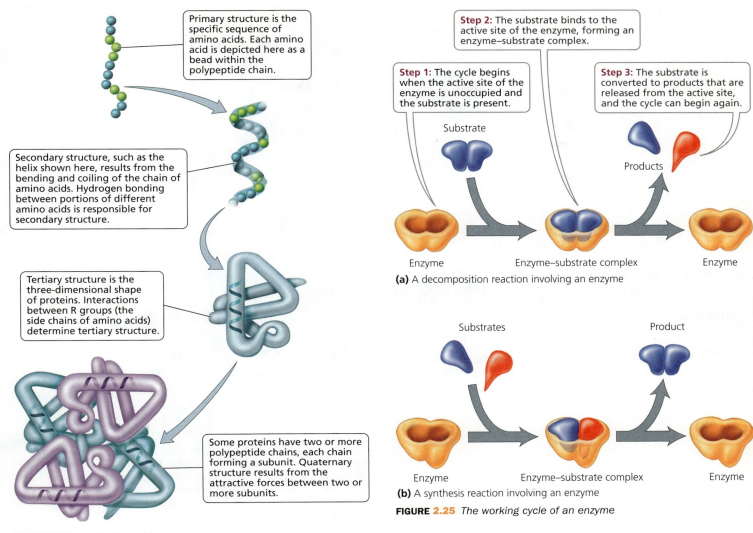

FIGURE 2.24 Levels of protein structure

FIGURE 2.25 The working cycle of an enzyme

Enzymes are very specific in their activity; each is capable of binding to and acting on only one or, at most, a few particular substrates. This specificity is due to the unique shape of each enzyme's active site. The enzyme's active site and the substrate fit together like pieces of a jigsaw puzzle.

Sometimes enzymes need *cofactors,* nonprotein substances that help them convert substrate to product. Some cofactors permanently reside at the enzyme's active site, whereas others bind to the active site at the same time as the substrate. Some cofactors are the organic (carbon-containing) substances we know as vitamins. Organic cofactors are called *coenzymes.* Other cofactors are inorganic (non-carbon-containing) substances, such as zinc or iron.

Enzyme deficiencies can affect our health. Lactase deficiency is one example. Lactase is the enzyme needed to digest the lactose in milk products, breaking it down to glucose and galactose. Infants and young children usually produce enough lactase, but many adults do not. For these adults, consumption of milk and milk products can lead to diarrhea, cramps, and bloating, caused by undigested lactose passing into the large intestine, where it feeds resident bacteria. The bacteria, in turn, produce gas and lactic acid, which irritate the bowels. The milk industry has responded to this problem (often called *lactose intolerance*) by marketing lactose-reduced milk. Tablets and caplets that contain the enzyme lactase are also available.

Nucleic Acids and Nucleotides

In our discussion of protein structure we mentioned that genes determine the protein's primary structure, which is the sequence of amino acids. Genes, our units of inheritance, are segments of long polymers called **deoxyribonucleic acid (DNA)**. DNA is one of the two types of nucleic acids.

Nucleotides The two nucleic acids in our cells are DNA and **ribonucleic acid (RNA)**. Both are polymers of smaller units called **nucleotides**, joined into chains through dehydration synthesis. Every nucleotide monomer consists of a five-carbon

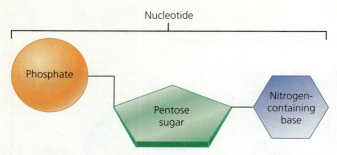

FIGURE 2.26 Structure of a nucleotide. Nucleotides consist of a five-carbon (pentose) sugar bonded to a phosphate molecule and one of five nitrogen-containing bases. Nucleotides are the building blocks of nucleic acids.

TABLE 2.4 Review of the Structural Differences between RNA and DNA

Characteristic	RNA	DNA
Sugar	Ribose	Deoxyribose
Bases	Adenine, guanine, cytosine, uracil	Adenine, guanine, cytosine, thymine
Number of strands	One	Two; twisted to form double helix

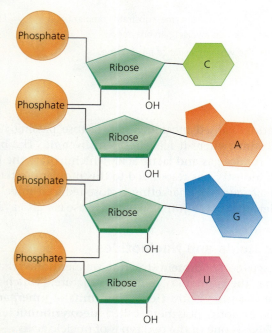

FIGURE 2.27 RNA is a single-stranded nucleic acid. It is formed by the linking together of nucleotides composed of the sugar ribose, a phosphate group, and the nitrogen-containing bases cytosine (C), adenine (A), guanine (G), and uracil (U).

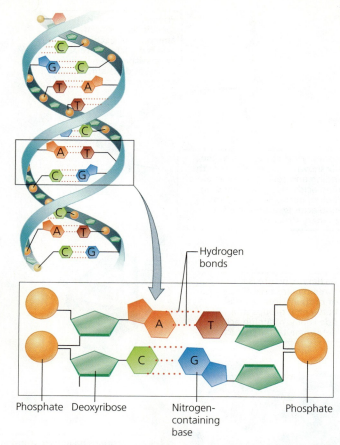

FIGURE 2.28 DNA is a nucleic acid in which two chains of nucleotides twist around one another to form a double helix. The two chains are held together by hydrogen bonds between the nitrogen-containing bases. Each nucleotide of DNA contains the pentose sugar deoxyribose, a phosphate group, and one of the following four nitrogen-containing bases: adenine (A), thymine (T), cytosine (C), or guanine (G).

(pentose) sugar bonded to one of five nitrogen-containing bases and at least one phosphate group (Figure 2.26). The five nitrogen-containing bases are adenine, guanine, cytosine, thymine, and uracil. The bases cytosine, thymine, and uracil have a single ring made of carbon and nitrogen atoms; adenine and guanine have two such rings. The sequence of bases in DNA and RNA determines the sequence of amino acids in a protein. DNA, as we said earlier, is the nucleic acid found in genes. RNA, in various forms, converts the genetic information found in DNA into proteins.

DNA and RNA Key differences in the structures of RNA and DNA are summarized in Table 2.4. RNA is a single strand of nucleotides (Figure 2.27). The five-carbon sugar in RNA is ribose. The nitrogen-containing bases in RNA are cytosine (C), adenine (A), guanine (G), and uracil (U). In contrast, DNA is a double-stranded chain (Figure 2.28). Its two parallel strands, held together by hydrogen bonds between the nitrogen-containing bases, twist around one another to form a double helix. The five-carbon sugar in DNA is deoxyribose. The nitrogen-containing bases in DNA are adenine (A), thymine (T), cytosine (C), and guanine (G).

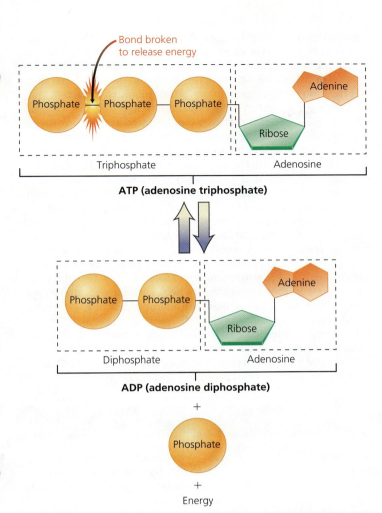

FIGURE 2.29 *Structure and function of adenosine triphosphate (ATP). This nucleotide consists of the sugar ribose, the base adenine, and three phosphate groups. The phosphate bonds of ATP are unstable. When cells need energy, the last phosphate bond is broken, yielding adenosine diphosphate (ADP), a phosphate molecule, and energy.*

ATP At this moment within your cells, many molecules of the nucleotide **adenosine triphosphate (ATP)** are each losing a phosphate group. As the phosphate group is lost, energy that the molecule stored by holding on to the phosphate group is released. Your cells trap that energy and use it to perform work. It is because of this activity that you are able to read this book. ATP consists of the sugar ribose, the base adenine, and three phosphate groups, attached to the molecule by phosphate bonds. It is formed from adenosine diphosphate (ADP) by covalent bonding of a phosphate group to the ADP in an energy-requiring reaction. The energy absorbed during the reaction is stored in the new phosphate bond. This high-energy phosphate bond is easily broken when the cell requires energy (Figure 2.29).

ATP is often described as the energy currency of cells. All energy from the breakdown of molecules, such as glucose, must be channeled through ATP before the body can use it.

looking ahead
In Chapter 2, we learned about the major molecules of life. In Chapter 3, we see how these molecules contribute to cell structure and function.

HIGHLIGHTING THE CONCEPTS

2.1 The Nature of Atoms (pp. 20–23)
- Atoms consist of subatomic particles called protons, neutrons, and electrons. Protons have a positive charge. Neutrons have no charge. Both are found in the nucleus and have atomic masses equal to 1. Electrons have a negative charge, weigh almost nothing, and are found around the nucleus in shells.
- Each element is made of atoms containing a certain number of protons. The atomic number of an element is the number of protons in one of its atoms. The atomic weight for any atom equals the number of protons plus the number of neutrons.
- Isotopes are atoms that have the same number of protons but different numbers of neutrons. Some isotopes, called radioisotopes, emit radiation.

2.2 Compounds and Chemical Bonds (pp. 23–25)
- Elements combine to form compounds. The characteristics of compounds are usually different from those of the elements that form them.
- When atoms join to form compounds, chemical bonds form between them. Covalent bonds form when atoms share electrons. Atoms that have lost or gained electrons have an electrical charge and are called ions. The attraction between oppositely charged ions is an ionic bond.

2.3 The Role of Water in Life (pp. 25–29)
- Sometimes the sharing of electrons in a covalent bond is unequal, resulting in a polar covalent bond and sometimes a polar molecule. Hydrogen bonds are weak attractive forces between the charged regions of polar molecules.
- Water is an important component of the human body because of its polarity (the tendency to have positive and negative regions) and hydrogen bonding. Its polarity makes water an excellent solvent. Its hydrogen bonds give water a high heat capacity, which helps the body's internal temperature remain constant, and a high heat of vaporization, which helps cool the surface of the body when perspiration evaporates.

- Acids increase the concentration of H+ in solution. Bases decrease the concentration of H+ in solution. The strengths of acids and bases are measured on the pH scale. Biological fluids must remain within a narrow pH range. Buffers prevent dramatic changes in pH.

2.4 Major Molecules of Life (pp. 29–39)

- A polymer is a large molecule made of many smaller molecules, called monomers. Polymers form through dehydration synthesis (removal of a water molecule) and are broken apart by hydrolysis (addition of a water molecule).
- Carbohydrates are sugars and starches that provide fuel for the human body. Monosaccharides are the smallest monomers of carbohydrates. Oligosaccharides are chains of a few monosaccharides. Disaccharides consist of two monosaccharides and are one type of oligosaccharide. Polysaccharides are complex carbohydrates formed by large numbers of monosaccharides linked together.
- Lipids, such as triglycerides, phospholipids, and steroids, are nonpolar molecules that do not dissolve in water. Triglycerides (fats and oils) are made of glycerol and three fatty acids and function in long-term energy storage. Phospholipids are important components of plasma membranes. Steroids include cholesterol, a component of the plasma membrane that also serves as the foundation for steroid hormones.
- Proteins are polymers made from a set of 20 amino acids linked together through dehydration synthesis to form chains. There are four levels of protein structure. Primary structure is the specific sequence of amino acids in a protein. Secondary structure results from the bending and coiling of the amino acid chain into pleated sheets or helices. Tertiary structure is the three-dimensional shape of the protein. Some proteins consist of subunits. Attractive forces between the subunits produce quaternary structure.
- Enzymes are proteins that speed up chemical reactions without being consumed in the process. The substrate binds to the enzyme at the active site, forming an enzyme–substrate complex, and is then converted to products that leave the active site.
- Deoxyribonucleic acid (DNA) and ribonucleic acid (RNA) are polymers of nucleotides. A nucleotide consists of a five-carbon sugar bonded to one of five nitrogen-containing bases and a phosphate group. Genes are stretches of DNA that determine the sequence of amino acids during protein synthesis. DNA is double stranded; its sugar component is deoxyribose; and its nitrogen-containing bases are adenine, guanine, cytosine, and thymine. RNA is a single-stranded molecule that also plays a major role in protein synthesis. The sugar in RNA is ribose, and the nitrogen-containing bases are adenine, guanine, cytosine, and uracil.
- Adenosine triphosphate (ATP), the energy currency of cells, is a nucleotide made of the sugar ribose, the base adenine, and three phosphate groups. When cells require energy, one of the high-energy phosphate bonds is broken, and energy is released.

RECOGNIZING KEY TERMS

chemistry p. 20
matter p. 20
atom p. 20
element p. 21
isotope p. 21
radioisotope p. 22
compound p. 23
covalent bond p. 23
molecule p. 24
ion p. 24
ionic bond p. 24
hydrogen bond p. 26
acid p. 28
base p. 28

pH p. 28
pH scale p. 28
buffer p. 28
macromolecule p. 29
polymer p. 30
monomer p. 30
dehydration synthesis p. 30
hydrolysis p. 30
carbohydrate p. 31
monosaccharide p. 31
oligosaccharide p. 31
disaccharide p. 31
polysaccharide p. 31
starch p. 31

glycogen p. 31
cellulose p. 31
lipid p. 31
triglyceride p. 32
fatty acid p. 32
phospholipid p. 32
hydrophobic p. 33
hydrophilic p. 33
steroid p. 33
protein p. 35
amino acid p. 35
peptide p. 36
polypeptide p. 36
primary structure p. 36

secondary structure p. 36
tertiary structure p. 36
denaturation p. 36
quaternary structure p. 36
enzyme p. 36
active site p. 36
enzyme–substrate complex p. 36
deoxyribonucleic acid (DNA) p. 37
ribonucleic acid (RNA) p. 37
nucleotide p. 37
adenosine triphosphate (ATP) p. 39

REVIEWING THE CONCEPTS

1. Define the following terms: *atom, element, compound,* and *molecule.* pp. 20–24
2. What is an isotope? p. 21
3. How is radiation used to diagnose or cure illness? pp. 22–23
4. What characteristics of water make it a critical component of the body? pp. 27–28
5. How are polymers formed and broken? Give three examples of polymers and their component monomers. pp. 30–39
6. Name two important energy-storage polysaccharides and one important structural polysaccharide. p. 31
7. Describe the structure of a phospholipid. What important roles do phospholipids play in the human body? pp. 32–33
8. Describe the structure and function of ATP. pp. 38–39
9. Compare and contrast the ways in which proteins and nucleic acids are used in the body. pp. 35, 38, 39
10. Choose the *incorrect* statement:
    a. Electrons are found outside the nucleus at certain energy levels (shells).
    b. For any atom, the number of protons plus the number of electrons equals the atomic weight.
    c. The atomic number equals the number of protons in the nucleus.
    d. Electrons have negligible mass.
11. Covalent bonds
    a. form when two or more atoms share electrons.
    b. result from the mutual attraction of oppositely charged ions.
    c. form between a hydrogen atom on one water molecule and an oxygen atom on another water molecule.
    d. involve the transfer of electrons.

12. Acids
    a. have a pH of 7.
    b. with a pH of 3 have 100 times the amount of H+ than do acids with a pH of 4.
    c. increase the concentration of OH+ in solution.
    d. release H+ when added to water.
13. Choose the *incorrect* statement:
    a. Peptide bonds are formed through hydrolysis.
    b. Vitamins sometimes help an enzyme convert substrate to product.
    c. The primary structure of a protein is the precise sequence of amino acids.
    d. Quaternary structure results from attractive forces between the subunits of a protein.
14. Hydrogen bonds
    a. are stronger than either ionic or covalent bonds.
    b. form between a slightly positively charged hydrogen atom and a slightly negatively charged atom nearby.
    c. maintain the shape of proteins and DNA.
    d. b and c
15. Water
    a. is a nonpolar molecule and therefore an excellent solvent.
    b. has a high heat capacity and therefore helps maintain a constant body temperature.
    c. has a low heat of vaporization and therefore helps prevent overheating of the body.
    d. makes up about 25% of the human body.
16. Carbohydrates
    a. consist of chains of amino acids.
    b. supply our cells with energy.
    c. contain glycerol.
    d. function as enzymes.
17. Triglycerides
    a. have one molecule of glycerol and three fatty acids.
    b. are poor sources of energy.
    c. are saturated when there are two or more double bonds linking carbon atoms.
    d. are major components of plasma membranes.
18. Enzymes
    a. speed up chemical reactions and are consumed in the process.
    b. function only in decomposition reactions.
    c. are usually nonspecific and therefore capable of binding to many different substrates.
    d. have locations, known as active sites, to which the substrate binds.
19. Choose the *incorrect* statement:
    a. The base uracil is found in RNA but not in DNA.
    b. The two chains of RNA are held together by hydrogen bonds between the bases.
    c. Nucleotides are the monomers of nucleic acids.
    d. DNA molecules contain genes, which specify the sequence of amino acids in proteins.
20. Choose the *incorrect* statement:
    a. Most biochemical reactions in the human body occur around pH 7.
    b. Buffers prevent dramatic changes in pH.
    c. HCl is an important buffering system in the blood.
    d. Burning of fossil fuels causes acid rain.
21. Changes in temperature or pH can cause a protein to lose its three-dimensional shape and become nonfunctional. This process is called _____.
22. _____ are arranged in a double layer (bilayer) that forms the plasma membrane of cells.

APPLYING THE CONCEPTS

1. A friend eyes your lunch and begins to lecture you on the perils of high-fat foods. You decide to acknowledge the health risks of eating a high-fat diet but also to point out to her the important roles that lipids play in your body. What will you say?
2. Bill claims that eating fruits and vegetables is overrated because humans lack the enzyme needed to digest cellulose and thus cellulose passes unchanged through our digestive tract. Is Bill correct that we gain nothing from eating food that contains cellulose?
3. For years your mother has been telling you to drink several large glasses of water each day. Having just read this chapter, you now understand the critical roles that water plays in your body. You vow to admit during your next phone call home that she has been right all along. Explain why dehydration might be dangerous.
4. Dental x-ray films are taken using small doses of radiation and are used to detect tooth decay, injuries to the roots of teeth, and problems with the bones supporting the teeth. A dental professional covers the patient with a lead apron that runs from the neck to the abdomen and then leaves the room to take the x-ray film. Why are such precautions necessary?

BECOMING INFORMATION LITERATE

In nutrition class, you and your fellow students are scheduled to debate the pros and cons of irradiating food. You have been assigned the position of supporting food irradiation. Use at least three reliable sources (books, scientific journals, Internet sites) to develop your arguments in favor of food irradiation and to gather information needed to refute potential arguments made by the opposing side. List the sources you used, and describe why you considered each source reliable.

MasteringBiology®

Go to MasteringBiology for practice quizzes, activities, eText, videos, current events, and more.

3 The Cell

Microscopes allow us to see cells and the diverse organelles inside them. Each organelle performs a specific function for the cell.

Did You Know?

- The average adult human is made up of 100 trillion cells.
- Bacterial cells far outnumber human cells in our bodies.

3.1 Eukaryotic Cells Compared with Prokaryotic Cells

3.2 Cell Size and Microscopy

3.3 Cell Structure and Function

3.4 Plasma Membrane

3.5 Organelles

3.6 Cytoskeleton

3.7 Cellular Respiration and Fermentation in the Generation of ATP

ENVIRONMENTAL ISSUE
The Deadly Interaction between Asbestos and Lysosomes

HEALTH ISSUE
Mitochondrial Diseases

In Chapter 2 we learned about the major molecules of life, which are carbohydrates, lipids, proteins, and nucleic acids. In this chapter, we see how these molecules contribute to cell structure, direct activities of cells, and serve as sources of cellular energy. We first compare two basic types of cells—eukaryotic cells and prokaryotic cells. Then we explore how eukaryotic cells work by examining the structures shared by all cells in the human body. Finally, we explore the ways that cells obtain the energy they need to do their work of running the body.

3.1 Eukaryotic Cells Compared with Prokaryotic Cells

The **cell theory** is a fundamental organizing principle of biology that guides the way biologists think about living things. The cell theory states that (1) a cell is the smallest unit of life; (2) cells make up all living things, from unicellular to multicellular organisms; and (3) new cells can arise only from preexisting cells.

As we mentioned, there are two basic types of cells—eukaryotic cells and prokaryotic cells. **Prokaryotic cells** are structurally simpler and typically smaller than eukaryotic cells. They are limited to bacteria and another group of microscopic organisms called Archaea. You are probably already aware of bacteria, some of which inhabit your body. Many bacterial inhabitants are harmless, but others can cause illness (see Chapters 13a and 15a). Archaea may be less familiar to you. They include species that inhabit extreme environments such as the high-saline Great Salt Lake or the hot sulfur springs of Yellowstone National Park. Most prokaryotic cells are surrounded by a rigid cell wall, as shown in Figure 3.1.

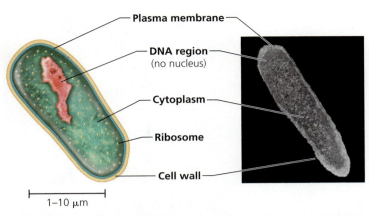

FIGURE 3.1 *Prokaryotic cells, such as a bacterium, lack internal membrane-bound organelles.*

The cells of plants, animals, and all other organisms except bacteria and archaea are eukaryotic. All the cells that make up your body, therefore, are eukaryotic. The difference between eukaryotic and prokaryotic cells relates to the presence or absence of membrane-bound organelles. An **organelle**, or "little organ," is a component within a cell that carries out specific functions. Some organelles have membranes, and others do not. Nonmembranous organelles, such as ribosomes and cytoskeletal elements, are found in both prokaryotic and eukaryotic cells. Unique to **eukaryotic cells**, however, are membrane-bound organelles, such as mitochondria and endoplasmic reticulum (Figure 3.2). Another of the membrane-bound organelles found in all typical eukaryotic cells is a well-defined nucleus containing DNA. Note that in prokaryotes, a membrane does not surround the DNA (refer, again, to Figure 3.1). Among

Q *The nucleus is a membrane-bound organelle. Look closely at the other organelles in the cell, and note their functions. Based on the structure and functions of the organelles shown, list those that you think are membrane-bound organelles and those that are nonmembranous organelles.*

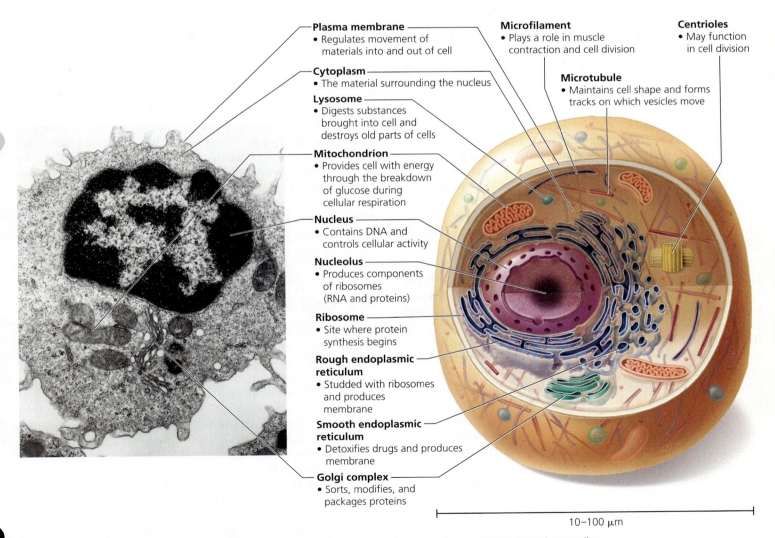

FIGURE 3.2 *Eukaryotic cells, such as the generalized animal cell shown here, have internal membrane-bound organelles.*

A *The membrane-bound organelles include the nucleus, rough endoplasmic reticulum, smooth endoplasmic reticulum, Golgi complex, mitochondrion, and lysosome. The nonmembranous organelles include ribosomes, microfilaments, centrioles, and microtubules.*

TABLE 3.1 Review of Features of Prokaryotic and Eukaryotic Cells

Feature	Prokaryotic Cells	Eukaryotic Cells
Organisms	Bacteria, archaea	Plants, animals, fungi, protists
Size	1–10 μm across	10–100 μm across
Membrane-bound organelles	Absent	Present
DNA form	Circular	Coiled, linear strands
DNA location	Cytoplasm	Nucleus
Internal membranes	Rare	Many
Cytoskeleton	Present	Present

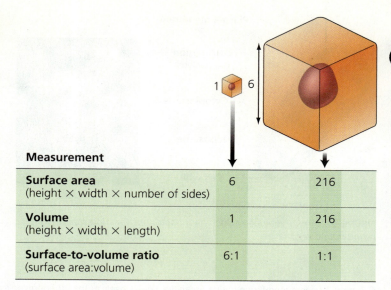

Measurement		
Surface area (height × width × number of sides)	6	216
Volume (height × width × length)	1	216
Surface-to-volume ratio (surface area:volume)	6:1	1:1

FIGURE 3.3 *Cells must remain small in size because the ratio of surface area to volume decreases rapidly as cell size increases.*

eukaryotes, plant cells have cell walls, but animal cells do not. Table 3.1 reviews the major differences between eukaryotic and prokaryotic cells.

3.2 Cell Size and Microscopy

Most eukaryotic and prokaryotic cells are so small that they are typically measured in micrometers (μm), which are equal to 1/1,000,000 meter (m). (An obvious exception is the chicken egg.) The small size of cells is dictated by a physical relationship known as the **surface-to-volume ratio**. As a cell gets larger, its surface area increases much more slowly than its volume (Figure 3.3). Nutrients enter a cell, and wastes leave a cell, at its surface. Therefore, a large cell would have difficulty moving all the nutrients it needs and all the wastes it produces across its inadequate surface and therefore would die. A small cell, in contrast, has sufficient surface for the uptake and removal of substances and would survive.

Because most cells are so small, you need a microscope to see them. Throughout this book you will see micrographs, which are photographs obtained using a microscope (Figure 3.4). Microscopic specimens can be imaged using beams of either light or electrons. Light microscopes, which are used in many classrooms, have the advantage of being relatively inexpensive and simple to operate. Electron microscopes, though more complex and expensive, have the capacity to reveal finer details because the wavelength of an electron beam is smaller than the wavelengths of visible light. Whether light or electrons are used, the beam can be either transmitted through a thinly sliced specimen or bounced off of the specimen's surface.

Figure 3.4a is a light micrograph showing three striated muscle cells that have been stained with biological dyes to increase the contrast between different cellular components. The nuclei visible in this picture are colored dark purple because the dye has an affinity for acidic components in the cell, such as DNA. Figure 3.4b is a transmission electron micrograph that shows the structure of striated muscle cells in more detail than is possible using light to image the tissue. In this case, the contrast between different cellular components is produced by staining the tissue with heavy metals.

Different components of the cells absorb different amounts of the metals. Components that readily absorb the metals differentially block the electron beam from passing through the sample. Figure 3.4c is a scanning electron micrograph produced by bouncing an electron beam off the surface of several striated muscle cells. The beam is scanned across the surface of the sample, and electrons that bounce off the surface are collected by a detector. For every point that is scanned, the number of electrons reaching the detector is used to calculate the relative brightness of that spot on the sample. This information is used to construct the image. Images produced with electron beams are not in color. The pictures shown in Figures 3.4b and 3.4c have been colored to highlight certain features, an improvement made possible by computer-assisted processing of images. Other micrographs in the text have also been colored.

3.3 Cell Structure and Function

The structure of a cell exquisitely reflects its functions. For example, few human cells are more specialized than sperm or eggs, the cells that carry genetic information and other materials needed to make a new individual of the next generation. A sperm is specialized for swimming to the egg and fertilizing it. As such, a sperm is streamlined and equipped with a whiplike tail. In the head of the sperm is an enzyme-containing sac that spills open to release enzymes that digest a path through the layers of cells surrounding the egg. In contrast, the egg is immobile and much larger than a typical cell because it is literally packed with nutrients and other materials needed to initiate development. A mature red blood cell is another example of a cell whose structure reflects its function. As the red blood cell matures, it extrudes its nucleus and most organelles, leaving more space for hemoglobin, the protein that transports oxygen. A mature

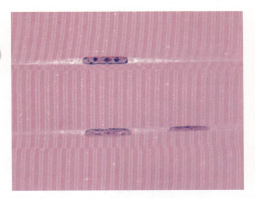
(a) Striated muscle cells viewed with a light microscope

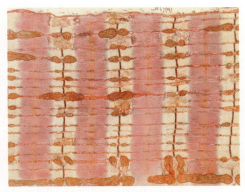
(b) Striated muscle cells viewed with a transmission electron microscope

(c) Striated muscle cells viewed with a scanning electron microscope

FIGURE 3.4 *Micrographs are photographs taken through a microscope. Here, striated muscle cells have been photographed using three different types of microscope. Electron microscopes use beams of electrons to produce images with finer details than those viewed with light microscopes.*

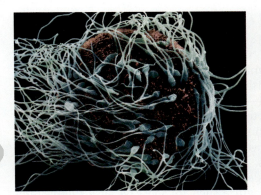
(a) A sperm is specialized to be highly mobile. In contrast, an egg is specialized to be large, immobile, and packed with material needed to initiate development.

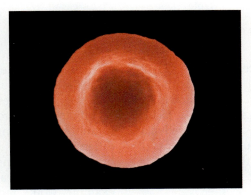
(b) A mature red blood cell, devoid of most organelles, is specialized for carrying oxygen.

(c) A cardiac muscle cell is specialized for contraction and for propagating the signal for contraction.

FIGURE 3.5 *A cell's structure reflects its specific function. These cell types from the human body illustrate the close tie between structure and function.*

red blood cell is thus an exception to the rule that eukaryotic cells have a well-defined nucleus and other membrane-bound organelles. Consider, also, a cardiac muscle cell. This cell is specialized for contraction and for propagating the signal for contraction from one muscle cell to the next. Thus it is filled with contractile proteins and is joined to adjacent cells by specialized junctions that strengthen cardiac tissue and promote rapid conduction of impulses throughout the heart. In each of these cases, careful study of the cell's structure provides excellent clues to its function, and vice versa (Figure 3.5).

3.4 Plasma Membrane

We begin our examination of the cell at its outer surface—the **plasma membrane**. This remarkably thin outer covering controls the movement of substances both into and out of the cell. Because the concentrations of substances in a cell's interior are critically balanced, molecules and ions are not permitted to move randomly in and out.

Both prokaryotic and eukaryotic cells have a plasma membrane, but only eukaryotic cells also contain internal membranes that divide the cell into many compartments. Each compartment contains its own assortment of enzymes and is specialized for particular functions. In general, the principles described for the plasma membrane also apply to the membranes inside the cell.

Plasma Membrane Structure

The plasma membrane is made of lipids, proteins, and carbohydrates. Recall from Chapter 2 that phospholipids are the major components of the plasma membrane. These molecules, with their hydrophilic ("water-loving") heads and hydrophobic ("water-fearing") tails, form a double layer—called the

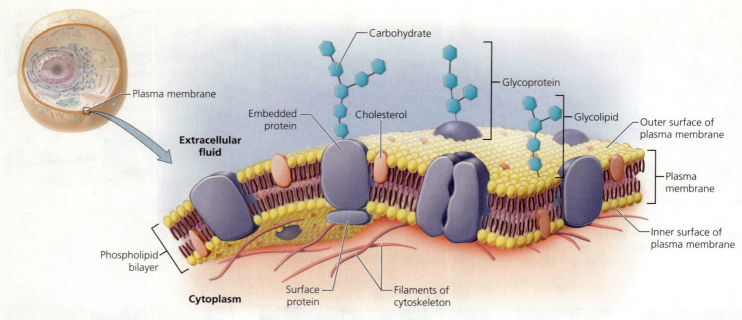

FIGURE 3.6 *The structure of the plasma membrane of a cell according to the fluid mosaic model*

phospholipid bilayer—at the surface of the cell (Figure 3.6). The hydrophilic heads facing outside the cell interact with the **extracellular fluid** (also known as *interstitial fluid*), which is the watery solution outside cells. The hydrophilic heads facing inside the cell interact with the **cytoplasm**, which is the jellylike solution inside the cell. The cytoplasm includes all contents of the cell between the plasma membrane and the nucleus. Within the phospholipid bilayer, the hydrophobic tails point toward each other and hold the plasma membrane together.

Interspersed in the phospholipid bilayer are proteins, as seen in Figure 3.6. Some proteins are embedded in the membrane, and some of these span the bilayer completely. Other proteins are simply attached to the inner or outer surface of the membrane. Molecules of cholesterol are also scattered throughout the bilayer.

As you can see in Figure 3.6, carbohydrates attach only to the outer surface of the plasma membrane. Most of these carbohydrates are attached to proteins, forming glycoproteins. Other carbohydrates are attached to lipids, forming glycolipids.

The structure of the plasma membrane is often described as a **fluid mosaic**. The proteins are interspersed among the lipid molecules like tiles of different colors within a mosaic. Many of the proteins are able to move sideways through the bilayer to some degree, giving the membrane its fluid quality.

Plasma Membrane Functions

The plasma membrane performs several vital functions for the cell. First, by imposing a boundary between the cell's internal and external environment, the plasma membrane maintains the cell's structural integrity. Second, the structure of the plasma membrane regulates the movement of substances into and out of the cell, permitting entry to some substances but not others. For this reason, the membrane is often described as being **selectively permeable**. You will read more about the transport of materials across the plasma membrane in the next section of this chapter.

The plasma membrane also functions in cell–cell recognition. Cells distinguish one type of cell from another by recognizing molecules—often glycoproteins—on their surface. Membrane glycoproteins differ from one species of organism to another and among individuals of the same species. Even different cell types within an individual have different membrane glycoproteins. This variation allows the body to recognize foreign invaders such as bacteria. Your own body, for example, would recognize such invaders because the bacteria lack the surface molecules found on your cells. The bacteria, in turn, "read" the different surface molecules of your cells to settle preferentially on some kinds of cells but not others.

Another important function of the plasma membrane is communication between cells. Such communication relies on *receptors*, specialized proteins in the plasma membrane (or inside the cell) that bind particular substances that affect cell activities. For example, hormones secreted by one group of cells may bind to receptors in the plasma membranes of other cells. The receptors then relay a signal to proteins inside the cell, which transmit the message to other nearby molecules. Through a series of chemical reactions, the hormone's "message" ultimately initiates a response by the recipient cell, perhaps causing it to release a certain chemical.

Finally, the plasma membrane plays an important role in binding pairs or groups of cells together. **Cell adhesion molecules (CAMs)** extend from the plasma membranes of most cells and help attach the cells to one another, especially during the formation of tissues and organs in an embryo. The functions of the plasma membrane are as follows:

- Maintain structural integrity of the cell
- Regulate movement of substances into and out of the cell
- Provide recognition between cells
- Provide communication between cells
- Stick cells together to form tissues and organs

stop and think

Of the five functions of the plasma membrane, which might explain the difficulty of transplanting tissues and organs successfully from one body to another? Why would rejection of such transplants occur? Under what circumstances might one body accept a tissue graft or organ from another?

Movement Across the Plasma Membrane

Recall that an important function of the plasma membrane is to control which substances move into and out of the cell. Substances cross the plasma membrane in several ways. These methods are described as either active (requiring the cell to expend energy) or passive (requiring no energy expenditure by the cell).

Simple diffusion Some materials cross the plasma membrane passively through **simple diffusion**, the random movement of a substance from a region of higher concentration to a region of lower concentration. *Concentration* is the number of molecules of a substance in a particular volume, and a **concentration gradient** is a difference in the relative number of molecules or ions of a given substance in two adjacent areas. The end result of simple diffusion is an equal distribution of the substance in the two areas; in other words, diffusion tends to eliminate the concentration gradient. Consider what happens when someone is cooking bacon in the kitchen. At first, the smell of bacon is localized in the kitchen. Soon, however, the smell permeates adjoining rooms, too, as odor molecules move from where they are more concentrated (the kitchen) to where they are less concentrated (other parts of the house). Eventually the odor molecules are equally distributed, but they still move randomly in all directions. Likewise, when a substance diffuses across a membrane from a region of higher concentration to a region of lower concentration, the movement of its molecules does not stop once the concentration has equalized. Instead, the molecules continue to move randomly back and forth across the membrane. The rate of movement in each direction, however, is now the same. Substances such as carbon dioxide and oxygen diffuse through the plasma membrane of our cells (Figure 3.7).

Facilitated diffusion Water-soluble substances are repelled by lipids, so they cannot move through the phospholipid bilayer by simple diffusion. If they are to cross a cell membrane, their transport must be assisted, or "facilitated," by certain proteins within the membrane. Some of these proteins, called carrier proteins, bind to a particular water-soluble substance. Such binding prompts a change in the protein's shape and has the effect of carrying the substance to the other side of the membrane. Other proteins form channels through which certain water-soluble substances can move. **Facilitated diffusion** is the movement of a substance from a region of higher concentration to a region of lower concentration with the aid of a membrane protein. Molecules of glucose, for example, enter fat cells by facilitated diffusion. In this example, a molecule of glucose in the extracellular fluid binds to a carrier protein in the plasma membrane, which helps to move the glucose molecule from outside to inside the fat cell (Figure 3.8). Facilitated diffusion does not require energy and is thus a form of passive transport.

Osmosis Osmosis is a type of diffusion in which water moves across a plasma membrane or any other selectively permeable membrane from a region of higher water concentration to a

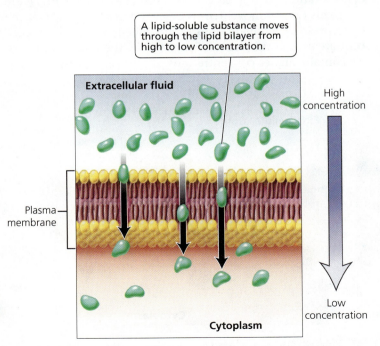

FIGURE 3.7 Simple diffusion is the random movement of a substance from a region of higher concentration to a region of lower concentration.

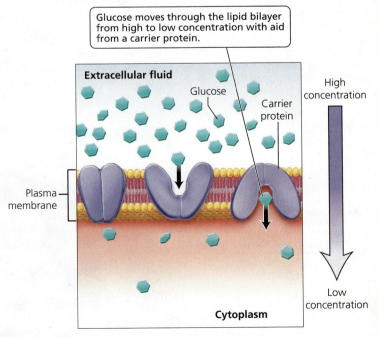

FIGURE 3.8 Facilitated diffusion is the movement through the plasma membrane of a substance from a region of higher concentration to a region of lower concentration with the aid of a membrane protein that acts as a channel or a carrier protein.

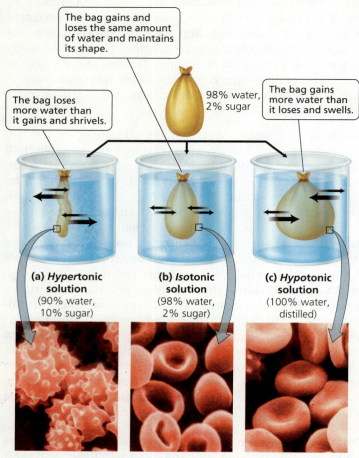

FIGURE 3.9 *Osmosis is the diffusion of water across a selectively permeable membrane. The drawings show what happens when a membranous bag through which water but not sugar can move is placed in solutions that are (a) hypertonic, (b) isotonic, or (c) hypotonic to the solution inside the bag. The width of the black arrows corresponds to the amount of water moving into and out of the bag. The photographs show what happens to red blood cells when placed in the three kinds of solutions. Red blood cells are normally shaped like flattened disks, as in part (b).*

region of lower water concentration. The movement of water occurs in response to a concentration gradient of a dissolved substance (solute). Consider what happens when a substance such as table sugar (in this case, our solute) is dissolved in water (our solvent) in a membranous bag through which water, but not sugar, can move. Keep in mind that when solute concentration is low, water concentration is high; and when solute concentration is high, water concentration is low. If the membranous bag is placed into a **hypertonic solution**, meaning a solution whose solute concentration is higher than that inside the bag, more water moves out of the bag than in, causing the bag to shrivel (Figure 3.9a). If, however, the bag is placed into an **isotonic solution**, one with the same solute (sugar) concentration as inside the bag, there is no net movement of water in either direction, and the bag maintains its original shape (Figure 3.9b). When the bag is placed into a **hypotonic solution**, in which the concentration of solute is lower than that inside the bag, more water moves into the bag than out, causing the bag to swell and possibly burst (Figure 3.9c). Osmosis does not require energy and is thus a form of passive transport.

Red blood cells behave the same way the bag in our example does, as shown at the bottom of Figure 3.9. Red blood cells move through a fluid, called plasma. As the figure illustrates, the shape of red blood cells responds to different levels of solute concentration in the plasma.

Active transport Active transport is a mechanism that moves substances across plasma membranes with the aid of a carrier protein and energy supplied by the cell (through the breakdown of ATP; see Chapter 2). So far in our discussion of movement across plasma membranes, we have described substances moving from regions of higher concentration to regions of lower concentration. However, in most cases of active transport, substances are moved from regions of lower concentration to higher concentration, as shown in Figure 3.10. This type of movement is described as going "against the concentration gradient" and occurs when cells need to concentrate certain substances. For example, the cells in our bodies contain higher concentrations of potassium ions (K+) and

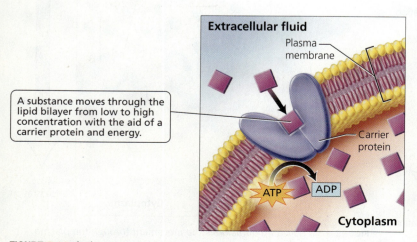
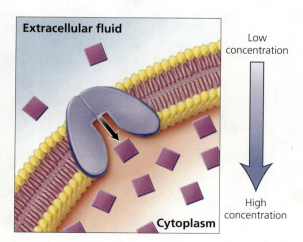

FIGURE 3.10 *Active transport is the movement of molecules across the plasma membrane, often from an area of lower concentration to one of higher concentration, with help from a carrier protein and energy, usually in the form of ATP.*

lower concentrations of sodium ions (Na+) than their surroundings. Through active transport, proteins in the plasma membrane help maintain these conditions, pumping potassium ions into the cell and sodium ions out of the cell. In this example, both potassium and sodium are moving from regions of lower concentration to regions of higher concentration.

Endocytosis Most small molecules cross the plasma membrane by simple diffusion, facilitated diffusion, or active transport. Large molecules, single-celled organisms such as bacteria, and droplets of fluid containing dissolved substances enter cells through endocytosis (Figure 3.11). In **endocytosis**, a region of the plasma membrane engulfs the substance to be ingested and then pinches off from the rest of the membrane; in this way it encloses the substance in a saclike structure called a **vesicle**. The vesicle then travels through the cytoplasm. Two types of endocytosis are phagocytosis ("cell eating")

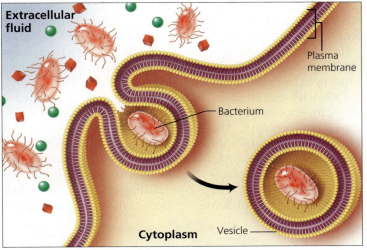

(a) Phagocytosis ("cell eating") occurs when cells engulf bacteria or other large particles.

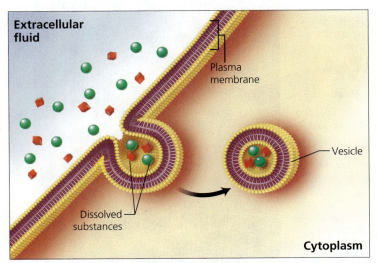

(b) Pinocytosis ("cell drinking") occurs when cells engulf droplets of extracellular fluid and the dissolved substances therein.

FIGURE 3.11 Endocytosis—phagocytosis or pinocytosis—occurs when a localized region of the plasma membrane surrounds a bacterium, large molecule, or fluid containing dissolved substances and then pinches inward to form a vesicle that moves into the cell.

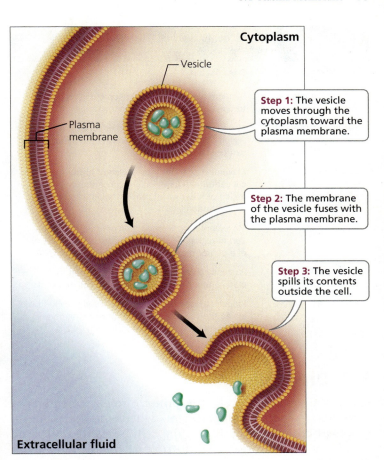

FIGURE 3.12 Cells package large molecules in membrane-bound vesicles, which then spill their contents by exocytosis.

and pinocytosis ("cell drinking"). In **phagocytosis**, cells engulf large particles or bacteria (Figure 3.11a). In **pinocytosis**, they engulf droplets of fluid (Figure 3.11b), thus bringing all of the substances dissolved in the droplet into the cell.

Exocytosis The process by which large molecules leave cells is **exocytosis**. In a cell that produces hormones, for example, the hormones are enclosed in membrane-bound vesicles that travel through the cell's cytoplasm toward the plasma membrane. When the vesicle reaches the plasma membrane, the vesicle membrane fuses with the plasma membrane, and then the vesicle opens up to release the hormone outside the cell. Nerve cells also release chemicals by exocytosis. Exocytosis is shown in Figure 3.12.

Table 3.2 reviews the ways in which substances move across the plasma membrane.

stop and think

People with kidney disorders may need dialysis to remove waste products from their blood. The blood is passed through a long, coiled tube submerged in a tank filled with dialyzing fluid. The tubing is porous, allowing small molecules to diffuse out of the blood into the dialyzing fluid. Urea is a waste product that must be removed from the blood. Glucose molecules, however, should remain in the blood. To achieve these goals, what should the concentrations of urea and glucose be like in the dialyzing fluid?

TABLE 3.2	Review of Mechanisms of Transport across the Plasma Membrane
Mechanism	Description
Simple diffusion	Random movement from region of higher concentration to region of lower concentration
Facilitated diffusion	Movement from region of higher concentration to region of lower concentration with the aid of a carrier or channel protein
Osmosis	Movement of water from region of higher water concentration (lower solute concentration) to region of lower water concentration (higher solute concentration)
Active transport	Movement, often from region of lower concentration to region of higher concentration, with the aid of a carrier protein and energy, usually from ATP
Endocytosis	Process by which materials are engulfed by plasma membrane and drawn into cell in a vesicle
Exocytosis	Process by which a membrane-bound vesicle from inside the cell fuses with the plasma membrane and spills contents outside the cell

3.5 Organelles

Inside the eukaryotic cell, the primary role of membranes is to create separate compartments where specific chemical processes critical to the life of the cell are carried out. The membrane-bound organelles distributed in the cells' cytoplasm have different functions—just like the different offices within a large company, some of which are responsible for production, some for purchasing, and others for shipping. The compartmentalization allows segregated combinations of molecules to carry out specific tasks (see Figure 3.2). Some organelles give directions for manufacturing cell products. Others make or modify the products or transport them. Still other organelles process energy or break down substances for use or disposal. Nonmembranous organelles also perform specific functions for the cell.

Nucleus

The cell **nucleus** contains almost all of the cell's genetic information (Figure 3.13). The DNA within the nucleus controls cellular structure and function because it contains a code for the production of proteins. All our cells contain the same genetic information. The characteristics of a particular cell—what makes it a muscle cell or a liver cell—are determined largely by the specific directions it receives from its nucleus.

A double membrane called the **nuclear envelope** surrounds the nucleus and separates it from the cytoplasm, as shown in Figure 3.13. Communication between the nucleus and cytoplasm occurs through openings in the envelope called **nuclear pores**. The traffic of selected materials across the nuclear envelope allows the contents of the cytoplasm to influence the nucleus and vice versa.

Genetic information within the nucleus is organized into **chromosomes**, threadlike structures made of DNA and associated proteins. The number of chromosomes varies from one species to another. For example, humans have 46 chromosomes (23 pairs), house mice have 40 chromosomes, and domestic dogs have 78. Individual chromosomes are visible with a light microscope during cell division, when they shorten and condense (Figure 3.14a). At all other times,

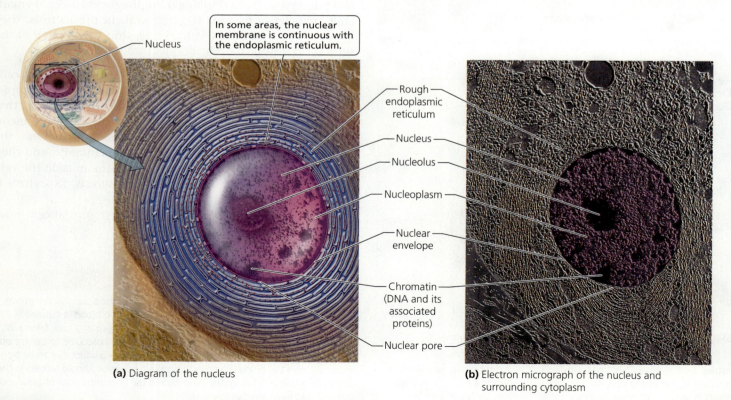

(a) Diagram of the nucleus

(b) Electron micrograph of the nucleus and surrounding cytoplasm

FIGURE 3.13 *The nucleus contains almost all the genetic information of a cell.*

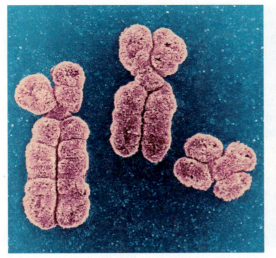
(a) Individual chromosomes are visible during cell division, when they shorten and condense.

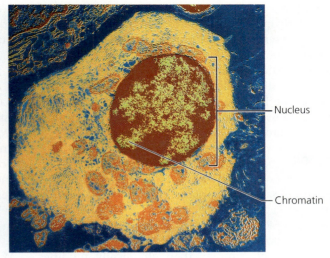
(b) At all other times, the genetic material is dispersed and called chromatin.

FIGURE 3.14 *Chromosomes are composed of DNA and associated proteins.*

however, the chromosomes are extended and not readily visible. In this dispersed state, the genetic material is called *chromatin* (Figure 3.14b). The chromatin and other contents of the nucleus constitute the *nucleoplasm*. We will discuss chromosomes and cell division in Chapter 19.

The **nucleolus**, a specialized region within the nucleus (see Figure 3.13), forms and disassembles during the course of the cell cycle (see Chapter 19). It is not surrounded by a membrane but is simply a region where DNA has gathered to produce a type of RNA called *ribosomal RNA (rRNA)*. Ribosomal RNA is a component of **ribosomes**, which are sites where protein synthesis begins. Ribosomes may be suspended in the cytoplasm (free ribosomes) or attached to the endoplasmic reticulum (bound ribosomes).

Endoplasmic Reticulum

The **endoplasmic reticulum (ER)** is part of an extensive network of channels connected to the nuclear envelope and certain organelles (Figure 3.15). In some regions, the ER is studded with ribosomes and because of this is called **rough endoplasmic reticulum (RER)**. The amino acid chains made by the attached ribosomes are threaded through the RER's membrane to its internal spaces. There the chains are processed and modified, enclosed in vesicles formed from the RER membrane, and transferred to the Golgi complex (discussed shortly) for additional processing and packaging. Proteins made by ribosomes bound to ER will be incorporated into membranes or eventually secreted by the cell. Proteins produced by free ribosomes will remain in the cell.

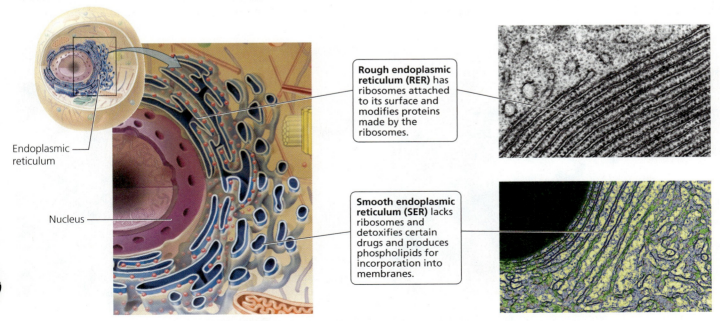

FIGURE 3.15 *The endoplasmic reticulum (ER) is continuous with the nuclear membrane and consists of two regions: rough ER and smooth ER.*

Smooth endoplasmic reticulum (SER) lacks ribosomes. The SER (particularly in liver cells) detoxifies alcohol and other drugs. Typically, enzymes of SER modify the drugs to make them more water soluble and easier to eliminate from the body. Another function of SER is the production of phospholipids. These phospholipids, along with proteins from the RER, are used to make the RER membrane. Because the RER membrane is continually used to form vesicles for shipping, it must be replenished constantly.

Golgi Complex

The **Golgi complex** consists of a series of interconnected, flattened membranous sacs. This organelle is the cell's protein processing and packaging center (Figure 3.16). Protein-filled vesicles from the RER arrive at the "receiving side" of the Golgi complex, fuse with its membrane, and empty their contents inside. The Golgi complex then chemically modifies many of the proteins as they move, by way of vesicles, from one membranous disk in the stack to the next. When the processing is finished, the Golgi complex sorts the proteins, much as a postal worker sorts letters, and sends them to their various destinations. Some of the proteins emerging from the "shipping side" are packaged in vesicles and sent to the plasma membrane for export from the cell or to become membrane proteins. Other proteins are packaged in lysosomes. Figure 3.17 summarizes the movement of protein-filled vesicles from the rough endoplasmic reticulum to the Golgi complex for processing and eventual release.

Lysosomes

How does the cell break down worn-out parts or digest bacteria that it takes in by phagocytosis? If it simply released digestive enzymes into its cytoplasm, for example, it would soon destroy itself. Instead, intracellular digestion occurs mainly within lysosomes. **Lysosomes** are roughly spherical organelles consisting of a single membrane packed with about 40 different digestive enzymes. The enzymes and membranes of lysosomes are made by the RER and then sent to the Golgi complex for additional processing. Eventually, enzyme-filled lysosomes bud and then pinch off from the Golgi complex (see Figure 3.17) and begin their diverse roles in digestion within the cell.

Consider, for example, what happens when a cell engulfs a bacterium. You can follow this process in Figure 3.18 (see pathway on right). During the process of phagocytosis (step 1), a vesicle encircles the bacterium. A lysosome released from the Golgi complex then fuses with the vesicle (step 2), and the lysosome's digestive enzymes break the bacterium down into smaller molecules. These molecules diffuse out of the vesicle into the cytoplasm, where they can be used by the cell (step 3). Indigestible residues may be expelled from the cell by exocytosis (step 4), or they may be stored indefinitely in vesicles inside the cell (step 5).

Lysosomes also break down obsolete parts of the cell itself. Worn-out organelles and macromolecules are broken down into smaller components that can be reused (see Figure 3.18, pathway on left). For example, an organelle called a mitochondrion (discussed later) lasts only about 10 days in a typical liver cell before being destroyed by lysosomes. After worn-out mitochondria are destroyed, their component monomers, such as amino acids, are returned to the cytoplasm for reuse. Such "housecleaning" keeps the cell functioning properly and promotes the recycling of essential materials.

The absence of a single kind of lysosomal enzyme can have devastating consequences. Molecules that would normally be broken down by the missing enzyme start to collect in the lysosomes and cause them to swell. Ultimately, the accumulating molecules interfere with cell function. These lysosomal storage diseases are inherited and progress with age.

Tay-Sachs disease is a lysosomal storage disease caused by the absence of the lysosomal enzyme hexosaminidase (Hex A), which breaks down lipids in nerve cells. When Hex A is missing, the lysosomes swell with undigested lipids. Infants with Tay-Sachs disease appear normal at birth but begin to deteriorate by about 6 months of age as abnormal amounts of lipid

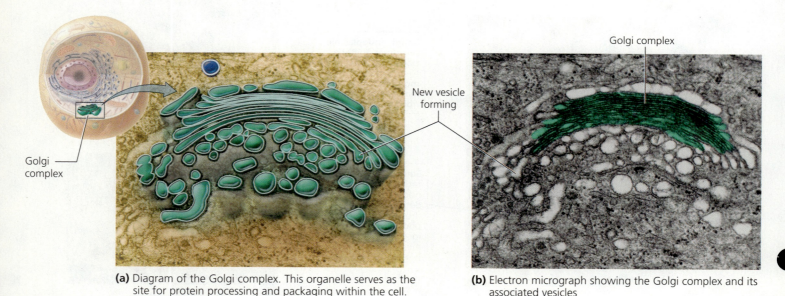

(a) Diagram of the Golgi complex. This organelle serves as the site for protein processing and packaging within the cell.

(b) Electron micrograph showing the Golgi complex and its associated vesicles

FIGURE 3.16 *The Golgi complex*

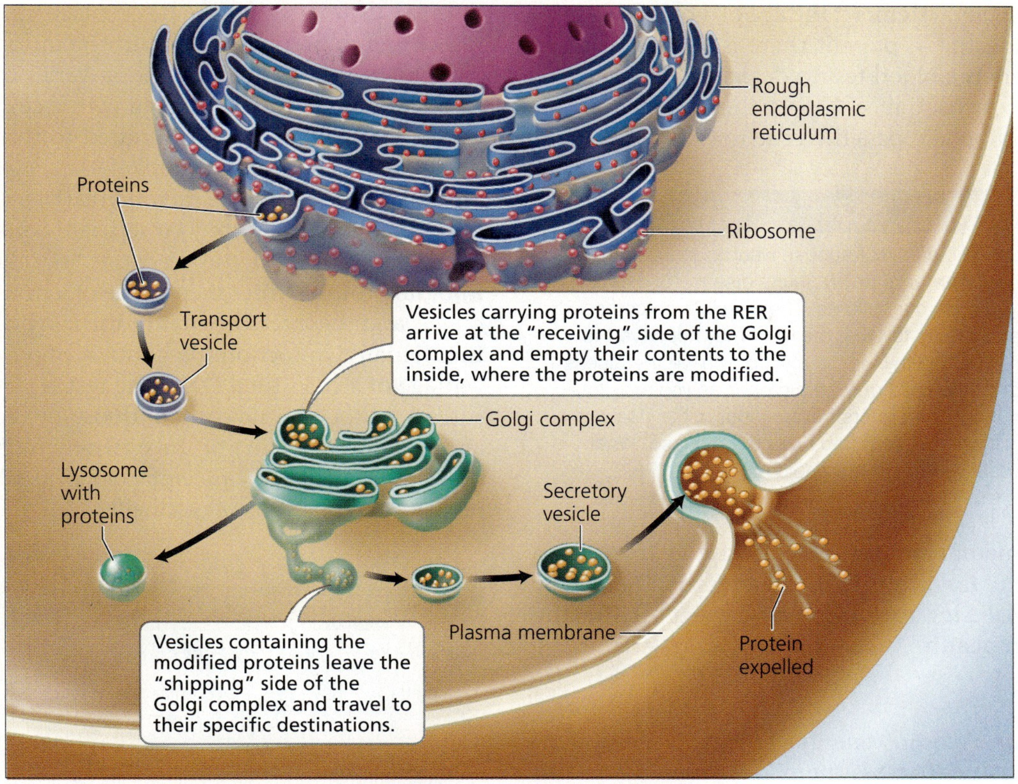

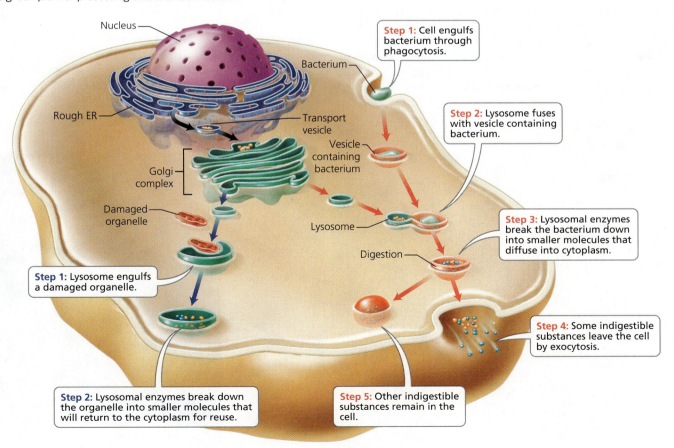

FIGURE 3.17 The route by which protein-filled vesicles from the rough endoplasmic reticulum travel to the Golgi complex for processing and eventual release.

FIGURE 3.18 Lysosome formation and function in intracellular digestion. Lysosomes, released from the Golgi complex, digest a bacterium engulfed by the cell (see pathway on right). Lysosomes also digest obsolete parts of the cell itself (see pathway on left).

accumulate in the nervous system. By the age of 4 or 5, Tay-Sachs causes paralysis and death. At present there is no cure for this disease. However, there is a blood test to detect individuals who carry the gene for Tay-Sachs. Called *carriers*, these individuals do not have the disease but could pass the gene to their offspring.

what would you do?

Imagine that you and your spouse want to start a family, but both of you are carriers of Tay-Sachs disease and could pass the gene to your children. The possible outcomes for any child you might conceive are as follows: the child may not have the gene for Tay-Sachs and may be healthy; or, the child may have the disease and die in early childhood; or, the child may be a carrier, as you are. Your parents urge adoption. Your spouse prefers not to adopt but to use prenatal screening to check whether your fetus has the disease. What would you do? On what would you base your choice?

Certain environmental factors cause disease by interfering with lysosomes. In the Environmental Issue essay, *The Deadly Interaction between Asbestos and Lysosomes,* we describe the impact of asbestos on health.

Mitochondria

Most cellular activities require energy. Energy is needed to transport certain substances across the plasma membrane and to fuel many of the chemical reactions that occur in the cytoplasm. Specialized cells such as muscle cells and nerve cells require energy to perform their particular activities. The energy that cells need is provided by **mitochondria** (singular, mitochondrion), the organelles where most of cellular respiration occurs. Cellular respiration, discussed later in the chapter, is a four-phase process in which oxygen and an organic fuel such as glucose are consumed and energy in the form of ATP is released. The first phase takes place in the cytoplasm. The remaining three phases occur in the mitochondria.

The number of mitochondria varies considerably from cell to cell and is roughly correlated with a cell's demand for energy. Most cells contain several hundred to thousands of mitochondria. Like the nucleus, but unlike other organelles, mitochondria are bounded by a double membrane (Figure 3.19). The inner and outer membranes create two separate compartments that serve as sites for some of the reactions in cellular respiration. The infoldings of the inner membrane of a mitochondrion are called *cristae,* and these are the sites of the last phase of cellular respiration. Finally, mitochondria contain ribosomes and a small percentage of a cell's total DNA (the rest is in the nucleus, as noted earlier). Mitochondria contain ribosomes and DNA because they are likely descendants of once free-living bacteria that invaded or were engulfed by ancient cells (see Chapter 22). Table 3.3 reviews the functions of organelles.

TABLE 3.3	Review of Major Organelles and Their Functions
Organelle	**Function**
Nucleus	Contains almost all the genetic information and influences cellular structure and function
Rough endoplasmic reticulum (RER)	Studded with ribosomes (sites where the synthesis of proteins begins); produces membrane
Smooth endoplasmic reticulum (SER)	Detoxifies drugs; produces membrane
Golgi complex	Sorts, modifies, and packages products of RER
Lysosomes	Digest substances imported from outside the cell; destroy old or defective cell parts
Mitochondria	Provide cell with energy through the breakdown of glucose during cellular respiration

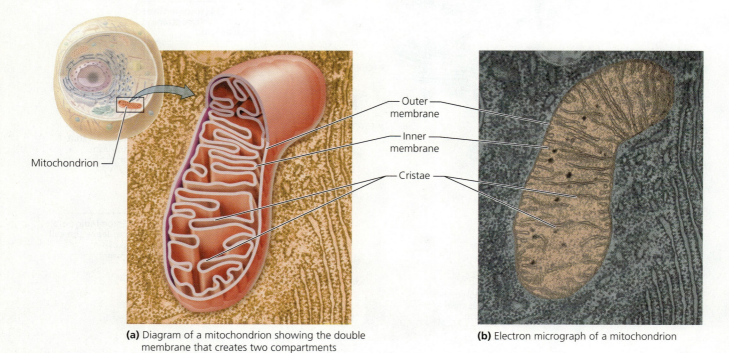

(a) Diagram of a mitochondrion showing the double membrane that creates two compartments

(b) Electron micrograph of a mitochondrion

FIGURE 3.19 *Mitochondria are sites of energy conversion in the cell.*

ENVIRONMENTAL ISSUE

The Deadly Interaction between Asbestos and Lysosomes

Asbestos is a fibrous silicate mineral, found in many forms in nature, that is strong, flexible, and resistant to heat and corrosion (Figure 3.A). Because of these properties, asbestos was used widely in construction—as an insulator on ceilings and pipes, for example, or to soundproof and fireproof the walls of schools.

The very properties that make asbestos an ideal building material—its fibrous nature and durability—also can make it deadly. For example, when fibers of asbestos insulation are dislodged, small, light particles become suspended in the air and can be inhaled into the lungs. There is some evidence that the different forms of asbestos differ in the time they persist in lung tissue. Nevertheless, particles of at least one form appear resistant to degradation and remain in the lungs for life (Figure 3.B).

Inhalation of asbestos particles can cause lung cancer and mesothelioma, a form of cancer specific to the lining of the lungs and chest cavity (pleura) and the lining of the abdomen (peritoneum). Asbestosis, a third condition, is the most common disease caused by exposure to asbestos. It results from the dangerous interaction between asbestos and lysosomes. Cells responsible for cleaning the respiratory passages engulf small particles of asbestos inhaled into the lungs; lysosomes inside the cleaning cells then fuse with the vesicles containing the asbestos particles. Unfortunately, the lysosomal enzymes cannot break down the asbestos particles. Instead, the particles destabilize the membranes of the lysosomes, causing massive release of enzymes, which destroy the cells of the respiratory tract. Irreversible scarring of lung tissue is the result, eventually interfering with the exchange of gases in the lungs. People with asbestos-damaged lungs experience chronic coughing and shortness of breath. These symptoms become more severe over time and may cause death from impaired respiratory function.

At present, there is no effective treatment for asbestosis. The focus, therefore, has been on prevention. In the United States, the use of asbestos for insulation and fireproofing, or for any new purposes, is banned. However, asbestos is still present in many buildings constructed before the ban went into effect. In these buildings, it is generally recommended and often required that exposed asbestos be removed, enclosed by other building materials, or covered with a sealant. Experts should determine which method is best in any given situation. Moreover, the sealing or removal of asbestos must be done by experts, because the greatest risk of asbestos exposure occurs when asbestos is handled improperly. Finally, workers at risk of exposure to asbestos—plumbers, electricians, insulation workers, and carpenters, to name a few—should insist upon frequent testing of the air in their workplaces.

FIGURE 3.A *The ideal, but deadly, building material asbestos*

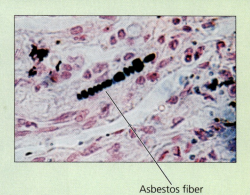

FIGURE 3.B *An asbestos fiber in lung tissue*

Questions to Consider

- Cigarette smoking worsens diseases caused by exposure to asbestos. If a worker who smokes is exposed over many years to asbestos in the workplace and subsequently develops lung cancer, then who is responsible for his developing cancer? Is the employer responsible, or does the worker bear some personal responsibility?
- What information would you consider when assessing responsibility?

---

stop and think

We have discussed the nucleus, endoplasmic reticulum (including rough and smooth), ribosomes, Golgi complex, lysosomes, and mitochondria. Assign each of these organelles to one of the following main functions within a cell: manufacturing, breakdown, or energy processing.

3.6 Cytoskeleton

Traversing the cytoplasm of the cell is a complex network of fibers called the **cytoskeleton**. The fibers are divided into three types: microtubules are the thickest; microfilaments are the thinnest; and intermediate filaments are the diverse group in between. Microtubules and microfilaments disassemble and reassemble, whereas intermediate filaments tend to be more permanent.

Microtubules are straight, hollow rods made of the protein tubulin. Some microtubules near the plasma membrane maintain cell shape. Microtubules also form tracks along which organelles or vesicles travel. Finally, microtubules play a role in the separation of chromosomes during cell division. A microtubule-organizing center located near the nucleus contains a pair of **centrioles**, each composed of nine sets of three microtubules arranged in a ring (Figure 3.20). Centrioles may function in cell division and in the formation of cilia and flagella.

Microtubules serve as the working parts of two types of cell extensions called cilia (singular, cilium) and flagella (singular, flagellum). **Cilia** are numerous, short extensions on a cell that move with the back-and-forth motion of oars. They are found, for example, on the surfaces of cells lining the respiratory tract (Figure 3.21a), where they sweep debris trapped in mucus away from the lungs. Smoking destroys these cilia and hampers

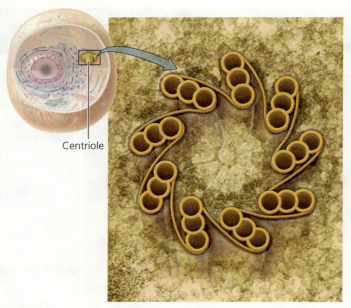

(a) Diagram of a centriole. Each centriole is composed of nine sets of triplet microtubules arranged in a ring.

Centriole

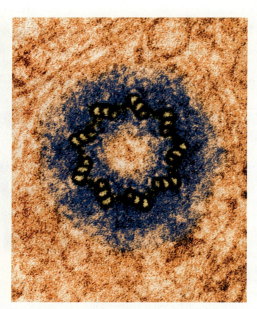

(b) Electron micrograph showing the microtubules of a centriole

FIGURE 3.20 *Centrioles may play a role in cell division.*

cleaning of respiratory surfaces. A **flagellum** resembles a whip and moves in an undulating manner. Flagella are much longer than cilia. The only cell with a flagellum in humans is the sperm cell (Figure 3.21b).

Cilia and flagella differ in length, number per cell, and pattern of movement. Nevertheless, they have a similar arrangement of microtubules, called a *9 + 2 pattern* (Figure 3.21c), at their core. This arrangement consists of nine pairs of microtubules arranged in a ring with two microtubules at the center.

Microfilaments are solid rods made of the protein actin. These fibers are best known for their role in muscle contraction, where they slide past thicker filaments made of the protein myosin. Microfilaments also play a role in cell division, forming a band that contracts and pinches the cell in two.

Intermediate filaments are a diverse group of ropelike fibers helping to maintain cell shape and anchoring certain organelles in place. Their protein composition varies from one type of cell to another.

3.7 Cellular Respiration and Fermentation in the Generation of ATP

Living requires work, and work requires energy. Logic tells us that, therefore, living requires energy.

We get our energy from the food we eat. Our digestive system (discussed in Chapter 15) breaks down complex

(a) Cilia on cells lining the respiratory tract

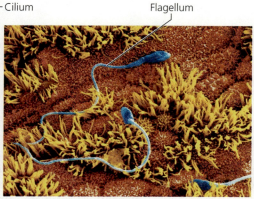

(b) Sperm cells in a fallopian tube. Each sperm cell has a single flagellum.

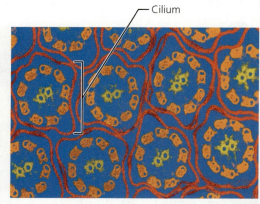

(c) Several cilia in cross section showing the 9 + 2 pattern of microtubules. Flagella (not shown) have a similar arrangement of microtubules.

FIGURE 3.21 *Microtubules are responsible for the movement of cilia and flagella.*

macromolecules such as carbohydrates, proteins, and fats into their simpler components, such as glucose, amino acids, and fatty acids. These simpler molecules are then absorbed into the bloodstream and carried to our cells, where some of the energy stored in the molecules' chemical bonds is used to make ATP, the energy-rich molecule that our cells use to do their work. (Some energy is also given off as heat.) Although carbohydrates, proteins, and fats are all sources of cellular energy, we will focus on carbohydrates. Cells have two ways of breaking glucose molecules apart for energy: cellular respiration and fermentation. Cellular respiration requires oxygen; fermentation does not.

All the chemical reactions that take place in a cell constitute its **metabolism**. These chemical reactions are organized into metabolic pathways. Each pathway consists of a series of steps in which a starting molecule is modified, eventually resulting in a particular product. Specific enzymes speed up each step of the pathway. Cellular respiration and fermentation are examples of *catabolic pathways*—pathways in which complex molecules, such as carbohydrates, are broken down into simpler compounds, releasing energy. *Anabolic pathways*, conversely, build complex molecules from simpler ones and consume energy in the process.

Cellular Respiration

Cellular respiration is the oxygen-requiring pathway by which cells break down glucose. It is an elaborate series of chemical reactions whose final products are carbon dioxide, water, and energy. In a laboratory beaker, glucose and oxygen can be combined to produce those products in a single step. However, under those circumstances, the glucose burns, and all the energy is lost as heat. The process used by cells, in which glucose is broken down in a series of steps, enables the cells to obtain much of the energy in a usable form—specifically, as a high-energy chemical bond in ATP. Recall from Chapter 2 that ATP is formed from ADP (adenosine diphosphate) and phosphate (here, abbreviated as P) in a process that requires energy.

Cellular respiration has four phases: (1) glycolysis, (2) the transition reaction, (3) the citric acid cycle, and (4) the electron transport chain. All four phases occur continuously within cells. Glycolysis takes place in the cytoplasm of the cell. The transition reaction, the citric acid cycle, and the electron transport chain take place in mitochondria. You will see that some of these phases consist of a series of reactions in which the **products** from one reaction become the **substrates** (raw materials) for the next reaction. You will also see that the transfer of electrons from one atom or molecule to another is a key feature of the process our cells use to capture energy from fuel. As the electrons are passed along a chain of intermediate compounds, their energy is used to make ATP.

Glycolysis The first phase of cellular respiration, called **glycolysis** (*glyco*, sugar; *lysis*, splitting), begins with glucose, a six-carbon sugar, being split into 2 three-carbon sugars. These three-carbon sugars are then converted into two molecules of **pyruvate** (Figure 3.22), another three-carbon com-

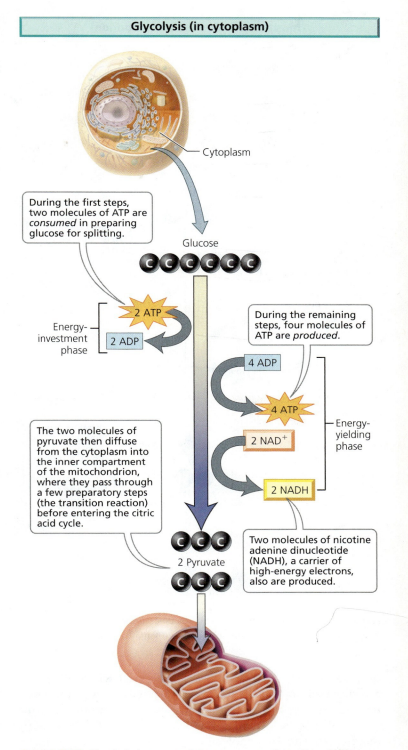

FIGURE 3.22 *Glycolysis is a several-step sequence of reactions in the cytoplasm. Glucose, a six-carbon sugar, is split into 2 three-carbon molecules of pyruvate.*

pound. Glycolysis occurs in several steps, each requiring a different, specific enzyme. During the first steps, two molecules of ATP are consumed because energy is needed to prepare glucose for splitting. During the remaining steps, four molecules of ATP are produced, for a net gain of two ATP. Glycolysis also produces two molecules of nicotine adenine

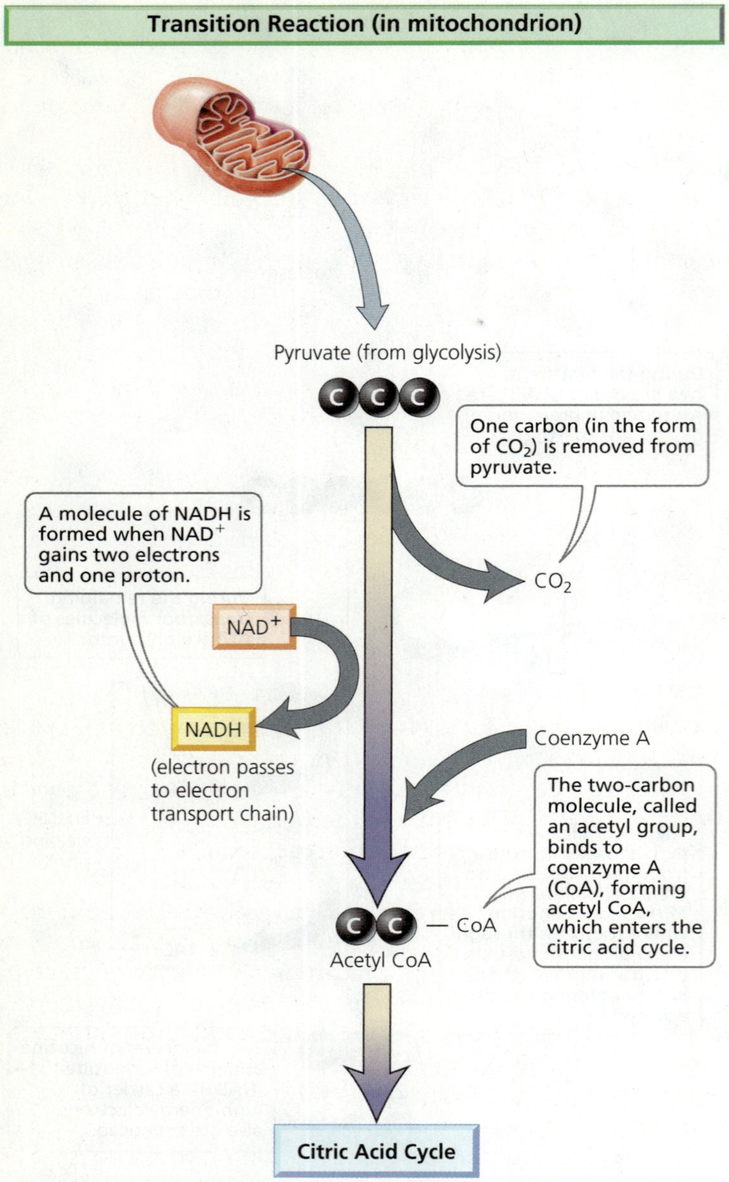

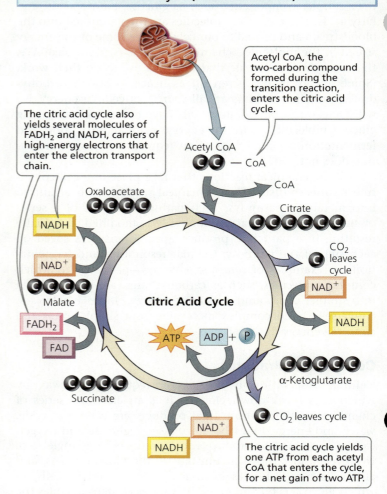

FIGURE 3.23 The transition reaction takes place inside the mitochondrion and is the link between glycolysis and the citric acid cycle.

FIGURE 3.24 The citric acid cycle is a cyclic series of eight chemical reactions that occurs inside the mitochondrion and yields two molecules of ATP and several molecules of NADH and $FADH_2$ per molecule of glucose.

dinucleotide (NADH), which are generated when electrons are donated to the coenzyme NAD^+. Glycolysis does not require oxygen and releases only a small amount of the chemical energy stored in glucose. Most of the energy remains in the two molecules of pyruvate. The pyruvate molecules move from the cytoplasm into the inner compartment of the mitochondrion.

Transition reaction Once inside the inner compartment of the mitochondrion, pyruvate reacts with a substance called coenzyme A (CoA) in a reaction called the *transition reaction*. The transition reaction results in the removal of one carbon (in the form of carbon dioxide, CO_2) from each pyruvate (Figure 3.23). The resulting two-carbon molecule, called an acetyl group, then binds to CoA to form acetyl CoA. A molecule of NADH is also produced from each pyruvate.

Citric acid cycle Still in the inner compartment of the mitochondrion, acetyl CoA reacts with a four-carbon compound in the first of a cyclic series of eight chemical reactions known as the **citric acid cycle**, named after the first product (citric acid, or citrate) formed along its route (Figure 3.24). The cycle is also called the *Krebs cycle*—after the scientist Hans Krebs, who described many of the reactions. Rather than considering each of the chemical reactions in the citric acid cycle, we will simply say that it completes the loss of electrons from glucose and yields two molecules of ATP (one from each acetyl CoA that enters the cycle) and several molecules of NADH and $FADH_2$ (flavin adenine dinucleotide). NADH and $FADH_2$ are carriers of high-energy electrons. The NADH and $FADH_2$ produced in glycolysis, in the transition reaction, and in the citric acid cycle enter the electron transport chain, the final phase of cellular respiration. The citric acid cycle also produces CO_2 as waste.

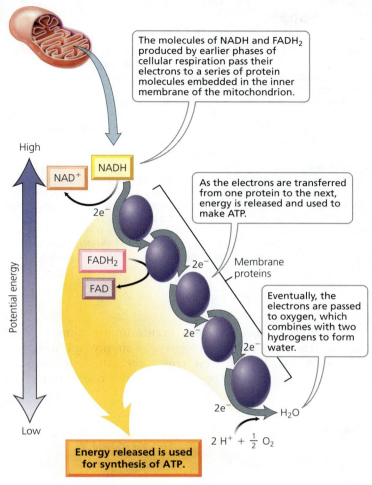

FIGURE 3.25 The electron transport chain is the final phase of cellular respiration. This phase yields 32 ATP molecules per molecule of glucose.

Electron transport chain During the final phase of cellular respiration, the molecules of NADH and $FADH_2$ produced by earlier phases pass their electrons to a series of carrier proteins embedded in the inner membrane of the mitochondrion. These proteins are known as the **electron transport chain** (Figure 3.25). (Recall that the inner membrane of the mitochondrion is highly folded, providing space for thousands of sets of carrier proteins.) During the transfer of electrons from one protein to the next, energy is released and used to make ATP. Eventually, the electrons are passed to oxygen, the final electron acceptor, which then combines with two hydrogen ions to form water. Oxygen has a critical role in cellular respiration. When oxygen is absent, electrons accumulate in the carrier proteins, halting the citric acid cycle and cellular respiration. But when oxygen is present, and accepts the electrons, respiration continues. The electron transport chain produces 32 molecules of ATP per molecule of glucose. In the Health Issue essay, *Mitochondrial Diseases*, we describe what happens when there is a deficiency of proteins in the electron transport chain.

Altogether, cellular respiration generally produces 36 molecules of ATP per molecule of glucose: 2 ATP from glycolysis, 2 ATP from the citric acid cycle, and 32 ATP from the electron transport chain. Basic descriptions of each phase can be found in Table 3.4. The results of cellular respiration are summarized in Figure 3.26.

Fermentation

As noted earlier, cellular respiration depends on oxygen as the final electron acceptor in the electron transport chain. Without oxygen, the transport chain comes to a halt, blocking the citric acid cycle and stopping cellular respiration. Is there a way for cells to harvest energy when molecules of oxygen are scarce? The answer is yes, and the pathway is fermentation.

Fermentation is the breakdown of glucose without oxygen. It begins with glycolysis, which occurs in the cytoplasm and does not require oxygen. From one molecule of glucose, glycolysis produces two molecules each of pyruvate, the electron carrier NADH, and ATP. The remaining fermentation reactions also take place in the cytoplasm, transferring electrons from NADH to pyruvate or a derivative of pyruvate. This transfer of electrons is critical because it regenerates NAD^+, which is essential for the production of ATP through glycolysis. Recall that in cellular respiration, oxygen is the final electron acceptor in the electron transport chain, whereas in fermentation it is pyruvate or a pyruvate derivative. Fermentation therefore nets only 2 molecules of

TABLE 3.4	Review of Cellular Respiration		
Phase	**Location**	**Description**	**Main Products**
Glycolysis	Cytoplasm	Several-step process by which glucose is split into 2 pyruvate	2 pyruvate 2 ATP 2 NADH
Transition reaction	Mitochondria	One CO_2 is removed from each pyruvate; the resulting molecules bind to CoA, forming 2 acetyl CoA	2 acetyl CoA 2 NADH
Citric acid cycle	Mitochondria	Cyclic series of eight chemical reactions by which acetyl CoA is broken down	2 ATP 2 $FADH_2$ 6 NADH
Electron transport chain	Mitochondria	Electrons from NADH and $FADH_2$ are passed from one protein to the next, releasing energy for ATP synthesis	32 ATP H_2O

HEALTH ISSUE

Mitochondrial Diseases

Outages in the electrical power grid can cause many services—air and rail transport, communications, water purification, heating and cooling—to slow and sometimes fail. A similar situation can happen in our bodies when mitochondria, our energy-processing organelles, fail. Mitochondria convert energy from food molecules into ATP for cells to use, and several steps of cellular respiration occur in mitochondria and require particular proteins to proceed. More than 40 illnesses, collectively known as *mitochondrial diseases*, can result from deficiencies in the proteins that function in energy metabolism within mitochondria, including proteins that are part of the electron transport chain. When mitochondria fail to function properly, less energy is generated, cell function is compromised, and cell death can result. This, in turn, can affect tissues, organs, and organ systems. Mitochondrial diseases cause the most damage to parts of the body that need the most energy, such as the brain, heart, lungs, kidneys, endocrine glands, and skeletal muscles.

Each year in the United States, an estimated 1000 to 4000 babies are born with mitochondrial diseases. Many physicians believe that mitochondrial diseases are underdiagnosed and frequently misdiagnosed, so estimates may be low. These diseases are caused by mutations (changes) in DNA that encodes for proteins critical in energy metabolism within mitochondria. Some of these proteins are encoded by mitochondrial DNA, but others are encoded by nuclear DNA and then imported into mitochondria. The mutations may be spontaneous or inherited. Children inherit nuclear DNA from their mother and father, but they inherit their mitochondrial DNA only from their mother. Environmental factors, such as infections and certain drugs, can also damage mitochondria.

People with a mitochondrial disease may experience fatigue and fail to gain weight. Symptoms of specific mitochondrial diseases are diverse because they reflect which cells and organs have compromised mitochondria. If cells in the brain are affected, then symptoms may include seizures, developmental delays, and dementia. When sensory organs, such as the eyes or ears are affected, then a decline or complete loss of vision or hearing may occur. Skeletal muscle cells with defective mitochondria can result in muscle weakness, cramps, and exercise intolerance. Sometimes multiple organ systems are affected.

ATP compared with the 36 molecules of ATP produced by cellular respiration (refer, again, to Figure 3.26). In short, fermentation is a very inefficient way for cells to harvest energy.

Lactic acid fermentation occurs in the human body. During strenuous exercise, the oxygen supply in our muscle cells runs low. Under these conditions, the cells increase lactic acid fermentation to ensure the continued production

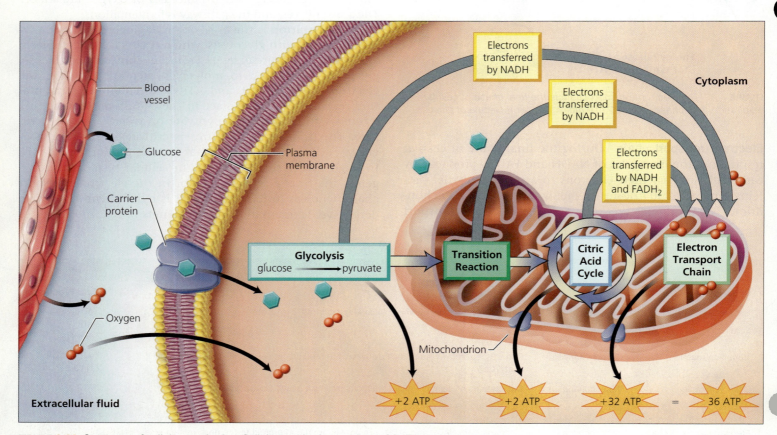

FIGURE 3.26 Summary of cellular respiration. Cellular respiration produces 36 ATP per molecule of glucose (2 ATP from glycolysis, 2 ATP from the citric acid cycle, and 32 ATP from the electron transport chain).

Additionally, cells typically contain hundreds of mitochondria, so a single cell can have some mitochondria that are normal and others that are defective (i.e., carry a mutation). The health of cells (and the severity of disease symptoms) will depend on the relative proportions of normal and defective mitochondria. Symptoms may vary dramatically even among members of the same family with an inherited mutation in mitochondrial DNA. Some of a woman's eggs may have large proportions of defective mitochondria while other eggs have mostly healthy mitochondria, so the severity of symptoms in the children will depend on the proportion of defective mitochondria passed to them by their mother. Finally, depending on the particular disorder, the onset of symptoms can occur in infancy, childhood, or adulthood. Adult onset may occur in a person who was born with the genetic mutation but whose symptoms did not appear until an environmental trigger, such as a severe illness, brought them on.

Given variation in the number of organ systems affected and in the timing and severity of symptoms, it is difficult to predict the course of a mitochondrial disease. Indeed, some people may experience relatively good quality of life, while others may have severe symptoms and not survive childhood. There are no known cures for mitochondrial diseases, so medical professionals focus on alleviating symptoms and slowing disease progression. For example, a dietician may help patients to maintain or gain weight, and physical therapy might benefit patients whose skeletal muscles are affected. Patients are also advised to avoid energetically stressful situations, such as fasting or extreme cold.

Mitochondrial dysfunction also may be linked to autism, aging, and many chronic conditions of adulthood. Clearly, much remains to be learned about the roles that mitochondria play in human health.

Questions to Consider

- A woman has an inherited, adult-onset mitochondrial disease that was diagnosed only after she had two sons and a daughter. Her symptoms are relatively mild, and her disease has been traced to a mutation in her mitochondrial DNA. Which, if any, of her children will inherit the mutation?
- What factors determine the extent of symptoms in any afflicted child? If all of her children reach adulthood and have families of their own, which of her children will pass on the mutation?

of ATP. The muscle pain we often experience after intense exercise is caused partly by the accumulation of lactic acid, a waste product of this type of fermentation. In time, the soreness disappears as the lactic acid moves into the bloodstream and is carried to the liver, where it is converted back to pyruvate.

looking ahead

In Chapter 3 we learned about the basic structure of cells. In Chapter 4 we describe how specialized cells form tissues, organs, and organ systems.

HIGHLIGHTING THE CONCEPTS

3.1 Eukaryotic Cells Compared with Prokaryotic Cells (pp. 42–44)

- There are two main types of cells. Prokaryotic cells, unique to Bacteria and Archaea, lack membrane-bound organelles. Eukaryotic cells, found in all other organisms, have membrane-bound organelles.

3.2 Cell Size and Microscopy (p. 44)

- As a cell grows, its volume increases more than its surface area; therefore, a cell that is too large will encounter problems caused by inadequate surface area. For that reason, most cells are very small.
- Because most cells are very small, they can be seen only with a microscope. Whereas light microscopes use beams of light to image specimens, electron microscopes use beams of electrons. Electron microscopes can reveal finer details because the wavelength of an electron beam is smaller than the wavelengths of visible light.

3.3 Cell Structure and Function (pp. 44–45)

- All eukaryotic cells have certain features in common, including a plasma membrane and membrane-bound organelles. Structural differences between eukaryotic cells often reflect differences in function. For example, a sperm has a tail and streamlined shape so that it can swim to the egg. The egg, in contrast, is large, immobile, and packed with materials needed to initiate development.

3.4 Plasma Membrane (pp. 45–50)

- The plasma membrane is made of phospholipids arranged in a bilayer with proteins and molecules of cholesterol interspersed throughout and carbohydrates attached to the outer surface.
- The plasma membrane maintains the cell's integrity, regulates movement of substances into and out of the cell, functions in cell–cell recognition, promotes communication between cells, and binds cells together to form tissues and organs.
- Substances cross the plasma membrane in several ways. Some cross by random movement from higher to lower concentration (simple diffusion); others need help from a carrier or channel protein (facilitated diffusion). Water also moves across the plasma membrane from higher to lower concentration (osmosis). Substances being concentrated by cells cross the plasma membrane from lower to higher concentration with the help of ATP and a carrier protein (active transport). Finally, cells may engulf outside materials by surrounding them with plasma membrane (endocytosis) or may release substances to the external environment by fusing internal vesicles with the plasma membrane and spilling their contents to the outside (exocytosis).

3.5 Organelles (pp. 50–55)
- Within a eukaryotic cell, membranes delineate specialized compartments within which specific processes occur. The nucleus contains almost all the genetic information and thus holds the code for the cell's structure and many of its functions. The nucleolus is a region within the nucleus that makes ribosomal RNA. Ribosomes are the sites where protein synthesis begins. Endoplasmic reticulum (ER) functions in membrane production and may be studded with ribosomes (rough endoplasmic reticulum, RER) or free of ribosomes (smooth endoplasmic reticulum, SER). The Golgi complex sorts, modifies, and packages products of the RER. Lysosomes digest bacteria engulfed by cells and break down old or defective cell components. Mitochondria process energy for cells.

3.6 Cytoskeleton (pp. 55–56)
- The cytoskeleton, a complex network of fibers throughout the cytoplasm of the cell, consists of three categories of fibers, all made of proteins: microtubules, microfilaments, and intermediate filaments. Microtubules are hollow rods of tubulin that function in cell movement (cilia and flagella), support, and the movement within cells of chromosomes, organelles, and vesicles. Microfilaments are rods of actin that function in muscle contraction and cell division. Intermediate filaments maintain cell shape and anchor organelles.

3.7 Cellular Respiration and Fermentation in the Generation of ATP (pp. 57–61)
- Cells require energy to work. We get this energy from the food we eat. Carbohydrates, fats, and proteins that we consume are each broken down by our digestive system into smaller units such as simple sugars, fatty acids, and amino acids. These simpler substances are absorbed into the bloodstream and carried to our cells, where the energy stored in the substances' chemical bonds is transferred for the cell's use to the chemical bonds of ATP. (Some energy is also given off as heat.)
- Cells use two catabolic pathways—cellular respiration and fermentation—to break down the carbohydrate glucose and store its energy as ATP. Cellular respiration requires oxygen and usually yields 36 molecules of ATP per molecule of glucose. Fermentation does not require oxygen and yields 2 ATP.
- Cellular respiration has four phases—glycolysis, the transition reaction, the citric acid cycle, and the electron transport chain—all occurring continuously within cells. Glycolysis occurs in the cytoplasm and splits glucose into pyruvate while producing NADH (a carrier of high-energy electrons) and a net gain of 2 ATP. The molecules of pyruvate move from the cytoplasm into the inner compartment of the mitochondrion, where they pass through a few preparatory steps, known as the transition reaction, before entering the citric acid cycle. The citric acid cycle completes the breakdown of glucose into carbon dioxide. It yields 2 ATP and carriers of high-energy electrons (NADH and $FADH_2$). In the electron transport chain, the carriers of high-energy electrons produced during glycolysis, the transition reaction, and the citric acid cycle pass their electrons to a series of proteins embedded in the inner membrane of the mitochondrion; oxygen is the final electron acceptor. Energy released during the transfer of electrons yields 32 ATP.
- Fermentation occurs in the cytoplasm and begins with glycolysis, the splitting of glucose into pyruvate. The remaining chemical reactions involve the transfer of electrons from NADH to pyruvate or a derivative of pyruvate. Compared with cellular respiration, fermentation is an inefficient way for cells to harvest energy.

RECOGNIZING KEY TERMS

cell theory p. 42
prokaryotic cell p. 42
organelle p. 43
eukaryotic cell p. 43
surface-to-volume ratio p. 44
plasma membrane p. 45
extracellular fluid p. 46
cytoplasm p. 46
fluid mosaic p. 46
selectively permeable p. 46
cell adhesion molecule (CAM) p. 46
simple diffusion p. 47
concentration gradient p. 47
facilitated diffusion p. 47
osmosis p. 47
hypertonic solution p. 48
isotonic solution p. 48
hypotonic solution p. 48
active transport p. 48
endocytosis p. 49
vesicle p. 49
phagocytosis p. 49
pinocytosis p. 49
exocytosis p. 49
nucleus p. 50
nuclear envelope p. 50
nuclear pore p. 50
chromosome p. 50
nucleolus p. 51
ribosome p. 51
endoplasmic reticulum (ER) p. 51
rough endoplasmic reticulum (RER) p. 51
smooth endoplasmic reticulum (SER) p. 52
Golgi complex p. 52
lysosome p. 52
mitochondrion p. 54
cytoskeleton p. 55
microtubule p. 55
centriole p. 56
cilia p. 55
flagellum p. 56
microfilament p. 56
intermediate filament p. 56
metabolism p. 57
cellular respiration p. 57
product p. 57
substrate p. 57
glycolysis p. 57
pyruvate p. 57
citric acid cycle p. 58
electron transport chain p. 59
fermentation p. 59
lactic acid fermentation p. 60

REVIEWING THE CONCEPTS

1. How do prokaryotic and eukaryotic cells differ? How are they the same? *pp. 42–44*
2. Describe the structure of the plasma membrane. *pp. 45–46*
3. List five functions of the plasma membrane. *p. 46*
4. What is the difference between simple and facilitated diffusion? Give an example of each. *p. 47*
5. Where are chromosomes found in the cell? What are chromosomes composed of? *pp. 50–51*
6. What are lysosomal storage diseases? *p. 52*
7. Name the three types of fibers that make up the cytoskeleton. Describe their structures and functions. *pp. 55–56*
8. What is the basic function of cellular respiration? *p. 57*

9. Choose the *incorrect* statement:
   a. Phagocytosis and pinocytosis are two types of endocytosis.
   b. Phagocytosis and pinocytosis are two types of exocytosis.
   c. Nerve cells release chemicals by exocytosis.
   d. Cells engulf bacteria by phagocytosis.
10. Lysosomes
    a. function in extracellular digestion.
    b. produce phospholipids to incorporate into membranes.
    c. have a double membrane.
    d. function in intracellular digestion.
11. Choose the *incorrect* statement:
    a. Cellular respiration requires oxygen.
    b. Glycolysis occurs in both cellular respiration and fermentation.
    c. Cellular respiration yields more ATP per molecule of glucose than does fermentation.
    d. Fermentation occurs in mitochondria.
12. Prokaryotic cells
    a. have internal membrane-bound organelles.
    b. are usually larger than eukaryotic cells.
    c. lack internal membrane-enclosed organelles.
    d. have linear strands of DNA within a nucleus.
13. The plasma membrane
    a. is selectively permeable.
    b. contains lipids that function in cell–cell recognition.
    c. has cell adhesion molecules that prevent cells from sticking together.
    d. is made of nucleic acids.
14. Facilitated diffusion is
    a. the random movement of a substance from a region of higher concentration to a region of lower concentration.
    b. the movement of water across the plasma membrane.
    c. the movement of molecules across the plasma membrane against a concentration gradient with the aid of a carrier protein and energy supplied by the cell.
    d. the movement of a substance from a region of higher concentration to a region of lower concentration with the aid of a membrane protein.
15. Almost all the genetic information of a cell is found in the
    a. endoplasmic reticulum.
    b. Golgi complex.
    c. nucleus.
    d. mitochondria.
16. Ribosomes
    a. are found on smooth endoplasmic reticulum.
    b. are sites where protein synthesis begins.
    c. process and modify proteins.
    d. break down foreign invaders and old organelles.
17. Mitochondria
    a. process energy for cells.
    b. lack ribosomes and DNA.
    c. are bounded by a single membrane.
    d. function in cell digestion.
18. Microtubules
    a. are found in eukaryotic cilia and flagella.
    b. are made of the protein actin.
    c. play a role in muscle contraction.
    d. pinch a cell in two during cell division.
19. Glycolysis
    a. occurs in the mitochondria.
    b. requires oxygen.
    c. splits glucose into pyruvate.
    d. nets 32 molecules of ATP per molecule of glucose.
20. _____ is the jellylike solution within a cell that contains everything between the nucleus and the plasma membrane.
21. _____ is the final electron acceptor in the electron transport chain.
22. Our muscle cells may switch from cellular respiration to _____ when oxygen is low.

APPLYING THE CONCEPTS

1. Given what you know about the composition of the plasma membrane, would you expect an anesthetic to be soluble or insoluble in lipids? Explain your answer.
2. Would you expect to find more mitochondria in muscle cells or bone cells? Explain your answer.
3. During cell division, the nucleus and cytoplasm of a cell split into two cells. Cancer is characterized by uncontrolled cell division and is often treated with chemotherapy. Some drugs used in chemotherapy halt cell division by affecting the cytoskeleton. Which of the cytoskeletal elements (microtubules, microfilaments, intermediate filaments) might be affected?

BECOMING INFORMATION LITERATE

Using at least three reliable sources, prepare a brochure on Tay-Sachs disease. Be sure to include the following sections: (1) symptoms; (2) causes; (3) transmission; (4) diagnosis (before and after birth); (5) treatment, if any; and (6) prognosis. List each source you considered, and explain why you chose the sources you used.

MasteringBiology®

Go to MasteringBiology for practice quizzes, activities, eText, videos, current events, and more.

4 Body Organization and Homeostasis

- By age 70, the average person has shed about 105 pounds of skin, at the rate of 600,000 particles every hour and about 1.5 pounds each year.
- People who use indoor tanning salons are at a much higher risk of developing skin cancer than people who have never tanned indoors.

4.1 From Cells to Organ Systems

4.2 Skin: An Organ System

4.3 Homeostasis

HEALTH ISSUES

Fun in the Sun?
Acne: The Miseries and Myths

Cells are arranged in tissues, and tissues, in turn, form organs. The skin, our largest organ, helps protect underlying tissues and helps to regulate body temperature.

In the previous chapter, we learned about cells. This chapter begins by describing the variety of cells and their function. It then describes the body's organization at four levels: cells, tissues, organs, and organ systems. It looks at the functions of the skin as an organ system and discusses how all the body's systems interact to maintain relatively constant internal conditions, when they can, at every organizational level.

4.1 From Cells to Organ Systems

Think for a moment about the multitude of functions taking place in your body at this very instant. Your heart is beating. Your lungs are taking in oxygen and eliminating carbon dioxide. Your eyes are forming an image of these words, and your brain is thinking about them. Your body can carry out these functions and more because its cells are specialized to perform specific tasks. But cell specialization is only the beginning. Specialized cells are organized into tissues, organs, and organ systems.

Tissues

A **tissue** is a group of cells of similar type that work together to serve a common function. Human tissues come in four primary types: epithelial tissue, connective tissue, muscle tissue, and nervous tissue. **Epithelial tissue** covers body surfaces, lines body cavities and organs, and forms glands. **Connective tissue** serves as a storage site for fat, plays an important role in immunity, and provides the body and its organs with protection and support. **Muscle tissue** is responsible for body movement and for movement of substances through the body. **Nervous tissue** conducts nerve impulses from one part of the body to another. As you read this chapter, you learn more about each of these types of tissue.

Epithelial tissue All epithelial tissues share two characteristics: a free surface and a basement membrane. The free surface may be specialized for protection, secretion, or absorption. The **basement membrane** is a noncellular layer that binds the epithelial cells to underlying connective tissue and helps the epithelial tissue resist stretching.

The three basic shapes of epithelial cells are suited to their functions. **Squamous** (skway'-mus) **epithelium** is made up of flattened, or scale-like, cells. These cells form linings—in the blood vessels or lungs, for instance—where their flattened shape allows oxygen and carbon dioxide to diffuse across the lining easily. In blood vessels, the smooth surface of the blood vessel lining reduces friction. **Cuboidal epithelium** is made up of cube-shaped cells. Cuboidal cells are found in many glands and in the lining of kidney tubules, where they provide some protection and are specialized for secretion and absorption. **Columnar epithelium**, consisting of elongated, column-shaped cells, is specialized for absorption and secretion. The small intestine is lined with columnar cells. These, like many examples of columnar cells, have numerous small, fingerlike folds on their exposed surfaces, which greatly increase the surface area for absorption. The goblet cells of this lining produce mucus to ease the passage of food and protect the cells of the lining.

Squamous, cuboidal, and columnar epithelium can be either simple (a single layer of cells) or stratified (multiple layers of cells). Stratified epithelium often serves a protective role, because its multiple layers provide additional thickness that makes it more difficult for molecules to pass through. Table 4.1 and Figure 4.1 summarize the types of epithelial tissue.

A **gland** is epithelial tissue that secretes a product. **Exocrine glands** secrete their products into ducts leading to body surfaces, cavities, or organs. Examples of exocrine glands include the glands that produce digestive enzymes, milk glands, and the oil and sweat glands of the skin. **Endocrine glands** (covered in more depth in Chapter 10) lack ducts and secrete their products, hormones, into spaces just outside the cells. Ultimately, hormones diffuse into the bloodstream and are carried throughout the body.

Connective tissue Connective tissue has many forms and functions. Sometimes described as the body's glue, its most common role is to bind (tendons and ligaments) and support the other tissues (cartilage and bone). However, certain connective tissues specialize in transport (blood) and energy storage (adipose tissue). Connective tissue is the most abundant and widely distributed tissue in the body.

All connective tissues contain cells embedded in an extracellular **matrix**. This matrix consists of protein fibers and a noncellular material called ground substance. The **ground substance** may be solid (as in bone), fluid (as in blood), or gelatinous (as in cartilage). It is secreted by the connective tissue cells themselves or by other cells nearby. Whereas all other types of tissue consist primarily of cells, connective tissue is made up mostly of its matrix. The cells are distributed in the matrix much like pieces of fruit suspended in a gelatin dessert.

The connective-tissue matrix contains three types of protein fibers in proportions that depend on the type of connective tissue. **Collagen fibers** are strong and ropelike and can withstand pulling because of their great tensile strength. **Elastic fibers** contain random coils and can stretch and recoil like a spring. They are common in structures where great elasticity is needed, including the skin, lungs, and blood vessels. **Reticular fibers** are thin strands of collagen[1] that branch extensively, forming interconnecting networks suitable for supporting soft tissues (for example, they support the liver and spleen).

All three types of protein fibers—collagen, elastic, and reticular—are produced by cells called **fibroblasts** in the connective tissue. Fibroblasts also repair tears in body tissues. For example, when skin is cut, fibroblasts move to the area of the wound and produce collagen fibers that help close the wound, cover the damage, and provide a surface upon which the outer layer of skin can grow back.

There are two broad categories of connective tissue—connective tissue proper (loose and dense connective tissue) and specialized connective tissue (cartilage, bone, and blood). Table 4.2 and Figure 4.2 group the many types of connective tissue and summarize the characteristics of each type. The characteristics of any specific connective tissue are determined more by its matrix than by its cells.

---

[1]Reticular fibers have the same subunits as collagen fibers, but they are assembled into a slightly different kind of structure.

TABLE 4.1 Review of Epithelial Tissue

Shape	Number of Layers	Example Locations	Functions
Squamous (flat, scale-like cells)	Simple (single layer)	Lining of heart and blood vessels, air sacs of lungs	Allows passage of materials by diffusion
	Stratified (more than one layer)	Linings of mouth, esophagus, and vagina; outer layer of skin	Protects underlying areas
Cuboidal (cube-shaped cells)	Simple	Kidney tubules, secretory portion of glands and their ducts	Secretes; absorbs
	Stratified	Ducts of sweat glands, mammary glands, and salivary glands	Protects underlying areas
Columnar	Simple	Most of digestive tract (stomach to anus), air tubes of lungs (bronchi), excretory ducts of some glands, uterus	Absorbs; secretes mucus, enzymes, and other substances
	Stratified	Rare; urethra, junction of esophagus and stomach	Protects underlying areas, secretes

SIMPLE EPITHELIUM

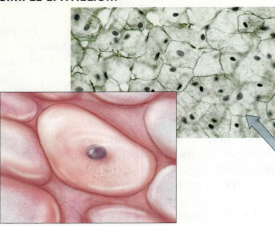

Simple squamous
- One layer of flattened cells
- Located in air sacs of lungs, heart and blood vessel linings
- Allows exchange of nutrients, gases, and wastes

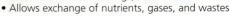

Simple cuboidal
- One layer of cube-shaped cells
- Located in linings of kidney tubules and glands
- Functions in absorption and secretion

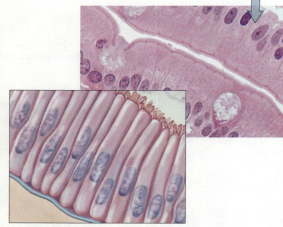

Simple columnar
- One layer of tall, slender cells
- Located in lining of gut and respiratory tract
- Functions in absorption and secretion

STRATIFIED EPITHELIUM

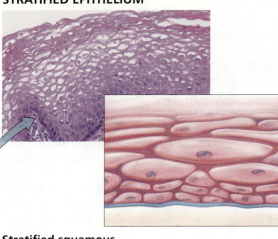

Stratified squamous
- Several layers of flattened cells
- Located on surface of skin, lining of mouth, esophagus, and vagina
- Provides protection against abrasion, infection, and drying out

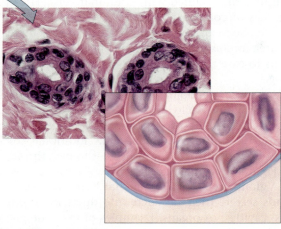

Stratified cuboidal
- Usually two layers of cube-shaped cells
- Located in ducts of mammary glands, sweat glands, and salivary glands
- Functions in protection

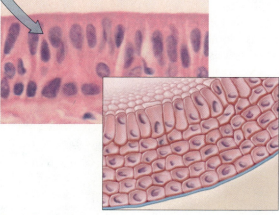

Stratified columnar
- Several layers of tall, slender cells
- Rare, located in urethra (tube through which urine leaves the body)
- Functions in protection and secretion

FIGURE 4.1 Types of epithelial tissue. These are named for the shape of the cell and the number of cell layers.

TABLE 4.2 Review of Connective Tissue

Type	Example Locations	Functions
Connective tissue proper		
Loose, areolar	Between muscles, surrounding glands, wrapping small blood vessels and nerves	Wraps and cushions organs
Loose, adipose (fat)	Under skin, around kidneys and heart	Stores energy, insulates, cushions organs
Dense	Tendons, ligaments	Attaches bone to bone (ligaments) or bone to muscle (tendons)
Specialized connective tissue		
Cartilage (semisolid)	Nose (tip); rings in respiratory air tubules; external ear	Provides flexible support, cushions
Bone (solid)	Skeleton	Provides support and protection (by enclosing) and levers for muscles to act on
Blood (fluid)	Within blood vessels	Transports oxygen and carbon dioxide, nutrients, hormones, and wastes; helps fight infections

Connective Tissue Proper Loose and dense connective tissues differ in the ratio of cells to extracellular fibers. **Loose connective tissue** contains many cells but has fewer and more loosely woven fibers than are seen in the matrix of dense connective tissue (Figure 4.2). One type of loose connective tissue, **areolar** (ah-ree'-o-lar) **connective tissue**, functions as a universal packing material between other tissues. Its many cells are embedded in a gelatinous matrix that is soft and easily shaped. Areolar connective tissue is found, for example, between muscles, where it permits one muscle to move freely over another. It also anchors the skin to underlying tissues and organs.

The second type of loose connective tissue is **adipose** (ad-ë'-pōs) **tissue**; it contains cells that are specialized for fat storage. Most of the body's long-term energy stores are fat. Fat also serves as insulation and, around certain organs, as a shock absorber.

Dense connective tissue forms strong bands because of its large amounts of tightly woven fibers. It is found in ligaments (structures that join bone to bone), tendons (structures that join muscle to bone), and the dermis (layer of skin below the epidermis).

Specialized Connective Tissue Specialized connective tissue, as shown in Figure 4.2, comes in three types: cartilage, bone, and blood. **Cartilage** (kär'tl-ĭj) is tough but flexible. It serves as cushioning between certain bones and helps maintain the structure of certain body parts, including the ears and the nose. The cells in cartilage (chondrocytes) sit within spaces in the matrix called *lacunae*. The protein fibers and somewhat gelatinous ground substance of cartilage are responsible for the tissue's resilience and strength. Cartilage lacks blood vessels and nerves, so nutrients reach cartilage cells by diffusion from nearby capillaries. Because this process is fairly slow, cartilage heals more slowly than bone, which is a tissue with a rich blood supply.

The human body has three types of cartilage:

- *Hyaline cartilage,* the most abundant, provides support and flexibility. It contains numerous cartilage cells in a matrix of collagen fibers and a bluish white, gel-like ground substance. Known commonly as gristle, hyaline cartilage is found at the ends of long bones (look carefully at your next drumstick), where it allows one bone to slide easily over another. It also forms part of the nose, ribs, larynx, and trachea.
- *Elastic cartilage* is more flexible than hyaline cartilage because of the large amounts of wavy elastic fibers in its matrix. Elastic cartilage is found in the external ear, where it provides strength and elasticity.
- *Fibrocartilage* contains fewer cells than either hyaline or elastic cartilage. Like hyaline cartilage, its matrix contains collagen fibers. Fibrocartilage forms a cushioning layer in the knee joint as well as the outer part of the shock-absorbing disks between the vertebrae of the spine. It is made to withstand pressure.

Bone, in combination with cartilage and other components of joints, makes up the skeletal system. To many people's surprise, bone is a living, actively metabolizing tissue with a good blood supply that promotes prompt healing. Bone has many functions: protection and support for internal structures; movement, in conjunction with muscles; storage of lipids (in yellow marrow), calcium, and phosphorus; and production of blood cells (in red marrow). The matrix secreted by bone cells is hardened by calcium, enabling bones to provide rigid support. Collagen fibers in bone also lend it strength. You will read more about the structure and function of bones in Chapter 5.

Blood is a specialized connective tissue consisting of a liquid matrix, called *plasma,* in which so-called formed elements (cells and cell fragments called platelets) are suspended (Figure 4.2). The "fibers" in blood are soluble proteins, visible only when the blood clots. An important function of blood is to transport various substances, many of which are dissolved in the plasma. One kind of formed element, the *red blood cell,* transports oxygen to cells and also carries some of the carbon dioxide away from cells. The other two kinds of formed elements are *white blood cells,* which help fight infection, and *platelets,* which help with clotting. Both white blood cells and platelets help protect the body. You will read more about blood in Chapter 11.

68 CHAPTER 4 Body Organization and Homeostasis

CONNECTIVE TISSUE PROPER

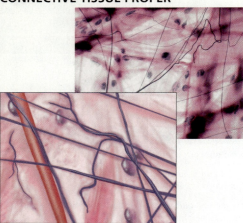

Areolar connective tissue
- Widely distributed; found under skin, around organs, between muscles
- Wraps and cushions organs

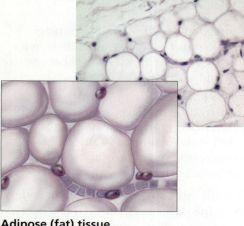

Adipose (fat) tissue
- Found under skin, around kidneys and heart
- Functions in energy storage and insulation; cushioning for organs

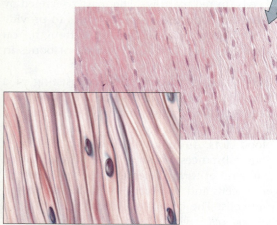

Dense connective tissue
- Found in tendons and ligaments
- Forms strong bands that attach bone to muscle or bone to bone

SPECIALIZED CONNECTIVE TISSUE

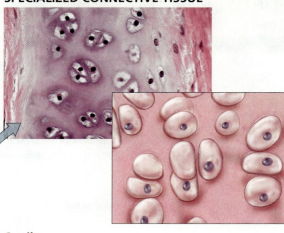

Cartilage
- Found in rings of respiratory air tubes, external ear, tip of nose
- Provides flexible support; cushions

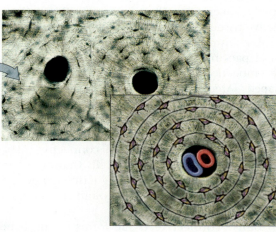

Bone
- Found in the skeleton
- Functions in support, protection (by enclosing organs), and movement

Blood
- Found within blood vessels
- Transports nutrients, gases, hormones, wastes; fights infections

FIGURE 4.2 *Types of connective tissue*

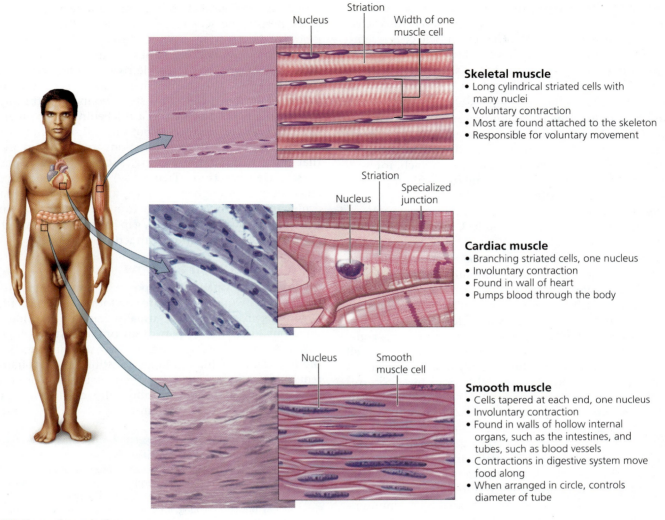

FIGURE 4.3 *Types of muscle tissue*

Muscle tissue Muscle tissue is composed of muscle cells (called *muscle fibers*) that contract when stimulated. As shown in Figure 4.3, there are three types of muscle tissue: skeletal, cardiac, and smooth. Their characteristics are summarized in Table 4.3. You will read more about muscle tissue in Chapters 6 and 12.

- **Skeletal muscle tissue** is so named because it is usually attached to bones. When skeletal muscle tissue contracts, therefore, it usually moves a part of the body. Because skeletal muscle is under conscious control, it is described as *voluntary muscle*. Skeletal muscle cells are long, cylinder-shaped cells, each containing several nuclei. In addition, skeletal muscle cells have striations, which are alternating light and dark bands visible under a light microscope. The striations are caused by the orderly arrangement of the contractile proteins actin and myosin, which interact to cause muscle contraction.

- **Cardiac muscle tissue** is found only in the walls of the heart, where its contractions are responsible for pumping blood to the rest of the body. Cardiac muscle contractions

TABLE 4.3	Review of Muscle Tissue		
Type	**Description**	**Example Locations**	**Functions**
Skeletal	Long, cylindrical cells; multiple nuclei per cell; obvious striations	Muscles attached to bones	Provides voluntary movement
Cardiac	Branching, striated cells; one nucleus; specialized junctions between cells	Wall of heart	Contracts and propels blood through the circulatory system
Smooth	Cells taper at each end; single nucleus; arranged in sheets; no striations	Walls of digestive system, blood vessels, and tubules of urinary system	Propels substances or objects through internal passageways

are not under conscious control; we cannot *make* them contract by thinking about them. Thus, cardiac muscle is considered *involuntary muscle*. Cardiac muscle cells resemble branching cylinders and have striations and typically only one nucleus. Special junctions at the plasma membranes of these cells strengthen cardiac tissue and promote rapid conduction of impulses throughout the heart.

- **Smooth muscle tissue** is involuntary and is found in the walls of blood vessels and airways, where its contraction reduces the flow of blood or air. Smooth muscle is also found in the walls of organs such as the stomach, intestines, and bladder, where it aids in mixing and propelling food through the digestive tract and in eliminating wastes. The cells of smooth muscle tissue taper at each end, contain a single nucleus, and lack striations.

Nervous tissue The final major type of tissue, nervous tissue, makes up the nervous system: brain, spinal cord, and nerves. Nervous tissue consists of two general cell types, neurons and accessory cells called neuroglia (Figure 4.4). **Neurons** generate and conduct nerve impulses, which they conduct to other neurons, muscle cells, or glands. Although neurons come in many shapes and sizes, most have three parts—the cell body, dendrites, and an axon. The cell body houses the nucleus and most organelles. Dendrites are highly branched processes that provide a large surface area for the reception of signals from other neurons. A neuron generally has one axon, a long extension that usually conducts impulses away from the cell body. Far more numerous than neurons, the **neuroglia** (or more simply, *glial cells*) support, insulate, and protect neurons. They increase the rate at which impulses are conducted by neurons and provide neurons with nutrients from nearby blood vessels. Recent studies indicate that glial cells communicate with one another and with neurons. You will read more about nervous tissue in Chapters 7 and 8.

Cell Junctions

In many tissues, especially epithelial tissue, the cell membranes have structures for forming attachments between adjoining cells. There are three kinds of junctions between cells: tight junctions, adhesion junctions, and gap junctions. Each type of junction suits the function of the tissue. In **tight junctions** (Figure 4.5a), the membranes of neighboring cells are attached so securely that they form a leakproof seal. Tight junctions are found in the linings of the urinary tract and intestines, where secure seals between cells prevent urine or digestive juices from passing through the epithelium. Less rigid than tight junctions, **adhesion junctions** (also called desmosomes; Figure 4.5b) resemble rivets holding adjacent tissue layers together. The plasma membranes of adjacent cells do not actually touch but are instead bound together by intercellular filaments attached to a thickening in the membrane. Thus, the cells are connected but can still slide slightly relative to one another. Adhesion junctions are common in tissues that must withstand stretching, such as the skin and heart muscle. **Gap junctions** (Figure 4.5c) connect the cytoplasm of adjacent cells through small holes, allowing certain small molecules and ions to flow directly from one cell into the next. In heart and smooth muscle cells, gap junctions help synchronize electrical activity and thus contraction.

Organs and Organ Systems

An **organ** is a structure composed of two or more different tissues that work together to perform a specific function. Organs themselves do not usually function as independent units but instead form part of an **organ system**—a group of organs with a common function. For example, organs such as the trachea, bronchi, and lungs constitute the respiratory system. The common function of these organs is to bring oxygen into the body and remove carbon dioxide. The human body includes 11 major organ systems.

what would you do?

When organs fail to function properly, the result can be illness and death. Certain organs can be transplanted from a living or a deceased donor. Most commonly, living donors give a kidney, but parts of lung, liver, and pancreas can now be donated. Would you consider being a living donor? If so, would you be willing to donate to a stranger? Are you willing to donate your organs after death? If so, have you signed an organ donor card and made your wishes known to family?

Body Cavities Lined with Membranes

Most of our organs are suspended in internal body cavities. These cavities have two important functions. First, they help protect the vital organs from being damaged when we walk or jump. Second, they allow organs to slide past one another and

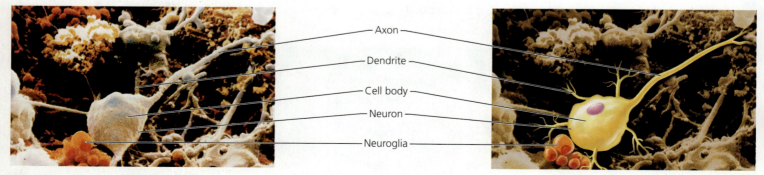

FIGURE 4.4 *Neurons and neuroglia*

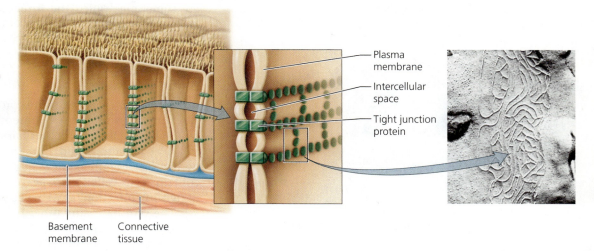

(a) Tight junction
- Creates an impermeable junction that prevents the exchange of materials between cells
- Found between epithelial cells of the digestive tract, where they prevent digestive enzymes and microorganisms from entering the blood

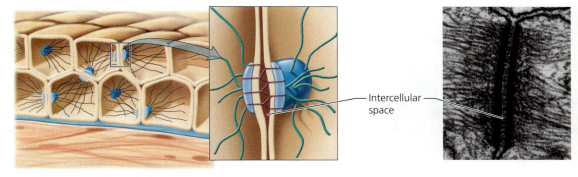

(b) Adhesion junction
- Holds cells together despite stretching
- Found in tissues that are often stretched, such as the skin and the opening of the uterus

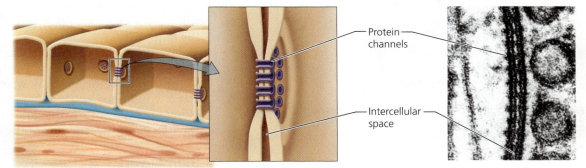

(c) Gap junction
- Allows cells to communicate by allowing small molecules and ions to pass from cell to cell
- Found in epithelia in which the movement of ions coordinates functions, such as the beating of cilia; found in excitable tissue such as heart and smooth muscle

FIGURE 4.5 Specialized cell junctions

change shape. Sliding and changing shape are important when the lungs fill with air, the stomach fills with food, the urinary bladder fills with urine, or when our bodies bend or stretch.

There are two main body cavities—the ventral and dorsal cavities—each of which is further subdivided. The ventral (toward the abdomen) cavity is divided into the *thoracic* (chest) *cavity* and the *abdominal cavity*. The thoracic cavity is subdivided again into the pleural cavities, which house the lungs, and the pericardial cavity, which holds the heart. The abdominal cavity contains the digestive system, the urinary system, and the reproductive system. A muscle sheet called the diaphragm separates the thoracic and abdominal cavities. The dorsal (toward the back) cavity is subdivided into the *cranial cavity,* which encloses the brain, and the *spinal cavity,* which houses the spinal cord.

Body cavities and organ surfaces are covered with membranes—sheets of epithelium supported by connective tissue. Membranes form physical barriers that protect underlying tissues. The body has four types of membranes.

- **Mucous membranes** line passageways that open to the exterior of the body, such as those of the respiratory, digestive, reproductive, and urinary systems. Some mucous membranes, including the mucous membrane of the small intestine, are specialized for absorption. Others, like those of the respiratory system, for instance, secrete mucus that traps bacteria and viruses that could cause illness.
- **Serous membranes** line the thoracic and abdominal cavities and the organs within them. They secrete a fluid that lubricates the organs within these cavities.

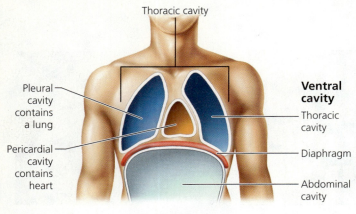

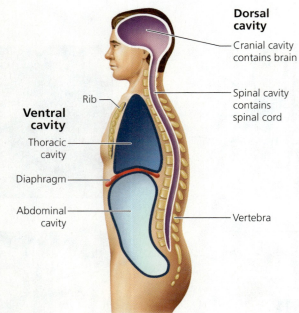

FIGURE 4.6 Body cavities. The internal organs are suspended in body cavities that protect the organs and allow organs to slide past one another as the body moves. Ventral means "toward the abdomen," and dorsal means "toward the back."

- **Synovial membranes** line the cavities of freely movable joints, such as the knee. These membranes secrete a fluid that lubricates the joint, easing movement.
- **Cutaneous membrane**, or skin, covers the outside of the body. Unlike other membranes, it is thick, relatively waterproof, and relatively dry.

The cavities of the human body are shown in Figure 4.6. We will consider the structure and function of the skin in the next section. Details about the structure and functions of the other organ systems, shown and described in Figure 4.7, are presented in subsequent chapters.

4.2 Skin: An Organ System

We have all been told that "beauty is only skin deep," but our skin does much more than just make us attractive. The skin and its derivatives—hair, nails, sweat glands, and oil glands—are sometimes called the **integumentary system** (an *integument* is an outer covering). It is considered an organ system because the skin and its derivatives function together to provide many services for the body.

Skin Functions

A major function of our skin is protection. It serves as a physical barrier that shields the contents of the body from invasion by foreign bacteria and other harmful particles, from ultraviolet (UV) radiation, and from physical and chemical insult. Besides offering this somewhat passive form of protection, skin contains cells called *macrophages* that have a more active way of fighting infection, as we will see in Chapter 13.

The skin has many other functions, as well. For example, because its outermost layer of cells contains the water-resistant protein keratin, the skin plays a vital role in preventing excessive water loss from underlying tissues. It plays a role in temperature regulation, too. Although we perspire (imperceptibly) through our skin almost constantly, during times of strenuous exercise or high environmental temperatures our sweat glands become active and increase their output of perspiration dramatically. The evaporation of this perspiration from the skin's surface helps rid the body of excess heat. Later we will see how changes in the flow of blood to the skin help regulate body temperature. The skin even functions in the production of vitamin D. Modified cholesterol molecules in the skin's outer layer are converted to vitamin D by UV radiation. The vitamin D then travels in the bloodstream to the liver and kidneys, where it is chemically modified to assume its role in stimulating the absorption of calcium and phosphorus from the food we eat.

In addition, the skin contains structures for detecting temperature, touch, pressure, and pain stimuli. These receptors—components of the nervous system—help keep us informed about conditions in our external environment. Keep these many functions of the integumentary system in mind as you read on about the structure of the skin and its derivatives.

Skin Layers

On most parts of your body, the skin is less than 5 mm (less than a quarter of an inch) thick, yet it is one of your largest organs. It represents about one-twelfth of your body weight and has a surface area of 1.5 to 2 m^2 (1.8 to 2.4 yd^2).

The skin has two principal layers, as shown in Figure 4.8. The thin, outer layer, the **epidermis**, forms a protective barrier against environmental hazards. The inner layer, the **dermis**, contains blood vessels, nerves, sweat and oil glands, and hair follicles. Beneath the skin is a layer of loose connective tissue called the *hypodermis* or *subcutaneous layer,* which anchors the skin to the tissues of other organ systems that lie beneath.

The epidermis The outermost layer of skin, the epidermis, is itself composed of several layers of epithelial cells. The outer surface of epidermis, the part you can touch, is made up of dead skin cells. Thus, when we look at another person, most of what we see on the person's surface is dead. These dead cells are constantly being shed, at a rate of about 30,000 to 40,000 each minute. In fact, much of the dust in any room consists of

Integumentary system
- Protects underlying tissues
- Provides skin sensation
- Helps regulate body temperature
- Synthesizes vitamin D

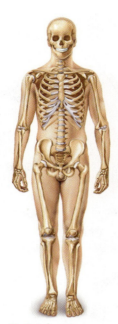

Skeletal system
- Attachment for muscles
- Protects organs
- Stores calcium and phosphorus
- Produces blood cells

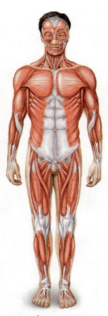

Muscular system
- Moves body and maintains posture
- Internal transport of fluids
- Generation of heat

Nervous system
- Regulates and integrates body functions via neurons

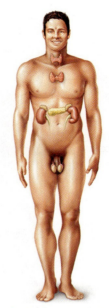

Endocrine system
- Regulates and integrates body functions via hormones

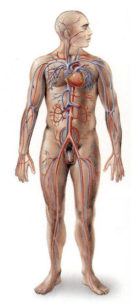

Cardiovascular system
- Transports nutrients, respiratory gases, wastes, and heat
- Transports immune cells and antibodies
- Transports hormones
- Regulates pH

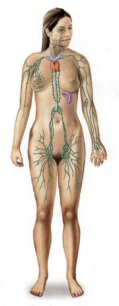

Lymphatic system
- Returns tissue fluids to bloodstream
- Protects against infection and disease

Respiratory system
- Exchanges respiratory gases with the environment

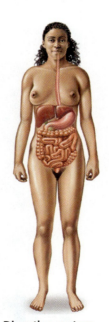

Digestive system
- Physical and chemical breakdown of food
- Absorbs, processes, stores food

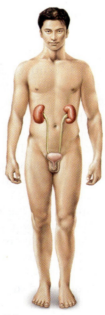

Urinary system
- Maintains constant internal environment through the excretion of nitrogenous waste

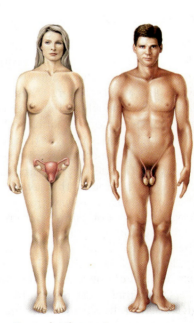

Reproductive system
- Produces and secretes hormones
- Produces and releases egg and sperm cells
- Houses embryo/fetus (females only)
- Produces milk to nourish offspring (females only)

FIGURE 4.7 *Major organ systems of the human body*

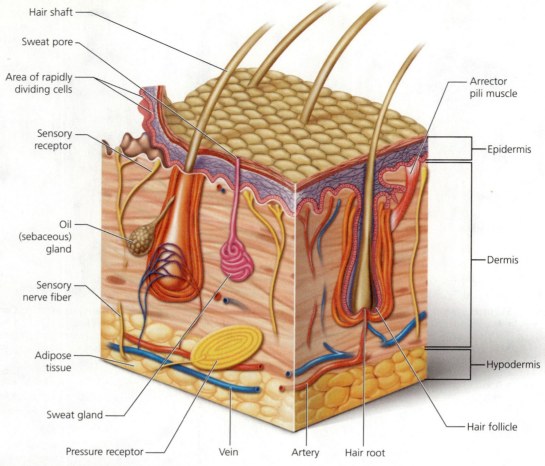

FIGURE 4.8 Structures of the skin and underlying hypodermis

dead skin cells. When you go swimming or soak in the bathtub for a long time, the dead cells on your skin's surface absorb water and swell, causing the skin to wrinkle. This is particularly noticeable where the layer of dead cells is thickest, such as on the palms of the hands and soles of the feet.

The skin does not get thinner as the dead cells are shed, because they are continuously replaced from below. The deepest layer of epidermis contains rapidly dividing cells. As new cells are produced in this layer, older cells are pushed toward the surface. On their way, they flatten and die because they no longer receive nourishment from the dermis. Along this death route, keratin—a tough, fibrous protein—gradually replaces the cytoplasmic contents of the cells. It is keratin that gives the epidermis its protective properties. About 2 weeks to a month pass from the time a new cell is formed to the time it is shed.

Drugs that must be continuously administered are often given across the skin (transdermally) using a drug-containing patch that adheres to the skin (Figure 4.9). Although the epidermis is a water-resistant protective barrier, lipid-soluble materials are able to cross the lipid cell membranes of the cells of the epidermis. Thus, if you dissolve a drug in a lipid solvent, it can cross the epidermis, diffuse into the underlying connective tissue, and be absorbed into the blood. Today, transdermal drug administration is commonly used to provide hormones for birth control or for treating menopausal symptoms, or to

FIGURE 4.9 A transdermal patch continuously delivers a drug across the skin. Here, a cigarette smoker uses a nicotine patch to help quit smoking.

provide nicotine to ease the urge to smoke while trying to quit, and to provide antiemetics to stop nausea from motion sickness.

what would you do?

To speed up the movement of a drug from a transdermal patch to the patient's blood supply, the patch is prepared with a higher concentration of the drug than would be found in a pill. In transdermal patches that deliver hormones for birth control, the blood levels of hormones may also be higher than would result with birth control pills. During the last few years, several young women using a transdermal patch for birth control have had heart attacks or strokes; some have died. There are warning labels with contraceptive patches informing women of these increased risks. Do you think that women or their families should be able to sue the patch manufacturer for their losses?

The dermis Over most parts of the body, the dermis is a much thicker layer than the epidermis. The dermis lies just beneath the epidermis and consists primarily of connective tissue. In addition, it contains blood vessels, hair follicles, oil glands, the ducts of sweat glands, sensory structures, and nerve endings. Unlike the epidermis, the dermis does not wear away. This durability explains why tattoos—designs created when tiny droplets of ink are injected into the dermal layer—are permanent (Figure 4.10). Because the dermis is laced with nerves and sensory receptors, getting a tattoo hurts. In the past, the only way to remove a tattoo was by surgical means, such as "shaving" (abrading) of the skin. Today, laser treatments that destroy the pigments of the tattoo can often be used for tattoo removal.

Blood vessels are present in the dermis but not in the epidermis. Nutrients reach the epidermis by passing out of dermal blood vessels and diffusing through tissue fluid into the layer above. Such tissue fluid is probably quite familiar to you. Where skin is traumatized by, for example, a burn or an ill-fitting shoe rubbing against your heel, this fluid accumulates

stop and think

Burns—tissue damage caused by heat, radiation, electric shock, or chemicals—can be classified according to how deeply the damage penetrates. First-degree burns are confined to the upper layers of epidermis, where they cause reddening and slight swelling. In second-degree burns, damage extends through the epidermis into the upper regions of the dermis. Blistering, pain, and swelling occur. Third-degree burns extend through the epidermis and dermis into underlying tissues. Severe burns, particularly if they cover large portions of the body, are life threatening. Given your knowledge of skin functions, what would you predict to be the immediate medical concerns for a patient with third-degree burns?

between the epidermis and dermis, separating the layers and forming blisters.

The lower layer of the dermis consists of dense connective tissue containing collagen and elastic fibers, a combination that allows the skin to stretch and then return to its original shape. The resilience of our skin also decreases as we age. The most pronounced effects begin in the late forties, when collagen fibers begin to stiffen and decrease in number and elastic fibers thicken and lose elasticity. These changes, combined with reductions in moisture and the amount of fat in the hypodermis, lead to wrinkles and sagging skin.

Certain wrinkles, such as frown lines, are caused by the contraction of facial muscles. A controversial and popular treatment for these wrinkles is to inject Botox, the toxin from the bacterium that causes botulism. When Botox is injected into facial muscles, they become temporarily paralyzed. The muscle contractions that form the wrinkles cannot occur, and the skin smoothes out. The muscles regain the ability to contract over the next several months, however, and the injection has to be repeated (Figure 4.11).

The hypodermis The **hypodermis**, a layer of loose connective tissue just below the epidermis and dermis, is not usually considered part of the skin. It does, however, share some of the skin's functions, including cushioning blows and helping to prevent extreme changes in body temperature, because it contains about half of the body's fat stores. In infants and toddlers, this layer of fat that lies under the skin—often called baby fat—covers the entire body, but as we mature, some of the fat stores are redistributed. In women, subcutaneous fat tends to accumulate in the breasts, hips, and thighs. In both sexes, it has a tendency to accumulate in the abdominal hypodermis, contributing to the all-too-familiar potbelly, and in the sides of the lower back, forming "love handles."

Liposuction, a procedure for vacuuming fat from the hypodermis, is a way to reshape the body. The physician makes a small incision in the skin above the area of unwanted fat, inserts a fine tube, and moves the tube back and forth to loosen the fat cells, which are then sucked into a container. Liposuction is not a way to lose a lot of weight, because only a small amount of fat—not more than a few pounds—can be removed. However, it is a way of sculpting the body and removing bulges. Furthermore, because liposuction removes the cells that store the fat, fat does not usually return to those areas. The procedure is generally safe, but it is not risk free.

FIGURE 4.10 *Tattoos—designs created when droplets of ink are injected into the dermis—are essentially permanent because, unlike the epidermis, the dermis is not shed.*

FIGURE 4.11 *Before and after Botox injections*

In some patients, it has produced blood clots that traveled to the lungs, causing death. People who are considering liposuction should choose their doctor carefully.

A newer procedure is to inject a drug combination, known as Lipodissolve, directly into the unwanted pockets of fat. The treatment consists of a series of injections containing two main ingredients: one that chemically digests fat molecules and another that causes fat cells to burst. Although the U.S. Food and Drug Administration (FDA) has approved both drugs for other purposes, it has not approved this procedure combining the drugs for the purpose of eliminating unwanted fat. There is no evidence that Lipodissolve is unsafe, and there is no scientific evidence that it is effective.

Skin Color

Two interacting factors produce skin color: (1) the quantity and distribution of pigment and (2) blood flow. The pigment, called **melanin**, is produced by cells called **melanocytes** at the base of the epidermis. These cells, with their spiderlike extensions, produce two kinds of melanin: a yellow-to-red form and the more common black-to-brown form. The melanin is then taken up by surrounding epidermal cells, thus coloring the entire epidermis.

All people have about the same number of melanocytes. Differences in skin color are due to differences in the form of melanin produced and the size and number of pigment granules. A person's genetic makeup determines the combination of the yellowish red or the brown form of melanin produced.

Circulation also influences skin color. When well-oxygenated blood flows through vessels in the dermis, the skin has a pinkish or reddish tint that is most easily seen in light-skinned people. Intense embarrassment can increase the blood flow, causing the rosy color to heighten, particularly in the face and neck. This response, known as blushing, is impossible to stop. Other intense emotions may cause color to disappear temporarily from the skin. A sudden fright, for example, can cause a rapid drop in blood supply, making a person pale. Skin color may also change in response to changing levels of oxygen in the blood. Compared to well-oxygenated blood, which is ruby red, poorly oxygenated blood is a much deeper red that gives the skin a bluish appearance. Poor oxygenation is what causes the lips to appear blue in extremely cold conditions. When it is cold, your body shunts blood away from the skin to the body's core, which conserves heat and keeps vital organs warm. This shunting reduces the oxygen supply to the blood in the small vessels near the surface of the skin. The oxygen-poor blood seen through the thin skin of the lips makes them look blue. When you do not get enough sleep, the amount of oxygen in your blood may be slightly lower than usual, causing the color to darken. In some people, the darker color of blood is visible through the thin skin under their eyes as dark circles.

Tanning is a change in skin color from exposure to the sun. Melanocytes respond to the UV radiation in sunlight by increasing the production of melanin. This is a protective response, because melanin absorbs some UV radiation, preventing it from reaching the lower epidermis and the dermis. The skin requires some UV radiation for the production of vitamin D, but too much can be harmful. See the Health Issue essay, *Fun in the Sun?*.

Hair, Nails, and Glands

The epidermis gives rise to many seemingly diverse structures: hair, nails, oil glands, sweat glands, and teeth. We will now consider the first four of these in view of their structure and roles in everyday life. (Teeth are discussed in Chapter 15.)

Hair Hair usually grows all over the body, except on a few areas such as the lips, palms of the hands, and soles of the feet. What functions do these dead cells serve? An important one is protection. Hair on the scalp protects the head from UV radiation. Hair in the nostrils and external ear canals keeps particles and bugs from entering. Likewise, eyebrows and eyelashes help keep unwanted particles and glare (and perspiration and rain) out of the eyes. Hair also has a sensory role: receptors associated with hair follicles are sensitive to touch.

A hair consists of a shaft and a root (see Figure 4.8). The shaft projects above the surface of the skin, and the root extends below the surface into the dermis or hypodermis, where it is embedded in a structure called the *hair follicle*. Nerve endings surround the follicle and are so sensitive to touch that we are aware of even slight movements of the hair shaft. (Try to move just one hair without feeling it.) Each hair is also supplied with an oil gland that opens onto the follicle and supplies the hair with an oily secretion that makes it soft and pliant. In the dermis, a tiny smooth muscle called the *arrector pili* is attached to the hair follicle. Contraction of this muscle—which pulls on the follicle, causing the hair to stand up—is associated with

HEALTH ISSUE

Fun in the Sun?

A beach crowded with people on a sunny day is hardly a scene we would equate with disfigurement and death. Nonetheless, that is a connection we should make, because skin cancers are increasing at an alarming rate, largely because of our exposure to the sun.

The ultraviolet (UV) radiation of sunlight causes the melanocytes of the skin to increase their production of the pigment melanin, which absorbs UV radiation before it can damage the genetic information of deeper layers of cells. Unfortunately, this protective buildup of melanin is not instantaneous. In skin cancer, UV radiation alters the genetic material in skin cells so that the cells grow and divide uncontrollably, forming a tumor. Some experts fear that the rates of skin cancer will increase dramatically if the ozone layer, which blocks some of the UV radiation before it reaches Earth, continues to become thinner (the ozone layer is discussed in Chapter 24). Three types of skin cancer are caused by overexposure to the sun (Figure 4.A):

- *Basal cell carcinoma,* the most common form of skin cancer, arises in the rapidly dividing cells of the deepest layer of epidermis.
- *Squamous cell carcinoma,* the second most common form of skin cancer, arises in the newly formed skin cells as they flatten and move toward the skin surface.
- *Melanoma* is the least common and most dangerous type of skin cancer. It arises in melanocytes, the pigment-producing cells of the skin. Unlike basal or squamous cell carcinomas, melanomas, when left untreated, often metastasize (spread rapidly) throughout the body, first infiltrating the lymph nodes and later the vital organs. The survival rate in persons whose melanoma is found before it has metastasized is about 90% but drops to about 14% if the cancerous cells have spread throughout the body.

You can catch melanomas at an early stage if you carefully examine your skin while applying the ABCD mnemonic of the American Cancer Society:

A—stands for "asymmetry." Most melanomas are irregular in shape.

B—stands for "border." Melanomas often have a diffuse, unclear border.

C—stands for "color." Melanomas usually have a mottled appearance and contain colors such as brown, black, red, white, and blue.

D—stands for "diameter." Growths with a diameter of more than 5 mm (about 0.2 in.) are threatening.

The best way to avoid getting skin cancer is to avoid prolonged exposure to the sun. If you must be out in the sun, wear a hat, long sleeves, and sunglasses. Use a sunscreen with a sun protection factor (SPF) of at least 15. Apply your sunscreen about 45 minutes before going out into the sun, allowing time for the skin to absorb it so that it is less likely to wash away with perspiration. Use sunscreen even when it is overcast, because UV rays can penetrate the clouds. Reapply it after swimming.

Always remember that sunscreens are not foolproof. Most block the higher-energy portion of the sun's UV radiation, known as UV-B, while providing only limited protection against the lower-energy portion, called UV-A. Whereas UV-B causes skin to burn, recent research suggests that exposure to UV-A weakens the body's immune system, possibly impairing its ability to fight melanoma. Ironically, by providing protection from sunburn, sunscreens have had the potentially devastating effect of enabling people to spend more time in the sun, possibly increasing their risk of developing melanoma.

Avoid tanning salons. For many years, tanning salons claimed to use "safe" wavelengths of UV radiation because they did not use skin-reddening UV-B. But these "safe" wavelengths are actually UV-A. Given the apparent link between UV-A and increased risk of melanoma, the potential danger of these "safe" wavelengths is now obvious.

Questions to Consider

- Do you think that tanning salons should be required to explain the dangers of UV-A to the public?
- Some physicians are now warning that excessive use of sunscreen can lower the body's ability to produce enough vitamin D, which can lead to osteoporosis (a loss of bone density). What steps might you take to balance the risk of developing skin cancer with the risk of vitamin D deficiency?

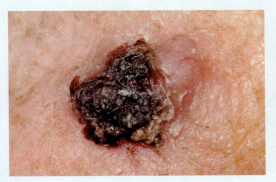
(a) Basal cell carcinoma

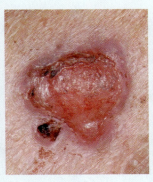
(b) Squamous cell carcinoma

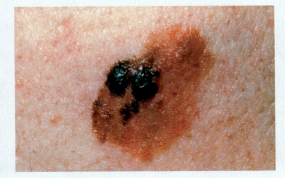
(c) Melanoma

FIGURE 4.A *Three skin cancers*

---

fear and with cold. The tiny mound of flesh that forms at the base of the erect hair is sometimes called a goose bump.

Nails Nails protect the sensitive tips of fingers and toes. Like hair, nails are modified skin tissue hardened by the protein keratin. Although the nail itself is dead and lacks sensory receptors, it is embedded in tissue so sensitive that we detect even the slightest pressure of an object on the nail. In this way, nails serve as sensory "antennas." They also help us manipulate objects, as when we undo a tight knot in a shoelace.

Glands Three types of glands—oil, sweat, and wax—are found in the skin. Although all three types develop from epidermal cells, they differ in their locations, structures, and functions.

HEALTH ISSUE

Acne: The Miseries and Myths

Acne and adolescence go hand in hand. In fact, about four out of five teenagers have acne, a skin condition that will probably annoy, if not distress, them well into their twenties and possibly beyond.

Simple acne is a condition that affects hair follicles associated with oil glands. During the teenage years, oil glands increase in size and produce larger amounts of oily sebum. These changes are prompted, in both males and females, by increasing levels of "male" hormones called *androgens* in the blood; the androgens are secreted by the testes, ovaries, and adrenal glands. The changes thus induced in the activity and structure of oil glands set the stage for acne.

It should come as no surprise, then, that acne occurs most often on areas of the body where oil glands are largest and most numerous: the face, chest, upper back, and shoulders.

Acne is the inflammation that results when sebum and dead cells clog the duct where the oil gland opens onto the hair follicle (Figure 4.B). A follicle obstructed by sebum and cells is called a whitehead. Sometimes the sebum in plugged follicles mixes with the skin pigment melanin, forming a blackhead. Thus, melanin, not dirt or bacteria, lends the dark color to these blemishes. The next stage of acne is pimple formation, beginning with the formation of a red, raised bump, often with a white dot of pus at the center. The bump occurs when obstructed follicles rupture and spew their contents into the surrounding epidermis. Such ruptures may occur naturally by the general buildup of sebum and cells or may be induced by squeezing the area. The sebum, dead cells, and bacteria that thrive on them then cause a small infection—a pimple or pustule—that will usually heal within a week or two without leaving a scar.

There are many misconceptions about the causes of acne. Eating nuts, chocolate, pizza, potato chips, or any of the other "staples" of the teenage diet does not cause acne. Also, acne is not caused by poor hygiene. Follicles plug from below, so dirt or oil on the skin surface is not responsible. (Most doctors do, however, recommend washing the face two or three times a day with hot water to help open plugged follicles.)

Questions to Consider
- Why do you think that there are so many misconceptions about the causes of acne?
- If a new medication for acne were marketed, how would you decide whether to use it?

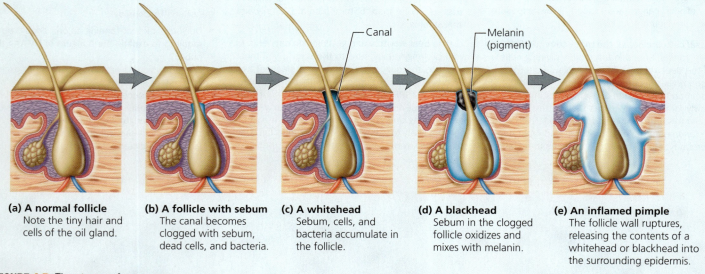

(a) A normal follicle Note the tiny hair and cells of the oil gland.

(b) A follicle with sebum The canal becomes clogged with sebum, dead cells, and bacteria.

(c) A whitehead Sebum, cells, and bacteria accumulate in the follicle.

(d) A blackhead Sebum in the clogged follicle oxidizes and mixes with melanin.

(e) An inflamed pimple The follicle wall ruptures, releasing the contents of a whitehead or blackhead into the surrounding epidermis.

FIGURE 4.B *The stages of acne*

Oil (sebaceous) glands are found virtually all over the body except on the palms of the hands and soles of the feet. They secrete sebum, an oily substance made of fats, cholesterol, proteins, and salts. The secretory part of these glands is located in the dermis, as shown in Figure 4.8. Sebum lubricates hair and skin and contains substances that inhibit growth of certain bacteria. Sometimes, however, the duct of an oil gland becomes blocked, causing sebum to accumulate and disrupt the gland's proper function. Then, bacteria can invade the gland and hair follicle, resulting in a condition called acne. See the Health Issue essay, *Acne: The Miseries and Myths*.

As their name implies, **sweat glands** produce sweat, which is largely water plus some salts, lactic acid, vitamin C, and metabolic wastes such as urea. Although some wastes are eliminated through sweating, the principal function of sweat is to help regulate body temperature by evaporating from the skin surface. Wax glands are modified sweat glands found in the external ear canal. As their name implies, they produce wax, which protects the ear by trapping small particles.

4.3 Homeostasis

To remain healthy, the organ systems of the body must constantly adjust their functioning in response to changes in the internal and external environment. We have already learned that the body's organ systems are interdependent, working together to provide the basic needs of all cells—water, nutrients, oxygen, and a normal body temperature. Just as city dwellers breathe the same air and drink the same city water, the cells of the body are surrounded by the same extracellular fluid. Changes in the makeup of that fluid will affect every cell.

One advantage of our body's multicellular, multiorgan-system organization is its ability to provide a controlled environment for the cells. Although conditions outside the body sometimes vary dramatically, our organ systems interact to maintain relatively stable conditions within. This ability to maintain a relatively stable internal environment despite changes in the surroundings is called **homeostasis** (meaning "to stay the same"). But conditions within the body never stay the same. As internal conditions at any level vary, the body's processes must shift to counteract the variation. Homeostatic mechanisms do not maintain absolute internal constancy, but they do dampen fluctuations around a set point to keep internal conditions within a certain range. Thus, homeostasis is not a static state but a dynamic one.

Illness can result if homeostasis fails. We see this in diabetes, a condition in which either the pancreas does not produce enough of the hormone insulin or the body cells are unable to use insulin. Normally, as a meal is digested and nutrients are absorbed into the bloodstream, the rising level of glucose in the blood stimulates the pancreas to release insulin. The general effect of insulin is to lower the blood level of glucose, returning it to a more desirable value. Without insulin, blood glucose can rise to a point that causes damage to the eyes, kidney, nerves, and blood vessels. A healthy diet, exercise, medication, and sometimes insulin injections can help people with diabetes regulate their blood glucose level.

Homeostasis depends on communication within the body. The nervous and endocrine systems are the two primary means of communication. The nervous system can bring about quick responses to changes in internal and external conditions. The endocrine system produces hormones, which bring about slower and longer-lasting responses to change.

stop and think
When you exercise, your breathing rate, blood pressure, and heart rate increase. Is this a violation of homeostasis?

Negative Feedback Mechanisms

Homeostasis is maintained primarily through **negative feedback mechanisms**—corrective measures that slow or reverse a variation from the normal value of a factor, such as blood glucose level or body temperature, and return the factor to its normal value. When the normal value is reached, the corrective measures cease; the normal value is the feedback that turns off the response. (In contrast, a positive feedback mechanism causes a change that promotes continued change in the same direction. Positive feedback mechanisms, which are described in Chapter 10, do not promote homeostasis.)

Homeostatic mechanisms have three components (Figure 4.12).

1. A *receptor* detects change in the internal or external environment. A receptor, in this context, is a sensor that monitors the environment. When the receptor detects a change in some factor or event—some variable—it sends that information to the control center, the second of the three components.

2. A *control center* determines the factor's set point—the level or range that is normal for the factor in question. The control center integrates information coming from all the pertinent receptors and selects an appropriate response. In most of the body's homeostatic systems, the control center is located in the brain.

3. An *effector,* often a muscle or gland, carries out the selected response.

Consider how a negative feedback mechanism controls the temperature in your home during frigid winter months. A thermostat in the heating system serves as both the temperature-sensing receptor and the control center. If it senses that the temperature inside your home has fallen below a certain programmed set point, the thermostat turns on the heating system (the effector). When the thermostat senses that the internal temperature has returned to the set point, it turns the heating system off. Thus, the temperature fluctuates around the set point but remains within a certain range. Now, let's apply these principles to see how homeostatic mechanisms regulate body temperature.

Hypothalamus and Body Temperature

The body's temperature control center is located in a region of the brain called the *hypothalamus.* Its set point is approximately 37°C (98.6°F), although it differs slightly from one person to the next. Body temperature must not vary too far from this mark, because even small temperature changes have dramatic effects on metabolism. Temperature is sensed at the body's outer surface by skin receptors and deep inside the body by receptors that sense the temperature of the blood. The hypothalamus receives input from both types of receptor. If the input indicates that body temperature is below the set point, the brain initiates mechanisms to increase heat production and conserve heat. When the input indicates that body temperature is above the set point, the brain initiates mechanisms that promote heat loss.

Let's consider what happens when we find ourselves in an environment where the temperature is above our set point, say 38°C (100.4°F). Thermoreceptors in the skin detect heat and activate nerve cells that send a message to the hypothalamus, which then sends nerve impulses to the sweat glands to increase their secretions. As the secretions (perspiration) evaporate and body temperature drops below 37°C (98.6°F), the signals from the brain to the sweat glands are discontinued. In this homeostatic system, the thermoreceptors in the skin are the receptors, the hypothalamus is the control center, and the sweat glands are the effectors. The system is a negative feedback mechanism because it produces an effect (cooling of the skin) that is relayed (feeds back) to the control center, shutting off the corrective mechanism when the desired change has been produced.

Q Stable blood calcium levels are important to many physiological processes. Calcitonin is a hormone from the thyroid gland that lowers blood calcium levels. Parathyroid hormone from the parathyroid glands is a hormone that raises blood calcium levels. Describe a negative feedback relationship involving these hormones as effectors that would maintain homeostasis by keeping blood calcium levels stable.

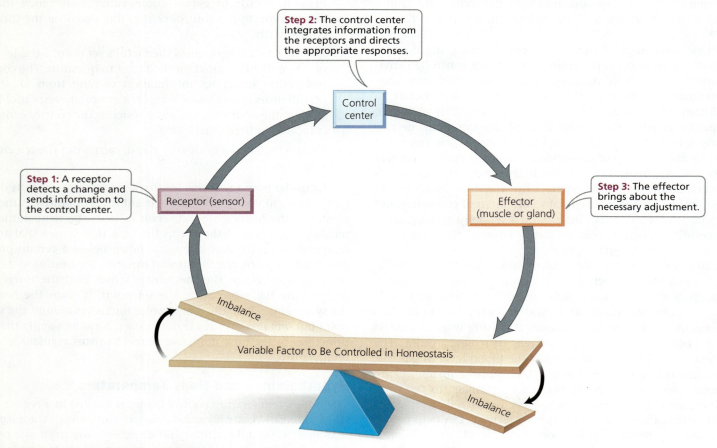

FIGURE 4.12 *The components of a homeostatic control system maintained by negative feedback mechanisms*

A An elevation in blood calcium level would be sensed by receptors. In response, a control center in the brain would stimulate the thyroid gland to increase production of calcitonin, which lowers blood calcium level. If the calcium level dropped too low, receptors would signal the control center in the brain. In response, the control center would stimulate the parathyroid glands to increase production of parathyroid hormone, which would raise blood calcium levels.

Other mechanisms that the brain may activate to lower body temperature include dilation of blood vessels in the dermis and relaxation of the arrector pili muscles attached to the hair follicles. The former response releases more heat to the surrounding air and explains the flushed appearance we get during strenuous exercise. The latter keeps damp, cooling hair lying close to the skin. The brain may also initiate behavioral responses to a high body temperature, such as seeking shade or removing a sweatshirt.

stop and think

As people age, the activity of their sweat glands declines. How does this explain why public authorities are less concerned about children than they are about the elderly developing a heat-related illness during a heat wave?

Now let's consider what happens when body temperature drops below the set point. Subtle drops in body temperature are detected largely by thermoreceptors in the skin, which send a message to the hypothalamus in the brain. The brain then sends nerve impulses to sweat glands, ordering a decrease in their activity, as well as messages to vessels in the dermis, telling them to constrict. This constriction reduces blood flow to the extremities, conserving heat for the internal organs and giving credence to the saying, "cold hands, warm heart." Another response to decreasing body temperature is contraction of arrector pili muscles, causing hairs to stand on end and thereby trapping an insulating layer of air near the body. This response, known as *piloerection,* is less effective in humans than in more heavily furred animals. The body also responds to cooling by increasing metabolic activity to generate heat and by the repeated contraction of skeletal muscles, known as shivering. Finally, behavioral responses, such as folding one's arms across one's chest, may be called upon to help combat a drop in body temperature. See Figure 4.13 for a summary of body temperature regulation.

Sometimes the mechanisms for lowering higher-than-normal temperatures fail, resulting in potentially deadly *hyperthermia*—abnormally elevated body temperature. Some marathon runners have died as a result of elevated core temperatures, as have people sitting in a hot tub heated to too high a temperature (say, up around 114°F). In both situations, perspiration could bring no relief; its evaporation would be prevented by high humidity in the case of the runners or by the surrounding water in the case of the hot-tubbers. Commonly called *heat stroke*, hyperthermia is marked by confusion and dizziness. If the core temperature reaches about 42°C (107°F), the heartbeat becomes irregular, oxygen levels in the blood drop, the liver ceases to function, and unconsciousness and death soon follow. Few people can survive core temperatures of 43°C (110°F).

If the body's temperature drops too far—to 35°C (95°F) or below—a condition called *hypothermia* results, disrupting nervous system function and temperature-regulating mechanisms. People suffering from hypothermia usually become giddy and confused at first. When their temperature drops to 33°C (91.4°F), they lose consciousness. Finally, at a body temperature of 30°C (86°F), blood vessels are completely constricted and temperature-regulating mechanisms are fully shut down. Death soon follows. Hypothermia can be treated if detected early enough. In severe cases, dialysis machines may be used to artificially warm the blood and pump it back into the body.

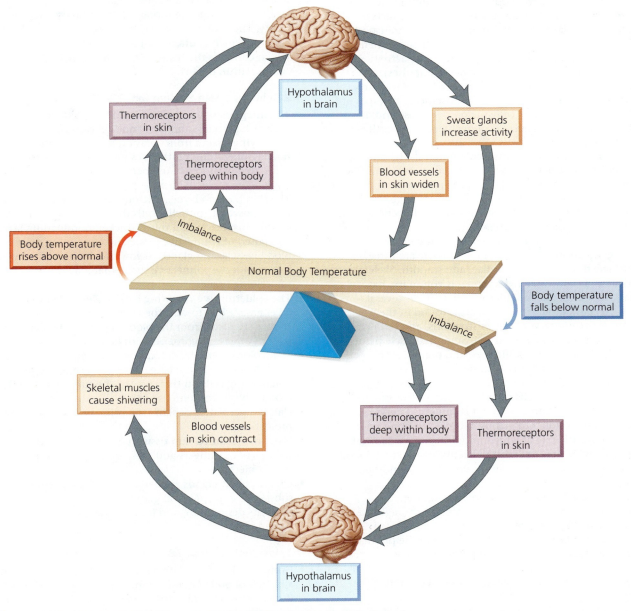

FIGURE 4.13 *Homeostatic regulation of body temperature by negative feedback mechanisms. In this homeostatic control system, thermoreceptors are the sensors; the hypothalamus is the control center; and sweat glands, blood vessels in the skin, and skeletal muscles are the effectors.*

stop and think

Frostbite is damage to tissues exposed to cold temperatures. Given what you know about the body's response to cold temperature, why are fingers and toes particularly susceptible to frostbite?

looking ahead

In this chapter we saw how cells form tissues, tissues form organs, and organs form organ systems. Next we will explore each organ system, beginning with Chapter 5, "The Skeletal System."

HIGHLIGHTING THE CONCEPTS

4.1 From Cells to Organ Systems (pp. 64–72)

- Tissues are groups of cells that work together to perform a common function. There are four main types of tissue in the human body: epithelial tissue (covers body surfaces, forms glands, and lines internal cavities and organs); connective tissue (acts as storage site for fat, plays a role in immunity and in transport, and provides protection and support); muscle tissue (generates movement); and nervous tissue (coordinates body activities through initiation and transmission of nerve impulses).
- All connective tissues contain cells in an extracellular matrix composed of protein fibers and ground substance. The types of connective tissue are connective tissue proper (loose and dense connective tissue) and specialized connective tissue (cartilage, bone, and blood).
- Blood consists of formed elements (red blood cells, white blood cells, and platelets) suspended in a liquid matrix (plasma). The protein fibers are normally dissolved in the plasma and play a role in blood clotting. Red blood cells transport oxygen and carbon dioxide; white blood cells aid in fighting infections; and platelets function in blood clotting.
- Muscle tissue is composed of muscle fibers that contract when stimulated, generating a mechanical force. There are three types of muscle tissue: skeletal, cardiac, and smooth. Skeletal muscle tissue is usually attached to bone, is voluntary, has cross-striations visible under a microscope, and has several nuclei in each cell. Cardiac muscle tissue is found in the walls of the heart, is involuntary, has cross-striations, and usually contains only one nucleus in each cell. Smooth muscle tissue is in the walls of blood vessels, airways, and organs. It is involuntary and lacks striations. A smooth muscle cell tapers at each end and has a single nucleus.
- Nervous tissue consists of cells called neurons and accessory cells called neuroglia. Neurons convert stimuli into nerve impulses that they conduct to glands, muscles, or other neurons. Neuroglia increase the rate at which impulses are conducted by neurons and provide neurons with nutrients from nearby blood vessels. Neuroglial cells can communicate with one another and with neurons.
- Three types of specialized junctions hold tissues together. Tight junctions prevent the passage of materials through the boundaries where cells meet. Adhesion junctions link cells by intercellular filaments attached to thickenings in the plasma membrane. Gap junctions have small pores that allow physical and chemical communication between cells.
- An organ is a structure that is composed of two or more different tissues and has a specialized function.
- Two or more organs that participate in a common function are collectively called an organ system. The human body has 11 major organ systems.
- Internal organs are located in body cavities. There are two main body cavities: (1) the dorsal cavity, which is subdivided into the cranial cavity, where the brain is located, and the spinal cavity, where the spinal cord is located; and (2) the ventral cavity, which is subdivided into the thoracic (chest) cavity and the abdominal cavity. The thoracic cavity is further divided into the pleural cavities, which contain the lungs, and the pericardial cavity, which contains the heart. Membranes line body cavities and spaces within organs.

4.2 Skin: An Organ System (pp. 72–78)

- The integumentary system includes the skin and its derivatives, such as hair, nails, and sweat and oil glands. It protects underlying tissues from abrasion and dehydration; regulates body temperature; synthesizes vitamin D; detects stimuli associated with touch, temperature, and pain; and initiates body defense mechanisms.
- The skin has two layers. The outermost layer, or epidermis, is composed of epithelial cells that die and wear away. The dermis, just below, is a much thicker, nondividing layer composed of connective tissue and containing nerves, blood vessels, and glands. Below the dermis is the hypodermis, a layer of loose connective tissue that anchors the skin to underlying tissues.
- The epidermis is a renewing barrier. Cells produced in the deepest layer are pushed toward the skin surface, flattening and dying as they move away from the blood supply of the dermis and replace their cytoplasmic contents with keratin.
- Skin color is partially determined by melanin, a pigment released by melanocytes at the base of the epidermis and taken up by neighboring cells on their way to the surface. Blood flow and blood oxygen content also influence skin color.
- Hair is a derivative of skin. The primary function of hair is protection.
- Nails are modified skin tissue hardened by keratin. They protect the tips of our fingers and toes and help us grasp and manipulate small objects.
- Oil and sweat glands are derivatives of skin. Sebum, the oily substance secreted by oil glands, lubricates the skin and hair, prevents desiccation, and inhibits the growth of certain bacteria.

4.3 Homeostasis (pp. 78–82)

- Homeostasis is the relative internal constancy maintained at all levels of body organization. It is a dynamic state, with small fluctuations occurring around a set point, and is sustained primarily through negative feedback mechanisms.
- Homeostatic mechanisms consist of receptors, a control center, and effectors.

RECOGNIZING KEY TERMS

tissue p. 64
epithelial tissue p. 64
connective tissue p. 64
muscle tissue p. 64
nervous tissue p. 64
basement membrane p. 65
squamous epithelium p. 65
cuboidal epithelium p. 65
columnar epithelium p. 65
gland p. 65
exocrine gland p. 65
endocrine gland p. 65
matrix p. 65

ground substance p. 65
collagen fibers p. 65
elastic fibers p. 65
reticular fibers p. 65
fibroblast p. 65
loose connective tissue p. 67
areolar connective tissue p. 67
adipose tissue p. 67
dense connective tissue p. 67
cartilage p. 67
bone p. 67
blood p. 67
skeletal muscle tissue p. 69

cardiac muscle tissue p. 69
smooth muscle tissue p. 70
neurons p. 70
neuroglia p. 70
tight junction p. 70
adhesion junction p. 70
gap junction p. 70
organ p. 70
organ system p. 70
mucous membrane p. 71
serous membrane p. 71
synovial membrane p. 72
cutaneous membrane p. 72

integumentary system p. 72
epidermis p. 72
dermis p. 72
hypodermis p. 75
melanin p. 76
melanocyte p. 76
oil gland p. 78
sweat gland p. 78
homeostasis p. 79
negative feedback mechanism
  p. 79

REVIEWING THE CONCEPTS

1. What are the four types of tissue found in the human body? p. 64
2. Contrast the organization of epithelial and connective tissues. How are differences in the matrix of different types of connective tissue related to their functions? pp. 65–68
3. Why does bone heal more rapidly than cartilage? p. 67
4. What type of tissue is blood? p. 67
5. Contrast skeletal, cardiac, and smooth muscle with respect to structure and function. pp. 69–70
6. What types of cells are found in nervous tissue? What are their functions? p. 70
7. Describe the body's homeostatic mechanisms for raising and lowering core body temperature. pp. 78–81
8. The four basic tissue types in the body are
   a. simple, cuboidal, squamous, columnar.
   b. neural, epithelial, muscle, connective.
   c. blood, nerves, bone, cartilage.
   d. fat, cartilage, muscle, neural.
9. A gland is composed of _____ tissue.
   a. epithelial
   b. connective
   c. muscle
   d. nervous
10. The lining of the intestine is composed primarily of
    a. epithelial cells.
    b. muscle cells.
    c. connective tissue cells.
    d. nerve cells.
11. Which of the following is *not* a function of epithelia?
    a. providing physical protection
    b. storing energy reserves
    c. producing specialized secretions
    d. absorption
12. _____ glands secrete hormones into the blood via tissue fluids.
    a. Endocrine
    b. Mixed
    c. Exocrine
    d. Unicellular
13. Which cells form disks that cushion the vertebrae?
    a. bone
    b. muscle
    c. epithelium
    d. cartilage
14. Which of the following is an example of dense connective tissue?
    a. tendon
    b. blood
    c. adipose (fat) tissue
    d. cartilage at the tip of the nose
15. Which type of tissue gets its characteristics from a matrix rather than cells?
    a. epithelial tissue
    b. muscle tissue
    c. connective tissue
    d. nerve tissue
16. Functions of connective tissue include
    a. establishing a structural framework for the body.
    b. transporting fluids and dissolved materials.
    c. storing energy reserves.
    d. all of the above
17. Which type of muscle cell propels substances or objects thrugh internal passageways?
    a. skeletal
    b. cardiac
    c. smooth
    d. striated
18. Which type of junction allows cells to communicate by allowing small molecules to pass from cell to cell?
    a. tight
    b. adhesion
    c. gap
19. Which is *not* a function of the integumentary system?
    a. production of blood cells
    b. protection of underlying tissues
    c. synthesis of vitamin D
    d. regulation of body temperature

20. Which type of membrane lines passageways that open to the exterior of the body?
    a. mucous
    b. serous
    c. synovial
    d. cutaneous

21. The effector in a negative feedback loop produces changes that
    a. are opposite to the change produced by the initial stimulus.
    b. increase the effect of the initial stimulus.
    c. are not related to the initial stimulus.

APPLYING THE CONCEPTS

1. Thor, a ski champion at his college, tore the cartilage in his knee in a ski accident. He asked the doctor whether he would be ready to compete in a month. Why would you expect the doctor's answer to be "No"?
2. Hannah has the flu. As her fever rises, she gets the chills. She shivers and covers herself in extra blankets. She takes some acetaminophen, and her fever breaks. As her body temperature returns to normal, she throws off the blankets, looks flushed, and perspires. Use the mechanisms of body temperature control to explain what is happening as Hannah's fever rises and falls.
3. Ehlers-Danlos syndrome (EDS) is a group of disorders caused by defects in genes that disrupt the production of collagen, which is a chief component of connective tissue. Explain why symptoms of EDS include joints that extend beyond their normal range and stretchy, saggy skin.

BECOMING INFORMATION LITERATE

Prepare a brochure or other presentation explaining ways to avoid skin cancer. Use at least three reliable sources (books, journals, websites). List each source you considered, and explain why you chose the three sources you used.

MasteringBiology®

Go to MasteringBiology for practice quizzes, activities, eText, videos, current events, and more.

The Skeletal System

5

- *You had more bones when you were born (perhaps as many as 300) than you do now (perhaps as few as 206), because some bones fuse as you age.*
- *The thighbone (femur) accounts for about 25% of your height.*

5.1 Bone Functions

5.2 Bone Structure

5.3 Bone as a Living Tissue

5.4 The Role of Fibroblasts and Osteoblasts in Repairing Bone Fractures

5.5 Bone Remodeling

5.6 Axial Skeleton

5.7 Appendicular Skeleton

5.8 Joints

HEALTH ISSUE

Osteoporosis: Fragility and Aging

In the previous chapter, we became familiar with our organ systems. In this chapter, we will consider the skeletal system and its role in determining our physical appearance. We will examine the structure of bone and see why it is able to support our bodies against gravity. We'll also consider our joints, places where bones meet, and see how they permit movement. In reading about the structure and properties of bone in this chapter, you may be surprised to learn that bone is a dynamic, living tissue and to discover the number of functions our skeletal system performs.

The 206 bones of the human body provide support against gravity and protect internal organs. The joints, places where bones meet, determine the type of movement that is possible.

5.1 Bone Functions

The **skeleton** is a framework of bones and cartilage that performs the following functions for the body:

1. **Support.** It provides a rigid framework that supports soft tissues. The leg bones and backbone hold our bodies upright, and the pelvic girdle supports the abdominal organs.
2. **Movement.** It provides places of attachment for muscles. Contraction of muscles allows bones to move at joints.
3. **Protection.** It shields our internal organs, such as the heart and lungs, which are enclosed within the chest cavity, and the brain, which lies within the skull.
4. **Storage of minerals.** It stores minerals, particularly calcium and phosphorus, that can be released to the rest of the body when needed.

5. **Storage of fat.** It stores energy-rich fat in yellow bone marrow (the soft tissue within some bones). The fat can be metabolized when the energy is needed.
6. **Blood cell production.** It produces blood cells in the red marrow of certain bones.

5.2 Bone Structure

The 206 bones of the human body come in a range of sizes and a variety of shapes. Most bones contain both compact and spongy bone in proportions that depend on the bone's size and shape.

Compact bone is very dense, with few internal spaces (Figure 5.1). It forms most of the shaft of long bones, such as those of the arms and legs. In the body, compact bone (except for the joint surfaces) is covered by a glovelike membrane, the **periosteum**, that nourishes the bone. The periosteum contains blood vessels and nerves as well as cells that function in bone growth and repair. The blood vessels pass through the periosteum and extend into the underlying bone. When a bone is bruised or fractured, most of the pain results from injury to the periosteum.

Spongy bone is a latticework of tiny beams and thin plates of bone with open areas between. This internal network braces the bone from within. Spongy bone is largely found in small, flat bones, such as most of the bones of the skull, and in the heads (enlarged ends) and near the ends of the shafts of long bones (refer to Figure 5.1). Some of the spongy bone in adults (for example, in the ribs, pelvis, backbone, skull, sternum, and long-bone ends) is filled with **red marrow**, where blood cells form. The cavity in the shaft of adult long bones is filled with **yellow marrow**, a fatty tissue used for energy storage.

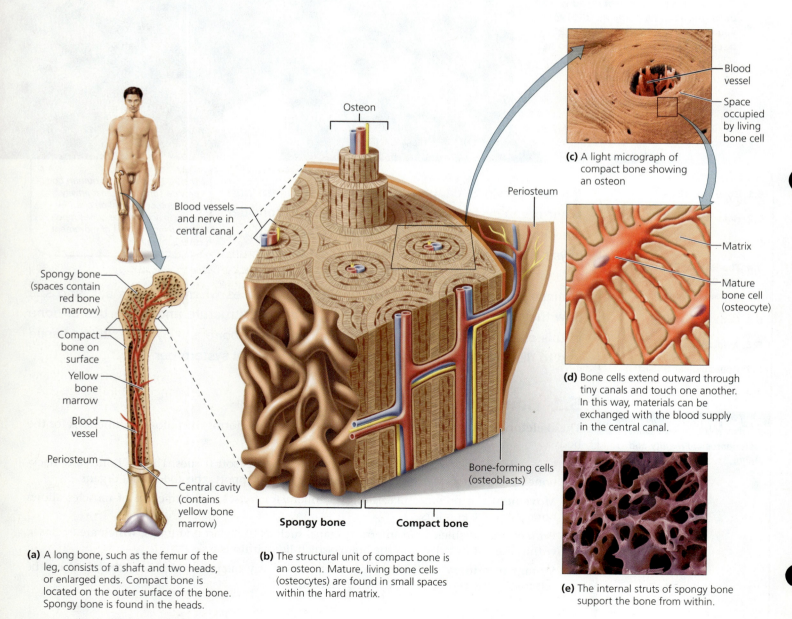

FIGURE 5.1 *The structure of bone*

what would you do?

Bone marrow transplants can save the life of a person with leukemia. In this procedure, a person's diseased bone marrow cells are removed and replaced with healthy blood marrow cells that will restore disease-fighting cells. The donor cells must be a close genetic tissue match to the recipient's cells, which usually means that the two people are relatives. If the best match is a minor, perhaps a younger sibling, should that child be forced to donate bone marrow? If there is no suitable donor, is it ethical for parents to conceive another child who might be a suitable donor to save the older child's life?

5.3 Bone as a Living Tissue

Compact bone is highly organized living tissue containing a microscopic, repeating structural unit called an **osteon** (shown in Figure 5.1). Each osteon consists of mature bone cells, called **osteocytes** (*osteo,* bone; *cyte,* cell), arranged in concentric rings around a central canal. Each osteocyte lies within a tiny cavity called a **lacuna** (plural, lacunae) in the hardened matrix. Tiny canals connect nearby lacunae and eventually connect with the central canal. Similarly, the osteocytes have projections that extend through the tiny canals to touch neighboring cells. Nutrients, oxygen, and wastes pass from cell to cell, traveling to or from the blood vessels in the central canal.

As noted, bone is a living tissue. But its most noticeable characteristics result from its nonliving component, the solid matrix, which makes the bone both hard and resilient. The hardness of the matrix comes from mineral salts, primarily calcium and phosphorus. The resilience comes from strands of the strong, elastic protein collagen that are woven throughout the matrix. Without the calcium and phosphorus salts, bone would be rubbery and flexible like a garden hose. Without the collagen, bone would be brittle and crumbly like chalk. In some disorders, bones do bend, causing bowlegs. An example is rickets, a disease due to greatly reduced amounts of calcium salts in the bones.

stop and think

Strontium-90 is a radioactive substance that enters the atmosphere after atomic explosions. Humans may ingest it in milk from cows that grazed on contaminated grass. The strontium-90 can then replace calcium in bone, killing nearby cells or altering their genetic information. Explain why exposure to strontium-90 can lead not just to bone cancer but also to disruption of blood cell formation.

Cartilage Model

During human embryonic development, most of the skeleton is first formed of cartilage, a strong yet flexible connective tissue (Figure 5.2). Unlike mature bone cells, which cannot divide because they are enclosed in a solid matrix, cartilage cells are able to divide and multiply quickly. Thus, the cartilage model can grow as rapidly as the fetus does. Beginning in the third month and continuing through prenatal development, the cartilage is gradually replaced by bone.

Q *It is recommended that young adult women consume 1100 to 1300 mg of calcium each day, but that requirement doubles during pregnancy. Explain why.*

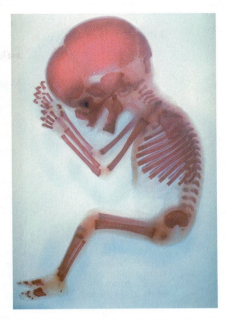

FIGURE 5.2 The fetal skeleton is first made of cartilage and gradually is replaced by bone. This image shows the cartilaginous model of the skeleton of a 16-week-old human fetus. The darker areas of this skeleton are regions that have already been replaced with bone.

A Calcium is needed for replacing the cartilage with bone in the skeleton of the fetus.

The transformation from cartilage to a mature long bone, such as an arm or leg bone, begins with the formation of a collar of bone around the shaft of the cartilaginous model. The collar is produced by bone-forming cells called **osteoblasts** (*osteo,* bone; *blast,* beginning or bud). Osteoblasts produce the matrix of bone around the shaft by secreting collagen (as well as other organic materials) and then depositing calcium salts on it. The bony collar supports the shaft as the cartilage within it breaks down, leaving the marrow cavity. Blood vessels then carry osteoblasts into the bone cavity, which in turn fill the cavity with spongy bone. Unlike cartilage cells, osteoblasts cannot undergo cell division. Once osteoblasts form the matrix around themselves, they are called osteocytes, which, as we have seen, are mature bone cells and the principal cells in bone tissue (Figure 5.3).

Near the time of birth, bone growth centers form in the ends of long bones, and spongy bone begins to fill them. Two kinds of structures made of cartilage will remain. One is a cap of cartilage over each end of the bone, where one bone glides over another in a joint. The second is a plate of cartilage, called an epiphyseal plate that separates each end of the bone from its shaft. The epiphyseal plate is commonly called the **growth plate**. Cartilage cells within the growth plate divide, forcing the end of the bone farther away from the shaft. As bone replaces the newly formed cartilage in the region closer to the shaft, the bone becomes longer. Note that the diameter of the bone also enlarges as the bone lengthens (Figure 5.4).

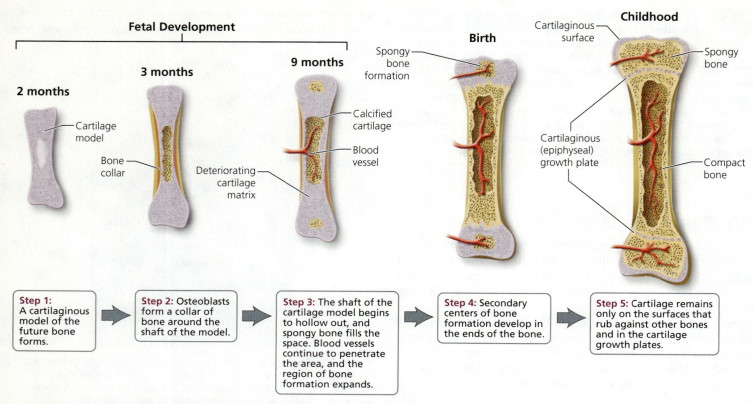

FIGURE 5.3 Steps of bone formation in long and short bones, from an embryo into childhood

Hormones and Bone Growth

During childhood, bone growth is powerfully stimulated by growth hormone, which is released by the anterior pituitary gland (see Chapter 10). Growth hormone prompts the liver to release growth factors that produce a surge of growth in the growth plate. Thyroid hormones modify the activity of growth hormone to ensure that the skeleton grows with the proper proportions.

At puberty, many children experience a growth spurt during which the length of pant legs seems to shrink almost weekly. These dramatic changes are orchestrated by the increasing levels of sex hormones (testosterone and estrogen) produced during puberty. Initially the sex hormones stimulate the cartilage cells of the growth plates into a frenzy of cell division. But this growth generally stops toward the end of the teenage years (about age 18 in females and 21 in males) because of later changes initiated by the sex hormones. The cartilage cells in the growth plates start dividing less frequently. Eventually, the growth plates become thinner as bone replaces cartilage. Finally, the ends of the bone where the growth plates were fuse with the bone in the shaft. Note that even though bones can no longer lengthen at this stage, they can continue to widen.

stop and think

We have considered the mechanism by which long bones grow and the role of growth hormone in that process. Why would it be ineffective for a short, middle-aged person to be treated with growth hormone to stimulate growth?

5.4 The Role of Fibroblasts and Osteoblasts in Repairing Bone Fractures

Despite their great strength, bones occasionally break. Bones break, or fracture, when the physical force applied to them is greater than the strength of the bone. Fractures are more common in children because their bones have not reached peak density and in older adults whose bones have lost density and

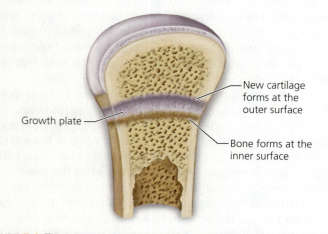

FIGURE 5.4 The lengthening of a bone. New cartilage forms on the outside edge of the growth plate, and bone forms on the inside of the growth plate.

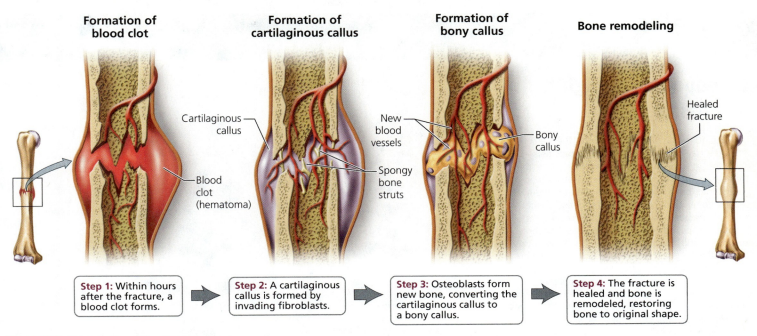

FIGURE 5.5 *The progress of healing in a bone*

become brittle. In a *nondisplaced fracture,* the bone may or may not break completely through, but the ends remain aligned. When the broken ends are not aligned, the break is described as a *displaced fracture.* The ends must be realigned for healing to occur. A fracture is called *closed* when the broken ends remain within the skin and *open* when a broken end punctures the skin. The broken bone in an open fracture may retreat beneath the skin, but the risk of bone infection is still higher than with a closed fracture.

Fortunately, bone tissue can heal. When a bone breaks, the first thing that happens is bleeding from blood vessels in the bone and periosteum, followed by formation of a clot (hematoma) at the break (Figure 5.5). Within a few days, connective tissue cells called *fibroblasts* grow inward from the periosteum and invade the clot. The fibroblasts secrete collagen fibers that form a mass called a *callus,* which links the broken surfaces of the bone. Some of the fibroblasts then transform into cartilage-producing cells that secrete cartilage into the callus.

Next, osteoblasts from the periosteum invade the callus and begin to transform the newly deposited cartilage into new bone material, changing it into a bony callus. During this transformation, the bony callus becomes thicker than the undamaged part of the bone and protrudes from it. In time, however, the extra material will be broken down, and the healed part of the bone returns to nearly normal size.

stop and think
Osteogenesis imperfecta is a condition in which the activities of osteoblasts and fibroblasts are abnormally low. What effects would you expect this to have on the skeletal system?

5.5 Bone Remodeling

Even after reaching our full height, we continue to undergo a lifelong process of bone deposition and breakdown called **remodeling**. Bone remodeling keeps bones strong by repairing tiny cracks in bones, such as might occur if you jump from a high place. We have seen that bone is deposited by osteoblasts. Another kind of bone cell, called an **osteoclast**, breaks down bone, releasing calcium and other minerals that are reabsorbed by the body. Thus, bone remodeling is also a mechanism for regulating blood calcium levels, which is important because calcium plays a role in the functioning of nerves and muscles as well as blood clotting.

Two hormones, calcitonin and parathyroid hormone, play an important part in both controlling bone remodeling and regulating blood levels of calcium. When blood levels of calcium are high, as might occur after a meal, the hormone **calcitonin**, which is released from the thyroid gland, removes calcium from the blood and causes it to be stored in bone. Calcitonin brings about these effects by stimulating the activities of osteoblasts while inhibiting those of osteoclasts. In contrast, when blood calcium levels are low, **parathyroid hormone (PTH)**, which is released by the parathyroid glands found embedded in the tissues of the thyroid gland, causes calcium to be released from bone and reabsorbed into the blood. PTH accomplishes this by stimulating the activities of osteoclasts. The interplay of these two hormones keeps blood calcium levels fairly steady. (The regulation of blood calcium levels is discussed in more detail in Chapter 10.)

In women, estrogen also plays a role in bone remodeling. It promotes the absorption of calcium from the digestive system, stimulates bone formation, and impairs the ability of osteoclasts to break down bone.

HEALTH ISSUE

Osteoporosis: Fragility and Aging

About 40 million Americans, most of them elderly white women, suffer from osteoporosis—a decrease in bone density that occurs when the breakdown of bone outpaces the formation of new bone during bone remodeling. The net breakdown causes bones to become thin, brittle, and susceptible to fracture. A person with osteoporosis becomes hunched and shorter as the vertebrae lose mass and compress (Figure 5.A). A fall, blow, or lifting action that would not bruise a person with healthy bones could easily cause a bone to fracture in a person with severe osteoporosis.

As we have seen, bone remodeling occurs throughout life. Until we reach about age 35, bone is formed faster than it is broken down. Our bones are strongest and densest during our mid-thirties, after which they begin to lose density. Peak bone density is influenced by several factors, including sex, race, nutrition (dietary levels of calcium and vitamin D), exercise (we have seen the importance of weight-bearing exercise), and overall health. In men, bone mass is generally 30% higher than in women. The bones of African Americans are generally 10% denser than those of Caucasians and Asians.

Women are at greater risk for developing osteoporosis than are men. This difference is not just because women have less bone mass at peak but also because their rate of bone loss is accelerated for several years after menopause (the time in a woman's life when she stops producing mature eggs or menstruating). The acceleration occurs because menopause is followed by a sharp decline in the hormone estrogen, and estrogen stimulates bone formation and is important in the absorption of calcium from the intestines.

Other factors have been implicated in the onset of osteoporosis. For example, height is important. Short people are at greater risk, perhaps because they generally start with less bone mass. People with a good supply of body fat are less at risk because fat can be converted to estrogen. Heavy drinkers are at higher risk, because alcohol interferes with estrogen function. Smoking is also bad for bones, because it can reduce estrogen levels. We have already noted that people who do not take in enough calcium, or who cannot absorb it because of insufficient vitamin D, have thinner bones because calcium is necessary for bone growth. (Calcium-rich foods include milk products as well as broccoli, spinach, shrimp, and soybean products.) Certain drugs, such as caffeine (a diuretic), tetracycline, and cortisone, can promote osteoporosis. Finally, sedentary people have thinner bones than do active people.

The degree to which our bones will become weakened as we age depends largely on how dense they were at their peak. We can expect osteoporosis and other afflictions associated with aging to become more common as life expectancy increases. Thus, it would seem prudent for each of us to start an early program of prevention. To build strong bones, eat a diet rich in calcium and vitamin D. As adults, we need 1100 to 1300 mg of calcium each day; so eat three to four servings of foods rich in calcium and vitamin D. Engage in regular weight-bearing exercise, such as walking or jogging, for 30 minutes every day. Avoid drinking alcohol and caffeinated beverages, and don't smoke cigarettes.

Questions to Consider

- Would diving or cross-country skiing on a regular basis be more likely to build bone density?
- If your mother were approaching menopause, what advice would you give her to help her maintain bone density?

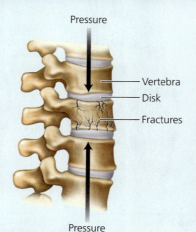
(a) Normal bending can place pressure on vertebrae that can cause small fractures.

(b) A loss of height and a stooped posture result as the weakened bones of the vertebrae become compressed.

FIGURE 5.A *Osteoporosis is a loss of bone density that occurs when bone destruction outpaces bone deposition during the continuous process of bone remodeling. The bones become brittle and are easily fractured.*

During bone remodeling, new bone forms along the lines of stress on the bone. Bone forms in response to stress and gets absorbed when it is not stressed. Weight-bearing exercise, such as walking or jogging, thickens the layer of compact bone tissue, leading to a stronger bone. Bones that are used frequently may actually change shape. For example, continual practice may enlarge the knuckles of pianists and the big toes of ballet dancers. Conversely, a few weeks without stress can cause a bone to lose nearly a third of its mass. This is a concern for astronauts living in the weightless environment of space as well as for a person who is using crutches with a leg in a cast.

If the breakdown process occurs faster than the deposition of new tissue, a bone becomes weak and easy to break. This imbalance occurs in **osteoporosis**, a condition in which there is a progressive loss in bone density. (See the Health Issue essay, *Osteoporosis: Fragility and Aging*.) Now that we are familiar with factors regulating bone growth and remodeling, we can understand why certain medications are used to treat osteoporosis. The bisphosphonates, such as Fosamax, Actonel,

5.6 Axial Skeleton

and Boniva, inhibit the osteoclasts that break down bone, but not the osteoblasts that build bone. Evista is an example of another class of drug, called selective estrogen receptor modulators (SERMs), that mimic the effects of estrogen on the bones. The thyroid hormone calcitonin can also be used to treat osteoporosis, because it increases the amount of calcium deposited in bones.

The human skeleton can be divided into two parts: the axial skeleton and the appendicular skeleton (Figure 5.6). The **axial skeleton** (shown in orange) includes the skull, the vertebral column (backbone), and the bones of the chest region (sternum and rib cage). The **appendicular skeleton** (shown in light

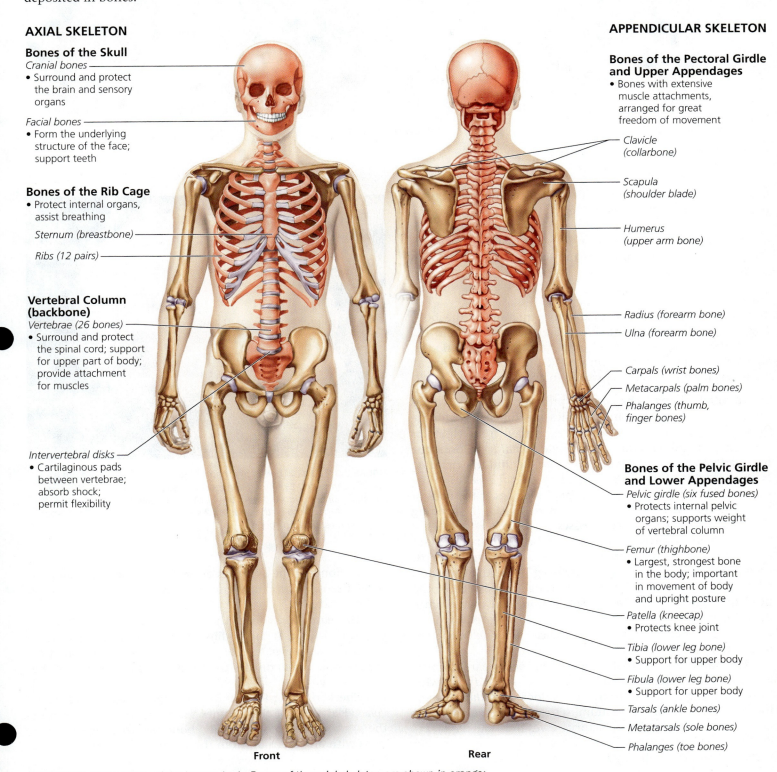

FIGURE 5.6 *Major bones of the human body. Bones of the axial skeleton are shown in orange; bones of the appendicular skeleton are shown in light brown. Cartilage is shown in light purple.*

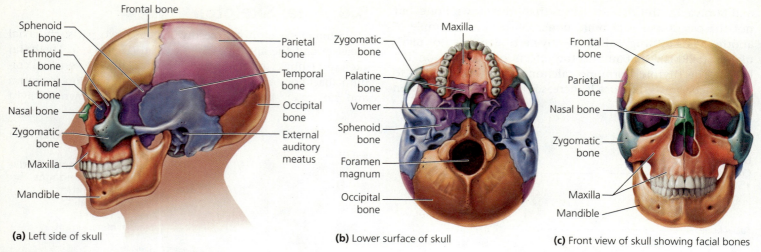

FIGURE 5.7 *Major bones of the skull and face*

brown) includes the pectoral girdle (shoulders), the pelvic girdle (pelvis), and the limbs (arms and legs).

In general, the axial skeleton protects and supports our internal organs. We will focus on only the major bones of the 80 that make up this portion of the skeleton, beginning with the bones of the skull.

Skull

The *skull* is the most complex bony structure in the body (Figure 5.7). Its principal divisions are the cranium and the face. Several bones of the skull contain air spaces called *sinuses*, which make the head lighter in weight and serve as resonating chambers for the voice, as discussed in Chapter 14.

The cranial bones The *cranium* protects the brain, houses the structures of hearing, and provides attachment sites for the muscles of the head and neck. It is formed from eight (or sometimes more) flat bones. The single *frontal bone* forms the forehead and the front of the brain case. Behind it, extending from either side of the midline, the two *parietal bones* form the top and sides of the skull. The *occipital bone* lies at the back of the head and surrounds the *foramen magnum*, the opening through which the spinal cord passes.

Before and shortly after birth, the bones of the cranium are connected by membranous areas called the *fontanels*, often referred to as "soft spots." During birth, the fontanels allow the skull to be compressed, easing the passage of the head through the birth canal. The fontanels also accommodate the rapid enlargement of the brain during fetal growth and infancy (Figure 5.8). They are replaced by bone by the age of 2 years.

The cranium also contains the *temporal bones,* part of which forms what we think of as our temples. The *sphenoid bone*, with its bow-tie shape, forms the cranium's floor. The *ethmoid bone*, the smallest bone in the cranium, separates the cranial cavity from the nasal cavity. The olfactory nerves from the nasal cavity, which are responsible for our sense of smell, communicate with the brain through tiny holes in the ethmoid.

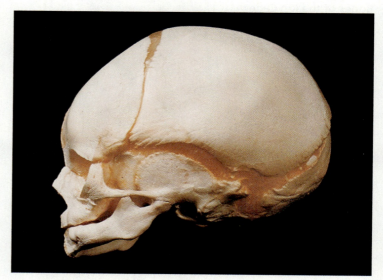

FIGURE 5.8 *The skull bones of a human newborn are not fused but are instead connected by fibrous connective tissue. These "soft spots" allow the skull bones to move during the birth process, easing the passage of the skull through the birth canal. By 2 years of age, the child's soft spots have been replaced by bone.*

Facial bones The 14 bones of the face (Figure 5.7c) support several sensory structures and serve as attachments for most facial muscles.

The *nasal bones* form the bridge of the nose. They are paired and fused at the midline. Inside the nose, a partition called the *nasal septum* (composed of the *vomer* and part of the ethmoid bone) divides the left and right chambers of the nasal cavity.

Cheekbones are formed largely from the paired *zygomatic bones.* Flat areas of these bones form part of the bottom of the eye sockets. Each zygomatic bone has an extension that joins with an extension from the temporal bone to form the *zygomatic arch* (the "cheekbone").

The smallest facial bones are the two *lacrimal bones,* located at the inner corners of the eyes, near the nose. A duct passes through each lacrimal bone and drains tears from the eyes into the nasal chambers; this connection explains why our nose runs when we cry.

The jaw is formed by two pairs of bones, the maxillae and the pair that forms the mandible. The *maxillae* form the upper jaw. Most of the other facial bones are joined to them, giving the upper jaw an important role in facial structure. The maxillae also form part of the hard palate, or roof of the mouth, the rest of which is formed by the two *palatine bones* that lie behind. When the maxillae fail to fuse to one another at the midline of the face below the nasal septum, the mouth cavity and nasal cavity do not fully separate, and a cleft palate results. This condition is easily corrected by surgery.

The lower jaw, called the *mandible,* is also formed from two bones connected at the midline. The mandible is connected to the skull at the temporal bone, forming a hinge called the *temporomandibular joint.* (Joints are discussed later in this chapter.) This joint allows the mouth to open and close. Emotional stress causes some people to clench or grind their teeth, sometimes unconsciously. This clenching can cause physical stress on the temporomandibular joint, causing headaches, toothaches, or even earaches. The condition is known as temporomandibular joint (or TMJ) syndrome.

Vertebral Column

The **vertebral column**, known more familiarly as the backbone or spine, is a series of bones through which the spinal cord passes as it descends the back (Figure 5.9). Each of these bones is called a **vertebra**. The vertebrae (plural of *vertebra*) are classified according to where they lie along the vertebral column. The vertebral column includes 26 vertebrae, organized as follows:

- 7 *cervical* (neck) vertebrae (C_1–C_7)
- 12 *thoracic* (chest) vertebrae (T_1–T_{12})
- 5 *lumbar* (lower back) vertebrae (L_1–L_5)
- 1 *sacrum* (formed by the fusion of five *sacral vertebrae*)
- 1 *coccyx* (or tailbone, formed by fusion of four vertebrae)

Scoliosis, which means "twisted disease," is an abnormal curvature of the spine to the left or right. The most common form has no known cause and affects over 1.5 million adolescents, primarily females. It usually begins during, and progresses through, the adolescent growth spurt. Treatment, if needed, may consist of a brace or surgery to straighten the spine.

The fusion of the sacral vertebrae into a single sacrum gives additional strength to the backbone in the region where the sacrum joins the pelvic girdle. This reinforcement is necessary because of the great stress placed on the sacrum by the weight of the vertebral column and the powerful movements of the leg. The coccyx, however, serves no function and is regarded as a vestigial tail (*vestigial* means "an evolutionary

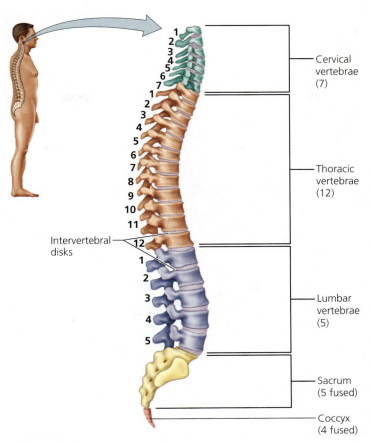

FIGURE 5.9 *A side view of the vertebral column*

relic") whose bones may have joined as a side effect of growth without movement. (The upper part of the coccyx, however, is supplied with nerves and can be extremely painful when injured.)

Above the sacrum, the vertebrae are separated from one another by **intervertebral disks**, pads of fibrocartilage that help cushion the bones of the vertebral column. The smooth, lubricated surfaces of these disks help give the column flexibility. As the intervertebral disks become compressed over the years, the person may become shorter. This compression, alone or coupled with osteoporosis, can have a pronounced effect on an aging person's height.

Excessive pressure on the disks, as might occur during improper lifting, can cause various problems. The term *slipped disk* is a misnomer, because a disk doesn't actually move out of place. Instead, it bulges. If a disk bulges inward, it can press against the spinal cord, interfering with muscle control and perception of incoming stimuli. Conversely, a disk that presses outward against a spinal nerve branching from the spinal cord can be a source of great pain. The sciatic nerve, a large nerve that extends down the back of the leg, is one of the nerves most frequently affected in this way. Sciatica, the resulting inflammation, can be cripplingly painful.

Lower back pain is a particularly mysterious human ailment that accounts for more missed workdays than does any

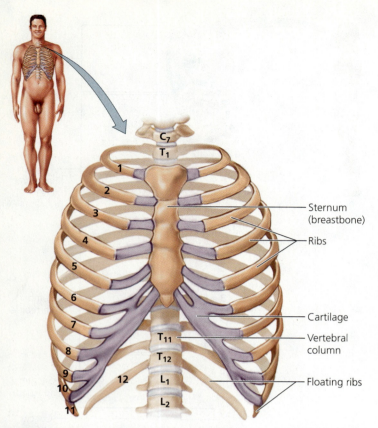

FIGURE 5.10 *The bones of the rib cage. Cartilage is shown in light purple.*

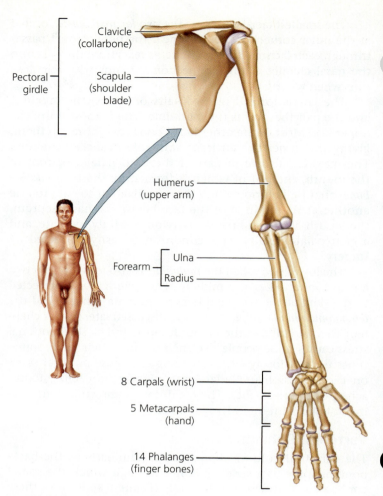

FIGURE 5.11 *The pectoral girdle and arm*

medical problem besides colds. The source of the pain is notoriously difficult to pinpoint. A person with a slipped disk in the region may or may not have pain, whereas a person whose back looks perfectly normal may complain of terrible pain. Part of the diagnostic problem seems to be that the pain may originate solely from muscle rather than bone, and muscles do not show up on x-ray films. Some cases may be due to weak abdominal muscles that cannot counteract the pull of the powerful back muscles and therefore cause the vertebrae to misalign.

Rib Cage

Twelve pairs of ribs attach at the back of the rib cage to the thoracic vertebrae (Figure 5.10). At the front, the upper 10 pairs of ribs are attached by cartilage either directly or indirectly to the *sternum* (breastbone). Their flexibility permits the ribs to take some blows without breaking and to move during breathing. The last two pairs of ribs do not attach to the sternum and are called floating ribs (although they *are* attached at the back to the thoracic vertebrae).

5.7 Appendicular Skeleton

The appendicular skeleton consists of the pectoral and pelvic girdles and the attached limbs. A *girdle* is a skeletal structure that supports the arms, in the case of the **pectoral girdle**, or the legs, in the case of the **pelvic girdle**. The pectoral girdle connects the arms to the rib cage. The pelvic girdle connects the legs to the vertebral column, enabling the body to move from one place to another.

Pectoral Girdle

The pectoral girdle is composed of the *scapulae* (shoulder blades) and the *clavicles* (collarbones; Figure 5.11). The clavicles are more curved in males than in females (one way to tell the sex of a skeleton), and each forms a relatively rigid bridge between a scapula and the sternum. One corner of the roughly triangular scapula has a socket into which fits one end of the *humerus*, the upper arm bone. The other end of the humerus joins, at the elbow, the *radius* and *ulna*—the bones of the lower arm, or forearm. The elbow is formed by the extension of the ulna past the junction with the humerus. The radius and ulna meet the eight *carpals* at the wrist, and these join with the five *metacarpals*, which are the bones of the hand. Each finger includes three *phalanges*, and the thumb includes two.

The carpal tunnel is a narrow opening through the carpal bones that form the wrist, and through it passes a nerve that controls sensations in the fingers and in some of the muscles in the hand. *Tendons,* bands of connective tissue that attach muscles to bones, also pass through the carpal tunnel. Repetitive motion in the hand or wrist can cause these tendons to become inflamed and press against the nerve, resulting in numbness or tingling in the hand and pain that may affect the wrist, hand, and fingers. This condition, known as carpal tunnel syndrome, has been experienced by barbers, cab drivers, and pianists. It is increasingly common as people spend more time operating computer keyboards and playing video games. The computer industry has responded by redesigning the keyboards and games. Employers have responded, too, by providing adjustable workstations and rest and exercise periods.

Pelvic Girdle

The pelvic girdle is much more rigid than the pectoral girdle (as you know if you have ever tried to shrug your hips). The two *pelvic bones* that make up the pelvic girdle (Figure 5.12) attach in back to the sacrum. The pelvic bones curve down and around to the front, where they join to a cartilage disk at the *pubic symphysis*. The male and female hips are easily distinguishable because the opening of the pelvic girdle in the female is wider to facilitate childbirth.

The *femur,* or thighbone, rotates within a socket in the pelvis. At the knee, the femur joins the *tibia* (the shinbone). The *fibula* is a smaller bone running down the side of the tibia. The junction where the tibia joins the femur is covered by a *patella* (kneecap). At the ankle, the lower leg bones meet the *tarsals,* or ankle bones. These are connected to the *metatarsals,* or foot bones, which in turn are connected to the toe bones, called phalanges (as with finger bones).

5.8 Joints

Joints are the places where bones meet. They can be classified as fibrous, cartilaginous, or synovial, depending on their components and structure. Some joints allow no movement; others permit slight movement; and still others are freely movable.

Fibrous joints are held together by fibrous connective tissue. They have no joint cavity, and most fibrous joints do not permit movement. For example, the joints between the skull bones of an adult, called **sutures**, are considered to be immovable because the bones are interlocked and held together tightly by fibrous connective tissue. In fact, the sutures can actually move ever so slightly. This ability is fortunate, because the movements serve as a shock absorber when you hit your head.

Cartilage, which is rather rigid, holds bones together in *cartilaginous joints.* Some cartilaginous joints are immovable, and others allow slight movement. We find cartilaginous joints between vertebrae; in the attachment of ribs to the sternum; and in the pubic symphysis, the joint between the two pelvic bones. In a pregnant woman, hormones loosen the cartilage of the pubic symphysis, allowing the pelvis to widen to ease childbirth.

Synovial Joints

Most of the joints of the body are freely movable **synovial joints**. Because of these joints, muscles can maneuver the body into thousands of positions. All synovial joints share certain common features. One is that the surfaces that move past one another in the joints have a thin layer of *cartilage.* The cartilage reduces friction, so the bones slide over one another without grating and grinding. In addition, a two-layered joint capsule surrounds synovial joints. The inner layer of the capsule, the *synovial membrane,* secretes viscous, clear fluid (synovial fluid) into the space, or *joint cavity,* between the two bones. The synovial fluid lubricates and cushions the joint. The outer layer of the capsule is continuous with the covering membranes of the bones forming the joint. The entire synovial joint is reinforced with **ligaments**, strong straps of connective tissue that hold the bones together, support the joint, and direct the movement of the bones.

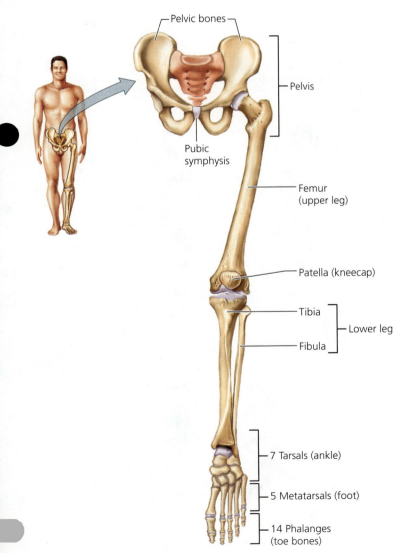

FIGURE 5.12 *The pelvic girdle and leg*

96 CHAPTER 5 The Skeletal System

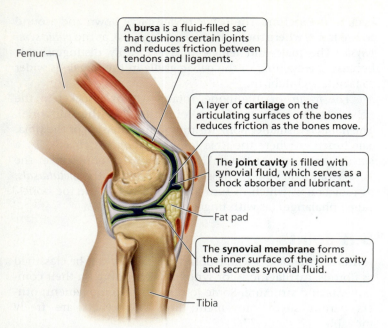

(a) Synovial joints, such as the knee shown here, permit a great range of movement.

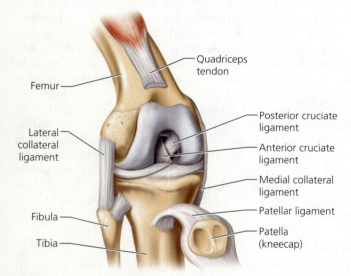

(b) Ligaments hold bones together, support the joint, and direct the movement of the bones.

FIGURE 5.13 *The knee is a synovial joint.*

All synovial joints share these features but may differ in the type and range of motion they permit (Figure 5.13). As the name implies, hinge joints, such as at the knee and elbow, resemble a hinge on a door in that they permit motion in only one plane. A ball-and-socket joint, such as the shoulder and hip, allows movement in all planes: the ball at the head of one bone fits into a socket on another bone. Notice that you can swing your arm around in a complete circle. Some of the ways that body parts move at synovial joints are shown in Figure 5.14.

Damage to Joints

Damage to a ligament is called a *sprain* and may range from slight, caused by overstretching, to serious, caused by tearing

Flexion
Motion that *decreases* the angle between the bones of the joint, bringing the bones closer together

Extension
Motion that *increases* the angle between the bones of the joint

Adduction
Movement of a body part *toward* the body midline

Abduction
Movement of a body part *away from* the body midline

Rotation
Movement of a body part around its own axis

Circumduction
Movement of a body part in a wide circle so that the motion describes a cone

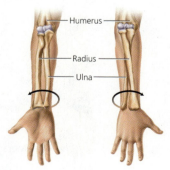

Supination
Rotation of the forearm so that the palm faces up

Pronation
Rotation of the forearm so that the palm faces down

FIGURE 5.14 *Types of movement at synovial joints*

FIGURE 5.15 *A common sports injury is a tear in the anterior cruciate ligament of the knee.*

of the ligament. A torn ligament results in swelling and enough pain to inhibit movement. Sometimes, because ligaments are covered with a concentration of pain receptors that are very sensitive to stretching and swelling, the injury is not as severe as the pain would suggest. As with most musculoskeletal swelling, the initial treatment is likely to be an ice pack to reduce the swelling. Ice constricts underlying blood vessels and reduces bleeding. Bleeding generally stops by the next day. Then, heat is applied to dilate blood vessels, allowing blood to carry in the nutrients, oxygen, and anti-inflammatory chemicals that assist healing. Like tendons, ligaments have few blood vessels and therefore heal slowly.

A common knee injury among athletes, especially gymnasts and football and soccer players, is a tear in the *anterior cruciate ligament* (ACL) (Figure 5.15). Why is this ligament so vulnerable? When the knee is bent, the ACL acts as a restraining wire that restricts front-to-back twisting movement between the thighbone (femur) and the shinbone (tibia). An external blow to a bent knee—as may occur during a tackle or a hard landing—can stretch the ACL. If the force applied to the two bones is greater than the strength of the ligament, the ACL can tear.

In synovial joints and certain other locations where movement might cause friction between moving parts, the body has its own "ball bearings" in the form of fluid-filled sacs called **bursae** (singular, bursa; meaning "pouch or purse"). Bursae are lined with synovial membranes. They are found in places where skin rubs over bone as the joint moves and between tendons and bones, muscles and bones, and ligaments and bones.

Joint injury or repeated pressure can cause bursae to become inflamed and swell with excess fluid, a condition called bursitis (–*itis* means "an inflammation"). For example, the repetitive movement of throwing a baseball or swinging a tennis racket may cause bursitis in the elbow. Bursitis is characterized by intense pain that becomes worse when the joint is moved and that cannot be relieved by resting in any position. Nonetheless, bursitis is not serious and usually subsides on its own within a week or two. In severe cases, a physician may drain some of the excess fluid to remove the pressure.

Arthritis

Arthritis is a general term referring to joint inflammation. Some of the more than 100 kinds of arthritis are far more serious than others. *Osteoarthritis* is a degeneration of the surfaces of a joint, caused by wear and tear. Eventually the slippery cartilage at the ends of the affected bones begins to disintegrate until the bones themselves come into contact and grind against each other, causing intense pain and stiffness. Any joint surface that undergoes friction is bound to wear down, but osteoarthritis is most likely to occur in weight-bearing joints, such as the hip, knee, and spine. It is also occasionally seen in the finger joints or wrist.

Prevention is always better than treatment. One tip for preventing osteoarthritis is to control your weight to avoid overburdening your knees and hips. Another tip is to exercise. Lifting weights will strengthen the muscles that help support your joints. Stretching will give you a greater range of movement.

A far more threatening form of arthritis is *rheumatoid arthritis*. Rheumatoid arthritis is marked by inflammation of the synovial membrane. The resulting accumulation of synovial fluid in the joint causes swelling, pain, and stiffness. Eventually, the constant irritation can destroy the cartilage, which may then be replaced by fibrous connective tissue that further impedes joint movement.

Rheumatoid arthritis differs from other types of arthritis in that it is an autoimmune disease. That is, the body mistakenly attacks its own synovial membranes just as it would some invasive foreign matter. Rheumatoid arthritis can vary in severity over time, but it is a permanent condition. It normally affects the joints of the fingers, wrist, knees, neck, ankles, and hips. Sometimes the only effective treatment is to replace the damaged joint with an artificial one.

looking ahead

In this chapter, we learned about our bony framework. In the next chapter, we will consider how muscles put flesh on our bones and attach to bones so that when our muscles contract, we can move.

HIGHLIGHTING THE CONCEPTS

d. osteocytes.
8. Which functions does bone perform?
   a. blood cell production
   b. support
   c. storage of fat
   d. all of the above
9. Where in a bone is fat stored?
   a. central canal
   b. red bone marrow
   c. periosteum
   d. yellow bone marrow
10. Which of the following hormones raises the blood level of calcium?

   c. parathyroid hormone
   d. all of the above
11. The _____ is a movable bone in the skull.
    a. mandible
    b. occipital bone
    c. frontal bone
    d. temporal bone
12. Which of the following is *not* a bone of the axial skeleton?

    b. vertebra
    c. femur
    d. sternum
13. What is the difference between compact and spongy bone?

    c. Spongy bone has larger cells than compact bone has.
    d. They have different arrangements of bone cells.

15. _____ is a decrease in bone density that causes a weakened bone.
    a. Osteomalacia
    b. Osteoblast
    c. Osteoporosis
    d. Osteon
16. Which bones comprise the pectoral girdle?
    a. carpals and metacarpals
    b. clavicle and scapula
    c. sternum and vertebrae
    d. all of the above
17. The action opposite to extension is
    a. abduction

18. Of the following persons, which one would you would expect to have the *densest* bones?

    b. a 16-year-old female who swims regularly
    c. a 33-year-old male who loves to drink milk and plays basketball regularly
    d. a 50-year-old male workaholic with a desk job

20. The cells that form bone are called _____.
21. A _____ is the overstretching or tear in a ligament that may cause pain after the injury.

APPLYING THE CONCEPTS

chair since she was 35 because of a car accident that left her legs paralyzed. Her leg bones are very thin and weak. Why?

2. Hannah and Becca are fraternal twins. They are both athletic

have denser bones?

femur (thighbone) and a tibia (lower leg bone). By measuring

could have belonged to a tall child or a short adult. What should

a child or short adult?

references to develop a plan for diagnosing the cause(s) of Mihoko's osteoporosis.

hypothesis. What questions would you ask Mihoko? What medical tests would you want to perform? List each source you considered, and explain why you chose the sources you used.

MasteringBiology®

Go to MasteringBiology for practice quizzes, activities, eText, videos, current events, and more.

6 The Muscular System

Did you know?

- Each time you take a step, you use as many as 200 muscles.
- The stapedius, the smallest muscle in your body, is located in the middle ear and is about the thickness of a cotton thread.
- Your muscles make up about 40% to 50% of your body weight.

6.1 Function and Characteristics of Muscles

6.2 Skeletal Muscles Working in Pairs

6.3 Contraction of Muscles

6.4 Voluntary Movement

6.5 Energy for Muscle Contraction

6.6 Slow-Twitch and Fast-Twitch Muscle Cells

6.7 Building Muscle

ETHICAL ISSUE
Anabolic Steroid Abuse

We have more than 600 skeletal muscles, most of which are attached to bones. Our muscles contract to move our body parts so that we can make our impact on the world.

In the previous chapter we learned about bones, which support us against gravity, and about joints, which allow movement. In this chapter, we will see how muscle contraction permits movement. We will explore muscle structure and the mechanism of contraction. Then, we will see how nerves control contractions and examine the energy sources that fuel muscle activity.

6.1 Function and Characteristics of Muscles

There are three kinds of muscles—*skeletal, cardiac,* and *smooth*. Each has distinct qualities and functions, but they all have four traits in common.

1. **Muscles are excitable.** They respond to stimuli.
2. **Muscles are contractile.** They have the ability to shorten (contract).
3. **Muscles are extensible.** They have the ability to stretch.
4. **Muscles are elastic.** They can return to their original length after being shortened or stretched.

This chapter focuses on skeletal muscle, the kind of muscle we usually think of when muscles are mentioned. Smooth and cardiac muscles are discussed in Chapters 4 and 12, respectively. Skeletal muscles allow you to smile with pleasure and scowl in anger. They allow you to move, some of us more gracefully than others. And they allow you to remain erect, maintaining your posture despite the pull of gravity. Besides these obvious functions of skeletal muscle, muscles forming the abdominal wall and the floor of the pelvic cavity provide support for the internal organs. Contraction of muscles also pushes against veins and lymphatic vessels, moving blood and lymph along. Importantly, contraction of skeletal muscles generates heat, which warms our bodies above most of the environmental temperatures we experience. (If we get too warm, other mechanisms come into play to cool our bodies, as described in Chapter 4.) Unlike cardiac and smooth muscle, skeletal muscle is under voluntary control—meaning we can contract it when we want to.

6.2 Skeletal Muscles Working in Pairs

Most muscles work in pairs or groups. Muscles that must contract at the same time to cause a certain movement are called *synergistic muscles*. Most muscles, however, are arranged in *antagonistic pairs* (Figure 6.1), which produce a movement when one of the pair contracts and the other relaxes. When a muscle in an antagonistic pair contracts, it shortens and pulls on one of the bones that meet at a given joint, causing movement of the bone in one direction. Whenever an antagonistic muscle contracts and pulls on a bone, its partner, which has an opposing action, must relax. For the bone to then move back to its former position, the first muscle must relax and the other muscle in the pair must contract, thus pulling the bone in the opposite direction. The biceps muscle (the muscle people like to show off), located on the top of the upper arm, cooperates in this way with the triceps muscle (the one we feel when we do push-ups), on the back of the arm. Contracting the biceps causes the arm to flex and bend at the elbow, but contracting the triceps causes the arm to extend and straighten at the elbow.

Most of the major muscles we use for locomotion, manipulation, and other voluntary movements are attached to bones. Each end of the muscle is attached to a bone by a **tendon**, which is a band of connective tissue. A muscle is often attached to two bones on opposite sides of a joint. The muscle's **origin** is the end attached to the bone that remains relatively stationary during a movement. The muscle's **insertion** is the end attached to the bone that moves. Thus, the bones act as levers in working with skeletal muscles to produce movement. The body has more than 600 skeletal muscles. The major ones—those most prominent in giving shape to the body—are shown in Figure 6.2.

In the preceding description, we said that a contracting muscle pulls a bone. In actuality, the contraction of a muscle pulls on the tendon that connects that muscle to a bone. Excessive stress on a tendon can cause it to become inflamed, a condition called *tendinitis*. Most tendinitis is probably due to overuse, misuse (as when lifting improperly), or age. Unfortunately, tendons heal slowly because they are poorly supplied with blood vessels. The most effective treatment is rest: if it hurts, do not use it.

The terms *muscle pull, muscle strain,* and *muscle tear* are used almost interchangeably to describe damage to a muscle or its tendons caused by overstretching. This common sports injury may also include damage to small blood vessels in the muscle, causing bleeding and pain. Treatment includes applying ice to the injured area to reduce swelling and keeping the muscle stretched.

"Text neck" is a stiff neck caused by damage to the neck muscles and ligaments from hunching over a cell phone while texting or using another mobile device excessively. This condition, which is increasingly common among the tech generation, can be avoided by taking frequent breaks, sitting upright, and rotating your shoulders with your arms by your side.

6.3 Contraction of Muscles

Now we'll look at the brawny bulk that gives shape to the body, gradually delving deeply into the fine structure forming the mechanism responsible for muscle contraction. An entire, intact muscle is formed from individual muscle fibers grouped in increasingly larger bundles, each wrapped in a connective tissue sheath. Each individual muscle, the biceps brachii for instance, is covered by a membrane made of fibrous connective tissue. The muscles themselves are formed of smaller bundles called **fascicles**, each wrapped in its own connective tissue sheath. The connective tissue sheaths merge and condense at the ends of the muscles to form the tendons that attach the muscle to bone.

Inside the fascicles are the **muscle fibers** themselves, the actual muscle cells. Skeletal muscle cells can be several

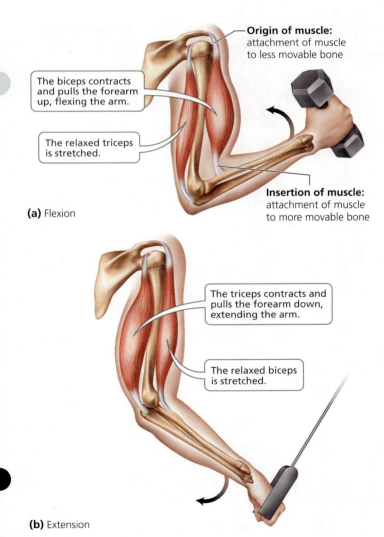

(a) Flexion

(b) Extension

FIGURE 6.1

102 CHAPTER 6 The Muscular System

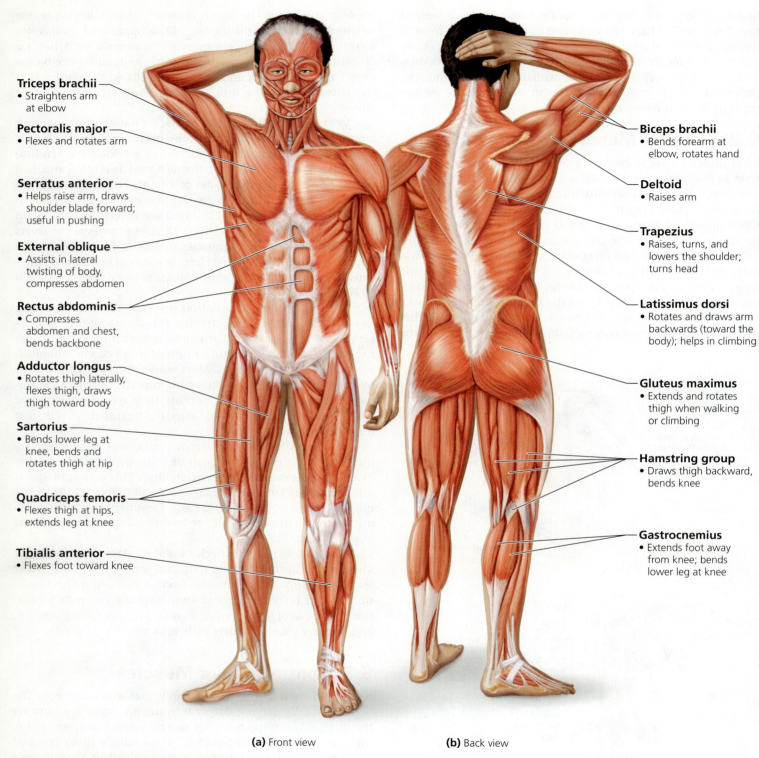

FIGURE 6.2 *Some major muscles of the body*

centimeters long, which is enormously long compared to most of the body's other cells. Muscle cells in the thigh can be 30 cm (1 ft) in length. The fine structure of these cells provides numerous clues to the mechanism responsible for muscle contraction.

Skeletal muscle is also called **striated** (striped) **muscle**, because under the microscope, the muscle cells have pronounced bands that look like stripes (Figure 6.3). The striations are caused by the orderly arrangement of many elongated **myofibrils**, which are specialized bundles of proteins within the muscle cell. Each myofibril contains two types of the long protein filaments called **myofilaments**: the thicker **myosin filaments** and a greater number of the thinner **actin filaments**. The bundles of myofilaments make up about 80% of the cell volume.

Along its length, each myofibril has tens of thousands of contractile units called **sarcomeres**. The ends of each sarcomere are marked by dark bands of protein, called *Z lines*. Within each sarcomere, the actin and myosin filaments are arranged in a specific manner. One end of each actin filament is attached to a Z line. Myosin filaments lie in the middle of the sarcomere, their ends partially overlapping with surrounding actin filaments. The degree of overlap increases when the muscle contracts.

Sliding Filament Model

According to the **sliding filament model** of muscle contraction, a muscle contracts when the actin filaments slide along the myosin filaments, increasing the degree of overlap between actin and myosin and thus shortening the sarcomere. When many sarcomeres shorten, the muscle as a whole contracts.

To understand the sliding of the actin and myosin filaments, let's look at their molecular structures. A thin, actin myofilament is made up of two chains of spherical actin molecules resembling two strings of beads twisted around each other to form a helix. A thick, myosin filament is composed of myosin molecules shaped like two-headed golf clubs. In a myosin filament, the "shafts" of several hundred myosin molecules lie along the filament's length, and the "heads" protrude along each end of the bundle in a spiral pattern. The club-shaped *myosin heads* are the key to the movement of actin filaments and, therefore, to muscle contraction (Figure 6.4).

The sliding filament model holds that muscle contraction results from the following cycle of interactions between myosin and actin:

- **Resting sarcomere.** At the start of each cycle, the myosin heads have already split a molecule of adenosine

(a) A section of a skeletal muscle

(b) A light micrograph of a longitudinal view of skeletal muscle cells

(c) A diagram and electron micrograph of a myofibril

(d) A sarcomere, the contractile unit of a skeletal muscle, contains actin and myosin myofilaments.

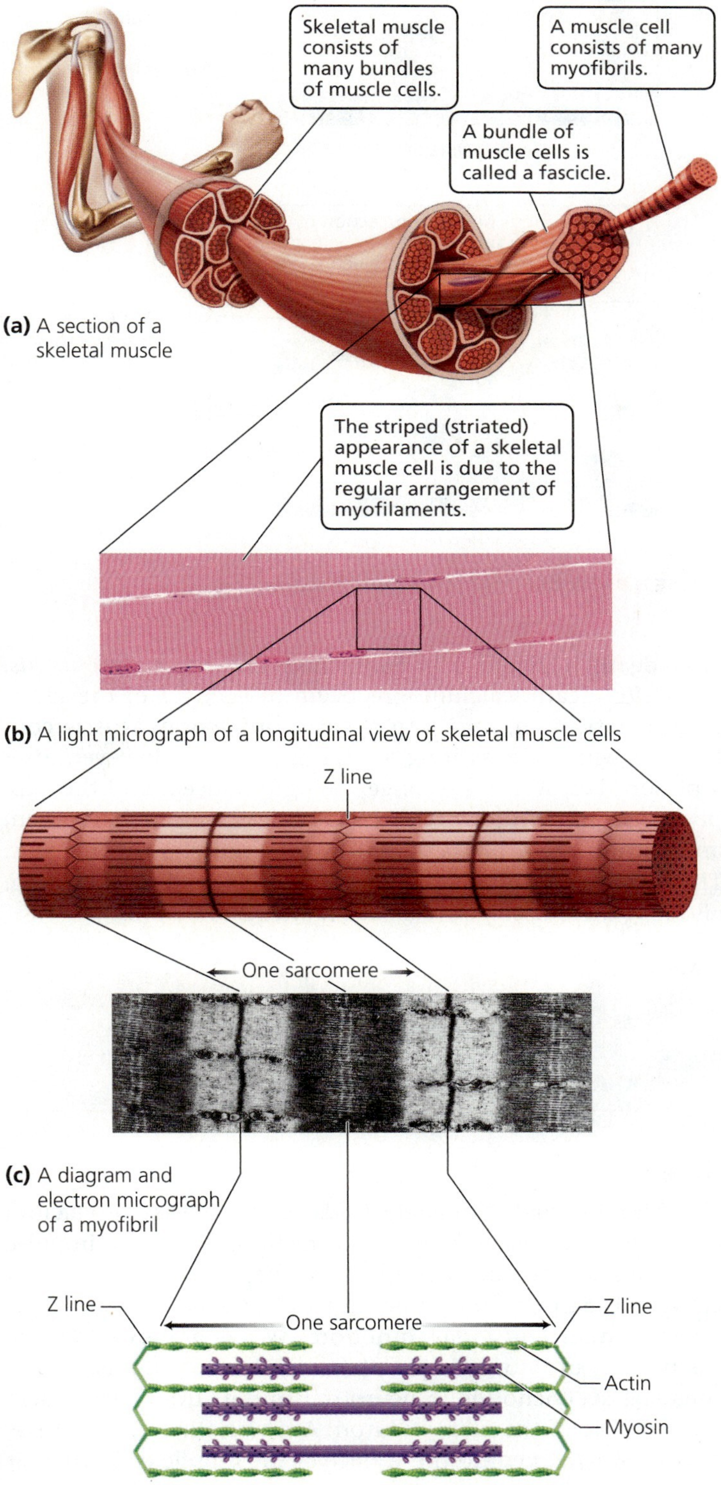

FIGURE 6.3 *The structure of a skeletal muscle*

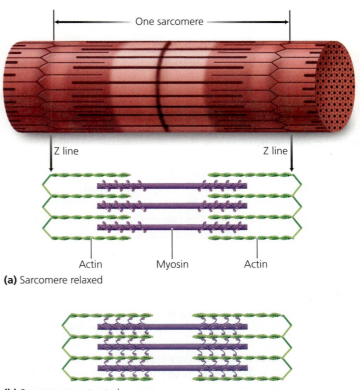

(a) Sarcomere relaxed

(b) Sarcomere contracted

FIGURE 6.4 *Each myofibril is packed with actin filaments and myosin filaments. When a muscle contracts, actin filaments slide past myosin filaments. Movements of the heads of myosin filaments pull actin filaments toward the center of a sarcomere.*

triphosphate (ATP) to adenosine diphosphate (ADP) and inorganic phosphate (P*i*). The energy released from splitting the ATP causes the myosin heads to swivel in a way that extends them toward the Z lines at the ends of the sarcomere. This step is analogous to cocking a pistol so that it is poised and ready to fire.

- **Step 1: Cross-bridge attachment.** The myosin heads attach themselves to the nearest actin filament. When the myosin head is bound to an actin molecule, it acts as a bridge between the thick and thin filaments. For this reason, myosin heads are also called **cross-bridges**.
- **Step 2: Bending of myosin head (the power stroke).** When myosin heads bind to actin, the energy previously stored from breaking down ATP to ADP and inorganic phosphate is released, causing the myosin heads to swing forcefully back to their bent positions. The actin filaments, still bound to the myosin heads, are therefore pulled toward the midline of the sarcomere. This so-called power stroke is analogous to pulling the trigger on a pistol. During the power stroke, the ADP and inorganic phosphate are released from the myosin head.
- **Step 3: Cross-bridge detachment.** New ATP molecules now bind to the myosin heads, causing the myosin heads to disengage from the actin.
- **Step 4: Myosin reactivation.** The myosin heads split the ATP into ADP and P*i* and store the energy, causing the contraction cycle to begin again.

This cycle of events is repeated hundreds of times in a second.

Calcium Ions and Regulatory Proteins

Muscle contractions are controlled by the availability of calcium ions. How? Muscle cells contain the proteins **troponin** and **tropomyosin**, which together form a troponin–tropomyosin complex (Figure 6.5). During muscle *relaxation,* the actin–myosin-binding sites where the myosin heads would otherwise attach to the actin filaments are covered over by the troponin–tropomyosin complex. Contraction occurs when calcium ions enter the sarcomere and bind to troponin, causing it to change shape. This change causes tropomyosin to shift position, which pulls the tropomyosin molecule away from the myosin-binding sites on the actin. The exposed binding sites enable actin to bind to myosin.

Where do the calcium ions come from? Calcium ions are stored in the **sarcoplasmic reticulum**, an elaborate form of smooth endoplasmic reticulum found in muscle cells. The sarcoplasmic reticulum can be pictured as a sleeve of thick lace surrounding each myofibril in a muscle cell. (Recall that a myofibril is a bundle of actin and myosin.) Also scattered through the cell are a number of **transverse tubules** (T tubules), which are tiny, cylindrical pockets in the muscle cell's plasma membrane. The T tubules carry signals from motor neurons deep into the muscle cell to virtually every sarcomere.

As we have seen, it takes an ATP molecule to break the cross-bridges so that new ones may be formed. This dependence explains why muscles become stiff within a few hours

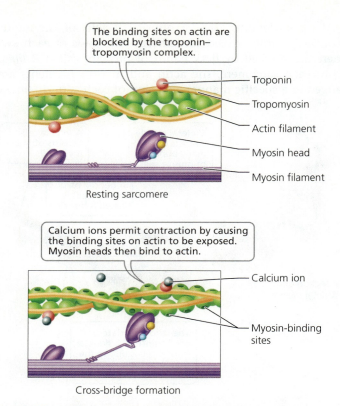

FIGURE 6.5 *Calcium ions initiate muscle contraction.*

after death, through a phenomenon known as *rigor mortis.* Soon after death, calcium ions begin to leak out of the sarcoplasmic reticulum, initiating muscle contraction. Muscle contraction will occur as long as ATP is present. However, after a person dies, ATP is no longer being produced, and the supply runs out. Without ATP, cross-bridges cannot be broken, and within 3 to 4 hours after death the muscles become stiff. The muscles' proteins actin and myosin gradually break down, allowing the muscles to relax again after 2 to 3 days.

stop and think

How would the degree of rigor mortis help a forensic scientist determine the approximate time of death of a corpse?

Role of Nerves

The stimulus that ultimately leads to the release of calcium ions, and therefore to muscle contraction, is a nerve impulse from a motor neuron (described in Chapter 7). The junction between the tip of a motor neuron and a skeletal muscle cell is called a **neuromuscular junction**. When a nerve impulse reaches a neuromuscular junction, it causes the release of the chemical acetylcholine (a neurotransmitter) from small packets in the tip of the motor neuron. Acetylcholine then diffuses across a small gap onto the surface of the muscle cell, where it binds to special receptors on the muscle cell membrane. Acetylcholine causes changes in the membrane's permeability, thus creating an electrochemical message similar to a nerve impulse. The message travels along the muscle cell's plasma

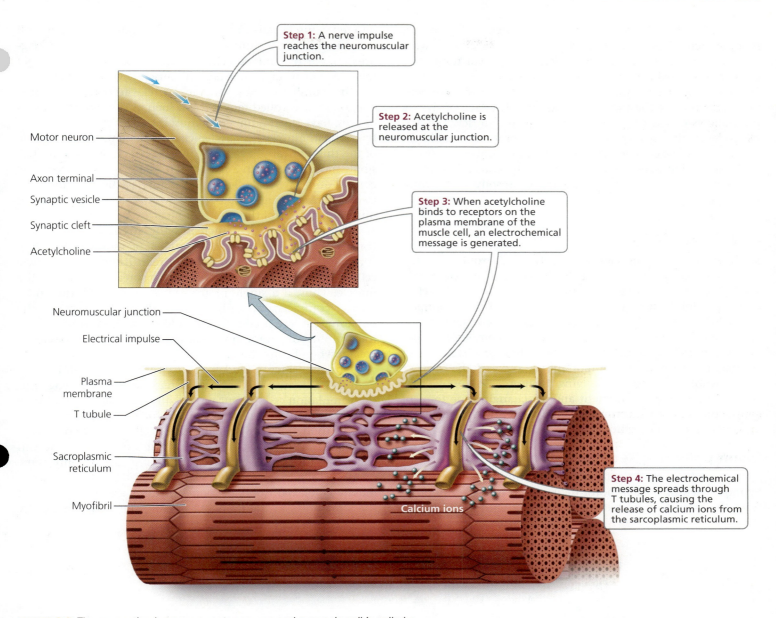

FIGURE 6.6 The connection between a motor neuron and a muscle cell is called a neuromuscular junction.

membrane, into the T tubules, and then to the sarcoplasmic reticulum, causing channels there to open and release calcium ions. The calcium ions then combine with troponin, the myosin-binding sites on actin are exposed, and the muscle contracts. This is the chain of events that must happen in order for you to absentmindedly scratch your head, and it happens a lot faster than it takes to describe (Figure 6.6).

When the nerve impulse stops, membrane pumps quickly clear the sarcomere of calcium ions, and the troponin–tropomyosin complexes move to where they again block the binding sites, causing the muscle to relax. The contraction of *other* muscles then stretches the sarcomere back to its original length.

Low blood levels of calcium, magnesium, or potassium are possible causes of *muscle cramps,* also called *muscle spasms,* which are forceful, involuntary muscle contractions. Muscle cramps can usually be relieved by stretching or gently rubbing the affected muscle.

stop and think

Curare is a poison used in South America on poison arrow darts. It prevents acetylcholine, the neurotransmitter released by motor neurons at neuromuscular junctions, from binding to muscle. Thus acetylcholine cannot cause its effects. The diaphragm is a voluntary skeletal muscle necessary for breathing. How do poison arrows cause death?

Muscular Dystrophy

We have seen that calcium ions must enter the cell for muscle contraction to occur. However, if too many calcium ions enter the cell, those ions can destroy proteins within the cell, causing it to die. If this cell death occurs on a large scale, muscles become increasingly damaged and weak. That is, in fact, what happens in *muscular dystrophy,* which is a general term for a group of inherited conditions. One of the most common forms, Duchenne muscular dystrophy, is caused by a defective gene for production of the protein dystrophin. The lack of dystrophin allows excess calcium ions to enter the muscle cell and rise to levels that destroy other important proteins and kill the cell. Dead muscle cells are replaced by fat and connective tissue, so the skeletal muscles become progressively weaker. (The mode of inheritance of Duchenne muscular dystrophy is described in Chapter 20.)

6.4 Voluntary Movement

Many of the same muscles are involved in strolling on the beach and kicking a soccer ball. However, the extent and strength of contraction is different in these two activities. How is this possible? Although an individual muscle cell contracts completely each time it is stimulated to do so, only some of the cells in a whole muscle contract at once. The combined output of these muscle cells determines the extent and strength of contraction. There are two important ways to vary the contraction of whole muscles: (1) changing the number of muscle cells contracting at a given moment and (2) changing the force of contraction in individual muscle cells by altering the frequency of stimulation. Let's see how this works.

Motor Units and Recruitment

As described earlier, skeletal muscles are stimulated to contract by motor neurons. Each motor neuron that brings an impulse from the brain to a muscle makes contact with a number of different muscle cells in that muscle. A motor neuron and all the muscle cells it stimulates are called a **motor unit** (Figure 6.7). All the muscle cells in a given motor unit contract together. On average there are 150 muscle cells in a motor unit, but this number is quite variable. Muscles responsible for precise, finely controlled movements, such as those of the fingers or eyes, have small numbers of muscle cells in each motor unit. In contrast, muscles for less precise movements, such as those of the thigh or calf, have many muscle cells in a motor unit. A single motor neuron in a motor unit in a tiny eye muscle may control only three muscle cells, making its action highly localized, whereas a single motor neuron in a motor unit in a calf muscle may control thousands of muscle cells.

The nervous system increases the strength of a muscle contraction by increasing the number of motor units being stimulated through a process called **recruitment**. The muscle cells of a given motor unit are generally spread throughout the muscle. Thus, if a single motor unit is stimulated, the entire muscle contracts, but only weakly. Although several of the same muscles are used to lift a table as are used to lift a fork, the number of motor units summoned in those muscles is greater when lifting the table.

Our movements tend to be smooth and graceful rather than jerky because the nervous system carefully choreographs the stimulation of different motor units, timing it so that they are not all active simultaneously or for the same amount of time. Although an individual muscle cell is either contracted or relaxed at any given instant, the muscle as a whole can consist of thousands of muscle cells, and usually only some of those cells are contracted at the same time. Even when a muscle is relaxed, some of its motor units are active (but not the same units all the time). As a result, the muscle is usually

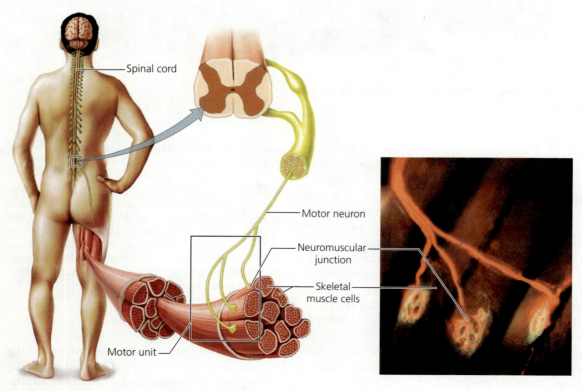

FIGURE 6.7 *A motor unit includes a motor neuron and the muscle cells it stimulates.*

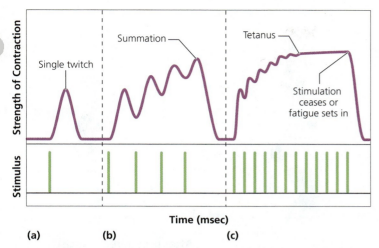

FIGURE 6.8 *Muscle contraction shown graphically: (a) muscle twitch, (b) summation, (c) tetanus*

firm and solid, even when it is not being used. This state of intermediate contraction of a whole skeletal muscle is called *muscle tone*. A muscle that lacks muscle tone is limp.

Muscle Twitches, Summation, and Tetanus

If a whole skeletal muscle is artificially stimulated briefly in the laboratory, some of the muscle cells will contract, causing a **muscle twitch** (Figure 6.8a). The interval between the reception of the stimulus and the time when contraction begins is called the latent period. The contraction phase is quite short and is followed by a longer relaxation phase as the muscle returns to its resting state. Muscle twitches are too brief to be part of normal movements (but they can generate heat to maintain body temperature when we shiver).

Most movements, including sipping a glass of water, require sustained contraction of muscles. Sustained contractions begin when the frequency of stimulation increases. If a second stimulus is given before the muscle is fully relaxed, the second twitch will be stronger than the first. This phenomenon is described as **summation**, because the second contraction is added to the first (Figure 6.8b). Summation occurs because an increasing number of muscle cells are stimulated to contract.

When stimuli arrive even more frequently, the contraction becomes increasingly stronger as each new muscle twitch is added. If the stimuli occur so frequently that there is no time for any relaxation before the next stimulus arrives, the muscle goes into a sustained, powerful contraction called **tetanus** (Figure 6.8c). Tetanus cannot continue indefinitely. Eventually, the muscle will become unable to produce enough ATP to fuel contraction, and lactic acid will accumulate (as discussed shortly). As a result, the muscle will stop contracting despite continued stimulation, a condition called *fatigue*.

6.5 Energy for Muscle Contraction

A single contracting muscle cell can require as much as 600 trillion ATP molecules per second simply to form and break the cross-bridges producing the contraction. When we consider that even small muscles contain thousands of muscle cells, it is clear that muscle contraction requires an enormous store of energy. So where does all the ATP come from?

Our muscles have various sources of ATP and typically use them in a certain sequence, depending on the duration and intensity of exercise. These sources include (1) ATP stored in muscle cells, (2) creatine phosphate stored in muscle cells, (3) anaerobic (without oxygen) metabolic pathways within cells, and (4) aerobic (with oxygen) respiration within cells (Figure 6.9). (Recall that cell respiration was discussed in Chapter 3.)

A resting muscle stores some ATP, but this reserve is used up quickly. During vigorous exercise, the ATP reserves in the active muscles are depleted in about 6 seconds. Earlier, when the muscles were resting, energy was transferred to another high-energy compound, called creatine phosphate, that is stored in muscle tissue. Creatine phosphate has a high-energy bond connecting the creatine and the phosphate parts of the molecule. A resting muscle contains about six times as much stored creatine phosphate as stored ATP and can release its energy when needed to convert ADP to ATP. This energy supports the next 10 seconds of exercise.

$$\text{Creatine phosphate} \xrightarrow{\text{ADP} \rightarrow \text{ATP}} \text{Creatine}$$

Activities such as diving, weight lifting, and sprinting, which require a short burst of intense activity, are powered entirely by ATP and creatine phosphate reserves.

Once the supply of creatine phosphate is diminished, ATP must be generated from either anaerobic metabolic pathways or aerobic respiratory pathways. The primary fuel for either pathway is glucose, and the glucose that fuels muscle contraction comes mainly from glycogen, which, as you learned in Chapter 2, is stored in muscle (and liver) cells and consists of a large chain of glucose molecules. When an active muscle cell runs short of ATP and creatine phosphate, enzymes begin converting glycogen to glucose. About 1.5% of a muscle cell's total weight is glycogen. Even so, long-term activity, as in an endurance sport, may deplete a muscle's glycogen reserves. The accompanying feeling of overwhelming fatigue is known to runners as "hitting the wall," to cyclists as "bonking," and to boxers as becoming "arm weary."

The cardiovascular system can supply enough oxygen for aerobic respiratory pathways to power *low* levels of activity, even if continued for a prolonged time. However, anaerobic pathways produce ATP 2.5 times faster than do aerobic pathways. Therefore, during strenuous activity lasting 30 or 40 seconds, anaerobic pathways supply the ATP that fuels muscle contraction. Indeed, burstlike activities such as used in tennis or soccer rely nearly completely on anaerobic pathways of ATP production. In anaerobic respiratory pathways, the pyruvic acid produced in glycolysis is converted to lactic acid.

$$\text{Glycogen} \rightarrow \text{Glucose} \xrightarrow[\text{No oxygen required}]{\text{ADP} \rightarrow \text{ATP}} \text{Pyruvic acid} \rightarrow \text{Lactic acid}$$

During more prolonged muscular activity, the body gradually switches back to aerobic pathways for producing ATP. The necessary oxygen can come from either of two sources. One is the oxygen bound to hemoglobin in the blood supply. As activity continues, the heart rate increases, and blood is pumped

Q *Which energy source would fuel walking up a single flight of stairs?*

6 seconds	10 seconds	30–40 seconds	End of exercise	After prolonged exercise
ATP stored in muscles	ATP formed from creatine phosphate and ADP	ATP generated from glycogen stored in muscles and broken down to form glucose		Oxygen debt paid back
		Oxygen limited • Glucose oxidized to lactic acid	Oxygen present • Heart beats faster to deliver oxygen more quickly • Myoglobin releases oxygen	Breathe heavily to deliver oxygen • Lactic acid used to produce ATP • Creatine phosphate restored • Oxygen restored to myoglobin • Glycogen reserves restored

FIGURE 6.9 *Energy sources for muscle contraction*

A *ATP stored in muscles*

more quickly; moreover, it is shunted to the neediest tissues. Another oxygen source is *myoglobin,* an oxygen-binding pigment within muscle cells. Aerobic pathways produce more than 90% of the ATP required for intense activity lasting more than 10 minutes. These pathways also produce nearly 100% of the ATP that powers a truly prolonged intense activity, such as running a marathon.

$$\text{Glycogen} \longrightarrow \text{Glucose} \xrightarrow[\text{Oxygen required}]{\text{ADP} \quad \text{ATP}} \text{Carbon dioxide + Water}$$

After prolonged exercise, a person continues to breathe heavily for several minutes. Taking in extra oxygen in this way relieves the **oxygen debt** that was created by the muscles' using more ATP than was provided by aerobic metabolism. Most of the oxygen is used to generate ATP to convert lactic acid back to glucose. In addition, the oxygen that was released by myoglobin is replaced, and glycogen and creatine phosphate reserves are restored.

stop and think

Some athletes take dietary creatine supplements to improve their performance. Creatine does seem to boost performance in sports that require short bursts of energy but not in those that require endurance. How might this difference be explained? Creatine does not increase muscle mass, yet it can enhance performance in sprint sports. How is this effect possible?

6.6 Slow-Twitch and Fast-Twitch Muscle Cells

There are two general types of skeletal muscle cells: slow-twitch and fast-twitch. **Slow-twitch muscle cells** contract slowly when stimulated, but with enormous endurance. These cells are dark and reddish because they are packed with the oxygen-binding pigment myoglobin and because they are richly supplied with blood vessels. Slow-twitch cells also contain abundant mitochondria, the organelles in which aerobic production of ATP occurs. Because they can produce ATP aerobically for a long time, slow-twitch cells are specialized to deliver prolonged, strong contractions.

In contrast, **fast-twitch muscle cells** contract rapidly and powerfully, but with far less endurance. Fast-twitch cells have a form of an enzyme that can split ATP bound to myosin more quickly than the same enzyme in slow-twitch cells. Because they can make and break cross-bridge attachments more quickly, fast-twitch cells can contract more rapidly. In addition, compared with their slow-twitch cousins, fast-twitch cells have a wider diameter because they are packed with more actin and myosin. This feature adds to their power. However, fast-twitch cells, rich in glycogen deposits, depend more heavily on anaerobic means of producing ATP. As a result, fast-twitch cells tire more quickly than slow-twitch cells.

The two kinds of cells are distributed unequally throughout the human body. The abdominal muscles do not need to contract rapidly, but they need to be able to contract steadily to hold our paunch in at the beach and to balance the powerful,

Q *Which type of muscle cell is best suited for cross-country skiing? Which type of muscle fiber is best suited for weight lifting?*

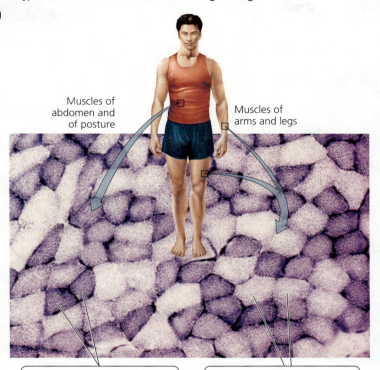

Slow-twitch muscle cells:
- Designed for endurance
- Contract slowly
- Strong, sustained contraction
- Steady supply of energy
  – Many mitochondria (structures for aerobic production of ATP)
  – Many capillaries
  – Packed with the oxygen-binding pigment myoglobin

Fast-twitch muscle cells:
- Designed for rapid, powerful response
- Contract rapidly
- Short, powerful contraction because there is more actin and myosin than in slow-twitch muscle cell
- Depend more heavily on anaerobic metabolic pathways to generate ATP, so fatigue rapidly

FIGURE 6.10 *Slow- and fast-twitch muscle cells*

A *Slow-twitch muscle cells for cross-country skiing; fast-twitch muscle cells for weight lifting*

slow-twitch back muscles so that we can stand upright. Fast-twitch cells are more common in the legs and arms, because our limbs may be called on to move quickly.

People vary in the relative amounts of slow- and fast-twitch muscle cells they possess. Whereas the leg muscles of endurance athletes, such as marathoners, are made up of about 80% slow-twitch cells, those of sprinters are about 60% fast-twitch cells. To some extent these differences are genetic, so if you are a "fast-twitch person," you can build a certain level of endurance, but only to a degree. Endurance runners, on the other hand, dread those times when, at the end of a long race, some well-trained competitor, genetically endowed with lots of fast-twitch cells, sprints past them at the finish line (Figure 6.10).

6.7 Building Muscle

Exercise can greatly influence the further development of muscle. The hormone testosterone also builds muscle (see the Ethical Issue essay, *Anabolic Steroid Abuse*). Different kinds of exercise produce different results. In *aerobic exercise,* such as walking, jogging, or swimming, enough oxygen is delivered to the muscles to keep them going for long periods. This kind of exercise fosters the development of new blood vessels that service the muscles and development of more mitochondria to facilitate energy usage. Aerobic exercise also increases muscle coordination, improves digestive tract movement, and increases the strength of the skeleton by exerting force on the bones. It brings cardiovascular and respiratory system improvements that help muscles to function more efficiently. For example, it enlarges the heart so that each stroke pumps more blood.

Aerobic exercises, however, generally do not increase muscle size. Muscular development—the kind that helps you look good on the beach—comes mostly from *resistance exercise,* for example, lifting heavy weights. To build muscle tone and mass, you have to make your muscles exert more than 75% of their maximum force. These exercises can be very brief. Only a few minutes of exercise every other day for a year are needed to build 50% more muscle and improve tone and strength. The bulk results from the existing muscle cells increasing in diameter, although some researchers suggest that heavy exercise splits or tears the muscle cells and that each of the parts then regrows to a larger size.

Nautilus and other muscle-building machines available at many gyms automatically adjust the resistance they offer to muscles during an exercise (Figure 6.11). This is a valuable feature because muscles are weaker at some parts of their range of motion than at others. The design of the exercise machine makes the lifting easier where the muscle is weaker and more difficult in the range the muscle can handle. "Free weights," such as barbells and dumbbells, require more training to use safely than does an exercise machine.

FIGURE 6.11 *Muscles get larger when they are repeatedly made to exert more than 75% of their maximum force.*

ETHICAL ISSUE

Anabolic Steroid Abuse

In recent years, allegations and confessions of anabolic steroid use have rocked professional sports. The prevalence of steroid use in professional sports is a reflection of the growing problem of steroid use in general, particularly among young adults. Although steroid abuse was at least 1% lower in 2011 than in 2000 among students in grades 8 through 12, a 2007 confidential survey of high school students funded by the National Institute on Drug Abuse revealed that 1.6% of eighth-grade students, 1.8% of tenth-grade students, and 2.7% of seniors had used steroids. Of the students who admitted steroid use, 57% said that steroid use by professional athletes influenced their decision.

Anabolic steroids are a class of drugs related to the steroid hormone testosterone. These synthetic hormones mimic testosterone's ability to stimulate the body to build muscle, often leading to a dramatic increase in strength. The steroids promote these improvements by stimulating protein formation in muscle cells and by reducing the amount of rest needed between workouts.

Anabolic steroids have more than 70 possible unwanted side effects that can range in severity from acne to liver cancer. Most common and most serious are the damage to the liver, cardiovascular system, and reproductive system. Cardiovascular problems, including heart attacks and strokes, may not show up for years, but effects on the reproductive system are more immediate. The testicles of male steroid users often become smaller, and the user may become sterile and impotent (unable to achieve an erection). These effects occur because anabolic steroids inhibit natural testosterone production. Female steroid abusers develop irreversible masculine traits, such as a deeper voice, growth of body hair, loss of scalp hair, smaller breasts, and an enlarged clitoris (during embryological development, the clitoris develops from the same structure that develops into the penis in a male). Among the other risks are injuries resulting from the intended effect of the drug—increased muscle strength—because that increase is not accompanied by a corresponding increase in the strength of tendons and ligaments. Injuries to tendons and ligaments may take a long time to heal.

Anabolic steroids are sometimes abused as an easy way to build muscle and strength.

Questions to Consider

- If you were coaching a high school football team and knew that some of your players were using anabolic steroids, would you ignore their actions?
- If you were the owner of a professional baseball team and suspected that one of your all-star players was using anabolic steroids, what would you do?

One difficulty with building muscle is that you have to keep at it. If you train in the spring only to look good in your summer clothes, you have an uphill battle because all the mass of muscle you built last year began to disappear 2 weeks after you stopped the training.

In addition to the benefits to the heart and blood vessels and an increase in strength, exercise can enhance your health in other ways. Exercise improves your mood. If you are having a stressful day, go for a walk or work out at the gym to calm down. Exercise even reduces depression and anxiety and improves the quality of sleep. Weight-bearing exercise strengthens bones. Are you concerned about your weight? Add exercise to your weight-loss program to burn calories faster.

Begin a fitness program slowly. If you have a medical condition, first seek medical advice. Warm up by walking and gently stretching to improve blood flow to your muscles. Increase the length of time you exercise gradually as you build stamina. While you are cooling down after exercise, walk and stretch as you did before exercise.

A day or two after intense exercise, muscles sometimes become tender and weak. This phenomenon, known as *delayed onset muscle soreness (DOMS)*, is especially common following exercises that force a muscle to lengthen while it is contracting. For example, DOMS is more likely to occur after you run downhill than after you run on level ground. Other activities that cause a muscle to lengthen while it is contracting include walking down stairs, climbing down a mountain, skiing, and horseback riding. The cause of DOMS isn't known for certain, but researchers suggest that microscopic muscle damage or an inflammatory response in the muscle may be to blame.

what would you do?

The International Olympics Committee and the World Anti-Doping Agency are concerned about the future possibility of creating genetically modified athletes. Gene therapy techniques (see Chapter 21) may soon be developed that ultimately might enable people to improve their athletic performance. For example, researchers have developed gene therapy that repairs and rebuilds muscles in old mice. Clinical trials of the technique are now underway on humans who have muscular dystrophy. If you were a competitive athlete, would you use gene therapy to repair damaged muscle tissue? Would you use gene therapy to build muscle tissue and enhance your performance? Do you think any of these uses of gene therapy should be illegal?

looking ahead

In this chapter, we learned about muscles, which contract and cause movement. In the next chapter, we will consider nerve cells and their messages that trigger muscle contraction or glandular secretion, as well as allowing communication among nerve cells.

HIGHLIGHTING THE CONCEPTS

6.1 Function and Characteristics of Muscles (pp. 100–101)
- There are three types of muscle: skeletal, smooth, and cardiac. All types of muscle cells are excitable, contractile, extensible, and elastic.

6.2 Skeletal Muscles Working in Pairs (p. 101)
- Many of the skeletal muscles of the body are arranged in antagonistic pairs so that the action of one opposes the action of the other. One member of such a pair usually causes flexion (bending at the joint) and the other, extension (straightening at the joint).
- Muscles are attached to bones by tendons. The muscle origin is the end attached to the bone that is more stationary during the muscle's movement, and the insertion is the end attached to the bone that moves.

6.3 Contraction of Muscles (pp. 101–106)
- A muscle cell is packed with myofibrils, which are composed of myofilaments (the contractile proteins actin and myosin).
- A muscle contracts when myosin heads in some of its cells bind to actin, swivel, and thereby pull the actin toward the midlines of sarcomeres, causing the actin and myosin filaments to slide past one another and increase their degree of overlap.
- Contraction is controlled by the availability of calcium ions. Calcium ions interact with two proteins on the actin filament—troponin and tropomyosin—that determine whether myosin can bind to actin. Calcium ions are stored in the sarcoplasmic reticulum of muscle cells and released when a motor nerve sends an impulse.
- Motor nerves contact muscle cells at neuromuscular junctions. When the nerve impulse reaches a neuromuscular junction, acetylcholine is released and causes a change in the permeability of the muscle cell's membrane that, in turn, causes calcium ions to be released from the sarcoplasmic reticulum.

6.4 Voluntary Movement (pp. 106–107)
- The extent and strength of contraction can be changed by altering the number of muscle cells contracting or by altering the strength of contraction of individual cells.
- A motor neuron and all the muscle cells it stimulates are collectively called a motor unit.
- The response of a muscle cell to a single brief stimulus in the laboratory is called a twitch. If a second stimulus arrives before the muscle has relaxed, the second contraction builds upon the first. This phenomenon is known as summation. Frequent stimuli cause a sustained contraction, called tetanus. Most body movements require muscle cells in tetanus.

6.5 Energy for Muscle Contraction (pp. 107–108)
- Each time an attachment between myosin and actin forms and is broken, two ATP molecules are used. The sources of ATP are (1) stored ATP, (2) creatine phosphate, (3) anaerobic metabolic pathways, and (4) aerobic respiration.

6.6 Slow-Twitch and Fast-Twitch Muscle Cells (pp. 108–109)
- Slow-twitch muscle cells contract slowly but with enormous endurance. Fast-twitch muscle cells contract rapidly and powerfully but with less endurance.

6.7 Building Muscle (pp. 109–110)
- Resistance exercise will increase muscle size. Forcing muscles to exert more than 75% of their maximal strength adds myofilaments to existing muscle cells and increases the cell diameter.

RECOGNIZING KEY TERMS

tendon *p. 101*
origin *p. 101*
insertion *p. 101*
fascicles *p. 101*
muscle fibers *p. 101*
striated muscle *p. 102*
myofibril *p. 102*
myofilament *p. 102*
myosin filament *p. 102*
actin filament *p. 102*
sarcomere *p. 103*
sliding filament model *p. 103*
cross-bridge *p. 104*
troponin *p. 104*
tropomyosin *p. 104*
sarcoplasmic reticulum *p. 104*
transverse tubules *p. 104*
neuromuscular junction *p. 104*
motor unit *p. 106*
recruitment *p. 106*
muscle twitch *p. 107*
summation *p. 107*
tetanus *p. 107*
oxygen debt *p. 108*
slow-twitch muscle cells *p. 108*
fast-twitch muscle cells *p. 108*

REVIEWING THE CONCEPTS

1. Why are most skeletal muscles arranged in antagonistic pairs? Give an example illustrating the roles of each member of an antagonistic pair of muscles. *p. 101*
2. Describe a skeletal muscle, including descriptions of muscle cells, myofibrils, and myofilaments. *pp. 102–103*
3. What is the sliding filament model of muscle contraction? *pp. 103–104*
4. What causes actin to move during muscle contraction? *p. 104*
5. Explain the roles of troponin, tropomyosin, and calcium ions in regulating muscle contraction. *p. 104*
6. Explain how muscle contraction results from the events that occur when an impulse from a motor nerve cell reaches a neuromuscular junction and calcium ions are released from the sarcoplasmic reticulum. *pp. 104–105*
7. Define *motor unit*. What are the consequences of the differences in sizes of motor units? *p. 106*
8. Define *muscle twitch, summation,* and *tetanus*. *p. 107*
9. List the sources of ATP for muscle contraction, and explain when each source is typically called on. *pp. 107–108*
10. Characterize the difference in function between slow- and fast-twitch muscle cells. *pp. 108–109*

11. What type of exercise can build muscles? *pp. 109–110*
12. A single motor neuron and all the muscle cells it stimulates is called
    a. the sarcoplasmic reticulum.
    b. a neuromuscular junction.
    c. a motor unit.
    d. summation.
13. In a muscle, energy is stored in the form of
    a. creatine phosphate.
    b. ADP.
    c. myosin.
    d. glucose.
14. The _____ are dark bands at the ends of each sarcomere.
    a. T-tubules
    b. myosin heads
    c. Z lines
    d. myofilaments
15. After heavy exercise, _____ help(s) repay the oxygen debt.
    a. breathing
    b. stretching
    c. myosin
    d. calcium ions
16. What causes the release of calcium ions from the sarcoplasmic reticulum?
    a. an increase in calcium ions within the T tubules
    b. the formation of cross-bridges
    c. an action potential
    d. contraction of a sarcomere
17. The thin filaments consist of
    a. actin.
    b. myosin.
    c. troponin.
    d. Z lines.
18. Myoglobin is
    a. the major component of the thick filament.
    b. an oxygen-binding molecule in muscle cells.
    c. released at the neuromuscular junction.
    d. the molecule that locks the actin-binding site.
19. What is the smallest functional unit of contraction?
    a. fiber
    b. sarcomere
    c. filament
    d. myofibril
20. What type of contraction is characterized by a rapid, jerky response to a single stimulus?
    a. summation
    b. treppe
    c. tonic
    d. twitch
21. The function of the transverse tubules (T tubules) is to
    a. distribute a supply of calcium ions through the muscle fiber.
    b. conduct action potentials to the interior of the muscle fiber.
    c. distribute a supply of glycogen throughout the muscle sarcoplasm.
    d. transport ATP molecules out of the mitochondria throughout the sarcoplasm.
22. Each myosin head has
    a. a binding site for an ATP molecule.
    b. a binding site for an actin molecule.
    c. the ability to swivel when powered by ATP.
    d. all of above
23. A muscle cell contracts when _____ filaments and _____ filaments slide past one another.
24. The contractile unit of a muscle is called a/an _____.

APPLYING THE CONCEPTS

1. Hakeem, who is 22 years old, ate some of his Aunt Sophie's canned tomatoes for dinner last night. This morning, his speech is slurred, and he is having trouble standing and walking, so his brother rushes him to the emergency room. The doctor tells Hakeem that he probably has botulism, a type of food poisoning produced by a bacterium that can grow in improperly sealed canned goods. The botulinum toxin prevents the release of acetylcholine at neuromuscular junctions. Explain how Hakeem's symptoms are related to this effect.
2. A new injury called "BlackBerry thumb" is becoming increasingly common as people send e-mails on their handheld devices and text messages on their cell phones. BlackBerry thumb is a form of tendinitis. What would cause BlackBerry thumb? What would the symptoms be? How would you treat it?
3. You are the CEO of a drug company. A research scientist approaches you with a plan to develop a new muscle-relaxing drug. The scientist explains that the drug works by flooding the muscle cell with calcium ions. Would you finance the development of this drug? Why or why not?
4. Chickens are ground-dwelling birds. They run to escape from predators, and they can fly only short distances. How do these activities explain the distribution of "white" and "dark" meat on a chicken? (Meat is skeletal muscle.)

BECOMING INFORMATION LITERATE

Use at least three reliable sources (books, journals, websites) to design an exercise program for yourself. Begin by deciding on your fitness goals. Next, plan a logical progression of activity that considers the type of activities you enjoy. List each source you considered, and explain why you chose the three sources you used.

MasteringBiology®

Go to MasteringBiology for practice quizzes, activities, eText, videos, current events, and more.

Neurons: The Matter of the Mind

7

Did youknow?

- It is possible to fit 30,000 neurons on the head of a pin.
- A "typical" neuron has 1000 to 10,000 synapses that communicate with other neurons.

7.1 Cells of the Nervous System

7.2 Structure of Neurons

7.3 Nerve Impulses

7.4 Synaptic Transmission

HEALTH ISSUE

Neurotransmitters and Disease

In the previous chapter, we learned how muscles contract when they receive messages from nerve cells. In this chapter, we will discover how nerve cells communicate with muscle cells and with one another. In Chapter 8, "The Nervous System," we will explore the parts and functions of the brain and spinal cord.

The brain cells we use to plot and plan our way to victory in a chess match have the same design and mode of functioning as the cells that tell our muscles to contract or that carry information from our sensory organs to our brain. They are neurons, the basic functional units of the nervous system.

7.1 Cells of the Nervous System

The *nervous system* integrates and coordinates all the body's varied activities. Its two primary divisions, which are the subject of Chapter 8, are (1) the central nervous system, consisting of the brain and spinal cord; and (2) the peripheral system, consisting of all the nervous tissue in the body outside the brain and spinal cord. Both these major divisions of the nervous system are composed of two types of specialized cells. **Neurons** (nerve cells) are excitable cells that generate and transmit messages. Outnumbering the neurons by about 10 to 1, **neuroglial cells** (also called *glial cells*) support and protect neurons.

Neuroglial Cells

The nervous system has several types of glial cells, each with a different job to do. Some glial cells provide structural support for the neurons of the brain and spinal cord. Glial cells also provide a steady supply of chemicals, called *nerve growth factors,* that stimulate nerve growth. Without nerve growth factors, neurons die. Other glial cells form insulating sheaths around axons that, as described shortly, are the long projections extending from certain neurons. This sheath, called the *myelin sheath,* has several important functions that are also described shortly. Scientists now know that glial cells can communicate with one another and with neurons.

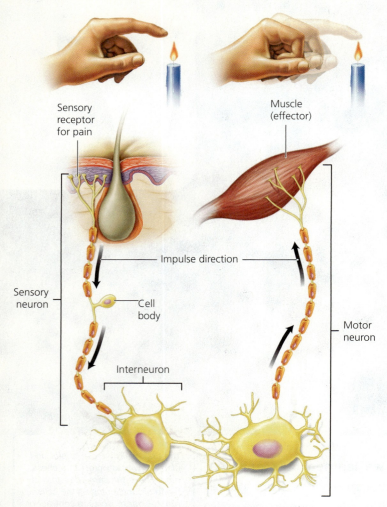

FIGURE 7.1 *Neurons may be sensory neurons, interneurons, or motor neurons. This diagram traces the pathway of an impulse from a sensory receptor to an interneuron and from there to a motor neuron and its effector. Sensory receptors detect changes in the external or internal environment. An interneuron usually receives input from many sensory neurons, integrates that information, and—if the input is appropriate—stimulates a motor neuron. The motor neuron then causes a muscle or a gland (an effector) to respond.*

Neurons

The basic unit of the nervous system is the neuron, or nerve cell. Neurons, which are responsible for an amazing variety of functions, can be grouped into the three general categories depicted in Figure 7.1.

- **Sensory** (or afferent) **neurons** conduct information *toward* the brain and spinal cord. These neurons generally extend from sensory receptors, which are structures specialized to gather information about the conditions within and around our bodies.
- **Motor** (or efferent) **neurons** carry information *away from* the brain and spinal cord to an **effector**—either a muscle, which will contract, or a gland, which will secrete its product—as a response to information from a sensory or interneuron.
- Association neurons, commonly called **interneurons**, are located between sensory and motor neurons. They are found only within the brain and spinal cord, where they integrate and interpret the sensory signals, thereby "deciding" on the appropriate response. Interneurons are by far the most numerous nerve cells in the body; they account for more than 99% of the body's neurons.

We can appreciate the specific roles of each type of neuron by considering the symptoms of a progressive disease called *amyotrophic lateral sclerosis (ALS)*, also known as *Lou Gehrig's disease*. In ALS, motor neurons throughout the brain and spinal cord die and stop sending messages to skeletal muscles. Without stimulation from motor neurons, the muscles gradually weaken, and the person loses control over arms, legs, and body. The cause of death is respiratory failure, because the muscles that control breathing (the diaphragm and rib muscles) eventually die. Sensory neurons and interneurons are not affected by ALS, so awareness and reasoning do not deteriorate.

7.2 Structure of Neurons

The shape of a typical neuron is specialized for communicating with other cells (Figure 7.2).

Axons and Dendrites

A neuron has many short, branching projections called **dendrites**, which provide a huge surface for receiving signals from other cells. Such signals travel toward an enlarged central region called the *cell body*, which has all the normal organelles, including a nucleus, for maintaining the cell. When a neuron responds to an incoming signal, it transmits its message along the **axon**, a single long extension of the neuron. The axon carries messages away from the cell body either to another neuron or to an effector, which can be a muscle or a gland. In some cases, the axon allows the neuron to communicate over very long distances. For example, a motor neuron that allows you to wiggle your big toe has its cell body in the spinal cord, and its axon runs all the way to the muscles of your toe. The end of the axon has many branches specialized to release a chemical, called a *neurotransmitter,* that alters the activity of the effector. The axon is the *sending* portion of the neuron, whereas the dendrites and cell body are typically the *receiving* portions.

To appreciate the dimensions of a neuron, imagine a "typical" motor neuron, one that carries a message from the spinal cord to a muscle. Now picture the cell body of this neuron as being the size of a tennis ball. At that scale, the axon of this neuron would be about 1 mile (1.6 km) long and only about one-half inch (1.3 cm) in diameter. At the same scale, the dendrites—the shorter but more numerous projections of the neuron—would fill an average-sized living room.

A **nerve** is a bundle of parallel axons, dendrites, or both arising from many neurons. Each nerve is covered with connective tissue and, depending on the type of neurons it contains, can be classified as sensory, motor, or mixed (a mixed nerve is made up of both sensory and motor neurons).

Myelin Sheath

Most of the axons outside the brain and spinal cord, and some of those within, have an insulating outer layer called a **myelin sheath** (Figure 7.3), which increases the rate of conduction of a nerve impulse and helps its repair. The myelin sheath is composed of the plasma membranes of glial cells. Outside of the

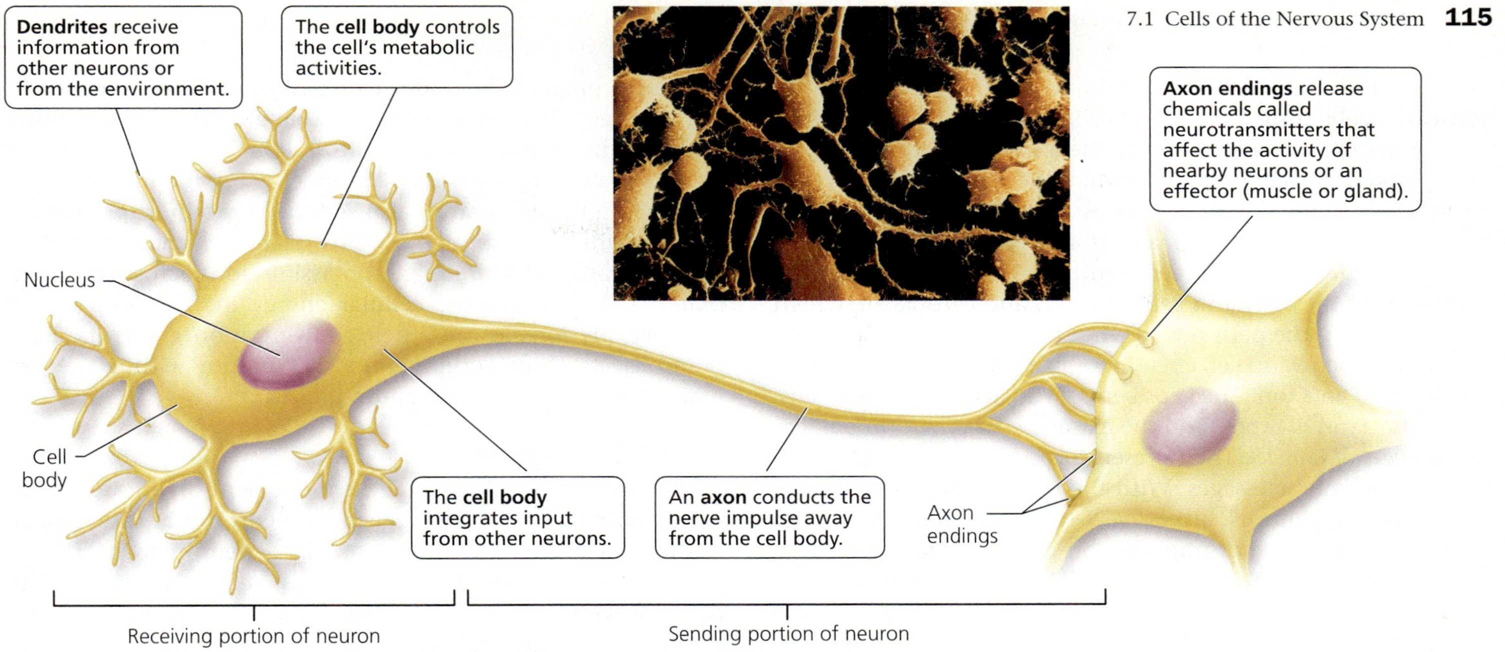

7.1 Cells of the Nervous System **115**

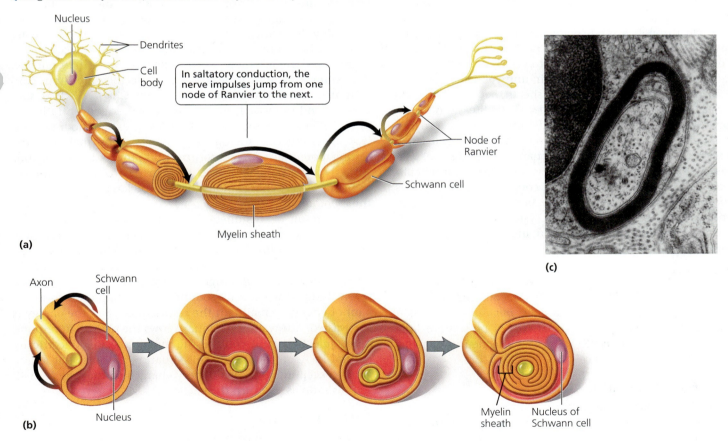

FIGURE 7.2 The structure of a neuron

Q Long axons are myelinated, but short axons may not be. Why?

FIGURE 7.3 The myelin sheath. (a) An axon protected by a myelin sheath. The Schwann cells that form the myelin sheath are separated by nodes of Ranvier—areas of exposed axon that allow for saltatory conduction. (b) The myelin sheath forms from multiple wrappings of Schwann cell plasma membranes. (c) An electron micrograph of the cut end of a myelinated axon.

A Saltatory conduction, which increases the speed of conduction, is possible only on myelinated axons. The increased speed is important when nerve impulses must travel long distances.

brain and spinal cord, for example, glial cells known as **Schwann cells** form neurons' myelin sheaths. A Schwann cell plasma membrane wraps around the axon many times to provide a covering that looks somewhat like a jelly roll. This covering, the myelin sheath, serves as a kind of living electrical tape, insulating individual axons and preventing messages from short-circuiting between neurons. The myelin sheath is kept alive by the Schwann cell's nucleus and cytoplasm, which are squeezed to the periphery as the sheath forms.

A single Schwann cell encloses only a small portion, about 1 mm long, of an axon. The gaps between adjacent Schwann cells, where the axon is exposed to the extracellular environment, are called *nodes of Ranvier*. Their presence is crucial to how rapidly a neuron transmits messages. With the myelin sheath in place, a nerve impulse "jumps" successively from one node of Ranvier to the next in a type of transmission called **saltatory conduction** (*saltare*, to jump), which is up to 100 times faster than signal conduction would be on an unmyelinated axon of the same diameter. Not surprisingly, the axons responsible for conducting signals over long distances are typically myelinated.

To get a sense of how this "jumping" mode of transmission increases the speed at which a message travels, think of the different ways a ball can be moved down the court during a basketball game. When only seconds are left in the game, dribbling the ball the length of the court would take too much time. Passing the ball through a series of players is faster. Likewise, an impulse passed from one node to the next, as occurs in myelinated nerves, moves faster than one traveling uniformly along the full length of the axon, as occurs in unmyelinated nerves.

The importance of the myelin sheath becomes dramatically clear in people with *multiple sclerosis,* a disease in which the myelin sheaths in the brain and spinal cord are progressively destroyed. The damaged regions of myelin become hardened scars called *scleroses* (hence the name of the disease) that interfere with the transmission of nerve impulses. Short-circuiting between normally unconnected conduction paths delays or blocks the signals going from one brain region to another. Depending on the part of the nervous system affected, the result can be paralysis or the loss of sensation, including loss of vision.

7.3 Nerve Impulses

The neuron membrane is specialized for communication. A nerve's message, which is called a *nerve impulse* or an *action potential,* is an electrochemical signal caused by sodium ions (Na+) and potassium ions (K+) crossing the neuron's membrane to enter and leave the cell.

Plasma Membrane of a Neuron

Like most living membranes, the plasma membrane of a neuron is selectively permeable; it allows some substances through but not others. The membrane contains many pores, called **ion channels**, that ions are able to pass through without using cellular energy. Each ion channel is designed to allow only certain ions to pass through. For example, sodium channels allow the passage of only sodium ions, and potassium channels allow the passage of only potassium ions. In this way, ion channels function as molecular sieves. Some channels are permanently open. Others are regulated by a "gate"—a protein that changes shape in response to changing conditions, either opening the channel, which allows ions to pass through, or closing it, which prevents ions from crossing the membrane.

The cell membrane also contains **sodium-potassium pumps**, which are special proteins in the cell membrane that actively transport sodium and potassium ions across the membrane. These pumps use cellular energy in the form of ATP to move the ions against their concentration gradients. Each pump ejects three sodium ions (Na+) from within the cell while bringing in two potassium ions (K+).

Resting Potential

It will be easier to understand the movement of ions during an action potential if we first consider a neuron that is not transmitting an action potential—that is, a neuron in its resting state. As we will see, however, *resting* is hardly the word to describe what is going on at this stage. The membrane of a resting neuron maintains a difference in the electrical charges near the two membrane surfaces (the surface facing inside the cell and the surface facing outside the cell), keeping the inside surface more negative than the outside one. This charge difference across the membrane, called the **resting potential**, results from the unequal distribution of ions across the membrane.

In a resting neuron, sodium and potassium ions are unequally distributed across the plasma membrane. There are more sodium ions outside the membrane than inside. Furthermore, there are more potassium ions inside than outside. Potassium ions tend to leak out because they are more concentrated inside the axon. (Recall from Chapter 3 that substances tend to move from an area of higher concentration to one of lower concentration.) To a lesser extent, sodium ions leak in. However, sodium-potassium pumps maintain the resting potential of a neuron by pumping out sodium ions while moving potassium ions back in. The cell also contains some negatively charged ions that are too large to pass through the membrane.

The result of this unequal distribution of ions is that the inner surface of a resting neuron's membrane is typically about 70 mV (millivolts) more negative than the outer surface. This voltage is the resting potential, and it is about 5% of the strength of a size AA flashlight battery.

Although the neuron's sodium-potassium pumps consume a lot of energy to maintain the resting potential, the energy is not wasted. The resting potential allows the neuron to respond more quickly than it could if the membrane were electrically neutral in its resting state. This situation is somewhat analogous to keeping your car's battery charged so that the car will start as soon as you turn the key.

Action Potential

Now we are ready to consider what happens when a neuron is stimulated—that is, when it receives some kind of excitatory signal. The **action potential**, or nerve impulse, resulting from such stimulation can be described briefly as a sudden reversal in the charge difference across the membrane, followed by the restoration of the original charge difference. Let's take a closer look at these two parts of a nerve impulse.

Resting Neuron
Plasma membrane is charged, with the inside negative relative to the outside.

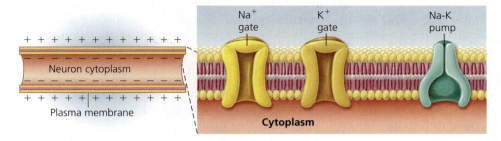

Action Potential
The charge difference across the membrane reverses and then is restored.

Step 1: The loss of the charge difference across the membrane (depolarization) occurs as sodium ions (Na$^+$) enter the axon.

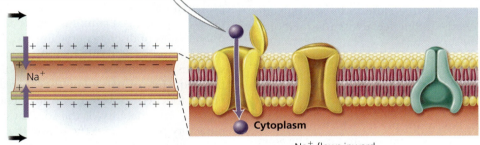

Na$^+$ flows inward

Step 2: The return of the membrane potential to near its resting value (repolarization) occurs as potassium (K$^+$) ions leave the axon.

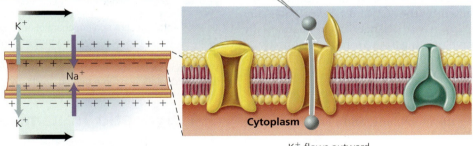

K$^+$ flows outward

Restoration of Original Ion Distribution
The sodium-potassium pump restores the original distribution of ions.

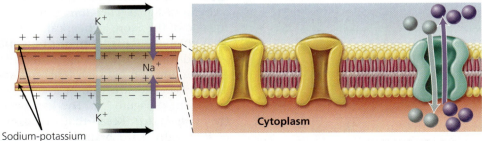

Na-K pump restores the original ion distribution

FIGURE 7.4 *The resting state and the propagation of an action potential along an axon. The sequential opening and closing of sodium-channel gates and potassium-channel gates produces the action potential.*

1. **Sodium ions (Na+) enter the axon.** An excitatory stimulus causes the gates on sodium channels to open. Sodium ions then enter the neuron, and their positive charge begins to reduce the negative charge within. Reduction of the charge difference across the membrane is called depolarization. The action potential begins when membrane depolarization reaches a certain value called the **threshold**. When the threshold is reached, the gates on more sodium channels open. Enough sodium ions enter through the open gates to create a net positive charge in that region (about +30 mV), as shown in Figure 7.4.

2. **Potassium (K+) ions leave the axon.** About halfway through the action potential, the gates on potassium channels open. Potassium ions now leave the cell. The exodus of potassium ions with their positive charge causes the interior of the neuron to become negative once again relative to the outside. The outward flow of potassium ions returns the membrane potential close to its resting value. Restoration of the charge difference across the membrane is called *repolarization*.

As noted, at the end of an action potential, the charge distribution across the membrane returns to the resting potential. However, there are slightly more sodium ions and slightly fewer potassium ions inside the cell than before. This alteration is corrected by the sodium-potassium pump, which restores the original distribution of sodium and potassium ions. The action of the sodium-potassium pump is slow. Therefore, it does not contribute directly to the events of the action potential.

To summarize, the action potential is a reversal of the charge difference across the membrane caused by the inward flow of sodium ions, followed immediately by restoration of the original charge difference caused by the outward flow of potassium ions (Figure 7.5 on p. 119). These changes occur sequentially along the axon, like a wave rippling away from the cell body.

The action potential is described as a wave of changes that travels down the neuron's plasma membrane because the events just described do not occur simultaneously along the entire length of the axon. Instead, as sodium ions enter the cell at one location along the membrane, and the charge inside the membrane becomes less negative in that region of the cell, the change in charge causes the opening of the sodium channel gates in an adjacent part of the membrane. As a result of this sequential opening of gates, the change in charge travels down the length of the axon. Once started, action potentials do not diminish—just like the last domino in a falling row falls with the same energy as the first. Moreover, the intensity of the nerve impulse does not vary with the strength of the stimulus that triggered it. If an action potential occurs at all, it is always of the same intensity as any other action potential. This "all-or-nothing" aspect of nerve cell conduction is similar to the firing of a gun in that the force of the bullet is not changed by how hard you pull the trigger.

Immediately after an action potential occurs, the neuron cannot be stimulated again for a brief instant called the **refractory period**. During the refractory period, the sodium channels are closed and cannot be reopened. Consequently, a new action potential cannot yet be generated. Because of the refractory period, a prolonged stimulus that is above threshold can cause only a series of discrete nerve impulses, not a single, larger, sustained impulse. For this reason, increasing the strength of a stimulus will increase the frequency of impulses. For example, the frequency of nerve impulses increases with the heat of an object touched. However, the frequency can increase only to a point. The inability of the sodium gates to open during the refractory period is also the reason that the nerve impulses cannot reverse and go backward toward the cell body.

stop and think

"Red tides" in the ocean are caused by proliferation of dinoflagellates. These single-celled marine algae contain a chemical called *saxitoxin (STX)*, extremely small concentrations of which prevent sodium channels in mammalian neurons from opening. The clams, scallops, and mussels that consume the dinoflagellates are insensitive to the toxin, but the STX accumulates in their tissues. What effect would you expect STX to have on nerve transmission in humans who accidentally consume tainted shellfish?

7.4 Synaptic Transmission

When a nerve impulse reaches the end of an axon, in almost all cases the message must be relayed to the adjacent cell across a small gap that the impulse cannot jump across. To transmit the message to the adjacent cell requires a brief change of the medium of communication from an electrochemical signal to a chemical signal. Therefore, when an action potential reaches the end of the axon, a chemical is released from the axon's tip. That chemical, called a **neurotransmitter**, diffuses across the gap and conveys the message to the adjacent cell.

The junction between a neuron and another cell is called a **synapse**. The structure of a synapse between two neurons is shown in Figure 7.6. The gap between the cells is called the *synaptic cleft*. Recall that the axon branches near the end of its length. Each branch ends with a small bulblike swelling called a *synaptic knob*. The neuron sending the message is the *presynaptic neuron* (meaning "before the synapse"). The neuron receiving the message is the *postsynaptic neuron* ("after the synapse").

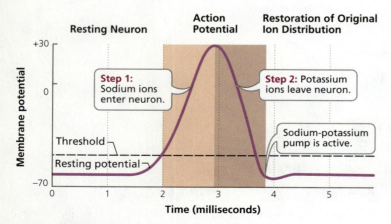

FIGURE 7.5 A graphic representation of an action potential. Voltage across the membrane can be measured by electrodes placed inside and outside the axon. The graph shows the changes in voltage that accompany an action potential.

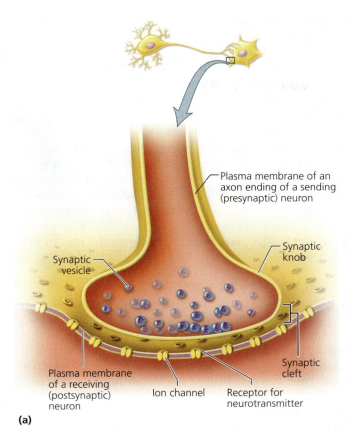

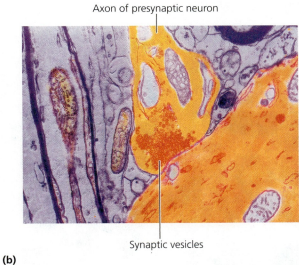

FIGURE 7.6 Structure of a synapse. (a) The synaptic knob at the end of the axon on the presynaptic neuron is separated from the dendrite or cell body of the postsynaptic neuron by a small gap called a synaptic cleft. Within the synaptic knob are small sacs, called synaptic vesicles, filled with neurotransmitter molecules. (b) An electron micrograph of a synapse.

Release of the Neurotransmitter and the Opening of Ion Channels

Let's consider the events that occur in the synapse as the message is sent from one neuron to the next (Figure 7.7).

1. **The nerve impulse reaches the axon ending of the presynaptic neuron.**

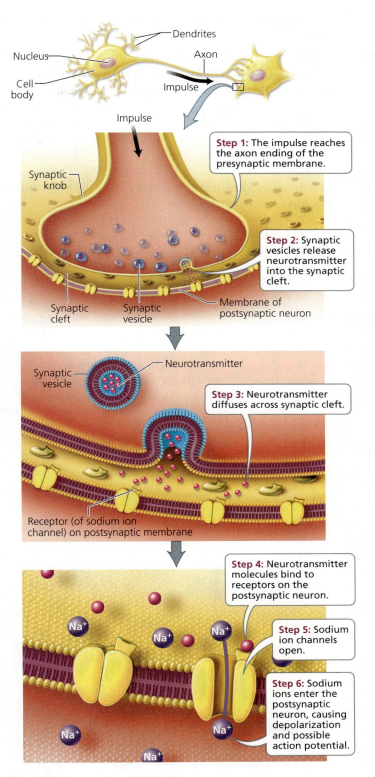

FIGURE 7.7 Transmission across an excitatory synapse

2. **Synaptic knobs release packets of neurotransmitter.** Within the synaptic knobs of the presynaptic neuron, the neurotransmitter is contained in tiny sacs called *synaptic vesicles*. When a nerve impulse reaches the synaptic knob, the gates of calcium ion channels in the membrane there open. Calcium ions move into the knob, causing the membranes of the synaptic vesicles to fuse with the plasma

membrane at the synaptic knob and to dump the enclosed neurotransmitter into the synaptic cleft.

3. **Neurotransmitter diffuses across the synaptic cleft.**
4. **Neurotransmitter binds with receptors on the membrane of the postsynaptic neuron.** A receptor is a protein that recognizes a particular neurotransmitter, much as a lock "recognizes" a key. The only cells a neurotransmitter can stimulate are cells that have receptors specific for that particular neurotransmitter. Thus, only certain neurons can be affected by a given neurotransmitter.
5. **When a neurotransmitter binds to its receptor, an ion channel is opened.** The binding of the neurotransmitter to a receptor causes the opening of an ion channel in the postsynaptic neuron. The response that is triggered as a result depends on the type of ion channel the receptor opens. It is the receptor that determines which ion channels will open and what the effect of a given neurotransmitter will be.

An *excitatory synapse* is one where the binding of the neurotransmitter to the receptor opens sodium channels, allowing sodium ions to enter and increasing the likelihood that an action potential will begin in the postsynaptic cell. In contrast, an *inhibitory synapse* is one where the binding of the neurotransmitter opens different ion channels, which decreases the likelihood that an action potential will be generated in the postsynaptic neuron. In this case, the cell's interior becomes more negatively charged than usual. As a result, the cell will require larger than usual amounts of an excitatory neurotransmitter in order to reach threshold.

Summation of Input from Excitatory and Inhibitory Synapses

A neuron may have as many as 10,000 synapses with other neurons at the same time (Figure 7.8). Some of these synapses will have excitatory effects on the postsynaptic membrane. Others will have inhibitory effects. The **summation** (combined effects) of excitatory and inhibitory effects on a neuron at any given moment determines whether an action potential is generated. This integration of input from large numbers of different kinds of synapses gives the nervous system fine control over neuronal responses, just as having both an accelerator and a brake gives you finer control over the movement of a car.

Removal of Neurotransmitter

After being released into a synapse, neurotransmitters are quickly removed, so their effects are temporary. If they were not removed, they would continue to excite or inhibit the postsynaptic membrane indefinitely. Depending on the neurotransmitter, disposal is accomplished in one of two ways. First, enzymes can deactivate a neurotransmitter. For example, the enzyme acetylcholinesterase removes the neurotransmitter **acetylcholine** from synapses where it has been released. Second, the neurotransmitter may be actively pumped back into the presynaptic knob.

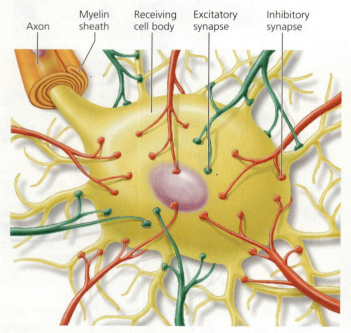 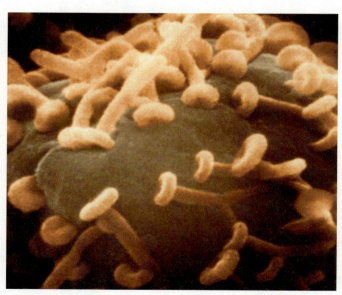

FIGURE 7.8 *A neuron may have as many as 10,000 synapses at which it receives input from other neurons. Some synapses have an excitatory effect on the membrane of the postsynaptic neuron and increase the likelihood that the neuron will fire. Other synapses have an inhibitory effect and reduce the likelihood that the postsynaptic neuron will fire. The net effect of all the synapses determines whether an action potential is generated in the postsynaptic neuron. (The shape of the synaptic knobs shown in this electron micrograph is distorted as a result of the preparation process.)*

HEALTH ISSUE

Neurotransmitters and Disease

Neurotransmitters affect our movements, memory, and emotions. It should not be surprising, then, that changes in neurotransmitter level can cause disorders.

Alzheimer's Disease and Acetylcholine

Alzheimer's disease is progressive and results in loss of memory, particularly for recent events, followed by sometimes severe personality changes. In Alzheimer's disease, the parts of the brain important in memory and intellectual functioning (hippocampus and cerebral cortex; see Chapter 8) lose large numbers of neurons. Some of the neurons in these regions use the neurotransmitter acetylcholine, which may decrease in level by as much as 90% in a person with Alzheimer's—a fact that may explain the loss of memory and mental capacity. In addition, a brain affected by Alzheimer's disease is pocked with clusters of proteins, some between the neurons (amyloid plaques) and others within the neurons (neurofibrillary tangles). The amyloid plaques and neurofibrillary tangles are the prime suspects for the cause of the death of acetylcholine-producing neurons.

In hypothesizing that the loss of acetylcholine is responsible for some of the Alzheimer's symptoms, researchers and physicians have attempted to treat Alzheimer's disease with drugs meant to raise or at least maintain acetylcholine levels. Although such drugs (Aricept, Exelon, and Reminyl) do improve the memory and intellectual ability of some people with Alzheimer's, they do not help in all cases. Moreover, any improvement is rapidly lost when a person stops taking the drugs.

Depression, Serotonin, Dopamine, and Norepinephrine

We all describe ourselves as feeling depressed at times. However, more than 19 million Americans experience depression that lasts for weeks, months, or years and interferes with their ability to function in daily life. This condition is considered *clinical depression,* and it can affect anyone. It is thought to be related in some way to insufficient levels of the neurotransmitter serotonin as well as of dopamine and norepinephrine. Signs of depression can include a loss of interest and pleasure in the activities and hobbies that were previously pleasurable; anxiety; sleep problems; decreased energy; and feelings of sadness, hopelessness, worthlessness, and guilt. Depression takes the joy out of life and complicates certain medical conditions such as heart disease, cancer, diabetes, epilepsy, and osteoporosis.

Depression can be treated successfully. Unfortunately, few of the millions of people suffering from depression recognize the symptoms and seek help. Antidepressant drugs affect the functioning of the neurotransmitters responsible for the problem: norepinephrine and serotonin. Older medications affected both neurotransmitters simultaneously. Newer ones, including Prozac, Zoloft, and Paxil, specifically affect serotonin functioning, increasing the level of serotonin in the synapse by reducing its rate of removal. Other antidepressants (such as Welbutrin) increase both norepinephrine and dopamine functioning by reducing their rates of removal from the synapse.

Parkinson's Disease and Dopamine

Actor Michael J. Fox and former heavyweight champion Muhammad Ali have Parkinson's disease (Figure 7.A), a progressive disorder that results from the death of dopamine-producing neurons deep in the brain's movement control center. A person with Parkinson's disease moves slowly, usually with a shuffling gait and a hunched posture, and may suffer from involuntary muscle contractions. The involuntary muscle contractions may cause tremors (involuntary rhythmic shaking) of the hands or head due to alternating contraction and relaxation of the muscles, or they may cause muscle rigidity due to continuous contraction of the muscles. This muscle rigidity may cause sudden "freezing" in the middle of a movement.

As the dopamine-producing neurons in the brain's movement control center die, dopamine levels begin to fall. Initially, the symptoms are subtle and are often written off as part of the aging process. By the time the Parkinson's diagnosis is apparent, 80% of the neurons in this small area of the brain may already have died.

Attempts to treat Parkinson's disease have focused on replacing dopamine or helping the brain get by with the remaining dopamine. Unfortunately, swallowing pills containing dopamine does not help, because dopamine is prevented from reaching the brain by the blood–brain barrier, which you will read about in Chapter 8. Instead, patients are given other substances that can reach the brain. The most common and effective treatment combines two drugs: L-dopa, an amino acid that the brain converts to dopamine; and carbidopa, which prevents dopamine from forming outside the brain and causing undesirable side effects. However, this treatment does not stop the steady loss of dopamine-producing neurons, so it loses effectiveness as the disease progresses. Some patients are treated with drugs that enhance their levels of dopamine by inhibiting the enzyme that breaks down dopamine.

FIGURE 7.A *Actor Michael J. Fox and former heavyweight champion Muhammad Ali have Parkinson's disease, a progressive debilitating disease caused by the death of dopamine-producing nerve cells in a movement control center of the brain. The Michael J. Fox Foundation for Parkinson's Research (www.michaeljfox.org) funds research on the early diagnosis and treatment of Parkinson's disease.*

Questions to Consider

- Few people would question the justifiability of providing drugs that elevate neurotransmitter function to someone who is depressed or suicidal in order to help the person live a normal life. However, researchers believe that the levels of key neurotransmitters also affect personality traits, such as shyness or impulsiveness. If so, we may someday be able to design our own personalities. Should minor personality problems be treated with drugs? Should a personality "flaw" such as shyness or impulsiveness be treated?
- Genetic tests can determine the likelihood of a person developing Alzheimer's disease. If an elderly member of the family is showing signs of dementia, is it ethical for other family members, who may become caretakers of this person, to demand that the elderly person be tested for genes that increase Alzheimer's risk?

what would you do?

Schizophrenia, a mental illness characterized by hallucinations and disordered thoughts and emotions, is caused by an imbalance between the neurotransmitters dopamine and glutamate in one part of the brain (the midbrain). To verify that a drug is useful in treating schizophrenia, controlled experiments must be performed. In some experiments, this means giving patients powerful antipsychotic drugs with unknown side effects. In other experiments, this means taking patients off medication to see whether they would suffer a psychotic relapse. Participation in such experiments now requires written, informed consent. Every detail of the experiment and its potential risks must be presented in writing. Do you think that someone with schizophrenia will understand the consent form? Should researchers be held accountable if they did not know about certain risks? Should the research be published if the participants did not give informed consent?

Roles of Different Neurotransmitters

As we have seen, neurotransmitters are the chemical means of communication within the nervous system. There are dozens of neurotransmitters, carrying messages among neurons and between neurons and muscles or glands. The activities of neurotransmitters produce our thoughts and feelings and enable us to interact appropriately with the world around us. Some neurotransmitters produce different effects on different types of cells.

Acetylcholine and norepinephrine are neurotransmitters that act in both the peripheral and the central nervous systems. Both have either excitatory or inhibitory effects, depending on where they are released. As we see in Chapter 8, most internal organs receive input from neurons that release acetylcholine and from neurons that release norepinephrine. Norepinephrine stimulates most organs but inhibits certain others. Whatever the effect of norepinephrine on any particular organ, acetylcholine will have the opposite effect.

Acetylcholine is also the neurotransmitter released at every neuromuscular junction (the junction of a motor neuron and a skeletal muscle cell), where it triggers contraction of voluntary (skeletal) muscles. We can see how important the nerve activation of muscle is whenever the interaction of nerve and muscle is interrupted. An example of such interruption is *myasthenia gravis,* an autoimmune disease in which the body's defense mechanisms attack the acetylcholine receptors at neuromuscular junctions. With any repeated movement, the amount of acetylcholine released with each nerve impulse decreases after the neurons have fired a few times in succession. The low number of acetylcholine receptors in people with myasthenia gravis makes them extremely sensitive to even the slightest decline in acetylcholine availability. As a result, people with myasthenia gravis have little muscle strength, and their repeated movements become feeble quite rapidly. Drugs that inhibit acetylcholinesterase are prescribed to prevent the breakdown of acetylcholine, elevating the level of acetylcholine in the neuromuscular junction.

About 50 neurotransmitters are used by the central nervous system for communication between the neurons in our brains. Why so many? One reason seems to be that different neurotransmitters are involved with different behavioral systems. Norepinephrine, for instance, is important in the regulation of mood, in the pleasure system of the brain, and in arousal. Norepinephrine is thought to produce an energizing "good" feeling. It is also thought to be essential in hunger, thirst, and the sex drive. Serotonin is thought to promote a generalized feeling of well-being. Dopamine helps regulate emotions. It is also used in pathways that control complex movements. A change in the level of a neurotransmitter affects the behaviors controlled by neurons that communicate using that neurotransmitter (see the Health Issue essay, *Neurotransmitters and Disease*).

looking ahead

In this chapter, we considered the structure and function of neurons. In the next chapter, we will see how neurons are organized to form a functional nervous system.

HIGHLIGHTING THE CONCEPTS

7.1 Cells of the Nervous System (pp. 113–114)

- The nervous system has two types of specialized cells: neurons (nerve cells) and neuroglial cells.
- Neuroglial cells outnumber neurons. They provide structural support for neurons, supply nerve growth factors, and form myelin sheaths around certain axons.
- There are three general categories of neurons. Sensory (or afferent) neurons conduct information from the sensory receptors toward the central nervous system. Motor (or efferent) neurons conduct information away from the central nervous system to an effector. Association neurons (interneurons) are positioned between sensory and motor neurons and are located in the central nervous system.

7.2 Structure of Neurons (pp. 114–116)

- Neurons are specialized for communicating with other cells. A typical neuron has a cell body containing the organelles that maintain the cell. Many branching fibers called dendrites conduct messages toward the cell body. A single long axon conducts impulses away from the cell body.
- An axon may be enclosed in an insulating layer called a myelin sheath. Neuroglia called Schwann cells form the myelin sheath by wrapping their plasma membranes repeatedly around the axon. The myelin sheath greatly increases the rate at which impulses are conducted along an axon; it also plays a role in the regeneration of cut axons in the peripheral nervous system.

7.3 Nerve Impulses (pp. 115–118)

- The message conducted by a neuron, called a nerve impulse or action potential, is caused by sodium ions (Na+) and potassium ions (K+) crossing the neuron's plasma membrane to enter and leave the cell.
- Ion channels in the plasma membrane are small pores through which ions move without the use of cellular energy. Ion channels are usually specific to one or a few types of ions.
- The sodium-potassium pump uses cellular energy in the form of ATP to pump sodium ions out of the cell and potassium ions into the cell against their concentration gradients.
- In the resting state, a neuron has an electrical potential difference, called the resting potential, across its plasma membrane. The resting potential is generated by an unequal distribution of charges across the membrane that makes the neuron more negative inside than outside; there are many negatively charged proteins inside the cell. Sodium ions are in greater concentration outside the neuron, and potassium ions are in greater concentration inside.
- The action potential begins when a region of the membrane suddenly becomes permeable to sodium ions. If enough sodium ions enter to reach a threshold, an action potential is generated. The gates on sodium channels open, and many sodium ions enter the cell, making its interior in that region of the membrane temporarily positive (depolarization). Potassium ions then leave the cell, making that region inside once again more negative than the outside (repolarization). The same events are repeated all along the axon in a wave of depolarization and repolarization called an action potential.
- At the end of an action potential, the sodium-potassium pump moves sodium ions out of the neuron and potassium ions into the neuron, restoring the original ion distribution.
- Once initiated, an action potential sweeps to the end of the axon without diminishing in strength.
- During the brief refractory period immediately following an action potential, the neuron cannot be stimulated.

7.4 Synaptic Transmission (pp. 118–122)

- The point where one neuron meets another is called a synapse. The neuron sending the message (the presynaptic neuron) and the neuron receiving the message (the postsynaptic neuron) are separated by a small gap called the synaptic cleft.
- The arrival of a nerve impulse at the axon's end causes calcium ions to enter the presynaptic cell there. These ions cause synaptic vesicles that store neurotransmitters to fuse with the plasma membrane of the presynaptic neuron and release their contents into the synaptic cleft. The neurotransmitter then diffuses across the gap and binds to receptors on the membrane of the postsynaptic neuron.
- If the synapse is excitatory, sodium ions enter the postsynaptic cell and increase the likelihood that it will generate a nerve impulse. If the synapse is inhibitory, the charge difference across the membrane of the receiving neuron is increased, reducing the likelihood that the postsynaptic neuron will generate a nerve impulse.
- Postsynaptic cells integrate excitatory and inhibitory input from many cells. If the threshold is reached, an action potential is generated in the postsynaptic cell.
- The neurotransmitter is quickly removed from the synapse either by enzymatic breakdown or by transport back into the presynaptic neuron.
- Acetylcholine, epinephrine, and norepinephrine are neurotransmitters used in both the peripheral and central nervous systems.
- Many neurotransmitters are found in the brain. Different ones are active in different behavioral systems. Disturbances in brain chemistry affect mood and behavior.

RECOGNIZING KEY TERMS

neuron *p. 113*
neuroglial cell *p. 113*
sensory neuron *p. 114*
motor neuron *p. 114*
effector *p. 114*
interneuron *p. 114*

dendrite *p. 114*
axon *p. 114*
nerve *p. 114*
myelin sheath *p. 114*
Schwann cell *p. 116*
saltatory conduction *p. 116*

ion channel *p. 116*
sodium-potassium pump *p. 116*
resting potential *p. 116*
action potential *p. 116*
threshold *p. 118*
refractory period *p. 118*

neurotransmitter *p. 118*
synapse *p. 118*
summation *p. 120*
acetylcholine *p. 120*

REVIEWING THE CONCEPTS

1. What are the functions of neuroglial cells? *p. 113*
2. List the three types of neurons, and give their general functions. *p. 114*
3. Draw a typical neuron and label the following: cell body, nucleus, dendrites, and axon. *p. 114*
4. Explain how a myelin sheath is formed. What are the functions of the myelin sheath? *pp. 114–116*
5. Describe the distribution of sodium ions and potassium ions during a neuron's resting state. Explain how the movements of these ions affect the charge difference across the membrane. *p. 116*
6. Why is the resting potential important? *p. 116*
7. What happens to sodium ions at the beginning of an action potential? *p. 118*
8. Describe the events that bring about the restoration of the charge difference across the membrane (repolarization). *p. 118*
9. How is the original ion distribution of sodium and potassium ions restored? *p. 118*

10. What is the refractory period? *p. 118*
11. Draw a synapse between two neurons. Label the following: presynaptic neuron, postsynaptic neuron, synaptic cleft, synaptic vesicles, neurotransmitter molecules, and receptors. *p. 118*
12. How do the events at an excitatory synapse differ from those at an inhibitory synapse? *p. 120*
13. How is the action of a neurotransmitter terminated? *p. 120*
14. Choose the *incorrect* statement:
    a. Neurotransmitters diffuse across the myelin sheath.
    b. In an inhibitory synapse, the neurotransmitter makes it less likely that an action potential will be generated in the postsynaptic (after the synapse) neuron.
    c. Neurotransmitters are stored in synaptic vesicles.
    d. Most interneurons are found in the central nervous system.
15. In botulism, a type of food poisoning, the poison produced by bacteria in the spoiled food prevents the person's synaptic vesicles from fusing with the neuron's membrane. You would expect that this effect would
    a. cause excessive destruction of neurotransmitter in the synaptic cleft.
    b. destroy myelin.
    c. prevent the message of the presynaptic cell from reaching the postsynaptic cell.
    d. cause neurotransmitters to clog the synaptic cleft.
16. In a resting neuron
    a. potassium ions are more concentrated outside the membrane than inside.
    b. the inside is more negative than the outside.
    c. sodium ions are more concentrated inside than outside.
    d. action potentials are being generated.
17. You are a neurophysiologist trying to identify the function of a particular nerve cell. You notice that this nerve cell fires immediately before a person's pinky finger bends. You correctly conclude that this axon
    a. may be part of a sensory neuron carrying information from the finger toward the brain.
    b. may be part of a motor neuron carrying information from the brain toward the pinky.
    c. is without a doubt part of an interneuron.
    d. is dead.
18. The synaptic cleft
    a. is a chemical that allows two neurons to communicate with one another.
    b. is a gap between two neurons.
    c. is a gap between two Schwann cells forming the myelin sheath.
    d. allows saltatory conduction.
19. Oubain is a drug that causes the axon to lose its membrane potential. (In other words, there is no separation of charges across the membrane.) What is the most likely effect that this would have on the neuron?
    a. It would lower the threshold and make it easier to generate action potentials.
    b. It would slow the rate at which action potentials move along the axon.
    c. Action potentials could only go halfway down the axon.
    d. Action potentials could not be generated.
20. Choose the *correct* statement:
    a. Neurotransmitters diffuse across the myelin sheath.
    b. An inhibitory neurotransmitter makes it less likely that an action potential will be generated in the *presynaptic* (before the synapse) neuron.
    c. Neurotransmitters are stored in synaptic vesicles.
    d. Most interneurons are found in the peripheral nervous system.
21. Multiple sclerosis is a disease in which
    a. patches of myelin are destroyed.
    b. neurotransmitters can no longer be produced.
    c. neurotransmitters can no longer be removed from the synapse.
    d. synaptic vesicles become clogged.
22. Choose the *incorrect* statement about ion channels:
    a. An ion channel is a small pore (opening) through the plasma membrane.
    b. An ion channel is usually specific for one type of ion.
    c. An ion channel uses cellular energy to pump sodium ions (Na^+) out of the cell and potassium ions (K^+) into the cell.
    d. The opening and closing of ion channels are important in the generation of an action potential.
23. What is the initial (first) event of an action potential?
    a. depolarization caused by potassium ions rushing to the inside
    b. repolarization due to rushing of potassium ions to the inside
    c. depolarization due to rushing of sodium ions to the inside
    d. repolarization due to departure of potassium ions from the axon
24. Repolarization
    a. establishes the threshold voltage.
    b. is due to movement of potassium ions out of the axon.
    c. is due to movement of sodium ions into the axon.
    d. occurs as the sodium-potassium pump moves potassium out of the axon.
25. The sodium-potassium pump
    a. uses energy to move sodium ions out of the cell and potassium ions into the cell.
    b. can move sodium ions only from an area where they are highly concentrated to an area where they are less concentrated.
    c. can move potassium ions (K^+) only toward a region that is positively charged.
    d. is part of the myelin sheath.
26. Saltatory conduction
    a. occurs only in myelinated nerve fibers.
    b. is slower than other types of nerve conduction.
    c. occurs only in the central nervous system.
    d. is the way an impulse gets from one neuron to another.
27. The _____ is an insulating layer formed by Schwann cells that increases the rate of an action potential.
28. The _____ uses cellular energy to move sodium ions out of the axon and potassium ions into the axon.

APPLYING THE CONCEPTS

1. Indira has taken the sedative Valium, a drug that has a molecular shape similar to that of the neurotransmitter GABA (an inhibitory neurotransmitter). When Valium binds to GABA receptors, it causes the same effects as the binding of GABA itself. Based on this information, explain why Valium has a calming effect on the nervous system.
2. A nerve gas (diisopropyl fluorophosphate) that was once used in chemical warfare blocks the action of acetylcholinesterase, the enzyme that breaks down acetylcholine in the synapse. What effects would you expect this gas to have at the synapses that use acetylcholine?
3. Mohammed has a kidney condition that raises the level of potassium ions in the extracellular fluid that bathes cells. What effect would you expect this condition to have on his ability to generate nerve impulses (action potentials)?
4. Kerry is experiencing a weakness in her legs. A physician tells her that she has Guillain-Barré syndrome, which is a progressive but reversible condition in which the myelin sheath is lost from parts of the nervous system. Explain why the loss of myelin would cause weakness in Kerry's legs.

BECOMING INFORMATION LITERATE

Attention deficit/hyperactivity disorder (ADHD) is caused by a neurotransmitter imbalance. Write a short article for the health section of a local newspaper describing the symptoms, causes, and treatment of ADHD. Use at least three reliable sources (books, journals, or websites). List each source you considered, and explain why you chose the three sources you used.

MasteringBiology®

Go to MasteringBiology for practice quizzes, activities, eText, videos, current events, and more.

8 The Nervous System

- Your brain cannot feel pain, because it has no pain receptors.
- You have almost 45 miles of nerves in your body.

The nervous system integrates sensory and motor information with memories, allowing us to perform highly skilled activities.

- 8.1 Organization of the Nervous System
- 8.2 The Central Nervous System
- 8.3 The Peripheral Nervous System
- 8.4 Disorders of the Nervous System

HEALTH ISSUE

Brain Injury: A Silent Epidemic

In the previous chapter we considered the structure and function of neurons, the cells of the nervous system that communicate with one another and with muscles or glands. In this chapter, we explore the organization of the nervous system and the structures responsible for its many functions. We also discuss some disorders of the brain and spinal cord and their effects on the human body and mind.

8.1 Organization of the Nervous System

If you were to view the nervous system apart from the rest of the body, you would see a dense mass of neural tissue where the head should be, with a cord of neural tissue extending downward from it where the middle of the back should be. These structures are the brain and spinal cord, respectively, and they constitute the **central nervous system (CNS)**, which integrates and coordinates all voluntary and involuntary nervous functions (Figure 8.1). Connected to the brain and spinal cord are many communication "cables"—the nerves that carry messages to and from the CNS. The nerves branch extensively, forming a vast network. Some of their cell bodies are grouped together in small clusters called **ganglia** (singular, ganglion). The nerves and ganglia are located outside of the CNS and make up the **peripheral nervous system (PNS)**. The PNS keeps the CNS in continuous contact with almost every part of the body.

The peripheral nervous system can be further subdivided on the basis of function into the somatic nervous system and the autonomic nervous system. The **somatic nervous system** consists of nerves that carry information to and from the CNS, resulting in sensations and voluntary movement. The **autonomic nervous system**, in contrast, governs the involuntary, subconscious activities that keep the body functioning properly. The autonomic nervous system has two parts that generally produce opposite effects on the muscles or glands they control.

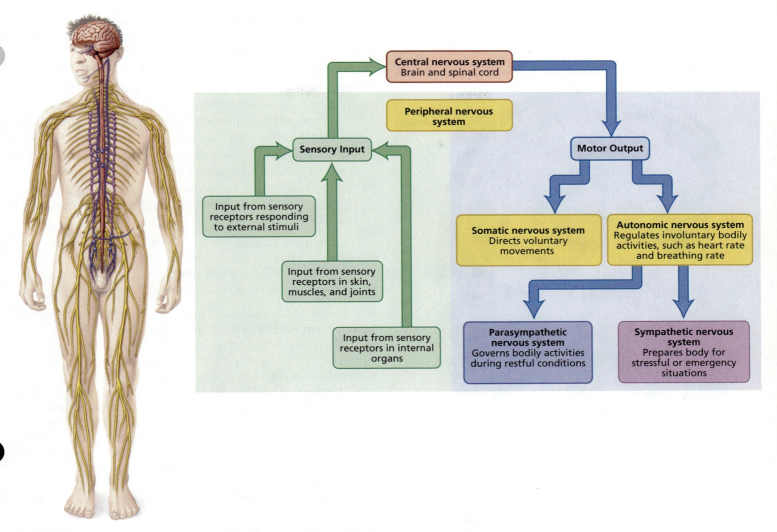

FIGURE 8.1 *An overview of the nervous system. The various parts of the human nervous system have special functions but work together as an integrated whole.*

One, the **sympathetic nervous system**, is in charge during stressful or emergency conditions. The other, the **parasympathetic nervous system**, adjusts bodily function so that energy is conserved during nonstressful times.

Although we have described the nervous system as having different parts and divisions, remember that all the parts function as a coordinated whole. Imagine for a moment that you are meditating in the park; your eyes are closed, and you are resting. While you are relaxing, the parasympathetic nervous system is ensuring that your life-sustaining bodily activities continue. Suddenly, someone grasps your hand. Sensory receptors in the skin (which are part of the somatic nervous system) respond to the pressure and warmth of the hand by sending messages over sensory nerves to the spinal cord. Neurons within the spinal cord relay the messages to the brain. The brain integrates incoming sensory information and "decides" on an appropriate response. For example, the brain may generate messages that cause your eyes to open. If the sight of the person holding your hand produces strong emotion, the sympathetic nervous system may speed up your heartbeat and perhaps even your breathing.

8.2 The Central Nervous System

The central nervous system includes the brain and spinal cord, which are made up of many closely packed neurons. Neurons are very fragile, and most cannot divide and produce new cells. Therefore, with few exceptions, a neuron that is damaged or dies cannot be replaced.

Protection of the Central Nervous System

The brain and spinal cord are protected by bony cases (the skull and vertebral column), membranes (the meninges), and a fluid cushion (cerebrospinal fluid).

The meninges The **meninges** are three protective connective tissue coverings of the brain and spinal cord (Figure 8.2). The outermost layer, the *dura mater,* is tough and leathery. Beneath the dura mater is the *arachnoid* (Latin, meaning "like a cobweb"). The arachnoid is anchored to the next-lower layer of meninges by thin, threadlike extensions that resemble a spider's web (hence the name of the layer). The innermost layer, the *pia mater,* is molded around the brain. Fitting like

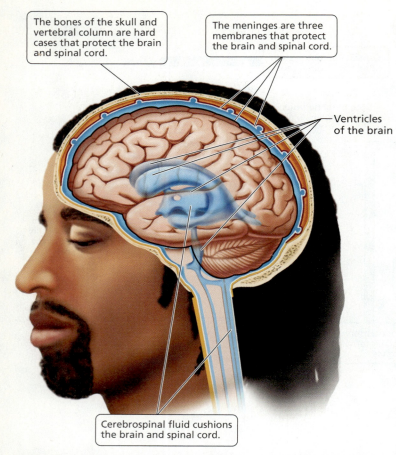

FIGURE 8.2 The central nervous system is protected by the meninges, the cerebrospinal fluid, and the bones of the skull and vertebral column.

a leotard, the pia mater dips into every indentation on the brain's surface.

Meningitis is an inflammation of the meninges. All cases of meningitis must be taken seriously because the infection can spread to the underlying nervous tissue and cause encephalitis (inflammation of the brain), which can be deadly. Many types of bacteria and certain viruses can cause meningitis. If bacteria are the cause, the person is treated with antibiotics. If a virus is the cause, treatment includes medicines to alleviate pain and fever while the body's immune system fights the virus.

Freshmen college students housed in dormitories are at increased risk of getting bacterial meningitis because of their close living quarters. Part of the reason is the means by which the bacteria are spread. People can carry the bacteria in their throat without having any symptoms of illness and can spread the infection through coughing, sneezing, or intimate kissing. Upperclassmen are less susceptible, perhaps because they have built up immune defenses against the bacteria. Vaccines are available against most, but not all, forms of meningitis. Some colleges are now requiring that incoming freshmen be vaccinated against some of the most common forms of meningitis.

Cerebrospinal fluid The **cerebrospinal fluid** fills the space between layers of the meninges as well as the internal cavities of the brain, called *ventricles*, and the cavity within the spinal cord, called the *central canal*. This fluid is formed in the ventricles and circulates from them through the central canal. Eventually, cerebrospinal fluid is reabsorbed into the blood.

Cerebrospinal fluid carries out several important functions:

- **Shock absorption.** Just as an air bag protects the driver of a car by preventing impact with the steering wheel, the cerebrospinal fluid protects the brain by cushioning its impact with the skull during blows or other head trauma.
- **Support.** Because the brain floats in the cerebrospinal fluid, it is not crushed under its own weight.
- **Nourishment and waste removal.** The cerebrospinal fluid delivers nutrients and chemical messengers and removes waste products.

The brain is, indeed, protected by the skull, meninges, and cerebrospinal fluid. Nonetheless, 7 million brain injuries occur annually in the United States. We discuss brain injury further in the Health Issue essay, *Brain Injury: A Silent Epidemic*.

The blood–brain barrier The CNS is also protected by the **blood–brain barrier**, a mechanism that selects the substances permitted to enter the cerebrospinal fluid from the blood. This barrier is formed by the tight junctions between the cells of the capillary walls that supply blood to the brain and spinal cord. Because the cells are held together much more tightly than are cells in capillaries in the rest of the body, substances in the blood are forced to pass through the cells of the capillaries instead of between the cells. Thus, the membranes of the capillary cells filter and adjust the composition of the filtrate by selecting the substances that can leave the blood. The plasma membranes of the capillary walls are largely lipid. So, lipid-soluble substances, including oxygen and carbon dioxide, can pass through easily. Certain drugs, including caffeine and alcohol, are lipid soluble, explaining why they can have a rapid effect on the brain. However, the blood–brain barrier prevents many potentially life-saving, infection-fighting, or tumor-suppressing drugs that are not lipid soluble from reaching brain tissue, which frustrates physicians.

Brain: Command Center

In a sense, your brain is more "you" than is any other part of your body, because it holds your emotions and the keys to your personality. Yet, if you were to look at your brain, you might not recognize it as yourself. The brain is the consistency of soft cheese and weighs less than 1600 g (3 lb), which is probably less than 3% of your body weight. Nevertheless, it is the origin of your secret thoughts and desires; it remembers your most embarrassing moment; and it keeps all your body systems functioning harmoniously while your conscious mind concentrates on other activities. Let's look at how its many circuits are organized to perform these amazing feats.

Cerebrum The **cerebrum** is the largest and most prominent part of the brain. It is, quite literally, your "thinking cap." Accounting for 83% of the total brain weight, the cerebrum gives you most of your human characteristics.

The many ridges and grooves on the surface of the cerebrum make it appear wrinkled. Some furrows are deeper than others. The deepest indentation is in the center and runs from front to back. This groove, called the *longitudinal fissure*, separates the cerebrum into two hemispheres. Each hemisphere receives sensory information from and directs the movements of the opposite side of the body. In addition, the hemispheres process information in slightly different ways and are, therefore, specialized for slightly different mental functions (Figure 8.3).

The thin outer layer of each hemisphere is called the **cerebral cortex**. (*Cortex* means "bark" or "rind.") The cerebral cortex consists of billions of neuroglial cells, nerve cell bodies, and unmyelinated axons and is described as **gray matter**. Although the cerebral cortex is only about 2.5 mm (about one-eighth inch) thick, it is highly folded. These folds, or convolutions, triple the surface area of the cortex.

Beneath the cortex is the cerebral **white matter**, which appears white because it consists primarily of myelinated axons. Recall from Chapter 7 that myelin sheaths increase the rate of conduction along axons and are, therefore, found on axons that conduct information over long distances. The axons of the cerebral white matter allow various regions of the brain to communicate with one another and with the spinal cord. A very important band of white matter, called the *corpus callosum*, connects the two cerebral hemispheres so they can communicate with one another.

Other grooves on the surface of the brain mark the boundaries of four lobes on each hemisphere: the *frontal, parietal, temporal,* and *occipital lobes* (Figure 8.4). Each of these lobes has its own specializations. Although the assignment of a specific function to a particular region of the cerebral cortex is imprecise, it is generally agreed that there are three types of functional areas: sensory, motor, and association.

Sensory Areas Our awareness of sensations depends on the sensory areas of the cerebral cortex. The various sensory receptors of the body send information to the cortex, where each sense is processed in a different region. If you stand on a street corner watching a parade go by, you hear the band play because information from your ears is sent to the auditory area in the temporal lobe. You see the flags wave because information from your eyes is sent to the visual area in the occipital lobe. When you catch a whiff of popcorn, information is sent from the olfactory (smell) receptors in your nose to the olfactory area in the temporal lobe of the cortex. As

Q *How do the left and right cerebral hemispheres communicate with one another?*

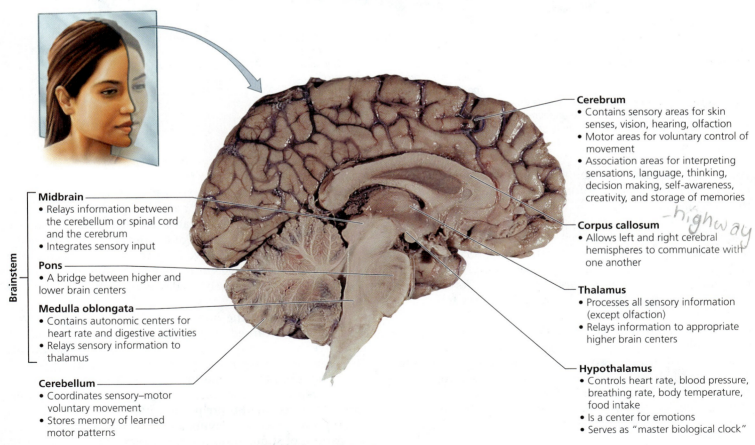

FIGURE 8.3 *A section through the brain from front to back, indicating the functions of selected structures*

A *The corpus callosum connects the hemispheres, allowing them to communicate.*

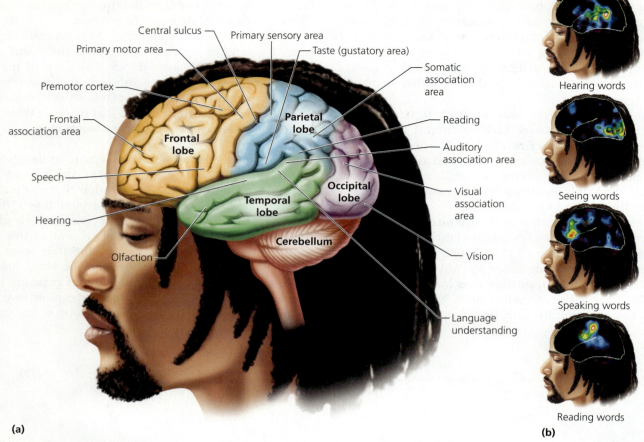

FIGURE 8.4 The cerebral cortex. (a) The cerebral cortex has four lobes. Some of the functions associated with each lobe are indicated. (b) These PET scans of the brain show regions of increased blood flow during different mental activities. The increased flow of blood shows which region becomes active when the cerebrum is engaged in hearing words, seeing words, speaking words, and reading words. Notice the relationship between active regions of the cerebral cortex during these tasks and the cortical areas for various language skills shown in part (a).

you eat that popcorn, you know it is too salty because information from the taste receptors is sent to gustatory areas in the parietal lobe.

Still watching the parade, you know that you are standing in the hot sun and that your belt is too tight because information from touch, pain, and temperature receptors in the skin and from receptors in the joints and skeletal muscles is sent to the **primary somatosensory area**. This region forms a band in the parietal lobes that stretches over the cortex from ear to ear (Figure 8.5). Sensations from different parts of the body are represented in different regions of the primary somatosensory area (of the hemisphere on the opposite side of the body). The greater the degree of sensitivity, the greater the area of cortex devoted to that body part. Thus, your most sensitive body parts, such as the tongue, hands, face, and genitals, have more of the cortex devoted to them than do less sensitive areas, such as the forearm.

Motor Areas If you decide to join the parade, the **primary motor area** (Figure 8.5) of the cerebral cortex will send messages to your skeletal muscles. This motor area controls voluntary movement. It also forms a band in the frontal lobe that stretches over the cortex, just anterior to the primary somatosensory area. The motor area is organized in a manner similar to the somatosensory area. Each point on its surface corresponds to the movement of a different part of the body. The parts of the body we have finer control over, such as the tongue and fingers, have greater representation on the motor cortex than do regions with less dexterity, such as the trunk of the body.

Just in front of the motor cortex is the *premotor cortex*. It coordinates learned motor skills that are patterned or repetitive, such as typing or playing a musical instrument. The premotor cortex coordinates the movement of several muscle groups at the same time. When a pattern of movement is repeated many times, the proper pattern of stimulation is stored in the premotor cortex. For example, as a guitar player practices playing a particular song many times, the pattern of stimulation needed to play that song is stored in the premotor cortex. Then, each time the song is played, the premotor cortex will stimulate the primary motor cortex in the pattern needed to play that song, without requiring

Q Our lips are more sensitive than is the skin on our forearm. We also have greater motor control of our lips than we do of our forearm. How is this difference in sensitivity and motor control represented on the cerebral cortex?

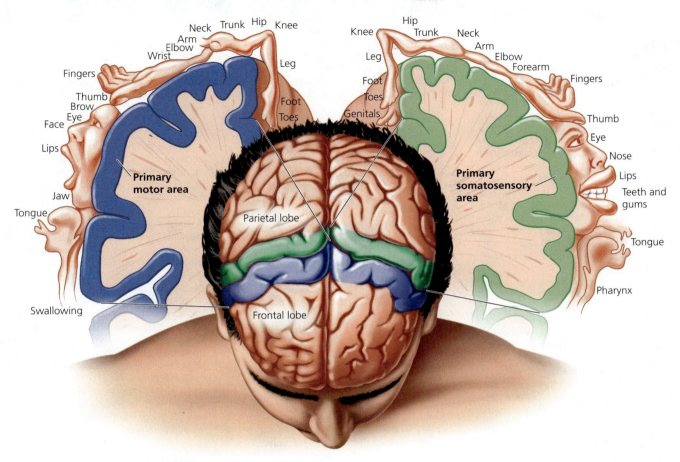

FIGURE 8.5 The primary motor and the primary somatosensory regions of the cerebral cortex are organized in such a way that each location on their surfaces corresponds to a particular part of the body. The general arrangement is similar in the two regions.

A Areas with greater sensitivity and motor control have a proportionally larger area of representation on the cortex than do areas with less sensitivity and motor control, such as the forearm.

the musician to think about where on the strings the fingers should be placed.

Association Areas Next to each primary sensory area is an association area. These communicate with the sensory and motor areas, and with other parts of the brain, to analyze and act on sensory input. In particular, each sensory association area communicates with the general interpretation area to recognize what the sensory receptors are sensing. The general interpretation area assigns meaning to sensory information by integrating the input from sensory association areas with stored sensory memories. For example, on a dark night your eyes may detect a small, moving object. If the object then rubs against your legs and purrs, your general interpretation area will assist you in recognizing it as the neighbor's friendly cat. However, if the object turns away from you and raises its tail, you will recognize it as a skunk.

Once the sensory input has been interpreted, the information is sent to the most complicated of all association areas, the **prefrontal cortex**. This most anterior part of the frontal lobe predicts the consequences of various possible responses to the information it receives and decides which response will be best for you in your current situation. The prefrontal cortex enables us to reason, plan for the long term, and think about abstract concepts. It also plays a key role in determining our personality.

Thalamus The cerebral hemispheres sit comfortably over the **thalamus** (see Figure 8.3). The thalamus is often described as the gateway to the cerebral cortex because all messages *to* the cerebral cortex must pass *through* the thalamus first. The thalamus functions in sensory experience, motor activity, stimulation of the cerebral cortex, and memory. Sensory input from every sense except smell and from all parts of the body is delivered to the thalamus. The thalamus sorts the information by function and relays it to appropriate regions of the cortex for processing. Some regions of the thalamus also integrate information from different sources rather than just relaying it.

HEALTH ISSUE

Brain Injury: A Silent Epidemic

Most of us take our brain for granted. We assume that this fragile control center is safe from harm, safely guarded by the thick bones of our skulls and cushioned by cerebrospinal fluid. The truth is that the brain is more vulnerable than we may think, and injuries to this vital organ are frighteningly common. Brain injury has been termed a "silent epidemic." It is silent because a brain-injured person doesn't have visible physical symptoms. It is an epidemic because it is so common. One in every 220 people in the United States is suffering from a brain injury. A brain injury occurs every 16 seconds; a death from head injury occurs every 12 minutes.

Brain injuries are categorized as either acquired or traumatic. *Acquired brain injury (ABI)* is caused by a disruption in oxygen flow to the brain. Examples of ABIs include strokes and aneurysms, heart attacks, brain tumors, anoxia, meningitis, seizure disorders, and substance abuse. There is a strong correlation between substance abuse and acquired brain injury because alcohol and other substances are neurotoxins that cause damage to the brain with repeated use. Furthermore, substance abuse is associated with poor nutrition, which can cause dehydration and ultimately wastes brain cells.

Traumatic brain injury, or *TBI*, is caused by an external force. There are two types of TBI: open and closed. An open head injury occurs when the scalp is cut through and the skull is broken, damaging the brain underneath. A closed head injury happens when the head suddenly changes motion, forcing the brain to follow the movement, such as when a car stops very suddenly. The brain is soft and jellylike, and it sits snugly within the skull. Sudden movement of the head can cause it to ricochet within the skull, damaging the millions of nerve fibers that run from one part of the brain to another. Also, the inside of the skull has many

To reduce your risk of traumatic brain injury, always wear a helmet when cycling.

ridges and sharp edges that can cut or bruise the brain. Common causes of TBI include motor vehicle accidents, gunshots, brawls, slip-and-fall accidents, and accidents related to sports such as skiing. Substance abuse is also associated with TBI, as the impairment caused by alcohol and drugs can lead to vehicular accidents and increase risk of falls and physical altercations. You can protect yourself against TBI by wearing a helmet when biking, skiing, or engaging in any other sport where a fall is likely. Motor vehicle accidents cause nearly half of all head injuries, so please buckle up!

TBI is getting a good deal of attention recently because nearly two-thirds of injured U.S. soldiers sent from Iraq to Walter Reed Medical Center have been diagnosed with traumatic brain injury. That percentage, thought to be higher than in any other past U.S. conflict, is said to be due to improved armor that allows soldiers to survive injuries that previously would have been fatal. Also, compared to other wars, fewer firearms and more improvised explosive devices (IEDs) are being used in combat; the intense vibrations from these explosives cause the brain to move within the soldiers' skulls.

Regardless of cause, no two brain injuries are the same. The symptoms are diverse and vary widely based on severity and location of injury as well as the individual's functioning before the accident. Symptoms frequently include cognitive and emotional limitations, including difficulties with memory, attention, and reasoning; depression; anxiety; and impulse control and anger management issues. Physical impairments are common and can range from weakness on one side of the body to paralysis.

Questions to Consider

- Many states have laws requiring motorcyclists and bicyclists to wear helmets. Do you think cyclists riding without helmets should be fined?
- Do you think that skiers and snowboarders should be allowed on the slopes without helmets?

---

 what would you do?

An increasing number of advertisers use neuromarketing to measure consumers' interest in their product. Neuromarketers use a brain scanner consisting of a cap containing electrodes to measure brain activity in all parts of a consumer's brain when he is presented with a product or advertisement. By noting the pattern of brain activity, a neuromarketer can determine a person's interest in, emotional response to, and perhaps even memories associated with a product. This information can then be used to manipulate desire for the product. Do you think neuromarketing is an invasion of privacy? Should it be legal?

At the thalamic level of processing, you have a general impression of whether the sensation is pleasant or unpleasant. If you step on a tack, for instance, you may experience pain by the time the messages reach the thalamus; however, you will not know where you hurt until after the message is directed to the cerebral cortex.

Hypothalamus Below the thalamus is the **hypothalamus** (*hypo*, under), a small region of the brain that is largely responsible for homeostasis—the body's maintenance of a stable environment for its cells (discussed in Chapter 4). The hypothalamus, shown in Figure 8.3, coordinates the activities of the nervous and endocrine (hormonal) systems through its influence on the pituitary gland. The hypothalamus also influences blood pressure, heart rate, digestive activity, breathing rate, and many other vital physiological processes. It keeps body temperature near the set point, and it regulates hunger and thirst and therefore the intake of food. Moreover, because the hypothalamus receives input from the cerebral cortex, it can make your heart beat faster when you so much as see or think of something exciting or dangerous—a rattlesnake about to strike, for instance.

The hypothalamus is part of the limbic system (discussed later in this chapter), so it is also part of the circuitry for emotions. Specific regions of the hypothalamus play a role in the sex drive and in the perception of pain, pleasure, fear, and anger.

Cerebellum The **cerebellum** (see Figure 8.3) is the part of the brain responsible for sensory–motor coordination. It acts as an automatic pilot that produces smooth, well-timed voluntary movements and controls equilibrium and posture. Sensory information concerning the position of joints and the degree of tension in muscles and tendons is sent to the cerebellum from all parts of the body. By integrating this information with input from the eyes and the equilibrium receptors in the ears, the cerebellum knows the body's position and direction of movement at any given instant.

The coordination of sensory input and motor output by the cerebellum involves two important processes: comparison and prediction. During every move you make, the cerebellum continuously compares the actual position of each part of the body with where it *ought* to be at that moment (in relation to the intended movement) and makes the necessary corrections. Try to touch the tips of your two index fingers together above your head. You probably missed on the first attempt. However, the cerebellum makes the necessary corrections, and you will likely succeed on the next attempt. At the same time, the cerebellum calculates future positions of a body part during a movement. Then, just before that part reaches the intended position, the cerebellum sends messages to stop the movement at a specific point. Therefore, when you scratch an itch on your cheek, your hand stops before slapping your face!

Brain stem The brain stem consists of the medulla oblongata, the midbrain, and the pons. The **medulla oblongata** is often called simply the medulla (see Figure 8.3). This marvelous inch of nervous tissue contains reflex centers for some of life's most vital physiological functions—including the pace of the basic breathing rhythm, the force and rate of heart contraction, and blood pressure. The medulla connects the spinal cord to the rest of the brain. Therefore, all sensory information going to the upper regions of the brain and all motor messages leaving the brain are carried by nerve pathways running through the medulla.

The **midbrain** processes information about sights and sounds and controls simple reflex responses to these stimuli. For example, when you hear an unexpected loud sound, your reflexive response is to turn your head and direct your eyes toward the source of the sound.

The **pons**, which means "bridge," connects lower portions of the CNS with higher brain structures. More specifically, it connects the spinal cord and cerebellum with the cerebrum, thalamus, and hypothalamus. In addition, the pons has a region that assists the medulla in regulating respiration.

Limbic system The **limbic system** is a collective term for a group of structures that help to produce emotions and memory (Figure 8.6). The limbic system is defined on the basis of function rather than anatomy, and it includes parts of several brain regions and the neural pathways that connect them.

The limbic system is our "emotional" brain. It allows us to experience countless emotions, including rage, pain, fear, sorrow, joy, and sexual pleasure. Emotions are important because they motivate behavior that will increase the chance of survival. Fear, for example, may have evolved to focus the mind on the threats in the environment so it can prepare the body to face them.

Connections between the cerebrum and the limbic system allow us to have *feelings* about *thoughts*. As a result, you may become excited at the thought of winning the lottery. Such connections also allow us to have thoughts about feelings, thus keeping us from responding to emotions, such as rage, in ways that would be unwise. The limbic system includes the

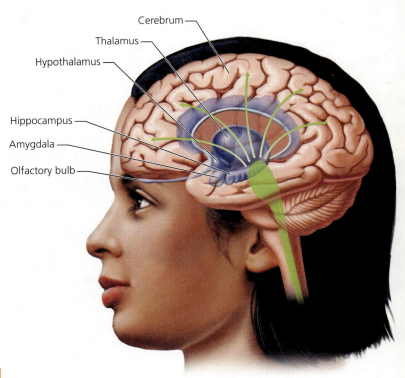

FIGURE 8.6 *The limbic system and reticular activating system. The diagram shows the limbic system in purple as a three-dimensional structure within the brain, viewed from the left side. The reticular activating system is shown in green. Note the upward arrows.*

stop and think

Why would a brain tumor that destroyed the functioning of nerve cells in the medulla lead to death more quickly than a tumor of the same size on the cerebral cortex?

hypothalamus, as Figure 8.6 shows. In addition, it is connected to lower brain centers, such as the medulla, that control the activity of internal organs. Therefore, we also have "gut" responses to emotions.

You wouldn't be you without your memories, and the limbic system plays a role in forming them. Memory, the storage and retrieval of information, takes place in two stages. The first is **short-term memory**, which holds a small amount of information for a few seconds or minutes, as when you look up a phone number and remember it only long enough to place the call. The second stage, **long-term memory**, stores seemingly limitless amounts of information for hours, days, or years. Not all short-term memories get consolidated into long-term memories, but when they do, the **hippocampus** plays an essential role. The *amygdala,* another part of the limbic system that functions in long-term memory, has widespread connections to sensory areas as well as to emotion centers. It associates memories gathered through different senses and links them to emotional states.

The olfactory bulb transmits information about odors from the nose to the limbic system. Thus, the limbic system is a center where emotions, memory, and our sense of smell meet. As a result, we often have emotional responses to odors. The association between odor and emotion is the basis of aromatherapy as well as the perfume and scented candle industries. The interaction of emotion, sense of smell, and memory explains why odors can bring back memories. For example, the smell of cinnamon rolls may be pleasant because it reminds you of your grandmother baking special treats.

Reticular activating system The **reticular activating system (RAS)** is an extensive network of neurons that runs through the medulla and projects to the cerebral cortex (shown in Figure 8.6 in green). The RAS functions as a net, or filter, for sensory input. Our brain is constantly flooded with tremendous amounts of sensory information, about 100 million impulses each second, most of them trivial. The RAS filters out repetitive, familiar stimuli—the sound of street traffic, paper rustling, the coughing of the person next to you, or the pressure of clothing. However, infrequent or important stimuli pass through the RAS to the cerebral cortex and, therefore, reach our consciousness. Because of the RAS, you can fall asleep with the television on but wake up when someone whispers your name.

In addition, the RAS is an activating center. Unless inhibited by other brain regions, the RAS activates the cerebral cortex, keeping it alert and "awake." Consciousness occurs only while the RAS stimulates the cerebral cortex. When sleep centers in other regions of the brain inhibit activity in the RAS, we sleep. In essence, then, the cerebrum "sleeps" whenever it is not stimulated by the RAS. Sensory input to the RAS results in stimulation of the cerebral cortex and an increase in consciousness, which explains why it is usually easier to sleep in a dark, quiet room than in an airport terminal. Conscious activity in the cerebral cortex can also stimulate the RAS, which in turn will stimulate the cerebral cortex. Therefore, thinking about a problem may keep you awake all night.

stop and think

When a boxer is hit very hard in the jaw, his head—containing his medulla and RAS—is twisted sharply. Why might this twisting result in a knockout, in which the boxer loses consciousness?

Spinal Cord: Message Transmission and Reflex Center

The other major component of the central nervous system besides the brain is the spinal cord. The **spinal cord** is a tube of neural tissue that is continuous with the medulla at the base of the brain and extends about 45 cm (17 in.) to just below the last rib. For most of its length, the spinal cord is about the diameter of your little finger. It becomes slightly thicker in two regions, just below the neck and at the end of the cord, because of the large group of nerves connecting these regions of the cord with the arms and legs. The central canal, filled with cerebrospinal fluid, runs the length of the spinal cord.

The spinal cord is encased in and protected by the stacked bones of the vertebral column (Figure 8.7). Pairs of spinal nerves (considered part of the peripheral nervous system) extend from the spinal cord through openings between the vertebrae to serve different parts of the body. The vertebrae are separated by disks of cartilage that act as cushions.

The spinal cord has two functions: (1) to transmit messages to and from the brain and (2) to serve as a reflex center. The transmission of messages is performed primarily by white matter, found toward the outer surface of the spinal cord. White matter in the spinal cord consists of myelinated axons grouped into tracts. Ascending tracts carry sensory information up to the brain. Descending tracts carry motor information from the brain to a nerve leaving the spinal cord.

The second function of the spinal cord is to serve as a reflex center. A reflex is an automatic response to a stimulus, prewired in a circuit of neurons called a **reflex arc**. The circuit consists of a receptor, a sensory neuron (which brings information from the receptors toward the CNS), usually at least one interneuron, a motor neuron (which brings information from the CNS toward an effector), and an effector (a muscle or a gland). The gray matter, which is located in the central region of the spinal cord, houses the interneurons and the cell bodies of motor neurons involved in reflexes.

Spinal reflexes are essentially "decisions" made by the spinal cord. They are beneficial when a speedy reaction is important to a person's safety. Consider, for example, the withdrawal reflex. When you step on a piece of broken glass, impulses speed toward the spinal cord over sensory nerves (Figure 8.8). Within the gray matter of the spinal cord, the sensory neuron synapses with an interneuron. The interneuron, in turn, synapses with a motor neuron that sends a message to the appropriate muscle to contract and lift your foot off the glass.

While the spinal reflexes were removing the foot from the glass, pain messages from the cut foot were sent to the brain through ascending tracts in the spinal cord. However, it takes longer to get a message to the brain than it does to get one to the spinal cord, because the distance and number of synapses

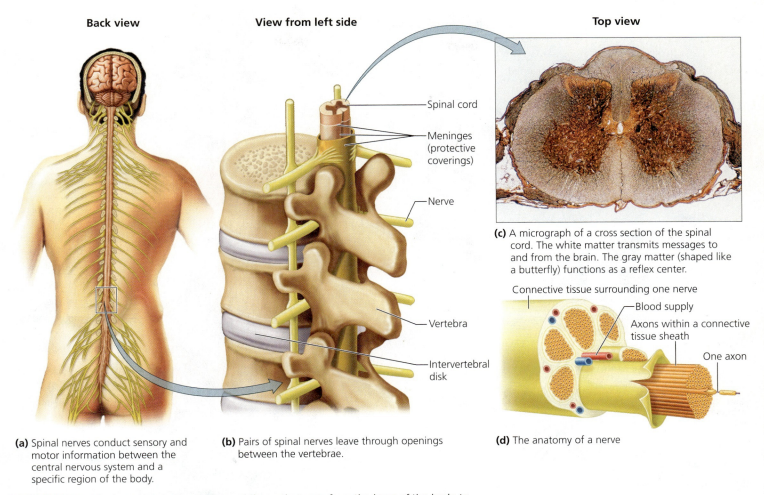

FIGURE 8.7 *The spinal cord is a column of neural tissue that runs from the base of the brain to just below the last rib. It is protected by the bones of the vertebral column.*

to be crossed are greater. Therefore, by the time pain messages reach the brain, you have already withdrawn your foot. Nonetheless, once the sensory information reaches the conscious brain, decisions can be made about how to care for the wound.

8.3 The Peripheral Nervous System

The nerves and ganglia of the PNS carry information between the CNS and the rest of the body. The PNS consists of spinal nerves and cranial nerves.

The body has 31 pairs of **spinal nerves**, each of which originates in the spinal cord and services a specific region of the body. One member of each pair serves a part of the right side of the body, and the other serves the corresponding part of the left side (Figure 8.9a). All spinal nerves carry both sensory and motor fibers. Fibers from the sensory neurons enter the spinal cord from the dorsal, or posterior, side, grouped into a bundle called the *dorsal root*. The cell bodies of these sensory neurons are located in a ganglion in the dorsal root. The axons of motor neurons leave the ventral (front side) of the spinal cord in a bundle called the *ventral root*. The cell bodies of motor neurons are located in the gray matter of the spinal cord. The dorsal and ventral roots join to form a single spinal nerve, which passes through the opening between the vertebrae.

The 12 pairs of **cranial nerves** (Figure 8.9b) arise from the brain and service the structures of the head and certain body parts, including the heart and diaphragm. Some cranial nerves carry only sensory fibers, others carry only motor fibers, and others carry both types of fibers.

Somatic Nervous System

The peripheral nervous system is subdivided into the somatic nervous system and the autonomic nervous system. The somatic nervous system carries sensory messages that tell us about the world around us and within us, and it controls movement. Sensory messages carried by somatic nerves result in conscious sensations, including light, sound, and touch. The somatic nervous system also controls our voluntary movements, allowing us to smile, stamp a foot, sing a lullaby, or frown as we sign a check.

Autonomic Nervous System

As part of the body's system of homeostasis, the autonomic nervous system automatically adjusts the functioning of our body organs so that the proper internal conditions are maintained and the body is able to meet the demands of the world around it. The somatic nervous system sends information about conditions within the body to the autonomic nervous

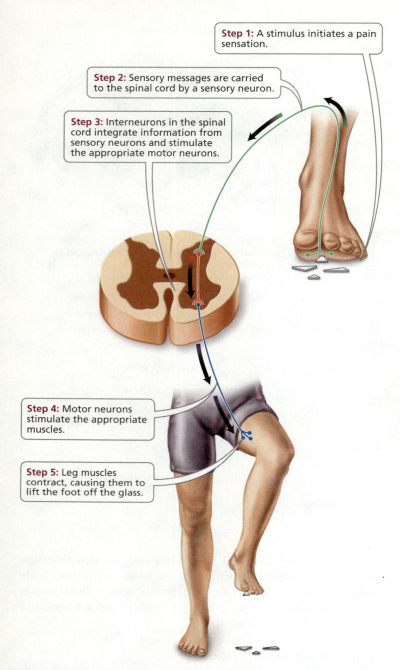

FIGURE 8.8 A reflex arc consists of a sensory receptor, a sensory neuron, usually at least one interneuron, a motor neuron, and an effector.

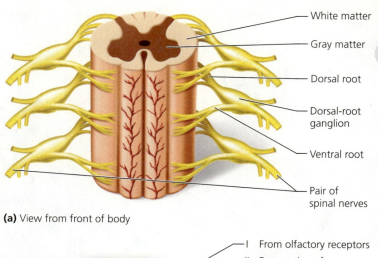

(a) View from front of body

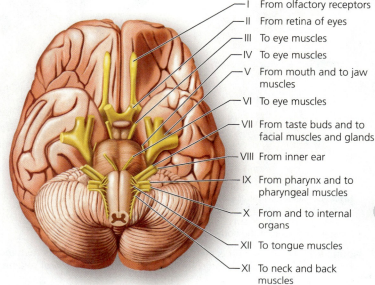

(b) View of underside of brain

FIGURE 8.9 (a) Spinal and (b) cranial nerves. The 12 pairs of cranial nerves can be seen in this view of the underside of the brain. Most cranial nerves service structures within the head, but some service organs lower in the body. The descriptions indicate whether the neuron carries sensory information (toward the brain) or motor information (away from the brain or both).

system. The autonomic nervous system then makes the appropriate adjustments. Its activities alter digestive activity, open or close blood vessels to shunt blood to areas that need it most, and alter heart rate and breathing rate.

Recall that the autonomic nervous system consists of two branches: the sympathetic and the parasympathetic nervous systems. The sympathetic nervous system gears the body to face an emergency or stressful situation, such as fear, rage, or vigorous exercise. Thus, the sympathetic nervous system prepares the body for fight or flight. In contrast, the parasympathetic nervous system adjusts body function so that energy is conserved during relaxation.

Both the parasympathetic and the sympathetic nervous systems send nerve fibers to most, but not all, internal organs (Figure 8.10). When both systems send nerves to a given organ, they have opposite, or antagonistic, effects on its function. If one system stimulates, the other system inhibits. The antagonistic effects are brought about by different neurotransmitters. Whereas sympathetic neurons release mostly norepinephrine at their target organs, parasympathetic neurons release acetylcholine at their target organs.

The sympathetic nervous system acts as a unified whole, bringing about all its effects at once. It is able to act in this way because its neurons are connected through a chain of ganglia. A unified response is exactly what is needed in an emergency. To meet a threat, the sympathetic nervous system increases breathing rate, heart rate, and blood pressure. It also

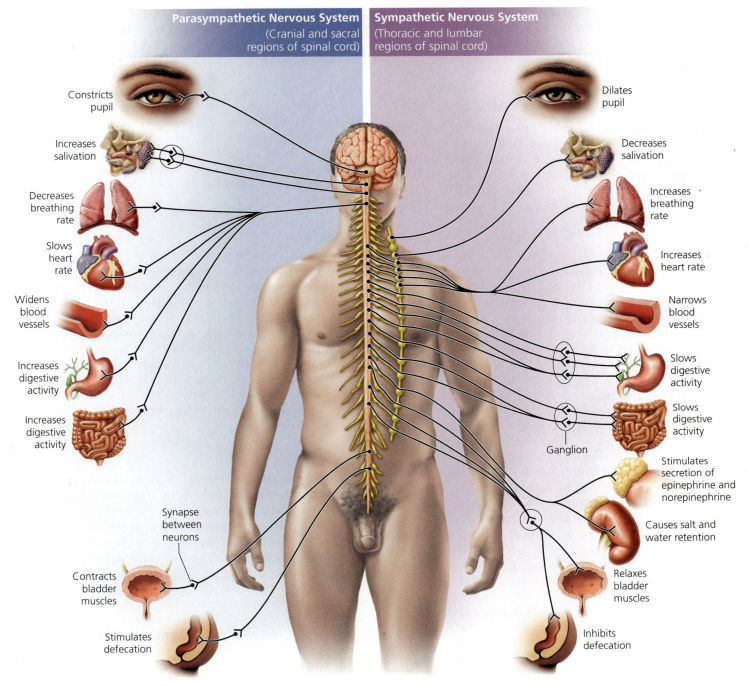

FIGURE 8.10 Structure and function of the autonomic nervous system. Most organs are innervated by fibers from both the sympathetic and the parasympathetic nervous systems. When this dual innervation occurs, the two branches of the autonomic nervous system have opposite effects on the activity level of that organ. A chain of ganglia links the pathways of the sympathetic nervous system, which therefore usually acts as a unit, with all its effects occurring together. In contrast, the ganglia of the parasympathetic nervous system are each near the organ they service, so parasympathetic effects are more localized.

increases the amount of glucose and oxygen delivered to body cells to fuel the response. In addition, it stimulates the adrenal glands to release two hormones, epinephrine and norepinephrine, into the bloodstream. These hormones back up and prolong the other effects of sympathetic stimulation. Lastly, the sympathetic nervous system inhibits digestive activity, because digesting the previous meal is hardly a priority during a crisis.

The effects of the parasympathetic nervous system occur more independently of one another. After the emergency, organ systems return to a relaxed state at their own pace. Organs can respond to the parasympathetic nervous system independently because the ganglia containing the parasympathetic neurons that stimulate each organ are located near the individual organs—not in a chain near the spinal cord, as they are in the sympathetic nervous system.

8.4 Disorders of the Nervous System

Disorders of the nervous system vary tremendously in severity and impact on the body. Some disorders, such as a mild headache, are often more of a nuisance than a health problem. Others, such as insufficient sleep, can cause more problems than a person might expect. Still other disorders, such as stroke, coma, and spinal cord injury, can have devastating effects on a person's well-being.

Headaches

Excessive exercise may make your muscles hurt. However, thinking too much cannot cause a headache. The brain has no pain receptors, so a headache is not a brain ache. Headaches can occur for almost any reason: they can be caused by stress or by relaxation, by hunger or by eating the wrong food, or by too much or too little sleep. The most common type of headache is a *tension headache,* affecting some 60% to 80% of people who suffer from frequent headaches. In response to stress, most of us unconsciously contract the muscles of our head, face, and neck. Therefore, the pain of a tension headache is usually a dull, steady ache, often described as feeling like a tight band around the head. *Migraine headaches* are usually confined to one side of the head, often centered behind one eye. A migraine headache typically causes a throbbing pain that increases with each beat of the heart. It is sometimes called a sick headache because it may cause nausea and vomiting. Some migraine sufferers experience an aura, a group of sensory symptoms, different for different people, that occurs just before an attack. The aura may include visual disturbances (a blind spot, zigzag lines, flashing lights), auditory hallucinations, or numbness. Though the causes of migraines are not entirely understood, some researchers believe that migraines are set off by an imbalance in the brain's chemistry. Specifically, the level of one of the brain's chemical neurotransmitters, serotonin, is low. With too little serotonin, pain messages flood the brain.

Strokes

A *stroke,* also called a *cerebrovascular accident,* is the death of nerve cells caused by an interruption of blood flow to a region of the brain. Neurons have a high demand for both oxygen and glucose. Therefore, when the blood supply to a portion of the brain is shut off, the affected neurons begin to die within minutes. The extent and location of the mental or physical impairment caused by a stroke depend on the region of the brain involved. If the left side of the brain is affected, the person may lose sensations in or the ability to move parts of the right side of his or her body because motor nerve pathways cross from one side of the brain to the other in the lower brain. Because the language centers are usually in the left hemisphere, the person may also have difficulty speaking. When the stroke damages the right rear of the brain, some people show what is called the neglect syndrome and behave as if the left side of things, even their own bodies, does not exist. The person may comb only the hair on the right side of the head or eat only the food on the right side of the plate.

Common causes of strokes include blood clots blocking a vessel, hemorrhage from the rupture of a blood vessel in one of the meninges, or the formation of fatty deposits that block a vessel. High blood pressure, heart disease, diabetes, smoking, obesity, and excessive alcohol intake increase the risk of stroke.

Coma

Although a comatose person seems to be asleep—with eyes closed and no recognizable speech—a coma is not deep sleep. A person in a *coma* is totally unresponsive to all sensory input and cannot be awakened. Although the cerebral cortex is most directly responsible for consciousness, damage to the cerebrum is rarely the cause of coma. Instead, coma is caused by trauma to neurons in regions of the brain responsible for stimulating the cerebrum, particularly those in the reticular activating system or thalamus. Coma can be caused by mechanical shock—as might be caused by a blow to the head—tumors, infections, drug overdose, or failure of the liver or kidney.

Spinal Cord Injury

The spinal cord is the pathway that allows the brain to communicate with the rest of the body. Therefore, damage to the spinal cord can impair sensation and motor control below the site of injury. The extent and location of the injury will determine how long these symptoms persist, as well as the degree of permanent damage. Depending on which nerve tracts are damaged, injury may result in paralysis, loss of sensation, or both. If the cord is completely severed, there is a complete loss of sensation and voluntary movement below the level of the cut.

Restoring the ability to function to people with spinal cord injuries is an active area of research. Some researchers are trying to reestablish neural connections by stimulating nerve growth through treatments with nerve growth factors. Others are exploring the potential use of stem cells for treatment (discussed in Chapter 19a). Stem cells retain the ability to develop into nerve cells, and they have been used successfully by researchers to restore some movement in laboratory mice with spinal cord injuries. Another approach to restoring the ability to move is to use computers to electronically stimulate specific muscles and muscle groups. The stimulation is delivered through wires that are either implanted under the skin or woven into the fabric of tight-fitting clothing. A small computer, usually worn at the wrist, directs the stimulation to the appropriate muscles. This technology has helped some people with a spinal cord injury to walk again. It has also helped some people by stimulating the diaphragm, a muscle important in breathing.

looking ahead

In Chapter 7, we learned that neurons communicate with one another using chemicals called neurotransmitters. Neurotransmitter molecules fit into receptors on the membrane of the receiving neuron and cause ion channels to open, either exciting or inhibiting the receiving neuron. Different neurotransmitters play roles in different behavioral systems. In this chapter, we learned that different parts of the nervous system are specialized for different functions. The limbic system of the brain is a "pleasure center." The sympathetic nervous system prepares the body for emergency situations.

Next, in Chapter 8a, "Special Topic: Drugs and the Mind," we will consider psychoactive drugs—those that affect a person's mental state. We will see that psychoactive drugs work by increasing or decreasing the effects of specific neurotransmitters and therefore affect specific regions of the brain.

HIGHLIGHTING THE CONCEPTS

8.1 Organization of the Nervous System (pp. 126–127)

- The nervous system is divided into the central nervous system (CNS), which includes the brain and spinal cord; and the peripheral nervous system (PNS), which includes all the neural tissue outside the CNS. The peripheral nervous system can be further subdivided into the somatic nervous system and the autonomic nervous system.

8.2 The Central Nervous System (pp. 127–135)

- The brain and the spinal cord are protected by the bony cases of the skull and vertebral column, by membranes (the meninges), and by a fluid cushion (cerebrospinal fluid).
- The meninges are three protective layers of connective tissue that cover the brain and spinal cord. Bacteria and viruses can cause inflammation of the meninges, resulting in the condition called meningitis.
- The cerebrospinal fluid, located between layers of the meninges, serves as a shock absorber for the brain, supports the brain, and provides nourishment to and removes waste from the brain.
- The blood–brain barrier is a filter that allows only certain substances to enter the cerebrospinal fluid from the blood, thus protecting the brain and spinal cord from many potentially damaging substances.
- The brain serves as the body's central command center, coordinating and regulating the body's other systems.
- The cerebrum is the thinking, conscious part of the brain. It consists of two hemispheres. Each hemisphere receives sensory impressions from and directs the movements of the opposite side of the body. The cerebrum has an outer layer of gray matter called the cerebral cortex and an underlying layer of white matter consisting of myelinated nerve tracts that allow communication between various regions of the brain.
- The cerebral cortex has three types of functional areas: sensory, motor, and association. Our awareness of sensation depends on the sensory areas of the cerebral cortex. Motor areas of the brain control the movement of different parts of the body. Association areas communicate with the sensory and motor areas to analyze and act on sensory input.
- The thalamus is an important relay station for all sensory experience except smell. It also plays a role in motor activity, stimulation of the cerebral cortex, and memory.
- The hypothalamus is essential in maintaining a stable environment within the body. It regulates many vital physiological functions, such as blood pressure, heart rate, breathing rate, digestion, and body temperature. The hypothalamus also coordinates the activities of the nervous and endocrine systems through its connection to the pituitary gland. As part of the limbic system, the hypothalamus is a center for emotions.
- The primary function of the cerebellum is sensory–motor coordination. It integrates information from the motor cortex and sensory pathways to produce smooth movements.
- The medulla oblongata regulates breathing, heart rate, and blood pressure. It also serves as a pathway for all sensory messages to higher brain centers and for motor messages leaving the brain.
- The pons connects lower portions of the CNS with higher brain structures. It connects the spinal cord and cerebellum to the cerebrum, thalamus, and hypothalamus.
- The limbic system, which includes several brain structures, is largely responsible for emotions. The hippocampus, which is part of the limbic system, is essential to converting short-term memory to long-term memory.
- The reticular activating system is a complex network of neurons that filters sensory input and keeps the cerebral cortex in an alert state.
- The spinal cord is a cable of nerve tissue extending from the medulla to approximately the bottom of the rib cage. The spinal cord has two functions: to conduct messages between the brain and the body and to serve as a reflex center.

8.3 The Peripheral Nervous System (pp. 135–138)

- The PNS consists of spinal nerves, each originating in the spinal cord and serving a specific region of the body, and cranial nerves, each arising from the brain and serving the structures of the head and certain body parts such as the heart and diaphragm.
- The PNS is divided into the somatic nervous system, which governs conscious sensations and voluntary movements, and the autonomic nervous system, which helps regulate our unconscious, involuntary internal activities.
- The autonomic nervous system can be divided into the sympathetic and parasympathetic nervous systems, two branches with antagonistic actions. The sympathetic nervous system gears the body to face stressful or emergency situations. The parasympathetic nervous system adjusts body functioning so that energy is conserved during restful times.

8.4 Disorders of the Nervous System (p. 138)

- Headaches can range from relatively mild (tension headaches) to severe (migraine).
- A stroke is caused by an interruption of blood flow leading to the death of nerve cells. The effects depend on the region of the brain affected.
- Coma, a condition in which a person is totally unresponsive to sensory input, can be caused by a blow to the head, tumors, infections, drugs, or failure of the liver or kidney.
- Because the spinal cord contains the pathways of communication between the brain and the rest of the body, damage to the spinal cord impairs functioning below the site of injury.

RECOGNIZING KEY TERMS

central nervous system p. 126
ganglia p. 126
peripheral nervous system p. 126
somatic nervous system p. 126
autonomic nervous system p. 126
sympathetic nervous system p. 127
parasympathetic nervous system p. 127
meninges p. 127
cerebrospinal fluid p. 128
blood–brain barrier p. 128
cerebrum p. 128
cerebral cortex p. 129
gray matter p. 129
white matter p. 129
primary somatosensory area p. 130
primary motor area p. 130
prefrontal cortex p. 131
thalamus p. 131
hypothalamus p. 132
cerebellum p. 133
medulla oblongata p. 133
midbrain p. 133
pons p. 133
limbic system p. 133
short-term memory p. 134
long-term memory p. 134
hippocampus p. 134
reticular activating system p. 134
spinal cord p. 134
reflex arc p. 134
spinal nerves p. 135
cranial nerves p. 135

REVIEWING THE CONCEPTS

1. Distinguish the central nervous system from the peripheral nervous system. List two components of each. p. 126
2. Describe three features that protect the brain and spinal cord. pp. 127–128
3. What are the functions of cerebrospinal fluid? p. 128
4. What forms gray matter? What is its function? What forms white matter? What is its function? p. 129
5. Describe the three types of functional areas of the cerebral cortex. pp. 129–131
6. In what way is the organization of the primary somatosensory area and that of the primary motor area of the cerebral cortex similar? In what way do these areas differ? pp. 129–130
7. List five functions of the hypothalamus. pp. 132–133
8. What is the function of the cerebellum? p. 133
9. Which functional system of the brain is responsible for emotions? pp. 133–134
10. You are cooking dinner and carelessly touch the hot burner on the stove. You remove your hand before you are even aware of the pain. Using the anatomy of a spinal reflex arc, explain how you could react before you were aware of the pain. pp. 134–135
11. What are the two divisions of the peripheral nervous system? What type of response does each control? pp. 135–138
12. The three protective membranes that surround the brain and spinal cord form the
    a. meninges.
    b. white matter.
    c. blood–brain barrier.
    d. reticular activating system.
    e. corpus callosum.
13. The autonomic nervous system
    a. cannot be affected by emotional states.
    b. controls conscious body movements.
    c. is entirely contained within the central nervous system.
    d. automatically adjusts the functioning of internal organs to suit conditions.
14. Which of the following lists the parts of a reflex arc in the correct sequence?
    a. receptor, sensory neuron, motor neuron, interneuron, effector
    b. effector, receptor, sensory neuron, motor neuron, interneuron
    c. effector, sensory neuron, receptor, interneuron, motor neuron
    d. receptor, sensory neuron, interneuron, motor neuron, effector
15. The area of the sensory cortex in the cerebrum devoted to each part of the body
    a. is exactly the same as the area of the motor cortex devoted to each region of the body.
    b. depends on how far away that region of the body is from the brain.
    c. is greater for regions of the body that are most sensitive to touch.
    d. is smaller on the left cerebral hemisphere than on the right cerebral hemisphere.
16. Choose the correct statement about the parasympathetic nervous system.
    a. It increases heart rate.
    b. Its actions back up those of the sympathetic nervous system; that is, usually its actions have the same effect as those of the sympathetic nervous system but start after a delay.
    c. It directs blood toward the skeletal muscles.
    d. none of the above
17. Your friend suffered a stroke and now has no feeling in her right arm and is unable to move it. The damage is most likely in the
    a. medulla.
    b. left cerebral hemisphere.
    c. right cerebral hemisphere.
    d. hypothalamus.
18. Which of the following is *not* a function of the hypothalamus?
    a. regulation of heart rate
    b. sensory–motor coordination
    c. appetite
    d. regulation of blood pressure
19. Subconscious, rather automatic, activities such as breathing and heart beat are regulated by the
    a. cerebrum.
    b. cerebellum.
    c. pons.
    d. medulla.
20. The brain stem consists of these structures:
    a. medulla oblongata, pons, and midbrain
    b. medulla oblongata, cerebellum, and pons
    c. pons, midbrain, and cerebellum
    d. pons, midbrain, and thalamus

21. When this region of a cat's brain is stimulated electrically, the cat behaves as if it is angry. It hisses, it arches its back, and its hair stands on end. This electrode is most likely placed in the
    a. hypothalamus.
    b. cerebellum.
    c. pons.
    d. medulla.
22. If someone were to yell, "Fire!" now and you suddenly began to smell smoke, which part of your nervous system would become activated to prepare your internal body systems to deal with the dangerous situation?
    a. somatic system
    b. parasympathetic system
    c. sympathetic system
    d. meninges
23. Which of the following is responsible for reasoning?
    a. amygdala
    b. basal nuclei
    c. hippocampus
    d. cerebral cortex
24. You are watching a football game with your friends. A wide receiver makes an incredible catch and then runs 20 yards, skillfully dodging defensive players to make a touchdown. Your friend Joe says, "Amazing! How *does* he do that?" The receiver's outstanding sensory–motor coordination is largely due to the actions of his
    a. cerebellum.
    b. medulla.
    c. reticular activating system.
    d. hypothalamus.
25. As you sit here studying, you are unlikely to be aware of the pressure of your clothes against your body and the rustling of paper as other students turn pages. Which part of the brain "decides" that these are unimportant stimuli?
    a. reticular activating system
    b. cerebellum
    c. hypothalamus
    d. medulla
26. Belinda was riding a bicycle without a helmet and was struck by a car. She hit the back of her head very hard in the fall. The physician is quite concerned because the medulla is located at the base of the skull. She explains to Belinda's parents that injury to the medulla could result in
    a. the loss of coordination, such that the child may never regain the motor skills needed to ride a bicycle.
    b. the loss of speech.
    c. amnesia (the loss of memory).
    d. death, because many life-support systems are controlled here.
27. The branch of the nervous system that prepares the body to respond to emergency situations is the _____.
28. The brain region responsible for intelligence and thinking is the _____.

APPLYING THE CONCEPTS

1. Joe and Henry were both in car accidents, and both suffered spinal cord damage. Joe's injury was in the lower back, and Henry's was in the neck region. The degree of injury to the spinal cord is similar in both Joe and Henry. Would the resulting problems be equal in severity? Explain. Describe some of the difficulties that you might expect Joe and Henry to have.
2. When you have a cold, you might take a decongestant to help you breathe. Some decongestants contain pseudoephedrine, which mimics the effects of the sympathetic nervous system. What side effects might you expect? Would you expect this medication to make you drowsy?
3. When you have dental work done, the dentist often administers a local anesthetic in the gums near the region that requires drilling. You are usually advised not to eat anything until the anesthetic wears off. This advice is given out of concern for your tongue, not your teeth. Why?
4. After Jorge's car accident he could remember events that took place before the accident, but he would quickly forget a conversation or a television show he just watched. What part of the brain was injured in the accident?

BECOMING INFORMATION LITERATE

When a person has a head injury, physicians may induce a coma by administering a high dose of barbiturates or sedative drugs. The intent of medically induced coma is to allow the brain to rest. It prevents additional injury by reducing blood flow, which reduces swelling. There are success stories. However, there are also risks—pneumonia or a blood clot in the lung—that could be fatal.

Use at least three reliable sources (books, journals, or websites) to gather information that would help you decide whether you would want a loved one who had sustained a severe head injury to be placed in a medically induced coma. Explain why you made your decision. List each source you considered, and explain why you chose the three sources you used.

MasteringBiology®

Go to MasteringBiology for practice quizzes, activities, eText, videos, current events, and more.

SPECIAL TOPIC 8a

Drugs and the Mind

Alcohol is a legal psychoactive drug.

- 8a.1 Psychoactive Drugs and Communication between Neurons
- 8a.2 Drug Dependence
- 8a.3 Alcohol
- 8a.4 Marijuana
- 8a.5 Stimulants
- 8a.6 Hallucinogens
- 8a.7 Opiates

In previous chapters, we learned how neurons function and form circuits that function as a nervous system. In this chapter, we will consider how drugs affect the nervous system and alter our state of mind. We will see why the use of some drugs leads to a need to continue using the drug. Then we will take a look at some of the more common mind-altering drugs.

8a.1 Psychoactive Drugs and Communication between Neurons

A drug that alters one's mood or emotional state is often described as a *psychoactive drug.* The mind-altering effects of these drugs result from their ability to cross the blood–brain barrier, a membranous structure that prevents most substances from reaching the brain. The drugs then interfere with the communication between nerve cells. As you learned in Chapter 7, neurons communicate with one another using neurotransmitters. Neurotransmitters are released by one neuron, diffuse across a small gap between neurons, and bind to specific receptors on another neuron, triggering changes in the activity of the second neuron. Under normal conditions, the action of the neurotransmitter is stopped almost immediately after its message is conveyed, because the neurotransmitter is either broken down by enzymes, reabsorbed into the cell that released it, or simply diffuses away.

A psychoactive drug may alter this communication between neurons in any of several ways (Figure 8a.1). A drug may stimulate the release of a neurotransmitter and enhance the response of the receiving neuron. It may inhibit the release of a neurotransmitter and thus dampen the response of the receiving neuron. Or it may increase and prolong the effect of a neurotransmitter by delaying its removal from the gap between neurons. If the drug is chemically similar to the natural neurotransmitter, the drug may bind to the receptor and affect the activity of the receiving neuron in the same manner as the neurotransmitter, increasing its effect. Alternatively, its binding to the receptor may prevent the neurotransmitter from acting at all.

One of the problems with using psychoactive drugs is that the user may develop some level of psychological or physical dependence on the drug. Before we consider the drugs themselves, let's consider the concepts of tolerance and dependence.

stop and think

Why would a drug that increases the amount of a particular neurotransmitter that is released and a drug that slows the removal of that neurotransmitter from the synapse have the same effects?

8a.2 Drug Dependence

Tolerance is a progressive decrease in the effectiveness of a drug in a given person. As tolerance to a particular drug develops, a person must take larger or more frequent doses to produce the same effect. Tolerance develops partly because the body naturally steps up its production of enzymes that break down the drug, disabling it from having a continued effect, and partly because of changes in the nerve cells themselves that make them less responsive to the drug. *Cross-tolerance* occurs when tolerance to one drug results in a lessened response to another, usually similar, drug. If a person abuses codeine, for instance, tolerance develops not only for codeine but also for other drugs that have similar effects on the nervous system, such as morphine and heroin.

A loose definition of *dependence* might be, "It is what causes a person to continue using a drug." More precisely, dependence is the state in which the drug is necessary for physical or psychological well-being. A person who is physically dependent on a drug experiences physical withdrawal symptoms when the drug use is stopped.

Certain drugs encourage users to continue using them because the drugs stimulate the "pleasure" centers in the limbic system of the brain (see Chapter 8). An animal with electrodes implanted in the pleasure center will quickly learn to press a lever to stimulate this brain region to the point of exhaustion; similarly, an animal will learn to press a lever to self-administer a dose of a drug that stimulates the pleasure center. Drugs that stimulate the pleasure center include cocaine, amphetamine, morphine, and nicotine.

In the following discussions of psychoactive drugs, we will consider their habit-forming potential as well as possible health risks associated with their use.

8a.3 Alcohol

In every alcoholic drink, the alcohol is *ethanol*. Ethanol is produced as a by-product of fermentation, when yeast cells break down sugar to release energy for their own use. The taste of the beverage is determined by the source of the sugar, which

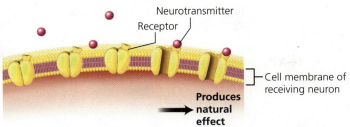

(a) The natural sequence of events: molecules of neurotransmitter released by one neuron diffuse across a gap and fit into receptors on the membrane of a receiving neuron, causing a response.

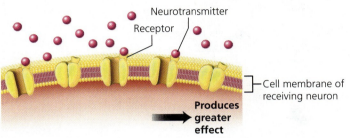

(b) A psychoactive drug may increase the number of neurotransmitter molecules released, increasing the response of the receiving neuron.

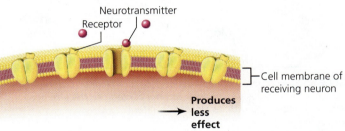

(c) A psychoactive drug may decrease the number of neurotransmitter molecules released, decreasing the response of the receiving neuron.

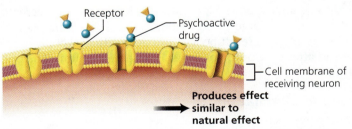

(d) A psychoactive drug may fit into the receptors for a neurotransmitter, causing a similar response by the receiving neuron.

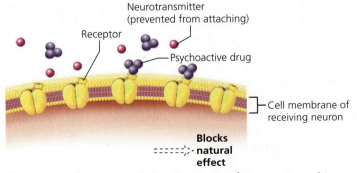

(e) A psychoactive drug may fit into the receptors for a neurotransmitter and prevent the neurotransmitter from entering the receptor, blocking the natural response by the receiving neuron.

FIGURE 8a.1 *Psychoactive drugs alter a person's mental state by affecting communication between neurons.*

comes from the fruit or vegetable that is fermented. For example, grapes are fermented to make wine, grain (barley, oats, rice, wheat, or malt) to make beer, and malt to make scotch.

The effects of alcohol on a person's behavior depend on the blood alcohol level, which is measured as the number of grams of alcohol in 100 ml of blood. One gram of alcohol in 100 ml of blood is a blood alcohol level of 1%. The blood alcohol level in a person depends on several factors, including how much alcohol is consumed; how rapidly it is consumed (over what period of time); and its rate of absorption, distribution, and metabolism.

A "drink" can mean different things to different people. For some, it is a can of beer; for others, a glass of wine; and for others, it is a scotch on the rocks. The amount of pure ethanol in the beverage varies tremendously. Natural fermentation, the process used to make beer and wine, cannot yield more than 15% alcohol, because the alcohol kills the yeast cells producing it. In contrast, liquor (distilled spirits) is produced by distillation, a process that concentrates the alcohol. The alcohol content of distilled spirits is measured as "proof." One degree of proof equals 0.5% alcohol. Most distilled spirits (vodka, gin, scotch, whiskey, rum, brandy, and cognac) are 80 proof, which means they contain 40% alcohol. But some brands are 90 proof (45% alcohol) or even 100 proof (50%). So, a small 3 oz[1] martini, which contains only distilled spirits and therefore about 1.2 oz of alcohol, is actually more intoxicating than a larger 8 oz mug of beer, which contains about 0.4 oz of alcohol.

The differences in the alcohol content of beverages are reflected in the standard amounts in which they are served as drinks. Generally, the standard drink contains 0.5 oz of pure ethanol. A standard drink is one 12 oz bottle or can of beer; a 5 oz glass of wine; or a jigger (1.5 oz) of 80-proof distilled spirits, such as gin, whiskey, vodka, or scotch (alone or in combination with another beverage) (Figure 8a.2).

Absorption and Distribution

The intoxicating effects of alcohol begin when it is absorbed from the digestive system into the blood and is delivered to the brain. As a rule, the absorption rate of alcohol depends on the concentration of alcohol in the drink: the higher the concentration, the faster the rate of absorption. So wine, beer, or distilled spirits diluted with a mixer will be absorbed more slowly than pure liquor. The choice of mixer also influences the rate of absorption. Carbonated beverages increase the absorption rate because of the pressure of the gas bubbles.

Although very few substances are absorbed across the walls of the stomach, about 20% of the alcohol consumed is absorbed there. The remaining alcohol is absorbed through the intestines. Because alcohol can be absorbed from the stomach, a person begins to feel the effects of a drink quickly, usually within about 15 minutes. The presence of food in the stomach slows alcohol absorption because it dilutes the alcohol, covers some of the stomach lining through which alcohol would be absorbed, and slows the rate at which the alcohol passes into the intestines.

Type of drink	Serving size	Caloric content
6 oz Mixer Fruit juice	1 glass	35–105 calories
8 oz Mixer Carbonated beverage	1 glass	~70–120 calories
1.5 oz Distilled spirits (80 proof gin, whiskey, vodka, scotch)	1 jigger	~100 calories
5 oz Wine (12% alcohol)	1 glass	110–200 calories
12 oz Most beer	1 bottle or can	140–150 calories

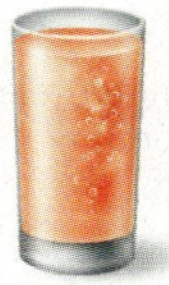

FIGURE 8a.2 *A standard drink contains 0.5 oz of alcohol. Different types of alcoholic beverages vary in their alcohol content, so the size of a standard drink varies with its alcoholic content.*

Ethanol is a small molecule that is soluble in both fat and water, so it is distributed to all body tissues. Therefore, overall body size affects one's blood alcohol level and the degree of intoxication. A large person would have a lower blood alcohol level and as a result be less intoxicated than would a small person after they both consumed the same amount of alcohol (Figure 8a.3).

Rate of Elimination

Ninety-five percent of the alcohol that enters the body is metabolized (broken down) before it is eliminated. Most of that metabolism occurs in the liver, which converts alcohol to carbon dioxide and water. The rate of metabolism is slow, about one-third of an ounce of pure ethanol per hour.

In practical terms, it takes slightly more than an hour for the liver to break down the alcohol contained in one standard drink—a can of beer or a glass of wine. Because alcohol cannot be stored anywhere in the body, it continues to circulate in the bloodstream until it is metabolized. Therefore, if more alcohol is consumed in an hour than is metabolized, both the blood alcohol level and the degree of intoxication increase along with consumption.

There is no way to increase the rate of alcohol metabolism by the liver and, therefore, no way to sober up quickly. The caffeine in a cup of coffee may slightly counter the drowsiness caused by alcohol, but it does not reduce the level

---

[1]In this chapter, the term *oz* (ounce) refers to a fluid ounce, which is equal to about 29 ml.

Q If a 140 lb man has three beers within 2 hours, in what range is his blood alcohol level? If a 180 lb man has three beers within 2 hours, in what range is his blood alcohol level?

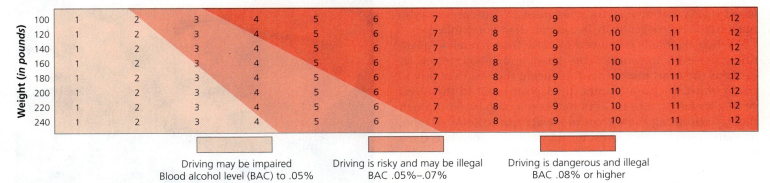

FIGURE 8a.3 Alcohol consumption can impair driving. The blood level of alcohol depends on both the number of drinks consumed and body size. After they consume the same amount of alcohol, a small person has a higher blood alcohol level than a large person does. The blood level of alcohol determines the effect on the nervous system.

A Probably over .05%; probably under .05%

of intoxication. Furthermore, because muscles do not metabolize alcohol, exercise does not help, either. Walking around the block will not make a person sober, nor will a cold shower.

A small amount of alcohol, about 5%, is eliminated from the body unchanged through the lungs or in the urine. Alcohol eliminated from the lungs is the basis of the breathalyzer test that may be administered by law enforcement officers who want an on-the-spot sobriety check.

Health-Related Effects

Because alcohol has a negative effect on virtually every organ of the body, people do not have to be alcoholics for alcohol to impair their health. In this section we look at some ways that alcohol affects the body.

Nervous system Many people believe that alcohol is a stimulant, but it is actually a depressant. In other words, it slows down the activity of all the neurons of the brain, beginning with the higher cortical, or "thinking," centers. Alcohol often is mistakenly thought to be a stimulant because it depresses the inhibitory neurons first, allowing the excitatory ones to take over. As alcohol removes the "brakes" from the brain, normal restraints on behavior may be lost. Release from inhibitory controls also tends to reduce anxiety and often creates a sense of well-being. However, discrimination, control of fine movements, memory, and concentration are gradually lost as well.

The brain centers for balance and coordination are affected next, causing a staggering gait. Numbed nerve cells send slower messages, resulting in slower reflexes. Eventually, the brain regions responsible for consciousness are inhibited, causing a person to pass out. Still higher concentrations of alcohol can cause coma and death from respiratory failure.

Liver Excessive alcohol consumption damages the liver, an organ that performs many vital functions in the body. Severe damage to the liver is a serious threat to life. Alcohol is metabolized in the liver before fats are, so fats therefore can accumulate in liver cells. Four or five drinks daily for several weeks are enough to cause fat to begin to accumulate. At this early stage, however, the liver cells are not yet harmed; and with abstinence, they can be restored to normal. With continued drinking, the accumulating fat causes liver cells to enlarge, sometimes so much that the cells rupture or form cysts. The accumulated fat in liver cells also reduces blood flow through the liver, causing inflammation known as *alcoholic hepatitis*. Signs of alcoholic hepatitis include fever and tenderness in the upper abdominal region. Gradually, fibrous scar tissue may form, a condition known as *cirrhosis*, which further impedes blood flow and impairs liver functioning. Cirrhosis can lead to intestinal bleeding, kidney failure, fluid accumulation, and eventually death, if drinking continues. Indeed, cirrhosis of the liver, the ninth leading cause of death in the United States, is most often caused by alcohol abuse.

Cancer A person who drinks heavily is at least twice as likely to develop cancer of the mouth, tongue, or esophagus than is a nondrinker. Evidence also exists that the combined risk of cancer from both drinking and smoking cigarettes is greater than the sum of the risks caused by either drinking or smoking alone.

Heart and blood vessels Here's the good news: *Moderate* amounts of alcohol can be good for the heart. Teetotalers are more likely to suffer heart attacks than are persons who drink moderately, say a drink a day. One reason may be that the relaxing effect of alcohol helps to relieve stress. But alcohol also seems to raise the levels of the "good" form of a cholesterol-carrying particle—high-density lipoprotein, or HDL—in the blood (discussed in Chapters 12a and 15). This form of cholesterol reduces the likelihood that fats in the blood will be deposited in the walls of blood vessels, clog the vessels, and reduce the blood supply to vital organs such as the heart or brain. In addition, moderate amounts of alcohol reduce the likelihood that blood clots will form when they should not. Such clots can cause a heart attack by blocking blood vessels that nourish the heart muscle. Thus, persons who imbibe moderately generally live longer than nondrinkers.

When alcohol is consumed in more than moderate quantities, however, it damages the heart and blood vessels. It weakens the heart muscle itself, reducing the heart's ability to pump blood. It also promotes the deposit of fat in the blood vessels, making the heart work harder to pump blood through them. Consuming more than moderate quantities of alcohol may also elevate blood pressure substantially. Together, these effects—damage to heart muscle, blood vessels clogged with fatty materials, and high blood pressure—can enlarge the heart to twice its normal size.

what would you do?

Fetal alcohol syndrome (FAS) is a pattern of growth abnormalities and birth defects common among children of women who drink during pregnancy. These adverse effects include mental retardation, growth deficiency, and characteristic facial features, although all the characteristics of FAS are not always present in any one infant. Recent studies suggest that the genes of a fetus may be permanently altered by exposure to alcohol during development. Considering these long-term effects on the child, do you think that there should be legal consequences for women who drink during pregnancy?

Alcoholism Alcohol is a legal drug, and alcoholism is America's number one drug problem. Alcoholics can be young or old, rich or poor, and of any race, economic status, or profession. Because there is no such thing as a typical alcoholic, alcoholism can be difficult to identify.

Different alcoholics have different drinking patterns. Some binge, and some chronically overindulge. But what they all share is a loss of control over their drinking. When an alcoholic takes the first sip of alcohol, he or she cannot predict how much or how long the drinking episode will continue.

8a.4 Marijuana

Marijuana is the most widely used illegal drug in the United States today. It consists of the leaves, flowers, and stems of the Indian hemp plant, *Cannabis sativa*. The principal psychoactive ingredient (the component that produces mind-altering effects) is delta-9-tetrahydrocannabinol, or THC.

The effects of marijuana depend on the concentration of THC and on the amount consumed. In small to moderate doses, THC produces feelings of well-being and euphoria. In large doses, THC can cause hallucinations and paranoia. Anxiety may even reach panic proportions at very high doses.

Marijuana is not addicting in the sense of producing severe, unpleasant withdrawal symptoms. Nonetheless, a withdrawal syndrome has been identified. It includes symptoms such as restlessness, irritability, mild agitation, insomnia, nausea, and cramping. Withdrawal symptoms are not common and, if they do occur, are usually relatively mild and short-lived.

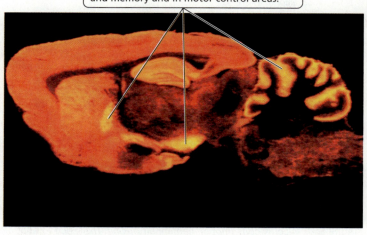

FIGURE 8a.4 *The areas with marijuana receptors appear yellow in this photograph showing a section of a rat's brain.*

THC Receptors in the Brain

Researchers are just beginning to understand how marijuana affects the brain. When THC binds to certain receptors on nerve cells, it triggers a cascade of events that ultimately leads to the "high" that the user experiences (Figure 8a.4).[2] After discovering the receptors, researchers quickly set about looking for the neurotransmitter that normally binds to them, reasoning that the receptors did not evolve millions of years ago just in case someone would someday decide to smoke marijuana. The researchers sorted through thousands of chemicals in pig brains to find one that would bind to THC receptors. The first one to be identified was a hitherto unknown chemical messenger they named *anandamide*, after the Sanskrit word meaning "internal bliss" (*ananda*). Anandamide functions as the brain's own THC. Its normal functions probably include the regulation of mood, memory, pain, appetite, and movement. In addition to mimicking anandamide, THC seems to stimulate the release of dopamine in the reward pathways of the brain.

Health-Related Effects of Long-Term Use

Marijuana is usually smoked. It should not be surprising, then, that the most clearly harmful effects of marijuana involve the respiratory system. However, the damage, it seems, is done by the residual materials in the smoke and not by the THC itself. Compared to the daily consumption of cigarettes by tobacco smokers, marijuana smokers smoke fewer joints (marijuana cigarettes) per day. However, compared with a regular cigarette, a joint has 50% more tar, which contains cancer-causing chemicals. Also, marijuana smoke is usually inhaled more deeply and held within the lungs longer. As a result, three times as much tar is deposited in the airways, and five times as much carbon monoxide is inhaled than with a normal cigarette. Carbon monoxide prevents red blood cells from carrying needed oxygen to the cells of the body. Like cigarette smoke, marijuana

---

[2]A slightly different type of THC receptor is found primarily on cells of the immune system.

smoke inflames air passages. However, a recent government-funded study revealed that long-term use of marijuana is less damaging to the lungs than is cigarette smoking.

Another known effect of smoking marijuana is that it makes the heart beat faster, sometimes double its normal rate. In some people, marijuana also increases blood pressure. Either change increases the heart's workload and could pose a threat to people with preexisting cardiovascular problems, such as high blood pressure or atherosclerosis (fatty deposits in the arteries). Thus, although the risk is small, marijuana increases one's risk of immediate heart attack.

Some studies have shown that THC can interfere with reproductive functions in both males and females. At least some of these disturbances may result from the structural similarity between THC and the sex hormone estrogen. Males who smoke marijuana often have lower levels of the sex hormone testosterone and produce fewer sperm than do nonusers. Testosterone and sperm levels return to normal when THC is cleared from the body.

The effects of marijuana on the female reproductive system are not clear. Although we know that THC interferes with ovulation in female monkeys, we know less about what it may do to human females. Some doctors, however, report menstrual problems and reproductive irregularities in women who smoke marijuana.

Medical Marijuana

Seventeen states and Washington DC have legalized marijuana for medical purposes, but controversy continues. How much research is enough, and how much credence should be given to anecdotal evidence? Some people claim that marijuana leads to more serious drugs, but so far, that has not been demonstrated. Other people oppose legalization of marijuana because of the health effects of smoking it. However, Sweden has approved use of a marijuana-based mouth spray to alleviate symptoms of multiple sclerosis, and a similar spray is waiting for approval in England. Do you think medical marijuana should be legalized in the United States?

8a.5 Stimulants

Stimulants are drugs that excite the central nervous system (CNS; see Chapter 8). Here we'll look at three types of stimulants: cocaine, the amphetamines, and nicotine.

Cocaine

Cocaine is extracted from the leaves of the coca plant (*Erythroxylon coca*), which grows naturally in the mountainous regions of South America. When cocaine powder extracted from the plant is inhaled into the nasal cavity ("snorted"), it reaches the brain within a few seconds and produces an effect almost as intense as when a solution of cocaine is injected. Smoking the drug is an even more effective delivery route. Forms of cocaine that can be smoked—specifically, freebase and crack—are obtained by further extraction and purification of the coca plant.

When ingested in one of these ways, cocaine brings about a rush of intense pleasure, a sense of self-confidence and

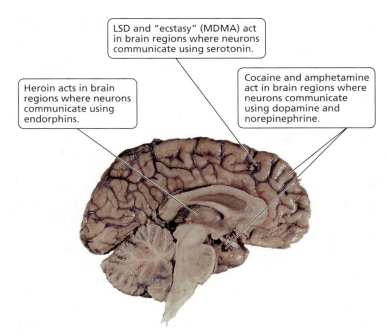

FIGURE 8a.5 *The effects of certain drugs on the brain*

power, mental alertness, and increased physical vigor. It does so by increasing the levels of two "feel-good" neurotransmitters: dopamine and norepinephrine (Figure 8a.5). The euphoria is caused primarily by cocaine's effect on dopamine, a neurotransmitter used by nerve cells in the pleasure centers of the brain. Normally, dopamine is almost immediately reabsorbed into the nerve cell that released it, and its effect soon ceases. Cocaine, however, interferes with the reuptake of dopamine, thus increasing and prolonging dopamine's effect. Cocaine also increases the effects of another neurotransmitter, norepinephrine. Norepinephrine stimulates the effects of the sympathetic nervous system, the part of the nervous system that prepares the body to face an emergency. Thus, cocaine also triggers the responses that ready the body for stress: increased heart rate and blood pressure, narrowing of certain blood vessels, dilation of pupils, a rise in body temperature, and a reduction of appetite.

The effects of cocaine are short-lived, lasting from 2 to 90 minutes depending on how the cocaine enters the body. When the high wears off, it is generally followed by a "crash," a period of deep depression, anxiety, and extreme fatigue. These uncomfortable feelings often produce a craving for more cocaine. The higher the high, the harder the crash, and therefore, the more intense the craving for more of the drug. Crack, which is estimated to be about 75% pure cocaine (by comparison, the purity of street cocaine is 10% to 35%) causes higher highs and lower crashes and therefore is extremely addicting.

Cardiovascular risks Cocaine may cause heart attack or stroke, emergencies in which an interruption of blood flow deprives heart cells or brain neurons of oxygen and nutrients. One way that cocaine blocks blood flow is by constricting arteries. Thus, one way it produces heart attack is by causing spasms in the arteries supplying the heart. Blood flow to the heart can also be disturbed by irregularities in heartbeat. By disabling the nerves that regulate heartbeat, cocaine can cause disturbed heartbeat rhythms that result in chest pain and heart palpitations (that uncomfortable feeling of being aware of your heart beating). Cocaine use may even cause the heart to stop beating. In addition, cocaine increases blood pressure, which may cause a blood vessel to burst.

Respiratory risks Stimulation of the CNS is always followed by depression. As the stimulatory effects of cocaine wear off, the respiratory centers in the brain that are responsible for breathing become depressed, or inhibited. Breathing may become shallow and slow and may stop completely. In short, cocaine use can cause respiratory failure that can lead to death.

Damage to the respiratory system resulting from cocaine use depends largely on how the drug is taken. If it is snorted, cocaine causes damage to the nerves, lining, and blood vessels of the nose. It can dry out the nose's delicate mucous membranes so that they crack and bleed almost continuously. Symptoms of a sinus infection—a perpetually runny nose and a dull headache spanning the bridge of the nose—are common. What is more alarming, the partition between the two nasal cavities may disintegrate. Because cocaine is a painkiller, considerable damage to the nose may occur before being discovered. Smoking crack damages the lungs and airways. The chronic irritation leads to bronchitis and may also cause lung damage from reduced oxygen flow into the lungs or blood flow through the lungs.

Amphetamines

Amphetamines are synthetically produced stimulants that closely resemble dopamine and norepinephrine (the same neurotransmitters whose levels are increased by cocaine). There are many forms of amphetamine, including dextroamphetamine and methamphetamine, also known as crystal meth. Methamphetamine is illegally sold as a powder that can be injected, snorted, swallowed, or smoked. All amphetamines are stimulants of the CNS. Like cocaine, when active, amphetamines make the user feel good—exhilarated, energetic, talkative, and confident. They suppress appetite and the need for sleep. Amphetamines are active for longer periods than cocaine—hours as opposed to minutes.

Amphetamines can be swallowed in a pill or injected intravenously. One crystalline form of methamphetamine is smoked to produce effects similar to those of crack cocaine. Methamphetamine can produce hazardous effects, including blood vessel spasm, blood clot formation, insufficient blood flow to the heart, and accumulation of fluid in the lungs.

Amphetamines bring about their physical and psychological effects by causing the release of certain neurotransmitters, such as norepinephrine and especially dopamine. As a result, the drugs elevate blood pressure, increase heart rate, and open up airways in the lungs.

Tolerance to amphetamines develops (accompanied by simultaneous tolerance to cocaine) along with both physical and psychological dependence. Withdrawal symptoms include extreme fatigue, depression, and increased appetite. Also, the pleasurable feelings caused by amphetamine use can lead to a compulsion to overuse the drug.

Amphetamines cross the placenta and can affect the developing fetus. Compared to nonusers, women who use amphetamines during pregnancy are more likely to give birth prematurely and to have a low-birth-weight baby.

Nicotine

Nicotine, the psychoactive ingredient in tobacco products, is a stimulant that activates acetylcholine receptors. In the peripheral nervous system, activation of acetylcholine receptors increases heart rate and blood pressure. In the central nervous system, activation of acetylcholine receptors, which facilitates the release of dopamine and serotonin, creates pleasurable feelings of relaxation. It is nicotine that causes smokers to become "hooked" on cigarettes, ensuring continued exposure to the other injurious substances in the smoke.

Although smoking can be relaxing, nicotine is actually a stimulant of the brain and heart and blood vessels. Under the influence of nicotine, the heart beats as many as 33 more beats per minute. At the same time, the blood vessels constrict, so the heart not only is beating faster, but also is forcing blood through a less receptive circulatory system. The result is an increase in blood pressure. Nicotine also affects the platelets in the blood (these are described in Chapter 11 as cell fragments containing chemicals that initiate clotting). Nicotine causes the platelets to become sticky, increasing the likelihood that abnormal clots may form, leading to heart attacks or strokes. A heart attack is the death of heart muscle cells, and a stroke is the death of nerve cells in the brain. Either can be caused when a clot blocks a blood vessel.

Nicotine is powerfully addictive. Ninety-five percent of smokers are physiologically dependent on it. Nicotine causes dopamine to be released in the brain's reward center, causing a pleasurable feeling. Drugs that stimulate the brain's reward center, such as nicotine, cocaine, and heroin, are highly addictive. Thus, smokers experience withdrawal symptoms when they try to quit. When a smoker tries to quit, nerve cells become hyperactive, causing withdrawal symptoms that include irritability, anxiety, headache, nausea, constipation or diarrhea, craving for tobacco, and insomnia. Most smokers continue smoking to avoid these unpleasant symptoms. Those who quit find that most of the withdrawal symptoms begin to lessen after a week without nicotine, but some symptoms may continue for weeks or even months. Certain symptoms, such as drowsiness, difficulty concentrating, and craving for a cigarette, seem to worsen about 2 weeks after quitting. Consequently, many smokers return to their habit and resume their exposure to nicotine and the other harmful substances in smoke.

8a.6 Hallucinogens

The drugs that are classified as *hallucinogenic* are grouped together because of their similar effects, even though they have very different structures. These effects include visual, auditory, or other distortions of sensation as well as vivid, unusual changes in thought and emotions.

The psychedelic drugs, including MDMA,[3] which is popularly known as "ecstasy," are thought to act by augmenting the action of the neurotransmitters serotonin, norepinephrine, or acetylcholine. For example, LSD, psilocybin, DMT (dimethyltryptamine), and bufotenin bind to the serotonin receptors in the brain, thereby mimicking the natural effects of serotonin. Ecstasy both binds to serotonin receptors and promotes the release of serotonin and dopamine. Mescaline, in contrast, is structurally similar to norepinephrine and binds to norepinephrine receptors.

The normal physiological reactions to psychedelic drugs are not especially harmful, but the distortions of reality that the drugs produce may lead to behavior that is quite dangerous. "Bad trip" is the term most often used to describe an unpleasant reaction to a psychedelic drug. The symptoms may include paranoia, panic, depression, and confusion.

Tolerance for psychedelic drugs develops quickly. Furthermore, cross-tolerance among these drugs is the rule; that is, a person who has become tolerant of one psychedelic drug will be tolerant to others as well. Craving and withdrawal reactions are unknown, however, and laboratory animals offered LSD will not take it voluntarily.

Ecstasy deserves additional consideration because its use is growing rapidly. It is sometimes called the "love drug" or "hug drug" because users say it puts them at peace with themselves and at ease with others. Also a stimulant, ecstasy increases heart rate, blood pressure, and body temperature—sometimes to dangerous levels. People using ecstasy feel energetic enough to dance all night, sometimes suffering dehydration and heat stroke, which occasionally lead to death. (Deaths are rare, however.) The release of serotonin from neurons causes a euphoric high, but the temporary depletion of serotonin in the days following ecstasy use can bring depression and anxiety. Ecstasy can also cause nausea, vomiting, and dizziness. Another danger is that ecstasy pills may contain drugs other than ecstasy. Among the drugs commonly found in these pills are the cough suppressant dextromethorphan, caffeine, ephedrine, and pseudoephedrine. Even the poison strychnine has been found.

8a.7 Opiates

The *opiates* are natural or synthetic drugs that affect the body in ways similar to morphine, the major pain-relieving agent in opium. These and related drugs have two faces. On one hand, they have a high potential for abuse because tolerance and physical dependence occur. On the other hand, they are medically important because they alleviate severe pain.

Morphine and codeine, which come from the opium poppy, were among the first opiates used. Today, prescription drugs containing oxycodone (oxycontin, Percocet, Percodan) are prescribed to alleviate pain. Unfortunately, they are frequently abused, causing addiction.

Heroin is a synthetic derivative of morphine that is more than twice as powerful. Heroin is usually injected intravenously, but there are now forms that can be smoked or inhaled into the nasal cavity. Regardless of how it enters the body, heroin reaches the brain very quickly, producing a feeling that is usually described in ecstatic or sexual terms.

The opiates, including heroin, exert their effects by binding to the receptors for the body's endogenous (natural, internally produced) opiates: compounds called *endorphins, enkephalins,* and *dynomorphins*. These are neurotransmitters whose functions include roles in the perception of pain and fear.

Heroin and other commonly abused opiates have effects similar to those of morphine: euphoria, pain suppression, and reduction of anxiety. They also slow the breathing rate. An overdose may cause the user to fall into a coma or stop breathing. Overdose is always a potential problem with opiates bought on the street for illegal use. The buyer has no idea of the strength of the drug. Street supplies are often diluted with sugar. If a heroin addict who is accustomed to a diluted drug injects heroin that is much more potent than he or she is used to, death due to overdose often occurs. Extremely constricted pupils in an unconscious person are a sign of heroin overdose.

Many of the problems associated with heroin use arise because heroin addicts often suffer from a general disregard for good health practices. Other problems are associated with the injections themselves. Frequent intravenous injection of any drug is associated with certain ailments. Because of the constant puncturing, veins can become inflamed. In addition, shared needles may spread disease-causing organisms, including the viruses that cause AIDS and hepatitis and the bacterium that causes syphilis.

looking ahead

In this chapter, we considered how psychoactive drugs affect communication among neurons. In the next chapter, we will examine how our sensory receptors detect external and internal stimuli.

---

[3] 3,4-methylenedioxymethamphetamine

9 Sensory Systems

We rely on our senses—sight, hearing, taste, touch, and so on—to identify the resources and conditions our bodies require and to recognize danger.

Did You Know?

- The prevalence of nearsightedness (myopia) in young adults in the United States has almost doubled in the last 30 years, probably because the amount of time spent on close work, such as using computers, has increased.
- A human has about 12 million olfactory receptors and can distinguish about 10,000 odors.

9.1 Sensory Receptors
9.2 Classes of Receptors
9.3 The General Senses
9.4 Vision
9.5 Hearing
9.6 Balance and the Vestibular Apparatus of the Inner Ear
9.7 Smell and Taste

ENVIRONMENTAL ISSUE

Noise Pollution

In Chapter 8, we learned how the brain integrates sensory information to direct appropriate responses, and in Chapter 8a we learned how psychoactive drugs may alter sensory integration. In this chapter, we explore the human body's general senses (such as touch, pressure, vibration, temperature, and pain) and our special senses (vision, hearing, balance, smell, and taste). We look closely at vision, hearing, and balance, examining how light, sound, and body position are reported by the intricate structures of the eye and ear. We also explore the relationship between smell and taste.

9.1 Sensory Receptors

Information about the external world and about the internal world of our bodies comes to us through our **sensory receptors**, structures that are specialized to detect and respond to changes in the environment, known as **stimuli**. If a stimulus is strong enough, these messages eventually become nerve impulses (action potentials) that are then conducted to the brain.

Sensation is an awareness of a stimulus. Whether a sensation is experienced as sight, sound, or something else depends on which part of the brain receives the nerve impulses. For example, photoreceptors respond best to light, but they can also respond to pressure. Regardless of the stimulus, nerve impulses from photoreceptors go to the visual cortex of the brain, and we see light. This is why, if you press gently on your closed eyelids, you will have the sensation of seeing spots of light. The pressure stimulates photoreceptors, which send nerve impulses to the visual cortex, and you sense light.

We use the word *perception* to describe the conscious awareness of sensations. Perception occurs when the cerebral cortex integrates sensory input (Figure 9.1). For example, light reflected

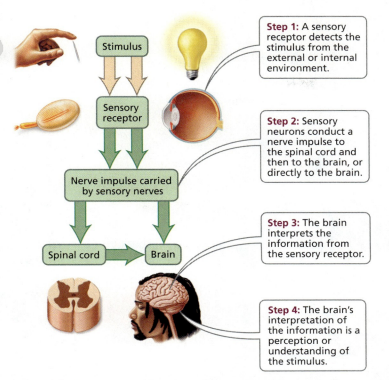

FIGURE 9.1 An overview of the steps involved in sensation and perception.

from a banana strikes the eye, stimulating some of the photoreceptors. The brain interprets the pattern of input from the photoreceptors, and we perceive a banana.

As we've seen, each type of sensory receptor responds best to one form of stimulus. The response of a sensory receptor is an electrochemical message (a change in the charge difference across the plasma membrane) that varies in magnitude with the strength of the stimulus. For instance, the louder a sound, the larger the change in the charge difference across the membrane—up to a point. When the change reaches a critical level, called threshold level, an action potential (nerve impulse) is generated. Most types of sensory receptors gradually stop responding when they are continuously stimulated. This phenomenon is called **sensory adaptation**. As receptors adapt in this way, we become less aware of the stimulus. For example, the musty smell of an antique store may be obvious to a person who just walked in, but the salesclerk working in the store no longer notices it. Some receptors, such as those for pressure and touch, adapt quickly. For this reason, we quickly become unaware of the feeling of our clothing against our skin. (Adaptation can also occur in the central nervous system. The reticular activating system in the brain filters stimuli, as discussed in Chapter 8.) Other receptors adapt more slowly or not at all. For instance, the receptors in muscles and joints that report on the position of body parts never adapt. Their continuous input is essential for coordinated movement and balance.

9.2 Classes of Receptors

Receptors are classified according to the stimulus they respond to. Several classes of receptors are traditionally recognized:

1. **Mechanoreceptors** are responsible for the sensations we describe as touch, pressure, hearing, and equilibrium. In addition, the body has mechanoreceptors that detect changes in blood pressure and others that indicate the body's position. Mechanoreceptors respond to distortions in the receptor itself or in nearby cells.

2. **Thermoreceptors** detect changes in temperature.

3. **Photoreceptors** detect changes in light intensity.

4. **Chemoreceptors** respond to chemicals. We describe the input from the chemoreceptors of the mouth as taste (gustation) and those from the nose as smell (olfaction). Other chemoreceptors monitor levels of specific substances such as carbon dioxide, oxygen, or glucose in our body fluids.

5. **Pain receptors** (or *nociceptors*) respond to very strong stimuli that usually result from physical or chemical damage to tissues. Pain receptors are sometimes classed with the chemoreceptors because they often respond to chemicals liberated by damaged tissue. These receptors are occasionally classed with the mechanoreceptors because they are stimulated by physical changes, such as swelling, in the damaged tissue.

We can sense stimuli both outside our bodies and within them. Receptors located near the body surface respond to stimuli in the environment. We are usually aware of these stimuli. Other receptors are inside the body and monitor conditions there. Although we often are unaware of the activity of internal receptors, they play a vital role in maintaining homeostasis. In fact, they are key components of the feedback loops that regulate blood pressure, blood chemistry, and breathing rate. Internal receptors also cause us to feel pain, hunger, or thirst, thereby prompting us to attend to our body's needs.

The **general senses**—touch, pressure, vibration, temperature, body and limb position, and pain—arise from receptors in the skin, muscles, joints, bones, and internal organs. Although we are not usually aware of the general senses, they are important because they provide information about body position and help keep internal body conditions within the limits optimal for health. The **special senses** are vision, hearing, the sense of balance or equilibrium, smell, and taste. These are what usually come to mind when we think of "the senses," largely because we are so dependent on them for perceiving and understanding the world. The receptors of the special senses are located in the head. Most of them reside within specific structures.

9.3 The General Senses

The receptors for general senses are distributed throughout the body. Some monitor conditions within the body; others provide information about the world around us. Some of the receptors are free nerve endings; in other cases the nerve endings are encapsulated. *Free nerve endings,* the tips of dendrites of sensory neurons, are not protected by an accessory structure. In contrast, an *encapsulated nerve ending* is one in which a connective tissue capsule encloses and protects the tips of the dendrites of sensory neurons (Figure 9.2).

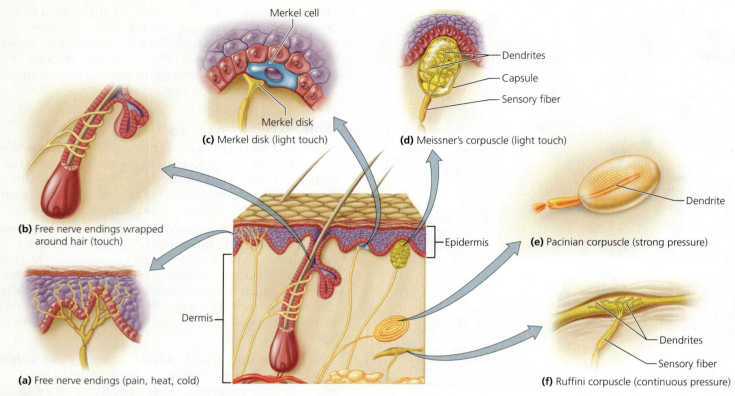

FIGURE 9.2 *General sense receptors of the skin allow us to feel touch, pressure, temperature, and pain.*

Touch, Pressure, and Vibration

As noted earlier, mechanoreceptors respond to touch and pressure—to any stimulus that stretches, compresses, or twists the receptor membrane. Throughout life, we actively use touch as a way of learning about the world and of communicating with one another. The messages we get from pressure can be equally important. Some pressure receptors inform us of the need to loosen our belt after a big meal. Other pressure receptors monitor internal conditions, including blood pressure.

Light touch, as when the cat's tail brushes your legs, is detected by several types of receptors. For example, free nerve endings wrapped around the base of the fine hairs on the skin detect any bending of those hairs. Free nerve endings and the special cells they end on (Merkel cells) form *Merkel disks*. When compressed, Merkel cells stimulate the free nerve endings in the associated Merkel disks to tell us that something has touched us. Merkel disks are found on both the hairy and hairless parts of the skin. *Meissner's corpuscles* are encapsulated nerve endings that tell us exactly where we have been touched. They are common on the hairless, very sensitive areas of skin, such as the lips, nipples, and fingertips.

The sensation of pressure generally lasts longer than does touch and is felt over a larger area. *Pacinian corpuscles,* which consist of onionlike layers of tissue surrounding a nerve ending, respond when pressure is first applied and quickly adapt. Therefore, they are important in sensing vibration. They are scattered in the deeper layers of skin and the underlying tissue. *Ruffini corpuscles* are encapsulated endings that respond to continuous pressure.

Temperature Change

Thermoreceptors respond to changes in temperature. In humans, thermoreceptors are specialized free nerve endings found just below the surface of the skin. One kind responds to cold, and another responds to warmth. They are widely distributed throughout the body but are especially numerous around the lips and mouth. You may have noticed that the sensation of hot or cold fades rapidly. This fading occurs because thermoreceptors are very active when temperature is changing but adapt rapidly when temperature is stable. As a result, the water in a hot tub may feel scalding at first, but very soon it feels comfortably warm.

Body and Limb Position

Whether you are at rest or in motion, the brain "knows" the location of all your body parts. It continuously scans the signals from muscles and joints to check body alignment and coordinate balance and movement. **Muscle spindles**—specialized muscle fibers wrapped in sensory nerve endings—monitor the length of a skeletal muscle. **Tendon organs**—highly branched nerve fibers located in tendons (connective tissue bands that connect muscles to bones)—measure the degree of muscle tension (Figure 9.3). The brain combines information from muscle spindles and tendon organs with information from the inner ear (as we see shortly) to coordinate our movements.

Pain

The receptors for pain are free nerve endings found in almost every tissue of the body. When tissue is damaged, cells

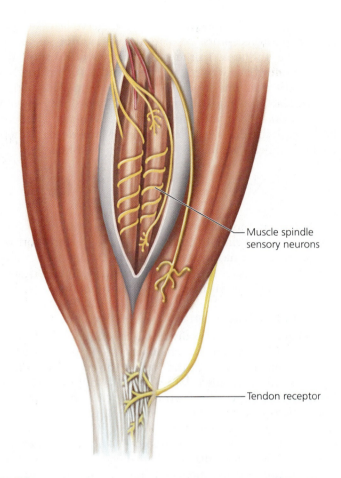

FIGURE 9.3 *Body and limb position are monitored by muscle spindles and tendon receptors. The muscle spindles respond when the muscle is stretched. When a muscle contracts, it increases tension on the tendon organs, which respond.*

release chemicals that alert the free nerve endings of the injury. The stimulated sensory neurons then carry the message to the brain, where it is interpreted as pain. Aspirin and ibuprofen reduce pain by interfering with the production of one of the released chemicals. Any stimulus strong enough to damage tissues, including heat, cold, touch, and pressure, can cause pain.

Many of our internal organs also have pain receptors. However, pain originating in an internal organ is sometimes perceived as pain in an uninjured region of the skin (Figure 9.4). This phenomenon is called **referred pain**. For example, the pain of a heart attack is often experienced as pain in the left arm. This pain probably occurs because sensory neurons from the internal organ and those from a particular region of the skin communicate with the same neurons in the spinal cord. Because the message is delivered to the brain by the same neurons, the brain interprets the input as coming from the skin.

Pain is an important mechanism that warns the body and protects it from further injury. For example, pain usually prevents a person with a broken leg from causing additional damage by moving the limb. Nonetheless, few of us appreciate the value of pain while we are experiencing it. Furthermore, pain that persists long after the warning is needed can be debilitating.

Q *If you had pain caused by a kidney stone, where would you experience the pain?*

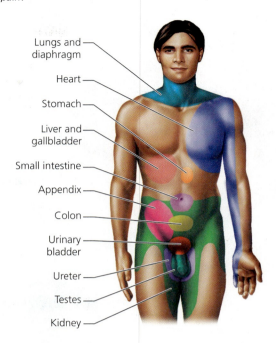

FIGURE 9.4 *Referred pain. Pain from certain internal organs is sensed as originating in particular regions of the skin.*

A *The lower body trunk and inner thigh.*

9.4 Vision

Humans are very visual creatures. We may not see detail as well as an eagle or movement as well as an insect, but we see far better than most other mammals and depend on vision in most of the activities that make up our daily lives.

Wall of the Eyeball

The human eyeball is an irregular sphere about 25 mm (1 in.) in diameter. As shown in Figure 9.5, the wall of the eyeball consists of three layers. The outermost layer is a tough, fibrous covering with two distinct regions: the sclera and the cornea. The **sclera**, often called the white of the eye, protects and shapes the eyeball and serves as an attachment site for the muscles that move the eye. In the front and center of the eye, the transparent **cornea** bulges slightly outward and provides the window through which light enters the eye.

The middle layer of the eye has three distinct regions—the choroid, the ciliary body, and the iris. The **choroid** is a layer containing many blood vessels that supply nutrients and oxygen to the tissues of the eye. The choroid layer also contains the brown pigment melanin, which absorbs light reflected from the light-sensitive layer inside the eye (the retina). This absorption of light results in sharper vision by helping to prevent excessive reflection of light within the eye.

Toward the front of the eye, the choroid becomes the **ciliary body**—a ring of tissue, primarily muscle—that encircles the **lens**, which focuses light on the retina. The ciliary body holds the lens in place and controls its shape, which is important for focusing light on the light-sensitive layer of the eye.

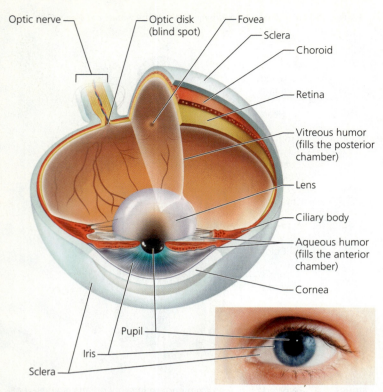

FIGURE 9.5 Structure of the human eye. Light enters the eye through the transparent cornea and then passes through the pupil. The lens focuses light on the light-sensitive retina, which contains the rods and cones.

The part of the choroid layer in front of the ciliary body is the iris. The **iris**, the colored portion of the eye that can be seen through the cornea, regulates the amount of light that enters the eye. The iris is shaped like a flat doughnut. The doughnut hole—the opening in the center of the iris through which light enters the eye—is called the **pupil**.

The iris contains smooth muscle fibers that automatically adjust the size of the pupil to admit the appropriate amount of light. The pupil becomes larger (dilates) in dim light and smaller (constricts) in bright light. Pupil size is also affected by emotions. The pupils dilate when you are frightened or when you are very interested in something. They constrict when you are bored. Candlelight creates a romantic setting partly because its dimness dilates the pupils, making the lovers appear more attentive and interested.

The innermost layer of the eye, the **retina**, contains almost a quarter-billion photoreceptors, the structures that respond to light by generating electrical signals. The retina contains two types of photoreceptors: rods and cones. The structure and function of rods and cones are discussed in more detail later in the chapter. The photoreceptors are most concentrated in a small region in the center of the retina called the **fovea**. Therefore, when we want to see fine details of an object, the light reflected from the object must be focused on the fovea. However, the fovea is only the size of the head of a pin. Consequently, at any given moment, only about a thousandth of our visual field is in sharp focus. Eye movements bring different parts of the visual field to the fovea. The **optic nerve** carries the message from the eye to the brain, where the message is interpreted. The region where the optic nerve leaves the retina has no photoreceptors. As a result, we cannot see an image that strikes this area. This area is called the **blind spot**. To find your blind spot, cover your left eye and focus on the X with your right eye. You should see the circle in your peripheral vision. Place this book in front of you, and slowly move your head toward it. As your head moves, the image of the circle becomes focused on different regions of the retina. When the circle disappears, its image is focused on the blind spot.

You are not usually aware of your blind spot because involuntary eye movements constantly move the position of the image on the retina, allowing the brain to "fill in" the missing parts as it processes the visual information. Table 9.1 summarizes the structures of the eye and their functions.

Fluid-Filled Chambers

The ciliary body and lens divide the interior of the eyeball into two fluid-filled cavities, or chambers (see Figure 9.5). The posterior chamber, located at the back of the eye between the lens and the retina, is filled with a jellylike fluid called **vitreous humor**. This fluid helps keep the eyeball from collapsing and holds the thin retina against the wall of the eye. The anterior chamber, located at the front of the eye between the cornea and the lens, is filled with a fluid called **aqueous humor**. This clear fluid supplies nutrients and oxygen to the cornea and lens and carries away their metabolic wastes. In addition, the aqueous humor creates pressure within the eye, helping to maintain the shape of the eyeball. Unlike the vitreous humor, which is produced during embryonic development and is never replaced, aqueous humor is replaced bit by bit about every 90 minutes. It is continuously produced from the capillaries of the ciliary body, circulates through the anterior chamber, and drains into the blood through a network of channels that surround the eye.

If the drainage of aqueous humor is blocked, the pressure within the eye may increase to dangerous levels. This condition, called *glaucoma,* is the second most common cause of blindness (after cataracts, which are discussed shortly). The accumulating aqueous humor pushes the lens partially into the posterior cavity of the eye, increasing the pressure there and compressing the retina and optic nerve. This pressure, in turn, collapses the tiny blood vessels that nourish the photoreceptors and the fibers of the optic nerve. Deprived of nutrients and oxygen, the photoreceptors and nerve fibers begin to die, and vision fades. Unfortunately, glaucoma is progressive but painless, making detection difficult. Late signs include blurred vision, headaches, and seeing halos around objects. Many people do not realize they have a problem until some vision has been lost.

Focusing and Sharp Vision

Sharp, clear vision requires that the light rays entering the eye converge so that their focal point, the point where they are in focus, is on the retina. The structures of the eye accomplish this focusing by bending the light rays to the necessary degree.

TABLE 9.1 A Review of the Structures of the Eye and Their Functions

Structure	Description	Function
Outer layer		
Sclera	Outer layer of the eye	Protects the eyeball
Cornea	Transparent dome of tissue forming the outer layer at the front of the eye	Refracts light, focusing it on the retina
Middle layer		
Choroid	Pigmented layer containing blood vessels	Absorbs stray light; delivers nutrients and oxygen to tissues of eye
Ciliary body	Encircles lens; contains the ciliary muscles	Controls shape of lens; secretes aqueous humor
Iris	Colored part of the eye	Regulates the amount of light entering the eye through the pupil
Pupil	Opening at the center of the iris	Opening for incoming light
Inner layer		
Retina	Layer of tissue that contains the photoreceptors (rods and cones); also contains bipolar and ganglion cells involved in retinal processing	Receives light and generates neural messages
Rods	Photoreceptor	Responsible for black-and-white vision and vision in dim light
Cones	Photoreceptor	Responsible for color vision and visual acuity
Fovea	Small pit in the retina that has a high concentration of cones	Provides detailed color vision
Other structures of the eye		
Lens	Transparent, semispherical body of tissue behind the iris and pupil	Fine focusing of light onto retina
Aqueous humor	Clear fluid found between the cornea and the lens	Refracts light and helps maintain shape of the eyeball
Vitreous humor	Gelatinous substance found within the chamber behind the lens	Refracts light and helps maintain shape of the eyeball
Optic nerve	Group of axons from the eye to the brain	Transmits impulses from the retina to the brain

Most bending of light occurs as the light passes through the curved surface of the cornea. Because of the way that the curved cornea bends the light rays, the image created on the retina is upside down and backward. The cornea has a fixed shape, so it always bends light to the same degree and cannot make the adjustments needed to focus on objects at varying distances.

The lens, in contrast, is elastic and can change shape to focus on both near and distant objects. Picture the lens as an underinflated round balloon. If you pull on the sides of such a balloon, it flattens. When you release its sides, the balloon assumes its usual, rounder shape. Similar pulling and releasing change the shape of the lens. Figure 9.6 shows that a rounder, thicker lens bends light to a greater degree, enabling the eye to

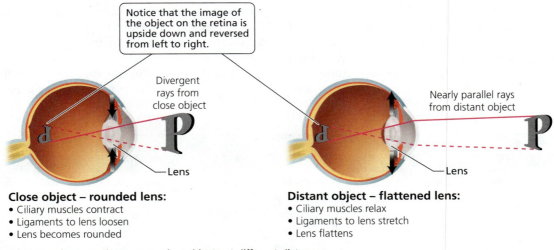

Close object – rounded lens:
- Ciliary muscles contract
- Ligaments to lens loosen
- Lens becomes rounded

Distant object – flattened lens:
- Ciliary muscles relax
- Ligaments to lens stretch
- Lens flattens

FIGURE 9.6 *The lens changes shape so the eye can view objects at different distances.*

FIGURE 9.7 A cataract is a lens that has become opaque.

focus on objects that are near. Changing the shape of the lens to change the bending of light is called **accommodation**.

These changes in lens shape are controlled by the ciliary muscle, which is attached to the lens by ligaments. Because the ciliary muscle is circular, its diameter becomes smaller when it contracts, much like lips when they are pursed. This contraction relaxes the tension on the ligaments, freeing the lens to assume the rounded shape needed to focus on nearby objects. Relaxation of the ciliary muscle increases its diameter, increasing the tension on the ligaments and the lens. Consequently, the lens flattens and focuses light from more distant objects on the retina. As we age, the lens becomes less elastic and does not round up to focus on nearby objects as easily. This is why we hold the newspaper farther away as we become older.

Cataracts A **cataract** is a cloudiness or opaqueness in the lens, usually a result of aging (Figure 9.7). Typically, the lens takes on a yellowish hue that blocks light on its way to the retina. In the beginning, cataracts may cause a person to see the world through a haze that can limit activities and cause automobile accidents, early retirement, and dangerous falls. As the lens becomes increasingly opaque, the fog thickens. Indeed, cataracts are the leading cause of blindness worldwide. Cataracts can be treated by surgically removing the clouded lens and replacing it with an artificial lens.

Cataracts are the most common eye problem affecting men and women older than 50. However, there are some things that everyone can do to minimize the risk of developing cataracts. Exposure to ultraviolet radiation and bright light can cause cataracts, so wear sunglasses in bright sunlight. Cigarette smoke also causes cataracts, so minimize your exposure.

Focusing problems The three most common visual problems that occur in individuals of all ages are problems in focusing. These include farsightedness, nearsightedness, and astigmatism. With each of these focusing problems, normal vision can usually be restored with corrective lenses. Roughly 60% of Americans are farsighted (Table 9.2). In *farsightedness,* distant objects are seen more clearly than nearby ones because the eyeball is too short or the lens is too thin, causing images of nearer objects to be focused behind the retina. Although distant objects can be seen clearly, the lens cannot become round enough to bend the light sufficiently to focus on nearby objects. Corrective lenses that are thicker in the middle than at the edges (convex) cause the light rays to converge a bit before they enter the eye. The lens can then focus the image on the retina.

People who are nearsighted can see nearby objects more clearly than those far away. *Nearsightedness* (called *myopia*) occurs when the eyeball is elongated or when the lens is too thick. This condition causes the image to focus in front of the retina. Nearsighted people see nearby objects clearly because the lens becomes round enough to focus the image on the retina. However, the lens simply cannot flatten enough to bring the focused image of distant objects to the retina, so those objects appear blurred. Lenses that are thinner in the middle than at the edges (concave) can correct nearsightedness. These lenses cause the light rays to diverge slightly before entering the eye.

Although genetics undoubtedly plays a role in the development of nearsightedness, frequent close work, such as reading or working at a computer, is also a cause. When you do a lot of close work, the frequent contractions of the ciliary muscles that change the shape of the lens increase the pressure within the eye. This pressure can cause the eye to stretch and elongate, resulting in nearsightedness. Eye specialists recommend that you look up from the page or away from the computer screen at frequent intervals—particularly if nearsightedness runs in your family.

Irregularities in the curvature of the cornea or lens will cause distortion of the image because they cause the light rays to converge unevenly. This condition is called *astigmatism*. Vision can almost always be restored to normal by corrective lenses that compensate for the asymmetrical bending of light rays (Figure 9.8).

Many people who are tired of depending on glasses or contact lenses have opted to undergo laser eye surgery. This

TABLE 9.2 Focusing Problems of the Eye

Problem	Description	Cause	Correction
Farsightedness	See distant objects more clearly than nearby objects	Eyeball too short or lens too thin; lens cannot become round enough	Convex lens; increases corneal curvature
Nearsightedness	See nearby objects more clearly than distant objects	Eyeball too long or lens too thick; lens cannot flatten enough	Concave lens; decreases corneal curvature
Astigmatism	Visual image is distorted	Irregularities in curvature of cornea or lens	Lenses that correct for the asymmetrical bending of light

(a) Normal eye
- Close and distant objects seen clearly
- Image focuses on retina

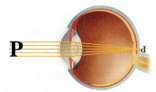

(b) Farsightedness
- Distant objects seen clearly
- Close objects out of focus
- Short eyeball causes image to focus behind retina

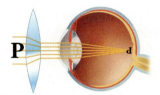

Convex lens
- Causes light rays to converge so that the image focuses on the retina

(c) Nearsightedness (myopia)
- Close objects seen clearly
- Distant objects out of focus
- Long eyeball causes image to focus in front of retina

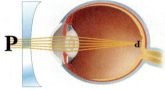

Concave lens
- Causes light rays to diverge so that the image focuses on the retina

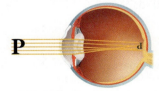

(d) Astigmatism
- Image blurred
- Irregular curvature of cornea or lens causes light rays to focus unevenly

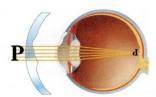

Uneven lens
- Focuses entire image on retina

FIGURE 9.8 *Focusing problems such as farsightedness, nearsightedness, and astigmatism result when the image of an object is not focused on the retina. These vision problems can be corrected with the use of specific lenses.*

procedure is popularly known as LASIK, which stands for "laser-assisted in situ keratomileusis." LASIK permanently changes the shape of the cornea. The procedure involves cutting a flap in the cornea, using pulses from a computer-controlled laser to reshape the middle layer of the cornea, and then replacing the flap. This treatment is used to correct nearsightedness, farsightedness, and astigmatism.

Another procedure for reshaping the cornea is LASEK (laser-assisted subepithelial keratomileusis). It differs from LASIK in that the surgeon creates a flap in the epithelium only, instead of in the entire cornea, eliminating some of the complications caused by the deeper corneal flaps. LASEK is used mostly for people who are poor candidates for LASIK because their corneas are thin or flat. In yet another procedure, photorefractive keratectomy (PRK), a surgeon uses short bursts of a laser beam to shave a microscopic layer of cells off the corneal surface and flatten it. A computer calculates and controls the laser exposure.

Light and Pigment Molecules

The eye contains a whopping 70% of all the sensory receptors in the body. The function of the eye's receptors, both rods and cones, is to respond to light by sending neural messages to the brain, where they are translated into images of our surroundings.

So how does vision work? First, light waves in the visible spectrum (the spectrum of wavelengths our eyes can detect) strike an object. At least some wavelengths are reflected off the object to the eye, where they are focused by the cornea and lens onto the retina and absorbed by photopigment molecules in the photoreceptors. When light strikes a photopigment, it causes the photopigment molecule's components to split apart. This change causes a series of reactions that reduce the photoreceptor's release of inhibitory neurotransmitter. The neurotransmitter normally inhibits the cells involved in processing the information (discussed shortly); so, by leading to reduced neurotransmitter levels, light increases the activity of these processing cells. The optic nerve carries these nerve signals from the retina to the thalamus of the brain, where some processing of the information occurs. The information is then sent to the visual cortex, located at the back of the brain, where a perception of the world outside takes shape.

The processing of electrical signals from the rods and cones begins before the signals leave the retina. The messages are first sent to *bipolar cells* in the retina. These cells are rare examples of neurons that have only two processes, an axon and a single dendrite, extending from opposite sides of the cell body. Bipolar neurons direct the information to interneurons called *ganglion cells* (Figure 9.9). Together, bipolar cells and ganglion cells convert the input from the retina into patterns, such as edges and spots.

Rods: Vision in Dim Light

The **rods** are the photoreceptors responsible for black-and-white vision (Figure 9.10). Rods, which are much more numerous than the cones, are exceedingly sensitive to light and are capable of responding to light measuring one 10-billionth of a watt—the equivalent of a match burning 50 mi (80.5 km) away

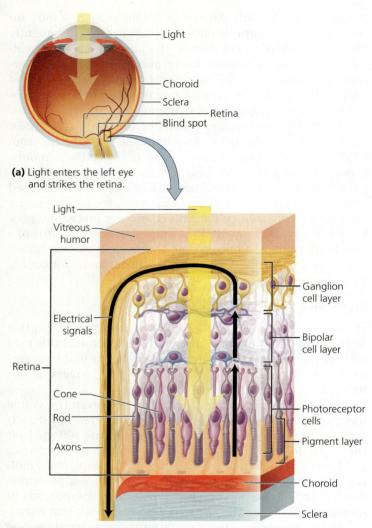

(a) Light enters the left eye and strikes the retina.

(b) When light is focused on the retina, it passes through the ganglion cell layer and bipolar cell layer before reaching the rods and cones. In response to light, the rods and cones generate electrical signals that are sent to bipolar cells and then to ganglion cells. These cells begin the processing of visual information.

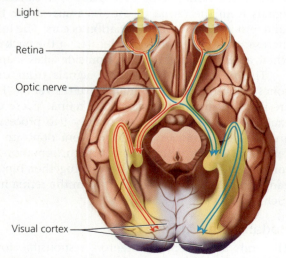

(c) The axons of the ganglion cells leave the eye at the blind spot, carrying nerve impulses to the brain (viewed from below) by means of the optic nerve.

FIGURE 9.9 *Neural pathways of the retina convert light to nerve signals.*

on a clear, pitch-dark night! (Of course, the ideal environmental conditions necessary to actually *see* a burning match at such a distance are impossible to attain.) The rods allow us to see in dimly lit rooms and in pale moonlight. By detecting changes in light intensity across the visual field, rods contribute to the perception of movements.

The pigment in rods, called **rhodopsin**, is packaged in membrane-bound disks that are stacked like coins in the outer segment of the rod. Light strikes a rod, splits the rhodopsin, and triggers events that reduce the permeability of the plasma membrane to sodium ions. This reaction eventually leads to changes in the activity of bipolar cells and ganglion cells. In the dark, rhodopsin is resynthesized. The reason you have difficulty seeing when you first walk into a dark room from a brightly lit area is that the rhodopsin in your photoreceptors has been split. As rhodopsin is resynthesized, your vision becomes sharper.

Cones: Color Vision

The **cones** are the photoreceptors responsible for color vision. Unlike the rods, the cones produce sharp images, because each cone usually sends information to only one bipolar cell for processing.

What makes you see red? Red, or any other color, is determined by wavelengths of light. White light, such as that emitted by an ordinary lightbulb or the sun, consists of all wavelengths. (Rainbows are created when the various wavelengths of white light are separated as they pass through tiny droplets of water in the air.) When light strikes an object, some of the wavelengths may be absorbed and others reflected; we see the wavelengths that are reflected. Thus, an apple looks red because it reflects mostly red light and absorbs most of the other wavelengths.

We see color because we have three types of cones, called blue, green, and red. The cones are named for the wavelengths they absorb best, not for their color. When light is absorbed, the cone is stimulated. Actually, each type of cone absorbs a range of wavelengths, and the ranges overlap quite a bit. As a result, colored light stimulates each type of cone to a different extent. For instance, when we look at a bowl of fruit, light reflected by a red apple stimulates red cones, light reflected by a ripe yellow banana stimulates both red and green cones, and light reflected from blueberries stimulates both blue and green cones. The brain then interprets color according to how strongly each type of cone is stimulated.

> **stop and think**
>
> Most bipolar cells receive input from several rods but from only one cone in the fovea. How might this difference explain why rods permit us to see at lower light intensities than do cones? How might it explain why the sharpest images are formed when the object is focused on the fovea?

Color blindness is a condition in which certain colors cannot be distinguished from each other. Most people who are color blind see some colors, but they tend to confuse certain colors with others. Thus, a standard test for color blindness is

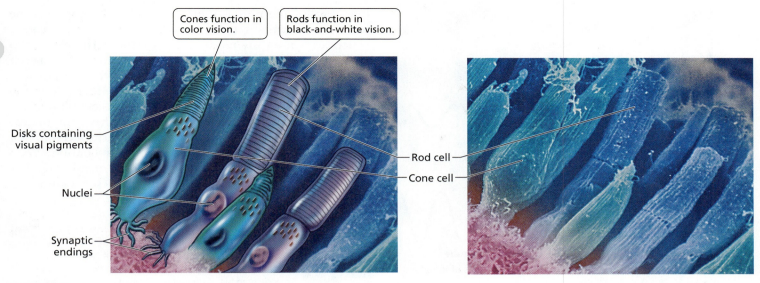

FIGURE 9.10 Rods and cones, the photoreceptors in the retina, are named for their shapes. Their function is to generate neural messages in response to light. The outer portion of each photoreceptor is packed with membrane-bound disks that contain pigment molecules.

to show the individual a color plate with dots of two colors (Figure 9.11). Dots of one color indicate a number. A person who is color blind cannot distinguish the number. A lack, or a reduced number, of one of the types of cones causes the confusion. A person who lacks red cones sees deep reds as black. In contrast, a person who lacks green cones sees deep reds but cannot distinguish between reds, oranges, and yellows. Absence of blue cones is extremely rare.

People who are color blind generally function normally in everyday life. They compensate for the inability to distinguish certain colors by using other cues, such as intensity, shape, or position. Ironically, red and green—the universal traffic colors for "stop" and "go"—are the most commonly confused colors. However, it is usually possible to pick out the brightest of the lights in a traffic signal, and all traffic signals are arranged the same way—red on top and green on the bottom. Schoolchildren are regularly tested for color blindness so that both students and teachers are aware of the condition. Color blindness may cause difficulty with class work, such as reading maps or graphs that are presented in color. If a teacher is aware of the condition, steps can be taken to avoid frustration and confusion for the student.

9.5 Hearing

From a social perspective, hearing is perhaps the most important of our senses because speech plays such an important role in the communication that binds society. Hearing can also make us aware of things in the external world that we cannot see—telling us of a car approaching from behind as we jog, for instance. Hearing can also add to our understanding of events that we *can* see, as when a mother determines by the sound of the cry whether her baby is hungry or in pain. And hearing can enrich the quality of our lives, as when we listen to music or hear waves crashing on the shore.

In every instance, what we hear are sound waves produced by vibration. Vibrating objects, such as guitar strings, the surface of the stereo speaker, or vocal cords, move rapidly back and forth. The vibrations push repeatedly against the surrounding air, creating sound waves, as shown in Figure 9.12. The loudness of sound is determined by the amplitude of the sound wave, represented graphically as the distance between the top of peaks and the bottom of troughs. The pitch is determined by the frequency, the number of cycles (repetitions of the wave) per second. The more cycles there are, the higher the pitch.

Form and Function of the Ear

The ear has three main parts—the outer ear, the middle ear, and the inner ear. The **outer ear** functions as a receiver. The part of the outer ear visible on the head consists of a fleshy

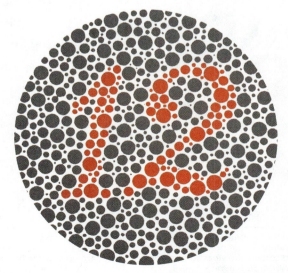

FIGURE 9.11 A standard test for color blindness. A person who lacks one of the three types of cones would not be able to distinguish the number 12 in the center of this circle.

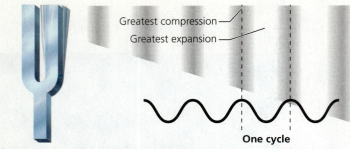

(a) Sound is caused by a vibrating object that causes pressure waves in the air (or water).

Low amplitude **Higher amplitude**

(b) The amplitude (height) of the wave determines the loudness of the sound. Loudness increases with amplitude.

Low frequency **Higher frequency**

(c) The frequency (cycles per second) of the waves determines the pitch of the sound. Pitch goes up as frequency increases.

FIGURE 9.12 *A sound wave and the effects of amplitude (height) and frequency (cycles per second)*

funnel called the **pinna** (Figure 9.13). The pinna gathers the sound and channels it into the **external auditory canal**, which leads from the pinna to the eardrum. The pinna also accentuates the frequencies of the most important speech sounds, making speech easier to pick out from background noise. The pinna's shape and placement at either side of the head help us determine the direction a sound comes from.

The tissue that separates the outer ear from the middle ear is the eardrum, or **tympanic membrane**. It is as thin as a sheet of paper and as taut as the head of a tambourine. When sound waves strike the eardrum, it vibrates at the same frequency as the sound waves and transfers these vibrations to the middle ear.

stop and think

As we age, the tissues of the eardrum thicken and become less flexible. How do these changes partially explain why sensitivity to high-frequency sounds is usually lost as we grow older?

The **middle ear** serves as an amplifier. It consists of an air-filled cavity within a bone of the skull and is spanned by the three smallest bones of the body: the **malleus** (hammer), **incus** (anvil), and **stapes** (stirrup)—named to indicate their respective shapes. Together, these bones function as a system of levers to

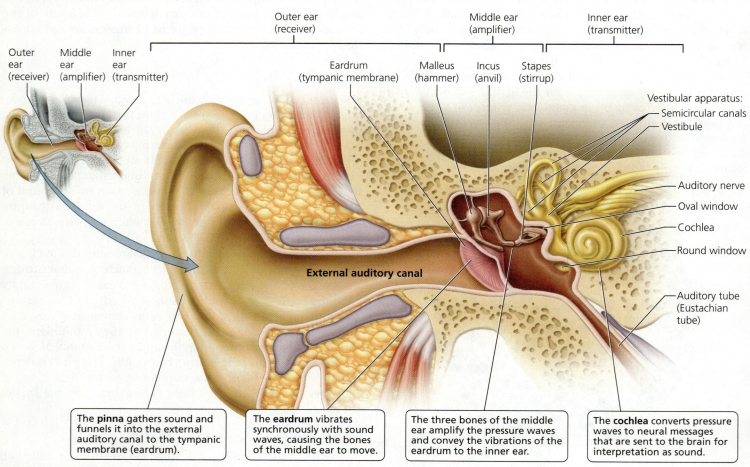

FIGURE 9.13 *Structure of the ear. The human ear has three parts: the outer ear (the receiver), the middle ear (the amplifier), and the inner ear (the transmitter).*

convey the airborne sound waves from the eardrum to the **oval window**, a sheet of tissue that forms the threshold of the inner ear. The malleus is attached to the inner surface of the eardrum. Therefore, the vibrations of the eardrum in response to sound cause the malleus to rock back and forth. The rocking of the malleus, in turn, causes the incus and then the stapes to move. The base of the stapes fits into the oval window.

The force of the eardrum's vibrations is amplified 22 times in the middle ear. The magnification of force is necessary to transfer the vibrations to the fluid of the inner ear. The amplification occurs because the eardrum is larger than the oval window. This size difference concentrates the pressure against the oval window.

The air pressure must be nearly equal on both sides of the eardrum for the eardrum to vibrate properly. If the pressure is not equal, the eardrum will bulge inward or outward and will be held in that position, causing pain and difficulty hearing. For instance, when a person ascends quickly to a high altitude, the atmospheric pressure is lower than the pressure in the middle ear, and the eardrum bulges outward. Unequal pressure usually is alleviated by the **auditory tube** (also known as the *Eustachian tube*), a canal that connects the middle ear cavity with the upper region of the throat. Most of the time, the auditory tube is closed and flattened. However, swallowing or yawning opens it briefly, allowing the pressure in the middle ear cavity to equalize with the air pressure outside the ear. The sensation of pressure suddenly equalizing is described as the ear "popping." It can occur whenever the pressure on the external eardrum changes quickly, as when you ride up or down in an elevator, take off or land in an airplane, or go scuba diving.

The **inner ear** is a transmitter. It generates neural messages in response to pressure waves caused by sound waves, and it sends the messages to the brain for interpretation. The inner ear contains two sensory organs, only one of which, the **cochlea** (kok'-le-ah), is concerned with hearing. The other sensory organ, the vestibular apparatus, is concerned with sensations of body position and movement and will be discussed in the next section.

The cochlea, which is about the size of a pea, is considered the true seat of hearing. It is a bony tube about 35 mm (1.4 in.) long that is coiled about two-and-a-half times and looks somewhat like the shell of a snail, as you can see in Figure **9.14**. (*Cochlea* is from the Latin for "snail.") The wider end of the tube, where the snail's head would be, has two membrane-covered openings. The upper opening is the oval window, into which the stapes fits. The lower opening, called the **round window**, serves to relieve the pressure created by the movements of the oval window.

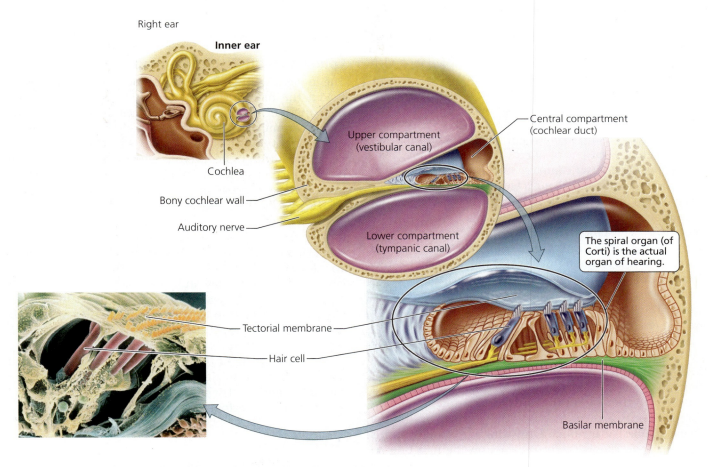

FIGURE 9.14 The cochlea houses the spiral organ (of Corti). The spiral organ consists of the hair cells and the overhanging gelatinous tectorial membrane. The spiral organ rests on the basilar membrane. The hairs are the receptors for hearing. In the micrograph, the hair cells are colored pink, and their "hairs" are colored yellow.

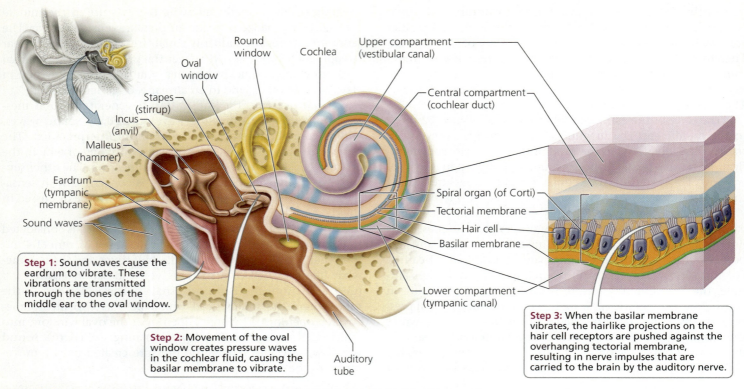

FIGURE 9.15 The sequence of events from sound vibration to a nerve impulse begins when sound enters the external auditory canal.

The internal structure of the cochlea is easier to understand if we imagine the cochlea uncoiled so that it forms a long straight tube. We would then see that two membranes divide the interior of the cochlea into three longitudinal compartments, each filled with fluid. The central compartment (the cochlear duct) ends blindly, like the finger on a glove, and does not extend completely to the end of the cochlea's internal tube. As a result, the upper and lower compartments (the vestibular canal and the tympanic canal) are connected at the end of the tube nearest to the tip of the coil. The **basilar membrane** is the floor of the central compartment. The **spiral organ** (of Corti), the portion of the cochlea most directly responsible for the sense of hearing, is supported on the basilar membrane (Figure 9.15). The spiral organ consists of the **hair cells** and the overhanging **tectorial membrane**. Each hair cell has about 100 "hairs," slender projections from its upper surface that project into the tectorial membrane. The hair cells are arranged on the basilar membrane in rows that resemble miniature picket fences.

When the stapes moves to and fro against the oval window, it sets up corresponding movements—pressure waves—in the fluid of the inner ear (Figure 9.15). Those pressure waves are then transmitted from the fluid of the cochlea's upper compartment to the fluid of the lower compartment, because the compartments are continuous. The movements of the fluid cause the basilar membrane to swing up and down; this swinging, in turn, causes the projections on the hair cells to be pressed against the tectorial membrane. The bending of the hairs ultimately alters the rate of nerve impulses in the auditory (cochlear) nerve, which arises within the cochlea and carries sound information to the brain. The structures of the ear and their functions are summarized in Table 9.3.

Loudness and Pitch of Sound

The louder the sound, the greater the pressure changes in the fluid of the inner ear and, therefore, the stronger the bending of the basilar membrane. Because hair cells have different thresholds of stimulation, more vigorous vibrations in the basilar membrane stimulate more hair cells. In addition, each hair cell bends more with more vigorous vibrations of the basilar membrane, which increases the number of nerve impulses. The brain interprets the increased number of impulses as louder sound.

How does the ear determine the pitch of a sound? Different regions of the basilar membrane vibrate in response to sounds of different pitch. In a sense, the cochlea is like a spiral piano keyboard. The basilar membrane varies in width and flexibility along its length. Near the oval window, the basilar membrane is narrow and stiff. Like the shorter strings on a piano, this region vibrates maximally in response to high-frequency sound, such as a whistle. At the tip of the tube in the cochlea, the basilar membrane is wider and floppier. Low-frequency sounds, such as a bass drum, cause maximum vibrations in this region of the basilar membrane. Thus, sounds of different pitches activate hair cells at different places along the basilar membrane. The brain then interprets input from hair cells in different regions as sounds of different pitch.

Hearing Loss

An estimated 28 million Americans have some degree of hearing loss, and 2 million of them are completely deaf. Hearing loss that is severe enough to interfere with social and job-related communication is among the most common chronic neural impairments in the United States.

TABLE 9.3 Review of the Structures of the Ear and Their Functions

Structure	Description	Function
Outer ear		
Pinna	Fleshy, funnel-shaped part of the ear protruding from the side of the head	Collects and directs sound waves
External auditory canal	Canal between pinna and tympanic membrane	Directs sound to the middle ear
Middle ear		
Eardrum (tympanic membrane)	Membrane spanning the end of the external auditory canal	Vibrates in response to sound waves
Malleus (hammer), incus (anvil), and stapes (stirrup)	Three tiny bones of the middle ear	Amplify the vibrations of the tympanic membrane and transmit vibrations to inner ear
Auditory tube (Eustachian tube)	A tube that connects the middle ear with the throat	Allows equalization of pressure in middle ear with external air pressure
Inner ear		
Cochlea	Fluid-filled, bony, snail-shaped chamber	Houses spiral organ (of Corti) and has openings called oval window and round window
Spiral organ (of Corti)	Contains hair cells	The organ of hearing
Oval window	Membrane between the middle and inner ear that the stapes presses against	Transmits the movements of the stapes to the fluid in the inner ear
Round window	Membrane at the end of the lower canal in cochlea	Relieves pressure created by the movements of the oval window
Vestibular apparatus	Fluid-filled chambers and canals	Monitors position and movement of the head
Vestibule (utricle and saccule)	Two fluid-filled chambers	Maintains static equilibrium (body and head stationary, information on position of head)
Semicircular canals	Three fluid-filled chambers oriented at right angles to one another	Maintain dynamic equilibrium (body or head moving)

There are two types of hearing loss: *conductive loss* and *sensorineural loss.* Conductive loss results when an obstruction anywhere along the route prevents sounds from being conducted through the external auditory canal to the eardrum and over the bones of the middle ear to the inner ear. For example, the external auditory canal can become clogged with wax or other foreign matter. In this case, inserting cotton-tipped swabs usually aggravates the situation by further clogging the canal and should therefore be avoided. Other possible causes for conductive hearing loss are thickening of the eardrum, which might occur with chronic infection, or perforation of the eardrum, which might occur as a result of trauma. Excess fluid in the middle ear, which often occurs with middle ear infections, can prevent movement of the eardrum and cause temporary conductive hearing loss.

Sensorineural deafness is caused by damage to the hair cells or to the nerve supply of the inner ear. A common cause is gradual loss of hair cells throughout life. Some of the loss is simply due to aging. After age 20, the loss of hair cells causes us to lose about 1 Hz of our total perceptive range of 20,000 Hz every day. However, much of the damage to hair cells is caused by exposure to loud noise (see the Environmental Issue essay, *Noise Pollution*). Neural damage can also be caused by infections, including mumps, rubella (German measles), syphilis, and meningitis. In addition, drugs such as those used for treating tuberculosis and certain cancers can cause neural damage.

Hearing can be damaged by exposure to noise that is loud enough to make conversation difficult. The louder a sound, the shorter the exposure time necessary to damage the ear. Even a brief, explosively loud sound is capable of damaging hair cells. More commonly, however, hearing loss results from prolonged exposure to volumes over 85 dB. At 110 dB, the average rock concert or stereo ear buds at full blast can damage your ears in as little as 30 minutes. If sounds seem muffled to you or if your ears are ringing after you leave a noisy area, you probably sustained some damage to your ears.

A surprising number of young people also have impaired hearing. The culprit is most likely noise—probably in the form of music. How can you protect yourself? Don't listen to loud music. Keep it tuned low enough that you can still hear other sounds. If you are listening with ear buds, no one else should be able to hear the sound from them. When you cannot avoid loud noise, as when you are mowing the lawn, vacuuming, or attending a rock concert, wear earplugs to protect your hearing. You can buy them at most drugstores, sporting goods stores, and music stores.

One of the first signs of sensorineural damage is the inability to hear high-frequency sounds. Because consonants, which are needed to decipher most words, have higher frequencies than do vowels, speech becomes difficult to understand. So beware if you find yourself asking, "Could you repeat that?"

The most effective way to deal with hearing loss is to wear a hearing aid. The basic job of a hearing aid is to amplify sound. One type of hearing aid presents amplified sound to the eardrum. Another type uses a vibrator or stimulator placed

ENVIRONMENTAL ISSUE

Noise Pollution

It is difficult to escape the din of modern life—noise from airports, city streets, loud appliances, mobile music players. Noise pollution threatens your hearing and your health. Exposure to excessive noise is to blame for the hearing loss of one-third of all hearing-impaired people. Loud noise damages the hairs on the hair cells of the inner ear. When the hairs are exposed to too much noise, they wear down, lose their flexibility, and can fuse together (Figure 9.A). Unfortunately, there is no way to undo the damage; you cannot grow spare parts for your ears.

The loudness of noise is measured in decibels (dB). The decibel scale is logarithmic. An increase of 10 dB generally makes a given sound twice as loud. The decibel ratings and effects of some familiar sounds are shown in Table 9.A. Most people judge sounds over 60 dB to be intrusive, over 80 dB to be annoying, and over 100 dB to be extremely bothersome. The federal Occupational Safety and Health Administration (OSHA) has set 85 dB as the safety limit for 8 hours of exposure. The threshold for physical pain is 140 dB.

Questions to Consider

- Do you think that local governments should have more regulations to control noise pollution?
- Do you think there should be regulations about how loudly the music at a concert can be played?

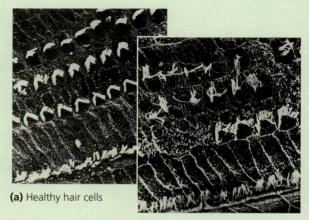

(a) Healthy hair cells

(b) Hair cells damaged by loud noise

FIGURE 9.A The hair cells of the inner ear can be permanently damaged by loud noise.

TABLE 9.A Loudness of Familiar Sounds

Sound Source	Loudness (dB)	Effect from Prolonged Exposure
Jet plane at takeoff	150	Eardrum rupture
Rock-and-roll band (at maximum volume)	130	Irreversible damage
Jet plane at 152 m (500 ft)	110	Loss of hearing
Electric blender	90	Annoyance
Traffic noise	70	
Quiet neighborhood (daytime)	50	
Soft background music	40	
Whisper	20	
Threshold of hearing	0	

in the bone of the skull behind the ear. Sound is then conducted through the bones of the skull to the inner ear.

Hearing aids often are unable to help profoundly deaf people, but cochlear implants can pierce the silence for some of them. A cochlear implant does not amplify sounds as a hearing aid does; instead, it compensates for parts of the inner ear that are not functioning properly. A cochlear implant consists of an array of electrodes that transforms sound into electrical signals that are delivered to the nerve cells near the cochlea. The implant is surgically inserted into the cochlea through the round window. A cochlear implant allows some people who were severely hearing impaired or totally deaf to converse with someone in person or on the phone. By providing exposure to sounds during the time when language is developing, cochlear implants also help deaf children learn to speak well.

what would you do?

Many deaf people consider themselves to be a cultural group, not a group with a medical problem. Some people argue, for instance, that communication is the real challenge faced by the deaf and that the solution should therefore be a social, not a medical, one. If medical technology is available, should we use it to "fix" conditions such as deafness? If you had a child who was born deaf, would you want a cochlear implant for that child? Why or why not? (Assume that the child is not yet able to make that decision for himself or herself.)

Ear Infections

An external ear infection is an infection in the ear canal. *Swimmer's ear,* the most common such infection, is precipitated by water trapped in the canal, which creates an environment favorable for bacterial growth. The first symptom of swimmer's ear is itching, usually followed by pain that can become intense and constant. Chewing food or touching the earlobe sharpens the pain. The usual treatment consists of antibiotic ear drops, heat, and pain medication. To prevent swimmer's ear, always be sure to drain any remaining water from the ear canal after swimming or bathing—if necessary, by shaking the head.

A *middle ear infection* (otitis media) usually results from an infection of the nose and throat that works its way through the auditory tubes connecting the throat and middle ear. At least half of all children get an ear infection at some time; some children get four or five ear infections a year. Middle ear infections are more common in children because the tubes are nearly horizontal, allowing infectious organisms to travel through them more easily. In adults, by contrast, the auditory tubes curve and tilt downward. The signs of a middle ear infection are a stabbing earache, impaired hearing, and a feeling of fullness in the ear, often accompanied by fever. A middle ear infection usually is treated with an antibiotic.

9.6 Balance and the Vestibular Apparatus of the Inner Ear

The **vestibular apparatus**, a fluid-filled maze of chambers and canals within the inner ear, is responsible for monitoring the position and movement of the head. The receptors in the vestibular apparatus are hair cells similar to those in the cochlea. Head movements or changes in velocity cause the hairs on these cells to bend, sending messages to the brain. The brain uses this input to maintain balance.

The vestibular apparatus consists of the semicircular canals and the vestibule. The **semicircular canals** are three canals in each ear that contain sensory receptors and help us stay balanced as we move. They monitor any sudden movements of the head, including those caused by acceleration and deceleration, to help us maintain our equilibrium when the body or head is moving (see the depiction of dynamic equilibrium in Figure 9.16). At the base of each semicircular canal is an enlarged region called the *ampulla*. Within the ampulla is a tuft of hair cells. The hairlike projections from these cells are embedded in a pointed "cap" of stiff, pliable, gelatinous material called the *cupula*. When you move your head, fluid in the canal lags a little behind, causing the cupula to bend the hair cells and stimulate them.

The three semicircular canals in each ear are oriented at right angles to one another. Head movements cause the fluid in the canals to move in the direction opposite the direction of movement, in the same way that sudden acceleration slams passengers back against the car seat. Whereas nodding your head to say yes will cause the fluid in the canals parallel to the sides of the head to swirl, shaking it from side to side to say no will cause fluid in the canals parallel to the horizon to move. Tilting your head to look under the bed will cause fluid in the canals that are parallel to your face to move back and forth. Even the most complex movement can be analyzed in terms of motion in three planes.

> **stop and think**
> When you stop moving, there is a time lag before the fluid in the semicircular canals stops swirling. How does this time lag account for the sensation of dizziness after you have been twirling around for a while?

The **vestibule**, the other part of the vestibular apparatus, is important for static equilibrium—the maintenance of balance when we are not moving. The vestibule consists of the utricle and the saccule, two fluid-filled cavities that literally let us know which end is up (Figure 9.16). These cavities tell the brain the position of the head with respect to gravity when the body is not moving. They also respond to acceleration and deceleration, but not to rotational changes as the semicircular canals do. Both the utricle and the saccule contain hair cells overlaid with a gelatinous material in which small chalklike granules of calcium carbonate are embedded. These granules, called *otoliths*, make the gelatin heavier than the surrounding material and thus make it slide over the hair cells whenever the head is moved. The movement of the gelatin stimulates the hair cells, which send messages to the brain, and the brain interprets the messages to determine the position of the head relative to gravity.

The utricle and saccule sense different types of movement. The utricle senses the forward tilting of the head, as well as forward motion, because its hair cells are on the floor of the chamber and oriented vertically when the head is upright. In contrast, the hair cells of the saccule are on the wall of the chamber, oriented horizontally when the head is upright. They respond when you move vertically, as when you jump up and down.

Motion sickness—that dreadful feeling of dizziness and nausea that sometimes causes vomiting—is thought to be caused by a mismatch between sensory input from the vestibular apparatus and sensory input from the eyes. For example, you may be one of the many people who get carsick from reading in the car. When looking down at a book, your eyes tell your brain that your body is stationary. However, as the car changes speed, turns, and hits bumps in the road, your vestibular system is detecting motion. Similarly, seasickness results when the vestibular apparatus tells the brain that your head is rocking back and forth, but the deck under your feet looks level. The brain is somehow confused by the conflicting information, and the result is motion sickness. Staring at the horizon, so you can see that you are moving, can sometimes relieve the feeling. Over-the-counter drugs to prevent motion sickness, such as Dramamine, work by inhibiting the messages from the vestibular apparatus.

9.7 Smell and Taste

Smell is perhaps the least appreciated of our senses. You could easily get along without it, but life would not be as interesting. For one thing, about 80% of what we usually think of as the flavor of a food is really detected by our sense of smell. This relationship explains why food often tastes bland when the nose is congested.

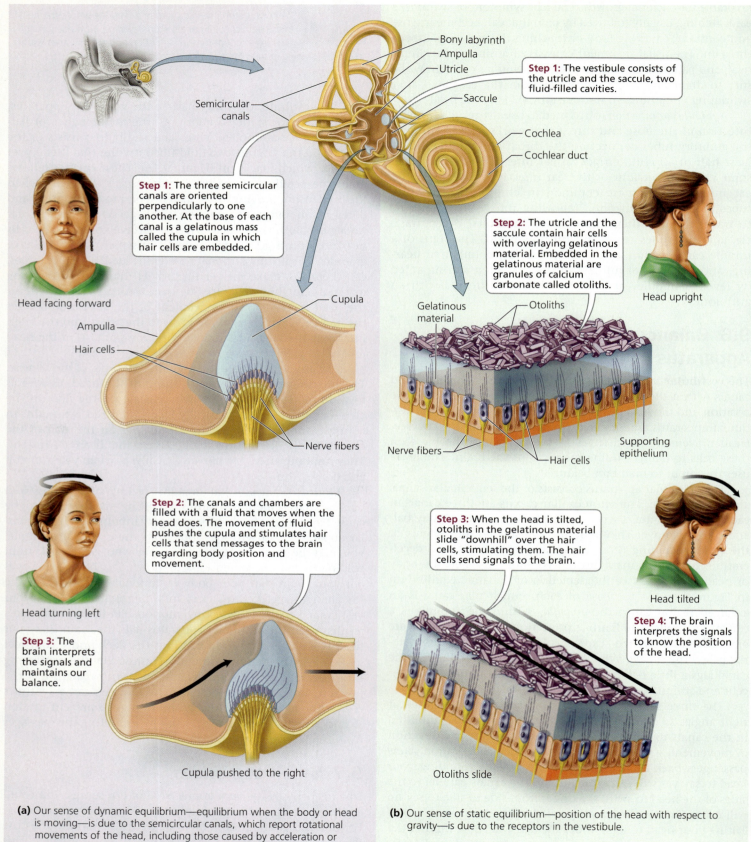

FIGURE 9.16 *Dynamic and static equilibrium*

You have millions of **olfactory (smell) receptors** located not in the nostrils but in a small patch of tissue the size of a postage stamp in the roof of each nasal cavity. These receptors are sensory neurons that branch into long olfactory hairs that project outward from the lining of the nasal cavity and are covered by a coat of mucus, which keeps them moist and is a solvent for odorous molecules. The hairs move, gently swirling the mucus. Olfactory receptors, by the way, are one of the few kinds of neuron known to be replaced during life—about every 60 days.

Odor molecules are carried into the nasal cavity in air, dissolve in the mucus, and bind to receptor sites on the hairs of the olfactory receptor cells, thereby stimulating the cells. If a threshold is reached, the message is carried to the two olfactory bulbs of the brain. The olfactory bulbs process the information from the olfactory receptors and pass it on to the limbic system and to the cerebral cortex (see Chapter 8), where it is interpreted. As you might recall, the limbic system is a center for emotions and memory. Thus, we rarely have neutral responses to odors. For example, the perfume industry makes a fortune from the association between scents and sexuality. Odors can also trigger a flood of long-forgotten memories. If your first kiss was near a blooming lilac bush, for instance, a simple sniff of lilac may take you back in time and place.

We have about 1000 types of olfactory receptors, with which we can distinguish about 10,000 odors. Each receptor responds to several odors. Thus, the brain relies on input from more than one type of receptor to identify an odor (Figure 9.17).

There are at least five primary tastes: sweet, salty, sour, bitter, and umami (oo-ma'-mee; savory flavor). These five tastes are enough to answer the important question about food or drink: Should we swallow it or spit it out? We are prompted to swallow if it tastes sweet, because sweetness implies a rich source of calories and thus energy. Salty tastes also prompt swallowing, because salts will replace those lost in perspiration. Sourness may pose a dilemma. Sour is the taste of unripe fruit that would have more food value later on. As fruit ripens, starches break down into sugars that create a sweet taste and mask the sourness. So it is often better to reject sour fruits and wait for them to ripen. However, some sour fruits, such as oranges, lemons, and tomatoes, are rich sources of vitamin C, an essential vitamin. Bitter is easy. We generally reject foods that taste bitter. Bitterness usually indicates that food is poisonous or spoiled. We swallow food if it tastes umami, because it indicates the food is high in protein.

The less widely known fifth taste, umami, is savory or brothy. *Umami* is derived from a Japanese word that means "delicious." The umami taste occurs when amino acids, especially glutamic acid, small proteins, and nucleotides (DNA and RNA), stimulate umami receptors. The brain interprets input from umami receptors, perhaps with simultaneous input from other receptors, as the taste umami. To experience the flavor umami, boil a cup of water with a tablespoon of dried shitake mushrooms. Let it cool and take a sip. Other umami foods include chicken broth, beef broth, and parmesan cheese.

We have about 10,000 **taste buds**, the structures responsible for our sense of taste. Most taste buds are on the tongue, but some are scattered on the inner surface of the cheeks, on the roof of the mouth, and in the throat. Most taste buds on the tongue are located in papillae, those small bumps that give the tongue a slightly rough feeling. Each papilla contains 100 to 200 taste buds. The cells of a taste bud are completely replaced about every 10 days, so you need not worry about permanently losing your sense of taste when you burn your mouth eating overly hot pizza, as almost always happens.

A taste bud is the meeting point between chemicals dissolved in saliva and the sensory neurons that will convey information about them to the brain. Each taste bud is a lemon-shaped structure containing about 40 modified epithelial cells. Some of these cells are the taste cells that respond to chemicals;

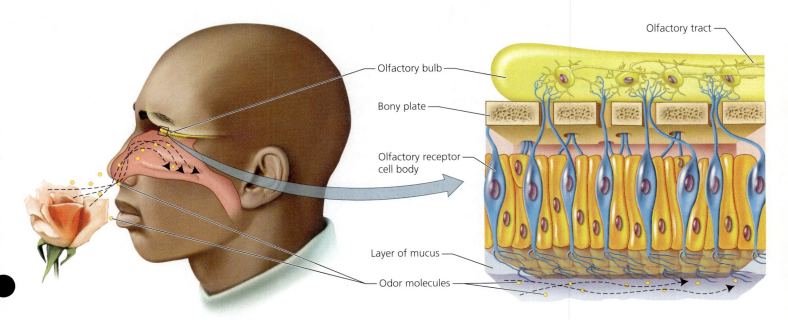

FIGURE 9.17 *Our sense of smell resides in a small patch of tissue in the roof of each nasal cavity. Within each cavity are some 5 million olfactory receptors.*

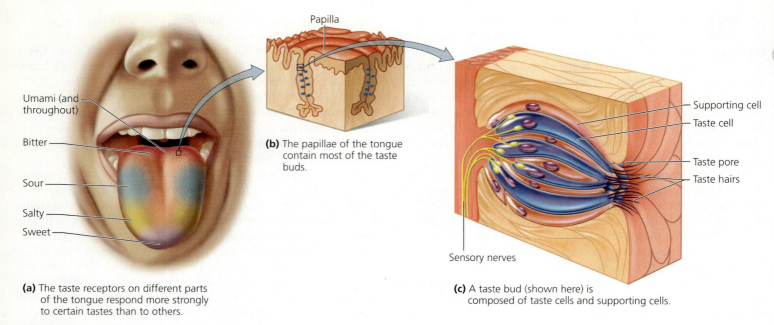

(a) The taste receptors on different parts of the tongue respond more strongly to certain tastes than to others.

(b) The papillae of the tongue contain most of the taste buds.

(c) A taste bud (shown here) is composed of taste cells and supporting cells.

FIGURE 9.18 Taste buds

others are supporting cells (Figure 9.18). The taste cells have long *taste hairs* that project into a pore at the tip of the taste bud. These taste hairs bear the receptors for certain chemicals found in food. When dissolved in saliva, food molecules enter the pore and stimulate the taste hairs. Although taste cells are not neurons, they generate electrical signals that are then sent to sensory nerve cells wrapped around the taste cell.

Each taste bud responds to all five basic tastes but is usually more sensitive to one or two of them than to the others. Because of the way the taste buds are distributed, different regions of the tongue have *slightly* different sensitivity to these tastes. In general, the tip of the tongue is most sensitive to sweet tastes, the back to bitter, and the sides to sour. Sensitivity to salty tastes is fairly evenly distributed on the tongue. Sensitivity to umami is thought to be distributed throughout the tongue, but it certainly exists at the back of the tongue.

looking ahead
In the previous chapters, we have seen how neurons and the nervous system function to create quick responses to stimuli. In the next chapter, we will consider the endocrine system, which produces hormones that cause slower, more prolonged responses within the body.

HIGHLIGHTING THE CONCEPTS

9.1 Sensory Receptors (pp. 150–151)
- Changes in external and internal environments stimulate sensory receptors, which in turn generate electrochemical messages that are converted to nerve impulses that are conducted to the brain.

9.2 Classes of Receptors (p. 151)
- There are five types of sensory receptors. Mechanoreceptors are responsible for touch, pressure, hearing, equilibrium, blood pressure, and body position. Thermoreceptors detect changes in temperature. Photoreceptors detect light. Chemoreceptors monitor chemical levels within the body and detect smells and tastes. Pain receptors respond to strong stimuli that cause physical or chemical damage to tissues.
- Some sensory receptors are located near the body surface and respond to environmental changes. Other sensory receptors are located inside the body and monitor internal conditions.

9.3 The General Senses (pp. 151–153)
- Touch, pressure, and vibration are monitored by mechanoreceptors: Merkel disks, Meissner's corpuscles, Pacinian corpuscles, and Ruffini corpuscles.
- Internal and external temperature is monitored by thermoreceptors, which are widely distributed throughout the body.
- Body and limb position is sensed by the mechanoreceptors called muscle spindles and tendon organs. Muscle spindles are specialized muscle fibers with sensory nerve endings wrapped around them, and tendon organs are highly branched nerve fibers in tendons.
- Pain receptors, located in almost every tissue of the body, respond to any strong stimulus that results in physical or chemical damage to tissues.

9.4 Vision (pp. 153–159)
- The wall of the eye has three layers. The sclera and the cornea make up the outer layer. The choroid, ciliary body, and iris make up the middle layer. The retina, which contains the photoreceptors (rods and cones), is the innermost layer.
- The cornea and the lens work together to focus images on the retina. A normal lens can accommodate (change in shape) to focus on near or distant objects.

- When pigment molecules in the rods or cones (photoreceptors) absorb light, a chemical change occurs in the pigment molecule. The resulting change in the permeability of the membrane of the photoreceptor generates a neural message that is carried by the optic nerve to the brain.
- Color vision depends on the cones. The three types of cones are named for the color of light they absorb best—green, red, or blue—and allow us to see colors.

9.5 Hearing (pp. 159–165)

- The ear is divided into three regions: the outer ear, the middle ear, and the inner ear. The outer ear, which functions as a receiver, consists of the pinna and the external auditory canal. The middle ear, which serves as an amplifier, consists of the tympanic membrane (eardrum), three small bones (malleus, incus, and stapes), and the auditory tube. The inner ear is a transmitter and consists of the cochlea and vestibular apparatus.
- Hearing is the sensing of sound waves caused by vibration. Sound enters the outer ear and vibrates the eardrum. These vibrations move the malleus, which moves the incus and the stapes. The stapes conveys these vibrations to the inner ear by means of the oval window.
- Pressure on either side of the eardrum is regulated by the auditory tube.
- The cochlea is a coiled tube enclosed in bone. Its interior is divided into three longitudinal tubes. The middle tube contains the spiral organ (of Corti), which is lined with hair cells and is the portion of the cochlea most responsible for hearing. Sensory hair cells extend from the basilar membrane, which forms the floor of the spiral organ. Vibrations in the oval window cause fluid in the cochlea to move. Movement of the fluid causes the basilar membrane to vibrate and push hair cells into the overlying tectorial membrane, bending hair cells and initiating nerve impulses.

9.6 Balance and the Vestibular Apparatus of the Inner Ear (pp. 165)

- Balance is controlled by the vestibular apparatus, which consists of the semicircular canals and the vestibule. The semicircular canals monitor sudden movements of the head. The vestibule is made of two components, the saccule and the utricle, which tell the brain the position of the head with respect to gravity.

9.7 Smell and Taste (pp. 165–168)

- Olfactory receptors are located in the nasal cavity. They are lined with cilia and coated with mucus. Odorous molecules dissolve in the mucus and bind to the receptors' hairs, stimulating the receptors. Information is passed to the olfactory bulbs and then to the limbic system and the cerebral cortex.
- Taste buds are responsible for our sense of taste. They are on the tongue, inside the cheeks, on the roof of the mouth, and in the throat. They sense the five primary tastes of sweet, salty, sour, bitter, and umami.

RECOGNIZING KEY TERMS

sensory receptors *p. 150*
stimuli *p. 150*
sensory adaptation *p. 151*
mechanoreceptor *p. 151*
thermoreceptor *p. 151*
photoreceptor *p. 151*
chemoreceptor *p. 151*
pain receptor *p. 151*
general senses *p. 151*
special senses *p. 151*
muscle spindle *p. 152*
tendon organ *p. 152*
referred pain *p. 153*
sclera *p. 153*

cornea *p. 153*
choroid *p. 153*
ciliary body *p. 153*
lens *p. 153*
iris *p. 154*
pupil *p. 154*
retina *p. 154*
fovea *p. 154*
optic nerve *p. 154*
blind spot *p. 154*
vitreous humor *p. 154*
aqueous humor *p. 154*
accommodation *p. 156*
cataract *p. 156*

rods *p. 157*
rhodopsin *p. 158*
cones *p. 158*
outer ear *p. 159*
pinna *p. 160*
external auditory canal *p. 160*
tympanic membrane *p. 160*
middle ear *p. 160*
malleus *p. 160*
incus *p. 160*
stapes *p. 160*
oval window *p. 161*
auditory tube *p. 161*
inner ear *p. 161*

cochlea *p. 161*
round window *p. 161*
basilar membrane *p. 162*
spiral organ *p. 162*
hair cells *p. 162*
tectorial membrane *p. 162*
vestibular apparatus *p. 165*
semicircular canal *p. 165*
vestibule *p. 165*
olfactory (smell) receptor *p. 167*
taste buds *p. 167*

REVIEWING THE CONCEPTS

1. Define *sensory adaptation*, and give an example. *p. 151*
2. What are the five classes of receptors? Which would give rise to the general senses? Which would give rise to special senses? *p. 151*
3. When stimulated, what do Merkel disks tell us? *p. 152*
4. What are the functions of the two distinct regions of the outer layer of the eye? *p. 153*
5. How is light focused on the retina? *p. 155*
6. How is light converted to a neural message? *p. 157*
7. How are sound waves produced? How are loudness and pitch determined? *p. 159*
8. What is the function of the eardrum (tympanic membrane)? *p. 160*
9. Why is it necessary for the force of the vibrations to be amplified in the middle ear? How is amplification accomplished? *pp. 160–161*
10. How does the basilar membrane respond to pitch? *p. 162*
11. What are the two types of hearing loss? Explain how they differ. *p. 163*
12. What are the two structures for equilibrium in the inner ear? Explain the structure of each and how it detects body position and motion. *p. 165*
13. What is believed to cause motion sickness? How can it be counteracted? *p. 165*
14. Where are the olfactory receptors located? Explain their structure as it is related to their function. *p. 167*

15. What are the structures responsible for taste? Where are they found? *p. 167*
16. What are the five primary tastes? *p. 167*
17. Describe the structure of a taste bud. *pp. 167–168*
18. Pacinian corpuscles sense
    a. pressure.
    b. light touch.
    c. warmth.
    d. pain.
19. The spiral organ (of Corti) is important in sensing
    a. body movement.
    b. sound.
    c. light.
    d. degree of muscle contraction.
20. The greatest concentration of cones is found in the
    a. sclera.
    b. lens.
    c. ciliary body.
    d. fovea.
21. Noah is a 4-year-old boy who has a middle ear infection. The cause of the infection could be
    a. bacteria that spread to the ear from a sore throat.
    b. excessive wax buildup in the ear canal.
    c. water that was trapped in the ear after bathing.
    d. noise pollution.
22. The blind spot of the eye is
    a. located in the cornea.
    b. the region where the optic nerve leaves the eye.
    c. the region of the eye where rods outnumber cones.
    d. the region where the ciliary muscle attaches.
23. The loudness or intensity of a sound wave is related to its
    a. amplitude.
    b. frequency.
    c. duration.
    d. pitch.
24. Sound waves are converted into mechanical movements by the
    a. bones of the middle ear.
    b. cochlea.
    c. oval window.
    d. round window.
25. Information about the position of arms, legs, and peripheral joints of the body in space comes from
    a. rods.
    b. cones.
    c. proprioceptors.
    d. semicircular canals.
26. The _____ determines the amount of light that enters the eye.
    a. retina
    b. cornea
    c. pupil
    d. fovea
27. A _____ can detect changes in hydrogen ion concentration.
    a. thermoreceptor
    b. chemoreceptor
    c. proprioceptor
    d. pain receptor
28. Receptors in the vestibular apparatus respond to changes in
    a. head position.
    b. position of limbs.
    c. brightness of light.
    d. taste.
29. The receptors responsible for color vision are the _____.
30. The _____ of the eye changes shape to focus.
31. The snail-shaped structure concerned with hearing is the _____.
32. The semicircular canals help us _____.

APPLYING THE CONCEPTS

1. Molly needs glasses for driving but not for reading. What is the name for her visual problem? Describe the shape of the lens she needs to correct her vision.
2. Sarah has a viral infection that is causing vertigo (dizziness). In which part of the ear is the infection?
3. You are at a baseball game. After watching the ball fly into the stands for a home run, you turn to your friend sitting next to you to ask for some peanuts. What changes occur in your eyes for your friend's face to come into focus?

BECOMING INFORMATION LITERATE

1. Computer visual syndrome (CVS) is a visual problem common among people who spend long hours in front of a computer. Use at least three reliable sources (books, journals, websites) to identify the causes, symptoms, treatment, and ways to avoid CVS. Cite your sources, and explain why you chose them.
2. Imagine that you are interested in gourmet cooking. Use at least three reliable sources (books, journals, or websites) to learn more about the fifth taste, umami. It not only is itself a taste, but also enhances the other flavors of foods. What natural sources of umami might you add to a dish to enhance its flavor? What beverages might you serve with it to create a true taste delight? Cite your sources, and explain why you chose them.

MasteringBiology®

Go to MasteringBiology for practice quizzes, activities, eText, videos, current events, and more.

The Endocrine System

10

Did You Know?

- *Laughter lowers blood levels of stress hormones, such as cortisol and epinephrine.*
- *Male hormones, called androgens, regulate sex drive in both males and females.*

10.1 Functions and Mechanisms of Hormones

10.2 Hypothalamus and Pituitary Gland

10.3 Thyroid Gland

10.4 Parathyroid Glands

10.5 Adrenal Glands

10.6 Pancreas

10.7 Thymus Gland

10.8 Pineal Gland

10.9 Locally Acting Chemical Messengers

ETHICAL ISSUE

Hormone Therapy

In Chapters 7, 8, and 9 we learned about the nervous system, which is used for rapid communication. In this chapter, we learn about the endocrine system, which is used for more leisurely communication. Recall that the nervous system sends its messages across synaptic clefts. The endocrine system, in contrast, sends its messages in the bloodstream. We consider the major endocrine glands and the hormones they secrete. These hormones initiate both long-term changes, such as growth and development, and more short-term changes, such as the fight-or-flight response. Because the nervous and endocrine systems share the common function of regulating and coordinating the activities of all body systems, some consider these two systems to be one—and call it the *neuroendocrine system*.

The endocrine system produces hormones, substances often described as the chemical messengers of the body. The messages they carry are varied but vital and include effects on body height, level of alertness, energy production, and fluid balance.

10.1 Functions and Mechanisms of Hormones

Our bodies contain two types of glands: exocrine glands (see Chapter 4) and endocrine glands. **Endocrine glands** (Figure 10.1) contain secretory cells that release their products, called *hormones*, to the fluid just outside the cells (extracellular fluid). Hormones diffuse from the extracellular fluid directly into the bloodstream. The **endocrine system** consists of endocrine glands and of organs that contain some endocrine tissue; these organs have other functions besides hormone secretion. The major endocrine glands are the pituitary gland, thyroid gland, parathyroid glands, adrenal glands, and pineal gland. Organs with some endocrine tissue include the hypothalamus, thymus, pancreas, ovaries, testes, heart, and placenta. Organs of the digestive and

Q Look closely at the structure of the endocrine gland. How does it differ from the structure of an exocrine gland, such as an oil gland?

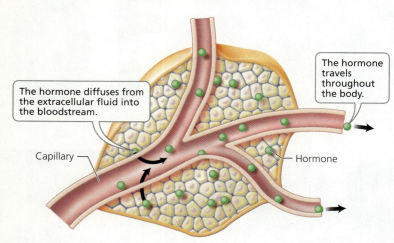

FIGURE 10.1 An endocrine gland. Cells of endocrine glands release their products, called hormones, into the extracellular fluid. The hormones then diffuse into the bloodstream to be transported throughout the body.

A Endocrine glands lack ducts. Rather than secreting their products into ducts that open to a surface, endocrine glands secrete hormones to the fluid just outside their cells, and from there the hormone moves into the bloodstream.

urinary systems, such as the stomach, small intestine, and kidneys, also have endocrine tissue.

Our discussion focuses on the major endocrine glands (Figure 10.2). As we describe individual glands, keep in mind that the main function of the endocrine system—like that of the nervous system—is to regulate and coordinate other body systems and thereby maintain homeostasis. We also examine three organs with endocrine tissue: hypothalamus, thymus, and pancreas. The other organs containing endocrine tissue are discussed in the chapters that cover the organs' other functions. For example, the kidneys are discussed in Chapter 16, which covers the urinary system, and the ovaries and testes are discussed in Chapter 17, which covers the reproductive systems.

Hormones as Chemical Messengers

Hormones are the chemical messengers of the endocrine system. They are released in very small amounts by the cells of endocrine glands and tissues and enter the bloodstream to travel throughout the body. Although hormones come into contact with virtually all cells, most affect only a particular type of cell, called a **target cell**. Target cells have *receptors*, protein molecules that recognize and bind to specific hormones. Once a hormone binds to its specific receptor, this hormone–receptor complex begins to exert its effects on the cell. Because cells other than target cells lack the correct receptors, they are unaffected by the hormone.

The mechanisms by which hormones influence target cells depend on the chemical makeup of the hormone. Hormones are classified as being either lipid soluble or water soluble. **Lipid-soluble hormones** include **steroid hormones**, a group of closely related hormones derived from cholesterol. The ovaries, testes, and adrenal glands are the main organs that secrete steroid hormones. Lipid-soluble hormones move easily through any cell's plasma membrane because it is a lipid bilayer (see Figure 10.3 on p. 174). Once inside a target cell, a steroid hormone combines with receptor molecules either in the cytoplasm or in the nucleus (only target cells have the proper receptors for a given hormone). If binding occurs in the cytoplasm, then the hormone–receptor complex moves into the nucleus of the cell. In the nucleus, the complex attaches to DNA and activates certain genes. Ultimately, such activation leads the target cell to synthesize specific proteins. (The precise steps involved in protein synthesis are described in Chapter 21.) These proteins may include enzymes that stimulate or inhibit particular metabolic pathways.

Water-soluble hormones, such as protein or peptide hormones, cannot pass through the lipid bilayer of the plasma membrane and therefore cannot enter target cells themselves. Instead, the hormone—which in this situation is called the **first messenger**—binds to a receptor on the plasma membrane of the target cell. This binding activates a molecule—called the second messenger—in the cytoplasm. **Second messengers** are molecules within the cell that influence the activity of enzymes, and ultimately the activity of the cell, to produce the effect of the hormone. Cyclic adenosine monophosphate (cAMP) is a common second messenger (illustrated in Figure 10.4 on p. 174).

As an example of how cAMP functions in its role as a second messenger, we consider the effects of the water-soluble hormone epinephrine on a liver cell. Binding of epinephrine to a receptor on the plasma membrane of a liver cell (the target cell) prompts the conversion of ATP to cAMP within the cell. Cyclic adenosine monophosphate then activates an enzyme within the cell (a protein kinase), which in turn activates another enzyme, and so on. The end result of this enzyme cascade is the activation of an enzyme that catalyzes the breakdown of glycogen to glucose within the liver cell. Thus, whereas lipid-soluble hormones stimulate the synthesis of proteins by a cell, water-soluble hormones such as epinephrine activate proteins that are already present in the cell. And water-soluble hormones do this without ever entering the cell.

We have presented the traditional dichotomy of steroid versus peptide hormones and their different mechanisms of action, but things are never so simple. For example, we describe steroid hormones as binding to receptors *inside* the cell and modifying gene expression and protein synthesis. These processes may take several hours or days to produce a response. In recent years, steroid hormones have been found to produce more rapid responses, on the order of seconds or a few minutes. Although not fully characterized, steroid effects that occur this rapidly probably do not involve modifications to gene expression and protein synthesis. Also, in these instances, steroid hormones seem to be interacting with membrane receptors rather than receptors inside the cell.

Feedback Mechanisms and Secretion of Hormones

Now that we have seen how hormones work at the cellular level, let's turn our attention to the factors that stimulate and regulate the release of hormones from endocrine glands. Stimuli that cause endocrine glands to manufacture and release

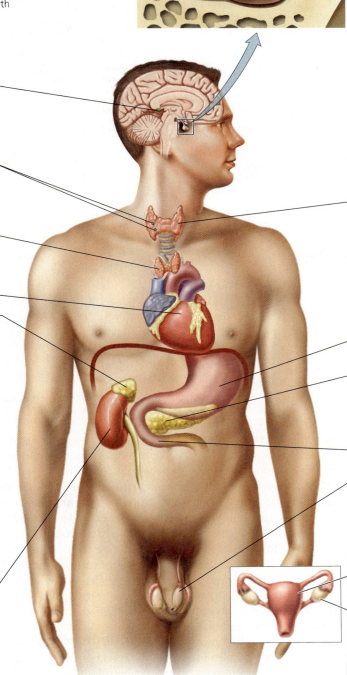

FIGURE 10.2 The endocrine system. The endocrine system is made up of endocrine glands and of organs that contain some endocrine tissue. Here, the hormones and their functions are listed under the endocrine gland or organ that produces them. (Hormones secreted by organs for which hormone secretion is a secondary function—the heart, stomach, small intestine, kidneys, testes, ovaries, and uterus with placenta—are discussed in chapters that cover the organs' other functions.)

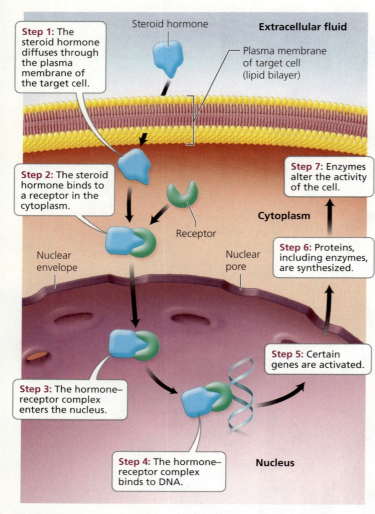

FIGURE 10.3 Mode of action of some lipid-soluble (steroid) hormones. Here, the hormone binds to its receptor in the cytoplasm. In other cases, binding occurs in the nucleus.

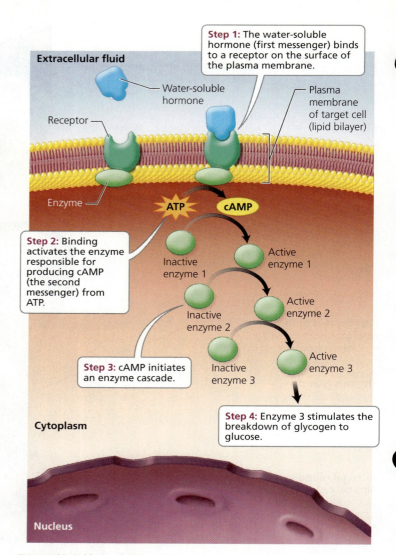

FIGURE 10.4 Mode of action of some water-soluble hormones: the second messenger system of cAMP

hormones include other hormones, signals from the nervous system, and changes in the levels of certain ions (such as calcium, Ca^{2+}) or nutrients (such as glucose) in the blood.

Recall from Chapter 4 that homeostasis keeps the body's internal environment relatively constant. Such constancy most often is achieved through **negative feedback mechanisms**, which, as we also saw in Chapter 4, are homeostatic mechanisms in which the outcome of a process feeds back to the system, shutting the process down. Negative feedback mechanisms regulate the secretion of most hormones. Typically, a gland releases a hormone, and then rising blood levels of that hormone inhibit its further release. In an alternative form of negative feedback, some endocrine glands are sensitive to the particular condition they regulate rather than to the level of the hormone they produce. For example, the pancreas secretes the hormone insulin in response to high levels of glucose in the blood. Insulin prompts the liver to store glucose, which in turn causes the blood level of glucose to decline. The pancreas senses the low glucose in the blood and stops secreting insulin.

Secretion of hormones is sometimes regulated by **positive feedback mechanisms**, in which the outcome of a process feeds back to the system and stimulates the process to continue. For example, during childbirth, the pituitary gland releases the hormone oxytocin (OT), which stimulates the uterus to contract. Uterine contractions then stimulate release of more oxytocin, which stimulates even more contractions (Figure 10.5). The feedback is described as positive because it acts to stimulate, rather than to inhibit, the release of oxytocin. Eventually, some change breaks the positive feedback cycle. In the case of childbirth, expulsion of the baby and placenta terminates the feedback cycle. When we discuss the various glands and their hormones in the sections that follow, we also describe the feedback mechanisms by which they are regulated.

Interactions between Hormones

Interactions between hormones may be antagonistic, synergistic, or permissive. When the effect of one hormone opposes that of another hormone, the interaction is described as *antagonistic*. For an example of an antagonistic interaction, consider glucagon and insulin, two hormones secreted by the pancreas. Whereas glucagon increases the level of glucose in the blood, insulin decreases the level of glucose in the blood.

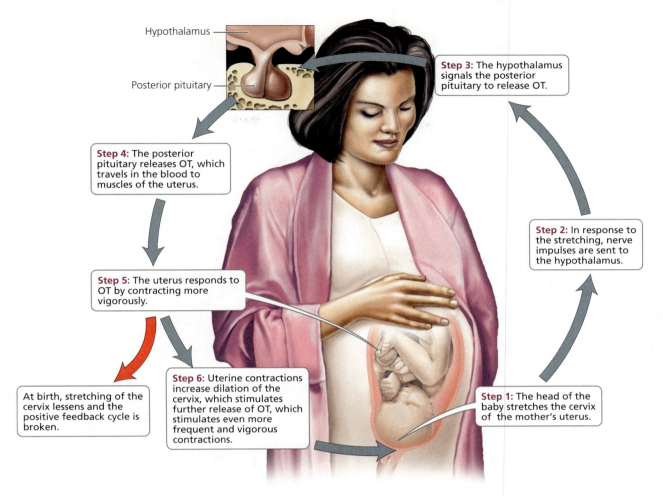

FIGURE 10.5 *The positive feedback cycle by which OT stimulates uterine contractions during childbirth*

During *synergistic* interactions, the response of a tissue to a combination of two hormones is much greater than its response to either individual hormone. For example, epinephrine (from the adrenal glands) and glucagon both prompt the liver to release glucose to the blood. When the two hormones act together, the amount of glucose released by the liver is greater than the combined amount released by each hormone acting alone. During *permissive* interactions, one hormone must be present for another hormone to exert its effects. For example, thyroid hormone must be present for the hormone aldosterone to stimulate reabsorption of sodium within the tubules of the kidneys.

Next we look at the individual endocrine glands, describing the location and general structure, hormones, and hormonal effects of each. We also consider disorders associated with each gland and its hormones.

10.2 Hypothalamus and Pituitary Gland

The **pituitary gland** is the size of a pea and is suspended from the base of the brain by a short stalk (see Figure 10.2). The stalk connects the pituitary gland to the **hypothalamus**, the area of the brain that regulates physiological responses such as body temperature, sleeping, and water balance. The pituitary gland consists of two lobes: the anterior lobe and the posterior lobe. These lobes differ in size and in their relationship with the hypothalamus. The two lobes release different hormones.

The anterior lobe is the larger one. A network of capillaries runs from the base of the hypothalamus through the stalk of the pituitary. The capillaries connect to veins that lead into more capillaries in the anterior lobe of the pituitary gland (Figure 10.6). This circulatory connection allows hormones of the hypothalamus to control the secretion of hormones from the anterior lobe of the pituitary. Specialized neurons in the hypothalamus synthesize and secrete hormones that travel by way of the bloodstream to the anterior lobe. These specialized neurons are called **neurosecretory cells** because they generate and transmit nerve impulses *and* make and secrete hormones. In effect, these cells function as neurons and as endocrine cells, providing a good example of the close relationship between the nervous and endocrine systems. Once these hormones from the hypothalamus reach the anterior pituitary, they stimulate or inhibit hormone secretion. Substances produced by the hypothalamus that stimulate hormone secretion by the anterior pituitary are called **releasing hormones**. Those

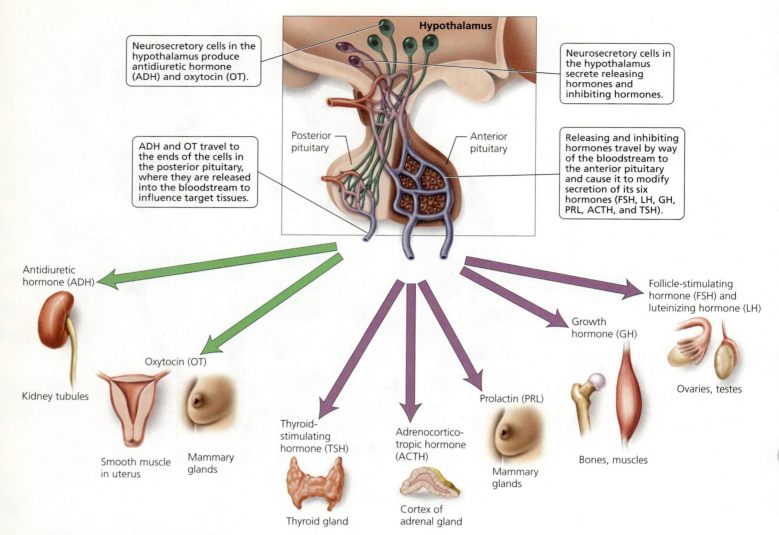

FIGURE 10.6 *The two lobes of the pituitary gland and the hormones they secrete*

that inhibit hormone secretion by the anterior pituitary are called **inhibiting hormones**. The anterior pituitary responds to releasing and inhibiting hormones from the hypothalamus by modifying its own synthesis and secretion of six hormones. These hormones are growth hormone (GH), prolactin (PRL), thyroid-stimulating hormone (TSH), adrenocorticotropic hormone (ACTH), follicle-stimulating hormone (FSH), and luteinizing hormone (LH).

The posterior lobe of the pituitary is very small, just larger than the head of a pin. It consists of neural tissue that releases hormones. In contrast to the *circulatory* connection between the hypothalamus and the anterior lobe, the connection between the hypothalamus and the posterior lobe is a *neural* one. As shown in Figure 10.6, neurosecretory cells from the hypothalamus project directly into the posterior lobe. These neurosecretory cells produce oxytocin (OT) and antidiuretic hormone (ADH). OT and ADH move down the axons to the axon terminals of these cells, which are located in the posterior pituitary. OT and ADH are stored in the posterior pituitary until their release into the bloodstream.

Anterior Lobe

As noted, the anterior lobe of the pituitary produces and secretes six major hormones. We begin with **growth hormone (GH)**, the primary function of which is to stimulate growth through increases in cell size and rates of cell division. The target cells of GH are quite diverse. Cells of bone, muscle, and cartilage are most susceptible to GH, but cells of other tissues are affected as well. Growth hormone also plays a role in glucose conservation by making fats more available as a source of fuel.

Two hormones of the hypothalamus regulate the synthesis and release of GH. Growth hormone–releasing hormone (GHRH) stimulates the release of GH. Growth hormone–inhibiting hormone (GHIH) inhibits the release of GH. Through the actions of these two hormones, levels of GH in the body are normally maintained within an appropriate range. However, excesses or deficiencies of the hormone can dramatically affect growth. For example, abnormally high production of GH in childhood, when the bones are still capable of growing in length, results in *gigantism*, a condition

FIGURE 10.7 Robert Wadlow, a pituitary giant, was born in 1918 at a normal size but developed a pituitary tumor as a young child. The tumor caused increased production of GH. Robert never stopped growing until his death at 22 years of age, by which time he had reached a height of 8 feet 11 inches.

(a) Age 9   (b) Age 16   (c) Age 33   (d) Age 52

FIGURE 10.8 Acromegaly. Excess secretion of GH in adulthood, when the bones can thicken but not lengthen, causes acromegaly, a gradual thickening of the bones of the hands, feet, and face. The disorder was not apparent in this female at ages 9 or 16, but it became apparent by age 33. The symptoms were even more obvious at age 52.

characterized by rapid growth and eventual attainment of height up to 8 or 9 feet (Figure 10.7). Increased production of GH in adulthood, when the bones can thicken but not lengthen, causes acromegaly (literally, "enlarged extremities"). *Acromegaly* is characterized by enlargement of the tongue and a gradual thickening of the bones of the hands, feet, and face (Figure 10.8). Both conditions are associated with decreased life expectancy. The excesses of GH that cause such conditions as gigantism and acromegaly may be caused by a tumor of the anterior pituitary. Tumors can be treated with surgery, radiation, or drugs that reduce GH secretion and tumor size. Insufficient production of GH in childhood results in *pituitary dwarfism*. Typically, pituitary dwarfs are sterile and attain a maximum height of about 4 feet (Figure 10.9). Administering GH in childhood can treat pituitary dwarfism but not other forms of dwarfism.

In the past, the use of GH for the treatment of medical conditions (such as pituitary dwarfism) was extremely limited because GH was scarce, given that the hormone had to be extracted from the pituitary glands of cadavers. Beginning in the late 1970s, however, GH could be made in the laboratory. With this greater availability came research on its potential uses in the treatment of aging in adults and below-average height in children. We examine these new uses of GH in the Ethical Issue essay, *Hormone Therapy*.

Prolactin (PRL), another hormone secreted by the anterior lobe of the pituitary gland, stimulates the mammary glands to

FIGURE 10.9 Pituitary dwarfism is caused by insufficient GH in childhood.

ETHICAL ISSUE

Hormone Therapy

Hormone therapy refers to medical treatment in which hormones are either given to a patient or blocked from acting in a patient's body. Therapy in which hormones are given may involve injections, pills, patches, or creams. For example, people with type 1 diabetes mellitus cannot produce the pancreatic hormone insulin. Thus, they need daily injections of insulin to survive. Because the hormone estrogen promotes the growth of some breast cancers, blocking the effects of estrogen or lowering estrogen levels may be part of a treatment plan for reducing the risk that breast cancer will return. Drugs, such as tamoxifen, can be used to block the effects of estrogen. Tamoxifen, taken daily as a pill, temporarily blocks the estrogen receptors on breast cancer cells. This action prevents estrogen from binding to the cells. In young women, the ovaries are the main source of estrogen. Thus, estrogen levels can be lowered in young women with breast cancer by surgically removing the ovaries. More commonly, drugs or synthetic hormones are used to shut down production of estrogen by the ovaries.

Using hormones to treat medical conditions such as diabetes or breast cancer is not controversial because the patient could die without medical intervention. However, some hormone therapies do provoke debate. For example, hormone therapy is sometimes used to treat declines in hormone secretion that occur as part of the normal aging process. Recall that growth hormone builds and maintains many tissues, including bone, muscle, and cartilage. Production of growth hormone by the anterior pituitary gland declines as we age. This decline has prompted some otherwise healthy adults to take synthetic human growth hormone to slow aging. In the United States, human growth hormone requires a doctor's prescription. Nevertheless, it can be purchased in various forms from foreign sources or over the Internet. Relatively few studies have monitored healthy adults taking growth hormone. Results to date indicate that growth hormone injections can increase muscle mass and decrease body fat. The increased muscle mass, however, does not increase strength. In fact, researchers found that strength training with weights was more effective than growth hormone therapy. Side effects of taking growth hormone included muscle, nerve, and joint pain, and swelling of the hands and feet. Elevated blood levels of glucose and cholesterol also have been reported.

Using synthetic growth hormone to treat idiopathic short stature (ISS) in children is quite controversial. ISS is formally defined as short stature without a known cause. It is informally described as "short, but otherwise normal." Studies monitoring children with ISS who were administered growth hormone typically find that such therapy can yield an additional 1.5 to 3 inches of adult height, although results for individual children vary considerably. So far, the data on ISS patients treated with growth hormone indicate that the treatment is safe. Nevertheless, parents and pediatricians remain concerned about the possible adverse effects that can happen long after the original use of a drug. Opponents of using growth hormone to treat ISS believe it is wrong to give a powerful hormone to healthy children for an essentially cosmetic reason. Rather than administering growth hormone to such children, opponents suggest working to increase societal acceptance of persons who are short. Opponents also point out that growth hormone therapy is invasive, requiring daily injections for several years. Also, growth hormone therapy for ISS is costly. Recent estimates indicate that each inch of height gained costs $35,000 to $52,000.

In the United States, growth hormone is approved for treating certain medical conditions, such as the muscle wasting that occurs with HIV/AIDS. In 2003, the U.S. Food and Drug Administration (FDA) approved the use of growth hormone to treat ISS in children whose heights are well below the average for their age and sex. The eligibility criteria established by the FDA made treatment with growth hormone possible for the shortest 1.2% of children. Growth hormone is not approved as an antiaging treatment.

Questions to Consider

- Are you in favor of taking synthetic hormones to replace hormones in our bodies that are naturally declining? Would you advise your parents or grandparents to take growth hormone to slow aging? Why or why not?
- Do you think that human growth hormone should be used to "treat" non-life-threatening conditions such as ISS in children? What would you do if you were the parent of a healthy child who was destined to be very short? What if your child were a pituitary dwarf; would you approve the use of human growth hormone then?

---

produce milk. (Oxytocin, a hormone secreted from the posterior pituitary, causes the ducts of the mammary glands to eject milk, as discussed later in the chapter.) PRL interferes with female sex hormones, which explains why most mothers fail to have regular menstrual cycles while nursing their newborn. (Lactation should not, however, be relied upon as a method for birth control, because the suppression of female hormones and ovulation lessens as mothers breast-feed their growing infants less frequently.)

Growth of a pituitary tumor may cause excess secretion of PRL, which may cause infertility in females, along with production of milk when birth has not occurred. In males, PRL appears to be involved in the production of mature sperm in the testes, but its precise role is not yet clear. Nevertheless, production of too much PRL, as might occur with a pituitary tumor, can cause sterility and impotence in men. Some hormones from the hypothalamus stimulate and others inhibit production and secretion of PRL.

The remaining hormones produced by the anterior lobe of the pituitary gland influence other endocrine glands. A hormone produced by one endocrine gland or organ that influences another endocrine gland is called a **tropic hormone**. Two such hormones secreted by the anterior lobe of the pituitary are thyroid-stimulating hormone and adrenocorticotropic hormone. **Thyroid-stimulating hormone (TSH)** acts on the thyroid gland in the neck to stimulate synthesis and release of thyroid hormones. **Adrenocorticotropic hormone (ACTH)**, also called corticotropin, controls the synthesis and secretion of glucocorticoid hormones from the outer portion (cortex) of the adrenal glands (see Figure 10.2).

Two other tropic hormones secreted by the anterior lobe of the pituitary gland influence the gonads (ovaries in the female and testes in the male). **Follicle-stimulating hormone (FSH)** promotes development of egg cells and secretion of the hormone estrogen from the ovaries in females. **Luteinizing hormone (LH)** causes ovulation, the release of a future egg cell by the ovary in females. LH also stimulates the ovaries to secrete estrogen and progesterone. These two hormones prepare the uterus for implantation of a fertilized ovum and the breasts for production of milk. In males, FSH promotes maturation of sperm, while LH stimulates cells within the testes to produce and secrete the hormone testosterone.

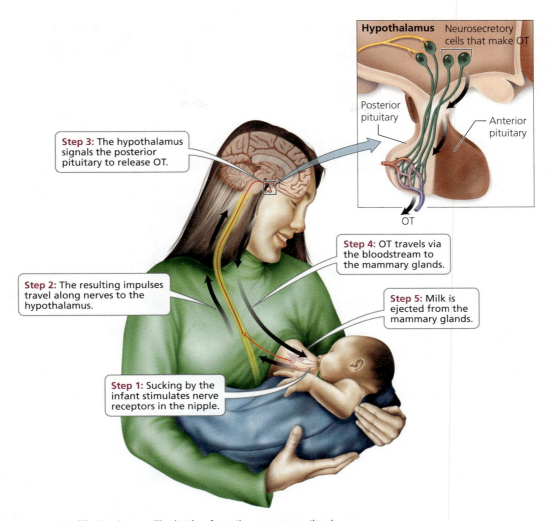

FIGURE 10.10 The steps by which OT stimulates milk ejection from the mammary glands

Posterior Lobe

The posterior pituitary does not produce any hormones of its own. However, neurosecretory cells of the hypothalamus manufacture antidiuretic hormone and oxytocin. These hormones travel down the axons of the neurosecretory cells into axon terminals in the posterior pituitary, where they are stored until their release to the bloodstream.

The main function of **antidiuretic hormone (ADH)** is to conserve body water by decreasing urine output. ADH accomplishes this task by prompting the kidneys to remove water from the fluid destined to become urine. The water is then returned to the blood. Alcohol temporarily inhibits secretion of ADH, causing increased urination following alcohol consumption. The increased output of urine causes dehydration and the resultant headache and dry mouth typical of many hangovers. ADH is also called *vasopressin*. This name comes from its role in constricting blood vessels and raising blood pressure, particularly during times of severe blood loss.

A deficiency of ADH may result from damage to either the posterior pituitary or the area of the hypothalamus responsible for the hormone's manufacture. Such a deficiency results in *diabetes insipidus*, a condition characterized by excessive urine production and resultant dehydration. Mild cases may not require treatment. Severe cases may cause extreme fluid loss, and death through dehydration can result. Treatment usually includes administration of synthetic ADH in a nasal spray. Diabetes insipidus (*diabetes*, overflow; *insipidus*, tasteless) should not be confused with *diabetes mellitus* (*mel,* honey). The latter is a condition in which large amounts of glucose are lost in the urine as a result of an insulin deficiency. Both conditions, however, are characterized by increased production of urine. (We briefly discuss diabetes mellitus later, when we describe the hormones of the pancreas. We describe it in more detail in Chapter 10a.)

Oxytocin (OT) is the second hormone produced in the hypothalamus and released by the posterior pituitary. The name *oxytocin* (*oxy*, quick; *tokos*, childbirth) reveals one of its two main functions: stimulating the uterine contractions of childbirth. As described earlier, the control of OT during labor is an example of a positive feedback mechanism (see Figure 10.5). Pitocin is a synthetic form of OT sometimes administered to induce and speed labor.

The second major function of oxytocin is to stimulate milk ejection from the mammary glands. Milk ejection occurs in response to the sucking stimulus of an infant (Figure 10.10). Recall that prolactin secreted by the anterior pituitary stimulates the mammary glands to produce, but not to eject, milk. Men also secrete OT, and there is some evidence that this hormone facilitates the transport of sperm in the male reproductive tract.

stop and think

Women who have just given birth are often encouraged to nurse their babies as soon as possible after delivery. How might an infant's suckling promote completion of, and recovery from, the birth process? Consider that the placenta (afterbirth) must still be expelled after the birth of the baby and that the uterus must return to an approximation of its prepregnancy form.

10.3 Thyroid Gland

The **thyroid gland** is a shield-shaped, deep red structure in the front of the neck, as shown in Figure 10.11a. (The color stems from its exceptional blood supply.) Within the thyroid are small, spherical chambers called follicles (Figure 10.11b). Cells line the walls of the follicles and produce thyroglobulin, the substance from which thyroxine (T_4) and triiodothyronine (T_3) are made. These two very similar hormones have different numbers of iodine molecules; as indicated by their abbreviations, thyroxine has four iodine molecules, and triiodothyronine has three. Thyroxine is usually produced in greater quantity than triiodothyronine, and most thyroxine is eventually converted to triiodothyronine. Because these two hormones are so similar, we will simply refer to them as **thyroid hormone (TH)**. Other endocrine cells in the thyroid, called *parafollicular cells* (because they occur near the follicles), secrete the hormone calcitonin (Figure 10.11b).

Nearly all body cells are target cells for TH. Therefore, it is not surprising that the hormone has broad effects. TH regulates the body's metabolic rate and production of heat. It also maintains blood pressure and promotes normal development and functioning of several organ systems. TH affects cellular metabolism by stimulating protein synthesis, the breakdown of lipids, and the use of glucose for production of ATP (Chapter 2). The pituitary gland and hypothalamus control the release of TH. Falling levels of TH in the blood prompt the hypothalamus to secrete a releasing hormone. The releasing hormone stimulates the anterior pituitary to secrete TSH, which, in turn, causes the thyroid to release more TH.

Iodine is needed for production of TH. A diet deficient in iodine can produce a *simple goiter*, that is, an enlarged thyroid gland (Figure 10.12a). When intake of iodine is inadequate, the level of TH is low, and the low level of TH in turn triggers secretion of TSH. TSH stimulates the thyroid gland to increase production of thyroglobulin. The lack of iodine prevents formation of TH from the accumulating thyroglobulin. In response to continued low levels of TH, the pituitary continues to release increasing amounts of TSH, which cause the thyroid to enlarge in a futile effort to filter more iodine from the blood. In the past, goiters were quite common, especially in parts of the Midwestern United States (dubbed the Goiter Belt), where iodine-poor soil and little access to iodine-rich shellfish led to diets deficient in iodine. The incidence of goi-

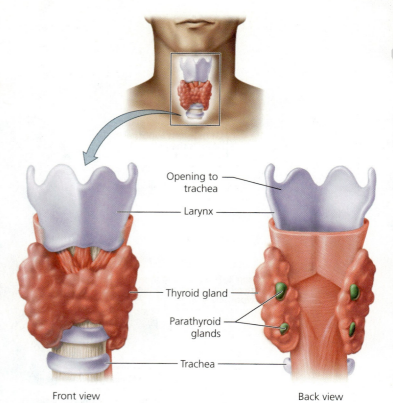

(a) The thyroid gland lies over the trachea, just below the larynx.

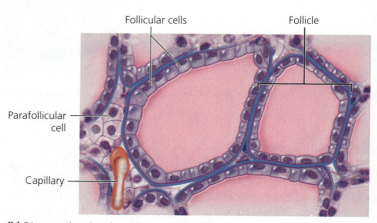

(b) Diagram showing thyroid tissue. Follicular cells produce the precursor to thyroid hormone, and the parafollicular cells produce calcitonin.

FIGURE 10.11 Location and structure of the thyroid gland and parathyroid glands

ter in the United States dramatically decreased once iodine was added to most table salt beginning in the 1920s. Simple goiter can be treated by iodine supplements or administration of TH.

Undersecretion of TH during fetal development or infancy causes *cretinism*, a condition characterized by dwarfism and delayed mental and sexual development (Figure 10.12b). If a pregnant woman produces sufficient TH, many of the symptoms of cretinism do not appear until after birth, when the deficient infant begins to rely solely on its own malfunctioning thyroid gland to supply the needed hormones. Oral doses of TH can prevent cretinism, so most infants in industrialized

(a) Simple goiter

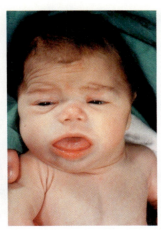
(b) Cretinism

(c) Exophthalmos

FIGURE 10.12 *Disorders of the thyroid gland*

The **calcitonin (CT)** secreted by the parafollicular cells of the thyroid helps regulate the concentration of calcium in the blood to ensure the proper functioning of muscle cells and neurons. Calcium ions bind to the protein troponin, leading to changes in other muscle proteins and eventually causing muscle contraction (Chapter 6). In addition, calcium causes the release of neurotransmitters into the synaptic cleft and therefore is critical in the transmission of messages from one neuron to the next (Chapter 7). When the level of calcium in the blood is high, CT stimulates the absorption of calcium by bone and inhibits the breakdown of bone, thereby lowering the level of calcium in the blood. CT also lowers blood calcium by stimulating an initial increase in the excretion of calcium in the urine. When the level of calcium in the blood is low, the parathyroid glands, which we discuss next, are prompted to release their hormone.

stop and think

Calcitonin is considered to be most important during childhood and possibly important at certain times in adulthood, such as during the late stages of a pregnancy. Why might calcitonin be important at these particular times?

10.4 Parathyroid Glands

The **parathyroid glands** are four small, round masses at the back of the thyroid gland (Figure 10.11a, back view). These glands secrete **parathyroid hormone (PTH)**, also called *parathormone*. As mentioned earlier, CT from the thyroid gland lowers the level of calcium in the blood. In contrast, PTH increases levels of calcium in the blood (see Chapter 5). Low levels of calcium in the blood stimulate the parathyroid glands to secrete PTH, which causes blood levels of calcium to rise. PTH exerts its effects by stimulating (1) bone-destroying cells called osteoclasts that release calcium from bone into the blood (Chapter 5); (2) the kidneys to reabsorb more calcium from the filtrate (the fluid inside the nephrons of kidneys, some of which will become urine) and return it to the blood; and (3) the rate at which calcium is absorbed into the blood from the gastrointestinal tract. PTH also inhibits bone-forming cells called osteoblasts and thereby reduces the rate at which calcium is deposited in bone. The feedback system by which CT and PTH together regulate levels of calcium in the blood is summarized in Figure 10.13.

Surgery on the neck or thyroid gland may damage the parathyroid glands. The resultant decrease in PTH causes decreased blood calcium that in turn produces nervousness and irritability (low calcium is associated with hyperexcitability of the membranes of neurons) and muscle spasms (recall that calcium is also important in muscle contraction). In severe cases, death may result from spasms of the larynx and paralysis of the respiratory system. PTH is difficult to purify, so deficiencies are not usually treated by administering the hormone. Instead, calcium is given either in tablet form or through increased dietary intake.

A tumor of the parathyroid gland can cause excess secretion of PTH. Oversecretion of PTH pulls calcium from bone

nations are now tested for proper thyroid function shortly after birth. In the United States, such testing reveals that incomplete development of the thyroid gland occurs in about 1 in every 3000 births. Undersecretion of TH in adulthood causes *myxedema,* a condition in which fluid accumulates in facial tissues. Other symptoms of TH undersecretion include decreases in alertness, body temperature, and heart rate. Oral administration of TH can prevent and treat these symptoms.

Oversecretion of TH causes *Graves' disease,* an autoimmune disorder in which a person's own immune system produces Y-shaped proteins called antibodies (discussed in Chapter 13) that in this case mimic the action of TSH. The antibodies stimulate the thyroid gland, causing it to enlarge and overproduce its hormones. Symptoms of Graves' disease include increased metabolic rate and heart rate, accompanied by sweating, nervousness, and weight loss. Many patients with Graves' disease also have *exophthalmos,* protruding eyes caused by the swelling of tissues in the eye orbits (Figure 10.12c). Graves' disease may be treated with drugs that block synthesis of thyroid hormones. Alternatively, thyroid tissue may be reduced through surgery or the administration of radioactive iodine. Because the thyroid gland accumulates iodine, ingestion of radioactive iodine (usually administered in capsules) selectively destroys thyroid tissue.

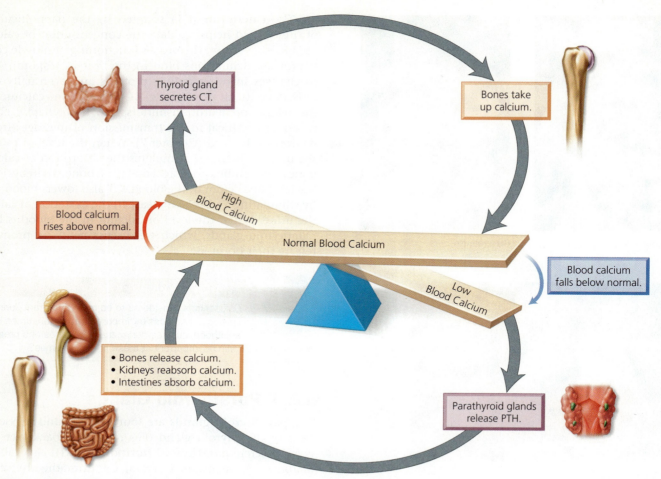

FIGURE 10.13 Regulation of calcium levels in the blood by CT from the thyroid gland (top) and by PTH from the parathyroid glands (bottom)

tissue, causing increased blood calcium and weakened bones. High levels of calcium in the blood may lead to kidney stones, calcium deposits in other soft tissue, and decreased activity of the nervous system.

10.5 Adrenal Glands

The body's two **adrenal glands** (*ad*, upon; *renal*, kidney), each about the size of an almond, are located at the tops of the kidneys. Each adrenal gland has an outer and an inner region. The outer region of the gland, the **adrenal cortex**, secretes more than 20 different lipid-soluble (steroid) hormones, generally divided into three groups: the gonadocorticoids, mineralocorticoids, and glucocorticoids (Figure 10.14). The inner region, called the **adrenal medulla**, secretes two water-soluble hormones, epinephrine (also known as *adrenaline*) and norepinephrine (also known as *noradrenaline*).

Adrenal Cortex

The **gonadocorticoids** are male and female sex hormones known as **androgens** and **estrogens**. In both males and females, the adrenal cortex secretes both androgens and estrogens. However, in normal adult males, androgen secretion by the testes far surpasses that by the adrenal cortex. Thus, the effects of adrenal androgens in adult males are probably insignificant. In females, the ovaries and placenta also produce estrogen, although during menopause, the ovaries decrease secretion of estrogen and eventually stop secreting it. The gonadocorticoids from the adrenal cortex may somewhat alleviate the effects of decreased ovarian estrogen in menopausal women. Menopause is discussed further in Chapter 17.

The **mineralocorticoids** secreted by the adrenal cortex affect mineral homeostasis and water balance. The primary mineralocorticoid is **aldosterone**, a hormone that acts on cells of the kidneys to increase reabsorption of sodium ions (Na^+) into the blood. This reabsorption prevents depletion of Na^+ and increases water retention. Aldosterone also acts on kidney cells to promote the excretion of potassium ions (K^+) in urine. *Addison's disease* is a disorder caused by the undersecretion of aldosterone and the glucocorticoid cortisol (see the following discussion). This disease appears to be an autoimmune disorder in which the body's own immune system perceives cells of the adrenal cortex as foreign and destroys them. The resulting deficiency of adrenal hormones causes weight loss, fatigue, electrolyte imbalance, poor appetite, and poor resistance to stress. A peculiar bronzing of the skin also is associated with Addison's disease (Figure 10.15). Addison's disease can be treated with hormone tablets.

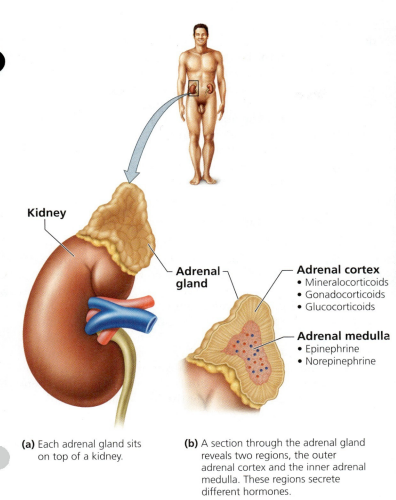

(a) Each adrenal gland sits on top of a kidney.

(b) A section through the adrenal gland reveals two regions, the outer adrenal cortex and the inner adrenal medulla. These regions secrete different hormones.

FIGURE 10.14 Location and structure of an adrenal gland

stop and think

High blood pressure can signal abnormal aldosterone secretion. Would high blood pressure be associated with the undersecretion or oversecretion of aldosterone?

The **glucocorticoids** are hormones secreted by the adrenal cortex that affect glucose levels. Glucocorticoids act on the liver to promote the conversion of fat and protein to intermediate substances that are ultimately converted into glucose. The glucocorticoids also act on adipose tissue to prompt the breakdown of fats to fatty acids that are released into the bloodstream, where they are available for use by the body's cells. Glucocorticoids further conserve glucose by inhibiting its uptake by muscle and fat tissue.

Glucocorticoids also inhibit the inflammatory response; such inhibition can be beneficial when the body is faced with the swelling and intense irritation associated with skin rashes such as that caused by poison ivy. One way glucocorticoids inhibit inflammation is to slow the movement of white blood cells to the site of injury. Another way is to reduce the likelihood that other cells will release chemicals that promote inflammation. Unfortunately, these activities of glucocorti-

FIGURE 10.15 John F. Kennedy suffered from Addison's disease, which is caused by undersecretion of cortisone and aldosterone from the adrenal cortex. JFK's complexion showed the peculiar bronzing of the skin characteristic of Addison's disease.

coids inhibit wound healing. Steroid creams containing synthetic glucocorticoids are therefore intended to be applied only to the surface of the skin and to be used for superficial rashes only. These creams should not be applied to open wounds. Some examples of glucocorticoids are cortisol, corticosterone, and cortisone.

Cushing's syndrome results from prolonged exposure to high levels of the glucocorticoid cortisol. Body fat is redistributed, and fluid accumulates in the face (Figure 10.16). Additional symptoms include fatigue, high blood pressure, and elevated glucose levels. A tumor on either the adrenal cortex or the anterior pituitary may cause the oversecretion of cortisol that leads to Cushing's syndrome. (Recall that the anterior pituitary secretes ACTH, which stimulates the release of hormones from the adrenal cortex.) Tumors are treated with radiation, drugs, or surgery. Cushing's syndrome also may result from glucocorticoid hormone treatment for asthma, lupus, or

(a) Patient with Cushing's syndrome   (b) Same patient after treatment

FIGURE 10.16 Cushing's syndrome. Prolonged exposure to cortisol causes fluid to accumulate in the face. Most often, Cushing's syndrome is caused by the administration of cortisol for allergies or inflammation.

rheumatoid arthritis. Treatment in medically induced cases of Cushing's syndrome typically entails a gradual reduction of the glucocorticoid dose, ideally to the lowest level necessary to control the existing disorder without prompting adverse affects.

Adrenal Medulla

As introduced earlier, the adrenal medulla produces **epinephrine** and **norepinephrine**. These hormones are critical in the **fight-or-flight response**, the reaction by the body's sympathetic nervous system to emergencies (Chapter 8). Imagine that you are walking home alone late at night and a stranger suddenly steps toward you from the bushes. Impulses received by your hypothalamus are sent by neurons to your adrenal medulla. These impulses cause cells in your adrenal medulla to increase output of epinephrine and norepinephrine. In response to these hormones, your heart rate, respiratory rate, and blood glucose levels rise. Blood vessels associated with the digestive tract constrict because digestion is not of prime importance during times of extreme stress. Vessels associated with skeletal and cardiac muscles dilate, allowing more blood, glucose, and oxygen to reach them. These substances also reach your brain in greater amounts, leading to the increased mental alertness needed for fleeing or fighting.

In contrast to the near instantaneous response of the sympathetic nervous system to a perceived threat, the hormonal response takes about 30 seconds to mount. This is because epinephrine and norepinephrine must be released by the adrenal medulla, travel in the bloodstream to all cells, bind to receptors on their target cells, and initiate changes in those cells. Even after the danger has passed, we feel the changes brought on by these hormones for a few additional minutes. Epinephrine and norepinephrine thus augment and prolong the response of the sympathetic nervous system to stress. The more leisurely onset and conclusion of the effects of epinephrine and norepinephrine highlight the differences between neural and hormonal systems of internal communication.

10.6 Pancreas

The **pancreas** is located in the abdomen just behind the stomach; it contains both endocrine and exocrine cells (Figure **10.17**). The exocrine cells secrete digestive enzymes and will be discussed in Chapter 15. The endocrine cells occur in small clusters called **pancreatic islets** (or *islets of Langerhans*). These clusters contain three types of hormone-producing cells. One type produces the hormone glucagon; a second produces the hormone insulin; and a third produces the hormone somatostatin. Somatostatin also is secreted by the digestive tract, where it inhibits secretions of the stomach and small intestine, and by the hypothalamus, where it inhibits secretion of growth hormone. The somatostatin

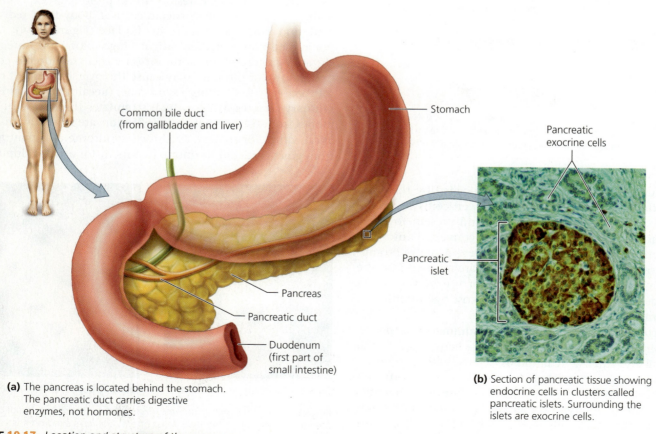

(a) The pancreas is located behind the stomach. The pancreatic duct carries digestive enzymes, not hormones.

(b) Section of pancreatic tissue showing endocrine cells in clusters called pancreatic islets. Surrounding the islets are exocrine cells.

FIGURE 10.17 *Location and structure of the pancreas*

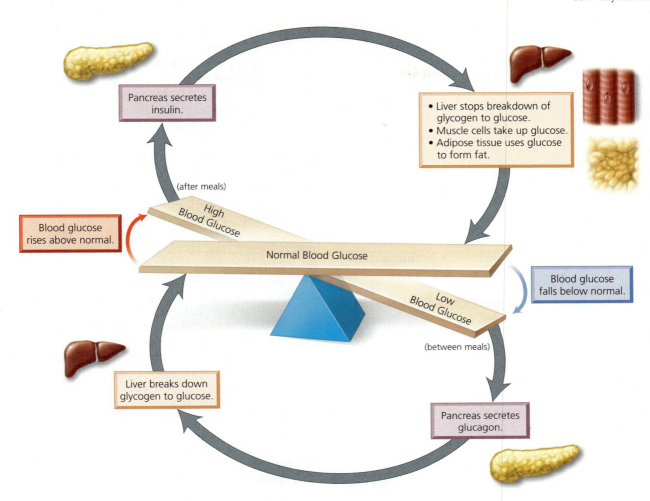

FIGURE 10.18 *Regulation of glucose level in the blood by insulin (top) and glucagon (bottom), both of which are secreted by the pancreas*

secreted by the pancreas may regulate the secretion of glucagon and insulin. However, the precise function of pancreatic somatostatin is not well understood, so we focus on glucagon and insulin.

Between meals, as the level of blood sugar declines, the pancreas secretes glucagon. **Glucagon** increases the level of blood sugar. It does so by prompting cells of the liver to increase conversion of glycogen (the storage polysaccharide in animals) to glucose (a simple sugar, or monosaccharide). Glucagon also stimulates the liver to form glucose from lactic acid and amino acids. The liver releases the resultant glucose molecules into the bloodstream, causing a rise in blood sugar level.

After a meal, as the level of blood sugar rises with the absorption of sugars from the digestive tract, the pancreas secretes insulin. In contrast to glucagon, **insulin** decreases glucose in the blood; insulin and glucagon thus have opposite, or antagonistic, effects. Insulin decreases blood glucose in several ways. First, insulin stimulates transport of glucose into muscle cells, white blood cells, and connective tissue cells. Second, insulin inhibits the breakdown of glycogen to glucose. Third, insulin prevents conversion of amino and fatty acids to glucose.

As a result of these actions, insulin promotes protein synthesis, fat storage, and the use of glucose for energy. Figure **10.18** summarizes the regulation of glucose in the blood by insulin and glucagon.

Diabetes mellitus is a group of metabolic disorders characterized by an abnormally high level of glucose in the blood. The high blood glucose levels are caused by problems with either insulin production or insulin function. We explore diabetes mellitus and the dramatic effects of insulin on our health in Chapter 10a.

10.7 Thymus Gland

The **thymus gland** lies just behind the breastbone, on top of the heart (see Figure 10.2). It is more prominent in infants and children than in adults because it decreases in size as we age. The hormones it secretes, such as **thymopoietin** and **thymosin**, promote the maturation of white blood cells called *T lymphocytes*. Precursor cells from bone marrow travel by way of the bloodstream to the thymus gland, where they mature into T lymphocytes, also known as *T cells*, to become part of the body's defense mechanisms (Chapter 13).

10.8 Pineal Gland

The **pineal gland** is a tiny gland at the center of the brain (see Figure 10.2). Its secretory cells produce the hormone **melatonin**. Levels of circulating melatonin are greater at night than during daylight hours, because of input the pineal gland receives from visual pathways. Neurons of the retina, stimulated by light entering the eye, send impulses to the hypothalamus and ultimately the pineal gland, where they inhibit secretion of melatonin.

Research in the past few decades has suggested diverse roles for melatonin. Melatonin may influence daily rhythms. Sleep and, for some people, seasonal changes in mood appear to be influenced by melatonin. Melatonin also may slow the aging process.

One disorder associated with too much melatonin is *seasonal affective disorder (SAD)*. This form of depression is associated with winter, when short day length and a decreased exposure to light results in overproduction of melatonin. Too much melatonin causes symptoms such as lethargy, long periods of sleep, low spirits, and a craving for carbohydrates. The symptoms usually appear around October and end about April in the Northern Hemisphere. Three-quarters of persons who suffer from SAD are female. Treatment of SAD often includes repeated exposure to very bright light for about an hour each day. The intense light inhibits melatonin production.

what would you do?

Scientific evidence indicates that melatonin helps to alleviate jet lag. Studies have shown that melatonin taken by mouth on the day of travel and continued for several days has the following effects in about half the people who take it to fight jet lag: (1) reduced fatigue during the day, (2) reduced time to fall asleep at night, and (3) more rapid development of a normal sleep pattern. Such benefits are usually most evident during eastward travel that crosses more than four time zones. In the United States, melatonin is sold as a dietary supplement, not as a drug. Thus, the regulations of the FDA that apply to medications do not apply to melatonin. Medications must be proven safe and effective for their intended use before they are made available to consumers. Dietary supplements do not require approval by the FDA before reaching the consumer. Instead, it is the manufacturer's responsibility for ensuring that a dietary supplement is safe. If you were scheduled to fly from California to New York for an important interview, would you feel safe taking melatonin to alleviate jet lag? If melatonin helped you during your trip and interview, would you feel comfortable recommending it to a friend?

10.9 Locally Acting Chemical Messengers

Now that we have surveyed the endocrine glands and their hormones, let's consider another group of chemical messengers—those that act locally. Once secreted by a cell, these **local signaling molecules** act near the site of their release, on adjacent target cells, within seconds or milliseconds. Communication via local signaling molecules occurs much more rapidly than the communication carried out by hormones, which travel to distant sites within the body (recall that the flight-or-fight response may take 30 seconds to initiate physiological changes). Neurotransmitters, discussed in Chapter 7, are examples of chemicals that rapidly convey messages from one cell (a neuron) to a neighboring cell (often another neuron). Prostaglandins, growth factors, and nitric oxide (NO) are other examples of local signaling molecules.

Prostaglandins are lipid molecules continually released by the plasma membranes of most cells. Different types of cells secrete different prostaglandins. At least 16 different prostaglandin molecules function within the human body. These molecules have remarkably diverse effects, influencing blood clotting, regulation of body temperature, diameter of airways to the lungs, and the body's inflammatory response. Prostaglandins also affect the reproductive system. Menstrual cramps are thought to be caused by prostaglandins released by cells of the uterine lining. These prostaglandins act on the smooth muscle of the uterus, causing muscle contractions and cramping. Anti-inflammatory drugs, such as aspirin and ibuprofen, inhibit the synthesis of prostaglandins and thus may lessen the discomfort of menstrual cramps. Prostaglandins also are found in semen. Once in the female reproductive tract, prostaglandins in semen cause the smooth muscles of the uterus to contract, perhaps helping the sperm continue their journey.

Other chemical messenger molecules, called **growth factors**, are peptides or proteins that, when present in the fluid outside target cells, stimulate those cells to grow, develop, and multiply. For example, one growth factor causes precursor cells in the bone marrow to proliferate and differentiate into particular white blood cells. Another growth factor prompts endothelial cells to proliferate and organize into tubes that eventually form blood vessels (see Chapter 12).

The gas **nitric oxide (NO)** functions in the cellular communication that leads to the dilation of blood vessels. Basically, endothelial cells of the inner lining of blood vessels make and release NO, which signals the smooth muscles in the surrounding (middle) layer to relax, allowing the vessel to dilate. NO aids in peristalsis, the rhythmic waves of smooth muscle contraction and relaxation that push food along the digestive tract. NO also functions as a neurotransmitter, carrying messages from one neuron to the next. Histamine, another local signaling molecule, is discussed in Chapter 13.

looking ahead

In Chapter 10 we learned about the hormones produced by endocrine glands and organs with some endocrine tissue. The pancreas is an organ with both endocrine and exocrine tissue. The endocrine cells make and secrete several hormones, among them insulin. In Chapter 10a we focus on diabetes mellitus, a group of diseases characterized by problems in insulin production or function. (The exocrine cells of the pancreas secrete digestive enzymes, which we describe in Chapter 15.)

HIGHLIGHTING THE CONCEPTS

10.1 Functions and Mechanisms of Hormones (pp. 171–175)

- Endocrine glands lack ducts and release their products, hormones, into the spaces just outside cells. The hormones then diffuse into the bloodstream. Endocrine glands and organs that contain some endocrine tissue constitute the endocrine system, which regulates and coordinates other organ systems and helps maintain homeostasis.
- Hormones, the chemical messengers of the endocrine system, contact virtually all cells within the body. However, hormones affect only target cells, those cells with receptors that recognize and bind specific hormones.
- Steroid hormones are lipid soluble. Steroids cross through the plasma membrane of target cells and combine with a receptor molecule inside the cell, forming a hormone–receptor complex. In the nucleus, the complex directs synthesis of specific proteins, including enzymes that stimulate or inhibit particular metabolic pathways.
- Water–soluble hormones, many of which are peptides and proteins, cannot pass through the lipid bilayer of the plasma membrane. Thus, they exert their effects indirectly by activating second messenger systems. The hormone, considered the first messenger, binds to a receptor on the plasma membrane. This event activates a molecule in the cytoplasm, considered the second messenger, which carries the hormone's message inside the cell, changing the activity of enzymes and chemical reactions. Thus, whereas lipid-soluble hormones prompt the synthesis of proteins, water-soluble hormones activate existing proteins.
- Endocrine glands are stimulated to manufacture and release hormones by chemical changes in the blood, hormones released by other endocrine glands, and messages from the nervous system. Hormone secretion is usually regulated by negative feedback mechanisms but sometimes by positive feedback mechanisms. The interactions between hormones may be antagonistic, synergistic, or permissive.

10.2 Hypothalamus and Pituitary Gland (pp. 175–180)

- The pituitary gland has an anterior lobe and a posterior lobe. The anterior lobe is influenced by the hypothalamus through a circulatory connection. Neurosecretory cells in the hypothalamus release hormones that travel by way of the bloodstream to the anterior lobe, where they stimulate or inhibit release of hormones. The anterior pituitary releases six hormones: growth hormone (GH), prolactin (PRL), thyroid-stimulating hormone (TSH), adrenocorticotropic hormone (ACTH), follicle-stimulating hormone (FSH), and luteinizing hormone (LH). Four (TSH, ACTH, FSH, LH) of the six hormones are tropic hormones, meaning they influence other endocrine glands.
- In contrast to the circulatory connection between the hypothalamus and the anterior lobe of the pituitary gland, the connection between the hypothalamus and the posterior lobe is neural. Neurosecretory cells in the hypothalamus make oxytocin (OT) and antidiuretic hormone (ADH); these two hormones travel down the axons of the cells to axon terminals in the posterior lobe, where they are stored and released.

10.3 Thyroid Gland (pp. 180–181)

- The thyroid gland, at the front of the neck, produces thyroid hormone (TH) and calcitonin (CT). Thyroid hormone—which includes two very similar hormones, thyroxine (T_4) and triiodothyronine (T_3)—has broad effects, including regulating metabolic rate, heat production, and blood pressure. CT decreases calcium in the blood.

10.4 Parathyroid Glands (pp. 181–182)

- The parathyroid glands, four small masses of tissue at the back of the thyroid gland, secrete parathyroid hormone (PTH, or parathormone), an antagonist to CT. As such, PTH is responsible for raising blood levels of calcium by stimulating the movement of calcium from bone and urine to the blood.

10.5 Adrenal Glands (pp. 182–184)

- Each of two adrenal glands sits on top of a kidney and has two regions. The adrenal cortex (outer region) secretes gonadocorticoids, mineralocorticoids, and glucocorticoids. The adrenal medulla (inner region) produces epinephrine (adrenaline) and norepinephrine (noradrenaline), which initiate the fight-or-flight response.

10.6 Pancreas (pp. 184–185)

- The pancreas secretes the hormones glucagon (increases glucose in the blood) and insulin (decreases glucose in the blood). Diabetes mellitus is a group of disorders characterized by problems with insulin production or function.

10.7 Thymus Gland (p. 185)

- The thymus gland lies on top of the heart and plays an important role in immunity. Its hormones influence the maturation of white blood cells called T lymphocytes.

10.8 Pineal Gland (p. 186)

- The pineal gland, at the center of the brain, secretes the hormone melatonin. Melatonin appears to be responsible for establishing biological rhythms and triggering sleep.

10.9 Locally Acting Chemical Messengers (p. 186)

- Some local chemical messengers convey information between adjacent cells, evoking rapid responses in target cells. Examples of local signaling molecules include neurotransmitters, prostaglandins, growth factors, and nitric oxide.

RECOGNIZING KEY TERMS

endocrine gland p. 171
endocrine system p. 171
hormone p. 172
target cell p. 172
lipid-soluble hormone p. 172
steroid hormone p. 172
water-soluble hormone p. 172
first messenger p. 172
second messenger p. 172
negative feedback mechanism p. 174
positive feedback mechanism p. 174
pituitary gland p. 175
hypothalamus p. 175
neurosecretory cell p. 175

releasing hormone p. 175
inhibiting hormone p. 176
growth hormone (GH) p. 176
prolactin (PRL) p. 177
tropic hormone p. 178
thyroid-stimulating hormone (TSH) p. 178
adrenocorticotropic hormone (ACTH) p. 178
follicle-stimulating hormone (FSH) p. 178
luteinizing hormone (LH) p. 178
antidiuretic hormone (ADH) p. 179
oxytocin (OT) p. 179
thyroid gland p. 180

thyroid hormone (TH) p. 180
calcitonin (CT) p. 181
parathyroid glands p. 181
parathyroid hormone (PTH) p. 181
adrenal glands p. 182
adrenal cortex p. 182
adrenal medulla p. 182
gonadocorticoids p. 182
androgen p. 182
estrogen p. 182
mineralocorticoids p. 182
aldosterone p. 182
glucocorticoids p. 183
epinephrine p. 184
norepinephrine p. 184

fight-or-flight response p. 184
pancreas p. 184
pancreatic islets p. 184
glucagon p. 185
insulin p. 185
thymus gland p. 185
thymopoietin p. 185
thymosin p. 185
pineal gland p. 186
melatonin p. 186
local signaling molecules p. 186
prostaglandin p. 186
growth factor p. 186
nitric oxide (NO) p. 186

REVIEWING THE CONCEPTS

1. Given that hormones contact virtually all cells in the body, why are only certain cells affected by a particular hormone? p. 172
2. How do lipid-soluble (steroid) and water-soluble hormones differ in their mechanisms of action? pp. 172, 174
3. Compare negative and positive feedback mechanisms with regard to regulation of hormone secretion. Provide an example of each. pp. 172, 174
4. How do the anterior and posterior lobes of the pituitary gland differ in size and relationship with the hypothalamus? pp. 175–176
5. Describe the feedback system by which calcitonin and parathyroid hormone regulate levels of calcium in the blood. pp. 181–182
6. What are the major functions of the glucocorticoids, mineralocorticoids, and gonadocorticoids secreted by the adrenal cortex? pp. 182–183
7. What is the fight-or-flight response? Which hormones are critical in initiating this response? p. 184
8. Explain the differences between diabetes insipidus and diabetes mellitus. p. 179
9. What is the basic function of hormones secreted by the thymus gland? p. 185
10. How do local signaling molecules differ from true hormones? p. 186
11. Choose the *incorrect* statement:
    a. Endocrine glands secrete their hormones through ducts.
    b. The endocrine system consists of endocrine glands and organs with endocrine tissue.
    c. Negative feedback mechanisms regulate the secretion of most hormones.
    d. During positive feedback, the outcome of a process feeds back to the system and stimulates the process to continue.
12. The interaction between glucagon and insulin is described as
    a. permissive.
    b. synergistic.
    c. antagonistic.
    d. tropic.
13. The posterior pituitary gland
    a. stores and releases antidiuretic hormone and oxytocin.
    b. produces and secretes growth hormone.
    c. does not produce any hormones of its own.
    d. a and c.
14. Choose the *incorrect* statement:
    a. The pancreas has endocrine and exocrine cells.
    b. Glucagon increases glucose in the blood.
    c. Insulin is an example of a local signaling molecule.
    d. Insulin stimulates the movement of glucose into cells.
15. Prostaglandins
    a. act more rapidly than hormones.
    b. act on nearby target cells.
    c. are examples of growth factors.
    d. a and b.
16. Choose the *incorrect* statement:
    a. The anterior pituitary releases oxytocin and antidiuretic hormone.
    b. The anterior pituitary has a circulatory connection to the hypothalamus.
    c. The anterior pituitary produces several tropic hormones.
    d. The anterior pituitary produces growth hormone and prolactin.
17. A diet deficient in iodine may produce
    a. cretinism.
    b. Graves' disease.
    c. Cushing's syndrome.
    d. goiter.
18. Which of the following does *not* characterize the adrenal medulla?
    a. inner region of the adrenal gland
    b. secretes epinephrine and norepinephrine
    c. secretes glucocorticoids
    d. secretes hormones involved in fight-or-flight response
19. Which of the following does *not* occur in a healthy person's body after meals?
    a. The pancreas secretes insulin.
    b. The liver stops breakdown of glycogen.
    c. The pancreas secretes glucagon.
    d. Muscle cells take up glucose.

20. Overproduction of melatonin by the pineal gland may cause
    a. seasonal affective disorder.
    b. diabetes insipidus.
    c. acromegaly.
    d. Addison's disease.

21. Oversecretion of growth hormone in childhood causes _____. Oversecretion in adulthood causes _____.
22. In males, androgens are produced by the testes and the _____.

APPLYING THE CONCEPTS

1. Mary has an itchy rash on the surface of her skin, and Rick has cut his finger on glass. Would either person benefit from applying a steroid cream containing cortisone? Why? Why not?
2. Which internal system of communication—the endocrine system or nervous system—would be responsible for the growth spurt that occurs at puberty? Which system would control the quick withdrawal of your foot when you step on a tack? Explain your answers.
3. It is winter in Massachusetts, and Theresa has felt "down" and lethargic since the fall. She has trouble getting out of bed in the morning, and once up, she craves carbohydrates. What might explain Theresa's symptoms? What might alleviate them, and why?
4. Velma tells her friend Carlos that he produces the female hormone estrogen. Is she correct? If yes, where is the estrogen produced in Carlos? Where is it produced in Velma?

BECOMING INFORMATION LITERATE

Recall that undersecretion of thyroid hormone in adulthood causes myxedema and that oversecretion causes Graves' disease. Prepare a brochure in which you describe the symptoms and treatments of these two conditions. Use at least three reliable sources (books, journals, or websites). List each source you considered, and explain why you chose the three sources you used.

MasteringBiology®

Go to MasteringBiology for practice quizzes, activities, eText, videos, current events, and more.

10a SPECIAL TOPIC
Diabetes Mellitus

- 10a.1 General Characterization and Overall Prevalence
- 10a.2 Type 1 and Type 2 Diabetes
- 10a.3 Gestational Diabetes
- 10a.4 Other Specific Types of Diabetes

In Chapter 10 we described how two hormones secreted by the pancreas—insulin and glucagon—act antagonistically to regulate the level of glucose in the blood to help maintain homeostasis. Insulin lowers blood glucose levels and glucagon raises blood glucose levels. In this chapter, we consider diabetes mellitus, a group of diseases characterized by problems in insulin production or insulin function. We introduce this group of diseases by describing the different forms of diabetes mellitus and their symptoms, complications, and treatments.

At present, type 1 diabetes cannot be prevented or cured. Type 2 diabetes, however, can be prevented or delayed by making key lifestyle changes.

10a.1 General Characterization and Overall Prevalence

We recognize four basic forms of diabetes mellitus: type 1, type 2, gestational, and "other specific types." Together, types 1 and 2 account for more than 95% of all diagnosed cases. All four forms are characterized by problems in glucose regulation. Glucose is a carbohydrate, specifically a monosaccharide (Chapter 2). Glucose is the main source of fuel for our cells, and we use it to make ATP (Chapter 3). We get glucose directly from the food we eat. After we have eaten a meal, our digestive system breaks down macromolecules, such as complex carbohydrates (polysaccharides), into their simpler components, such as glucose. Glucose is then absorbed into the bloodstream for transport to cells. In response to high levels of glucose in the blood, the pancreas secretes insulin. Insulin promotes the movement of glucose into target cells by stimulating increases in the number of glucose transport proteins in target cell membranes. These proteins then transport glucose into cells by facilitated diffusion (Chapter 3). By stimulating the movement of glucose from the blood into cells, insulin lowers blood glucose levels (Figure 10a.1). As mentioned, all forms of diabetes mellitus are characterized by defects in the production or function of insulin. If there is not

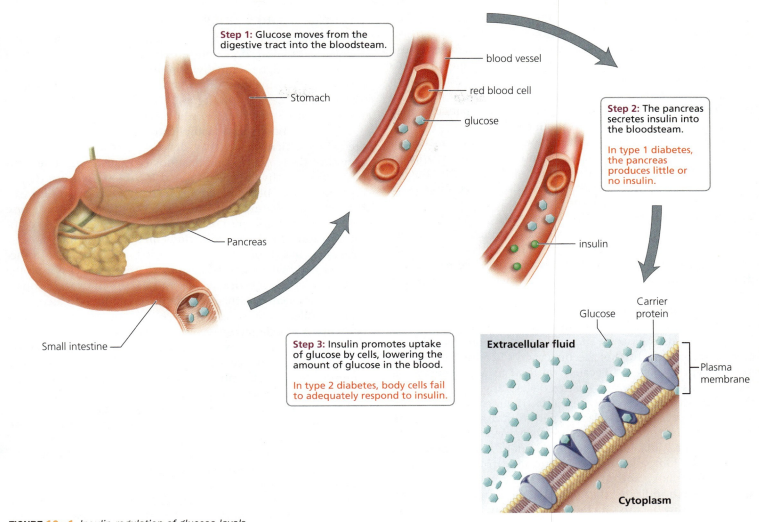

FIGURE 10a.1 *Insulin regulation of glucose levels*

enough insulin, or if target cells do not adequately respond to insulin, then glucose cannot move from the bloodstream into cells. As a result, glucose reaches abnormally high levels in the blood. We will see that many of the long-term health complications faced by people with diabetes mellitus result from chronic exposure to high levels of blood glucose.

Diabetes mellitus is a major cause of death and disability worldwide. In the United States, diabetes is the seventh leading cause of death, ranking behind heart disease, cancer, chronic lower respiratory disease, stroke, accidents, and Alzheimer's disease. About 25.8 million Americans alive today (about 8.3% of the population) have diabetes. In the next 25 years, the total number of Americans with diabetes is predicted to double. Much of the predicted increase in number of cases concerns type 2 diabetes, a form associated with a lifestyle of overeating and inactivity (see the following discussion).

10a.2 Type 1 and Type 2 Diabetes

Now that we understand the basics of diabetes mellitus and its general prevalence in the population, let's examine the different forms of diabetes. We begin with type 1 and type 2. These two diseases differ from one another in many ways, but they also have some things in common, so we will consider them together and point out their similarities and differences.

Characterization and Risk Factors

Type 1 diabetes used to be known as *insulin-dependent diabetes* because taking insulin is always part of treatment. It also was called *juvenile-onset diabetes*, because it usually develops in people younger than 25 years of age. Type 1 represents about 5% to 10% of all diagnosed cases of diabetes. In this autoimmune disorder, a person's own immune system attacks the cells of the pancreas (called *beta cells*) responsible for insulin production. A virus may trigger the attack. Having a family history of type 1 diabetes is considered a risk factor for development of the disease. Thus, a person has a genetic predisposition for type 1 diabetes, and then an environmental factor, such as a virus, triggers the autoimmune response. Given the destruction of beta cells, people with type 1 diabetes must have insulin delivered by injection or pump (see the discussion on treatments).

Type 2 diabetes used to be known as *non-insulin-dependent diabetes* because treatment did not require insulin. This name was dropped because about 40% of people with this disease do need insulin injections. It was formerly called *adult-onset diabetes* as well, because it usually develops after age 40, although it recently has begun showing up in younger people who are overweight and sedentary. Type 2 diabetes is also associated with high blood pressure. Type 2 diabetes accounts for 90% to 95% of diabetes cases. It is characterized by insulin resistance, a condition in which the body's cells fail to adequately respond to insulin. Over time, the pancreas may reduce the amount of insulin it produces.

Obesity is a primary culprit in type 2 diabetes, but the reasons are not fully understood at the molecular level. It appears that increased adipose (fat) tissue interferes with glucose metabolism. One line of research implicates free fatty acids released by fat cells when they break down triglycerides (a type of lipid; see Chapter 2). These free fatty acids travel in the bloodstream, and cells take them up as a source of fuel. Some cells can more easily metabolize free fatty acids than glucose. As a result, they use free fatty acids in preference to glucose, thus disrupting glucose metabolism. Other lines of research implicate other molecules released by adipose tissue, such as hormones and inflammation-promoting proteins. Their links to insulin resistance are the focus of active research.

Exercise uses glucose and makes cells more sensitive to insulin, so failure to exercise (inactivity) is another risk factor for type 2 diabetes. Because we tend to gain weight and exercise less as we age, the risk for developing type 2 diabetes increases with age, especially after 45. Having a parent or sibling with type 2 diabetes places a person at risk for developing it. This could be due to genetic or environmental factors. Additionally, type 2 diabetes disproportionately affects some populations, including Hispanic Americans, African Americans, American Indians, and Alaska Natives (Figure 10a.2). In summary, research indicates that type 2 diabetes results from several interacting factors, including obesity, insulin resistance, and failure of the beta cells of the pancreas to fully compensate for the decreased responsiveness to insulin.

The characteristics of type 1 and type 2 diabetes are compared in Table 10a.1.

Non-Hispanic whites	7.1%
Asian Americans	8.4%
Hispanics	11.8%
Non-Hispanic blacks	12.6%
American Indians and Alaska Natives	16.1%

FIGURE 10a.2 *Percentages of people in the United States with diabetes by race and ethnicity, 2007–2009. These data are age-adjusted.*
Source: Centers for Disease Control and Prevention. National diabetes fact sheet: national estimates and general information on diabetes and prediabetes in the United States, 2011. Atlanta, GA: U.S. Department of Health and Human Services, CDC, 2011.

Symptoms and Complications

Physical symptoms of type 1 diabetes include increased thirst, frequent urination, dry mouth, extreme hunger, unexplained weight loss, fatigue, blurry vision, and sores that are slow to heal. Frequent infections of the gums, skin, bladder, and vagina also occur. A person with type 2 diabetes may have none of these physical symptoms or only some of the symptoms.

Diabetes has serious complications. Some complications are acute, which means they develop over a relatively short period of time. Acute complications of diabetes include high blood glucose (hyperglycemia) and low blood glucose (hypoglycemia). High blood glucose is treated by taking insulin (see the following discussion). Taking too much insulin may cause blood glucose to drop to dangerously low levels. Initial symptoms of low blood glucose include anxiety, sweating, hunger, weakness, and disorientation. Because brain cells fail to function properly when starved of glucose, these initial symptoms may be followed by convulsions and unconsciousness. Consequences associated with severe depletion of blood glucose are known collectively as *insulin shock*. Insulin shock can prove fatal unless blood glucose levels are raised.

Diabetics also may experience acute, life-threatening biochemical imbalances. These imbalances are much more common in type 1 diabetics than in type 2 diabetics. One such imbalance is *diabetic ketoacidosis (DKA)*. When insulin is unavailable to stimulate the movement of glucose from the blood into cells, the cells become starved of energy. These cells turn to breaking down lipids. Increased breakdown of lipids results in increases in free fatty acids in the blood. The liver metabolizes these fatty acids, producing ketones as a by-product. Because ketones are acids, they dissociate in solution and release hydrogen ions (Chapter 2). If produced in excess, ketones can overwhelm the body's buffering abilities and cause blood pH to drop. If left untreated, such decreases in blood pH can disrupt normal body functions and cause coma, abnormal heart rhythms, and death. Common triggers of DKA include insufficient home insulin therapy (for example, missed insulin treatments), illnesses or infections (for example, pneumonia or urinary tract infections), stress, physical or emotional trauma, and abuse of alcohol or drugs.

Symptoms of DKA usually develop over a few hours and may include nausea, vomiting, excessive thirst and urination, shortness of breath, weakness, fatigue, and confusion. Sometimes DKA leads to a diagnosis of type 1 diabetes because the symptoms are severe enough that they prompt an undiagnosed person to seek medical attention. Even diagnosed diabetics may fail to recognize DKA because some of its symptoms are typical of other illnesses, such as the flu or foodborne infections (Chapter 15a). To help diabetics recognize this potentially fatal condition, there is a specific over-the-counter test for ketones in urine (ketones in the blood eventually appear in urine, where they can be easily detected with a test strip). Inability to lower blood glucose levels through home insulin therapy is another key feature of DKA. Blood glucose rises because the liver produces glucose in response to the energy crisis (liver cells can make glucose from noncarbohydrates, such as amino acids), but the glucose cannot move into the cells of the body without

TABLE 10a.1 A Comparison of Type 1 and Type 2 Diabetes Mellitus

Characteristic	Type 1	Type 2
Percentage of all diagnosed cases of diabetes	5–10%	90–95%
Previous names	Juvenile-onset diabetes Insulin-dependent diabetes	Adult-onset diabetes Non-insulin-dependent diabetes
Typical age of onset	<25 years	>40 years, but now appearing at younger ages
Cause	Autoimmune reaction destroys the beta cells of the pancreas	Body cells become resistant to insulin
Risk factors	Family history Viral infection	Family history Obesity Inactivity Member of a high risk population
Percentage of patients requiring insulin	100%	~40%
Development of ketoacidosis	Likely if undiagnosed or if treatment is compromised	Rare
Treatment	Insulin; management of diet; exercise	Some need insulin; some take oral antidiabetic medications; management of diet; exercise
Prognosis	Cannot be prevented	Can be delayed or prevented

insulin. People with blood glucose levels consistently higher than 300 mg/dl (milligrams per deciliter), ketones in their urine, and other symptoms of ketoacidosis should seek immediate medical attention. (For reference, normal blood glucose levels are between 70 and 150 mg/dl; over the course of a day, levels fluctuate within this range in response to factors such as size and nutritional composition of meals, exercise, etc.)

Treatment for DKA requires hospitalization and includes intravenous insulin therapy (because of the failure of home insulin therapy to lower blood glucose) and fluid replacement (needed to restore body fluids lost through vomiting and excessive urination). Also, because insufficient insulin is associated with decreases in electrolytes (minerals in the blood, such as sodium, potassium, and chloride, that carry an electrical charge), these, too, must be replenished to ensure proper functioning of nerve and muscle cells, including those of the heart.

Other complications of diabetes take years or decades to develop. Over time, high blood glucose is especially damaging to blood vessels, so diabetes can seriously impact the cardiovascular system (heart and blood vessels; Chapter 12). Diabetics are at increased risk for high blood pressure and atherosclerosis, which is the buildup of fatty deposits in the arteries that can lead to heart attack and stroke. Damage to blood vessels affects more than the cardiovascular system. Because blood brings oxygen and nutrients to all cells of the body and removes wastes from them, damage to blood vessels can affect many different tissues, organs, and organ systems. One organ system at risk is the urinary system. High blood glucose can damage the millions of tiny clusters of blood vessels (known as glomeruli) that filter blood in the kidneys. Damage to these vessels can cause kidney failure, necessitating either a kidney transplant or hemodialysis (use of artificial devices to cleanse the blood, Chapter 16). High blood glucose also affects the nervous system and sense organs (Chapters 8 and 9). Many diabetics eventually experience impaired sensation in the hands and feet, when the tiny blood vessels that supply nerves become damaged. Poor circulation and problems with nerves in the lower legs may necessitate amputation of the lower limbs. This is a very common outcome: diabetics account for more than half of nontraumatic lower limb amputations in the United States. High blood glucose also can damage tiny blood vessels in the retina of the eye; these vessels rupture and proliferate abnormally, eventually causing blindness. Other complications that may develop over time include gum disease and skin infections. Finally, diabetics are at increased risk for developing depression. This may reflect, in part, the daily challenges associated with managing this chronic disease (see the following discussion).

Diagnosis

Given the very serious acute and long-term complications, it is important to diagnose and treat diabetes early. In the case of type 2 diabetes, the goal is to prevent it when possible (type 2 diabetes can be prevented or delayed; type 1 diabetes cannot be actively prevented at this time). Several diagnostic blood tests are available. One, called the *fasting blood glucose test*, involves collecting a blood sample after an overnight fast. For this test, a blood glucose level of 70 mg/dl to 99 mg/dl is considered normal; a value of 126 mg/dl or higher on two separate tests indicates diabetes. Readings of 100 mg/dl to 125 mg/dl indicate prediabetes, a condition in which blood glucose levels are higher than normal but not high enough for a diagnosis of type 2 diabetes. (Note: The American Diabetes Association provides these blood sugar levels as general guidelines.) Another test, the *random (nonfasting) blood glucose test*, assesses glucose in blood collected at a random time.[1]

Testing for type 1 and type 2 diabetes now includes the glycated hemoglobin A1c test. This blood test indicates a person's average blood glucose level for the past 2 or 3 months.

[1] © 2010 American Diabetes Association. From Diabetes Care®, Vol. 33, 2010; S62-S69. Adapted with permission from The American Diabetes Association.

FIGURE 10a.3 A diabetic uses a glucometer to measure the amount of glucose in a drop of blood.

Hemoglobin is the oxygen-carrying pigment found in red blood cells. This test measures the percentage of hemoglobin molecules that have glucose attached. Basically, within red blood cells, glucose reacts with hemoglobin to form glycated hemoglobin. Once a hemoglobin molecule is glycated, it remains that way for the life of the red blood cell, which is about 120 days. Thus, the percentage of glycated hemoglobin molecules indicates the average blood glucose level for the past few months. People with diabetes have a higher percentage of glycated hemoglobin than those without diabetes. Specifically, diabetes is indicated by glycated hemoglobin values of 6.5% or higher on two separate tests; for comparison, people without diabetes have readings of 4% to 6%. Prediabetes is indicated by glycated hemoglobin values of 6% to 6.5%.

Treatments

Treatment of type 1 and type 2 diabetes has several components. First, it is essential to maintain a diet that emphasizes foods that are high in nutrition and low in fat and calories (Chapter 15). Many physicians advise diabetic patients to consult a registered dietician who can help them develop healthy and consistent meal plans. Second, physical activity is critical to lowering levels of blood glucose and increasing sensitivity to insulin. After consulting their physician to make sure that an exercise program is safe, patients should aim for 30 minutes of aerobic exercise, such as walking or swimming, most days of the week. Healthy meal plans and exercise also are recommended for people diagnosed with prediabetes.

Monitoring blood glucose levels is key to treating type 1 and type 2 diabetes. Depending on the particular treatment plan, glucose levels may need to be checked several times a day or less frequently. Target glucose levels also depend on the particular treatment plan. Diabetics use glucometers for self-testing blood glucose levels (Figure 10a.3). A glucometer measures the amount of glucose in a drop of blood obtained after a diabetic pricks his or her finger with a lancet. Physicians may recommend glycated hemoglobin A1c testing every few months to assess overall success of the treatment plan.

Patients with type 1 diabetes require insulin to survive. Some patients with type 2 diabetes also need insulin. Insulin cannot be taken by mouth because stomach enzymes break it down. Insulin must be injected. This can be accomplished using a needle, syringe, and vial of insulin; an insulin pen (a device that resembles a pen except that the cartridge is filled with insulin rather than ink); or an insulin pump (Figure 10a.4). An insulin pump is worn outside the body. Inside the pump is a reservoir of insulin. A tube connects this reservoir to a catheter inserted under the skin of the abdomen. The device does not measure glucose levels. Instead, it is programmed to dispense specific amounts of insulin, which can be adjusted according to meals, activity, and so on.

Type 2 diabetics and prediabetics at high risk of developing type 2 diabetes may be prescribed oral antidiabetic medications (most type 1 diabetics do not take oral antidiabetic medications). Some of these medications make tissues more sensitive to insulin, and some inhibit production and release of glucose from the liver. Finally, diabetics may take medications to prevent or treat certain long-term complications of diabetes. For example, they may take low-dose aspirin therapy and cholesterol-lowering drugs to help prevent cardiovascular disease.

Medical management of diabetes is clearly a time-consuming but necessary endeavor for patients. Management can also impose a significant financial burden, especially for diabetics who are uninsured or underinsured.

Lifestyle Changes and Key Recommendations after Diagnosis

A diagnosis of type 1 or type 2 diabetes necessitates many lifestyle changes. These changes include checking blood glucose regularly, taking insulin or oral medication, exercising, planning and eating healthy meals, and being prepared for health emergencies. Successful diabetes management requires a strong commitment from patients. Many patients begin by establishing a relationship with a diabetic educator. Diabetic educators help patients develop daily routines regarding diet, exercise, glucose monitoring, and administration of insulin. They also help when new questions or problems arise in the course of treatment. If a diabetic smokes cigarettes, then quitting is essential because smoking also negatively impacts the cardiovascular system (Chapter 12a). Indeed, smokers who have diabetes are three times more likely to die of cardiovascular disease than are nonsmokers with diabetes. Also, if someone with diabetes chooses to consume alcohol, then it should occur on an occasional basis, with food, and only when his or her diabetes and blood glucose are well controlled. In moderation, alcohol can raise blood glucose. In excess, alcohol can cause blood glucose to drop to dangerously low levels. Stress typically causes higher blood glucose levels and thus is especially problematic for diabetics. Stress also can make it difficult to maintain the daily routines that are critical to managing diabetes. For both of these reasons, diabetics are advised to reduce or eliminate stress whenever possible. It also is recommended that diabetics

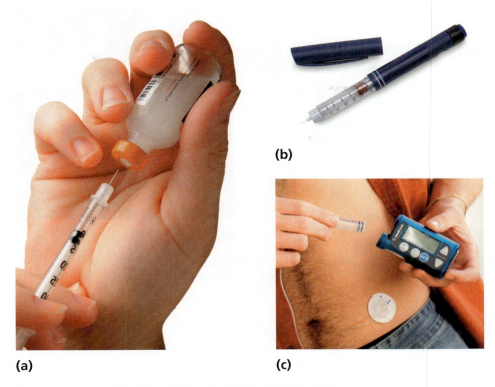

FIGURE 10a.4 There are several methods for self-administering insulin. (a) Some diabetics use a syringe, fine needle, and vial of insulin. (b) Other diabetics use an insulin pen, which contains a cartridge of insulin. (c) Still other diabetics use an insulin pump, which has a reservoir of insulin inside. A tube connects the reservoir to a catheter inserted under the skin of the abdomen.

manage unavoidable stress through relaxation techniques, exercise, and enjoyable activities.

Health care professionals encourage their diabetic patients to follow some key recommendations. One is to wear a diabetic identification tag or bracelet (Figure 10a.5). Such identification can immediately inform medical personnel or others of the existing health condition. This can be critical if a diabetic is unable to communicate, as in the case of insulin shock or very young children who may be unable to describe their condition. Also, anyone who uses insulin should have an emergency kit that contains glucagon. Recall that glucagon is the pancreatic hormone that acts in opposition to insulin to raise blood glucose. If a diabetic loses consciousness because of severely low blood glucose, then a family member or friend can inject glucagon to raise glucose levels. Diabetics who use insulin will often encounter less dramatic episodes of low blood glucose that still require care. For such episodes, a diabetic can drink orange juice or consume glucose tablets to quickly raise blood glucose to acceptable levels. Finally, given all of the long-term complications of diabetes, it is important for diabetics to have regular physical, dental, and eye exams.

Prognoses

At this time, type 1 diabetes cannot be prevented, and there is no cure. Nevertheless, strict control of blood glucose can dramatically delay or prevent the cardiovascular, kidney, nerve, and eye complications associated with this disease. Scientists are working to identify genes and environmental factors (e.g., viruses, toxins, and diet) that contribute to type 1 diabetes. Such knowledge may help us understand how to prevent or reverse the autoimmune destruction of insulin-producing cells in the pancreas. Other approaches include transplants of either a whole pancreas or pancreatic islets (clusters of cells that include the beta cells). Finally, researchers are working to develop an artificial pancreas that would monitor blood glucose levels and deliver appropriate amounts of insulin at the right times.

Type 2 diabetes can be prevented or delayed through lifestyle interventions, specifically weight loss and exercise. Without these lifestyle changes, prediabetes can progress to type 2 diabetes in as little as 10 years. Through strict control of blood glucose and blood pressure, patients with type 2 diabetes can reduce their risks of developing long-term complications. Scientists are looking for susceptibility genes for type 2 diabetes. They also are striving to better understand the molecular link between obesity and development of insulin resistance.

what would you do?

Medical expenses for diabetics are more than twice those of nondiabetics. Type 1 diabetes cannot be prevented. Type 2 diabetes can be prevented or delayed by losing weight. Should health insurance companies charge overweight people more? Should they charge type 1 diabetics more? What do you think?

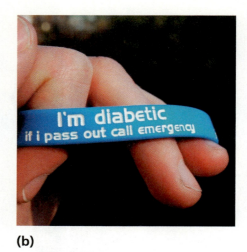

FIGURE 10a.5 *It is recommended that diabetics wear a medical ID so that they get appropriate care if they are unable to communicate: (a) medical identification necklace for a diabetic; (b) medical identification bracelet for a diabetic.*

10a.3 Gestational Diabetes

When a pregnant woman develops diabetes mellitus, the disorder is called *gestational diabetes*. This condition is diagnosed in 2% to 10% of pregnancies in the United States (these percentages are expected to increase as a result of new diagnostic criteria for gestational diabetes). It is more prevalent in certain populations, including African Americans, Hispanic Americans, and American Indians. Gestational diabetes usually begins in the second half of pregnancy. It is characterized by progressive insulin resistance. The placenta is the organ that supplies the growing fetus with nutrients and oxygen and carries away wastes and carbon dioxide (Chapter 18). The placenta also produces hormones, some of which make the mother's cells more resistant to insulin. Normally, the mother's pancreas responds by producing more insulin to overcome this resistance. Sometimes, however, her production of insulin is insufficient, and gestational diabetes results. The disease is progressive because the placenta initially increases in size as pregnancy proceeds. This means that production of placental hormones, including those that cause insulin resistance, also increases. In addition, increases in adipose tissue during pregnancy contribute to development of insulin resistance. Risk factors for developing gestational diabetes include family history (having a parent or sibling with type 2 diabetes), personal history (having gestational diabetes in a previous pregnancy), age over 25 years, being overweight before pregnancy, and being a member of a minority or ethnic group with high prevalence for gestational diabetes.

Gestational diabetes typically resolves after delivery because the placenta is expelled soon after the baby is born. Nevertheless, there can be serious health consequences for mother and baby during the pregnancy, at birth, and even thereafter. Some women with gestational diabetes develop preeclampsia during their pregnancy. This is a potentially life-threatening condition characterized by high blood pressure, excess protein in the urine, and fluid retention. The intimate circulatory connection between the mother and developing fetus means that gestational diabetes also has health consequences for the fetus. Recall that insulin resistance means that the mother's cells fail to adequately respond to insulin. This results in high levels of glucose in the blood of the mother. A major function of the placenta is to allow nutrients, such as glucose, to diffuse from the mother's blood into fetal blood. Thus, the extra glucose in maternal blood moves into fetal blood. Unlike glucose, insulin from the mother cannot cross the placenta to help lower blood glucose in the fetus. By about three months gestation, however, the pancreas of the fetus can respond to the high blood glucose, and it does so by producing insulin in excess. Because fetal insulin prompts the movement of glucose from the blood into fetal cells, its overproduction promotes excessive growth in the fetus. As a result, babies born to mothers with gestational diabetes often exhibit excessive birth weight. Excessive fetal growth is associated with difficult deliveries and may necessitate a cesarean section (procedure in which the baby and placenta are removed from the uterus through an incision in the abdominal wall and uterus). After birth, babies born to mothers with gestational diabetes may have low blood glucose (hypoglycemia) because their own production of insulin is so high. In addition to these immediate complications, such babies have an increased risk for developing obesity and type 2 diabetes in childhood and adulthood. Women who have had gestational diabetes are at risk for developing it in a subsequent pregnancy and of developing type 2 diabetes later in life.

Women with gestational diabetes may not experience any physical symptoms or may experience some or all of the symptoms of type 1 and type 2 diabetes. Fortunately, screening for gestational diabetes is usually a part of routine prenatal care. Between 24 and 28 weeks of pregnancy (earlier if a woman is considered at high risk), an expectant mother takes a glucose challenge test in which she drinks a glucose solution and then has blood drawn 1 hour later. If blood glucose is above a certain level (usually 140 mg/dl), then another test is scheduled to confirm gestational diabetes. In this second test, the woman fasts overnight. In the morning, she has blood drawn, after which she drinks a more concentrated glucose solution, and has her blood glucose checked every hour for 3 hours. At least two of the four values (fasting; 1 hour post challenge; 2 hours post challenge; 3 hours post challenge) must be abnormal for the test to be considered positive for gestational diabetes. If gestational diabetes is confirmed, then she works with her physician to develop a treatment plan, which typically includes testing blood glucose levels, eating a healthy diet, and engaging in regular physical activity. If diet and exercise are insufficient to control blood glucose, then some women may need to take insulin or medication. Finally, because women diagnosed with gestational diabetes are at increased risk for developing gestational diabetes in subsequent pregnancies and type 2 diabetes later in life, it is important that they continue to have their blood screened for diabetes and that they make all efforts to prevent development of the disease.

stop and think

Is gestational diabetes more similar to type 1 diabetes or type 2 diabetes with respect to causes, risk factors, and treatments?

10a.4 Other Specific Types of Diabetes

"Other specific types" of diabetes represent only 1% to 2% of all diagnosed cases. These types include insulin deficiencies resulting from damage to the pancreas from disease, infection, drugs, or trauma. As an example, consider the case of Tre F. Porfirio (Figure 10a.6). This 21-year-old serviceman was shot three times in the back by an insurgent in Afghanistan. The bullets damaged large portions of Porfirio's gastrointestinal tract, necessitating the removal of his gallbladder, portions of his small and large intestines, and a large section of his pancreas. Following these surgeries at combat hospitals, Porfirio was flown back to the United States for further treatment. Surgeons at the Walter Reed Army Medical Center in Washington, D.C., determined that the part of his pancreas that remained would have to be removed, because it was leaking digestive enzymes. (In addition to having endocrine cells that produce hormones, the pancreas has exocrine cells that produce digestive enzymes.) Complete removal of the pancreas would

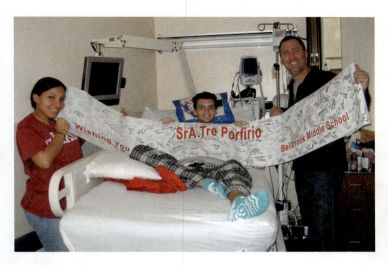

FIGURE 10a.6 Serviceman Tre F. Porfirio received an autotransplant of pancreatic islets cells to his liver after his pancreas was removed following traumatic injury.

eliminate all insulin-producing cells and render Porfirio a severe diabetic. To prevent this, surgeons at Walter Reed removed the remaining section of his pancreas and shipped it to the Diabetes Research Institute at the University of Miami Miller School of Medicine. At the institute, a team of doctors extracted and purified thousands of islets from the pancreas. The islets were then sent to Walter Reed, where they were injected into Porfirio's liver. The islets became established in the liver and began producing insulin. Porfirio died about a year later, but the transplant made it possible for him to witness the birth of his son. Tre Porfirio's case is believed to be the first pancreatic autotransplantation (the transplantation of tissues from one part of the body to another within the same individual) following a traumatic injury.

looking ahead

In Chapter 10a, we discussed diabetes mellitus, a group of diseases primarily diagnosed through checking levels of glucose in the blood. Blood transports substances other than glucose and has functions beyond transport. In Chapter 11, we examine the functions and composition of blood, as well as disorders of red and white blood cells.

11 Blood

Did You Know?
- Each year, 4.5 million Americans will need a blood transfusion.
- Every second, 2 million red blood cells die.

Blood donations save lives.

11.1 Functions of Blood
11.2 Composition of Blood
11.3 Blood Cell Disorders
11.4 Blood Types
11.5 Blood Clotting

ENVIRONMENTAL ISSUE
Lead Poisoning

In the previous several chapters we discussed the nervous and endocrine systems, which control and coordinate body activities. Here and in upcoming chapters, we consider systems that maintain the human body. In this chapter, we turn our attention to blood—a critical life-sustaining fluid. As we consider the functions and composition of blood, many of the reasons why it is so vital will become clear. In addition, we discuss various blood disorders, the classification of blood into blood types, and the significance of blood types. Finally, we consider how blood forms clots to prevent blood loss from a wound.

11.1 Functions of Blood

Blood is sometimes referred to as the river of life. The comparison is apt because, like many a river, blood serves as a transportation system. It carries vital materials to all the cells of the body and carries away the wastes that cells produce. But blood does more than passively move its cargo. Its white blood cells help protect us against disease-causing organisms, and its clotting mechanisms help protect us from excessive blood loss when a vessel is damaged. In addition, buffers in the blood help regulate the acid–base balance of body fluids. Blood also helps regulate body temperature by absorbing heat produced in metabolically active regions and distributing it to cooler regions and to the skin, where the heat is dissipated. So the diverse functions of blood can be grouped into three categories: transportation, protection, and regulation.

11.2 Composition of Blood

Blood *is* thicker than water. The reason is that blood contains cells suspended in its watery fluid. In fact, a single drop of blood contains more than 250 million blood cells. You may recall from Chapter 4 that blood is classified as a connective tissue because it contains cellular elements suspended in a matrix. The liquid matrix is called *plasma,* and the cellular elements are collectively called the *formed elements* (Figure 11.1).

Plasma

Plasma is a straw-colored liquid that makes up about 55% of blood. Plasma serves as the medium in which materials are transported by the blood. Almost every substance that is transported by the blood is dissolved in the plasma. These include nutrients (such as simple sugars, amino acids, lipids, and vitamins), ions (such as sodium, potassium, and chloride), dissolved gases (including carbon dioxide, nitrogen, and a small amount of oxygen), and every hormone. In addition to transporting materials to the cells, the plasma carries away cellular wastes. For example, urea and uric acid are carried to the kidneys, where they can be removed from the body. Blood also transports carbon dioxide from the cells where it is produced to the lungs for release. Indeed, if wastes were not removed, the cells would die.

Despite the amount and variety of substances transported by the blood, most of the dissolved substances (solutes) in the blood are **plasma proteins**, which make up 7% to 8% of plasma. Plasma proteins help balance water flow between the blood and the cells. You may recall from Chapter 3 that water moves by osmosis across biological membranes from an area of lesser solute concentration to an area of greater solute concentration. Without the plasma proteins, water would be drawn out of the blood by the proteins in cells. As a result, fluid would accumulate in the tissues, causing swelling.

Most of the 50 or so types of plasma proteins fall into one of three general categories: albumins, globulins, and clotting proteins. The albumins make up more than half of the plasma proteins. They are most important in the blood's water-balancing ability. The globulins have a variety of functions. Some globulins transport lipids, including fats and some cholesterol, as well as fat-soluble vitamins. Other globulins are antibodies, which provide protection against many diseases. An example of the third category of plasma protein—the clotting proteins—is fibrinogen, whose role we discuss later in the chapter.

Formed Elements

Among the substances transported by the plasma are the **formed elements**—platelets, white blood cells, and red blood cells. These cell fragments and cells perform some of the key functions of the blood. Descriptions and functions of the formed elements of the blood are summarized in Table 11.1.

Red bone marrow fills the cavities within many bones and is the birthplace and nursery for the formed elements. Its spongelike framework supports fat cells, but it also supports the

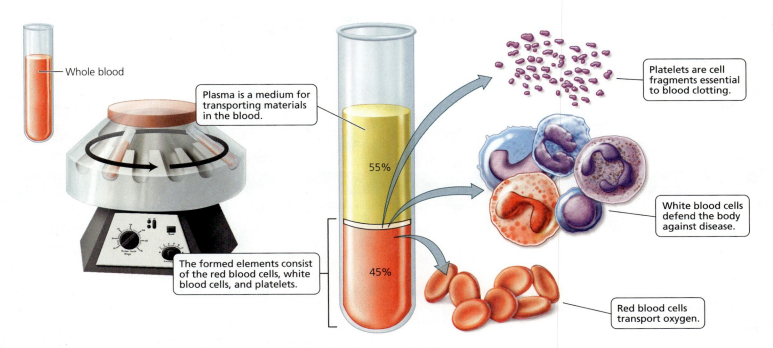

FIGURE 11.1 *To separate blood into its major components, it is placed in a test tube with a substance that prevents coagulation, and then it is spun in a centrifuge. Whole blood consists of a straw-colored liquid, called plasma, in which cellular elements, called formed elements, are suspended. After separation, the uppermost layer in the test tube consists of plasma. The formed elements are found in two layers below the plasma. Just below the plasma is a thin layer consisting of platelets and white blood cells (leukocytes). The red blood cells (erythrocytes) are packed at the bottom of the test tube.*

TABLE 11.1 The Formed Elements of Blood

Type of Formed Element		Cell Function	Description	No. of Cells/mm³	Life Span
Platelets		Play role in blood clotting	Fragments of a megakaryocyte; small, purple-stained granules in cytoplasm	250,000–500,000	5–10 days
White Blood Cells (WBCs; leukocytes)					
Granulocytes					
Neutrophils		Consume bacteria by phagocytosis	Multilobed nucleus, clear-staining cytoplasm, inconspicuous granules	3000–7000	6–72 hours
Eosinophils		Consume antibody–antigen complex by phagocytosis; attack parasitic worms	Large, pink-staining granules in cytoplasm, bilobed nucleus	100–400	8–12 days
Basophils		Release histamine, which attracts white blood cells to the site of inflammation and widens blood vessels	Large, purple-staining cytoplasmic granules; bilobed nucleus	20–50	3–72 hours
Agranulocytes					
Monocytes		Give rise to macrophages, which consume bacteria, dead cells, and cell parts by phagocytosis	Gray-blue cytoplasm with no granules; U-shaped nucleus	100–700	Several months
Lymphocytes		Attack damaged or diseased cells or disease-causing organisms; produce antibodies	Round nucleus that almost fills the cell	1500–3000	Many years
Red Blood Cells (RBCs; erythrocytes)					
		Transport oxygen and carbon dioxide	Biconcave disk, no nucleus	4–6 million	About 120 days

undifferentiated cells called blood **stem cells** that divide and give rise to all the formed elements (Figure 11.2).

Platelets

Platelets, sometimes called *thrombocytes* (*thromb–*, clot; *–cyte*, cell), are essential to blood clotting. They are actually fragments of larger cells called *megakaryocytes* and are formed in the red bone marrow when megakaryocytes break apart. The fragments are released into the blood at the astounding rate of about 200 billion a day. They then mature during the course of a week, after which they circulate in the blood for about 5 to 10 days. Platelets contain several substances important in stopping the loss of blood through damaged blood vessels. This vital function of platelets is considered later in the chapter.

stop and think

Thrombocytopenia is a medical term for the presence of relatively few platelets in the blood. Why would this condition cause unusual bruising and bleeding?

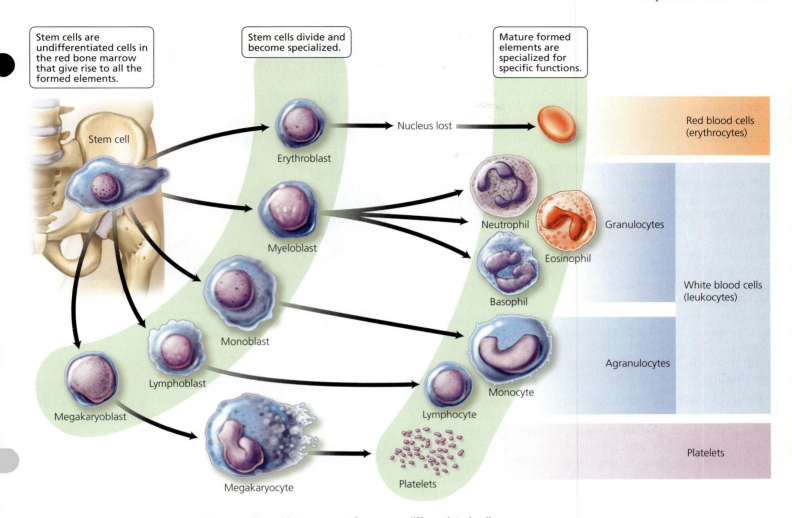

FIGURE 11.2 *All formed elements originate in the red bone marrow from an undifferentiated cell called a stem cell. Stem cells divide and differentiate, giving rise to the various types of blood cells.*

White Blood Cells and Defense against Disease

White blood cells (WBCs), or **leukocytes** (*leuk–*, white; *–cyte*, cell), perform certain mundane housekeeping duties—such as removing wastes, toxins, and damaged or abnormal cells—but they also serve as warriors in the body's fight against disease (see Chapter 13). Although leukocytes represent less than 1% of whole blood, we simply could not live without them. We would succumb to the microbes that surround us. Because the number of white blood cells increases when the body responds to microbes, white blood cell counts are often used as an index of infection.

Like the other formed elements, white blood cells are produced in the red bone marrow.[1] Unlike platelets and red blood cells, however, white blood cells are nucleated. Moreover, although they circulate in the bloodstream, they are not confined there. By squeezing between neighboring cells that form the walls of blood vessels, white blood cells can leave the circulatory system and move to a site of infection, tissue damage, or inflammation (Figure 11.3). Having slipped out of the capillary into the fluid bathing the cells, white blood cells roam through

FIGURE 11.3 *White blood cells can squeeze between the cells that form the wall of a capillary. They then enter the fluid surrounding body cells and, attracted by chemicals released by microbes or damaged cells, gather at the site of infection or injury.*

---

[1] One type, the lymphocytes, also may be produced in the lymph nodes and other lymphoid tissues.

the tissue spaces. Chemicals released by invading microbes or damaged cells attract the white blood cells and cause them to gather in areas of tissue damage or infection. Certain types of white blood cells may then engulf the "offender" in a process called **phagocytosis** (*phago–*, to eat; *–cyt*, cell; see Chapter 3). Our consideration of white blood cells in this chapter is brief; Chapter 13 contains a more detailed discussion of the many tactics that white blood cells use in defending our bodies.

There are two groups of white blood cells. **Granulocytes** have granules in their cytoplasm. The granules are actually sacs containing chemicals that are used as weapons to destroy invading pathogens, especially bacteria. **Agranulocytes** lack cytoplasmic granules or have very small granules. These two groups comprise a total of five types of white blood cells.

Granulocytes All white blood cells are colorless, and thus they are often stained to make them visible under a microscope. Depending on the color of the granules after they have been stained for microscopic study, the granulocytes are classified as neutrophils, eosinophils, or basophils. Neutrophils have small granules that don't stain, either with the acidic red stain eosin or with a basic blue stain. The granules in eosinophils pick up a pink color from the stain eosin. Basophils have purple-staining granules (see Table 11.1).

- **Neutrophils**, the most abundant of all white blood cells, are the blood cell soldiers on the front lines. Arriving at the site of infection before the other types of white blood cells, neutrophils immediately begin to engulf microbes by phagocytosis, thus curbing the spread of the infection. After engulfing a dozen or so bacteria, a neutrophil dies. But even in death it helps the body's defense by releasing chemicals that attract more neutrophils to the scene. Dead neutrophils, along with bacteria and cellular debris, make up pus, the yellowish liquid we usually associate with infection.
- **Eosinophils** contain substances that are important in the body's defense against parasitic worms, such as tapeworms and hookworms. They also lessen the severity of allergies by engulfing antibody–antigen complexes and inactivating inflammatory chemicals.
- **Basophils** release histamine, a chemical that attracts other white blood cells to the site of infection and causes blood vessels to dilate (widen), thereby increasing blood flow to the affected area. They also play a role in some allergic reactions.

Agranulocytes The agranulocytes, which lack visible granules in the cytoplasm, are classified as monocytes or lymphocytes.

- **Monocytes**, the largest of all formed elements, leave the bloodstream and enter various tissues, where they develop into macrophages. *Macrophages* are phagocytic cells that engulf invading microbes, dead cells, and cellular debris.
- **Lymphocytes** are classified into two types: B lymphocytes and T lymphocytes. The *B lymphocytes* give rise to plasma cells, which, in turn, produce antibodies. *Antibodies* are proteins that recognize specific molecules—called *antigens*—on the surface of invading microbes or other foreign cells. After recognizing the foreign cell by its antigens, the antibodies help prevent it from harming the body. There are several types of *T lymphocytes*, specialized white blood cells that play roles in the body's defense mechanisms. We will discuss lymphocytes further in Chapter 13.

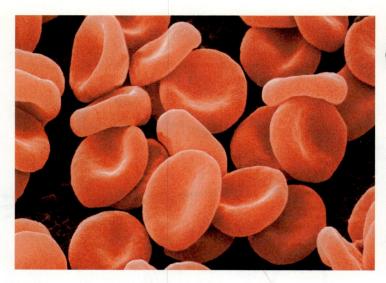

FIGURE 11.4 Red blood cells serve to ferry oxygen from the lungs to the needy tissues. Each red blood cell is packed with the oxygen-binding pigment hemoglobin. A red blood cell is a small disk that is indented on both sides. This design maximizes the surface area for gas exchange.

Red Blood Cells and Transport of Oxygen

Red blood cells (RBCs), also called **erythrocytes** (*erythro–*, red; *–cyte*, cell), pick up oxygen in the lungs and ferry it to all the cells of the body. Red blood cells also carry about 23% of the blood's total carbon dioxide, a metabolic waste product. They are by far the most numerous cells in the blood. Indeed, they number 4 million to 6 million per cubic millimeter (mm^3) of blood and constitute approximately 45% of the total blood volume.

Red blood cells and hemoglobin The shape of red blood cells, as shown in Figure 11.4, is marvelously suited to their function of picking up and transporting oxygen. Red blood cells are quite small, and each is shaped like a biconcave disk—a flattened disk indented on each side. This shape maximizes the surface area of the cell. Because of the greater surface area, oxygen can enter the red blood cell more rapidly than if the disk did not have an indent. A red blood cell is unusually flexible and thus able to squeeze through capillaries (the smallest blood vessels, those in which gas exchange occurs), even those with a diameter much smaller than red blood cells. Each red blood cell is packed with **hemoglobin**, the oxygen-binding pigment that is responsible for the cells' red color. As a red blood cell matures in the red bone marrow, it loses its nucleus and most organelles. Thus, it is scarcely more than a sac of hemoglobin molecules. Each red blood cell is packed with approximately 280 million molecules of hemoglobin. As can be seen in Figure 11.5, each hemoglobin molecule is made up

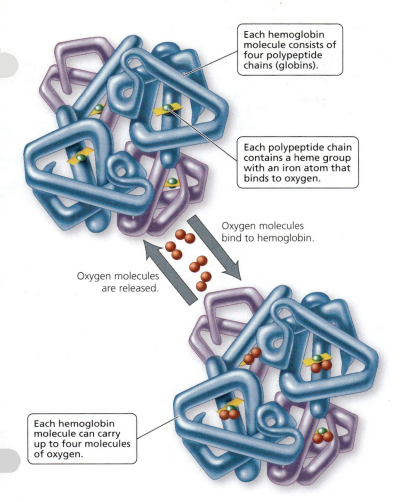

FIGURE 11.5 *The structure of hemoglobin, the pigment in red blood cells that transports oxygen from the lungs to the cells*

of four subunits. Each subunit consists of a polypeptide chain and a heme group. The heme group includes an iron ion that actually binds to the oxygen. Therefore, each hemoglobin molecule can carry up to four molecules of oxygen. The compound formed when hemoglobin binds with oxygen is called, logically enough, *oxyhemoglobin*. Body cells use the oxygen to boost the amount of energy extracted from food molecules in cellular respiration. As oxygen is used, carbon dioxide is produced. Most of the carbon dioxide travels to the lungs as a bicarbonate ion, but some of it binds to hemoglobin (at a site other than that where the iron atom binds oxygen).

As wonderfully adapted as it is for carrying oxygen, the hemoglobin molecule binds 200 times more readily to carbon monoxide, a product of the incomplete combustion of any carbon-containing fuel. In other words, if concentrations of carbon monoxide and oxygen were identical in inhaled air, for every 1 molecule of hemoglobin that binds to oxygen molecules, 200 molecules of hemoglobin would bind to carbon monoxide molecules. This is the reason carbon monoxide can be deadly. When carbon monoxide binds to the oxygen-binding sites on hemoglobin, it blocks oxygen from binding to it, preventing the blood from carrying life-giving oxygen to the cells. Carbon monoxide is a particularly insidious poison because it is odorless and tasteless. Its primary source is automobile exhaust, but it can also come from indoor sources, including improperly vented heaters and leaky chimneys. Thus, indoor carbon monoxide detectors can save lives.

Life cycle of red blood cells The creation of a red blood cell, which takes about 6 days, entails many changes in the cell's activities and structure. First, the very immature cell becomes a factory for hemoglobin molecules. And as noted earlier, after the cell is packed with hemoglobin, its nucleus is pushed out. Then a structural metamorphosis occurs, culminating in a mature red blood cell with a typical biconcave shape. Once this change in shape takes place, the cell leaves the bone marrow and enters the bloodstream. Red marrow produces roughly 2 million red blood cells a second, throughout the life of an individual, for a cumulative total of more than half a ton in a lifetime.

A red blood cell lives for only about 120 days. During that time, it travels through approximately 100 km (62 mi) of blood vessels—being bent, bumped, and squeezed along the way. Its life span is probably limited by the lack of a nucleus that would otherwise maintain it and direct needed repairs. Without a nucleus, for instance, protein synthesis needed to replace key enzymes cannot take place, so the cell becomes increasingly rigid and fragile.

The liver and spleen are the "graveyards" where worn-out red blood cells are removed from circulation. The old, inflexible red blood cells tend to become stuck in the tiny circulatory channels of these organs. Macrophages then engulf and destroy the dying cells. The demolished cells release their hemoglobin, which the liver degrades into its protein (globin) component and heme. The protein is digested to amino acids, which can be used to make other proteins. The iron from the heme is salvaged and sent to the red marrow for recycling.

The remaining part of the heme is degraded to a yellow pigment, called *bilirubin*, which is excreted by the liver in bile. Bile is released into the small intestine, where it assists in the digestion of fats. It is then carried along the digestive system to the large intestine with undigested food and becomes a component of feces. The color of feces is partly due to bilirubin that has been broken down by intestinal bacteria.

Products formed by the chemical breakdown of heme also create the yellowish tinge in a bruise that is healing. A bruise, or black-and-blue mark, results when tiny blood vessels or capillaries are ruptured and blood leaks into the surrounding tissue. As the tissues use up the oxygen, the blood becomes darker and, viewed through the overlying tissue, looks black or blue. Gradually, the red blood cells degenerate, releasing hemoglobin. The breakdown products of hemoglobin then make the bruise appear yellowish.

stop and think

Hepatitis is an inflammation of the liver that can be caused by certain viruses or exposure to certain drugs. It impairs the liver's ability to handle bilirubin properly. A symptom of hepatitis is jaundice, a condition in which the skin develops a yellow tone. Why does hepatitis cause jaundice?

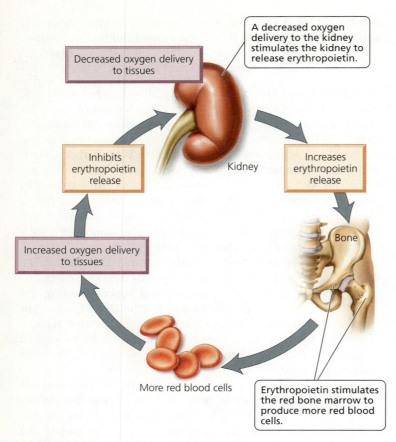

FIGURE 11.6 *The production of red blood cells is regulated by a negative feedback relationship between the oxygen-carrying capacity of the blood and the production of erythropoietin.*

Disorders of Red Blood Cells

Anemia, a condition in which the blood's ability to carry oxygen is reduced, can result from too little hemoglobin, too few red blood cells, or both. The symptoms of anemia include fatigue, headaches, dizziness, paleness, and breathlessness. In addition, an anemic person's heart often beats faster to compensate for the blood's decreased ability to carry oxygen. The accelerated pumping can cause heart palpitations—the uncomfortable awareness of one's own heartbeat. Although anemia is not usually life threatening, it can affect the quality of life because lack of energy and low levels of productive activity often go hand in hand.

Worldwide, the most common cause of anemia is an insufficiency of iron in the body, which leads to inadequate hemoglobin production. *Iron-deficiency anemia* can be caused by a diet that contains too little iron, by an inability to absorb iron from the digestive system, or from blood loss—such as might occur because of menstrual flow or peptic ulcers. Treatment for iron-deficiency anemia includes dealing with the cause of the iron depletion, as well as restoring iron levels to normal by eating foods that are rich in iron—such as meat, leafy green vegetables, and fortified cereals—or by taking pills that contain iron.

Blood loss will obviously lower red blood cell counts, but so will any condition that causes the destruction of red blood cells at an amount that exceeds the production of red blood cells. For example, in hemolytic anemias, red blood cells are ruptured because of infections, defects in the membranes of red blood cells, transfusion of mismatched blood, or hemoglobin abnormalities. *Sickle-cell anemia* is an example of a hemolytic anemia caused by abnormal hemoglobins. This abnormal hemoglobin (hemoglobin S) causes the red blood cells to become deformed to a crescent (or sickle) shape when the blood's oxygen content is low. The misshapen cells are fragile and rupture easily, clogging small blood vessels and promoting clot formation. These events prevent oxygen-laden blood from reaching the tissues and can cause episodes of extreme pain.

Red blood cell numbers also drop when the production of red blood cells is halted or impaired, as occurs in *pernicious anemia.* The production of red blood cells depends on a supply of vitamin B_{12}. The small intestine absorbs vitamin B_{12} from the diet with the aid of a chemical called intrinsic factor, which is produced by the stomach lining. People with pernicious anemia do not produce intrinsic factor and are therefore unable to absorb vitamin B_{12}. They are treated with injections of B_{12}. Lead poisoning can also cause anemia. (See the Environmental Issue essay, *Lead Poisoning.*)

A negative feedback mechanism regulates red blood cell production according to the needs of the body, especially the need for oxygen (Figure 11.6). Most of the time, red blood cell production matches red blood cell destruction. However, there are circumstances—blood loss, for instance—that trigger a homeostatic mechanism that speeds up the rate of red blood cell production. This mechanism is initiated by a decrease in the oxygen supply to the body's cells. Certain cells in the kidney sense the reduced oxygen, and they respond by producing the hormone **erythropoietin**. Erythropoietin then travels to the red marrow, where it steps up both the division rate of stem cells and the maturation rate of immature red blood cells. When maximally stimulated by erythropoietin, the red marrow can increase red blood cell production tenfold—to 20 million cells per second! The resulting increase in red blood cell numbers is soon adequate to meet the oxygen needs of body cells. The increased oxygen-carrying capacity of the blood then inhibits erythropoietin production.

11.3 Blood Cell Disorders

Disorders of red and white blood cells have many different causes. The problems associated with each disorder depend on the type of blood cells affected, because red and white blood cells have different functions.

Disorders of White Blood Cells

Infectious mononucleosis ("mono") is a viral disease of the lymphocytes. It is caused by the Epstein-Barr virus, which is common in humans but usually asymptomatic. In mononucleosis, the infection causes an increase in lymphocytes that have an atypical appearance. Because mono often is spread from person to person by oral contact, it is sometimes called the "kissing disease"; however, it can also be spread by sharing eating utensils or drinking glasses. Mono is most common among teenagers and young adults, particularly those living in dormitories. It often strikes at stressful times, such as during final exams, when immune resistance is low.

ENVIRONMENTAL ISSUE

Lead Poisoning

What do children in old buildings in America today have in common with the ancient upper-class Romans? Lead poisoning. Today, some children are poisoned by the lead in paint chips that they unintentionally ingest as aging paint peels (or, in very young children, when they chew or bite objects that have been treated with lead paint). Centuries ago, Romans were poisoned by lead in their eating and drinking vessels.

The types of cells that are most sensitive to lead are nerve cells and the bone marrow cells that give rise to red blood cells. The best-known effects of lead on human health are probably those on the nervous system—mental retardation, lowered IQs, reading disabilities, irritability, hyperactivity, and even death. However, the effects of lead on the blood, including anemia and changes in blood enzymes, can also be devastating. Indeed, a blood test is the only way to positively diagnose lead poisoning.

Lead launches a two-pronged attack on the blood's ability to carry oxygen. First, it interferes with the absorption of iron from the digestive system. Second, it inhibits one of the essential enzymes leading to hemoglobin synthesis.

As our example of ancient Romans suggests, lead has been a common pollutant for centuries, encountered in a surprising number of substances. The major causes of lead poisoning in the United States are as follows:

1. **House paint** Although lead has not been added to house paint for years, this remains the most prevalent cause of poisoning.
2. **Drinking water** One out of every five Americans drinks tap water containing excess levels of lead, according to estimates by the U.S. Environmental Protection Agency (EPA).
3. **Air and soil pollution** The primary source of lead in the atmosphere has been leaded gasoline. Leaded gasoline is no longer sold in the United States for use in automobiles, although it is permissible in farm equipment, race cars, and other off-road vehicles.
4. **Lead solder on food cans** The lead in the solder used to seal the seams in food and beverage cans can leach into the contents, especially if the contents are acidic foods such as tomatoes or citrus juices. Although the use of lead solder on food cans has been banned in the United States, it is still found on some imported food cans.
5. **Lead-containing dishware** China, ceramic, and earthenware dishes may be coated with lead-containing glaze. When the glazes are not properly fired, lead can leach into the food or beverage contained in the dish.
6. **Fruits and vegetables grown in lead-contaminated soil** Soil can be contaminated by lead from the atmosphere or by leakage from lead-containing solid wastes, such as lead batteries in automobiles that are dumped into landfills. Important sources of atmospheric lead are the combustion of leaded motor fuel, the smelting of ores and other industrial processes, and the incineration of refuse.

Although it is true that blood levels of lead are at an all-time low in the United States, the lead problem has not been eliminated. In 2010 the EPA passed a new Renovation Repair and Painting rule to protect children from lead dust caused by lead paint removal. An important problem now is deciding who should be financially responsible for keeping us safe.

Questions to Consider

- Who do you think should bear the financial responsibility of removing lead paint from old apartment buildings? The residents? The landlords? The city? The federal government?
- Do you think the residents of buildings with lead paint should be compensated for any lead-related health problems they experience? If so, who do you think should pay—the landlords? The original contractors? The government?

Children often put objects into their mouths. They may unintentionally ingest lead paint chips if the object has been treated with lead.

---

 what would you do?

Blood doping is a practice that boosts red blood cell numbers and, therefore, the blood's ability to carry oxygen. In one method, blood is drawn from the body and stored, prompting the body temporarily to boost its production of red blood cells. The stored blood is then returned to the body. Some athletes who compete in aerobic events, such as swimming or cycling, have used blood doping to increase their endurance and speed. Although blood doping is considered unethical and is prohibited in Olympic competition, it is difficult to detect. One sign is an exceptionally high number of red blood cells; another is the presence in the urine of a hormone (erythropoietin) that stimulates red blood cell production—but tests can detect only about 50% of the athletes who practice blood doping. Considering the large percentage of blood-doped athletes who will escape detection even if they are tested, do you think that those who are caught should be punished?

The initial symptoms of mono are similar to those of influenza: fever, chills, headache, sore throat, and an overwhelming sense of being ill. Within a few days, the lymph nodes (commonly called "glands") in the neck, armpits, and groin become painfully swollen. There is no treatment, so mono must simply run its course. The major symptoms generally subside within a few weeks, but fatigue may linger for several months or longer.

Leukemia is a cancer of the white blood cells that causes their uncontrolled multiplication, so that their number increases greatly. These cancerous cells—all descendants of a single abnormal cell—remain immature and are therefore unable to defend the body against infectious organisms. Because they divide more rapidly than do normal cells, the abnormal cells "take over" the bone marrow, preventing the development of all normal blood cells, including red blood cells, white blood cells, and platelets.

Symptoms of leukemia generally result either from the insufficient number of normal blood components or from the invasion of organs by abnormal white blood cells. The increased number of white blood cells crowds out the other formed elements. Insufficient numbers of platelets lead to inadequate clotting, which causes gum bleeding and frequent bruising. Reduced levels of red blood cells lead to anemia, which in turn causes chronic fatigue, breathlessness, and pallor. Because their white blood cells do not function properly, leukemia patients may suffer from repeated respiratory or throat infections, herpes, or skin infections. Also, people with leukemia may experience bone tenderness because the immature white blood cells pack the red marrow. Headaches, another symptom of leukemia, may be caused by anemia or by the effects of abnormal white blood cells in the brain.

Treatment of leukemia usually includes radiation therapy and chemotherapy to kill the rapidly dividing cells. In addition, transfusions of red blood cells and platelets may be given to alleviate anemia and prevent excessive bleeding. Today, many patients with acute leukemia are cured with bone marrow transplants. Someone, most often a family member whose tissue type closely matches that of the patient, must be located and agree to serve as a donor. Then the bone marrow in the leukemia patient must be destroyed by irradiation and drugs. Next, the donor's bone marrow is given to the patient intravenously, just like a blood transfusion. The marrow cells find their way to the patient's marrow and begin to grow. Treatment with stem cells from umbilical cord blood seems to be as effective as bone marrow transplantation in helping some patients with leukemia go into remission (the signs and symptoms of leukemia go away). Stem cells from umbilical cord blood are discussed in Chapters 19a, "Special Topic: Stem Cells—A Repair Kit for the Body."

11.4 Blood Types

Human blood is classified into different **blood types** according to the presence or absence of certain molecules—mostly proteins—on the surface of a person's red blood cells. Each of your body cells is labeled as "self" (that is, as belonging to your body) by proteins on its surface. If a cell that lacks these self markers enters the body, the body's defense system recognizes that the foreign cell is "nonself" and does not belong. The foreign cell will have different proteins on its surface. To the body's defense system, these proteins are antigens, identifying the cell as foreign and marking it for destruction. As was mentioned earlier, one way the body attacks the foreign cell is to produce proteins called antibodies that specifically target the antigen on the foreign cell's surface. Let's consider the role of antigens and antibodies in blood types and transfusions.

ABO Blood Types

When asked about your blood type, you are probably used to responding by indicating one of the types in the ABO series: A, B, AB, or O. Red blood cells with only the antigen A on their surface are type A. When only the B antigen is on the red blood cell surface, the blood is type B. Blood with both A and B antigens on the red blood cell surface is designated type AB. When neither A nor B antigens are present, the blood is type O.

Normally, a person's plasma contains antibodies against those antigens that are *not* on his or her own red blood cells. Thus, individuals with type A blood have antibodies against the B antigen (anti-B antibodies[2]), and those with type B blood have antibodies against A (anti-A antibodies). Because individuals with type AB blood have both antigens on their red blood cells, they have neither antibody. Those with type O blood have neither antigen, so they have both anti-A and anti-B antibodies in their plasma.

In a typical test for blood type, technicians mix a drop of a person's blood with a solution containing anti-A antibodies and mix another drop of the blood with a solution containing anti-B antibodies. If clumping occurs in one of the mixtures, it means the antigen corresponding to the antibody in that mixture is present (Figure 11.7).

Similarly, when a person is given a blood transfusion with donor blood containing foreign antigens, the antibodies in the recipient's blood will cause the donor's cells to clump, or **agglutinate**. This clumping of the donor's cells is damaging and perhaps even fatal. The clumped cells can get stuck in small blood vessels and block blood flow to body cells. Or they may break open, releasing their cargo of hemoglobin. The hemoglobin then clogs the filtering system in the kidneys, causing death.

It is important to be sure that the blood types of a donor and recipient are compatible, which means that the recipient's blood does not contain antibodies to antigens on the red blood cells of the donor. The plasma of the donor's blood may contain antibodies against antigens on the recipient's red blood cells, but these will be diluted as they enter the recipient's circulation. Therefore, the donor's antibodies are not a major problem. The questions to ask in this case are (1) what, if any, antigens are on the donor's cells and (2) what, if any, antibodies are in the recipient's blood. For example, if a person with blood type A is given a transfusion of blood type B or of type AB, the naturally occurring anti-B antibodies in the recipient's blood will cause the donor's red blood cells to clump because they have the B antigen. The transfusion relationships among blood types in the ABO series are shown in Table 11.2.

Rh Factor

The A and B antigens are not the only important antigens found on the surface of red blood cells. The presence or absence of an **Rh factor** is also an important component of blood type. The name *Rh* comes from the beginning of the name of the rhesus monkey, in which an Rh antigen was first discovered. People who have the Rh antigens on their red blood cells are considered Rh-positive (Rh+). When Rh antigens are missing from the red blood cell surface, the individual is considered Rh-negative (Rh-).

An Rh-negative person will not form anti-Rh antibodies unless he or she has been exposed to the Rh antigen. For this reason, an Rh-negative individual should be given only Rh-negative blood in a transfusion. If he or she is mistakenly given Rh-positive blood, it will stimulate the production of anti-Rh antibodies. A transfusion reaction will not occur after the first

---

[2] It is not certain why these antibodies form without exposure to red blood cells bearing the foreign antigen. It may be that either the bacteria that invade our bodies or the foods we eat contain a small amount of A and B antigens—enough to stimulate antibody production.

11.4 Blood Types

Q What blood type will agglutinate to each serum when mixed separately with sera containing anti-A, anti-B, and anti-Rh antibodies?

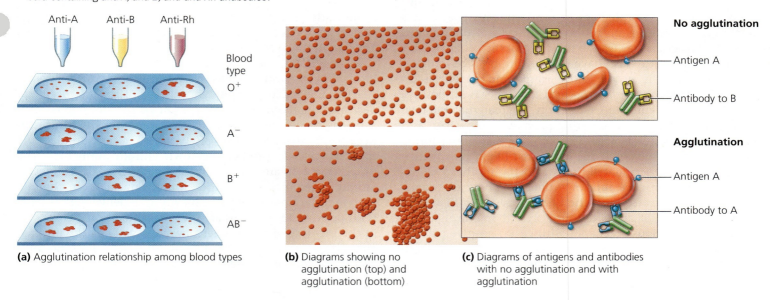

(a) Agglutination relationship among blood types

(b) Diagrams showing no agglutination (top) and agglutination (bottom)

(c) Diagrams of antigens and antibodies with no agglutination and with agglutination

FIGURE 11.7 Blood is typed by mixing it with serum known to contain antibodies specific for a certain antigen. If blood containing that antigen is mixed with the serum, the blood will agglutinate (clump). Thus, one drop of blood is mixed with serum containing anti-A and another drop with serum containing anti-B. A third drop of blood is mixed with serum containing antibodies to Rh. Agglutination in response to an antibody reveals the presence of the antigen.

A AB+

TABLE 11.2 Transfusion Relationships among Blood Types

Blood Types	Antigens on Red Blood Cells	Antibodies in Plasma	Blood Types (RBCs) That Can Be Received in Transfusions	Incidence of Blood Type in United States
A	A	Anti-B	A, O	Caucasian, 40%
				African American, 26%
				Asian, 27.3%
				Hispanic 31%
				Native American, 8%
B	B	Anti-A	B, O	Caucasian, 11%
				African American, 19%
				Asian, 25.5%
				Hispanic 10%
				Native American, 1%
AB	A and B	None	A, B, AB, O	Caucasian, 4%
				African American, 4%
				Asian, 7.1%
				Hispanic 2.2%
				Native American, 0%
O	None	Anti-A, Anti-B	O	Caucasian, 45%
				African American, 51%
				Asian, 40%
				Hispanic 57%
				Native American, 91%

such transfusion, because it takes time for the body to start making anti-Rh antibodies. After a second transfusion of Rh-positive blood, however, the antibodies in the recipient's plasma will react with the antigens on the red blood cells of the donated blood. This reaction may lead to the death of the patient.

The Rh factor can also have medical importance for pregnancies in which the mother is Rh-negative and the fetus is Rh-positive, a situation that may occur if the father is Rh-positive (see Chapter 20; Figure 11.8). Ordinarily, the maternal and fetal blood supplies do not mix during pregnancy. However, as a result of blood vessel damage, some mixing may occur during a miscarriage or delivery. If the baby's red blood cells, which bear Rh antigens, accidentally pass into the mother's bloodstream, she will produce anti-Rh antibodies. There are usually no ill effects associated with the first introduction of the Rh antigen. However, if antibodies are present in the maternal blood from a previous pregnancy with an Rh-positive child or from a transfusion of Rh-positive blood, the anti-Rh antibodies may pass into the blood of the fetus during a subsequent pregnancy. This transfer can occur because anti-Rh antibodies, unlike red blood cells, can cross the placenta (a structure that forms during pregnancy to allow the exchange of selected substances between the maternal and fetal circulatory systems). These anti-Rh antibodies may destroy the fetus's red blood cells. As a result, the child may be stillborn or very anemic at birth. This condition is called *hemolytic disease of the newborn*.

The incidence of hemolytic disease of the newborn has decreased in recent years as a result of the development of a means of destroying any Rh-positive fetal cells in an Rh-negative mother's blood supply before they can stimulate the mother's cells to produce her own anti-Rh antibodies. The Rh-positive cells are killed by injecting RhoGAM, a serum containing antibodies against the Rh antigens, at about the seventh month of pregnancy and shortly after delivery if the baby is Rh-positive. Rh antigens are thus prevented from being "set" in the memory of the mother's immune system. The injected antibodies disappear after a few months. Therefore, no antibodies linger to affect the fetus in a subsequent pregnancy.

Blood Donation

We have seen that blood is indeed a life-saving fluid, and someone in the United States or Canada needs donated blood every 2 seconds. If you are a healthy person at least 17 years old (16 years in some states), weigh at least 110 lb, and are willing to spend 10 to 20 minutes, you may be eligible to give the gift of life. When you donate, you may feel a slight twinge as the needle is inserted, but the procedure is generally painless. Each donation is a pint of blood. The donated plasma is replaced within hours, and the cells are replaced within weeks. The donated blood can be separated into its components: red blood cells to increase the oxygen-carrying capacity of the recipient's blood, platelets to help the recipient's blood clot, and plasma to help the recipient control bleeding.

11.5 Blood Clotting

When a blood vessel is cut, several responses are triggered to stop the bleeding. To understand the process of clotting, imagine how you might respond if the garden hose you are using springs a leak. Your initial response might be to squeeze the hose in hopes of stopping the water flow. Likewise, the body's immediate response to blood vessel injury is for the vessel to constrict (squeeze shut).

The next response is to plug the hole (Figure 11.9). Your thumb might do the job on your garden hose; in an injured blood vessel, platelets form a plug that seals the leak. The *platelet plug* is formed when platelets cling to cables of collagen, a protein fiber on the torn blood vessel surface. When the platelets attach to collagen, they swell, form many cellular extensions, and stick together. Platelets also produce a chemical that attracts other platelets to the wound and makes them stick together even more. Aspirin prevents the formation of this chemical and, therefore, inhibits clot formation. For this reason, a daily dose of aspirin is sometimes prescribed to prevent the formation of blood clots that could block blood vessels nourishing heart tissue and thus cause the death of heart cells (a heart attack). It is also why aspirin can cause excessive bleeding.

An Rh+ man and an Rh− woman could have an Rh+ baby.

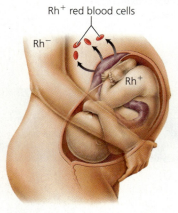
First pregnancy: At birth some of the Rh+ blood of the fetus may enter the mother's circulation.

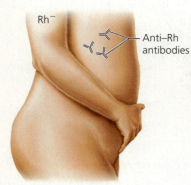
After delivery: The mother forms anti-Rh antibodies over the next few months.

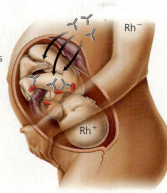
Second pregnancy with an Rh+ fetus: Anti-Rh antibodies may pass into the fetus's blood, causing its blood cells to burst.

FIGURE 11.8 *Rh incompatibility can result when an Rh-negative (Rh−) woman is pregnant with an Rh-positive (Rh+) baby if the woman has been previously exposed to Rh+ blood.*

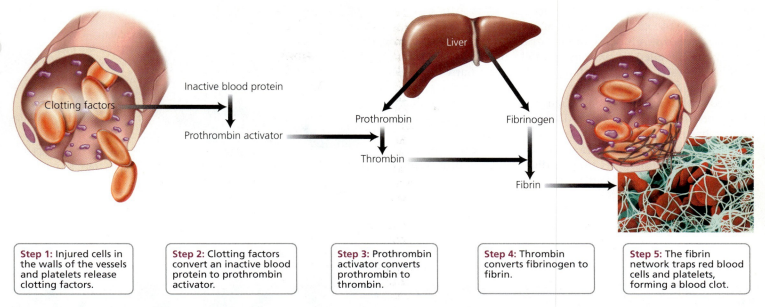

FIGURE 11.9 Selected steps in the blood-clotting process

Step 1: Injured cells in the walls of the vessels and platelets release clotting factors.

Step 2: Clotting factors convert an inactive blood protein to prothrombin activator.

Step 3: Prothrombin activator converts prothrombin to thrombin.

Step 4: Thrombin converts fibrinogen to fibrin.

Step 5: The fibrin network traps red blood cells and platelets, forming a blood clot.

The next stage in stopping blood loss through a damaged blood vessel is the formation of the clot itself. There are more than 30 steps in the process of clot formation, but here we will describe only the key events. Clot formation begins when clotting factors are released from injured tissue and from platelets. At the site of the wound, the clotting factors convert an inactive blood protein to **prothrombin activator**,[3] which then converts **prothrombin**—a plasma protein produced by the liver—to an active form, **thrombin**. Thrombin then causes a remarkable change in another plasma protein produced by the liver, **fibrinogen**. The altered fibrinogen forms long strands of **fibrin**, a protein that makes a web that traps blood cells and forms a clot. The clot is a barrier that prevents additional blood loss through the wound in the vessel.

stop and think

Thromboplastin, a chemical important in initiating clot formation, is released from both damaged tissue and activated platelets. How does this fact explain why a scrape, which causes a great deal of tissue damage, generally stops bleeding more quickly than a clean cut, such as a paper cut or one that might occur with a razor blade?

If even one of the many factors needed for clotting is lacking, the process can be slowed or completely blocked. Vitamin K is needed for the liver to synthesize prothrombin and three other clotting factors. Thus, without vitamin K, clotting does not occur. We have two sources of vitamin K. One is our diet. Vitamin K is found in leafy green vegetables, tomatoes, and vegetable oils. The second source is bacteria living in our intestines and producing vitamin K, some of which we absorb for our own use. Body tissues rapidly use vitamin K, so both sources are needed for proper blood clotting. Antibiotic treatment for serious bacterial infections can kill gastrointestinal bacteria and lead to a vitamin K deficiency in as few as 2 days.

Hemophilia is an inherited condition in which the affected person bleeds excessively owing to a fault in one of the genes involved in producing clotting factors. As noted earlier, there are many steps in the clotting process, and so there are many clotting factors. Although a loss of any of the clotting factors could cause hemophilia, the two most common types of hemophilia are caused by insufficient amounts of clotting factors VIII or IX. Because of the way hemophilia is inherited, the condition usually occurs in males (see Chapter 20). Symptoms appear when the affected child first becomes physically active. Crawling, for instance, causes bruises on the elbows and knees, and cuts tend to bleed longer than is normal. Excessive internal bleeding can damage nerves or, when it occurs in joints, permanently cripple the hemophiliac.

Treatment for hemophilia involves restoring the missing clotting factor. Bleeding episodes can be controlled with repeated transfusions of fresh plasma or with injections of concentrated clotting factor. Concentrated clotting factor is made by combining the plasma donations of many people. In fact, each injection requires 2000 to 5000 individual donations. Fortunately, two of the clotting factors (factors VIII and IX) are now being manufactured using recombinant DNA technology.

The failure of blood to clot can shorten the life of a person with hemophilia. But in people without hemophilia, the formation of unnecessary blood clots can have much more

---

[3] Prothrombin activator is an enzyme called *prothrombinase*.

immediate health consequences because clots can disrupt blood flow. A blood clot lodged in an unbroken blood vessel is called a *thrombus*. A blood clot that drifts through the circulatory system is called an *embolus*. Emboli can drift through the circulatory system until they become stuck in a narrow vessel. When the tiny vessels that nourish the heart or brain become clogged with a clot, the consequences can be severe—a heart attack or stroke, which can be disabling or even fatal. After a wound has healed, clots are normally dissolved by an enzyme called **plasmin**, which is formed from an inactive protein, *plasminogen*. Plasmin dissolves clots by digesting the fibrin strands that form the framework of the clot.

stop and think

Heparin is a drug that inactivates thrombin. It is sometimes administered to patients for the purpose of inhibiting the clotting response. How would heparin act to achieve these ends?

looking ahead

In this chapter, we looked at the composition and functions of blood. In the next chapter, we will consider the blood vessels through which blood travels and the heart that pumps the blood.

HIGHLIGHTING THE CONCEPTS

11.1 Functions of Blood (p. 198)
- Blood transports vital material to cells and carries wastes away from cells. White blood cells defend against disease. Blood clotting prevents excessive blood loss. Blood also helps regulate body temperature.

11.2 Composition of Blood (pp. 199–204)
- Blood is a type of connective tissue that contains formed elements in a liquid (plasma) matrix.
- Plasma is the liquid portion of the blood, consisting of water and certain dissolved components, mostly proteins. (The substances transported by the blood are also dissolved in the plasma. These include nutrients, ions, gases, hormones, and waste products.) Plasma proteins aid in the balance of water flow between blood and cells. These proteins are grouped into three categories: albumins, globulins, or clotting proteins (such as fibrinogen).
- Blood stem cells are undifferentiated cells that divide and give rise to all of the formed elements—namely, platelets, white blood cells, and red blood cells.
- Platelets play an important role in blood clotting. They are fragments of a larger cell called a megakaryocyte.
- White blood cells (leukocytes) help the body fight off disease and help remove wastes, toxins, and damaged cells.
- There are five types of leukocytes: neutrophils, eosinophils, basophils, monocytes, and lymphocytes. Neutrophils are the most abundant and immediately phagocytize foreign microbes. Eosinophils help the body defend against parasitic worms and play a role in allergic reactions. Basophils can increase the flow of blood by releasing histamines. Monocytes are the largest leukocytes and, after maturing into macrophages, actively fight chronic infections. Lymphocytes take part in body defense responses.
- Red blood cells (erythrocytes) are flexible cells packed with hemoglobin, an oxygen-binding pigment. Red blood cell production is controlled by erythropoietin, a hormone produced by the kidney in response to low oxygen. Worn and dead red blood cells are removed from circulation and broken down in the liver and spleen.

11.3 Blood Cell Disorders (pp. 204–206)
- Anemia is a reduction in the blood's ability to carry oxygen. There are several forms, including iron-deficiency anemia; hemolytic anemias, such as sickle-cell anemia; and pernicious anemia, caused by the stomach's failure to produce intrinsic factor, which is needed to absorb the vitamin necessary for red blood cell production.
- Infectious mononucleosis is a viral disease of lymphocytes. Leukemia is a cancer of the white blood cells. Although the numbers of white blood cells are high, they do not function properly to defend against infectious agents.

11.4 Blood Types (pp. 206–208)
- Blood types are determined by the presence of certain proteins (antigens) on the surface of red blood cells: the ABO and Rh groups. The plasma contains antibodies against A and B antigens if the antigens are not present on the red blood cells. Antibodies to Rh antigens are formed only after exposure to Rh-positive blood. If blood containing foreign antigens is introduced during a transfusion, a reaction between the antibodies and antigens can cause red blood cells to clump and clog the recipient's bloodstream, which can lead to death.
- Rh incompatibility becomes important to an Rh-negative woman who has been previously exposed to Rh-positive blood and who is pregnant with an Rh-positive fetus. Antibodies to the Rh antigens can cross the placenta and destroy the red blood cells of the fetus; this condition is called hemolytic disease of the newborn.

11.5 Blood Clotting (pp. 208–210)
- The prevention of blood loss involves three mechanisms: blood vessel constriction, platelet plug formation, and clotting. Clotting is initiated when platelets and damaged tissue release clotting factors that lead to the production of a chemical called prothrombin activator, which converts the blood protein prothrombin to thrombin. Thrombin then converts the blood protein fibrinogen to fibrin. Strands of fibrin make a mesh that traps red blood cells and forms the clot. Hemophiliacs are missing an important clotting factor and, therefore, bleed excessively.

RECOGNIZING KEY TERMS

plasma *p. 199*
plasma protein *p. 199*
formed elements *p. 199*
stem cell *p. 200*
platelet *p. 200*
white blood cell (WBC) *p. 201*
leukocyte *p. 201*
phagocytosis *p. 202*
granulocyte *p. 202*
agranulocyte *p. 202*
neutrophil *p. 202*
eosinophil *p. 202*
basophil *p. 202*
monocyte *p. 202*
lymphocyte *p. 202*
red blood cell (RBC) *p. 202*
erythrocyte *p. 202*
hemoglobin *p. 202*
erythropoietin *p. 204*
blood type *p. 206*
agglutinate *p. 206*
Rh factor *p. 206*
prothrombin activator *p. 209*
prothrombin *p. 209*
thrombin *p. 209*
fibrinogen *p. 209*
fibrin *p. 209*
plasmin *p. 210*

REVIEWING THE CONCEPTS

1. What is plasma? What are its functions? *p. 199*
2. What are the three categories of plasma proteins? *p. 199*
3. List the three types of formed elements, and describe the function of each. *pp. 199–204*
4. Compare the size, structure, and numbers of leukocytes with the size, structure, and numbers of erythrocytes. *pp. 201–202*
5. White blood cells function in defending the body against foreign invaders. List the five types of white blood cells, and describe the role each plays in body defense. *p. 202*
6. Describe the characteristics of red blood cells that make them specialized for delivering oxygen to the tissues. *p. 202*
7. Describe the function of hemoglobin. *p. 202*
8. Where are red blood cells produced? How is the production of red blood cells controlled? *p. 203*
9. Describe what happens to worn or damaged red blood cells and their hemoglobin. *p. 203*
10. Why would pernicious anemia be treated with regular injections of vitamin B_{12} instead of by dietary supplements of this vitamin? *p. 204*
11. What determines a person's blood type? *p. 206*
12. What happens if a person is given a blood transfusion with blood of an incompatible type? *p. 206*
13. What blood type(s) can a person with type B blood receive? Explain. *pp. 206–207*
14. What is hemolytic disease of the newborn? What causes it? *p. 208*
15. After a blood vessel is cut, what mechanisms prevent excessive blood loss? *pp. 208–209*
16. Describe the steps involved in blood clotting. *pp. 208–209*
17. What is the difference between the blood clotting that occurs after an injury and the clumping that occurs after a mismatched transfusion? *p. 209*
18. Type B blood contains which antibodies?
    a. anti-A
    b. anti-B
    c. both anti-A and anti-B
    d. neither anti-A nor anti-B
19. An important function of white blood cells is
    a. blood clotting.
    b. transportation of oxygen.
    c. fighting infection.
    d. maintaining blood pressure.
20. A condition in which the oxygen-carrying capacity of the blood is low is
    a. infection.
    b. anemia.
    c. leukemia.
    d. a kidney problem.
21. A person whose blood type is AB-negative can
    a. receive any blood type in moderate amounts except that with the Rh antigen.
    b. donate to all blood types in moderate amounts.
    c. receive types A, B, and AB, but not type O.
    d. donate to types A, B, and AB, but not to type O.
22. Platelets
    a. stick to the damaged area of a blood vessel and help seal the break.
    b. have a life span of about 120 days.
    c. are the precursors of leukocytes.
    d. have multiple nuclei.
23. The _____ regulates red blood cell production.
    a. kidney
    b. spleen
    c. liver
    d. red bone marrow
24. Hemolytic disease of the newborn will *never* happen to the child of an Rh-negative mother *if*
    a. the child is type O-positive.
    b. the child is Rh-positive.
    c. the father is Rh-positive.
    d. the father is Rh-negative.
25. Choose the *incorrect* statement about white blood cells.
    a. They carry oxygen to the body cells.
    b. They can leave the capillaries and move through body tissues.
    c. They can help the body defend against disease.
    d. They are formed in red bone marrow.
26. The primary function of red blood cells is to _____.
27. _____ is a protein in red blood cells that transports oxygen.
28. The protein that forms a net that traps red blood cells and platelets and forms blood clots is _____.

APPLYING THE CONCEPTS

1. Erin has leukemia, and her white blood cell count is elevated. Why does Erin have an elevated risk of infection?
2. José is a runner for his college track team from New York City. The championship meet is in Denver, which is called the Mile High City because of its elevation above sea level. There is less oxygen available for breathing as elevation increases. Because this meet is so important, José plans to arrive in Denver several weeks before the meet. Why? Would you expect José's red blood cell count to be higher or lower than normal when he returns home to New York City, which is at sea level?
3. Indira is a 25-year-old woman who has uterine polyps that cause heavy vaginal bleeding. She complains that she tires easily with physical activity and always feels fatigued. The doctor orders that a blood test be done to determine what percentage of her blood is red blood cells. Why? Would you expect the percentage to be higher or lower than normal? The doctor suggests that she take iron supplements. Why?
4. Raul is in a car accident and is taken to the emergency room. He has type AB blood. Which blood types can he receive?
5. Elizabeth has type Rh-negative blood. She is pregnant for the second time. What information would the doctor want to know about the first and second children to know whether to expect a problem with this pregnancy?
6. Sarala was given an antibiotic that caused her platelet count to fall. What symptoms would be expected?

BECOMING INFORMATION LITERATE

The umbilical cord is the lifeline for a growing fetus that connects the fetus to the mother through the placenta. Within the umbilical blood (and blood from the placenta) are blood stem cells that retain the ability to develop into red blood cells, white blood cells, and platelets. There is a growing interest in collecting umbilical cord blood when a baby is born and storing it for future use.

Use at least three reliable sources (books, journals, websites) to answer the following questions. Cite your sources, and explain why you chose them.

1. What are potential uses for stem cells from umbilical cord blood?
2. What are advantages of using umbilical cord blood over a bone marrow transplant as a source of stem cells?
3. Based on what you have learned about umbilical cord blood, do you think that there should be umbilical cord blood banks? Why? If so, do you think use of the cord blood should be restricted to the family whose infant donated the blood, or should it be made available to anyone?

MasteringBiology®

Go to MasteringBiology for practice quizzes, activities, eText, videos, current events, and more.

The Cardiovascular and Lymphatic Systems

12

 **Did you know?**

- *Laughter is good medicine. It protects your heart and blood vessels. When you start to laugh, your blood pressure increases, but then it drops to levels below normal. Laughter also gives your heart a good workout and can increase blood flow by 20%. LOL.*
- *Even when you are sleeping, your heart is working twice as hard as your leg muscles do when you sprint to catch the bus.*
- *It takes 10 capillaries to equal the thickness of one human hair.*

12.1 Cardiovascular System

12.2 Blood Vessels

12.3 Heart

12.4 Blood Pressure

12.5 Lymphatic System

HEALTH ISSUE

Benefits of Cardiovascular Exercise

In Chapter 11, we learned about blood—its components and functions. In this chapter, we consider how the blood circulates throughout the body through blood vessels pumped by the heart. We will also learn about the vessels and other structures that make up the lymphatic system. Together, the heart and blood vessels and the lymphatic system constitute the circulatory system.

The cardiovascular system consists of a pump—the heart—and a loop of blood vessels. All the organs of the body require a continuous supply of blood to deliver oxygen and nutrients and to remove waste. This blood supply is particularly important for the heart—the hardest-working muscle in the body. If the blood vessels of the heart become clogged, steps must be taken to restore adequate blood flow.

12.1 Cardiovascular System

The **cardiovascular system** consists of the **heart**—a muscular pump that contracts rhythmically and provides the force that moves the blood—and the blood vessels—a system of tubules through which blood flows (Figure 12.1). The blood delivers a continuous supply of oxygen and nutrients to the cells of the body and carries away metabolic waste products so that they cannot poison the cells.

Why is the cardiovascular system so critical to survival? It is the body's transportation network, similar in some ways to the highways within a country. Our bodies are too large and complex for diffusion alone to distribute materials efficiently. The cardiovascular system provides a means to distribute vital chemicals from one part of the body to another quickly enough to sustain life. The cardiovascular system is more than just a passive system of pipelines, however. The heart rate and the diameter of certain blood vessels are continually being adjusted in prompt response to the body's changing needs.

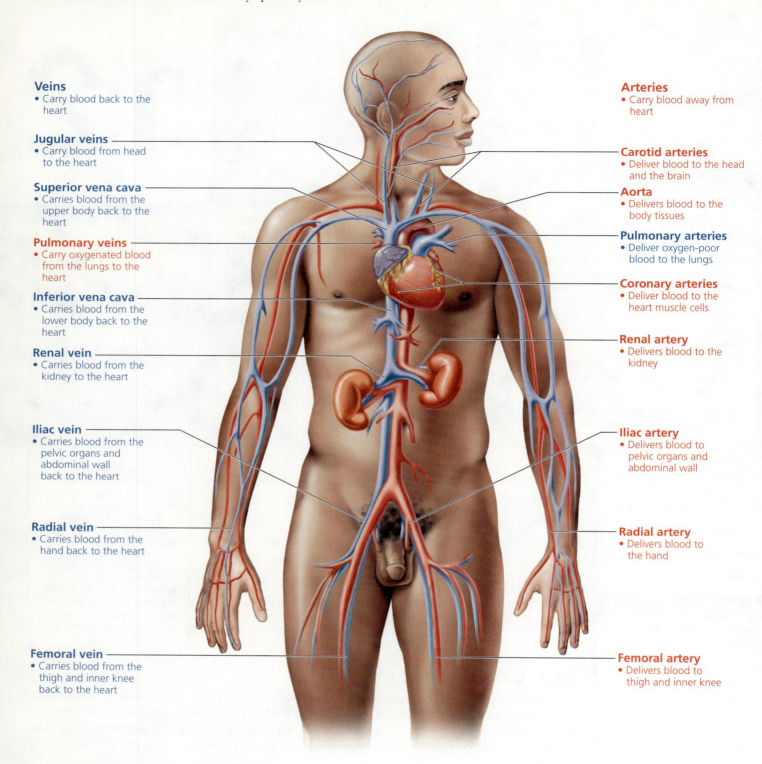

FIGURE 12.1 *A diagrammatic view of the cardiovascular system (heart and blood vessels). Throughout this chapter, red indicates blood that is high in oxygen, and blue indicates blood that is low in oxygen.*

12.2 Blood Vessels

Once every minute, or about 1440 times each day, the blood moves through a life-sustaining circuit. The circuit of blood vessels is extensive. Indeed, if all the vessels in an average adult's body were placed end to end, they would stretch about 100,000 km (60,000 mi), long enough to circle Earth's equator more than twice!

The blood vessels do not form a single long tube. Instead, they are arranged in branching networks. With each circuit through the body, blood is carried away from the heart in *arteries*, which branch to give rise to narrower vessels called *arterioles*. Arterioles lead into networks of microscopic vessels called *capillaries*, which allow the exchange of materials between the blood and body cells. The capillaries eventually merge to form *venules*, which in turn join to form larger tubes called *veins*. The venules and veins return the blood to the heart.

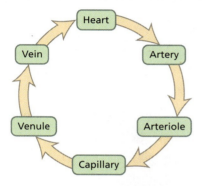

All blood vessels share some common features, but each type also has its own traits and is marvelously adapted for its specific function (Figure 12.2). The hollow interior of a blood vessel, through which the blood flows, is called the *lumen*. The inner lining that comes into contact with the blood flowing through the lumen is composed of simple squamous epithelium (the flattened, tight-fitting cells we encountered in Chapter 4). This lining, called the *endothelium*, provides a smooth surface that minimizes friction so that the blood flows easily. The lumen and endothelium are characteristic of all blood vessels.

Arteries

Arteries are muscular tubes that transport blood *away* from the heart, delivering it rapidly to the body tissues. As noted, the innermost layer of an arterial wall is the endothelium. Immediately outside the endothelium is a middle layer that contains elastic fibers and circular layers of smooth muscle. The elastic fibers allow an artery to stretch and then return to its original shape. The smooth muscle enables the artery to contract. The outer layer of an arterial wall is a sheath of connective tissue that contains elastic fibers and collagen. This layer adds strength to the arterial wall and anchors the artery to surrounding tissue.

The elastic fibers in the middle layer of an artery have two important functions: (1) They help the artery tolerate the pressure shock caused by blood surging into it when the heart contracts, and (2) they help maintain a relatively even pressure within the artery, despite large changes in the volume of blood passing through it. Consider, for instance, what happens when the heart contracts and sends blood into the **aorta**, the body's main artery. Each beat of the heart causes 70 ml (about one-fourth cup) of blood to pound against the wall of the aorta like a tidal wave. A rigid pipe could not withstand the repeated pressure surges, but the elastic walls of the artery stretch with each wave of blood and return to their original size when the surge has moved past; this results in a continuous stream of blood rather than intermittent waves.

The alternate expansion and recoil of arteries create a pressure wave, called a **pulse**, that moves along the arteries with each heartbeat. Thus, the pulse rate is the same as the heart rate. You can feel the pulse by slightly compressing with your fingers any artery that lies near the body's surface, such as the one at the wrist or the one under the angle of the jaw.

The middle layer of an artery wall also contains smooth muscle that enables the artery to contract. When this circular muscle contracts and the diameter of the lumen becomes narrower, a process called *vasoconstriction*, blood flow through the artery is reduced. Conversely, when the smooth muscle relaxes and the arterial lumen increases in diameter, a process called *vasodilation*, blood flow through the artery increases. The smooth muscle is best developed in small- to medium-sized arteries. These arteries serve to regulate the distribution of blood, adjusting flow to suit the needs of the body.

We see the importance of the strength of an artery wall when it becomes weakened, as might occur as a result of disease, inflammation, injury, or a birth defect. When the arterial wall becomes weakened, the pressure of the blood flowing through the weakened area may cause the wall to swell outward like a balloon, forming an *aneurysm*. Most aneurysms do not cause symptoms, but the condition can be threatening just the same. The primary risk is that the aneurysm will burst, causing blood loss. The tissues serviced by that vessel will then be deprived of oxygen and nutrients, a situation that can be fatal. Even if the aneurysm does not rupture, it can cause the formation of life-threatening blood clots. A clot can break free from the site of formation and float through the circulatory system until it lodges in a small vessel, where it can block blood flow and cause tissue death beyond that point. In some cases, an aneurysm can be repaired surgically. Nicotine in tobacco products can increase the risk of blood clots and aneurysms.

The smallest arteries, called **arterioles**, are barely visible to the unaided eye. Their walls have the same three layers found in arteries, but the middle layer is primarily smooth muscle with only a few elastic fibers, and the outer layer is much thinner.

Arterioles have two extremely important regulatory roles. First, they are the prime controllers of blood pressure, which is the pressure of blood against the vessel walls (discussed later in this chapter). When the muscle in arteriole walls contracts, blood pressure increases. The greater the number of arterioles contracted, the higher the blood pressure. Relaxation of arteriole walls lowers blood pressure. Second, arterioles serve as gatekeepers to the capillary networks. A capillary network can be open or closed, depending on whether the smooth muscle in the walls of the arteriole leading to it allows blood through. In this way, arterioles can regulate the amount of blood sent to cells based on those cells' immediate needs. Arterioles are constantly responding to input from hormones, the nervous

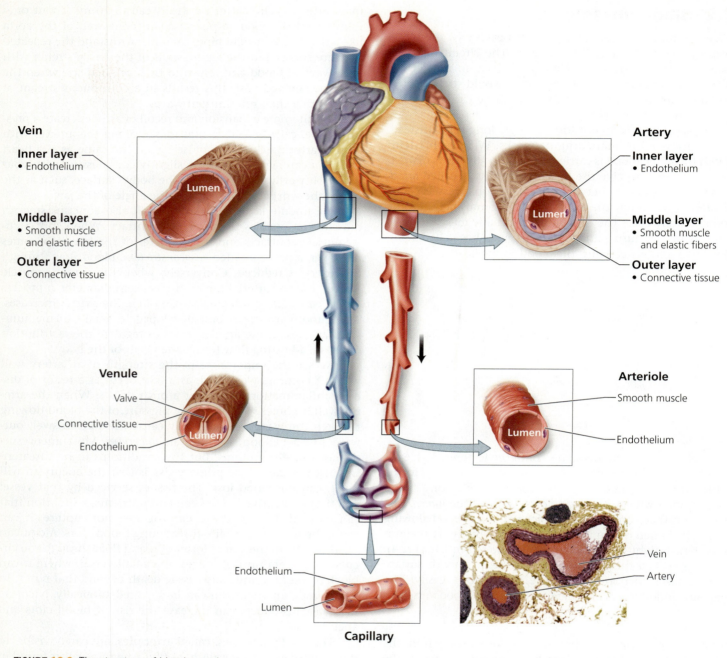

FIGURE 12.2 *The structure of blood vessels*

system, and local conditions, modifying blood pressure and flow to meet the body's changing needs.

Capillaries

Capillaries are microscopic blood vessels that connect arterioles and venules. The capillaries are well suited to their primary function: the exchange of materials between the blood and the body cells (Figure 12.3). Capillary walls are only one cell layer thick, so substances move easily between the blood and the fluid surrounding the cells outside the capillary. The plasma membrane of the capillary's endothelial cells is an effective and selective barrier that determines which substances can cross. Some substances that cross the capillary walls do not pass *through* the endothelial cells. Instead, these substances filter through small slits *between* adjacent endothelial cells. The slits between the cells are just large enough for some fluids and small dissolved molecules to pass through. Blood pressure filters fluid and nutrients out of the capillary at the arterial end. At the venous end, the presence of proteins in the blood causes fluid and wastes to be drawn back into the capillary by osmosis.

The design of the capillary networks allows blood flow through capillaries to be adjusted to deliver the necessary

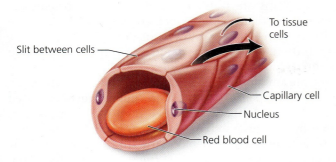

(a) Substances are exchanged between the blood and tissue fluid across the plasma membrane of the capillary cells or through slits between capillary cells.

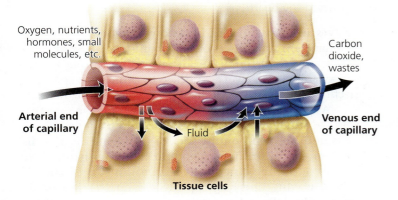

(b) At the arterial end of a capillary, blood pressure forces fluid out of the capillary to the fluid surrounding tissue cells. At the venous end, fluid is drawn back into the capillary by osmotic pressure.

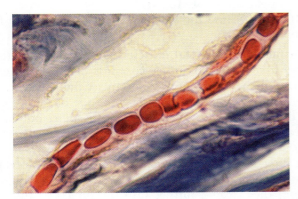

(c) Capillaries are so narrow that red blood cells must travel through them in single file.

FIGURE 12.3 *Capillaries are the sites where materials are exchanged between the blood and the body cells.*

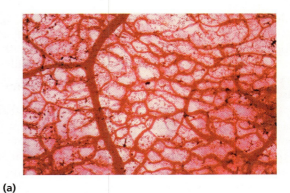

(a)

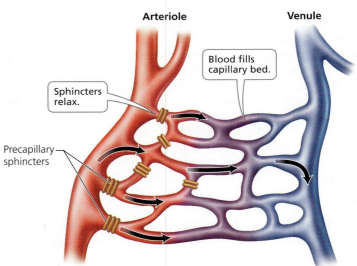

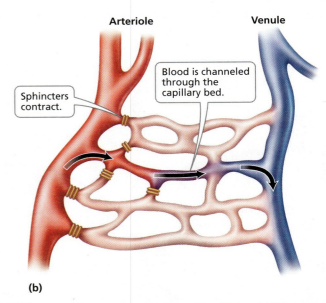

(b)

FIGURE 12.4 *(a) A capillary bed is a network of capillaries. (b) The entrance to each capillary (the arterial side) is guarded by a ring of muscle called a precapillary sphincter. Blood flow through a capillary bed is regulated according to the body's metabolic needs.*

amount of oxygen and nutrients to meet the needs of particular regions of the body. The network of capillaries servicing a particular area is called a **capillary bed** (Figure 12.4). The number of capillaries in a bed generally ranges from 10 to 100, depending on the type of tissue. A ring of smooth muscle called a **precapillary sphincter** surrounds the capillary where it branches off the arteriole and regulates blood flow into it.

The precapillary sphincters act as valves that open and close the capillary beds. Contraction of the precapillary sphincter squeezes the opening to the capillary shut. For instance, while you are resting on the beach after finishing a picnic lunch, the capillary beds servicing the digestive organs will be open, and nutrients from the food will be absorbed. If you dive into the water and start swimming, the capillary beds of the digestive organs will close down, and those in the skeletal muscles will open.

Collectively, the capillaries provide a tremendous surface area for the rapid exchange of materials between body and blood. Capillary beds bring capillaries very close to nearly every cell. Your fingernails provide windows that allow you to appreciate the efficiency with which capillary networks reach all parts of the body. You may have noticed that the tissue beneath a fingernail normally has a pink tinge. The color results from blood flowing through numerous capillaries there. Gentle pressure on the nail causes the tissue to turn white as the blood is pushed from those capillaries.

A capillary is so narrow that red blood cells must squeeze through single file. Despite their size, there are so many capillaries that their *combined* cross-sectional area is enormous, much greater than that of the arteries or veins. Because of the large cross-sectional area of the capillaries, the blood flows much more slowly through them than through the arteries or veins. The slower rate of flow in the capillaries provides more time for the exchange of materials (Figure 12.5).

Veins

After the capillary bed, capillaries merge to form the smallest kind of vein, a **venule**. Venules then join to form larger veins. **Veins** are blood vessels that return the blood to the heart.

Although veins share some structural features with arteries, there are also some important differences. The walls of veins have the same three layers found in arterial walls, but the walls of veins are thinner and the lumens of veins are larger than those of arteries of equal size (see Figure 12.2). The thin walls and large lumens allow veins to hold a large volume of blood. Indeed, veins serve as blood reservoirs, holding up to 65% of the body's total blood supply.

The same amount of blood that is pumped out of the heart must be conducted back to the heart, but it must be moved through the veins without assistance from the high pressure generated by the heart's contractions. In the head and neck, of course, gravity helps move blood toward the heart. But how is it possible to move blood against the force of gravity—from the foot back to the heart, for instance (unless, by chance, your foot was in your mouth)?

Three mechanisms move blood from lower parts of the body toward the heart:

1. **Valves in veins prevent backflow of blood.** Veins often contain valves that act as one-way turnstiles, allowing blood to move toward the heart but preventing it from flowing backward. These valves are pockets of connective tissue projecting from the lining of the vein, as shown in Figure 12.6a.

   A simple experiment can demonstrate the effectiveness of venous valves. Allow your hand to hang by your side until the veins on the back of your hand become distended. Place two fingertips from the other hand at the end of one of the distended veins nearest to the knuckles. Then, leaving one fingertip pressed on the end of the vein, move the other toward the wrist, pressing firmly and squeezing the blood from the vein. Lift the fingertip near the knuckle, and notice that blood immediately fills the vein. Repeat the procedure, but this time lift the fingertip near the wrist. You will see that the vein remains flattened, because the valves prevent the backward flow of blood.

2. **Contraction of skeletal muscle squeezes veins.** Virtually every time a skeletal muscle contracts, it squeezes nearby veins. This pressure pushes blood past the valves toward the heart. The mechanism propelling the blood is similar to the one that causes toothpaste to squirt out of the uncapped end of the tube regardless of where the tube is squeezed; valves in veins ensure that blood flows in only one direction. When skeletal muscles relax, any blood that moves backward fills the valves. As the valves fill with blood, they extend further into the lumen of the vein, closing the vein and preventing the flow of blood from reversing direction (Figure 12.6b). Thus, the skeletal muscles are always squeezing the veins and driving blood toward the heart.

3. **Breathing causes pressure changes that move blood toward the heart.** The thoracic (chest) cavity increases in size when we inhale (see Chapter 14). The expansion reduces pressure within the thoracic cavity and at the same time increases pressure in the abdominal cavity. Blood naturally moves toward regions of lower pressure. Thus, the reduced pressure in the thoracic cavity that comes with each breath pulls blood back toward the heart. In addition, the increased pressure in the abdominal cavity squeezes veins, also forcing blood back toward the heart.

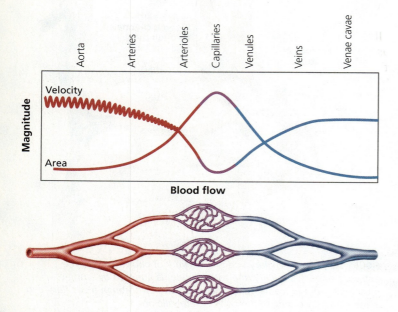

FIGURE 12.5 The capillaries are so numerous that their total cross-sectional area is much greater than that of arteries or veins. Thus, blood pressure drops and blood flows more slowly as it passes through a capillary bed. The slower rate of flow allows time for the exchange of materials between the blood and the tissues.

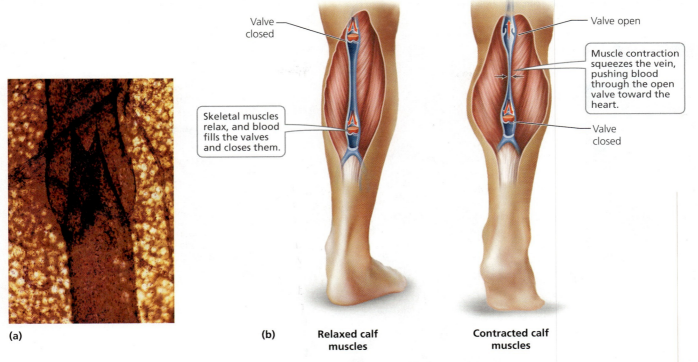

(a) (b) Relaxed calf muscles | Contracted calf muscles

FIGURE 12.6 (a) A micrograph of a vein showing a valve. (b) Pocketlike valves on the inner surface of veins prevent backflow, thereby assisting the return of blood to the heart against gravity.

stop and think

If an artery is cut, blood is lost in rapid spurts. In contrast, blood loss through a cut vein has an even flow. What accounts for these differences?

12.3 Heart

The heart is about the size of a fist, but it is an incredible muscular pump that generates the force needed to circulate blood. It beats about 72 times a minute, every hour of every day—although this rate varies with age, physical fitness, and current physical exertion. To appreciate the work done by the heart, alternately clench and relax a fist 70 times a minute. How many minutes does it take before the muscles of your hand are too tired to continue? In contrast, the healthy heart does not fatigue. It beats more than 100,000 times each day, which adds up to about 2 billion beats over a lifetime. The volume of blood pumped by the heart is equally remarkable. It pumps slightly less than 5 liters (10 pt) of blood a minute through its chambers, which adds up to more than 9400 liters (2500 gal) per day.

The heart has three layers, each one contributing to the heart's ability to function as a pump. The wall of the heart, called the **myocardium**, is mostly cardiac muscle tissue. The myocardium's contractions are responsible for the heart's incredible pumping action. The **endocardium** is a thin lining in the cavities of the heart. By reducing friction, the endocardium's smooth surface lessens the resistance to blood flow through the heart. The **pericardium** is a thick, fibrous sac that holds the heart in the center of the chest (thoracic) cavity and slides over the surface of the heart without hampering its movements, even when they are vigorous.

Although the heart appears to be a single structure, it actually has two halves, with the right and left halves functioning as two separate pumps. As we will see shortly, the right side of the heart pumps blood to the lungs, where it picks up oxygen. The left side pumps the blood to the body cells. The two pumps are physically separated by a partition called a **septum**. Each half of the heart consists of two chambers: an upper chamber, called an **atrium** (plural, atria), and a lower chamber, called a **ventricle** (Figure 12.7). The two atria function as receiving chambers for the blood returning to the heart. The two ventricles function as the main pumps of the heart. Contraction of the ventricles forces blood out of the heart under great pressure. When we think about the work of the heart, we are in fact thinking about the work of the ventricles. It should not be surprising, then, that the ventricles are much larger chambers than the atria and have thicker, more muscular walls.

Two pairs of valves ensure that the blood flows in only one direction through the heart. The first pair is the **atrioventricular (AV) valves**, each leading from an atrium to a ventricle, as shown in Figure 12.8. The AV valves are connective tissue flaps, called cusps, anchored to the wall of the ventricle by strings of connective tissue called the *chordae tendineae*—the heartstrings. These strings prevent the AV valves from flapping back into the atria under the pressure developed when the ventricles contract.

Q When the ventricles contract, which valves open, and which valves close?

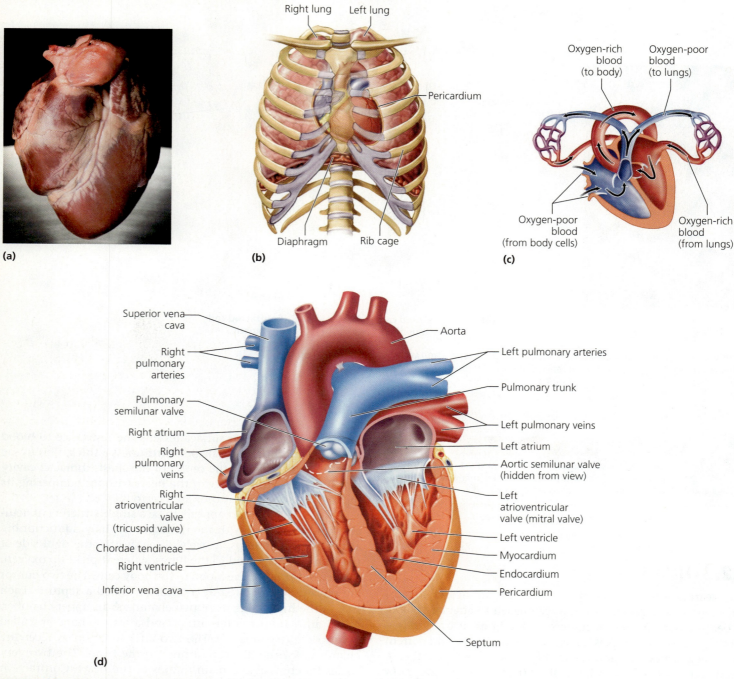

FIGURE 12.7 (a) The human heart. (b) The heart is located in the thoracic (chest) cavity. (c) Blood flows through the heart from the atria to the ventricles. (d) This diagram of a human heart shows the four chambers, the major vessels connecting to the heart, and the two pairs of heart valves.

▲ The semilunar valves open, and the atrioventricular (AV) valves close.

The AV valve on the right side of the heart has three flaps and is called the *tricuspid valve*. The AV valve on the left side of the heart has two flaps and is called the *bicuspid*, or *mitral, valve*.

Each of the second pair of valves, the **semilunar valves**, is located between a ventricle and its connecting artery—either the aorta or the pulmonary artery. The cusps of the semilunar valves are small pockets of tissue attached to the inner wall of the respective artery. When the pressure in the arteries becomes greater than the pressure in the ventricles, these valves fill with blood much like a parachute filling with air. In this way, the semilunar valves prevent the backflow of blood into the ventricles from the aorta or pulmonary artery.

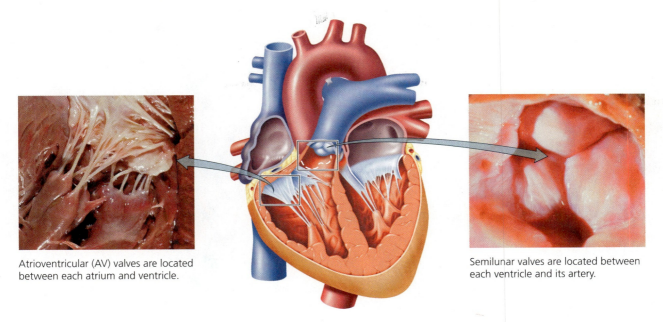

Atrioventricular (AV) valves are located between each atrium and ventricle.

Semilunar valves are located between each ventricle and its artery.

FIGURE 12.8 *The valves of the heart keep blood flowing in one direction.*

Two Circuits of Blood Flow

As we mentioned, the left and right sides of the heart function as two separate pumps, each circulating the blood through a different route, as shown in Figure 12.9. Note that in both circuits the blood moves through the arteries, arterioles, capillaries, and venules before returning to the heart via the veins. The right side of the heart pumps blood through the **pulmonary circuit**, which transports blood to and from the lungs. The left side of the heart pumps blood through the **systemic circuit**, which transports blood to and from body tissues. This arrangement prevents oxygenated blood (blood rich in oxygen) from mixing with blood that is low in oxygen.

The pulmonary circuit begins in the right atrium, as veins return oxygen-poor blood from the systemic circuit. (You can trace the flow of blood through the heart in the pulmonary and systemic circuits in Figure 12.9 as you read the following description.) The blood then moves from the right atrium to the right ventricle. Contraction of the right ventricle pumps poorly oxygenated blood to the lungs through the pulmonary trunk (main pulmonary artery), which divides to form the left and right *pulmonary arteries*. In the lungs, oxygen diffuses into the blood, and carbon dioxide diffuses out. The now oxygen-rich blood is delivered to the left atrium through four *pulmonary veins*, two from each lung. (Note that the pulmonary circulation is an *exception* to the general rule that arteries carry oxygen-rich blood and veins carry oxygen-poor blood. Exactly the opposite is true of vessels in the pulmonary circulation.) The pathway of blood pumped through the pulmonary circuit by the right side of the heart is

Right atrium ⟶ AV valve (tricuspid) ⟶
Right ventricle ⟶ Pulmonary semilunar valve ⟶
Pulmonary trunk ⟶ Pulmonary arteries ⟶ Lungs ⟶
Pulmonary veins ⟶ Left atrium

The systemic circuit begins when oxygen-rich blood enters the left atrium (see Figure 12.9). Blood then flows to the left ventricle. When the left ventricle contracts, oxygenated blood is pushed through the largest artery in the body, the aorta. The aorta arches over the top of the heart and gives rise to the smaller arteries that eventually feed the capillary beds of the body tissues. The venous system collects the oxygen-depleted blood and eventually culminates in veins that return the blood to the right atrium. These veins are the *superior vena cava*, which delivers blood from regions above the heart, and the *inferior vena cava*, which returns blood from regions below the heart. Thus, the pathway of blood through the systemic circuit pumped by the left side of the heart is

Left atrium ⟶ AV (bicuspid or mitral) valve ⟶
Left ventricle ⟶ Aortic semilunar valve ⟶ Aorta ⟶
Body tissues ⟶ Inferior vena cava or superior
vena cava ⟶ Right atrium

The familiar sounds of the heart, which are often described as "lub-dup," are associated with the closing of the valves. The first heart sound ("lub") is produced by the turbulent blood flow when the AV valves snap shut as the ventricles begin to contract. The higher-pitched second heart sound ("dup") is produced by the turbulent blood flow when the semilunar valves close and ventricular relaxation begins.

Heart murmurs, which are swooshing heart sounds other than lub-dup, are created by disturbed blood flow. Although heart murmurs are sometimes heard in normal, healthy people, they can also indicate a heart problem. For instance, malfunctioning valves often disturb blood flow through the heart, causing the swishing or gurgling sounds of heart murmurs. Several conditions can cause valves to malfunction. In some cases, thickening of the valves narrows the opening and impedes blood flow. In other cases, the valves do not close properly and,

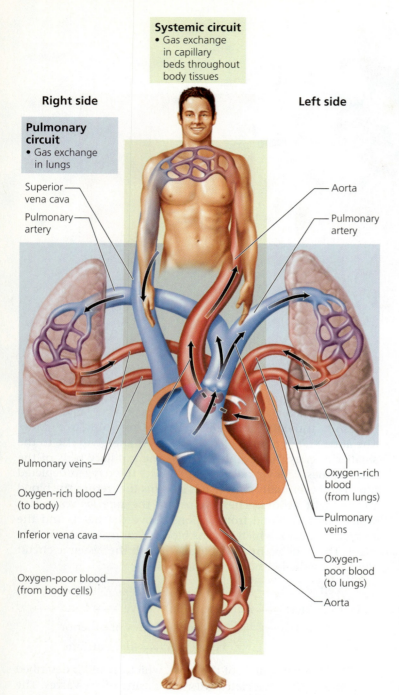

FIGURE 12.9 Circuits of blood flow. The right side of the heart pumps blood through the pulmonary circuit, which carries blood to and from the lungs. The left side of the heart pumps blood through the systemic circuit, which conducts blood to and from the body tissues.

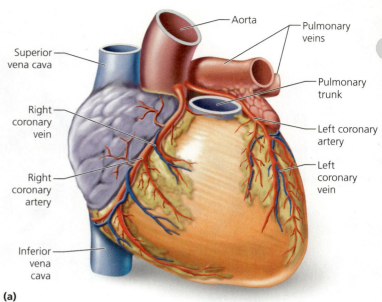

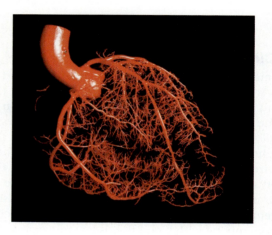

FIGURE 12.10 Coronary circulation. (a) The coronary vessels deliver a rich supply of oxygen and nutrients to the heart muscle cells and remove the metabolic wastes. (b) This cast of the coronary blood vessels reveals the complexity of the coronary circuit.

therefore, allow the backflow of blood. In either case, the heart is strained because it must work harder to move the blood.

Coronary Circulation

Cells of the heart muscle themselves obtain little nourishment from blood flowing through the heart's chambers. Instead, an extensive network of vessels, known as the **coronary** circulation, services the tissues of the heart. The first two arteries that branch off the aorta are the **coronary arteries** (Figure 12.10). These arteries give rise to numerous branches, ensuring that the heart receives a rich supply of oxygen and nutrients. After passing through the capillary beds that nourish the heart tissue, blood enters cardiac veins and eventually flows into the right atrium.

Cardiac Cycle

Although the two sides of the heart pump blood through different circuits, they work in tandem. The two atria contract simultaneously, and then the two ventricles contract simultaneously.

We see, then, that a heartbeat is not a single event. Each beat involves contraction, which is called **systole** (sis'-to-lee),

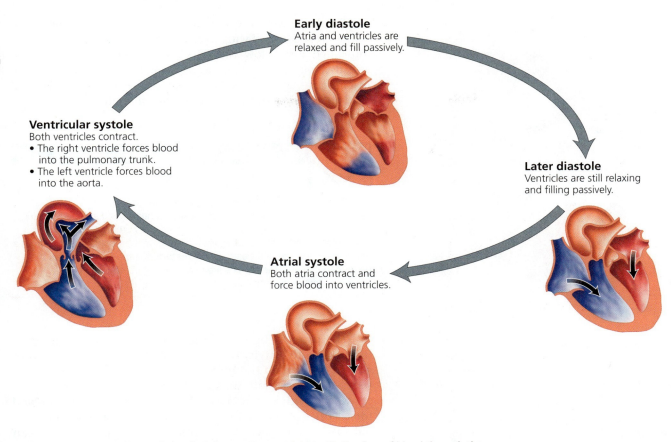

FIGURE 12.11 *The cardiac cycle is all of the events associated with the flow of blood through the heart during each heartbeat. The atria contract together, and the ventricles contract together. Red indicates blood high in oxygen. Blue indicates blood low in oxygen.*

and relaxation, which is called **diastole** (di-as'-to-lee). All the events associated with the flow of blood through the heart chambers during a single heartbeat are collectively called the **cardiac cycle**, as illustrated in Figure 12.11. First, all chambers relax (diastole), and blood passes through the atria and enters the ventricles. When the ventricles are about 70% filled, the atria contract (atrial systole) and push their contents into the ventricles. The atria then relax (atrial diastole), and the ventricles begin their contraction phase (ventricular systole). Upon completion of this contraction, the whole heart again relaxes. If we were to add the contraction time of the heart during a day and compare it with the relaxation time during a day, the heart's workday might turn out to be equivalent to yours. In 24 hours, the heart spends a total of about 8 hours working (contracting) and 16 hours relaxing. However, unlike your workday, the heart's day is divided into repeating cycles of work and relaxation.

Internal Conduction System

If a human heart is removed, as in a transplant operation, and placed in a dish, it will continue to beat, keeping a lonely and useless rhythm until its tissues die. In fact, if a few cardiac muscle cells are grown in the laboratory, they, too, will beat on their own, each twitch a reminder of the critical role the intact organ plays. Clearly, then, the heart muscle does not require outside stimulation to beat. Instead, the tendency is intrinsic, within the heart muscle itself.

Another remarkable observation has been made of heart muscle cells grown in a laboratory dish. Although isolated heart cells twitch independently of each other, if two cells should touch, they will begin beating in unison. This, too, is an inherent property of the cells, but it is partly due to the type of connections between heart muscle cells. The cell membranes of adjacent cardiac muscle cells interweave with one another at specialized junctions called *intercalated disks*. Cell junctions in the intercalated disks mechanically and electrically couple the connected cells. Adjacent cells are held together so tightly that they do not rip apart during contraction but instead transmit the pull of contraction from one cell to the next. At the same time, the junctions permit electrical communication between adjacent cells, allowing the electrical events responsible for contraction to spread rapidly over the heart by passing from cell to cell. Yet even though heart cells contract automatically, they still need some outside control to contract at the proper rate.

The tempo of the heartbeat is set by a cluster of specialized cardiac muscle cells, called the **sinoatrial (SA) node**, located in the right atrium near the junction of the superior vena cava

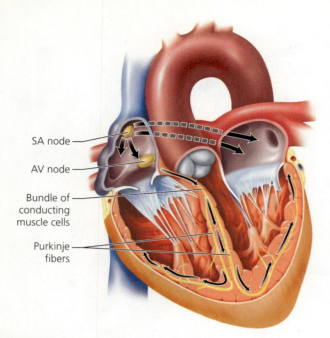

FIGURE 12.12 The conduction system of the heart consists of specialized cardiac muscle cells that speed electrical signals through the heart. The sinoatrial (SA) node serves as the heart's internal pacemaker and determines the heart rate. Electrical signals from the SA node spread through the walls of the atria, causing them to contract. The signals then stimulate the atrioventricular (AV) node, which in turn sends the signals along the AV bundle to its forks and finally to the many Purkinje fibers that penetrate the ventricular walls. The Purkinje fibers distribute the signals to the walls of the ventricles, causing them to contract.

(Figure 12.12). Because the SA node sends out the impulses that initiate each heartbeat, it is often referred to as the **pacemaker**. About 70 to 80 times a minute, the SA node sends out an electrical signal that spreads through the muscle cells of the atria, causing them to contract. The signal reaches another cluster of specialized muscle cells called the **atrioventricular (AV) node**, located in the partition between the two atria, and stimulates it. The AV node then relays the stimulus by means of a bundle of specialized muscle fibers, called the *atrioventricular bundle,* that runs along the wall between the ventricles. The bundle forks into right and left branches and then divides into many other specialized cardiac muscle cells, called *Purkinje fibers,* that penetrate the walls of the ventricles. The rapid spread of the impulse through the ventricles ensures that they contract smoothly.

When the heart's conduction system is faulty, cells may begin to contract independently. Such cellular independence can result in rapid, irregular contractions of the ventricles, called *ventricular fibrillation,* which render the ventricles useless as pumps and stop circulation. With the brain no longer receiving the blood it needs to function, death will occur unless an effective heartbeat is restored quickly. A method for stopping ventricular fibrillation is to subject the heart to an electric shock; in many cases, the SA node will once again begin to function normally. Although costly, an implantable defibrillator (a device that electrically shocks the heart) can provide a life-saving "jump start" when needed.

Problems with the conduction system of the heart can sometimes be treated with an artificial pacemaker, a small device implanted just below the skin that monitors the heart rate and rhythm and responds to abnormalities if they occur. For instance, if the heart rate becomes too slow, the pacemaker will send an electrical stimulus to the heart through an electrode.

> **what would you do?**
>
> In a heart transplant operation, a person's diseased heart is replaced with a healthy heart from a person who recently died. However, in a controversial procedure, doctors removed the hearts from three infants whose hearts had stopped beating when they were taken off respirators. The hearts were transplanted into other infants and restarted. Legally, a heart can be transplanted only from a donor who has died. If a heart can be restarted in the recipient, was the donor dead, or was the donor's life ended by organ removal? Is this procedure legal? Is it ethical?

The pace or rhythm of the heartbeat changes constantly in response to activity or excitement. The autonomic nervous system and certain hormones make the necessary adjustments so that the heart rate suits the body's needs. During times of stress, the sympathetic nervous system increases the rate and force of heart contractions. As part of this response, the adrenal medulla produces the hormone epinephrine, which can prolong the sympathetic nervous system's effects. In contrast, when restful conditions prevail, the parasympathetic nervous system dampens heart activity, in keeping with the body's more modest metabolic needs.

Electrocardiogram

The electrical events that spread through the heart with each heartbeat actually travel throughout the body, because body fluids are good conductors. Electrodes placed on the body surface can detect these electrical events, transmitting them so that they cause deflections (movements) in the tracing made by a recording device. An **electrocardiogram** (ECG or EKG) is an image of the electrical activities of the heart, generated by such a recording device.

A typical ECG consists of three distinguishable deflection waves, as shown in Figure 12.13. The first wave, called the *P wave,* accompanies the spread of the electrical signal over the atria and the atrial contraction that follows. The next wave, the *QRS wave,* reflects the spread of the electrical signal over the ventricles and ventricular contraction. The third wave, the *T wave,* represents the return of the ventricles to the electrical state that preceded contraction (ventricular repolarization). Because the pattern and timing of these waves are remarkably consistent in a healthy heart, abnormal patterns can indicate heart problems.

12.4 Blood Pressure

We hear a lot about blood pressure, usually when someone worries about having a high blood pressure reading or brags about having a low one. **Blood pressure** is the force exerted by the blood against the walls of the blood vessels. When the

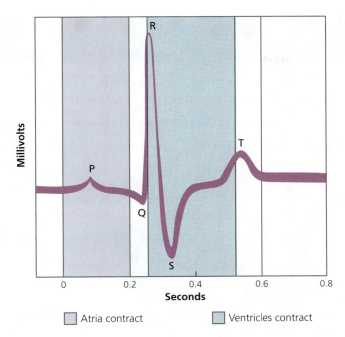

FIGURE 12.13 *The electrical activity that accompanies each heartbeat can be visualized in an ECG tracing. The P wave is generated as the electrical signals from the SA node spread across the atria and cause them to contract. The QRS wave represents the spread of the signal through the ventricles and ventricular contraction. The T wave occurs as the ventricles recover and return to the electrical state that preceded contraction.*

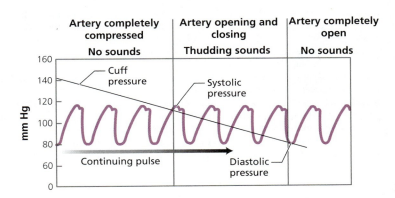

FIGURE 12.14 *Blood pressure is measured with a sphygmomanometer, which consists of an inflatable cuff and a means of measuring the pressure within the cuff. The cuff is placed around the upper arm and inflated so that it compresses the brachial artery. The pressure in the cuff is slowly released, and as it descends it reaches a point where blood is able to spurt through the constricted artery only at the moments of highest blood pressure. This pressure, at which "tapping" sounds are first heard, is the systolic pressure, the blood pressure when the heart is contracting. As the pressure in the cuff continues to drop, a point is reached where the sounds disappear. The blood is now flowing continuously through the brachial artery. The pressure in the cuff when the sounds first disappear is the diastolic pressure, the blood pressure when the heart is relaxing.*

ventricles contract, they push blood into the arteries under great pressure. This pressure is the driving force that moves blood through the body, but it also pushes outward against vessel walls. Ideally, a person's blood pressure should be great enough to circulate the blood but not so great that it stresses the blood vessels and the heart, as we see in Chapter 12a. Many factors influence blood pressure, including gender, age, time of day, physical activity, stress, and lifestyle.

Blood pressure in the arteries varies predictably during each heartbeat. It is highest during the contraction of the ventricles (ventricular systole), when blood is being forced into the arteries. In a typical, healthy adult, the optimal **systolic pressure**, the highest pressure in the artery during each heartbeat, is 110 mm to 120 mm of mercury (mm Hg).[1] Blood pressure is lowest when the ventricles are relaxing (diastole). In a healthy adult, the optimal lowest pressure, or **diastolic pressure**, is 70 mm to 80 mm Hg. A person's blood pressure is usually expressed as two values—the systolic followed by the diastolic. For instance, optimal adult blood pressure is said to be less than 120/80. (Do you know what your blood pressure is?)

Blood pressure is measured with a device called a *sphygmomanometer* (sfig-mo-mah-nom'-e-ter), which consists of an inflatable cuff that wraps around the upper arm and is attached to a device that can measure the pressure within the cuff. Figure 12.14 shows how a manually operated sphygmomanometer uses the easily measured pressure of the air pumped into the cuff to measure the blood pressure in the brachial artery (which runs along the inner surface of the arm).

12.5 Lymphatic System

The **lymphatic system** consists of **lymph**, which is a fluid identical to interstitial fluid (the fluid that bathes all the cells of the body); of **lymphatic vessels**, through which the lymph flows; and of various lymphoid tissues and organs scattered throughout the body.

---

[1]Pressure is measured as the height to which that pressure could push a column of mercury (Hg).

HEALTH ISSUE

Benefits of Cardiovascular Exercise

What would you say if you were told there is a simple way to reduce your risk of heart attack, stroke, diabetes, and cancer while controlling your weight, strengthening your bones, relieving anxiety and tension, and improving your memory? "Impossible!" you might say, "What's the catch?" There is none. This key to life is regular aerobic exercise that uses large muscle groups rhythmically and continuously and elevates heart rate and breathing rate for at least 15 to 20 minutes.

Although exercise has many beneficial effects on the body, here we will consider only the benefits to the cardiovascular system. Exercise benefits the heart by making it a more efficient pump, thus reducing its workload. A well-exercised heart beats more slowly than the heart of a sedentary person—during both exercise and rest. The lower heart rate gives the heart more time to rest between beats. Yet at the same time, the well-exercised heart pumps more blood with each beat.

Exercise also increases the oxygen supply to the heart muscle by widening the coronary arteries, thereby increasing blood flow to the heart. Moreover, because the capillary beds within the heart muscle become more extensive with regular exercise, oxygen and nutrients can be delivered to the heart cells and wastes can be removed more quickly.

Furthermore, exercise helps to ensure continuous blood flow to the heart. One way it accomplishes this benefit is by increasing the body's ability to dissolve blood clots that can lead to heart attacks or strokes. Exercise stimulates the release of a natural enzyme that prevents blood clotting and remains effective for as long as 90 minutes after you stop exercising. Besides this, exercise stimulates the development of collateral circulation, that is, additional blood vessels that provide alternative pathways for blood flow. As a result, blood flows continuously through the heart, even if one vessel becomes blocked.

Exercise affects the blood in ways that allow more oxygen to be delivered to the cells. The amount of hemoglobin, the oxygen-binding protein in red blood cells, increases. In addition, the blood volume and the numbers of red blood cells increase.

Exercise lowers the risk of coronary artery disease by lowering blood pressure and by shifting the balance of lipids in the blood. High-density lipoproteins (HDLs), which are the "good" form of cholesterol-carrying particles that remove cholesterol from the arterial walls, increase with exercise.

To reap cardiovascular benefits, you must exercise hard enough, long enough, and often enough. The exercise must be vigorous enough to elevate your heart rate to the so-called target zone, which is between 70% and 85% of your maximal attainable heart rate. The target zone can be determined by subtracting your age in years from 222 beats per minute. The exercise must continue for at least 20 minutes and be performed at least 3 days a week, with no more than 2 days between sessions.

Doing something active on a regular basis can improve your quality of life. Moderate activity helps you feel better emotionally and physically. The difference has been likened to traveling first class instead of coach.

Questions to Consider

- Do you plan to exercise regularly? What factors will you consider in making this decision?
- Do you think that promoting exercise programs now would result in future savings in medical costs?

The functions of the lymphatic system are as diverse as they are essential to life:

1. **Return excess interstitial fluid to the bloodstream.** The lymphatic system maintains blood volume by returning excess interstitial fluid to the bloodstream. Only 85% to 90% of the fluid that leaves the blood capillaries and bathes the body tissues as interstitial fluid is reabsorbed by the capillaries. The rest of that fluid (as much as 3 liters a day) is absorbed by the lymphatic system and then returned to the circulatory system. This job is important. If the surplus interstitial fluid were not drained, it would cause the tissue to swell; the volume of blood would drop to potentially fatal levels; and the blood would become too viscous (thick) for the heart to pump.

   A dramatic example of the importance of returning fluid to the blood is provided by *elephantiasis*, a condition in which parasitic worms block lymphatic vessels (Figure 12.15). The blockage can cause a substantial buildup of fluid in the affected body region, followed by the growth of connective tissue. Elephantiasis is so named because it results in massive swelling and the darkening and

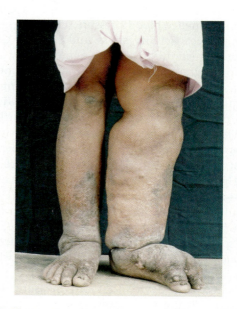

FIGURE 12.15 The leg of a person with elephantiasis. In this condition, parasitic worms plug lymphatic vessels and prevent the return of fluid from the tissues to the circulatory system.

thickening of the skin in the affected region, making the region resemble the skin of an elephant. Elephantiasis is a tropical disease that is transmitted by mosquitoes.

2. **Transport products of fat digestion from the small intestine to the bloodstream.** The products of fat digestion are too large to be absorbed into the capillaries in the small intestine. Instead, these products enter a lymphatic vessel, called a *lacteal,* and travel in the lymphatic system to be returned to the blood circulatory system.

3. **Help defend against disease-causing organisms.** The lymphatic system helps protect against disease and cancer. We will learn more about its role in body defense in Chapter 13.

The structure of the lymphatic vessels is central to their ability to absorb the interstitial fluid not carried away by capillaries. The extra fluid enters a branching network of microscopic tubules, called *lymphatic capillaries,* that penetrate between the cells and the capillaries in almost every tissue of the body (except teeth, bones, bone marrow, and the central nervous system; Figure 12.16). The lymphatic capillaries differ from the blood capillaries in two ways. First, unlike blood capillaries, which form continuous networks, lymphatic capillaries end blindly, like the fingers of a glove. In essence, lymphatic capillaries serve as drainage tubes. Fluid enters at the "fingertips" and moves through the system in only one direction. Second, lymphatic capillaries are much more permeable than blood capillaries, a feature that is crucial to their ability to absorb the digestive products of fats as well as excess interstitial fluid. The lymphatic capillaries drain into larger lymphatic vessels, which merge into progressively larger tubes with thicker walls. Lymph is eventually returned to the blood through one of two large ducts that join with the large veins at the base of the neck.

With no pump to drive it, lymph flows slowly through the lymphatic vessels, propelled by the same forces that move blood through the veins. That is, the contractions of nearby skeletal muscles compress the lymphatic vessels, pushing the lymph along. One-way valves similar to those in veins prevent backflow. And as with the flow of blood back to the heart from the lower part of the body, pressure changes in the thorax (chest cavity) that accompany breathing also help pull the lymph upward from the lower body. Gravity assists the flow from the upper body.

The lymphatic vessels are studded with **lymph nodes**, small bean-shaped structures that cleanse the lymph as it slowly filters through. The lymph nodes contain macrophages and lymphocytes, white blood cells that play an essential role in the body's defense system. Macrophages engulf bacteria, cancer cells, and other debris, clearing them from the lymph. Lymphocytes serve as the surveillance squad of the immune system. They are continuously on the lookout for specific disease-causing invaders, as we will see in Chapter 13. When lymphocytes detect bacteria or viruses, the lymphocytes are stimulated to divide. The increased number of lymphocytes causes the lymph nodes to swell. Thus, swollen and painful lymph nodes (commonly called "glands") are a symptom of infection.

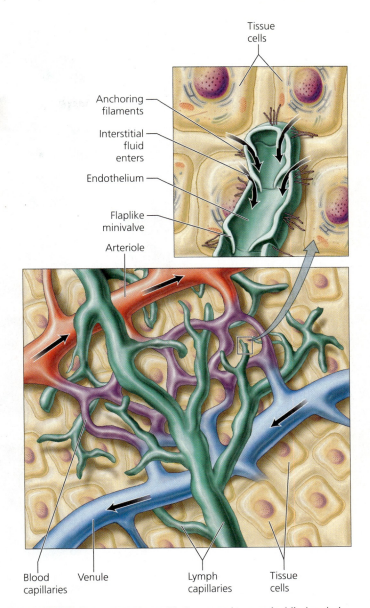

FIGURE 12.16 *The lymphatic capillaries are microscopic, blind-ended tubules through which surplus tissue fluid enters the lymphatic system to be returned to the bloodstream.*

Besides the lymph nodes, there are several other *lymphoid organs* (Figure 12.17). Among these is the **spleen**, which is the largest lymphoid organ. The red pulp of the spleen contains blood-filled sinuses. Macrophages roam the red pulp, clearing the blood of microorganisms as well as old and damaged red blood cells and platelets. After it is cleansed, blood is stored in the red pulp and can be released when a boost in oxygen-carrying capacity is needed. The white pulp functions as a meeting place for lymphocytes and disease-causing organisms.

The **thymus gland**, located in the chest, is another lymphoid organ. It plays its part during early childhood by helping the maturation of certain lymphocytes that protect us from specific disease-causing organisms. Within the thymus, lymphocytes learn to distinguish "self" (normal body cells) from "nonself."

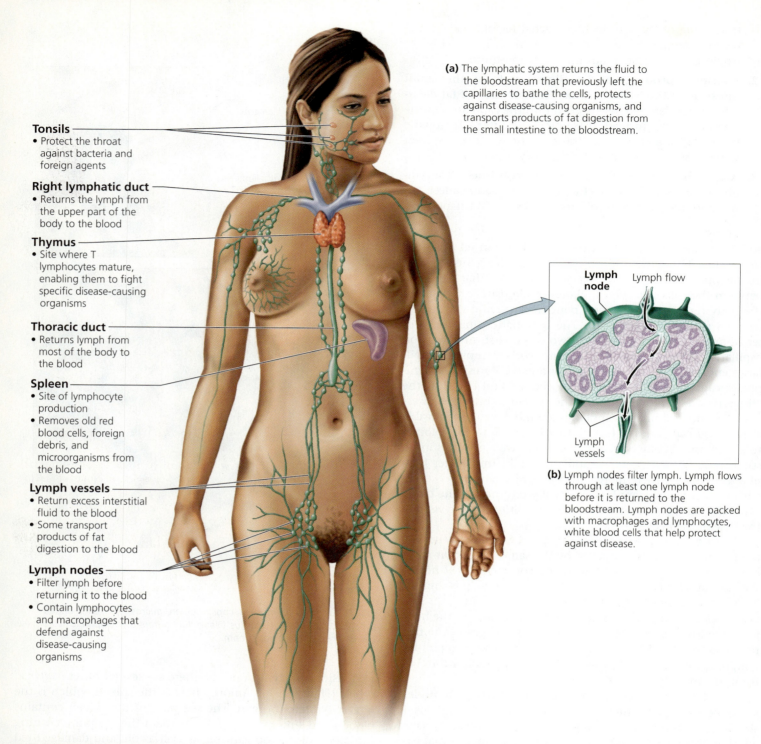

FIGURE 12.17 *The lymphatic system is a system of lymphatic vessels containing a clear fluid, called lymph, and various lymphatic tissues and organs located throughout the body. Green indicates lymphatic vessels and nodes.*

Other lymphoid tissues include the tonsils, Peyer's patches, and the red bone marrow. The *tonsils* form a ring around the entrance to the throat, where they help protect against disease organisms that are inhaled or swallowed. Isolated clusters of lymph nodules along the small intestine, known as *Peyer's patches*, keep bacteria from breaching the intestinal wall. Finally, the *red bone marrow*, where white blood cells and other formed elements are produced, is a lymphoid organ found in the ends of long bones, ribs, sternum, and vertebrae.

stop and think

Cancer cells that break loose from their original site in a process called *metastasis* have easy access to the highly permeable lymphatic capillaries. The lymphatic vessels then provide a route by which the cancer cells may spread to nearly every part of the body. Why are the lymph nodes often examined to determine whether cancer has spread? Why are the lymph nodes near the original cancer site often removed?

looking ahead

In this chapter, we learned about the structure and function of the cardiovascular and lymphatic systems as they are in a healthy person. In the next chapter, we will consider some common disorders of the cardiovascular system.

HIGHLIGHTING THE CONCEPTS

12.1 Cardiovascular System (pp. 213–214)
- The heart serves as a pump that pushes blood through blood vessels.

12.2 Blood Vessels (pp. 215–219)
- Blood circulates through a branching network of blood vessels in a path that travels from the heart to arteries to arterioles to capillaries to venules to veins and then back to the heart.
- Arteries are elastic, muscular tubes that carry blood away from the heart. By stretching and then returning to their original shape, they withstand the high pressure of blood as it is pumped from the heart. These changes help maintain a relatively even blood pressure within the arteries despite large changes in the volume of blood within them.
- The pressure change along an artery as it expands and returns to its original size is called a pulse.
- Arteries branch to form narrower tubules called arterioles. Arterioles are important in regulating blood pressure, and they regulate blood flow through capillary beds.
- The exchange of materials between the blood and tissues takes place across the thin walls of capillaries. Capillaries are arranged in highly branched networks that provide a tremendous surface area for the exchange. Each network of capillaries is called a capillary bed. Rings of muscle called precapillary sphincters determine whether blood flows into a capillary bed or is channeled through it.
- Capillaries merge to form venules; these, in turn, merge to form veins, which conduct blood back to the heart. Blood is returned to the heart against gravity by nearby skeletal muscles that contract and push the blood along within the veins. Valves within the veins prevent blood from flowing backward when the skeletal muscles are relaxed. Pressure differences generated by breathing also help draw blood toward the heart from the lower torso.

12.3 Heart (pp. 219–224)
- Every minute, the heart beats about 72 times and moves slightly less than 5 liters (10 pt) of blood through its chambers.
- Most of the wall of the heart is composed of cardiac muscle and is called the myocardium. The endocardium is a thin inner lining. The heart is enclosed in a fibrous sac called the pericardium, which allows the heart to beat while still confining it near the midline of the thoracic cavity.
- The right and left halves of the heart function as two separate pumps. Each side consists of two chambers: an upper chamber called the atrium, and a lower chamber called the ventricle. The smaller, thin-walled atria function primarily as receiving chambers that accept blood returning to the heart and pump it a short distance to the ventricles. When the larger, thick-walled ventricles contract, they push blood through the arteries to all parts of the body.
- Blood circulates in one direction through the heart as a result of the action of two pairs of valves. The atrioventricular (AV) valves are located between each atrium and ventricle. The semilunar valves are located between each ventricle and its connecting artery. The heart sounds, lub-dup, are caused by blood turbulence associated with the closing of the heart valves.
- The right side of the heart pumps blood to the lungs through a loop of vessels called the pulmonary circuit. The left side of the heart pumps blood to all parts of the body except the lungs through a loop of vessels called the systemic circuit.
- The heart has its own network of vessels, called the coronary circuit, that services the heart tissue itself.
- Each heartbeat consists of contraction (systole) and relaxation (diastole). The atria contract in unison, and then the ventricles do the same. The events associated with each heartbeat are collectively called the cardiac cycle.
- The rhythmic contraction of the heart is produced by its internal conduction system. A cluster of specialized cardiac muscle cells, called the sinoatrial (SA) node, usually sets the tempo of the heartbeat and is therefore called the pacemaker. When the electrical signal reaches another cluster of specialized muscle cells, called the atrioventricular (AV) node, the stimulus is quickly relayed along the atrioventricular bundle that runs through the wall between the two ventricles and then fans out into the ventricular walls through the Purkinje fibers.
- The rate of heartbeat changes constantly to suit the body's level of activity.
- An electrocardiogram (ECG or EKG) is a recording of the electrical events associated with each heartbeat.

12.4 Blood Pressure (pp. 224–225)
- Blood pressure, the force created by the heart to drive the blood around the body, is measured as the force of blood against the walls of blood vessels. The blood pressure in an artery peaks when a ventricle contracts. This is called the systolic pressure. In contrast, the lowest blood pressure of each cardiac cycle, the diastolic pressure, occurs while the heart is relaxing between contractions. Blood pressure is measured with a device called a sphygmomanometer.

12.5 Lymphatic System (pp. 225–229)
- The lymphatic system consists of lymph, lymphatic vessels, lymphoid tissue, and lymphoid organs.
- Three vital functions of the lymphatic system are to return interstitial fluid to the bloodstream, to transport products of fat digestion from the digestive system to the bloodstream, and to defend the body against disease-causing organisms or abnormal cells.

- Tissue fluid enters lymphatic capillaries—microscopic tubules that end blindly and are more permeable than blood capillaries. The fluid, then called lymph, is moved along larger lymphatic vessels by contraction of nearby skeletal muscles. The lymphatic vessels have valves to prevent the backflow of lymph.
- Lymph nodes filter lymph and contain cells that actively defend against disease-causing organisms.
- Lymphoid organs include the red bone marrow, lymph nodes, tonsils, thymus gland, spleen, and Peyer's patches of the small intestine.

RECOGNIZING KEY TERMS

cardiovascular system *p. 213*
heart *p. 213*
artery *p. 215*
aorta *p. 215*
pulse *p. 215*
arteriole *p. 215*
capillary *p. 216*
capillary bed *p. 217*
precapillary sphincter *p. 217*
venule *p. 218*
vein *p. 218*

myocardium *p. 219*
endocardium *p. 219*
pericardium *p. 219*
septum *p. 219*
atrium *p. 219*
ventricle *p. 219*
atrioventricular (AV) valve *p. 219*
semilunar valves *p. 220*
pulmonary circuit *p. 221*
systemic circuit *p. 221*

coronary circulation *p. 222*
coronary arteries *p. 222*
systole *p. 222*
diastole *p. 223*
cardiac cycle *p. 223*
sinoatrial (SA) node *p. 223*
pacemaker *p. 224*
atrioventricular (AV) node *p. 224*
electrocardiogram (ECG or EKG) *p. 224*

blood pressure *p. 224*
systolic pressure *p. 225*
diastolic pressure *p. 225*
lymphatic system *p. 225*
lymph *p. 225*
lymphatic vessel *p. 225*
lymph nodes *p. 227*
spleen *p. 227*
thymus gland *p. 227*

REVIEWING THE CONCEPTS

1. What is a pulse? *p. 215*
2. What are two important functions of arterioles? *p. 215*
3. What determines whether blood flows through any particular capillary bed? *p. 217*
4. Compare the structure of arteries, capillaries, and veins. Explain how the structure is suited to the function of each type of vessel. *pp. 215–218*
5. Explain how blood is returned to the heart from the lower torso against the force of gravity. *p. 218*
6. Describe the structure of the heart. Explain how it functions as two separate pumps. *p. 219*
7. Trace the path of blood from the left ventricle to the left atrium, naming each major vessel associated with the heart and the heart chambers in the correct sequence. *p. 221*
8. Describe the cardiac cycle. *p. 223*
9. Explain how clusters or bundles of specialized cardiac muscle cells coordinate the contraction associated with each heartbeat. *pp. 223–224*
10. List three important functions of the lymphatic system. *pp. 226–227*
11. What is the function of lymph nodes? *p. 227*
12. The semilunar valves prevent the backflow of blood from the
    a. arteries to the ventricles.
    b. veins to the atria.
    c. ventricles to the atria.
    d. arteries to the atria.
13. If you cut all the nerves to the heart but kept the heart alive,
    a. the heart would stop beating.
    b. the heart would continue beating.
    c. only systole would occur.
    d. only diastole would occur.

For questions 14–16, imagine that you have been miniaturized and are riding through the circulatory system using a red blood cell as a life raft.

14. You are in the big toe traveling toward the heart. The last vessel you pass through before entering the heart is the
    a. aorta.
    b. inferior vena cava.
    c. coronary artery.
    d. pulmonary vein.
15. You are nearly deafened by the first heart sound, which is caused by
    a. the opening of the atrioventricular valves.
    b. the closing of the atrioventricular valves.
    c. the opening of the semilunar valves.
    d. the closing of the semilunar valves.
16. You are wearing sensitive equipment that may be damaged by high pressure. The blood pressure would be highest when you are in the aorta during
    a. atrial systole.
    b. ventricular systole.
    c. atrial diastole.
    d. ventricular diastole.
17. Heart tissue is nourished by:
    a. the SA node.
    b. the blood passing through the chambers of the heart.
    c. the chordae tendineae.
    d. blood in the coronary blood vessels (i.e., through the coronary arteries to capillaries).
18. If you were to cut all the nerves to the heart and keep the heart alive,
    a. the heart would stop beating.
    b. the heart would continue beating.
    c. only systole would occur.
    d. only diastole would occur.
19. The second heart sound (dup) is the sound of the
    a. pacemaker initiating its contraction.
    b. blood entering the aorta.
    c. the closing of the semilunar valves.
    d. the closing of the atrioventricular (AV) valves.
20. Oxygen and nutrients move through the walls of _____ to reach the body cells.

APPLYING THE CONCEPTS

1. Abnormally short (or long) chordae tendineae of the mitral (bicuspid) valve can cause a condition known as mitral valve prolapse, in which the mitral valves do not close properly. Why would mitral valve prolapse cause the heart to produce abnormal heart sounds?
2. Nicotine in cigarette smoke causes vasoconstriction (narrowing of certain blood vessels) and increases heart rate. Explain how these effects lead to high blood pressure. Explain why cigarette smokers are more likely to die of cardiovascular diseases than are nonsmokers.
3. Amelia is a friend of yours. When you call her to ask about dinner plans, she tells you that she is not feeling well. She aches all over and has swollen "glands" in her neck. You explain that these are not really glands, because they do not secrete anything. What are they? Why are they swollen?

BECOMING INFORMATION LITERATE

Use at least three reliable sources (books, journals, websites) to plan a program for yourself to increase your cardiovascular fitness. Begin by deciding which aspects of your lifestyle you might change to improve cardiovascular fitness. Examples might include, but are not limited to, weight loss, change in diet, increased physical activity, and stress reduction. Next, plan a logical progression of changes that considers the type of activities and the foods that you enjoy. List each source you considered, and explain why you chose the three sources you used.

MasteringBiology®

Go to MasteringBiology for practice quizzes, activities, eText, videos, current events, and more.

12a SPECIAL TOPIC: Cardiovascular Disease

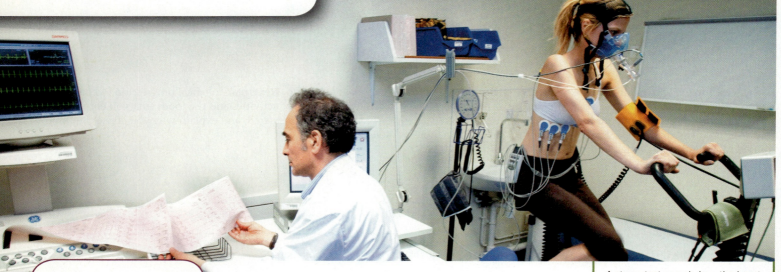

A stress test reveals how the heart responds to exertion. This test can help you develop a healthful exercise program.

- 12a.1 The Prevalence of Cardiovascular Disease
- 12a.2 Blood Clots
- 12a.3 Problems with Blood Vessels
- 12a.4 Heart Attack and Heart Failure
- 12a.5 Cardiovascular Disease and Cigarette Smoking
- 12a.6 Preventing Cardiovascular Disease

In Chapter 12, you examined the heart and blood vessels as they function in health. In this chapter, you will consider the problems that arise in some common cardiovascular diseases.

12a.1 The Prevalence of Cardiovascular Disease

Cardiovascular disease is the single biggest killer of men and women in the United States. It affects slightly more men than women because, until menopause, women receive some natural protection from cardiovascular disease through the action of the sex hormone estrogen. Ironically, although slightly fewer women than men have heart attacks, women who do have heart attacks are twice as likely to die within the following weeks as are men. Because heart attacks are commonly thought of as a male problem, women and their physicians often fail to recognize the symptoms, thus delaying treatments that could be lifesaving.

12a.2 Blood Clots

A *thrombus* is a stationary blood clot that the forms along the wall of a blood vessel or within the heart that may obstruct blood flow. The rough surface of a fatty deposit in an artery (atherosclerotic plaque, discussed shortly) can damage platelets and trigger clot formation. A thrombus can also form in the deep veins of the legs or pelvis when blood flow is sluggish, which can result from long periods of inactivity, such as during travel or after surgery or illness.

An *embolism* is a blockage of a blood vessel, which is often caused by a thrombus that breaks free of its site of formation and floats through the circulatory system until it lodges in a small vessel, where it can block blood flow and cause tissue death beyond that point. If the blockage is in the brain, nerve tissue can die, causing a stroke. Blockage in the heart causes the death of heart muscle cells, which is a heart attack. The clot can also travel to the lungs (a pulmonary embolism), where it blocks blood flow and prevents adequate oxygenation of the blood.

We see, then, that abnormal blood clots can threaten life. Blood-thinning drugs, such as warfarin and heparin, reduce the likelihood of clot formation. When blood clots are already lodged in coronary (heart), cerebral (brain), or pulmonary (lung) blood vessels and block blood flow, it is essential to dissolve the clot quickly. One clot-dissolving drug is *tissue plasminogen activation (t-PA)*. You may recall from Chapter 11 that plasmin is an enzyme formed from plasminogen in the blood that dissolves blood clots; t-PA activates plasminogen to form plasmin. It should be administered within 6 hours of the problem to minimize damage.

12a.3 Problems with Blood Vessels

As we learned in Chapter 12, blood vessels conduct blood from the heart to body cells and the lungs. Here we will examine some serious consequences resulting from problems with the blood vessels.

High Blood Pressure

In Chapter 12, we learned that blood pressure is the outward pressure exerted by blood against the vessel walls. It must be high enough to circulate blood, but not so high that it stresses blood vessels. Recall that blood pressure is usually reported with two numbers. The first value—the *systolic* value—is the pressure when the heart is contracting. The second value, called the *diastolic* pressure, is the pressure when the heart is relaxing.

High blood pressure, or *hypertension,* is often called the silent killer. It is *silent* because it does not produce any telltale symptoms. It is a *killer* because it can cause fatal problems, usually involving the heart, brain, blood vessels, or kidneys. Hypertension damages the heart in a number of ways, primarily by causing the heart to work harder to keep the blood moving. In response, the heart muscle thickens, and the heart enlarges. The enlarged heart works less efficiently and has difficulty keeping up with the body's needs. At the same time, the increased workload increases the heart's need for oxygen and nutrients. If these cannot be delivered rapidly enough, a heart attack can result.

High blood pressure can also damage the kidneys, reducing the blood flow through them. In response, the kidneys make matters worse by secreting renin, a chemical that leads to further increases in blood pressure in an ever-escalating cycle.

Although about 90% of the cases of hypertension have no *known* cause, many contributing factors have been identified. When a cause can be identified, the kidneys are sometimes to blame. If kidneys have an impaired ability to handle sodium, the resulting fluid retention increases blood pressure by increasing blood volume. In other cases of hypertension, the sympathetic nervous system reacts too strongly to stress, constricting the blood vessels and increasing heart rate. Thus, more blood per minute is pumped through vessels that provide a greater resistance to flow.

Most physicians agree that a blood pressure of 160/90 is high and should be treated (Figure 12a.1). But uncertainty

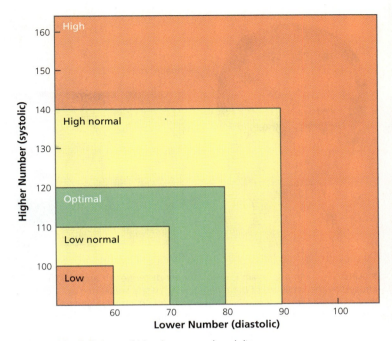

FIGURE 12a.1 *Values of blood pressure in adults*

clouds the treatment issue when the person's diastolic pressure is between 80 and 89—the higher end of normal. Although drug treatment may help in borderline cases, it usually must be continued for life. A diagnosis of high blood pressure may influence other aspects of a person's life, such as life insurance premiums. So, sometimes, only lifestyle changes are recommended for high normal values of blood pressure.

When the diagnosis of hypertension is clear, one or more of various kinds of drugs can be prescribed, each type combating a different mechanism that contributes to high blood pressure. The diuretics, for instance, decrease blood volume by increasing the excretion of sodium and fluids, thereby reducing blood pressure by reducing blood volume. Other drugs cause the blood vessels to dilate (become wider), reducing hypertension in instances where is it caused by overly constricted vessels.

Aneurysm

When the wall of an artery becomes weakened, as may be caused by disease, inflammation, injury, or a congenital defect, the pressure of the blood flowing through the weakened area may cause it to swell outward like a balloon, forming an aneurysm (Figure 12a.2). Common locations for aneurysms include arteries in the brain and the aorta, the major artery leaving the left ventricle that carries blood to body cells. An aneurysm does not always cause symptoms, but it can be threatening just the same. The primary risk is that the aneurysm will burst, causing blood loss and depriving tissues of oxygen and nutrients, a situation that can be fatal. Even if it does not rupture, an aneurysm can cause life-threatening blood clots to form. An aneurysm can be detected with either an MRI or ultrasound. Treatment often includes surgical removal or applying support with a coil or a stent.

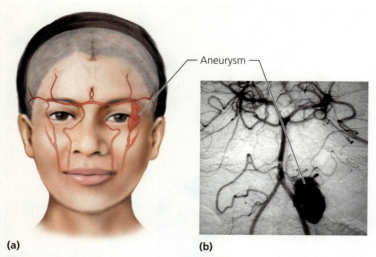

FIGURE 12a.2 An aneurysm is a permanent distention in an artery wall caused by the pressure of the blood flowing past a weakened area of that wall. (a) A diagram of a cerebral aneurysm. (b) An x-ray image of a cerebral aneurysm. Rupture of this aneurysm would cause a stroke.

Atherosclerosis

Atherosclerosis (*ather-*, yellow, fatty deposit; *sclerosis*, a hardening) is a buildup of fatty substances in the walls of arteries, fueled by an inflammatory response. In some cases, the deposits narrow the artery (Figure 12a.3). Such narrowing causes problems because it reduces blood flow through the vessel, choking off the vital supply of oxygen and nutrients to the tissues served by that vessel. But contrary to the beliefs held just a few years ago, atherosclerosis is more than just a plumbing problem, akin to a clog in a passive pipeline.

The inflammatory response thought to cause atherosclerosis is the same process that wards off infection when you scrape your knee (discussed in Chapter 13). In this case, it begins with an injury to the wall of an artery. The injury may be caused by some kind of bloodborne irritant (such as the chemicals inhaled in cigarette smoke), by cholesterol deposits in the artery lining, or by infection. Perhaps the excessively rapid or turbulent blood flow caused by high blood pressure can also produce such arterial damage. The damaged cells begin to pick up low-density lipoproteins (LDLs), the so-called bad form of cholesterol. This accumulation of LDLs is most likely to occur when the LDL concentration in the blood is high. The LDLs undergo chemical changes that stimulate the cells of the arterial lining to enlist the body's defense responses: inflammatory chemicals and defense cells. Growth factors produced by defense cells stimulate smooth muscle cells in the arterial wall to divide, thereby thickening the wall. Other defense cells engulf LDLs, become enlarged, and form fatty streaks on the arterial lining. As these defense cells continue to scavenge lipids, the fatty streaks increase in size and forms plaque, a bumpy, fatty layer in the artery wall. The plaque can bulge into the artery channel, blocking the blood flow, or it can expand outward into the artery wall. Although the plaque has a fibrous cap that initially keeps pieces from breaking away, the fat-filled cells secrete inflammatory substances that weaken the cap. A small break in the cap can allow the plaque to rupture, triggering the formation of a blood clot. In any case, restriction of the blood flow can starve and kill the cells lying downstream. Such an occurrence in the heart (leading to a heart attack) or brain (leading to a stroke) can be fatal. Because plaques do not necessarily bulge into the artery and cause symptoms of atherosclerosis before they rupture, heart attacks often occur in patients without previous symptoms.

There are several approaches to treating atherosclerosis. A healthy lifestyle promotes healthy arteries, so control your weight, engage in regular aerobic exercise, and eat a heart-healthy diet (see Chapter 15a). Medications to treat atherosclerosis include drugs to lower blood pressure, drugs to reduce blood cholesterol levels, and drugs to prevent unwanted blood clot formation. Some physicians recommend a daily dose of a low-strength aspirin, because aspirin reduces the risk of blood clot formation.

stop and think

C-reactive protein (CRP) is an inflammatory chemical released by injured cells in the artery lining. Why might CRP be a better predictor of atherosclerosis than blood cholesterol level?

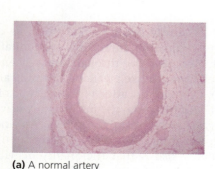

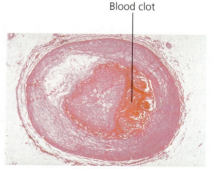

(a) A normal artery
(b) An artery partially obstructed with plaque
(c) An artery completely obstructed with plaque and a blood clot

FIGURE 12a.3 Atherosclerosis, a low-level inflammatory response in the wall of an artery, is associated with the formation of fat-filled plaques. Plaque can obstruct blood flow through the artery, thus depriving the cells that would be fed life-sustaining blood by the artery. Plaque can also rupture, causing a blood clot to form. The clot may then completely clog the vessel and cause the death of tissue downstream.

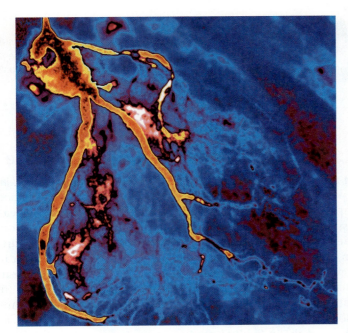

FIGURE 12a.4 *A coronary angiograph. A dye is injected into the coronary arteries, making them visible on an x-ray film.*

what would you do

Although C-reactive protein is an excellent predictor of atherosclerosis, it cannot be modified by drugs or changes in behavior. With this in mind, do you think that patients who would like to have this test should pay for it themselves?

Coronary artery disease is a condition in which the fatty deposits associated with atherosclerosis form within coronary arteries, the arteries that nourish the heart muscle. Coronary artery disease is the underlying cause of the vast majority of heart attacks.

A temporary shortage of oxygen to the heart is accompanied by angina pectoris—chest pain, usually experienced in the center of the chest or slightly to the left. The name *angina* comes from the Latin word *angere,* meaning "to strangle." The name is apt, because the pain of angina is often described as suffocating, viselike, or choking. Typically, the pain begins during physical exertion or emotional stress, when the demands on the heart are increased and the blood flow to the heart muscle can no longer meet the needs. The pain stops after a period of rest. Angina serves as a warning that part of the heart is receiving insufficient blood through the coronary arteries, but it does not cause permanent damage to the heart. The warning should be taken seriously, however, because each year up to 15% of people who have angina later die from a heart attack.

Although coronary artery disease is usually diagnosed from the symptoms of angina and a physical examination, a procedure called *coronary angiography* may be used to spot areas in the coronary arteries that have become narrowed by atherosclerosis. In this procedure, a contrast dye that is visible in x-ray images is released in the heart, allowing the coronary vessels to be seen on film. A catheter (a slender, flexible tube) is inserted into an artery in the arm or leg and then threaded through the blood vessels until it reaches the heart. The dye is then squirted into the openings of the coronary arteries, and its movement through the arteries is recorded in a series of high-speed x-rays (Figure 12a.4).

Coronary artery disease can be treated with medicines or surgery. Among the medicines commonly used are some that dilate (widen) blood vessels, such as nitroglycerin. Certain other drugs specifically dilate only the coronary arteries. Wider blood vessels make it easier for the heart to pump blood through the circuit. When the coronary arteries dilate, more blood is delivered to the heart muscle. Also used are drugs that dampen the heart's response to stimulation from the sympathetic nervous system, thus decreasing its need for oxygen.

Two surgical operations for treating coronary artery disease are balloon angioplasty and coronary artery bypass. In *angioplasty,* the channel of an artery narrowed by soft, fatty plaque is widened by inflating a tough plastic balloon inside the artery (Figure 12a.5). First, the tiny, uninflated balloon

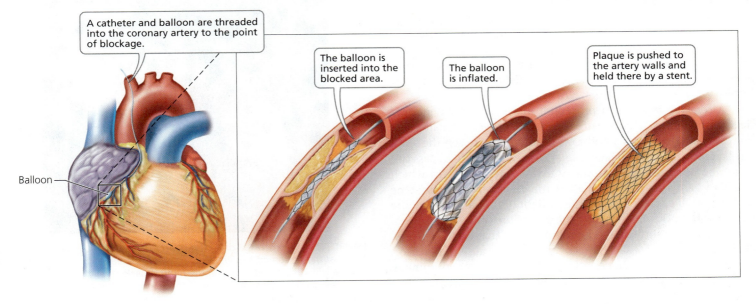

FIGURE 12a.5 *Balloon angioplasty opens a partially blocked artery.*

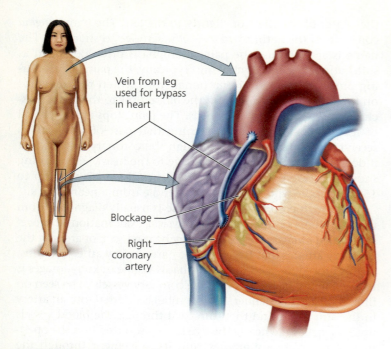

FIGURE 12a.6 In coronary bypass surgery, a section of a leg vein is removed. One end of the vein is attached to the heart's main artery, the aorta, and the other to a coronary artery, bypassing the obstructed region. The grafted vein provides a pathway through which blood can reach the previously deprived region of heart muscle.

is attached to the end of a catheter and inserted through an artery in the arm or upper thigh. It is then pushed to the blocked spot in a coronary artery; high-speed x-rays are used to track its progress. After it reaches the blockage, the balloon is inflated under pressure, stretching the artery and pressing the soft plaque against the wall to widen the lumen, the hollow central part of the vessel that blood flows through.

After angioplasty, physicians commonly insert a metal-mesh tube called a *stent* into the treated arteries. The stent prevents the arteries from collapsing and keeps loose pieces of plaque from being swept into the bloodstream. Although stents boost the percentage of arteries that stay open, they sometimes trigger an inflammatory response, and some arteries become clogged again, even with a stent in place. Some stents slowly release drugs that further reduce the risk that the artery will become blocked again.

A *coronary bypass* is a procedure in which a segment of a leg vein is removed and grafted so that it provides an alternate pathway that bypasses a point of obstruction between the aorta and a coronary artery (Figure 12a.6). Today, coronary bypass can often be done using a minimally invasive technique in which small incisions are made that allow the insertion of "arms." One arm carries a camera and another a light. Other arms allow the surgeon to work while viewing the procedure on a video screen. Sometimes, the surgeon guides robotic arms that do the surgery. The advantages are a shorter recovery time (because the incision is smaller) and less chance of infection. However, minimally invasive surgery usually takes longer and, therefore, involves increased risks.

stop and think

In a coronary bypass operation, why is it important for the surgeon to insert the grafted vein in the correct orientation? What would happen if the piece of vein were inserted backward?

12a.4 Heart Attack and Heart Failure

In a heart attack, technically known as a *myocardial infarction,* a part of the heart muscle dies because of an insufficient blood supply. (*Myocardial* refers to heart muscle; *infarct* refers to dead tissue.) Heart muscle cells begin to die if they are cut off from their essential blood supply for more than 2 hours. Depending on the extent of damage, the effects of a heart attack can spread quickly throughout the body. As a result of a heart attack, the brain receives insufficient oxygen; the lungs fill with fluid; and the kidneys fail. Within a short time, white blood cells swarm in to remove the damaged heart tissue. Then—if the individual survives—over the next 8 weeks or so, scar tissue replaces the dead cardiac muscle (Figure 12a.7). Because scar tissue cannot contract, part of the heart permanently loses its pumping ability.

Heart attacks can be caused in many ways. The most common type of heart attack is a *coronary thrombosis,* which means the attack is caused by a blood clot blocking a coronary artery. Coronary thrombosis is unlikely to happen unless the artery already contains plaques of atherosclerosis, as occurs in coronary artery disease. In some instances, the blood clot is formed elsewhere in the body but is swept along in the bloodstream until it lodges in a coronary artery. In other instances of heart attacks, the blockage is temporary, caused by constriction of a coronary artery, called a *coronary artery spasm.*

Chest pain is a common indication that a heart attack is in progress, especially in men. Unlike angina, the pain is not

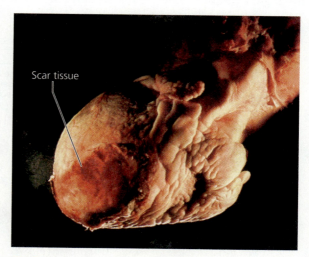

FIGURE 12a.7 The ravages of a prior heart attack are visible as scar tissue at the bottom of this lifeless heart. Scar tissue replaced cardiac muscle when the blood supply to the heart muscle was shut down. Because scar tissue cannot contract, that part of the life-sustaining pump becomes ineffective.

necessarily brought on by activity, and it doesn't go away with rest. In some cases, the victim feels a severe, crushing pain that begins in the center of the chest and often spreads down the inside of one or both arms (most commonly the left one) as well as up to the neck and shoulders. Although the pain is usually severe enough to cause the victim to stop whatever he or she is doing, it is not always so strong that it is recognized as a sign of heart attack. A heart attack may also cause nausea and dizziness, which can prompt the victim to interpret the symptoms as an upset stomach. Oddly, persons who experience severe pain may be the lucky ones, because they are more likely to realize they are having a heart attack and seek immediate help. Doubt about the cause of the symptoms is unfortunate because it often delays treatment, and treatment within the first few hours after a heart attack can make the difference between life and death.

If a large enough section of heart muscle is damaged by a heart attack, the heart may not be able to continue pumping blood to the lungs and the rest of the body at an adequate rate. A condition in which the heart is no longer an efficient pump is known as *heart failure*. Symptoms of heart failure include shortness of breath, fatigue, weakness, and fluid accumulation in the lungs or limbs. Although the heart can never be restored to its former health, the symptoms of heart failure can be treated with drugs. For instance, digitalis can increase the strength of heart contractions, and diuretics reduce fluid accumulation, thus lessening the heart's workload. Other drugs relax constricted arteries, thereby reducing resistance to blood flow and lowering the blood pressure. Together, these drugs help a weakened heart pump more efficiently.

Sometimes, drugs and bypass surgery cannot halt progressive heart failure. In this case, heart transplant surgery may provide hope for some patients, but generally only for those younger than 80 years of age. First, a donor heart must be found whose tissue is an acceptable match with that of the recipient. During transplant surgery, a heart-lung machine takes over the circulation while the weakened heart is removed, and the new heart is sewn in place. The patient is then treated with drugs that lessen the chances of organ rejection (discussed in Chapter 13).

12a.5 Cardiovascular Disease and Cigarette Smoking

When we think about the hazards of tobacco smoke, lung cancer generally leaps to mind. But the increased risk of cardiovascular (heart and blood vessel) disease is even more significant. Each year, cardiovascular disease kills many more people than does lung cancer, and smokers have a twofold to threefold increase in the risk of heart disease (Figure 12a.8). The American Heart Association estimates that about 25% of all fatal heart attacks are caused by cigarette smoke. This translates to roughly 200,000 heart attacks a year in the United States that could have been prevented by not smoking.

Nicotine and carbon monoxide in cigarette smoke stress the heart and blood vessels in many ways. Nicotine makes the heart beat faster and constricts blood vessels, raising blood

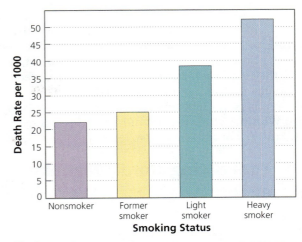

FIGURE 12a.8 *Death rates due to heart disease among nonsmokers and smokers. Notice that the death rate from heart disease increases with the number of cigarettes smoked per day. People who smoke over a pack per day (heavy smokers) have more than twice the risk of death due to heart disease than do people who have never smoked. In any case, a smoker who successfully quits is much less likely to die of heart disease than if he or she continues to smoke.*

pressure. In addition, nicotine makes platelets, the cell fragments responsible for blood clotting, stick together, increasing the risk of formation of abnormal blood clots. The resulting clots may break loose from the site where they form and travel through the bloodstream until they lodge in a small vessel and block the blood flow. These blockages may result in a heart attack or stroke.

The amount of carbon monoxide in cigarette smoke is 1600 ppm (parts per million), which greatly exceeds the 10 ppm considered dangerous in industry. Furthermore, carbon monoxide from smoking a cigarette lingers in the bloodstream for up to 6 hours. You may recall from Chapter 11 that carbon monoxide is a poison that prevents red blood cells from transporting oxygen. In fact, carbon monoxide from smoking a cigarette lowers the oxygen-carrying capacity of the blood by about 12%, reducing oxygen delivery to every part of the body, including the brain and heart. Thus, the heart must work harder to deliver oxygen to the cells. The diminished oxygen supply to the brain can impair judgment, vision, and attentiveness to sounds. For these reasons, smoking can be hazardous for drivers.

A less immediate, but no less important, way that smoking leads to cardiovascular disease is by increasing the risk of developing atherosclerosis. Smoking influences atherosclerosis in two ways. First, smoking decreases the levels of protective cholesterol-transport particles, called *high-density lipoproteins (HDLs)*; HDLs carry cholesterol to the liver, and perhaps even remove cholesterol from cells, so that it can be eliminated from the body. With fewer HDLs, more cholesterol begins to clog the arteries. A second way that smoking promotes cholesterol deposits is by raising blood pressure. The elevated blood pressure causes rapid, turbulent blood flow that damages the walls of the arteries, making them more susceptible to cholesterol deposits. The cholesterol deposits cause inflammation, which leads to atherosclerosis. The narrowing of blood vessels caused by atherosclerosis leads to starvation of tissue downstream and an increase in blood pressure. When atherosclerotic deposits form in the arteries that supply blood to the heart, as they often do, the blood supply to the heart may be reduced or shut down completely, causing heart cells to die.

When a person quits smoking, the cardiovascular system benefits. Within 5 years, the risk of heart disease is reduced by 61% and the risk of stroke by 42%.

12a.6 Preventing Cardiovascular Disease

A number of changes in lifestyle are recommended to treat or prevent cardiovascular disease.

1. **Control weight.** Maintaining normal body weight can help control blood pressure. Many overweight people with high blood pressure benefit from shedding just a few extra pounds. The best way to lose weight is to eat a moderate, balanced diet; reduce fat intake; and increase physical activity.
2. **Exercise regularly.** Aerobic exercise, such as brisk walking, jogging, swimming, or cycling, performed for at least 20 minutes at least three times a week, helps lower blood pressure and keep it low.
3. **Do not smoke.** Cigarette smoke contains nicotine, a drug that increases heart rate and constricts blood vessels; both of those effects increase blood pressure.
4. **Eat a heart-healthy diet.** Limit saturated fats, which come from animal sources, and *trans* fats, which are found in many commercial baked goods and fried food. These fats raise the blood level of low-density lipoprotein, or LDL ("bad"), cholesterol. Choose low-fat protein sources, such as fish and legumes. Eat plenty of fats and vegetables.
5. **Limit dietary salt.** Some people with hypertension can lower their blood pressure by lowering the amount of salt in their diet. Salt can affect fluid retention and, therefore, blood volume.

looking ahead

In this chapter, we learned about cardiovascular disease. In the next chapter, we will consider how the body defends itself against infectious disease.

Body Defense Mechanisms

13

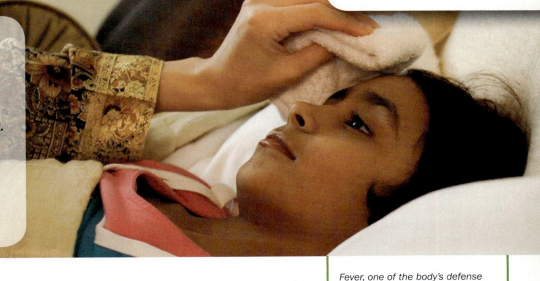

Did you Know?

- The red, itchy bump on your skin following a mosquito bite indicates that your body defenses are working.
- When you are fighting an infection, your lymph nodes can grow to four times their normal size.

Fever, one of the body's defense mechanisms, helps us fight bacterial infections.

13.1 The Body's Defense System

13.2 Three Lines of Defense

13.3 Distinguishing Self from Nonself

13.4 Antibody-Mediated Responses and Cell-Mediated Responses

13.5 Steps of the Adaptive Immune Response

13.6 Active and Passive Immunity

13.7 Monoclonal Antibodies

13.8 Problems of the Immune System

HEALTH ISSUE
Rejection of Organ Transplants

In the previous two chapters we learned about the structure and function of the cardiovascular and lymphatic systems. We were introduced to white blood cells and their roles in body defenses. In this chapter, we study how the body reacts to the invasion of disease-causing organisms and substances that it perceives as threats. We see that there are three lines of defense. We also learn that the body can acquire long-lasting resistance to a microbe by becoming ill or by being immunized. Finally, we consider some potential problems caused by the immune system.

13.1 The Body's Defense System

Your body generally defends you against anything that it does not recognize as being part of or belonging inside you. Common targets of your defense system include organisms that cause disease or infection and body cells that have turned cancerous.

The bacteria, viruses, protozoans, fungi, parasitic worms, and prions (infectious proteins) that cause disease are called **pathogens** (discussed further in Chapter 13a). Note that this term does not apply to most of the microorganisms we encounter. Many bacteria, for example, are actually beneficial. Some bacteria flavor our cheese and produce other foods, such as yogurt, beer, and pickles; other bacteria help rid the planet of corpses through decomposition, thereby recycling nutrients to support new life; and some bacteria help keep other, potentially harmful bacteria in check within our bodies.

Cancerous cells also threaten our well-being. A cancer cell was once a normal body cell, but because of changes in its genes, it can no longer regulate its cell division. If left unchecked, these renegade cells can multiply until they take over the body, upsetting its balance, choking its pathways, and ultimately causing great pain and sometimes death.

13.2 Three Lines of Defense

The body has three strategies for defending against foreign organisms and molecules or cancer cells.

1. **Keep the foreign organisms or molecules out of the body in the first place.** This is accomplished by the first line of defense— *chemical and physical surface barriers.*
2. **Attack any foreign organism or molecule or cancer cell inside the body.** The second line of defense consists of *internal cellular and chemical defenses* that become active if the surface barriers are penetrated.
3. **Destroy a specific type of foreign organism or molecule or cancer cell inside the body.** The third line of defense is the *adaptive immune response,* which destroys *specific* targets (usually disease-causing organisms) and remembers those targets so that a quick response can be mounted if they enter the body again.

Thus, the first and second lines of defense consist of nonspecific mechanisms that are effective against *any* foreign organisms or substances. We are born with these defense mechanisms, so they are described as *innate responses*. We acquire the third line of defense, the *immune response,* which is an adaptive, specific mechanism of defense. We acquire adaptive immunity when we are exposed to chemicals and organisms that are not recognized as belonging in the body. The three lines of defense against pathogens are summarized in Figure 13.1.

First Line of Innate Defense: Physical and Chemical Barriers

The skin and mucous membranes that form the first line of defense are physical barriers that help keep foreign substances from entering the body (Figure 13.2). In addition, they produce several protective chemicals.

Physical barriers Like a suit of armor, unbroken skin helps shield the body from pathogens by providing a barrier to foreign substances. A layer of dead cells forms the tough outer layer of skin. These cells are filled with the fibrous protein keratin, which waterproofs the skin and makes it resistant to the disruptive toxins (poisons) and enzymes of most would-be invaders. Some of the strength of this barrier results from the tight connections binding the cells together. What is more, the dead cells are continuously shed and replaced, at the rate of about a million cells every 40 minutes. As dead cells flake off, they take with them any microbes that have somehow managed to latch on. Another physical barrier, the mucous membranes lining the digestive and respiratory passages, produces sticky mucus that traps many microbes and prevents them from fully entering the body. The cells of the mucous membranes of the upper respiratory airways have cilia—short, hairlike structures that beat constantly. This beating moves the contaminated mucus to the throat. We eliminate the mucus in the throat by swallowing, coughing, or sneezing.

Chemical barriers The skin also provides chemical protection against invaders. Sweat and oil produced by glands in the skin wash away microbes. Moreover, the acidity of the secretions slows bacterial growth, and the oils contain chemicals that kill some bacteria.

Other chemical barriers include the lining of the stomach, which produces hydrochloric acid and protein-digesting enzymes that destroy many pathogens. Beneficial bacteria in a woman's vagina create an acidic environment that discourages the growth of some pathogens. The acidity of urine slows bacterial growth. (Urine also works as a physical barrier, flushing microbes from the lower urinary tract.) Saliva and tears contain an enzyme called *lysozyme* that kills some bacteria by disrupting their cell walls.

stop and think
Harmful bacteria within the digestive system often cause diarrhea. How might this be a protective response of the body?

Second Line of Innate Defense: Defensive Cells and Proteins, Inflammation, and Fever

The second line of defense consists of nonspecific internal defenses against any pathogen that breaks through the physical and chemical barriers and enters the body. This second line of defense includes defensive cells and proteins, inflammation, and fever (Table 13.1).

Defensive cells Specialized "scavenger" cells called **phagocytes** (*phag-*, to eat; *-cyte,* cell) engulf pathogens, damaged tissue, or dead cells by the process of phagocytosis (Chapter 3). This class of white blood cells serves not only as the front-line soldiers in the body's internal defense system but also as janitors that clean up debris. When a phagocyte encounters a foreign particle, cytoplasmic extensions flow from the phagocytic cell, bind to the particle, and pull it inside the cell.

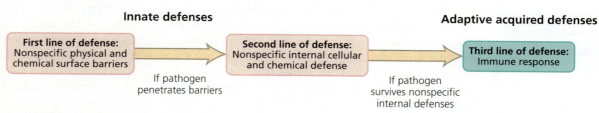

FIGURE 13.1 *The body's three lines of defense against pathogens*

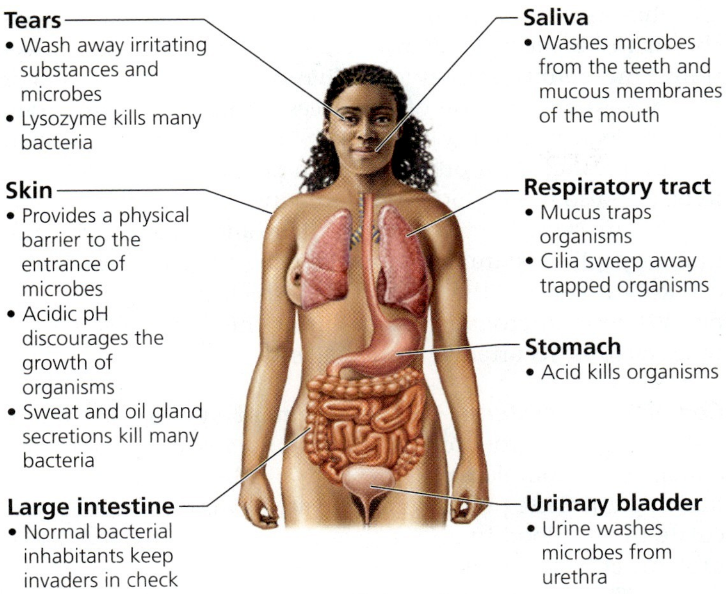

Tears
- Wash away irritating substances and microbes
- Lysozyme kills many bacteria

Skin
- Provides a physical barrier to the entrance of microbes
- Acidic pH discourages the growth of organisms
- Sweat and oil gland secretions kill many bacteria

Large intestine
- Normal bacterial inhabitants keep invaders in check

Saliva
- Washes microbes from the teeth and mucous membranes of the mouth

Respiratory tract
- Mucus traps organisms
- Cilia sweep away trapped organisms

Stomach
- Acid kills organisms

Urinary bladder
- Urine washes microbes from urethra

FIGURE 13.2 The body's first line of defense consists of physical and chemical barriers that serve as innate, nonspecific defenses against any threats to our well-being. Collectively, they prevent many invading organisms and substances from entering the body or confine them to a local region, kill them, remove them, or slow their growth.

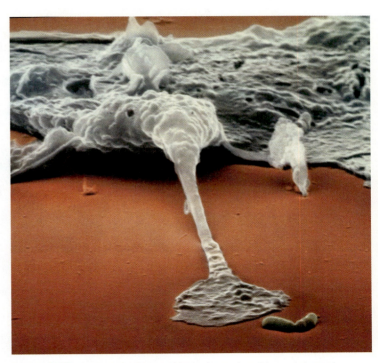

FIGURE 13.3 A macrophage ingesting a bacterium (the rod-shaped structure). The bacterium will be pulled inside the cell within a membrane-bound vesicle and quickly killed.

Once inside the cell, the particle is enclosed within a membrane-bound vesicle and quickly destroyed by digestive enzymes.

The body has several types of phagocytes. One type, *neutrophils*, arrives at the site of attack before the other types of white blood cells and immediately begins to consume the pathogens, especially bacteria, by phagocytosis. Other white blood cells (monocytes) leave the vessels of the circulatory system and enter the tissue fluids, where they develop into large **macrophages** (*macro-*, big; *-phage*, to eat). Macrophages have hearty and less discriminating appetites than neutrophils do, and they attack and consume virtually anything that is not recognized as belonging in the body—including viruses, bacteria, and damaged tissue (Figure 13.3).

A second type of white blood cell, *eosinophils*, attacks pathogens that are too large to be consumed by phagocytosis, such as parasitic worms. Eosinophils get close to the parasite and discharge enzymes that destroy the organism. Macrophages then remove the debris.

Natural killer cells A third type of white blood cell, called **natural killer (NK) cells**, roams the body in search of abnormal cells and quickly orchestrates their death. In a sense, NK cells function as the body's police walking a beat. They are not seeking a specific villain. Instead, they respond to any suspicious character, including a cell whose cell membrane has been altered by the addition of proteins that are unfamiliar to the NK cell. The prime targets of NK cells are cancerous cells and

TABLE 13.1 The Second Line of Defense—Innate, Nonspecific Internal Defenses

Defense	Example	Function
Defensive cells	Phagocytic cells, such as neutrophils and macrophages	Engulf invading organisms
	Eosinophils	Kill parasites
	Natural killer cells	Kill many invading organisms and cancer cells
Defensive proteins	Interferons	Slow the spread of viruses in the body
	Complement system	Stimulates histamine release; promotes phagocytosis; kills bacteria; enhances inflammation
Inflammation	Widening of blood vessels and increased capillary permeability, leading to redness, heat, swelling, and pain	Brings in defensive cells and speeds healing
Fever	Abnormally high body temperature	Slows the growth of bacteria; speeds up body defenses

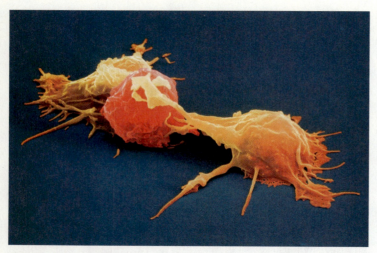

FIGURE 13.4 *Natural killer, or NK, cells (shown in orange) attacking a leukemia cell (shown in red). NK cells patrol the body, bumping and touching other cells as they go. When NK cells contact a cell with an altered cell surface, such as a cancer cell or a virus-infected cell, a series of events is immediately initiated. The NK cell attaches to the target cell and releases proteins that create pores in the target cell, making the membrane leaky and causing the cell to burst.*

cells infected with viruses. Cancerous cells routinely form but are quickly destroyed by NK cells and prevented from spreading (Figure 13.4).

As soon as it touches a cell with an abnormal surface, the NK cell attaches to the abnormal cell and delivers a "kiss of death" in the form of proteins that create many pores in the target cell. The pores make the target cell "leaky," so that it can no longer maintain a constant internal environment and eventually bursts.

Defensive proteins The second line of defense also includes defensive proteins. We will discuss two types of defensive proteins: interferons, which slow viral reproduction, and the complement system, which assists other defensive mechanisms.

Interferons A cell that has been infected with a virus can do little to help itself. But cells infected with a virus can help cells that are not yet infected. Before certain virally infected cells die, they secrete small proteins called **interferons** that act to slow the spread of viruses already in the body. Thus, interferons interfere with viral activity.

Interferons function in a manner somewhat similar to a fire alarm in a public building, which both summons firefighters to extinguish the fire and warns occupants to take the necessary precautions. Interferons also mount a two-pronged attack. First, they help rid the body of virus-infected cells by attracting macrophages and NK cells that destroy the infected cells immediately. Second, interferons warn cells that are not yet infected with the virus to take protective action. When released, interferons diffuse to neighboring cells and stimulate them to produce proteins that prevent viruses from replicating in those cells. Because viruses cause disease by replicating inside body cells, preventing replication curbs the disease. Interferons help protect uninfected cells from *all* strains of viruses, not just the one responsible for the initial infection.

Pharmaceutical preparations of interferons have been shown to be effective against certain cancers and viral infections. Interferons inhibit cell division of cancer cells. For instance, interferons are often successful in combating a rare form of leukemia (hairy cell leukemia) and Kaposi's sarcoma, a form of cancer that often occurs in people with AIDS. Interferon has also been approved for treating the hepatitis C virus, which can cause cirrhosis of the liver and liver cancer; the human papillomavirus (HPV), which causes genital warts and cervical cancer; and the herpes virus, which causes genital herpes. When interferons are taken during the first attack of multiple sclerosis, recurrences are less likely.

Complement system The **complement system**, or simply *complement*, is a group of at least 20 proteins whose activities enhance, or complement, the body's other defense mechanisms. Until these proteins are activated by infection, they circulate in the blood in an inactive state. Once activated, these proteins enhance both nonspecific and specific defense mechanisms. The effects of complement include the following:

- **Destruction of pathogen.** Complement can act *directly* by punching holes in a target cell's membrane (Figure 13.5) so that the cell is no longer able to maintain a constant internal environment. Just as when NK cells secrete proteins that make a target cell's membrane leaky, water enters the cell, causing it to burst.

- **Enhancement of phagocytosis.** Complement enhances phagocytosis in two ways. First, complement proteins attract macrophages and neutrophils to the site of infection to remove the foreign cells. Second, one of the complement proteins binds to the surface of the microbe, making it easier for macrophages and neutrophils to "get a grip" on the intruder and devour it.

- **Stimulation of inflammation.** Complement also causes blood vessels to widen and become more permeable. These changes provide increased blood flow to the area and increased access for white blood cells.

Inflammation When body tissues are injured or damaged, a series of events called the **inflammatory response** or *inflammatory reaction* occurs. This response destroys invaders and helps repair and restore damaged tissue. The four cardinal signs of inflammation that occur at the site of a wound are redness, heat (or warmth), swelling, and pain. These signs announce that certain cells and chemicals have combined efforts to contain infection, clean up the damaged area, and heal the wound. Let's consider the causes of the cardinal signs and how they are related to the benefits of inflammation.

- **Redness.** Redness occurs because blood vessels dilate (widen) in the damaged area, causing blood flow in this area to increase. The dilation is caused by **histamine**, a substance that is also released during allergic reactions (discussed later in the chapter). Histamine is released by small, mobile connective tissue cells called **mast cells** in response to chemicals from damaged cells.

The increased blood flow to the site of injury delivers phagocytes, blood-clotting proteins, and defensive proteins,

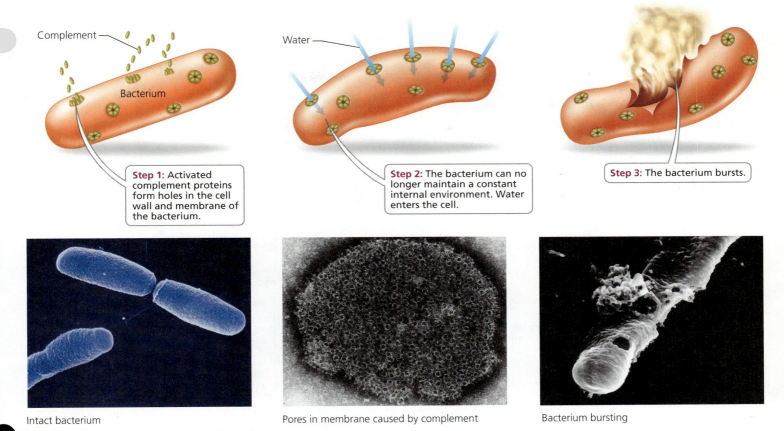

FIGURE 13.5 *Complement has a direct destructive effect on pathogens.*

including complement and antibodies. At the same time, the increased blood flow washes away dead cells and toxins produced by the invading microbes.

- **Heat.** The increased blood flow also elevates the temperature in the area of injury. The elevated temperature increases the metabolic rate of the body cells in the region and speeds healing. Heat also increases the activities of phagocytic cells and other defensive cells.

- **Swelling.** The injured area swells because histamine also makes capillaries more permeable, or leakier, than usual. Fluid seeps into the tissues from the bloodstream, bringing with it many beneficial substances. Blood-clotting factors enter the injured area and begin to wall off the region, thereby helping to protect surrounding areas from injury and preventing excessive loss of blood. The seepage also increases the oxygen and nutrient supply to the cells. If the injured area is a joint, swelling can hamper movement—an effect that might seem to be an inconvenience, but that permits the injured joint to rest and recover.

- **Pain.** Several factors cause pain in an inflamed area. For example, the excessive fluid into the tissue presses on nerves and contributes to the sensation of pain. Some soreness might be caused by bacterial toxins, which can kill body cells. Injured cells also release pain-causing chemicals, such as prostaglandins. Pain usually prompts a person to protect the area to avoid additional injury.

Because of the wider blood vessels and increased capillary permeability that bring about the inflammatory response, phagocytes begin to swarm to the injured site, attracted by chemicals released when tissue is damaged. Within minutes, the neutrophils squeeze through capillary walls into the fluid around cells and begin engulfing pathogens, toxins, and dead body cells. Soon macrophages arrive and continue the body's counterattack for the long term. Macrophages are also important in cleaning debris, such as dead body cells, from the damaged area. As the recovery from infection continues, dead cells (including microbes), body tissue cells, and phagocytes may begin to ooze from the wound as pus (Figure 13.6).

Fever A *fever* is an abnormally high body temperature (Figure 13.7). Fevers are caused by *pyrogens* (*pyro-*, fire; *-gen*, producer), chemicals that raise the "thermostat" in the brain (the hypothalamus) to a higher set point. Bacteria release toxins that sometimes act as pyrogens. It is interesting to note, however, that the body produces its own pyrogens as part of its defensive strategy. Regardless of the source, pyrogens have the same effect on the hypothalamus, raising the set point so that physiological responses, such as shivering, are initiated to raise body temperature (as discussed in Chapter 4). Thus, we have the chills while the fever is rising. When the set point is lowered, the fever breaks, and physiological responses such as perspiring reduce the body temperature until it reaches the new set point.

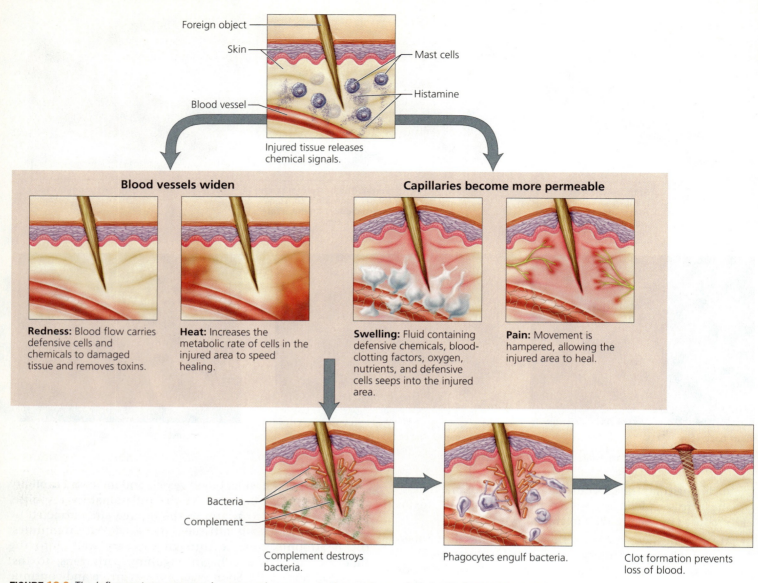

FIGURE 13.6 *The inflammatory response is a general response to tissue injury or invasion by foreign microbes. It serves to defend against pathogens and to clear the injured area of pathogens and dead body cells, allowing repair and healing to occur. The four cardinal signs of inflammation are redness, heat, swelling, and pain.*

A mild or moderate fever helps the body fight bacterial infections by slowing the growth of bacteria and stimulating body defense responses. Bacterial growth is slowed because a mild fever causes the liver and spleen to remove iron from the blood, and many bacteria require iron to reproduce. Fever also increases the metabolic rate of body cells; the higher rate speeds up defensive responses and repair processes. However, a very high fever (over 105°F, or 40.6°C) is dangerous. It can inactivate enzymes needed for biochemical reactions within body cells.

Third Line of Defense: Adaptive Immune Response

When the body's first and second lines of defense fail to stop a pathogen, cancer cell, or foreign molecule from entering the body, the third line of defense, the **adaptive immune response**, takes over. The adaptive immune response provides

FIGURE 13.7 *Although a fever might make us feel uncomfortable, it can help the body fight disease.*

the specific responses and memory needed to target the invader. The organs of the lymphatic system (see Chapter 12) are important components of the immune system because they produce the various cells responsible for immunity. The immune system is not an organ system in an anatomical sense. Instead, the immune system is defined by its *function*: recognizing and destroying specific pathogens or foreign molecules.

There are several important characteristics of an adaptive immune response. First, it is directed at a particular pathogen. For example, the adaptive immune responses of a child infected with measles recognize the measles virus as a foreign substance (not belonging in the body) and then act to immobilize, neutralize, or destroy it. An effective immune response will enable the child to recover from the illness. Second, the adaptive immune response has memory. If the same child is again exposed to the same pathogen years later, the immune system remembers the pathogen and attacks it so quickly and vigorously that the child will not become ill with measles a second time.

13.3 Distinguishing Self from Nonself

To defend against a foreign organism or molecule, the body must be able distinguish it from a body cell and recognize it as foreign. This ability depends on the fact that each cell in your body has special molecules embedded in the plasma membrane that label the cell as *self*. These molecules serve as flags declaring the cell as a "friend." The molecules are called **MHC markers**, named for the *major histocompatibility complex* genes that code for them. The self labels on your cells are different from those of any other person (except an identical twin) as well as from those of other organisms, including pathogens. The immune system uses these labels to distinguish what is part of your body from what is not (Figure 13.8). It doesn't attack cells that are recognized as self.

A nonself substance or organism that triggers an immune response is called an **antigen**. Because an antigen is not recognized as belonging in the body, the immune system directs an attack against it. Typically, antigens are large molecules, such as proteins, polysaccharides, or nucleic acids. Often, antigens are found on the surface of an invader—embedded in the plasma membrane of an unwelcome bacterial cell, for instance, or part of the protein coat of a virus. However, pieces of invaders and chemicals secreted by invaders, such as bacterial toxins, can also serve as antigens. Each antigen is recognized by its shape.

Certain white blood cells, called lymphocytes, are responsible for both the specificity and the memory of the adaptive immune response. There are two principal types of lymphocytes: **B lymphocytes**, or more simply *B cells,* and **T lymphocytes**, or *T cells*. Both types form in the bone marrow, but they mature in different organs of the body. It is thought that B cells mature in the bone marrow. The T cells mature in the thymus gland, which overlies the heart.

As the T lymphocytes mature, they develop the ability to distinguish cells that belong in the body from those that do not. The T cells must be able to recognize the specific MHC self markers of that person and *not* respond vigorously to cells bearing that MHC self marker. T cells that do respond to cells with those self markers are destroyed. Once they are mature, T lymphocytes circulate through the body, bumping into other cells and checking to be sure those cells have the correct self (MHC) marker. Cells with proper MHC markers are passed by.

In addition, both T and B lymphocytes are programmed during development to recognize one particular type of antigen. This recognition is the basis of the specificity of the adaptive immune response. Each lymphocyte develops its own particular receptors—molecules having a unique shape—on its surface. Thousands of *identical* receptor molecules pepper the surface of each lymphocyte, and they are unlike the receptor molecules on other lymphocytes. When an antigen fits into a lymphocyte's receptors, much like a key into a lock, the body's defenses target that particular antigen. Because of the tremendous diversity of receptor molecules, each type occurring on a different lymphocyte, a few of the billions of lymphocytes in your body are able to respond to each of the thousands of different antigens that you will be exposed to in your lifetime. Thus lymphocytes are prepared to respond to each antigen when it is first encountered.

How do we build an effective force of lymphocytes to defend against a particular antigen? When an antigen is detected, B cells and T cells bearing receptors able to respond to that particular invader are stimulated to divide repeatedly, forming two lines of cells. One line of descendant cells is made up of **effector cells**, which carry out the attack on the enemy. Effector cells generally live for only a few days. Thus, after the invader has been eliminated from the body, the number of effector cells declines. The other line of descendant cells is composed of **memory cells**, long-lived cells that "remember" that particular invader and mount a rapid, intense response to it if it should ever appear again. The quick response of memory cells is the mechanism that prevents you from getting ill from the same pathogen twice.

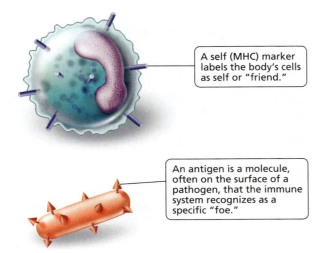

FIGURE 13.8 *All nucleated cells in the body have molecular MHC markers on their surface that label them as self. Foreign substances, including potential disease-causing organisms, have molecules on their surfaces that are not recognized as belonging in the body. Foreign molecules that are capable of triggering an adaptive immune response are called antigens.*

13.4 Antibody-Mediated Responses and Cell-Mediated Responses

An analogy can be made between the body's adaptive immune defenses and a nation's military defense system. The military has scouts who look for invaders. If an invader is found, the scout alerts the commander of the military forces and provides an exact description of the enemy. The scout must also provide the appropriate password so that the commander knows he or she is not a spy planting misinformation. The body also has scouts, called macrophages, that are part of the nonspecific defenses. Macrophages roam the tissues, looking for any invader. The cells that act as the immune system's commander are a subset of T cells called helper T cells. When macrophages properly alert helper T cells, they respond by calling out the body's specific defensive forces, and the adaptive immune responses begin.

A nation's military may have two (or more) branches. For example, it may consist of an army and a navy. Specialized to respond in slightly different ways to enemy invasion, each branch is armed with certain types of weapons. Either branch can be activated to combat a particular threat, say little green people with purple hair. The navy may be called into action if the enemy is encountered at sea, whereas the army will come to the defense if the enemy is on land.

The body, similarly, has two types of specific defenses. These specific defenses recognize the same antigens and destroy the same invaders, but they do so in different ways.

- *Antibody-mediated immune responses* defend primarily against antigens found traveling freely in intercellular and other body fluids—for example, toxins or extracellular pathogens such as bacteria or free viruses. The warriors of this branch of immune defense are the effector B cells (also called *plasma cells*), and their weapons are Y-shaped proteins called *antibodies,* which neutralize and remove potential threats from the body. Antibodies are programmed to recognize and bind to the antigen posing the threat; they help eliminate the antigen from the body. We discuss how this works in greater detail later in the chapter.

- *Cell-mediated immune responses* protect against cellular pathogens or abnormal cells, including body cells that have become infected with viruses or other pathogens and cancer cells. The lymphocytes responsible for cell-mediated immune responses are a type of T cell called *cytotoxic T cell* (discussed at greater length later in the chapter). Once activated, cytotoxic T cells quickly destroy the cellular pathogen, infected body cells, or cancerous cells by causing them to burst.

Now that we have introduced the various defenders, let's see how they work together to produce your body's highly effective adaptive immune response. Table 13.2 summarizes the functions of the cells participating in the adaptive immune response, and Table 13.3 summarizes the steps in the adaptive immune response.

TABLE 13.2 Cells Involved in the Adaptive Immune Response

Cell	Functions
Macrophage, dendritic cell, or B cell	**An antigen-presenting cell** • Engulfs and digests pathogen or invader • Places a piece of digested antigen on its plasma membrane • Presents the antigen to a helper T cell • Activates the helper T cell
T Cells	
Helper T cell	**The "on" switch for both lines of immune response** • After activation by macrophage, the helper T cell divides, forming effector helper T cells and memory helper T cells • Helper T cells activate B cells and T cells
Cytotoxic T cell (effector T cell)	**Responsible for cell-mediated immune responses** • When activated by helper T cells, the cytotoxic T cell divides to form effector cytotoxic T cells and memory cytotoxic T cells • Destroys cellular targets, such as infected body cells, bacteria, and cancer cells
Suppressor T cell	**The "off" switch for both lines of immune responses** • Suppresses the activity of the B cells and T cells after the foreign cell or molecule has been successfully destroyed
B Cells	**Involved in antibody-mediated responses** • When activated by helper T cells, the B cell divides to form plasma cells and memory cells
Plasma Cell	**Effector in antibody-mediated response** • Secretes antibodies specific to extracellular antigens, such as toxins, bacteria, and free viruses
Memory Cells	**Responsible for memory of immune system** • Generated by B cells or any type of T cell during an immune response • Enable quick and efficient response on subsequent exposures of the antigen • May live for years

TABLE 13.3 Steps in the Adaptive Immune Response

Step 1: Threat	Foreign cell or molecule enters the body.
Step 2: Detection	• Macrophage detects foreign cell or molecule and engulfs it.
Step 3: Alert	• Macrophage puts antigen from the pathogen on its surface and finds the helper T cell with correct receptors for that antigen. • Macrophage presents antigen to the helper T cell. • Macrophage alerts the helper T cell that there is an invader that "looks like" the antigen. • Macrophage activates the helper T cell.
Step 4: Alarm	Helper T cell activates both lines of defense to fight that specific antigen.
Step 5: Building specific defenses (clonal selection)	• Antibody-mediated defense—B cells are activated and divide to form plasma cells that secrete antibodies specific to the antigen. • Cell-mediated defense—T cells divide to form cytotoxic T cells that attack cells with the specific antigen.
Step 6: Defense	• Antibody-mediated defense—antibodies specific to the antigen eliminate the antigen. • Cell-mediated defense—cytotoxic T cells cause cells with the antigen to burst.
Step 7: Continued surveillance	Memory cells formed when helper T cells, cytotoxic T cells, and B cells were activated remain to provide swift response if the antigen is detected again.
Step 8: Withdrawal of forces	Once the antigen has been destroyed, suppressor T cells shut down the immune response to that antigen.

13.5 Steps of the Adaptive Immune Response

Although the cell-mediated immune response and the antibody-mediated immune response use different mechanisms to defend against pathogens or foreign molecules (nonself), the general steps in these responses are the same (Figure 13.9).

Threat The adaptive immune response begins when a molecule or organism (an antigen) lacking the self (MHC) marker manages to evade the first two lines of defense and enters the body (Figure 13.10).

Detection Recall that macrophages are phagocytic cells that roam the body, engulfing any foreign material or organisms they may encounter. Within the macrophage, the engulfed material is digested into smaller pieces.

Alert The macrophage then alerts the immune system's commander, a helper T cell, that an antigen is present. The macrophage accomplishes this task by transporting some of the digested pieces to its own surface, where they bind to the MHC self markers on the macrophage membrane. On the one hand, the self marker acts as a secret password that identifies the macrophage as a "friend." On the other hand, the antigen bound to the self marker functions as a kind of wanted poster, telling the lymphocytes that there is an invader and revealing how the invader can be identified. The displayed antigens trigger the adaptive immune response. Thus, the macrophage is an important type of **antigen-presenting cell (APC)**. (B cells and dendritic cells—cells with long extensions found in lymph nodes—are two other kinds of antigen-presenting cells.)

The macrophage presents the antigen to a **helper T cell**, the kind of T cell that serves as the main switch for the entire adaptive immune response. However, the macrophage must alert the *right* kind of helper T cell—a helper T cell bearing

Q Why are helper T cells critical to the adaptive immune response?

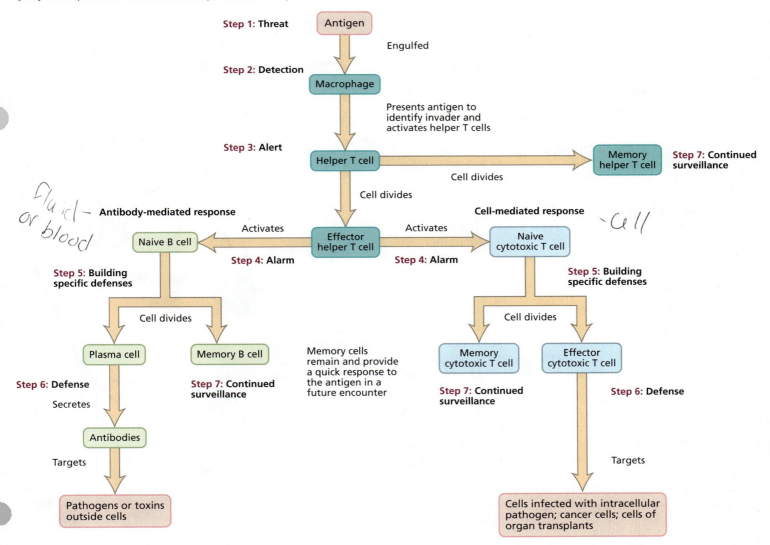

FIGURE 13.9 An overview of the adaptive immune response

A Helper T cells activate both naive cytotoxic T cells and naive B cells. Thus, helper T cells turn on both the cell-mediated and the antibody-mediated adaptive immune responses.

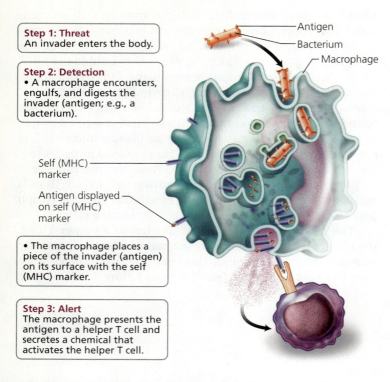

FIGURE 13.10 A macrophage is an important antigen-presenting cell. It presents the antigen, which is attached to a self (MHC) marker, to a helper T cell, and it activates the helper T cell.

receptors that recognize the specific antigen being presented. These specific helper T cells constitute only a tiny fraction of the entire T cell population. Finding the right helper T cell is like looking for a needle in a haystack. The macrophage wanders through the body until it literally bumps into an appropriate helper T cell. The encounter most likely occurs in one of the lymph nodes, because these bean-shaped structures, discussed in Chapter 12, contain huge numbers of lymphocytes of all kinds. When the antigen-presenting macrophage meets the appropriate helper T cell and binds to it, the macrophage secretes a chemical that activates the helper T cell.

Alarm Within hours, an activated helper T cell begins to secrete its own chemical messages. The helper T cell's message calls into active duty the appropriate B cells and T cells—those with receptors that recognize the particular antigen that triggered the response, which is the antigen displayed by the antigen-presenting cell.

Building Specific Defenses When the appropriate "naive"[1] B cells or T cells are activated, they begin to divide repeatedly. The result is a clone (a population of genetically identical cells) that is specialized to protect against the particular target antigen.

The process by which this highly specialized clone is produced, called **clonal selection**, underlies the entire adaptive immune response (Figure 13.11). We have seen that each lymphocyte is equipped to recognize an antigen of a specific

[1]A "naive" cell is one that has been programmed to respond to a particular antigen but has not been previously activated to respond.

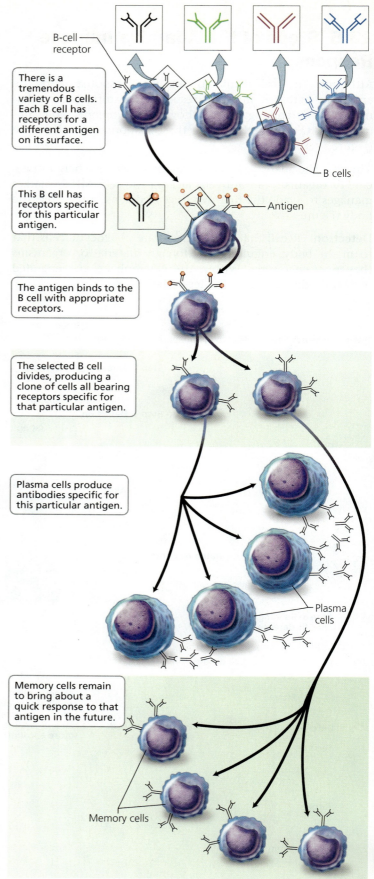

FIGURE 13.11 Clonal selection is the process by which an adaptive immune response to a specific antigen becomes amplified. This figure shows clonal selection of B cells, but a similar process occurs with T cells.

shape. Any antigen that enters the body will be recognized by only a few lymphocytes at most. By binding to the receptors on a lymphocyte's surface, an antigen *selects* a lymphocyte that was preprogrammed during its maturation with receptors able to recognize that particular antigen. That particular lymphocyte is then stimulated to divide and produces a clone of millions of identical cells able to recognize that same antigen.

The following analogy may be helpful for understanding clonal selection. Consider a small bakery with only sample cookies on display. A customer chooses a particular cookie and places an order for many cookies of that type. The cookies are then prepared especially for that person. The sample cookies do not take a lot of space, so a wide selection can be on display for other customers to select. The baker does not waste energy making cookies that have not been specifically requested. Your body prepares samples of many kinds of lymphocytes; a given lymphocyte responds to only one antigen. When an antigen selects the appropriate lymphocyte, the body produces many additional copies of the lymphocyte chosen by that particular antigen.

stop and think

A primary target of HIV, the human immunodeficiency virus that leads to AIDS, is the helper T cell. Why does the virus's preference for the helper T cell impair the immune system more than if another type of lymphocyte were targeted?

We have already mentioned that when building specific defenses, the body produces two types of cells: memory cells and effector cells. Before turning to the role of memory cells, let's look more closely at exactly *how* the effector cells protect us.

Defense—The Antibody-Mediated Response In the **antibody-mediated immune response**, activated B cells divide. The effector cells they produce through clonal selection, which are called **plasma cells**, secrete antibodies into the bloodstream to defend against antigens free in the blood or bound to a cell surface (Figure 13.12). **Antibodies** are Y-shaped proteins that recognize a specific antigen by its shape. Each antibody is specific for one particular antigen. The specificity results from the shape of the proteins that form the tips of the Y (Figure 13.13). Because of their shapes, the antibody and antigen fit together like a lock and a key. Each antibody can bind to two identical antigens, one at the tip of each arm on the Y.

Antibodies can bind only to antigens that are free in body fluids or attached to the surface of a cell. Their main targets are toxins and extracellular microbes, including bacteria, fungi, and protozoans. Antibodies help defend against these pathogens in several ways that can be remembered with the acronym PLAN:

- **Precipitation.** The antigen–antibody binding causes antigens to clump together and precipitate (settle out of

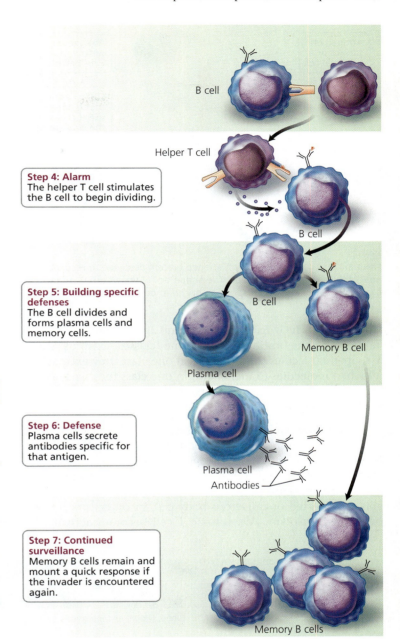

Step 4: Alarm
The helper T cell stimulates the B cell to begin dividing.

Step 5: Building specific defenses
The B cell divides and forms plasma cells and memory cells.

Step 6: Defense
Plasma cells secrete antibodies specific for that antigen.

Step 7: Continued surveillance
Memory B cells remain and mount a quick response if the invader is encountered again.

FIGURE 13.12 Antibody-mediated immune response

solution), enhancing phagocytosis by making the antigens easier for phagocytic cells to capture and engulf.
- **Lysis (bursting).** Certain antibodies activate the complement system, which then pokes holes through the membrane of the target cell and causes it to burst.
- **Attraction of phagocytes.** Antibodies also attract phagocytic cells to the area. Phagocytes then engulf and destroy the foreign material.
- **Neutralization.** Antibodies bind to toxins and viruses, neutralizing them and preventing them from causing harm.

There are five classes of antibodies, each with a special role to play in protecting against invaders. Antibodies are also called **immunoglobulins** (Ig), and each class is designated

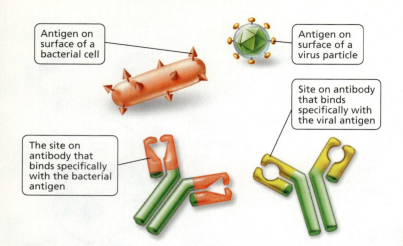

FIGURE 13.13 An antibody is a Y-shaped protein designed to recognize an antigen having a specific shape. The shape of the tips of the Y in the antibody molecule allows a specific antigen to be recognized.

with a letter: IgG, IgA, IgM, IgD, and IgE. As you can see in Table 13.4, the antibodies of some classes exist as single Y-shaped molecules (monomers); in one class they exist as two attached molecules (dimers); and in one class they exist as five attached molecules (pentamers) radiating outward like the spokes of a wheel.

Defense—The Cell-Mediated Response The **cytotoxic T cells** are the effector T cells responsible for the **cell-mediated immune response** that destroys antigen-bearing cells. Each cytotoxic T cell is programmed to recognize a particular antigen bound to MHC markers on the surface of a cellular pathogen, an infected or cancerous body cell, or on cells of a tissue or organ transplant. A cytotoxic T cell becomes activated to destroy a target cell when two events occur simultaneously, as shown in Figure 13.14. First, the cytotoxic T cell must encounter an antigen-presenting cell, such as a macrophage. Second, a helper T cell must release a chemical to activate the cytotoxic T cell. When activated, the cytotoxic T cell divides, producing memory cells and effector cytotoxic T cells.

An effector cytotoxic T cell releases chemicals called **perforins**, which cause holes to form in the target cell membrane. The holes are large enough to allow some of the cell's contents to leave the cell so that the cell disintegrates. The cytotoxic T cell then detaches from the target cell and seeks another cell having the same type of antigen.

stop and think

Rejection of an organ transplant occurs when the recipient's immune system attacks and destroys the cells of the transplanted organ. Why would this attack occur? Which branch of the immune system would be most involved?

Continued Surveillance The first time an antigen enters the body, only a few lymphocytes can recognize it. Those lymphocytes must be located and stimulated to divide in order to produce an army of lymphocytes ready to eliminate that particular antigen. As a result, the *primary response*, the one that occurs during the body's first encounter with a particular antigen, is relatively slow. A lapse of several days occurs before the antibody concentration begins to rise, and the concentration does not peak until 1 to 2 weeks after the initial exposure to the antigen (Figure 13.15).

Following subsequent exposure to the antigen, the *secondary response* is strong and swift. Recall that when naive B cells

TABLE 13.4	Classes of Antibodies			
Class	Structure	Location	Characteristics	Protective Functions
IgG	Monomer	Blood, lymph, and intestines	Most abundant of all antibodies in body; involved in primary and secondary immune responses; can pass through placenta from mother to fetus and provides passive immune protection to fetus and newborn	Enhances phagocytosis; neutralizes toxins; triggers complement system
IgA	Dimer or monomer	Present in tears, saliva, and mucus as well in secretions of gastrointestinal system and excretory systems; present in breast milk	Levels decrease during stress, raising susceptibility to infection	Prevents pathogens from attaching to epithelial cells of surface lining
IgM	Pentamer	Attached to a B cell, where it acts as a receptor for antigens; free in blood and lymph	First Ig class released by plasma cell during primary response	Powerful agglutinating agent (10 antigen-binding sites); activates complement
IgD	Monomer	Surface of many B cells; blood and lymph	Life span of about 3 days	Thought to be involved in recognition of antigen and in activating B cells
IgE	Monomer	Secreted by plasma cells in skin, mucous membranes of gastrointestinal and respiratory systems	Binds to surface of mast cells and basophils	Involved in allergic reactions by triggering release of histamine and other chemicals from mast cells or basophils

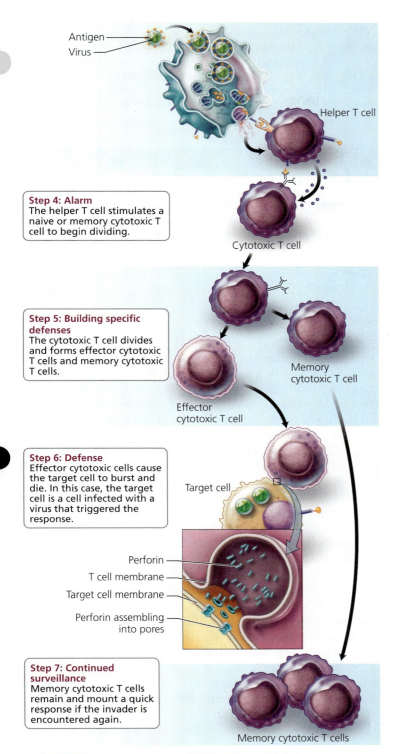

FIGURE 13.14 Cell-mediated immune response

Step 4: Alarm
The helper T cell stimulates a naive or memory cytotoxic T cell to begin dividing.

Step 5: Building specific defenses
The cytotoxic T cell divides and forms effector cytotoxic T cells and memory cytotoxic T cells.

Step 6: Defense
Effector cytotoxic cells cause the target cell to burst and die. In this case, the target cell is a cell infected with a virus that triggered the response.

Step 7: Continued surveillance
Memory cytotoxic T cells remain and mount a quick response if the invader is encountered again.

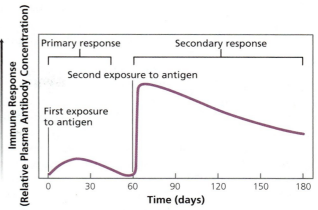

FIGURE 13.15 *The primary and secondary immune responses. In the primary response, which occurs after the first exposure to an antigen, there is a delay of several days before the concentration of circulating antibodies begins to increase. It takes 1 to 2 weeks for the antibody concentration to peak because the few lymphocytes programmed to recognize that particular antigen must be located and activated. (The T cells show a similar pattern of response.) The secondary response following a subsequent exposure to an antigen is swifter and stronger than the primary response. The difference is due to the long-lived memory cells produced during the primary response; these are a larger pool of lymphocytes programmed to respond to that particular antigen.*

cells and additional memory cells specific for that antigen. Therefore, the number of effector cells rises quickly during the secondary response and within 2 or 3 days reaches a higher peak than it did during the primary response.

Withdrawal of Forces As the immune system begins to conquer the invading organism and the level of antigens declines, another type of T cell, the **suppressor T cell**, releases chemicals that dampen the activity of both B cells and T cells. Suppressor T cells turn off the adaptive immune response when the antigen no longer poses a threat. This may be a mechanism that prevents the immune system from overreacting and harming healthy body cells.

13.6 Active and Passive Immunity

There are two types of immunity: In **active immunity**, the body actively defends itself by producing memory B cells and T cells following exposure to an antigen. Active immunity happens naturally whenever a person gets an infection. Fortunately, active immunity can also develop through *vaccination* (also known as *immunization*), a procedure that introduces a harmless form of an antigen into the body to stimulate adaptive immune responses against that antigen. Today, some vaccines, such as the vaccine for hepatitis B, are prepared using yeast that are genetically modified to produce a protein from the pathogen. Because only the protein (antigen) is injected, rather than the actual virus, the vaccine can't cause disease. In some kinds of vaccination—those for whooping cough and typhoid fever, for instance—the microbe is killed before the vaccine is prepared. Other vaccines must be made from live organisms in order to be effective. In these cases, the microbes are first weakened so that they can no longer cause disease. The microbes are weakened by transferring them repeatedly

and T cells are stimulated to divide, they produce not only effector cells that actively defend against the invader, but also memory cells. These memory B cells, memory cytotoxic T cells, and memory helper T cells live for years or even decades. As a result, the number of lymphocytes programmed to respond to that particular antigen is much greater than it was before the first exposure. When the antigen is encountered again, each of those memory cells divides and produces new effector in tissue culture, which allows unpredictable mutations to occur. Still other vaccines, including the one against smallpox, are prepared from microbes that cause related but milder diseases.

Because it leads to the production of memory cells, active immunity—occurring naturally or via vaccination—is relatively long-lived. The first dose of a vaccine causes the primary immune response, and antibodies and some memory cells are generated. In certain cases, especially when inactivated antigens are used in the vaccine, the immune system may "forget" its encounter with the antigen after a time. A booster is administered periodically to make sure the immune system does not forget. The booster results in a secondary immune response and enough memory cells to provide for a quick response should a potent form of that pathogen ever be encountered.

Vaccinations have saved millions of lives. In fact, they have been so effective in preventing diseases such as whooping cough and tetanus that many people mistakenly think those diseases have been eliminated. However, most of the diseases that vaccines prevent still exist, so vaccinations are still important. For example, measles was declared to be eliminated in 2000, but several outbreaks in the United States in 2011 caused more cases of measles than occurred during the previous fifteen years. The frequency of whooping cough also increased dramatically in 2012. Children should be immunized (given vaccines) on a recommended schedule. (The safety of vaccinations is addressed in Chapter 18a.)

Passive immunity is protection that results when a person receives antibodies produced by another person or animal. For instance, some antibodies produced by a pregnant woman can cross the placenta and give the growing fetus some immunity. These maternal antibodies remain in the infant's body for as long 3 months after birth, at which point the infant is old enough to produce its own antibodies. Antibodies in breast milk also provide passive immunity to nursing infants, especially against pathogens that might enter through the intestinal lining. The mother's antibodies are a temporary yet critical blanket of protection, because most of the pathogens that would otherwise threaten the health of a newborn have already been encountered by the mother's immune system.

People can acquire passive immunity medically by being injected with antibodies produced in another person or animal. In this case, passive immunity is a good news–bad news situation. The good news is that the effects are immediate. Gamma globulin, for example, is a preparation of antibodies used to protect people who have been exposed to diseases such as hepatitis B or who are already infected with the microbes that cause tetanus, measles, or diphtheria. Gamma globulin is often given to travelers before they visit a country where viral hepatitis is common. The bad news is that the protection is short-lived. The borrowed antibodies circulate for 3 to 5 weeks before being destroyed in the recipient's body. Because the recipient's immune system was not stimulated to produce memory cells, protection disappears with the antibodies.

stop and think
The viruses that cause influenza (the flu) mutate rapidly, so the antigens in the protein coat continually change. Why does this characteristic make it difficult to develop a flu vaccine that will be effective for several consecutive years?

what would you do?
There is now a vaccine against the human papillomavirus, a sexually transmitted virus that is also the most important cause of cervical cancer. Health officials recommend the vaccine for girls and boys 11 or 12 years of age, but it can be given to girls as young as 9 years and to women as old as 26 years. Some people fear that use of this vaccine will encourage vaccinated teenagers to be sexually active. If you were (or are) a parent, would you have your daughter vaccinated? Do you think that the government should mandate that young girls be vaccinated?

13.7 Monoclonal Antibodies

Suppose you wanted to determine whether a particular antigen was present in a solution, tissue, or even somewhere in the body. An antibody specific for that antigen would be just the tool you would need. Because of its specificity, any such antibody would go directly to the target antigen. If a label (such as a radioactive tag or a molecule that fluoresces) were attached to the antibody, the antibody could reveal the location of the antigen. You can see that for a test of this kind, it is desirable to have a supply of identical antibodies that react with a specific antigen. Groups of identical antibodies that bind to one specific antigen are called **monoclonal antibodies**.

Monoclonal antibodies have many uses. Home pregnancy tests contain monoclonal antibodies produced to react with a hormone (human chorionic gonadotropin; see Chapter 18) secreted by membranes associated with the developing embryo. Monoclonal antibodies have also proved useful in screening for certain diseases, including Legionnaire's disease, hepatitis, certain sexually transmitted diseases, and certain cancers, including those of the lung and prostate. Some monoclonal antibodies are also used in cancer treatment. The radioactive material or chemical treatment to combat the cancer is attached to a monoclonal antibody that targets the tumor cells but has little effect on healthy cells.

13.8 Problems of the Immune System

The immune system protects us against myriad threats from agents not recognized as belonging in the body. However, sometimes the defenses are misguided. In autoimmune disease, the body's own cells are attacked. Allergies result when the immune system protects us against substances that are not harmful. Tissue rejection following organ transplant is also

HEALTH ISSUE

Rejection of Organ Transplants

Each year, tens of thousands of people receive a gift of life in the form of a transplanted kidney, heart, lungs, liver, or pancreas. Although these transplants seem almost commonplace today, they have been performed for only about 30 years. Before organ transplants could be successful, physicians had to learn how to prevent the effector T cells of the immune system from attacking and killing the transplanted tissue because it lacked appropriate self markers. When transplanted tissue is killed by the host's immune system, we say that the transplant has been rejected.

The success of a transplant depends on the similarity between the host tissues and the transplanted tissues. The most successful transplants are those in which tissue is taken from one part of a person's body and transplanted to another part. In cases of severe burns, for example, healthy skin from elsewhere on the body can replace badly burned areas of skin.

Another way to increase the likelihood that a transplant will be accepted is to use cells from the person's body to grow the transplant in a laboratory. Today, it is possible to grow some organs—urinary bladders, for instance—in the laboratory. Cells are taken from the defective organ and grown in tissue culture. When there are enough cells, they are placed on a three-dimensional model of the organ. Then the cell-covered mold is incubated until the new organ is formed. We discuss laboratory-grown organs in more detail in Chapter 19a.

Because identical twins are nearly genetically identical, their cells have the same self markers, and organs can be transplanted from one twin to another with little fear of tissue rejection. But few of us have an identical twin. The next best source for tissue for a transplant, and the most common, is a person whose cell surface markers closely match those of the host. Usually the transplanted tissue comes from a person who has recently died. Typically, the donor is someone who is brain-dead, but whose heart is kept beating by life-support equipment. Some organs—primarily kidneys—can be harvested from someone who has died and whose heart has stopped beating. In some cases, living people can donate organs; one of two healthy kidneys can be donated to a needy recipient, as can sections of liver.

Unfortunately, the waiting list of patients in need of an organ from a suitable donor has outpaced the supply. Some researchers believe that in the future, organs from nonhuman animals may fill the gap between the supply of organs and the demand. So far, however, attempts to transplant animal organs into people have failed. The biggest obstacle is hyperacute rejection. Within minutes to hours after transplant, the animal organ dies because its blood supply is choked off by the human immune system. Today, researchers are hopeful that genetic modification of animals for organ transplants will reduce the chances of rejection.

Other dangers may remain, even if the rejection problem is solved. Animals carry infectious agents that are harmless to their hosts but that might "jump species" and then gain the ability to spread from the transplant recipient to another person. If that were possible, we would have to ask whether it is ethical to expose a third party to risk.

Questions to Consider

- If you were a tissue match for someone who needed a kidney or a bone marrow transplant, how would you decide whether to be a donor?
- Do you think that family members who are a tissue match should be obligated to donate a kidney or bone marrow?
- Do you think that people should be able to buy a kidney or bone marrow from a suitable donor?

---

caused by the immune system (see the Health Issue essay, *Rejection of Organ Transplants*).

Autoimmune Disorders

Autoimmune disorders occur when the immune system fails to distinguish between self and nonself and attacks the tissues or organs of the body. If the immune system can be called the body's military defense, then autoimmune disease is the equivalent of friendly fire.

As we have seen, during their development, lymphocytes are programmed to attack a specific foreign antigen while still tolerating self antigens. The body usually destroys lymphocytes that do not learn to make this distinction. Unfortunately, some lymphocytes that are primed to attack self antigens escape destruction. These cells are like time bombs ready to attack the body's own cells at the first provocation. For example, if these renegade lymphocytes are activated by a virus or bacterium, they may direct their attack against healthy body cells as well as the invading organism.

Autoimmune disorders are often classified as organ-specific or non-organ-specific. As the name implies, organ-specific autoimmune disorders are directed against a single organ. Organ-specific autoimmune disorders are usually caused by T cells that have gone awry. The thyroid gland, for example, is attacked in *Hashimoto's thyroiditis*. Type 1 diabetes is an autoimmune disorder in which the pancreatic cells that produce insulin are attacked (see Chapter 10a). In contrast, non-organ-specific autoimmune disorders are generally caused by antibodies produced by B cells gone awry and tend to have effects throughout the body. In *systemic lupus erythematosus*, for instance, connective tissue is attacked. Because connective tissue can be found throughout the body, almost any organ can be affected. Lupus can cause skin lesions or rashes, most notably a butterfly-shaped rash centered on the nose and spreading to both cheeks. It may affect the heart (pericarditis), joints (arthritis), kidneys (nephritis), or nervous system (seizures).

A number of autoimmune disorders occur because portions of disease-causing organisms resemble proteins found on normal body cells. If the immune system mistakes the body's antigens for the foreign antigens, it may attack them. For instance, the body's attack on certain streptococcal bacteria that cause a sore throat may result in the production of antibodies that target not only the streptococcal bacteria but also similar molecules that are found in the valves of the heart and joints. The result is an autoimmune disorder known as *rheumatic fever*.

Treatment of autoimmune disorders is usually two-pronged. First, any deficiencies caused by the disorder are

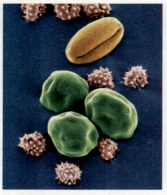
Pollen grains

Dust mite

FIGURE 13.16 *Common causes of allergies are pollen grains and the feces of dust mites, such as the mite shown here.*

corrected. Second, drugs are administered to depress the immune system.

Allergies

An **allergy** is an overreaction by the immune system to an antigen, in this case called an *allergen*. The immune response in an allergy is considered an overreaction because the allergen itself usually is not harmful to the body. The most common allergy is hay fever—which, by the way, is not caused by hay and does not cause a fever. *Hay fever* is more correctly known as *allergic rhinitis* (*rhin-*, nose; *-itis*, inflammation of). The symptoms of hay fever—sneezing and nasal congestion—occur when an allergen is inhaled, triggering an immune response in the respiratory system. Mucous membranes of the eyes may also respond, causing red, watery eyes. Common causes of hay fever include pollen, mold spores, animal dander, and the feces of dust mites—microscopic creatures that are found throughout your home (Figure 13.16). The same allergens, however, can trigger asthma. During an asthma attack, the small airways in the lung (bronchioles) constrict, making breathing difficult. In food allergies, the immune response occurs in the digestive system and may cause nausea, vomiting, abdominal cramps, and diarrhea. Food allergies can also cause hives, a skin condition in which patches of skin temporarily become red and swollen.

An immediate allergic response begins when a person is exposed to an allergen and a primary immune response is launched (Figure 13.17). Soon, plasma cells churn out the antibody IgE, which binds to either basophils or mast cells. In subsequent exposures to that allergen, the allergen binds to IgE antibodies on the surface of basophils or mast cells and causes granules within the cells to release their contents: histamine.

Histamine then causes the swelling, redness, and other symptoms of an allergic response. The blood vessels widen, slowing blood flow and causing redness. At the same time, the blood vessels become leaky, allowing fluid to flow from the vessels into spaces between tissue cells, swelling the tissues. Histamine also causes the release of large amounts of mucus, so the nose begins to run. In addition, histamine can cause smooth muscles of internal organs to contract. Thus, if the allergen

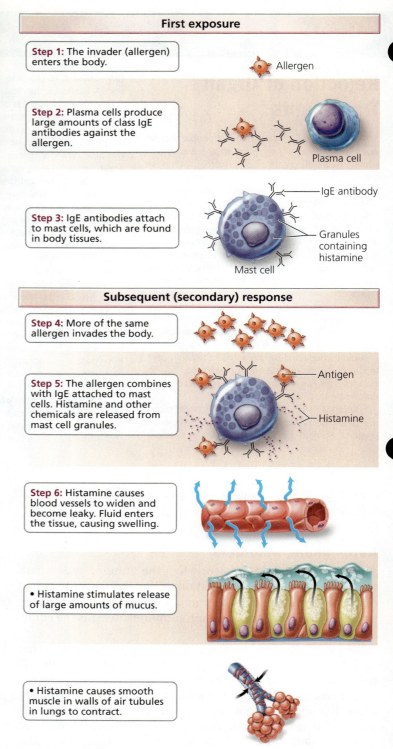

FIGURE 13.17 *Steps in an allergic reaction*

is in the respiratory system, histamine can trigger an asthma attack by causing the air tubules to contract. If the allergen moves from the area where it entered the body, these effects can be widespread. The result can be anaphylactic shock.

Anaphylactic shock is an extreme allergic reaction that occurs within minutes after exposure to the substance a person is allergic to. It can cause pooling of blood in capillaries, which leads to dizziness, nausea, and sometimes unconsciousness as well as extreme difficulty in breathing. Anaphylactic shock can be fatal, but most people survive. Allergies that are common triggers of anaphylactic shock include certain foods; medicines, including antibiotics such as penicillin; and insect stings, especially stings from bees, wasps, yellow jackets, and hornets.

People with allergies often know which substances cause their problems. When the culprits are not known, doctors can identify them using a crude but effective technique in which small amounts of suspected allergens are injected into the skin. If the person is allergic to one of the suspected allergens, a red welt will form at the site of injection.

If you know you have an allergy, the simplest way to avoid the miseries of an allergic reaction is to avoid exposure to the substances that cause problems. During pollen season, spend as much time as possible indoors, using an air conditioner to filter pollen out of the incoming air. Unfortunately, spores from molds growing in air conditioners and humidifiers are also common triggers of allergies. Some common allergy-causing foods—for instance, strawberries or shellfish—may be easy to avoid. Others, such as peanut oil, can show up in some unlikely dishes, including stew, chili, baked goods, or meat patties.

Certain drugs may reduce allergy symptoms. As their name implies, antihistamines block the effects of histamine. Antihistamines are most effective if they are taken before the allergic reaction begins. Unfortunately, allergies tend to become less susceptible to antihistamines over time, and most antihistamines cause drowsiness, which can impair performance on the job or in school and can make driving a car extremely hazardous.

Some allergies can be treated by gradually desensitizing the person to the offending allergens. Allergy shots containing gradually increasing amounts of a known allergen are injected into the person's bloodstream. During this treatment, the allergen causes the production of another class of antibodies—IgG. Afterward, when the allergen enters the body, IgG antibodies bind to it and prevent it from binding to IgE antibodies on mast cells and triggering an allergic reaction.

looking ahead

In this chapter, we learned about the mechanisms that protect us against harmful organisms and substances. In the next chapter, we consider some infectious organisms that cause disease.

HIGHLIGHTING THE CONCEPTS

13.1 The Body's Defense System (p. 239)
- The targets of the body's defense system include anything that is not recognized as belonging in the body, such as disease-causing organisms and cancerous cells. These foreign agents that cause illness are called pathogens.

13.2 Three Lines of Defense (pp. 240–245)
- The first line of defense is innate—nonspecific physical barriers, such as skin and mucous membranes, and chemical barriers, such as sweat, oil, tears, and saliva, all of which prevent entry of pathogens.
- The second line of innate defense includes defensive cells and proteins, inflammation, and fever. Defensive cells include phagocytes, eosinophils, and natural killer cells. Two types of defensive proteins are antiviral interferons and complement proteins, which cause cells to burst.
- The inflammatory response occurs as a result of tissue injury or invasion by foreign microbes. It begins when mast cells in the injured area release histamine, which increases blood flow by dilating blood vessels to the region and by increasing the permeability of capillaries there. Increased blood flow causes redness and warmth in the region. Fluid leaking from the capillaries causes swelling.
- Fever, an abnormally high body temperature, helps the body fight invading microbes by enhancing several body defense mechanisms and slowing the growth of many pathogens.
- The third line of defense, the adaptive immune response, targets specific pathogens. The immune system has memory.

13.3 Distinguishing Self from Nonself (p. 245)
- All body cells are labeled with proteins called major histocompatibility complex (MHC) proteins, which serve as self markers. Cells that lack self markers (MHC) are considered nonself and are attacked. A nonself substance or organism triggers an immune response and is called an antigen.
- Lymphocytes are white blood cells that are responsible for immune responses. Both B lymphocytes (B cells) and T lymphocytes (T cells) develop in the bone marrow. The B cells are thought to mature in the bone marrow, but the T cells mature in the thymus gland. During maturation, B cells and T cells develop receptors on their surfaces that allow each of those cells to recognize an antigen of a different shape.
- When an antigen is detected, B cells and T cells with receptors that respond to that antigen divide repeatedly, forming effector cells that destroy the antigen and forming memory cells that remain in the body over years or even decades to provide a quick response on subsequent exposure to that antigen.

13.4 Antibody-Mediated Responses and Cell-Mediated Responses (p. 246)
- The antibody-mediated immune response and the cell-mediated immune response simultaneously defend against the same antigen.

13.5 Steps of the Adaptive Immune Response (pp. 247–251)
- Macrophages are an example of antigen-presenting cells. They are phagocytic cells that engulf any foreign material or organism they encounter. After engulfing the material, the macrophage places a part of the destroyed substance on its own surface to serve as an antigen that alerts lymphocytes to the presence of an invader and reveals what the invader looks like. Macrophages also have molecular (MHC) markers on their membranes that identify them as belonging in the body, that is, as self.

- A macrophage then presents the antigen to a helper T cell, which serves as the main switch to the entire immune response. When this encounter occurs, the macrophage secretes a chemical that activates the helper T cell. The helper T cell, in turn, secretes a chemical that activates the appropriate B cells and T cells (those specific for the antigen that the macrophage engulfed).
- B cells are responsible for antibody-mediated immune responses, which defend against antigens that are free in body fluids, including bacteria, free virus particles, and toxins. When called into action by a helper T cell, a B cell divides repeatedly, forming two lines of descendant cells: effector cells that transform into plasma cells and memory B cells. Plasma cells secrete Y-shaped proteins called antibodies into the bloodstream. Antibodies bind to the particular antigen and inactivate it or help remove it from the body.
- Cytotoxic T cells are responsible for cell-mediated immune responses, which are effective against cellular threats, including infected body cells and cancer cells. When a T cell is activated, it divides, forming two lines of descendant cells: effector cells, called cytotoxic T cells, and memory T cells. Cytotoxic T cells secrete perforins that poke holes in the foreign or infected cell, causing it to burst and die.
- After the first encounter with a particular antigen, the primary response is initiated, which may take several weeks to become effective against the antigen. However, because of memory cells, a subsequent exposure to the same antigen triggers a quicker response, called a secondary response.
- Suppressor T cells dampen the activity of B cells and T cells when antigen levels begin to fall.

13.6 Active and Passive Immunity (pp. 251–252)
- In active immunity, the body actively participates in forming memory cells to defend against a particular antigen. Active immunity may occur when an antigen infects the body, or it may occur through vaccination, a procedure that introduces a harmless form of an antigen into the body.
- Passive immunity results when a person receives antibodies that were produced by another person or animal. Passive immunity is short-lived.

13.7 Monoclonal Antibodies (p. 252)
- Monoclonal antibodies are identical antibodies that bind to a specific antigen. They are useful in research and in the diagnosis and treatment of diseases.

13.8 Problems of the Immune System (pp. 252–255)
- Autoimmune disorders occur when the immune system mistakenly attacks the body's own cells.
- An allergy is a strong immune response against an antigen (called an allergen). An allergy occurs when the allergen binds to IgE antibodies on the surface of mast cells or basophils, causing them to release histamine. Histamine, in turn, causes the redness, swelling, itching, and other symptoms of an allergic response.

RECOGNIZING KEY TERMS

pathogen *p. 239*
phagocyte *p. 240*
macrophage *p. 241*
natural killer (NK) cells *p. 241*
interferon *p. 242*
complement system *p. 242*
inflammatory response
   p. 242
histamine *p. 242*
mast cell *p. 242*

adaptive immune response
   p. 244
MHC marker *p. 245*
antigen *p. 245*
B lymphocyte *p. 245*
T lymphocyte *p. 245*
effector cell *p. 245*
memory cell *p. 245*
antigen-presenting cell (APC)
   p. 247

helper T cell *p. 247*
clonal selection *p. 248*
antibody-mediated immune
   response *p. 249*
plasma cell *p. 249*
antibody *p. 249*
immunoglobulin *p. 249*
cytotoxic T cell *p. 250*
cell-mediated immune
   response *p. 250*

perforins *p. 250*
suppressor T cells *p. 251*
active immunity *p. 251*
passive immunity *p. 252*
monoclonal antibody *p. 252*
autoimmune disorders
   p. 253
allergy *p. 254*

REVIEWING THE CONCEPTS

1. Explain the difference between innate nonspecific and adaptive specific defense mechanisms. *p. 240*
2. List seven types of nonspecific defense mechanisms. Explain how each type helps protect us against disease. *pp. 240–244*
3. How does a natural killer cell kill its target cell? *pp. 241–242*
4. What are interferons? What type of cell produces them? How do they help protect the body? *p. 242*
5. What is the complement system? Explain how it acts directly and indirectly to protect the body against disease. *p. 242*
6. Signs of inflammation include redness, warmth, swelling, and pain. What causes each of these symptoms? How does inflammation help defend against infection? *pp. 242–243*
7. What does an antigen-presenting cell do? How do other cells recognize the antigen-presenting cell as a "friend"? *p. 245*
8. What cells are responsible for antibody-mediated immune responses? What are the targets of antibody-mediated immune responses? *p. 246*
9. Describe an antibody. How do antibodies inactivate or eliminate antigens from the body? *p. 249*
10. What is responsible for cell-mediated immune responses? What are the targets of cell-mediated immune responses? *pp. 246, 250*
11. How does a natural killer cell differ from a cytotoxic T cell? *pp. 241–242, 250–251*

12. Why does a secondary response occur more quickly than the primary response? *pp. 250–251*
13. Differentiate active immunity from passive immunity. *pp. 251–252*
14. What are monoclonal antibodies? What are some medical uses for them? *p. 252*
15. What is an autoimmune disorder? *pp. 253–254*
16. What is an allergy? What causes the symptoms? *pp. 254–255*
17. Indicate the *correct* statement:
    a. An antibody is specific to one particular antigen.
    b. Antibodies are held within the cell that produces them.
    c. Antibodies are produced by macrophages.
    d. Antibodies can be effective against viruses that are inside the host cell.
18. What is an antigen?
    a. a cell that produces antibodies
    b. a receptor on the surface of a lymphocyte that recognizes invaders
    c. a memory cell that causes a quick response to an invader when it is encountered a second time
    d. a large molecule on the surface of an invader that triggers an immune response
19. Which of the following pairings of cell type and function is *incorrect*?
    a. helper T cell—serves as "main switch" that activates both the cell-mediated immune responses and the antibody-mediated immune responses
    b. cytotoxic T cell—presents antigen to the helper T cell
    c. macrophage—roams the body looking for invaders, which are engulfed and digested when they are found
    d. suppressor T cell—shuts off the immune response when the invader has been removed
20. When the doctors say they are looking for a suitable donor for a kidney transplant, they are looking for someone
    a. whose tissues have self markers similar to those of the recipient.
    b. who lacks antibodies to the recipient's tissues.
    c. who has suppressor T cells that will suppress the immune response against the donor kidney.
    d. who lacks macrophages.
21. The piece of the antigen displayed on the surface of a macrophage
    a. stimulates the suppressor T cells to begin dividing.
    b. attracts other invaders to the cell, causing them to accumulate and making it easier to kill the invaders.
    c. informs the other cells in the immune system of the exact nature of the antigen they should be looking for (what the antigen "looks like").
    d. has no function in the immune response.
22. A cytotoxic T cell could attack all of the following *except*
    a. transplants of foreign tissue.
    b. cells infected with viruses.
    c. cancerous cells.
    d. viruses that are free in the bloodstream.
23. Which of the following is *not* a function of the inflammatory response?
    a. preventing the injurious agent from spreading to nearby tissue
    b. replacing injured tissues with connective tissue
    c. disposing of cellular debris and pathogens
    d. setting the stage for repair processes
24. In clonal selection of B cells, which substance is responsible for determining which cells will eventually become cloned?
    a. antigen
    b. interferon
    c. antibody
    d. complement
25. Innate immune system defenses include which of the following?
    a. B cells
    b. T cells
    c. plasma cells
    d. phagocytosis
26. Fever
    a. is a higher-than-normal body temperature that is always dangerous.
    b. decreases the metabolic rate of the body to conserve energy.
    c. results from the actions of chemicals that reset the body's thermostat to a higher setting.
    d. causes the liver to release large amounts of iron, which seems to inhibit bacterial replication.
27. A cell that kills any unrecognized cell in the body and is part of the nonspecific body defenses is a(n) _____.
28. _____ is a chemical released by mast cells and basophils that produces most of the symptoms of an allergy.
29. Antibodies are produced by _____.
30. _____ are important antigen-presenting cells.

APPLYING THE CONCEPTS

1. After being exposed to the hepatitis B virus, Barbara goes to the doctor and asks to be vaccinated against it. Instead, the doctor gives her an injection of gamma globulin (a preparation of antibodies). Why wasn't she given the vaccine?
2. More than 100 viruses can cause the common cold. How does this fact explain why you can catch a cold from Raymond immediately after recovering from a cold you caught from Jessica?
3. HIV is a virus that kills helper T cells. This virus is not the direct cause of death in people who are infected with it. Instead, people die of diseases caused by organisms that are common in the environment. Explain why HIV-infected persons are susceptible to these diseases.
4. Ira finds a deer tick attached to the back of his leg. He knows that deer ticks can transmit the bacterium that causes Lyme

disease and that untreated Lyme disease can cause arthritis and fatigue. He immediately goes to the doctor to get tested, which would involve drawing blood to look for antibodies to the bacterium. The doctor refuses to test Ira for Lyme disease. Why?

5. Rashon has leukemia, a cancer in which the number of white blood cells increases dramatically. The doctors decide that a bone marrow transfer might help by replacing defective bone stem cells with healthy ones. His girlfriend offered to be a donor, but the doctors chose his brother instead. Why? Why was Rashon given drugs to suppress his immune system after the transplant?

BECOMING INFORMATION LITERATE

1. Vaccination has reduced or eliminated many diseases that once were killers, including diphtheria, typhoid, polio, and smallpox. The Centers for Disease Control and Prevention (CDC) has a recommended schedule for childhood immunization. Yet some parents don't have their children vaccinated.

    If you were or are a parent, would you have your child vaccinated according to the recommended schedule? Write a few paragraphs that include the benefits and potential risks of vaccination. Explain how and why you made your decision.

    Use at least three reliable sources (books, journals, or websites) to gather information that would help you decide. List each source you considered, and explain why you chose the three sources you used.

2. AIDS is an immune deficiency disease that is occurring in epidemic proportions in Africa. Write a few paragraphs describing the African AIDS crisis and strategies for relieving the crisis.

    Use at least three reliable sources (books, journals, or websites) to gather information. List each source you considered, and explain why you chose the three sources you used.

MasteringBiology®

Go to MasteringBiology for practice quizzes, activities, eText, videos, current events, and more.

SPECIAL TOPIC
Infectious Disease
13a

Vaccinations have helped eliminate certain infectious diseases.

- 13a.1 Pathogens
- 13a.2 Spread of a Disease
- 13a.3 Infectious Diseases as a Continued Threat

In the previous chapter, we learned about the ways that our bodies protect us against pathogens. In this chapter, we discuss the most important categories of pathogens, the disease-causing organisms introduced in Chapter 13. We explore how pathogens cause harm, how they are transmitted from person to person, and how they are studied so that steps may be taken to hold them in check.

13a.1 Pathogens

As we explained in Chapter 13, *pathogens* are disease-causing organisms. There are different types of pathogens and a wide range of differences within each type. As a result, each pathogen has specific effects on the body, and some pathogens are a greater menace than others. In this chapter, we look at bacteria, viruses, protozoans, fungi, parasitic worms, and prions. We consider the general means these different types of pathogens use to attack the body and cause symptoms. Keep in mind, however, as we do so, that some of the symptoms are caused not by the pathogen itself but by the immune responses our body uses to protect us (Chapter 13).

Virulence is the relative ability of a pathogen to cause disease. Some factors contributing to this ability are how easily the pathogen invades tissues and the degree and type of damage it does to body cells. An organism that always causes disease—the typhoid bacterium, for instance—is highly virulent. In contrast, the yeast *Candida albicans,* which *sometimes* causes disease, is moderately virulent.

Bacteria

Bacterial cells differ from the cells that make up our bodies. Recall from Chapter 3 that our bodies are made up of eukaryotic cells, which contain a nucleus and membrane-bound organelles. Bacteria, in contrast, are prokaryotes, which means they lack a nucleus and other membrane-bound organelles. Nearly all bacteria have a semirigid cell wall composed of a strong

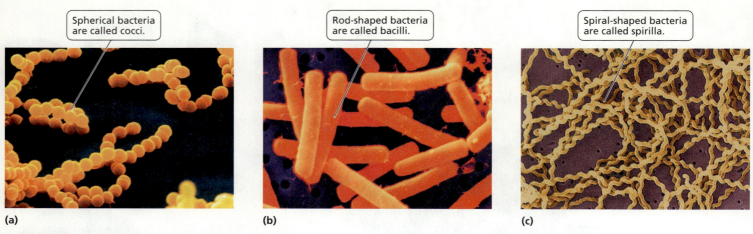

FIGURE 13a.1 Bacteria have three basic shapes: (a) spherical (coccus), (b) rod-shaped (bacillus), and (c) corkscrew-shaped (spirillum). All bacteria are prokaryotic cells, meaning that they lack a nucleus and membrane-bound organelles.

mesh of peptidoglycan, a type of polymer consisting of sugars and amino acids. The cell wall endows most types of bacteria with one of three common shapes: a sphere (a spherical bacterium is called a *coccus*), which can occur singly, in pairs, or in chains; a rod (called a *bacillus*), which usually occurs singly; or a spiral or corkscrew shape (such bacteria are called *spirilla*) (Figure 13a.1).

Bacteria can reproduce rapidly. This rapid growth rate is a matter of concern because the greater the number of bacteria, the greater harm they can potentially do. Rapid reproduction is possible because bacteria reproduce asexually in a type of cell division called *binary fission*. In binary fission, the bacterial genetic material (DNA) is copied, the cell is pinched in half, and each new cell contains a complete copy of the original genetic material. Under ideal conditions, certain bacteria can divide every 20 minutes. Thus, if every descendant lived, a single bacterium could result in a massive infection of trillions of bacteria within 24 hours. If a percentage of the descendant bacteria die before dividing, the population of bacteria will begin to grow more slowly than does a population in which all descendants survive, but the populations will eventually have the same growth rate.

Bacteria have defenses or other adaptive mechanisms that affect their virulence. Some bacteria have long, whiplike structures called *flagella* that allow them to move and spread through tissues to new areas where they can cause infection. Bacteria may also have filaments, called *pili*, that help them attach to the cells they are attacking. Outside the bacterial cell, there is often a capsule that provides a means of adhering to a surface and prevents scavenger cells of the immune system (phagocytes; see Chapter 13) from engulfing them.

Beneficial bacteria Many bacteria are beneficial. For instance, certain bacteria are important in food production, especially of dairy products such as cheese and yogurt. Other bacteria are important in the environment, serving as decomposers or driving the cycling of nitrogen, carbon, and phosphorus between organisms and the environment (see Chapter 23). Yet other bacteria are important in genetic engineering (see Chapter 21). Some bacteria are normal residents in the body that keep potentially harmful microorganisms in check, and bacteria in our intestines produce vitamin K.

Bacterial enzymes and toxins Destructive enzymes and toxins (poisons) are among the offensive mechanisms that certain bacteria use to spread and to attack. Some of these bacteria secrete enzymes that directly damage tissue and cause lesions, allowing the bacteria to push through tissues like a bulldozer. An example is *Clostridium*, the bacterium that causes gas gangrene, a condition in which tissue dies because its blood supply is shut off. The bacterium secretes an enzyme that dissolves the material holding muscle cells together, permitting the bacteria to spread with ease. When this bacterium digests muscle cells for energy, a gas is produced that presses against blood vessels and shuts off the blood supply. In addition, *Clostridium* causes anemia by secreting an enzyme that bursts red blood cells.

Most bacteria, however, do their damage by releasing toxins (poisons) into the bloodstream or the surrounding tissues. If the toxins enter the bloodstream, they can be carried throughout the body and disturb body functions.

The disease symptoms depend on which body tissues are affected by the toxin. Thus, the bacteria that cause various types of food poisoning have different effects. *Staphylococcus* is often the culprit responsible for contaminating poultry, meat and meat products, and creamy foods such as pudding or salad dressing. These bacteria multiply when food is undercooked or unrefrigerated. The toxins they produce stimulate cells in the immune system to release chemicals that result in inflammation, vomiting, and diarrhea. Another type of food poisoning is caused by *Salmonella*, often encountered in undercooked contaminated chicken or eggs. In this case, the toxin causes changes in the permeability of intestinal cells, leading to diarrhea and vomiting. One type of *Escherichia coli* (*E. coli*) food poisoning is often caused by contaminated meat, particularly ground meat. Besides vomiting and diarrhea, *E. coli* toxin can cause kidney failure in children and the elderly. The toxin that causes botulism, a type of food poisoning often brought on by eating improperly canned food, is one of the most toxic substances known. Produced by the bacterium *Clostridium botulinum*, it interferes with nerve functioning, especially motor nerves that cause muscle contraction. Death occurs because muscle paralysis prevents breathing. If enough of it is consumed, this toxin is almost always fatal.

Antibiotics Fortunately, bacteria can be killed. As we learned in Chapter 13, the human body has its own array of defenses against foreign invaders. But when the body needs outside help, we can call on *antibiotics,* chemicals that inhibit the growth of microorganisms. Antibiotics work to reduce the number of bacteria or slow the growth rate of the population, allowing time for body defenses to conquer the bacteria. Some antibiotics kill bacteria directly by preventing the synthesis of bacterial cell walls, causing them to burst. Recall that our body cells lack cell walls (see Chapter 3). Thus, our cells are unaffected by antibiotics that target cell walls. Some antibiotics block protein synthesis by bacteria but do so without interfering with protein synthesis in human body cells. This selective action is possible because the structure of ribosomes, the organelles on which proteins are synthesized, is slightly different in bacteria and humans.

When antibiotics were introduced during the 1940s, they were considered miracle drugs. For the first time, there was a cure for such devastating bacterial diseases as pneumonia, bacterial meningitis, tuberculosis, and cholera. Today, there are more than 160 antibiotics. These lifesaving drugs have become so commonplace that we take them for granted.

Unfortunately, antibiotics are losing their power. Infections that were once easy to cure with antibiotics can now turn deadly as bacteria gain resistance to the drugs. Several bacterial species capable of causing life-threatening illnesses have produced strains that are resistant to every antibiotic available today.[1]

Contradictory as it may seem, the use of antibiotics can actually increase antibiotic resistance in a strain of bacteria. When a strain of bacteria is exposed to an antibiotic, the bacteria that are susceptible die. The more resistant bacteria, however, may survive and multiply. If the bacteria are exposed to the antibiotic again, the selection process is repeated. With each exposure to the drug, the resistant bacteria gain a stronger foothold. Making matters worse, antibiotics kill beneficial bacteria along with the harmful ones. Normally, the beneficial bacterial strains help keep the harmful strains in check. Loss of the "good" bacteria can allow the harmful ones to dominate.

The overuse and misuse of antibiotics are largely to blame for the resistance problem. Prescribing antibiotics for illnesses that are viral, such as a cold or flu, is an example of overuse. Antibiotics have no effect on viruses, so they are unnecessary for treating such illnesses. Patients misuse antibiotics when they stop taking their medicine as soon as they feel better, instead of completing the full course of treatment. By stopping too early, they may be leaving the bacteria with greater resistance alive. Hospitals use antibiotics heavily, so it is not surprising that they are breeding grounds for antibiotic-resistant bacteria. The resistant bacteria survive, outgrow susceptible strains, and spread from person to person. Indeed, most infections by antibiotic-resistant bacteria occur in hospitals. An example is *Staphylococcus aureus,* which can cause many types of infections, including blood poisoning, pneumonia, skin infections, heart infections, and nervous system infections. The strain of *S. aureus* called MRSA (**m**ethicillin-**r**esistant **S**taphylococcus **a**ureus) is actually resistant to many antibiotics. For many years, MRSA existed only in hospitals, but it is now found in the community at large. In the last several years, outbreaks of MRSA have occurred in public places, including schools, locker rooms, fitness centers and doctors' offices. For a time, vancomycin was the only antibiotic that remained effective against MRSA. Unfortunately, a vancomycin-resistant *S. aureus* (VRSA) strain has arisen. An antibiotic-resistant strain of another bacterium, *Clostridium difficile* (often called "C. diff"), is more dangerous than other strains, because it produces more toxin. Outbreaks of *C. difficile* are spreading in health care facilities, such as hospitals and nursing homes, because antibiotics are used heavily there. Hospital-acquired *C. difficile* infections cause 18,000 to 20,000 deaths a year in the United States, mostly among the elderly.

More than 40% (by mass) of the antibiotics used in the United States are given to livestock to promote growth and ensure health. Farmers also spray crops with antibiotics to control or prevent bacterial infections in the crops. These practices also contribute to antibiotic resistance, and the antibiotic-resistant bacteria can then infect humans. For this reason, the U.S. Food and Drug Administration (FDA) has recently banned the use of antibiotics in animal feed.

What can you do to slow the spread of drug-resistant bacteria? Use antibiotics responsibly. Do not insist on a prescription for antibiotics against your doctor's advice. Take antibiotics exactly as prescribed, and be sure to complete the recommended treatment. Also, reduce your risk of getting an infection that might require antibiotic treatment: wash your hands frequently, rinse fruits and vegetables before eating them, and cook meat thoroughly.

Viruses

Viruses are responsible for many human illnesses. Some viral diseases, such as the common cold, are usually not very serious. Other viral diseases, such as yellow fever, can be deadly.

Most biologists do not consider a virus to be a living organism because, on its own, it cannot perform any life processes (see Chapter 1 for a review on the basic characteristics of life). To copy itself, a virus must enter a host cell. The virus exploits the host cell's nutrients and metabolic machinery to make copies of itself that then infect other host cells.

Viruses are much smaller than bacteria. A virus consists of a strand or strands of genetic material, either DNA or RNA, surrounded by a coat of protein, called a capsid. The genetic material carries the instructions for making new viral proteins. Some of these proteins become structural parts of the new viruses. Some of them serve as enzymes that help carry out biochemical functions important to the virus. Some are regulatory proteins, such as the proteins that trigger specific viral genes to become active under certain sets of conditions or the proteins that convert the host cell into a virus-producing factory.

---

[1]Bacteria resistant to all antibiotics available today include some strains of *Staphylococcus aureus* (which cause skin infection and pneumonia), *Mycobacterium tuberculosis* (the cause of tuberculosis), *Enterococcus faecalis* (which causes intestinal infections), and *Pseudomonas aeruginosa* (which causes many types of infections).

Q Which part of a virus would have to change for it to be able to infect a new type of tissue?

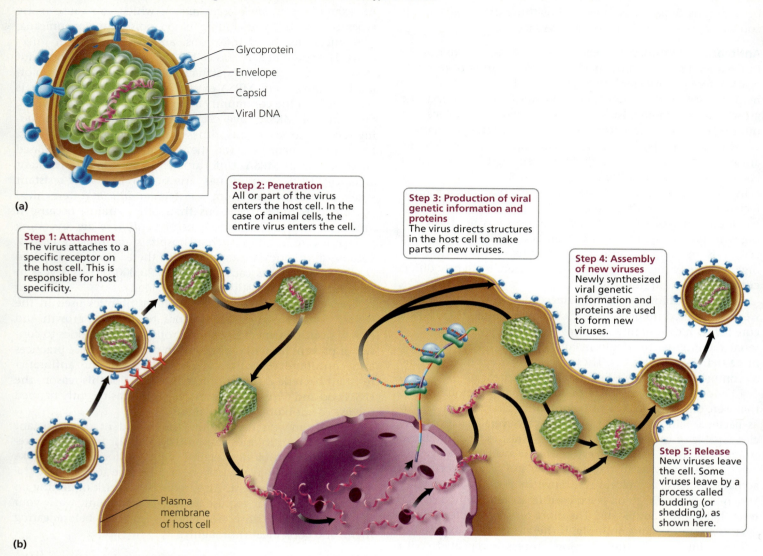

FIGURE 13a.2 (a) The structure of a typical virus. A coat, called a capsid, made of protein surrounds a core of genetic information made of DNA or RNA. Some viruses have an outer membranous layer, called the envelope, from which glycoproteins project. (b) Steps in viral replication.

A The glycoprotein on its surface. It is the fit between the glycoprotein and the host cell receptor that determines whether the virus can infect the cell.

Some viruses have an envelope, an outer membranous layer studded with glycoproteins. In some viruses, the envelope is actually a bit of plasma membrane from the previous host cell that became wrapped around the virus as it left the host cell. The envelope of certain other viruses—those in the herpes family, for instance—comes from a previous host cell's nuclear membrane. In any case, the virus produces the glycoproteins on the envelope. Some glycoproteins are important for attachment of the virus to the host cell.

A virus can replicate (make copies of itself) only when its genetic material is inside a host cell. Figure 13a.2 illustrates the general steps in the replication of viruses that infect animal cells:

1. **Attachment.** The virus gains entry by binding to a receptor (a protein or other molecule of a certain configuration) on the host cell surface. Such binding is possible because the viral surface has molecules of a specific shape (glycoproteins or capsids) that fit the host's receptors. The host cell receptors play a role in normal cell functioning. However, a molecule on the surface of the virus has a shape that is similar to the chemical that would normally bind to the receptor. Viruses generally attack only certain kinds of cells in certain species, because a particular virus can infect only cells bearing a receptor the virus can bind to. For example, the virus that causes the common cold infects only cells in the respiratory system, and the virus that causes hepatitis infects only liver cells.

2. **Penetration.** After a virus has bound to a receptor on a host cell, the entire virus enters the host cell, often by phagocytosis by the host cell. Once inside, the virus loses its capsid, leaving only its genetic material intact.

3. **Production of viral genetic information and proteins.** Viral genes then direct the host cell machinery to make thousands of copies of viral DNA or RNA. Next, viral genes direct the synthesis of viral proteins, including coat proteins and enzymes.

4. **Assembly of new viruses.** Copies of the viral DNA (or RNA) and viral proteins then assemble to form new viruses.

5. **Release.** Some viruses leave the cell through *budding*, or shedding. In this process, the newly formed viruses push through the host cell's plasma membrane and become wrapped in this membrane, which forms an envelope. Budding need not kill the host cell. Other virus types do not acquire an envelope, but rather cause the host cell membrane to rupture, releasing the newly formed viruses and killing the host cell.

Viruses can cause disease in several ways, as summarized in Table 13a.1. Some viruses cause disease when they kill the host cells or cause the cells to malfunction. The host cell dies when viruses leave it so rapidly that it lyses (bursts). In such cases, disease symptoms will depend on which cells are killed. However, if viruses are shed slowly, the host cell may remain alive and continue to produce new viruses. Slow shedding causes *persistent infections* that can last a long time. Some viruses can produce *latent infections,* in which the viral genes remain in the host cell for an extended period without harming the cell. At any time, however, the virus can begin replicating and cause cell death as new viruses are released.

An example of a virus that can act in all of these ways is the herpes simplex virus that causes fever blisters ("cold sores") on the mouth. The virus is spread by contact (discussed shortly) and enters the epithelial cells of the mouth, where it actively replicates. Rapid shedding kills the host cells, causing fever blisters. Slow shedding may not cause outward signs of infection, but the virus can still be transmitted. When the blisters are gone, the virus remains in a latent form within nerve cells without causing symptoms. However, stress can activate the virus. It then follows nerves to the skin and begins actively replicating, causing new blisters in the same region of the mouth.

Certain viruses can also cause cancer. Some do this when they insert themselves into the host chromosome near a gene that regulates cell division and, in so doing, alter the functioning of that gene. Still other viruses bring cancer-causing genes with them into the host cell.

Unfortunately, viruses are not as easy to destroy as bacteria. It is difficult to attack viruses inside their host cells without killing the host cell itself. Most attempts to develop antiviral drugs have failed for this reason. Nonetheless, some drugs are now available to slow viral growth, and others are being developed. Most of the antiviral drugs available today, including those against the herpes virus and HIV, work by blocking one of the steps necessary for viral replication. As mentioned in Chapter 13, interferons are proteins produced by virus-infected cells that protect neighboring cells from all strains of viruses. Interferons are not as useful as originally hoped, but they have been used for certain viral infections, including hepatitis C and the human papillomavirus that causes genital warts.

Because of these obstacles to treatment, the best way to fight viral infections is to prevent them with vaccines (discussed in Chapter 13).

TABLE 13A.1	Possible Effects of Animal Virus on Cells
Lytic infection	Rapid release of new viruses from infected cell causes cell death. Symptoms of the disease depend on which cells are killed.
Persistent infection	Slow release of new viruses causes cell to remain alive and continue to produce new viruses for a prolonged period of time.
Latent infection Primary infection  Latent period  Secondary infection 	Delay between infection and symptoms. Virus is present in the cell without harming the cell. Symptoms begin when the virus begins actively replicating, and new viruses exiting the host cell can cause cell death.
Transformation to cancerous cell 	Certain viruses insert their genetic information into host cell chromosomes. Some carry oncogenes (cancer-causing genes) that are active in the host cell. Some disrupt the functioning of the host cell's genes that regulate cell division, causing the cell to become cancerous.

stop and think

How do the structure and replication cycle of viruses explain why antibiotics are not effective against viral diseases?

Protozoans

Protozoans are single-celled eukaryotic organisms with a well-defined nucleus. They can cause disease by producing toxins or by releasing enzymes that prevent host cells from functioning normally. Protozoans are responsible for many diseases, including malaria, sleeping sickness, amebic dysentery, and giardiasis. Giardiasis is a diarrheal disease that can last for weeks. Outbreaks of giardiasis occur frequently in the United States, most of them resulting from water supplies contaminated with

FIGURE 13a.3 *Giardia is a protozoan that is commonly found in lakes and streams used as sources of drinking water, even those in pristine areas. It causes severe diarrhea that lasts for weeks and can be especially dangerous for children.*

human or animal feces. Even clear and seemingly clean lakes and streams in the wilderness can contain *Giardia* (Figure 13a.3). Fortunately, drugs are available to treat protozoan infections. Some of these drugs work by preventing protozoans from synthesizing proteins.

Fungi

Like the protozoans, fungi are also eukaryotic organisms with a well-defined nucleus in their cells. Some fungi exist as single cells. Others are organized into simple multicellular forms, with not much difference among the cells. There are more than 100,000 species of fungi, but fewer than 0.1% cause human ailments. Fungi obtain food by infiltrating the bodies of other organisms—dead or alive—secreting enzymes to digest the food, and absorbing the resulting nutrients. If the fungus is growing in or on a human, body cells of the human are digested, causing disease symptoms. Some fungi cause serious lung infections, such as histoplasmosis and coccidioidomycosis. Other, less-threatening fungal infections occur on the skin and include athlete's foot and ringworm. Most fungal infections can be cured. Fungal cell membranes have a slightly different composition from those of human cells. As a result, the membrane is a point of vulnerability. Some antifungal drugs work by altering the permeability of the fungal cell membrane. Others interfere with membrane synthesis by fungal cells. Fungal infections of the skin, hair, and nails can be combated with a drug that prevents the fungal cells from dividing.

Parasitic Worms

Parasitic worms are multicellular animals that benefit from a close, prolonged relationship with their hosts while harming, but usually not killing, their hosts. They include flukes, tapeworms, and roundworms, such as hookworms and pinworms. They can cause illness by releasing toxins into the bloodstream, feeding off blood, or competing for food with the host. Parasitic worms cause many serious human diseases, including ascariasis, schistosomiasis, and trichinosis.

Ascariasis is caused by a large roundworm, *Ascaris*, that is about the size of an earthworm. People become infected with *Ascaris* when they consume food or drink contaminated with *Ascaris* eggs. The eggs develop into larvae (immature worms) in the person's intestine. The larvae then penetrate the intestinal wall, enter the bloodstream, and travel to the lungs. After developing further, the worms are coughed up and swallowed, thus returning to the intestine. Within 2 to 3 months, they mature into male and female worms, which live for about 2 years. During those years, female worms can produce more than 200,000 eggs a day.

As much as 25% of the world population is infected with *Ascaris,* particularly in tropical regions. Up to 50% of the children in some parts of the United States (mostly rural areas in the Southeast) are infected. Many people with ascariasis have no symptoms. However, the worms can cause lung damage and severe malnutrition. When many worms are present, they can block or perforate the intestines, leading to death.

Prions

Prions (pree'-ons) are infectious particles of proteins—or, more simply, infectious proteins. They are misfolded versions of a harmless protein normally found on the surface of nerve cells. If a prion is present, it somehow causes the host protein to change its shape to the abnormal form. Prions cause a group of diseases called *transmissible spongiform encephalopathies (TSEs)*, which are associated with degeneration of the brain. The misshapen proteins clump together and accumulate in the nerve tissue of the brain. These clumps of prions may damage the plasma membrane or interfere with molecular traffic. Spongelike holes develop in the brain, causing death.

Transmissible spongiform encephalopathies are progressive and fatal. Prions cannot be destroyed by heat, ultraviolet light, or most chemical agents. Currently, there is no treatment for any disease they cause. Several of the TSEs are animal infections, notably mad cow disease, scrapie in sheep, and chronic wasting disease (CWD), which affects deer and elk.

Prions also cause a human neurological disorder called *Creutzfeldt-Jakob disease (CJD)*. Indeed, the prion responsible for mad cow disease is thought to cause one form of CJD. The incubation period for CJD can be months to decades. Symptoms include sensory and psychiatric problems. Once the symptoms begin, death usually occurs within a year.

How does an animal become infected with prions? In the case of mad cow disease, it appears that cattle become infected when they eat prions in contaminated food. For example, prions have been passed along in the protein supplements fed to cattle to increase their growth and milk production. Those protein supplements had been prepared from the carcasses of animals considered unfit for human consumption, a practice banned in the United States in 1997. A broader ruling prohibiting the use of any high-risk animal parts in *any* animal feed went into effect in 2009. Any protein supplements prepared from animals infected with mad cow disease would have contained prions. The prions pass through the intestinal wall, enter the lymphatic system, and are then transported by nerves to the brain and spinal cord. In contrast, CWD can apparently be spread by animal-to-animal contact, including contact with body fluids such as urine or feces from infected animals. Scientists also think that the prions responsible for CWD may remain in the soil or water for years. As a result, healthy animals may become infected from living in a region previously occupied by diseased animals. Humans can become infected with prions by eating contaminated substances, through tissue transplant, or through contaminated surgical instruments.

what would you do?

Mad cow disease is spread in cattle when they consume contaminated food. There are laws to prevent feeding cattle food that might be contaminated. When violations of the laws occur, who should be held responsible: the food manufacturers or the farmers?

13a.2 Spread of a Disease

Obviously, you catch a disease when the pathogen enters your body. But how do diseases travel from person to person or enter the body in the first place? The answer to this question varies with the type of pathogen.

- **Direct contact.** One means of transmission is direct contact of an infected person with an uninfected person, as might occur when shaking hands, hugging and kissing, or being sexually intimate. For example, sexually transmitted diseases (STDs)—including chlamydia, gonorrhea, syphilis, genital herpes, and human papillomavirus (HPV)—are spread when a susceptible body surface touches an infected body surface (see Chapter 17a). The organisms that cause STDs generally cannot remain alive outside the body for very long, so direct intimate contact is necessary. A few disease-causing organisms—HIV and the bacterium that causes syphilis, for instance—can spread across the placenta from a pregnant woman to her growing fetus.

- **Indirect contact.** Indirect contact, the transfer from one person to another without their touching, can spread other diseases. Most respiratory infections, including the common cold, are spread by indirect contact (see Chapter 14). When an infected person coughs or sneezes, airborne droplets of moisture full of pathogens are carried through the air (Figure 13a.4). The infected droplets may be inhaled or land on nearby surfaces. When another person touches an affected surface, the organisms are transmitted. In this way, droplet infection spreads pathogens on contaminated inanimate objects, including doorknobs, drinking glasses, and eating utensils.

- **Contaminated food or water.** Certain diseases are transmitted in contaminated food or water (see also Chapter 15a). You have read that spoiled food can cause food poisoning. Another disease transmitted by food or water is hepatitis A, an inflammation of the liver caused by a certain virus. *Legionella,* the bacterium that causes a severe respiratory infection known as Legionnaires' disease, is a common inhabitant of the water in condensers of large air conditioners and cooling towers. The disease-causing bacteria are spread through tiny airborne water droplets. Coliform bacteria come from the intestines of humans and are, therefore, an indicator of fecal contamination of water. Their numbers are monitored in drinking and swimming water. To be safe, drinking water should not have any coliform bacteria.

- **Animal vectors** Another means of transmission is by *vector,* an animal that carries a disease from one host to another. The most common vector-borne disease in the United States is Lyme disease. It is caused by a bacterium transmitted by the deer tick (the vector), which is about the size of the head of a pin (Figure 13a.5). The tick larva picks up the infectious agent when it bites and sucks blood from an infected animal. When the tick subsequently feeds on a human or other mammalian host, the bacteria gradually move from the tick's gut to its salivary glands and then are passed to its victim. The incubation period, during which there are no symptoms, can be as long as 6 to 8 weeks. Early symptoms include a headache, backache, chills, and fever. Often, a rash resembling a bull's eye develops, with an intense red center and border. Over a period of weeks, the circle increases in diameter. Weeks to months later, unless the disease is treated promptly, pain, swelling, and arthritis may develop. Cardiovascular and nervous system problems may follow the arthritis.

FIGURE 13a.4 *Pathogens can be spread through the air in droplets of moisture when an infected person sneezes or coughs.*

✺ Actual size

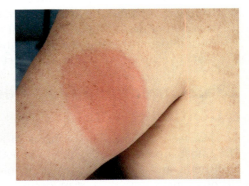

FIGURE 13a.5 *A tiny tick, the deer tick, is a vector that transmits the bacterium responsible for Lyme disease. One characteristic sign of Lyme disease is a red bull's-eye rash surrounding the tick bite. The rash gradually increases in diameter.*

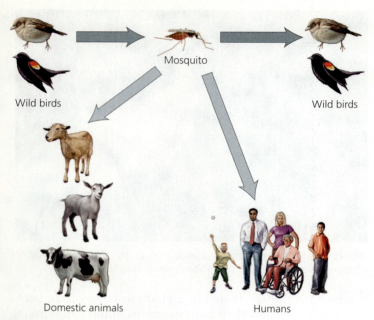

FIGURE 13a.6 *The mosquito is the vector that transmits West Nile virus.*

Emerging Diseases and Reemerging Diseases

An *emerging disease* is a disease with clinically distinct symptoms whose incidence has increased, particularly over the last two decades. Among these diseases are HIV, SARS, H1H5 influenza, and H1N1 influenza. Other diseases have reemerged that were thought to have been conquered. A *reemerging disease* is a disease that has reappeared after a decline in incidence. For example, because new drug-resistant strains of bacteria have emerged, tuberculosis is once again a global problem. We consider three factors that play important roles in the emergence and reemergence of disease.

1. **Development of new organisms that can infect humans and development of drug-resistant organisms.** Most of the time, a pathogen infects only one type or a few types of organisms. Mutations are changes in genetic information that occur randomly. Some mutations allow the pathogen to "jump species" from its original host and infect another type of organism. Recall that a virus can penetrate a cell only if the virus has the appropriate molecule on its surface—one that will fit into a receptor on the host cell. Another mechanism that could allow an animal virus to infect humans is *antigenic shift*, the mixing of genetic information of an animal virus and a human virus, which might occur if both viruses infected the same cell. This is how the H1N1 virus that causes swine flu developed: a person passed human influenza A viruses to a pig with influenza A. When the viruses infected the same cell, pieces of the viruses' genetic material were mixed and created a new strain of virus.

   Pathogens can also undergo changes in their response to drugs. We have seen that certain bacteria have acquired resistance to antibiotics, for example. As a result, some diseases that were once easily cured by antibiotics are now much more difficult to treat. Improper antibiotic treatment during the reemergence of tuberculosis (TB) has led to antibiotic-resistant strains of TB that, in turn, make TB more difficult to treat. Infections with multi-drug-resistant strains of *Mycobacterium tuberculosis*, the bacterium that causes TB, are increasing at an alarming rate. The World Health Organization coined a new term to describe drug resistance in a new strain of the tuberculosis bacterium—*XDR*, which stands for *extensively drug resistant*. The new XDR strain causes a tuberculosis infection that is nearly impossible to treat.

Mosquitoes transmit the West Nile virus, which can cause both meningitis (inflammation of the meninges, the protective coverings of the central nervous system) and encephalitis (brain inflammation; Figure 13a.6). The first reported cases of West Nile virus in North America were in New York City in 1999. Since then, the disease has spread to nearly every state. The virus can infect certain vertebrates, including humans, horses, birds, and occasionally dogs and cats. Testing mosquitoes and dead birds, especially crows and starlings, for the virus is one way to track its spread. Because the symptoms are similar to those of the flu (fever, headache, and muscle and joint pain) many people who become infected are unaware of it—and most infected people under the age of 50 have few symptoms or none at all. However, older people have weaker immune systems. If they become infected, they are more likely to develop meningitis or encephalitis, either of which can cause brain damage, paralysis, or death.

You can protect yourself from West Nile virus and other mosquito-borne viral infections such as Eastern equine encephalitis by avoiding wet and humid places that harbor mosquitoes. If you must enter an area where mosquitoes are likely to be, wear light-colored clothing that covers your body, and use insect repellent.

13a.3 Infectious Diseases as a Continued Threat

An *epidemic* is a large-scale outbreak of an infectious disease. The most notorious epidemics—bubonic plague, cholera, diphtheria, and smallpox—have happened in the distant past, although new outbreaks may occur sporadically. However, outbreaks of serious new diseases continue to present problems. We discuss some of these modern-day plagues elsewhere in the text.

what would you do?

The number of people with drug-resistant TB is rising, especially in poorer counties. Inadequate treatment is one reason for this rise. When new drugs to combat drug-resistant strains become available, they are extremely expensive. People in poor countries cannot afford the drugs, and cases of XDR increase, which puts us all at risk. Clearly it would be globally beneficial to decrease the prevalence of XDR tuberculosis by using effective drugs. Who do you think should foot the bill?

2. **Environmental change.** Changes in local climate—the annual amount of rainfall and the average temperature—can affect the distribution of organisms and change the size of the geographical region where certain organisms can live. Global warming makes the redistribution of pathogens a growing concern.

3. **Population growth.** Another important factor in the emergence or reemergence of diseases is the increase of the human population in association with the development and growth of cities. Swelling human populations in cities cause people to move out of the city into surrounding areas, creating suburbs. If the surrounding areas were previously undeveloped, the move brings more people into contact with animals and insects that might carry infectious organisms. Indeed, wild animals serve as reservoirs for more than a hundred species of pathogens that can affect humans. The development of suburbs also destroys populations of predators, such as foxes and bobcats. In some regions of New York, the loss of predators has led to an increase in numbers of tick-carrying mice and an increase in the incidence of Lyme disease.

Population density and mobility also enable infectious diseases to spread more easily today than in the past. Densely populated cities allow diseases to begin spreading quickly, and air travel enables them to spread over great distances (Figure 13a.7).

Global Trends in Emerging Infectious Diseases

Emerging infectious diseases are a concern because of economic costs and public health issues. These diseases are not evenly distributed throughout the world. The most important factors determining where new infectious disease will emerge are (1) the rate of human population growth and the density of the human population, and (2) the number of species of wild mammals. Most pathogens responsible for emerging infectious diseases are spread to humans by animals—wildlife, pets and livestock, and vectors. As we have seen, the development of drug resistance in some pathogens has also led to emergent infectious diseases.

Epidemiology

Epidemiology is the study of patterns of disease, including rate of occurrence, distribution, and control. Most diseases can be described as having one of the following four patterns:

- *Sporadic diseases* occur only occasionally at unpredictable intervals. They affect a few people within a restricted area.
- *Endemic diseases* are always present in a population and pose little threat. The common cold provides an example.
- An *epidemic disease* occurs suddenly and spreads rapidly to many people. Outbreaks of smallpox and cholera are examples of epidemics.
- A *pandemic* is a global outbreak of disease. HIV/AIDS is considered to be a pandemic.

Epidemiologists are "disease detectives" who try to determine why a disease is triggered at a particular time and place. The first step in answering this question is to verify that there is indeed a disease outbreak, defined as more than the expected number of cases of individuals with similar symptoms in a given area. Next, epidemiologists try to identify what causes the disease; whether it can be transmitted to other people; and, if it can be, how the disease is transmitted. To identify the cause of an infectious disease, epidemiologists try to isolate the same infectious agent from all people showing symptoms of the condition. They also try to identify factors—including age, sex, race, personal habits, and geographic location—shared by people with symptoms of the condition. These factors might provide a clue as to whether the condition can be transmitted and how.

FIGURE 13a.7 *Air travel is one reason that new diseases can spread rapidly.*

looking ahead

In this chapter, we considered infectious diseases—the pathogens that cause them, the methods by which they spread, reasons for emerging and reemerging diseases, and the epidemiologists who track the causes. In Chapter 14, we examine the respiratory system, which brings life-giving oxygen into the body and rids the body of carbon dioxide.

14 The Respiratory System

Did you Know?

- The average person takes 17,280 to 23,040 breaths each day.
- The world record holder for hiccupping hiccupped for 68 years.
- When you sneeze, the air can travel 100 miles per hour.

Your respiratory system brings in oxygen-laden air and removes carbon dioxide from the body. A healthy respiratory system enhances the quality of life.

- 14.1 Structures of the Respiratory System
- 14.2 Mechanism of Breathing
- 14.3 Transport of Gases between the Lungs and the Cells
- 14.4 Respiratory Centers in the Brain
- 14.5 Respiratory Disorders

HEALTH ISSUES

*Surviving a Common Cold
Smoking and Lung Disease*

In the previous chapters, we learned about the circulatory system and cardiovascular disease. We learned that the circulatory system transports oxygen to the cells and carbon dioxide to the lungs. In this chapter, we learn about the role the respiratory system plays in obtaining oxygen and ridding the body of carbon dioxide. We follow the course of inhaled air to the lungs and describe the mechanics of breathing. We then consider the transport of oxygen and carbon dioxide between the lungs and the cells and examine the control of respiration. Finally, we discuss several disorders of the respiratory system.

14.1 Structures of the Respiratory System

Without oxygen, we would die within a few minutes. Why? To stay alive, our cells need energy, and oxygen plays an essential role in extracting energy from food molecules (see Chapter 3). We store the extracted energy by producing a molecule called *ATP (adenosine triphosphate)*, which then releases the energy as needed to do the work of the cell. Our cells can make a little ATP without oxygen, but it is not enough to supply the body's energy needs. Cells can make 18 times more ATP if oxygen is present.

The same chemical reactions that require oxygen for the production of ATP produce carbon dioxide as a by-product. In solution—for example, in water or blood—carbon dioxide forms carbonic acid, which can be harmful to cells.

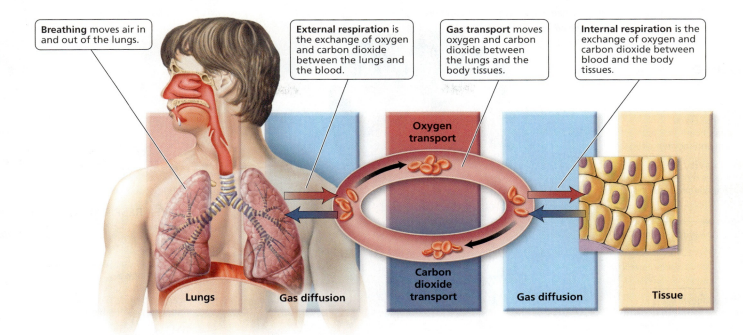

FIGURE 14.1 *An overview of respiration*

The function of the **respiratory system** is to provide the body with oxygen and to dispose of carbon dioxide, an exchange that also regulates the acidity of body fluids. Four processes play a part in respiration (Figure 14.1).

- **Breathing (also called *ventilating*).** Bringing oxygen-rich air into the lungs and moving air laden with carbon dioxide out of the lungs.
- **External respiration.** The exchange of oxygen and carbon dioxide between the lungs and the blood. Oxygen moves from the lungs into the blood, and carbon dioxide moves from the blood into the lungs.
- **Gas transport.** Transport of oxygen from the lungs to the cells and of carbon dioxide from the cells to the lungs.
- **Internal respiration.** The exchange of oxygen and carbon dioxide between the blood and tissue cells. Oxygen moves from the blood to the cells, where it is used in cellular respiration to produce ATP and carbon dioxide. Carbon dioxide produced by the cells moves into the blood.

We begin our exploration of how humans obtain oxygen and dispose of carbon dioxide by following the path of air from the nose to the lungs. The structures the air passes along the way are identified and described in Figure 14.2 and Table 14.1 on page 271. The path the air travels is summarized in Figure 14.3. The respiratory system is generally divided into upper and lower regions. The nose (nasal cavities) and pharynx make up the *upper respiratory system*. The *lower respiratory system* consists of the larynx, epiglottis, trachea, bronchi, bronchioles, and lungs.

Nose

However large someone's nose might seem from the outside, the inside is not as roomy as you might imagine. One reason is that a thin partition of cartilage and bone called the *nasal septum* divides the inside of the nose into two **nasal cavities**. In addition, much of the space within the nasal cavities is taken up by three convoluted, shelflike bones. These bones increase the surface area inside the nasal cavities and divide each cavity into three narrow passageways through which the air flows. Moist mucous membrane covers the entire inner surface of the nasal cavities.

We all know what a nose looks like, but what does a nose do? Your nose has three important functions.

- **Filtration and cleansing.** The nose helps clear particles from the air that moves through its passages in a variety of ways. Hairs inside the nose filter out the largest particles. In addition, certain cells in the membrane lining the surface of the nasal cavities and air tubules produce mucus, a sticky substance that catches dust particles. Cilia, tiny projections extending from the membranous lining, then sweep the mucus, trapped dirt particles, and bacteria toward the throat. The trapped particles can then either be swallowed and subsequently destroyed by digestive enzymes or coughed up. Particles that do not become trapped in the nasal cavities or the air tubules are deposited in the lungs. Many of the particles deposited in the lungs are engulfed and removed by macrophages, large irregularly shaped cells that wander across the surfaces of the lungs. However, if too many particles are inhaled or if the mechanisms for removing them fail, the particles may accumulate in the lungs and cover some of the gas exchange surfaces,

Q What is the path of oxygen from air entering the nose to the structures in the lungs where oxygen enters the blood supply?

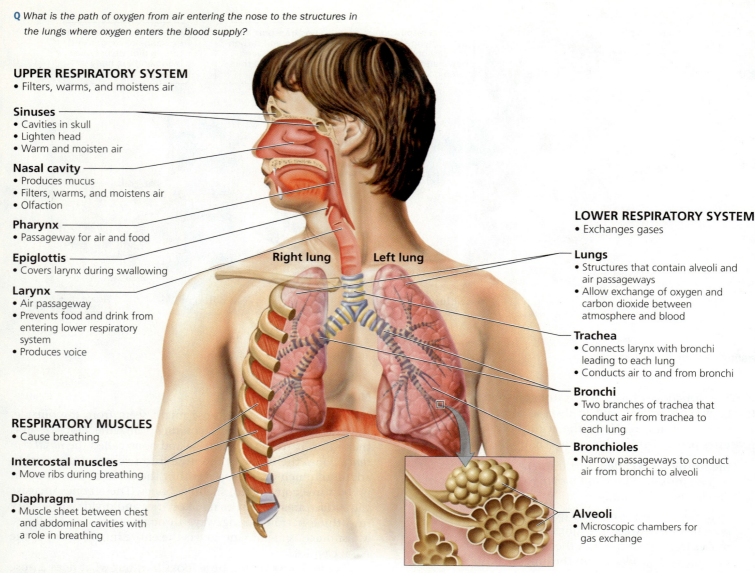

FIGURE 14.2 The respiratory system

A Nasal cavity, pharynx, trachea, bronchi, bronchioles, alveoli

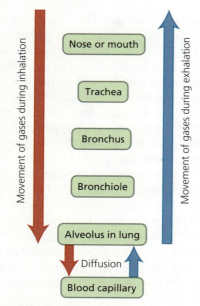

FIGURE 14.3 The path of air during inhalation and exhalation

reducing their efficiency and setting the stage for infection (Figure 14.4).

- **Conditioning the air.** The blood in the extensive capillary system of the mucous membrane lining the nasal cavity warms and moistens incoming air. The profuse bleeding that follows a punch to the nose is evidence of the rich supply of blood in these membranes. Warming the air before it reaches the lungs is extremely important in cold climates because frigid air can kill the delicate cells of the lung. Moistening the inhaled air is also essential because oxygen cannot cross dry membranes. Mucus helps moisten the incoming air so that lung surfaces do not dry out.

- **Olfaction.** Our sense of smell is due to the olfactory receptors located on the mucous membranes high in the nasal cavities behind the nose. The sense of smell is discussed in Chapter 9.

TABLE 14.1 Review of Structures of the Respiratory System

Structure	Description	Function
Upper respiratory system		
Nasal cavity	Cavity within the nose, divided into right and left halves by nasal septum; has three shelflike bones	Filters and conditions (moistens and warms incoming air); olfaction (sense of smell)
Sinuses	Large, air-filled spaces in the bones of the face	Lessen the weight of the head; warm and moisten inhaled air
Pharynx (throat)	Chamber connecting nasal cavities to esophagus and larynx	Common passageway for air, food, and drink
Epiglottis	Flap of tissue reinforced with cartilage	Covers the glottis during swallowing
Larynx	Cartilaginous, boxlike structure between the pharynx and trachea that contains the vocal cords and the glottis	Allows air but not other materials to pass to the lower respiratory system; source of the voice
Lower respiratory system		
Trachea	Tube reinforced with C-shaped rings of cartilage that leads from the larynx to the bronchi	The main airway; conducts air from larynx to bronchi
Bronchi (primary)	Two large branches of the trachea reinforced with cartilage	Conduct air from trachea to each lung
Bronchioles	Narrow passageways leading from bronchi to alveoli	Conduct air to alveoli; adjust airflow in lungs
Lungs	A pair of elastic structures within the thoracic (chest) cavity containing surfaces for gas exchange	Exchange oxygen and carbon dioxide between blood and air
Alveoli	Microscopic sacs within lungs, bordered by extensive capillary network	Provide immense, internal surface area for gas exchange

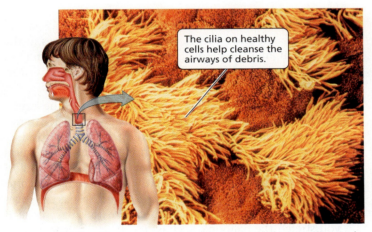

(a) The cilia are yellow in this color-enhanced electron micrograph. The cells without cilia secrete mucus.

(b) Cigarette smoke first paralyzes and then destroys the cilia. As a result, hazardous materials can accumulate on the surfaces of the air passageways.

FIGURE 14.4 *The respiratory passageways are lined with clumps of short hairlike structures, called cilia, interspersed between mucus-secreting cells.*

stop and think

Very cold temperatures can slow the action of the cilia in the nasal cavities. Why does the loss of ciliary action sometimes cause a runny nose on a very cold day?

Sinuses

Connected to the nasal cavities are large air-filled spaces in the bones of the face. These spaces are called the **sinuses**. Because these air spaces occupy space that would otherwise be occupied by heavy bone, one advantage of the sinuses is to make the head lighter. The sinuses also help warm and moisten the air we breathe because they, too, are lined with mucous membranes and some incoming air does pass through them. In addition, the sinuses are part of the resonating chamber that affects the quality of the voice. When you have a cold, your voice becomes muffled because the mucous membranes of the sinuses swell and produce excess fluid.

Because the air spaces of the sinuses are connected with those of the nasal cavities, any excess mucus and fluids drain from the sinuses into the nasal cavities. However, when the mucous membranes of the sinuses become inflamed, as they do in *sinusitis* (*–itis*, inflammation of), the swelling can block the connection between the nasal cavities and the sinuses, preventing the sinuses from draining the mucous fluid they produce. The pressure caused by the accumulation of fluids in the sinuses causes pain over one or both eyes or in the cheeks or jaws; this condition is usually called a *sinus headache*. Sinusitis may be caused by the virus responsible for a cold or by a subsequent bacterial infection. Decongestant nasal sprays reduce the swelling in the tubes that connect the sinuses with the nasal cavity, allowing the sinuses to drain more easily—but such sprays should be used only as directed, because you may become dependent on them to breathe freely.

Pharynx

The **pharynx**, commonly called the *throat*, is the space behind the nose and mouth. It is a passageway for air, food, and drink. Small, narrow passages, called the *auditory (Eustachian) tubes*, connect the upper region of the pharynx with the middle ear. These passages help equalize the air pressure in the middle ear with that of the pharynx.

Larynx

After moving through the pharynx, the air next passes through the **larynx**, which is commonly called the *voice box* or *Adam's apple.* The larynx is a boxlike structure composed primarily of cartilage (Figure 14.5).

The larynx has two main functions. It is a traffic director for materials passing through the structures in the neck, allowing air, but not other materials, to enter the lower respiratory system. The larynx is also the source of the voice. Let's consider these two functions in more detail.

1. **A selective entrance to the lower respiratory system.** The larynx provides a selective opening to the trachea (windpipe) and lower respiratory system: It can be opened to allow air to pass into the lungs and closed to prevent other matter, such as food, from entering the lungs. Because the esophagus (the tube leading to the stomach) is behind the larynx, food and drink must pass over the opening to the larynx to reach the digestive system. If solid material such as food were to enter the lower respiratory system, it could lodge in one of the tubes conducting air to the lungs and prevent airflow. Fluid entering the lungs is equally dangerous because it can cover the respiratory surfaces, decreasing the area available for gas exchange. Normally, foreign material is prevented from entering the lower respiratory system during swallowing because the larynx rises and causes a flap of cartilage called the **epiglottis** to move downward and form a lid over the **glottis**, the opening in the larynx through which air passes. You can feel the larynx moving if you put your fingers on your Adam's apple while swallowing. Because of this movement, you cannot breathe and swallow at the same time. (Try it!)

   If food or drink accidentally enters the trachea, we usually cough and expel it. However, if food lodges in the trachea, it may block airflow. The **Heimlich maneuver** can be used to remove the blockage and restore airflow (Figure 14.6).

2. **Production of voice.** The voice is generated in the larynx by the vibration of the **vocal cords**, two thick strands of tissue stretched over the opening of the glottis (see Figure 14.5). When you speak, muscles stretch the vocal cords across the air passageway, narrowing the opening of the glottis. Air passing between the stretched vocal cords causes them to vibrate and produce a sound, just as the edges of the neck of an inflated balloon vibrate and make noise if you stretch the balloon's neck while allowing air to escape. The vibrations of the vocal cords set up sound waves in the air spaces of the nose, mouth, and pharynx. This resonation is largely responsible for the tonal quality of your voice.

   The pitch of the voice depends on the tension of the vocal cords. When the cords are stretched, becoming thinner and tighter, the pitch of the sound when they vibrate is higher. You can demonstrate the relationship between thickness and pitch for yourself by plucking a rubber band stretched between your thumb and forefinger. The more the rubber band is stretched, the higher the pitch of the twang.

   When you suffer from *laryngitis,* an inflammation of the larynx, the vocal cords become swollen and thick. As a result, they cannot vibrate freely, and the voice becomes deeper and huskier. When the vocal cords are very inflamed, a person can hardly speak at all because the cords cannot vibrate in that condition.

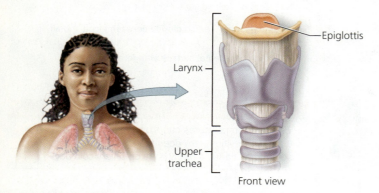

(a) The epiglottis is open during breathing but covers the opening to the larynx during swallowing to prevent food or drink from entering the trachea.

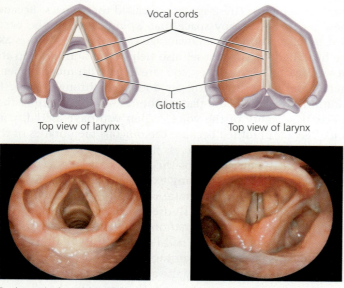

During quiet breathing, the vocal cords are near the sides of the larynx, and the glottis is open.

During speech, the vocal cords are stretched over the glottis and vibrate as air passes through them, producing the voice.

(b) The vocal cords are the folds of connective tissue above the opening of the larynx (the glottis) that produce the voice.

FIGURE 14.5 *The larynx, commonly called the voice box or Adam's apple, is an adjustable entryway to the trachea and the source of the voice.*

Trachea

The **trachea**, or windpipe, is a tube that conducts air between the outside of the body and the lungs. It is held open by rings

14.1 Structures of the Respiratory System

A person who is choking cannot speak or breathe and needs immediate help. The **Heimlich maneuver** is a procedure intended to force a large burst of air out of the lungs and dislodge the object blocking air flow.

Step 1: Stand behind the choking person with arms around the waist.

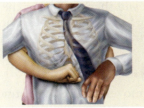

Step 2: Make a fist and place the thumb of the fist beneath the victim's rib cage about midway between the navel (belly button) and the breastbone.

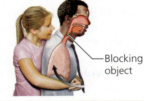

Step 3: Grasp the fist with your other hand and deliver a rapid "bear hug" up and under the rib cage with the clenched fist. Be careful not to press on the ribs or the breastbone because doing so could cause serious injury.

Step 4: Repeat until the object is dislodged.

FIGURE 14.6 *The Heimlich maneuver can be performed on a choking person who is standing or sitting. If a choking victim is lying on the ground, the same lifesaving pressure changes can be generated by pushing inward and upward on the upper part of the victim's abdomen. If you begin to choke and there is no one to perform the Heimlich maneuver on you, it may be possible to dislodge the obstruction in your trachea by throwing your upper abdominal region against a table, chair, or other stationary object.*

of cartilage that give it the general appearance of a vacuum cleaner hose. These rings of cartilage are C-shaped; the open ends of the rings face the side of the trachea next to the esophagus, which allows the esophagus to expand and compress the trachea when a large mass of food is swallowed. You can feel these rings of cartilage in your neck, just below the larynx.

The support rings are necessary in the trachea and its branches to prevent these airways from collapsing during each breath when the rapid flow of air into the lungs creates a drop in pressure. Air (or fluid) passing rapidly over a surface causes a lower pressure, experienced as a "pull," on that surface. Maybe you have noticed that when you get into the shower and turn on the water, the shower curtain is drawn in toward you. The curtain moves inward because the moving water lowers the air pressure, just as the rapid movement of air through the respiratory tubules does. If the trachea were not supported open by cartilage rings, the rapid flow of air during breathing would cause it to collapse or flatten.

Bronchial Tree

The trachea divides into two air tubes called primary **bronchi**; each bronchus (singular) conducts air from the trachea to one of the lungs. The bronchi branch repeatedly within the lungs, forming progressively smaller air tubes. The smallest bronchi divide to form yet smaller tubules called **bronchioles**, which finally terminate in *alveoli*, sacs with surfaces specialized for gas exchange (discussed shortly).

The repeated branching of air tubules in the lung is reminiscent of a branching tree. In fact, the resemblance is so close that the system of air tubules is often called the **bronchial tree** (Figure 14.7). All the bronchi are held open by cartilage, just as occurs in the trachea. However, the amount of cartilage decreases with the diameter of the tube. The bronchioles have no cartilage, but their walls contain smooth muscle, which is controlled by the autonomic nervous system so that airflow can be adjusted to suit metabolic needs (see Chapter 8).

Although the contraction of the muscle in bronchial walls is usually closely attuned to the body's needs, sometimes the bronchial muscles go into spasms that severely obstruct the flow of air. Such is the case with *asthma*, a chronic condition characterized by recurring attacks of wheezing and difficulty breathing. The difficult breathing is worsened by persistent inflammation of the airways. An allergy to substances such as pollen, dog or cat dander (skin particles), and the feces of tiny mites in household dust often trigger asthma attacks. However, a cold or respiratory infection, certain drugs, inhaling irritating substances, vigorous exercise, and psychological stress can also cause an attack. Some attacks start for no apparent reason. Certain inhalants prescribed to treat asthma attacks work by relaxing the bronchial muscles. Other inhalants contain steroids that reduce the inflammation of the air tubules that occurs in asthma.

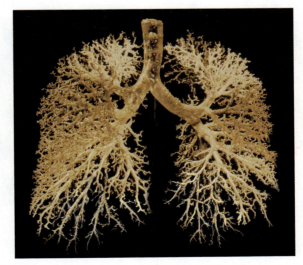

FIGURE 14.7 *A resin cast of the bronchial tree of the lungs. In the body, this branching system of air tubules is hollow and serves as a passageway for the movement of air between the atmosphere and the alveoli, where gas exchange takes place.*

Alveoli

Each bronchiole ends either with an enlargement called an **alveolus** (plural, alveoli) or, more commonly, with a grapelike cluster of alveoli. Each alveolus is a thin-walled, rounded chamber surrounded by a dense network of capillaries (Figure 14.8). Oxygen diffuses from the alveoli into the blood, which delivers the oxygen to cells. Carbon dioxide produced by the cells diffuses from the blood into the alveolar air to be exhaled.

Most of the lung tissue is composed of alveoli, making the structure of the lung much more like foam rubber than like a balloon, the image sometimes used to describe a lung. The surface area inside a simple, hollow balloon the same size as our lungs would be roughly 0.01 m² (about 0.2 yd²). However, each of our lungs contains approximately 300 million alveoli, whose total surface area is about 70 to 80 m² (about 84 to 96 yd²). In other words, the alveoli increase the surface area of the lung about 8500 times.

For the alveoli to function properly as a surface for gas exchange, they must be kept open. Phospholipid molecules called **surfactant**, which coat the alveoli, act to keep them open. Moist membranes, such as those of the alveolar walls, are attracted to one another because of an attraction between water molecules called *surface tension*. If this attraction were not disrupted by surfactant, it would pull the alveolar walls together, collapsing the air chambers.

Surfactant production usually begins during the eighth month of fetal life, so enough surfactant is present to keep the alveoli open when the newborn takes his or her first breath. Unfortunately, some premature babies have not yet produced enough surfactant to overcome the attractions between the alveolar walls. As a result, their alveoli collapse after each breath. This condition, called *respiratory distress syndrome* (RDS), makes breathing difficult for the premature newborn. Some newborns with RDS die as a result. However, many are saved by the use of mechanical respirators and artificial surfactant to keep them alive until their lungs mature.

stop and think
Pneumonia is a lung infection that results in an accumulation of fluid and dead white blood cells in the alveoli. Why might this result in lower blood levels of oxygen?

Lungs

We see, then, that the paired lungs consist of the bronchi, bronchioles, alveoli, and blood vessels. Each lung differs slightly in size and shape. Whereas the right lung has three lobes, the left lung has only two lobes. The smaller left lung has a notch in its central side to make room for the heart, which points to the left.

14.2 Mechanism of Breathing

Air moves between the atmosphere and the lungs in response to pressure gradients. It moves into the lungs when the pressure in the atmosphere is greater than the pressure in the lungs, and it moves out when the pressure in the lungs is greater than the pressure in the atmosphere.

The pressure changes in the lungs are created by changes in the volume of the thoracic cavity, a relationship explained by the characteristics of the *pleural membrane*. Each lung is enclosed in a double-layered sac of pleural membrane. One layer of membrane adheres to the wall of the thoracic cavity; the other adheres to the lung. Fluid between the layers of membrane lubricates the membrane layers and holds them together by surface tension. As a result, a change in the volume

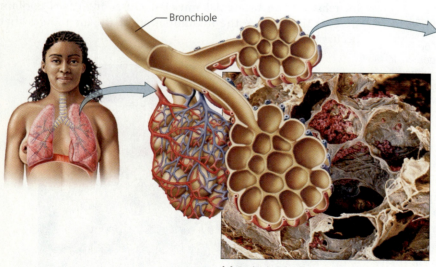

(a) Each alveolus is a small air-filled sac. In this section, some of the alveoli have been cut open and you can see into them.

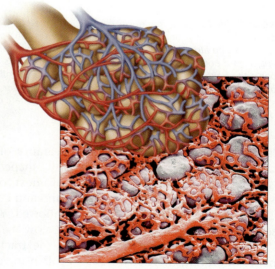

(b) Much of the surface of each alveolus is covered with capillaries. The interface provides a vast surface area for the exchange of gases between the alveoli and the blood.

FIGURE 14.8 *Alveoli in the lungs create a huge surface area where oxygen and carbon dioxide are exchanged between the lungs and the blood. Oxygen diffuses from the alveoli into the blood, and carbon dioxide diffuses from the blood to the alveoli.*

of the thoracic cavity causes a similar change in the volume of the lungs. Let's consider how the changes in the size of the thoracic cavity are brought about.

Inhalation

Air moves into the lungs when the size of the thoracic cavity increases; this increase causes the pressure in the lungs to drop below atmospheric pressure. The increase is due to the contraction of both the **diaphragm**, a broad sheet of muscle that separates the abdominal and thoracic cavities, and the muscles of the rib cage, called the **intercostal muscles** (*costa*, rib; Figure **14.9a**). The air pressure in the lungs decreases, and air rushes into the lungs. This process is called *inhalation* or **inspiration**. The intercostals lie between the ribs, so that when those muscles contract, they pull the rib cage upward and outward. By placing your hands on your rib cage while you inhale, you can feel the rib cage move up and out. Raising the rib cage increases the size of the thoracic cavity from the front to the back. Meanwhile, the contraction of the diaphragm lengthens the thoracic cavity from top to bottom. This lengthening occurs because the diaphragm is dome shaped when relaxed, and it flattens when contracted, as shown in Figure 14.9a.

Exhalation

The process of breathing out, called *exhalation* or **expiration**, is usually passive. In other words, it doesn't require work but occurs when the muscles of the rib cage and the diaphragm relax. The lungs are elastic; that is, after stretching they return to their former size. When the elastic tissues of the lung recoil, the rib cage falls back to its former lower position, and the diaphragm bulges into the thoracic cavity (Figure **14.9b**). The pressure within the lungs increases as the volume of the lungs decreases. When the pressure within the lungs exceeds atmospheric pressure, air moves out.

If it is necessary to exhale more air than usual, as in heavy breathing or coughing, other muscles assist the process. For example, there is another layer of muscles between the ribs. When the rib muscles in this layer contract, they pull the rib cage even further down and inward, increasing the pressure on the lungs. In addition, the muscles of the abdomen can be contracted. Abdominal contractions push the organs in the abdomen against the diaphragm, causing it to bulge even further into the thorax.

stop and think

In Victorian times, a woman often wore a corset containing whalebone that formed a band around her waist and lower chest. Corsets were laced tightly to create a wasplike waistline, and women who wore them frequently fainted. What is the most likely cause of these fainting spells?

The Volume of Air Moved Into or Out of the Lungs during Breathing

The volume of air moved during each breath varies from person to person, depending largely on the person's sex, age, and height. During quiet breathing, about 500 ml, or roughly

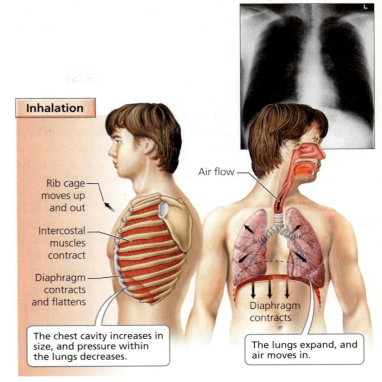

(a)

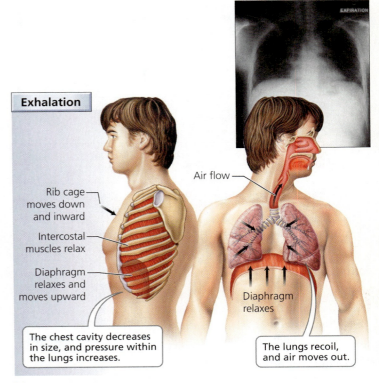

(b)

FIGURE 14.9 Changes in the volume of the thoracic cavity bring about inhalation and exhalation. The x-ray images show the actual changes in lung volume during inhalation and exhalation.

1 pint, of air moves in and out with each breath. The amount of air inhaled or exhaled during a normal breath is called the **tidal volume** (Figure 14.10).

If, after inhaling normally, you were to inhale until you could not take in any more air, you would probably bring another 1900 to 3300 ml of air into your lungs. This volume of air is nearly four to seven times the volume moved during quiet breathing. The additional volume of air that can be brought into the lungs after normal inhalation is called the **inspiratory reserve volume**.

After you have exhaled normally, you can still force about 1000 ml of additional air from the lungs. This additional volume of air that can be expelled from the lungs after the tidal volume is called the **expiratory reserve volume**. Decreases in expiratory reserve volume are characteristic of obstructive lung diseases, such as bronchitis and asthma. New York City firefighters and other rescue workers at the scene of the collapse of the World Trade Center on September 11, 2001, experienced a significant decrease in expiratory reserve volume in the year following the disaster. The decline in lung function among rescue workers, presumably due to the inhalation of toxic dust, was equivalent to that expected from 12 years of aging. Those who were on the scene when the towers fell, or shortly afterward, suffered the most damage.

If you were to take the deepest breath possible and exhale until you could not force any more air from your lungs, you would be demonstrating your **vital capacity**, the maximum amount of air that can be moved into and out of the lungs during forceful breathing. The vital capacity, therefore, equals the sum of the tidal volume, the inspiratory reserve, and the expiratory reserve. Although average values for college-aged people are about 4800 ml in men and 3400 ml in women, the values can vary tremendously with a person's health and fitness. One illness that affects vital capacity is pneumonia, which causes fluid to accumulate within the alveoli and takes up space that would normally be occupied by air.

The lungs can never be completely emptied, even with the most forceful expiration. The amount of air that remains in the lungs after exhaling as much air as possible, called the **residual volume**, is roughly 1100 to 1200 ml of air. As we will see later in the chapter, emphysema is a lung condition in which the alveolar walls break down, creating larger air spaces that are more difficult to empty. Thus, the residual volume increases. This residual air is lower in oxygen than is inhaled air, so a person with emphysema feels short of breath.

Because some air is always left in the lungs, the vital capacity is not a measure of the total amount of air that the lungs can hold. The **total lung capacity**, the total volume of air contained in the lungs after the deepest possible breath, is calculated by adding the residual volume to the vital capacity. This volume is approximately 6000 ml in men and 4500 ml in women.

14.3 Transport of Gases between the Lungs and the Cells

We have seen that breathing brings air into the lungs and expels air from the lungs. Recall that three other processes then play roles in delivering the oxygen to the cells and disposing of carbon dioxide from the cells. External respiration occurs in the alveoli of the lungs; there, oxygen diffuses into the blood, and carbon dioxide diffuses from the blood. Gas transport is accomplished by the blood, which carries oxygen to the cells and carbon dioxide away from the cells. Internal respiration occurs in the various tissues; there, oxygen diffuses out of the blood and into the cells, and carbon dioxide diffuses out of the cells and into the blood (Figure 14.11).

Oxygen Transport and Hemoglobin

Oxygen is carried from the alveoli throughout the body by the blood. Almost all—about 98.5%—of the oxygen that reaches the cells is bound to hemoglobin, a protein in the red blood cells. Hemoglobin bound to oxygen is called **oxyhemoglobin** (HbO_2). The remaining 1.5% of the oxygen delivered to the cells is dissolved in the plasma. Whole blood, which consists of cells as well as plasma, carries 70 times more oxygen than an equal amount of plasma alone.

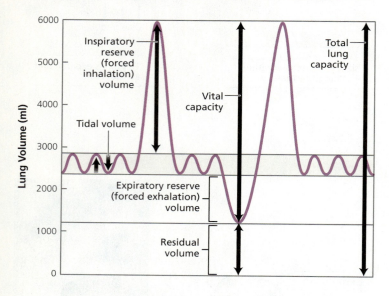

Tidal volume (~500 ml)	Amount of air inhaled or exhaled during an ordinary breath
Inspiratory reserve volume (~1900–3300 ml)	Amount of air that can be inhaled in addition to a normal breath
Expiratory reserve volume (~1000 ml)	Amount of air that can be exhaled in addition to a normal breath
Vital capacity (~3400–4800 ml)	Maximum amount of air that can be inhaled or exhaled in a single forced breath
Residual volume (~1100–1200 ml)	Amount of air remaining in the lungs after maximum exhalation
Total lung capacity (4500–6000 ml)	Total amount of air in the lungs after maximal inhalation (vital capacity + residual volume)

FIGURE 14.10 A spirometer is used to measure the volumes of air in the lungs.

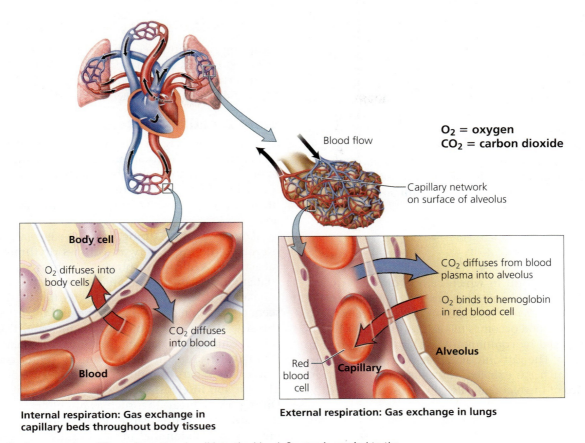

FIGURE 14.11 In the lungs, oxygen diffuses from the alveoli into the blood. Oxygen is carried to the cells in red blood cells. At the cells, oxygen diffuses from the blood to the body cells, which use the oxygen and produce carbon dioxide in the process. Carbon dioxide diffuses into the blood and is carried back to the lungs, where it diffuses from the blood into an alveolus and is exhaled.

Hemoglobin picks up oxygen at the lungs and releases it at the cells. But what determines whether hemoglobin will bind to oxygen or release it? The most important factor deciding this question is the partial pressure of oxygen, which is directly related to its concentration. In a mixture of gases, each gas contributes only part of the total pressure of the whole mixture of gases. The pressure exerted by one of the gases in a mixture is called its *partial pressure*.

Recall from Chapter 3 that substances always diffuse from regions of higher concentration or pressure to regions of lower concentration or pressure. In the alveoli of the lungs, where the concentration of oxygen is high, hemoglobin in the red blood cells in nearby capillaries picks up oxygen. The oxygen is then released near the cells, where the oxygen concentration is low.

Carbon Dioxide Transport and Bicarbonate Ions

The carbon dioxide produced by cells as they use oxygen is removed by the blood. Carbon dioxide transport occurs in three principal ways:

1. **Dissolved in blood plasma.** Between 7% and 10% of the carbon dioxide is transported dissolved in the plasma as molecular carbon dioxide.

2. **Carried by hemoglobin.** Hemoglobin molecules in red blood cells carry slightly more than 20% of the transported carbon dioxide. When carbon dioxide combines with hemoglobin, it forms a compound called **carbaminohemoglobin**.

3. **As a bicarbonate ion.** By far the most important means of transporting carbon dioxide is as bicarbonate ions dissolved in the plasma. About 70% of the carbon dioxide is transported this way. Carbon dioxide (CO_2) produced by cells diffuses into the blood and into the red blood cells. In both the plasma and the red blood cells, it reacts with water (H_2O) and forms carbonic acid (H_2CO_3). Carbonic acid quickly dissociates to form hydrogen ions (H^+) and bicarbonate ions (HCO_3^-). The process of bicarbonate ion formation in the capillaries of the tissues is represented by the following formula:

$$\underset{\substack{\text{Carbon} \\ \text{dioxide}}}{CO_2} + \underset{\text{Water}}{H_2O} \xrightarrow{\text{Carbonic anhydrase}} \underset{\substack{\text{Carbonic} \\ \text{acid}}}{H_2CO_3} \rightarrow \underset{\substack{\text{Bicarbonate} \\ \text{ion}}}{HCO_3^-} + \underset{\substack{\text{Hydrogen} \\ \text{ion}}}{H^+}$$

Although these reactions occur in the plasma as well as in red blood cells, they occur hundreds of times faster in red blood cells. The higher rate of reaction is caused by the enzyme **carbonic anhydrase**, which is found within red blood cells but not in the plasma. The hydrogen ions produced by the reaction combine with hemoglobin. In this way, hemoglobin acts as a buffer, and the acidity of the blood changes only slightly as it passes through the tissues. The bicarbonate ions diffuse out of the red blood cells into the plasma and are transported to the lungs.

In the lungs, the process is reversed. When the blood reaches the capillaries of the lungs, carbon dioxide diffuses from the blood into the alveoli because the concentration (partial pressure) of carbon dioxide is comparatively low in the alveoli. Because the concentration of carbon dioxide in the blood is higher than in the alveoli, the chemical reactions we have just described reverse direction. The bicarbonate ions rejoin the hydrogen ions to form carbonic acid. In the presence of carbonic anhydrase within the red blood cells, carbonic acid is converted to carbon dioxide and water. The carbon dioxide then leaves the red blood cells, diffuses into the alveolar air, and is exhaled. The reactions in the lungs are summarized below.

$$HCO_3^- + H^+ \xrightarrow{\text{Carbonic anhydrase}} H_2CO_3 \rightarrow CO_2 + H_2O$$

Bicarbonate ion + Hydrogen ion → Carbonic acid → Carbon dioxide + Water

Besides being the form in which carbon dioxide is transported by the blood, bicarbonate ions are an important part of the body's *acid–base buffering system*. They help neutralize acids in the blood. If the blood becomes too acidic, the excess hydrogen ions are removed by combining with bicarbonate ions to form carbonic acid. The carbonic acid then forms carbon dioxide and water, which are exhaled. (Bicarbonate ions and the acid–base balance of the blood are also discussed in Chapter 2.)

stop and think
Carbon monoxide is a poisonous gas that binds to hemoglobin much more readily than does oxygen, and it binds in the same site as oxygen. Thus, when carbon monoxide is bound to hemoglobin, oxygen cannot bind. Why is carbon monoxide poisoning potentially fatal?

14.4 Respiratory Centers in the Brain

Breathing rate influences the amount of oxygen that can be delivered to cells and the amount of carbon dioxide that can be removed from the body. Neural and chemical controls adjust breathing rate to meet the body's needs.

Basic Breathing Pattern

As you sit reading your text, your breathing is probably rather rhythmic, with about 12 to 15 breaths a minute. The basic rhythm is controlled by a breathing (respiratory) center located in the medulla of the brain (Figure 14.12). Within the breathing center are an inspiratory area and an expiratory area.

During quiet breathing, when you are calm and breathing normally, the inspiratory area shows rhythmic bouts of neural activity. While the inspiratory neurons are active, impulses that stimulate contraction are sent to the muscles involved in inhalation (the diaphragm and the intercostals). As we have seen, contraction of the diaphragm and the intercostals causes the size of the thoracic cavity to increase, thus moving air into the lungs. After about 2 seconds of inhalation, the activity of the neurons in the inspiratory center ceases for about 3 seconds. When inspiratory neurons cease activity, the diaphragm and intercostals relax, and passive exhalation occurs. During heavy breathing, such as might occur during exercise, the expiratory center causes contraction of other intercostal muscles and abdominal muscles, quickly pushing air out of the lungs.

Most of the time we breathe without giving it a thought. However, we can voluntarily alter our pattern of breathing through impulses originating in the cerebral cortex (the "conscious" part of the brain). We control breathing when we speak

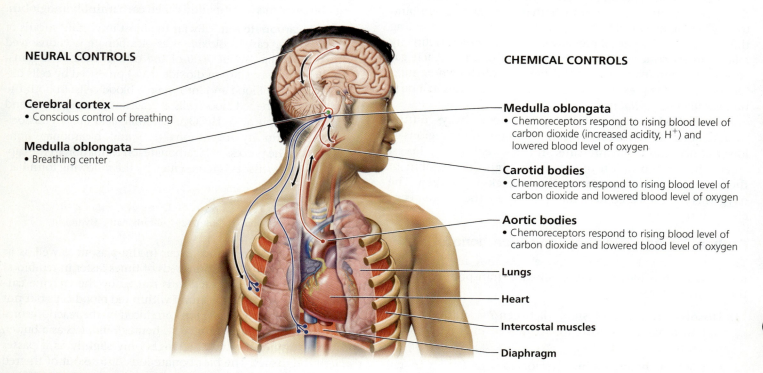

FIGURE 14.12 *Neural and chemical controls of breathing*

or sigh, and we can voluntarily pant like a dog. Holding our breath while swimming underwater is obviously a good idea. At certain other times, holding our breath can protect us from inhaling smoke or irritating gases.

During forced breathing, as might occur during strenuous exercise, stretch receptors in the walls of the bronchi and bronchioles throughout the lungs prevent the overinflation of the lungs. When a deep breath greatly expands the lungs and stretches these receptors, they send impulses over the vagus nerve that inhibit the breathing center, permitting exhalation. As the lungs deflate, the stretch receptors are no longer stimulated.

Chemoreceptors

The purpose of breathing is to control the blood levels of carbon dioxide and oxygen. We now consider how the levels of these gases control the breathing rate, which in turn influences the levels of the gases (Figure 14.13).

Carbon dioxide The most important chemical influencing breathing rate is carbon dioxide. The mechanism by which carbon dioxide regulates breathing depends on the hydrogen ions produced when carbon dioxide goes into solution and forms carbonic acid:

$$CO_2 + H_2O \rightarrow H_2CO_3 \rightarrow H^+ + HCO_3^-$$
Carbon dioxide, Water, Carbonic acid, Hydrogen ion, Bicarbonate ion

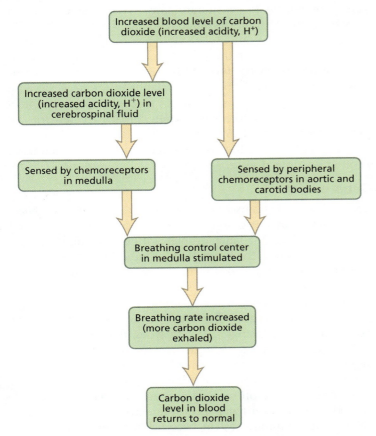

FIGURE 14.13 The role of carbon dioxide in controlling the breathing rate

Three groups of chemoreceptors respond to the changing levels of hydrogen ions in the blood (see Figure 14.12). Central chemoreceptors are located in a region of the brain called the medulla, near the breathing center. Peripheral chemoreceptors are located in the aortic bodies and the carotid bodies, small structures associated with the main blood vessel leaving the heart and going to the body as well as with the main blood vessels to the head. The chemoreceptors in the aortic bodies and carotid bodies also respond when the blood level of oxygen is low.

The chemoreceptors in the medulla are near its surface, where they are bathed in cerebrospinal fluid. Carbon dioxide diffuses from the blood into the cerebrospinal fluid, where it raises the hydrogen ion concentration by forming carbonic acid. When the rising hydrogen ion concentration stimulates the chemoreceptors in the medulla, breathing rate is increased and more carbon dioxide is exhaled, causing a decrease in the blood level of carbon dioxide.

Oxygen Because oxygen, not carbon dioxide, is essential to survival, it may be somewhat surprising to learn that oxygen does not influence the breathing rate unless its blood level falls dangerously low. Only then do oxygen-sensitive chemoreceptors in the aortic bodies and carotid bodies send a message that the blood oxygen level is at a critical point, initiating a last-minute call to the medulla to increase the breathing rate and raise oxygen levels. If the oxygen level falls much more, the neurons in the inspiratory area die from a lack of oxygen and do not respond well to impulses from the chemoreceptors. As a result, the inspiratory area begins to send fewer impulses to the muscles of inspiration, the breathing rate decreases, and breathing may even cease completely.

14.5 Respiratory Disorders

Even on a restful day, you move more than 86,000 liters of air into and out of your lungs. The inhaled air may contain disease-causing organisms or noxious chemicals and particles. The respiratory system clears most of these materials out before they can cause harm. However, some of the disease-causing organisms, including viruses and bacteria, and other harmful substances remain and cause problems.

Common Cold

Any one of more than 200 viruses can cause a cold. Not surprisingly, some 30 million Americans have a cold at this moment. (The "common cold" is indeed common.) Because there are so many cold-causing viruses, you can, and usually do, get several colds a year, each from a different virus. Most people get their first cold before they are one year old.

Typically, a cold begins with a runny nose, possibly a sore throat, and sneezing. In the beginning, the nasal discharge is thin and watery, but it becomes thicker as it fills the nasal cavity. Almost any part of the respiratory system can be affected. Sneezing and a stuffy nose indicate that the infection is in the upper respiratory system. When the pharynx is affected, a sore throat results. The infection may spread to the bronchi, causing a cough, or to the larynx, making your voice hoarse.

As miserable as you feel with a cold, you can take comfort from knowing that your suffering will not last forever (see the

Health Issue essay, *Surviving a Common Cold*). A cold is self-limiting, lasting only 1 to 2 weeks. Furthermore, colds are seldom fatal, except occasionally among the very young or very old or in those people already seriously ill with another malady.

Colds are spread when the causative virus is transmitted from an infected person. The viruses are plentiful in nasal secretions. However, transmission of the virus is not usually the result of direct inhalation of infected droplets from a cough or sneeze. It is more likely to be the result of handling an object that is contaminated with the virus. Cold viruses may remain capable of causing an infection on the skin or on an object for several hours, waiting for an unsuspecting person to touch them and contaminate his or her fingers. Subsequently, when the virus-laden fingers are touched to the mucous membranes of the nose, the transfer is completed. The best ways to prevent a cold are to wash your hands frequently and avoid being around people who have colds.

Flu

Flu is an abbreviation of *influenza,* another viral disease. The kinds of viruses that cause the flu are few compared to the number of different viruses that can cause a cold. In fact, the viruses that cause the flu in humans are all variants of three major types—A, B, and C (but there are hundreds of variants of these three basic types). Influenza of the A type is often more serious than B in that it is more frequently accompanied by severe complications, and it more frequently results in death. Influenza C causes a mild illness with cold symptoms.

Symptoms of the flu are similar to those of a cold, but they appear more suddenly and are more severe. (Though the onset of the flu may seem sudden, actually, by the time flu symptoms emerge, the disease has been incubating for several days.) A typical flu begins with chills and a high fever, about 103°F (39°C) in adults and perhaps higher in children. Many flu victims experience aches and pains in the muscles, especially in the back. Other common symptoms include a headache, sore throat, dry cough, weakness, pain and burning in the eyes, and sensitivity to light. When the flu hits, you usually feel sick enough to go to bed. The flu generally lasts for 7 to 10 days, but an additional week or more may pass before you are completely back on your feet.

The flu is often complicated by secondary infections, which set in when other disease-causing organisms take advantage of the body's weakened state. The most common complication is pneumonia, an inflammation of the lungs (described next). Possible secondary infections caused by bacteria include bronchitis, sinusitis, and ear infections.

One way to prevent the flu is to get a flu shot—a vaccine made from the strains of viruses that scientists anticipate will cause the next outbreaks of the illness. Flu shots are only about 60% to 70% effective because the viruses they target mutate rapidly, causing new strains to appear. The new strains are not recognized by the immune system defenses that were programmed by the latest vaccine. Because each flu season brings new strains of flu viruses, the effectiveness of the vaccine lasts only as long as that season's most prevalent strains. As a result, new vaccines must be developed continually to protect us, and flu shots must be repeated each year.

Pneumonia

Pneumonia is an inflammation of the lungs that causes fluid to accumulate in the alveoli, thus reducing gas exchange. It also causes the bronchioles to swell and narrow, making breathing difficult. Pneumonia is usually caused by infection with bacteria or viruses, but fungi and protozoans can also cause it. Many cases of pneumonia develop after a common cold or influenza. Radiation, chemicals, and allergies can also bring it on.

Symptoms of pneumonia often begin suddenly. They include fever and chills, chest pain, cough, and shortness of breath. The severity of pneumonia varies from mild to life threatening. Treatment depends on the cause of the illness.

Strep Throat

Strep throat, a sore throat caused by *Streptococcus* bacteria, is a problem mainly in children 5 to 15 years old. The soreness is usually accompanied by swollen glands (lymph nodes) and a fever.

Although the pain of strep throat may be so mild that a doctor is never consulted, ignoring a strep infection can have serious consequences. If untreated, the *Streptococcus* bacteria can spread to other parts of the body and cause rheumatic fever or kidney problems. The main symptoms of rheumatic fever are swollen, painful joints and a characteristic rash. About 60% of rheumatic fever sufferers develop disease of the heart valves. Another possible consequence of a streptococcal infection is kidney disease (glomerulonephritis). The kidney damage is due to a reaction from the body's own protective mechanisms. The body produces antibodies that destroy the bacteria, but if these antibodies persist after the bacteria have been killed, they can cause the kidneys to become inflamed. The inflamed kidneys may be unable to filter the blood, and blood may leak into the urine.

Symptoms of strep throat include a sore throat and two or three of the following: fever of 101°F (38.33°C), white or yellow coating on tonsils, or swollen glands in the neck. Because many viruses can cause sore throats that look like strep infections, the only way to identify strep throat is to test for the causative organism. If *Streptococcus* bacteria are found, an antibiotic, usually penicillin, is prescribed to prevent rheumatic fever and kidney disease.

Tuberculosis

Tuberculosis (TB) is caused by a rod-shaped bacterium, *Mycobacterium tuberculosis*. It is spread when the cough of an infected person sends bacteria-laden droplets into the air and the bacteria are inhaled into the lungs of an uninfected person. Because the bacteria are inhaled, the lungs are usually the first sites attacked, but the bacteria can spread to any part of the body, especially to the brain, kidneys, or bone.

As a defense against the bacteria, the body forms fibrous connective tissue casings, called *tubercles,* that encapsulate the bacteria (hence, the name of the disease). Although the formation of tubercles slows the spread of the disease, it does not actually kill the bacteria. The immune system destroys at least some of the walled-off bacteria and may, in fact, kill them all. But pockets of bacteria may persist undetected for many years. Later, if the immune system becomes weakened, the disease may progress to the secondary stage as pockets of bacteria

HEALTH ISSUE

Surviving a Common Cold

The only thing more common than the cold is advice on how to treat it. Here we will examine the validity of some frequently suggested treatments for a cold.

1. **Take large doses of vitamin C?** Vitamin C will not prevent a cold unless you are malnourished or under extreme physical stress. Nonetheless, some people who take vitamin C may experience less severe or shorter colds than do people who do not take vitamin C.

2. **Suck on a zinc lozenge?** Some evidence exists that zinc slows viral replication, prevents cold viruses from adhering to nasal membranes, and boosts the immune system.

3. **Take echinacea?** Commonly known as the purple coneflower, echinacea has been used for centuries by cold sufferers. However, scientific studies examining whether the herbal medication is effective in preventing or treating the common cold have produced conflicting results.

4. **Take an antibiotic?** Antibiotics are *not* effective against viruses and cannot cure a cold. Unnecessary use of an antibiotic may cause side effects such as diarrhea and can lead to the development of bacterial resistance to the drug, as discussed in Chapter 13a.

5. **Go to bed?** Bed rest enables the body to muster its resources and fight secondary infections. Staying at home with a cold is also socially responsible, because it helps prevent the spread of the virus. Bed rest will not, however, cure your cold or shorten its duration.

6. **Have some chicken soup?** Grandmothers have long prescribed chicken soup to treat a cold, and doctors finally agree that the advice has some merit. You should always consume plenty of fluids when you have a cold. They help loosen secretions in the respiratory tract and thus reduce congestion, allowing you to breathe more easily. Hot fluids, such as chicken soup, are more effective than cold ones for increasing the flow of nasal mucus.

Questions to Consider

- Do you think that it should be illegal to market cold remedies that have not been proven effective?
- A cold is not usually a serious illness. Should employees be prohibited from taking sick days for a cold?

---

become reactivated and multiply. Furthermore, bacteria may escape from the tubercles and be carried by the bloodstream to other parts of the body. Whenever the patient becomes weak, ill, or poorly nourished, bacteria can become active and multiply, causing the disease to flare up.

The initial symptoms of tuberculosis, if they occur, are similar to those of the flu. In the secondary stage, the patient usually develops a fever, loses weight, and feels tired. If the infection is in the lungs, as is usual, it causes a dry cough that eventually produces pus-filled and blood-streaked phlegm. In this stage the bacteria multiply rapidly and destroy the cells of the affected organ. The second stage of TB can be fatal, especially if it is caused by a multi-drug-resistant strain of bacteria, and resistant strains are becoming increasingly common (discussed in Chapter 13a).

what would you do?

DOTS (Directly Observed Therapy, short course) is the World Health Organization's recommended treatment for tuberculosis. DOTS, which mandates that someone witness the TB patient swallowing medication each day, is in place in most major cities in the United States. The treatment ranges from 6 months to 2 years (the latter, if a multi-drug-resistant strain is present). Patients' failure to complete TB treatment is what has led to the drug-resistant strains of the TB bacterium. Is DOTS a fair balance of protection of the public with personal rights? What do you think?

Cystic Fibrosis

Cystic fibrosis is an inherited disorder of the lungs and digestive system. A mutation in a single gene causes thick, sticky mucus that clogs air passageways and traps bacteria that can cause serious lung infections. (The inheritance of cystic fibrosis is discussed in Chapter 20.) Treatment for the respiratory effects of cystic fibrosis is two-pronged: preventing and controlling lung infections and loosening and removing mucus.

Bronchitis

Viruses, bacteria, or chemical irritation may cause the mucous membrane of the bronchi to become inflamed—a condition called *bronchitis*. The inflammation results in the production of excess mucus, which triggers a deep cough that produces greenish yellow phlegm.

There are two types of bronchitis: acute and chronic. *Acute bronchitis*, which often follows a cold, is usually caused by the cold virus itself, but it may be caused by bacteria that take advantage of the body's lowered resistance and invade the trachea and bronchi. An antibiotic will hasten recovery if the cause is bacterial.

When a cough that brings up phlegm is present for at least 3 months in each of 2 consecutive years, the condition is called *chronic bronchitis*, a more serious problem that is usually associated with cigarette smoking or air pollution. Some people with chronic bronchitis may lack an enzyme that normally protects the air passageways from such irritants. As the disease progresses, breathing becomes increasingly difficult, partly because the linings of the air tubules thicken, narrowing the passageway for air. Contraction of the muscles in bronchiole walls and excessive secretion of mucus further obstruct the air tubules.

Chronic bronchitis can have serious consequences. The degenerative changes in the lining of the air tubules make removal of mucus more difficult. As a result, the patient is more likely to develop lung infections such as pneumonia, which can be fatal, and degenerative changes—such as emphysema—in the lungs.

Emphysema

Emphysema is a common consequence of smoking, although this condition can have other causes as well. In *emphysema*,

(a) Normal alveoli

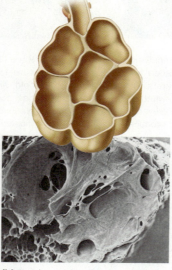
(b) Emphysema causes breakdown of alveolar walls.

FIGURE 14.14 *A comparison of (a) normal alveoli and (b) alveoli in an individual with emphysema. Notice that in emphysema, the alveolar walls rupture. As a result, the surface area for gas exchange decreases, the dead air space increases, and the alveolar walls thicken.*

the walls of the alveoli break down and merge, thereby making the alveoli fewer and larger (Figure 14.14). This change has two major effects: a reduction in the surface area available for gas exchange and an increase in the volume of residual, or "dead," air in the lungs. Exhalation, you may recall, is a passive process that depends on the elasticity of lung tissue. In emphysema, the lungs lose that elasticity, and air becomes trapped inside them. As the dead air space increases, adequate ventilation requires more forceful inhalation. Forcing the air in causes more alveolar walls to rupture, further increasing the dead air space. Lung size gradually increases as the residual volume of air becomes greater, giving a person with emphysema a characteristic barrel chest, but gas exchange continues to become more difficult because surface area is reduced, despite the increase in lung size. To get an idea of what poor lung ventilation caused by increased dead air space feels like, take a deep breath, then exhale only slightly, and repeat this process several times. Notice how quickly you feel an oxygen shortage if you continue taking very shallow breaths that keep the lungs almost completely filled with air.

Shortness of breath, the main symptom of emphysema, is a result of both the decreased alveolar surface area and the increased dead air space. As the disease progresses, gas exchange becomes even more difficult because the alveolar walls thicken with fibrous connective tissue. The oxygen that does reach the alveoli has difficulty crossing the connective tissue to enter the blood. Thus, a person with emphysema constantly gasps for air.

Emphysema can be treated, but it cannot be cured. Cigarette smoking is the most common cause for emphysema, so the first step in treatment is usually to quit smoking. Medicines to widen the respiratory air tubules can be prescribed to make airflow to and from the lungs easier. In addition, supplemental oxygen may be administered to increase the amount of oxygen reaching the patient's body cells, which will relieve some of the symptoms of emphysema.

Lung Cancer

Between 85% and 90% of all cases of lung cancer are caused by smoking and are, therefore, preventable. Lung cancer usually has no symptoms until it is quite advanced. Therefore, it is often not detected in time for a cure.

The progression to lung cancer begins with chronic inflammation of the lungs and is marked by changes in the cells of the airway linings. These cell changes are often caused by inhaled carcinogens, including those found in tobacco smoke (more chemicals found in tobacco smoke—and the respiratory problems they cause—are discussed in the Health Issue essay, *Smoking and Lung Disease*.)

In a nonsmoker, the lining of the air passageways has a basement membrane underlying basal cells and a single layer of ciliated columnar cells. In a smoker, one of the first signs of damage is an increase in the number of layers of basal cells. Next, the ciliated columnar cells die and disappear. The nuclei of the basal cells then begin to change as mutations accumulate, and the cells become disorganized. This is the beginning of cancer. Eventually, the uncontrolled cell division forms a tumor (Figure 14.15). When cancer cells break through the basement membrane, they can spread to other parts of the lung and to the rest of the body through a process called *metastasis* (see Chapter 21a).

looking ahead

In this chapter, we considered how the respiratory system obtains the oxygen we need to survive and rids our bodies of carbon dioxide. In Chapter 15, we will examine another function vital to life—digestion.

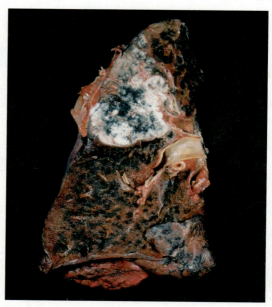
FIGURE 14.15 *Lung cancer. The tumor is the light-colored solid mass shown in the upper region of the lung.*

HEALTH ISSUE

Smoking and Lung Disease

Smoking is the greatest single preventable cause of disease, disability, and death in our society. In fact, every cigarette pack and cigarette advertisement in the United States must bear a warning from the Surgeon General stating the dangers of tobacco smoke. Yet tobacco, which causes bodily harm when used exactly as intended, is legal to sell to anyone at least 18 years old.

Because the objective of smoking is to bring smoke into the lungs, we can expect some of its most damaging effects to occur there. The damage to the respiratory system of smokers is gradual and progressive. It begins as the smoke hampers the actions of two of the lungs' cleansing mechanisms—cilia and macrophages. Even the first few puffs from a cigarette slow the movement of the cilia, making them less effective in sweeping debris from the air passageways. Smoking an entire cigarette prevents the cilia from moving for an hour or longer. With continued smoking, the nicotine and sulfur dioxide in the smoke paralyze the cilia, and the cyanide destroys the ciliated cells.

Cigarette smoke causes a smoker's lungs to be chronically inflamed. The inflammation summons macrophages, the wandering cells that engulf foreign debris, to the lungs for the purpose of cleaning the lung surfaces. But, just as smoke paralyzes the cilia, it also paralyzes the macrophages, further hampering the cleansing efforts. As the cilia and macrophages become less effective, greater quantities of tar and disease-causing organisms remain within the respiratory system. As a result, cigarette smokers spend more time sick in bed and lose more workdays each year than do nonsmokers.

At the same time that the cilia and macrophages are being slowed, the smoke stimulates the mucus-secreting cells in the linings of the respiratory passageways. Consequently, the smaller airways become plugged with mucus, making breathing more difficult. At this point, if not before, "smoker's cough" begins. Coughing is a protective reflex, and initially the smoker coughs simply because smoke irritates air passageways. However, as smoking continues and the cilia become less and less able to remove mucus and debris, the only way to remove the material from the passageways is to cough. The cough is generally worse in the morning, when the body attempts to clear away the mucus that accumulated overnight.

Gradually, the inflammation and congestion within the lungs, along with the constant irritation from smoke, lead to chronic bronchitis, a disease characterized by a persistent deep cough that brings up mucus. The air passageways become narrow due to the thickening of their linings caused by repeated infection, accumulation of mucus, and contraction of the smooth muscle in their walls. Airflow is restricted, resulting in breathlessness and wheezing.

Emphysema, in which the walls of alveoli are destroyed, is often the next stage in the progressive damage to the lungs. Airways and alveolar walls lose elasticity as tar causes body defense cells to secrete destructive enzymes. Consequently, the lung tissues can no longer absorb the increase in pressure that accompanies a cough, and the delicate alveolar walls break like soap bubbles. With more and more alveoli destroyed, the surface area for gas exchange is reduced, so less oxygen is delivered to the body. Besides all these negative effects on the respiratory system, smoking is the single major cause of lung cancer, and it causes other cancers as well. In fact, smoking is responsible for 30% of all cancer deaths.

The health benefits if you quit smoking are enormous. Much of the damage caused by smoke is reversible once you quit. For example, the risk of lung cancer drops, though it never falls as low as the level for people who have never smoked.

Questions to Consider

- Do you think that a ban on smoking in buildings violates smokers' rights?
- Many states are drastically increasing the tax on cigarettes. This action increases revenues and discourages an unhealthy behavior. Do you think that such a tax is unfair because it targets a particular segment of the population?
- What steps might you take to help a loved one quit smoking?

HIGHLIGHTING THE CONCEPTS

14.1 Structures of the Respiratory System (pp. 268–274)

- The role of the respiratory system is to exchange oxygen and carbon dioxide between the air and the blood. The oxygen that we breathe is needed to maximize the number of energy-storing ATP molecules formed from food energy. Exhaling carbon dioxide, a waste product formed by the same reactions, helps regulate the acid–base balance of body fluids.
- The first structure that inhaled air usually passes through is the nose, which serves to clean, warm, and moisten the incoming air.
- The sinuses are air-filled spaces in the facial bones that also help warm and moisten the air.
- After leaving the nose, the inhaled air passes through the pharynx, or throat, and then the larynx, or voice box. Movements of the larynx during swallowing prevent food from entering the airways and lungs.
- The air passageways include the trachea, which branches to form bronchi, which branch various times in the lungs and eventually form progressively smaller branching tubules called bronchioles. The bronchioles terminate at the lungs' gas-exchange surfaces, the alveoli. Each alveolus is a thin-walled air sac encased in a capillary network.

14.2 Mechanism of Breathing (pp. 274–276)

- Pressure changes within the lungs, caused by changes in the size of the thoracic cavity, move air into and out of the lungs. Inspiration occurs when the size of the thoracic cavity increases, causing the pressure in the lungs to drop below atmospheric pressure. Expiration occurs when the size of the thoracic cavity decreases and pressure in the lungs rises above atmospheric pressure.

14.3 Transport of Gases between the Lungs and the Cells (pp. 276–278)

- Oxygen and carbon dioxide are exchanged between the alveolar air and the capillary blood by diffusion along their concentration (partial pressure) gradients. Oxygen diffuses from the alveoli into the blood, where it binds to hemoglobin within the red blood cells and is delivered to the body cells. A small amount of carbon dioxide is carried to the lungs dissolved in the blood plasma or bound to hemoglobin. Most, however, is transported to the lungs as bicarbonate ions.

14.4 Respiratory Centers in the Brain (pp. 278–279)

- The basic rhythm of breathing is controlled by the inspiratory area within the medulla of the brain. The neurons within this

center undergo spontaneous bouts of activity. When they are active, messages are sent, causing contraction of the diaphragm and the muscles of the rib cage. As a result, the thoracic cavity increases in size, and air is drawn into the lungs. When the inspiratory neurons are inactive, the diaphragm and rib cage muscles relax, and exhalation occurs passively. Also in the medulla is an expiratory area that causes forceful exhalation during heavy breathing.
- The most powerful stimulant to breathing is an increased number of hydrogen ions in the blood, formed from carbonic acid when carbon dioxide dissolves in plasma. Extremely low levels of oxygen also increase breathing rate.

14.5 Respiratory Disorders (pp. 279–282)
- The common cold and the flu are caused by viruses. Pneumonia is an inflammation of the lungs (caused by bacteria or viruses) that causes fluid to fill the alveoli and narrows bronchioles. Strep throat is a sore throat caused by *Streptococcus* bacteria. Acute bronchitis is caused by either bacteria or a virus. Chronic bronchitis is a persistent irritation of the bronchi. Emphysema is a breakdown of the alveolar walls and thus a reduction in the gas-exchange surfaces. Chronic bronchitis and emphysema are usually caused by smoking or air pollution. The primary cause of lung cancer is cigarette smoking.

RECOGNIZING KEY TERMS

respiratory system *p. 269*
nasal cavities *p. 269*
sinuses *p. 271*
pharynx *p. 272*
larynx *p. 272*
epiglottis *p. 272*
glottis *p. 272*
Heimlich maneuver *p. 272*

vocal cords *p. 272*
trachea *p. 272*
bronchi *p. 273*
bronchioles *p. 273*
bronchial tree *p. 273*
alveolus *p. 274*
surfactant *p. 274*
diaphragm *p. 275*

intercostal muscles *p. 275*
inspiration *p. 275*
expiration *p. 275*
tidal volume *p. 276*
inspiratory reserve volume *p. 276*
expiratory reserve volume *p. 276*

vital capacity *p. 276*
residual volume *p. 276*
total lung capacity *p. 276*
oxyhemoglobin *p. 276*
carbaminohemoglobin *p. 277*
carbonic anhydrase *p. 277*

REVIEWING THE CONCEPTS

1. Why must we breathe oxygen? *p. 268*
2. Trace the path of air from the nose to the lungs. *p. 269*
3. How are most particles and disease-causing organisms removed from the inhaled air before it reaches the lungs? *pp. 269–270*
4. Explain why food does *not* usually enter the lower respiratory system when you swallow. *p. 272*
5. How is human speech produced? *p. 272*
6. What is the function of the cartilage rings in the trachea? What would happen to the trachea without these rings? *p. 273*
7. What is the bronchial tree? *p. 273*
8. How are the pressure changes in the thoracic cavity that are responsible for breathing created? *pp. 274–275*
9. Which is a larger volume of air: tidal volume or vital capacity? Explain. *p. 276*
10. How is most oxygen transported to the body cells? *pp. 276–277*
11. How is most carbon dioxide transported from the cells to the lungs? *p. 277*
12. Which region of the brain causes the basic breathing rhythm? *p. 278*
13. Explain how blood carbon dioxide levels regulate the breathing rate. *p. 279*
14. What are the causes of the shortness of breath experienced by people with emphysema? *p. 282*
15. Choose the correct statement.
    a. During quiet breathing, expiration does not usually involve the contraction of muscles.
    b. Expiration occurs when the diaphragm and the rib muscles contract.
    c. Expiration occurs as the chest (thoracic) cavity enlarges.
    d. The larynx acts like a suction pump to pull air into the lungs.
16. You should be able to hold your breath longer than normal after you hyperventilate (breathe rapidly for a while) because hyperventilating
    a. decreases blood oxygen levels.
    b. decreases blood carbon dioxide levels.
    c. increases blood oxygen levels.
    d. increases blood carbon dioxide levels.
17. Which structure is specialized to produce the sound of your voice?
    a. trachea
    b. larynx
    c. bronchiole
    d. epiglottis
18. As a molecule of oxygen enters the body and is delivered to the cells, it passes through many structures. Which sequence shows the correct pathway?
    a. nose, pharynx, larynx, alveolus, bronchus, trachea
    b. pharynx, bronchiole, larynx, alveolus, bronchus, trachea
    c. larynx, pharynx, alveolus, trachea, bronchus
    d. nose, pharynx, trachea, bronchus, bronchiole, alveolus
19. What is the most important cue that prompts you to breathe again after holding your breath?
    a. the decrease in carbon dioxide levels
    b. the increase in carbon dioxide levels
    c. the increase in oxygen levels
    d. the decrease in oxygen levels
20. What is the reason why we must breathe oxygen?
    a. Oxygen is broken down to yield energy.
    b. Oxygen allows us to get more energy (in the form of ATP) out of the food molecules we break down for energy.
    c. Oxygen stimulates enzyme activity.
    d. Oxygen is the chemical detected by chemoreceptors that keeps us breathing.

21. Choose the correct statement:
    a. Cartilage rings keep the alveoli open.
    b. The temperature and humidity of the air are adjusted as the air flows through the nasal cavities.
    c. The larynx changes shape as we speak, creating resonance chambers of different sizes.
    d. The center of each lung contains a large hollow area in which most of the gas exchange occurs.
22. Choose the *incorrect* statement about the production of sound called the voice:
    a. Speaking sounds are produced when the epiglottis vibrates.
    b. Thicker vocal cords produce deeper voice sounds.
    c. Laryngitis results when the vocal cords become swollen and thick and cannot vibrate easily.
    d. There are only two true vocal cords.
23. When you swallow food, why does it *not* usually enter the respiratory system?
    a. The Heimlich maneuver prevents it from doing so.
    b. The bronchioles produce mucus.
    c. The diaphragm contracts.
    d. The epiglottis covers the opening to the respiratory system.
    e. The vocal cords block the opening to the respiratory system.
24. In a healthy person, most of the particles that are inhaled into the respiratory system
    a. are trapped in the mucus and moved by cilia to the pharynx (toward the digestive system).
    b. pass through the alveoli into the circulatory system, where they are engulfed by white blood cells.
    c. are caught on the vocal cords.
    d. are trapped in the sinuses.
25. In emphysema,
    a. the number of alveoli is reduced.
    b. cartilage rings in the trachea break down.
    c. the diaphragm is paralyzed.
    d. the epiglottis becomes less mobile
26. The _____ is the flap that covers the trachea to prevent food from entering during swallowing.
27. The enzyme in red blood cells that reversibly converts carbonic acid to bicarbonate ions and hydrogen ions is _____.

APPLYING THE CONCEPTS

1. Tatyana is a young woman with iron-deficiency anemia, so her blood does not carry enough oxygen. Would you expect this condition to affect her breathing rate or tidal volume? Why or why not?
2. Cigarette smoke destroys the cilia in the respiratory system. Explain why the loss of these cilia is a reason that cigarette smokers tend to lose more workdays because of illness than do nonsmokers.
3. Rosa has a 4-year-old son, Juan, who threatens to hold his breath until she gives him a candy bar. Should she be worried about Juan's holding his breath? Why?
4. Vincent is having an asthma attack. During an asthma attack, the bronchioles constrict (get narrower in diameter). Does Vincent have more difficulty inhaling or exhaling? Why?

BECOMING INFORMATION LITERATE

Use at least three reliable sources (books, journals, or websites) to answer the following questions. List each source you considered, and explain why you chose the three sources you used.
1. The incidence of childhood asthma has increased steadily during the last few decades. What hypotheses have been proposed (at least three) to explain this increase in asthma? What evidence supports each hypothesis?
2. Studies have shown that secondhand cigarette smoke can pose health risks to nonsmokers nearby. Many state and local governments have laws preventing smoking in public places so that nonsmokers are not exposed to secondhand smoke. Present at least two arguments for and two against laws that ban smoking in public places. Then take a stand for or against such laws and defend your position.
3. Many of the first responders to the site of the collapse of the World Trade Center on September 11, 2001, have developed respiratory problems. Describe some of these respiratory problems. What is being done to treat these problems? Is there financial assistance for the affected people to help pay for their treatment?

MasteringBiology®

Go to MasteringBiology for practice quizzes, activities, eText, videos, current events, and more.

15 The Digestive System and Nutrition

Did you Know?

- Of the 11.5 liters of digested food, liquids, and digestive juices that flow through your digestive system each day, only 100 ml are lost in feces.
- Your stomach replaces its mucous lining every 2 weeks.

15.1 The Gastrointestinal Tract
15.2 Specialized Compartments for Food Processing
15.3 Nerves and Hormones in Digestion
15.4 Planning a Healthy Diet
15.5 Nutrients
15.6 Food Labels
15.7 Energy Balance
15.8 Obesity
15.9 Weight-Loss Programs
15.10 Eating Disorders

HEALTH ISSUE
Peptic Ulcers

In the previous chapter, we learned how we obtain oxygen and rid our bodies of carbon dioxide. In this chapter, we learn how we obtain necessary nutrients from the food we eat. We learn that food usually travels in one direction through the digestive system. We see how the digestive system is organized for breaking down food molecules and making them available for use as a source of energy or as raw material for the growth and repair of cells. Finally, we consider the nutrients in a balanced diet—proteins, fats, vitamins, minerals, and water. We see how the body uses these nutrients and examine ways to combine the foods we eat to promote health and fitness.

The digestive system breaks food into its component subunits that are small enough to absorb. The subunits are then used for energy or as building blocks to make new molecules.

15.1 The Gastrointestinal Tract

There is some truth to the saying, "You are what you eat." Rest assured, however, that no matter how many hamburgers you eat, you will never become one. Instead, the hamburger becomes *you*. This transformation is largely due to the activities of the digestive system. Like an assembly line in reverse, the digestive system takes the food we eat and breaks the complex organic molecules into their chemical subunits. The subunits are molecules small enough to be absorbed into the bloodstream and delivered to body cells, where they provide either materials for growth and repair of the body or energy for daily activities. Imagine that you just ate a cheeseburger on a bun. The starch in the bun may fuel a jump for joy; the protein in the beef may be used to build muscle; and the fat in the cheese may become myelin sheaths that insulate nerve fibers. The digestive system breaks food into molecules small enough to be absorbed and delivered to the cells that use them. Table **15.1** identifies the structures of the digestive system and describes their roles in both mechanical and chemical digestion. You can refer to the table as you read about each of the following digestive structures.

TABLE 15.1	Review of Structures of the Digestive System		
Structure	**Description/Functions**	**Mechanical Digestion**	**Chemical Digestion**
Mouth	Receives food; contains teeth and tongue; tongue manipulates food and monitors quality	Teeth tear and crush food into smaller pieces	Digestion of carbohydrates begins
Pharynx	Passageway for both food and air	None	None
Esophagus	Tube that transports food from mouth to stomach	None	None
Stomach	J-shaped muscular sac for food storage	Churning of stomach mixes food with gastric juice, creating liquid chyme	Protein digestion begins
Small intestine	Long tube where digestion is completed and nutrients are absorbed	Segmental contractions mix food with intestinal enzymes, pancreatic enzymes, and bile	Carbohydrate, protein, and fat digestion are completed
Large intestine	Final tubular region of GI tract; absorbs water and ions; houses bacteria; forms and expels feces	None	Some digestion is carried out by bacteria
Anus	Terminal outlet of digestive tract	None	None

The **digestive system** consists of a long, hollow tube, called the **gastrointestinal (GI) tract**, into which various accessory glands release their secretions (Figure 15.1). The GI tract begins at the mouth and continues to the pharynx, esophagus, stomach, small intestine, and large intestine. The hollow area of the tube through which food and fluids travel is called the *lumen*. Along most of its length, the walls of the GI tract have four basic layers.

- **Mucosa.** The innermost layer of the GI tract is the moist, mucus-secreting layer called the **mucosa**. The mucus helps lubricate the lumen, allowing food to slide through easily. Mucus also helps protect the cells in the lining of the GI tract from rough substances in the food and from digestive enzymes. In some regions of the digestive system, cells in the mucosa also secrete digestive enzymes. In addition, the mucosa in some parts of the digestive tract is highly folded, which increases the surface area for absorption.
- **Submucosa.** The next layer, the **submucosa**, consists of connective tissue containing blood vessels, lymph vessels, and nerves. The blood supply maintains the cells of the digestive system and, in some regions, picks up and transports the products of digestion. The nerves are important in coordinating the contractions of the next layer.
- **Muscularis.** The next layer, the **muscularis**, is responsible for movement of materials through the GI tract and for mixing ingested materials with digestive secretions. In most sections of the GI tract, the muscularis is a double layer of smooth muscle. (The stomach, as you will read later, is an exception; it has three layers of muscle.) In the inner, "circular" layer of muscle, the muscle cells encircle the tube, causing a constriction when they contract. In the outer, longitudinal layer, the muscle cells are arranged parallel to the GI tract. Longitudinal muscles shorten the GI tract when they contract. The muscle layers churn the food until it is liquefied, mix the resulting liquid with enzymes, and propel the food along the GI tract in a process called *peristalsis*, which is discussed later in this chapter. Local muscle contractions in random regions of the GI tract, especially in the small intestine, are called segmental contractions. Segmental contractions mix intestinal contents—food and digestive juices (fluids containing enzymes and other fluids that aid in digestion). Segmental contractions also help the body absorb digested food by moving intestinal contents over the intestinal wall.
- **Serosa.** The **serosa**, a thin layer of epithelial tissue supported by connective tissue, wraps around the GI tract. It secretes a fluid that lubricates the outside of the GI tract to reduce friction with surfaces of the intestine and other abdominal organs that come into contact with the GI tract (Figure 15.2 on page 289).

15.2 Specialized Compartments for Food Processing

As food moves along the GI tract, it passes through the mouth, pharynx, esophagus, stomach, small intestine, and large intestine. The salivary glands, liver, and pancreas add secretions along the way. Most nutrients are absorbed from the small intestine. Additional water is absorbed in the large intestine. Undigested and indigestible materials pass out the anus.

Mouth

The entryway to the digestive system and the first stop on food's journey through it is the *mouth*, also called the *oral cavity*. The roof of the mouth is called the *palate*. The region of the palate closest to the front of the mouth, the hard palate, is reinforced with bone. Toward the back of the mouth is the soft palate, which consists only of muscle and prevents food from entering the nose during swallowing. The mouth serves several functions: (1) it begins mechanical and, to some extent, chemical digestion; (2) it monitors food quality; and (3) it moistens and manipulates food so that it can be swallowed. The teeth, salivary glands, and tongue all contribute to these functions.

Teeth and mechanical digestion As we chew, our teeth break solid foods into smaller fragments that are easier to swallow and digest. The sharp, chisel-like incisors in the front of the mouth (Figure 15.3a) slice the food as we bite into it. At the same time, the pointed canines to the sides of the incisors tear the food.

288 CHAPTER 15 The Digestive System and Nutrition

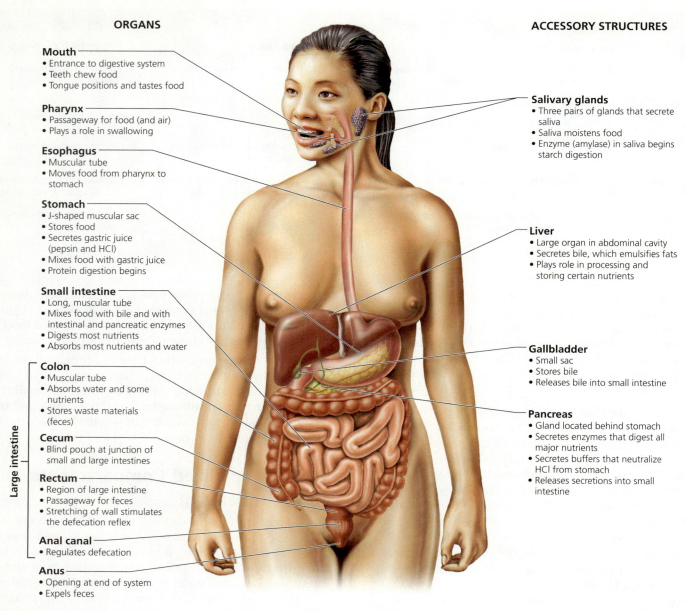

FIGURE 15.1 The digestive system consists of a long tube, called the gastrointestinal tract, into which accessory structures release their secretions.

Then the food is ground, crushed, and pulverized by the premolars and molars, which lie along the sides of the mouth.

Teeth are alive. In the center of each tooth is the pulp, which contains the tooth's life-support systems—blood vessels that nourish the tooth and nerves that sense heat, cold, pressure, and pain (Figure 15.3b). Surrounding the pulp is a hard, bonelike substance, called *dentin*. The crown of the tooth (the part visible above the gum line) is covered with enamel, a nonliving material that is hardened with calcium salts. The root of the tooth (the part below the gum line) is covered with a calcified, yet living and sensitive connective tissue called cementum. The roots of the teeth fit into sockets in the jawbone. Blood vessels and nerves reach the pulp through a tiny tunnel through the root, called the root canal.

Tooth decay is caused by acid produced by bacteria living in the mouth. When you eat, food particles become trapped between the teeth, in the small spaces where the teeth meet the gums, and in the hollows of molars. The sugar in these food particles nourishes bacteria in the mouth. As bacteria digest the sugar, acid is produced that erodes the enamel and causes a cavity to form. Blood vessels in the pulp widen in response to this erosion, allowing greater numbers of white bloods cells to reach the area and fight infection. The widened blood vessels may press on nerves, causing a toothache. When the enamel has been penetrated, bacteria can invade the softer dentin beneath. If a dentist does not fill the cavity, the bacteria can infect the pulp. *Plaque,* an invisible film of bacteria, mucus, and food particles, promotes tooth decay because it holds the acid against the enamel. Daily brushing and flossing helps remove plaque, reducing the chance of tooth decay.

Gum disease, which affects two of three middle-aged people in the United States, is a major cause of tooth loss

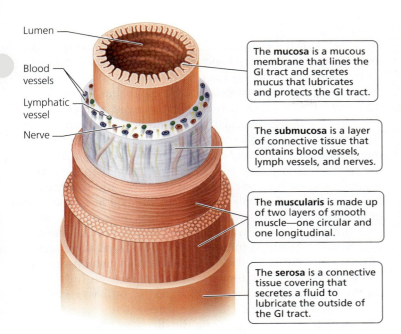

FIGURE 15.2 Along most of its length, the wall of the GI tract has four basic layers: the mucosa, the submucosa, the muscularis, and the serosa.

in adults. *Gingivitis* (*gingiv–*, the gums; *–itis*, inflammation of), an early stage of gum disease, occurs when plaque that has formed along the gum line causes the gums to become inflamed and swollen. The swollen gums can bleed and do not fit as tightly around the teeth as they should. The pocket that forms between the tooth and the gum traps additional plaque. The bacteria in the plaque can then attack the bone and soft tissues around the tooth, a condition called *periodontitis* (*peri–*, around; *dont*, teeth; *–itis*, inflammation of). As the tooth's bony socket and the tissues that hold the tooth in place are eroded, the tooth becomes loose.

Salivary glands and chemical digestion Three pairs of salivary glands—the sublingual (below the tongue), submandibular (below the jaw), and parotid (in front of the ears)—release their secretions, collectively called **saliva**, into the mouth (Figure 15.4). As we chew, food is mixed with saliva. Water in saliva moistens food, and mucus binds food particles together, making it easier for the food to pass through the GI tract.

Saliva also contains an enzyme, called **salivary amylase**, that begins to chemically digest starches into shorter chains of sugar. You will notice the result of salivary amylase activity if you chew a piece of bread for several minutes: The bread will begin to taste sweet. Try it.

Tongue: Taste and food manipulation The tongue is a large skeletal muscle studded with taste buds. Our ability to control the position and movement of the tongue is critical to both speech and the manipulation of food within the mouth. Once

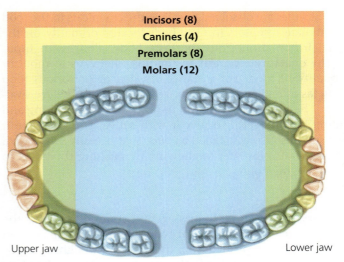

(a) The teeth slice, tear, and grind food until it can be swallowed.

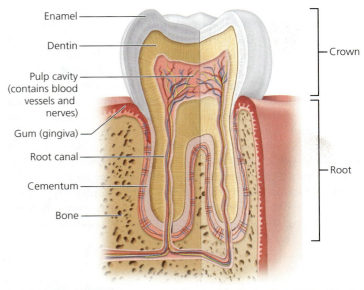

(b) The structure of the human tooth is suited for its function of breaking food into smaller pieces.

FIGURE 15.3 Adult human teeth

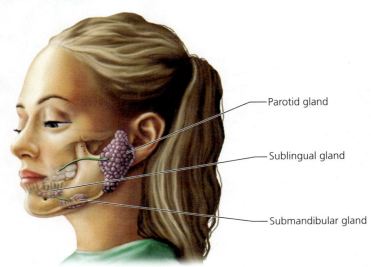

FIGURE 15.4 Three pairs of salivary glands release their secretions into the mouth. These secretions, collectively called saliva, make food easier to swallow, dissolve substances so they can be tasted, and begin the chemical digestion of starch.

food molecules are dissolved in saliva, the chemicals in the food can stimulate receptors in taste buds located primarily on the tongue. Information from the taste buds, along with input from the olfactory receptors in the nose, helps us to monitor the quality of food. For instance, spoiled or poisonous food usually tastes and smells bad, so we can spit it out before swallowing. The tongue moves food to position it for crushing and grinding by the teeth, to mix it with saliva, and to shape it into a small, soft mass, called a *bolus,* that is easily swallowed. The tongue also initiates swallowing by pushing the bolus to the back of the mouth.

Pharynx

The **pharynx**, which is the passageway commonly called the *throat,* is shared by the respiratory and digestive systems. When we swallow, food is pushed from the mouth, through the pharynx, and into the **esophagus**, the tube that connects the pharynx to the stomach.

Swallowing consists of a voluntary component followed by an involuntary one. When a person begins to swallow, the tongue pushes the bolus of softened and moistened food into the pharynx (Figure 15.5). Once food is in the pharynx, it is too late to change one's mind about swallowing. Sensory receptors in the wall of the pharynx detect the presence of food and stimulate the involuntary swallowing reflex. Reflex movements of the soft palate prevent food from entering the nasal cavities. Other involuntary muscle contractions push the larynx (the voice box, commonly called the *Adam's apple*) upward. As we learned in Chapter 14, the movement of the larynx causes a cartilaginous flap called the *epiglottis* to move, covering the opening to the airways of the respiratory system (the glottis). The movement of the epiglottis prevents food from entering the airways. Instead, food is pushed into the esophagus.

Esophagus

The esophagus is a muscular tube that conducts food from the pharynx to the stomach. Food is moved along the esophagus and all the rest of the GI tract by rhythmic waves of muscle contraction called **peristalsis**. In the esophagus, small intestine, and large intestine, peristalsis is produced by the two layers of muscle in the muscularis. The muscles of the inner layer circle the tube, causing a constriction when they contract. The muscles in the outer layer run lengthwise, causing a shortening of the region where they contract. The presence of food stretches the walls in one region of the tube, and this triggers the contraction of circular muscles in the region of the tube immediately behind the food mass. When these circular muscles contract, that region of the tube pinches inward, pushing food forward. The food then stretches the next adjacent region of the tube, again stimulating contraction of circular muscles behind it. At the same time, longitudinal muscles in front of the food contract, shortening this region and widening its walls to receive the food. We see, then, that gravity is not important in moving food along the digestive tract. It is possible, therefore, to swallow while standing on your head or in the weightless conditions of outer space (Figure 15.6).

Stomach

The **stomach** is a muscular sac that is well designed to carry out its three important functions: (1) storing food and regulating the release of food to the small intestine, (2) liquefying food, and (3) carrying out the initial chemical digestion of proteins. The structure of the stomach is shown in Figure 15.7.

Storage of food and regulation of the release of food to the small intestine Like any good storage compartment, the stomach is expandable and has openings that can close to seal the contents within as well as open for filling and emptying.

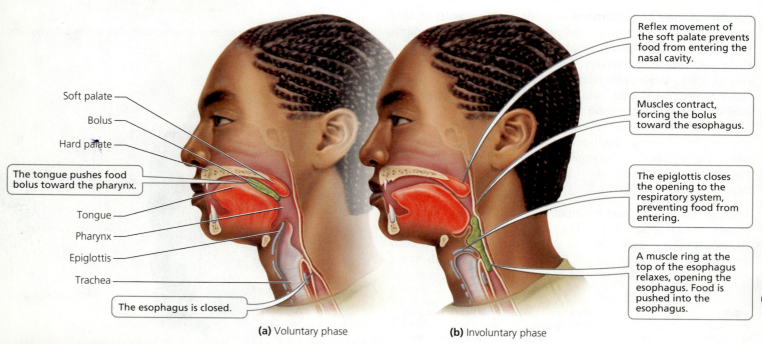

(a) Voluntary phase   **(b)** Involuntary phase

FIGURE 15.5 Swallowing consists of (a) voluntary and (b) involuntary phases.

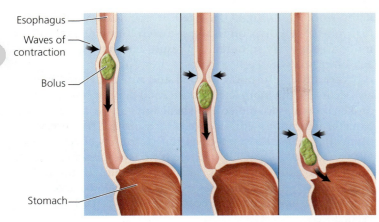

FIGURE 15.6 *Peristalsis is a wave of muscle contraction that pushes food along the esophagus and the entire remaining GI tract. When circular muscles contract, the tube is narrowed, and food is pushed forward. The longitudinal muscles in front of the bolus contract, shortening that region.*

When empty, the stomach is a small, J-shaped sac that can hold only about 50 ml (a quarter of a cup) without stretching. However, the wall of the empty stomach has folds that can spread out, allowing the stomach to expand as it fills. When fully expanded, as after a large meal, the stomach can hold several liters of food. Bands of circular muscle called *sphincters* guard the openings at each end of the stomach and regulate the release of food to the small intestine. Contraction of a sphincter closes the opening, and relaxation of a sphincter allows material to pass through.

Heartburn occurs when the pressure of the stomach contents overwhelms the sphincter at the lower end of the esophagus, causing a burning sensation behind the breastbone. In some people, this sphincter is weak and unable to keep stomach contents out of the esophagus. A person with a normal esophageal sphincter will experience heartburn when the stomach contents exert a greater pressure than usual against the sphincter, as might occur after a large meal, during pregnancy, when lying down, or when constipated.

People who have chronic heartburn—at least one attack per week—have an increased risk of developing cancer of the esophagus. The risk increases with the frequency of attacks and the number of years of experiencing heartburn. Those who have one attack per week have a risk eight times higher than normal. Esophageal cancer is particularly aggressive and has become more common in recent years. It is detected using a test called endoscopy, in which a thin, lighted tube is snaked down the throat. A more recently developed means of detecting potentially harmful changes in the esophagus is the PillCam. The patient swallows a camera that is about the size of a vitamin pill. As the camera moves through the esophagus, it takes thousands of pictures and sends them to a small recording device worn by the patient.

Liquefaction of food Food is generally stored and processed within the stomach for 2 to 6 hours. The stomach wall has three layers of smooth muscle, each oriented in a different direction. The coordinated contractions of these layers twist, knead, and compress the stomach contents, physically

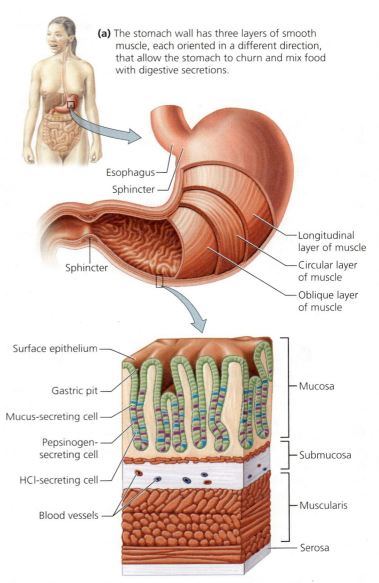

(a) The stomach wall has three layers of smooth muscle, each oriented in a different direction, that allow the stomach to churn and mix food with digestive secretions.

(b) Gastric glands in the wall of the stomach produce gastric juice, a mixture of hydrochloric acid and pepsin.

(c) The holes seen in this electron micrograph are the gastric pits, openings in the stomach wall through which gastric glands release their secretions.

FIGURE 15.7 *The structure of the stomach is well suited to its functions of churning food with digestive secretions, storing food, and beginning protein digestion.*

breaking food into smaller pieces. This additional mechanical digestion occurs as the food is churned and mixed with secretions produced by the glands of the stomach until it is a soupy mixture called **chyme**.

Initial chemical digestion of proteins Chemical digestion in the stomach is limited to the initial breakdown of proteins. The lining of the stomach has millions of gastric pits, within which are **gastric glands** containing several types of secretory cells. Certain secretory cells produce hydrochloric acid (HCl), which kills most of the bacteria swallowed with food or drink. Hydrochloric acid also breaks down the connective tissue of meat and activates pepsinogen, which is secreted by other cells in the gastric glands. Once activated by HCl, pepsinogen becomes **pepsin**, a protein-digesting enzyme. When the mixture of pepsin and HCl, called *gastric juice*, is released into the stomach, the pepsin begins the chemical digestion of the protein in food. Still other cells within the gastric glands secrete mucus, which helps protect the stomach wall from the action of gastric juice. Although not related to the digestive function of the stomach, a very important material secreted by the gastric glands is *intrinsic factor*, a protein necessary for the absorption of vitamin B_{12} from the small intestine.

The stomach wall is composed of the same materials that gastric juice is able to attack, so various protections are in place to keep the stomach from digesting itself. One, mentioned previously, is the presence of mucus. Mucus forms a thick, protective coat that prevents gastric juice from reaching the cells of the stomach wall. The alkalinity of mucus helps to neutralize the HCl. Another protection is that pepsin is produced in an inactive form (pepsinogen) that cannot digest the cells that produce it. In addition, neural and hormonal reflexes regulate the production of gastric juice so that little is released unless food is present to absorb and dilute it. Finally, if the stomach lining is damaged, it is quickly repaired. The high rate of cell division in the stomach lining replaces a half million cells every minute. As a result, you have a new stomach lining every 3 days!

Very little absorption of food and other ingested materials occurs in the stomach because food simply has not been broken down into molecules small enough to be absorbed. Notable exceptions are alcohol and aspirin. The absorption of alcohol from the stomach is the reason its effects can be felt very quickly, especially if there is no food present to dilute it. The absorption of aspirin can cause bleeding of the wall of the stomach, which is the reason aspirin should be avoided by people who have stomach ulcers (see the Health Issue essay, *Peptic Ulcers*).

Small Intestine

The next region of the digestive tract, the **small intestine**, has two major functions: chemical digestion and absorption. As food moves along this twisted tube, it passes through three specialized regions: the duodenum, the jejunum, and the ileum. Chyme from the stomach enters the **duodenum**, the first region of the small intestine, in squirts, so that only a small amount enters the small intestine at one time. Digestive juices also enter the duodenum from the pancreas and liver. However, most chemical digestion and absorption occur in the *jejunum* and the *ileum*.

Chemical digestion within the small intestine Within the small intestine, a battery of enzymes completes the chemical digestion of virtually all the carbohydrates, proteins, fats, and nucleic acids in food. Although both the small intestine and the pancreas contribute enzymes, most of the digestion that occurs in the small intestine is actually performed by pancreatic enzymes (Table **15.2**).

TABLE 15.2 Major Digestive Enzymes

Enzyme	Site of Production	Site of Action	Substrate and Products
Carbohydrate digestion			
Salivary amylase	Salivary glands	Mouth	Polysaccharides into shorter molecules
Amylase	Pancreas	Small intestine	Polysaccharides into disaccharides
Maltase	Small intestine	Small intestine	Maltose into glucose units
Sucrase	Small intestine	Small intestine	Sucrose into glucose and fructose
Lactase	Small intestine	Small intestine	Lactose into glucose and galactose
Protein digestion			
Pepsin	Stomach	Stomach	Proteins into protein fragments (polypeptides)
Trypsin	Pancreas	Small intestine	Proteins and polypeptides into smaller fragments
Chymotrypsin	Pancreas	Small intestine	Proteins and polypeptides into smaller fragments
Carboxypeptidase	Pancreas	Small intestine	Polypeptides into amino acids
Lipid digestion			
Lipase	Pancreas	Small intestine	Triglycerides (fats) into fatty acids and glycerol

Fats present a special digestive challenge because they are insoluble in water. You have observed this when the oil (a fat) quickly separates from the vinegar (a water solution) in your salad dressing. In water, droplets of fat tend to coalesce into large globules. This poses a problem for digestion because lipase, the enzyme that chemically breaks down fats, is soluble in water and not in fats. As a result, lipase can work only at the surface of a fat globule. Large fat globules have less combined surface area than do smaller droplets, so their digestion by lipase proceeds more slowly.

Bile, a mixture of water, ions, cholesterol, bile pigments, and bile salts, plays an important role in the mechanical digestion of fats, which assists lipase in chemically digesting fats. The bile salts emulsify fats; that is, they keep fats separated into small droplets that disperse in liquid. This separation exposes a larger combined surface area to lipase, making the chemical digestion and absorption of fats faster and more complete. Bile is produced by the liver, is stored in the gallbladder, and acts in the small intestine.

We have more to say about the pancreas, liver, and gallbladder shortly.

Structure of the small intestine The small intestine, the primary site of absorption in the digestive system, is extremely effective at its task because it is long and has several structural specializations that vastly increase its surface area (Figure 15.8). First, the entire lining of the small intestine is pleated, like an accordion, into circular folds. These circular folds increase the surface area for absorption and cause chyme to flow through the small intestine in a spiral pattern. The spiral flow helps mix the chyme with digestive enzymes and increases its contact with the absorptive surfaces. Covering the entire lining surface are tiny 1 mm projections called **villi** (singular, villus). The villi give the lining a velvety appearance and, like the pile on a bath towel, increase the absorptive surface. Indeed, the villi increase the surface area of the small intestine tenfold. In addition, the absorptive epithelial cells covering the surface of each villus contain thousands of microscopic projections, called *microvilli,* that increase the surface area of the small intestine by another 20 times. The microvilli form a fuzzy surface, known as a *brush border.* The circular folds, villi, and microvilli create a surface area of 300 to 600 m²—greater than the size of a tennis court!

Q List three structural adaptations of the small intestine that increase its surface area for absorbing digested nutrients.

FIGURE 15.8 The small intestine features several structural modifications that increase its surface area, making it especially effective in absorbing nutrients. (a) Its wall contains accordion-like pleats called circular folds. (b) This electron micrograph shows the intestinal villi, the numerous fingerlike projections on the intestinal lining (the mucosa in part c). (d) The surface of each villus bristles with thousands of microscopic projections (of cell membranes) called microvilli. In the center of each villus is a network of blood capillaries, which carries away absorbed products of protein and carbohydrate digestion as well as ions and water. This network of capillaries surrounds a lacteal (a small vessel of the lymphatic system) that carries away the absorbed products of fat digestion. (e) An electron micrograph of the microvilli that cover each villus.

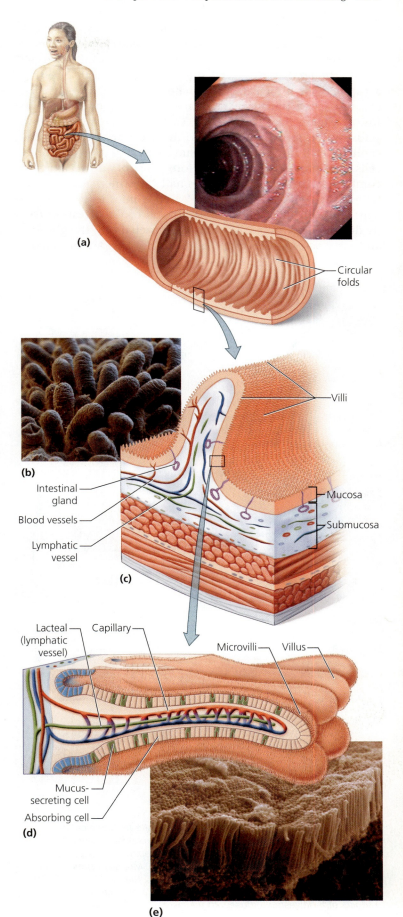

▲ Circular folds, villi, and microvilli.

The core of each villus is penetrated by a network of capillaries and a **lacteal**, which is a lymphatic vessel. As substances are absorbed from the small intestine, they cross only two cell layers: the epithelial cells of the villi and the wall of either a capillary or a lacteal. Most materials enter the epithelial cells by active transport, facilitated diffusion, or simple diffusion (Figure 15.9; see Chapter 3). Monosaccharides, amino acids, water, ions, vitamins, and minerals diffuse across the capillary wall into the bloodstream and are delivered to body cells. The products of fat digestion, glycerol and fatty acids, combine with bile salts in the small intestine, creating particles called *micelles* (mī-selz´). When a micelle contacts an epithelial cell of a villus, the products of fat digestion easily diffuse into the cell. Within an epithelial cell, the glycerol and fatty acids are reassembled into triglycerides, are mixed with cholesterol and phospholipids, and are coated with special proteins, thus becoming part of a complex known as a *chylomicron* (kī-lō-mī´krän). The protein coating makes the fat soluble in water, allowing it to be transported throughout the body. The chylomicrons leave the epithelial cell by exocytosis (see Chapter 3). Chylomicrons are too large to pass through capillary walls. However, they easily diffuse into the more porous lacteal and enter the lymphatic system, which carries them to the bloodstream.

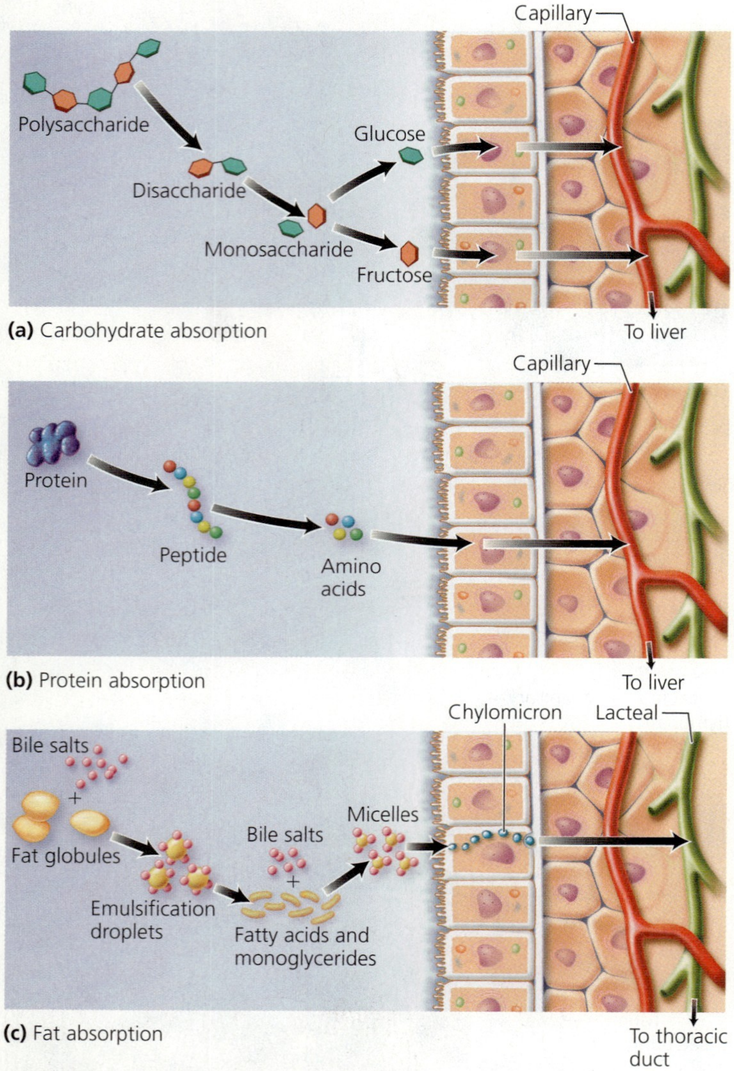

Accessory Organs: Pancreas, Liver, and Gallbladder

The pancreas, liver, and gallbladder are not part of the GI tract. However, they play vital roles in digestion by releasing their secretions into the small intestine. The functions of the secretions of the accessory organs are summarized in Table 15.3.

Pancreas The **pancreas** is an accessory organ that lies behind the stomach, extending from the small intestine toward the left side of the body. Pancreatic juice drains from the pancreas into the pancreatic duct, which fuses with the common bile duct from the liver just before entering the duodenum of the small intestine (Figure 15.10). In addition to enzymes, pancreatic juice contains water and ions, including bicarbonate ions that are important in neutralizing the acid in chyme when it emerges from the stomach. Neutralization is essential for optimal enzyme activity in the small intestine.

Collectively, the pancreatic enzymes and intestinal enzymes break nutrients into their component building blocks: proteins to amino acids, carbohydrates to monosaccharides, and triglycerides (a type of lipid) to fatty acids and glycerol.

FIGURE 15.9 The small intestine is the primary site for chemical digestion and absorption. The digestive products, such as monosaccharides, amino acids, fatty acids, and glycerol, enter absorptive epithelial cells of villi by active transport, facilitated diffusion, or diffusion. Monosaccharides and amino acids, along with water, ions, and vitamins, then enter the capillaries within the villus and are carried to body cells by the bloodstream. The products of fat digestion diffuse into the lacteal within the villus.

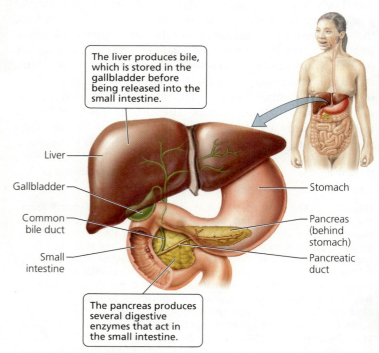

FIGURE 15.10 The pancreas, liver, and gallbladder are accessory organs of the digestive system.

TABLE 15.3 Review of Accessory Structures of the Digestive System

Structure	Secretions/Functions	Site of Action of Chemical Secretions
Salivary glands (sublingual, submandibular, parotid)	Secrete saliva, a liquid that moistens food and contains an enzyme (amylase) for digesting carbohydrates	Mouth
Pancreas	Digestive secretions include bicarbonate ions that neutralize acidic chyme and enzymes that digest carbohydrates, proteins, fats, and nucleic acids	Small intestine
Liver	Digestive function is to produce bile, a liquid that emulsifies fats, making chemical digestion easier and facilitating absorption	Small intestine
Gallbladder	Stores bile and releases it into small intestine	Small intestine

Liver The nutrient-laden blood from the capillaries in the villi travels through the hepatic portal vein to the **liver**, the largest internal organ in the body, which has a variety of metabolic and regulatory roles. A portal system consists of the blood vessels that link two capillary beds. The hepatic portal system delivers blood from a capillary bed in the small intestine to a second capillary bed in the liver (Figure 15.11).

We have already seen that the liver's primary role in digestion is to produce bile. One of its other roles is to control the glucose level of the blood, either removing excess glucose and storing it as glycogen or breaking down glycogen to raise blood glucose levels. Thus, the liver keeps the glucose levels of the blood within the proper range. The liver also packages lipids with protein carrier molecules to form lipoproteins, which transport lipids in the blood. After the liver adjusts the blood composition, the blood is returned to the general circulation through the hepatic veins. In addition, the liver stores iron and vitamins A, D, E, K, B_{12}, and folate.

The liver also removes poisonous substances, including lead, mercury, and pesticides, from the blood and, in some cases, breaks them down into less harmful chemicals. What's more, the liver converts the breakdown products of amino acids into urea, which can then be excreted by the kidney. These are but a few of the approximately 500 functions of the liver.

Considering the liver's many vital functions, it is no surprise that diseases of the liver can be serious and life threatening. *Cirrhosis* is a condition in which the liver becomes fatty and gradually deteriorates, its cells eventually being replaced by scar tissue. Cirrhosis is sometimes caused by prolonged, excessive alcohol use (see Chapter 8a).

Hepatitis is inflammation of the liver. It is most commonly caused by one of six viruses, designated as A, B, C, D, E, and G. Although all the hepatitis viruses attack the liver and destroy liver cells, there are differences in their means of transmission and their symptoms, as well as their severity. All forms of hepatitis do have one symptom in common: liver cells injured by hepatitis viruses stop filtering bilirubin from the blood. Bilirubin is a yellowish pigment produced by the breakdown of red blood cells; healthy liver cells remove it from the bloodstream and use it to make bile. In a person with hepatitis, the accumulating bilirubin is deposited in the skin and the whites of the eyes, giving them a yellowish tint. This condition, called *jaundice,* is characteristic of any disease that damages the liver.

Currently, about 4 million people in the United States have hepatitis C, and most of them have no idea that they are infected. For years, the disease has spread silently because it has no outward warning signs or very mild symptoms—vague fatigue, or flu-like muscle and joint pain. The virus is spread primarily through contaminated blood. Hundreds of thousands of intravenous drug users have been infected by sharing contaminated needles. Hepatitis C can also be spread through contaminated needles used in body piercing or tattooing. Before there was a way to test for the virus, many people became infected when they received transfusions of contaminated blood.

Gallbladder After it is produced by the liver, bile is stored, modified, and concentrated in a muscular, pear-shaped sac called the **gallbladder**. When chyme enters the small intestine, a hormone causes the gallbladder to contract, squirting bile through the common bile duct into the duodenum of the small intestine.

Bile is rich in cholesterol. Sometimes, if the balance of dissolved substances in bile becomes upset, a tiny crystalline particle precipitates out of solution. Cholesterol and other

Q Where are the two capillary beds in the hepatic portal system? What does the liver do in this system?

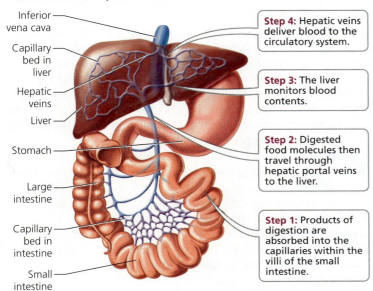

FIGURE 15.11 A portal system transports blood from one capillary bed to another. In the hepatic portal system, the hepatic portal vein carries blood from the capillary network of the villi of the small intestine to the capillary beds of the liver. The liver monitors blood content and processes nutrients before they are delivered to the bloodstream.

A The capillary beds are in the small intestine and the liver. The liver monitors and adjusts blood content.

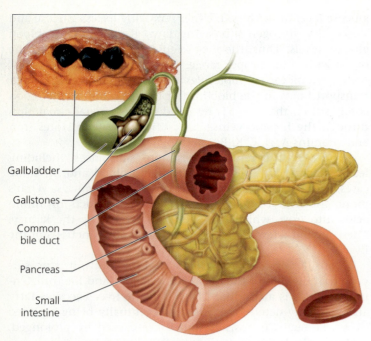

FIGURE 15.12 *Gallstones consist primarily of cholesterol that has precipitated from bile during storage in the gallbladder. A gallstone can intermittently or continuously block the ducts that drain bile into the small intestine. The photograph shows a gallbladder and several gallstones.*

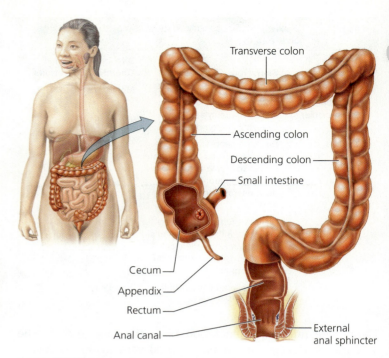

FIGURE 15.13 *The large intestine consists of the cecum, colon, rectum, and anal canal. It absorbs water from undigested material, forming the feces, and houses bacteria.*

substances can then build up around the particle to form a *gallstone* (Figure 15.12). Many people develop several gallstones, which can cause problems if they block the flow of bile and may necessitate surgical removal of the gallbladder.

Large Intestine

Now that we have discussed the accessory organs that assist digestion and absorption in the small intestine, we go back to following the movement of ingested substances through the GI tract. The material that was not absorbed in the small intestine moves into the final major structure of the digestive system, the **large intestine**. The principal functions of the large intestine are (1) to absorb most of the water remaining in the indigestible food residue, thereby adjusting the consistency of the waste material, or feces; (2) to store the feces; and (3) to eliminate them from the body. The large intestine is home to many types of bacteria, some of which produce vitamins that may be absorbed for use by the body.

Regions of the large intestine The large intestine has four regions: the cecum, colon, rectum, and anal canal, as shown in Figure 15.13. The **cecum** is a pouch that hangs below the junction of the small and large intestines. Extending from the cecum is another slender, wormlike pouch, called the **appendix**. The appendix has no digestive function. Some scientists believe the appendix plays a role in the immune system, which protects the body against disease.

Each year, about 1 in 500 people develops *appendicitis*, inflammation of the appendix. Appendicitis is usually caused by an infection that arises in the appendix after it becomes blocked by a piece of hardened stool from the cecum, by a piece of food, or by a tumor. After the blockage occurs, bacteria that are normally present in the appendix can multiply and cause an infection. At first, the person with appendicitis usually experiences vague bloating, indigestion, and a mild pain in the region of the navel (bellybutton). As the condition worsens, the pain becomes more severe and is localized in the region of the appendix, the lower right abdomen. The pain is typically accompanied by fever, nausea, and vomiting. When appendicitis is diagnosed, antibiotics are administered, and the infected appendix is surgically removed. Untreated infection in the appendix usually causes the appendix to rupture, allowing its infected contents to spill into the abdominal cavity. Spillage from the appendix generally leads to infection and inflammation throughout the abdomen (a condition called *peritonitis*), which is potentially fatal.

The largest region of the large intestine, the **colon**, is composed of the ascending colon on the right side of the abdomen, the transverse colon across the top of the abdominal cavity, and the descending colon on the left side (see Figure 15.13). Although much of the water that is originally in chyme is absorbed in the small intestine, the material entering the colon is still quite liquid. The colon absorbs 90% of the remaining water and sodium and potassium ions. The material left in the large intestine after passing through the colon is called *feces* and consists primarily of undigested food, sloughed-off epithelial cells, water, and millions of bacteria. The brown color of feces comes from bile pigments.

The bacteria in the colon do not normally cause disease and, in fact, are beneficial. Intestinal bacteria produce several vitamins that we are unable to produce on our own, including vitamin K and some of the B vitamins. Some of these

HEALTH ISSUE

Peptic Ulcers

At some point in their lives, nearly 13% of Americans experience a failure of the mechanisms that protect the stomach and duodenum (the first region of the small intestine) from their acidic contents, so that the lining of some region in the GI tract becomes eroded. The resulting sore resembles a canker sore of the mouth and is called a *peptic ulcer* (Figure 15.A). Although a peptic ulcer may form in the esophagus or the stomach, the most common site is the duodenum of the small intestine. Ulcers are usually between 10 and 25 mm (0.33 and 1 in.) in diameter and may occur singly or in multiple locations.

The symptoms of an ulcer are variable. A common symptom is abdominal pain, which can be quite severe. Vomiting, loss of appetite, bloating, indigestion, and heartburn are other common symptoms. However, some people with ulcers, especially those who are taking nonsteroidal anti-inflammatory drugs (NSAIDs), have no pain. Unfortunately, the degree of pain is a poor indicator of the severity of ulceration. Often, people with no symptoms are unaware of their ulcers until serious complications develop. Gastric juice can erode the lining of the GI tract until it bleeds. In some cases, the ulcer can eat a hole completely through the wall of the GI tract (a condition called *perforated ulcer*). Recurrent ulcers can cause scar tissue to form, which may narrow or block the lower end of the stomach or duodenum.

Although acidic gastric juice is the direct cause of peptic ulcers, factors that interfere with the mechanisms that normally protect the lining of the GI tract from the acid are considered to be the real causes. For example, NSAIDs, which include aspirin, ibuprofen, and naproxen, can cause ulcers because they slow the production of chemicals called *prostaglandins*, which normally help protect the lining of the GI tract from damage by acid.

The leading cause of peptic ulcers, however, is an infection with the bacterium *Helicobacter pylori*. More than 80% of persons with ulcers in the stomach or duodenum are infected with *H. pylori*. This bacterium produces an alkaline compound that neutralizes the stomach acid. These corkscrew-shaped bacteria live in the layer of mucus that protects the lining of the GI tract. Here, the bacteria attract body defense cells—specifically, macrophages and neutrophils—that cause inflammation, leading to ulcer formation. Toxic chemicals produced by the bacteria also contribute to ulcers.

H. pylori infections may last for years. These bacteria affect more than a billion people throughout the world and approximately 50% of the people in the United States who are over 60 years of age. For some reason, however, only about 10% to 15% of those who are infected actually develop peptic ulcers. Besides ulcers, an *H. pylori* infection is a risk factor for esophageal and stomach cancer. However, it may soon be possible to vaccinate children against this bacterium, thereby reducing the risk of both peptic ulcers and stomach cancer.

Questions to Consider

- An endoscopy is a procedure in which a long flexible tube with a camera is used to view your esophagus or stomach. If you were going to have an endoscopy, would you request that the physician test for the presence of *H. pylori* at the same time?
- The Korean American and Japanese American populations have an exceptionally high rate of stomach cancer. Would you recommend that people in these populations be routinely screened for the presence of *H. pylori*?

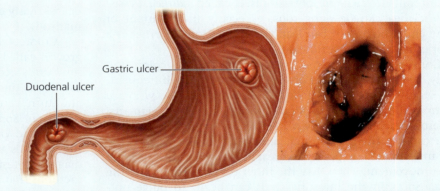

FIGURE 15.A *A peptic ulcer is an open sore that forms when gastric juice erodes the lining of the esophagus, stomach, or, most commonly, the duodenum. The ulcer shown in this stomach wall is bleeding. The most common symptom is abdominal pain that occurs when the stomach is empty.*

---

vitamins are then absorbed from the colon for our own use. Roughly 50 species of bacteria, including the well-known *Escherichia coli* (*E. coli*), live in the healthy colon. Bacteria are nourished by undigested food and by material that we are unable to digest, including certain components of plant cells. When the intestinal bacteria use the undigested and indigestible food for their own nutrition, their metabolic processes liberate gas that sometimes has a foul odor. Although most of the gas is absorbed through the intestinal walls, the remaining gas can produce some embarrassing moments when released as flatus.

People who are *lactose intolerant* lack the enzyme lactase, which is normally produced by and acts in the small intestine. Lactase breaks down lactose, the primary sugar in milk, into its component monosaccharides. Without lactase, then, lactose moves intact into the colon, where it provides a nutritional bonanza for the bacteria living there. As a result, when people who are lactose intolerant consume milk products, the intestinal bacteria ferment the lactose and produce the gases carbon dioxide and methane that, in turn, produces bloating, gas, and abdominal discomfort. Although lactose intolerance is common in adults, it is not dangerous. People with lactose intolerance can usually avoid these problems by swallowing capsules or tablets of lactase or by avoiding dairy products in their diet.

 stop and think

Certain foods—beans, for instance—are notorious for producing intestinal gas. Beans contain large amounts of certain short-chain carbohydrates that our bodies are unable to digest. How is the nutritional content of beans related to flatulence?

Periodic peristaltic contractions move material through the large intestine, but these contractions are slower than the contractions in the small intestine. Slower contractions allow for adequate water absorption as material moves through the large intestine. Eventually, the feces are pushed into the **rectum**, stretching the rectal wall and initiating the *defecation reflex*. Nerve impulses from the stretch receptors in the rectal wall travel to the spinal cord, which sends motor impulses back to the rectal wall, stimulating muscles there to contract and propel the feces into the **anal canal**. Two rings of muscles, called sphincters, must relax to allow *defecation,* the expulsion of feces. The internal sphincter relaxes automatically as part of the defecation reflex. The external sphincter is under voluntary control, allowing the person to decide whether to defecate. Conscious contraction of the abdominal muscles can increase abdominal pressure and help expel the feces.

Disorders of the colon The water absorption that occurs in the colon adjusts the consistency of feces. When material passes through the colon too rapidly, as might occur when colon contractions are stimulated by toxins from microorganisms or by excess food or alcoholic drink, too little water is absorbed. As a result, the feces are very liquid. This condition, which results in frequent loose stools, is called *diarrhea*. Persistent diarrhea can be dangerous, especially in an infant or young child, because it can lead to dehydration. Diarrhea is a major cause of death worldwide.

Conversely, if material passes through the colon too slowly, too much water is absorbed, resulting in infrequent, hard stools—a condition called *constipation*. People who are constipated may need to strain during bowel movements. Straining increases pressure within veins in the rectum and anus, causing them to stretch and enlarge. The wall of the large intestine also experiences great pressure during a strained bowel movement. Then, like the inner tube of an old tire, the weaker spots in the intestinal wall can begin to bulge outward, forming small pouches called *diverticula* (singular, diverticulum; Figure 15.14). Diverticula are very common in people older than age 50. When diverticula are present but do not cause problems or symptoms, the condition is called *diverticulosis*. But if the diverticula become infected with bacteria and inflamed, the condition is then called *diverticulitis*, which can cause abrupt, cramping abdominal pain, a change in bowel habits, fever, and rectal bleeding. Treatments for diverticulitis include changing the diet (for example, increasing fiber intake and avoiding nuts), taking drugs that reduce muscle spasms, or having surgery.

stop and think

Stimulant laxatives enhance peristalsis in the large intestine but do not affect the small intestine. Some people with eating disorders use laxatives to speed the movement of food through the digestive system, thinking that the calories in the food will not be absorbed. Are they correct in their thinking? After purging in this way, the person may weigh less on the bathroom scale. What accounts for the weight loss? Could laxative use help a person lose body fat?

Cancers of the colon or rectum are common and can be deadly. *Colorectal cancer* is the second leading cause of cancer deaths. Early detection and treatment cut the risk of death dramatically.

Colorectal cancer begins with a small, noncancerous growth called a *polyp*. If the polyp continues to grow, genetic mutations accumulate in it that can transform a cell into a cancerous tumor. Polyps may take as long as 10 years to grow and turn cancerous, generally allowing plenty of time for detection. Sometimes polyps bleed, so one sign of colorectal polyps and perhaps cancer is blood in the stool. However, the blood is not usually visible and must be detected with a diagnostic test. Because many polyps do not bleed, screening methods that allow a direct view of the wall of the rectum and colon are generally more effective. A long, flexible fiber-optic tube is threaded through the first third of the colon in a procedure known as a *sigmoidoscopy,* or through the entire colon in a procedure called a *colonoscopy*. If a polyp is detected, it can be removed and biopsied to determine whether it is cancerous (see Chapter 21a).

15.3 Nerves and Hormones in Digestion

As we have seen, ingested material moves along the GI tract, stopping for specific kinds of treatment along the way. For digestion to occur, enzymes must be present in the right place at the right time. However, because the body is composed of many of the same substances found in food, digestive enzymes should not be released until food is present (otherwise the enzymes might start digesting the digestive system). Both nerves and hormones play a role in orchestrating the release of digestive secretions, timing the release of each to the presence of food at each stop.

Food spends little time in the mouth; so to be effective, saliva must be secreted quickly. Because nervous stimulation is faster than hormonal stimulation, it is not surprising to learn that the nervous system controls salivation. Some saliva is released before food even enters the mouth, which

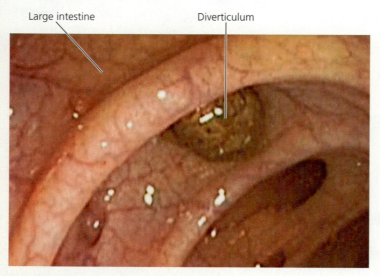

FIGURE 15.14 *A diverticulum is a small pouch that forms in the wall of the large intestine and is usually caused by repeated straining during bowel movements. A high-fiber diet results in softer, bulkier stools that are easier to pass, thus making it less likely that diverticula will form.*

TABLE 15.4 Examples of Neural Controls on Digestive Activity	
Stimulus	Effect
Sight of food, thought of food, presence of food in mouth	Release of saliva from salivary glands
Chewing food	Release of gastric juice (enzymes from stomach and HCl) and mucus from cells of stomach lining
Presence of acidic chyme in small intestine	Release of enzymes from small intestine and pancreas into the small intestine; release of bile from gallbladder into small intestine; increased motility in small intestine

may begin to "water" simply at the *thought* of food—and certainly begins at the sight or smell of food. The major trigger for salivation, however, is the presence of food in the mouth—its flavor and pressure. Salivary juices continue to flow for some time after the food is swallowed, helping to rinse out the mouth.

While food is still being chewed, neural reflexes stimulate the stomach lining to begin secreting gastric juice and mucus. Distention of the stomach by swallowed food, along with the presence of partially digested proteins, stimulates cells in the stomach lining to release the hormone *gastrin*. Gastrin enters the bloodstream and circulates throughout the body and back to the stomach, where it increases the production of gastric juice.

The presence of acidic chyme in the small intestine, which triggers local nerve reflexes, is the most important stimulus for the release of enzymes from both the small intestine and the pancreas, as well as bile from the gallbladder. Acid chyme also causes the small intestine to release several hormones that, in turn, are responsible for the release of digestive enzymes and bile. For instance, one hormone, *vasoactive intestinal peptide*, is released from the small intestine into the bloodstream and is carried back to the small intestine, where it causes the release of intestinal juices. At the same time, the small intestine releases a second hormone, *secretin*, which stimulates the release of sodium bicarbonate from the pancreas into the small intestine to help neutralize the acidity of chyme. A third hormone from the small intestine is *cholecystokinin* (kō′lĭ-sĭs-tə-kī-nĭn), which causes the pancreas to release its digestive enzymes and causes the gallbladder to contract and release bile. The neural and hormonal controls of the digestive system are summarized in Table 15.4 and Table 15.5, respectively.

15.4 Planning a Healthy Diet

Planning a healthy diet requires more than simply eating certain foods to avoid deficiencies of particular nutrients. Choosing the right balance of foods can help improve health and reduce the risk of serious chronic diseases, such as heart disease, cancer, and diabetes.

MyPlate (www.choosemyplate.gov) is a food guide released by the U.S. Department of Agriculture (USDA) to help any person plan a well-balanced diet. MyPlate follows the recommendations in the USDA's Dietary Guide for Americans (http://www.dietaryguidelines.gov/DGAs2010-PolicyDocument.htm). This guide promotes a healthy lifestyle that includes wise decisions about food and physical activity. As you can see in Figure 15.15, MyPlate is simple to understand because it shows the proportions of different food groups that we should eat on a plate—where we usually see food. Whereas MyPyramid described food in terms of classes of nutrients, MyPlate describes food in more familiar terms: half of your plate should consist of fruits and vegetables; grains and proteins should each be 25% of your plate; and you should have three servings of fat-free or 1% fat dairy products each day. You should adjust the size of portions of each food group according to the number of calories you use in a day. The SuperTracker tool, which is found on the Choose MyPlate website, can help you plan your diet and level of physical activity.

15.5 Nutrients

What does the body do with the food you eat? Food provides fuel, building blocks, molecules needed to carry out chemical processes in the body, and water.

- Food is needed as an energy source for all cellular activities. (Energy is measured in a unit called a *calorie*, which is the amount needed to raise the temperature of 1 g of water 1°C. In discussions of biochemical reactions, gains and losses of energy are usually reported in kilocalories. A

TABLE 15.5 Examples of Hormonal Control on Digestive Activity				
Hormone	Stimulus	Origin	Target	Effects
Gastrin	Distention of stomach by food; presence of partially digested proteins in stomach	Stomach	Stomach	Release of gastric juice (enzymes from stomach and HCl)
Vasoactive intestinal peptide	Presence of acidic chyme in small intestine	Small intestine	Small intestine	Release of enzymes from small intestine
Secretin	Presence of acidic chyme in small intestine	Small intestine	Pancreas	Release of sodium bicarbonate from pancreas into small intestine to neutralize acidic chyme
Cholecystokinin	Arrival of chyme-containing lipids	Small intestine	Pancreas	Release of enzymes from pancreas
			Gallbladder	Contraction of gallbladder and release of bile

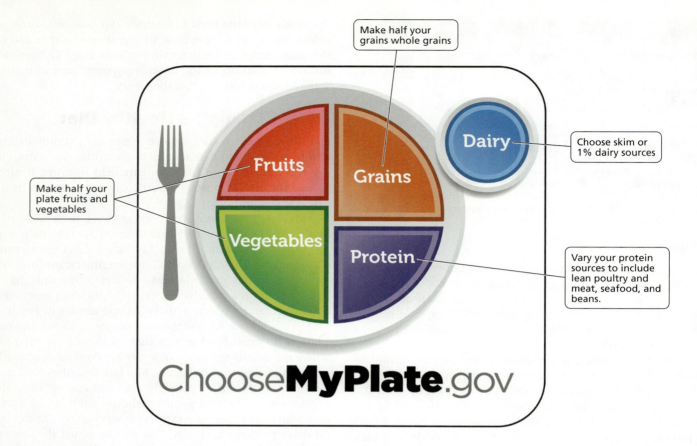

FIGURE 15.15 *MyPlate helps you plan a well-balanced diet, even if you are not an expert nutritionist.*
U.S. Department of Agriculture and U.S. Department of Health and Human Services, Dietary Guidelines for Americans, 2010 (7th ed.), Washington, DC: U.S. Government Printing Office, December 2010.

kilocalorie is 1000 calories of energy. In nonscientific discussions, however, the "kilo" is dropped, and kilocalories are referred to simply as "calories." We will follow this tradition because it is the way most nutritional values are reported to the public.[1]

- Building blocks are needed for cell division, maintenance, and repair.
- Molecules, such as vitamins, are needed to coordinate life's processes.
- Water is necessary for maintaining the proper cellular environment and for certain cellular reactions.

We have seen that the digestive system breaks the complex molecules of carbohydrates, proteins, fats, and nucleic acids into their component subunits. Most cells of the body, especially those of the liver, are able to use these subunits as building blocks by joining them together and sometimes by

---

[1] In technical writing, kilocalories are written as Calories (with a capital *C*).

converting one type of molecule into another. Your body, however, cannot synthesize all of the required kinds of amino and fatty acids, at least not in quantities sufficient to meet bodily needs. These substances that the body cannot synthesize are called *essential amino acids* and *essential fatty acids*. *Essential* means that the substances must be included in the diet because the body cannot produce them.

A **nutrient** is a substance in food that provides energy; becomes part of a structure; or performs a function in growth, maintenance, or repair. Three nutrients—fats (lipids), carbohydrates, and proteins—can provide energy. Although proteins are included in this list, they are usually used to build cell structures or regulatory molecules, such as enzymes or certain hormones. Vitamins, minerals, water, and fiber do not provide energy, but they are essential to cellular functioning, as we'll see later.

Lipids

Lipid is the more technical name for what we have been calling "fat." There are several types of lipids. They include fats, oils, and cholesterol. Even so, 95% of the lipids found in food are triglycerides—the fats that we commonly think of when we hear the term **fat**. A triglyceride is a molecule made from three fatty acids (hence, *tri–*) attached to a molecule of glycerol (hence, *–glyceride*). The fatty acids in the triglyceride give the molecule its characteristics.

An important way that fatty acids can differ is in their degree of saturation, or the extent to which each carbon in the fatty acid is bonded to as many hydrogen atoms as possible. Recall from Chapter 2 that a saturated fatty acid contains all the hydrogen it can hold. A polyunsaturated fatty acid can hold four or more additional hydrogens, and a monounsaturated fatty acid can hold two more hydrogens. In general, saturated fats are solid at room temperature, and they usually come from animal sources. In contrast, unsaturated fats are liquid at room temperature. Oils are unsaturated fats. The oils in our diet usually come from plant sources.

We have a biological need for lipids. Certain lipids, including **cholesterol**, are essential components of all cell membranes; some are used in the construction of myelin sheaths that insulate nerve fibers. Lipids are also needed for the absorption of the fat-soluble vitamins A, D, E, and K. These vitamins are absorbed from the intestines along with the products of fat digestion. Lipids then carry these vitamins in the bloodstream to the cells that use them. Glands in our skin produce oils that keep the skin soft and prevent dryness. Cholesterol is the structural basis of the steroid hormones, including the sex hormones.

The daily need for dietary fat is a mere tablespoon, yet the average American adult consumes 6 to 8 tablespoons of fat (78 to 196 g) a day, which can quickly add inches to the waistline. Obesity is associated with health problems such as high blood pressure and increased risk of diabetes. In addition, high consumption of fat is related to certain cancers, including cancer of the colon, prostate gland, lungs, and perhaps the breast.

The blood level of cholesterol, which is partly influenced by the amount and kinds of fat in the diet, affects the risk of developing atherosclerosis, a condition in which fatty deposits form in the walls of blood vessels. The deposits promote an inflammatory response in the artery wall, thereby increasing the risk of heart attack and stroke (see Chapter 12a). The risk of atherosclerosis increases with the blood level of cholesterol. In general, blood cholesterol levels under 200 mg/dl[2] are recommended for adults. The bad news is that the average blood cholesterol level for a middle-aged adult in the United States is 215 mg/dl. The good news is that people who lower their blood cholesterol levels can slow, or even reverse, atherosclerosis and, therefore, their risk of heart attack.

Blood cholesterol comes from one of two sources: the diet or the liver. Of the two, cholesterol production by the liver is more significant because most of the cholesterol in blood comes from the liver and not from the food we eat. Diet is still important, and, surprisingly, saturated fat in the diet raises blood levels of cholesterol more than dietary cholesterol does. Moreover, different people's bodies handle cholesterol differently depending on their genetic makeup (as well as the amount of cholesterol consumed in a meal). As a result, the relationship between dietary cholesterol and blood cholesterol is not clear.

Total blood cholesterol provides only a partial picture of a person's risk of atherosclerosis. Before cholesterol can be transported in the blood or lymph, it is combined with protein and triglycerides to form a lipoprotein, which makes it soluble in water. **Low-density lipoproteins (LDLs)** are considered to be a damaging, or bad, form of cholesterol. Although LDLs bring cholesterol to the cells that need it to sustain life, they also deposit cholesterol in artery walls. In contrast, **high-density lipoproteins (HDLs)** carry cholesterol from the cells, perhaps even from the artery walls, to the liver for elimination. Because HDLs are protective against heart disease, they are considered to be a good form of cholesterol.

Because of the different roles the two kinds of lipoproteins play in transporting cholesterol, the proportion of HDLs and LDLs in the blood is considered a more important indicator of risk than is the total blood cholesterol (HDL + LDL) level alone. The ratio of total cholesterol to HDLs should not be greater than 4:1. HDL levels higher than 60 mg/dl are considered to reduce the risk of heart disease.

The types of fat in the diet can influence blood cholesterol level and, more importantly, the ratios of LDLs and HDLs. Saturated fats and *trans* fats are bad for the heart. Saturated fat consistently boosts blood levels of harmful LDL cholesterol, both by directly stimulating the liver to step up its production of LDLs and by slowing the rate at which LDLs are cleared from the bloodstream. Foods high in saturated fat include meat (especially red meat), butter, cheese, whole milk, and other dairy products. *Trans* fatty acids are formed when hydrogens are added to unsaturated fats (oils) in such a way as to stabilize them or to solidify them, as when margarine is formed from

---

[2]dl = deciliter (100 milliliters).

FIGURE 15.16 *Types of dietary fats.* The total fat content of the diet should be moderate. The types of fats in the diet can influence the risk of heart attack and stroke.

vegetable oil. *Trans* fats are also found in many packaged foods (Figure 15.16). *Trans* fatty acids may behave like saturated fatty acids and raise LDL levels. In addition, they lower good HDLs.

But not all fats are bad. Monounsaturated fats and polyunsaturated fats, such as omega-3 fatty acids and omega-6 fatty acids,[3] are good fats; they lower total blood cholesterol and LDLs. Monounsaturated fats are found in olive, canola, and peanut oils and in nuts. Omega-3 fatty acids are found in the oils of certain fish, such as Atlantic mackerel, lake trout, herring, tuna, and salmon. The most important omega-6 fatty acid is linoleic acid, an essential fatty acid found in corn and safflower oils.

 what would you do?

Statins are cholesterol-lowering drugs. Some scientists believe that statins lower cholesterol equally in men and women. Other scientists believe that statins are not effective in women, and they issue warnings about the drug's side effects. If your mother, sister, or aunt had high cholesterol and her physician suggested that she take statins, what would you advise her to do? What information would you want to have before making that decision? How would you find that information?

[3]The number in omega-3 and omega-6 refers to the location of the first double bond in the carbon chain of the fatty acid.

Carbohydrates

Carbohydrates in the diet come primarily from plant sources and include sugars, starches, and roughage, or **dietary fiber** (Figure 15.17). The basic unit of a carbohydrate molecule is a monosaccharide. Sugars are simple carbohydrates, meaning they are generally monosaccharides or disaccharides (two monosaccharides linked together). Sugars taste sweet and are naturally present in whole foods such as fruit and milk. In whole foods, sugars are present with other nutrients, including vitamins and minerals. However, refined sugars have been removed from their plant sources and concentrated, so they are no longer mixed with other nutrients. The calories from refined sugars are therefore described as "empty." They provide energy but have no other nutritive value. Refined sugars are found in candies, cookies, cakes, pies, and sodas.

Starches and fiber are complex carbohydrates, which means they are polysaccharides. Plants store energy in starches—long, sometimes branched chains of hundreds or thousands of linked molecules of the monosaccharide glucose. Common sources of starches include wheat, rice, oats, corn, potatoes, and legumes. Fiber is the indigestible part of edible plants. Complex carbohydrates are usually accompanied by other nutrients, but even complex carbohydrates are not all healthful, because they differ in the way they affect blood glucose.

The most important function of carbohydrates is to provide fuel for the body. Recall that the digestive system breaks down all carbohydrates (except fiber) to simple sugars, primarily glucose, which is absorbed into the bloodstream. Glucose is the carbohydrate that cells use for fuel most of the time. A measure called the **glycemic response** describes how quickly a serving of food is converted to blood sugar and how much the level of blood sugar is affected. The glycemic index is a numerical ranking of carbohydrates based on their glycemic response. The scale of the glycemic index ranges from 0 to 100, with pure glucose serving as a reference point of 100. Sugar

FIGURE 15.17 *Types of carbohydrates.* You should minimize your consumption of simple carbohydrates, especially refined sugars. Most of the carbohydrates in the diet should be complex carbohydrates.

and starchy foods such as white bread, potatoes, and white rice have a high value on the glycemic index because they cause the blood sugar level to rise sharply. Foods with a low value on the glycemic index, including whole fruit, whole-grain foods, brown rice, and barley, cause a more modest and gradual increase in blood sugar. Generally, foods with a low value on the glycemic index are high in fiber.

The glycemic response is important because it influences how the body reacts to different foods. After you eat a carbohydrate with a high glycemic index value, your blood glucose rises. As discussed in Chapter 10, a rise in blood glucose level causes the pancreas to release more of the hormone insulin, which lowers blood glucose (thus inducing hunger) by converting excess glucose to fat. In short, in a healthy person, consumption of a high-glycemic food is followed by fat formation and by increased hunger. In people with diabetes mellitus, rising blood sugar levels due to high-glycemic foods may cause medical problems because these people are unable to produce or to use insulin (see Chapter 10a).

Nutritionists recommend that you limit your intake of simple carbohydrates, especially refined sugars, and instead choose sources of carbohydrates that contain more than just calories. For example, fruits, vegetables, grains, and milk are food groups that have low glycemic index values and supply many other nutrients besides carbohydrates.

Dietary fiber is found in all plants that are eaten for food, including fruits, vegetables, dried beans, and whole grains. Fiber is a form of carbohydrate that humans cannot digest into its component monosaccharides. Because only monosaccharides can be absorbed from the small intestine, dietary fiber is passed along to the large intestine. Some of the fiber is digested by the bacteria living in the large intestine, and the remaining fiber gives bulk to feces.

Although fiber cannot be digested or absorbed, it is still an important part of a healthful diet. Fiber is good for the heart and blood vessels, because it lowers LDLs but does not lower the beneficial HDLs. Intestinal disorders such as constipation and hemorrhoids improve when the amount of fiber in the diet is increased (discussed in Chapter 15). Fiber can absorb an amazing amount of water, thereby softening stools and making them easier to pass.

Proteins

A **protein** consists of one or more chains of amino acids. Human proteins contain 20 different kinds of amino acids. The protein in the food you eat is digested into its component amino acids, which are then absorbed into the bloodstream and delivered to the cells, creating a pool of available amino acids. Your cells then draw the amino acids needed to build proteins in your body from those available in the pool. In addition, the body is able to synthesize 11 of the amino acids from nitrogen and molecules derived from carbohydrates, fats, or other amino acids. The 9 remaining amino acids that the body cannot synthesize—called **essential amino acids**—must be supplied by the diet.

When your body makes a protein, the amino acids are strung together in a specific order, depending on which protein is being made. If a particular amino acid is needed for a

Complete proteins contain all the essential amino acids and usually come from animal sources.

Incomplete proteins lack one or more of the essential amino acids and usually come from plant sources.

FIGURE 15.18 *Types of protein in the diet*

certain protein but is not available, then that protein cannot be synthesized. Consider this analogy: A sign maker with a bag containing only one copy of the letter *R* and 100 copies of the other 25 letters in the alphabet could make only one NO PARKING sign. There is simply no way to substitute for the limiting letter, *R*. The same principle applies to protein synthesis; if a needed amino acid is lacking, protein production stops, or the body breaks down existing proteins to get that amino acid.

In short, the pool of amino acids available for protein synthesis must always contain sufficient amounts of all the essential amino acids. Dietary proteins described as **complete proteins** contain ample amounts of all the essential amino acids. Animal proteins are generally complete. Plant proteins are generally incomplete proteins and are low in one or more of the essential amino acids. Eating certain combinations of **incomplete proteins** from two or more plant sources ensures that the pool of amino acids available for protein synthesis contains ample amounts of all the essential amino acids (Figure 15.18). Such combinations are called **complementary proteins** because if the combinations are correct, they supply enough of all the essential amino acids (Figure 15.19). A vegetarian must be sure to consume complementary proteins.

Sources of animal protein usually contain a lot of fat, whereas plant protein sources usually contain less fat and are good sources of carbohydrates, too. The tendency in the United States to make meat the focal point of every meal is the main reason that the typical diet here contains excess fat. For a healthier diet, choose lean, low-fat, or fat-free sources of protein. Eating a variety of plant proteins will supply all the essential amino acids and will help reduce the percentage of fat calories in the diet as well as boost the percentage of calories from complex carbohydrates. When choosing among animal sources of protein, reduce the amount of fatty red meat in favor of chicken (with the fatty skin removed) or fish.

Vitamins

A **vitamin** (*vita*, life) is an organic (carbon-containing) compound that, although essential for health and growth, is needed only in minute quantities—milligrams or micrograms. All the vitamins you need in a day would fill only an eighth of a teaspoon. These tiny amounts are sufficient because vitamins are

- Hummus (chickpeas and sesame seeds)
- Tofu and cashew stir-fry
- Trail mix (roasted soybeans and nuts)
- Tahini (sesame seeds) and peanut sauce

Nuts and seeds

Legumes

- Rice and beans
- Black-eyed peas and corn bread
- Bean burrito in corn tortilla
- Peanut butter on bread
- Rice and tofu
- Rice and lentils

Grains

FIGURE 15.19 *Complementary proteins are combinations of two or more incomplete proteins that together supply all the essential amino acids.*

not broken down or destroyed during use. Most function as coenzymes, which are nonprotein molecules necessary for certain enzymes to function. Enzymes and coenzymes are continuously recycled and, therefore, can be used repeatedly by the body.

There are two categories of vitamins: water-soluble vitamins, which dissolve in water, and fat-soluble vitamins, which are stored in fat. Of the 13 vitamins known to be needed by humans, 9 are water soluble (C and the various B vitamins), and 4 are fat soluble (A, D, E, and K). Table 15.6 lists the vitamins, their functions, good sources for them, and the problems associated with deficiencies or excesses.

Except for vitamin D, our cells cannot make vitamins, so we must obtain them in our food. A varied, balanced diet is the best way to ensure an adequate supply of all vitamins. No one food contains every vitamin, but most contain some. Vitamins are often more easily available for absorption when the foods containing them are cooked. Cooked carrots, for instance, are a better source of vitamin A than are raw carrots. However, water-soluble vitamins are likely to be lost if the vegetables containing them are cooked by being boiled in water (because much of the vitamin content ends up in the water). Steaming those vegetables is a better way to preserve their vitamin content.

Certain vitamins have to be consumed in adequate amounts every day. Folic acid, one of the B vitamins that is particularly abundant in dark leafy greens, plays a role in preventing birth defects, such as spina bifida, that involve the brain and spinal cord. It now seems that folic acid, along with vitamins B_6 and B_{12}, may also help prevent heart disease. Five daily servings of fruits and vegetables should provide enough of these vitamins to protect the heart. Antioxidant vitamins, including vitamin C, vitamin E, and beta-carotene, protect against cell damage due to oxidation. Some researchers believe that antioxidants slow the aging process and protect against cancer, atherosclerosis, and macular degeneration (the leading cause of irreversible blindness in people over age 65). These, too, should be ingested every day. Spinach, collard greens, and carrots are good sources of antioxidant vitamins.

stop and think

Some weight-loss medications work by preventing fats from being absorbed from the digestive system. Why might these drugs lead to deficiencies in vitamins A, D, E, and K?

Minerals

The **minerals** needed in our diet are inorganic substances essential to a wide range of life processes. We need fairly large amounts, although not what would be described as "megadoses," of seven minerals: calcium, phosphorus, potassium, sulfur, sodium, chloride, and magnesium. In addition, we need trace amounts of about a dozen others. Table 15.7 on page 306 lists selected minerals, their functions, good sources for them, and the problems associated with deficiencies or excesses.

We can obtain the necessary minerals from the foods we eat, as long as we prepare them in ways that do not reduce their mineral content. Like certain vitamins, many minerals are water soluble and can be lost during food preparation.

Sodium, a component of table salt, is essential to health, but most Americans consume too much of it. High salt intake causes high blood pressure in some people. The Dietary Recommendations for Americans advises that salt intake should not exceed 2400 mg (slightly more than 1 teaspoon) a day. Processed foods are especially high in salt, which is added to preserve food and enhance the taste. Salt is found in nearly every processed food product, including canned vegetables, cheese, bread, and processed meats. To reduce your salt intake, use salt sparingly when cooking fresh food, and read the labels on prepared food.

Water

Water is perhaps the most important nutrient; it transports materials through our bodies, lubricates and cushions organs, helps in temperature regulation, and provides a medium for many vital chemical reactions. We can live without food for about 8 weeks, but without water for only about 3 days. Nutritionists recommend that we consume eight 8-ounce glasses (2 quarts) of water a day. Water in fruits and vegetables makes up about half of that requirement for the average adult. The rest of the requirement does not have to be consumed as plain water, but it should not come from carbonated sweet drinks, caffeinated beverages, or alcoholic beverages. Carbonation interferes with water absorption, and sugar adds empty calories. Caffeine and alcohol increase water loss in urine.

15.6 Food Labels

Grocery stores generally offer several choices for essentially the same product. Food labels and the knowledge you now have about nutrients can help you make healthy choices.

TABLE 15.6 Vitamins

Vitamin	Good Sources	Function	Effects of Deficiency	Effect of Excess
Fat-soluble vitamins				
A	Liver, egg yolk, fat-containing and fortified dairy products; formed from beta-carotene (found in deep yellow and deep green leafy vegetables)	Components of rhodopsin, the eye pigment responsible for black-and-white vision; maintains skin and mucous membranes; cell differentiation	Night blindness; dry, scaly skin; dry hair; skin sores; increased respiratory, urogenital, and digestive infections; xerophthalmia (the leading cause of preventable blindness worldwide); most common vitamin deficiency in world	Drowsiness; headache; dry, coarse, scaly skin; hair loss; itching; brittle nails; abdominal and bone pain
D	Fortified milk, fish liver oil, egg yolk; formed in skin when exposed to ultraviolet light	Increases absorption of calcium; enhances bone growth and calcification	Bone deformities in children; rickets, bone softening in adults	Calcium deposits in soft tissues, kidney damage, vomiting, diarrhea, weight loss
E	Whole grains, dark green vegetables, vegetable oils, nuts, seeds	May inhibit effects of free radicals; helps maintain cell membranes; prevents oxidation of vitamins A and C in gut	Rare; possible anemia and nerve damage	Muscle weakness, fatigue, nausea
K	Primary source from bacteria in large intestine; leafy green vegetables, cabbage, cauliflower	Important in forming proteins involved in blood clotting	Easy bruising, abnormal blood clotting, severe bleeding	Liver damage and anemia
Water-soluble vitamins				
C (ascorbic acid)	Citrus fruits, cantaloupe, strawberries, tomatoes, broccoli, cabbage, green pepper	Collagen synthesis; may inhibit free radicals; improves iron absorption	Scurvy, poor wound healing, impaired immunity	Diarrhea, kidney stones; may alter results of certain diagnostic lab tests
Thiamin (B_1)	Pork, legumes, whole grains, leafy green vegetables	Coenzyme in energy metabolism; nerve function	Water retention in tissues, nerve changes leading to poor coordination, heart failure, beriberi	None known
Riboflavin (B_2)	Dairy products such as milk; whole grains, meat, liver, egg whites, leafy green vegetables	Coenzyme used in energy metabolism	Skin lesions	None known
Niacin (B_3)	Nuts, green leafy vegetables, potatoes; can be formed from tryptophan found in meats	Coenzyme used in energy metabolism	Contributes to pellagra (damage to skin, gut, nervous system)	Flushing of skin on face, neck, and hands; possible liver damage
B_6	Meat, poultry, fish, spinach, potatoes, tomatoes	Coenzyme used in amino acid metabolism	Nervous, skin, and muscular disorders; anemia	Numbness in feet, poor coordination
Pantothenic acid	Widely distributed in foods, animal products, and whole grains	Coenzyme in energy metabolism	Fatigue, numbness and tingling of hands and feet, headaches, nausea	Diarrhea, water retention
Folic acid (folate)	Dark green vegetables, orange juice, nuts, legumes, grain products	Coenzyme in nucleic acid and amino acid metabolism	Anemia (megaloblastic and pernicious), gastrointestinal disturbances, nervous system damage, inflamed tongue, neural tube defects	High doses mask vitamin B_{12} deficiency
B_{12}	Poultry, fish, red meat, dairy products except butter	Coenzyme in nucleic acid metabolism	Anemia (megaloblastic and pernicious), impaired nerve function	None known
Biotin	Legumes, egg yolk; widely distributed in foods; bacteria of large intestine	Coenzyme used in energy metabolism	Scaly skin (dermatitis), sore tongue, anemia	None known

TABLE 15.7 Selected Minerals

Mineral	Good Sources	Function	Effects of Deficiency	Effects of Excess
Major minerals				
Calcium	Milk, cheese, dark green vegetables, legumes	Hardness of bones, tooth formation, blood clotting, nerve and muscle action	Stunted growth, loss of bone mass, osteoporosis, convulsions	Impaired absorption of other minerals; kidney stones
Phosphorus	Milk, cheese, red meat, poultry, whole grains	Bone and tooth formation; components of nucleic acids, ATP, and phospholipids; acid–base balance	Weakness, demineralized bone	Impaired absorption of some minerals
Magnesium	Whole grains, green leafy vegetables, milk, dairy products, nuts, legumes	Component of enzymes	Muscle cramps, neurological disturbances	Neurological disturbances
Potassium	Available in many foods, including meats, fruits, vegetables, and whole grains	Body water balance, nerve function, muscle function, role in protein synthesis	Muscle weakness	Muscle weakness, paralysis, heart failure
Sulfur	Protein-containing foods, including meat, legumes, milk, and eggs	Component of body proteins	None known	None known
Sodium	Table salt	Body water balance, nerve function	Muscle cramps, reduced appetite	High blood pressure in susceptible people
Chloride	Table salt, processed foods	Formation of hydrochloric acid in stomach, role in acid–base balance	Muscle cramps, reduced appetite, poor growth	High blood pressure in susceptible people
Trace minerals				
Iron	Meat, liver, shellfish, egg yolk, whole grains, green leafy vegetables, nuts, dried fruit	Component of hemoglobin, myoglobin, and cytochrome (transport chain enzyme)	Iron-deficiency anemia, weakness, impaired immune function	Liver damage, heart failure, shock
Iodine	Marine fish and shellfish, iodized salt, dairy products	Thyroid hormone function	Enlarged thyroid	Enlarged thyroid
Fluoride	Treated drinking water, tea, seafood	Bone and tooth maintenance	Tooth decay	Digestive upsets, mottling of teeth, deformed skeleton
Copper	Nuts, legumes, seafood, drinking water	Synthesis of melanin, hemoglobin, and transport chain components; collagen synthesis; immune function	Rare; anemia, changes in blood vessels	Nausea, liver damage
Zinc	Seafood, whole grains, legumes, nuts, meats	Component of digestive enzymes; required for normal growth, wound healing, and sperm production	Difficulty in walking; slurred speech, scaly skin, impaired immune function	Nausea, vomiting, diarrhea, impaired immune function
Manganese	Nuts, legumes, whole grains, leafy green vegetables	Role in synthesis of fatty acids, cholesterol, urea, and hemoglobin; normal neural function	None known	Nerve damage

Figure 15.20 shows a food label from a box of baked whole-wheat crackers. It provides some useful general pointers for reading food labels. First, note the serving size described on the label, and remember that the amount you actually eat, whether larger or smaller than what is described on the label, will be what determines the number of calories and amount of nutrients you consume.

Next, notice the number of calories reported per serving and the number of calories from fat. In the example shown in Figure 15.20, 40 of the 120 calories in a serving, or nearly 34% of the calories, come from fat. Dietary guidelines recommend that you consume less than 30% of your total calories as fat. However, if you crave these crackers, you can make up for the amount of fat in them, to some extent, by eating low-fat food at another time during the day. Also note the amount of saturated fat and *trans* fat reported on the label, because you should minimize these kinds of fat in your diet.

In considering the Percent Daily Values of various nutrients reported on the label, keep two things in mind. First, these values are based on a 2000-calorie diet. Your own diet may require more or fewer calories. Second, as becomes clear from the daily values provided at the bottom of the label, you should be attempting to keep dietary fat and sodium consumption *below* those amounts.

Food labels can also help you increase your intake of nutrients that are important to consume. Be sure to get enough vitamin A, vitamin C, calcium, iron, and phosphorus. The daily values reported for carbohydrates and dietary fiber will also help you consume enough of those substances.

abolic rate for a while afterward. When you determine your caloric needs, the calories you use in exercise are added to those needed to maintain your BMR.

Exercise helps keep the body in good working order. The Dietary Guide for Americans encourages adults (18 to 64 years) to engage in at least 2½ hours of moderate-intensity physical activity a week. Aerobic exercise reduces the risk of diseases of the heart and blood vessels and lowers blood pressure (discussed in Chapter 12). Weight-bearing exercise reduces the risk of osteoporosis, a loss of bone density (discussed in Chapter 5). In general, regular physical exercise reduces stress and the risk of certain chronic diseases, including diabetes. Regular exercise also manages body weight, helping to prevent the unhealthy weight gain that can occur in adulthood.

15.8 Obesity

Although *overweight* and *obese* are both terms used to describe people who have excess body weight, they do not have exactly the same meaning. An **obese** person is overweight because of excess fat. An **overweight** person weighs more than the ideal on a height and weight chart. An athletic person whose muscles are well developed may weigh more than the weight listed as desirable on height–weight tables, but such a person is not obese. However, most people who are overweight have too much body fat. *Overweight* and *obese* generally refer to ranges of weight that are considered to be unhealthy.

The body mass index (BMI) is a number that provides a reliable indicator of body fat because it evaluates your weight in relation to your height (Figure 15.21). A BMI greater than 30 is generally considered unhealthy and an indication of obesity.

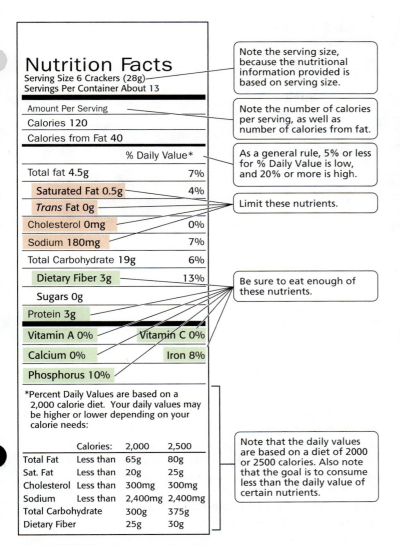

FIGURE 15.20 Tips for reading food labels

15.7 Energy Balance

The body requires energy for maintaining basic body functions, for physical activity, and for processing the food that is eaten. The energy that is needed strictly for maintenance is called the **basal metabolic rate (BMR)**; it is the minimum energy needed to keep an awake, resting body alive, and it generally represents between 60% and 75% of the body's energy needs. A male usually has a higher metabolic rate than does a female because a male's body has more muscle and less fat than a female's. Muscles use more energy than fat does. So, while a man and a woman of equal size sit on the couch and watch television together, he burns 10% to 20% more calories than she does. As you age, muscle mass and metabolic rate both decline. Together, these factors reduce caloric needs. If the person makes no other adjustments in lifestyle to compensate, these metabolic changes can add extra pounds each year after age 35. Besides gender and age, a person's BMR is influenced by age, health, food intake, and genetics.

The second largest use of energy is physical activity. Exercise is an excellent way to burn calories. It not only boosts your energy needs during the activity but also speeds up met-

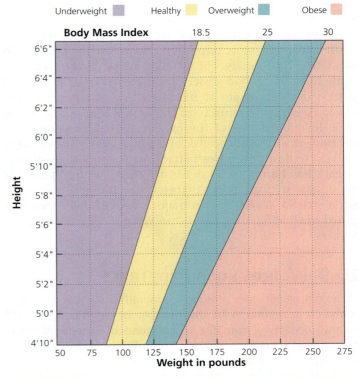

FIGURE 15.21 The body mass index (BMI) evaluates body weight relative to height. A BMI over 30 is usually considered to be unhealthy and a sign of obesity.

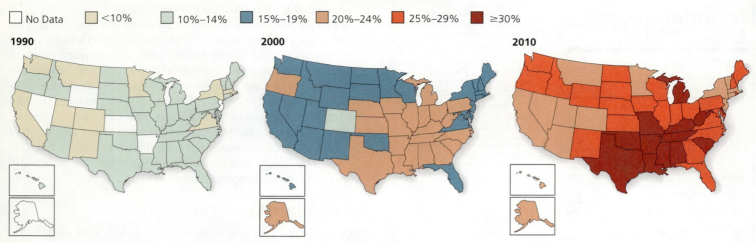

FIGURE 15.22 Obesity trends among adults in the United States. Obesity is defined as a high amount of body fat relative to total body weight. The number of obese people has increased alarmingly over the last decade. Centers for Disease Control and Prevention.

However, just as it is possible for a very muscular person to have a BMI above 30 and not be considered obese, it is possible for a person in the healthy weight range to have too much fat and little muscle. The Centers for Disease Control and Prevention (CDC) estimates that 68% of Americans are overweight and that almost 34% are obese. The number of obese people in the United States has been steadily rising since 1985, but the rate of increase is slowing (Figure 15.22).

Although most dieters are motivated to slim down for cosmetic reasons, more important reasons are the health risks associated with obesity. For instance, obesity leads to disease of the heart and blood vessels. Even individuals who are just slightly overweight are at increased risk of having a heart attack. Obesity also raises total cholesterol levels in the blood while lowering levels of the beneficial HDL cholesterol. In addition, it increases the risk of high blood pressure, which can lead to death from heart attack, stroke, or kidney disease. Obesity has harmful effects besides those on the heart and blood vessels. It can induce diabetes, which results in elevated blood glucose levels; it is a major cause of gallstones (discussed in Chapter 10a); and it can worsen degenerative joint diseases.

15.9 Weight-Loss Programs

Successful weight-loss programs generally have three components: (1) a reduction in the number of calories consumed, without departing from the recommended nutritional guidelines; (2) an increase in energy expenditure; and (3) behavior modification. Gradual changes in eating habits are most likely to lead to permanent lifestyle changes.

To determine the number of calories needed each day to maintain a desirable weight, a person must take activity level and age into account. A pound of fat contains approximately 3500 calories, so to lose 1 pound a week, a person should reduce calorie consumption by 500 calories a day (500 calories × 7 days = 3500 calories), or increase calorie use by 500 calories a day, or any equivalent combination. The lowest daily caloric intake recommended is 1200 calories for an adult female and 1500 calories for an adult male, unless they are in a medically supervised program.

The easiest way to reduce calorie intake while continuing to eat healthily is to cut back on fatty foods, especially those containing saturated fat, because fat contains more than twice as many calories as an equivalent weight of carbohydrate or protein. In addition, it is easier for the body to store the unused fat from foods as body fat than as protein or carbohydrate. Indeed, most of us do end up wearing the fat we eat.

Another recommended way of reducing calories is to avoid sugar. As we have seen, foods that taste sweet are packed with sugar and, therefore, calories. Recall that sugar is a high-glycemic carbohydrate.

A third healthful diet tip is to increase the amount of fiber in the diet. High-fiber foods, such as fruits and vegetables, tend to be low in calories and fat but also high in vitamins and minerals. Because fiber is bulky, these foods are also filling.

Approximately 60% to 90% of dieters who lose weight will later regain all the weight they lost. In many cases, the reason is that they had achieved the weight loss by drastically cutting back calories, which can be unhealthy and is very difficult to sustain for long periods. As old eating habits return, so do the pounds. When the determination to shed some pounds returns, the diet begins again. This process is commonly known as "the yo-yo effect."

What causes the yo-yo effect? When calories are severely restricted, as occurs in the typical crash diet, the body adopts a calorie-sparing defense that reduces the resting metabolic rate by as much as 45%. This response conserves energy and evolved long ago to help our ancestors survive during times of food scarcity. Today, it makes weight loss from successive diets progressively more difficult. Furthermore, with each diet, a person usually loses both lean and fat tissue, especially if increased exercise is not part of the weight-loss program. When the weight is regained, most of it is fat. Fat is less metabolically active than lean muscle tissue, so the person's metabolic rate drops even lower. Thus, repeated crash dieting may be a "no-lose situation."

what would you do?
Genetics and social environment are important causes of obesity. Do you think it is ethical for the weight-loss industry to place responsibility for obesity on the individual?

15.10 Eating Disorders

Obesity can be considered an eating disorder associated with overeating. People who are obese or overweight would be wise to follow a nutritionally sound weight-loss program. It is important to remember, however, that dieting can be taken too far. Most people with the eating disorders anorexia nervosa and bulimia nervosa began their descent into these disorders by dieting. **Anorexia nervosa** is a deliberate self-starvation. A person whose body weight is 85% or less than expected for his or her height is considered to be anorexic. In contrast, **bulimia nervosa** is marked by binge eating large quantities of food and then purging by self-induced vomiting, enemas, laxatives, diuretics, or excessive exercise.

The changes in eating habits associated with these eating disorders are thought to be the result of psychological, social, and physiological factors. Both disorders are associated with a preoccupation with body size and shape. Anxious depression also seems to play a role.

The behavior patterns associated with anorexia nervosa and bulimia differ, but both result in a severe deficit in calories and nutrients. A person with anorexia nervosa eats very little food and, therefore, consumes few calories. But *anorexia*, which means "lack of appetite," is misnamed. Although people with anorexia nervosa deny feeling hunger, their refusal to eat stems from an intense fear of becoming fat—not from lack of appetite. Excessive exercise is also typical in anorexia nervosa. No matter how much weight the person with anorexia loses, it is never enough, because the person has a distorted body image. He or she perceives the body as fat even when emaciated (Figure 15.23). In contrast, a person with bulimia eats a huge amount of food at one time but then eliminates it from the body. During a bulimic binge, which may last as long as 8 hours, the person may consume as many as 20,000 calories. Each binge is followed by attempts to purge the body of the calories, usually by self-induced vomiting or by laxatives.

Eating disorders have many negative effects on the body. Negative health effects of bulimia include esophageal injuries, tooth decay, and gum disease resulting from frequent vomiting, as well as dehydration, constipation, and electrolyte imbalance. One major side effect of anorexia nervosa is a severe decrease in bone health. Although the excessive exercise engaged in by most anorexic people may have a slight strengthening effect on bones, many other factors work to weaken the bones. For example, amenorrhea (cessation of menstruation), malnutrition, and low body weight, particularly low body fat, are all consequences of anorexia nervosa, and they can contribute to poor bone health.

The effects of anorexia nervosa are not so different from those of starvation. In the early phase of the illness, an anorexic person typically chooses a diet that is low in energy-dense foods but rather high in proteins and other essential nutrients. Dietary protein, combined with the high activity levels characteristic of a person with anorexia nervosa, has a nitrogen-sparing effect. As a result, the initial weight loss is almost entirely due to loss of fat tissue. However, when fat reserves are exhausted and refusal of food becomes more severe, the body begins to break down its own proteins to use as an energy source. The primary sources of these proteins are skeletal and heart muscle; therefore,

FIGURE 15.23 *Anorexia is a form of self-starvation. No matter how emaciated the person becomes, the individual still perceives the body as being fat.*

skeletal and heart muscle mass decreases. At the same time, water loss is accelerated, especially from within the body cells. This water loss leads to disturbances in metabolism and electrolyte balance.

Without treatment, up to 20% of people with serious eating disorders die. Even with treatment, 2% to 3% die. Heart problems are the most common cause of death in people with anorexia nervosa. Starvation, dehydration, and electrolyte disturbances cause the heartbeat to slow (bradycardia) and blood pressure to fall (hypotension). A potential cause of death associated with either anorexia nervosa or bulimia is hypoglycemia, an abnormally low blood glucose level. Because the brain depends entirely on glucose for its metabolism, hypoglycemia can cause unconsciousness and death.

Help for people with anorexia nervosa and bulimia nervosa is available. Treatment usually involves the family and centers on psychotherapy to help the person develop a healthier body image.

looking ahead

In this chapter, we have seen how the digestive system is organized to break down food molecules for energy or building materials. We also learned about our body's use of specific nutrients and considered how what we eat affects our health. In Chapter 15a, we will consider the causes, treatments, and prevention of foodborne illnesses.

HIGHLIGHTING THE CONCEPTS

15.1 The Gastrointestinal Tract (pp. 286–287)

- The digestive system consists of a long tube called the gastrointestinal (GI) tract, which starts at the mouth and continues through the pharynx, esophagus, stomach, small intestine, and large intestine. The digestive system also includes several accessory organs (salivary glands, pancreas, liver, and gallbladder).
- The GI tract has four layers. Moving from the inside out, they are the mucosa, submucosa, muscularis, and serosa.

15.2 Specialized Compartments for Food Processing (pp. 287–298)

- The mouth serves several functions. Teeth tear and grind food, making it easier to swallow. The salivary glands produce salivary amylase, which is released into the mouth to begin the chemical breakdown of starches. Taste buds help monitor the quality of food. Finally, the tongue manipulates food so that it can be swallowed.
- The pharynx is shared by the digestive and respiratory systems. Food swallowed moves from the mouth through the pharynx and into the esophagus.
- The esophagus is a tube that leads to the stomach. When we swallow, food is pushed from the mouth, and waves of muscle contraction called peristalsis push the food along the esophagus.
- The stomach stores food and regulates its release to the small intestine, liquefies it by mixing it with gastric juice, begins the chemical digestion of proteins, and regulates the release of chyme into the small intestine. Gastric juice consists of hydrochloric acid (HCl) and pepsin, a protein-splitting enzyme. Pepsin is produced in an inactive form, called pepsinogen, that is activated by HCl.
- The small intestine is the primary site of digestion and absorption. Enzymes produced by the small intestine and pancreas work to chemically digest carbohydrates, proteins, and fats into their component subunits.

- Bile emulsifies fat (breaks it into tiny droplets) in the small intestine, thereby increasing the combined surface area of fat droplets. Bile's action makes fat digestion by water-soluble lipase faster and more complete.
- The small intestine's surface area for absorption is increased by circular folds in its lining, fingerlike projections called villi, and microscopic projections covering the villi, called microvilli.
- Products of digestion are absorbed into the epithelial cells of villi by active transport, facilitated diffusion, or simple diffusion. Most materials, including monosaccharides, amino acids, water, and ions, then enter the capillary blood network in the center of each villus. However, fatty acids and glycerol are resynthesized into triglycerides, combined with cholesterol and phospholipids, and covered in protein, forming droplets called chylomicrons. The chylomicrons diffuse into a lymphatic vessel called a lacteal in the core of each villus. They are then delivered to the bloodstream by way of lymphatic vessels.
- The pancreas, liver, and gallbladder are accessory organs that aid in digestion and absorption within the small intestine. The pancreas secretes enzymes to digest most nutrients. The liver produces bile, which is then stored in the gallbladder.
- The large intestine consists of the cecum, colon, rectum, and anal canal. The large intestine absorbs water, ions, and vitamins. It is home to millions of beneficial bacteria that live on undigested material that has passed from the small intestine. The bacteria produce several vitamins, some of which we then absorb for our own use.
- Material left in the large intestine after passing through the colon is called feces. Feces consist of undigested or indigestible material, bacteria, sloughed-off cells, and water. Defecation is the discharge of feces from the rectum through the anus.

15.3 Nerves and Hormones in Digestion (pp. 298–299)

- Neural and hormonal mechanisms regulate the release of digestive secretions. Neural reflexes trigger the release of saliva, initiate the secretion of some gastric juice, and are the most important factors regulating the release of intestinal secretions. Gastrin, vasoactive intestinal peptide, secretin, and cholecystokinin are hormones that regulate digestive activities.

15.4 Planning a Healthy Diet (p. 299)

- MyPlate is a useful tool for planning a healthy diet. Make half your plate fruits and vegetables. Make half of your grains whole grains. Choose varied, lean sources of protein and fat-free or 1% fat sources of dairy foods. Balance caloric intake with energy expenditure.

15.5 Nutrients (pp. 299–304)

- A nutrient is a substance in food that provides energy; becomes part of a structure; or performs a function in growth, maintenance, or repair.
- Lipids, which provide 9 calories per gram, include fats, oils, and cholesterol. We should reduce the amount of saturated fat (fat that is solid at room temperature) and *trans* fats in our diet in favor of unsaturated fats or oils (fats that are liquid at room temperature) and omega-3 and omega-6 fats. Elevated blood levels of cholesterol increase one's risk of heart disease.
- A carbohydrate provides 4 calories per gram. A healthy diet should limit simple carbohydrates and maximize complex carbohydrates.
- Proteins are chains of amino acids. A protein provides 4 calories per gram. Essential amino acids are the 9 amino acids out of 20 that the body cannot synthesize and must be obtained in the diet. A complete protein, which generally comes from animal sources, contains all the essential amino acids. An incomplete protein, which comes from plant sources, lacks one or more of the essential amino acids. Complementary proteins are combinations of incomplete proteins that together supply all of the essential amino acids.
- A vitamin is a nutrient that is needed in very small amounts and serves as an enzyme or a coenzyme. Whereas vitamin C and the B vitamins are water soluble, vitamins A, D, E, and K are fat soluble.
- Minerals are inorganic substances needed for a variety of life processes.
- Water is a nutrient that it transports materials through our bodies, lubricates and cushions organs, helps in temperature regulation, and provides a medium for many vital chemical reactions.

15.6 Food Labels (pp. 304–307)

- Food labels can help you choose foods that will help you increase nutrients that are important to consume and limit those that could be harmful.

15.7 Energy Balance (p. 307)

- The basal metabolic rate (BMR) is the minimal energy needed to keep an awake, resting person alive. The sum of your BMR and the number of calories you use in physical activity is the number of calories you should consume to maintain a stable weight.

15.8 Obesity (pp. 307–308)

- Obesity, weighing more than height–weight charts recommend because of too much body fat, is harmful to your health.

15.9 Weight-Loss Programs (pp. 308–309)

- To maintain a stable weight, balance the number of calories consumed with those used. To lose weight, eat fewer calories than you use. Physical activity is an important part of weight-loss programs.

15.10 Eating Disorders (pp. 309–310)

- Anorexia nervosa is an eating disorder that is a form of self-starvation, and bulimia is a disorder in which a person consumes a huge amount of food and then purges it from the body.

RECOGNIZING KEY TERMS

digestive system p. 287
gastrointestinal (GI) tract p. 287
mucosa p. 287
submucosa p. 287
muscularis p. 287
serosa p. 287
saliva p. 289
salivary amylase p. 289
pharynx p. 290
esophagus p. 290
peristalsis p. 290
stomach p. 290
chyme p. 292

gastric glands p. 292
pepsin p. 292
small intestine p. 292
duodenum p. 292
bile p. 293
villi p. 293
lacteal p. 294
pancreas p. 294
liver p. 295
gallbladder p. 295
large intestine p. 296
cecum p. 296
appendix p. 296
colon p. 296

rectum p. 298
anal canal p. 298
nutrient p. 301
lipid p. 301
fat p. 301
cholesterol p. 301
low-density lipoproteins (LDLs) p. 301
high-density lipoproteins (HDLs) p. 301
carbohydrates p. 302
dietary fiber p. 302
glycemic response p. 302

protein p. 303
essential amino acids p. 303
complete proteins p. 303
incomplete proteins p. 303
complementary proteins p. 303
vitamin p. 303
minerals p. 304
basal metabolic rate (BMR) p. 307
obese p. 307
overweight p. 307
anorexia nervosa p. 309
bulimia nervosa p. 309

REVIEWING THE CONCEPTS

1. List the structures of the GI tract, in the order that food passes through them. *p. 287*
2. Describe how food is processed in the mouth. What are the functions of the teeth and tongue? *pp. 287–290*
3. Describe the structures of a tooth that are involved in tooth decay. What causes tooth decay? *pp. 287–289*
4. What are the three primary functions of the stomach? What are the functions of gastric juice? *pp. 290–292*
5. How does bile assist in the digestion and absorption of fats? *p. 293*
6. Where are carbohydrates digested? Proteins? Fats? *pp. 292–293*
7. Describe the structural features that increase the surface area for absorption in the small intestine. *p. 293*
8. What are the functions of the large intestine? *pp. 296–297*
9. Describe the neural or hormonal mechanisms that regulate the release of digestive juices from each structure involved in digestion. *pp. 298–299*
10. Explain how MyPlate helps you plan a healthful diet. *p. 299*
11. List the nutrients, and describe their functions. *pp. 301–304*
12. How does being overweight differ from being obese? *p. 307*
13. The villi in the wall of the small intestine function to
    a. increase the surface area for absorption.
    b. help mix the food with digestive enzymes.
    c. secrete bile.
    d. secrete digestive enzymes.
14. Why is it is possible to swallow while standing on one's hands?
    a. Valves in the digestive system keep food from moving backward.
    b. Peristalsis pushes food along the digestive tract in the right direction.
    c. Bacteria clog the digestive tube and prevent food from moving in the wrong direction.
    d. A wave of muscle contraction called emulsification prevents food from moving backward.
15. What is the most important function of the stomach?
    a. absorption of nutrients
    b. chemical digestion
    c. mucus secretion
    d. storage of food
16. You go out with your friends to celebrate your birthday and share a sausage pizza. Where does the digestion of the oil *begin*?
    a. in the mouth
    b. in the esophagus
    c. in the stomach
    d. in the small intestine
17. You are a pediatrician. A woman brings in her 1-year-old son, who has severe diarrhea. You would be most immediately concerned about the child's loss of _____ due to the diarrhea.
    a. water
    b. fiber
    c. fat
    d. protein
18. For the child in question 17, you prescribe an antidiarrhea medicine, which
    a. speeds up peristalsis in the small intestine.
    b. speeds up peristalsis in the large intestine.
    c. kills all bacteria in the large intestine.
    d. increases the amount of water absorbed from the large intestine.
19. What is the longest part of the gastrointestinal tract?
    a. esophagus
    b. stomach
    c. small intestine
    d. large intestine
20. Eating which of the following is most helpful in lowering your blood level of LDL cholesterol?
    a. fiber
    b. protein
    c. saturated fats
    d. carbohydrates
21. After a meal of greasy fries, which of the following digestive secretions would you expect to be most active?
    a. salivary amylase
    b. bile and lipase
    c. pepsin
    d. HCl
22. Which of the following is *not* a function of saliva?
    a. begins the chemical breakdown of starches
    b. helps clean the mouth
    c. contains fluoride to harden the tooth enamel
    d. moistens food and helps stick it together to make it easier to swallow
23. Which of the following soups contains a complete protein?
    a. lentil
    b. chunky chicken
    c. tomato
    d. cream of mushroom
24. Which of the following statements about vegetarian diets is *incorrect*?
    a. A vegetarian diet is likely to contain fewer calories than is a diet that contains meat.
    b. A vegetarian diet is likely to be lower in saturated fats than is a diet that contains meat.
    c. Vegetarians must consume milk products for a source of complete proteins.
    d. A vegetarian diet is likely to be high in complex carbohydrates.
25. A fast-food hamburger
    a. is deficient in protein.
    b. is deficient in fats.
    c. is high in salt and fats.
    d. supplies all the vitamins you need.
26. To reduce the amount of fat in your diet, it would be best to reduce your consumption of
    a. baked potatoes.
    b. hamburgers.
    c. spaghetti.
    d. bread.
27. _____ is a liquid mixture of food and gastric juices found in the stomach.
28. The _____ is the organ that produces bile.
29. Saliva contains an enzyme that begins the chemical digestion of _____.
30. _____ is a hormone produced by the small intestine that causes the gallbladder to contract and release bile.

APPLYING THE CONCEPTS

1. Barbara is a 20-year-old woman with cystic fibrosis, a genetic disease in which the body produces abnormally thick mucus. The thick mucus sometimes blocks the pancreatic duct that allows pancreatic juice to enter the small intestine. Explain why cystic fibrosis has caused Barbara to be malnourished.
2. Dehydration occurs when too much body fluid is lost. It can be a serious, even fatal, condition. Doctors worry about dehydration if a person, especially an infant, has severe diarrhea lasting several days. Why?
3. A 35-year-old man sporting a tattoo visits the dermatologist to have the tattoo removed. Noticing that the patient's skin has a yellowish tint, the dermatologist strongly recommends that the patient visit his primary care physician. Why? What is the dermatologist concerned about?
4. You spend the day with your Aunt Sally on her 55th birthday. She has a salad for lunch and suffers no ill effects. However, she indulges in fried chicken and fries for dinner. A few hours later, she is rushed to the emergency room with a severe pain on the right side of her abdomen. She reports that she has had bouts of pain over the last month or so. What do you think is Aunt Sally's problem? Why isn't the pain continuous?

BECOMING INFORMATION LITERATE

The PillCam is a small camera that is swallowed by the patient and takes pictures of the GI tract as it moves through the body. Write a short article for the health section of a local newspaper, describing the PillCam in more detail and answering the following questions in easily understood words. How does a PillCam differ from a traditional endoscope? What conditions and diseases can be diagnosed using the PillCam? Compared with endoscopy, what are the advantages and disadvantages of the PillCam?

Use at least three reliable sources (books, journals, or websites). List each source you considered, and explain why you chose the three sources you used.

MasteringBiology®

Go to MasteringBiology for practice quizzes, activities, eText, videos, current events, and more.

15a SPECIAL TOPIC
Food Safety and Defense

Careful food selection, preparation, and storage can help prevent foodborne illness.

- 15a.1 Foodborne Illnesses
- 15a.2 Keeping Food Safe at International and National Levels
- 15a.3 Food Defense and Bioterrorism
- 15a.4 Personal Food Safety

In Chapter 15 we described the digestive system and nutrition. In this chapter, we examine some of the illnesses caused by consuming contaminated food or water. Typical symptoms of these illnesses include vomiting and diarrhea, which can cause dehydration and disrupt body homeostasis. We examine how these illnesses are diagnosed and treated. We also look at the agencies that protect our food supply and what we can do as individuals to make sure that the food we eat and the water and beverages we drink are safe.

15a.1 Foodborne Illnesses

Foodborne illnesses result from ingesting contaminated food or water. Disease-causing agents such as bacteria, viruses, prions (infectious proteins), and parasitic protozoans (single-celled eukaryotic organisms) or worms can contaminate food; collectively, these agents are called *pathogens*. There are more than 250 different foodborne illnesses, including some caused by harmful chemicals such as those found in pesticides or those found naturally in certain mushrooms or fish (Figure 15a.1). Foodborne illnesses caused by pathogens are described as *infections*, whereas those caused by chemicals or toxins are considered *poisonings*. The general public usually ignores this distinction and simply refers to all foodborne illnesses as "food poisoning." (Food allergies, described in Chapter 13, are also a type of foodborne illness.)

General Symptoms, Diagnosis, and Treatment

Despite their different causes and classifications, all foodborne illnesses have symptoms that first appear in the gastrointestinal (GI) tract. These early symptoms may include nausea, vomiting, abdominal cramps, and diarrhea. Some pathogens do not move beyond the GI tract. Other pathogens, including some bacteria, produce toxins that are absorbed into the bloodstream, and

(a) The death cap mushroom contains lethal toxins that cannot be destroyed through cooking, drying, or freezing.

(b) Pufferfishes contain tetrodotoxin, which causes death if consumed. Despite this danger, pufferfish is a highly prized dish in Japan. Known as *fugu*, the dish is prepared by licensed chefs who carefully remove the poisonous parts of the fish.

FIGURE 15a.1 *Naturally occurring poisons can cause foodborne illness and death.*

still others directly invade other body tissues. For example, the bacterium *Clostridium botulinum,* found in improperly canned foods, produces a toxin that blocks communication between nerve and muscle cells, causing paralysis of muscles and making it difficult or impossible to breathe. Because new canning procedures have virtually eliminated the risk of botulism from commercially canned foods, foods canned improperly at home are the typical source of *C. botulinum* contamination (see Chapter 13a for discussion of how bacteria and viruses cause disease).

Many foodborne illnesses go undiagnosed and unreported to public health officials. Some ill people do not seek medical care, perhaps because they do not realize that they have a foodborne illness or because they can't afford to see a doctor. For those who do seek medical attention, diagnostic tests may not be performed to confirm the particular illness and its causative agent; or, if tests are performed, the results may not be communicated to public health officials. The latter situation can arise because each state decides which diseases should be monitored in that state, so some diseases are not reportable in the first place. Some estimates suggest that the actual number of cases of mild foodborne illnesses is 20 to almost 40 times the numbers reported. Even foodborne illnesses with very severe symptoms are underreported, though to a much lesser degree than those with milder symptoms. In 1996, the Centers for Disease Control and Prevention (CDC), the Food Safety and Inspection Service (FSIS), the Food and Drug Administration (FDA), and selected state health departments developed the Foodborne Diseases Active Surveillance Network (FoodNet), a collaborative monitoring system that collects data on pathogens commonly transmitted through food. This program tracks seven bacterial and two parasitic foodborne illnesses through laboratory testing of patients.

A diagnosis of foodborne illness requires a physical exam and additional research to obtain a detailed history of recently consumed foods and beverages. A stool sample may be collected for specific laboratory tests. When a bacterial illness is suspected, stool samples are cultured, and the bacteria that grow on the culture medium are identified. If parasites are suspected, then the stool sample is also examined with a microscope. Because parasites often have several life stages, stool samples are examined for the presence of adults or younger stages, such as eggs or larvae. In the case of the protozoan parasite *Cryptosporidium parvum* (a major cause of waterborne diarrheal illness in the United States), trained laboratory personnel look for oocysts, the infectious early stage in the life history of this organism. During the oocyst stage, the parasite is enclosed in a protective capsule that makes it resistant to chlorine-based disinfectants. *Cryptosporidium parvum* is commonly transmitted through the swallowing of recreational water, such as that found in pools, hot tubs, and ponds. Tests for viruses usually involve examining the stool sample for the genetic markers that indicate the presence of a specific virus. Patients may have a blood sample collected for examination.

According to the CDC, the symptoms associated with diarrheal illness that warrant consultation with a health care professional are the following:

- Temperature over 101.5°F (measured orally)
- Blood in stools
- Prolonged vomiting (especially the inability to keep liquids down)
- Dizziness when standing up, dry mouth and throat, decreased urination (all indicate dehydration)
- Diarrhea lasting more than 3 days

Recall from Chapter 15 that water performs important functions in our body, such as serving as the main transport medium and helping to prevent dramatic changes in body temperature. The vomiting and diarrhea associated with foodborne illness function to expel the offending pathogens or toxins from the GI tract, but they also rob the body of water it needs to function properly. Dehydration from diarrhea and vomiting can be deadly if not treated. Patients treated for foodborne illnesses are usually given clear, clean fluids to prevent dehydration. When foodborne illness is caused by bacteria, antibiotics may be prescribed as well. Antiparasitic drugs can be used to kill parasites.

Common Foodborne Infections

The most common foodborne infections in the United States are caused by the bacteria *Campylobacter jejuni*, *Salmonella*, and *Escherichia coli* O157:H7, along with a group of viruses known as caliciviruses. *Campylobacter jejuni* typically enters the body during the handling of raw poultry or the consumption of raw or undercooked poultry (chickens often carry the bacteria but show no signs of illness). The resulting illness is known as campylobacteriosis.

Species of bacteria within the genus *Salmonella* most often enter our bodies when we consume food contaminated with animal feces. The food, which is often—but not always—animal in origin (meat, milk, or eggs), usually looks and smells normal. The illness that results is salmonellosis.

Most strains of the bacterium *Escherichia coli* do not cause illness; in fact, *E. coli* bacteria live harmlessly in the intestines of most people. *Escherichia coli* O157:H7, however, is not harmless (the letters and numbers in its name refer to specific markers found on its surface that allow scientists to differentiate this bacterium from other strains of *E. coli*). This strain lives in the intestines of healthy farm animals such as cattle, goats, and sheep, and it is most commonly passed to humans when they consume undercooked ground beef that contains the pathogen. *Escherichia coli* O157:H7 also occurs in animals at petting zoos, where it has been found on the animals' fur as well as on the ground, on railings, and in feed bins. Once a person is infected with *Escherichia coli* O157:H7, good hygiene is necessary to prevent its spread to others. If hand washing is inadequate, *Escherichia coli* O157:H7 in diarrhea can be passed from person to person.

Finally, viral gastroenteritis, commonly called the "stomach flu," is caused by many viruses, including a group known as caliciviruses. Individuals can become infected by consuming food or beverages contaminated with caliciviruses. Viral gastroenteritis is contagious and is spread by close contact with an infected person.

These four common foodborne illnesses—campylobacteriosis, salmonellosis, infection with *Escherichia coli* O157:H7, and viral gastroenteritis—have similar symptoms: diarrhea, nausea, vomiting, and abdominal cramps. Fever is present in some cases. *Campylobacter* and *Salmonella* infections that spread beyond the GI tract to the bloodstream are more serious, and infection with *Escherichia coli* O157:H7 occasionally causes kidney failure. All but caliciviruses can cause life-threatening illness. People most at risk include those whose immune system is compromised or incompletely developed, such as the very old, the very young, and people already suffering from a disease that reduces their immune function. However, even healthy people can die from foodborne illnesses if they are exposed to a very high dose of the pathogen or if they do not receive proper treatment.

How Does Food Become Contaminated?

Food can become contaminated during any of the many steps that typically occur as it moves from "farm to fork" (Figure 15a.2). At early stages, bacteria normally present in the intestines of food animals, such as cows or chickens, can contaminate carcasses at slaughterhouses and get into meat and poultry products. *Salmonella* can infect the ovaries and oviducts of chickens, which then produce eggs containing the bacterium. This route of infection, called the *transovarian route*, is thought to be the predominant way that eggs become contaminated. A less common route, called the *trans-shell route*, occurs when bacteria in the environment, such as those on nesting material or in shipping containers, penetrate the eggshell after the egg has been laid.

Oysters and other filter-feeding shellfish concentrate bacteria naturally found in seawater along with the bacteria dumped into our oceans in human sewage. Oysters also concentrate toxins and pollutants. Large individuals of some top predatory fish species in our oceans, such as king mackerel, act as bioconcentrators and store potentially dangerous levels of heavy metals (such as mercury) in their tissues (see Chapter 23). In freshwater and on land, parasitic worms, such as roundworms, tapeworms, and flukes, release eggs into the environment, where they may be picked up by fish, pigs, cattle, and other animals. The eggs of the parasitic worms hatch into larvae in these animals, and the larvae can survive for some time after the animals have been killed for food.

Certain foods are more likely than others to be associated with foodborne illness. As evident from our discussion, raw foods originating from animals are most likely to be contaminated. In addition, foods that contain the products of many animals, such as ground beef, are especially dangerous because only one of the many contributing animals need contain the pathogen for the meat to be contaminated. To put this in perspective, it is estimated that a single hamburger patty may contain meat from hundreds of cattle and that a glass of milk may contain milk from hundreds of cows!

Contamination is not restricted to food from animals. Indeed, fresh fruits and vegetables can become contaminated with bacteria, viruses, and protozoan parasites when they are irrigated or washed with water containing animal (including human) waste. For example, fresh spinach contaminated with *E. coli* O157:H7 killed three people and sickened about 200 others in an outbreak that spanned 26 states in the fall of 2006. Investigators were able to trace the contaminated spinach back to at least one field and farm in California, where tests of soil, water, and animal manure showed that *E. coli* O157:H7 was more widespread in the environment than previously believed. Washing fruits and vegetables may not protect against foodborne illness. Contaminated water can introduce pathogens, and washing with clean water has been shown to decrease but not to completely eliminate contamination. Contamination also can occur after harvesting. In 2011, more than 140 people from 28 states were hospitalized after eating cantaloupe infected with strains of the bacterium

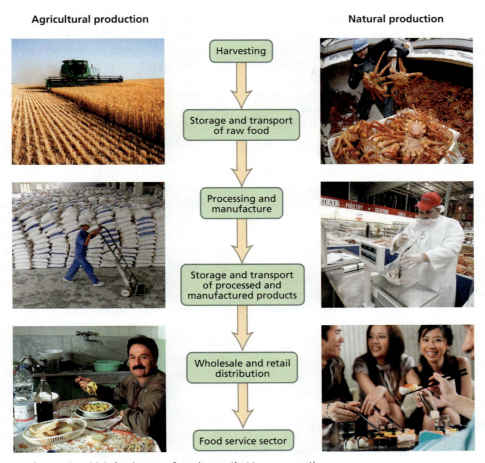

FIGURE 15a.2 *A typical pattern of steps by which food moves from harvesting to consumption*

Listeria monocytogenes; 30 people died. The contamination was traced to the cantaloupe-packing facility.

Food also can be contaminated during processing (refer, again, to Figure 15a.2). For example, from fall 2008 through spring 2009, peanut butter and peanut paste contaminated with *Salmonella* sickened more than 700 people in 46 states and may have contributed to 9 deaths. The contamination was traced to a peanut-processing plant in Georgia and may have resulted from improper roasting or subsequent processing. Finally, food also can be contaminated during preparation. People handling food with unwashed hands can introduce pathogens, and pathogens can be transferred from one food to another when utensils and cutting boards are used for different foods without being washed in between.

 what would you do?

Most egg-laying hens in the United States are housed in small battery cages in which they are unable to spread their wings or display species-typical behavior. Free-range hens, in contrast, have access to the outdoors. Studies on whether free-range hens are less likely than conventionally housed hens to become infected with *Salmonella* have produced conflicting results. Eggs from free-range hens are more expensive than eggs from battery hens. Would you be willing to spend more money on eggs from free-range hens based on animal welfare concerns alone? Why or why not?

Methods of Combating Food Contamination

Once food is contaminated, the way it is handled becomes critical. Large numbers of bacteria are usually required to cause foodborne illness, so leaving contaminated food in conditions conducive to bacterial reproduction is especially dangerous. Bacteria reproduce extremely rapidly, particularly under warm, moist conditions (temperatures between 40°F and 140°F are considered the danger zone because they represent conditions under which bacteria are most likely to reproduce). In fact, a single bacterium reproducing by binary fission (dividing itself in half) can produce millions of new cells in only a few hours (see Chapter 13a).

Prompt refrigeration can slow reproduction by many bacteria, as can acidic conditions. Most microorganisms do not thrive at a pH of less than or equal to 5. To create such acidic conditions, vinegar (which is dilute acetic acid) is often added to such foods as cucumbers and peppers during a process called pickling. High levels of salt or sugar also can be used to inhibit microbial growth in food; these solutes reduce the amount of water available to microorganisms. Fruits are typically preserved by adding sugar (to make jams and jellies), and meats and fish are often preserved with salt. Drying also reduces water available to microorganisms, and this method is used to preserve fruits, vegetables, meat, fish, eggs, and milk.

With few exceptions, heat kills parasites, bacteria, and viruses; the microorganisms die because very high temperatures cause their macromolecules, such as DNA and proteins, to lose their structure and function. In this way, heating contaminated food to sufficiently high temperatures can render it safe to eat. A food thermometer, placed in the thickest part of the food item, should reach at least 160°F to ensure that the food item is safe to eat. Toxins produced by bacteria vary in whether or not they are destroyed by heat. Prions are not destroyed by heat alone. However, research in the last few years has shown that the combination of high temperature (275°F) and short bursts of very high pressure inactivates prions in processed meats, such as hot dogs. Prions also are inactivated by treatment with certain very strong chemicals, such as sodium hydroxide (a strong base) and household bleach, but such treatments cannot be applied to food.

stop and think
Sponges used to clean kitchen countertops have a reputation for being notoriously contaminated items. What characteristics associated with sponges and their use lead to such high levels of contamination? What could we do to prevent or reduce their contamination?

15a.2 Keeping Food Safe at International and National Levels

Our food supply today is global: most of us consume food grown and processed in other countries as well as that produced nearer to home. The nature of our food supply, along with the extreme mobility of the population, makes the regulation of food safety and the tracking of foodborne illnesses extremely challenging. To this end, several agencies regulate, monitor, and investigate the safety of our food.

International Oversight

The World Health Organization (WHO) is a United Nations agency whose primary mandate is to protect public health. One of its major tasks is to fight diseases, especially those that are communicable (able to be passed from one person to another). Various functions of WHO regarding food safety include coordinating international efforts to effectively monitor and respond to outbreaks of foodborne illness and implementing international health regulations. These regulations relate to international travelers and the import and export of contaminated foods. WHO also sponsors research on foodborne illnesses and programs to prevent and treat these illnesses.

National Oversight

Several agencies within the United States are charged with overseeing food safety. The Food and Drug Administration (FDA) is responsible for the safety of about 80% of our food supply. More specifically, the FDA oversees all domestic and imported food except meat and poultry. In addition, the FDA oversees eggs (in the shell), bottled water, and wines containing less than 7% alcohol. The Bureau of Alcohol, Tobacco, Firearms, and Explosives (ATF) oversees all other alcoholic beverages. The Food Safety and Inspection Service (FSIS) oversees domestic and imported meat and poultry, including products that contain meat or poultry, such as frozen foods. Processed egg products (such as liquid egg whites) are also overseen by the FSIS. The Environmental Protection Agency (EPA) regulates the use of pesticides in agriculture and establishes standards for safe public drinking water. The National Oceanic and Atmospheric Administration (NOAA) inspects fishing vessels and seafood-processing plants. Finally, the U.S. Customs Service works with federal regulatory agencies to make sure the imported foods meet U.S. regulations.

The Centers for Disease Control and Prevention (CDC) is not a regulatory agency, in the sense that it does not have the power to enforce federal rules. Nevertheless, it plays a critical role in keeping our food safe by working with state and local health officials to monitor the occurrence of foodborne illnesses (as well as other diseases). Another major function of the CDC is to investigate disease outbreaks. Working with local food safety officials, the CDC determines whether a cluster of similar cases suggests an outbreak of foodborne illness in a particular area. When an outbreak is suspected, the CDC begins an investigation to determine the particular pathogen involved and its source. Such investigations typically involve searching for more cases; characterizing the outbreak by time, place, and people involved; identifying the pathogen through laboratory tests performed on stool or blood samples; and determining through extensive interviews the source of the outbreak. The CDC also develops methods to prevent foodborne illnesses and continually assesses the effectiveness of prevention efforts. Because the CDC does not restrict its oversight activities to particular foods, it effectively oversees all foods.

15a.3 Food Defense and Bioterrorism

So far we have considered unintentional threats to our food and water supplies, such as pathogens accidentally entering food or people mistakenly eating food that contains toxins. However, intentional threats also exist. For example, there is considerable concern that terrorists could introduce biological, chemical, or radiological agents into food or water supplies to generate fear, economic losses, and human casualties. Whereas food safety prevents the *unintentional* contamination of food, food defense prevents *deliberate* contamination. Here we consider food defense, which is overseen by the Department of Homeland Security (DHS) and many of the agencies previously described as overseeing food safety. The DHS has many food defense responsibilities, for example, developing screening procedures for food entering the United States, assessing the vulnerabilities of the food and beverage industries, and coordinating efforts by the private sector and federal, state, and local governments to protect these industries.

Bioterrorism is the use of biological agents (bacteria and their toxins, viruses, or parasites) to intimidate or attack societies or governments. The CDC classifies biological agents that could be used in terrorist attacks into three categories (A, B, and C) based on characteristics of the biological agent, such as availability, ease of dissemination (distribution), severity of health effects produced, and degree of preparedness required by public health agencies to respond to an attack.

Biological agents in Category A are the highest priority because they are easily disseminated or highly transmissible from person to person, produce high mortality rates, and require significant planning by public health agencies. Included in Category A is the toxin produced by the bacterium *Clostridium botulinum*. This toxin is the most poisonous substance known. Although the resulting illness (botulism) is not spread from person to person, the agent is placed in this category because of its extreme potency and lethality and because it can so easily be produced and transported. We mentioned earlier in this chapter that the toxin might enter food unintentionally as a result of improper canning, but it could also intentionally be placed in food. It is unlikely to be placed in water because standard water treatment processes, such as chlorination, inactivate it.

Biological agents in Category B are moderately easy to disseminate and result in moderate rates of morbidity (sickness) and low rates of mortality. Included in Category B are several species of bacteria that could be intentionally added to food, including *Salmonella* and *Escherichia coli* O157:H7. The main threats to water safety also are included in Category B. These include the bacterium *Vibrio cholerae*, which causes the acute diarrheal illness cholera, and the single-celled parasite *Cryptosporidium parvum*, discussed earlier as a major cause of waterborne illness.

Biological agents in Category C include emerging pathogens that could be used in the future because they are easy to produce and disseminate and are associated with high rates of morbidity and mortality.

15a.4 Personal Food Safety

Regardless of whether a risk is intentional or unintentional, systems are in place to limit exposure to contaminated food or water. Although we may be uneasy about the many steps between the production of food and its arrival on our plate, each of us can do some things to make sure that the food we consume is as safe as possible.

Good personal food safety practices begin with a commonsense approach and the awareness that you should practice food safety from the moment you select an item in a market until you prepare it at home and store the leftovers.

Food Selection

When selecting any food, you should always look at "sell by" dates and carefully examine the packaging for damage or tampering. Bacterial contamination and the possible growth of bacteria on food surfaces of fresh meat, poultry, seafood, or other food products rich in animal proteins is a very real concern. To limit bacterial growth, we rely primarily on cold temperatures in store display cases, rapid (and cool) transport of food from the store to a home refrigerator, and consistently cold temperatures in home refrigerators (always below 40°F). It is best to select refrigerated and frozen foods last when shopping and to put them into a refrigerator or freezer within 1 hour of purchase.

Meat A good rule of thumb when buying meat products is that whole cuts, such as steaks or roasts, are far less likely to carry bacterial contamination than are ground meat products. This is so because the surface area of the food serves as the site for bacterial contamination, and ground meat products have many times the surface area of an equivalent weight of steak. Recall, also, that meat from many individual animals goes into ground meat, raising the chances that pathogens may be present in the first place.

Poultry Fresh poultry is susceptible to bacterial contamination, particularly by *Salmonella*. Processors now deliver many fresh poultry items to markets packaged in trays or strong bags that limit the contact and potential cross-contamination by bacteria from one item to another. That said, it is impossible to eliminate all risk, and you should not assume that the outside of the package does not harbor bacteria. When you select poultry from a market case, it is a good idea to place it apart from other foods—particularly vegetables or fruits that will be eaten raw—and to clean your hands before handling other food products. Always carry poultry products in bags separate from the rest of the groceries.

Seafood Shopping for seafood poses some special concerns. For example, some seafood, such as fresh oysters, is traditionally eaten raw, so we cannot rely on high temperatures during cooking to kill pathogens or to destroy their toxins (Figure 15a.3). Also, as noted earlier, oysters, because they are filter feeders, concentrate local microorganisms, toxins, and pollutants in their tissues. If you choose to eat raw oysters, then you should ask where the oysters came from. Even when you "know" their source, you must still be willing to accept the risk that the source may not be the pristine natural environment you might assume it is. When you are selecting fresh fish, keep

FIGURE 15a.3 Eating raw oysters poses certain risks because oysters are filter feeders that concentrate bacteria, toxins, and pollutants in their tissues.

in mind that firmness, color, and odor are paramount. Reject any fish with a strong fishy odor. (Fresh saltwater fish smells like the sea.) You may also want to ask whether the fish has been refreshed with any solutions. Such "refreshing" refers to washing the fish with a dilute bleach solution to eliminate bacterial contamination on the surface—not a very appetizing way to treat any food, and not even necessary if the fish has been handled carefully up to that point. (A quick sniff will often reveal the presence of bleach.)

A final concern when buying fresh fish—even fish that will be cooked to high enough temperatures to ensure safety from typical bacterial contamination—relates to the species of fish you are selecting. As mentioned earlier, large individuals of some top predator species concentrate and store potentially dangerous levels of mercury in their tissues. To avoid consuming dangerous levels of mercury, be aware of the current safety concerns related to individual species of fishes, and do not exceed the recommended weekly consumption rates for those species, especially if you are pregnant or nursing. Further information on mercury contamination is available at the FDA's Food Safety website (www.fda.gov/Food/FoodSafety/Product-SpecificInformation/Seafood/default.htm), and information regarding the safety of locally caught fish can be found at the EPA's Fish Advisory website (www.epa.gov/waterscience/fish).

Produce When you're shopping for fresh vegetables, raw prepacked, prewashed lettuce and salad mixes vie for your attention against whole lettuces and other vegetables. Should you buy the time-saving prepacked versions, whose labels often promise they have been washed twice? From a food safety standpoint, there is no one clear answer. Bacterial contamination of fresh vegetables can happen at many points between farm and fork. Wash these products as thoroughly as possible, even those that are packaged as ready to eat, before consumption, while recognizing that there is always some risk.

Conventionally produced fruits and vegetables are treated with pesticides, exposed to herbicides, and depend upon chemically produced fertilizers derived from finite resources, such as fossil fuels. Washing such produce with running water (and a brush if the fruit or vegetable is firm), peeling fruits whenever possible, and discarding the outer leaves of lettuce and cabbage can reduce your exposure to pesticides and herbicides. Some consumers, concerned with chemical residues in fruits and vegetables, select certified organic foods, which are produced without certain synthetic pesticides and fertilizers (some synthetic pesticides have been approved for use on organic farms and are sometimes used). At present, there is no evidence that organic foods are safer or nutritionally superior to conventionally produced foods. Each consumer will have to decide whether to accept fruits and vegetables grown in a conventional manner or to pay somewhat more for foods produced organically. A deciding factor for many is that foods that are certified to have been produced organically may be less harmful to natural environments than are conventionally produced foods. Organically grown products are often produced closer to home, leading to lower oil consumption during transport.

In addition to concerns about pesticide and herbicide exposure, there are concerns related to genetically modified (GM) foods and to foods that result from the cloning of animals; all of these issues ensure that many basic questions about food safety and the environment will continue far into the future (see Chapter 21 for a discussion of genetically modified food). As a responsible consumer, it is important that you stay informed and make conscientious decisions about your personal tolerance for food safety risks and the future of foods and the environment.

Food Handling and Storage

At home, safe food-handling practices can be sorted into four basic categories: (1) cleanliness, (2) separation, (3) cooking, and (4) chilling (Figure 15a.4). *Cleanliness* refers to both hand

Q Explain the different effects that chilling and cooking have on bacteria in food. Also, is it best to thaw food in the refrigerator or on the counter in your kitchen?

FIGURE 15a.4 This logo from the Be Food Safe program of the USDA and Partnership for Food Safety Education emphasizes the four basic ways to keep food safe at home.

A Placing food in the refrigerator or freezer does not kill bacteria; it simply stops bacteria from increasing in number. Cooking food to an appropriate temperature kills bacteria. Food should always be thawed in the refrigerator because room temperatures allow bacteria to multiply on the surface of the food, even if the interior is still frozen.

TABLE 15a.1 Storage Times for Refrigerated Foods

Foods	Storage Times	Foods	Storage Times
Eggs		**Hot dogs and luncheon meats**	
Fresh, in shell	3–5 weeks	Hot dogs	Unopened package, 2 weeks
Raw yolks, whites	2–4 days		Opened package, 1 week
Hard-cooked	1 week	Luncheon meats	Unopened package, 2 weeks
Liquid pasteurized eggs, egg substitutes	Unopened, 10 days Opened, 3 days	**Bacon and sausage**	
Cooked egg dishes	3–4 days	Bacon	7 days
Mayonnaise, commercial, opened	2 months	Sausage, raw, from meat or poultry	1–2 days
Deli and vacuum-packed products		Smoked breakfast links, patties	7 days
Store-prepared (or homemade) egg, chicken, tuna, ham, and macaroni salads	3–5 days	Summer sausage labeled "Keep Refrigerated"	Unopened, 3 months Opened, 3 weeks
Prestuffed pork, lamb chops, and chicken breasts	1 day	Hard sausage (such as pepperoni)	2–3 weeks
		Cooked meat, poultry, and fish leftovers	
Store-cooked dinners and entrees	3–4 days	Pieces and cooked casseroles	3–4 days
Commercial brand vacuum-packed dinners with USDA seal, unopened	2 weeks	Gravy and broth, patties, and nuggets	1–2 days
		Soups and stews	3–4 days
Raw hamburger, ground meat, and stew meat		**Fresh meat (beef, veal, lamb, and pork)**	
Ground beef, turkey, veal, pork, lamb	1–2 days	Steaks, chops, roasts	3–5 days
Stew meats	1–2 days	Variety meats (tongue, kidneys, liver, heart, chitterlings)	1–2 days
Ham, corned beef			
Ham, canned, labeled "Keep Refrigerated"	Unopened, 6–9 months Opened, 3–5 days	**Fresh poultry**	
		Chicken or turkey, whole	1–2 days
Ham, fully cooked, whole	7 days	Chicken or turkey, parts	1–2 days
Ham, fully cooked, half	3–5 days	Giblets	1–2 days
Ham, fully cooked, slices	3–4 days	**Fresh fish and shellfish**	
Corned beef in pouch with pickling juices	5–7 days	Fresh fish and shellfish	1–2 days

Source: USDA Food Safety and Inspection Service, *Fact Sheets: Safe Food Handling,* www.fsis.usda.gov.

washing by the food preparer (and subsequently the food consumer) and cleanliness of surfaces that will come into contact with raw food, such as cutting boards, knives, and storage containers. Keeping the kitchen workplace clean is an obvious and basic way to improve food safety at home. *Separation* refers to keeping foods apart from one another and avoiding cross-contamination. For example, keeping fresh meat out of direct contact with cooked foods is an essential part of kitchen hygiene that can limit exposure to bacteria. *Cooking* refers to thoroughly cooking food products to temperatures that will kill most bacteria, which generally do not survive temperatures greater than 160°F. Proper and prompt *chilling* of both fresh and cooked foods helps to stop bacterial growth. Nevertheless, even refrigerated food is not safe to eat indefinitely (Table 15a.1). Freezing to 0°F inactivates bacteria in food, but it does not kill them. Indeed, once the food is thawed, the bacteria will begin to reproduce at room temperature. When in doubt about the freshness of refrigerated food, throw it out.

In the end, personal food safety practices in shopping, preparing, and storing foods at home play a major role in ensuring that your food is safe. Keep these principles in mind when you plan, shop for, and prepare safer meals.

stop and think

When freezing a large amount of food, it is best to divide the food into smaller portions and freeze each portion in a freezer-safe container. Why would freezing multiple, small portions be better than freezing one large portion?

looking ahead

In this chapter we learned how pathogens cause foodborne illnesses, which are associated with vomiting, diarrhea, and dehydration. In Chapter 16 we focus on how the urinary system helps to regulate fluid balance in the body.

16 The Urinary System

Did you Know?

- The nephrons of one kidney, when placed end to end, would stretch about 5 miles.
- As many as eight lives can be saved by one organ donor.

Proper fluid balance is critical to homeostasis. We challenge the balance of fluids in our bodies through diet and exercise. In the face of such challenges, our kidneys help to achieve the correct balance by filtering the blood and regulating its volume and solute concentration.

- 16.1 Eliminating Waste
- 16.2 Components of the Urinary System
- 16.3 Kidneys and Homeostasis
- 16.4 Dialysis and Transplant Surgery
- 16.5 Urination
- 16.6 Urinary Tract Infections

HEALTH ISSUE

Urinalysis

ETHICAL ISSUE

Kidney Donation and Trafficking

Homeostasis is the ability to maintain a relatively stable internal environment in the face of changing internal and external conditions. In Chapter 15a we learned that foodborne illnesses disrupt body homeostasis by causing dehydration. In this chapter, we focus on the urinary system and maintenance of fluid balance in the body. Specifically, we consider the critical roles that kidneys play in maintaining homeostasis by removing wastes from the body and regulating blood volume and blood pressure. The kidneys also stimulate the production of red blood cells, regulate the concentration of solutes in plasma, and help stabilize blood pH. Finally, the kidneys transform vitamin D into its active form, which promotes the body's use of calcium and phosphorus. Clearly, kidneys have multiple roles in maintaining body homeostasis.

16.1 Eliminating Waste

Excretion is the elimination of wastes and excess substances from the body. Metabolic wastes include carbon dioxide, water, heat, excess ions, and nitrogen-containing wastes such as ammonia, urea, uric acid, and creatinine. Nitrogen-containing wastes are produced during the normal breakdown of proteins. This breakdown initially produces ammonia. Because ammonia is toxic, the liver converts it to urea, a less harmful waste. Uric acid is formed during the recycling of nitrogen-containing bases of nucleotides (see Chapter 2). In excess, uric acid may form crystals in the synovial fluid of joints (particularly in the big toe), causing the painful condition called *gout*. Creatinine is generated in skeletal muscle from the breakdown of *creatine phosphate,* a compound that serves as an alternative energy source for muscle contraction. Excess ions—such as hydrogen

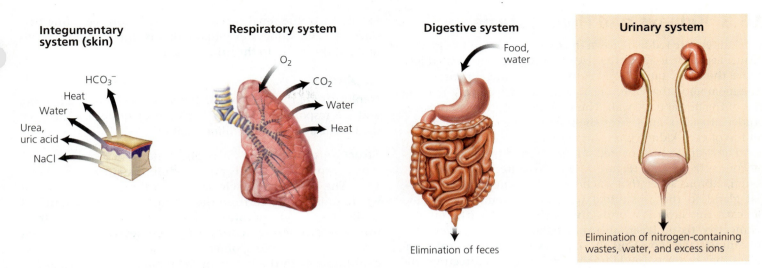

FIGURE 16.1 *Organs from several systems eliminate wastes of different kinds from the body. The kidneys, organs of the urinary system, play a major role in the excretion of nitrogen-containing wastes, water, and excess ions.*

(H^+), sodium (Na^+), potassium (K^+), and chloride (Cl^-)—come from ingested food and the breakdown of nutrients.

The organs that help to maintain homeostasis by eliminating metabolic wastes and excess ions are shown in Figure 16.1. Lungs and skin eliminate heat, water, and carbon dioxide (as the gas CO_2 from the lungs and as bicarbonate ions, HCO_3^-, from the skin). Skin also excretes salts (mostly NaCl) and small amounts of urea and uric acid. Minor amounts of other substances, such as alcohol, are also eliminated from the skin and lungs. (Alcohol excretion by the lungs forms the basis of the breathalyzer test.) Organs of our gastrointestinal tract eliminate solid wastes and some metabolic wastes. Defecation is the elimination of feces from the digestive tract. Feces contain undigested food, bacteria, water, bile pigments, and sloughed-off epithelial cells (see Chapter 15).

Our focus in this chapter is the kidneys, which are the organs of the urinary system that form urine. **Urine**, a yellowish fluid, is a mix of water and solutes. Through urine, the body excretes water, excess ions, urea, uric acid, and creatinine.

16.2 Components of the Urinary System

The **urinary system** consists of two kidneys, two ureters, one urinary bladder, and one urethra (Figure 16.2). The main function of this system is to regulate the volume, pressure, and composition of the blood. The **kidneys** are the organs of the urinary system that accomplish this task by regulating the amount of water and dissolved substances that are removed from and returned to the blood. Substances not returned to the blood form urine.

Urine from the kidneys travels down muscular tubes called **ureters** to the urinary bladder. Each ureter is 25 to 30 cm in length. Peristaltic (wavelike) contractions begin where the ureters leave the kidneys and travel down the ureters, pushing urine toward the urinary bladder. Each ureter enters the urinary bladder through a slit-like opening. The **urinary bladder**

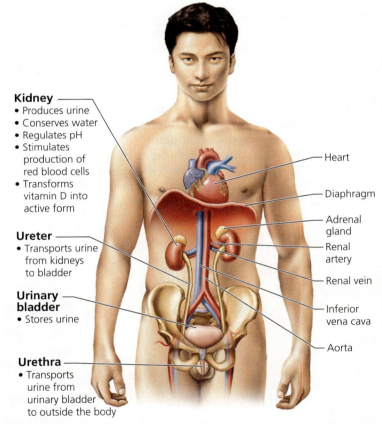

FIGURE 16.2 *Organs of the urinary system (labeled on the left side) and their relation to major blood vessels (identified on the right side)*

is a muscular, expandable organ that temporarily stores urine until it is excreted from the body. Urine leaves the body through the **urethra**, a tube that transports urine from the urinary bladder to the outside of the body through an external opening. The female urethra transports only urine. The male urethra, however, carries urine and reproductive fluids (but not simultaneously). The reproductive function of the male urethra is discussed in Chapter 17.

16.3 Kidneys and Homeostasis

Our kidneys are reddish brown in color and shaped like kidney beans. Each one is about the size of a fist. They are located just above the waist, sandwiched between the parietal peritoneum (the membrane that lines the abdominal cavity) and the muscles of the dorsal body wall. The slightly indented, or concave, border of each kidney faces the midline of the body. Perched on top of each kidney is an adrenal gland.

The kidneys are covered and supported by several layers of connective tissue (Figure 16.3). The outermost layer is a tough, fibrous layer that anchors each kidney and its adrenal gland to the abdominal wall and surrounding tissues. Beneath this layer is a protective cushion of fat (the middle connective tissue layer, also called the *adipose capsule*). The innermost layer covering the kidneys is a layer of collagen fibers that protects the kidneys from trauma. Protecting the kidneys is critical, given their somewhat peripheral location in the body.

Structure of the Kidneys

One ureter leaves each kidney at a notch in the concave border, as shown in Figure 16.3. This notch is also the area where blood vessels enter and exit the kidney. The renal arteries branch off the aorta and carry blood to the kidneys. The renal veins carry filtered blood away from the kidneys to the inferior vena cava, which transports the blood to the heart.

Each kidney has three regions: an outer region, the **renal cortex**; a region enclosed by the cortex, the **renal medulla**; and an inner chamber, the **renal pelvis** (see Figure 16.3). The renal cortex begins at the outer border of the kidney, and portions of it, called renal columns, extend inward between the pyramid-shaped subdivisions (*renal pyramids*) of the renal medulla. The narrow end of each renal pyramid joins a cuplike extension of the renal pelvis. As we will soon see, urine produced by the kidneys eventually drains into the renal pelvis and out the ureter to the urinary bladder.

Nephrons

Nephrons are the microscopic functional units of the kidneys and are responsible for the formation of urine. Each kidney contains 1 million to 2 million nephrons.

Structure of nephrons A nephron has two basic parts: the renal corpuscle and the renal tubule, as shown in Figure 16.4. The **renal corpuscle** is the portion of the nephron where fluid is filtered from the blood. It consists of a tuft of capillaries, the **glomerulus**, and a surrounding cuplike structure, the **glomerular capsule** (sometimes called *Bowman's capsule*). Within the glomerular capsule is a space that is continuous with the lumen of the renal tubule. Blood enters a glomerulus by an afferent (incoming) arteriole. Inside the glomerular capillaries, water and small solutes move from the blood into the space within the glomerular capsule and then into the renal tubule, where they are considered filtrate (a fluid similar to blood plasma but typically lacking proteins). The blood then leaves the glomerulus by an efferent (outgoing) arteriole.

The **renal tubule** is the site where substances are removed from and added to the filtrate. This tubule has three sections: the *proximal convoluted tubule*, the *loop of the nephron* (sometimes called the *loop of Henle*), and the *distal convoluted tubule*. The **proximal convoluted tubule**, whose name reflects its location nearest the glomerular capsule, has cells with many tiny projections called *microvilli*. Microvilli allow the efficient removal (reabsorption) of useful substances from the filtrate, which are eventually returned to the blood (see later in this discussion). The **loop of the nephron** resembles

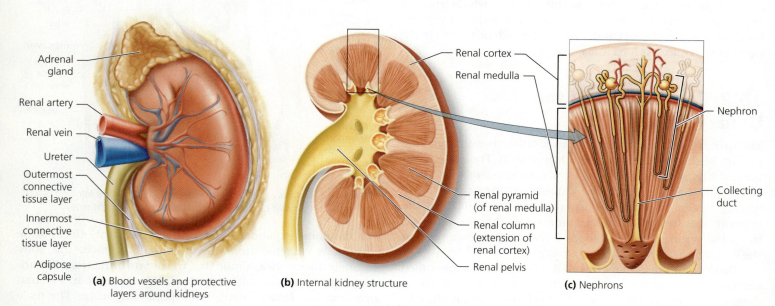

(a) Blood vessels and protective layers around kidneys
(b) Internal kidney structure
(c) Nephrons

FIGURE 16.3 Structure of a kidney

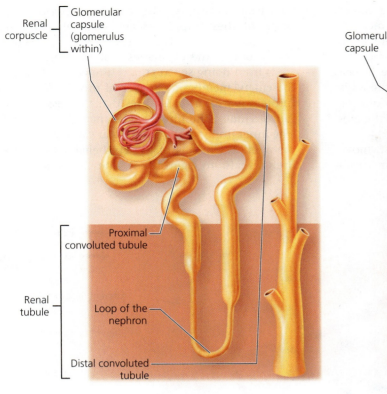
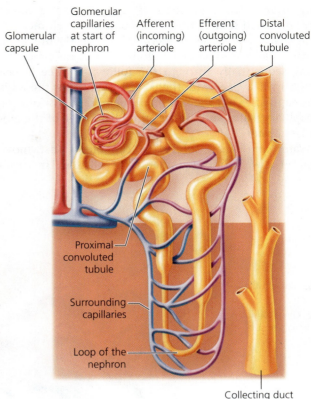

(a) Simplified view of a nephron, showing the basic structural components but not the associated capillaries

(b) A nephron and its blood supply

FIGURE 16.4 *Structure of a nephron*

a hairpin turn, with a descending limb and an ascending limb. Next comes the **distal convoluted tubule**, whose name reflects this section's more distant location from the glomerular capsule. The loop of the nephron and the distal convoluted tubule have cells with few or no microvilli. The distal convoluted tubules of many nephrons empty into a single **collecting duct**; many collecting ducts eventually drain into the renal pelvis. The renal pelvis is connected to the ureter. Urine exits the kidney by way of the ureter and moves into the urinary bladder.

About 80% of the nephrons in our kidneys are confined almost entirely to the renal cortex; they have short loops that dip only a short distance into the renal medulla. The remaining 20% have long loops that extend from the cortex deep into the renal medulla. Once in the medulla, the loops of these nephrons turn abruptly upward, back into the cortex, where they lead into distal convoluted tubules. As we will see, the nephrons whose loops extend deep into the medulla play an important role in water conservation.

Functions of nephrons The kidneys are crucial for maintaining homeostasis. For one thing, they filter wastes and excess materials from the blood. Indeed, over the course of one day, two healthy kidneys filter all the blood in the body 30 times! They also assist the respiratory system in the regulation of blood pH. Finally, the kidneys maintain fluid balance by regulating the volume and composition of blood and urine.

To understand what occurs in the kidneys, we must examine the work of nephrons. Nephrons perform three functions: (1) glomerular filtration, (2) tubular reabsorption, and (3) tubular secretion.

We can compare these functions to the steps you might take during a selective cleaning of small items from your bedroom closet. First, you might remove almost all of the small items—the valuable along with the unwanted. This activity is analogous to glomerular filtration, which removes from the blood all materials small enough to fit through the pores of the kidney's filter (discussed next). The next step in cleaning your closet might be to look through the various items you have removed and put back "the good stuff," the items worth saving. Returning valuable materials to the closet is analogous to tubular reabsorption, which returns useful materials to the blood. The final step in cleaning the closet might be to once again scan what you have in your closet and to selectively remove items, such as those in excess. This last step is analogous to tubular secretion, in which wastes and excess materials are removed from the blood and added to the filtrate that will eventually leave the body as urine. Secretion also removes from the blood substances not naturally found in the body, such as pesticides and certain drugs. Here are the three steps in more detail.

1. **Glomerular filtration** occurs as blood pressure forces water and small solutes from the blood in the glomerulus to the space inside the glomerular capsule (Figure 16.5). Reaching the space inside the glomerular capsule requires passing sequentially through the following three layers: (1) the single layer of endothelial cells that forms the walls of the capillaries, (2) the basement membrane just outside the capillary walls, and (3) the epithelial lining of the glomerular capsule.

   Water and small solutes in the blood first move through the walls of the glomerular capillaries. The capillary walls consist of a single layer of endothelial cells; the walls of these capillaries are fenestrated, meaning they have many pores. The pores allow some substances to move out of the capillaries, but they prevent red blood cells from doing so. Next, the filtered substances cross the basement membrane, a layer of protein fibers in a glycoprotein matrix. This layer restricts the passage of large proteins. Finally, the filtered substances pass through slits in the inner lining of the glomerular capsule; these slits occur between extensions of the epithelial cells of the capsule (refer, again, to Figure 16.5). These substances are known

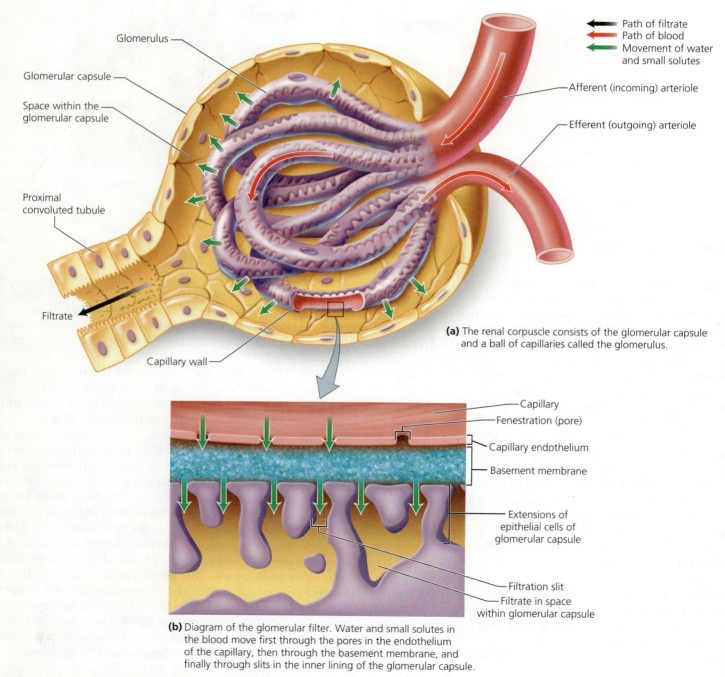

(a) The renal corpuscle consists of the glomerular capsule and a ball of capillaries called the glomerulus.

(b) Diagram of the glomerular filter. Water and small solutes in the blood move first through the pores in the endothelium of the capillary, then through the basement membrane, and finally through slits in the inner lining of the glomerular capsule.

FIGURE 16.5 *The renal corpuscle is the site of glomerular filtration.*

collectively as *glomerular filtrate*. The concentrations of the molecules dissolved in the glomerular filtrate are approximately the same as in the blood plasma.

Several things can change the rate of filtration by the glomerulus. An increase in the diameter of afferent (incoming) arterioles brings more blood into the glomerulus and produces higher pressure in the glomerular capillaries. Higher pressure results in higher filtration rates (more filtrate produced by the kidneys each minute). A reduction in the diameter of efferent (outgoing) arterioles also produces higher pressure in glomerular capillaries and higher filtration rates. General (systemic) increases in blood pressure can also produce higher filtration rates.

2. **Tubular reabsorption** is the process that removes useful materials from the filtrate and returns them to the blood. This process occurs in the renal tubule, primarily in the proximal convoluted tubule. The characteristics of the cells lining the proximal convoluted tubule make it an ideal location for reabsorption (Figure 16.6). As mentioned, these epithelial cells have numerous microvilli (projections of the plasma membrane) that reach into the lumen of the tubule. Similar in function to the microvilli in the small intestine, these microvilli dramatically increase the surface area for the reabsorption of materials. Reabsorption returns water, essential ions, and nutrients to the blood (Table 16.1). Remarkably, as the glomerular filtrate passes through the renal tubule, about 99% of it is returned to the blood in the surrounding capillaries. Thus, only about 1% of the

TABLE 16.1 Reabsorption by Nephrons of Some Substances

Substance	Amount Filtered per Day (in liters or grams)	% Reabsorbed (Removed from Filtrate and Returned to Blood)
Water	180 L	99
Glucose	180 g	100
Urea	52 g	50
Creatinine	1.6 g	0
Sodium ions (Na$^+$)	620 g	99
Potassium ions (K$^+$)	30 g	93
Bicarbonate (HCO$_3^-$)	275 g	100

glomerular filtrate is eventually excreted as urine. Put another way, of the approximately 180 liters (48 gal) of filtrate that enter the glomerular capsule each day, between 178 and 179 liters (47 gal) are returned to the blood by reabsorption. The remaining 1 to 2 liters (about 0.5 to 1 gal) are excreted as urine. Imagine how much water and food we would have to consume if we did not have reabsorption to offset the losses from glomerular filtration! Some wastes are not reabsorbed at all, and they will eventually be excreted; others, such as urea, are partially reabsorbed. We will see that antidiuretic hormone (ADH), manufactured by the hypothalamus and released by the posterior pituitary gland, regulates the amount of water reabsorbed in parts of the renal tubule and collecting ducts.

3. **Tubular secretion** removes additional wastes and excess ions from the blood. For example, hydrogen ions (H$^+$), potassium ions (K$^+$), and ammonium ions (NH$_4^+$) in the blood are actively transported into the renal tubule, where they become part of the filtrate to be excreted. Tubular secretion also removes foreign substances from the blood, including pesticides and drugs such as penicillin, cocaine, and marijuana. These substances are added to the filtered fluid that will become urine. Tubular secretion occurs along the proximal and distal convoluted tubules and collecting duct.

By the end of glomerular filtration, tubular reabsorption, and tubular secretion, blood leaving the kidneys contains most of the water, nutrients, and essential ions that it contained upon entering the kidneys. Wastes, foreign substances, and excess materials have been removed, leaving the blood cleansed. This purified blood now moves from the capillaries surrounding the nephron into small veins that eventually join the renal vein.

The final filtrate, now called *urine*, contains all of the materials that were filtered from the blood and not reabsorbed, plus substances that were secreted from the blood. Urine empties from the distal convoluted tubules into collecting ducts. There, more water may be reabsorbed, and some additional excess substances, such as hydrogen and potassium ions, are removed. From the collecting ducts, urine moves into the renal pelvis and leaves each kidney through a ureter. It then travels down the ureters to the urinary bladder,

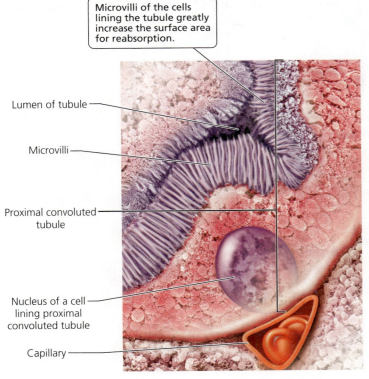

FIGURE 16.6 *The proximal convoluted tubule is the site of tubular reabsorption. Substances move from the filtrate through the cells of the proximal convoluted tubule and into the surrounding fluid. Eventually, the substances move into the capillaries nearby.*

328 CHAPTER 16 The Urinary System

Q How does the structure of glomerular capillaries aid filtration?

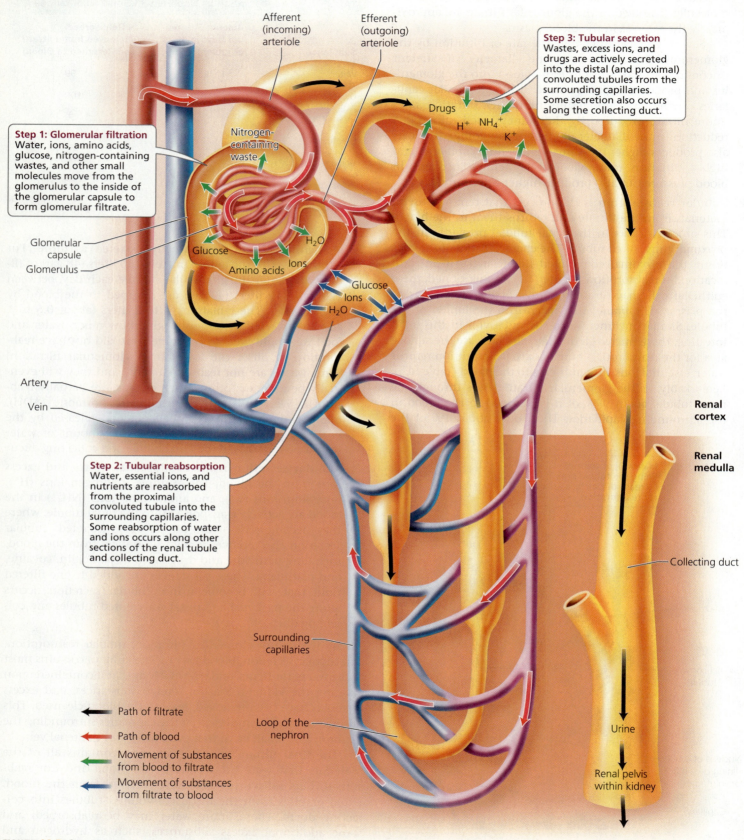

FIGURE 16.7 Overview of glomerular filtration, tubular reabsorption, and tubular secretion along the nephron

A The walls of glomerular capillaries have pores that allow many substances (but not red blood cells) to move out of the capillaries.

HEALTH ISSUE

Urinalysis

The kidneys are our body's filtering system, so the urine they produce contains substances that originate from almost all of our organs. A detailed analysis of our urine, therefore, tells us not only how the organs of our urinary tract are functioning but also about the general health of our other organs. In other words, a typical urinalysis, which assesses the physical and chemical properties of urine and evaluates whether microorganisms are present, provides an overview of a person's basic health.

Healthy urine exiting the body typically contains no microorganisms. The presence of bacteria in a properly collected urine sample usually signals infection of the urinary system. Bacteria found in a sample may be cultured to determine their identity, which can help in diagnosing the infection. Urine also may be screened for fungi or protozoans that cause inflammation within the urinary tract.

Sometimes a urine sample is contaminated because of improper collection. For example, if the perineal area has not been cleaned and the collection cup or urine is allowed to contact that area, bacteria may be introduced. (In males, this is the area between the anus and scrotum; in females, it is between the anus and vulva.) It is therefore important to be careful when collecting a urine sample.

The following physical characteristics are routinely checked in urinalysis: color; turbidity (cloudiness); pH; and specific gravity, or density, of urine (the ratio of its weight to the weight of an equal volume of distilled water). The yellow color of urine comes from the pigment *urobilin*, which is made from pigments derived from bilirubin. *Bilirubin*, you may recall, is the yellow pigment produced as a waste product during the breakdown of hemoglobin from aging red blood cells (see Chapter 11). The color of urine varies somewhat according to diet. Beets, for example, lend urine a red color, and asparagus causes a green tinge. The color of urine also varies with concentration. More concentrated urine, such as urine collected first thing in the morning, is darker than more dilute urine. An abnormal color of urine—particularly red, when followed by microscopic confirmation of the presence of red blood cells—can indicate trauma to urinary organs.

Freshly voided urine is usually transparent. Cloudy, or turbid, urine may indicate a urinary tract infection, particularly if white blood cells are detected. Healthy urine has a pH of about 6, although considerable variation may occur in response to diet. High-protein diets produce acidic urine, and vegetarian diets produce alkaline urine. An alkaline pH also is associated with some bacterial infections. Specific gravity is a measure of the concentration of solutes in urine and, therefore, of the concentrating ability of the nephrons within our kidneys. When urine becomes highly concentrated for long periods, substances such as calcium and uric acid may precipitate out and form kidney stones.

Quantitative chemical tests are run to assess the levels of specific chemical constituents of urine. The major constituent is water, making up about 95% of the total volume. The remaining 5% consists of solutes from the metabolic activities of our cells or from outside sources, such as drugs. Tests run on urine samples look for certain abnormal constituents of urine (such as glucose, red blood cells, or white blood cells) and for normal constituents present in abnormal amounts (such as higher-than-normal amounts of protein). Drugs (legal and illegal) and their metabolites may pass through the glomerular filter if they are small enough. Because drugs are foreign substances, most are not actively reabsorbed and returned to the blood; thus, they are detectable in urine samples. Additionally, some drugs are actively secreted into the filtrate along the renal tubule and collecting ducts; these, too, will show up in urine.

Questions to Consider

- Why would it be important for a woman to inform her doctor that she is menstruating at the time of urine collection?
- Urinalysis is often preferred over blood testing in workplace drug-testing programs. What might explain this preference?

---

which stores the urine until it is eliminated from the body through the urethra. The precise composition of urine can be examined in a medical laboratory. See the Health Issue essay, *Urinalysis*.

The regions of the nephron and their roles in filtration, reabsorption, and secretion are shown in Figure **16.7**. When reviewing these, keep in mind the directions that substances move in each of the three processes: (1) during glomerular filtration, substances move from the blood into the nephron to form filtrate; (2) during tubular reabsorption, useful substances move from the filtrate within the nephron back into the blood; (3) during tubular secretion, drugs and substances in excess move from the blood into the filtrate.

Acid–Base Balance

In addition to removing wastes and regulating the volume and solute concentration of blood plasma, the kidneys help regulate the pH of blood. Recall from Chapters 2 and 14 that blood pH must be regulated precisely for proper functioning of the body. This precise regulation is achieved through the actions of the kidneys, through buffer systems in the blood, and through respiration. Buffer systems regulate pH by picking up hydrogen ions (H^+) when their concentrations are high and releasing hydrogen ions when their concentrations are low. Chapter 2 described the importance of carbonic acid as such a buffer in the blood. The role of the kidneys in maintaining pH is twofold. First, by secreting hydrogen ions into the urine, the kidneys remove excess hydrogen ions from the blood, thereby increasing blood pH. Second, the kidneys help sustain the carbonic acid buffer system by returning bicarbonate to the blood. Because of their role in regulating blood pH, the kidneys ultimately influence breathing rate. Recall from Chapter 14 that chemoreceptors in the medulla of the brain respond to changes in the pH of blood (and cerebrospinal fluid) by adjusting breathing rate.

Water Conservation

By producing concentrated urine, our kidneys enable us to conserve water. Because of their role in water conservation, the kidneys participate in the maintenance of cardiac output and blood pressure. Production of concentrated urine is performed by the 20% of nephrons with long loops that dip deep into the renal medulla. The special ability of these nephrons to concentrate urine derives from an increasing concentration of solutes in the interstitial fluid (the fluid that fills the spaces between cells) from the cortex to the medulla of the kidneys. The most important of these solutes are sodium chloride ($NaCl$) and urea. Let's explore this mechanism for urine concentration by

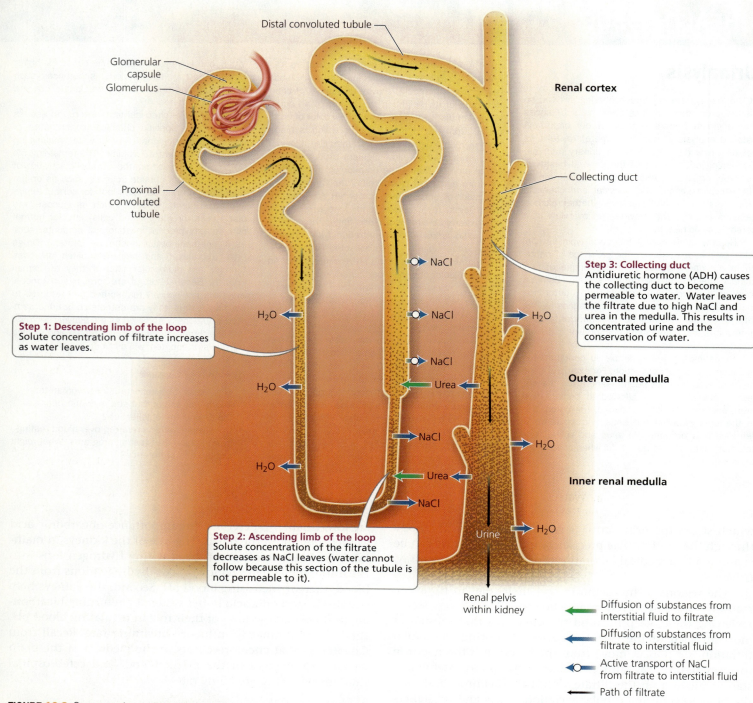

FIGURE 16.8 Some nephrons have loops that extend deep into the medulla. These nephrons are responsible for water conservation. The steps by which these nephrons concentrate urine and conserve water are shown here. Stippling indicates solute concentration of the filtrate within the nephron and collecting duct. The color gradient behind the nephron and collecting duct indicates solute concentration in the interstitial fluid of different regions of the kidney (renal cortex and outer and inner renal medulla; darker is more concentrated).

tracing the path of the filtrate as it flows through these long loops (Figure 16.8).

The solute concentration of the filtrate passing from the glomerular capsule to the proximal tubule is about the same as that of blood. As the filtrate moves through the proximal convoluted tubule, large amounts of water *and* salt are reabsorbed (recall from our earlier discussion of reabsorption that nutrients are also reabsorbed at this time). This reabsorption produces dramatic reductions in the volume of filtrate but little change in its solute concentration (because both water and salt are reabsorbed). However, when the filtrate enters the descending limb of the loop of the nephron, major changes in solute concentration begin (see Figure 16.8). This path takes the filtrate from the cortex to the medulla.

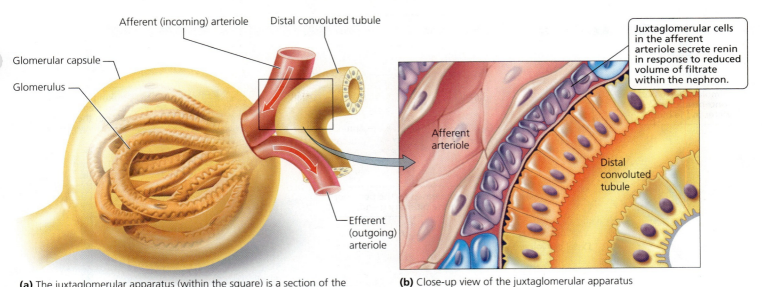

(a) The juxtaglomerular apparatus (within the square) is a section of the nephron where the distal convoluted tubule contacts the afferent arteriole. The nearby renal corpuscle is shown in ghosted view to reveal its components, the glomerular capsule and the glomerulus.

(b) Close-up view of the juxtaglomerular apparatus

FIGURE 16.9 *The juxtaglomerular apparatus*

Along the descending limb, water leaves the filtrate by osmosis (a special type of diffusion described in Chapter 3). The departure of water creates an increase in the concentration of solutes, including sodium chloride, within the filtrate. The concentration of salt in the filtrate peaks at the curve of the loop, setting the stage for the next step in the process of urine concentration. Now the filtrate moves up the ascending limb of the loop. As it does so, large amounts of sodium chloride are actively transported out of the filtrate into the interstitial fluid of the medulla. Water, however, remains in the filtrate because the ascending limb is not permeable to water. When the filtrate reaches the distal convoluted tubule in the cortex, it is quite dilute. In fact, the filtrate is hypotonic to body fluids. (Recall from Chapter 3 that *hypotonic* means having a lower solute concentration than another fluid. *Hypertonic* means having a greater solute concentration than another fluid. *Isotonic* means having the same solute concentration as another fluid.) The filtrate then moves into a collecting duct and begins to descend once again toward the medulla. This pathway is one of increasing salt concentration in the interstitial fluid because of all the salt that was transported out of the filtrate as it ascended the loop. Collecting ducts are permeable to water but not to salt (antidiuretic hormone increases the permeability of the collecting duct to water; see Figure 16.8). Thus, as the filtrate encounters increasing concentrations of salt in the fluid of the inner medulla, water leaves the filtrate by osmosis. With this departure of large amounts of water, urea is now concentrated in the filtrate. In the lower regions of the collecting duct, some of the urea moves into the interstitial fluid of the medulla. This leakage of urea contributes to the high solute concentration of the inner medulla and thus aids in concentrating the filtrate. The remaining urea is excreted.

At its most concentrated, urine is hypertonic to blood and interstitial fluid from any other part of the body except the inner medulla, where it is isotonic. Together, the loop of the nephron and collecting duct maintain extraordinarily high solute concentrations in the interstitial fluid of the kidneys, making it possible for the kidneys to concentrate urine and conserve water.

Hormones and Kidney Function

Our health depends on our keeping the salt and water levels in our body near certain optimum values. This, as we have seen, is an important job of the kidneys. It is also a challenging job, because our activities produce constant fluctuations in those levels. For example, on a hot day or after exercise, we may lose body water and salts through perspiration. In contrast, eating a tub of salted popcorn at the movies can boost our salt intake. The kidneys must deal with these challenges and adjust the concentration of solutes in the urine and in the blood to keep water and salt levels in our body relatively constant.

Three hormones—aldosterone, ADH, and atrial natriuretic hormone—adjust kidney function to meet the body's needs. **Aldosterone** is released by the adrenal cortex. It increases reabsorption of sodium by the distal convoluted tubules and collecting ducts. Aldosterone accomplishes this task by stimulating the synthesis of sodium ion pumps and sodium channels in cell membranes along the distal convoluted tubules and collecting ducts; at these locations, sodium ions are reabsorbed, usually in exchange for potassium ions. Sodium reabsorption is important because water follows sodium. As more sodium is transported out of the nephron into the capillaries, increased amounts of water go with it, resulting in increased blood volume and pressure and the production of small amounts of concentrated urine.

What stimulates the release of aldosterone? Ultimately, it is the blood pressure in the afferent (incoming) arteriole carrying blood to the glomerulus. The blood pressure in this arteriole is monitored by a part of the nephron called the **juxtaglomerular apparatus**, a group of cells located where the distal convoluted tubule contacts the afferent arteriole (Figure 16.9). When the blood pressure in the afferent arteriole drops, so does the glomerular filtration rate. In turn, this drop

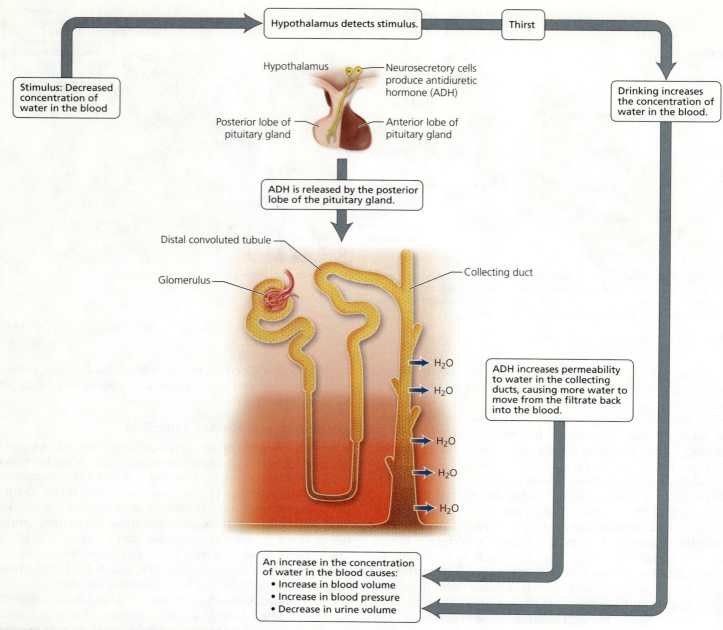

FIGURE 16.10 Regulation of blood volume and blood pressure by ADH

in filtration rate reduces the volume of filtrate within nephrons. Cells within the juxtaglomerular apparatus respond by releasing the enzyme renin. **Renin** converts angiotensinogen, a protein produced by the liver, into another protein, *angiotensin I*. Angiotensin I is then converted by yet another enzyme into angiotensin II. *Angiotensin II* is the active form of the protein that stimulates the adrenal gland to release aldosterone. Aldosterone increases reabsorption of sodium and water by the distal convoluted tubules and collecting ducts of nephrons, resulting in increased blood volume and pressure. These changes increase the filtration rate within the glomerulus, resulting in an increased volume of filtrate within the nephron.

Antidiuretic hormone (ADH) is manufactured by the hypothalamus and then travels to the posterior pituitary for storage and release. This hormone regulates the amount of water reabsorbed by the collecting ducts. The hypothalamus responds to changes in the concentration of water in the blood by increasing or decreasing secretion of ADH. Decreases in the concentration of water in the blood stimulate thirst and increased secretion of ADH, as shown in Figure 16.10. Higher levels of ADH in the bloodstream then increase the collecting ducts' permeability to water; as a result, more water is reabsorbed from the filtrate. The movement of increased amounts of water from the filtrate back into the blood results in increased blood volume and pressure and production of small amounts of concentrated urine.

Just the opposite occurs when the concentration of water in the blood increases. In this case, the release of ADH is inhibited, thereby reducing water reabsorption from the filtrate.

TABLE 16.2 Review of Some Hormones That Influence Kidney Function

Hormone	Effect on Water and Solute Reabsorption in Nephron	Effect on Blood Volume and Pressure	Urine Produced
Aldosterone	Increases reabsorption of Na^+ by distal convoluted tubules and collecting ducts, resulting in more water following Na^+ as it moves from filtrate to blood	Increases	Concentrated
Antidiuretic hormone (ADH)	Increases permeability to water of collecting ducts, resulting in more water moving from filtrate to blood	Increases	Concentrated
Atrial natriuretic hormone (ANH)	Decreases reabsorption of Na^+ by distal convoluted tubules and collecting ducts, resulting in more Na^+ and water remaining in filtrate	Decreases	Dilute

Reduced water reabsorption causes reduced blood volume and pressure and the production of large amounts of dilute urine. Alcohol, too, inhibits the secretion of ADH, causing reduced water reabsorption by the kidneys and production of large amounts of dilute urine. Thus, it makes little sense to try to quench your thirst and restore body fluids by drinking an alcoholic beverage on a hot day. Substances, such as alcohol, that promote urine production are called **diuretics**. *Diabetes insipidus*, a disease characterized by excretion of large amounts of dilute urine, is caused by a deficiency of ADH (see Chapter 10).

A final hormone that influences kidney function is **atrial natriuretic hormone (ANH)**, released by cells in the right atrium of the heart. The cells release ANH in response to stretching of the heart caused by increased blood volume and pressure. Atrial natriuretic hormone decreases water and Na^+ reabsorption by the kidneys, causing declines in blood volume and pressure and the production of large amounts of dilute urine. ANH also influences kidney function by inhibiting secretion of aldosterone and renin.

Table 16.2 summarizes the actions of aldosterone, ADH, and ANH.

stop and think

Estrogens are sex hormones that are chemically similar to aldosterone. As a result, the effects of estrogens on the distal convoluted tubules and collecting ducts of the kidneys are similar to those of aldosterone. How might this similarity explain the water retention many women experience as their estrogen levels rise during the menstrual cycle?

Red Blood Cells and Vitamin D

The kidneys have two additional functions that are important to homeostasis but are not directly related to the urinary system. First, the kidneys release **erythropoietin**, a hormone that travels to the red bone marrow, where it stimulates production of red blood cells. Second, the kidneys have an effect on vitamin D. Vitamin D is a substance provided by certain foods in our diet or produced by the skin in response to sunlight. The kidneys transform vitamin D into its active form, *calcitriol*, which promotes the body's absorption and use of calcium and phosphorus.

16.4 Dialysis and Transplant Surgery

Renal failure is a decrease or complete cessation of glomerular filtration. In other words, the kidneys stop working. The failure can be acute or chronic. *Acute renal failure* is an abrupt, complete—or nearly complete—cessation of kidney function. It typically develops over a few hours or days and is characterized by little output of urine. Causes of acute renal failure include nephrons damaged by severe inflammation, certain drugs, and poison. Acute renal failure also may be caused by low blood volume due to profuse bleeding or obstruction of urine flow by kidney stones. Kidney stones are small, hard crystals that form in the kidneys when substances such as calcium or uric acid precipitate out of urine because of higher-than-normal concentrations. Some kidney stones are surgically removed, others are dissolved with drugs, and still others are pulverized with high-energy shock waves, which break the stones into tiny pieces that can be passed painlessly in the urine.

Chronic renal failure is a progressive and often irreversible decline in the rate of glomerular filtration over a period of months or years. Kidney disease may destroy nephrons and cause a progressive decline of this type. Nephrons lost to kidney disease cannot be replaced. *Polycystic kidney disease*, for example, is an inherited and progressive condition in which fluid-filled cysts and tiny holes form throughout kidney tissue. Few symptoms are apparent at the beginning of chronic renal failure because the remaining nephrons enlarge and take over for those that have been destroyed. With time, however, the loss of nephrons becomes so severe that symptoms of decreased glomerular filtration rate appear. For example, increased levels of nitrogen-containing wastes are found in the blood. By end-stage renal failure, about 90% of the nephrons have been lost, making necessary a kidney transplant or the use of an artificial kidney machine.

Renal failure, whether acute or chronic, has many consequences. These include (1) acidosis, a decrease in blood pH caused by the inability of the kidneys to excrete hydrogen ions; (2) anemia, low numbers of red blood cells caused by the failure of damaged kidneys to release erythropoietin; (3) edema, the buildup of fluid in the tissues because of water and salt retention; (4) hypertension, an increase in blood pressure caused by failure of the renin–angiotensin system and salt and water retention; and (5) accumulation of nitrogen-containing wastes in the blood. In short, failure of the kidneys severely disrupts homeostasis. Untreated kidney failure will lead to death within a few days.

stop and think

Why would inadequate water consumption increase your chances of developing kidney stones?

Dialysis

A common way of coping with failure or severe impairment of the kidneys is **hemodialysis**, the use of artificial devices to cleanse the blood of wastes and excess fluid. Often this is done with an artificial kidney machine (Figure 16.11). A tube is inserted into an artery of the patient's arm. Blood flows through the tube and into the kidney machine. The machine pumps the blood into a large canister (called the *hemodialyzer*), where it is filtered and then returned to the patient's body. Within the hemodialyzer, the blood flows through tubing made of a selectively permeable membrane and immersed in a dialysis solution called the *dialysate*. The selectively permeable membrane of the tubing permits wastes and excess small molecules to move from the blood into the dialysate. At the same time, the membrane prevents passage into the dialysate of blood cells and most proteins because they are too large. Nutrients are sometimes provided in the dialysate for absorption into the blood. The composition of the dialysate is precisely controlled to maintain proper concentration gradients between the dialysate and the blood. Once the blood has completed the circuit through the tubing, it is returned, free of wastes, to a vein in the arm. Patients requiring hemodialysis typically undergo the procedure about three times a week at a dialysis clinic.

Another option for removing wastes from the blood is *continuous ambulatory peritoneal dialysis (CAPD)*, a procedure that is normally performed by the patient at home. In this procedure the peritoneum, one of the body's own selectively permeable membranes, is used as the dialyzing membrane. The peritoneum lines the abdominal cavity and covers the internal organs. Dialyzing fluid held in a plastic container suspended over the patient flows down a tube inserted into the abdomen. As the fluid bathes the peritoneum, wastes move from the blood vessels that line the abdomen across the peritoneum and into the solution. The solution is then returned to the plastic container and discarded. Typically, fluid is passed into the abdomen, left for a few hours, and then removed and replaced with new fluid. CAPD requires about three or four fluid changes each day. However, the patient is free to move around between fluid changes while the dialysis proceeds internally, which tends to be less disruptive than traditional hemodialysis.

Because CAPD is performed daily over long periods of time, it is gentler in its waste removal than hemodialysis, resulting in a steadier physical condition in the patient. CAPD is also less costly than hemodialysis. However, CAPD requires that the patient perform several daily changes of dialysis fluid, and each change requires hooking up a new container of dialysate. Thus, CAPD provides more opportunities for bacteria to move down the tube and into the abdomen, where they may cause peritonitis (inflammation of the peritoneum).

In some cases of acute renal failure, dialysis may be needed for a short time, until kidney function returns. Patients with chronic renal failure, however, will need dialysis for the rest of their lives, unless they receive a kidney transplant.

Kidney Transplant Surgery

The ultimate hope for many people whose kidneys fail is to receive a healthy kidney from another person. The kidney was one of the first organs to be successfully transplanted. The availability of dialysis was a crucial part of this success, because dialysis keeps people alive until a suitable donor organ can be found.

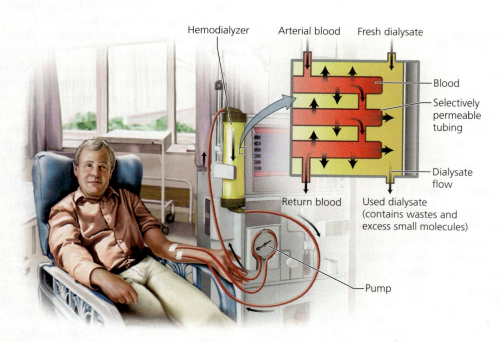

FIGURE 16.11 *An artificial kidney machine can be used to cleanse the blood when the kidneys fail.*

What makes a suitable donor organ? The main obstacle to successful acceptance of a transplanted organ is the rejection of foreign tissue by the patient's own immune system (see Chapter 13). The most suitable kidney would come from a patient's identical twin. In fact, the first successful transplant occurred in 1954 between identical twins. A kidney from a close relative, such as a parent or sibling, would be the next best choice. Removal of a healthy kidney is a safe operation, and a healthy donor who gives a kidney away can get along fine with the remaining kidney (of course, all surgeries involve risks, and in the United States about 3 in 10,000 people die from kidney donation). About 90% of kidneys donated by close relatives are still functioning 2 years after transplantation. However, most donated kidneys come from individuals who are unrelated to the patient but who agreed before their death (usually a sudden death in an accident) to donate their organs. About 75% of kidneys donated from unrelated individuals, matched as closely as possible to the patient's tissue and blood type, are still functioning 2 years after transplantation. Even if a transplant fails after a few years, successful second and third transplants may extend a patient's life.

In most situations, the kidneys of the patient needing a transplant are not removed, because survival after surgery is higher when the recipient's own kidneys are left in place. The donor kidney is simply transplanted into a protected area within the pelvis (Figure 16.12). The ureter of the donated kidney is attached to the recipient's bladder, and blood vessels of the donated kidney are attached to the recipient's vessels. The recipient's own kidneys would be removed if they were infected or if they were causing problems such as high blood pressure.

The high success rate of kidney transplants is linked to the use of drugs that suppress the recipient's immune system. Typically, transplant recipients must take two or more of these medications for the rest of their lives. Recently, a team of researchers developed a way to perform kidney transplants without the recipients' needing to permanently remain on immune-suppressing drugs. The suppressing drugs were given to recipients after transplant surgery. Then, the recipients received several radiation treatments, which were directed at reducing the number of cells in their immune system capable of attacking the new organ. Finally, the recipients received blood stem cells from the person who donated the transplanted kidney. These new stem cells moved to the recipient's bone marrow and generated new blood cells and immune cells. In essence, each recipient now had a hybrid immune system—a mix of his or her own cells and those of the donor. The hybrid immune systems did not recognize the transplanted kidneys as foreign, and recipients were gradually weaned off the immune-suppressing medications.

Despite major advances in kidney transplantation, there still is room for improvement. For example, donor kidneys can be kept alive and healthy for only about 48 to 72 hours (Figure 16.13). This short window of opportunity necessitates rapid location of a recipient, shipment of the donor kidney, and transplantation. Perhaps most important, kidneys available for transplantation are always in short supply. The constant shortage of donor organs has led some people to suggest that we need a new approach to organ donation. The shortage also has led to the illegal sale of kidneys in some countries. See the Ethical Issue essay, *Kidney Donation and Trafficking*.

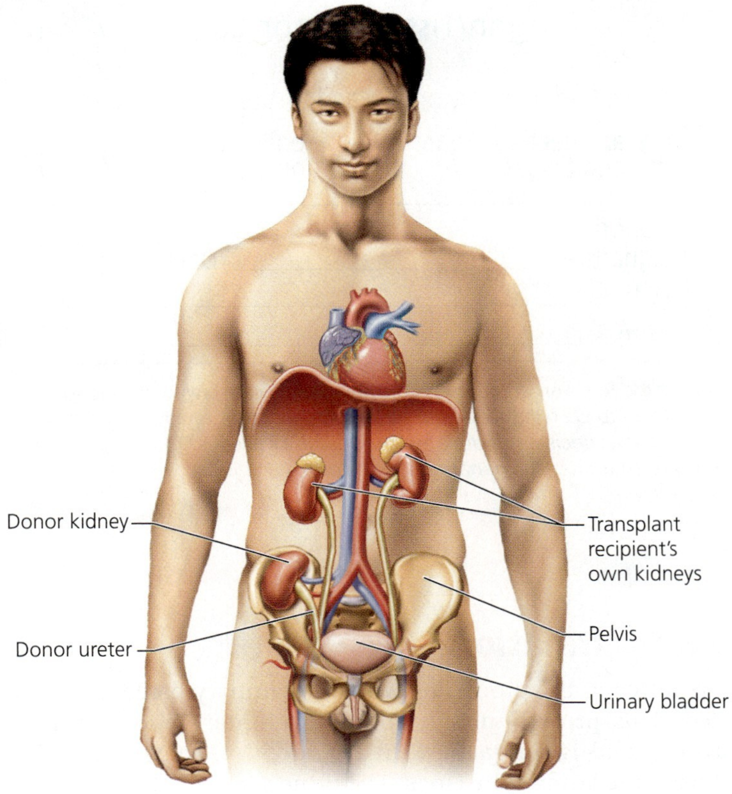

FIGURE 16.12 *During a kidney transplant, the donor organ is located in a safe region within the pelvis, and its ureter is attached to the recipient's bladder. Blood vessels from the transplanted kidney are attached to the recipient's vessels. Typically, the recipient's own kidneys are left in place.*

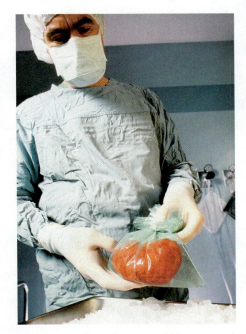

FIGURE 16.13 *Before transplantation, donor kidneys must be kept in a cool salt solution under sterile conditions. Even under such conditions, kidneys will deteriorate after about 2 or 3 days.*

ETHICAL ISSUE

Kidney Donation and Trafficking

Kidneys can be transplanted from living or deceased donors, but in many countries there is a long waiting list for organs. Living kidney donors come from several sources. Many are close relatives of the person needing the transplant. Others donate a kidney to someone who is not well known to them. In the United States it is illegal to sell organs, so donors are not paid for their kidneys. In Iran, however, some live donors legally receive monetary compensation. In the Iranian system, a person needing a transplant must first seek a willing donor from among family members. If this fails, then he or she must wait up to 6 months for a kidney from an appropriate deceased donor. Only if this second option fails can the person apply for a kidney from a list of living donors; people on this list are compensated when they donate one of their kidneys. Iran began this legal compensation system in 1988, and its waiting list for kidneys was eliminated by 1999. About 70% of transplants performed in Iran are compensated; the service is not available to foreign nationals. A final way to obtain a kidney from a living donor is to illegally purchase one. Some patients from developed countries buy kidneys from donors in developing countries. When donors are coerced or deceived into giving up one of their kidneys, the practice is called *kidney trafficking* (Figure 16.A). The growing illegal trade in kidneys has prompted calls to consider legal compensation of living donors.

For deceased organ donation, the United States currently employs the opt-in system, in which a person must give permission to use his or her organs after death. If a potential donor is incapable of giving such permission, then medical personnel may consult close relatives. In recent years, people in the United States concerned about organ shortages have suggested changing the law so that people would have to provide written proof that they did *not* want their organs made available for transplantation after death. In other words, without anything in writing, the revised law would presume that people have no objection to organ donation—this is the opt-out system, also known as *presumed consent*.

The total number of candidates on the waiting list for organs in the United States is updated minute by minute. At 9:05 A.M. on April 28, 2012, that number totaled 114,180. Most of these candidates were waiting for kidneys. Many patients wait years for a new kidney. Perhaps you have considered carrying an organ donor card that says you wish to donate organs or tissues at the time of your death (Figure 16.B). If you haven't already signed an organ donor card, you might want to reconsider it after you have finished reading this essay: Your kidney could save someone else's life—or perhaps someone else's kidney could save you.

Questions to Consider

- Do you support legal compensation of living donors?
- Should the United States change from the opt-in to the opt-out system of deceased organ donation?

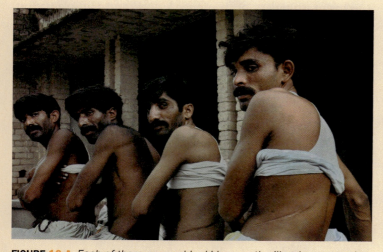

FIGURE 16.A Each of these men sold a kidney on the illegal organ market.

FIGURE 16.B An organ donor card available through www.organdonor.gov. In addition to carrying this card, it is important to let family members know of your decision to donate your organs or tissues. You can also indicate your intent to donate on your driver's license.

what would you do?

Kidneys available for transplantation are extremely scarce. Some people in need of a kidney transplant have mental disabilities. In some cases, mental disability is accompanied by health conditions such as heart disease and a compromised immune system. Should mental disability disqualify a patient from receiving a kidney? What if additional health conditions accompanied a patient's mental disability? If you were in charge of allocating kidneys for transplantation, what would you do?

16.5 Urination

Urination is the process by which the urinary bladder is emptied. This process includes both involuntary and voluntary actions and is summarized in Figure 16.14. The kidneys produce urine around the clock. The urine trickles into the ureters and down to the urinary bladder, where it is temporarily stored. When at least 200 ml (0.42 pt) of urine have accumulated, stretch receptors in the wall of the bladder send impulses along sensory neurons to the lower part of the spinal cord.

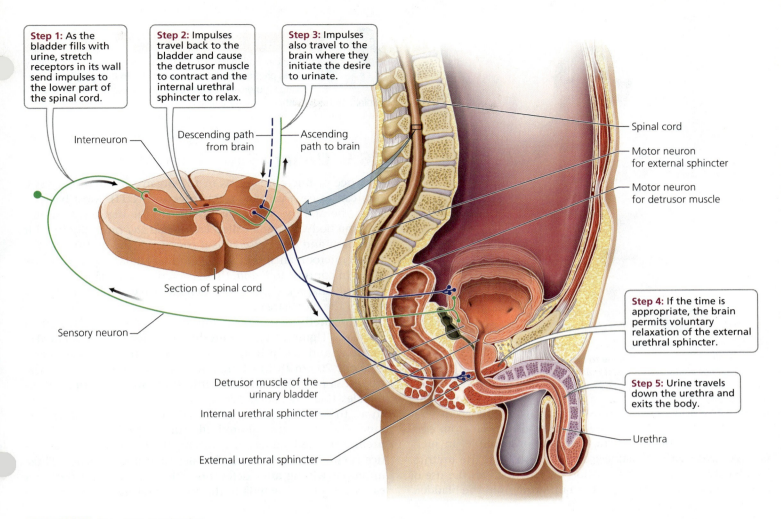

FIGURE 16.14 The steps in urination

From there, impulses are sent along motor neurons back to the bladder, where they cause a smooth muscle in the wall of the bladder, called the *detrusor muscle,* to contract. The impulses also cause the *internal urethral sphincter,* a thickening of smooth muscle located at the junction of the bladder and urethra, to relax. Upon arrival of sensory impulses in the lower spinal cord, information also travels up to the brain and initiates a desire to urinate. If the circumstances are appropriate, the brain permits voluntary relaxation of the *external urethral sphincter,* a small band of skeletal muscle farther down the urethra. When the external sphincter relaxes, urine exits the body. If the circumstances are not appropriate for urination, the brain does not permit the external sphincter to relax, and the bladder continues to store the urine until a better time. With normal functioning, urine will exit the body later, when the person consciously allows the external sphincter to relax.

Thus, although urination is a reflex—a relatively rapid response to a stimulus that is mediated by the nervous system—it can nevertheless be started and stopped voluntarily because of the conscious control exerted by the brain over the external urethral sphincter. However, not everyone can control his or her external urethral sphincter. Lack of voluntary control over urination is called *urinary incontinence*. Incontinence is the norm for infants and children younger than 2 or 3 years of age, because nervous connections to the external urethral sphincter are incompletely developed. As a result, infants and young children void whenever their bladder fills with enough urine to activate its stretch receptors (Figure 16.15). Toilet training occurs when toddlers learn to bring urination under conscious control. This step is made possible by the development of complete neural connections to the external sphincter.

Damage to the external sphincter may cause incontinence in adults. In men, such damage may occur during surgery on the prostate gland. The prostate gland surrounds the male urethra just below the urinary bladder and contributes substances to semen (see Chapter 17). Incontinence also may occur when bladder muscles contract before the bladder is full. This condition is often described as an *overactive* or *spastic bladder*. Infection of the urinary system (see the next discussion) can also cause incontinence, in any age group.

Mild incontinence, especially the kind called stress incontinence, is common in adults. *Stress incontinence* is characterized by the escape of small amounts of urine when sudden increases in abdominal pressure force urine past the external sphincter. Laughing, sneezing, or coughing may cause these sudden increases in pressure.

Urinary retention is the failure to expel urine from the bladder to a normal degree. This condition may result from a lack

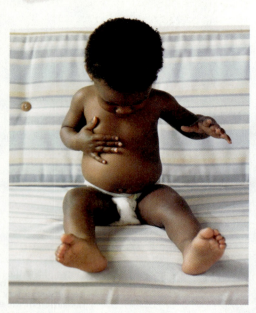

FIGURE 16.15 *Conscious control over urination is usually acquired by the age of 3, when neural connections to the external urethral sphincter are fully developed. Before that age, urination occurs as an involuntary reflex.*

of the sensation that one must urinate, as might occur temporarily after general anesthesia. It may also result from contraction or obstruction of the urethra. For example, in men, enlargement of the prostate gland may obstruct the urethra. Immediate treatment for retention usually involves the use of a tube called a *urinary catheter* to drain urine from the bladder.

stop and think

Stress incontinence is more common in women, particularly late in pregnancy and after childbirth. In late pregnancy, the enlarging uterus puts pressure on the bladder. What might explain stress incontinence after childbirth?

16.6 Urinary Tract Infections

Presence of microorganisms in the organs of the urinary system can cause a **urinary tract infection (UTI)**. Most bacteria enter the urinary system by moving up the urethra from outside the body. Bacteria called *Escherichia coli* normally inhabit the colon and can cause UTIs if they enter the urethra. Sometimes, microorganisms called *Chlamydia* and *Mycoplasma* cause UTIs. In contrast to *E. coli,* these bacteria are sexually transmitted (see Chapter 17a). Some bacteria arrive at the kidneys by way of the bloodstream.

The urethras of males and females differ in length, as shown in Figure 16.16. The urethra in females is 3 to 4 cm (1.5 in.) long and lies in front of the vagina. The urethra in males is about 20 cm (8 in.) long and opens to the outside at the tip of the penis. Urinary tract infections are more common in females than in males, perhaps because of their shorter urethras. In females, the bacteria need to travel only a short distance from the external urethral orifice (opening) through the urethra to the bladder. In addition, the external urethral orifice is closer to the anus in females than it is in males. Thus, improper wiping after defecation (that is, back to front) can easily carry fecal bacteria to the female urethra. (Proper wiping

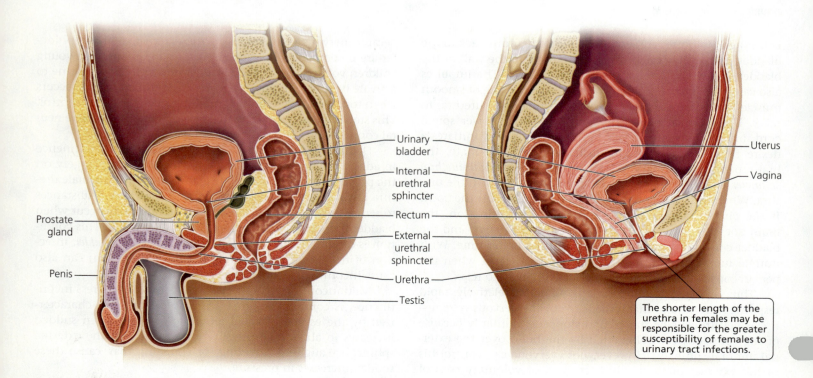

FIGURE 16.16 *The urinary bladder and urethra in a male (left) and a female (right)*

is front to back.) Bacteria also may enter the urethra during sexual intercourse. Women are advised to urinate soon after sex to flush bacteria from the lower urinary tract.

Symptoms of urinary tract infections include fever, blood in the urine, and painful and frequent urination. Bed wetting may be a symptom in young children. In addition, there may be lower abdominal pain (if the bladder is involved) or back pain (if the kidneys are involved). Physicians usually diagnose such infections by having the urine checked for bacteria and blood cells. People who have recurrent UTIs may wish to purchase nonprescription dipsticks to check for infection. The dipstick is held in the stream of the first morning urine and changes color when it detects nitrite, a compound made by the bacteria. These dipsticks can detect about 90% of UTIs.

Urinary tract infections are treated with antibiotics. Infections of the lower urinary tract (the urethra and bladder) should be treated immediately and the prescribed antibiotics taken for their full term. These steps are critical to prevent the spread of infection to the kidneys, where very serious damage can occur.

looking ahead

In Chapter 16, we learned that the urethra is a component of the urinary system—it is the tube in males and females that transports urine from the urinary bladder to outside the body. In Chapter 17, we discover that the urethra also is considered part of the male reproductive system because it separately transports semen, the fluid containing sperm and secretions from accessory glands.

HIGHLIGHTING THE CONCEPTS

16.1 Eliminating Waste (pp. 322–323)

- Several organs from different body systems remove metabolic wastes and excess ions from our bodies. These include lungs, skin, and organs of the digestive system and urinary system.

16.2 Components of the Urinary System (p. 323)

- Two kidneys, two ureters, the urinary bladder, and the urethra make up the urinary system. The kidneys filter excess materials and wastes from the blood and produce urine. Urine travels down the ureters to the urinary bladder, where it is stored until excreted from the body through the urethra. The male urethra transports reproductive fluid in addition to urine (but not at the same time). The female urethra transports only urine.

16.3 Kidneys and Homeostasis (pp. 324–333)

- The kidneys are located just above the waist against the back wall of the abdominal cavity. Each kidney has an outer portion (the renal cortex), an inner region (the renal medulla), and an inner chamber (the renal pelvis).
- Properly functioning kidneys are critical to maintaining homeostasis. Kidneys filter wastes and excess materials from the blood, help regulate blood pH, and maintain fluid balance by regulating the volume and composition of blood and urine.
- Nephrons are tiny tubules responsible for the formation of urine. There are two basic parts to a nephron: (1) the renal corpuscle, which consists of a tuft of capillaries called the glomerulus and a surrounding cuplike structure called the glomerular capsule, and (2) the renal tubule. The renal tubule has three sections: the proximal convoluted tubule, the loop of the nephron, and the distal convoluted tubule.
- The distal convoluted tubules of several nephrons empty into a single collecting duct. Urine in collecting ducts moves into a large chamber within the kidneys called the renal pelvis. From the renal pelvis, urine leaves the kidneys by traveling down the ureters to the urinary bladder.
- Glomerular filtration is the process by which only certain substances are allowed to pass out of the blood and into the nephron. High pressure within glomerular capillaries forces water and small solutes in the blood across a filter into the space within the glomerular capsule to form filtrate.
- From the glomerular capsule, the filtrate moves into the renal tubule of the nephron. As it passes through the renal tubule, about 99% of it is returned to the blood through tubular reabsorption. Almost all the water, ions, and nutrients in the filtrate are returned to the blood. Wastes are either partially reabsorbed or not reabsorbed at all. Most reabsorption occurs in the proximal convoluted tubule.
- During tubular secretion, drugs and any wastes or excess essential ions that may have escaped glomerular filtration are removed from the blood in the capillaries surrounding nephrons and added to the filtered fluid that will become urine. Secretion occurs along the proximal and distal convoluted tubules and collecting ducts.
- The kidneys work with the lungs and buffer systems to regulate pH. Tubular reabsorption returns bicarbonate to the blood (for use in buffering), and tubular secretion removes excess hydrogen ions from the blood.
- About 20% of nephrons have long loops that extend from the cortex deep into the renal medulla. These nephrons are responsible for the extreme water-conserving ability of the kidneys because they maintain high solute concentrations in the interstitial fluid within the kidneys. Maintenance of this gradient between the interstitial fluid and the filtrate enables water to move out of collecting ducts by osmosis so that the body can conserve it.
- Several hormones influence kidney function. Antidiuretic hormone (ADH) is produced by the hypothalamus and released by the posterior pituitary gland. ADH increases water reabsorption in the collecting ducts of nephrons. Aldosterone, secreted by the adrenal cortex, increases sodium reabsorption by the distal convoluted tubules and collecting ducts. Because water follows sodium, aldosterone increases water reabsorption by nephrons. Atrial natriuretic hormone (ANH) is released by cells of the right atrium of the heart. ANH decreases water and sodium reabsorption by the kidneys.
- The kidneys release erythropoietin, a hormone necessary for red blood cell production. The kidneys also convert vitamin D into its active form that promotes the body's absorption of calcium and phosphorus.

16.4 Dialysis and Transplant Surgery (pp. 333–336)

- Renal failure is a decrease or complete cessation of glomerular filtration. Failure of the kidneys severely disrupts homeostasis, resulting in conditions such as acidosis, anemia, edema, and hypertension.
- Treatments for renal failure include hemodialysis, continuous ambulatory peritoneal dialysis, and kidney transplantation.

16.5 Urination (pp. 336–338)

- Urination is a reflex initiated by stretching of the bladder as it fills with urine. We can start or stop urinating because the brain has conscious control over the external urethral sphincter.

16.6 Urinary Tract Infections (pp. 338–339)

- Urinary tract infections (UTIs) are more common in females than in males, possibly because bacteria can more readily move from outside the body up the shorter urethra of females to the bladder. Infections may occur in the urethra, bladder, and kidneys.

RECOGNIZING KEY TERMS

excretion *p. 322*
urine *p. 323*
urinary system *p. 323*
kidney *p. 323*
ureter *p. 323*
urinary bladder *p. 323*
urethra *p. 323*
renal cortex *p. 324*
renal medulla *p. 324*
renal pelvis *p. 324*

nephron *p. 324*
renal corpuscle *p. 324*
glomerulus *p. 324*
glomerular capsule *p. 324*
renal tubule *p. 324*
proximal convoluted tubule *p. 324*
loop of the nephron *p. 324*
distal convoluted tubule *p. 325*

collecting duct *p. 325*
glomerular filtration *p. 326*
tubular reabsorption *p. 327*
tubular secretion *p. 327*
aldosterone *p. 331*
juxtaglomerular apparatus *p. 331*
renin *p. 332*
antidiuretic hormone (ADH) *p. 332*

diuretic *p. 333*
atrial natriuretic hormone (ANH) *p. 333*
erythropoietin *p. 333*
renal failure *p. 333*
hemodialysis *p. 334*
urination *p. 336*
urinary tract infection (UTI) *p. 338*

REVIEWING THE CONCEPTS

1. Name four nitrogen-containing wastes produced by the human body. Describe the processes that generate them, and name the organs primarily responsible for excreting them. *pp. 322–323*
2. List the components of the urinary system and their functions. *p. 323*
3. How do nephrons contribute to the regulation of blood pH? *p. 329*
4. How do the kidneys conserve water, and why is this important? *pp. 329–331*
5. Identify three hormones that play a significant role in kidney function, and explain how each hormone works to help the kidneys meet the body's needs on a daily basis. *pp. 331–333*
6. Does the solute concentration in interstitial fluid increase, decrease, or stay the same from the renal cortex to the renal medulla? Explain. *pp. 329–331*
7. Describe the process and mechanisms involved in urination. *pp. 336–337*
8. How do the urethras of males and females differ? What are the clinical implications of these differences? *pp. 338–339*
9. Select the correct order of organs for the path urine takes from its formation to its leaving the body:
    a. bladder; urethra; ureter; kidney
    b. bladder; kidney; ureter; urethra
    c. kidney; ureter; bladder; urethra
    d. kidney; urethra; bladder; ureter
10. Choose the *incorrect* statement:
    a. During tubular reabsorption, useful substances move from the filtrate within the nephron back into the blood.
    b. During tubular reabsorption, useful substances move from the blood into the filtrate within the nephron.
    c. During tubular secretion, excess substances move from the blood into the filtrate within the nephron.
    d. During glomerular filtration, substances move from the blood into the nephron to form filtrate.
11. Glomerular filtration
    a. occurs at the renal corpuscle.
    b. returns useful substances to the blood.
    c. occurs along the renal tubule.
    d. is the last step of three in cleansing the blood.
12. Which of the following substances would normally *not* pass through the glomerular filter?
    a. water
    b. ions
    c. red blood cells
    d. glucose
13. Microvilli dramatically increase surface area for reabsorption. In the nephron, microvilli characterize the
    a. proximal convoluted tubule.
    b. collecting duct.
    c. distal convoluted tubule.
    d. glomerulus.

14. Which of the following would raise blood pH?
    a. reabsorption of H^+
    b. secretion of HCO_3^-
    c. secretion of H^+
    d. reabsorption of glucose
15. Which of the following hormones results in production of large amounts of dilute urine?
    a. atrial natriuretic hormone (ANH)
    b. erythropoietin
    c. aldosterone
    d. antidiuretic hormone (ADH)
16. Urinary incontinence is
    a. lack of voluntary control over the internal urethral sphincter.
    b. found only in infants and young children.
    c. lack of voluntary control over urination.
    d. failure to expel urine from the bladder to a normal degree.
17. Which of the following substances has the lowest percent reabsorption by nephrons?
    a. glucose
    b. water
    c. bicarbonate
    d. urea
18. _____ is the use of artificial devices to cleanse the blood.
19. The _____ urethral sphincter is made of _____ muscle and is involuntary. The _____ urethral sphincter is made of _____ muscle and allows voluntary control over urination.

APPLYING THE CONCEPTS

1. Rachel has noticed that since she gave birth, a small amount of urine escapes whenever she sneezes. What condition might Rachel have? What would explain this condition?
2. Miguel has noticed that after he drinks beer, his urine output increases. Why does this happen?
3. Li has polycystic disease, an inherited and progressive condition that is destroying her kidneys. What will be the health consequences of her chronic renal failure? What are her options for restoring kidney function? What are the advantages and disadvantages of each option?
4. Varian has proteinuria, a condition in which an abnormal amount of protein is present in his urine. His physician suspects that hypertension caused the proteinuria. Which region of the nephron did hypertension likely damage to produce proteinuria? How might proteinuria be diagnosed?

BECOMING INFORMATION LITERATE

Write a paper on some of the major changes that occur in the urinary system across the life span. Include in your paper the general changes to kidney function caused by age-related declines in number of nephrons, rates of glomerular filtration, and sensitivity of nephrons to ADH. Keep a log of the sources (books, journal articles, websites) that you look at during your search for information. Next to each source, indicate whether you found it helpful and accurate.

MasteringBiology®

Go to MasteringBiology for practice quizzes, activities, eText, videos, current events, and more.

17 Reproductive Systems

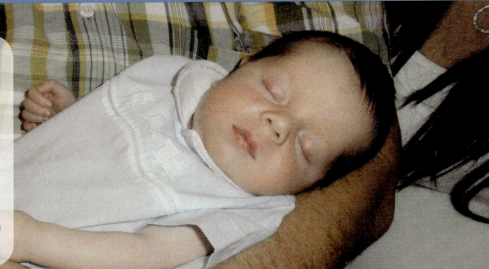

This beautiful baby boy seems comfortable in his new surroundings, although they can hardly be as sheltering and secure as his mother's uterus.

 **Did you Know?**

- *A sexually mature male produces more than 100 million sperm each day, whereas a female produces about 400 mature eggs in her lifetime.*
- *The largest cell in the human body is the female egg, and the smallest is the male sperm.*
- *A human sperm cell can travel 75 times its own length in a minute.*

17.1 Gonads

17.2 Male and Female Reproductive Roles

17.3 Form and Function of the Male Reproductive System

17.4 Form and Function of the Female Reproductive System

17.5 Disorders of the Female Reproductive System

17.6 Stages of the Human Sexual Response

17.7 Birth Control

HEALTH ISSUE
Breast Cancer

In the previous few chapters, we learned about the systems that maintain our bodies. In this chapter, we look at the structures of the reproductive systems and how they function. We also consider the sex hormones and the ways in which they regulate reproductive processes.

17.1 Gonads

In both males and females, the **gonads**—testes or ovaries—are the most important structures in the reproductive system. The gonads serve two important functions: (1) they produce the **gametes**, meaning the eggs and sperm—the cells that will fuse and develop into a new individual; and (2) they produce the sex hormones. The male gonads are the **testes**, and the gametes they produce are the **sperm**. The testes also produce the sex hormone **testosterone**. The **ovaries** are the female gonads. The gametes they produce are the **eggs**. The sex hormones they produce are **estrogen** and **progesterone**.

17.2 Male and Female Reproductive Roles

Males and females make an equal genetic contribution to the next generation by contributing one copy of each chromosome to their offspring. However, they have different "reproductive strategies" to help ensure that their DNA (packaged in their chromosomes) is passed along to a new generation. A male's strategy is to produce millions of sperm, deliver them to the female reproductive system, and hope that one sperm reaches an egg. In contrast, a female usually produces only one egg approximately once a month. If sperm delivery is appropriately timed with egg production, a sperm and egg may fuse in a process called **fertilization**. If conditions are right, the cell created by fertilization, called a zygote, will develop into a new individual (see Chapter 18). A male's role in the reproductive process ends with delivery of sperm, but a female's role continues after egg production; she nourishes and protects the offspring until birth. The egg is packed with nutrients to provide for early development, and the woman's uterus serves as a nourishing, protective environment for the developing offspring.

The development of eggs and sperm involves two types of cell division. The first is *mitosis,* a process in which a cell with 23 pairs of chromosomes replicates its chromosomes and divides into two identical cells, each with 23 pairs of chromosomes. The alterations in genetic content occur during the second type of cell division, called *meiosis,* which begins with a cell that has two copies of each of the 23 kinds of chromosome (called a *diploid* cell) and ends with up to four cells that each contain only one of each kind of chromosome (called *haploid* cells). Meiosis entails two rounds of cell division (called meiosis I and meiosis II). (The details of mitosis and meiosis are described in Chapter 19.)

Each cell in your body contains 23 pairs of chromosomes (for a total of 46 chromosomes). One member of each pair came from your father's sperm; the other member of each pair came from your mother's egg. Each of the 23 kinds of chromosomes contains a different portion of the instructions for making and maintaining your body, and two of each kind are necessary for health and normal development. It is important, therefore, that only one member of each pair be present in a gamete.

17.3 Form and Function of the Male Reproductive System

The male reproductive system consists of the testes, a system of ducts through which the sperm travel; the penis; and various accessory glands. Accessory glands produce secretions that help protect and nourish the sperm as well as provide a transport medium that aids the delivery of sperm to the outside of the male's body. The structure of the male reproductive system is summarized in Table 17.1.

TABLE 17.1 Review of Male Reproductive System

Structure	Function
Testes	Produce sperm and testosterone
Epididymis	Location of sperm storage and maturation
Vas deferens	Conducts sperm from epididymis to urethra
Urethra	Tube through which sperm or urine leaves the body
Prostate gland	Produces secretions that make sperm mobile and that counteract the acidity of the female reproductive tract
Seminal vesicles	Produce secretions that make up most of the volume of semen
Bulbourethral glands	Produce secretions just before ejaculation; may lubricate; may rinse urine from urethra
Penis	Delivers sperm to female reproductive tract

Testes

The male reproductive system has two testes (singular, testis; Figure 17.1). Each is located externally to the body in a sac of skin called the **scrotum**. The temperature in the scrotum is several degrees cooler than the temperature within the abdominal cavity. The lower temperature is important in the production of healthy sperm.

Reflexes in the scrotum help keep the temperature within the testes fairly stable. Cooling of the scrotum, such as might occur when a man jumps into frigid water, triggers contraction of a muscle that pulls the testes closer to the warmth of the body. In a hot shower, however, the muscle relaxes, and the testes hang low, away from the heat of the body. The skin of the scrotum is also amply supplied with sweat glands that help cool the testes.

Beginning at puberty (the transition to sexual maturity), which may occur as young as age 10, a healthy male produces more than 100 million sperm each day. The microscopic sites of sperm production are the **seminiferous tubules** (Figure 17.2 on page 345). We will consider the development of sperm in more detail later in this chapter.

The *interstitial cells* are located between the seminiferous tubules of the testis. They produce the steroid sex hormones, collectively called *androgens*. The most important androgen is *testosterone,* which is needed for sperm production and the maintenance of male reproductive structures.

Testicular cancer is the most common form of cancer among men between the ages of 15 and 35 years, although it affects only 1% of all men. Testicular cancer is more likely to occur in men whose testes did not descend into the scrotum or descended after 6 years of age. Because this cancer does not usually cause pain, it is important for every man to examine his testes each month to feel for a lump or a change in consistency. The cure rate for tumors caught in the early stages is nearly 100%.

Duct System

Sperm produced in the seminiferous tubules next enter a highly coiled tubule called the **epididymis**, where the sperm mature and are stored (see Figure 17.1). Sperm that enter the epididymis look mature, but they cannot yet function as mature sperm. During their stay in the epididymis, the sperm become capable of fertilizing an egg and of moving on their own, although they do not yet do so.

The tube that conducts sperm from the epididymis to the urethra is called the **vas deferens**. Some sperm may be stored in the part of the vas deferens closest to the epididymis. When a man reaches sexual climax, rhythmic waves of muscle contraction propel the sperm along the vas deferens.

The *urethra* conducts urine (from the urinary bladder) or sperm (from the vas deferens) out of the body through the *penis,* the male organ of sexual intercourse and urination. Sperm and urine do not pass through the urethra at the same time. A circular muscle contracts and pinches off the connection to the urinary bladder during sexual excitement.

Accessory Glands

Semen, the fluid released through the urethra at sexual climax, contains sperm and is composed of the secretions of the accessory glands: the prostate gland, the paired seminal vesicles,

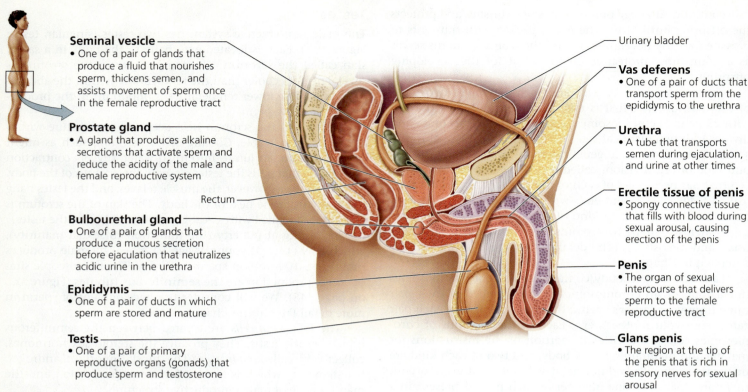

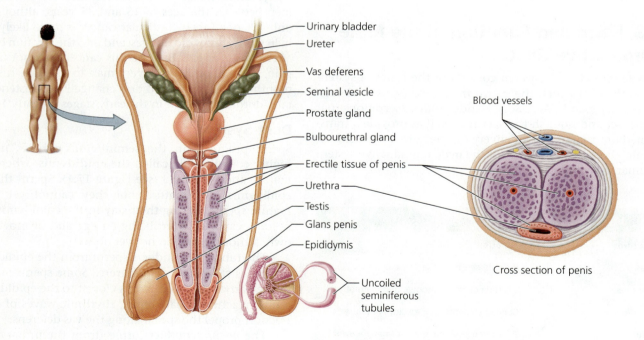

FIGURE 17.1 *The male reproductive system*

and paired bulbourethral glands. Sperm make up very little of the volume of semen.

About the size of a walnut, the **prostate gland** surrounds the upper portion of the urethra, just beneath the urinary bladder. Prostate secretions are slightly alkaline and serve both to activate the sperm, making them fully capable of moving, and to counteract the acidity of the female reproductive tract.

The prostate often begins enlarging when a man reaches middle age. The enlarged prostate may squeeze the urethra and restrict urine flow, making urination difficult. This benign enlargement of the prostate gland that accompanies aging is not related to prostate cancer.

Prostate cancer is an important cause of cancer deaths among men. Unlike testicular cancer, prostate cancer usually affects older men. There are two ways of detecting prostate cancer: a

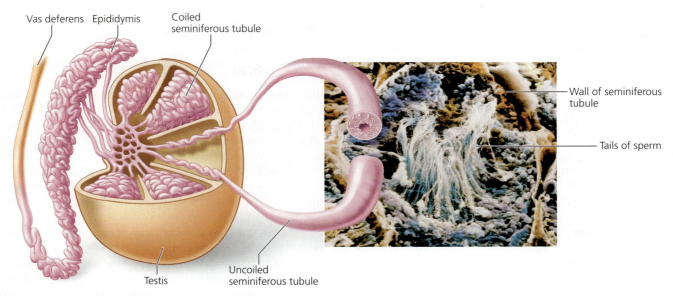

FIGURE 17.2 *The internal structure of the testis and epididymis*

rectal exam and a blood test. Most physicians recommend that both tests be used in the diagnosis. During a rectal exam for prostate cancer, a physician inserts a gloved finger into the rectum and feels the prostate through the rectum's wall (see Figure 17.1). A cancerous prostate is firm or may contain a hard lump. The blood test for prostate cancer measures the amount of a protein called *prostate-specific antigen (PSA)*. Because this protein is produced only by the prostate gland, the amount of PSA in the blood reflects the size of the prostate. As a tumor within the prostate grows, the blood level of PSA usually rises.

The secretions of the **seminal vesicles** contain citric acid, fructose, amino acids, and prostaglandins, which are chemicals secreted by one cell that alter the activity of nearby cells. Fructose is a sugar that provides energy for the sperm's long journey to the egg. Some of the amino acids thicken the semen. This thickening helps keep the sperm within the vagina and protects the sperm from contacting the acidic environment of the vagina. The prostaglandins serve to cut the viscosity of female cervical mucus (which could otherwise slow the movement of sperm) and to cause uterine contractions that assist the movement of sperm.

The **bulbourethral glands** release a clear, slippery liquid immediately before ejaculation. This fluid may serve to rinse the slightly acidic urine remnants from the urethra before the sperm pass through.

Penis

The **penis** is a cylindrical organ whose role in reproduction is to deliver the sperm to the female reproductive system. The tip of the penis is enlarged, forming a smooth, rounded head known as the *glans penis*, which has many sensory nerve endings and is important in sexual arousal. When a male is born, a cuff of skin called the *foreskin* covers the glans penis. The foreskin can be pulled back to expose the glans penis. The surgical removal of the foreskin is called *circumcision*.

The transfer of sperm to the female reproductive system during sexual intercourse usually requires the penis to be erect. An erection consists of increases in the length, width, and firmness of the penis that are due to changes in the blood supply to the organ. Within the penis are three columns of spongy *erectile tissue*, which is a loose network of connective tissue with many empty spaces (see Figure 17.1). During sexual arousal, the arterioles that pipe blood into the spongy tissue dilate (widen), and the spongy tissue fills with blood, causing the penis to become larger. At the same time, the expanding spongy tissue squeezes shut the veins that drain the blood from the penis. As a result, the blood flows into the penis faster than it can leave, causing the spongy tissue to fill with blood and press against a connective tissue casing. This makes the penis firm, larger, and erect.

Erectile dysfunction (*ED*, also called *impotence*) is a male's inability to achieve or maintain an erection (and thus to have sexual intercourse). It is not unusual for a man to experience ED at some point in his life. This condition has some psychological causes, including worry, stress, a quarrel with the partner, and depression. However, it also has many physical causes. Nerve damage, which often accompanies chronic alcoholism and sometimes diabetes, can be responsible. Because an erection depends on adequate blood supply, fatty deposits in the arteries serving the penis (as in atherosclerosis) can also cause ED. Medications—especially certain drugs used to treat high blood pressure, antihistamines, antinausea and antiseizure drugs, antidepressants, sedatives, and tranquilizers—may cause the problem. Cigarette smoking, heavy alcohol consumption, or marijuana use can also cause ED.

The first step in treating ED is to eliminate the cause of the problem, if possible. Should impotence continue, drugs for erectile dysfunction—Viagra, Levitra, Cialis—may be prescribed. These drugs can help a man achieve and maintain an erection when he is sexually aroused. They work by prolonging the effect of nitric oxide, a chemical that is released when the man becomes sexually aroused and that causes the widening of the arterioles in the penis. As the arterioles widen, blood flow increases, and an erection results.

stop and think

Why is it that Viagra, Levitra, or Cialis cannot cause an erection in a man who is not sexually aroused?

Sperm Development

The sequence of events within the seminiferous tubules that leads to the development of sperm is called *spermatogenesis*. This process reduces the number of chromosomes in the resulting gametes to one member of each pair and changes the sperm cells' shape and functioning to make the sperm efficient chromosome-delivery vehicles.

The process of spermatogenesis (Figure 17.3) begins in the outermost layer of each seminiferous tubule, where undifferentiated diploid cells called *spermatogonia* (singular, spermatogonium) develop. Each spermatogonium divides by mitosis to produce two new diploid spermatogonia. One of these spermatogonia pushes deeper into the wall of the tubule, where it will enlarge to form a *primary spermatocyte*, which is also a diploid cell. Primary spermatocytes undergo the two divisions of meiosis to form *secondary spermatocytes* (after meiosis I), which mature into *spermatids* (after meiosis II). Although the spermatids are haploid and have the correct set of chromosomes for joining with the egg during fertilization, numerous structural changes must still occur to create cells capable of swimming to the egg and fertilizing it. These changes convert spermatids to streamlined *spermatozoa*, or sperm. Spermatogonia mature into spermatozoa in about 2 months.

The mature sperm cell has three distinct regions: the head, the midpiece, and the tail (Figure 17.4). The head of the sperm is a flattened oval that contains little else besides the 23 densely packed chromosomes (remember that the sperm is haploid). Positioned like a ski cap on the head of the sperm is the **acrosome**, a membranous sac containing enzymes. A few hours after the sperm have been deposited in the female reproductive system, the membranes of the acrosomes break down. The enzymes then spill out and digest through the layers of cells surrounding the egg, assisting fertilization. Within the midpiece, mitochondria are arranged in a spiral. Remember that the mitochondria are the powerhouses of the cell. They provide energy in the form of ATP to fuel the movements of the tail. The tail contains contractile filaments. The whiplike movements of the tail propel the sperm during its long journey through the female reproductive system.

FIGURE 17.3 The stages of spermatogenesis in the wall of a seminiferous tubule. As the cells that will become sperm develop, they are pushed from the outer wall of the tubule to the lumen, or central canal.

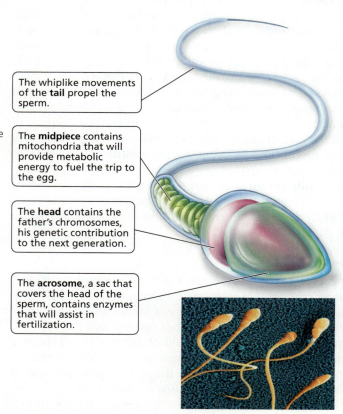

FIGURE 17.4 The structure of a mature sperm (spermatozoon)

Hormones

Testosterone, secreted by the interstitial cells of the testes, is important for sperm production as well as the development of male characteristics. At puberty, testosterone turns boys into men (literally). It is responsible for the growth spurt that occurs when a male reaches puberty, stimulating the growth of the long bones of the arms and legs. It also makes the male reproductive organs, including the testes and penis, larger. In addition, testosterone is responsible for the development and maintenance of the male secondary sex characteristics, which are features associated with "masculinity" but not directly related to reproductive functioning. For example, the growth of muscles and the skeleton tends to result in wide shoulders and narrow hips. Pubic hair develops, as does hair under the arms. A beard appears, perhaps accompanied by hair on the chest. Meanwhile, the voice box enlarges and the vocal cords thicken, causing a deepening of the male voice. Another important role of testosterone is that it is responsible for sex drive.

The level of testosterone in the body remains relatively steady because its production is regulated by a negative feedback loop (see Chapter 10) involving hormones from the hypothalamus, the anterior pituitary gland, and the testes. The hormones that regulate male reproductive processes are summarized in Figure 17.5 and Table 17.2. Testosterone levels rise when the hypothalamus, a region of the brain, releases **gonadotropin-releasing hormone (GnRH)**. This hormone stimulates the anterior pituitary gland, a pea-sized endocrine gland on the underside of the brain, to secrete **luteinizing hormone (LH)**. In turn, LH stimulates the production of testosterone by the interstitial cells of the testis. (For this reason, LH is sometimes called *interstitial cell–stimulating hormone*, or *ICSH*.) The rising testosterone level inhibits the release of GnRH from the hypothalamus. As a result, the amount of LH produced by the anterior pituitary drops. This decrease in the LH level then lowers the amount of testosterone secretion, removing the inhibition on the hypothalamus. These interactions keep testosterone levels constant.

Sperm production is regulated by another negative feedback loop. Besides secreting LH, the anterior pituitary gland produces **follicle-stimulating hormone (FSH)**. Follicle-stimulating hormone stimulates sperm production by making the cells that will become sperm more sensitive to the stimulatory effects of testosterone. FSH works by causing certain cells within the seminiferous tubules to secrete a protein that binds and concentrates testosterone.

When sperm numbers are high, the seminiferous tubules produce a hormone called *inhibin* in addition to sperm. Inhibin is named for its ability to *inhibit*, or lessen, the production of FSH from the anterior pituitary. It also may inhibit the hypothalamic secretion of GnRH. Thus, rising levels of inhibin cause a decline in testosterone level and sperm production. As the sperm count falls, so does the level of inhibin. Released from inhibition, the anterior pituitary then increases production of FSH, and the hypothalamus produces GnRH, which increases sperm production again.

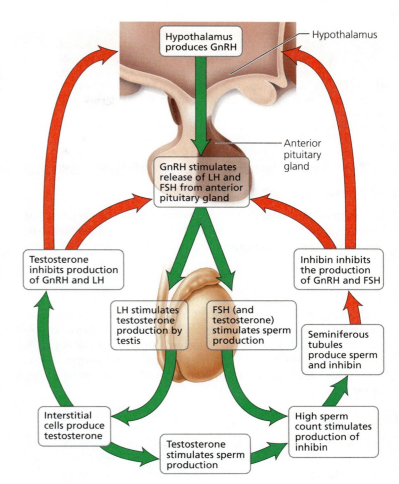

FIGURE 17.5 *The feedback relationships among the hypothalamus, anterior pituitary, and testes control the production of both sperm and testosterone.*

TABLE 17.2 Hormones Important in Regulating Male Reproductive Processes

Hormone	Source	Effects
Testosterone	Interstitial cells in testes	Sperm production; development and maintenance of male reproductive structures, male secondary sex characteristics; sex drive
Gonadotropin-releasing hormone (GnRH)	Hypothalamus (in brain)	Stimulates the anterior pituitary gland to release LH
Luteinizing hormone (LH)	Anterior pituitary gland (in brain)	Stimulates interstitial cells of testis to produce testosterone
Follicle-stimulating hormone (FSH)	Anterior pituitary gland (in brain)	Enhances sperm formation
Inhibin	Seminiferous tubules in testes	Inhibits FSH secretion by anterior pituitary gland, causing a decrease in sperm production and testosterone production

stop and think

Some male athletes take anabolic steroids to build muscles. A side effect of steroid abuse is a reduction in testis size. Considering that the anabolic steroids mimic the effect of testosterone, how can the shrinking testis size be explained?

17.4 Form and Function of the Female Reproductive System

The female reproductive system consists of the ovaries, the oviducts, the uterus, the vagina, and the external genitalia. Its structure is shown in Figure 17.6, and its functions are summarized in Table 17.3.

Ovaries

Each ovary is about twice the size of an almond. There is an ovary on each side of the uterus. The ovaries have two important functions: (1) to produce eggs in a process called *oogenesis* and (2) to produce the hormones estrogen and progesterone. These two functions are intimately related and occur in cycles, as we will see.

Oviducts

Two **oviducts**, also known as *fallopian tubes* or *uterine tubes*, extend from the uterus to the ovaries—although they do not attach to the ovaries. They transport the immature egg, called an *oocyte,* from the ovary to the uterus. The end of each oviduct nearest the ovary is open and funnel shaped. Its many ciliated, fingerlike projections drape over the ovary but are rarely in direct contact with it. Shortly before the egg is released from the ovary, the projections from the tubes begin to wave. The currents they create, and those caused by the cilia lining the tubes, help draw the oocyte into the oviduct. If fertilization occurs, it usually takes place in the oviduct, near the ovary. The resulting zygote begins its development into an embryo in the oviduct. The beating cilia and the rhythmic, muscular contractions of the oviduct sweep the egg or the early embryo along the oviduct toward the uterus.

TABLE 17.3 Review of Female Reproductive System	
Structure	**Function**
Ovary	Produces eggs and the hormones estrogen and progesterone
Oviducts	Transport ovulated oocyte (or embryo if fertilization occurred) to the uterus; the usual site of fertilization
Uterus	Receives and nourishes embryo
Vagina	Receives penis during intercourse; serves as birth canal
Clitoris	Contributes to sexual arousal
Breasts	Produce milk

Uterus

The **uterus** is a hollow, muscular organ that before pregnancy is about the size and shape of an inverted pear. The uterus receives and nourishes the developing baby (first called an *embryo,* but after 8 weeks called a *fetus*). During pregnancy, as the fetus grows, the uterus expands to about 60 times its original size. After childbirth, the uterus never quite returns to its prepregnancy size.

The wall of the uterus has two main layers: a muscular layer and a lining called the **endometrium** (*endo-*, within; *metr-*, uterus; *-ium*, region). The smooth muscle of the uterine wall contracts rhythmically in waves during childbirth and forces the infant out. The thickness of the endometrium varies over a cycle of approximately 1 month (this cycle is discussed later in this chapter). If an embryo forms, it implants (embeds) in the endometrium, which provides nourishment during early pregnancy. If an embryo does not form, the endometrium completes its cycle by being lost as menstrual flow.

If an embryo implants in an area other than the uterus, the condition is described as an *ectopic pregnancy* (*ect-*, outside). The most common type of ectopic pregnancy is a *tubal pregnancy,* in which the embryo implants in an oviduct. A tubal pregnancy cannot be carried to term. It must be surgically terminated because it endangers the mother's life. If the embryo is permitted to continue growing, it will eventually rupture the oviduct, and the rupture can cause the mother to bleed to death internally.

The narrow neck of the uterus, called the **cervix**, projects into the **vagina**, a muscular tube that opens to the outside of the body. The vagina receives the penis during sexual intercourse. Sperm that are deposited in the vagina can enter the uterus through an opening in the cervix and then swim to the oviduct to meet the egg. The vagina is also the birth canal. The infant is pushed through the cervix and then the vagina on the way to greet the world.

External Genitalia

The female reproductive structures that lie outside the vagina are collectively known as the *external genitalia* or the **vulva**. Two hair-covered folds of skin surround the vaginal opening. The *labia majora* ("big lips") enclose two thinner skin folds, the *labia minora* ("little lips"). The anterior portions of the labia minora form a hood over the *clitoris*. Like the penis, the clitoris has many nerve endings sensitive to touch. During tactile stimulation, the erectile tissue in the clitoris becomes swollen with blood and contributes to a woman's sexual arousal.

Breasts

The *breasts,* or *mammary glands,* are present in both sexes, but they produce milk to nourish a newborn only in females. Inside the female breast are 15 to 25 groups of milk-secreting glands. A milk duct drains each group through the nipple. Interspersed around the glands and ducts is fibrous connective tissue that supports the breast. Most of the breast consists of

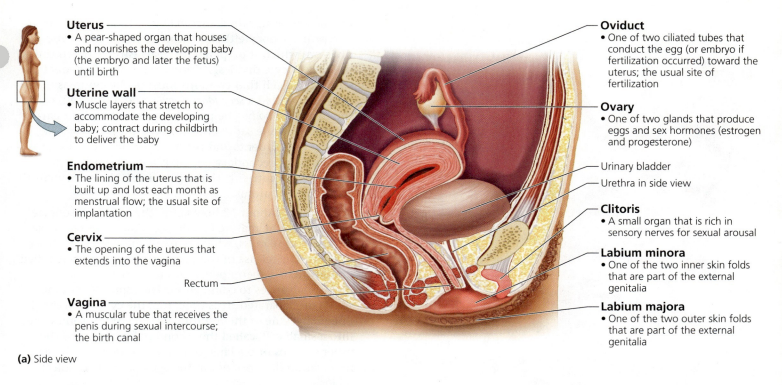

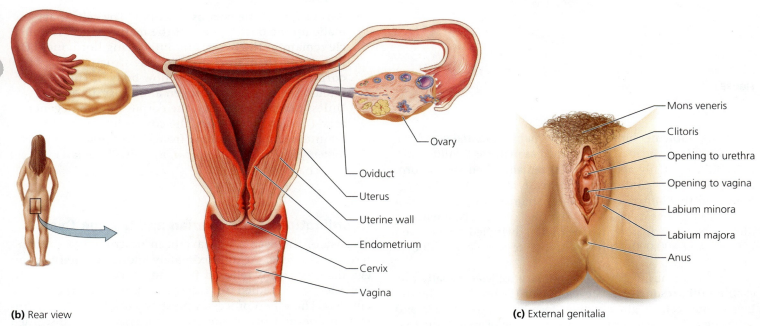

FIGURE 17.6 *The female reproductive system*

fatty tissue (Figure 17.7). Breast cancer is the second leading cause of cancer deaths among women (see the Health Issue essay, *Breast Cancer*).

Ovarian Cycle

The changes in the ovary that produce the egg occur in a cycle about 1 month long called the **ovarian cycle**. Although the ovaries do not produce mature eggs until a female reaches puberty, the preparations for egg production begin before she is born. Recall that the egg is a haploid cell formed by meiosis, the type of cell division that forms gametes. During fetal development, *oogonia*, which are diploid cells in the ovaries, enlarge and begin to store nutrients. Some begin meiosis by making a copy of their chromosomes (but the copies of each chromosome remain attached to one another). The cells that prepare for meiosis are called *primary oocytes*. (They are immature eggs.) A single layer of flattened cells, called *follicle cells*, surrounds each primary oocyte. The entire structure is called a primary **follicle**.

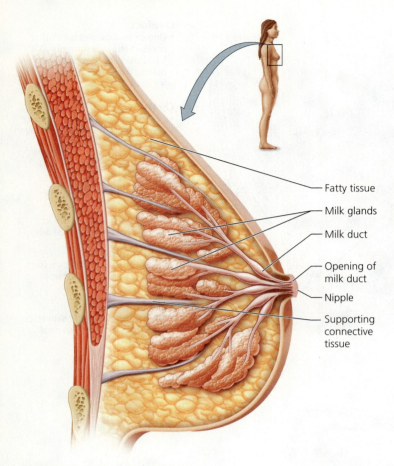

FIGURE 17.7 Breast structure

Long before her birth, all of a woman's potential eggs have formed. In fact, she may have formed as many as 2 million primary follicles. Only about 700,000 remain when she is born, and the number continues to decrease after birth. The eggs remain in this immature state until she reaches puberty, usually at 10 to 14 years of age. By the time of puberty, a female's lifetime supply of potential eggs has dwindled to between 200,000 and 400,000. Only 400 to 500 of these eggs will ever mature.

Beginning at puberty, some primary follicles, usually one each month, resume their development (Figure 17.8). The follicle's outer cells begin dividing, forming layers of cells and secreting a fluid that contains estrogen. The ovarian cycle that converts a primary follicle into an egg each month consists of the following steps:

- **Follicle maturation.** The follicle cells continue dividing, and fluid begins to accumulate between them. As the fluid accumulates, the wall of follicle cells splits into two layers. The inner layer of follicle cells directly surrounds the primary oocyte. The outer layer forms a balloonlike sphere enclosing the fluid and the oocyte. This structure grows rapidly.
- **Formation of a mature follicle.** Within about 10 to 14 days after its development began, the follicle assumes its mature form, called a *secondary* or *Graafian follicle*. The primary oocyte then completes the first meiotic division, which it prepared for years earlier. When the primary oocyte divides, it forms two cells of unequal size: a large cell that contains most of the cytoplasm and nutrients, called a *secondary oocyte,* and a tiny cell, called the first *polar body.* The polar body is essentially a garbage bag for one set of chromosomes. It has very few cellular constituents and plays no further role in reproduction. The secondary oocyte is a much larger cell packed with nutrients that will nourish the embryo until it reaches the uterus.
- **Ovulation.** About 12 hours after the secondary oocyte has formed, the mature follicle pops, like a blister, releasing the oocyte mass, which is about the size of the head of a pin. The release of the oocyte from the ovary is called **ovulation**. If a sperm penetrates the secondary oocyte, meiosis advances to completion: The replicate chromosomes are separated. This time, too, the cytoplasm divides unequally. One of the resulting sets of chromosomes goes into a small cell, called the second polar body, and the other set goes into a large cell, the mature *ovum,* or egg. If fertilization does not occur, the egg does not complete meiosis.
- **Formation of the corpus luteum.** The cells that made up the outer sphere of the mature Graafian follicle remain in the ovary, and luteinizing hormone (LH) transforms them into an endocrine structure called the **corpus luteum** (meaning "yellow body"). The corpus luteum secretes both estrogen and progesterone. Unless pregnancy occurs, the corpus luteum degenerates. If pregnancy occurs, the corpus luteum will be maintained by a hormone from the embryo called *human chorionic gonadotropin,* as we will see later in this chapter.

Coordination of the Ovarian and Uterine Cycles

Because of the events we have been describing, a woman's fertility is cyclic. At approximately monthly intervals, an egg matures and is released from an ovary. Simultaneously, the uterus is readied to receive and nurture the young embryo. These two processes must be coordinated. If fertilization does not occur, the uterine provisions are discarded as menstrual flow. The ovaries and uterus will prepare for fertilization again during the next cycle. The events in the ovary, known as the *ovarian cycle,* must be closely coordinated with those in the uterus, known as the **uterine cycle** (or the **menstrual cycle**). The length of the uterine cycle may vary from cycle to cycle and from woman to woman. The timing of events within a uterine cycle will vary with the cycle's length. We will describe the events as they might occur in a 28-day cycle.

A female's fertility is governed by hormones (Table 17.4). Events in both the uterus and the ovary are coordinated by interactions between hormones from the anterior pituitary

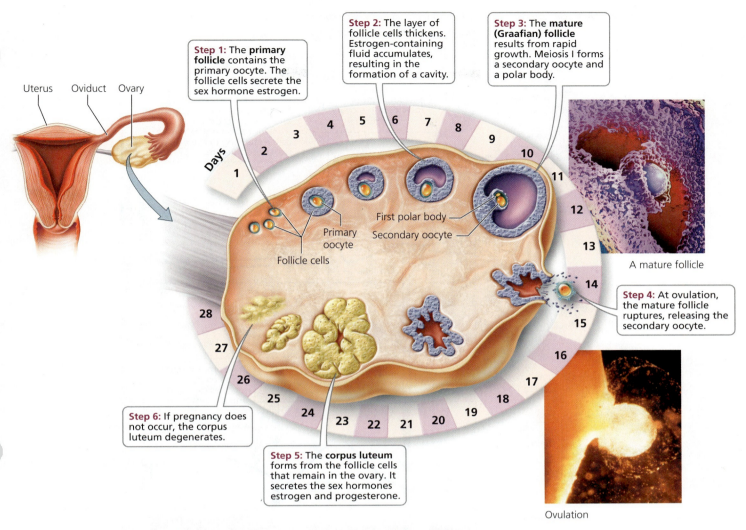

FIGURE 17.8 The ovarian cycle. A follicle does not move around the ovary during its development, as depicted here. The steps indicate the sequence of events that occur in a single place in the ovary during the approximately 28-day cycle.

gland and from the ovary. The anterior pituitary gland produces follicle-stimulating hormone (FSH) and luteinizing hormone (LH). As in the male, the release of these hormones is regulated by a hormone from the hypothalamus called GnRH. In the female, FSH and LH cause the ovary to release its hormones, estrogen and progesterone. Progesterone and estrogen (except at a very high level) exert negative feedback on the anterior pituitary, which causes the levels of FSH and LH to decline. As a result, the levels of estrogen and progesterone decline, reducing the inhibition on the anterior pituitary gland—and so, the hormones cycle.

TABLE 17.4 Hormones Involved in Regulating Female Reproductive Processes

Hormone	Source	Effects
Estrogen	Ovaries (follicle cells and corpus luteum)	Maturation of the egg; development and maintenance of female reproductive structures, secondary sex characteristics; thickens endometrium of uterus in preparation for implantation of embryo; cell division in breast tissue
Progesterone	Ovaries (corpus luteum)	Further prepares uterus for implantation of embryo; maintains endometrium
Follicle-stimulating hormone (FSH)	Anterior pituitary gland (in brain)	Stimulates development of a follicle in the ovary
Luteinizing hormone (LH)	Anterior pituitary gland (in brain)	Triggers ovulation; causes formation of the corpus luteum

stop and think

Fertility clinics may administer FSH to women who are having difficulty getting pregnant. Why would FSH increase the likelihood of pregnancy? Why would women treated with FSH be more likely to have multiple births?

352 CHAPTER 17 Reproductive Systems

Q Couples who are trying to conceive a child can purchase kits to help them predict when ovulation will occur, so that intercourse can be timed appropriately to increase the chance that conception will occur. Which hormone would be the best predictor of ovulation? Why?

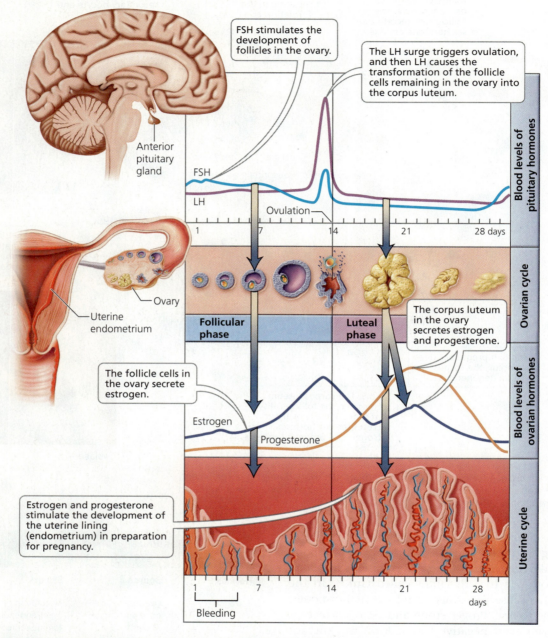

FIGURE 17.9 *The ovarian and uterine cycles are coordinated by the interplay of hormones from the anterior pituitary gland and the ovary. The timing of events shown is for a 28-day cycle.*

A LH, because the surge in LH levels triggers ovulation.

Menstruation Traditionally, the first day of menstrual flow is considered day 1 of the uterine cycle because it is the most easily recognized event in the cycle (Figure 17.9). The bleeding usually lasts an average of 5 days. During menstruation, the ovarian hormones, estrogen and progesterone, are at their lowest levels, allowing the anterior pituitary gland to produce its hormones, especially FSH. In turn, FSH causes a new egg follicle in the ovary to develop and produce estrogen. (Hence its name, *follicle-stimulating hormone*.) Thus, even as the uterus loses the endometrial lining it prepared during the previous cycle in which no egg was fertilized, the egg that will be released in the next cycle is developing.

Endometrium thickens As the egg follicle develops, it produces an increasing amount of estrogen. Estrogen causes the cells in the endometrial lining of the uterus to divide. These cells store nutrients to nourish a future embryo during its early stages of development. In addition, estrogen inhibits the release of FSH through a negative feedback cycle (see Chapter 10). When the egg and follicle are nearly mature, the estrogen level rises rapidly and causes a sudden and spectacular release of LH (and FSH) from the anterior pituitary gland.[1]

Ovulation and formation of corpus luteum (in ovary) The LH surge causes several important events. It causes the egg to undergo its first meiotic division. Next, LH triggers ovulation, and the egg bursts out of the ovary to begin its journey along the oviduct. Continued secretion of LH then transforms the remaining follicle cells into the corpus luteum. The corpus luteum continues the estrogen secretion begun by the follicle cells and, importantly, also secretes progesterone.

Endometrium further prepared for implantation Together, estrogen and progesterone make the endometrial lining of the uterus a hospitable place for the embryo. The blood supply to the endometrium increases. Uterine glands develop and then secrete a mucous material that can nourish the young embryo when it arrives in the uterus.

The rising estrogen and progesterone levels inhibit pituitary secretion of FSH and LH. As FSH levels decline, the development of new follicles is inhibited.

Corpus luteum degenerates If fertilization does not occur, the corpus luteum will degenerate within about 2 weeks (14 days plus or minus 2 days, regardless of the length of the menstrual cycle). The degeneration of the corpus luteum results in falling levels of estrogen and progesterone.

Menstruation begins again Progesterone is essential to the maintenance of the endometrium. As progesterone levels drop because of the degeneration of the corpus luteum, the blood vessels nourishing the endometrial cells collapse. The cells of the endometrium then die and are sloughed off, along with mucus and blood, as menstrual flow. No longer inhibited by estrogen and progesterone, FSH and LH levels begin to climb. Thus, the cycle begins again (Table 17.5).

If the egg is fertilized, a hormone from the embryo, called *human chorionic gonadotropin* (*HCG*), maintains the corpus luteum. If implantation occurs, HCG will prevent the degeneration of the corpus luteum, keeping estrogen and progesterone levels high enough to prevent endometrial shedding. HCG is detectable in the mother's blood within 7 to 9 days after implantation and in her urine less than 2 weeks after implantation, which is usually before her first missed period. Between the second and third month of development, the placenta has developed sufficiently to take over the production of estrogen and progesterone during the rest of the pregnancy. The corpus luteum then degenerates.

what would you do?

It is clear that we have polluted our environment with chemicals that mimic the effects of estrogen. Sex hormones choreograph so many biological activities—development, anatomy, physiology, and behavior—that environmental estrogens can play havoc with human reproduction in myriad ways. It is also clear that this pollution is already affecting various animal species. What is not clear is how great a threat these chemicals are to humans. The International Union of Pure and Applied Chemistry (IUPAC) recommends research and monitoring of environmental estrogens. Implementing these recommendations would be costly. As a taxpayer, are you willing to pay for these precautions?

Menopause

A woman's fertility usually peaks when she is in her twenties and then gradually declines. By the time she reaches 45 to 55 years of age, few of the potential eggs prepared before birth remain in the ovaries. Those that do remain become increasingly less responsive to FSH and LH, the hormones that cause egg development and ovulation. The ovaries, therefore, gradually stop producing eggs, and the levels of estrogen and progesterone fall. During this time, menstrual cycles become increasingly irregular. Eventually ovulation and menstruation stop completely, at a stage in a woman's life called **menopause**.

The drop in estrogen levels has some physiological effects that can range in severity from annoying to life threatening. At one end of the scale is the loss of a layer of fat that was formerly promoted by estrogen. This loss results in a reduction in breast size and the appearance of wrinkles. Estrogen also plays an important role in regulating a woman's body thermostat. So, as estrogen levels fall, many women experience hot flashes—waves of warmth spreading upward from the trunk to the face. In addition, the absence of estrogen can cause vaginal dryness that can make sexual intercourse painful. Furthermore, without estrogen, the male hormones produced by the adrenal gland predominate and can cause facial hair to grow.

[1] Above a critical level, estrogen exerts positive feedback on the brain and anterior pituitary gland.

TABLE 17.5 Ovarian and Uterine Cycles

Ovarian Cycle		Uterine Cycle	
Approximate Timing in 28-Day Cycle	**Events**	**Approximate Timing in 28-Day Cycle**	**Events**
Days 1–13	Follicle develops, caused by FSH; Follicle cells produce estrogen	Day 1	Onset of menstrual flow (breakdown of endometrium)
Day 14	Ovulation is triggered by LH surge	Day 6	Endometrium begins to get thicker
Days 15–21	Corpus luteum forms and secretes estrogen and progesterone	Days 15–23	Endometrium is further prepared for implantation of the embryo by estrogen and progesterone
Days 22–28	Corpus luteum degenerates, causing estrogen and progesterone levels to decline	Days 24–28	Endometrium begins to degenerate owing to declining maintenance by progesterone

HEALTH ISSUE

Breast Cancer

Breast cancer usually begins with abnormal growth of the cells lining the milk ducts of the breast, but it sometimes begins in the milk glands themselves. Some types of breast cancer aggressively invade surrounding tissues. Typically, cancerous cells begin to spread when the tumor is about 20 mm (about 3/4 in.) in diameter. At this point, they break through the membranes of the ducts or glands where they initially formed and move into the connective tissue of the breast. They may then move into the lymphatic vessels or blood vessels permeating the breast or into both; the vessels may transport the cells throughout the body.

Detecting Breast Cancer

Early detection is a woman's best defense against breast cancer. A monthly breast self-exam (BSE) is helpful in detecting a lump early (Figure 17.A). If a woman begins doing regular breast self-exams in early adulthood, she becomes familiar with the consistency of her breast tissue. With this experience, it is easier to notice changes that might be signs of breast cancer. Mammograms, which are x-ray exams of breast tissue, can also help detect early breast cancer because they can reveal a tumor too small to be felt as a lump. A tumor large enough to be felt contains a billion or more cells—a few of which may already have spread from the tumor to other tissues of the body. After cancer cells spread, the woman's chance of survival decreases dramatically. An added benefit of mammograms is that they can detect tumors that are small enough to be removed by a type of surgery called *lumpectomy*, which removes the lump but spares the breast. At later stages of breast cancer, the entire breast may have to be removed in a type of surgery called *mastectomy*.

Risk Factors

Many of the factors that increase the risk of breast cancer cannot be altered. The most important factor is gender. Although men can develop breast cancer, it is much more common in women. The genes she inherited from her parents also alter breast cancer risk. Mutations in at least two known genes, *BRCA1* and *BRCA2*, increase a woman's risk of breast cancer. Although only 5% to 10% of all breast cancers are related to these genes, the women who inherit either of them have up to an 85% chance of developing breast cancer at some point in life.

Exposure to estrogen is a common thread running through the tapestry of risk factors associated with breast cancer. During each menstrual cycle, estrogen stimulates breast cells to begin dividing in preparation for milk production, in case the egg is fertilized. Excessive estrogen exposure, therefore, may be a factor that pushes cell division to a rate characteristic of cancer.

The number of times a woman ovulates during her lifetime alters breast cancer risk, because estrogen is produced by both the maturing ovarian follicle and the corpus luteum. The number of times a woman ovulates is, in turn, affected by factors such as the following:

1. **Age when menstruation begins.** Ovulation usually occurs in each menstrual cycle. Thus, the younger a woman is when menstruation begins, the more opportunities there are for ovulation and the greater her exposure to estrogen.
2. **Menopause after age 55.** The older a woman is at menopause, the more menstrual cycles she is likely to have experienced. Estrogen levels are low and ovulation ceases after menopause.
3. **Childlessness and late age at first pregnancy.** Ovulation does not occur during pregnancy. Thus, pregnancy gives the ovaries a rest. Furthermore, the hormonal patterns of pregnancy appear to transform breast tissue in a way that protects against cancer. As a result, delaying pregnancy until after age 30 or remaining childless increases a woman's risk of developing breast cancer later in life.
4. **Breast-feeding.** Nursing may guard against breast cancer by blocking ovulation and, therefore, the monthly changes in estrogen level. Another possible reason is that during breast-feeding, the breast tissue is differentiated and is not undergoing mitosis. Women who breast-feed their infants have a 20% lower risk of developing breast cancer at a young age—before they reach menopause—than do women who bottle-feed their infants.
5. **Exercise.** Even moderate exercise can suppress ovulation (and thus estrogen levels) in adolescents and women in their twenties (but, unfortunately, exercise is not as likely to suppress ovulation in older women).

Other factors may also influence estrogen levels. For instance, obese women have higher estrogen levels than do thin ones because estrogen is produced in fat cells as well as by the ovaries. Indeed, obesity—especially if the fat is carried above the waist—translates to a threefold elevated risk of breast cancer.

Questions to Consider

- Federal guidelines, which are based on a cost/benefit analysis, no longer recommend routine

---

More serious problems associated with the estrogen deficit that accompanies menopause include an increased risk of diseases of the heart and blood vessels. Estrogen offers some protection against atherosclerosis, a condition in which fatty deposits clog the arteries. Thus, after menopause, this protective effect on the heart is lost. Another serious problem stems from the role estrogen plays in the body's ability to absorb calcium from the digestive system and to deposit it in bone. Without calcium, bone becomes weak and porous—a condition known as *osteoporosis* (discussed in Chapter 5).

Until 2002, doctors routinely prescribed hormone replacement therapy (HRT), which contains estrogen and progesterone, for postmenopausal women. HRT reduces hot flashes and vaginal dryness and protects against osteoporosis. The routine recommendation ended with a study showing that HRT elevates a woman's risk of stroke. Today, doctors help postmenopausal women weigh the benefits and risks of HRT, so a woman can make a wise decision on whether to use HRT.

17.5 Disorders of the Female Reproductive System

Premenstrual syndrome (PMS) is a collection of symptoms that appear in some women 7 to 10 days before their period begins. These symptoms include depression, irritability, fatigue, and headaches.

Some researchers suggest that a progesterone deficiency is to blame for the symptoms of PMS. Progesterone has a calming effect on the brain. It also decreases fluid retention. As progesterone levels plummet in the days just before menstruation, the nervous system may be stimulated and fluids may be retained. Fluid retention causes an uncomfortable, bloated feeling.

Treatments for PMS are varied. For women whose symptoms are severe, drugs that elevate the levels of serotonin—a neurotransmitter in the brain (Chapter 7)—are administered. Some women with milder symptoms are helped by changes in diet. Caffeine, alcohol, fat, and sodium should be avoided.

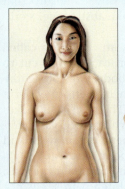 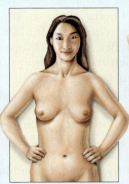  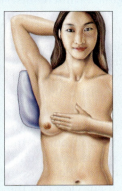 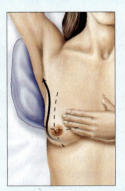 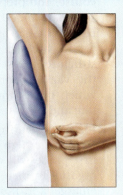

Step 1: Stand in front of the mirror and look at each breast to see if there is a lump, a depression, a difference in texture, or any other change in appearance.

Step 2: Get to know how your breasts look. Be especially alert for any changes in the nipples' appearance.

Step 3: Raise both arms and check for any swelling or dimpling in the skin of your breasts.

Step 4: Lie down with a pillow under your shoulder and put your arm behind your head. Perform a manual breast examination. With the nipple as the center, divide your breast into imaginary quadrants.

Step 5: With the pads of the fingers, make firm circular movements over each quadrant, feeling for unusual lumps or areas of tenderness. When you reach the upper, outer quadrant of your breast, continue toward your armpit. Press down in all directions.

Step 6: Feel your nipple for any change in size and shape. Squeeze your nipple to see if there is any discharge. Repeat from step 4 on the other breast.

FIGURE 17.A *A monthly breast self-exam can help a woman find lumps in her breast before they have spread to surrounding tissue. It also makes make a woman familiar with the consistency of her breast tissue and therefore able to detect changes in consistency that might indicate cancer.*

mammograms for women over 40. The American Cancer Society says that routine mammograms reduce the number of mastectomies and deaths from breast cancer. Would you recommend that your female family members and friends who are over 40 get routine mammograms? Why or why not?

- Some women who carry the mutant genes that increase the risk of breast cancer decide to have their breasts removed to reduce the risk developing breast cancer. Would you recommend that your female family members and friends who have the genes for breast cancer have preventative mastectomies? Why or why not?

Conversely, women should increase their consumption of foods high in calcium, potassium, manganese, and magnesium. Aerobic exercise for at least half an hour a day stimulates the release of enkephalins and endorphins (brain chemicals that produce pleasure), which may provide some relief.

Prostaglandins, chemicals used in communication between cells in many parts of the body, are the primary cause of *menstrual cramps*. Endometrial cells produce prostaglandins that, among other things, make the smooth muscle cells of the uterus contract, causing cramps. High levels of prostaglandins can cause sustained contractions, called *muscle spasms,* of the uterus. These muscle spasms may cut down blood flow—and therefore, the oxygen supply to the uterine muscles—and result in pain.

Endometriosis is a condition in which tissue from the lining of the uterus is found outside the uterine cavity—commonly in the oviducts, on the ovaries, or on the outside surface of the uterus, the bladder, or the rectum. In these cases, the endometrial tissue has moved out the open ends of the oviducts to the abdominal cavity. Endometrial tissue, wherever it is, grows and breaks down with the hormonal changes that occur during each menstrual cycle. These cyclic changes can cause extreme pain. Moreover, women with endometriosis often have difficulty becoming pregnant because endometrial tissue may block the oviduct or coat the ovary.

17.6 Stages of the Human Sexual Response

In both men and women, sexual arousal and sexual intercourse involve two basic physiological changes: certain tissues fill with blood (vasocongestion), and certain muscles undergo sustained or rhythmic contractions (myotonia). However, men and women differ in which tissues fill with blood and which muscles contract. The sequence of events that accompanies sexual arousal and intercourse is called the *sexual response cycle.* Let's consider the four stages of the human response cycle in men and in women.

1. **Excitement.** During the excitement phase of the sexual response cycle, sexual arousal increases. This stage occurs during foreplay, the activities that precede sexual intercourse, and prepares the penis and vagina for intercourse. In a male, blood fills the spongy tissue of the penis, which becomes erect, as described earlier. In a female, blood flow to the breasts, nipples, labia, and clitoris causes these structures to swell. Partly due to an increase in blood flow, the vaginal walls seep fluid that serves as a lubricant, which makes the insertion of the penis easier. Breathing rate and heart rate increase.

2. **Plateau.** During the plateau stage, sexual arousal is maintained at a high level. The changes that began during the first phase will continue through the plateau stage until orgasm. In men, increased blood flow causes the testes to enlarge, and muscle contraction pulls them closer to the body. In women, increased blood flow causes the outer third of the vagina to enlarge and the clitoris to retract under the clitoral hood. In both men and women, breathing rate and heart rate continue to increase.

3. **Orgasm.** The peak of the sexual response cycle, orgasm, is brief but usually intensely pleasurable. In both sexes, orgasm is characterized by muscle contractions of the reproductive structures. In men, ejaculation occurs during orgasm. Ejaculation takes place in two stages. In the first stage, contractions of glands and ducts in the reproductive system force sperm and the secretions of the accessory glands to the urethra. During the second stage of ejaculation, the urethra contracts and semen is forcefully expelled from the penis. Orgasm in women involves rhythmic muscle contractions in the vagina and uterus. Some women experience orgasm rarely or not at all, and others can experience orgasm repeatedly. The absence of orgasm does not affect a woman's ability to become pregnant.

4. **Resolution.** During the resolution stage, the body slowly returns to its normal level of functioning. Resolution usually takes longer in women than it does in men. Many women are capable of becoming aroused again during the resolution phase, but men usually experience an interval during the resolution stage when it is impossible for them to achieve another erection or orgasm.

17.7 Birth Control

Sexual activity carries with it the risk of both unwanted pregnancy and the possibility of contracting a sexually transmitted disease (STD; many STDs are discussed in Chapter 17a). The method couples use (if any) to avoid pregnancy influences the chances of contracting STDs. For this reason, as we discuss various means of contraception, we will also consider whether they reduce the risk of spreading sexually transmitted infections.

Abstinence

Abstinence—not having sexual contact at all—is the most reliable way to avoid both pregnancy and the spread of STDs. However, abstinence is not always used as a means of contraception.

Sterilization

Well before the end of their reproductive life spans, most people have had all the children they want. Other than abstinence, *sterilization* is the most effective way to ensure that pregnancy does not occur. However, unlike abstinence, sterilization offers no protection against STDs.

Sterilization in men usually involves an operation called a *vasectomy*, in which the vas deferens on each side is cut to prevent sperm from leaving the man's body. The procedure usually takes about 20 minutes and can be performed in a physician's office under local anesthesia. The physician makes small openings in the scrotum through which each vas deferens can be pulled and cut; a small segment is removed, and at least one end is sealed shut. Because sperm make up only a small percentage of the semen, the volume of semen the man ejaculates is not noticeably reduced. His interest in sex is not lessened, because testosterone, which is responsible for the sex drive, is still released from the interstitial cells of the testis and carried around the body in the blood.

The most common method of female sterilization, *tubal ligation*, involves blocking the oviducts to prevent the egg and sperm from meeting. Commonly, the tubes are cut and the ends seared shut or mechanically blocked with clips or rings. Tubal ligation is frequently done using a procedure called *laparoscopy*, in which two small incisions are made in the abdominal cavity and the operation is visualized through a telescopic lens inserted through one of them. Laparoscopy is generally performed in a hospital under general anesthesia. Because the abdominal cavity is opened, the risk of infection is greater after tubal ligation than it is after a vasectomy. A woman continues to menstruate after a tubal ligation, because the ovarian hormones continue to be produced in a cyclic manner.

A newer method of sterilization for women does not require an incision. Instead, a small wire coil or a flexible insert is inserted into the oviducts through the vagina, cervix, and uterus. During the next 3 months, scar tissue forms around the coil and blocks the passage of sperm.

Sterilization should be considered permanent even though it is sometimes possible to reverse the procedure. At best, the reversal procedure is expensive and requires that the surgeon have special training in microsurgical techniques. Even then, success is not guaranteed.

Hormonal Contraception

Hormonal contraception is currently available only to females. There are two basic types: the methods that combine estrogen and progesterone and the progesterone-only means of contraception.

Combination estrogen and progesterone contraception

Several forms of birth control combine synthetic forms of estrogen and progesterone. Each of these methods works by mimicking the effects of natural hormones that would ordinarily be produced by the ovaries. Among these effects is the suppression of the release of FSH and LH from the anterior pituitary gland. Without these pituitary hormones, the egg does not mature and is not released from the ovary.

When most people speak about "the pill," they are referring to the combination birth control pill, so named because it contains synthetic forms of both estrogen and progesterone. A hormone-containing "combination" birth control pill is taken daily for 3 weeks, followed by a week of daily pills without hormones.

Over the past few years, additional methods of combined hormonal contraception have been developed. There is a skin patch (Ortho Evra) and a vaginal ring (NuvaRing), each of which slowly releases hormones for the 3 weeks it is in place (Figures 17.10a and 17.10b). When a woman uses the birth control pill, the vaginal ring, or the patch, she menstruates during the week in which hormones are not administered. Women who have difficulty remembering to take a pill each day may prefer these newer options. In addition, there are regimens of hormonal contraception (Seasonale and Seasonique) in which estrogen and progesterone pills are taken for 84 days, followed by a week of placebo pills or low-dose estrogen pills. A woman who is using Seasonale or Seasonique menstruates only four times a year, while taking the placebo pills. Lybrel is a regimen of pills that are taken continuously for one year, so that a woman menstruates only once a year.

For most healthy, nonsmoking women younger than 35 years of age, combination hormonal methods are extremely safe means of contraception. Nonetheless, a small number of hormonal contraceptive users do die each year from a complication, so the possibility of such complications should be recognized. The risk increases with age and with cigarette smoking. Hormonal contraception and cigarette smoking can both increase blood pressure and the risk of abnormal blood clot formation. Together, they have a greater effect than either does alone. Problems with the circulatory system are the most important of the serious (sometimes fatal) complications of hormonal contraception use. Most pill-associated deaths are caused by heart attack or stroke that results when the blood supply to the heart or a region of the brain is blocked. Hormonal contraception also increases the risk of abnormal blood clot formation.

Women who use combination hormonal contraceptives have a greater risk of catching certain STDs from an infected partner than do women who use other forms of birth control or no birth control at all. One way the pill increases the likelihood of transmission of these diseases is by making the vaginal environment more alkaline, a condition that favors the growth of bacteria that cause gonorrhea and chlamydia. Another reason that women on the pill have a greater risk of getting STDs is that it makes the user's cervix more vulnerable to disease-causing organisms; the pill causes delicate cells that line the cervical canal to migrate to the exposed surface of the cervix.

Progesterone-only contraception The most popular form of progesterone-only contraception is an injection (Depo-Provera) given every 3 months. Progesterone injections are an extremely effective means of contraception. There is also a progesterone-only pill (POP; also called the *minipill*). As when using the combined birth control pill, a woman must take the minipill faithfully every day. POPs are generally less effective than the combined pill. A progesterone-releasing intrauterine device (IUD; discussed shortly) is also available. Two types of progesterone-containing implants, Implanon and Nexplanon, are available. Each is a slender, flexible silicone rod about the size of a matchstick. When inserted under the skin in a woman's upper arm, the progesterone-containing implant is barely visible and provides extremely effective protection for 3 years as the progesterone slowly diffuses out of the capsules.

Progesterone-only contraceptives can prevent pregnancy in several ways. All types may prevent ovulation, but they vary in their ability to do so. They all cause thickening of cervical mucus, which hampers the passage of sperm to reach the egg. In addition, because estrogen is not used in this contraceptive, the endometrium is not prepared properly for the implantation of an embryo should fertilization occur. An important drawback is that progesterone-only contraceptives provide no protection against sexually transmitted diseases.

Intrauterine Devices

An *intrauterine device (IUD)* is a small device that is inserted into the uterus by a physician to prevent pregnancy (Figure 17.10c). It can be left in place for several years and should be removed by

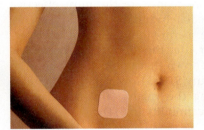
(a) Hormone-containing skin patch

(b) Vaginal contraceptive ring

(c) Intrauterine device (IUD)

(d) Diaphragm and spermicidal cream or jelly

(e) Male latex condom

(f) Female polyurethane condom

FIGURE 17.10 Selected methods of birth control

a physician. An IUD is highly effective in preventing pregnancy. It can interfere with both fertilization and implantation. However, IUDs offer no protection against the spread of sexually transmitted diseases. In the future, there may be a male equivalent of an IUD, called an Intra Vas Device, which is a plug inserted into the vas deferens to block the passage of sperm.

A risk associated with using an IUD is pelvic inflammatory disease, which is a general term for any bacterial infection of the pelvic organs. The risk is greatest in the weeks following IUD insertion, because disease-causing organisms, usually sexually transmitted organisms, can enter the uterus from the vagina at this time. The risk of pelvic infection due to an IUD is low—less than 1%. However, if it does occur, it can lead to sterility or ectopic pregnancy.

Barrier Methods

Barrier methods of contraception include the diaphragm, cervical cap, contraceptive sponge, and male and female condoms. Methods in this category work, as their name suggests, by creating a barrier between the egg and sperm.

A *diaphragm* is a dome-shaped, soft rubber cup containing a flexible ring. It is inserted into the vagina before intercourse so that it covers the cervix, preventing sperm from passing through (Figure 17.10d). A *cervical cap* (FemCap) is smaller than a diaphragm and fits snugly over the cervix. Before the diaphragm or cervical cap is inserted, spermicidal cream or jelly should be added to the inner surface. Because a diaphragm covers the cervix and the top of the vagina, it offers some, though limited, protection to the woman against important STDs, including chlamydia and gonorrhea. By protecting the cervix from the virus that causes genital warts, many types of which cause cervical cancer, a diaphragm also lessens the risk of cervical cancer. A cervical cap, being smaller, offers no protection against STDs.

Unlike a diaphragm or cervical cap, a contraceptive sponge or condom can be purchased without a prescription. A *contraceptive sponge* is a small sponge that contains a sperm-killing chemical. One side of the sponge has an indentation that the cervix fits into. The other side has a strap that makes removal easier. To activate the spermicide, the sponge must be thoroughly moistened before use. It is then inserted and offers protection against pregnancy for the next 24 hours.

The *male condom* is a thin sheath of latex, polyurethane, or natural membranes ("skin") that is rolled onto an erect penis, where it fits like a glove (Figure 17.10e). The sperm are trapped within the condom and cannot enter the vagina. The effectiveness of condoms in preventing pregnancy depends largely on how consistently they are used.

Other than complete abstinence from sexual activity, the latex condom is the best means of preventing the spread of STDs available today. The skin condoms may offer greater sensitivity, but they do allow some microorganisms, particularly viruses, to pass through. Latex has no pores, so microorganisms cannot pass through it. Health advisors highly recommend that people who might be exposed to STDs use a latex condom for disease protection, even if they are already using another means of birth control to prevent pregnancy. Keep in mind, however, that a condom can prevent disease transmission only between the body surfaces it separates. Diseases can still be spread by contact between other, unprotected body surfaces that are vulnerable to infection.

The *female condom* is a loose sac of polyurethane, a clear plastic that resembles the type used in a food-storage bag. At each end of the sac is a flexible ring that helps hold the device in place (Figure 17.10f). The female condom does reduce the risk of spreading sexually transmitted infections. Female condoms offer a barrier against STDs and prevent pregnancy for women who cannot count on their male partner to use condoms.

stop and think

A latex condom provides an impenetrable barrier to disease-causing organisms. Yet, studies indicate that women who use a diaphragm or vaginal sponge routinely have lower rates of STDs than do women who rely on their male partners to use a condom. What might explain this seemingly illogical difference?

Spermicidal Preparations

Spermicidal preparations consist of a sperm-killing chemical (nonoxynol-9) in some form of carrier, such as foam, cream, jelly, film, or tablet. When spermicide is used without any other means of contraception, foams are the most effective option because they act immediately and disperse more evenly than other preparations. The sperm-killing effect of any spermicide lasts for about 1 hour after the product has been activated. Laboratory tests have shown that nonoxynol-9 kills the organisms responsible for many STDs. However, this chemical also damages the cells lining the vagina, and this damage could increase a woman's susceptibility to STDs.

Fertility Awareness

Fertility awareness, which also goes by the names of *natural family planning* and the *rhythm method*, is a way to reduce the risk of pregnancy by avoiding intercourse on all days on which sperm and egg might meet. This sounds easier than it is. Sperm can live in the female reproductive tract for 2 to 5 days, but the lower value is most likely. An egg lives only 12 to 24 hours after ovulation. Consequently, there are only 4 days in each cycle during which fertilization might occur. But which 4 days? Therein lies the problem. It is difficult enough to pinpoint when ovulation occurs, much less predict it several days in advance. The situation is not entirely hopeless, however. There are several ways for a woman to track her menstrual cycle: the calendar, body temperature, and cervical mucus. All of these methods require training to be used effectively.

Emergency Contraception

Emergency contraception, or the so-called morning-after pill, is a means of contraception that can actually be used in the *first few days* after unprotected intercourse. Doctors have prescribed birth control pills and inserted IUDs as emergency contraception for decades, but only recently have two pills specifically marketed for this use become available in the United States. One type (Preven) combines estrogen and progesterone. The second type (Plan B) contains only progesterone. With either type, the first dose should be taken within

3 to 5 days of unprotected intercourse and the second 12 hours later. Scientists do not fully understand the precise mechanisms of action; possible explanations for the pill's efficacy include inhibition or delay of ovulation, prevention of fertilization, thickening of cervical mucus, and alteration of the endometrium, making it an inhospitable place for implantation of the young embryo. Plan B can be purchased over the counter in a pharmacy by women 18 years of age or older.

looking ahead

In this chapter, we have seen how the male and female reproductive systems function to produce offspring. Unfortunately, during sexual intimacy, the vulnerable surfaces of certain parts of the reproductive system come into contact, potentially allowing sexually transmitted organisms to spread. We explore this topic further in Chapter 17a.

HIGHLIGHTING THE CONCEPTS

17.1 Gonads (p. 342)

- The gonads (testes and ovaries) are the reproductive structures that produce the gametes (sperm and eggs), as well as sex hormones. Male testes produce the sex hormone testosterone. Female ovaries produce the sex hormones estrogen and progesterone. A sperm and egg fuse at fertilization, producing a cell called a zygote that under the right circumstances will develop into a new individual.

17.2 Male and Female Reproductive Roles (pp. 342–343)

- A male produces a large number of sperm and delivers them to the female reproductive system. A female usually produces only one nutrient-filled egg during each approximately month-long interval. The female also nourishes and protects the developing baby (the embryo and later the fetus) in her uterus.

17.3 Form and Function of the Male Reproductive System (pp. 343–348)

- The male reproductive system is composed of the testes, a series of ducts (the epididymis, vas deferens, and urethra), accessory glands (the prostate, seminal vesicles, and bulbourethral glands), and the penis.
- The testes are located outside the body cavity in a sac called the scrotum, which helps regulate the temperature of the testes to ensure proper sperm development.
- Within the seminiferous tubules of the testes, the production of gametes, called spermatogenesis, is a continuous process. Sperm are haploid gametes produced by meiosis.
- After leaving the seminiferous tubules, sperm mature and are stored in the epididymis. At ejaculation, sperm travel through the vas deferens and the urethra to leave the body.
- Semen is the fluid released when a man ejaculates. Most of the semen consists of the secretions of the accessory glands.
- The penis transfers sperm to the female during sexual intercourse. Erectile dysfunction is the inability to achieve or maintain an erection. There are several options for treatment.
- Each sperm has three distinct regions. The head of the sperm contains the male's genetic contribution to the next generation. The midpiece is packed with mitochondria, which produce ATP to power movement. The whiplike tail propels the sperm to the egg.
- In males, the hormone testosterone is produced by the interstitial cells, which are located within the testes between the seminiferous tubules.
- Male reproductive processes are regulated by an interplay of hormones from the anterior pituitary gland in the brain (LH and FSH), from the hypothalamus (GnRH), and from the testes (testosterone and inhibin).

17.4 Form and Function of the Female Reproductive System (pp. 348–354)

- The female reproductive system consists of the ovaries, oviducts, uterus, vagina, and external genitalia.
- The ovaries produce eggs and the sex hormones estrogen and progesterone.
- The oviducts transport the immature egg, zygote, and then the early embryo to the uterus. Fertilization usually occurs in the oviduct.
- The uterus supports the growth of the developing baby (embryo and then fetus). The embryo implants in the lining of the uterus, the endometrium, which provides nourishment during development. The muscular wall of the uterus allows the uterus to stretch as the fetus grows, and contractions of the muscle force the baby out of the uterus during childbirth. The cervix is the opening of the uterus.
- The female breasts produce milk to nourish the baby after childbirth.
- The events of the ovarian cycle lead to the release of an egg. A female is born with a finite number of primary follicles. A primary follicle is a primary oocyte (an immature egg) surrounded by a single layer of follicle cells. The primary oocytes will remain in this state until puberty, when (usually) one each month will resume development and be ovulated as a secondary oocyte.
- Hormones from the pituitary gland (FSH and LH) and from the ovary (estrogen and progesterone) regulate the ovarian cycle (which prepares an egg for fertilization) and the uterine, or menstrual, cycle (which prepares the endometrium of the uterus for implantation of the embryo). However, if the egg is fertilized, the young embryo produces human chorionic gonadotropin (HCG), which maintains the corpus luteum. HCG is the hormone that pregnancy tests detect.
- At menopause, which usually occurs between the ages of 45 and 55, menstruation and ovulation stop. As a result, the levels of estrogen and progesterone drop. Menopause has many psychological and physiological effects on a woman's body.

17.5 Disorders of the Female Reproductive System (pp. 354–355)

- Premenstrual syndrome (PMS), a collection of symptoms that occur in some women several days before their period begins, may be caused by low levels of progesterone or of the brain chemical serotonin. Menstrual cramps are caused by prostaglandins, which make muscles contract.
- Endometriosis is a condition in which endometrial tissue is found outside the uterus.

17.6 Stages of the Human Sexual Response (pp. 355–356)

- The human sexual response is the sequence of events that occur during sexual intercourse. This cycle consists of four phases: excitement (increased arousal), plateau (continued arousal), orgasm (climax), and resolution (return to a normal level of functioning).

17.7 Birth Control (pp. 356–359)

- Abstinence (refraining from intercourse) and sterilization (vasectomy or tubal ligation) are the most effective ways to prevent pregnancy. Hormonal contraception (combined estrogen plus progesterone or progesterone-only methods) interferes with the regulation of reproductive processes. An intrauterine device interferes with fertilization and implantation. Barrier methods of contraception (the diaphragm, cervical cap, contraceptive sponge, and male or female condom) prevent the union of sperm and egg. Spermicidal preparations kill sperm. Natural family planning (the rhythm method) consists of abstinence at times when fertilization could occur.
- Morning-after pills are emergency contraception that can reduce the risk of an unwanted pregnancy resulting from unprotected intercourse.

RECOGNIZING KEY TERMS

gonad p. 342
gamete p. 342
testes p. 342
sperm p. 342
testosterone p. 342
ovary p. 342
egg p. 342
estrogen p. 342
progesterone p. 342
fertilization p. 342

scrotum p. 343
seminiferous tubule p. 343
epididymis p. 343
vas deferens p. 343
semen p. 343
prostate gland p. 344
seminal vesicles p. 345
bulbourethral gland p. 345
penis p. 345

acrosome p. 346
gonadotropin-releasing hormone (GnRH) p. 347
luteinizing hormone (LH) p. 347
follicle-stimulating hormone (FSH) p. 347
oviduct p. 348
uterus p. 348
endometrium p. 348

cervix p. 348
vagina p. 348
vulva p. 348
ovarian cycle p. 349
follicle p. 349
ovulation p. 350
corpus luteum p. 350
uterine cycle (menstrual cycle) p. 350
menopause p. 353

REVIEWING THE CONCEPTS

1. Name the male and female gonads. What are the functions of these organs? *p. 342*
2. How is the temperature maintained in the testes? Why is temperature control important? *p. 343*
3. Trace the path of sperm from their site of production to their release from the body, naming each tube the sperm pass through. *p. 343*
4. Name the male accessory glands, and give their functions. *pp. 343–345*
5. What is the function of the penis in reproduction? Describe the process by which the penis becomes erect. *p. 345*
6. Name and describe the functions of the three regions of a sperm cell. *p. 346*
7. List the hormones from the hypothalamus, the anterior pituitary gland, and the testes that are important in the control of sperm production. Explain the interactions between these hormones. *p. 347*
8. What are the two main layers in the wall of the uterus? *p. 348*
9. List the major structures of the female reproductive system, and give their functions. *pp. 348–349*
10. What is an ectopic pregnancy? Why is it dangerous to the mother's health? *p. 348*
11. Describe the structure of female breasts. What is their function? *pp. 348–349*
12. Describe the ovarian cycle. Include in your description primary oocytes, primary follicles, mature Graafian follicles, and the corpus luteum. *pp. 349–350*
13. Describe the interplay of hormones from the anterior pituitary and from the ovaries that is responsible for the menstrual cycle. *pp. 350–353*
14. What is menopause? Why does menopause lower estrogen levels? What are some effects of lowered estrogen levels? *pp. 353–354*
15. What are the stages of the human sexual response? *pp. 355–356*
16. Describe a vasectomy and a tubal ligation. *p. 356*
17. How does the combination birth control pill reduce the chances of pregnancy? *pp. 356–357*
18. What health risks are associated with use of the birth control pill? *p. 357*
19. How do progesterone-only means of contraception reduce the risk of pregnancy? *p. 357*
20. What is an IUD? *pp. 357–358*
21. How do the diaphragm, male condom, and female condom prevent pregnancy? *p. 358*
22. The interstitial cells
    a. are found in the seminal vesicles.
    b. produce a secretion that makes up most of the volume of semen.
    c. secrete testosterone.
    d. store sperm.
23. Choose the *incorrect* statement about semen.
    a. It helps lubricate passageways in the male reproductive system, so sperm travel through them more easily.
    b. Sperm cells make up most of the volume of semen.
    c. It helps reduce the acidity of the female reproductive system, thereby increasing sperm survival.
    d. It contains nourishment for the sperm.
24. After sperm are released in the female reproductive system, they can move to the egg and fertilize it. Which is the correct order of structures through which the *sperm* will pass?
    a. vagina, cervix, body of uterus, oviduct
    b. ovary, cervix, body of uterus, oviduct
    c. endometrium, cervix, oviduct, ovary
    d. ovary, oviduct, endometrium, cervix

25. An injection of LH might
    a. trigger ovulation.
    b. cause menstruation.
    c. cause breast development.
    d. cause a miscarriage.
26. Which of the following structures is *incorrectly* paired with its function?
    a. prostate gland: adds alkaline fluid to sperm
    b. seminiferous tubules: secrete testosterone
    c. epididymis: provides for storage and maturation of sperm
    d. ovary: secretes estrogen and progesterone
27. A zygote
    a. is the cell formed when the sperm fertilizes the egg.
    b. will develop into a gamete.
    c. contains one complete set of chromosomes.
    d. usually forms in the endometrium.
28. Choose the *incorrect* statement about the endometrium.
    a. It is the part of the uterus that is lost each month as part of the menstrual flow.
    b. It is where fertilization usually takes place.
    c. It nourishes the young embryo.
    d. It is where the embryo usually implants.
29. Which of the following is the correct pairing of a structure with its function?
    a. endometrium: the usual site of fertilization
    b. corpus luteum: production of estrogen and progesterone
    c. epididymis: production of testosterone
    d. seminal vesicles: storage of sperm while they mature
30. The controversial pill mifepristone (formerly called RU486) works to terminate a pregnancy by preventing progesterone from acting. This effect would cause an abortion because progesterone is needed to
    a. trigger ovulation.
    b. cause the formation of the corpus luteum.
    c. maintain the endometrium.
    d. increase the levels of LH.
31. Sperm are stored and mature in the _____.
32. Fertilization usually occurs in the _____.

APPLYING THE CONCEPTS

1. You are a health care provider in a family planning clinic. Your first client is a woman who is 22 years old and in excellent health. She is about to marry her high school sweetheart. She and her fiancé have never had sexual intercourse with anyone. They want a very effective means of contraception because they would like to wait until he finishes graduate school to start a family. She does not smoke. Which means of birth control would you recommend for this client? Why?
2. Endometriosis, a condition in which endometrial tissue implants on pelvic organs outside the uterus, causes pain in the pelvic area. Why would the pain be greatest around the time of menstruation?
3. A male friend of yours and his wife would like to have a baby. They are having difficulty conceiving a child. The doctor told him that the reason for this difficulty is a low sperm count. Since further tests are costly and would take time to complete, the doctor told your friend that as a first step, he might try wearing loose clothing and switching to boxer shorts, rather than his customary tight briefs. Why do you think the doctor made this suggestion?

BECOMING INFORMATION LITERATE

Use at least three reliable sources (journals, newspaper, websites) to answer the following questions. Explain why you chose those sources.

1. Three extended-use hormonal contraceptives are available for women: Lybrel (used for one year), Seasonale, and Seasonique (both used for 90 days). How do these contraceptives prevent pregnancy? Are they safe? What are the pros and cons of using them?
2. There is contradictory evidence that environmental estrogens are affecting human reproductive systems. What are some effects that some researchers attribute to environmental estrogen? Discuss the evidence for and against a link between environmental estrogens and at least one of these effects.

MasteringBiology®

Go to MasteringBiology for practice quizzes, activities, eText, videos, current events, and more.

17a SPECIAL TOPIC
Sexually Transmitted Diseases and AIDS

HIV/AIDS is a global pandemic.

- 17a.1 Long-Lasting Effects of STDs and STIs
- 17a.2 STDs Caused by Bacteria
- 17a.3 STDs Caused by Viruses
- 17a.4 HIV/AIDS

In the previous chapter, we discussed the structure and function of the reproductive systems. We also discussed birth control. In that discussion we mentioned the effectiveness of each method of birth control in preventing pregnancy and in preventing the spread of sexually transmitted diseases (STDs). In this chapter, we first describe three STDs that are caused by bacteria: chlamydia, gonorrhea, and syphilis. Then we describe three STDs that are caused by viruses: genital herpes, genital warts, and HIV/AIDS. HIV is discussed in this chapter because sexual contact is the most common means of transmission, although the virus can be transmitted in other ways as well.

17a.1 Long-Lasting Effects of STDs and STIs

Sexually transmitted diseases (STDs) take a serious toll on humanity, both in their direct effects on the infected individuals and in their cost to society in general. Nineteen million infections occur each year in the United States. Two-thirds of the U.S. residents infected are younger than 25. The cost of treating their infections is in the billions of dollars.

STDs are most common among adolescents and young adults (aged 14–26) because they are the most sexually active segment of the population. Indeed, 26% of female adolescents 14 to 19 years old have at least one STD, and 15% have more than one STD. Furthermore, 20.1% of females who have had only one lifetime sexual partner (so far) have contracted at least one STD. And 50% or more of female adolescents with three or more partners has an STD. The most common STD is genital infections with the human papillomavirus (HPV).

One reason STDs are so rampant is that people are often unaware that they have been infected. Many cases of STDs lack symptoms. Also, the symptoms of some STDs, such as syphilis, disappear without treatment, leading the person to believe—mistakenly—that he or she is cured. For these reasons, some authorities now favor using the term *sexually transmitted infection (STI)* rather than *sexually transmitted disease*.

We should also mention STDs are sexist. They affect women more severely than men, because women expose more vulnerable surface area, mucous membranes, than men do and because women are less likely to develop symptoms that would prompt treatment.

17a.2 STDs Caused by Bacteria

We consider three STDs that are caused by bacteria: chlamydia, gonorrhea, and syphilis.

Chlamydia and Gonorrhea

Chlamydia is the most frequently reported infectious disease in the United States. It is caused by a very small bacterium, *Chlamydia trachomatis*, which cannot grow outside a human cell. *Gonorrhea,* one of the oldest known sexually transmitted diseases, is caused by the bacterium *Neisseria gonorrhoeae*.

Chlamydial infections are rapidly becoming epidemic because they are highly contagious and because they do not necessarily cause noticeable symptoms that would prompt the infected person to seek treatment (Figure 17a.1). In fact, most people who have chlamydia do not have symptoms, and many people learn that they have it only because a responsible partner diagnosed with the infection informs them. Any symptoms that do develop may take weeks or months to appear. In the meantime, even without outward signs of infection, the disease can be passed along to others.

Chlamydia and gonorrhea primarily infect the mucous membranes of the genital or urinary tract, throat, or anus. The bacteria that cause these STDs cannot survive long if exposed to air, so they are generally transferred by direct contact between an infected mucous membrane and an uninfected one during sexual intimacy. The bacteria can infect the cells of mucous membranes, including those of the urethra, vagina, cervix, oviducts, throat, anus, and eyes. The most common sites of infection are the mucous membranes of the urethra or the reproductive structures, because these are most likely to be in contact during sexual intimacy. Chlamydia and gonorrhea can also spread to the lining of the anus during anal intercourse or to the throat during oral sex. Furthermore, they can infect the eyes if an eye is touched by a finger that has just touched an infected area. In the case of infants, the eyes can become infected as the baby passes through the vagina or cervix during the birth process.

The most common symptoms of chlamydia or gonorrhea are those of urinary tract infection. Inflammation of the urethra causes a burning sensation during urination, itching or burning around the opening of the urethra, and a discharge from the urethra (Figure 17a.2). In males, the most common site of chlamydial infection is the urethra, because this is the mucous membrane most likely to be exposed to the bacteria during sexual intimacy.

In women, chlamydia and gonorrhea are more likely to affect the pelvic organs than the urethra, because the vagina and cervix are a woman's primary sites of contact during sexual intimacy. The general term for an infection of the pelvic organs is *pelvic inflammatory disease (PID)*. The symptoms of PID—abdominal pain or tenderness, lower-back pain, pain during intercourse, abnormal vaginal bleeding or discharge, fever, or chills—can come on gradually or suddenly and vary in intensity from mild to severe.

Pelvic inflammatory disease can be caused by a number of organisms, but the most common culprits are the sexually transmitted bacteria that cause chlamydia and gonorrhea. The bacteria ascend from the vagina through the cervix to the uterus. There, they may infect the endometrium and then spread to the wall of the uterus and to the oviducts. Near the ovary, the oviduct opens to the abdominal cavity, allowing the infection to spread into the abdominal cavity. The infection can be curbed by treatment with antibiotics, which kill the bacteria. Unfortunately, pelvic infections are less likely to produce symptoms than are urinary tract infections, and the untreated infection may permanently damage a female's reproductive system.

Because a man is more likely to experience noticeable symptoms of chlamydia and gonorrhea, men are more likely to seek diagnosis and treatment before the infections have caused lasting damage. However, undiagnosed chlamydia and gonorrhea can have long-term reproductive consequences for both men and women, so it is important that a man who is diagnosed with chlamydia or gonorrhea inform everyone he has been sexually intimate with. If left untreated, chlamydia and

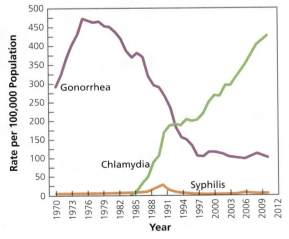

FIGURE 17a.1 The relative rates of chlamydia, gonorrhea, and syphilis from 1984 to 2010. The bacteria that cause these STDs are shown at the right. Centers for Disease Control and Prevention

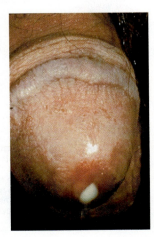

FIGURE 17a.2 A yellowish white discharge from the urethra caused by gonorrhea

gonorrhea can cause scar tissue to form in the tubes through which gametes travel. If the vas deferens or oviducts are completely blocked by scar tissue, the man or woman becomes sterile. Women are more likely to become sterile than are men, because women are less likely to have symptoms that prompt treatment. In a woman, partial blockage of the oviduct may allow the tiny sperm to reach the egg and fertilize it, but the blockage will not allow the much larger embryo to move past the scar tissue to reach the uterus (Figure 17a.3). The embryo may then implant in the oviduct, resulting in an ectopic pregnancy that places the woman's life in danger.

Chlamydia and gonorrhea are of special concern to pregnant women. Newborns can be infected with either chlamydia or gonorrhea as they pass through an infected cervix or vagina at birth. Chlamydial or gonorrheal infections picked up during birth can affect the mucous membrane of the eye (the conjunctiva), the throat, vagina, rectum, or lungs. These infections are sometimes fatal to an infant. Chlamydial infections during pregnancy place the fetus at risk as well. Specifically, chlamydia can cause the protective membranes around the fetus to rupture early, which can kill the fetus.

Chlamydia and gonorrhea can now be diagnosed in a physician's office with quick, accurate tests. Urine tests detect the DNA of *Chlamydia* or the gonococcal bacterium. Gonorrhea can also be diagnosed by observing the bacterium in a smear of cells taken from the infected area or grown in a laboratory. Both chlamydia and gonorrhea can be cured with antibiotics. Some strains of the gonococcal bacteria, however, are now antibiotic resistant. Therefore, every patient undergoing treatment for gonorrhea must be retested for the bacterium when the antibiotic regimen has been completed to make sure the infection is cured.

FIGURE 17a.3 Chlamydia and gonorrhea can cause pelvic inflammatory disease, which can cause scar tissue to form in the oviducts.

Syphilis

Syphilis rates decreased steadily during the 1990s, reaching their lowest point in 2000. Since then, the frequency of syphilis has risen slightly, primarily among men who have sex with men.

Syphilis is caused by a corkscrew-shaped bacterium (a spirochete) called *Treponema pallidum*. As terrible as these bacteria are, they are extremely delicate and cannot survive drying or even minor temperature changes. A person can contract syphilis only by direct contact with an infected sexual partner. The bacteria can invade any mucous membrane or enter through a break in the skin. If a pregnant woman is infected, the bacteria can also cross the placenta and infect the growing fetus.

If untreated, syphilis progresses through three stages. The first stage is characterized by a painless bump, called a *chancre* (shang'-ker), that forms at the site of contact, usually (but not always) the genitals. The chancre normally appears within 2 to 8 weeks of the initial contact (Figure 17a.4) as a hard, reddish brown bump with raised edges that make it resemble a crater. Unfortunately, a chancre is not always noticed, because it can

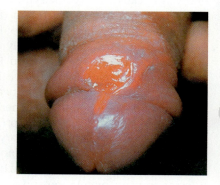

Step 1: The first stage of syphilis is characterized by a chancre, a hard, painless, crater-shaped bump at the place in the body where the bacteria entered, usually the genitals.

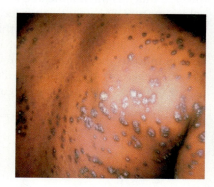

Step 2: A reddish-brown rash covers the entire body, including the palms of the hands and the soles of the feet.

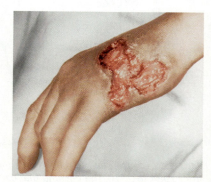

Step 3: Lesions, called gummas, shown here on the hand, are characteristic of the third stage of syphilis. These lesions can also form on the blood vessels, the central nervous system, and the bones.

FIGURE 17a.4 Untreated syphilis goes through three stages.

form in places that are difficult to see and because it may be mistaken for something else. The chancre lasts for one to a few weeks. During that time it ulcerates, becomes crusty, and disappears. The disease, however, has not disappeared.

During the primary stage, syphilis is diagnosed by identifying the bacterium in the discharge from a chancre. At this stage, syphilis can be treated and cured with an antibiotic, such as penicillin.

If syphilis is not detected and treated in the first stage, it can progress to the second stage, which is characterized by a reddish brown rash covering the entire body, including the palms of the hands and soles of the feet. The rash usually appears within a few weeks to a few months after the disappearance of the chancre. The rash does not hurt or itch. Each bump eventually breaks open, oozes fluid, and becomes crusty. The ooze contains millions of bacteria. Therefore, the secondary stage is the most contagious stage of syphilis.

Syphilis in this stage is diagnosed by a blood test to detect antibodies for *Treponema* bacteria. Second-stage syphilis *can* be cured with antibiotics such as penicillin, but the disease becomes increasingly difficult to treat as it progresses. Again, the symptoms go away whether or not they are treated. However, in some people, the symptoms recur periodically. Other people may have no outward signs of syphilis for years.

The third stage brings the drama to a grisly conclusion. At this stage, lesions called *gummas* may appear on the skin or certain internal organs. Gummas often form on the aorta, the major artery that delivers blood from the heart to the rest of the body. As a result, the artery wall may be weakened and may burst, causing the person to bleed to death internally. The bacteria that cause syphilis can also infect the nervous system, damaging the brain and spinal cord, so that the person may have difficulty walking, become paralyzed, or become insane. The disease may also cause blindness by affecting the optic nerve, the iris, and other parts of the eye.

In its third stage, syphilis is very difficult to treat. Treatment requires massive doses of antibiotics over a prolonged period. The damage that has already been done to the body cannot be repaired.

Characteristics of bacterial STDs (their symptoms, diagnosis, treatment, and effects) are summarized in Table 17a.1.

TABLE 17a.1 Overview of Bacterial STDs

Disease	Symptoms	Diagnosis and Treatment	Effects
Chlamydia	First symptoms occur 7–21 days after contact Up to 75% of women and 50% of men show no symptoms Women: Vaginal discharge Vaginal bleeding between periods Pain during urination and intercourse Abdominal pain accompanied by fever and nausea Men: Urethral discharge Pain during urination	Diagnosis: Urine test for chlamydial DNA Treatment: Antibiotics	Long-term reproductive consequences, such as sterility Infection can pass to infant during childbirth Can cause rupture of the protective membrane surrounding the fetus
Gonorrhea	First symptoms occur 2–21 days after contact About 30%–40% of men and women show no symptoms Women: Vaginal discharge Pain during urination and bowel movement Cramps and pain in lower abdomen More pain than usual during menstruation Men: Thick yellow or white discharge from penis Inflammation of the urethra Pain during urination and bowel movements	Diagnosis: Examination of penile discharge or cervical secretions Urine test for DNA of the bacterium that causes gonorrhea Cell culture Treatment: Antibiotics	Can cause long-term reproductive consequences, such as sterility Infection can pass to infant during childbirth Can cause heart trouble, arthritis, and blindness
Syphilis	Stage 1: Occurs 2–8 weeks after contact Chancre forms at site of contact Lymph nodes in groin area swell Stage 2: Occurs 6 weeks to 6 months after contact Reddish brown rash appears anywhere on the body Flu-like symptoms present Ulcers or warty growths may appear Patches of hair may be lost Stage 3: Lesions appear on skin and internal organs May affect nervous system Blindness Brain damage	Diagnosis: Identification of the bacterium from a chancre Blood test to detect antibodies to the bacterium that causes syphilis Treatment: Large doses of antibiotics over a prolonged period of time	Infection can pass to fetus during pregnancy Can cause heart disease, brain damage, blindness, and death

17a.3 STDs Caused by Viruses

Unlike STDs caused by bacteria, viral STDs cannot be cured with antibiotics. One can treat the symptoms, but one can never be certain that the virus has been eliminated. Therefore, it is always important to take precautions not to pass the virus on to others.

Genital Herpes

Genital herpes is caused by *herpes simplex viruses (HSVs)*. There are actually two types of HSV. They cause similar sores but tend to be active in different parts of the body. Type 1 (HSV-1) is most commonly found above the waist, where it causes fever blisters or cold sores. Type 2 (HSV-2) is more likely to be found below the waist, where it causes genital herpes on the genitals, buttocks, or thighs. As a result of oral–genital sex with an infected person, however, HSV-1 can cause sores on the genitals, and HSV-2 can cause cold sores (Figure 17a.5).

Herpes simplex viruses are quite contagious and can be spread by direct contact with viruses that are being shed from an infected surface or that are in the fluid on the blisters. Mucous membranes are most susceptible. Skin is a good barrier, unless there is a cut, abrasion, burn, acne, eczema, or other break.

The first hints of genital herpes infection begin about 2 to 20 days (an average of 6 days) after the initial contact. The initial bout may be severe, often accompanied by fever, aching muscles, and swollen glands in the groin. Soon blisters appear, accompanied by local swelling, itching, and possibly burning, especially if the blisters get wet during urination. The blisters form at the site of contact and last about 2 days before they ulcerate, leaving small, painful sores. At this point individuals infected with genital herpes are at an increased risk of acquiring an HIV infection as well, if they are exposed to the HIV virus.

During dormant periods, the virus retreats to ganglia (clusters of nerve cells) near the spinal cord. Then, at times of emotional or physical stress, the virus may be reactivated and blisters may re-form.

Most people with herpes apparently never develop symptoms, or the symptoms are so mild that they go unnoticed. Nonetheless, these asymptomatic individuals can unwittingly transmit the virus. In fact, *most* cases of genital herpes are transmitted by partners who never had symptoms or whose symptoms were atypical and undiagnosed! If they had known, they would likely have avoided unprotected sexual contact.

Genital herpes is most contagious when active sores are present, so sexual contact should be avoided during an outbreak. Unprotected contact should be avoided at all times, because genital herpes can also be spread when there are no symptoms.

Most women with herpes have successful pregnancies and normal deliveries. In rare cases, however, the infection spreads to the fetus as it is growing in the uterus and can cause miscarriage or stillbirth. HSV can also be transmitted to the newborn during vaginal delivery, especially if active sores are present. To avoid exposing the baby to HSV, physicians often recommend delivery by cesarean section (surgical removal from the uterus) if the mother has active sores or is shedding the virus.

Clinicians diagnose herpes by examining the sores or by testing the fluid from sores for the presence of the virus. It is also possible to identify the DNA of the herpes virus in the material wiped off when the genital area is gently swabbed. Blood tests can also detect antibodies to the herpes viruses. However, such blood tests may not be useful during an initial attack, because the antibodies may not show up for several months.

Although there is no cure for HSV, there are three antiviral drugs—Zovirax, Famvir, and Valtrex—that help ease symptoms during the initial outbreak and reduce the frequency of recurrences. Unfortunately, strains of HSV that are resistant to antiviral drugs are beginning to crop up.

HPV and Genital Warts

Genital warts may be caused by any of several *human papillomaviruses (HPVs)*. These are not the same viruses that cause warts on the hands and feet. Although chlamydia holds the title of being the most common of all STDs in the United States, HPV is the most common of the viral STDs. Roughly 50% of sexually active people will acquire an infection with HPV at some point in their lives. The body's defense mechanisms usually eliminate HPV without the virus causing serious health problems.

Most people with genital HPV are not aware of the infection, because they don't have any symptoms. However, HPV can be spread by genital contact with an infected person, even when there are no symptoms. Recall from Chapter 13a that viruses replicate in a host cell, and new viruses can be released from the cell slowly, without causing symptoms. The newly released viruses can cause genital HPV on uninfected areas of the same person or on another person if genital contact is made.

There are many strains of HPV. Genital infections with some strains of HPV are considered low-risk infections; these cause genital warts. The warts can be flat or raised and can occur singly or in groups. Without treatment, they often grow in size. Although warts can form in a visible location (Figure 17a.6), they often form in locations where they are not likely to be discovered—the vagina, cervix, or anus. Genital warts are usually

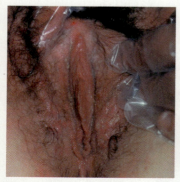 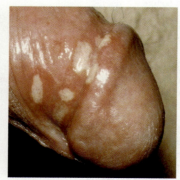

(a) Genital herpes, shown here on the external genitalia of a female, is usually caused by HSV-2.

(b) Genital herpes, shown here on the penis, is usually caused by HSV-2.

FIGURE 17a.5 *Herpes simplex viruses cause blisters to form at the site of infection.*

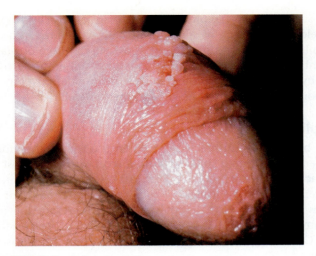

FIGURE 17a.6 *Genital warts on the penis*

diagnosed on the basis of their appearance. However, flat warts are not usually visible unless they are painted with a vinegar-like solution. In women, a Pap test, which looks for precancerous cells on the cervix, is also helpful. A Pap test is a painless test in which the cervix is gently swabbed to collect cells that are then examined for abnormalities under a microscope.

Treatments for genital warts are intended to kill the cells that contain the virus. Methods for removing genital warts include (1) freezing (cold cautery), (2) burning with an electrical instrument (hot cautery), (3) laser (high-intensity light), (4) surgery, and (5) podophyllin (a chemical that is painted onto the warts and washed off after the prescribed time period, before it burns the skin). Also available are creams that the patient applies to shrink genital warts. The creams work by boosting the body's defense mechanisms against the virus. These treatments may destroy visible warts, but HPV may remain nearby in normal-looking tissue and can cause new warts to form weeks or months after old ones have been destroyed.

Other strains of HPV cause high-risk genital infections, which can persist for long periods of time and are closely linked to both cervical cancer in women and penile cancer in men. HPV is thought to be responsible for one-third of all cases of penile cancers in the United States, and it can be isolated in 90% of women with cervical cancer. Therefore, women should have Pap tests regularly, and those who have ever had HPV should have a Pap test at least once a year. HPV can also cause oral cancer if spread by oral sex and anal cancer if spread by anal intercourse.

There are now vaccines, Gardasil and Cervarix, that are effective against several types of HPV, including two strains that are responsible for most cases of cervical cancer. Health advisors recommend that girls be vaccinated when they are 10 to 12 years old. Health advisors also recommend that boys aged 11 to 12 years be vaccinated with Gardasil to protect against genital warts and HPV-related cancers.

The characteristics of genital herpes and genital HPV, both of which are caused by viruses, are summarized in Table 17a.2.

 what would you do?

Do you think that you should be permitted to file a lawsuit against someone for transmitting an STD? Why or why not? If you were a juror in such a lawsuit, what factors would influence your decision?

17a.4 HIV/AIDS

In the past several decades, *acquired immune deficiency syndrome,* or *AIDS,* has left its mark on many aspects of society, including medicine, science, law, economics, and education. The name of the condition is apt, because the symptoms are those associated with a damaged immune system. Unlike many other immune deficiencies, however, AIDS is not inherited but acquired. A syndrome is a set of symptoms that tend to occur together, and people with AIDS experience a devastating set of symptoms.

We now know that a virus, the *human immunodeficiency virus (HIV),* causes AIDS. However, as we will learn shortly, AIDS is actually the final stage in an HIV infection during which the immune system is slowly weakened. A primary target of HIV is helper T cells, which serve as the main switch for the entire immune response (see Chapter 13). Once

TABLE 17a.2 Overview of Viral STDs

Disease	Symptoms	Diagnosis and Treatment	Effects
Genital herpes	First symptoms appear 2–20 days after contact Many people have no symptoms Flu-like symptoms present Small, painful blisters that can leave painful ulcers appear Blisters go away, but the virus remains Symptoms recur periodically	*Diagnosis:* Examination of blisters Laboratory test on the fluid from the sore to detect the presence of the virus Blood test for antibodies *Treatment:* Antiviral drugs can ease symptoms	Cannot be cured Recurrences of blisters Infection can pass to fetus, causing miscarriage or stillbirth Can cause brain damage in newborns
HPV genital infection	First symptoms appear 1–6 months after exposure; presence of symptoms depends on type of HPV Small warts appear on sex organs May cause itching, burning, irritation, discharge, bleeding	*Diagnosis:* Appearance of growth In women, Pap test may help *Treatment:* For removal: freezing, burning, laser surgery	Formation of additional warts Closely associated with cervical cancer and penile cancer Infection can pass to infant during childbirth

HIV enters a helper T cell, the T cell stops functioning well, although this cessation of function is not immediately apparent. HIV will eventually kill the infected helper T cell. The infection and eventual death of helper T cells cripple the immune system, giving disease-causing organisms that routinely surround us the opportunity to cause infection. These are the opportunistic infections that characterize AIDS and, in time, cause death.

Global Pandemic

According to the World Health Organization (WHO), toward the end of 2010, an estimated 34 million people globally were living with an HIV infection. Africa is currently the area hardest hit by HIV: two-thirds of the people living with HIV are in sub-Saharan Africa. There are now more people living with HIV than ever before, but because of antiretroviral therapy, the number of new HIV infections has dropped since 2001. Unfortunately, because of insufficient global funding, the United Nations' goal to stop and reverse HIV infections by 2015 may not be reached.

Form of HIV

The genetic material in HIV is RNA. The RNA, along with several virus-specified enzymes, is encased in a protein coat. The RNA and protein coat constitute the core of the virus (Figure 17a.7).

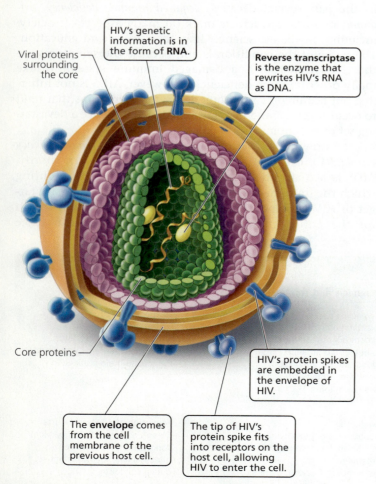

FIGURE 17a.7 *The structure of HIV (human immunodeficiency virus)*

Surrounding the protein coat is an outer covering, or envelope, consisting of spikes of protein embedded in a lipid membrane. The lipid membrane is actually a piece of plasma membrane stolen from the previous host cell and altered for use by the virus. The protein spikes of the viral envelope are responsible for binding the virus to the host cell.

Replication of HIV

HIV binds to an uninfected cell when the spherical region at the end of a protein spike fits into a receptor on the host cell surface—like a key in a lock. The host cell receptor is a surface protein called *CD4*. Helper T cells are the predominant cell types with these CD4 receptors and are therefore the most common targets of HIV. When an HIV spike is properly docked in a CD4 receptor, the contents of the virus enter the host cell (Figure 17a.8).

Once within the host cell, the RNA of HIV undergoes a process called *reverse transcription*. During this process, the viral RNA is rewritten as double-stranded DNA. Because going from RNA to DNA is the reverse of the usual genetic information transfer, viruses that work this way are known as *retroviruses* (*retro–*, backward). The backward copying of genetic information from RNA into DNA is performed by an enzyme called *reverse transcriptase*, which is inserted into the host cell along with viral RNA.

The newly formed viral DNA, which contains all the instructions necessary for producing thousands of new viruses, is then spliced into the host DNA. Once HIV's DNA is integrated into the host cell chromosome, that DNA is called the HIV *provirus*. After they have been incorporated into the host cell DNA, the host cell treats the HIV genes as it would its own. Each time the cell reproduces, the viral DNA is copied along with the cell's DNA. When HIV is transmitted to another person, it is usually in the form of a latent—that is, inactive—HIV provirus when the infected cell enters the other person's body.

The HIV genes can reside in a helper T cell chromosome for years, until the cell is activated to respond to some foreign antigen, such as another kind of virus, a fungus, or a parasite. At this point, the virus begins making copies of itself instead of allowing the cell to fight the invader. The viral genome is activated and turns the cell into a virus factory. Some of the newly produced HIV RNA will become genetic material for new viruses, and some will be used to produce virus-specified proteins. The viral components gather at the cell membrane and bud off from the host cell. The HIV RNA and proteins then self-assemble into new, mature viruses, which move through the bloodstream to infect new cells.

Transmission of HIV

HIV is found in bodily fluids—blood, semen, vaginal secretions, breast milk, saliva, tears, urine, cerebrospinal fluid, and amniotic fluid. However, HIV is not easily transmitted. Only the first four of the bodily fluids listed contain HIV concentrations high enough to cause infection in another person.

HIV *cannot* be transmitted by casual contact. The major modes of transmission are as follows:

- **Unprotected sexual activity.** HIV can be transmitted through unprotected sexual activity of any kind—that is,

Q Why is the insertion of HIV DNA into the host cell chromosome so important to HIV's replication?

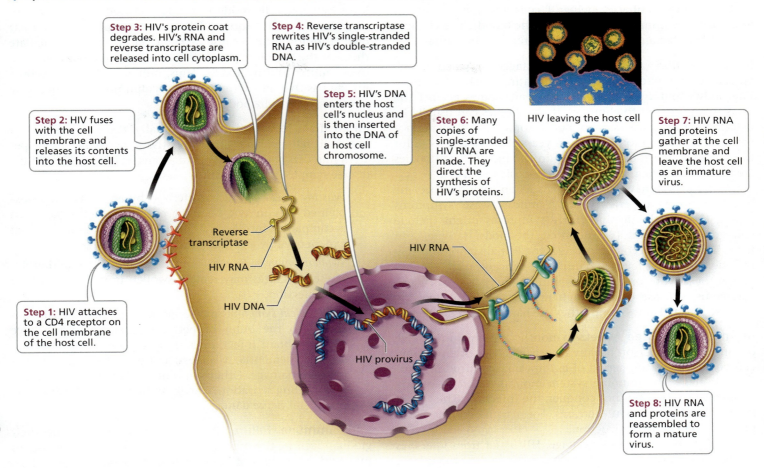

FIGURE 17a.8 The life cycle of HIV

▲ If HIV DNA is not inserted into the host cell chromosome, it cannot produce HIV RNA or HIV-specified proteins.

activity in which broken skin or mucous membranes of the vagina, vulva, penis, mouth, or anus of the partners are not separated by a barrier (i.e., a latex condom) that the virus cannot penetrate.

- **Intravenous drug use.** Contact with infected blood is another way that HIV can be transmitted. Since 1985, all blood donated to blood banks is tested for HIV; this testing has essentially eliminated the risk of getting HIV from a blood transfusion. However, intravenous drug users who share needles *do* have an increased risk of HIV infection.
- **Infected mother to offspring before, during, and after birth.** HIV can be transmitted from an infected mother to her offspring across the placenta during prenatal development, during delivery, or in breast milk during nursing. An HIV-infected woman who is not treated for HIV during pregnancy and does not breast-feed her infant has about a 25% chance of spreading HIV to her newborn. However, if the mother is treated with drugs that slow replication of HIV (antiretroviral drugs), the chance of transmitting the virus to her newborn is less than 2%.

stop and think

Hospitals do not isolate patients with HIV unless they have another infectious disease. Why is this a safe practice?

Sites of HIV Infection

HIV can infect any cell that has a CD4 receptor. The most important of these cells is the helper T cell, which turns on the entire immune response (see Chapter 13). The hallmark of an HIV infection is a progressive decline in helper T cell numbers. This loss is devastating because it leaves the body increasingly defenseless against other infections.

HIV can also infect the brain, killing nerve cells. When the brain is infected, the symptoms can include forgetfulness, impaired speech, inability to concentrate, depression, seizures, and personality changes. Roughly 60% of people with AIDS have signs of dementia.

Stages of HIV Infection

An HIV infection usually progresses through a series of stages: the initial infection, an asymptomatic stage, initial disease symptoms, early immune failure, and AIDS. The progress of an

HIV infection is usually monitored by following both the decline in the number of helper T cells and the increase in viral load (the number of HIV free in the blood). The changes in T cell number and viral load are summarized in Figure 17a.9.

Initial infection During the initial stages of infection, the virus actively replicates, and the circulating level of HIV rises. The body's immune system produces antibodies against the virus in an attempt to eliminate it. Antibodies can usually be detected within 8 weeks or so, but in some cases they are not detected for many months or even years. An HIV test looks for the presence of antibodies to HIV in the blood. If antibodies are found, the person is said to be HIV-positive (HIV+).

Many people have no symptoms when they first become infected; others experience some mild disease symptoms during the initial infection. Early symptoms might include enlarged lymph nodes throughout the body, fatigue, and fever. If the brain becomes infected, the initial symptoms may include headaches, fever, and difficulty concentrating, remembering, or solving problems. This initial stage is particularly dangerous, because even though many people are unaware they are infected at this point, they are particularly contagious.

Asymptomatic stage Weeks to months after the initial stage of infection, the person usually feels well again, often for several years. This is the period of asymptomatic infection. During this stage, the immune system mounts a defense strong enough to control, but not conquer, the infection. The rate of helper T cell production increases. For a while, helper T cell production matches destruction.

During the asymptomatic stage, HIV is far from idle. It is "hiding out" in certain lymphatic organs, including the spleen, tonsils, and adenoids, but most significantly in the lymph nodes, where it replicates. Consequently, the spaces between cells in the lymph nodes become packed with virus particles. Most helper T cells reside in the lymph nodes, and many of them have the DNA form of HIV genes incorporated into their own chromosomes. Because helper T cells frequently circulate through the lymph nodes, this is an opportune place for HIV to encounter healthy T cells. Thus, millions of T cells become infected as they move through the lymph nodes on their way to other parts of the body.

Initial disease symptoms Eventually, the immune system begins to falter as the virus gains the upper hand. As helper T cell numbers gradually drop, symptoms set in again. The initial disease symptoms fall into three classes.

1. *Wasting syndrome* is characterized by an otherwise unexplained loss in body weight of more than 10%, often accompanied by diarrhea. The weight loss is similar to that experienced by a cancer patient.

2. *Swollen lymph nodes* occur in the neck, armpits, and groin. These structures may remain swollen until later stages of HIV infection.

3. *Neurological symptoms* may appear, either because HIV has spread to the brain or because other organisms have caused a brain infection. Among the neurological symptoms are dementia, weakness, and paralysis caused by spinal cord damage. Other symptoms include pain, burning, or a tingling sensation, usually in hands or feet, caused by peripheral nerve damage.

Early immune failure As T cell numbers continue their gradual decline, the body becomes increasingly vulnerable to infection. Early signs of immune failure include thrush (a fungal infection in the mouth) and shingles (a painful rash caused by the same virus that causes chickenpox).

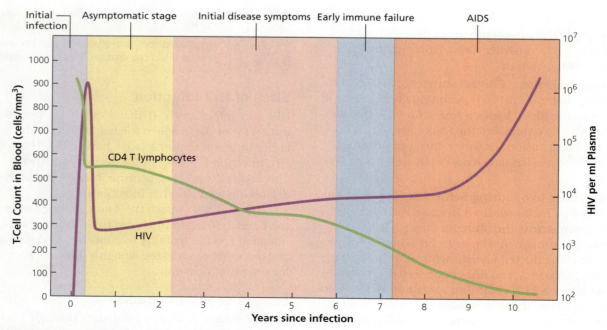

FIGURE 17a.9 An HIV infection can be monitored by following the decline in the number of T cells and the rise in viral load (number of viruses in the bloodstream).

AIDS AIDS is the final stage of an HIV infection. The time from HIV infection to diagnosis of AIDS can be 10 or more years, depending on the number of other infections the person is exposed to and the health of the immune system before infection with HIV. Now, although the body has fought long and hard, it is beginning to lose the battle. The immune system is crippled, making the person vulnerable to certain characteristic fungal, protozoan, bacterial, and viral infections, as well as cancers.

A diagnosis of AIDS is made when an HIV-positive person develops one of the following conditions: (1) a helper T cell count below 200 cells per mm^3 of blood; (2) one of 26 opportunistic infections, the most common of which are *Pneumocystis jiroveci* pneumonia and Kaposi's sarcoma, a cancer of connective tissue that primarily affects the skin; (3) a greater than 10% loss of body weight (wasting syndrome); or (4) dementia.

Treatments

Death in a person with AIDS is usually caused by one of the opportunistic infections. Thus, treating opportunistic infections, or preventing their onset, can improve the quality of life and extend the length of life for people with AIDS. Treatment for HIV infection is also aimed at slowing the progress of the infection, both before and after the diagnosis of AIDS. The current prevailing strategy is to slow the rate at which HIV can make new copies of itself. At present, there are five classes of drugs that slow viral replication: reverse transcriptase inhibitors, protease inhibitors, fusion inhibitors, HIV entry inhibitors, and inhibitors of HIV DNA integration into the host cell chromosome. Each class of drug prevents HIV from replicating by blocking a different step in the virus replication cycle.

The original class of antiviral drugs used to slow the progression of an HIV infection consisted of drugs that act on reverse transcriptase (AZT, ddI, ddC). As discussed earlier, reverse transcriptase is the enzyme necessary for converting the RNA of HIV to DNA that can then be inserted into the DNA of the host cell chromosome.

A second class of antiviral drugs consists of protease inhibitors. Recall that after the HIV genetic information has been inserted into the DNA of the host cell, it can begin producing proteins needed to form new copies of HIV. The proteins initially produced are too big to be used and must be cut to the proper size by an enzyme called a *protease*. Protease inhibitors block the action of this enzyme, preventing a necessary step in the preparation of proteins for the assembly of new viruses.

There are newer classes of antiviral drugs. Fusion inhibitors block the proteins on the surface of HIV that bind to the CD4 receptor on host cells, thereby blocking HIV from entering the host cell. Integrase inhibitors prevent HIV DNA from inserting itself into the host chromosome.

A treatment called *highly active antiretroviral therapy (HAART)*, also known as a *drug cocktail*, consists of combinations of drugs from at least two classes of antiretroviral drugs. The idea behind this drug combination is that a combination of drugs will minimize the development of drug resistance. Combination drug treatments are prolonging the lives of HIV-positive individuals. HAART can reduce the viral load to undetectable levels, but the treatment is not a cure. If treatment is stopped, HIV can leave the latent state and begin actively replicating again.

stop and think

The drugs used to combat HIV effectively stop or slow HIV replication, but they are not effective against HIV when it is incorporated into the host cell chromosome. How does this explain why the treatment must be continued for life?

Several changes in HIV drug treatment are prolonging lives. Treatment regimens may be as simple as one pill a day and newer antiretroviral drugs have fewer side effects, which increase the likelihood that a person will adhere to the treatment regimen. Treatment can be started earlier—before symptoms appear—to reduce initial damage to the immune system. These changes in treatment protocol help reduce the viral load to undetectable levels, reducing (but not eliminating) the possibility of sexual transmission.

A variety of other approaches to combating an HIV infection are under active investigation. Clearly, though, the best way to fight the devastation caused by HIV is to prevent it from establishing the initial infection. Currently, the best way to do this is to avoid high-risk behaviors such as unprotected sex or intravenous drug use. Theoretically, prevention of HIV infection could also be accomplished by developing a vaccine similar to those for other viral infections such as smallpox and measles. Recall from Chapter 13 that vaccines work by causing a person's body to produce antibodies against the organisms that cause a disease, without actually causing the disease. There are many vaccines in clinical trials, but HIV has several characteristics that so far have thwarted efforts to develop a vaccine:

- The mutation rate in HIV is very high, especially for the proteins in its outer coat, which are the proteins that an antibody would have to recognize.
- There are two types of HIV and many strains of each type. Antibodies would have to recognize the exact strain entering the body.
- An HIV infection is latent when its genetic material has become incorporated into the host DNA (as a provirus). Antibodies cannot attack the virus while it is hidden within the host cell.
- HIV can pass directly from the interior of an infected cell to the interior of an adjacent uninfected cell. Antibodies can attack viruses outside the cell only; they cannot prevent the virus from spreading by this means of transmission.

looking ahead

In Chapter 17 we learned about reproductive systems and in this chapter we considered some of the organisms that have the potential to spread through sexual intimacy. In the next chapter we will consider the events of human development and aging from fertilization to old age.

18 Development throughout Life

- Each of us spent about 30 minutes as a single cell at the start of life.
- Eyelids develop by about 10 weeks' gestation, close a few weeks later, and don't reopen until the 28th week.

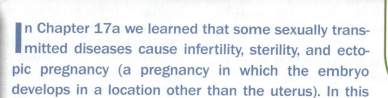

18.1 Periods of Development in Human Life

18.2 Prenatal Period

18.3 Birth

18.4 Birth Defects

18.5 Milk Production by Mammary Glands

18.6 Postnatal Period

ETHICAL ISSUE

Making Babies

HEALTH ISSUE

Disparities in Health and Health Care at All Life Stages

In Chapter 17a we learned that some sexually transmitted diseases cause infertility, sterility, and ectopic pregnancy (a pregnancy in which the embryo develops in a location other than the uterus). In this chapter, we describe the major milestones of human development, from the beginning of pregnancy (fertilization) to the transition into old age. We also consider treatments for infertility and sterility.

Human development begins with fertilization and continues until death. This couple is viewing ultrasound images of their fetus, who is a few months into a life that may last 70 to 80 years.

18.1 Periods of Development in Human Life

A newborn enters the world outside his mother's uterus and rests (Figure 18.1). How did this tiny human develop? Consider that he—like you—began as a fertilized egg, no bigger than the period at the end of this sentence. The miracle of human development begins with **fertilization**, the union of a sperm and an egg, and the early stages take place within the female reproductive system (see Chapter 17). Birth occurs about 266 days after fertilization, marking the transition to development outside the mother's body.

The period of development before birth is the prenatal period. The period after birth is the postnatal period. Relative to prenatal development, postnatal development is a lengthy process, taking from 20 to 25 years for the person to reach adulthood. Bodily changes do not cease once adulthood is reached, however. Our bodies continue to change throughout our lives as we age.

18.2 Prenatal Period

We start our discussion of human development at the moment it all begins—when sperm meets egg. In this section, which presents the major events of prenatal development, we consider what

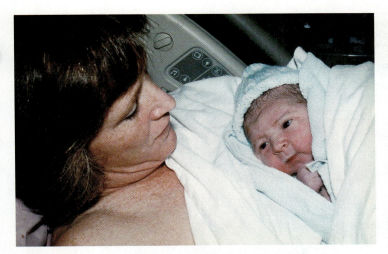

FIGURE 18.1 *A newborn and his mother. Birth is the transition from prenatal to postnatal development.*

TABLE 18.1	Review of Major Events during Prenatal Development
Period	**Major Events**
Pre-embryonic period (fertilization to week 2)	Fertilization Cleavage Formation and implantation of the blastocyst Beginning of formation of extraembryonic membranes and placenta
Embryonic period (weeks 3–8)	Gastrulation Formation of tissues, organs, and organ systems
Fetal period (week 9 to birth)	Continued differentiation and growth of tissues and organs Increase in length as measured from head to rump Increase in weight

normally happens and what can happen when things go wrong. The major events that occur during prenatal development are summarized by developmental period in Table 18.1.

The human prenatal period is divided into three periods: (1) the **pre-embryonic period**, from fertilization through the second week (during which the developing human is called a **pre-embryo**); (2) the **embryonic period**, from week 3 through week 8 (during which the developing human is called an **embryo**); and (3) the **fetal period**, from week 9 until birth (during which the developing human is called a **fetus**). *Gestation* refers to the time a pre-embryo, embryo, or fetus is carried in the female reproductive tract. Several stages of prenatal human development are shown in Figure 18.2.

Pre-embryonic Period

The pre-embryonic period begins with fertilization, the union between the nucleus of an egg and the nucleus of a sperm. Fertilization takes about 24 hours and usually occurs in a widened portion of the oviduct, not far from the ovary, as shown in Figure 18.3. Fertilization is a beautifully orchestrated performance by egg and sperm, aided by the actions of the oviducts and uterus.

Fertilization At ovulation, the ovary releases the egg, still a secondary oocyte (see Chapter 17). The egg is swept into the oviduct by fingerlike projections called *fimbriae* (singular, fimbria; see Figure 18.3). Cilia and waves of peristalsis move the oocyte slowly along the oviduct toward the uterus (*peristalsis* is a general term for the rhythmic waves of contraction and relaxation of smooth muscles in the walls of tubular organs, such as oviducts or digestive organs, that push the contents through the tubes).

The trip made by sperm is a considerably more competitive and precarious venture than the movements of the typically lone oocyte. During sexual intercourse, from 200 million to 600 million sperm are deposited in the vagina and on the cervix, the neck of the uterus that extends into the vagina. Many sperm become trapped at the boundary of the vagina and cervix. Indeed, fewer than 1% of deposited sperm actually enter the uterus. Sperm that escape entrapment use their whiplike tails (recall from Chapters 3 and 17 that each sperm has a flagellum) to move into the uterus and eventually the oviduct. Along the way, the sperm are aided by small uterine contractions stimulated by chemicals (prostaglandins) in the semen. Of the vast number of sperm originally

(a) A zygote

(b) A blastocyst implanting into the wall of the uterus

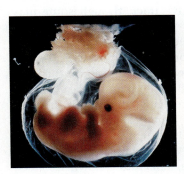
(c) An embryo about 7 weeks old

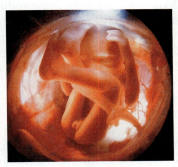
(d) A fetus about 5 months old

FIGURE 18.2 *The developing human*

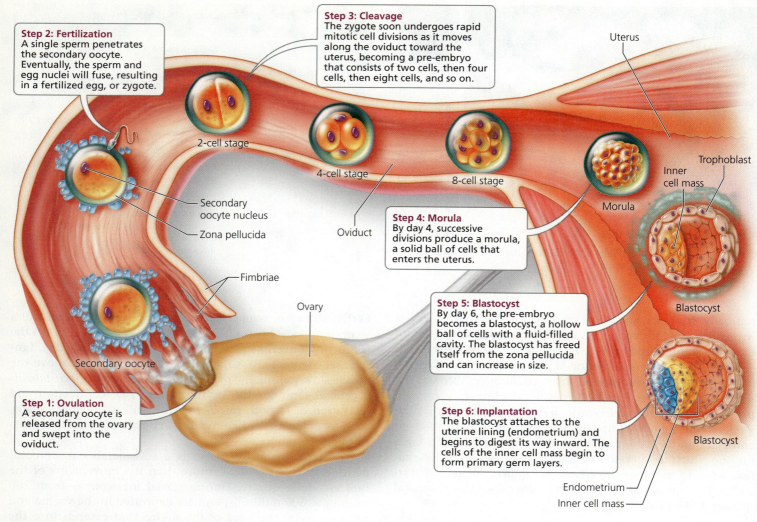

FIGURE 18.3 Early stages in the reproductive process

deposited in the female reproductive tract, only about 200 reach the site of fertilization in the widened portion of the oviduct.

Typically, an egg lives only 12 to 24 hours after its release from the ovary. For most eggs that get fertilized, the encounter with the sperm occurs within 12 hours of the egg's release. In addition, most sperm survive no more than 2 days in the female reproductive tract, though some may survive 5 days.

Two critical processes must precede fertilization (Figure 18.4). First, secretions from the uterus or oviducts must alter the surface of the acrosome, the enzyme-containing cap on the head of a sperm. These changes destabilize the sperm's plasma membrane. The second process occurs when sperm with unstable membranes contact the *corona radiata*, a layer of cells surrounding the secondary oocyte. Once such contact occurs, perforations develop in the sperm's weakened plasma membrane and outer acrosomal membrane. Enzymes (shown as green dots in Figure 18.4) then spill out of the acrosome and digest the attachments between cells of the corona radiata. Digestion of these attachments enables sperm to pass between the cells of the corona radiata to the layer below called the *zona pellucida*, the thick noncellular layer that immediately surrounds the secondary oocyte. Enzymes released from the acrosome also break apart the zona pellucida, creating a pathway to the oocyte.

Several sperm may arrive at the secondary oocyte at about the same time, although usually only one sperm crosses the zona pellucida and reaches the plasma membrane of the secondary oocyte. The plasma membrane of the successful sperm and that of the secondary oocyte fuse, and the sperm enters the egg cytoplasm. Fusion of the plasma membranes triggers two important events. First, enzymes (shown as red dots in Figure 18.4) released by granules near the plasma membrane of the secondary oocyte cause the zona pellucida to quickly harden and thereby prevent passage of other sperm. This block to entry by additional sperm ensures equal genetic contributions from each parent and prevents abnormal numbers of chromosomes in the resulting embryo, which would make it incapable of normal development.

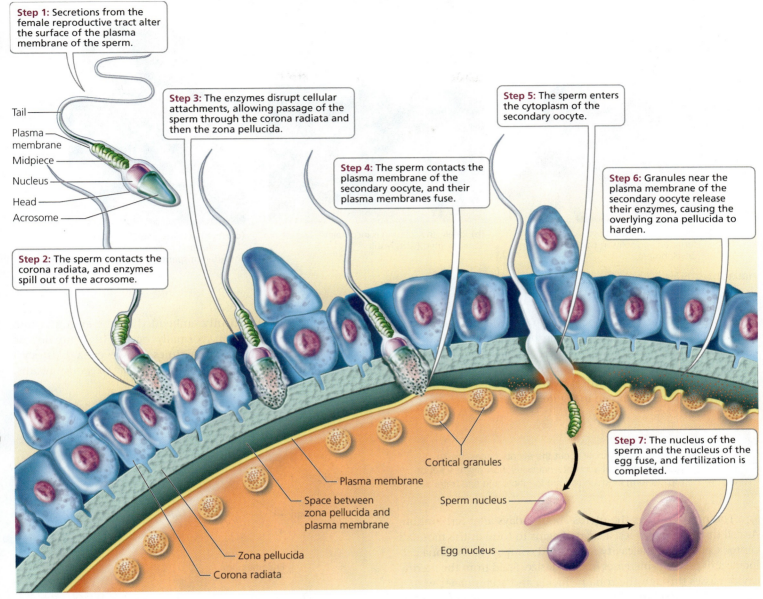

FIGURE 18.4 Fertilization

Second, the oocyte undergoes its second meiotic division (Chapter 17) and is now considered an ovum. The nucleus of the sperm (located in the sperm's head) and the nucleus of the ovum fuse (Figure 18.4). The remaining parts of the sperm (midpiece and tail) degenerate. The fertilized ovum is called a **zygote**. Just visible to the unaided eye, the zygote contains genetic material from both the mother (23 chromosomes) and father (23 chromosomes).

Cleavage About 1 day after fertilization, the zygote undergoes **cleavage**, a rapid series of mitotic cell divisions. During cleavage, the single-celled zygote becomes a pre-embryo, first consisting of two cells, then four cells, then eight cells, and so on. Cleavage occurs as the pre-embryo moves along the oviduct toward the uterus (see Figure 18.3). The cleaving pre-embryo becomes a solid ball of cells by about day 3. By day 4, the pre-embryo is a *morula*, a solid ball of 12 or more cells produced by successive divisions of the zygote. These early cell divisions do not result in an overall increase in size; such increases are prevented by the tight-fitting zona pellucida that still covers the morula. Instead, the cells within the ball become progressively smaller as divisions occur.

Sometimes during an early stage of cleavage, the mass of cells splits, and two pre-embryos are formed. *Identical twins*, also called *monozygotic twins* ("from one zygote"), develop in this way. Such twins are always the same gender and have nearly identical genetic material (see Chapter 18a for a discussion of twin studies). In rare cases, the splitting of the pre-embryo is incomplete, and *conjoined twins* result. Such twins may be surgically separated after birth once doctors have evaluated which structures are shared and analyzed the likelihood of successful separation. *Fraternal twins*

(a) Identical twins result when a single fertilized egg splits in two very early in development.

(b) Rarely, the splitting is incomplete, and conjoined twins result.

(c) Fraternal twins result from the fertilization of two oocytes by different sperm.

FIGURE 18.5 *Twins*

occur when two secondary oocytes are released from the ovaries and fertilized by different sperm. Such twins also are called *dizygotic twins* ("from two zygotes"). Fraternal twins may or may not be the same gender and are no more genetically similar than siblings who are not twins. Identical twins (separate and conjoined) and fraternal twins are shown in Figure 18.5.

stop and think
Given what you know about the events necessary to produce identical and fraternal twins, how might a set of triplets form in which two are identical and one fraternal?

The morula enters the uterus about 5 days after fertilization. As cell division continues, a cavity begins to form at the morula's center. Cells lining the cavity flatten and compact as the zona pellucida still prevents increases in overall size. Fluid from the uterine cavity passes through the zona pellucida and accumulates within the forming cavity, and the morula is converted into a **blastocyst**, a ball of cells with an inner fluid-filled cavity. As shown in Figure 18.3, the blastocyst is made up of two parts: an inner cell mass and a trophoblast. The **inner cell mass** is a group of cells that will become both the embryo proper and some of the extraembryonic membranes that will extend from or surround the embryo. The **trophoblast** is a thin layer of cells that will give rise to the extraembryonic membrane that is the embryo's contribution to the placenta. The **placenta** is the organ that delivers oxygen and nutrients to the embryo and carries carbon dioxide and other wastes away. Before the blastocyst becomes implanted in the uterine wall and the placenta develops, the blastocyst floats freely in the uterus. Increases in overall size of the blastocyst become possible when it frees itself from the degenerating zona pellucida.

Implantation The blastocyst attaches to the endometrium (the lining of the uterus) about 6 days after fertilization. The trophoblast rapidly proliferates and differentiates into two cell layers, one of which begins to invade the endometrium. During this process, called **implantation**, the blastocyst digests its way inward until it is firmly embedded in the endometrium (Figure 18.6). Implantation normally occurs high up on the back wall of the uterus. Sometimes, however, a blastocyst implants outside the uterus, and an *ectopic pregnancy* results.

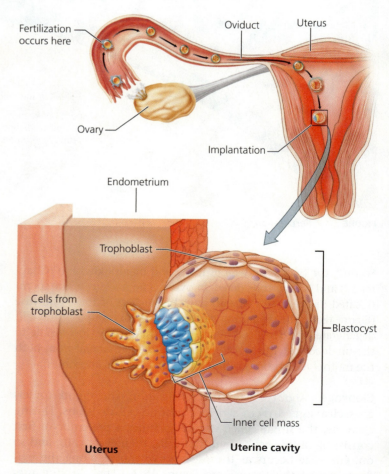
FIGURE 18.6 *Implantation. About 6 days after fertilization, the blastocyst attaches to the endometrium of the uterus and begins to digest its way inward.*

Most implantations outside the uterus occur in the oviducts, usually because passage of the dividing pre-embryo through the oviduct is impaired in some way. Factors that hinder passage of the pre-embryo through the oviduct include structural abnormalities and scar tissue from surgery or pelvic inflammatory disease (see Chapters 17 and 17a). In such instances, the embryo is surgically removed (and hence the pregnancy terminated) because the oviduct cannot support a pregnancy, and because rupture of the oviduct by a growing embryo and the resulting hemorrhage can be fatal to the mother.

An estimated one-third to one-half of all zygotes fail to become blastocysts and fail to implant. The failed zygotes and early pre-embryos are either reabsorbed by cells of the endometrium or expelled from the uterus in an early spontaneous abortion (such an end to a pregnancy is described as *spontaneous* because it was not medically induced). Causes of early spontaneous abortion include chromosomal abnormalities in the zygote and an inhospitable uterine environment for implantation. The latter might result from the presence of an intrauterine device (IUD) or inadequate production of the hormones estrogen and progesterone by the corpus luteum. (Recall from Chapter 17 that the corpus luteum is a glandular structure that forms from the ovarian follicle after ovulation.) These early spontaneous abortions are included in the term *miscarriage*, which describes pregnancy loss occurring before 20 weeks' gestation. (After 20 weeks, the fetus is considered potentially viable, with medical support, and such deliveries are described as *preterm* or *premature births* when they occur before 37 weeks' gestation; this topic is discussed in more detail later in this chapter).

When conditions are right, the blastocyst completes implantation by the end of the second week of the pre-embryonic period. During implantation, cells within the blastocyst produce **human chorionic gonadotropin (HCG)**, a hormone that enters the mother's bloodstream and is excreted in her urine. Many pregnancy tests screen for the presence of this hormone in the mother's urine. Enough HCG is produced by the end of the second week to be detected by the pregnancy test and yield a positive result. The physiological function of HCG is to maintain the corpus luteum and stimulate it to continue producing progesterone. Progesterone is essential for maintaining the endometrium; without an adequate supply of progesterone, the endometrium would be shed, as during menstruation.

Infertility is the inability of a female to conceive (become pregnant) or of a male to cause conception. Implantation is a major hurdle in the series of steps leading to a successful pregnancy. The Ethical Issue essay *Making Babies* describes the options available to infertile couples wishing to have children.

Extraembryonic membranes Toward the end of the pre-embryonic period, four membranes—the amnion, yolk sac, allantois, and chorion—begin to form around the pre-embryo (Figure 18.7). Formation of these membranes extends into the embryonic period, when the developing human is called an embryo. The membranes lie outside the embryo and are called **extraembryonic membranes**. These membranes protect and

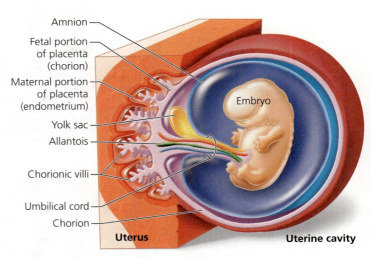

FIGURE 18.7 Extraembryonic membranes. The amnion, yolk sac, chorion, and allantois begin to form during the second to third week after fertilization.

nourish the embryo and later the fetus. The **amnion** surrounds the entire embryo, enclosing it in a fluid-filled space called the *amniotic cavity*. Amniotic fluid forms a protective cushion around the embryo that later can be examined as part of prenatal testing in a procedure known as *amniocentesis* (see Chapter 20). The **yolk sac** is the primary source of nourishment for embryos in many species of vertebrates. In embryonic humans, however, it remains quite small and does not provide nourishment. Human embryos receive nutrients from the placenta. The yolk sac in humans is a site of early blood cell formation. It also contains *primordial germ cells* that migrate to the embryo's gonads (testes or ovaries), where they differentiate into immature cells that will eventually become sperm or oocytes. The **allantois** is a small membrane whose blood vessels become part of the **umbilical cord**, the ropelike connection between the embryo and the placenta. The umbilical cord consists of blood vessels and supporting connective tissue. Finally, the **chorion** is the outermost membrane, which rests against the uterine cavity. The chorion develops largely from the trophoblast and becomes the embryo's major contribution to the placenta. The amnion, yolk sac, and allantois develop from the inner cell mass.

The placenta The placenta orchestrates all interactions between the mother and the fetus. The placenta forms from the chorion of the embryo and a portion of the endometrium of the mother (specifically, the endometrium in the area where implantation occurred).

A major function of the placenta is to allow oxygen and nutrients to diffuse from maternal blood into embryonic blood. Wastes such as carbon dioxide and urea diffuse from embryonic blood into maternal blood. The placenta also produces hormones such as HCG, estrogen, and progesterone. Placental hormones, as mentioned earlier, are essential for the continued maintenance of pregnancy. High levels of HCG from the placenta may be responsible for *morning sickness*, the nausea and vomiting experienced by some women early in pregnancy. (Morning sickness is not restricted to the morning, by the way.)

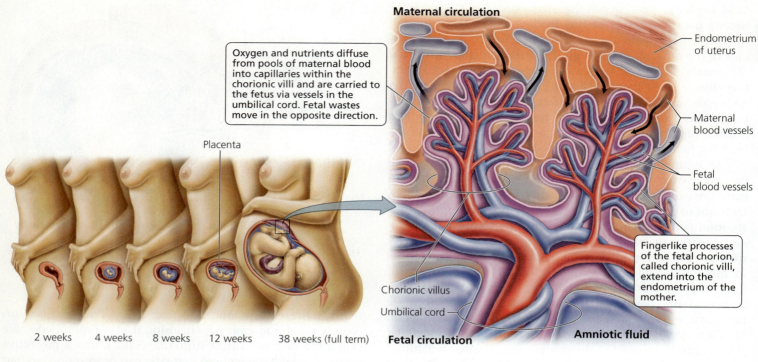

(a) The placenta begins to form at about 2 weeks (shortly after implantation) and is fully developed by about 12 weeks.

(b) The internal structure of the placenta

FIGURE 18.8 *The placenta is formed from the chorion of the embryo and the endometrium of the mother.*

Formation of the placenta begins shortly after implantation, when cells derived from the trophoblast rapidly divide and invade the endometrium. Cavities form that fill with blood from maternal capillaries severed by the invading cells. Soon, fingerlike processes of the chorion, called **chorionic villi**, grow into the endometrium. These chorionic villi grow, divide, and continue their invasion of maternal tissue, causing ever-larger cavities and pools of maternal blood to form. The placenta is fully developed by the third month of pregnancy (Figure 18.8a). At this time, it weighs about 680 g (1.5 lb).

Chorionic villi contain blood vessels connected to the developing embryo. Oxygen and nutrients in the pools of maternal blood diffuse through the capillaries of the villi into the umbilical vein and travel to the embryo. Wastes leave the embryo by the umbilical arteries, move into capillaries within the chorionic villi, and diffuse into maternal blood. Thus, the chorionic villi provide exchange surfaces for diffusion of nutrients, oxygen, and wastes (Figure 18.8b). Some chorionic villi also anchor the embryonic portion of the placenta to maternal tissue.

Under normal circumstances, there is no direct mixing of maternal and fetal blood; all exchanges occur across capillary walls. Nevertheless, the closeness of the maternal and fetal blood vessels within the placenta causes the fetus to share in maternal nutrition, habits, and lifestyle. In addition to beneficial nutrients and oxygen, harmful substances can cross the placenta from the mother to the fetus, including some drugs, alcohol, caffeine, toxins in cigarette smoke, and HIV (human immunodeficiency virus, the virus that causes AIDS; see Chapter 17a).

Usually, implantation occurs high up on the back wall of the uterus, so the placenta forms in the upper portions of the uterus. Sometimes, however, implantation occurs lower in the uterus, and the developing placenta grows to cover the cervix, the neck of the uterus that projects into the vagina; this condition is called *placenta previa*. Placenta previa may cause premature birth or maternal hemorrhage. Most women with diagnosed placenta previa have their baby delivered by cesarean section before labor begins. **Cesarean section**, often shortened to *C-section*, is the procedure by which the fetus and placenta are removed surgically through an incision in the abdominal wall and uterus.

Embryonic Period

The embryonic period extends from the third to the eighth week of development. It is a time of great change, beginning with the formation of three distinct germ layers from which all tissues and organs develop. We first describe the general process by which germ layers form, and then we consider development of the central nervous system and reproductive system as examples of tissue and organ formation during the embryonic period. By the end of the embryonic period, all organs have formed, and the embryo has a distinctly human

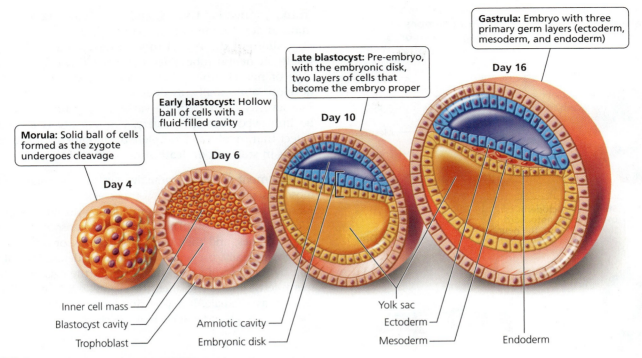

FIGURE 18.9 Early stages of development in cross section

appearance. Three interrelated processes produce this tiny human: cell division (which continues from the pre-embryonic period); **cell differentiation** (the process by which cells become specialized with respect to structure and function); and **morphogenesis** (the development of overall body organization and shape).

Gastrulation   Morphogenesis begins during the third week after fertilization. Just before the start of morphogenesis, during implantation, the inner cell mass moves away from the surface of the blastocyst, and the amniotic cavity forms. The amniotic cavity is filled with amniotic fluid and lined by the amnion, one of the four extraembryonic membranes. The inner cell mass, which now becomes a flattened, platelike structure called the *embryonic disk* (Figure 18.9), is destined to become the embryo proper. But first, the cells within the embryonic disk must differentiate and migrate, forming three **primary germ layers** known as ectoderm, mesoderm, and endoderm. The cell movements that establish the primary germ layers are called **gastrulation**, and the embryo during this period is called a *gastrula* (Figure 18.9).

All tissues and organs develop from the primary germ layers. **Ectoderm** covers the surface of the embryo and forms the outer layer of skin and its derivatives, such as hair, nails, oil glands, sweat glands, and mammary glands. Ectoderm also forms the nervous system. Another germ layer, the **endoderm**, turns inward to form the lining of the digestive, urinary, and respiratory tracts, as well as some organs and glands (for example, the pancreas, liver, thyroid gland, and parathyroid glands). **Mesoderm** fills some of the space between ectoderm and endoderm and gives rise to muscle; bone and other connective tissue; and various organs, including the heart, kidneys, ovaries, and testes. Because gastrulation initiates the processes by which the developing embryo's body takes shape and becomes organized, it is considered a key part of morphogenesis.

As gastrulation proceeds, a flexible rod of mesoderm tissue called the **notochord** develops where the vertebral column will form. The notochord defines the long axis of the embryo and gives the embryo some rigidity. Vertebrae eventually form around the notochord, and the notochord degenerates. The pulpy, elastic material in the center of intervertebral disks (pads that help cushion the bones of the vertebral column) is all that remains of our notochord.

Development of the central nervous system   A major milestone in embryonic development is formation of the central nervous system (brain and spinal cord) from ectoderm. First, the ectoderm overlying the notochord thickens. This thickened region of ectoderm above the notochord is called the *neural plate*. The neural plate folds inward, forming a groove that extends the length of the embryo, along its back surface. The raised sides of the groove, known as *neural folds*, grow upward and eventually meet and fuse to form the **neural tube**, a fluid-filled tube that will become the central nervous system. The process by which the neural tube is formed is called **neurulation**, and the embryo during this period is called a *neurula*. The anterior portion of the neural tube develops into the brain, and the posterior portion forms the spinal cord. Alongside the neural tube, mesoderm cells organize into blocks called **somites**. Somites eventually form skeletal muscles of the neck and

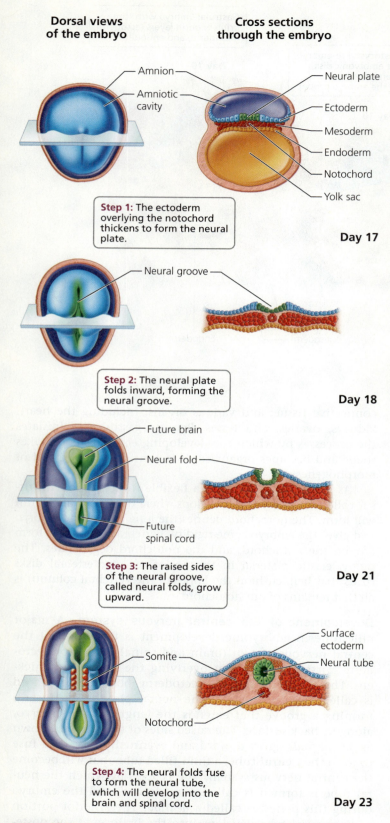

FIGURE 18.10 *Formation of the central nervous system from ectoderm. Dorsal views of a human embryo (left) and corresponding cross-sectional views or parts of such views (right) are shown at four different days during development of the brain and spinal cord.*

trunk, connective tissues, and vertebrae. Figure 18.10 summarizes development of the central nervous system.

Failure of the neural tube to develop and close properly results in neural tube defects. *Spina bifida* ("split spine") is a type of neural tube defect in which part of the spinal cord develops abnormally, as does the adjacent area of the spine. The severity of spina bifida varies greatly, and some cases can be improved through surgery. *Anencephaly* is a neural tube defect that involves incomplete development of the brain and results in stillbirth or death shortly after birth.

Development of the reproductive system Among the 23 chromosomes provided by the mother's egg is an X chromosome. Among the 23 chromosomes provided by the father's sperm is an X or a Y chromosome. The X and Y chromosomes are called sex chromosomes, and they determine the sex of human embryos. More specifically, the sex of an embryo is determined at fertilization by the type of sperm that fertilizes the egg. If an X-bearing sperm fertilizes the egg, then the zygote is XX and will normally develop as a female. If a Y-bearing sperm fertilizes the egg, then the zygote is XY and will normally develop as a male.

what would you do?

Methods are available to separate X-bearing sperm from Y-bearing sperm. During the procedure, sperm are treated with a fluorescent dye that binds to DNA. X-bearing and Y-bearing sperm glow differently because there is almost 3% more DNA in X chromosomes than Y chromosomes. Thus X- and Y-bearing sperm can be sorted by an automated machine. Such "sperm sorting" makes it possible to select a child's sex. Couples could decide whether to use X-bearing sperm to produce a female baby or Y-bearing sperm to produce a male baby. Some couples want to select the sex of their child to avoid X-linked genetic disorders; others want to balance their families with respect to numbers of sons and daughters. The chosen sperm are deposited within the woman's vagina or cervix at a time when pregnancy is likely to occur.

Should parents be allowed to choose the sex of their children? What implications would sex selection have for families and for society? If you were given the responsibility of deciding whether such technology should continue to be made available to parents, what would you do?

Male and female embryos have the same external and internal anatomy for the first few weeks of development; their reproductive organs have not yet differentiated. The embryos at this time are described as being "sexually indifferent." However, about 6 weeks after fertilization, a region of the Y chromosome—called SRY for "sex-determining region of the Y chromosome"—initiates development of testes in XY embryos. The testes soon begin producing testosterone, the hormone that directs development of male reproductive organs. Female embryos lack the Y chromosome; in its absence, ovaries develop. We focus on development of the external genitalia, but keep in mind that internal reproductive organs also are differentiating.

Male and female embryos of about 6 weeks of age have a small bud of tissue between their legs (Figure 18.11). Just below

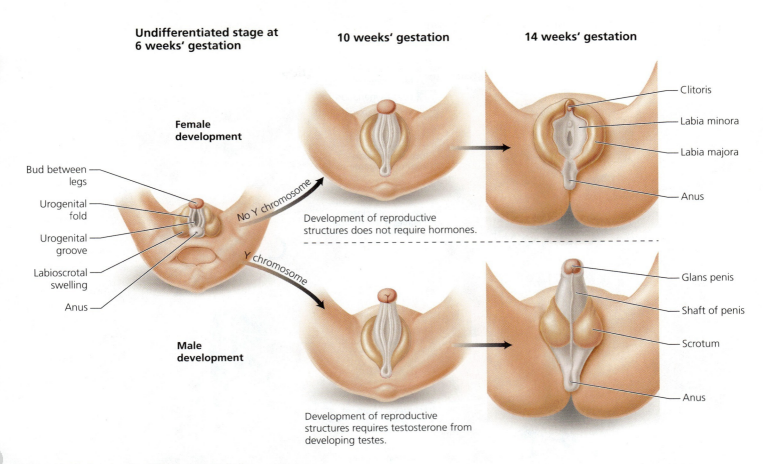

FIGURE 18.11 *Development of external genitalia*

this bud is a shallow depression called the *urogenital groove*. The urogenital groove is surrounded by urogenital folds, which in turn are surrounded by labioscrotal swellings. In the male embryo, testosterone produced by the newly differentiated testes causes the urogenital groove to elongate and completely close. The bud then becomes the glans penis; the urogenital folds develop into the shaft of the penis; and the labioscrotal swellings become the scrotum (into which the testes will later descend). In the female, the urogenital groove also elongates, but instead of closing, as in the male, it remains open. The bud becomes the clitoris in females, and the urogenital folds develop into the labia minora. The labioscrotal swellings of females become the labia majora. The formation of male and female reproductive structures is typically completed by the end of the third month of gestation.

Fetal Period

The fetal period extends from the ninth week after fertilization until birth. Although tissues and organs continue to differentiate during this period, the most notable change in most body parts is rapid growth, made possible by the placenta. The placenta completes its development early in the fetal period.

Growth Growth during the fetal period is extremely rapid, as reflected in phenomenal increases in weight (from 8 g to 3400 g, or from about 0.3 oz to 120 oz) and substantial increases in length (the length as measured from the head to the rump increases from 50 to 360 mm, or from about 2 in. to 14 in.), when the pregnancy continues for the full 38 weeks. One striking change during the fetal stage is that the growth rate of the head slows relative to the growth rate of other regions of the body, altering the way the body is proportioned (Figure 18.12). Difference in the relative rates of growth of various parts of the body is called **allometric growth**. Such growth continues after birth and helps to shape developing humans and other organisms.

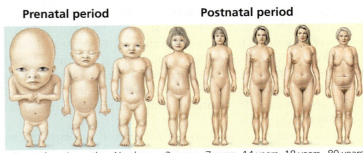

FIGURE 18.12 *Allometric growth. Changes in body proportions occur throughout prenatal and postnatal development. For comparison, all stages are drawn to the same total height.*

Q Which two structures in the fetus allow blood to bypass the pulmonary circulation?

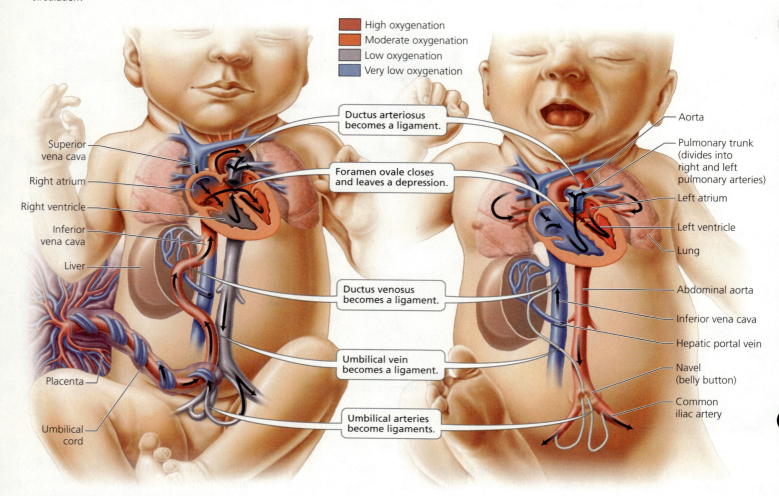

(a) Fetal circulation is characterized by a connection to the placenta (the umbilical cord) and several bypasses around organs that do not yet perform their postnatal functions.

(b) At birth, the umbilical cord is tied off and cut, leaving the navel (belly button). The bypasses close, allowing more blood to reach the now functional organs of the newborn.

FIGURE 18.13 *Fetal circulation and changes at birth*

A In the fetus, the foramen ovale (between the right and left atria) and the ductus arteriosus (on the pulmonary trunk) divert blood away from the lungs, which are not yet functional.

Fetal circulation The fetal circulatory system differs from circulation after birth because several organs—the lungs, kidneys, and liver, to name a few—do not perform their postnatal functions in the fetus. Before birth, most blood is shunted past these organs through temporary vessels or openings. Recall that the fetus receives its oxygen and nutrients from maternal blood and gives up its carbon dioxide and wastes to maternal blood at the placenta.

Fetal blood must travel to the placenta to rid itself of carbon dioxide and wastes and to pick up oxygen and nutrients. It does so by flowing through umbilical arteries that branch off large arteries in the legs of the fetus. The two umbilical arteries run through the umbilical cord to the placenta. Once exchanges between fetal and maternal blood have occurred at the placenta, blood rich in oxygen and nutrients travels back to the fetus in the umbilical vein (within the umbilical cord; Figure 18.13a). Some of this blood flows to the fetal liver, which produces red blood cells but does not yet function in digestion. Most of the blood from the placenta, however, bypasses the liver by means of a temporary vessel called the *ductus venosus* and heads to the heart.

Like the liver, fetal lungs are not yet functional and need only a small amount of blood for growth and removal of wastes from their cells. Once blood enters the right atrium of the heart, some of it passes to the right ventricle and on to the lungs, as it does in circulation after birth. Most of the blood, however, moves into the left atrium through a small hole in the wall between the atria, called the *foramen ovale*. The blood in the left atrium moves into the left ventricle and out to the body of the fetus. Another shunt, called the *ductus arteriosus*, connects the pulmonary trunk to the aorta and also functions to divert blood away from the lungs.

Thus, blood flowing out of the right ventricle into the pulmonary trunk is shunted away from the lungs and into the aorta, where it travels to all areas of the fetus. Another glance at Figure 18.13a reveals that much of the blood traveling through the fetus is moderate to low in oxygen. Indeed, the umbilical vein is the only fetal vessel that carries fully oxygenated blood.

At birth—when the lungs, liver, and other organs begin their postnatal functions—fetal circulation converts to the postnatal pattern (Figure 18.13b). Blood flow to the placenta ceases when the umbilical cord is tied and cut off. The scar left by the cord becomes the baby's navel, or belly button. Within the infant, the umbilical arteries, umbilical vein, ductus venosus, and ductus arteriosus constrict, shrivel, and form ligaments. The foramen ovale normally closes shortly after birth, leaving a small depression in the wall between the two atria.

stop and think
In some newborns, the foramen ovale fails to close. These infants, called *blue babies*, have a bluish appearance. What might explain their appearance?

18.3 Birth

Birth, also called **parturition**, usually occurs about 38 weeks after fertilization. The process by which the fetus is expelled from the uterus and moved through the vagina to the outside world is called **labor**. Uterine contractions during labor occur at regular intervals and are usually painful. Recall from Chapter 10 that the hormone oxytocin, which is released from the posterior pituitary gland, causes contractions of the smooth muscle of the uterus. These contractions, in turn, stimulate further release of oxytocin in a positive feedback loop. As labor proceeds, contractions become more intense, and the interval between them decreases.

Labor can be divided into three stages (Figure 18.14). The first stage, known as the *dilation stage*, begins with the onset of regular contractions and ends at the point when the cervix has fully dilated to 10 cm (4 in.). This stage usually lasts about 6 to 7 hours, although it can be shorter in women who previously have given birth. During the dilation stage, the amniotic sac usually ruptures, releasing amniotic fluid. Known commonly as *breaking water*, rupture of the amniotic sac may be done deliberately by a doctor or midwife if it does not happen spontaneously. The second stage, the *expulsion stage*, begins with full dilation of the cervix and ends with delivery of the baby. This stage may last 1 or more hours. Some physicians make an incision to enlarge the vaginal opening, just before passage of the baby's head. This procedure, an *episiotomy*, facilitates delivery and avoids ragged tearing, which is generally slower to heal. Once delivery of the baby is complete, the umbilical cord is clamped and cut. The newborn takes his or her first breath, and the conversion of fetal circulation to the postnatal pattern begins. Labor, however, is not over. The third (and final) stage of labor is the *placental stage*. It begins with delivery of the newborn and ends when the afterbirth, consisting of the placenta and fetal membranes, is expelled from the mother's body about 15 minutes after the baby. Continuing and powerful uterine contractions help expel the placenta and constrict maternal blood vessels torn during delivery. Any vaginal incision or tearing is sutured once labor has ended.

Most babies are born head first, facing the vertebral column of their mother. Some babies, however, are born buttocks first, and their delivery is called *breech birth*. Breech births are associated with difficult labors and umbilical cord accidents, such as compression or looping of the cord around the baby's neck. Thus, many physicians attempt to turn breech babies into the headfirst position before delivery. The baby may be delivered by cesarean section if turning is not successful.

Babies born at least 38 weeks after fertilization are called *full-term infants*. These babies weigh, on average, 3400 g (7.5 lb). Some babies are born 1 or more weeks late.

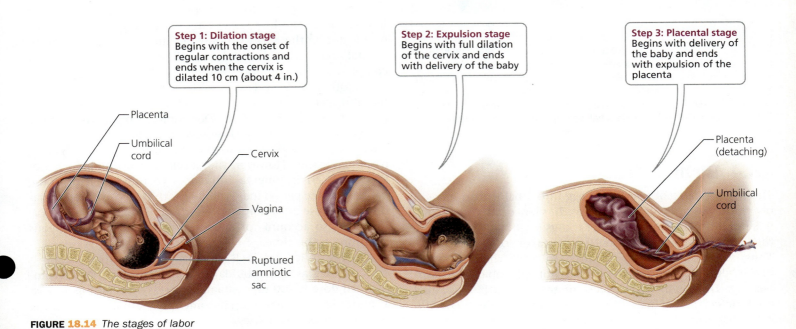

FIGURE 18.14 *The stages of labor*

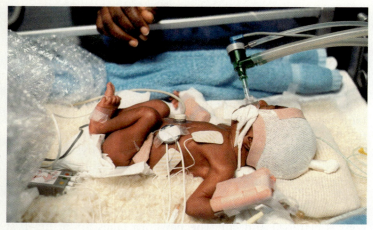

FIGURE 18.15 Many premature infants require intensive care because their organs are not yet functional.

Because the placenta begins to deteriorate with time, labor is typically induced if a pregnancy continues for 41 or more weeks. *Premature infants* are those born before 37 weeks of gestation. Their chances for survival increase with length of time spent in the female reproductive tract. Doctors do not usually attempt to save infants born at or before 22 weeks' gestation because even with intensive medical care, major disabilities are almost inevitable. Such babies weigh about 630 g (1.4 lb). Amazingly, infants born at 25 weeks' gestation and weighing about 1 kg (2.2 lb) have a 50% to 80% chance of surviving, provided they receive intensive care. Intensive medical care is needed because the organ systems of these infants have not matured sufficiently to take over the functions normally performed by the mother's body. Care for premature infants often includes blood transfusions, tube feeding, and maintenance on a respirator (Figure 18.15). Even with excellent intensive care, health problems may persist well beyond infancy.

 what would you do?

Babies born before 25 weeks' gestation typically have a low chance of survival, even with excellent medical care. In addition, they have a very high probability of suffering from severe disabilities, if they do survive. Should efforts be made to save all such infants? Or should criteria be used to decide which infants the doctors should try to save and which they should let die? If you were given the responsibility of developing such criteria, what would you suggest? More to the point, if your infant were born before 25 weeks, what would you ask the doctor to do?

18.4 Birth Defects

The development of a fertilized egg into a baby is an incredibly wondrous yet complex process. In view of its complexity, it is not difficult to imagine that things might go wrong along the way. Indeed, the potential for mistakes seems enormous. Thus, we should not be surprised to learn that not every zygote develops into a baby, and not all babies are born healthy. Developmental defects present at birth are called **birth defects**. They may involve structure, function, behavior, or metabolism.

The causes of birth defects may be genetic or environmental. Genetic causes include mutant genes or changes in the number or structure of chromosomes. Environmental causes include drugs, chemicals, radiation, deficiencies in maternal nutrition, and certain viruses, such as herpes simplex and rubella (the cause of German measles). Birth defects resulting from genetic causes will be discussed in Chapter 19. Some of the birth defects associated with psychoactive drugs were considered in Chapter 8a.

Environmental agents that disrupt development have their greatest effects during periods of rapid differentiation, when cells and tissues acquire specific functions and form organs. Recall that these events occur largely during the embryonic period (week 3 through week 8). Exposure to disruptive agents during the embryonic period can cause major birth defects (Figure 18.16). Such defects, characterized by major structural abnormalities, occur in 2% to 3% of newborns. Spina bifida is an example of a major birth defect. Disruptive agents produce more minor defects during the fetal period, when most organs are growing rather than differentiating. Minor birth defects, such as having a single horizontal crease across the upper palm of the hand (instead of two creases), occur in about 4% to 5% of newborns.

Even though all organs and organ systems develop primarily during the embryonic period, different organs develop at somewhat different times and rates. Because of such variation, each organ system has a critical period in which it is most susceptible to disruption by environmental agents. The CNS has the longest critical period, extending from week 3 through week 16—beyond the embryonic period and into the fetal period. During this time, the developing nervous system is highly sensitive, and disruptive agents can have major consequences. (By comparison, exposure to such agents after week 16 tends to induce relatively minor defects.) This long critical period may explain why the brain is the organ in which defects in structure and function are most commonly found.

The critical period for development of the upper and lower limbs is relatively short, extending from about week 4 to week 6. It is during this short window of time that certain environmental agents cause major abnormalities, such as tiny limbs or complete absence of limbs.

Another look at Figure 18.16 reveals that exposure to harmful agents during the first 2 weeks after fertilization is not known to cause birth defects. Development during this period (the pre-embryonic period) centers on formation of structures outside the embryo, such as the extraembryonic membranes described earlier. However, the developing pre-embryo is not immune to environmental influences. Disruptive agents in the pre-embryonic period may interfere with cleavage or implantation and cause a spontaneous abortion. Chromosomal abnormalities also account for a large proportion of spontaneous abortions at this time.

Taking folic acid, a B vitamin, can help prevent birth defects such as spina bifida and anencephaly. The recommended dose is 400 micrograms every day, beginning at least 1 month before the woman becomes pregnant and continuing throughout the pregnancy.

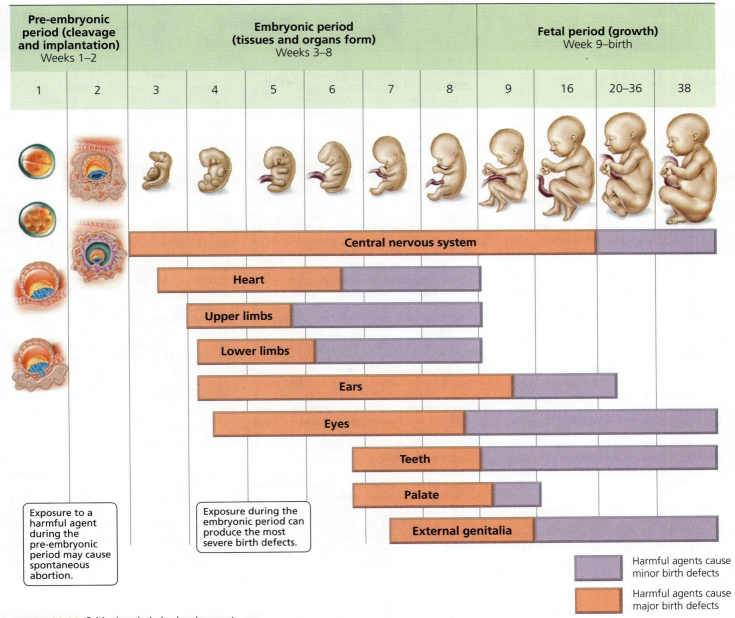

FIGURE 18.16 *Critical periods in development*

18.5 Milk Production by Mammary Glands

Lactation is the production and ejection of milk from the mammary glands. Recall from Chapter 10 that prolactin from the anterior pituitary gland promotes milk production. Recall also that oxytocin from the posterior pituitary gland stimulates milk ejection. The structure of the breast was described in Chapter 17.

Prolactin levels increase during pregnancy and reach their peak at birth. Milk production does not begin during pregnancy, because high levels of estrogen and progesterone inhibit the actions of prolactin. Thus, prolactin cannot initiate milk production until birth, when levels of estrogen and progesterone decline. Milk usually is available about 3 days later. In the interim, infants consume *colostrum*, a cloudy yellowish fluid produced by the breasts that has a different composition from that of breast milk. After birth, the infant's sucking on the breast stimulates the release of prolactin, promoting and maintaining the production of milk. At weaning, when sucking on the breast ends, so, too, does milk production.

Which is better—bottle or breast? Most people today believe that breast-feeding is better for infants. Advocates of breast-feeding point out that compared with commercial formulas—many of them based on cow's milk—breast milk is more digestible and much less likely to cause an allergic reaction, constipation, or obesity in infants. In addition, maternal milk and colostrum contain antibodies and special proteins that boost the immune system. Finally, obtaining milk from a breast requires more effort than sucking on a bottle. Health professionals think this greater effort promotes optimum development of the infant's jaws, teeth, and facial muscles.

ETHICAL ISSUE

Making Babies

The inability to become pregnant or to cause a pregnancy is called infertility. Couples are considered infertile if conception does not occur after 1 year of unprotected sexual intercourse. Infertility strikes about one in six American couples. In some of these couples, the woman is infertile. Hormonal imbalances may disrupt ovulation or implantation, or scarring from disease may block the oviducts and prevent passage of gametes. Some women are sterile (they can never conceive and carry a child) because they have had their uterus (and sometimes their ovaries) removed for medical reasons. In other couples, it is the man who is infertile. Male infertility often is caused by production of few or sluggish sperm. Some men are sterile (they can never cause a pregnancy) as a result of conditions such as cryptorchidism (failure of the testes to descend from the abdomen to the scrotum). Also, in a strange twist of fate, otherwise compatible couples may be incompatible at the cellular level—some women produce antibodies that kill their partner's sperm. For some couples, the cause of infertility is unknown.

Can anything be done to help those who are infertile? For roughly half of the couples seeking help the answer is yes, although the road to reproduction may be long, expensive, and filled with emotional ups and downs. The procedures to treat infertility are collectively called *assisted reproductive techniques (ARTs)*.

Administering hormones can treat some cases of female infertility. Hormones that trigger ovulation may be given to women whose ovaries fail to release eggs properly. Women who tend to miscarry (spontaneously abort) may be given progesterone to enhance the receptivity of their uterine lining. However, some causes of infertility, such as scarred oviducts in women and low sperm counts in men, do not respond to hormone therapy. What can be done in these cases?

Artificial insemination is one option available to couples whose infertility is caused by a low sperm count. In this procedure, sperm that have been donated and stored at a sperm bank are deposited (usually with a syringe) in the woman's cervix or vagina at about the time of ovulation. Sperm from the male member of the couple may be concentrated and then used. Alternatively, couples may use semen from an anonymous donor. Another possibility for couples in which the man has few sperm, or sperm that lack the strength or enzymes necessary to penetrate an egg, is *intracytoplasmic sperm injection (ICSI)*. In ICSI, a tiny needle is used to inject a single sperm into an egg. If fertilization occurs, the embryo is transferred to the uterus.

Some cases of infertility can be treated with *in vitro fertilization (IVF)*. IVF involves placing eggs and sperm together in a laboratory dish. First, the woman is treated with hormones to trigger superovulation—the ovulation of many eggs. Next, the eggs are removed from the woman's ovaries and placed into a dish that contains some of the man's sperm. If fertilization occurs, the zygotes are transferred to a solution that will support further development. Eventually one or more of the pre-embryos is transferred to the woman's uterus, which has been primed with hormones to support implantation. Women with blocked oviducts may conceive with IVF because the technique bypasses the oviducts altogether. If the woman lacks a uterus, the couple can hire a surrogate mother to gestate their baby. The egg and sperm may come from the couple or from other individuals.

IVF is sometimes unsuccessful because pre-embryos fail to implant. The problem seems to be the rather violent squirt of the pre-embryo into a reproductive tract that is already traumatized from hormone treatments and retrieval of eggs. Two other procedures avoid the problem of the pre-embryo's abrupt arrival in the uterus. In *gamete intrafallopian transfer (GIFT)*, eggs and sperm are collected from a couple and inserted into the woman's oviduct, where fertilization may occur. Afterward, any resulting pre-embryos drift naturally (and gently) into the uterus. The second procedure is *zygote intrafallopian transfer (ZIFT)*. In ZIFT, eggs and sperm are collected and brought together in a laboratory dish. If fertilization occurs, the resulting zygotes are inserted into the woman's oviducts, where they continue to develop and travel on their own to the uterus. Normal, healthy oviducts are needed for GIFT and ZIFT to work.

Success rates of ARTs range from about 20% to 28% live births per egg retrieved. Although costs vary somewhat, each attempt typically costs between $10,000 and $20,000, and multiple attempts often are needed. What about the couples who do not conceive with ARTs? Is there hope that some day they might also conceive? The answer is yes—advances in reproductive research and technology are rapidly providing potential treatments for infertility.

Questions to Consider

- If you donated eggs or sperm for use by infertile couples, should you have a claim (legal or otherwise) to any offspring produced?
- If you were charged with establishing policies at an infertility clinic, how would you decide what should happen to pre-embryos created at the clinic that turned out to be "extra"? Do they have a right to life? And who gets the pre-embryos when couples divorce or die?

Breast-feeding also benefits the mother. Nursing helps the mother's uterus return to nearly its prepregnant size. When oxytocin is released in response to an infant's sucking, it helps the uterus shrink. Also, because milk production is energetically expensive, breast-feeding helps with maternal weight loss in the weeks and months after giving birth. Finally, breast-feeding, when done for at least 1.5 years, seems to offer some protection against breast cancer.

Despite its benefits, breast-feeding is not for everyone. Some women simply prefer bottle-feeding, and others have difficulty breast-feeding. Still others are warned against breast-feeding because the medications they take pass into breast milk. Finally, mothers who are HIV-positive should not breast-feed because the virus can be transmitted in breast milk.

18.6 Postnatal Period

The postnatal period—the period of growth and development after birth—includes several stages, all of which involve biological, social, and psychological changes. We consider the following stages: infancy (from birth to 12 months), childhood (from 13 months to 12 or 13 years), adolescence (from puberty to late teens), and adulthood (generally reached by around 20 or 21 years of age).

Infancy is a time of rapid growth, though not as rapid as during prenatal development. Physical milestones reached during the first year include (in approximate chronological order) rolling over, sitting, crawling, and standing; by the end of the first year, some infants are walking. Vocalizations also change: cooing is replaced by babbling, which in turn is replaced by single words in some 1-year-olds. An infant's responsiveness to social stimuli increases over the 12 months. Childhood is a time of continued growth during which gross and fine motor skills improve and coping skills develop. With the exception of the reproductive system, organ systems become fully functional.

Adolescence begins with *puberty,* the period of sexual maturation that typically occurs from 12 to 15 years in girls and from 13 to 16 years in boys. A period of rapid growth

TABLE 18.2 Changes in Organ Systems as We Age

Organ System	Some Changes That Occur with Aging
Integumentary	Wrinkles appear as skin becomes thinner and less elastic.
	Sweat glands decrease in number, making regulation of body temperature more challenging.
	Hair thins owing to death of hair follicles and turns gray as pigment-producing cells die.
Skeletal	Bones become lighter and more brittle, especially in women after menopause.
	About 7.6 cm (3 in.) of height are lost as the intervertebral disks deteriorate and the vertebrae move closer together.
	Joints become stiff and painful owing to decreased production of synovial fluid.
Muscular	Muscle mass decreases owing to loss of muscle cells and decreased size of remaining muscle cells.
Nervous	Brain mass decreases.
	Movements and reflexes slow as conduction velocity of nerve fibers decreases and release of neurotransmitters slows.
	Hearing becomes less acute as hair cells in the inner ear are lost.
	Ability of the eye to focus declines as the lens of the eye stiffens.
	Smell and taste become less acute.
Endocrine	In women, production of estrogen and progesterone decreases with menopause.
	In men, production of testosterone decreases.
	In both sexes, production of growth hormone decreases.
Cardiovascular	Cardiac output decreases as walls of the heart stiffen.
	Blood pressure rises as arteries become less elastic and clogged by fatty deposits.
Respiratory	Lung capacity decreases as alveoli break down and lung tissue becomes less elastic.
Digestive	Basal metabolic rate declines.
	Ability of the liver to detoxify substances declines.
Urinary	In both sexes, kidney mass declines, as does the rate of filtration of the blood by nephrons.
	Particularly in women, the external urethral sphincter weakens, causing incontinence.
	In men, the prostate gland (part of the reproductive system) enlarges, causing painful and frequent urination.
Reproductive	In men, fewer viable sperm are produced.
	In women, ovulation and menstruation cease at menopause.

(the growth spurt) occurs in both sexes. Growth slows for a few years, and adolescence ends with the cessation of growth in the late teens or early twenties. The start of adulthood is imprecise because of the many factors—physical, behavioral, societal—that influence how people define an adult. **Aging**, the normal and progressive decline in the structure and function of the bodies of adults, begins only a few years after growth has ceased.

Observable characteristics of old age—graying and thinning hair, wrinkles and sagging skin, reduced muscle mass, and stooped posture—are familiar to us. However, we also grow old on the inside. Table 18.2 summarizes some of the changes to organ systems that typically occur with aging. For example, bones weaken, and blood vessels stiffen. Nephrons, the functional units of the kidneys, decline in number and efficiency, ultimately challenging the kidneys' efforts to maintain the balance of fluids in our bodies. Aging also strikes our nervous system and sense organs, as evidenced by our deteriorating ability to remember, see, hear, taste, and smell. Of course, we do not suddenly become "old." Changes to body systems occur gradually, many of them beginning in middle age. For example, declines in basal metabolic rate will result in weight gain after about age 35 unless lifestyle changes are made to reduce caloric intake, increase activity, or both. Most people in their forties begin to experience a decreasing ability to focus on close objects, necessitating more frequent visits to the eye doctor and new prescriptions for lenses. Also, around age 50, our ability to hear high-frequency sounds starts to gradually decline, reflecting a slow but continual loss of hair cells in the inner ear that begins in our twenties!

Even so, you might wonder why we discuss aging in a textbook geared toward college students. Our reasoning is simple; you can do a lot right now to ensure a high quality of life during old age (Figure 18.17). Also, as described in the Health Issue essay *Disparities in Health and Health Care at All Life Stages*, significant health and health care disparities exist among populations in the United States, so being aware of these disparities and making your health a key focus while you are young can lead to a much better old age.

FIGURE 18.17 *Wise and informed lifestyle decisions made when young can lead to a healthy old age.*

HEALTH ISSUE

Disparities in Health and Health Care at All Life Stages

Developing good health care habits as a young adult will have lasting positive impacts on the quality of your adult life, aging, and longevity. Unfortunately, significant differences exist in access to health care in the United States, and in the effectiveness of the care received, so you need to be proactive. The Centers for Disease Control and Prevention, the Institute of Medicine, and the U.S. Department of Health and Human Services consistently report differences in health, health care access, and outcomes of care received among certain populations, including those based on age, race, ethnicity, gender, socioeconomic status, sexual orientation, and place of residence (for example, geographic region). Here, we focus on differences related to race and ethnicity.

Some health disparities occur around birth, some in childhood, and others in adulthood. For example, when compared with non-Hispanic whites, African Americans have 2.3 times the infant mortality rate. Asthma is much more prevalent among African American children than among non-Hispanic white children, and among 5- to 24-year-olds, African Americans are four to six times more likely to die from asthma than are whites. Health disparities in mid-to-late adulthood include chronic diseases, such as diabetes: African American and Hispanic adults are almost twice as likely as non-Hispanic white adults to develop diabetes. Further, the rates of cancer, cardiovascular disease, and HIV/AIDS are higher among minority populations than among non-Hispanic whites.

Racial and ethnic disparities extend to access to health care and the quality of health care received at all life stages. The percentage of live births in which the mother received prenatal care starting in the first trimester (months 1 to 3) is 76.1% for African Americans and 88.1% for non-Hispanic whites. Hispanic children between 19 and 35 months of age are less likely than white and African American children to receive all recommended immunizations. Hispanic women over 40 are less likely than white women of this age group to receive recommended screening for breast cancer, and minorities are more likely to have late-stage cancer when the disease is diagnosed. Some health outcomes also are worse for minorities: African Americans and Hispanics with late-stage diabetes are more likely than whites to have their lower legs and feet amputated. With few exceptions, the pattern is that racial and ethnic minorities have higher incidences of chronic diseases (such as asthma, diabetes, and heart disease), higher mortality rates, and lower-quality health care.

Although racial and ethnic disparities in health and health care are well documented, the causes of the disparities have been a challenge to identify, which makes it difficult to solve the problem. A major complication concerns the complex influences of environments—both medical and social—on health problems. How much do disparities reflect unequal treatment by health care professionals? How much do disparities reflect aspects of the social environment, such as income, diet, stress, and other aspects of lifestyle? Do disparities reflect differences in access to education needed for people to make informed choices about their own care? Most people who study health disparities suggest that improvements are necessary both within the health care system and within minority communities (for example, improving the availability of healthy food options and health care providers). Toward this end, Healthy People 2010, an initiative of the U.S. Department of Health and Human Services, lists elimination of health disparities as one of its two major goals (the other is to increase the quality and years of healthy life). Healthy People 2010 emphasizes prevention and builds on similar initiatives over the last few decades (Healthy People 2020 is now being developed). Backed by the best scientific information available, based on broad consultation with health agencies and organizations, and open to input from the public, the initiatives provide 10-year national health objectives that can be used by states and communities to develop programs to improve health. These comprehensive, prevention-based approaches may help to eliminate disparities in health and health care.

Questions to Consider

- If you were charged with improving minority access to health care, where would you begin? Would you increase insurance coverage, availability of providers, health literacy, or something else?
- How might communication between minority patients and their providers be improved?

Possible Causes of Aging

Scientists do not agree about what causes our bodies to age. Some suggest that aging results from changes in critical body systems. For example, aging might be prompted by a decline in function of the immune system or by changing levels of certain hormones. Recall from Chapter 10 that the production of some hormones, such as growth hormone, declines as we age. Perhaps such declines induce the changes in bodily structure and function that characterize aging. Other scientists seek explanations for the causes of aging at the cellular and molecular levels.

Cessation of cell division Ongoing cell division is necessary to replace cells that die. Without new cells, most tissues and organs could not continue to function effectively. Yet cell division appears to slow in aging animals. Cells grown in culture in the laboratory do not divide indefinitely, supporting the idea that cessation of cell division is genetically programmed. In fact, they divide only a certain number of times. The number of divisions seems to be correlated with the age of the individual who donated the cells (whether the donor is a roundworm, mouse, or human) and the life span of the particular species. Telomeres, protective pieces of DNA at the tips of chromosomes, may help cells keep track of how many times they have divided (see Chapter 21a). A tiny piece of each telomere is sliced off each time the DNA is copied before cell division. After a certain number of cell divisions, the telomere is gone, and the chromosome is no longer protected from the slicing. Chromosomes that sustain such damage can no longer participate in cell division. Cells grown in culture demonstrate this phenomenon, yet a clear link to human longevity has not been established.

Damage to DNA and other macromolecules Other researchers postulate that highly reactive molecules known as free radicals disrupt cell processes and lead to aging. Free radicals are by-products of normal cellular activities. They have an unpaired electron and readily combine with and damage DNA, proteins, and lipids. Aging also is associated with a decline in the ability of cells to repair damaged DNA. Such a decline might lead to an accumulation of gene mutations and ultimately to a decline in cell function.

Finally, some scientists have suggested that glucose—our main source of fuel—is the culprit. Glucose changes proteins such as collagen by causing cross-linkages to form between their molecules. This shackling of protein molecules may cause the stiffening of connective tissue and heart muscle associated with aging.

Is there a primary cause of aging? Most scientists believe that aging is not caused by a single factor, but rather by several processes that interact to produce our eventual deterioration and death.

High-Quality Old Age

The news today is full of stories about antiaging products. Aging, however, is a normal biological process that, at present, cannot be slowed, stopped, or reversed. Today, the maximum documented life span for humans is 122 years, a record established by Madame Jeanne Calment of France (Figure 18.18). Most of us will likely never reach her age. In fact, life expectancy for babies born in the United States today is about 77 years. While this news might depress you, especially after reading the long list of declines in organ systems outlined in Table 18.2, the good news is that we can stave off much of the disease and disability associated with aging. The human life span appears to be determined by genes, environment, and lifestyle. Physicians are now able to treat many conditions of old age. For example, worn-out hip or knee joints can be replaced with artificial joints. The clouding of vision caused by cataracts is now treated readily by replacing the old lens of the eye with an artificial lens. Drugs are being tested that may break the cross-links that glucose creates in proteins and thereby restore elasticity to our arteries. In addition, research is under way to determine whether antioxidants (substances that inhibit the formation of free radicals) are effective in delaying certain aspects of aging.

Lifestyle is the factor over which we have the most personal control. Some components of a healthy lifestyle include proper nutrition, plenty of exercise and sleep, refraining from smoking, and routine medical checkups. Indeed, the lifestyle choices we make when we are young can delay some aspects of aging. For

FIGURE 18.18 *The longest documented human life span is 122 years; this record was established by Madame Jeanne Calment, shown here on her last birthday. She died in 1997.*

example, we can delay wrinkling and aging of the skin by avoiding excessive exposure to the sun. The importance of a healthy lifestyle continues into our later years, when exercise, social and intellectual stimulation, and good nutrition can help to prevent the physical and mental declines associated with aging. Although it is never too late to change lifestyle habits, the earlier we begin to develop healthy habits, the better. Finally, aging not only is about declining organ systems, but also has positive aspects as well. Not the least of these, as the years pass, are the gains we make in experience, perspective, and wisdom.

looking ahead

In this chapter, we described the major milestones of human development, with a particular focus on the prenatal period. In the next chapter, we consider the developmental milestones of early childhood and how careful monitoring of these milestones can be used to diagnose autism spectrum disorder. We also describe the symptoms and possible causes of this lifelong disorder as well as available treatments.

HIGHLIGHTING THE CONCEPTS

18.1 Periods of Development in Human Life (p. 372)
- The period of development before birth is the prenatal period. The period of development after birth is the postnatal period.

18.2 Prenatal Period (pp. 372–383)
- The prenatal period can be subdivided into the pre-embryonic period (from fertilization through week 2), embryonic period (from week 3 through week 8), and fetal period (from week 9 until birth).
- The pre-embryonic period is characterized by formation and implantation of the blastocyst. The extraembryonic membranes and placenta also begin to form at this time. The embryonic period is characterized by gastrulation and the formation of organs and organ systems. Intense growth occurs during the fetal period.
- Fertilization is the union of an egg and a sperm to form a zygote.

- Cleavage is a rapid series of mitotic cell divisions that transforms the zygote into a morula (a solid ball of cells) and then a blastocyst (a hollow ball of cells with a fluid-filled cavity). Within the blastocyst are the inner cell mass, which will become the embryo and some of the extraembryonic membranes, and the trophoblast, which will form part of the placenta.
- Toward the end of the first week, the blastocyst begins implantation, the process by which the embryo becomes embedded in the endometrium of the mother's uterus.
- Extraembryonic membranes—the amnion, yolk sac, chorion, and allantois—lie outside the embryo and begin to form a few weeks after fertilization.
- The placenta, formed from the chorion of the embryo and the endometrium of the mother, delivers nutrients and oxygen to the embryo (and later the fetus) and carries wastes away. The placenta also produces hormones such as estrogen, progesterone, and human chorionic gonadotropin (HCG).
- Gastrulation includes the cell movements by which the primary germ layers—ectoderm, mesoderm, and endoderm—are established.
- Ectoderm forms the nervous system and the epidermis and its derivatives. Mesoderm forms muscle, bone, other connective tissue, and organs such as the heart, ovaries, and testes. Endoderm forms organs such as the liver, some endocrine glands, and the lining of the urinary, respiratory, and digestive tracts.
- Neurulation is the process by which the neural tube forms from ectoderm. The anterior portion of the neural tube forms the brain, and the posterior portion forms the spinal cord.
- Somites, blocks of mesoderm organized alongside the neural tube, eventually form vertebrae, connective tissue, and skeletal muscles of the neck and trunk.
- The sex of a human embryo is determined at fertilization by the type of sperm (X-bearing or Y-bearing) that fertilizes the egg (which is X-bearing). An XX zygote will develop into a female, and an XY zygote into a male. Internal reproductive organs and external anatomy begin to differentiate about 6 weeks after fertilization, when a region on the Y chromosome initiates development of testes in male embryos. Absence of the Y chromosome results in the development of a female.
- Growth during the fetal period is extremely rapid. The growth rate of the head slows relative to that of other parts of the body, producing changes in the way the body is proportioned. Allometric growth is change in the relative growth rates of different parts of the body.
- Fetal circulation differs from circulation after birth in that most blood is shunted through temporary vessels or openings past organs such as the lungs and liver, which do not yet perform their postnatal functions. At birth, when organs begin their postnatal functions, the bypasses of fetal circulation close.

18.3 Birth (pp. 383–384)

- Labor is the process by which the fetus is expelled from the uterus and moved through the vagina into the outside world. Parturition (birth) usually occurs about 38 weeks after fertilization and marks the transition from the prenatal period of development to the postnatal period.

18.4 Birth Defects (pp. 385–386)

- Birth defects are developmental defects present at birth. Genetic factors or environmental agents cause them. Disruptive agents have their greatest effects on the developing embryo during periods of rapid differentiation, when organs and organ systems are forming.

18.5 Milk Production by Mammary Glands (p. 386)

- Lactation is the production and ejection of milk from the mammary glands. At birth, when levels of estrogen and progesterone drop, prolactin initiates milk production. Oxytocin causes milk ejection.

18.6 Postnatal Period (pp. 386–389)

- Stages in postnatal development include infancy, childhood, adolescence, and adulthood. Growth stops in adulthood, and a few years later aging begins.
- Aging is the progressive decline in the structure and function of the body. Changes occur in all organ systems. Possible causes of aging include genetically programmed cessation of cell division; damage to DNA, protein, and lipids caused by free radicals; the inability of cells to repair damaged DNA; alteration of proteins by glucose such that cross-linkages form and stiffen tissues; and declines in the functioning of key organ systems.

RECOGNIZING KEY TERMS

fertilization p. 372
pre-embryonic period p. 373
pre-embryo p. 373
embryonic period p. 373
embryo p. 373
fetal period p. 373
fetus p. 373
zygote p. 375
cleavage p. 375
blastocyst p. 376
inner cell mass p. 376
trophoblast p. 376
placenta p. 376
implantation p. 376
human chorionic gonadotropin (HCG) p. 377
infertility p. 377
extraembryonic membranes p. 377
amnion p. 377
yolk sac p. 377
allantois p. 377
umbilical cord p. 377
chorion p. 377
chorionic villi p. 378
cesarean section p. 378
cell differentiation p. 379
morphogenesis p. 379
primary germ layers p. 379
gastrulation p. 379
ectoderm p. 379
endoderm p. 379
mesoderm p. 379
notochord p. 379
neural tube p. 379
neurulation p. 379
somite p. 379
allometric growth p. 381
parturition p. 383
labor p. 383
birth defect p. 385
lactation p. 386
aging p. 387

REVIEWING THE CONCEPTS

1. List the three stages of prenatal development. What ages and major developmental milestones are associated with each stage? p. 373
2. What prompts destabilization of the acrosome on the head of a sperm? p. 374
3. How is fertilization of an oocyte by more than one sperm prevented? p. 374
4. Describe implantation. Where does it usually occur? What happens if it occurs elsewhere? pp. 376–377
5. What are the functions of the four extraembryonic membranes? p. 377
6. What is the placenta, and how does it form? Explain how nutrients, oxygen, and wastes are exchanged between fetal and maternal blood. pp. 377–378
7. What is gastrulation? What tissues and organs are formed from each of the three primary germ layers? p. 379
8. Describe formation of the central nervous system. pp. 379–380
9. How is the sex of a human embryo determined? pp. 380–381
10. How does fetal circulation differ from circulation after birth? pp. 382–383
11. What are the three stages of labor, and what happens during each stage? p. 383
12. Explain the relationship between embryonic age and the effects of environmental agents that disrupt development. pp. 385–386
13. When do a mother's breasts begin producing milk? What explains the timing? p. 386
14. What are some possible causes of aging? What can you do today to improve the quality of your old age? pp. 388–389
15. Fertilization
    a. typically occurs in the uterus.
    b. often involves more than one sperm uniting with an egg.
    c. is the process by which the zygote becomes embedded in the uterine lining.
    d. typically occurs in an oviduct.
16. The yolk sac
    a. is a primary source of nourishment for human embryos.
    b. is the embryo's major contribution to the placenta.
    c. contains primordial germ cells.
    d. encloses the embryo in a fluid-filled sac.
17. Which of the following does *not* characterize the placenta?
    a. It produces estrogen, progesterone, and human chorionic gonadotropin.
    b. It is the site where fetal and maternal blood directly mix.
    c. It usually forms in upper portions of the uterus.
    d. It is expelled after birth.
18. Neurulation
    a. is the process by which the neural tube forms.
    b. is the process by which the notochord forms.
    c. occurs just before gastrulation.
    d. involves endoderm.
19. The fetal period
    a. is the time when tissues and organs form.
    b. is characterized by rapid growth.
    c. runs from week 2 to week 8.
    d. is when exposure to harmful environmental agents is most likely to produce major birth defects.
20. Which of the following statements regarding childbirth is *true*?
    a. Most babies are born feet first.
    b. Delivery of the baby follows delivery of the placenta.
    c. Full-term babies are those born at 32 weeks' gestation.
    d. Infants born prematurely with very low birth weight require intensive care after birth.
21. Within the blastocyst, which of the following becomes part of the placenta?
    a. inner cell mass
    b. yolk sac
    c. chorion derived from trophoblast cells
    d. amnion
22. Soon after fertilization, the zygote undergoes a series of rapid mitotic divisions called
    a. gastrulation.
    b. neurulation.
    c. cleavage.
    d. cell differentiation.
23. Which of the following statements about the notochord is *false*?
    a. The notochord forms from mesoderm.
    b. The notochord is part of the nervous system.
    c. The notochord provides rigidity and defines the long axis of the embryo.
    d. The vertebral column replaces the notochord, which degenerates, leaving remnants in intervertebral disks.
24. Which of the following statements about gastrulation is *false*?
    a. Gastrulation occurs after implantation.
    b. Gastrulation occurs during the fetal period.
    c. Gastrulation establishes three germ layers (ectoderm, mesoderm, and endoderm).
    d. Gastrulation is a key part of morphogenesis.
25. Which hormone stimulates milk ejection?
    a. prolactin
    b. human chorionic gonadotropin
    c. progesterone
    d. oxytocin
26. Highly reactive molecules, produced by normal cellular metabolism, that damage DNA are called
    a. antioxidants.
    b. telomeres.
    c. collagen.
    d. free radicals.

APPLYING THE CONCEPTS

1. Juan was described as a "blue baby" at birth and underwent corrective surgery. What did the surgeons correct, and why?
2. Why might exposure to thalidomide, a drug known to disrupt development of the limbs, during weeks 4 to 6 of gestation produce more severe birth defects than would a similar level of exposure toward the end of gestation?
3. Dora is expecting her baby in about a month. She is deciding whether to breast-feed or bottle-feed. You are her physician. First describe how milk is produced and ejected from the mammary glands. Then, describe the advantages of breast-feeding, and explain the circumstances in which you would recommend bottle-feeding to Dora instead.
4. Your grandmother is taking vitamin E and explains to you that it is a potential antioxidant. What is an antioxidant, and what "condition" is she hoping to treat? Is the vitamin E likely to work?

BECOMING INFORMATION LITERATE

Human growth hormone (HGH) is often touted as having antiaging effects. Prepare a position paper that recommends the use of HGH for this purpose. Use at least three reliable sources (books, journals, or websites). List each source you considered, and explain why you chose the three sources you used.

MasteringBiology®

Go to MasteringBiology for practice quizzes, activities, eText, videos, current events, and more.

SPECIAL TOPIC
Autism Spectrum Disorder

18a

This boy has autism spectrum disorder, a developmental disorder characterized by impaired communication, poor social skills, and unusual patterns of behavior.

18a.1 Characterization and Prevalence

18a.2 Diagnosis

18a.3 Possible Causes

18a.4 Treatment and Therapy

18a.5 Fear That Vaccines Cause Autism Spectrum Disorder

In Chapter 18 we examined human development, with a focus on the prenatal period. In this chapter, we examine autism spectrum disorder, a lifelong neurodevelopmental disorder diagnosed in early childhood. We consider how autism spectrum disorder is defined, diagnosed, and treated. We also look at possible causes. Finally, we explore the fear that childhood vaccines are linked to the onset of the disorder.

18a.1 Characterization and Prevalence

Autism spectrum disorder (ASD) refers to a family of neurodevelopmental disorders characterized by deficits in social communication and interaction and by the performance of repetitive and restricted patterns of behavior. Symptoms must be present in early childhood and impair daily functioning. ASD impairs a child's daily functioning, which can impact parents, siblings, friends, teachers, social workers, and others who care for the child and make decisions regarding special services. For these reasons, ASD can be an emotionally charged and far-reaching issue.

Definitions of mental and neurodevelopmental disorders are based on the *Diagnostic and Statistical Manual of Mental Disorders (DSM)*, a handbook widely used by psychologists and psychiatrists to classify these disorders. The handbook is published and regularly updated by the American Psychiatric Association. The fifth edition (*DSM-5*) is expected in May 2013, and it will contain significant changes from the Fourth Edition, Text Revision of the *DSM* (*DSM-IV-TR*). Four previously distinct neurodevelopmental disorders—*autistic disorder* (also called *autism*), *Asperger's disorder* (sometimes called *Asperger's syndrome*), *pervasive developmental disorder not otherwise specified (PDD-NOS)*, and *childhood disintegrative disorder*—now are subsumed under the single category ASD. The American Psychiatric Association drafted new diagnostic

(a) Failure to make eye contact is an example of an impairment in nonverbal communication that may occur in someone with ASD.

(b) Social withdrawal occurs in some people with ASD.

(c) Restricted and repetitive behaviors are also common in people with ASD. Here, an autistic teen continually touches his ears.

FIGURE 18a.1 *Autism spectrum disorder (ASD) is characterized by deficits in social communication and interaction and by the performance of unusual behaviors.*

criteria for ASD (see below) and posted them on their dedicated website, DSM-5 Development. Public review and comment continued through the spring of 2012.

Using and understanding nonverbal communication pose problems for people with ASD (Figure 18a.1). In some cases, their facial expressions are disconnected from their words; for example, they may smile when describing an extremely sad or frightening experience. Some have difficulty making eye contact, and others stand too close to the person with whom they are speaking. The ability to notice and understand facial expressions and body language of others also may be poorly developed. People with ASD find the give-and-take of normal conversation difficult and may interrupt or continue talking too long on a subject of particular interest to them.

Impairments in communication go hand in hand with impairments in social skills. For example, an individual who has trouble listening to others and engaging in the give-and-take of conversation will likely experience difficult social interactions. Some people with ASD may not want to interact with other people at all. Others may pursue social interactions but have difficulty forming and maintaining social relationships because of their impaired ability to communicate and to interact.

As noted, repetitive movements, such as flapping the hands, can occur with ASD. In children, repeated actions may involve toys (for example, spinning the blades of a toy helicopter close to the eyes) or other objects (for example, repetitively turning lights on and off or opening and closing a door). These repetitive motor activities are called *self-stimulatory activities*, and they, too, make social interaction difficult.

Some people with ASD develop unusual rituals, such as touching several objects in succession and in precise order before leaving home. Whereas these rituals may seem odd or unnecessary to observers, they can be very important to the person performing them. Indeed, disrupting or preventing these rituals can cause extreme frustration and tantrums. Preoccupation with all-consuming interests also characterizes ASD. Some people with ASD may fixate on certain topics, such as train or tide timetables, and talk about them incessantly. Under- or over-reactions to sensory stimuli such as pain or cold are also typical.

Some ASD symptoms can be present in people with other medical conditions; for example, people with Fragile X syndrome may show some symptoms of ASD (see Chapter 20). However, some symptoms can be present in people who do *not* have ASD or another medical condition. For example, some normally developing toddlers may go through a brief period during which they flap their hands when excited or frustrated—a behavior that is typical of children with ASD. Normally developing toddlers tend to flap their hands because they are incapable of fully expressing themselves through speech; when they become more fluent, they stop flapping their hands. In contrast, the stereotyped behaviors such as hand flapping characteristic of some individuals with ASD often extend well beyond the toddler years and can occur with such regularity and for such prolonged periods that they interfere with other activities.

stop and think

Tantrums are not unusual in normally developing toddlers. What would you predict regarding timing and severity of tantrums in children with ASD?

Autism spectrum disorder is quite common: the Centers for Disease Control and Prevention (CDC) estimates that about 1 in 88 children in the United States has ASD. ASD crosses racial, ethnic, and socioeconomic lines, but it is four times more common in boys than girls.

18a.2 Diagnosis

Specific diagnostic criteria for ASD proposed for inclusion in *DSM-5* are shown in Table 18a.1. Deficits in communication and social interaction, which *DSM-IV-TR* lists as separate areas of impairment for autistic disorder, are collapsed into a single area of impairment called *social/communication deficits,* and several diagnostic criteria have been combined and streamlined in this new category. Language delays are no longer considered a diagnostic criterion because such delays are not unique to children with ASD and are not found in all individuals with ASD. *DSM-5* also includes diagnostic criteria—such as overreacting or underreacting to sensory experiences, and unusual interest in sensory aspects of the environment—not present in *DSM-IV-TR* for the subsumed disorders. Despite changes in specific diagnostic criteria over the years, physicians generally begin to consider a diagnosis of ASD when a child fails to meet age-specific developmental milestones.

Children differ in the pace at which they develop. Nevertheless, there are age-specific developmental milestones to which their progress can be compared to identify delays (Figure 18a.2 and Table 18a.2). Identifying developmental delays is important

TABLE 18a.1 Diagnostic Criteria for Autism Spectrum Disorder Proposed for *DSM-5*.

Diagnosis would require impairment in Criteria A (all three deficits) and B (at least two symptoms) as well as timing of onset and limited functioning specified in Criteria C and D.*

Criteria	Deficits and Symptoms
A. Persistent deficits in social communication and social interaction across contexts, not accounted for by general developmental delays	1. Deficits in social-emotional reciprocity 2. Deficits in nonverbal communicative behaviors used for social interaction 3. Deficits in developing and maintaining relationships appropriate to developmental level (beyond those with caregivers)
B. Restricted, repetitive patterns of behavior, interests, or activities	1. Stereotyped or repetitive speech, motor movements, or use of objects 2. Excessive adherence to routines, ritualized patterns of verbal or nonverbal behavior, or excessive resistance to change 3. Highly restricted, fixated interests that are abnormal in intensity or focus 4. Hyper- or hypo-reactivity to sensory input or unusual interest in sensory aspects of the environment
C. Symptoms must be present in early childhood (but may not become fully manifest until social demands exceed limited capacities)	
D. Symptoms together limit and impair everyday functioning	

*Modified from Proposed Revision to *DSM-5*, January 26, 2011.

(a) One milestone, typically attained by the end of the first year, is the ability to pull up to a standing position.

(b) The motor skills needed to build towers of blocks or other toys are usually attained after the first year.

(c) Some milestones concern social skills and behavior; these include showing independence and defiance.

FIGURE 18a.2 *The pace at which a young child develops can be viewed in the context of age-specific developmental milestones. Shown here is the same child at 10 months, 20 months, and nearly 3 years of age.*

TABLE 18a.2 Some Developmental Milestones Used to Assess a Young Child's Progress in the Areas of Communication, Social Behavior, and Motor Skills

Age	Developmental Milestones		
	Communication	Social Behavior	Motor Skills
By the end of the first year	Babbles with changes in tone Tries to imitate words Uses simple gestures and single words	Is shy with strangers Repeats gestures for attention Tests parental responses to behavior	Reaches sitting position without help Crawls forward on belly Pulls self to standing position
By the end of the second year	Uses simple phrases and several single words Points to named objects Repeats words heard in conversation	Imitates behavior of others Shows independence and defiance Shows some self-awareness	Walks independently and begins to run Kicks a ball Pulls or carries a toy while walking
By the end of the third year	Recognizes and identifies common objects Uses sentences with four or five words Speaks such that strangers can understand most words	Spontaneously shows affection Takes turns in games Separates from parents with relative ease	Runs with ease Pedals a tricycle Builds a tower of more than six blocks

*More complete lists of developmental milestones can be found at www.cdc.gov/ncbddd/autism/actearly/screening.

because early diagnosis of ASD can lead to early intervention, which can benefit the child. To this end, pediatricians and nurses routinely conduct developmental screening of infants and young children at periodic physical examinations. In addition to asking the parents about their child's development, these health care professionals directly interact with the child to see how the child reacts, speaks, and moves (Figure 18a.3). If delays are noted, then the parents may be advised to seek the opinion of a developmental pediatrician or psychologist, and additional screening may be performed. This screening typically involves several of the following diagnostic tools: extensive interviews of the parents regarding their child's developmental history; detailed records from the pediatrician; clinical observations of the child's behavior; psychological testing; speech, language, and hearing assessments; physical and neurological examinations; and reviews of family history.

Several diagnostic checklists and rating scales are available. Results from screening can be used by a trained clinician, together with other diagnostic tools, to consider a diagnosis of ASD.

18a.3 Possible Causes

Early thinking about neurodevelopmental disorders now subsumed under ASD regarded aberrant parenting as the cause. According to the "Refrigerator Mother" hypothesis, a cold, uncaring parent (usually the mother) caused the child to retreat into ASD. It was not until the mid-1960s that this hypothesis was refuted and replaced with the understanding that ASD is a neurodevelopmental disorder with a genetic basis. ASD cannot be linked to any particular parenting style.

Although we know that aberrant parenting does not cause ASD, we still do not know the precise causes of ASD. Most researchers agree that there are probably multiple causes and that both genes and environment play a role.

Some research focuses on neurobiological bases of ASD. Neuroimaging studies and studies of brain tissue from people with ASD indicate fundamental differences from controls in brain growth and organization. Many of these differences probably originate in the prenatal period. For example, abnormal migration of cells during formation of the nervous system may affect brain structure and neural circuitry, and ultimately influence thinking and behavior. Imaging the brain during cognitive tasks or response to visual or auditory stimulation suggests that people with ASD use different cognitive strategies and may process certain types of information in different brain areas. A particularly consistent finding of neuroimaging studies is that the part of the brain involved in face recognition is underactivated in people with ASD. Despite these advances in identifying structural and functional differences in the brains of people with ASD, a reliable marker has yet to be identified, and, for this reason, routine neuroimaging is not recommended as a diagnostic tool.

Two main types of studies indicate that ASD has a strong genetic component. First, studies show higher-than-expected rates of ASD among family members. For example, the rate of ASD in siblings of ASD individuals is 2% to 6%, which is 10 to 60 times greater than the prevalence of ASD in the general population. Second, studies of twins demonstrate that ASD has a genetic basis.

Recall from Chapter 18 that dizygotic (fraternal) twins occur when two secondary oocytes are fertilized by different sperm; these twins are no more genetically similar than siblings who are not twins. Monozygotic twins result from the splitting of a pre-embryo and are nearly genetically identical. (We say "nearly" here because new research has shown that although identical twins have very similar DNA, they do not have identical DNA. Identical twins may differ, for example, in the number of copies of a particular gene. Prior to this discovery, scientists knew that identical twins experience epigenetic changes, which are changes in gene expression induced by environmental factors, such as diet. Although epigenetic changes affect the way genes are expressed, they do not reflect changes in the underlying DNA sequence, so this new information about identical twins is very exciting.)

Researchers studying the genetics of ASD have compared its incidence in monozygotic and dizygotic twins. If one monozygotic twin has ASD, then there is a 75% chance that the other twin will have ASD. In contrast, if one dizygotic twin has ASD, then there is a 3% chance that the other twin will have ASD (note that this falls within the 2% to 6% rate of recurrence

Q Why does developmental screening play such an important role in diagnosing ASD? Is there a medical test for this disorder?

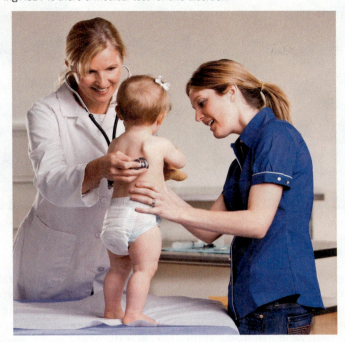

FIGURE 18a.3 Routine developmental screening at periodic visits to the pediatrician can help identify developmental delays or losses.

A There is no medical test (e.g., blood test or scan of the brain) for ASD; instead, doctors rely on several tools to assess behavioral symptoms. Typically, delays detected during routine developmental screening are further evaluated through parent interviews, psychological testing, and assessments of speech, language, and hearing.

among non-twin siblings mentioned previously). These comparative findings indicate a genetic basis for ASD because the incidence among twins is higher when they are nearly genetically identical. However, data for monozygotic twins also suggest an environmental component because although they are nearly genetically identical, it is not always the case that both twins will have ASD. These findings suggest that one twin was exposed to environmental factors during the prenatal and/or postnatal period that, in combination with a susceptible genetic background, led to the expression of ASD.

It is unlikely that a single gene is responsible for the genetic component of ASD. Indeed, current models suggest that several genes, perhaps as many as 15, are involved and that when these genes are present in combination, they lead to an increased vulnerability to ASD that may be triggered by environmental factors. Research is under way to identify candidate genes and environmental triggers. Finally, older fathers may be more likely than younger fathers to have a child with ASD. An older father has had more time for random genetic mutations to occur (mutations are changes in DNA; see Chapter 21). And, the more mutations a child inherits, the more likely that some will increase ASD risk. A mother's age did not contribute to this genetic risk.

Little is known about environmental risk factors for ASD. Some research focuses on maternal effects (specifically the intrauterine environment) and complications during pregnancy and birth. Prenatal exposure to certain prescription drugs, such as thalidomide or valproic acid (an anticonvulsant and mood stabilizer), is linked to an increased risk of ASD. Some evidence links maternal illness, such as rubella (German measles), during pregnancy to ASD. Recent studies indicate an association between infertility treatments (specifically, drugs used to induce ovulation) and ASD, and the association appears to strengthen with duration of treatment. In other words, the longer a woman is treated for infertility, the higher are her chances of having a child with ASD. The challenge now is to sort out the roles of many factors—premature delivery, low birth weight, twinning, and maternal age—that are associated with both ASD and infertility treatments. One now discredited environmental variable—early childhood vaccines—is discussed later in the chapter.

18a.4 Treatment and Therapy

There are no cures for ASD, but educational interventions can help children with these neurodevelopmental disorders reach their full potential. Successful educational programs provide structure, organization, and direction for the child. Such programs directly teach children with ASD the social skills that other children learn by other means. Given the enormous individual variation in manifestation of ASD, it is imperative to design educational programs for the individual child; to this end, school-aged children with ASD typically have an Individualized Educational Program (IEP). An IEP is developed and periodically reviewed by a team that typically includes the parents (or guardians), teachers, social workers, school psychologists, and representatives of the school district. An IEP is federally guaranteed under the Individuals with Disabilities Education Act (IDEA) and ensures that a child with ASD will receive the most appropriate education available. In addition to classroom accommodations, such programs may include occupational therapy (designed to improve social, fine motor, and self-help skills) and sensory integration therapy (designed to facilitate development of the nervous system's ability to process external stimuli, such as touch, sound, and sight, in more typical ways). Dogs or horses are sometimes included in sensory integration activities (Figure 18a.4).

FIGURE 18a.4 Sensory integration therapy helps children with ASD process sensory stimuli in more typical ways. Here, a dog is part of the therapy.

To date, no therapeutic drug is known to improve core symptoms of ASD, which are impairments in communication and social behavior coupled with repetitive or unusual behaviors. Ongoing research holds promise. One candidate drug is oxytocin, a hormone normally released by the posterior pituitary gland (see Chapter 10), delivered to patients as a nasal spray. Oxytocin promotes uterine contractions during childbirth, prompts milk letdown during lactation, and enhances social bonding. A small study published in 2010 showed that adults with ASD who had been treated with oxytocin were more likely to focus on the eyes when viewing pictures of human faces. Other measures of social function also improved. Several drugs are used with limited to moderate success to treat behavioral features that accompany ASD. For example, the U.S. Food and Drug Administration approved the use of the antipsychotic drug risperidone in children with ASD who display severe tantrums, aggression, or self-injurious behavior. Other drugs may be prescribed to treat hyperactivity, impulsivity, and anxiety often present in children with ASD.

The prognosis for individuals with ASD depends on many factors. In general, for children, better outcomes are associated with better cognitive abilities, less severe symptoms overall, and early identification of the disorder and appropriate intervention programs. The outcomes for adults with ASD correlate best with cognitive abilities; those with at least normal intelligence have the best outcomes.

18a.5 Fear That Vaccines Cause Autism Spectrum Disorder

A major medical success story of the last century has been the development of vaccines to prevent potentially deadly diseases, including measles, mumps, rubella, diphtheria, tetanus, and polio. Indeed, vaccine programs have reduced, and in some cases even eradicated, serious infectious diseases and saved innumerable lives. These programs have been so successful that many people today may not know what these diseases even look like or appreciate how severe they can be. Despite the success of immunization programs, however, concerns about a possible link between certain vaccines and ASD have received much attention over the last decade (Figure 18a.5). The debate continues, even though there is no scientific evidence of a causal relationship between childhood vaccines and ASD.

The first hypothesis that such a link existed came from a now discredited 1998 study of the records of 12 children who had both ASD and irritable bowel disease (a condition thought to be caused by a persistent measles-virus infection in the intestines). The parents or pediatricians of nine of the children suspected that the MMR (measles, mumps, and rubella) vaccine the children had received had caused the irritable bowel disease and contributed to the children's ASD. The idea, which has not been supported by subsequent research, included the following suppositions: (1) The live virus in the MMR vaccine could cause a chronic measles infection in certain susceptible children; (2) this, in turn, could damage the bowels of these children; (3) the damaged bowels could allow certain

FIGURE 18a.5 *Actress Jenny McCarthy has been outspoken in her belief that vaccines are an environmental trigger for ASD.*

peptides that are normally broken down by the gut to leak into the bloodstream before being completely broken down; and (4) once in the bloodstream, the peptides could damage the developing brain. Because there were too few cases in the study to establish a causal link between the MMR vaccine and ASD, 10 of the 13 authors of the paper later retracted the interpretation of the data that had been presented in the paper. Other scientists criticized the study for lacking controls and using a biased sample of subjects.

Subsequent studies explored the possible link between the MMR vaccine and ASD. The Institute of Medicine (IOM) of the National Academies reviewed all of these studies and issued reports in 2001 and 2004 concluding that there is no link between the MMR vaccine and ASD. Studies published since 2004 have supported this conclusion. Indeed, a study published in 2008 found no evidence of association between ASD and persistent measles infection in the intestines or exposure to the MMR vaccine. In 2010, the editors of the journal that had published the 1998 study retracted the paper from the published record, citing its claims as false.

Another hypothesis suggested that a mercury-containing organic compound called thimerosal might cause ASD. Thimerosal was used to prevent the growth of microorganisms in certain vaccines from the 1930s until the early 2000s. (Thimerosal was never used in the MMR vaccine, the other focus of contention.) Exposure to mercury is harmful to the nervous system, so it is perhaps reasonable to think that thimerosal might cause problems in the developing nervous system that might result in ASD. In 2004, an IOM report concluded that it is biologically plausible for the cumulative amount of mercury in the series of recommended vaccinations for infants and toddlers to exceed maximum federal safety standards. However, the standards pertain to *methylmercury*, whereas thimerosal contains *ethylmercury*, which is handled differently in the body. Further, although the IOM did not dispute that mercury-containing compounds, such as thimerosal, might damage the nervous system, there was no evidence that such damage was associated with ASD. Thus, the IOM concluded in 2004 that the bulk of evidence rejects a causal relationship between thimerosal-containing vaccines and ASD. To date, there is no convincing scientific evidence to refute this conclusion.

Despite the lack of evidence that thimerosal-containing vaccines cause ASD, the Public Health Service agencies, the American Academy of Pediatrics, and vaccine manufacturers agreed in 1999 that, as a precautionary measure, thimerosal should be reduced or eliminated in vaccines. Although it took time and effort to develop thimerosal-free vaccines, it was feasible to remove mercury from vaccines and thereby reduce an individual's total exposure to mercury. It is far more difficult to control an individual's exposure to other sources of environmental mercury (forest fires, landfills, incinerators, and certain industrial processes). Currently, vaccines routinely recommended for children under the age of 6 do not contain thimerosal, except for some formulations of the inactivated influenza (flu) vaccine. Vaccines for adults and for children 7 years of age and older are increasingly available in formulations without thimerosal.

Current scientific evidence does not support a causal association between either the MMR vaccine or thimerosal-containing vaccines and ASD. Even so, a recent survey of parents of children with ASD showed that 54% believe immunizations caused their child's ASD.

what would you do?

If you were a young pediatrician just establishing a practice and a parent refused to vaccinate her children, would you exclude the family from your practice? Why or why not?

looking ahead

In Chapter 18a, we discussed autism spectrum disorder, which has a strong genetic component. In Chapter 19, we take a closer look at chromosomes, which are the physical basis of heredity.

19 Chromosomes and Cell Division

Did you Know?

- If the DNA from the chromosomes in a single cell were unwound, as it is during interphase, and attached end to end, it would be more than 5 feet long but only 50 trillionths of an inch in diameter.
- You developed from one of more than 8.3 million possible genetically different gametes that your mother might have produced.

Down syndrome, which results from an error in cell division, is the most frequent inherited cause of mild to moderate retardation.

- 19.1 Two Types of Cell Division
- 19.2 Form of Chromosomes
- 19.3 The Cell Cycle
- 19.4 Mitosis: Creation of Genetically Identical Diploid Body Cells
- 19.5 Cytokinesis
- 19.6 Karyotypes
- 19.7 Meiosis: Creation of Haploid Gametes

ETHICAL ISSUE
Trisomy 21

In the previous few chapters, we considered reproduction and development. In this chapter, we examine the role of two types of cell division, mitosis and meiosis, in the human life cycle. We consider the physical basis of heredity—the chromosomes—and how the chromosomes are parceled out during mitosis and meiosis. We finish the chapter by examining why it is important for each cell to have the correct number of chromosomes.

19.1 Two Types of Cell Division

We begin life as a single cell called a *zygote,* formed by the union of an egg and a sperm. By adulthood, our bodies consist of trillions of cells. What happened in the intervening years? How did we go from a single cell to the multitude of cells that make up the tissues of a fully functional adult? The answer is cell division, which happened over and over again as we grew. Even in adults, many cells continue to divide for growth and repair of body tissues. With very few exceptions, each of those cells carries the same genetic information as its ancestors. The type of nuclear division that results in identical body cells is called *mitosis.*

In Chapter 17 you learned that males and females produce specialized reproductive cells called gametes (eggs or sperm). You'll recall that *meiosis* is a special type of nuclear division that gives rise to gametes. In females, meiosis occurs in the ovaries and produces eggs. In males, meiosis occurs in the testes and produces sperm. Meiosis is important because through it the gametes end up with half the amount of genetic information (half the number of chromosomes) in the original cell. When the nuclei of an egg and sperm unite (fertilization), the chromosome number is restored to that of the original cell. As a result, the number of chromosomes in body cells remains constant from one generation to the next.

The roles of mitosis (which produces new body cells) and meiosis (which forms gametes) are summarized in the diagram of the human life cycle in Figure 19.1. You will learn more about both mitosis and meiosis later in this chapter.

19.2 Form of Chromosomes

A **chromosome** is a tightly coiled combination of a DNA molecule (which contains genetic information for the organism) and specialized proteins called *histones*. Chromosomes are found in the cell nucleus. The information contained in the DNA molecules in chromosomes directs the development and maintenance of the body. The histones combined with the DNA are for support and control of gene activity. A **gene** is a specific segment of the DNA that directs the synthesis of a protein, which in turn plays a structural or functional role within the cell. By coding for a specific protein, a gene determines the expression of a particular characteristic, or trait. Each chromosome in a human cell contains a specific assortment of genes. Like beads on a string, genes are arranged in a fixed sequence along the length of specific chromosomes.

In the human body, **somatic cells**—that is, all cells except for eggs or sperm—have 46 chromosomes. Those 46 chromosomes are actually 23 pairs of chromosomes. One member of each pair came from the mother's egg, and another member of each pair came from the father's sperm. Thus, each cell contains 23 **homologous chromosome pairs,** a pair being two chromosomes (one from the mother and one from the father) with genes for the same traits. Homologous pairs are called *homologues* for short. Any cell with two of each kind of chromosome is described as being **diploid** (annotated as $2n$, with n representing the number of each kind of chromosome). In diploid cells, then, genes also occur in pairs. The members of each gene pair are located at the same position on homologous chromosomes.

One of the 23 pairs of chromosomes consists of the **sex chromosomes** that determine whether a person is male or female. There are two types of sex chromosomes, X and Y. A person who has two X chromosomes is described as XX and is genetically female; a person who has an X and a Y chromosome is described as XY and is genetically male. The other 22 pairs of chromosomes are called the **autosomes**. The autosomes determine the expression of most of a person's inherited characteristics.

19.3 The Cell Cycle

In **mitosis**, one nucleus divides into two daughter nuclei containing the same number and kinds of chromosomes. But mitosis is only one phase during the life of a dividing cell. The

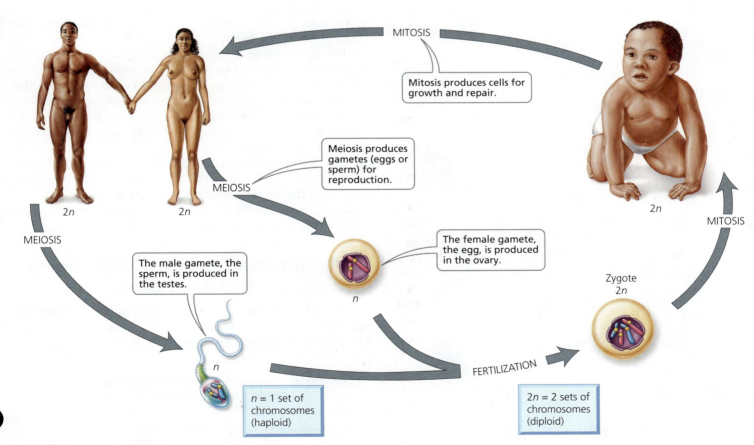

FIGURE 19.1 The human life cycle

entire sequence of events that a cell goes through, from its origin in the division of its parent cell through its own division into two daughter cells, is called the **cell cycle** (Figure 19.2). The cell cycle consists of two major phases: interphase and cell division.

Interphase

Interphase is the period of the cell cycle between cell divisions. It accounts for most of the time that elapses during a cell cycle. During active growth and divisions (depending on the type of cell), an entire cell cycle might take about 16 to 24 hours to complete, and only 1 to 2 hours are spent in division. Interphase is not a "resting period," as once thought. Instead, interphase is a time when the cell carries out its functions and grows. If the cell is going to divide, interphase is a time of intense preparation for cell division. During interphase, the DNA and organelles are duplicated. These preparations ensure that when the cell divides, each of its resulting cells, called *daughter cells,* will receive the essentials for survival.

Interphase consists of three parts: G_1 (first "gap"), S (DNA synthesis), and G_2 (second "gap"). All three parts of interphase are times of cell growth, characterized by the production of organelles and the synthesis of proteins and other macromolecules. There are, however, some events specific to certain parts of interphase:

- G_1: A time of major growth before DNA synthesis begins
- S: The time during which DNA is synthesized (replicated)
- G_2: A time of growth after DNA is synthesized and before mitosis begins

The details of DNA synthesis (replication) are described in Chapter 21. Our discussion in this chapter introduces some basic terminology pertaining to the cell cycle.

Throughout interphase, the genetic material is in the form of long, thin threads that are often called *chromatin* (Figure 19.3). They twist randomly around one another like tangled strands of yarn. In this state, DNA can be synthesized (replicated), and genes can be active. At the start of interphase, during G_1, each chromosome consists of a DNA molecule and proteins. When the chromosomes are being replicated during the S phase, the chromosome copies remain attached. The two copies, each an exact replicate of the original chromosome, stay attached to one another at a region called the **centromere**. As long as the replicate copies remain attached, each copy is called a **chromatid**. The two attached chromatids are genetically identical and are called *sister chromatids.*

Division of the Nucleus and the Cytoplasm

Body cells divide continually in the developing embryo and fetus. Such division also plays an important role in the growth and repair of body tissues in children. In the adult, specialized cells, such as most nerve cells, lose their ability to divide. Late in G_1 of interphase, these cells enter what is called the G_0 stage: they are carrying out their normal cellular activities but do not divide. Other adult cells, such as liver cells, stop dividing but retain the ability to undergo cell division should the need for tissue repair and replacement arise. Still other cells actively divide throughout life. For example, the ongoing cell division in skin cells in adults serves to replace the enormous numbers of cells worn off each day.

We see, then, that the cell cycle requires precise timing and accuracy. Proteins monitor the environment within the cell to ensure that it is appropriate for cell division and that the DNA has been accurately replicated. Healthy cells will not divide unless these two conditions are met. However, as we will see in Chapter 21a, cancer cells escape this regulation and divide uncontrollably.

The division of body cells (after interphase) consists of two processes that overlap somewhat in time. The first process, division of the nucleus, is called *mitosis.* The second process is *cytokinesis,* which is the division of the cytoplasm that occurs toward the end of mitosis (Figure 19.4).

19.4 Mitosis: Creation of Genetically Identical Diploid Body Cells

For the purpose of discussion, mitosis is usually divided into four stages: prophase, metaphase, anaphase, and telophase. The major events of each stage are depicted in Figure 19.5 (pp. 404–405).

- **Prophase.** Mitosis begins with **prophase**, a time when changes occur in the nucleus as well as the cytoplasm. In the nucleus, the chromatin condenses and forms chromosomes as DNA wraps around histones. The DNA then loops and twists to form a tightly compacted structure (see Figure 19.3). When DNA is in this condensed state, it cannot be replicated, and gene activity is shut down. In

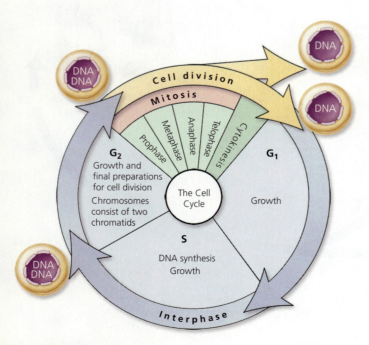

FIGURE 19.2 *The cell cycle*

19.4 Mitosis: Creation of Genetically Identical Diploid Body Cells 403

Q Describe the difference in the structure of a chromosome between the start of interphase and at the end of interphase.

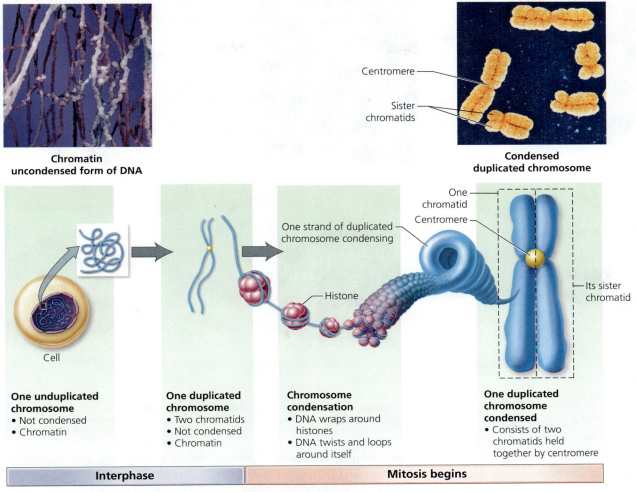

FIGURE 19.3 Changes in chromosome structure due to DNA replication during interphase and preparation for nuclear division in mitosis

A At the start of interphase, a chromosome is a single strand of DNA. At the end of interphase, a chromosome consists of two sister chromatids that are replicate copies of the original strand of DNA held together by a centromere.

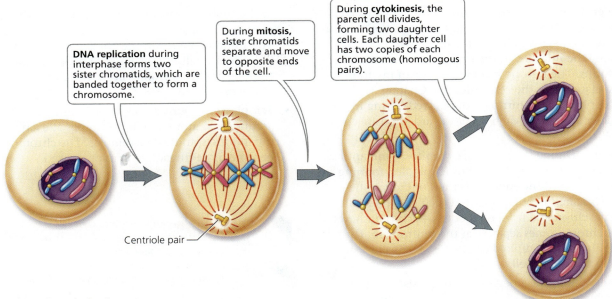

FIGURE 19.4 An overview of mitosis

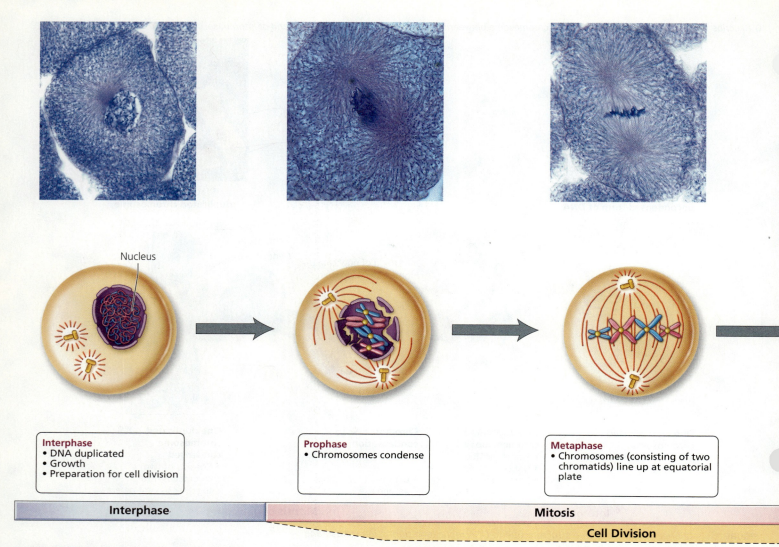

FIGURE 19.5 *The stages of cell division (mitosis and cytokinesis) captured in light micrographs and depicted in schematic drawings*

this condensed state, the sister chromatids are easier to separate without breaking. At about this time, the nuclear membrane also begins to break down.

Outside the nucleus, in the cytoplasm, the mitotic spindle forms. The mitotic spindle is made of microtubules associated with the centrioles (see Chapter 3). During prophase, the centrioles, duplicated during interphase, move away from each other toward opposite ends of the cell.

- **Metaphase.** During the next stage of mitosis, **metaphase**, the chromosomes attach to the mitotic spindles, forming a line at what is called the equator (center) of the mitotic spindles. This alignment ensures that each daughter cell receives one chromatid from each of the 46 chromosomes when the chromosomes separate at the centromere. Thus, each daughter cell is a diploid cell that is genetically identical to the parent cell.
- **Anaphase.** **Anaphase** begins when the sister chromatids of each chromosome begin to separate, splitting at the centromere. Now separate entities, the sister chromatids are considered chromosomes in their own right. The spindle fibers pull the chromosomes toward opposite poles of the cell. By the end of anaphase, equivalent collections of chromosomes are located at the two poles of the cell.
- **Telophase.** During **telophase**, a nuclear envelope forms around each group of chromosomes at each pole, and the mitotic spindle disassembles. The chromosomes also become more threadlike in appearance.

stop and think

Cancer cells divide rapidly and without end. One type of drug used in cancer chemotherapy inhibits the formation of spindle fibers. Why can this be an effective anticancer treatment?

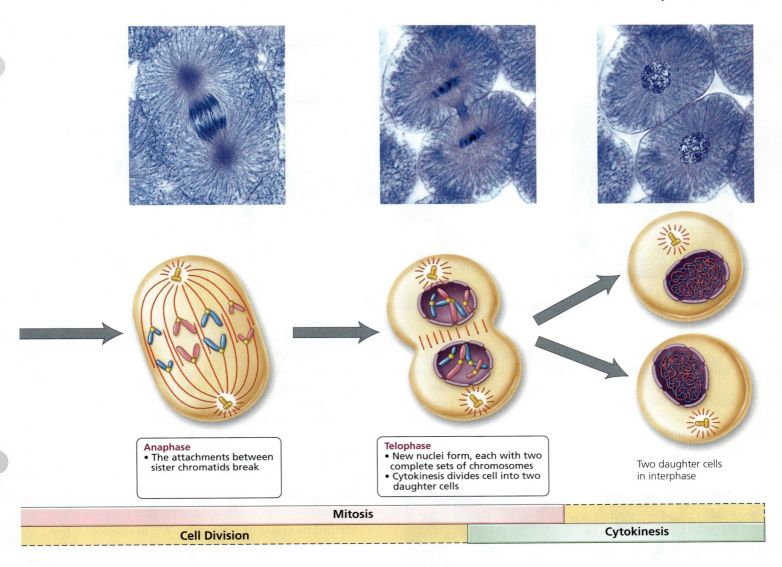

19.5 Cytokinesis

Cytokinesis—division of the cytoplasm—begins toward the end of mitosis, sometime during telophase. During this period, a band of microfilaments in the area where the chromosomes originally aligned contracts and forms a furrow, as shown in Figure 19.6. The furrow deepens, eventually pinching the cell in two.

stop and think

What would happen if a cell completed mitosis but did not complete cytokinesis?

19.6 Karyotypes

As we have seen, a major feature of cell division is the shortening and thickening of the chromosomes. In this state, the chromosomes are visible with a light microscope and can be used for diagnostic purposes, such as when potential parents want to check their own chromosomal makeup for defects. One often-used method takes white blood cells from a blood sample and grows them for a while in a nourishing medium. The culture then is treated with a drug that destroys the mitotic spindle, thus preventing separation of the chromosomes and halting cell division at metaphase. Next the cells are fixed, stained, and photographed so that the images of the chromosomes can be arranged in pairs based on physical characteristics, such as location of the centromere and overall length. The chromosomes are numbered from largest to smallest, in an arrangement called a **karyotype** (Figure 19.7). Karyotypes can be checked for irregularities in number or structure of chromosomes.

19.7 Meiosis: Creation of Haploid Gametes

We have seen that the somatic cells contain a homologous pair of each type of chromosome, one member of each pair from the father and one member of each pair from the mother.

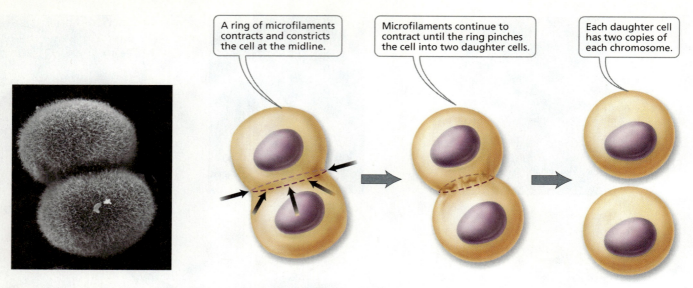

FIGURE 19.6 Cytokinesis is the division of the cytoplasm to form two daughter cells.

Recall that a cell with homologous pairs of chromosomes is described as being diploid, $2n$. The gametes—eggs or sperm—differ from somatic cells in that they are **haploid**, indicated by n, meaning that they have only one member of each homologous pair of chromosomes. As you read earlier in the chapter, gametes are produced by a type of cell division called **meiosis**, which is actually two divisions that result in up to four haploid daughter cells. When a sperm fertilizes an egg, a new cell—the zygote—is created. Because the egg and sperm both contribute a set of chromosomes to the zygote, it is diploid. After many mitotic cell divisions, a zygote and its descendant cells can eventually develop into a new individual.

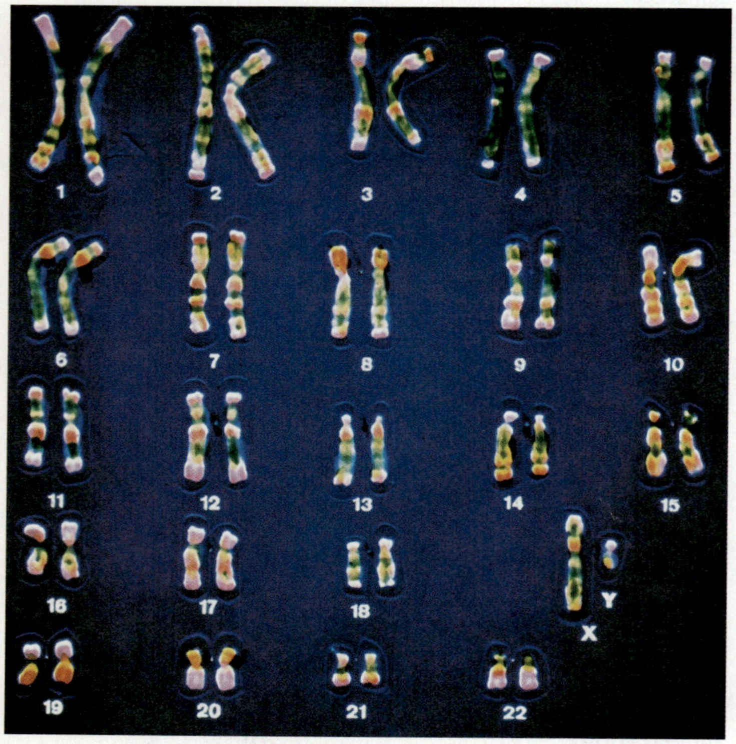

Functions of Meiosis

Meiosis serves two important functions in sexual reproduction:

- Meiosis keeps the number of chromosomes in a body cell constant from generation to generation.
- Meiosis increases genetic variability in the population.

Meiosis keeps the number of chromosomes in a body cell constant over generations because it creates haploid gametes (sperm and eggs) with only one member of each homologous pair of chromosomes. If gametes were produced by mitosis, they would be diploid; each sperm and egg would contain 46 chromosomes instead of 23. Then, if a sperm containing 46 chromosomes fertilized an egg with 46 chromosomes, the zygote would have 92 chromosomes. The zygote of the next generation would have 184 chromosomes, having been formed by an egg and sperm each containing 92 chromosomes. The next generation would have 368 chromosomes in each cell, and the next one 736—and so on. You can see that the chromosome number would quickly become unwieldy and, more importantly, alter the amount of genetic information in each cell. As we will see toward the chapter's end, even one extra copy of a single chromosome usually causes an embryo to die.

Meiosis also increases genetic variability in the population. Later in this chapter we consider the mechanisms by which it accomplishes this increase. Genetic variability is important because it provides the raw material through which natural selection can act, leading to the changes described collectively as evolution. The relationship between genetic variability and evolution is discussed in Chapter 22.

FIGURE 19.7 Chromosomes in dividing cells can be examined for defects in number or structure. A karyotype is constructed by arranging the chromosomes from photographs based on size and centromere location.

Two Meiotic Cell Divisions: Preparation for Sexual Reproduction

First, let's consider how meiosis keeps the chromosome number constant. The stages in meiosis are summarized in Figure 19.8. Meiosis and mitosis begin the same way. Both are preceded by the same event—the replication of chromosomes. Unlike mitosis, however, meiosis involves *two* divisions. In the first division, the chromosome number is reduced, because the two homologues of each pair of chromosomes (each replicated into two chromatids attached by a centromere) are separated into two cells so that each cell has one member of each homologous pair of chromosomes. In the second division, the replicated chromatids of each chromosome are separated. We see, then, that meiosis begins with one diploid cell and, two divisions later, produces four haploid cells. The orderly movements of chromosomes during meiosis ensure that each haploid gamete produced contains one member of each homologous pair of chromosomes. Although not shown in the summary figure, each of the two meiotic divisions has four stages similar to those in mitosis: prophase, metaphase, anaphase, and telophase.

Meiosis I The first meiotic division—meiosis I—produces two cells, each with 23 chromosomes. Note that the daughter cells do not contain a random assortment of any 23 chromosomes. Instead, each daughter cell contains one member of each homologous pair, with each chromosome consisting of two sister chromatids.

It is important that each daughter cell receive one of each kind of chromosome during meiosis I. If one of the daughter cells had two of chromosome 3 and no chromosome 6, it would not survive. Although there would still be 23 chromosomes present, part of the instructions for the structure and function of the body (chromosome 6) would be missing. The separation of homologous chromosomes occurs reliably during meiosis I because, during prophase I (the *I* indicates this phase takes place during meiosis I), members of homologous pairs line up next to one another by a phenomenon called **synapsis** ("bringing together"). For example, the chromosome 1 that was originally from your father would line up with the chromosome 1 originally from your mother. Paternal chromosome 2 would pair with maternal chromosome 2, and so on. During metaphase I, matched homologous pairs become positioned at the midline of the cell and attach to spindle fibers. The pairing of homologous chromosomes helps ensure that the daughter cells will receive one member of each homologous pair. Consider the following analogy. By pairing your socks before putting them in a drawer, you are more likely to

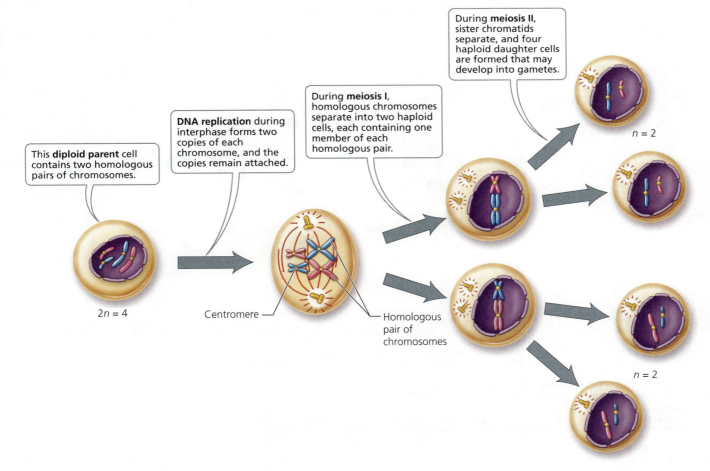

FIGURE 19.8 *Overview of meiosis. Meiosis reduces the chromosome number from the diploid number to the haploid number. Meiosis involves two cell divisions.*

put matching socks on your feet than if you randomly pulled out two socks.

Next, during anaphase I, the members of each homologous pair of chromosomes separate, and each homologue moves to opposite ends of the cell. During telophase I, cytokinesis begins, resulting in two daughter cells, each with one member of each chromosome pair. Each chromosome in each daughter cell still consists of two replicated sister chromatids. Telophase I is followed by *interkinesis,* a brief interphase-like period. Interkinesis differs from mitotic interphase in that there is no replication of DNA during interkinesis.

Meiosis II During the second meiotic division—meiosis II—each chromosome lines up in the center of the cell independently (as occurs in mitosis), and the sister chromatids (attached replicates) making up each chromosome separate.

Separation of the sister chromatids occurs in both daughter cells that were produced in meiosis I. This event results in four cells, each containing one of each kind of chromosome. The events of meiosis II are similar to those of mitosis, except that only 23 chromosomes are lining up independently in meiosis II compared with the 46 chromosomes aligning independently in mitosis. Figure 19.9 depicts the events of meiosis. Table 19.1 and Figure 19.10 compare mitosis and meiosis.

Subsequent changes in the shape and functioning of the four haploid cells result in functional gametes. In males, one diploid cell can result in four functional sperm. In contrast, in females, meiotic divisions of one diploid cell result in only one functional egg and up to three nonfunctional cells called *polar bodies.* As a result, the egg contains most of the nutrients found in the original diploid cell, and these nutrients will nourish the early embryo (Figure 19.11, p. 411).

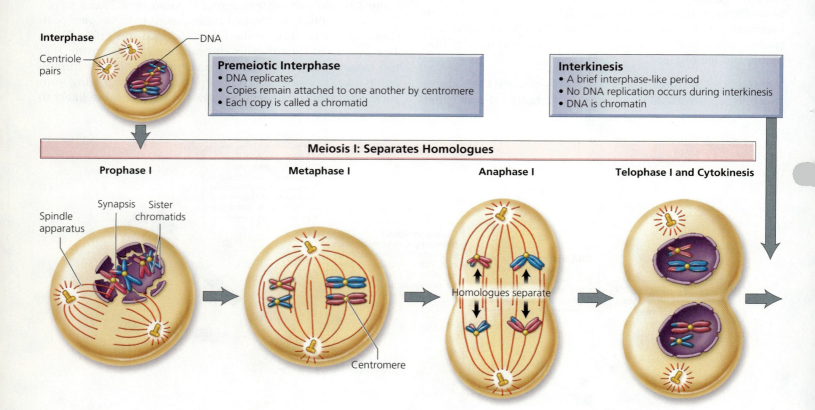

FIGURE 19.9 *Stages of meiosis*

stop and think

If you were examining dividing cells under a microscope, how could you determine whether a particular cell was in metaphase of mitosis or metaphase I of meiosis?

Genetic Variability: Crossing Over and Independent Assortment

At the moment of fertilization, when the nuclei of an egg and a sperm fuse, a new, *unique* individual is formed. Although certain family characteristics may be passed along, each child bears its own assortment of genetic characteristics (Figure 19.12).

Genetic variation arises largely because of the shuffling of maternal and paternal forms of genes during meiosis. One

TABLE 19.1 Mitosis and Meiosis Compared

Mitosis	Meiosis
Involves one cell division	Involves two cell divisions
Produces two diploid cells	Produces up to four haploid cells
Occurs in somatic cells	Occurs only in ovaries and testes during the formation of gametes (egg and sperm)
Results in growth and repair	Results in gamete (egg and sperm) production
No exchange of genetic material	Parts of chromosomes are exchanged in crossing over
Daughter cells are genetically similar	Daughter cells are genetically dissimilar

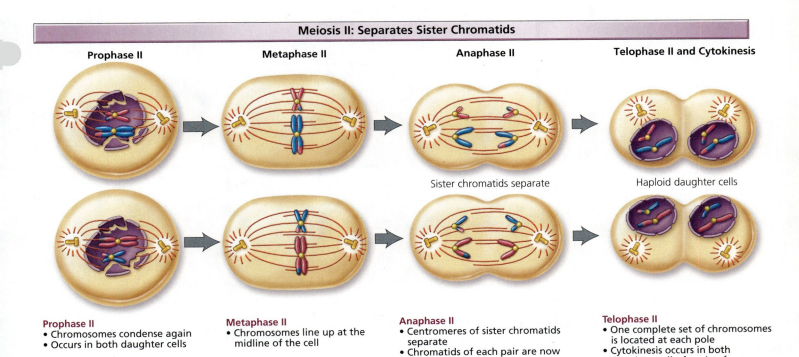

Meiosis II: Separates Sister Chromatids

Prophase II
- Chromosomes condense again
- Occurs in both daughter cells

Metaphase II
- Chromosomes line up at the midline of the cell

Anaphase II
- Centromeres of sister chromatids separate
- Chromatids of each pair are now called chromosomes
- Chromosomes move to opposite poles

Telophase II
- One complete set of chromosomes is located at each pole
- Cytokinesis occurs in both daughter cells, forming four haploid daughter cells

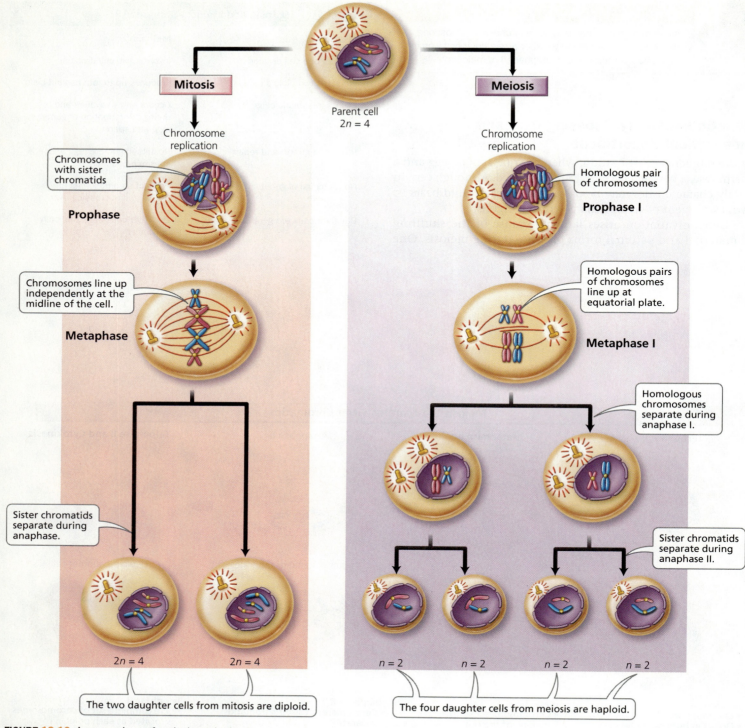

FIGURE 19.10 *A comparison of meiosis and mitosis*

way this mixing occurs is through a process called **crossing over**, in which corresponding pieces of chromatids of maternal and paternal homologues (nonsister chromatids) are exchanged during synapsis when the homologues are aligned side by side. After crossing over, the affected chromatids have a mixture of DNA from the two parents. Because the homologues align gene by gene during synapsis, the exchanged segments contain genetic information for the same traits. However, because the genes of the mother and those of the father may direct different expressions of the trait—attached or unattached earlobes, for instance—the chromatids affected by crossing over have a new, novel combination of genes. Thus, crossing over increases the genetic variability of gametes (Figure 19.13).

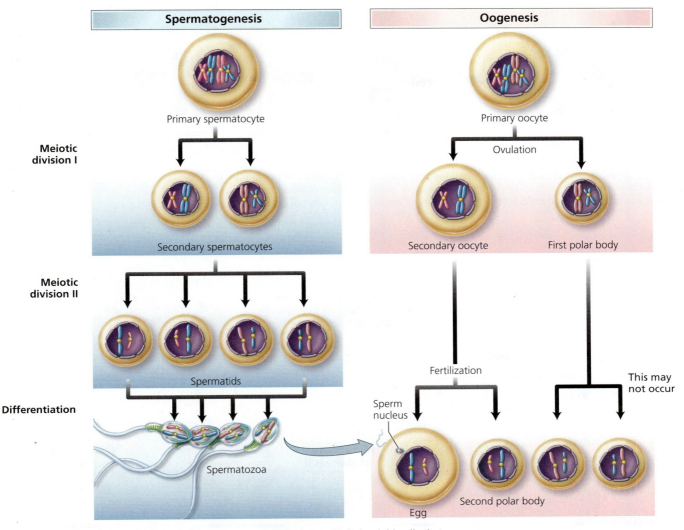

FIGURE 19.11 *Comparison of spermatogenesis and oogenesis. Meiosis results in haploid cells that differentiate into mature gametes. Spermatogenesis produces four sperm cells that are specialized to transport the male's genetic information to the egg. Oogenesis produces up to three polar bodies and one ovum that is packed with nutrients to nourish the early embryo.*

Independent assortment is a second way that meiosis provides for the shuffling of genes between generations (Figure 19.14). Recall that the homologous pairs of chromosomes line up at the equator (midpoint) of the mitotic spindles during metaphase I. However, the orientation of the members of the pair is random with respect to which member is closer to which pole. Thus, like the odds that a flipped coin will come up heads, there is a fifty-fifty chance that a given daughter cell will receive the maternal chromosome from a particular pair. Each of the 23 pairs of chromosomes orients independently during metaphase I. The orientations of all 23 pairs will determine the assortments of maternal and paternal chromosomes in the daughter cells. Thus, each child (other than identical siblings) of the same parents has a unique genetic makeup.

Extra or Missing Chromosomes

Most of the time, meiosis is a precise process that results in the even distribution of chromosomes to gametes. But meiosis is

FIGURE 19.12 *Each child inherits a unique combination of maternal and paternal genetic characteristics due to the shuffling of chromosomes that occurs during meiosis. This photograph shows Eric and Mary Goodenough with their four sons: Derick, Stephen, David, and John.*

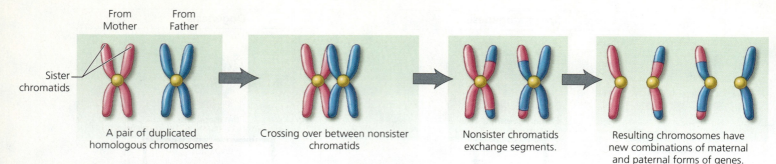

FIGURE 19.13 Crossing over. During synapsis, when the homologous chromosomes of the mother and the father are closely aligned, corresponding segments of nonsister chromatids are exchanged. Each of the affected chromatids has a mixture of maternal and paternal genetic information.

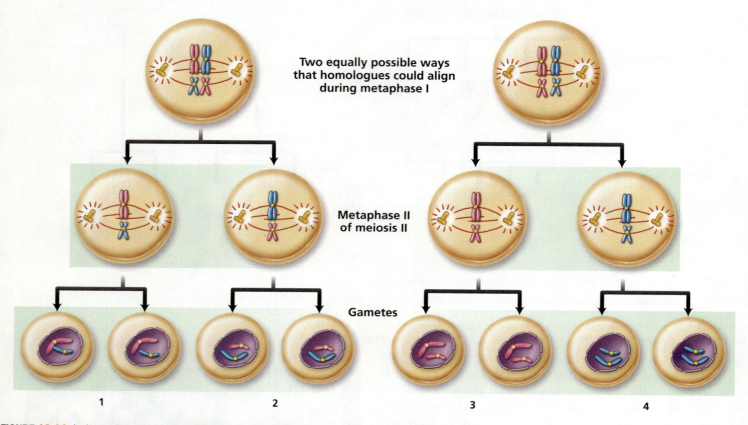

FIGURE 19.14 Independent assortment. The relative positioning of homologous maternal and paternal chromosomes with respect to the poles of the cell is random. The members of each homologous pair orient independently of the other pairs. Notice that with only two homologous pairs, there are four possible combinations of chromosomes in the resulting gametes.

not foolproof. A pair of chromosomes or sister chromatids may adhere so tightly to one another that they do not separate during anaphase. As a result, both go to the same daughter cell, and the other daughter cell receives none of this type of chromosome (Figure 19.15). The failure of homologous chromosomes to separate during meiosis I or of sister chromatids to separate during meiosis II is called **nondisjunction**.

What happens if nondisjunction creates a gamete with an extra or a missing chromosome and that gamete is then united with a normal gamete during fertilization? The resulting zygote will have an excess or deficit of chromosomes. For instance, if the abnormal gamete has an extra chromosome, the resulting zygote will have three of one type of chromosome and two of the rest. This condition, in which there are three representatives of one chromosome, is called **trisomy**. If, however, a gamete that is missing a representative of one type of chromosome joins with a normal gamete during fertilization, the resulting zygote will have only one of that type of chromosome, rather than the normal two chromosomes. The condition in which there is only one

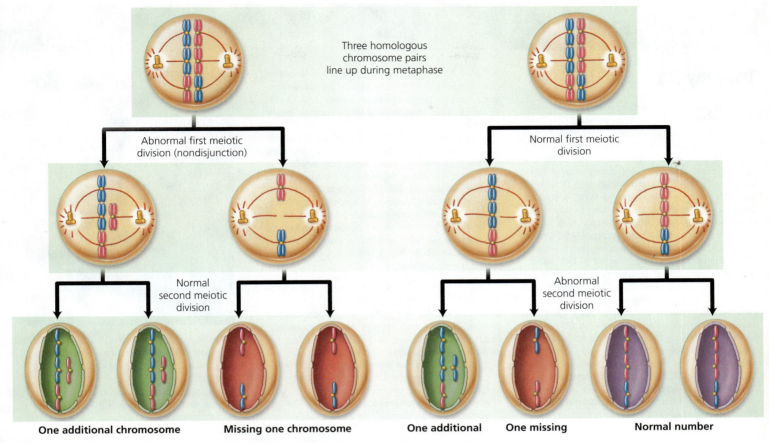

FIGURE 19.15 *Nondisjunction is a mistake that occurs during cell division in which homologous chromosomes or sister chromatids fail to separate during anaphase. One of the resulting daughter cells will have three of one type of chromosome, and the other daughter cell will be missing that type of chromosome.*

representative of a particular chromosome in a cell is called **monosomy**. The imbalance of chromosome numbers usually causes abnormalities in development. Most of the time, the resulting malformations are severe enough to cause the death of the fetus, which will result in a miscarriage. Indeed, in about 70% of miscarriages, the fetus has an abnormal number of chromosomes.

When a fetus inherits an abnormal number of certain chromosomes—for instance, chromosome 21 or the sex chromosomes—the resulting condition is usually not fatal (see the Ethical Issue essay, *Trisomy 21*). The upset in chromosome balance does, however, cause a specific syndrome. (A *syndrome* is a group of symptoms that generally occur together.)

Like autosomes, sex chromosomes may fail to separate during anaphase. This error can occur during either egg or sperm formation. A male is chromosomally XY, so when the X and Y separate during anaphase, equal numbers of X-bearing and Y-bearing sperm are produced. However, if nondisjunction of the sex chromosomes occurs during sperm formation, half of the resulting sperm will carry both X and Y chromosomes, whereas the other resulting sperm will not contain any sex chromosome. A female is chromosomally XX, so each of the eggs she produces should contain a single X chromosome. When nondisjunction of sex chromosomes occurs, however, an egg may contain two X chromosomes or none at all. When a gamete with an abnormal number of sex chromosomes is joined with a normal gamete during fertilization, the resulting zygote has an abnormal number of sex chromosomes (Figure 19.16 on page 414).

Turner syndrome occurs in individuals who have only a single X chromosome (XO). Approximately 1 in 5000 female infants is born with Turner syndrome, but this represents only a small percentage of the XO zygotes that are formed. Most of these XO zygotes are lost as miscarriages. A person with Turner syndrome has the external appearance of a female. The only hint of Turner syndrome may be a thick fold of skin on the neck. As she ages, however, she generally is noticeably shorter than her peers. Her chest is wide, and her breasts underdeveloped. In 90% of the women with Turner syndrome, the ovaries are also poorly developed, leading to infertility. Pregnancy may be possible through in vitro fertilization (see Chapter 18), in which a fertilized egg from a donor is implanted in her uterus.

Klinefelter syndrome is observed in males who are XXY. Although the extra X chromosome can be inherited as a result of nondisjunction during either egg or sperm formation, it is twice as likely to come from the egg. Increased maternal age may increase the risk slightly.

ETHICAL ISSUE

Trisomy 21

One in every 700 infants is born with three copies of chromosome 21 (trisomy 21), a condition known as *Down syndrome*. Symptoms of Down syndrome include moderate to severe mental retardation, short stature or shortened body parts due to poor skeletal growth, and characteristic facial features (Figure 19.A). Individuals with Down syndrome typically have a flattened nose, a forward-protruding tongue that forces the mouth open, upward-slanting eyes, and a fold of skin at the inner corner of each eye. Approximately 50% of all infants with Down syndrome have heart defects, and many of them die as a result of this defect. Blockage in the digestive system, especially in the esophagus or small intestine, is also common and may require surgery shortly after birth.

The risk of having a baby with Down syndrome increases with the mother's age. A 30-year-old woman is twice as likely to give birth to a child with Down syndrome as is a 20-year-old woman. After age 30, the risk rises dramatically. At age 45, a mother is 45 times as likely to give birth to a Down syndrome infant as is a 20-year-old woman.

Today, people with Down syndrome live longer and with a higher quality life than they did in the past. These improvements are due to better health care, more effective teaching approaches, and a greater range of opportunities. Life expectancy is now approaching 60 years in many countries.

Prenatal screening for Down syndrome is common and usually recommended for pregnant women aged over 30 years. Approximately 95% of the "positive" screening tests are wrong. Nonetheless, *all* women who initially test positive for carrying a fetus with Down syndrome are encouraged to undergo more invasive tests, and 1% to 2% of the pregnancies tested by these procedures result in miscarriage. As a result, prenatal screening for Down syndrome poses a risk to 700,000 pregnancies each year.

Questions to Consider

Down Syndrome International is encouraging reviews of screening policies and public debate about the acceptance of genetic screening for mental and physical disabilities.

- If you or a loved one were pregnant, would you advocate for prenatal screening for Down syndrome? Why or why not?
- Who should pay for prenatal screening? The person? Health insurer? The government?
- Do you agree that genetic screening for mental and physical disabilities should be recommended?

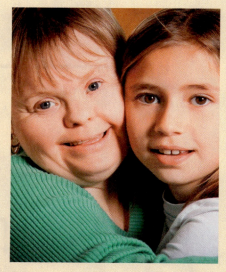

FIGURE 19.A *A person with Down syndrome is moderately to severely mentally retarded and has a characteristic appearance.*

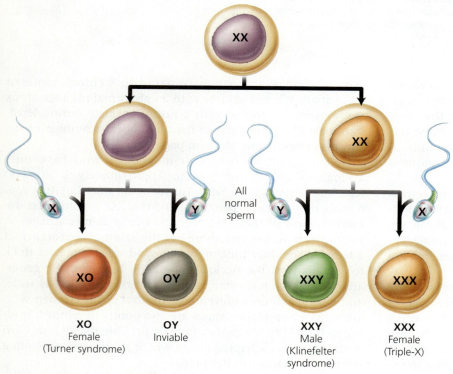

(a) Nondisjunction during egg development

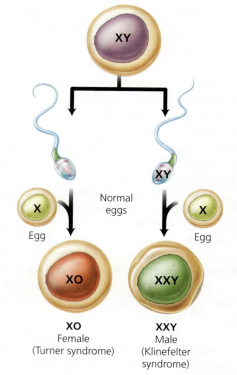

(b) Nondisjunction during sperm development

FIGURE 19.16 *The sex chromosomes may fail to separate during formation of a gamete. Here, an egg with an abnormal number of sex chromosomes joins a normal sperm in fertilization; the resulting zygote has an abnormal number of sex chromosomes. Imbalances of sex chromosomes upset normal development of reproductive structures.*

Klinefelter syndrome is fairly common. Approximately 1 in 500 to 1 in 1000 of all newborn males is XXY. However, not all XXY males display the symptoms of having an extra X chromosome. In fact, some of them live their lives without ever suspecting that they are XXY. When there are signs that a male has Klinefelter syndrome, they do not usually show up until puberty. During the teenage years, the testes of an XY male gradually increase in size. In contrast, the testes of many XXY males remain small and do not produce an adequate amount of the male sex hormone, testosterone. As a result of the testosterone insufficiency, these males may grow taller than average but remain less muscular. Secondary sex characteristics, such as facial and body hair, may fail to develop fully. The breasts may also develop slightly. The penis is usually of normal size, but the testes may not produce sperm; so, men with Klinefelter syndrome may be sterile.

Nondisjunction can also result in a female with three X chromosomes (XXX, triple-X syndrome) or a male with two Y chromosomes (XYY, Jacob syndrome, produced when the chromatids of a replicated Y chromosome fail to separate).

Most women with triple-X syndrome (XXX) have normal sexual development and are able to conceive children. Some triple-X females have learning disabilities and delayed language skills. Males with two Y chromosomes (XYY) are often taller than normal, and some have slightly lower than normal intelligence.

what would you do?
If you had a son with Klinefelter syndrome, would you want him to have testosterone treatments after puberty?

looking ahead
In this chapter we considered cell division: mitosis, which gives rise to new body cells for growth and repair; and meiosis, which gives rise to the gametes (eggs and sperm). In Chapter 19a, we consider mitosis further and explore stem cells, which are unspecialized cells that can divide continuously and develop into different tissue types.

HIGHLIGHTING THE CONCEPTS

19.1 Two Types of Cell Division (pp. 400–401)
- The human life cycle requires two types of nuclear division—mitosis and meiosis. Mitosis creates cells that are exact copies of the original cell. Mitosis occurs in growth and repair. Meiosis creates cells with half the number of chromosomes of the original cell. Gamete production requires meiosis.

19.2 Form of Chromosomes (p. 401)
- A chromosome contains DNA and proteins called histones. A gene is a segment of DNA that codes for a protein that plays a structural or functional role in the cell. Genes are arranged along a chromosome in a specific order. Each of the 23 different kinds of chromosomes in human cells contains a specific sequence of genes.
- Somatic cells (all cells except for eggs and sperm) are diploid; that is, they contain pairs of chromosomes, one member of each pair from each parent. Homologous chromosomes carry genes for the same traits. In humans, the diploid number of chromosomes is 46—or 23 homologous pairs. One pair of chromosomes, the sex chromosomes, determines gender. Males are XY, and females are XX. The other 22 pairs of chromosomes are called autosomes. Eggs and sperm are haploid; they contain only one set of chromosomes.

19.3 The Cell Cycle (pp. 401–402)
- The cell cycle consists of two major phases: interphase and cell division. Interphase is the period between cell divisions.
- During interphase, DNA and organelles become replicated in preparation for the cell to divide and produce two identical daughter cells. Somatic cell division consists of mitosis (division of the nucleus) and cytokinesis (division of the cytoplasm).

19.4 Mitosis: Creation of Genetically Identical Diploid Body Cells (pp. 402–404)
- In mitosis, the original cell, having replicated its genetic material, distributes it equally between its two daughter cells. There are four stages of mitosis: prophase, metaphase, anaphase, and telophase.

19.5 Cytokinesis (p. 405)
- Cytokinesis, division of the cytoplasm, usually begins sometime during telophase. A band of microfilaments at the midline of the cell contracts and forms a furrow. The furrow deepens and eventually pinches the cell in two.

19.6 Karyotypes (p. 405)
- A karyotype is an arrangement of chromosomes based on their physical characteristics, such as length and position of the centromere.

19.7 Meiosis: Creation of Haploid Gametes (pp. 405–415)
- Meiosis, a special type of nuclear division that occurs in the ovaries or testes, begins with a diploid cell and produces up to four haploid cells that will become gametes (eggs or sperm).
- Meiosis is important because it halves the number of chromosomes in gametes, thereby keeping the chromosome number constant between generations. When a sperm fertilizes an egg, a diploid cell called a zygote is created. After many successful mitotic divisions, the zygote and its descendant cells may develop into a new individual.
- Before meiosis begins, the chromosomes are replicated, and the copies remain attached to one another by centromeres. The attached replicated copies are called sister chromatids.
- There are two cell divisions in meiosis. During the first meiotic division (meiosis I), members of homologous pairs are separated.

Thus, the daughter cells contain only one member of each homologous pair (although each chromosome still consists of two replicated sister chromatids). During the second meiotic division (meiosis II), the sister chromatids are separated.
- Genetic recombination during meiosis results in variation among offspring from the same two parents. One cause of genetic recombination is crossing over, in which corresponding segments of DNA are exchanged between maternal and paternal homologues, creating new combinations of genes in the resulting chromatids.
- A second cause of genetic recombination is the independent assortment of maternal and paternal homologues into daughter cells during meiosis I. The orientation of the members of the pair relative to the poles of the cell determines whether a daughter cell will receive the maternal or the paternal chromosome from a given pair. Each pair aligns independently of the others.
- Nondisjunction is the failure of homologous chromosomes or sister chromatids to separate during cell division. It results in an abnormal number of chromosomes in the resulting gametes, and in zygotes created by fertilization involving these gametes, which generally results in death of the fetus. Nondisjunction of chromosome 21 can result in Down syndrome.

RECOGNIZING KEY TERMS

chromosome p. 401
gene p. 401
somatic cells p. 401
homologous chromosome pair p. 401
diploid p. 401
sex chromosomes p. 401
autosomes p. 401
mitosis p. 401
cell cycle p. 402
interphase p. 402
centromere p. 402
chromatid p. 402
prophase p. 402
metaphase p. 404
anaphase p. 404
telophase p. 404
cytokinesis p. 405
karyotype p. 405
haploid p. 406
meiosis p. 406
synapsis p. 407
crossing over p. 410
independent assortment p. 411
nondisjunction p. 412
trisomy p. 412
monosomy p. 413

REVIEWING THE CONCEPTS

1. Explain the relationship between genes and a chromosome. p. 401
2. Define *mitosis* and *cytokinesis*. pp. 401–405
3. Why is meiosis important? p. 406
4. Describe the alignment of chromosomes at the midline during meiosis I and meiosis II. Explain the importance of these alignments in creating haploid gametes from diploid cells. pp. 407–408
5. Explain how crossing over and independent assortment result in genetic recombination that causes variability among offspring (aside from identical twins) from the same two parents. pp. 409–411
6. Define *nondisjunction*. Explain how nondisjunction can result in abnormal numbers of chromosomes in a person. p. 412
7. What causes Down syndrome? What are the usual characteristics of the condition? p. 412
8. The process of mitosis results in
   a. two haploid cells.
   b. two diploid cells.
   c. four haploid cells.
   d. four diploid cells.
9. DNA is synthesized (replicated) during
   a. interphase.
   b. prophase.
   c. metaphase.
   d. anaphase.
10. Crossing over occurs during which stage of meiosis?
    a. prophase I
    b. metaphase I
    c. prophase II
    d. metaphase II
11. Which of the following expresses an important difference between spermatogenesis and oogenesis?
    a. A sperm is haploid, and a mature ovum is diploid.
    b. During spermatogenesis, four functional sperm are produced; during oogenesis, one mature ovum and up to three polar bodies are formed.
    c. Spermatogenesis involves mitosis and meiosis, but oogenesis involves only meiosis.
    d. Crossing over occurs during spermatogenesis but not during oogenesis.
12. What kind of chromatid is attached at the centromere?
    a. sister
    b. mother
    c. daughter
    d. programmed
13. After mitosis, the number of chromosomes in a daughter cell is _____ those in the parent cell.
    a. one-half
    b. the same as
    c. twice
14. After meiosis, the number of chromosomes in a daughter cell is _____ those in the parent cell.
    a. one-half
    b. the same as
    c. twice
15. An abnormal number of chromosomes can result during meiosis because of
    a. crossing over.
    b. recombination.
    c. nondisjunction.
    d. synapsis

16. During meiosis, the processes of _____ and _____ increase genetic diversity.
17. _____ chromosomes carry genes for the same traits.
18. _____ is the pairing of chromosomes during meiosis.
19. The stage of mitosis during which sister chromatids separate is _____.
20. The stage of meiosis during which sister chromatids separate is _____.

APPLYING THE CONCEPTS

1. A cell biologist is studying the cell cycle. She is growing the cells in culture, and they are actively dividing mitotically. One particular cell has half as much DNA as most of the other cells. Which stage of mitosis is this cell in? How do you know?
2. What would happen if the spindle fibers failed to form during mitosis?
3. What condition is indicated by the following karyotype?

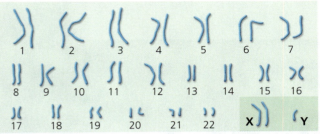

sex chromosomes

BECOMING INFORMATION LITERATE

Several genetic disorders are caused by too many or too few chromosomes. Use at least three reliable sources (books, journals, websites) to describe at least one such disorder other than Down syndrome, Turner syndrome, and Klinefelter syndrome. Indicate which chromosomes are extra or missing in the disorder, and note the symptoms of the disorder. List each source you considered, and explain why you chose the three sources you used.

MasteringBiology®

Go to MasteringBiology for practice quizzes, activities, eText, videos, current events, and more.

19a SPECIAL TOPIC
Stem Cells—A Repair Kit for the Body

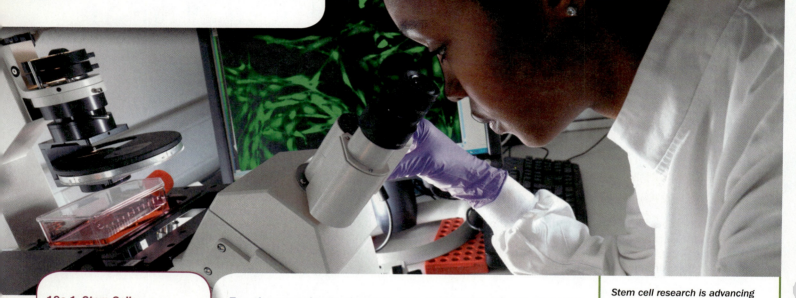

Stem cell research is advancing rapidly.

- 19a.1 Stem Cells: Unspecialized Cells
- 19a.2 Sources of Human Stem Cells
- 19a.3 Potential Uses for Stem Cells

In the previous chapter, we considered cell division for growth and repair. In this chapter, we learn that, unlike most cells, stem cells are able to divide continuously and without becoming a specialized cell type. We will consider where they are found and the promises they may hold for the future.

19a.1 Stem Cells: Unspecialized Cells

Imagine growing new heart cells to repair damage from a heart attack. Think about curing Parkinson's disease by restoring the dopamine-producing neurons that are lacking in such patients. What about generating insulin-producing pancreatic islet cells to cure diabetes? Can we find a cure for multiple sclerosis, Alzheimer's disease, Lou Gehrig's disease, and spinal cord injuries? How? These things may be possible by using stem cells—cells that continually divide and retain the ability to develop into many types of cells—to produce the needed type of cell. You might think of stem cells as a repair kit for the body.

The common thread among these disorders is that they are caused by too little cell division or defective cell function. When there is too little cell division, the body cannot repair damaged tissue; for example, spinal cord injury cannot be repaired, because neurons do not usually divide. Other disorders result from defective cell function; examples include Parkinson's disease, which is characterized by a lack of neurons that produce the neurotransmitter dopamine, and hemophilia, which is characterized by a lack of production of certain blood clotting factors. If the damaged or defective cells could be replaced by healthy, functional cells, the disorders would be cured. Stem cells may be a source of functional cells to bring about those cures.

Most of the trillions of cells in your body have become specialized to perform a particular job. Muscle cells are specialized to contract, and neurons are specialized to conduct nerve impulses. The specialization of cells during embryonic development is directed by biological cues in each cell's immediate environment. The cues are usually provided by neighboring cells and include growth factors, surface proteins, salts, and contact with other cells. Because all cells within the same body have the same genetic makeup, the specialization of different cells to perform different functions means that particular genes in each type of cell have been turned on or off. Recall from Chapter 19 that specialized cells usually do not divide again.

Stem cells, in contrast, are relatively unspecialized cells that divide continually, creating a pool of undifferentiated cells for possible use. If stem cells are given the correct signals—exposure to a particular growth factor or hormone, for example—they can be coaxed into differentiating into a particular specialized cell type. The specialized cell types can then be used to treat diseases or regenerate injured tissues. As we will see, stem cells from different sources differ in the variety of specialized cell types they can develop into.

Stem cells are categorized by the degree of flexibility in their developmental path (Figure 19a.1). A fertilized egg can develop into all the types of cells in the body and is therefore considered to be *totipotent*. After about a week of development, the embryo is a source of embryonic stem cells, which can develop into nearly every type of cell. Because the developmental path of embryonic stem cells is so flexible, they are described as being *pluripotent*.[1] In an adult, some stem cells remain able to differentiate into several types of cells and are described as being *multipotent*. Still other adult stem cells are described as *unipotent*, because they can differentiate into only one type of cell. Thus, embryonic stem cells are more versatile than are adult stem cells.

The source of stem cells used for therapeutic purposes influences the kinds of cells they can become as well as the likelihood of their rejection in the body of the person being treated. Each person's cells have self markers (MHC markers) identifying the cells as belonging in that body (see Chapter 13). These self markers are genetically determined, so the self markers of a relative's cells will be more similar than those of unrelated individuals. Cells that are not recognized as belonging in the body, because of the difference in the self markers, are usually attacked and killed by cells of the body's defense system. The self markers develop during prenatal development and early childhood. Thus, stem cells from an embryo or fetus have less developed self markers and are less likely to be rejected than are adult stem cells. Likewise, adult stem cells from the person being treated or a close relative are less likely to be rejected than are cells from an unrelated person.

19a.2 Sources of Human Stem Cells

There are several sources of stem cells, including certain adult tissues, umbilical cord blood, and early embryos.

Adult Stem Cells: Unipotent and Multipotent

Adults have stem cells, but they are more difficult to locate than those in an embryo are. Waiting in the bone marrow, brain, skin, liver, and other organs, adult stem cells remain ready to generate new cells for repair or replacement of old

[1]Totipotent cells can give rise to all cells of the body and those of all four extraembryonic membranes (see Chapter 18). Pluripotent cells can give rise to all cells of the body but cannot give rise to cells of the extraembryonic membrane that will contribute to the placenta.

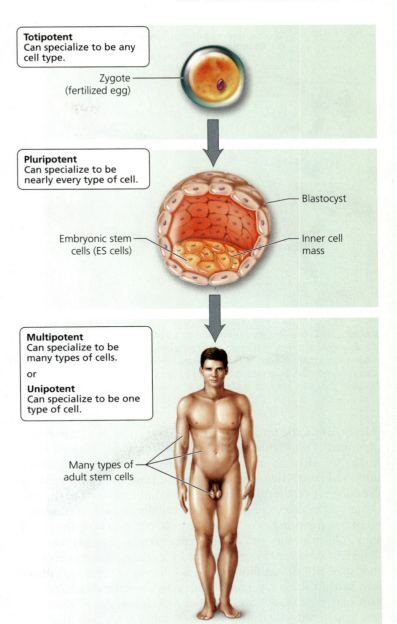

FIGURE 19a.1 *Stem cells at different stages in human development.*

ones in many parts of the body. Diseased or dying cells provide biological cues that summon stem cells and prompt them to differentiate into the needed type of cell. When a stem cell divides in an adult, one cell usually specializes to be a particular type of cell, and the other remains a stem cell (Figure 19a.2).

Adult stem cells do not seem to be as versatile as stem cells from an embryo. Whereas embryonic stem cells can transform into any cell type in the body, adult stem cells usually form one type of cell or form cells of a particular lineage only. However, we now know that adult stem cells are more versatile than was once believed.

Umbilical Cord and Placental Stem Cells: Multipotent

Blood from the umbilical cord and the embryonic part of the placenta (the chorion) are also good sources of stem cells. Stem

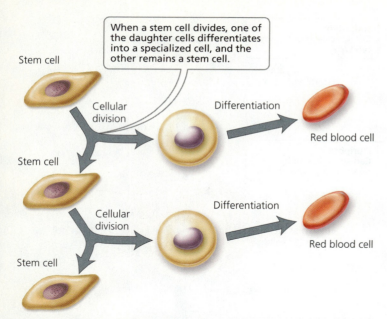

FIGURE 19a.2 *Two important characteristics of stem cells are that they divide continually and that their daughter cells can differentiate into one or more specialized stem cells. Here, a blood stem cell divides, producing one daughter cell that becomes specialized to perform a function and another daughter cell that remains a stem cell to replenish the stem cell supply.*

cells do not settle into an infant's bone marrow until a few days after birth. Before that, during fetal development, they circulate in the bloodstream and therefore travel through vessels in the umbilical cord when circulating to and from the placenta. After the baby is born, it no longer needs its umbilical cord or the placenta, so the cord and placenta are usually discarded. However, an increasing number of parents are choosing to have the blood stem cells from the cord and placenta saved in a storage bank.

Most of the stem cells in cord blood are blood stem cells that give rise to red blood cells, white blood cells, and platelets. Within the body, blood stem cells produce about 260 billion new cells (about an ounce of blood) every day. However, umbilical cord blood also contains some stem cells that can turn into other kinds of cells—bone, cartilage, heart muscle, nerve tissue, and liver tissue.

Stem cells from umbilical cord blood and the placenta have some advantages over adult blood stem cells taken from bone marrow. Blood stem cells can be harvested from bone marrow, but the procedure is done in a hospital under general anesthesia. Stem cells are more easily extracted from umbilical cord blood because the cord is removed from the infant's body. Stem cells from cord blood are usually used to treat childhood diseases because the cord blood has a limited number of stem cells. However, the placenta has several times more stem cells and, when added to stem cells from cord blood, can be used to treat adult illnesses. Another advantage of using these stem cells is that they are less likely to be rejected by the recipient than are adult stem cells from bone marrow. The self markers on the cord blood stem cells or placental stem cells have not yet fully developed, so the match does not have to be as close as with adult stem cells from bone marrow. Finally, stem cells from cord blood or the placenta are less likely to carry infections.

Because blood stem cells are able to produce an endless supply of blood cells, they hold promise for treating a host of blood conditions. For instance, some patients with the inherited disorders sickle-cell anemia and beta-thalassemia, in which abnormal forms of hemoglobin are produced, have already been treated successfully with stem cell therapy. Stem cells that produce normal hemoglobin can be transplanted into a person with the inherited anemia. Normal red blood cells then begin replacing the abnormal ones. Leukemia patients have also benefited from umbilical cord stem cells. Traditionally, leukemia has been treated with bone marrow transplants from family members or from donors from the National Marrow Donor Program who have close tissue types. However, most patients in need of a bone marrow transplant do not find donors with matching tissue types. Because the tissue markers on umbilical cord stem cells are immature, the cells are less likely to be rejected by the recipient.

Most of the medical benefits that may be achieved using umbilical cord or placental blood have yet to be realized. Nonetheless, there are already banks that store umbilical cord blood and placentas from newborn babies. Some parents pay a setup fee and an annual service fee to store their baby's cord blood and placenta just in case the child needs that blood or those stem cells some day. Government-owned banks collect and store thousands of samples of cord blood in hopes of having a tissue match for most individuals in the population. Researchers have developed ways to increase the number of umbilical cord blood stem cells by growing them in the laboratory.

The existence of umbilical cord blood banks raises certain social issues. Should blood stored in private blood banks be reserved only for the possible future need of the donor, or should it be made available for anyone in need whose tissue type matches? Should cord blood banks be privately owned or funded by the government? These are just a few of the many questions remaining to be addressed.

Embryonic Stem Cells: Pluripotent

Embryonic stem cells can divide continually and specialize into nearly any cell type. They come from several sources.

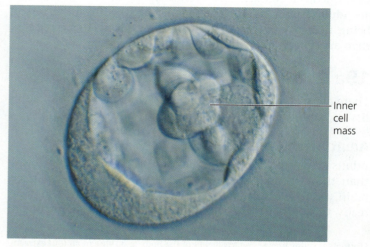

FIGURE 19a.3 *Embryonic stem cells come from the inner cell mass, a cluster of 20 to 30 cells, from a 6- to 7-day-old embryo. At this age, the embryo is about the size of a head of a pin.*

Unused embryos from fertility clinics Most of the stem cells used for current research come from embryos that were created for reproductive purposes but were not used. Only a few days old, these embryos left in fertility clinics were destined for destruction. When the early embryo is about 6 to 7 days old, it is a rich source of stem cells because the cells it consists of are able to produce all the cells of the new individual. When stem cells are extracted from an early embryo, the embryo is about the size of the head of a pin and is called a *blastocyst* (see Chapter 18; see also Figure 19a.3). It is a sphere of cells containing a cluster of about 20 to 30 cells, called the *inner cell mass,* adhering to one side of the sphere. The stem cells make up the inner cell mass. Embryonic stem cells are easy to harvest: the cells of the inner cell mass are extracted and cultured in the laboratory. They will continue to divide for years, creating a stem cell line.

what would you do?
If you were a parent with an embryo stored in a fertility clinic, would you consent to using that embryo for stem cell research?

Somatic cell nuclear transfer In *somatic cell nuclear transfer (SCNT)*, the nucleus from a somatic body cell is transferred to an egg from which the egg nucleus has been removed. The newly created cell is then stimulated to begin embryonic development. All the cells of the embryo are genetically identical. Thus, somatic cell nuclear transfer is a form of *cloning*—producing genetically identical copies of a cell or organism.

Dolly, born on July 5, 1996, was the first sheep—indeed, the first mammal—to be cloned from an adult cell (Figure 19a.4). She was created using somatic cell nuclear transfer, illustrated in Figure 19a.5. In Dolly's case, the nucleus of a mammary cell from an adult sheep's udder was placed into an egg from which the nucleus had been removed. (Dolly was named after Dolly Parton, a country singer known for her mammary "cells" as well as for her voice.) The goal of Dr. Ian Wilmut, the Scotsman whose work resulted in Dolly, was to develop techniques that would

FIGURE 19a.4 *Dolly, the first animal cloned from an adult cell, was created using a technique called somatic cell nuclear transfer.*

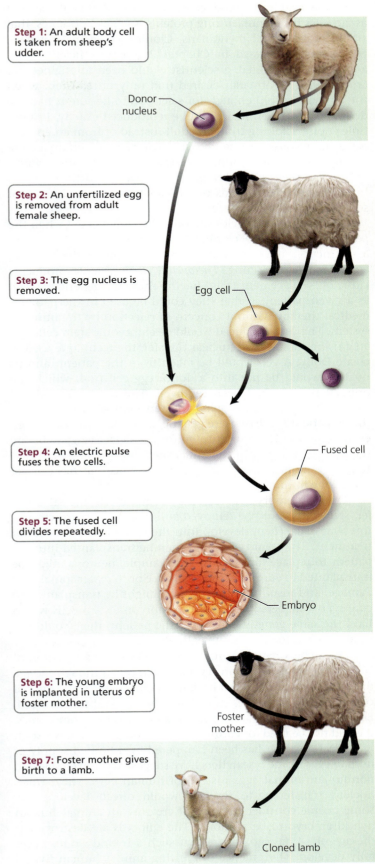

FIGURE 19a.5 *Somatic cell nuclear transfer used for reproductive cloning.*

eventually lead to the production of animals that could be used as factories for manufacturing proteins, such as hormones, that would be beneficial to humans. Cloning could make genetic engineering (discussed in Chapter 21) more efficient. If an adult could be cloned, a scientist would need to engineer an animal with a particular desired trait only once. Then, when the animal was old enough for scientists to be sure it had the desired trait, the animal could be reproduced exactly, in multiple copies. Cloning from an adult instead of from an embryo is advantageous because the scientist can be certain before cloning that the animal has the desired traits. In other words, "What you see is what you get." Since Dolly's birth, scientists have cloned several kinds of animals using the technique of somatic cell nuclear transfer.

Discussions about cloning often become heated when the possibility of human cloning is raised. In these discussions, a distinction is often drawn between reproductive cloning and therapeutic cloning. *Therapeutic cloning* would produce a clone of "replacement cells" having the same genetic makeup as a given patient so that they could be used in the patient's medical therapy without concern of rejection by the immune system. The cells produced would be embryonic stem cells created using somatic cell nuclear transfer, the technique used to create Dolly. A cell would be taken from the patient, and its nucleus would be put into a donor egg cell from which the original nucleus had been removed. The newly created cell would then begin developing into an embryo (Figure 19a.6). (If the patient's disease were genetic, a normal form of the relevant gene would be substituted for the defective gene before the nucleus was put in the enucleated egg. Replacement of a specific gene has been accomplished in mice but has not yet been done in humans.)

Once an embryo is created by somatic cell nuclear transfer, stem cells could be removed and treated with growth factors that make them develop into the type of cells needed for treatment. The stem cells could then be transplanted into the patient to replace faulty cells. For example, neurons might be transplanted to cure Parkinson's disease or to repair spinal cord damage, and insulin-producing cells might be transplanted to cure type 1 diabetes. Because the embryonic stem cells would have the same genetic makeup as the patient, they would not be attacked as foreign by the patient's immune responses.

Researchers have recently created cells using a variation of SCNT to combine a human egg and a skin cell. Although the egg did reprogram the skin cell back to an embryonic state, the cell contained chromosomes from both the egg and the skin cell. Because the resulting stem cell contained excess chromosomes, it could not be used for curing diseases. Nonetheless, it is a start.

SCNT research has been hampered by a limited supply of human eggs. Some researchers have suggested using enucleated nonhuman animal eggs instead of human eggs. The human nucleus transferred to the egg would direct development. Some people consider the idea of creating an animal–human hybrid embryo to be immoral. The embryos are destroyed at the age of 14 days to ensure that such a hybrid would never fully develop. The practice of creating animal–human hybrid embryos for use in stem cell research has been permitted in the United Kingdom since 2008.

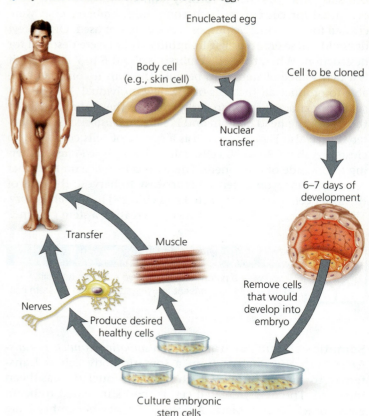

Q Why is the nucleus removed from the donor egg?

FIGURE 19a.6 *Potential method for human therapeutic cloning using somatic cell nuclear transfer.*

A The egg nucleus is removed so that the only genetic information directing development will come from the adult cell nucleus. The donor nucleus has genes for the desired trait.

In contrast to therapeutic cloning, *reproductive cloning* produces a new individual with a known genetic makeup. The cloning of nonhuman animals (like Dolly) is reproductive cloning. Currently, reproductive cloning is an extremely inefficient process. Hundreds of cloned embryos must be used to obtain one newborn mammal. When cloned embryos do proceed to develop into a new individual, they grow faster and larger than normal. There is also a suspicion that cloned mammals age prematurely. For example, there are indications that Dolly was genetically older than her birth age: one is that Dolly had arthritis, a painful joint condition usually found only in older animals. For reasons like these, experts agree that scientists are still far from being able to produce a cloned human baby.

If reproductive cloning of humans does become possible someday, what might this technology mean to our society? How would it affect human life? One thing it would *not* do is give us the power to create an exact copy of a specific person. Environment *and* genes (also often referred to as "nurture" and "nature") determine personality characteristics. Although cloning could replicate the exact genetic makeup of a specific person, there would be no way to replicate all the experiences in the life of the person being cloned; so, the clone would be a different person from the original one. Thus we could not replace a loved one who has died or achieve our own immortality in the body of a clone. If such technology is ever successful, however, we could produce an identical twin of a person.

Induced Pluripotent Stem Cells

Suppose one could turn back the clock and stimulate an adult cell to revert to a pluripotent stem cell. In 2006, researchers announced that they had done just that—they caused adult mouse skin cells to revert to pluripotent stem cells that could differentiate into any type of tissue. A short time later, in 2007, scientists created induced pluripotent stem cells (iPSCs) using human skin cells. Both groups of researchers used retroviruses to insert four genes into the skin cells. This is an exciting breakthrough—it creates pluripotent stem cells without the need to destroy an embryo, and it could produce patient-specific cell lines without the possibility of rejection. Two concerns with pluripotent stem cells created in this way are that two of the inserted genes are cancer-producing genes and that using retroviruses to deliver the genes carries a risk. Retroviruses integrate themselves directly into the cell's DNA, which allows the genes delivered by the virus to become permanently active in the cell. However, depending on where the virus inserts itself, this incorporation can cause the cell to become cancerous.

Researchers have addressed these concerns. They created pluripotent stem cells from adult skin cells using two genes that are not linked to cancer and then used a cocktail of drugs to revert the skin cells to pluripotent stem cells. Adenoviruses, which normally cause the human cold, have been modified to serve as the gene-delivery system. These viruses persist in the cell for only a short time.

This research brings stem cells created from adult cells closer to clinical use. Researchers have used iPSCs to rescue vision in rats. The rats used in these experiments have a gene mutation that causes blindness by killing the photoreceptor cells. The research team surgically inserted iPSCs of retinal epithelial cells into the retina of rats before the photoreceptors degenerated, and vision was retained. But don't expect human cures in the near future; it will take years before iPSCs can be put to practical use in humans.

19a.3 Potential Uses for Stem Cells

The use of stem cells is just beginning to blossom. Let's consider some of their potential uses.

Replacement for Damaged Cells

As of this writing, there are no established stem cell therapies in the United States—except for those that use cells extracted from bone marrow, circulating blood, or umbilical cord blood. There are, however, more than 2000 clinical trials in progress or looking for recruits. For example, clinical trials are underway to determine the effectiveness of umbilical cord stem cells to repair and replace nerve cells in children with cerebral palsy, a condition in which neurons are damaged at birth. Certain other countries, China and India for example, are offering stem cell treatments. Although these treatments are unproven by U.S. scientific standards, some people are traveling abroad for treatment. To date, most of the studies on stem cells as a source of replacement cells or as factories to produce proteins or hormones that the body is not making have been done on nonhuman animals. Insulin-producing stem cells have eliminated symptoms of type 1 diabetes in mice. In rats, bone marrow stem cells move toward heart muscle with damage similar to that occurring in a heart attack. In the damaged heart muscle, stem cells promote heart muscle strength.

Growing New Organs

Regenerative medicine is still in its infancy, but it is growing rapidly. Its aim is to rebuild or repair damaged organs by coaxing stem cells to grow and fill in tissue scaffolds that were engineered in the laboratory. In 2004, a German man had his jaw removed because of cancer. Doctors created a mesh mold in the shape of the jawbone that had been removed, and they implanted the mold into the muscle below the man's shoulder blade, along with some cow-derived bone mineral, a growth factor to stimulate bone growth, and blood extracted from the man's bone marrow, which contained stem cells (Figure 19a.7). The mold was left in place on the man's shoulder until bone

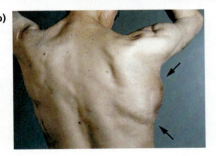

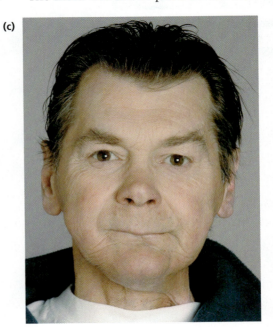

FIGURE 19a.7 Adult stem cells from bone marrow were used to grow a new jaw to replace one that had been removed because of cancer. (a) A mold in the shape of the missing jaw was created out of mesh and seeded with bone marrow cells and bone growth factor. (b) The mold was placed in the muscle of the man's back until bone filled the mold. (c) The new jaw was then implanted to replace the missing jaw.

scans showed that new bone had formed around the mold. The new jaw then was removed, along with some muscle and blood vessels, and implanted in the correct location. Four weeks later, the man could chew solid food—a bratwurst sandwich!

Since then, other organs have been grown and transplanted. Doctors have grown new urinary bladders using the patients' own cells. In those cases, doctors extracted muscle and bladder cells from the patient and grew them in a Petri dish. As the cells formed layers of tissue, the tissue was shaped into new bladders, which were then implanted into the patients. Within a few weeks, the bladders grew to normal size and performed the required functions. New windpipes (tracheas) have been built from patients' own stem cells. In early cases, doctors removed cells from a cadaver's trachea, but more recently a scaffold of the trachea was built from plastic microfibers. Researchers then seeded the tracheal scaffolds with stem cells from the patient. The new trachea grew from the patient's cells and was, therefore, not rejected.

An exciting application of regenerative medicine is "regrowing" limbs of wounded soldiers returning from Afghanistan (Figure 19a.8). Many soldiers are so badly mutilated by explosives that limbs must be amputated. Now there is hope that some limbs can be saved. The secret is an application of a powder created from pig bladders, which is nicknamed "magic pixie dust." The pig bladders contain an extracellular matrix that contains the protein collagen. Researchers think that the powder works by attracting stem cells in the body and giving these cells the chemical signals to make new tissues. The new limbs have all the appropriate tissue types, including skin, muscle, and nerve.

stop and think
Why would chemical signals be necessary for the development of new tissues?

Testing New Drugs

Before a new drug can go to clinical trials, it must be shown to be safe for animals and human cells cultured in Petri dishes. The problem is that not all tissues will survive in tissue culture. For example, heart cells, brain cells, and liver cells do not. Stems cells may be a way to test new drugs for safety and effectiveness before testing them on humans.

Researchers are also using iPSCs to create cells that display the characteristics of cells with a particular disease. Cell cultures of these cells could then be used to study the nature of the disease or to develop drugs to treat the disease. Skin samples from patients with Alzheimer's disease are being collected with an interest in converting them into pluripotent stem cells for such purposes.

FIGURE 19a.8 *Corporal Isais Hernandez's leg was severely injured by a mortar round. New tissue was grown in the wound using a powder created from pulverized pig bladders.*

looking ahead
In Chapter 19 we considered how cells copy their genetic material and undergo mitosis, creating cells for growth and repair, or undergo meiosis, creating gametes for reproduction. In this chapter, we considered stem cells, which divide endlessly. In the next chapter, we will relate what we learned about cell division to genetic inheritance.

Genetics and Human Inheritance

20

Did you Know?

- You have fewer genes than corn has. Scientists believe that humans have about 25,000 genes and that corn has about 59,000 genes.
- Your traits are determined by only 2% of your genetic material.

Identical twins develop from a single zygote, so they have identical genetic origins. Nonetheless, they have slight differences in appearance.

20.1 Principles of Inheritance

20.2 Breaks in Chromosomes

20.3 Detecting Genetic Disorders

ETHICAL ISSUE

Gene Testing

In the Chapter 19 we considered how chromosomes and the genes they carry are distributed during cell division. We learned that mitosis is necessary for growth and repair of body tissues. We also learned that in humans, meiosis is necessary to prepare the gametes needed for reproduction. In Chapter 19a we considered stem cells, which divide endlessly and hold promise for curing conditions that are caused by too little cell division or defective cells. In this chapter, we will see that the genes we receive at the moment of conception influence all the biochemical reactions taking place inside our cells, our susceptibility to disease, our behavior patterns, and even our life span. Our environment is also an important influence, but our genes provide the basic blueprint for our possibilities and limitations. In this chapter, you will learn more about the genetic foundation that has been so important in shaping who you are. You will learn how heritable traits are passed to new generations and how to predict the distribution of traits from one generation to the next.

20.1 Principles of Inheritance

The understanding of meiosis we gained in Chapter 19 helps us answer important questions about inheritance: Why is your brother the only sibling with Mom's freckles and widow's peak (a hairline that comes to a point on the forehead)? How can you have blue eyes when both your parents are brown eyed? Will you be bald at 40, like Dad? Let's delve more deeply into how chromosomes, meiosis, and heredity are related.

Before beginning, you may want to review some of the terms that were introduced in Chapter 19 and that are summarized in Figure 20.1. Recall that somatic (body) cells have a pair of

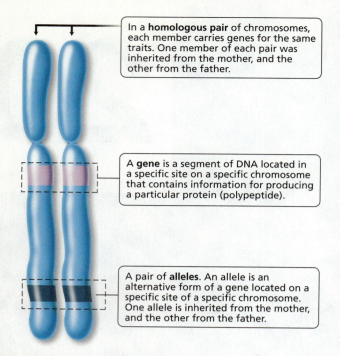

FIGURE 20.1 *Important terms in genetics*

every chromosome. Thus, human cells have 23 pairs of chromosomes. One member of each pair was inherited from the female parent, and the other from the male parent. The chromosomes that carry genes for the same traits are a *homologous pair of chromosomes,* or homologues. Chromosomes are made of DNA and protein. Certain segments of the DNA of each chromosome function as genes. A **gene** directs the synthesis of a specific protein that can play either a structural or a functional role in the cell.[1] (In some cases, the gene directs the synthesis of a polypeptide that forms part of a protein.) In this way, the gene-determined protein can influence whether a certain **trait**, or characteristic, will develop. For instance, the formation of your brother's widow's peak was directed by a protein coded for by a gene that he inherited from Mom. Genes for the same trait are found at the same specific location on homologous chromosomes.

Different forms of a gene are called **alleles**. Alleles produce different versions of the trait they determine. For example, there is a gene that determines whether freckles will form. One allele of this gene causes freckles to form; the other does not. Somatic cells carry two alleles for each gene, one allele on each homologous chromosome. If a person has at least one allele for freckles, melanin will be deposited, and freckles will form. If neither homologue bears the freckle allele that leads to melanin deposition, for example, the person will not have freckles.

Individuals with two copies of the same allele of a gene are said to be **homozygous** (*homo,* same; *zygo,* joined together) for that trait. Individuals with different alleles of a given gene are said to be **heterozygous** (*hetero,* different; *zygo,* joined together). When the effects of a certain allele can be detected regardless of whether an alternative allele is also present, the allele is described as a **dominant allele**. Dimples and freckles are human traits dictated by dominant alleles. An allele whose effects are masked in the heterozygous condition is described as a **recessive allele**. Because of this masking, only homozygous recessive alleles are expressed. Some disorders result from recessive alleles; examples include cystic fibrosis, in which excessive mucus production impairs lung and pancreatic function; and albinism, in which the pigment melanin is missing in the hair, skin, and eyes. It is customary to designate a dominant allele with a capital letter and a recessive allele with a lowercase letter—*A* and *a,* for example.

We observe the dominant form of the trait whether the individual is homozygous dominant (*AA*) or heterozygous (*Aa*) for that trait. As a consequence, we cannot always tell exactly which alleles are present. It is important to remember that an individual's genetic makeup is not always revealed by the individual's appearance. A **genotype** is the precise set of alleles a person possesses for a given trait or traits. It tells us whether the individual is homozygous or heterozygous for a given gene. A **phenotype**, in contrast, is the *observable* physical trait or traits of an individual. Figure 20.2 shows the genotype or genotypes for several phenotypes in humans (for example, the freckled phenotype has two genotypes: *FF* and *Ff*). Table 20.1 summarizes terms commonly used in genetics as they apply to the inheritance of freckles, which is determined by a dominant allele.

Gamete Formation

Recall from Chapter 19 that the members of each homologous pair of chromosomes segregate during meiosis I, with each homologue going to a different daughter cell. Thus, an egg or a sperm has only one member of each homologous pair. Consider what this fact means for inheritance. Because the alleles for each gene segregate during gamete formation, half the gametes bear one allele, and half bear the other. This principle, known as the *law of segregation,* explains how, for every gene in our chromosomes, one of the alleles comes from our mother and one comes from our father.

Furthermore, each pair of homologous chromosomes lines up at the midline of the cell during meiosis I independently of the other pairs. The orientation of the paternal and maternal homologues relative to the poles of the cell (that is, with regard to which homologue is closer to which pole) is entirely random. What this fact means for inheritance is that pairs of alleles for genes on different chromosomes segregate into gametes independently. This principle, known as the *law of independent assortment,* explains why the mixture of alleles that came from the mother and alleles that came from the father is different in every gamete.

Mendelian Genetics

During the nineteenth century, Gregor Mendel, a monk who grew up in a region of what was then Austria and is now part of the Czech Republic, worked out much of what we know today about the laws of heredity by performing specific crosses

---

[1]This is a simplified definition of *gene*. As we will see in Chapter 21, some genes contain regulatory regions of DNA within their boundaries. Also, some genes code for RNA molecules that are needed for the production of a protein but are not part of it.

20.1 Principles of Inheritance   **427**

Freckles: *FF* or *Ff*

No freckles: *ff*

Widow's peak: *WW* or *Ww*

Straight hairline: *ww*

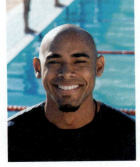
Unattached earlobes: *EE* or *Ee*

Attached earlobes: *ee*

Tongue rolling: *TT* or *Tt*

FIGURE 20.2 *Genotypes for selected human phenotypes*

TABLE 20.1 Review of Common Terms in Genetics		
Genotypes: The Alleles That Are Present	**Description**	**Phenotype: The Observable Trait (Examples)**
FF	Homozygous dominant: • Two dominant alleles present • Dominant phenotype expressed	Freckles
Ff	Heterozygous: • Different alleles present • Dominant phenotype expressed	Freckles
ff	Homozygous recessive: • Two recessive alleles present • Recessive phenotype expressed	No freckles

of pea plants. Although Mendel knew nothing about chromosomes, his ideas about the inheritance of traits are consistent with what we now know about the movement of chromosomes during meiosis. Mendel's ideas are used today to predict the outcome of hereditary crosses.

One-trait crosses   We begin our discussion by using Mendel's ideas to consider the inheritance of a single trait, using the example of freckles, a characteristic determined by a dominant allele. Suppose a freckled female who is homozygous dominant (*FF*) had a child with a homozygous recessive male with no freckles (*ff*; Figure **20.3**). The following sequence of steps allows us to predict the degree of likelihood that their child will have freckles.

1. **Identify the possible gametes that each parent can produce.**   We know that alleles segregate during meiosis, but in this example, both parents are homozygous. Therefore, each can produce only one type of gamete (as far as freckles are concerned). The female produces gametes (eggs) with the dominant allele (*F*), and the male produces gametes (sperm) with the recessive allele (*f*).

2. **Use a Punnett square to determine the probable outcome of the genetic cross.**   A *Punnett square* is a diagram that is used to predict the genetic makeup of the offspring of individuals of particular genotypes. In a Punnett square, columns are set up and labeled to represent each of the possible gametes of one parent, let's say the father (Figure **20.4**). Remember, the alleles of each gene segregate during meiosis. In this case, then, there would be two columns to represent the gametes of the male without freckles, each labeled with a recessive allele, *f*. In a similar manner, rows are established across the columns and labeled to represent all the possible gametes formed by the other parent, in this case, the mother with freckles. There would be two rows in this Punnett square, each labeled *F*. Each square in the resulting table is then filled in by combining the labels of the corresponding rows and columns. The squares represent the possible offspring of these parents. In this first case, all the children would have freckles and would be heterozygous for the trait (*Ff*).

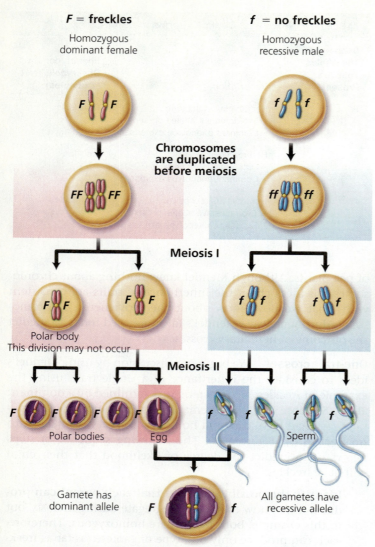

FIGURE 20.3 Gamete formation by a female who is homozygous dominant for freckles (FF), a male who is homozygous recessive for no freckles (ff), and the heterozygous (Ff) individual resulting from the union of these gametes.

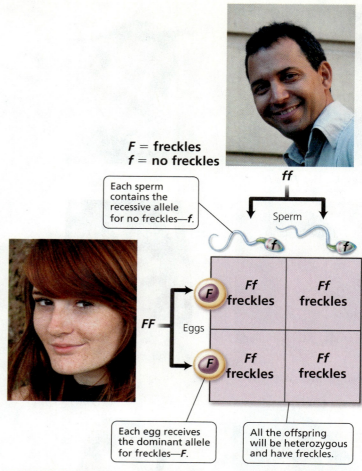

FIGURE 20.4 This Punnett square illustrates the probable offspring from a cross between a homozygous dominant female with freckles (FF) and a homozygous recessive male without freckles (ff). The columns are labeled with the possible gametes the male could produce (one gamete per column). The rows are labeled with the possible gametes the female could produce (one gamete per row). Combining the labels on the corresponding rows and columns yields the genotype of possible offspring.

Now let's consider why it is possible for parents who are both heterozygous for the freckle trait (Ff) to have a child without freckles. The F and f alleles segregate during meiosis. So half the gametes bear the F allele, and half bear the f allele (Figure 20.5a). This time the two rows in the Punnett square are labeled with F and f, and so are the two columns (Figure 20.5b). After filling the boxes according to the labels on rows and columns, we see that the probable genotypes of the children would be homozygous dominant (FF), heterozygous (Ff), and homozygous recessive (ff), in a genotypic ratio of 1 FF : 2 Ff : 1 ff. The ratio of phenotypes of children with freckles (FF and Ff) compared with children without freckles (ff) would be 3 : 1. This ratio means that each child born to this couple will have a 75% (3/4) chance of having freckles. But there is a 1 in 4, or 25%, chance that each child will not have freckles. A cross in which both parents are heterozygous for one trait of interest is called a *monohybrid cross*.

stop and think

At the beginning of the chapter, we raised a question about why only one sibling, your hypothetical brother, inherited your mother's freckles. Implicit in the way we phrased the question was that there is at least one other sibling (you), who, like your father, does not have freckles. Use a Punnett square to explain why the mother in this example must be heterozygous for the freckle trait.

Two-trait crosses We use the same steps to predict the probable outcome for inheritance of two traits of interest that we used for inheritance of a single trait—as long as the genes for the two traits are on different chromosomes.

1. **Identify the possible gametes that each parent can produce.** Like freckles, a widow's peak is controlled by a dominant allele. What would the children look like if the parents were a woman who is homozygous for both freckles

and a widow's peak (*FFWW*) and a man with no freckles and a straight hairline (*ffww*)? To answer that question, we begin by determining the possible gametes each mate can produce. Because they are homozygous for both traits, the woman can produce only gametes bearing dominant alleles (*FW*), and the man can produce only gametes with the recessive alleles (*fw*).

2. **Use a Punnett square to determine the probable outcome of the genetic cross.** In this case, each parent can produce only one type of gamete, so all the children will have the same genotype and the same phenotype. All the children will be heterozygous for each trait (*FfWw*) and will have both freckles and a widow's peak.

Suppose one of these children mates with someone who is also heterozygous for freckles and a widow's peak. This would be a *dihybrid cross*, a mating of individuals who are both heterozygous for two traits of interest. Would it be possible for this couple to have a child who had neither freckles nor a widow's peak? It *would* be possible, but the chances are only 1 in 16. Let's see why.

First, determine the possible gametes that each parent could produce (Figure 20.6). Keep in mind that alleles for a gene segregate and that genes for different traits that are located on different chromosomes will segregate independently of one another. There are, therefore, four possible allele combinations in gametes produced by a person who is heterozygous for these two different genes. The gametes are *FW*, *Fw*, *fW*, and *fw*. In this instance, we need to construct a Punnett square with *four* columns, because this time there are four possible male gametes, and four rows labeled with the possible female gametes. Each type of gamete has an equal chance of joining with any other type at fertilization. The squares are filled by combining the labels of the corresponding columns and rows, giving us the possible genotypes of the offspring. Notice in Figure 20.7 that the relative numbers of phenotypes possible for the next generation are 9 freckled, widow's peak; 3 freckled, straight hairline; 3 no freckles, widow's peak; 1 no freckles, straight hairline. We see, then, that the expected phenotypic ratio resulting from a dihybrid cross is always 9 : 3 : 3 : 1.

Pedigrees

It is sometimes important to find out the genotype of a human with a dominant phenotype for a particular trait. Because there are two possible genotypes for a dominant trait (homozygous dominant and heterozygous dominant), we cannot know genotype based on phenotype alone. However, we can often deduce the unknown genotype by looking at the expression of the trait in the person's ancestors or descendants. A chart showing the genetic connections between individuals in a family is called a **pedigree**. Family or medical records are used to fill in the pattern of expression of the trait in question for as many family members as possible. Pedigrees are useful not just in determining an unknown genotype but also in predicting the chances that one's offspring will display the trait. Pedigrees for the inheritance of genetic disorders caused by a dominant

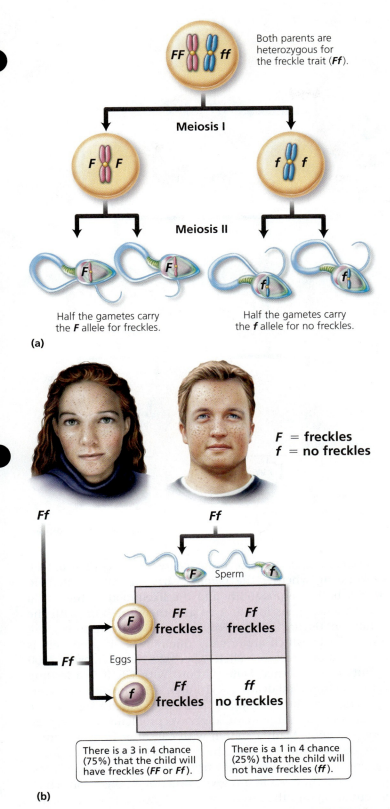

FIGURE 20.5 *(a)* Gamete formation by a person who is heterozygous for the freckle trait (*Ff*). *(b)* A Punnett square showing the probable outcome of a mating between two people who are heterozygous for the freckle trait (*Ff*).

Q *If a parent were homozygous recessive for freckles and heterozygous for widow's peak, how many types of gametes could form?*

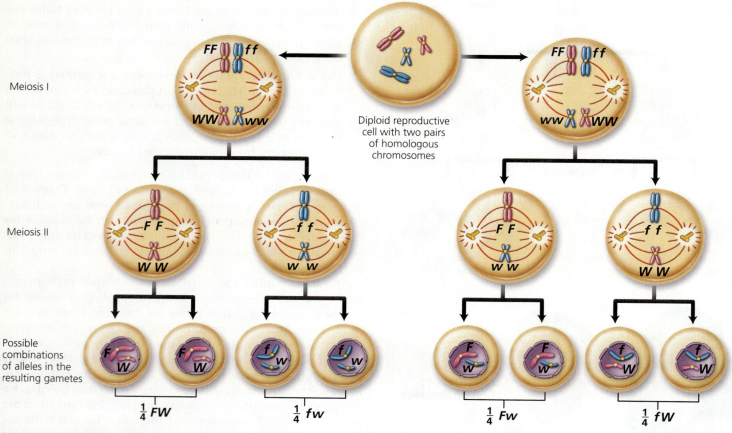

FIGURE 20.6 *A person who is heterozygous for two genes that are located on different chromosomes can produce four different types of gametes.*

▲ *Two types: fW and fw.*

allele on one of the non-sex chromosomes (called *autosomal dominant*) and of genetic disorders that are caused by recessive alleles on a pair of non-sex chromosomes (called *autosomal recessive*) are shown in Figure **20.8**. We will discuss genes on sex chromosomes later in the chapter.

Marfan syndrome is an autosomal dominant disorder. Marfan syndrome is a connective tissue disorder in which a dominant allele for the production of an elastic connective tissue protein, fibrillin, produces a nonfunctional protein. Because connective tissue is widespread in the body, so are the symptoms of Marfan syndrome. Connective tissue is an important component in blood vessel walls, heart valves, tendons, ligaments, and cartilage. A serious problem caused by Marfan syndrome is weakness in walls of large blood vessels, because they can tear or burst.

stop and think

If the pedigree in Figure 20.8a were depicting the inheritance of Marfan syndrome, what would be the genotype of the males in the third generation who have Marfan syndrome? How do you know?

Genetic disorders are often caused by recessive alleles, so knowing whether the parents carry the allele will help predict both the possibility and the likelihood of that child being homozygous for the trait and therefore born with the condition (Figure 20.8b). For example, cystic fibrosis (CF), a disorder in which abnormally thick mucus is produced, is controlled by a recessive allele. Approximately 1 in every 2000 infants in the United States is born with CF. It is a leading cause of childhood death; for children with the disorder, average length of life is 24 years. The three primary signs of CF are abnormally salty sweat, digestive problems, and respiratory problems. Many children with CF suffer from malnutrition because mucus clogs the pancreatic ducts, preventing pancreatic digestive enzymes from reaching the small intestine, where they would otherwise function in digestion. Thick mucus also plugs the respiratory passageways, making breathing difficult and increasing susceptibility to lung infections such as pneumonia (Figure **20.9**).

Because CF is inherited as a recessive allele, children with CF are usually born to normal, healthy parents who had no idea they carried the trait. A **carrier** is someone who displays the dominant phenotype but is heterozygous for a trait and can, therefore, pass the recessive allele to descendants.

Approximately 1 in 22 Caucasians in the United States is a carrier for CF. (CF is less common among Asian Americans and is rare among African Americans.)

stop and think

If the pedigree in Figure 20.8b were depicting the inheritance of CF, what would be the genotype of the male in the third generation who has CF? How do you know? Which individuals are carriers for the trait? How do you know?

The U.S. Surgeon General urges all Americans to create a pedigree of their family health history. The Surgeon General's Health Initiative website (www.hhs.gov/familyhistory) provides a computerized tool for creating a pedigree, but it can also be done without a computer. To create your own pedigree, collect information from your relatives about their own and their ancestors' health at your next family gathering. On the pedigree, indicate which individuals have experienced high blood pressure, a heart attack or stroke, cancer, diabetes, or any other disease common in your family. This health history may help predict your risk of developing certain diseases and encourage you to take preventive action before problems develop.

Dominant and Recessive Alleles

You may wonder what makes an allele dominant or recessive. In many cases, the dominant allele produces a normal, functional protein, but the recessive allele either produces the protein in an altered form that does not function properly,

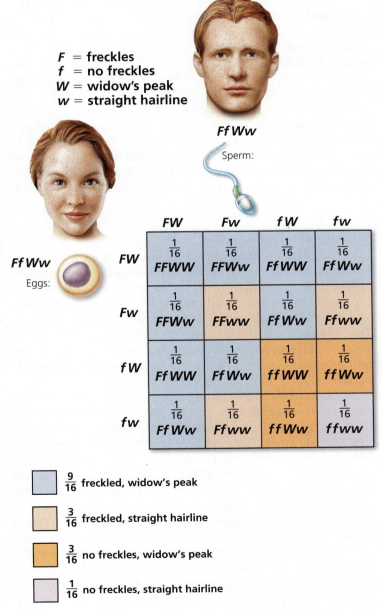

FIGURE 20.7 A dihybrid cross is a mating of individuals heterozygous for two traits, each governed by a gene on a different chromosome. Analyzed on a Punnett square, this cross illustrates the law of independent assortment—that is, each allele pair is inherited independently of others found on different chromosomes. The phenotypes of the offspring of a dihybrid cross would be expected to occur in a 9 : 3 : 3 : 1 ratio.

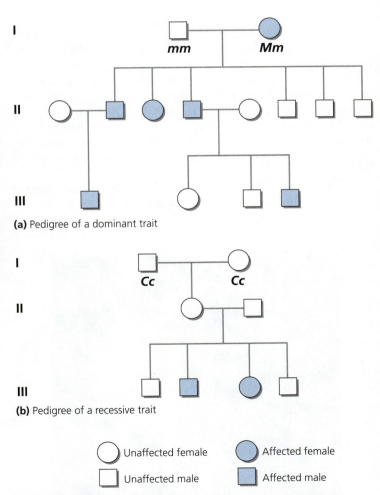

(a) Pedigree of a dominant trait

(b) Pedigree of a recessive trait

FIGURE 20.8 Pedigrees showing the inheritance of (a) a dominant autosomal trait and (b) a recessive autosomal trait. A pedigree is constructed so that each generation occupies a different horizontal line, numbered from top to bottom, with the most ancestral at the top. Males are indicated as squares, and females as circles. A horizontal line connects mating partners. An affected individual is indicated with a colored symbol.

FIGURE 20.9 Cystic fibrosis, controlled by a recessive gene, is a condition in which abnormally thick mucus is produced, causing serious digestive and respiratory problems. This child with cystic fibrosis is using a therapeutic toy to enhance airflow. When she exhales with enough force, the ribbons wave.

or it does not produce any protein. Consider the inheritance of the most common form of albinism, the inability to produce the brown pigment melanin that normally gives color to the eyes, hair, and skin. Because of the lack of melanin, a person with albinism has pale skin and white hair (Figure 20.10). A child with the trait has pink eyes, but the eye color darkens to blue in an adult. Because there is no melanin in the skin to protect against sunlight's ultraviolet rays, a person with albinism is quite vulnerable to sunburn and skin cancer.

The ability to produce melanin depends on the enzyme tyrosinase. The dominant allele that results in normal skin pigmentation produces a functional form of tyrosinase. A single copy of the dominant allele can produce all of the necessary amount of this enzyme. The recessive allele that causes albinism produces a nonfunctional form of tyrosinase. If there are two copies of the recessive allele, melanin cannot be formed, and albinism results.

Codominant Alleles

The examples we have described so far represent **complete dominance**, a situation in which a heterozygous individual exhibits the trait associated with the dominant allele but not that of the recessive allele. In other words, the dominant allele produces a functional protein, and the protein's effects are apparent, but the recessive allele produces a less functional protein or none at all, and its effects are not apparent. However, complete dominance is not the only possibility for a heterozygous genotype. In some cases, both alleles produce functional proteins. In this situation, the effects of *both* alleles are apparent in the heterozygous phenotype.

The inheritance of type AB blood is an example of **codominance**. Two alleles, I^A and I^B, result in the production of two different polysaccharides on the surface of red blood cells. In persons with type AB blood, both alleles are expressed, and their red blood cells have both A and B polysaccharides on their surface. (We return to the example of blood groups shortly.)

Incomplete Dominance

In **incomplete dominance**, the expression of a trait in a heterozygous individual is somewhere between the expression of the trait in a homozygous dominant individual and the expression of the trait in a homozygous recessive individual. The dominant allele produces a functional protein product. The recessive allele does not produce that product. As a result, a heterozygous person has only one "dose" of the protein product—half the amount in a homozygous dominant person.

Sickle-cell hemoglobin provides an example of incomplete dominance (Figure 20.11). Recall from Chapter 11 that hemoglobin is the pigment in red blood cells that carries oxygen. A red blood cell filled with normal hemoglobin (Hb^A) is a biconcave disk. The allele for sickling hemoglobin (Hb^S) produces an abnormal form of hemoglobin that is less efficient in binding oxygen. In the homozygous sickling condition (Hb^SHb^S), called *sickle-cell anemia*, the red blood cells contain only the abnormal form of hemoglobin. When the oxygen content of the blood drops below a certain level, as might occur during excessive exercise or respiratory difficulty, these red blood cells with abnormal hemoglobin become sickle-shaped and tend to clump together. The clumped cells can break open and clog capillaries, causing great pain. Vital organs may be damaged by lack of oxygen. People who are homozygous for sickle-cell anemia usually die at a young age.

The normal allele for the sickle-cell gene shows incomplete dominance, so people who are heterozygous Hb^AHb^S

FIGURE 20.10 A person with albinism lacks the brown pigment melanin in the skin, hair, and irises of the eyes.

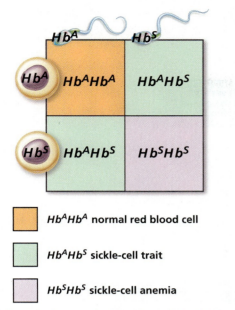

FIGURE 20.11 The inheritance of sickle-cell trait (Hb^AHb^S) is an example of incomplete dominance.

have the *sickle-cell trait*. They have only one "dose" of normal hemoglobin (Hb^A) instead of the normal two doses needed for healthy red blood cells. Thus, people with sickle-cell trait are generally healthy, but sickling and clumping of red cells may occur if there is a prolonged drop in the oxygen content of the blood, as might happen during travel at high elevations. (Sickle-cell anemia is discussed in more detail in Chapter 11.)

stop and think

Straight hair shows incomplete dominance over curly hair. A homozygous dominant person has straight hair; a heterozygous person has wavy hair; and a homozygous recessive person has curly hair. What is the probability that a curly-haired person and a wavy-haired person would have a child with wavy hair?

Pleiotropy

Besides providing an example of incomplete dominance, sickle-cell anemia is an example of **pleiotropy**: one gene leading to many effects. As you can see in Figure 20.12, the sickling of red blood cells caused by the abnormal hemoglobin has effects throughout the body. The sickled cells can break down, clog blood vessels, and accumulate in the spleen. These effects can affect the heart, brain, lungs, kidneys, and muscles and joints.

Multiple Alleles

Many genes have more than two alleles. When three or more forms of a given gene exist, they are referred to as **multiple alleles**. Keep in mind, however, that one individual has only two alleles for a given gene (one on each homologue), even if multiple alleles exist in the population.

The ABO blood types (discussed in Chapter 11) provide an example of multiple alleles. Blood type is determined by the presence of certain polysaccharides (sugars) on the surface of red blood cells. Type A blood has the A polysaccharide; type B has the B polysaccharide; type AB has both A and B polysaccharides; and type O has neither. A specific enzyme directs the synthesis of each kind of polysaccharide: one enzyme produces A, and a different enzyme produces B. Each enzyme is specified by a different allele of the gene.

The gene controlling ABO blood types therefore has three alleles, I^A, I^B, and i. Alleles I^A and I^B specify the A and B polysaccharides, respectively. When both these alleles are present, both polysaccharides are produced. I^A and I^B are, therefore, codominant. Allele i, which produces no enzyme, is recessive to both I^A and I^B. The possible combinations of these alleles and the resulting blood types are shown in Table 20.2.

stop and think

A man who has blood type AB is named as the father of a child with blood type O. The mother has blood type B. Is it possible for the named man to be the father of this child? (*Hint:* What are the possible genotypes of the man, woman, and child? Given each possible genotype of this man and woman, what gametes could each produce? Use Punnett squares to determine whether any combination of these genotypes could produce a child with the same genotype as this child.)

Polygenic Inheritance

So far, we have discussed traits governed by single genes, although the genes may have multiple alleles. When a single gene controls a trait, the trait usually either is present or is not, even though incomplete dominance or environmental factors may modify its expression. In the case of multiple alleles, there may be several distinct classes of phenotypes, as in A, B, AB, and O blood types.

The expression of most traits is much more variable, however. Many traits, including height, skin color, and eye color, vary almost continuously from one extreme to another. Environment can play a role in creating such a smooth continuum. For instance, diet and disease influence adult height, and exposure to sunlight darkens skin color. But even when all environmental factors are equal, there is still considerable variation in the expression of certain traits. Such variation results from **polygenic inheritance**—that is, the involvement of two or more genes, often on different chromosomes, in producing a trait. The more genes involved, the smoother the gradations and the greater the extremes of trait expression.

Although human height is probably controlled by more than three genes, for our purposes we will imagine it is determined by only three, partly to simplify our discussion but also

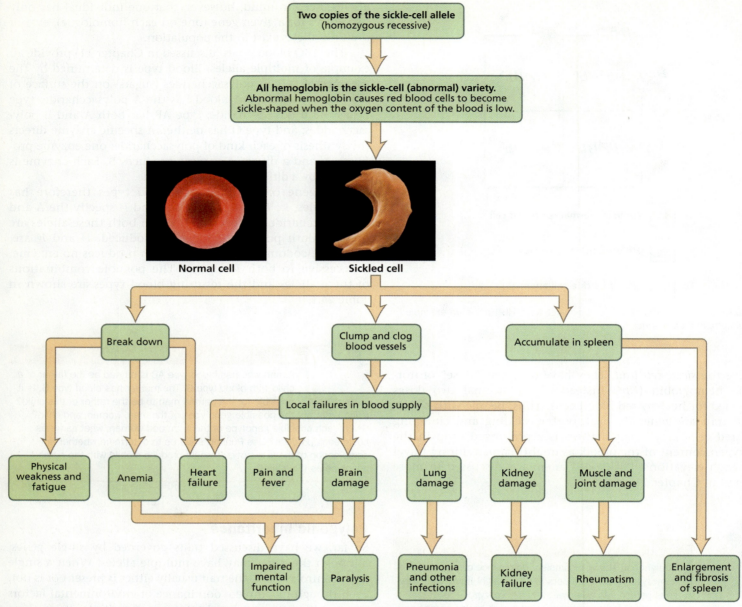

FIGURE 20.12 Sickle-cell anemia is an example of pleiotropy, a condition in which a single gene has many effects.

to show how much variation in expression is possible even with as few as three genes—*A*, *B*, and *C*—interacting to determine a trait. Let's assume that the dominant alleles (*A*, *B*, *C*) of each gene add height, and the recessive alleles (*a*, *b*, *c*) do not.

How tall would we expect the children to be if both parents were of medium height and heterozygous for all three genes? As you can see in Figure 20.13, there would be seven genetic height classes, ranging from very short to very tall. The probability of the children having either extreme of stature is slim, 1/64. The most likely probability (a 20/64 chance) is that they will be of medium height, like their parents.

Skin color is also determined by several genes. The allele for the kind of albinism described previously prevents melanin production, so if a person is homozygous recessive for this allele, no melanin can be deposited in the skin. In addition, there are probably at least four other genes involved in determining the amount of melanin deposited in the skin. Two alleles for each of four genes would create nine classes of skin color, ranging from pale to dark.

TABLE 20.2 The Relationship between Genotype and ABO Blood Types

Genotype	Blood Type
$I^A I^A$, $I^A i$	A
$I^B I^B$, $I^B i$	B
$I^A I^B$	AB
ii	O

Genes on the Same Chromosome

Scientists estimate that there are about 20,000 to 25,000 human genes distributed among the 23 pairs of chromosomes. Thus, each chromosome bears a great number of genes. Genes on the same chromosome tend to be inherited together because an entire chromosome moves into a gamete as a unit. Genes that tend to be inherited together are described as being *linked*. We see, then, that linked genes *usually* do not assort independently. *Usually* is emphasized here because there is a mechanism that can unlink genes on the same chromosome: crossing over (discussed in Chapter 19).

Sex-Linked Genes

Recall from Chapter 19 that one pair of chromosomes consists of the sex chromosomes. There are two kinds of sex chromosomes, X and Y. They are not truly homologous, because the Y chromosome is much smaller than the X chromosome, and they do not carry all of the same genes.[2] The Y chromosome carries very few genes, but it is important in determining gender. If a particular gene found only on the Y chromosome is present, an embryo will develop as a male. Without that gene (whether because there is no Y chromosome or because the gene is missing from the Y chromosome), an embryo will develop as a female. In contrast, most of the genes on an X chromosome have nothing to do with sex determination. For instance, genes for certain blood-clotting factors and for the pigments in cones (the photoreceptors responsible for color vision) are found on the X chromosome but not on the Y chromosome. Furthermore, the X chromosome has about as many genes as a typical autosome does, but the Y chromosome has relatively few. Thus, most genes on the X chromosome have no corresponding alleles on the Y chromosome and are known as **X-linked genes.**

Because most X-linked genes have no homologous allele on the Y chromosome, their pattern of inheritance is different from that of autosomes. A male is XY and therefore will express virtually all of the alleles on his single X chromosome, *even alleles that are recessive*. A female, in contrast, is XX, so she does not always express recessive X alleles. As a result, the recessive phenotype of X-linked genes is much more common in males than in females (Figure 20.14). Furthermore, a son cannot inherit an X-linked recessive allele from his father. To be male, a child must have inherited his father's Y chromosome, not his father's X chromosome. Consequently, a son can inherit an X-linked recessive allele only from his mother. A daughter, however, can inherit an X-linked recessive allele from either parent. She must be homozygous for the recessive allele to show the recessive phenotype. If she is heterozygous for the trait, she will have a normal phenotype but be a carrier for that trait. Among the disorders caused by X-linked recessive alleles are red-green color blindness, two forms of hemophilia, and Duchenne

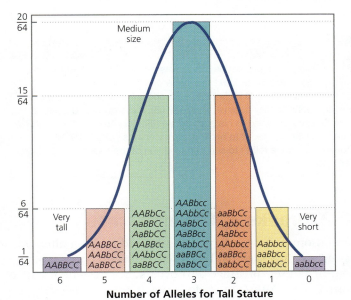

FIGURE 20.13 *Human height varies along a continuum. (a) One reason is that height is determined by more than one gene (polygenic inheritance). This figure shows the distribution of alleles for tallness in children of two parents of medium height, assuming that three genes are involved in the determination of height. The top line of boxes shows the parental genotypes, and the second line of boxes indicates the possible genotypes of the offspring. Alleles for tallness are indicated with dark squares. (b) Students organized according to height.*

[2]X and Y are considered to be a homologous pair because they each have a small region at one end that carries some of the same genes. During meiosis, the tiny homologous region on X and Y will pair in synapsis. As a result, they segregate into gametes in the same way that autosomes do.

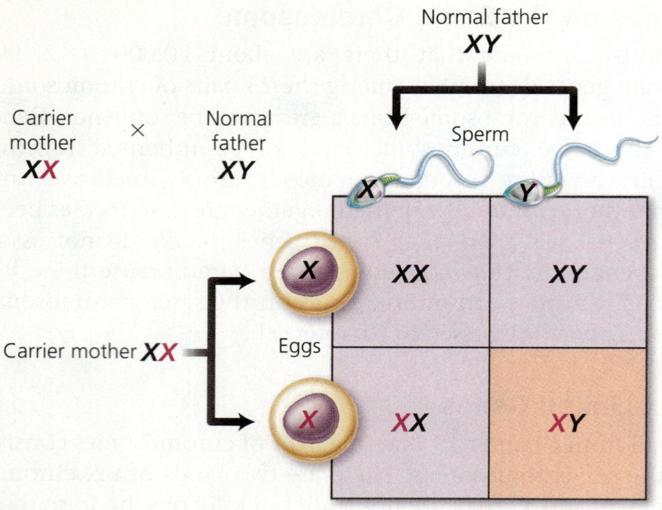

FIGURE 20.14 Genes that are X linked have a different pattern of inheritance from genes on autosomes, as seen in this cross between a carrier mother and a father who is normal for the trait. The recessive allele is indicated in red. Notice that each son has a 50% chance of displaying the recessive phenotype. All daughters will appear normal, but each daughter has a 50% chance of being a carrier.

muscular dystrophy. Red-green color blindness, the inability to distinguish red and green, is discussed in Chapter 9. Hemophilia (discussed in Chapter 11) is a bleeding disorder caused by a lack of a blood-clotting factor: hemophilia A is due to lack of blood-clotting factor VIII, and hemophilia B is due to lack of clotting factor IX. Duchenne muscular dystrophy is discussed in Chapter 6.

stop and think

Why is Duchenne muscular dystrophy inherited from one's mother but usually expressed only in sons? (*Hint:* Consider its mode of inheritance.)

Sex-Influenced Genes

The expression of certain autosomal genes, those that are not located on the sex chromosomes, is powerfully influenced by the presence of sex hormones, so their expression differs in males and females. These traits are described as *sex-influenced* traits.

Male pattern baldness, premature hair loss on the top of the head but not on the sides, is an example of a sex-influenced trait. Male pattern baldness is much more common in men than in women because its expression depends on both the presence of the allele for baldness and the presence of testosterone, the male sex hormone. The allele for baldness, then, acts as a dominant allele in males because of their high level of testosterone and as a recessive allele in females because females have a much lower testosterone level. A male with the allele will develop pattern baldness whether he is homozygous or heterozygous for the trait. However, only women who are homozygous for the trait will develop pattern baldness. When a woman does develop pattern baldness, it usually appears later in life than it does in a man. The allele is expressed in women because the adrenal glands produce a small amount of testosterone. After menopause, when the supply of estrogen declines, adrenal testosterone may cause the expression of the baldness gene. However, in many women, balding may be merely thinning of hair.

stop and think

Male pattern baldness was passed from father to son through at least four generations of the Adams family. John Adams (1735–1826), the second U.S. president, passed it to his son, John Quincy Adams (1767–1848), the sixth U.S. president. He, in turn, passed the gene to his son, Charles Francis Adams (1807–1886), a diplomat, who passed it to his son, Henry Adams (1838–1918), a historian. Why does father-to-son transmission of the trait rule out X-linked inheritance?

20.2 Breaks in Chromosomes

Chromosomes can break, which may lead to other alterations in structure. Breakage can be caused by certain chemicals, radiation, or viruses. Breakage also occurs as an essential part of crossing over. Although it does not happen often, chromosomes can be misaligned when crossing over occurs. Then, when the pieces reattach, one chromatid will have lost a segment, and the other will have gained a segment.

The loss of a piece of chromosome is called a **deletion**. The most common type of deletion occurs when the tip of a chromosome breaks off and then, during cell division, does not move into the same daughter cell as the rest of the chromosome. Deletion of more than a few genes on an autosome is usually lethal, and the loss of even small regions causes disorders.

In humans, the most common deletion, the loss of a small region near the tip of chromosome 5, causes cri-du-chat syndrome (meaning "cry of the cat"). An infant with this syndrome has a high-pitched cry that sounds like a kitten meowing. The unusual sound of the cry is caused by an improperly developed larynx (voice box). Infants with this syndrome have a round face; wide-set, downward-sloping eyes with a fold of skin at the corner of each eye; and misshapen ears (Figure 20.15). Although the condition is not usually fatal, it does cause severe mental retardation.

The addition of a piece of chromosome is called a **duplication**. The effects of a duplication depend on its size and position. In general, however, a small duplication is less harmful than a deletion of comparable size. A small region of chromosome 9 is sometimes duplicated, resulting in cells containing three copies of this segment. The result is mental retardation, accompanied by facial characteristics that may include a bulbous nose; wide-set, squinting eyes; and a lop-sided grin.

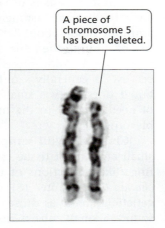

FIGURE 20.15 *Occurring in 1 out of every 50,000 live births, cri-du-chat syndrome is the most common genetic deletion found in humans. It is caused by the loss of a small region near the tip of chromosome 5.*

Genetic disorders also occur when certain sequences of DNA are duplicated multiple times. Fragile X syndrome provides an example. The syndrome is so named because an abnormally long sequence of repeats caused by duplication makes the X chromosome fragile and easily broken. Besides making the chromosome fragile, the repeated DNA can shut down the activity of the entire chromosome. Fragile X syndrome is the most common form of inherited mental retardation, affecting roughly 1 in 1250 males and 1 in 2500 females. It is not known, however, exactly how Fragile X syndrome causes the retardation. Other characteristics may include attention deficit, hyperactivity, autistic symptoms, large ears, long face, and flat feet (Figure 20.16).

20.3 Detecting Genetic Disorders

More than 4000 disorders are known to have their roots in our genes. Tests are now available to look for predispositions to many of these genetic disorders. Some tests can even confirm the presence of a suspected disease-related allele in a particular person. Knowledge about the presence or absence of a faulty gene can be very helpful to couples who are planning a family. Normal, healthy parents can carry recessive alleles for disorders such as CF and Tay-Sachs disease (a disorder of lipid metabolism that causes death, usually between the ages of 1 and 5; discussed in Chapter 3), which they may pass on to their children. Testing is available for both of these recessive alleles. You can read more about the pros and cons of genetic testing in the Ethical Issue essay, *Gene Testing*.

When specific tests are unavailable, pedigree analysis can help prospective parents determine whether they *might* be carriers of recessive alleles. It cannot answer the question with certainty, but it does allow people to weigh the risks of passing a lethal allele to their children versus the possibility of having a child who is not affected.

Prenatal Genetic Testing

Prenatal testing of a fetus is usually recommended when a defective gene runs in the family or when the mother is older

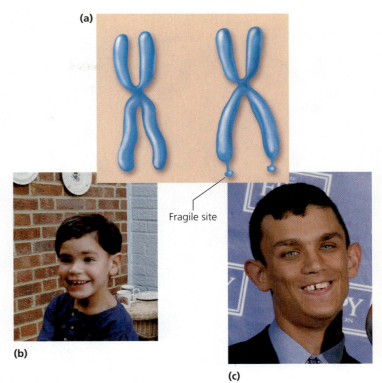

FIGURE 20.16 *Fragile X syndrome. (a) Duplication of a region on the X chromosome makes the chromosome fragile and easily broken. (b) A child with Fragile X syndrome appears normal. (c) Characteristics of an adult with Fragile X syndrome include a long face and large ears. In addition, Fragile X syndrome causes mental retardation.*

than 35 (because age increases the risk of problems due to failure of the homologous chromosomes to separate; see Chapter 19). There are two available procedures for diagnosing genetic problems in the fetus: amniocentesis and chorionic villi sampling (Figure 20.17). Although it is possible to look for more than 100 disorders with these procedures, tests are run only for those disorders that are common as well as any that are of particular concern in that pregnancy. Reassuring results from either form of prenatal testing cannot guarantee a healthy baby, even though these tests are highly accurate in detecting genetic disorders.

In **amniocentesis,** a needle is inserted through the lower abdomen into the uterus, and a small amount of amniotic fluid—10 to 20 ml (about 2 to 4 tsp)—is withdrawn. (Ultrasound is performed first, to find the safest spot for insertion: away from the fetus, umbilical cord, and placenta.) Floating in the amniotic fluid are living cells that were sloughed off the fetus. These cells are grown in the laboratory for a week or two and then examined for abnormalities in the number of chromosomes and for the presence of certain alleles that are likely to cause specific diseases. Biochemical tests are also done on the fluid to look for certain chemicals that indicate problems. For example, a high level of alpha-fetoprotein, a substance produced by the fetus, suggests a problem with the development of the central nervous system (a neural tube defect).

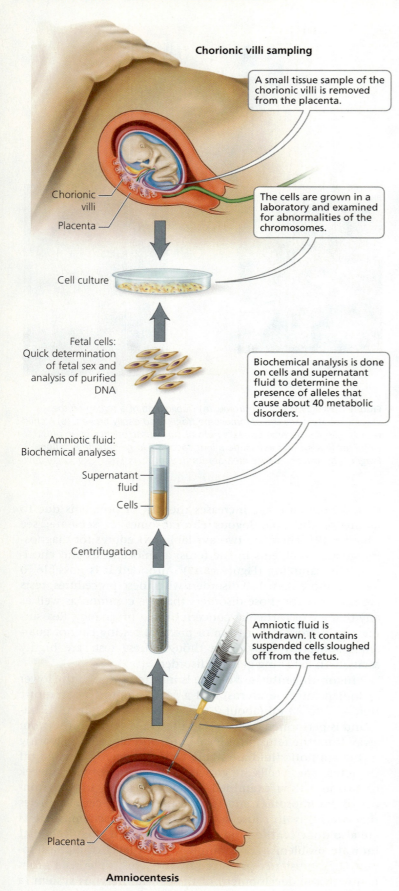

FIGURE 20.17 *Amniocentesis and chorionic villi sampling (CVS) are procedures available for prenatal genetic testing.*

Amniocentesis is usually done between 15 and 20 weeks after the woman's last menstrual period, when there is enough amniotic fluid—about 250 ml (about 1 cup)—to minimize the risk of injuring the fetus. (A few medical centers are now able to perform amniocentesis at 12 weeks of pregnancy.) Amniocentesis is generally safe for both the mother and the fetus, but it does carry a small risk of triggering a miscarriage or of needle injury to the mother and subsequent infection or bleeding.

Chorionic villi sampling (CVS) removes and analyzes a small amount of tissue containing chorionic villi, the small, fingerlike projections of the part of the placenta called the *chorion* (see Chapter 18). Cells of the chorion have the same genetic material as those of the fetus. With the help of ultrasound, a small tube is inserted through the vagina and cervix to where the villi are located. Gentle suction is then used to remove a small tissue sample, which can be analyzed for genetic abnormalities.

There are pros and cons to CVS. Advantages are that it can be performed several weeks earlier in the pregnancy than amniocentesis can, and the results are available within a few days. More than 95% of the high-risk women who opt for prenatal tests receive good news, and an early diagnosis saves weeks of worry over the health of the fetus. Also, if there is a genetic problem in the fetus and the couple wishes to terminate the pregnancy, the procedure can be performed earlier in the pregnancy, when it is safer for the mother. When abortion is not chosen, early diagnosis allows more time to plan the safest time, location, and method of delivery. A disadvantage of CVS is that it has a slightly greater risk of triggering miscarriage than does amniocentesis.

If detected early enough, certain birth defects can be prevented. Congenital adrenal hyperplasia (an overgrowth of the adrenal glands), for instance, will cause a female fetus to develop abnormal genitalia, unless she is treated with hormones from week 10 to week 16 of gestation. Early diagnosis of this disorder through CVS can tell a physician whether hormone treatment is needed.

Newborn Genetic Testing

A simple blood test is now used routinely to screen newborns for phenylketonuria (PKU), an inherited metabolic disorder. People with PKU produce a defective enzyme that prevents them from converting phenylalanine (an amino acid in food) to tyrosine. As a result, they have too much phenylalanine in their bodies and too little tyrosine, an imbalance that somehow causes brain damage. Although nothing can be done to correct the enzyme, diagnosis of the genetic disorder allows doctors and parents to prevent brain damage by keeping the infant on a strict diet that excludes most phenylalanine. Nearly all proteins contain phenylalanine, so it is often necessary to substitute a specially prepared mixture of amino acids for most protein-containing foods, such as meat. It is also necessary to avoid foods containing the artificial sweetener NutraSweet, which contains aspartame, consisting of the amino acids phenylalanine and aspartic acid. When aspartame is digested, phenylalanine is separated from aspartic acid and can reach dangerous levels.

ETHICAL ISSUE

Gene Testing

Genetic screening is the practice of testing people who have no symptoms to determine whether they carry genes that will influence their chances of developing certain genetic diseases. Genetic screening technologies are advancing rapidly, and their use is gaining popularity. However, genetic screening raises many ethical questions.

Among the advantages of genetic testing is that it enables people who discover they are at risk for a treatable or preventable condition to take steps to reduce their risk. By informing people that they carry a recessive allele they were unaware of or a dominant allele that is not expressed until late in life, genetic screening can also help reduce the incidence of serious genetic disorders in future generations. Consider Tay-Sachs disease, an autosomal recessive disorder that causes the death of children, usually by the age of 5. Tay-Sachs disease is especially prevalent in descendants of Jewish people from eastern Europe. As a result of voluntary screening programs, the number of children born with Tay-Sachs disease has decreased tenfold in many communities.

Genetic testing also has a dark side. The psychological consequences of test results can be devastating. Many genetic disorders cannot be prevented or treated. How does a person who may have one of these disorders prepare for the consequences of knowing *now* what will cause his or her death? Huntington's disease, for example, is caused by a dominant allele that provides no hints of its existence until relatively late in life, usually past child-bearing years. About 60% of the people with Huntington's disease are diagnosed between the ages of 35 and 50. The gene causes degeneration of the brain, leading to muscle spasms, personality disorders, and death, usually within 10 to 15 years. Because Huntington's disease is caused by a dominant allele, a bearer has a 50% chance of passing it to his or her children. Thus, a person whose parent died of Huntington's disease might well be tested and receive the good news that the test did not detect the allele. But it is equally likely that the allele *will* show up in the test. Many persons at risk for Huntington's disease prefer to live without knowing their possible fate.

There is also concern that the results of gene tests will not remain private information but instead be used by employers as well as life and health insurers. If you were an employer who had genetic information about prospective employees, would you choose to invest time and money in training a person who carried an allele that increased the risk of cancer, heart disease, Alzheimer's disease, or alcoholism? As an insurer, would you knowingly cover such a carrier?

The results of gene testing can have both positive and negative consequences for individuals being tested and for their families. Who, then, should decide whether screening should be done, for which genes, on whom, and in which communities? At first blush, one might be tempted to say, "There oughta be a law!" Should we leave ethical issues to judges and legislators? Should moral matters be decided by society or clergy? Or should they be personal decisions? These are not easy questions to answer, or even to think about—yet if we do not take part in the debate, we will be allowing others to decide these crucial issues for us.

Questions to Consider

- If gene testing *is* done, should the person being tested be told the results no matter what? If the affected person is an infant, should the parents always be told the results, even if the condition is poorly understood? How do we balance helping such children with the possibility of stigmatizing them?
- We live in a world of limited resources. In addition to deciding who should be tested, we must decide who should pay the bill. Both testing and treatment are expensive. Should testing be done only when treatment or preventive measures are available? How much say should the agent that pays for the procedure have in who is tested and who receives medical treatment?

Adult Genetic Testing

Many predictive genetic tests for adults are now available or being developed. These tests identify people who are at risk of getting a disease but do not yet have symptoms. They are simple tests that can usually be done with a small blood sample. When steps can be taken to prevent the disease, predictive gene tests can be lifesaving. For instance, colon cancer will develop in nearly everyone who has the alleles for familial adenomatous polyposis, a condition in which thousands of benign polyps grow in the intestine. If the disorder is diagnosed, a person can be routinely inspected for colon polyps. Any polyps found can be removed, and cancer can be prevented. Genetic counselors help people understand their risks of developing inherited disorders and the medical consequences of such disorders.

Other predictive gene tests look for alleles that might predispose a person to a disorder. A protein that transports cholesterol in the blood, called ApoE, comes in three forms, each specified by a different allele. Having two alleles for ApoE-2, one form of the protein, causes catastrophically high blood cholesterol levels, which can lead to heart attack and stroke. Knowing that a person has this genetic makeup, a physician could prescribe medication to lower blood cholesterol.

what would you do?

Having two copies of another *ApoE* allele, *ApoE-4*, increases a person's risk of heart disease by 30% to 50%. It also nearly guarantees that the person will develop Alzheimer's disease by 80 years of age. Alzheimer's disease is an untreatable condition in which brain tissue degenerates, gradually robbing the person of memories, of the ability to function normally in society, and eventually of life itself (see Chapter 7). Suppose a gene test that is performed out of concern for a person's risk of heart disease reveals the presence of two copies of *ApoE-4*. In your opinion, is a physician with this knowledge morally obligated to tell the patient about the inevitability of Alzheimer's disease? If your genes were tested and found to have two copies of *ApoE-4*, would you want to be told? Why or why not?

looking ahead

In this chapter, we learned about the chromosomal basis of human inheritance. In the next chapter, we will look more closely at DNA and the mechanisms by which genes affect cellular activity and development.

HIGHLIGHTING THE CONCEPTS

20.1 Principles of Inheritance (pp. 425–436)

- Different forms of a gene are called alleles. An individual who has two of the same alleles is said to be homozygous. An individual with two different alleles for a gene is said to be heterozygous. The allele that is expressed in the heterozygous condition is described as dominant. The allele that is masked in the heterozygous condition is described as recessive.
- The genotype for one or more traits consists of the specific alleles present in an individual. The phenotype refers to the observable expression of the trait or traits.
- The alleles for each gene separate during gamete formation so that half the gametes receive one allele and the other half receive the other allele.
- Each pair of alleles located on one kind of chromosome separate into gametes independently of the alleles of a gene pair located on a different kind of chromosome.
- If a homozygous dominant individual (*AA*) mates with a homozygous recessive one (*aa*), the genotypes of all the offspring would be heterozygous (*Aa*) and the dominant phenotype. If heterozygous individuals (*Aa*) mate, the probable genotypes of children will be 1 in 4 (25%) homozygous dominant (*AA*) : 2 in 4 (50%) heterozygous dominant (*Aa*) : 1 in 4 (25%) homozygous recessive (*aa*). Their phenotypes will be 3 dominant : 1 recessive.
- Pedigrees, which are diagrams constructed to show the genetic relationships among individuals in an extended family, are often useful in determining the unknown genotypes of humans showing dominant phenotypes.
- In many cases, the dominant allele produces a functional protein, and the recessive allele produces a nonfunctional protein or no protein at all.
- In complete dominance, the dominant allele completely masks the recessive allele. When two alleles are codominant, both are apparent in the phenotype. In incomplete dominance, the expression of the trait in a heterozygous individual is somewhere in between the expression of the trait in a homozygous dominant individual and the expression of the trait in a homozygous recessive individual.
- Three or more alleles for a particular gene in a population are called multiple alleles. ABO blood types are determined by three alleles: I^A, I^B, and i. Blood type refers to the presence of certain polysaccharides on the surface of red blood cells.
- Many traits, including height, skin pigmentation, and eye color, are determined by more than one gene (polygenic inheritance). Such traits show a wide range of variation. Moreover, when many genes are involved in determining a trait, the variation in expression can be continuous.
- An X-linked gene is one that is located on the X chromosome and that has no corresponding allele on the Y chromosome. A recessive X-linked allele will always be expressed in a male, but in a female it will be expressed only in the homozygous condition.
- The expression of a sex-influenced trait depends on both the presence of the allele and the presence of sex hormones. Therefore, the expression of the allele depends on the person's sex.

20.2 Breaks in Chromosomes (pp. 436–437)

- The loss of a piece of a chromosome is called a deletion. The gain of a piece of chromosome is called a duplication. Either chromosome abnormality can cause a genetic disorder.

20.3 Detecting Genetic Disorders (pp. 437–439)

- Tests such as amniocentesis and chorionic villi sampling (CVS) are available to determine whether a fetus is likely to develop a genetic disease. In amniocentesis, a sample of amniotic fluid is taken. The fluid and the fetal cells in the fluid are analyzed for genetic problems. CVS samples cells from the chorion of the placenta and analyzes them for genetic disorders.

RECOGNIZING KEY TERMS

gene *p. 426*
trait *p. 426*
allele *p. 426*
homozygous *p. 426*
heterozygous *p. 426*
dominant allele *p. 426*

recessive allele *p. 426*
genotype *p. 426*
phenotype *p. 426*
pedigree *p. 429*
carrier *p. 430*
complete dominance *p. 432*

codominance *p. 432*
incomplete dominance *p. 432*
pleiotropy *p. 433*
multiple alleles *p. 433*
polygenic inheritance *p. 433*

X-linked genes *p. 435*
deletion *p. 436*
duplication *p. 436*
amniocentesis *p. 437*
chorionic villi sampling (CVS) *p. 438*

REVIEWING THE CONCEPTS

1. What is the ratio of genotypes and phenotypes in the offspring resulting from a cross between a homozygous dominant individual and a homozygous recessive one? *p. 428*
2. What is a pedigree? What can a family pedigree reveal about the inheritance of a trait? *pp. 429–431*
3. Explain what is meant by codominance by using an example. *p. 432*
4. Differentiate multiple alleles from polygenic inheritance. *pp. 433–434*
5. What are linked genes? Why are they usually inherited together? What can cause such genes to become unlinked? *p. 435*
6. Explain why the pattern of inheritance for recessive X-linked genes is different from the pattern for recessive autosomal alleles. *p. 435*
7. What two procedures are used for prenatal genetic testing? How do they differ? *pp. 437–438*
8. Which of the following crosses could produce offspring with the recessive phenotype?
   a. *AA* × *aa*
   b. *Aa* × *Aa*
   c. *AA* × *Aa*
   d. *AA* × *AA*

9. All of the following crosses have a 50% probability of producing heterozygous offspring *except*
   a. AA × aa
   b. Aa × Aa
   c. AA × Aa
   d. Aa × aa
10. Which of the following is an example of a phenotype?
    a. a man with hemophilia
    b. a female carrier for cystic fibrosis
    c. XY
    d. a heterozygote
11. Huntington's disease is caused by a dominant allele. If a mother has Huntington's disease but the father does not carry the dominant allele, what is the probability that the first child will be normal?
    a. 0%   c. 50%
    b. 25%   d. 100%
12. In the previous example, what is the probability that the second child will be normal?
    a. 0%   c. 50%
    b. 25%   d. 100%
13. How many different gametes could a person with the genotype *Aabbcc* form? (The genes are on different chromosomes.)
    a. 2   c. 16
    b. 4   d. 64
14. How many different gametes could a person with the genotype *AaBbCc* form? (The genes are on different chromosomes.)
    a. 2   c. 16
    b. 4   d. 64
15. Cystic fibrosis is caused by a recessive allele, *c*. What is the probability that two people who are carriers for cystic fibrosis will have a child who is also a carrier?
    a. 0%   c. 50%
    b. 25%   d. 100%
16. A trait controlled by many genes is described as being _____.
17. The _____ of an individual is the physical expression of one or more genes of interest, and the _____ is the set of alleles the person possesses for the gene or genes of interest.
18. The genotype of a person with two copies of the same allele is _____, and the genotype of a person with two different alleles for a trait is _____.
19. Genes for different traits that are located on the same chromosome are described as being _____.

APPLYING THE CONCEPTS

1. Klaus has red-green color blindness, which is a sex-linked recessive trait. His wife, Helen, is homozygous for normal color vision. (Normal color vision is dominant to red-green color blindness.) What is the probability of their having a color-blind daughter? What is the probability of their having a color-blind son?
2. George and Sue have an infant, Sammy, who is lethargic, is vomiting, and has liver disease. Sammy is diagnosed with galactosemia, an autosomal recessive disorder in which the affected individuals are unable to metabolize the milk sugar galactose. Sammy is placed on a diet free of lactose and galactose and slowly recovers. George and Sue would like to have a second child. What are the chances that the second child will have galactosemia? (*Hint:* What are the genotypes of George and Sue? Use a Punnett square to determine the expected results of a cross with those genotypes.)
3. Explain why the offspring of first cousins are more likely to have harmful recessive traits than are offspring of unrelated individuals.
4. A woman with blood type AB names a man with blood type O as the father of her child. The child has blood type AB. Could the man be the father? Why or why not?
5. Straight hair shows incomplete dominance over curly hair. A homozygous dominant person has straight hair; a heterozygote has wavy hair; and a homozygous recessive person has curly hair. The first child of a curly-haired person and a wavy-haired person has wavy hair. What is the probability that a second child would have wavy hair?

BECOMING INFORMATION LITERATE

Use at least three reliable sources (books, journals, websites) to answer the following questions. Cite your sources, and explain why you choose them.

1. A "savior sibling" is a child born to save the life of a sibling. The older sibling's illness is usually a genetic blood disease, such as leukemia or Fanconi's anemia. It is hoped that the savior sibling will be a tissue match for the older sibling, so a bone marrow transplant from the younger sibling may save the older sibling's life. The embryos for a potential savior sibling are often created in a fertility clinic and tested for tissue compatibility before being implanted in the mother's uterus. If you were the parent of a child with leukemia and there were no family members who were tissue matches, would you consider conceiving a savior sibling? Identify the arguments on both sides of the issue, and defend your decision.
2. Identify at least two disorders that are caused by chromosome deletions and two that are caused by chromosome duplications. Which chromosomes are affected? What are the symptoms of the disorders?

MasteringBiology®

Go to MasteringBiology for practice quizzes, activities, eText, videos, current events, and more.

21 DNA and Biotechnology

Did you Know?

- The DNA in each of your cells consists of about 3 billion chemical "letters," called nucleotides.
- If you unwound all the DNA in all your cells and attached the strands end to end, it would reach to the sun and back more than 600 times.

This child was rescued from tsunami debris and claimed by nine sets of parents. She was returned to her real parents following a DNA test.

- 21.1 Form of DNA
- 21.2 Replication of DNA
- 21.3 Gene Expression
- 21.4 Mutations
- 21.5 Regulating Gene Activity
- 21.6 Genetic Engineering
- 21.7 Genomics

ENVIRONMENTAL ISSUE
Environment and Epigenetics

HEALTH ISSUE
Genetically Modified Food

ETHICAL ISSUE
Forensic Science, DNA, and Personal Privacy

In Chapter 20, we learned about the chromosomal basis of inheritance. In this chapter, we become familiar with the structure and function of DNA and discover how this molecule is able to serve as the basis of our genetic inheritance as well as the source for the diversity of life on Earth. We learn that the importance of DNA on a personal level is that it directs the synthesis of specific polypeptides (proteins) that play structural or functional roles in our bodies. We then consider the technology that our understanding of DNA has already made available and what possibilities such technology may hold for the future.

21.1 Form of DNA

DNA is sometimes referred to as the thread of life—and a very slender thread it is. When DNA is unwound, it measures a mere 50-trillionths of an inch in diameter. If all the DNA strands in a single cell were fastened together end to end, the thread would stretch more than 5 feet in length. DNA might also be considered the thread that ties all life together, because the DNA of organisms ranging from bacteria to humans is built from the same kinds of subunits. The order of these subunits encodes the information needed to make the proteins that build and maintain life.

Deoxyribonucleic acid, or **DNA**, is a double-stranded molecule resembling a ladder that is gently twisted to form a spiral called a *double helix,* as shown in Figure **21.1**. Each side of the ladder, including half of each rung, is made from a string of repeating subunits called *nucleotides.* You may recall from Chapter 2 that a nucleotide is composed of three subunits, including one sugar (deoxyribose, in DNA), one phosphate, and one nitrogenous base. DNA contains four

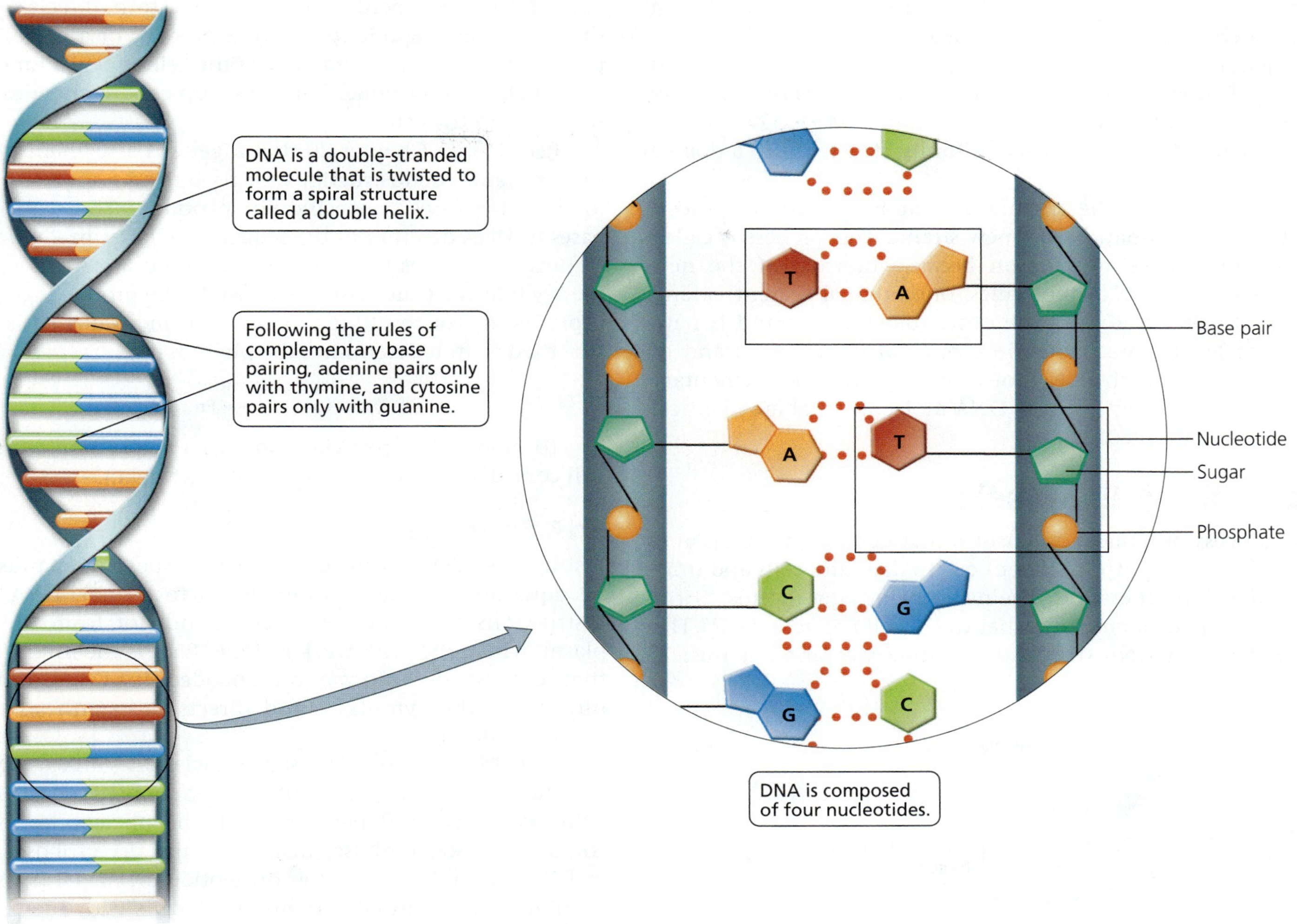

FIGURE 21.1 *DNA is a double-stranded molecule that twists to form a spiral structure called a double helix.*

- DNA is a double-stranded molecule that is twisted to form a spiral structure called a double helix.
- Following the rules of complementary base pairing, adenine pairs only with thymine, and cytosine pairs only with guanine.
- DNA is composed of four nucleotides.

Labels: Base pair, Nucleotide, Sugar, Phosphate

types of nitrogenous bases: adenine (A), guanine (G), thymine (T), and cytosine (C). The sides of the ladder are composed of alternating sugars and phosphates; the rungs consist of paired nitrogenous bases. The bases attach to each other according to the rules of **complementary base pairing**: adenine pairs *only* with thymine (creating an A–T pair), and cytosine pairs *only* with guanine (creating a C–G pair). Each base pair is held together by weak hydrogen bonds. The pairing of complementary bases is specific because of the shapes of the bases and the number of hydrogen bonds that can form between them. You may also recall from Chapter 2 that a molecule formed by the joining of nucleotides is called a *nucleic acid*. Thus, DNA is a nucleic acid.

Because base pairing is so specific, the bases on one strand of DNA are *always* complementary to the bases on the other strand. Thus, the order of bases on one strand determines the sequence of bases on the other strand. For instance, if the sequence of bases on one strand were CATATGAG, what would the complementary sequence be? Remember, cytosine (C) always pairs with guanine (G), and adenine (A) always pairs with thymine (T). As a result, the complementary sequence on the opposite strand would be GTATACTC.

The DNA within each human cell has an astounding 3 billion base pairs. Although the pairing of adenine with thymine and cytosine with guanine is specific and does not vary, the sequence of bases throughout the length of different DNA molecules can vary in myriad ways. As we will see, genetic information is encoded in the exact sequence of bases.

21.2 Replication of DNA

For DNA to be the basis of inheritance, its genetic instructions must be passed from one generation to the next. Moreover, for DNA to direct the activities of each cell, its instructions must be present in every cell. These requirements dictate that DNA be copied before both mitotic and meiotic cell division (see Chapter 19). It is important that the copies be exact. The key to the precision of the copying process is that the bases are complementary.

The copying process, or **DNA replication**, begins when an enzyme breaks the weak hydrogen bonds that hold together the paired bases that make up nucleotide strands of the double helix, thereby "unzipping" and unwinding the strands. As a result, the nitrogenous bases on the separated regions of each strand are temporarily exposed. Free nucleotide bases, which are always present within the nucleus, can then attach to complementary bases on the open DNA strands. Enzymes called *DNA polymerases* link the sugars and phosphates of the newly attached nucleotides to form a new strand. As each of the new double-stranded DNA molecules forms, it twists into a double helix.

Each strand of the original DNA molecule serves as a template for the formation of a new strand. This process is called **semiconservative replication** because in each of the new double-stranded DNA molecules, one original (parent) strand is saved (conserved) and the other (daughter) strand is new. Look at Figure 21.2. Notice the original (parental) strand of nucleotides in each new molecule of DNA. Complementary base pairing creates two new DNA molecules that are identical to the parent molecule.

21.3 Gene Expression

The process of replication ensures that genetic information is passed accurately from a parent cell to daughter cells and from generation to generation. The next obvious question is, "How does DNA issue commands that direct cellular activities?" The answer is that DNA directs the synthesis of another nucleic acid—**ribonucleic acid**, or **RNA**. RNA, in turn, directs the synthesis of a polypeptide (a part of a protein) or a protein. The protein may be a structural part of the cell or play a functional role, such as an enzyme that speeds up certain chemical reactions within the cell.

Recall from Chapter 20 that a **gene** is a segment of DNA that contains the instructions for producing a specific protein (or in some cases, a specific polypeptide).[1] The sequence of bases in DNA determines the sequence of bases in RNA, which in turn determines the sequence of amino acids of a protein. We say that the gene is expressed when the protein it codes for is produced. The resulting protein is the molecular basis of the inherited trait; it determines the phenotype.

Gene expression:
DNA ⟶ RNA ⟶ Protein

To more fully appreciate how gene expression works, we will consider each step in slightly greater detail.

RNA Synthesis

Just as the CEO of a major company issues commands from headquarters instead of from the factory floor, DNA issues instructions from the cell nucleus and not from the cytoplasm, where the cell's work is done. RNA is the intermediary that carries the information encoded in DNA from the nucleus to the cytoplasm and directs the synthesis of the specified protein.

Like DNA, RNA is composed of nucleotides linked together, but there are some important differences between DNA and RNA, as shown in Table 21.1. First, the nucleotides of RNA contain the sugar ribose, instead of the deoxyribose found in DNA. Second, in RNA the nucleotide uracil (U) pairs with adenine, whereas in DNA thymine (T) pairs with adenine (A). Third, most RNA is single stranded. Recall that DNA is a double-stranded molecule.

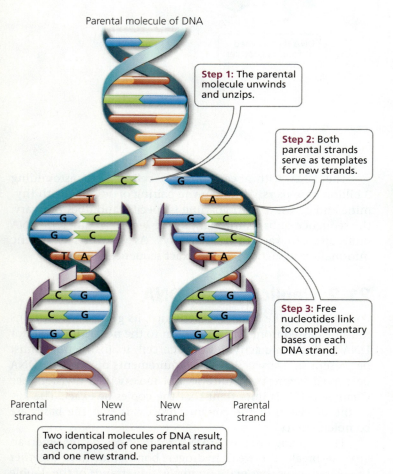

FIGURE 21.2 *DNA replication is called semiconservative because each daughter molecule consists of one "parental" strand and one "new" strand.*

TABLE 21.1	Comparisons of DNA and RNA	
	DNA	**RNA**
Similarities	Are nucleic acids	
	Are composed of linked nucleotides	
	Have a sugar-phosphate backbone	
	Have four types of bases	
Differences	Is a double-stranded molecule	Is a single-stranded molecule
	Has a sugar deoxyribose	Has a sugar ribose
	Contains the bases adenine, guanine, cytosine, and thymine	Contains the bases adenine, guanine, cytosine, and uracil (instead of thymine)
	Functions primarily in the nucleus	Functions primarily in the cytoplasm

[1]We will develop this concept as the chapter proceeds. Some genes code for a polypeptide that is only part of a functional protein. A gene can also code for RNA that forms part of a ribosome or that transports amino acids during protein synthesis.

The first step in converting the DNA message to a protein is to copy the message as RNA, by a process called **transcription**.

$$\text{DNA} \xrightarrow{\text{transcription}} \text{RNA} \longrightarrow \text{Protein}$$

Three types of RNA are produced in cells. Each plays a different role in protein synthesis (Table 21.2). **Messenger RNA (mRNA)** carries DNA's instructions for synthesizing a particular protein from the nucleus to the cytoplasm. The order of bases in mRNA specifies the sequence of amino acids in the resulting protein, as we will see. Each **transfer RNA (tRNA)** molecule is specialized to bring a specific amino acid to where it can be added to a polypeptide that is under construction. **Ribosomal RNA (rRNA)** combines with proteins to form ribosomes, which are the structures on which protein synthesis occurs.

Transcription begins with the unwinding and unzipping of the specific region of DNA to be copied; these actions are performed by an enzyme. The DNA message is determined by the order of bases in the unzipped region of the DNA molecule. One of the unwound strands of the DNA molecule serves as the template during transcription. RNA nucleotides present in the nucleus pair with their complementary bases on the template—cytosine with guanine and uracil with adenine (Figure 21.3). The signal to start transcription is given by a specific sequence of bases on DNA, called the **promoter**. An enzyme called **RNA polymerase** binds with the promoter on DNA and then moves along the DNA strand, opening up the DNA helix in front of it and then aligning the appropriate RNA nucleotides and linking them together; the region of DNA that has been transcribed zips again after RNA polymerase passes by. Another sequence of bases on the DNA signals RNA polymerase to stop transcription. After transcription ceases, the newly formed strand of RNA, called the *RNA transcript*, is released from the DNA.

Messenger RNA usually undergoes certain modifications before it leaves the nucleus (Figure 21.4). Most stretches of DNA between a promoter and the stop signal include regions that do not contain codes that will be translated into protein. These unexpressed regions of DNA are called *introns*, short for *intervening sequences*. The regions of mRNA corresponding to the introns are snipped out of the newly formed mRNA strand by enzymes before the strand leaves the nucleus. The remaining segments of DNA or mRNA, called *exons* for *expressed sequences*, splice together to form the sequence that directs the synthesis of a protein.

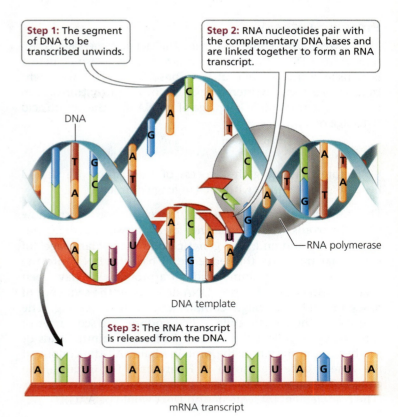

FIGURE 21.3 Transcription is the process of producing RNA from a DNA template.

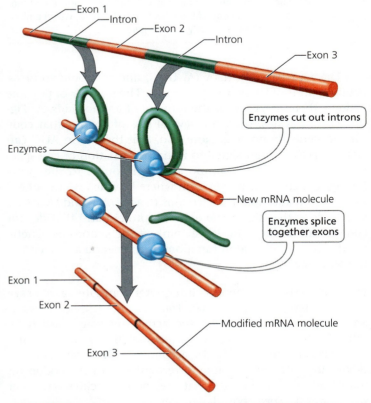

FIGURE 21.4 Newly formed messenger RNA is modified before it leaves the nucleus. Noncoding regions of DNA called introns are snipped out of the corresponding regions of mRNA molecule. Segments of mRNA that code for protein are then spliced together.

TABLE 21.2	Review of the Functions of RNA
Molecule	**Functions**
Messenger RNA (mRNA)	Carries DNA's information in the sequence of its bases (codons) from the nucleus to the cytoplasm
Transfer RNA (tRNA)	Binds to a specific amino acid and transports it to be added, as appropriate, to a growing polypeptide chain
Ribosomal RNA (rRNA)	Combines with protein to form ribosomes (structures on which polypeptides are synthesized)

Protein Synthesis

The newly formed mRNA carries the genetic message (transcribed from DNA) from the nucleus to the cytoplasm, where it is translated into protein at the ribosomes. Just as we might translate a message written in Spanish into English, **translation** converts the nucleotide language of mRNA into the amino acid language of a protein.

$$\text{DNA} \xrightarrow{\text{transcription}} \text{RNA} \xrightarrow{\text{translation}} \text{Protein}$$

Before examining the process of translation, we should become more familiar with the language of mRNA.

The genetic code To use any language, you must know what the words are and what they mean, as well as where sentences begin and end. The **genetic code** is the "language" of genes that translates the sequence of bases in DNA into the specific sequence of amino acids in a protein. We have seen that the sequence of bases in DNA determines the sequence of bases in mRNA through complementary base pairing. The "words" in the genetic code, called **codons**, are sequences of three bases on mRNA that specify 1 of the 20 amino acids or the beginning or end of the protein chain. All the codons of the genetic code are shown in Figure 21.5. For instance, the codon UUC on mRNA specifies the amino acid phenylalanine. (The complementary sequence on DNA would be AAG.)

> **stop and think**
> Look at the mRNA transcript in Figure 21.3. Notice that the codon at the end of the mRNA strand is GUA. Which amino acid does this specify? (Use Figure 21.5.)

The four bases in RNA (A, U, C, and G) could form 64 combinations of three-base sequences. The number of possible codons, therefore, exceeds the number of amino acids. As Figure 21.5 indicates, there are several sets of codons that code for the same amino acid. Note, too, that the codon AUG can either serve as a start signal to initiate translation or can specify the addition of the amino acid methionine to the growing protein chain, depending on where it occurs in the mRNA molecule. In addition, three codons (UAA, UAG, and UGA) are stop codons that signal the end of a protein and that do not code for an amino acid. If we think of the codons as genetic words, then a stop codon functions as the period at the end of the sentence.

Transfer RNA A language interpreter translates a message from one language to another. Transfer RNA (tRNA) serves as an interpreter that converts the genetic message carried by mRNA into the language of protein, which is a particular sequence of amino acids. To accomplish this conversion, a tRNA molecule must be able to recognize both the codon on mRNA and the amino acid that the codon specifies—in other words, it must speak both languages.

There are many kinds of tRNA—at least one for each of the 20 amino acids. Each type of tRNA molecule binds to a particular amino acid. Enzymes ensure tRNA binds with the correct

Q *If the sequence of bases following a start signal were AACUCAGCC, which amino acids would be specified?*

First base	Second base U	Second base C	Second base A	Second base G	Third base
U	UUU Phenylalanine	UCU Serine	UAU Tyrosine	UGU Cysteine	U
U	UUC Phenylalanine	UCC Serine	UAC Tyrosine	UGC Cysteine	C
U	UUA Leucine	UCA Serine	UAA *stop*	UGA *stop*	A
U	UUG Leucine	UCG Serine	UAG *stop*	UGG Tryptophan	G
C	CUU Leucine	CCU Proline	CAU Histidine	CGU Arginine	U
C	CUC Leucine	CCC Proline	CAC Histidine	CGC Arginine	C
C	CUA Leucine	CCA Proline	CAA Glutamine	CGA Arginine	A
C	CUG Leucine	CCG Proline	CAG Glutamine	CGG Arginine	G
A	AUU Isoleucine	ACU Threonine	AAU Asparagine	AGU Serine	U
A	AUC Isoleucine	ACC Threonine	AAC Asparagine	AGC Serine	C
A	AUA Isoleucine	ACA Threonine	AAA Lysine	AGA Arginine	A
A	AUG **(start)** Methionine	ACG Threonine	AAG Lysine	AGG Arginine	G
G	GUU Valine	GCU Alanine	GAU Asparagine	GGU Glycine	U
G	GUC Valine	GCC Alanine	GAC Asparagine	GGC Glycine	C
G	GUA Valine	GCA Alanine	GAA Glutamic acid	GGA Glycine	A
G	GUG Valine	GCG Alanine	GAG Glutamic acid	GGG Glycine	G

FIGURE 21.5 The genetic code. Each sequence of three bases on the mRNA molecules, called a codon, specifies a specific amino acid, a start signal, or a stop signal.

A Asparagine, serine, alanine

amino acid. The tRNA then ferries the amino acid to the correct location along a strand of mRNA (Figure 21.6).

How does the tRNA know the correct location along mRNA? The location is determined by a sequence of three nucleotides on the tRNA called the **anticodon**. In a sense, the anticodon "reads" the language of mRNA by binding to a codon on the mRNA molecule according to the complementary base-pairing rules. When the tRNA's anticodon binds to the mRNA's codon, the specific amino acid attached to the tRNA is brought to the growing polypeptide chain. For example, a

21.3 Gene Expression **447**

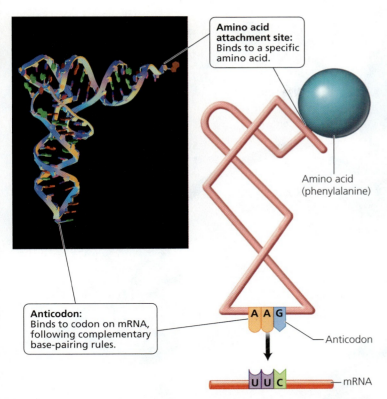

FIGURE 21.6 *A tRNA molecule is a short strand of RNA that twists and folds on itself. The job of tRNA is to ferry a specific amino acid to the ribosome and insert it in the appropriate position in the growing peptide chain.*

tRNA molecule with the anticodon AAG binds to the amino acid phenylalanine and ferries it to the mRNA molecule, where the codon UUC is presented for translation. Phenylalanine will then be added to the growing amino acid chain.

Ribosomes Ribosomes function as the workbenches on which proteins are built from amino acids. A ribosome consists of two subunits (small and large), each composed of ribosomal RNA (rRNA) and protein. The subunits form in the nucleus and are shipped to the cytoplasm. They remain separate except during protein synthesis. The role of the ribosome in protein synthesis is to bring the tRNA bearing an amino acid close enough to the mRNA to interact. As you can see in Figure 21.7, when the two subunits fit together to form a functional ribosome, a groove for mRNA is formed. Two binding sites position tRNA molecules so that an enzyme in the ribosome can cause bonds to form between their amino acids.

Protein synthesis Translation—essentially, protein synthesis—can be divided into three stages: initiation, elongation, and termination.

1. During initiation, the major players in protein synthesis (mRNA, tRNA, and ribosomes) come together (Figure 21.8).
   - Step 1: The small ribosomal subunit attaches to the mRNA strand at the start codon, AUG.
   - Step 2: The tRNA with the complementary anticodon pairs with the start codon. The larger ribosomal subunit then joins the smaller one to form a functional, intact

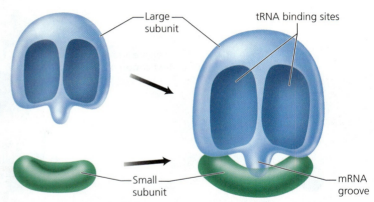

FIGURE 21.7 *A ribosome consists of two subunits of different sizes. When the two subunits join together to form a functional ribosome, a groove for mRNA is formed. The ribosome has two binding sites for tRNA molecules. It also contains an enzyme that promotes the formation of a peptide bond between the amino acids that are attached to the tRNAs in the binding sites.*

ribosome with mRNA positioned in a groove between the two subunits.

2. **Elongation of the protein occurs as additional amino acids are added to the chain** (Figure 21.9).
   - *Step 1: Codon recognition.* With the start codon positioned in one binding site, the next codon is aligned in the other binding site.
   - *Step 2: Peptide bond formation.* The tRNA bearing an anticodon that will pair with the exposed codon slips into place at the binding site, and the amino acid it carries forms a peptide bond with the previous amino acid with the assistance of enzymes.
   - *Step 3: Ribosome movement.* The tRNA in the first binding site leaves the ribosome. The ribosome moves along the mRNA molecule, carrying the growing peptide chain and the remaining tRNA with its amino acid to the first binding site. This movement positions the next codon

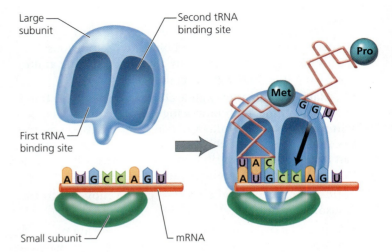

Step 1: The small ribosomal subunit joins to mRNA at the start codon, AUG.

Step 2: A tRNA with complementary anticodon pairs with the start codon. Ribosomal subunits join to form a functional ribosome.

FIGURE 21.8 *Initiation of translation*

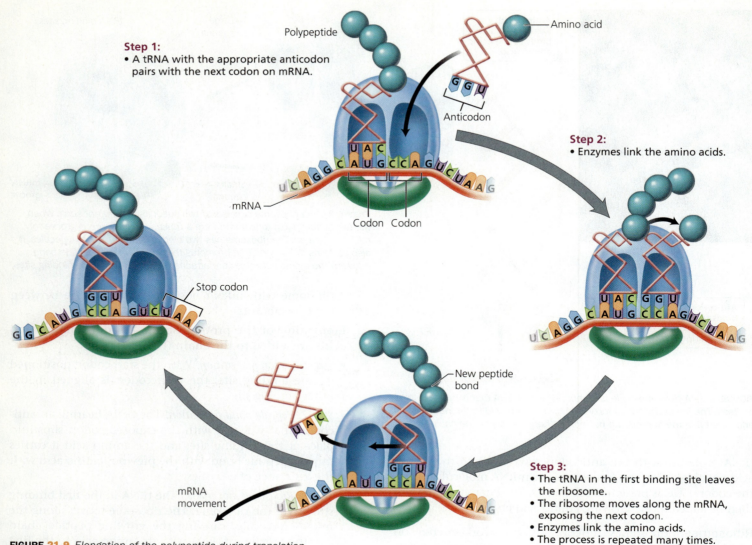

FIGURE 21.9 Elongation of the polypeptide during translation

in the open site. An appropriate tRNA slips into the open site, and its amino acid binds to the previous one. This process is repeated many times, adding one amino acid at a time to the growing polypeptide chain.

Many ribosomes may glide along a given mRNA strand at the same time, each producing its own copy of the protein directed by that mRNA (Figure 21.10). As soon as one ribosome moves past the start codon, another ribosome can attach. A cluster of ribosomes simultaneously translating the same mRNA strand is called a **polysome**.

3. **Termination occurs when a stop codon moves into the ribosome** (Figure 21.11).
   - *Step 1: Stop codon moves into ribosome.* There are no tRNA anticodons that pair with the stop codons, so when a stop codon moves into the ribosome, protein synthesis is terminated.
   - *Step 2: Parts disassemble.* The newly synthesized polypeptide, the mRNA strand, and the ribosomal subunits then separate from one another.

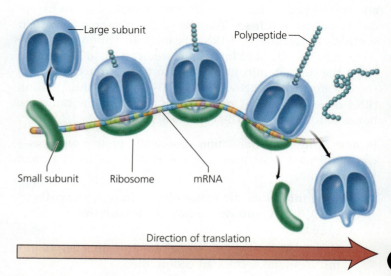

FIGURE 21.10 A polysome is a group of ribosomes reading the same mRNA molecule.

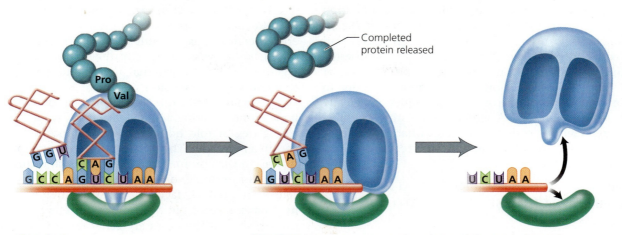

Step 1: The stop codon moves into the ribosome.

Step 2: Release factors cause the release of the newly formed polypeptide and the separation of the ribosomal subunits and the mRNA.

FIGURE 21.11 *Termination of translation*

stop and think

Streptomycin is an antibiotic, a drug taken to slow the growth of invading bacteria and allow body defense mechanisms more time to destroy them. Streptomycin works by binding to the bacterial ribosomes and preventing an accurate reading of mRNA. Why would this process slow bacterial growth?

21.4 Mutations

DNA is remarkably stable, and the processes of replication, transcription, and translation generally occur with amazing precision. However, sometimes DNA is altered, and the alterations can change its message. Changes in DNA are called **mutations**. One type of mutation occurs when whole sections of chromosomes are duplicated or deleted, as discussed in Chapter 20. Now that we are familiar with the chemical structure of DNA and how it directs the synthesis of proteins, we can consider another type of mutation—a gene mutation. A gene mutation results from changes in the order of nucleotides in DNA. Although a gene mutation can occur in any cell, it can be passed on to offspring only if it is present in a cell that will become an egg or a sperm. A mutation that occurs in a body cell can affect the functioning of that cell and the subsequent cells produced by that cell, sometimes with disastrous effects, but it cannot be transmitted to a person's offspring.

One type of gene mutation is the replacement of one nucleotide pair by a different nucleotide pair in the DNA double helix. During DNA replication, bases may accidentally pair incorrectly. For example, adenine might mistakenly pair with cytosine instead of thymine. Repair enzymes normally replace the incorrect base with the correct one. However, sometimes the enzymes recognize that the bases are incorrectly paired but mistakenly replace the original base (the one on the old strand) rather than the new, incorrect one. The result is a complementary base pair consisting of the wrong nucleotides (Figure 21.12).

Other types of gene mutations are caused by the insertion or deletion of one or more nucleotides. Generally, a mutation

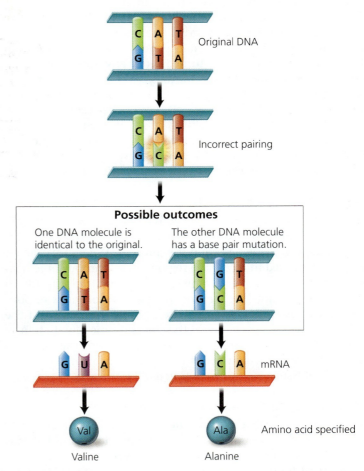

FIGURE 21.12 *A base-pair substitution is a DNA mutation that results when a base is paired incorrectly. This may change the amino acid specified by the mRNA and alter the structure of the protein.*

of this kind has more serious effects than does a mutation caused by substitution of one base pair for another. Recall that the mRNA is translated in units of three nucleotides (a unit called a codon). If one or two nucleotides are inserted or deleted, *all* the triplet codons that follow the insertion or deletion are likely to change. Consequently, mutations due to the insertion or deletion of one or two nucleotides can greatly change the resulting protein. A sentence consisting of three-letter words (representing codons) illustrates what can happen. Deleting a single letter from the sentence "The big fat dog ran" renders the sentence nonsensical:

Original: THE BIG FAT DOG RAN
After deletion of the E in THE: THB IGF ATD OGR AN

21.5 Regulating Gene Activity

At the time of your conception, you received one set of chromosomes from your father and one set from your mother. The resulting zygote then began a remarkable series of cell divisions—some of which continue in many of your body cells to this day. With each division, the genetic information was faithfully replicated, and exact copies were parceled into the daughter cells. Thus, every nucleated cell you possess, except gametes, contains a complete set of identical genetic instructions for making every structure and performing every function in your body.

How, then, can liver, bone, blood, muscle, and nerve cells look and act so differently from one another? The answer is deceptively simple: Only certain genes are active in a certain type of cell; most genes are turned off in any given cell, which leads to specialization for specific jobs. The active genes produce specific proteins that determine the structure and function of that particular cell. Indeed, as cells become specialized for specific jobs, the timing of the activity of specific genes is critical.

But what controls gene activity? The answer to this question is a bit more complex, because gene activity is controlled in several ways. Genes are regulated on several levels simultaneously.

Gene Activity at the Chromosome Level

At the chromosome level, gene activity is affected by the coiling and uncoiling of the DNA. When the DNA is tightly coiled, or condensed, the genes are not expressed. When a particular protein is needed in a cell, the region of the chromosome containing the necessary gene unwinds, allowing transcription to take place. Presumably, the uncoiling allows enzymes responsible for transcription to reach the DNA in that region of the chromosome. Other regions of the chromosome remain tightly coiled and therefore are not expressed. Indeed, the environment may affect which genes are turned on or off (see Environmental Issue essay, *Environment and Epigenetics*).

Regulating the Transcription of Genes

Some regions of DNA regulate the activity of other regions. As we have seen, a promoter is a specific sequence of DNA that is located adjacent to the gene it regulates. When regulatory proteins called *transcription factors* bind to a promoter, RNA polymerase can bind to the promoter, which begins transcription of the regulated genes.

Transcription factors can also bind to enhancers, segments of DNA that increase the *rate* of transcription of certain genes and, therefore, the amount of a specific protein that is produced. Enhancers also specify the timing of expression and a gene's response to external signals and developmental cues that affect gene expression.

You may recall from Chapter 10 that one of the ways certain hormones bring about their effects is by turning on specific genes. Steroid hormones, for instance, bind to receptors within a target cell. The hormone–receptor complex then finds its way to the chromatin in the nucleus and turns on specific genes. For example, one such complex turns on the genes in cells that produce facial hair—explaining why your father may have a beard but your mother probably does not, even though she has the necessary genes to grow one. In this case, the sex hormone testosterone binds to a receptor and turns on hair-producing genes. Facial hair follicle cells of both men and women have the necessary testosterone receptors. However, women usually do not produce enough testosterone to activate the hair-producing genes, so bearded women are rare.

stop and think

Why do female athletes who inject themselves with testosterone to stimulate muscle development sometimes develop increased facial hair?

21.6 Genetic Engineering

The manipulation of genetic material for human purposes, a practice called **genetic engineering**, began almost as soon as scientists started to understand the language of DNA. Genetic engineering is part of the broader endeavor of **biotechnology**, a field in which scientists make controlled use of living cells to perform specific tasks. Genetic engineering has been used to produce pharmaceuticals and hormones, improve diagnosis and treatment of human diseases, increase food production from plants and animals, and gain insight into the growth processes of cells.

Recombinant DNA

The basic idea behind genetic engineering is to put a gene of interest—in other words, one that produces a useful protein or trait—into another piece of DNA to create **recombinant DNA**, which is DNA combined from two or more sources. The recombinant DNA, carrying the gene of interest, is then placed into a rapidly multiplying cell that quickly produces many copies of the gene. The final harvest may consist of large amounts of the gene product or many copies of the gene itself. Let's take a closer look at the procedure one step at a time.

1. **The gene of interest is sliced out of its original organism and spliced into vector DNA.** Both the DNA originally containing the gene of interest and the vector DNA, which receives the transferred genes and transports it to a new cell, are cut at specific sequences that are recognized by a **restriction enzyme**. This is a type of

ENVIRONMENTAL ISSUE

Environment and Epigenetics

Your lifestyle may influence the health of your great grandchild. How is this possible? It can occur through epigenetics, which involves a stable alteration in gene expression *without* changes in DNA sequence. In other words, it regulates how genes are expressed without changing the proteins they encode. We will consider two epigenetic processes: DNA methylation and histone acetylation. These processes alter gene expression by affecting how tightly packaged the DNA molecule is. DNA is packaged with proteins to form chromosomes. DNA methylation (adding a methyl group to the cytosine bases in DNA) turns off the activity of a gene by bringing in proteins that act to compact DNA into a tighter form. Histone acetylation, in contrast, makes the DNA less tightly coiled and gene expression easier.

We now know that these processes can be affected by the environment and that the pattern of DNA methylation is dynamic and changes over time. DNA methylation patterns can be affected by environmental factors, cause disease, be transmitted through generations, and, potentially, influence evolution. DNA is sensitive to the environment, so what we eat and the chemicals we are exposed to, including pesticides, tobacco smoke, hormones, and nutrients, may influence our health by affecting our gene expression patterns. For example, maternal nutrition during pregnancy can cause epigenetic changes in gene activity in the fetus that may increase susceptibility to obesity, type-2 diabetes, heart disease, and cancer. The quantity of food consumed during pregnancy alters the offspring's susceptibility to cardiovascular disease. Epigenetics is also thought to play a role in human behavioral disorders, such as autism spectrum disorders (discussed in Chapter 18a), Rett syndrome (a developmental disorder that affects the nervous system), and Fragile-X syndrome (an inherited form of mental impairment). For example, there is some evidence that a gene needed to respond to oxytocin (a hormone important in social bonding) is turned off in some people with autism. As we will see in Chapter 21a, cancer development is controlled by cancer-inhibiting and cancer-promoting genes. If cancer-inhibiting genes are turned off or cancer-promoting genes turned on, cancer can result. Changes in the pattern of gene expression are found in cancers of the cervix, prostate, breast, stomach, and colon.

Although DNA methylation patterns are considered to be stable, some studies suggest that methylation can be reversed in adulthood. Foods such as broccoli, onions, and garlic may reduce methylation, allowing genes to be expressed. Researchers are actively looking for drugs that will alter the pattern of methylation and cure cancer.

Questions to Consider

- Do you think that epigenetics increases or decreases a person's responsibility for their own behavior?
- Researchers may someday develop an "epigenetic diet" that favors positive changes in gene activity. Would you follow that diet? Do you think that pregnant women should be required to follow that diet?

---

enzyme that makes a staggered cut between specific base pairs in DNA, leaving several unpaired bases on each side of the cut. There are many kinds of restriction enzymes; each kind recognizes and cuts a different sequence of DNA. The stretch of unpaired bases produced on each side of the cut is called a *sticky end* because of its tendency to pair with the single-stranded stretches of complementary base sequences on the ends of other DNA molecules that were cut with the same restriction enzyme (Figure 21.13).

The sticky ends are the secret to splicing the gene of interest and the vector DNA. The sticky ends of DNA from different sources will be complementary and stick together as long as they have been cut with the same restriction enzyme. The initial attachment between sticky ends is temporary, but the ends can be "pasted" together permanently by another enzyme, DNA ligase. The resulting recombinant DNA contains DNA from two sources.

2. **The vector is used to transfer the gene of interest to a new host cell.** Biological carriers that ferry the recombinant DNA to a host cell are called **vectors**. A common vector is bacterial **plasmids**, which are small, circular pieces of self-replicating DNA that exist separately from the bacterial chromosome.[2] As previously described, the source DNA (that is, the source of the gene of interest) and the plasmid (vector) DNA are both treated with the same restriction enzyme. Afterward, fragments of source DNA, some of which will contain the gene of interest, will be incorporated into plasmids when their sticky ends join. The recombinant DNA is mixed with bacteria in a test tube. Under the right conditions, some of the bacterial cells will then take up the recombined plasmids (Figure 21.14).

Although the basic strategy is usually the same, there are many variations on this theme of transporting a gene into a new host. For instance, the gene of interest is sometimes combined with viral DNA. The viruses are then used as vectors to insert the recombinant DNA into a host cell. Cells other than bacteria, including yeast or animal cells, can also be used as vectors.

3. **The recombinant organism containing the gene of interest is identified and isolated from the mixture of recombinants.** When plasmids are used as vector DNA, each recombinant plasmid is introduced into a single bacterial cell, and each cell is then grown into a colony. Each colony contains a different recombinant plasmid. The bacteria containing the gene of interest must be identified and isolated.

4. **The gene is amplified through bacterial cloning or by use of a polymerase chain reaction.** After the colony containing the gene of interest has been identified, researchers usually amplify (that is, replicate) the gene, producing numerous copies. Gene amplification is accomplished using one of two techniques: bacterial cloning or a polymerase chain reaction.

---

[2] Plasmids seem to have evolved as a means to move genes between bacteria. A plasmid can replicate itself and pass, with its genes, into another bacterium.

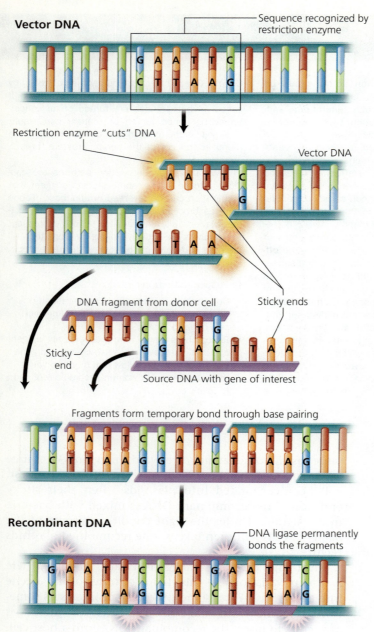

FIGURE 21.13 DNA from different sources can be spliced together using a restriction enzyme to make cuts in the DNA. A restriction enzyme makes a staggered cut at a specific sequence of DNA, leaving a region of unpaired bases on each cut end. The region of single-stranded DNA at the cut end is called a sticky end, because it tends to pair with the complementary sticky end of any other piece of DNA that has been cut with the same restriction enzyme, even if the pieces of DNA came from different sources.

Cloning Bacteria containing the plasmid with the gene of interest can be grown in huge numbers by cloning. Each bacterium divides many times to form a colony. Thus, each colony constitutes a clone—a group of genetically identical organisms all descended from a single cell. In this case, all the members of the clone carry the same recombinant DNA. Later, the plasmids can be separated from the bacteria, a process that partially purifies the gene of interest. The plasmids then can be taken up by other bacteria that will thus become capable of

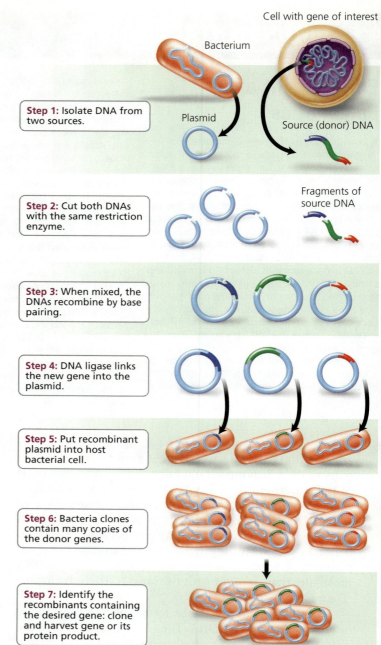

FIGURE 21.14 An overview of genetic engineering using plasmids.

performing a service deemed useful by humans. Alternatively, the plasmids can be transferred into plants or animal cells, creating transgenic organisms—organisms containing genes from another species.

Polymerase chain reaction (PCR) In the **polymerase chain reaction (PCR)** (Figure 21.15), the DNA of interest is unzipped, by gentle heating, to break the hydrogen bonds and form single strands. The single strands, which will serve as templates, are then mixed with *primers*—special short pieces of nucleic acid—one primer with bases complementary to each strand. The primers serve as start tags for DNA replication. Nucleotides and a special heat-resistant DNA polymerase,

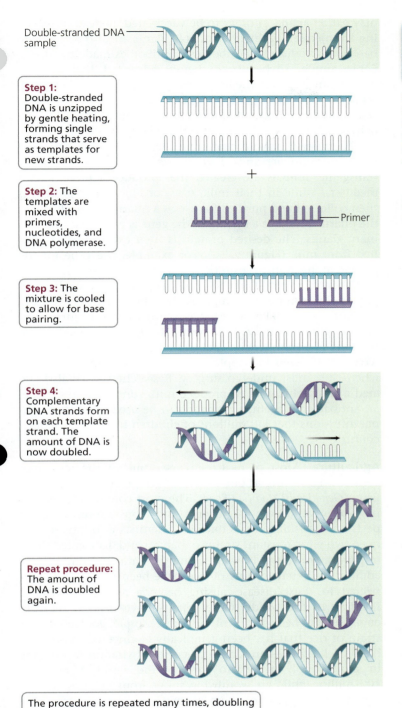

Double-stranded DNA sample

Step 1: Double-stranded DNA is unzipped by gentle heating, forming single strands that serve as templates for new strands.

Step 2: The templates are mixed with primers, nucleotides, and DNA polymerase.

Primer

Step 3: The mixture is cooled to allow for base pairing.

Step 4: Complementary DNA strands form on each template strand. The amount of DNA is now doubled.

Repeat procedure: The amount of DNA is doubled again.

The procedure is repeated many times, doubling the amount of DNA with each round.

FIGURE 21.15 *The polymerase chain reaction (PCR) rapidly produces a multitude of copies of a single gene or of any desired segment of DNA. PCR amplifies DNA more quickly than does bacterial cloning. It has many uses besides genetic engineering, including DNA fingerprinting.*

which promotes DNA replication, are also added to the mixture, which is then cooled to allow base pairing. Through base pairing, a complementary strand forms for each single strand. The procedure is then repeated many times, and each time the number of copies of the DNA of interest is doubled. In this way, billions of copies of the DNA of interest can be produced in a short time.

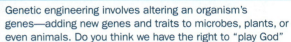

what would you do?

Genetic engineering involves altering an organism's genes—adding new genes and traits to microbes, plants, or even animals. Do you think we have the right to "play God" and alter life-forms in this way? The U.S. Supreme Court has approved patenting of genetically engineered organisms, first of microbes and now of mammals, such as pigs, that are genetically modified for use in organ transplant. If you were asked to decide whether it is ethical to patent a new life-form, how would you respond?

Applications of Genetic Engineering

Genetic engineering has been used in two general ways.

- **Genetic engineering provides a way to produce large quantities of a particular gene product.** The useful gene is transferred to another cell, usually a bacterium or a yeast cell, that can be grown easily in large quantities. The cells are cultured under conditions that cause them to express the gene, after which the gene product is harvested. For example, genetically engineered bacteria have been used to produce large quantities of human growth hormone (Figure 21.16). Treatment with growth hormone allows children with an underactive pituitary gland to grow to nearly normal height.

- **Genetic engineering allows a gene for a trait considered useful by humans to be taken from one species and transferred to another species.** The transgenic

FIGURE 21.16 *Genetic engineering is used to produce large quantities of a desired protein or to create an organism with a desired trait. This boy has an underactive pituitary gland. Its undersecretion of growth hormone would have caused him to be very short, even as an adult. However, growth hormone from genetically engineered bacteria has helped him grow to an almost normal height.*

organism then exhibits the desired trait. For example, scientists have endowed salmon with a gene from an eel-like fish. This gene causes the salmon to produce growth hormone year-round (something they do not normally do). As a result, the salmon grow faster than normal.

Environmental applications Genetic engineering also has environmental applications. For example, in sewage treatment plants, genetically engineered microbes lessen the amount of phosphate and nitrate discharged into waterways. Phosphate and nitrate can cause excessive growth of aquatic plants, which could choke waterways and dams, and of algae, which can produce chemicals that are poisonous to fish and livestock. Microorganisms are also being genetically engineered to modify or destroy chemical wastes or contaminants so that they are no longer harmful to the environment. For instance, oil-eating microbes that can withstand the high salt concentrations and low temperatures of the oceans have proven useful in cleaning up marine oil spills.

Livestock Genetic engineering has also been used on livestock. Genetically engineered vaccines have been created to protect piglets against a form of dysentery called scours, sheep against foot rot and measles, and chickens against bursal disease (a viral disease that is often fatal). Genetically engineered bacteria produce bovine somatotropin (BST), a hormone naturally produced by a cow's pituitary gland that enhances milk production. Injections of BST can boost milk production by nearly 25%.

Transgenic animals have been created by injecting a fertilized egg with the gene of interest in a Petri dish. The goals of creating transgenic animals include making animals with leaner meat, sheep with softer wool, cows that produce more milk, and animals that mature more quickly.

Pharmaceuticals Genes have been put into a variety of cells, ranging from microbes to mammals, to produce proteins for treating allergies, cancer, heart attacks, blood disorders, autoimmune disease, and infections.

Genetically engineered bacteria have also been used to create vaccines for humans. You may recall from Chapter 13 that a vaccine typically uses an inactivated bacterium or virus to stimulate the body's immune response to the active form of the organism. The idea is that the body will learn to recognize proteins on the surface of the infectious organism and mount defenses against any organism bearing those proteins. Because the organism used in the vaccine was rendered harmless, the vaccine cannot trigger an infection. Scientists produce genetically engineered vaccines by putting the gene that codes for the surface protein of the infectious organism into bacteria. The bacteria then produce large quantities of that protein, which can be purified and used as a vaccine. The vaccine cannot cause infection, because only the surface protein is used instead of the infectious organism itself.

Plants have also been used to produce therapeutic proteins. Engineered bananas that produce an altered form of the hepatitis B virus surface protein are being developed as an edible vaccine against the liver disease hepatitis B. Someday, you may eat a banana in order to be vaccinated, instead of receiving an injection against hepatitis B. Plants are also being engineered to produce "plantibodies," antibodies made by plants. For example, soybeans are being cultivated that contain human antibodies to the herpes simplex virus that causes genital herpes. A human gene for an antibody that binds to tumor cells has been transplanted into corn. The antibodies can then deliver radioisotopes to cancer cells, selectively killing them.

Pharming is a word that comes from the combination of the words *farming* and *pharmaceuticals*. In gene pharming, transgenic animals are created that produce a protein with medicinal value in their milk, eggs, or blood. The protein is then collected and purified for use as a pharmaceutical. When the pharm animal is a mammal, the gene is expressed in mammary glands. The desired protein is then extracted and purified from milk (Figure **21.17**). For example, the gene for the protein alpha-1-antitrypsin (AAT) has been inserted into sheep that then secrete AAT in their milk. People with an inherited, potentially fatal form of emphysema (a lung disease) take AAT as a drug. It is also being tested as a drug to prevent lung damage in people with cystic fibrosis. The first drug made from the milk of a transgenetic goat was an anticlotting drug called ATryn. It is given to people with a blood-clotting deficiency when they must undergo surgery. Researchers have also created a transgenic goat to produce milk containing lysozyme, an antibacterial agent. Lysozyme can be used to treat intestinal infections that kill millions of children in underdeveloped countries.

Agriculture Most of us experience some of the results of genetic engineering at our dinner tables (see the Health Issue Essay, *Genetically Modified Food*). The most common traits that have been genetically engineered into crops are resistance to pests and resistance to herbicides. Scientists also have developed two virus-resistant strains of papaya and distributed them to papaya growers in Hawaii, saving the industry from ruin. In addition, different strains of rice have been genetically engineered to resist disease-causing bacteria and to withstand flooding of the paddy. Other plants have been genetically engineered to be more nutritious. For example, golden rice is a strain of rice that has been genetically engineered to produce high levels of beta-carotene, a precursor of vitamin A, which is in short supply in certain parts of the world. More than 100 million children worldwide suffer from vitamin A deficiency, and 500,000 of them go blind every year because of that deficiency. Although golden rice cannot supply a complete recommended daily dose of vitamin A, the amount it contains could be helpful to a person whose diet is very low in vitamin A. Other crops have also been created that grow faster, produce greater yields, and have longer shelf lives.

Gene Therapy

The problems associated with many genetic diseases arise because a mutant gene fails to produce a normal protein product. The goal of **gene therapy** is to cure genetic diseases by putting normal, functional genes into the body cells that were affected by the mutant gene. The functional gene would then produce the needed protein.

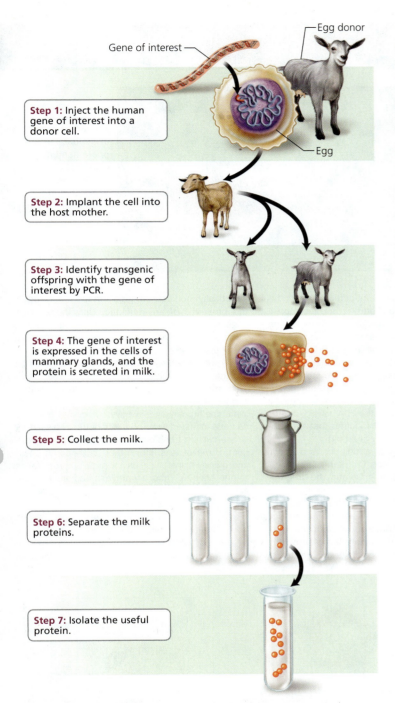

FIGURE 21.17 The procedure for creating a transgenic animal that will produce a useful protein in its milk.

Methods of delivering a healthy gene One way a healthy gene can be transferred to a target cell is by means of viruses. Viruses generally attack only one type of cell. For instance, an adenovirus, which causes the common cold, typically attacks cells of the respiratory system. A virus consists largely of genetic material, commonly DNA, surrounded by a protein coat (see Chapter 13a). Once inside a cell, the viral DNA uses the cell's metabolic machinery to produce viral proteins. If a healthy gene is spliced into the DNA of a virus that has first been rendered harmless, the virus will deliver the healthy gene to the host cell and cause the desired gene product to be produced (Figure 21.18).

Another type of virus used in gene therapy is a retrovirus, a virus whose genetic information is stored as RNA rather than DNA. Once inside the target cell, a retrovirus rewrites its genetic information as double-stranded DNA and inserts the viral DNA into a chromosome of the target cell.

Gene therapy results More than 4000 human diseases have been traced to defects in single genes. Although the Food and Drug Administration has not yet approved a gene therapy for any of these conditions, hundreds of clinical trials of gene therapies are currently under way, including trials of possible therapies for cystic fibrosis and cancer (discussed in Chapter 21a).

The first condition to be treated experimentally with gene therapy is a disorder referred to as *severe combined immunodeficiency disease (SCID)*. In children with SCID, the immune system is nonfunctional, leaving them vulnerable to infections. The cause of the problem is a mutant gene that prevents the production of an enzyme called adenosine deaminase (ADA). Without ADA, white blood cells never mature—they die while still developing in the bone marrow. The first gene therapy trial began in 1990, when white blood cells of a 4-year-old SCID patient, Ashanthi DeSilva, were genetically engineered to carry the ADA gene and then returned to her tiny body. Her own gene-altered white blood cells began producing ADA, and her body defense mechanisms were strengthened. Ashanthi's life began to change. She was not ill as often as she had been before. She could play with other children. However, the life span of white blood cells is measured in weeks, and when the number of gene-altered cells declined, new gene-altered cells had to be infused. Ashanthi is now in her twenties and has a reasonably healthy immune system. However, she still needs repeated treatments.

French scientists believe they have cured 10 children with X-SCID by using gene therapy. X-SCID is a severe combined immunodeficiency syndrome caused by a mutant gene on the X chromosome. It is still too soon to know for certain whether all 10 children will require treatment in the future. However, four children in the French studies who had shown improvement in symptoms of X-SCID developed leukemia from the therapy.

The results of a recent study on gene therapy to treat HIV infections allay some concerns that gene therapy can cause cancer, especially leukemia. Researchers modified the DNA in the patients' own immune cells (T cells; see Chapter 13) to kill cells infected with HIV and injected the modified cells into patients. The patients are healthy, and the cells are still present more than a decade later.

A form of muscular dystrophy has recently been treated with gene therapy. Researchers packed a virus with the normal form of the protein that is deficient in this form of muscular dystrophy. The virus was then injected directly into the muscles. The level of the protein and the gene expression of the protein remained high for several months, restoring some muscle function to the patients.

HEALTH ISSUE

Genetically Modified Food

From dinner tables to diplomatic circles, people are discussing genetically modified (GM) food. This relatively recent interest is somewhat ironic, considering that people in the United States have been eating GM food since the mid-1990s. More than 70% of processed foods sold in the United States contain genetically modified ingredients. Yet many people vehemently object to GM food.

Why is something as common as GM food controversial? The concerns about it can generally be divided into three categories: health issues, social issues, and environmental issues. Let's explore these categories one at a time.

The larger salmon in the back of the photo has been genetically modified to grow faster than the normal salmon in front.

Health Concerns

A panel of the National Academy of Sciences (NAS) has issued a report saying that genetically engineered crops do not pose health risks that cannot also be caused by crops created by conventional breeding. However, because genetic engineering could produce unintended harmful changes in food, the NAS panel recommends scrutiny of GM foods before they can be marketed. Currently, the U.S. Department of Agriculture, the Food and Drug Administration (FDA), and the Environmental Protection Agency regulate genetically modified foods. The NAS panel concluded that the GM foods already on the market are safe.

A common safety concern is that GM foods may contain allergens (substances that cause allergies). After a protein is produced, the cell modifies it in various ways. The protein may be modified in the genetically modified plant differently from the way it would in an unmodified cell, and the modification could produce an allergen. Rigorous testing can reduce the likelihood that this may occur. Most known allergens share certain properties. They are proteins, relatively small molecules, and are resistant to heat, acid, and digestion in the stomach. If a protein produced by a GM plant has any of the properties typical of an allergen or is structurally similar to a known allergen, the FDA considers it to be a potential allergen and requires that the protein undergo additional allergy testing.

Bacterial resistance to antibiotics is a major threat to public health (see Chapter 13a). When bacteria are resistant to an antibiotic, the drug will not kill them and thus will no longer cure the human disease for which they are the cause. Some people worry about the scientific practice of putting genes for resistance to an antibiotic into GM crops as markers to identify the plants with the modified genes. Plant seedlings thought to be genetically modified are grown in the laboratory in the presence of an antibiotic. Only those seedlings with the gene for resistance will survive. Because of the way they were engineered, the surviving plants also contain the "useful" gene.

What worries some people is that the genes for resistance to antibiotics could be transferred to bacteria, making the bacteria resistant to antibiotics. The receiving bacteria might be those that normally live in the human digestive system, or they might be bacteria ingested with food. It is not known whether genes can be transferred from a plant to a bacterium. However, it is known that bacteria can easily and quickly transfer genes for antibiotic resistance to one another. Thus, a harmless bacterium in the gut could transfer the gene for antibiotic resistance to a disease-causing bacterium.

The transfer of antibiotic-resistance genes from GM plants to bacteria could have serious consequences. For this reason, antibiotic-resistance marker genes are being phased out in favor of

Step 1: Incorporate a healthy form of the gene into the virus.

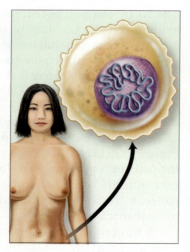

Step 2: Remove bone marrow stem cells from the patient.

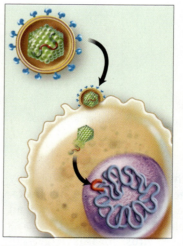

Step 3: Infect the patient's stem cells with the virus that is carrying the healthy form of the gene.

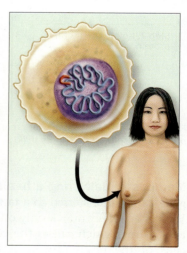

Step 4: Return the genetically engineered stem cells to the patient. The gene is expressed to produce the needed protein.

FIGURE 21.18 *Gene therapy using a virus. In gene therapy, a healthy gene is introduced into a patient who has a genetic disease caused by a faulty gene.*

other marker genes, such as a green fluorescent protein. Scientists also have developed a way to inactivate the antibiotic-resistance gene if it were to be transferred to bacteria.

Environmental Concerns

Proponents of GM foods argue that herbicide-resistant and pesticide-resistant crops reduce the need for spraying with herbicides and pesticides. So far, experience has shown that the validity of this argument depends on the crop. Pest-resistant cotton has substantially reduced the use of pesticides, but pest-resistant corn probably has not. Farmers who grow herbicide-resistant crops still spray with herbicides, but they change the type of herbicide they use to a type that is less harmful to animals.

Unfortunately, engineered crops containing insecticides could have undesirable effects. Genetically engineering insecticides into plants could hasten the development of insect resistance to that insecticide, making the insecticide ineffective—not just for the genetically modified crop, but for all crops.

Another concern is that genetically modified organisms could harm other organisms. For example, pollen from pest-resistant corn has been shown to harm monarch butterfly caterpillars. Fortunately, monarch caterpillars rarely encounter enough pollen to be harmed, and most of the pest-resistant corn grown in the United States today does not produce pollen that is harmful to monarch butterflies. However, farmers are increasing their use of certain herbicides to kill weeds, because their crops are genetically modified to resist the herbicides. The herbicides are killing milkweeds, which are essential for monarch butterfly survival.

A second example of a genetically modified organism that has the potential to harm other organisms is the salmon that produce more growth hormone and grow several times faster than their wild relatives do. These salmon are grown on fish farms. If the FDA grants approval, the gene-altered salmon could dramatically cut costs for fish farmers and consumers. However, when the genetically modified salmon are grown in tanks with ordinary salmon and food is scarce, the genetically modified salmon eat most of the food and some of their ordinary companions. What would happen if the genetically modified salmon escaped from their pens on the fish farm? They might mate with the wild salmon and create less healthy offspring or outcompete wild salmon for food, which could eventually cause the extinction of the wild salmon. Escape is possible. During the past few years, hundreds of thousands of fish have escaped from fish farms when floating pens were ripped apart by storms or sea lions. To minimize the risk that genetically modified salmon could destroy the population of wild salmon, scientists plan to breed the fish inland, sterilize the offspring, and ship only sterile fish to coastal pens. The sterilization procedure is effective in small batches of fish, but it is not known whether the procedure is completely effective in large batches of fish.

Critics of GM foods also point out that pollen from crops genetically engineered to resist herbicides can cause their wild relatives growing nearby to become "superweeds" that are resistant to many existing chemicals. In North Dakota, GM canola plants that are resistant to herbicides are growing along roadsides. Canola can hybridize with at least two wild weeds. The two original strains of GM canola were each resistant to a different herbicide. As a result of cross-pollination, some of the canola plants found in the wild are resistant to *both* herbicides, which suggests that the GM traits are stable in the wild and are evolving.

Social Concerns

Proponents of GM food claim that GM food can assist in the battle against world hunger. We have seen that genetic engineering can produce crops that resist pests and disease. It can also produce crops with greater yields and crops that will grow in spite of drought, depleted soil, or excess salt, aluminum, or iron. Foods can also be genetically modified to contain higher amounts of specific nutrients.

Critics of using GM food to battle world hunger argue that the problem of hunger has nothing to do with an inability to produce enough food. The problem, they say, is a social one of distributing food so that it is available to the people who need it.

Some developing countries are resisting the use of GM seeds. Part of the resistance stems from lingering health concerns. However, many farmers in developing countries also object to GM seeds because the GM plants do not seed themselves. The need to buy seeds each year places a financial burden on poor farmers.

Questions to Consider

- Do you consider GM food to be a blessing or a danger to the world? What are your reasons?
- If a genetically modified organism that was created for food begins to cause environmental problems, who should be held accountable?
- Do you think that foods containing genetically modified components should be labeled as such?

what would you do?

Without treatment, children with X-SCID die at a young age. The gene therapy treatment for X-SCID that caused leukemia in four French boys does appear to have cured this deadly disorder in other patients. If you had a child with X-SCID, would you want him to have this gene therapy treatment? Why or why not? What factors would you consider in making your decision?

21.7 Genomics

A **genome** is the entire set of genes carried by one member of a species—in our case, one person. **Genomics** is the study of entire genomes and the interactions of the genes with one another and the environment.

Human Genome Project

One goal of genomics is to determine the location and sequences of genes. Researchers have developed supercomputers that automatically sequence (determine the order of bases in) DNA. The supercomputers were put to use on a massive scale in the Human Genome Project, a worldwide research effort, completed in 2003, to sequence the human genome. As a result, we now have some idea of the locations of genes along all 23 pairs of human chromosomes and the sequence of the estimated 3 billion base pairs that make up those chromosomes. Although the exact number of human genes is still not known for certain, scientists now estimate that the human genome consists of 20,000 to 25,000 genes, not 100,000 as originally thought. One reason for the smaller number of actual genes is that many gene families have related or redundant functions and therefore are able to share certain genes, so that fewer are needed to carry out all the body's functions. A second reason is that many genes are now known to code for parts of more than one protein.

In addition, researchers have identified and mapped to specific locations on specific chromosomes genes for more than 1400 genetic diseases. It is hoped that this information will give scientists a greater ability to diagnose, analyze, and eventually treat many of the 4000 diseases of humans that have a known genetic basis. Researchers have already cloned the genes responsible for many genetic diseases, including Duchenne muscular dystrophy, retinoblastoma, cystic fibrosis, and neurofibromatosis. These isolated genes can now be used to test for the presence of the same disease-causing genes in specific individuals. As we saw in Chapter 20, some gene tests can be used to identify people who are carriers for certain genetic diseases such as cystic fibrosis, allowing families to make choices based on known probabilities of bearing an affected child; similar tests can be used for prenatal diagnosis and for diagnosis before symptoms of the disease begin. After a disease-related gene has been identified, scientists can study it to learn more about the protein it codes for and perhaps discover ways to correct the problem.

We have also learned from the Human Genome Project that humans are identical in 99.9% of the sequences of their genes. As scientists gain greater understanding of the 0.1% of DNA that differs from person to person, they expect to learn more about why some people develop heart disease, cancer, or Alzheimer's disease and others do not.

Microarray Analysis

A second goal of genomics is to understand the mechanisms that control gene expression. More than 95% of human DNA does not code for protein; however, some of these noncoding DNA sequences function as regulatory regions that determine when, where, and how much of certain proteins are produced. Because gene activity plays a role in many diseases, the study of how these regions turn genes on or off may lead to advances in diagnosis and treatment.

One of the tools researchers use in this effort is the *microarray*, which consists of thousands of DNA sequences stamped onto a single glass slide called a *DNA chip*. Researchers use microarrays to monitor large numbers of DNA segments to discover which genes are active and which are turned off under different conditions, such as in different tissue types, in different stages of development, or in health and disease. For example, they may use microarrays to identify genes that are active in cancerous cells but not in healthy cells (Figure 21.19). Presumably, the genes that are active in cancerous cells play a role in the development of cancer.

Besides identifying gene activity in health and disease, microarray analysis is useful in identifying genetic variation in the members of a population. Some of these genetic differences are in the form of single-nucleotide polymorphisms (SNPs, or *snips*). These are DNA sequences that can vary by one nucleotide from person to person, and the differences in their protein products are thought to influence how we respond to stress and diseases, among other things. As researchers learn more about SNPs, they may be able to develop treatments tailored to the genetic makeup of each individual. These are the kinds of discoveries that can open the door for gene therapy. Whereas individualized gene therapy may be useful in the future, we already use identification of individual differences in DNA on a regular basis in DNA fingerprinting (see the Ethical Issue essay, *Forensic Science, DNA, and Personal Privacy*).

what would you do?

Some people worry that once we know the location and function of every gene and have perfected the techniques of gene therapy, we will no longer limit gene manipulation to repairing faulty genes but will begin to modify genes to enhance human abilities. Should people be permitted to design their babies by choosing genes that they consider superior? What do you think? Where should the line be drawn? Who should draw that line? Who should decide which genes are "good?"

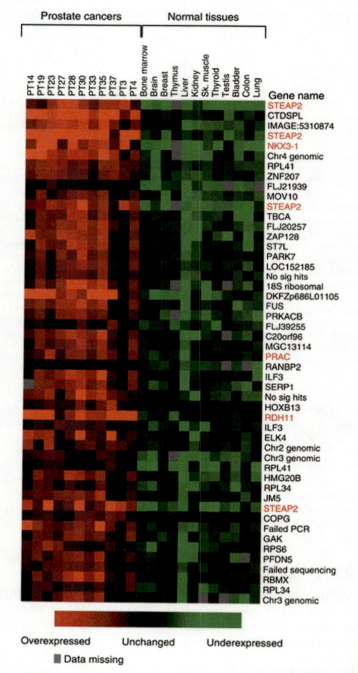

FIGURE 21.19 *A comparison of microarrays showing the pattern of gene activity in prostate cancer and in normal tissue. The genes that are active in tissue with prostate cancer but not in normal tissue probably play a role in the development of cancer.*

ETHICAL ISSUE

Forensic Science, DNA, and Personal Privacy

So-called DNA fingerprints, like the more conventional prints left by fingers, can help identify the individuals they belong to out of a large population. *DNA fingerprinting* refers to techniques of identifying individuals on the basis of unique features of their DNA. DNA fingerprints are possible because many regions of DNA are composed of small, specific sequences of DNA that are repeated many times. Most commonly used are repeated units of one to five bases, which are called *short tandem repeats* (STRs). The number of times these sequences are repeated varies considerably from person to person, from a few to 100 repeats. Because of these differences, the segments can be used to match a sample of DNA to the person whose cells produced the sample.

The first step in preparing a DNA fingerprint is to extract DNA from a tissue sample. The type of tissue does not matter, and one type of tissue can be successfully compared to another type. Commonly used sources include blood, semen, skin, and hair follicles, because they are not too painful to remove, they are readily available, or they are left at a crime scene.

First, the amount of DNA is greatly increased using PCR, as described in Figure 21.15. The primers used are sequence specific for the regions on either side of the repeating region. This produces many copies of the repeating region, which are then analyzed to determine the number of repeats present.

The FBI uses 13 STRs as a core set for forensic analysis. The resulting DNA fingerprint is unique to the person who produced the DNA. Moreover, any DNA sample taken from the same person would always be identical. But the fingerprint profiles resulting from the DNA of *different* people are always different (except perhaps for identical siblings), because the number and sizes of the fragments are determined by the unique sequence of bases in each person's DNA.

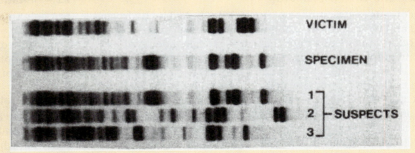

FIGURE 21.A *The pattern of banding in a DNA fingerprint is determined by the sequence of bases in a person's DNA and is, therefore, unique to each person. A match between DNA fingerprints can identify the source of a tissue sample from a crime scene with a high degree of certainty. Which suspect's DNA fingerprint matches this specimen found at a crime scene?*

DNA fingerprinting has many applications, but the most familiar is probably its use in crime investigations. In these cases, the DNA fingerprint is usually created from a sample of tissue, such as blood or hair follicles, collected at the crime scene. A fingerprint can be produced from tissue left at the scene years before. This fingerprint is then compared with the DNA fingerprints of various suspects. A match reveals, with a high degree of certainty, the person who was the source of the sample from the crime scene (Figure 21.A).

Of course, the degree of certainty of a match between DNA fingerprints depends on how carefully the analysis was done. Because DNA fingerprints are being used as evidence in an increasing number of court cases each year, it is important that national standards be set to ensure the reliability of these molecular witnesses. It is generally easier to declare with certainty that two DNA fingerprints do *not* match than it is to be sure that they do. In the United States, more than 200 convicts have been found innocent through DNA testing during the last decade.

Has DNA testing gone too far? All states in the United States collect DNA samples from people convicted of sex crimes and murder. Several other states also collect DNA samples from people convicted of other felonies, such as robbery. About 30 states collect DNA samples from people accused of misdemeanors, including loitering, shoplifting, or vandalism. In many cases, the DNA is stored in a database, even if the person is found innocent of the crime. People who are simply cooperating with the investigation may also provide DNA samples. These, too, are added to a national database.

Questions to Consider

- Do you think everyone arrested for a crime should have the right to DNA fingerprinting to prove his or her innocence? If so, who should pay for the process?
- Is the creation of a national database of DNA fingerprints an invasion of privacy? Is it any more so than a database of actual fingerprints or mug shots?
- Under what conditions do you think DNA samples should be obtained?

When SNPs are located near one another on a chromosome, they tend to be inherited together. A group of SNPs in a region of a chromosome is called a *haplotype*. The International HapMap Project is a scientific consortium whose purpose is to describe genetic variation between populations. Researchers collaborating on this project compare haplotype frequencies in groups of people who have a certain disease to those of a group without the disease, hoping to identify genes associated with the disease.

Comparison of Genomes of Different Species

The DNA of certain widely studied organisms, including the mouse, the fruit fly, a roundworm, yeast, slime mold, and the honeybee, have also been mapped. From these genomes, geneticists hope to gain some insight into basic biology, including basic principles of the organization of genes within the genome, gene regulation, and molecular evolution. Humans share many genes with other organisms. For example, we share 50% of our genes with the fruit fly and 90% of our genes with the mouse. These genetic similarities are evidence of our common evolutionary past. The genes and genetic mechanisms we share with other organisms are likely to be important in determining body form as well as influencing development and aging.

looking ahead

In Chapter 19, we learned about the cell cycle. In Chapters 20 and 21, we learned about genes, their inheritance, and their regulation, and we also considered how mutations affect gene functions. In Chapter 21a, we use this information to understand cancer, a family of diseases in which mutations in genes that regulate the cell cycle cause a loss of control over cell division.

HIGHLIGHTING THE CONCEPTS

21.1 Form of DNA (pp. 442–443)

- DNA consists of two strands of nucleotides linked by hydrogen bonds and twisted together to form a double helix. Each nucleotide consists of a phosphate, a sugar called deoxyribose, and one of four nitrogenous bases: adenine, thymine, cytosine, or guanine. The sugar and phosphate components of the nucleotides alternate along the two sides of the molecule. Pairs of bases meet and form hydrogen bonds in the interior of the double helix; these pairs resemble the rungs on a ladder.
- According to the rules of complementary base pairing, adenine binds only with thymine, and cytosine binds only with guanine.

21.2 Replication of DNA (pp. 443–444)

- DNA replication is semiconservative. Each new double-stranded DNA molecule consists of one old and one new strand. The enzyme DNA polymerase "unzips" the two strands of a molecule (the parent molecule), allowing each strand to serve as a template for the formation of a new strand. Complementary base pairing ensures the accuracy of replication.

21.3 Gene Expression (pp. 444–448)

- Genetic information is transcribed from DNA to RNA and then translated into a protein.
- Transcription is the synthesis of RNA by means of base pairing on a DNA template. RNA differs from DNA in that the sugar ribose replaces deoxyribose, and the base uracil replaces thymine. Most RNA is single stranded.
- Messenger RNA (mRNA) carries the DNA genetic message to the cytoplasm, where it is translated into protein. The genetic code is read in sequences of three RNA nucleotides; each triplet is called a codon. Each of the 64 codons specifies a particular amino acid or indicates the point where translation should start or stop.
- Transfer RNA (tRNA) interprets the genetic code. At one end of the tRNA molecule is a sequence of three nucleotides called the anticodon that pairs with a codon on mRNA in accordance with base-pairing rules. The tail of the tRNA binds to a specific amino acid.
- Each of the two subunits of a ribosome consists of ribosomal RNA (rRNA) and protein. A ribosome brings tRNA and mRNA together for protein synthesis.
- Translation of the genetic code into protein begins when the two ribosomal subunits and an mRNA assemble, with the mRNA sitting in a groove between the ribosome's two subunits. The mRNA attaches to the ribosome at the mRNA's start codon. Then the ribosome slides along the mRNA molecule, reading one codon at a time. Molecules of tRNA ferry amino acids to the mRNA and add them to a growing protein chain. Translation stops when a stop codon is encountered. The protein chain then separates from the ribosome.

21.4 Mutations (pp. 449–450)

- A point mutation is a change in one or a few nucleotides in the sequence of a DNA molecule. When one nucleotide is mistakenly substituted for another, the function of the resulting protein may or may not affect the function of the protein. The insertion or deletion of a nucleotide always changes the resulting protein.

21.5 Regulating Gene Activity (p. 450)

- Gene activity is regulated at several levels. Usually, most of the DNA is folded and coiled. For a gene to be active, the region of DNA in which it is located must be uncoiled. Gene activity can be affected by other segments of DNA. Regions of DNA called enhancers can increase the amount of RNA produced. Chemical signals such as regulatory proteins or hormones can also affect gene activity.

21.6 Genetic Engineering (pp. 450–457)

- Genetic engineering is the purposeful manipulation of genetic material by humans. It can be used to produce large quantities of a particular gene product or to transfer a desirable genetic trait from one species to another or to another member of the same species.
- Genetic engineering uses restriction enzymes to cut the source DNA, which contains the gene of interest, and the vector DNA at specific places, creating sticky ends composed of unpaired complementary bases that allow the cut segments to recombine. The recombinant vector is then used to transfer the recombinant DNA to a host cell. Common vectors include bacterial plasmids and viruses. The host cell is often a type of cell that reproduces rapidly, such as a bacterium or yeast cell. Each time a host cell divides, both daughter cells receive a copy of the gene of interest. Introducing a gene from one species into a different species results in a transgenic plant or animal.
- Genetic engineering has had many applications in plant and animal agriculture, environmental science, and medicine.
- In gene therapy, a healthy form of a gene is introduced into body cells to correct problems caused by a defective gene.

21.7 Genomics (pp. 457–459)

- A genome consists of all the genes in a single organism. Genomics is the study of genomes and the interaction between genes and the environment. It is now believed that the human genome consists of 20,000 to 25,000 genes.
- Scientists use microarrays to analyze gene activity under different conditions. It uses DNA chips—glass slides with thousands of DNA segments stamped on them. The information may be helpful in treating genetic diseases. Microarray analysis also allows scientists to discover small differences in the gene sequences of people. Scientists hope to use this information to develop individualized treatments.
- Large portions of our DNA are the same as the DNA in other organisms. The closer the evolutionary relationship, the greater portion of DNA we have in common.

RECOGNIZING KEY TERMS

deoxyribonucleic acid (DNA) p. 442
complementary base pairing p. 443
DNA replication p. 443
semiconservative replication p. 444
ribonucleic acid (RNA) p. 444
gene p. 444
transcription p. 445
messenger RNA (mRNA) p. 445
transfer RNA (tRNA) p. 445
ribosomal RNA (rRNA) p. 445
promoter p. 445
RNA polymerase p. 445
translation p. 446
genetic code p. 446
codon p. 446
anticodon p. 446
polysome p. 448
mutation p. 449
genetic engineering p. 450
biotechnology p. 450
recombinant DNA p. 450
restriction enzyme p. 450
vector p. 451
plasmid p. 451
polymerase chain reaction (PCR) p. 452
gene therapy p. 454
genome p. 457
genomics p. 457

REVIEWING THE CONCEPTS

1. Describe the structure of DNA. pp. 442–443
2. Explain why complementary base pairing is crucial to exact replication of DNA. p. 443
3. Why is DNA replication described as semiconservative? p. 444
4. Explain the roles of transcription and translation in converting the DNA message to a protein. p. 444
5. In what ways does RNA differ from DNA? p. 444
6. What roles do mRNA, tRNA, and rRNA play in the synthesis of protein? pp. 445–448
7. Define *codon*. What role do codons play in protein synthesis? p. 446
8. What is the role of an anticodon? pp. 446–447
9. Describe the events that occur during the initiation of protein synthesis, the elongation of the protein chain, and the termination of synthesis. pp. 447–448
10. Why does a deletion have such a major effect on a cell? pp. 449–450
11. How is gene activity regulated? p. 450
12. Define *genetic engineering*. Explain the roles of restriction enzymes and vectors in genetic engineering. pp. 450–451
13. Describe some of the ways in which genetic engineering has been used in farming and medicine. p. 454
14. What is gene therapy? pp. 454–455
15. What is the complementary base for thymine?
    a. adenine
    b. cytosine
    c. guanine
    d. uracil
16. Although the amount of any particular base in DNA will vary among individuals, the amount of guanine will always equal the amount of
    a. thymine.
    b. adenine.
    c. cytosine.
    d. uracil.
17. A codon is located on
    a. DNA.
    b. mRNA.
    c. tRNA.
    d. rRNA.
18. Translation produces
    a. a polypeptide chain.
    b. mRNA complementary to a template strand of DNA.
    c. tRNA complementary to mRNA.
    d. rRNA.
19. Choose the *incorrect* statement about restriction enzymes.
    a. They are used to create recombinant plasmids.
    b. They are necessary for translation.
    c. They are used to produce DNA fingerprints.
    d. They cut DNA at specific sequences.
20. Which codon on mRNA is the start signal for translation?
    a. CUG
    b. ATC
    c. AUG
    d. UAA
21. The polymerase chain reaction is a technique used to
    a. attach amino acids to a growing polypeptide chain.
    b. increase the number of copies of a small sample of DNA.
    c. splice DNA into a plasmid.
    d. regulate the transcription of genes.
22. What is a genome?
    a. a technique for creating recombinant DNA
    b. a technique for determining which genes are active under certain conditions
    c. the coding sequence of a gene
    d. the entire set of genes carried by one member of a species
23. What is microarray analysis?
    a. a technique for creating recombinant DNA
    b. a technique for determining which genes are active under certain conditions
    c. a technique used to identify the coding sequence of a gene
    d. the entire set of genes carried by one member of a species
24. The anticodon is located on a molecule of _____.
25. In RNA, the nucleotide _____ binds with adenine.
26. In genetic engineering, the staggered cuts in DNA that allow genes to be spliced together are made by _____.

APPLYING THE CONCEPTS

Use the genetic code in Figure 21.5 (page 446) to answer questions 1 and 2.

1. What would be the amino acid sequence in the polypeptide resulting from a strand of mRNA with the following base sequence?

    AUG ACA UAU GAG ACG ACU

2. The following are base sequences in four mRNA strands: one normal and three with mutations. (Keep in mind that translation begins with a start codon and ends with a stop codon.) Which of the mutated sequences is likely to have the most severe effects? Why? Which would have the least severe effects? Why?

Normal mRNA:	AUG ACA UAU GAG ACG ACU
Mutation 1:	AUG ACC UAC GAA ACG ACC
Mutation 2:	AUG ACU UAA GAG ACG ACA
Mutation 3:	AUG ACG UAU GAG ACG ACG

3. You are a crime scene investigator testifying at a murder trial. The DNA fingerprints shown on the right are those of a bloodstain at the murder scene (not the victim's blood) and those of seven suspects (numbered 1 through 7). Which suspect's blood matches the bloodstain from the crime scene?

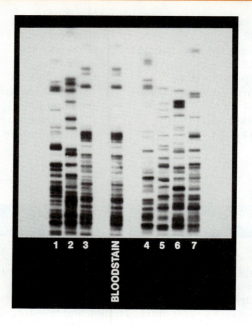

BECOMING INFORMATION LITERATE

Describe some of the possible medical uses of information gained from the Human Genome Project. What ethical issues are raised by the project? Use at least three reliable sources (journals, books, websites) to answer these questions. Explain why you chose those sources.

MasteringBiology®

Go to MasteringBiology for practice quizzes, activities, eText, videos, current events, and more.

SPECIAL TOPIC
Cancer

21a

This child is entertaining himself while undergoing chemotherapy for cancer.

- 21a.1 Uncontrolled Cell Division
- 21a.2 Development of Cancer
- 21a.3 Multiple Mutations
- 21a.4 Cancer Stem Cell Hypothesis
- 21a.5 Known Causes of Cancer
- 21a.6 Reducing the Risk of Cancer
- 21a.7 Diagnosing Cancer
- 21a.8 Treating Cancer

In the previous few chapters, we learned about cell division, genes, and gene function. In this chapter, we consider how cancer cells escape the normal controls over cell division. We then learn about some causes of cancer and how we can reduce our risk of developing the disease. Finally, we identify some means of diagnosing and treating cancer.

Cancer, the "Big C," is perhaps the disease that people in the industrialized world dread the most—and with good reason. Cancer touches the lives of nearly everyone. One of every three people in the United States will develop cancer at some point in life, and the other two are likely to have a friend or relative with cancer.

21a.1 Uncontrolled Cell Division

All forms of cancer share one characteristic—uncontrolled cell division. Cancer cells may behave like aliens taking over our bodies, but they are actually traitorous cells from within our own bodies.

Benign or Malignant Tumors

An abnormal growth of cells can form a mass of tissue called a *tumor* or *neoplasm* (meaning "new growth"). However, not all tumors are cancerous: tumors can be either benign or malignant. A *benign tumor* is an abnormal mass of tissue that is surrounded by a capsule of connective tissue and that usually remains at the site where it forms. Its cells do not invade surrounding tissue or spread to distant locations, although they can and do grow. In most cases, a benign tumor does not threaten life, because it can be removed completely by surgery.

Benign tumors can be harmful when they press on nearby tissues enough to interfere with the functioning of those tissues. If a "benign" tumor of that type is also inoperable, as may occur with a tumor in the brain, it can be life threatening. Even so, only *malignant tumors,* tumors that can invade surrounding tissue and spread to multiple locations throughout the body, are properly called cancerous. The spread of cancer cells from one part of the body to another is called *metastasis.*

Stages of Cancer Development

Cells on their way to becoming cancerous are accumulating genetic damage. As a result, precancerous cells typically look different from normal cells. *Dysplasia* is the term used to describe the changes in shape, nuclei, and organization within tissues of precancerous cells (Figure 21a.1). Their ragged edges give precancerous cells an abnormal shape. Their nuclei become unusually large and atypically shaped and may contain increased amounts of DNA. We can see the differences by comparing the chromosomes of a normal cell with those from a cancer cell. Notice in Figure 21a.2 that the cancer cell has extra copies of some chromosomes, is missing parts of some chromosomes, and has extra parts of other chromosomes. In a group, precancerous cells form a disorganized clump and, significantly, have an unusually high percentage of cells in the process of dividing.

Eventually, the tumor will reach a critical mass consisting of about a million cells. Although the tumor is still only a millimeter or two in diameter (smaller than a BB), the cells in the interior cannot get a sufficient supply of nutrients, and their own waste is poisoning them. This tiny mass is now called *carcinoma in situ,* which literally means "cancer in place."

If a tumor this size is to continue growing, it must attract a blood supply. Some of the tumor cells will start to secrete chemicals that cause blood vessels to invade the tumor. This process marks an ominous point of transition, because the tumor cells now have supply lines bringing in nutrients to support continued growth and carry away waste (Figure 21a.3). Of equal importance, the tumor cells have an escape route; they can enter the blood or nearby lymphatic vessels and travel throughout the body. Like cancerous seeds, the cancer cells that spread, or metastasize, can begin to form tumors in their new locations.

As long as a tumor stays in place, it can grow quite large, and a surgeon would still be able to remove it (depending on the location). However, once cancer cells leave the original tumor, they usually spread to so many locations that a surgeon's scalpel is no longer an effective weapon. At this point, chemotherapy or radiation is generally used to kill the cancer cells wherever they are hiding. (These treatments are discussed later in the chapter.) The original tumor is rarely a cause of death. Instead, the tumors that form in distant sites

Q *Which chromosomes have extra copies in the karyotype of the cancer cell? Which have extra pieces in the karyotype of the cancer cell? Which are missing pieces in the karyotype of the cancer cell?*

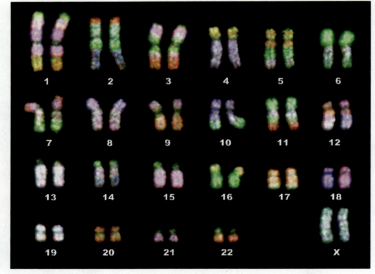

(a)

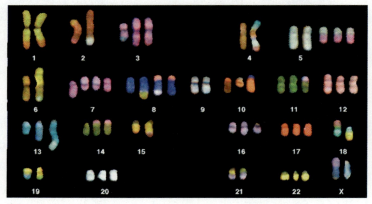

(b)

FIGURE 21a.2 *A comparison of karyotypes from (a) a normal cell and (b) a cancerous cell shows that cancer cells contain extra chromosomes and chromosomes with extra pieces. (A karyotype is an arrangement of photographed chromosomes in their pairs, identified by physical features.)*

A *Chromosomes 3, 5, 7, 8, 10, 11, 12, 13, 14, 16, 17, 20, and 22 have extra copies. Chromosomes 2, 4, 6, and 13 have extra pieces. Chromosomes 3, 4, 5, 6, 7, 8, 10, 13 14, and 18 are missing pieces.*

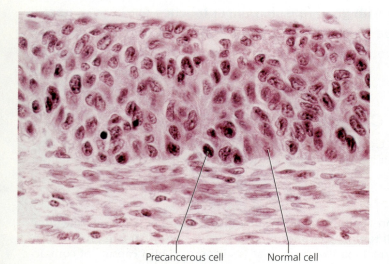

FIGURE 21a.1 *Cancer cells have an abnormal appearance. In dysplasia, precancerous cells have large, irregularly shaped nuclei that contain increased amounts of DNA.*

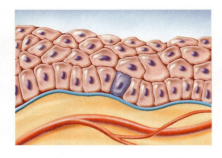
Initial tumor cell: One cell acquires mutations, causing loss of control of cell division.

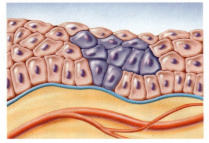
Cell divides more frequently than others.

Carcinoma in situ: Tumor remains at its site of origin.

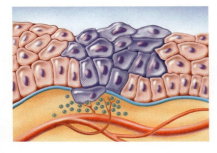
Cells of the tumor release growth factors to attract a blood supply.

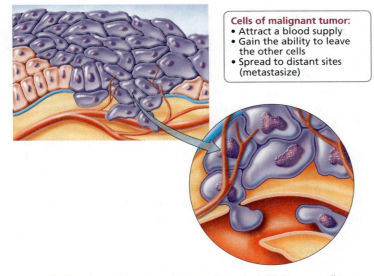
Cells of malignant tumor:
- Attract a blood supply
- Gain the ability to leave the other cells
- Spread to distant sites (metastasize)

FIGURE 21a.3 *The progression of cancer from the initial tumor cell to a malignant tumor.*

in the body are responsible for 90% of the deaths of people with cancer.

Once cancer cells have separated from the original tumor, they usually enter the cardiovascular or lymphatic system, which carries the renegade cells to distant sites. Thus, circulatory pathways in the body often explain the patterns of metastasis. For example, cancer cells escaping from tumors in most parts of the body, including the skin, encounter the next capillary bed in the lungs. Consequently, many cancers spread to the lungs. However, blood leaving the intestine travels directly to the liver, so colon cancer typically spreads to the liver.

The simplest explanation of how cancer causes death is that it interferes with the ability of body cells to function normally. For instance, cancer cells are greedy. They deprive normal cells of nutrients and thereby weaken the cells, sometimes to the point of death. Cancer cells can also prevent otherwise healthy cells from performing their usual functions. In addition, tumors can block blood vessels or air passageways in the respiratory system or press on vital nerve pathways in the brain.

21a.2 Development of Cancer

The 30 trillion to 50 trillion cells in the human body generally work cooperatively, much like the members of any organized society. There are "rules," or controls, that tell a cell when and how often to divide, when to self-destruct, and when to stay in place. However, cancer cells are outlaws. They evade the many controls that would normally maintain order in the body. Let's discuss some of the normal systems of checks and balances that regulate healthy cells and see how cancer cells are able to get around these safeguards to divide indefinitely and spread out of control.

Recall from Chapter 19 that the cell cycle is the life cycle of the cell. During the cell cycle, there are checkpoints at which the cell determines whether conditions are favorable for moving on to the next stage. This is the cell's system of damage control. If a healthy cell detects damage, such as a mutated gene, it stops the cell cycle, assesses the damage, and begins repair (Figure 21a.4). If the repair is successful, the cell cycle resumes. If the damage is recognized as too severe to repair, a program of cell death is initiated, as will be discussed shortly. Unfortunately, if the damage repair is unsuccessful or incomplete, genetic damage accumulates and can lead to cancer.

Tumor-suppressor genes are an important part of the cell's system of damage control. Some tumor-suppressor gene products detect damaged DNA.[1] When they do, other tumor-suppressor gene products serve as "managers of the cell's repair shop." These gene products assess the damage and coordinate the activities of other genes, whose products serve as "mechanics" and repair the damage. If the damage turns out to be too extensive, the manager activates still other genes whose products cause cell death. A particularly important tumor-suppressor gene is *p53*. We consider some of the activities of the p53 protein as we discuss the relationship between genes and cancer.

Lack of Restraint on Cell Division

When genes that regulate cell division are mutated, they usually do not function properly, and the cell loses control over cell division. Cancer is fundamentally a disease in which certain genes mutate and produce proteins that malfunction or produce proteins in abnormal amounts or in inappropriate locations. We now know that gene activity can be turned on or

---

[1] Recall that genes exert their effect through the proteins they code for. Thus, it is really the proteins that are acting.

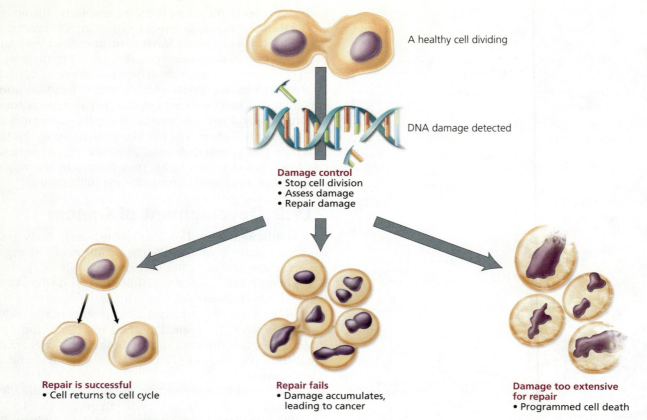

FIGURE 21a.4 *Steps in controlling DNA damage during the cell cycle.*

off by changes in how tightly the DNA is coiled without changes in the DNA sequence. Thus, cancer can also result if tumor-suppressor genes are turned off or if oncogenes (cancer-promoting genes) are turned on.

Two types of genes usually regulate cell division: proto-oncogenes and the tumor-suppressor genes mentioned earlier. Proto-oncogenes stimulate cell division in a variety of ways, for example, by producing growth factors or affecting their function or by producing proteins that affect the activity of certain genes. In contrast, tumor-suppressor genes inhibit or stop cell division. Thus, tumor-suppressor gene products act like brakes on cell division. The combined activities of these two types of genes allow the body to control cell division to divide and develop normally, repair defective cells, and replace dead cells.

When mutations affect the functioning of these gene products, the normal system of checks and balances that regulates cell division goes awry, and the disruption can result in the unrestrained cell division that characterizes cancer. A mutation in a tumor-suppressor gene can promote cancer by taking the brakes off cell division. The tumor-suppressor gene *p53* produces a protein that regulates another gene whose job it is to produce a protein that keeps cells in a nondividing state. However, when *p53* mutates, cell division is no longer curbed. Mutant *p53* seems to be an important factor in more than half of all cancers. Mutation in another tumor-suppressor gene, called *RB,* causes retinoblastoma, which is a rare form of childhood eye cancer. The normal product of *RB* turns off a proto-oncogene that stimulates cell division. When *RB* mutates, the activity of the proto-oncogene is no longer inhibited, and cell division continues because the cell cannot exit the cell cycle. Two tumor-suppressor gene products that play a role in breast cancer—BRCA1 and BRCA2 proteins—initiate DNA repair. When these two genes mutate, damaged DNA is not repaired. Mutations then accumulate, which may set the cell on a course toward cancer.

A mutation in a proto-oncogene can also destroy the regulation of cell division. The mutated gene is called an *oncogene,* and it increases the stimulus for cell division or promotes cell division without a stimulus. An oncogene does to cell division what a stuck accelerator would do to the speed of a car. The proteins produced by many proto-oncogenes are growth factors or receptors for growth factors. When a proto-oncogene mutates and becomes an oncogene, the transformation often causes too much protein to be produced or makes the protein more active than usual. For example, the product of the *ras* gene normally signals the presence of a growth factor, which stimulates cell division. The *ras* oncogene protein is hyperactive and stimulates cell division even in the absence of growth factors. The *ras* oncogene is thought to be important in the development of most pancreatic and colon cancers, as well as some lung cancers. Other oncogenes play a role in leukemia and many of the most deadly forms of breast and ovarian cancer.

When normal cells are grown in tissue culture, they divide until they form a single layer of cells. If healthy cells contact a neighbor, they stop dividing. This phenomenon is called *contact inhibition*. But cancer cells do not exhibit contact inhibition. Instead, they continue to divide, pile up on one another, and form a tumor.

DNA Damage and Cell Destruction

When the genes that regulate cell division are faulty, backup systems normally swing into play to protect the body from the renegade cell. One such system is programmed cell death, also called *apoptosis,* in which cells activate a genetic suicide program in response to a biochemical or physiological signal. Activation of the so-called death genes prompts cells to manufacture proteins that then kill the cells. Often, the condemned cells go through a predictable series of physical changes that indicate the cell will die. During apoptosis, the outer membrane of the condemned cell produces bulges, called blebs, that are pinched off the cell (Figure 21a.5). The blebs are an indication that the cell will break down into membrane-enclosed fragments that are engulfed and removed by other cells.

Cancer cells often fail to trigger apoptosis. Although it is not the only way that cancer cells evade this safeguard, a faulty *p53* tumor-suppressor gene is often at least partly responsible. Besides producing a protein that inhibits cell division, the p53 protein normally prevents the replication of damaged DNA. If damage is detected, the p53 protein halts cell division until the DNA can be mended. If the damage is beyond repair, the p53 protein triggers the events that lead to programmed cell death.

In a cancer cell, however, a faulty p53 protein fails to initiate the events leading to cellular self-destruction, so the cells are free to divide in spite of genetic damage. Tumors containing cells with damaged *p53* genes grow aggressively and spread easily and quickly to new locations in the body. Cancer cells containing mutations in *p53* are difficult to kill with radiation or chemotherapy, because these techniques are intended to damage the DNA of the cancer cell and trigger programmed self-destruction. Because of mutations in *p53,* many cancer cells are simply unable to self-destruct in response to DNA damage.

Unlimited Cell Division

Healthy cells have yet another safeguard against unrestrained cell division: a mechanism that limits the number of times a cell can divide during its lifetime. When grown in the laboratory, most human cells divide only about 50 or 60 times before entering a nondividing state called *senescence*. Like sand running through an hourglass, cell division has a predetermined end.

How does a cell "count" the number of times it has divided? The answer might lie in telomeres—pieces of DNA at the tips of chromosomes that protect the ends of the chromosomes much like the plastic pieces on the ends of shoelaces protect the shoelace—or in telomerase, the enzyme that constructs the telomeres (Figure 21a.6). Soon after an embryo is fully developed, most cells stop producing telomerase, putting an end to the maintenance of telomere length. Each time DNA is copied in preparation for cell division, a tiny piece of every telomere in the cell is shaved off, shortening the chromosomes slightly. When the telomeres are completely gone, the chromosome tips can fuse together, disrupting the genetic message and causing the cell to die. Telomeres, then, may be the cell's way of limiting the number of times division can occur. When the telomeres are gone, time is up for that cell. Thus, telomere length may serve both as a gauge of a cell's age and as an indicator of how long that line of cells will continue to divide.

The "fountain of youth" that bestows immortality on cancer cells and allows them to divide indefinitely is their unceasing production of telomerase. This enzyme reconstructs the telomeres after each cell division, stabilizing telomere

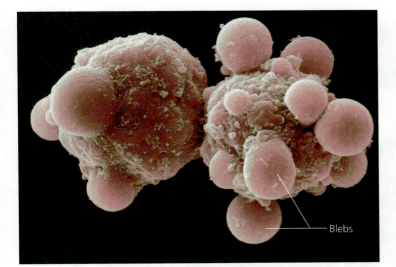

FIGURE 21a.5 *Programmed cell death is a backup system that protects the body from a cell in which the genes regulating cell division have been damaged. DNA damage that is too extensive to repair normally triggers a genetic suicide program that causes the cell to self-destruct, as these cells are doing. As the cell goes through programmed cell death, its plasma membrane forms bulges called blebs. Cancer cells are able to evade this protective mechanism.*

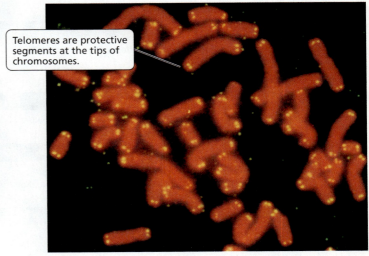

FIGURE 21a.6 *Telomeres (shown here in yellow) may serve as molecular counting mechanisms that limit the number of times a cell can divide. Cancer cells retain the ability to construct new telomeres to replace the bits that are shaved off.*

length and protecting the important genes at chromosome tips. Although most types of mature human cells can no longer produce telomerase, the genes for telomerase apparently become turned on in most cancer cells. Telomerase is present in nearly 90% of biopsies of human cancerous tumors.

stop and think

Why would telomerase activity serve as a tissue marker for cancerous cells?

Blood Supply to Cancer Cells

We have seen that cancer cells have escaped the normal cellular controls on cell division and are unable to issue the orders that would lead to their own death. Instead, they multiply and form a tumor.

As mentioned earlier, when a tumor reaches a critical size of about a million cells, its growth will stop unless it can attract a blood supply to deliver the nutrients it needs to support its growth and to remove waste. Cancer cells release special growth factors that cause capillaries to invade the tumor. These tiny blood vessels become the tumor's lifeline, removing wastes and delivering fresh nutrients and additional growth factors that will spur tumor growth. They also serve as pathways by which the cancer cells are able to leave the tumor and spread to other sites in the body.

In a healthy person, blood vessel formation is uncommon and is usually limited to the repair of cuts or other wounds. Abnormal invasion of blood vessels into tissues can cause damage. For instance, when blood vessels invade the eye's light-sensitive retina, blindness can result. When vessels invade joints, they can cause arthritis. To avoid such damage, cells produce a protein that prevents new blood vessels from forming in tissues. The gene that normally produces this protein is by now familiar—*p53*. Mutations in *p53* can block the production of the protein that prevents the attraction of blood vessels, allowing blood vessels to invade the tumor.

Adherence to Neighboring Cells

With access to blood vessels, the cancer cells can begin to spread. Their ability to travel through the body is yet another example of their freedom from normal cellular control mechanisms. Normal cells are "glued" in place by special molecules on their surfaces called *cellular adhesion molecules (CAMs)*. When most normal cells become "unglued" from other cells, they stop dividing, and their genetic program for self-destruction is activated.

Cancer cells must become "unglued" from other cells to travel through the body. One way cancer cells break loose is by secreting enzymes that break down the CAMs that hold them and their neighbors in place. In this way, their anchors are broken, and mechanical barriers, such as basement membranes, that would prevent metastasis are breached. Unanchored cancer cells can continue dividing and evade self-destruction because their oncogenes send a false message to the nucleus saying that the cell is properly attached. Table 21a.1 presents a review of the control mechanisms that can fail, resulting in cancer.

Body Defense Cells

Despite all the safeguards that prevent cells from becoming cancerous, cancer cells develop in our bodies every day. Fortunately, certain body defense cells—natural killer cells and cytotoxic T cells (see Chapter 13)—usually kill those cancer cells. The processes that lead to the creation of cancer cells also produce new and slightly different proteins on cancer cell membranes. The defense cells recognize these proteins as nonself and destroy the cancer cells. But sometimes cancer cells evade destruction. Some types of cancer cells actively inhibit

TABLE 21a.1 Review of Control Mechanisms That Fail in Cancer	
Mechanisms That Protect Cells from Cancer	**Method of Evasion Used by Cancer Cells**
Genetic controls on cell division	
Proto-oncogenes stimulate cell division through effects on growth factors and certain other cell-signaling mechanisms	Oncogenes promote cell division
Tumor-suppressor genes inhibit cell division	Mutations in tumor-suppressor genes take the "brakes" off cell division
Programmed cell death	
A genetic program that initiates events that lead to the death of the cell when damaged DNA is detected or another signal is received	Mutations in tumor-suppressor genes: Mutant gene *p53* no longer triggers cell death when damaged DNA is detected
Limitations on the number of times a cell can divide	
Telomeres protect the ends of chromosomes, but a fraction of each is shaved off each time the DNA is copied; when the telomeres are gone, the chromosome tips can stick together, causing the cell to die	Genes to produce telomerase, the enzyme that reconstructs telomeres, are turned on in cancer cells so telomere length is stabilized
Controls that prevent the formation of new blood vessels	
These controls are normally in effect except in a few instances, such as wound healing	Cancer cells produce growth factors that attract new blood vessels and proteins that counter the normal proteins that inhibit blood vessel formation
Controls that keep normal cells in place	
Cellular adhesion molecules (CAMs) hold cells in place; unanchored cells stop dividing and self-destruct	Cancer cells' oncogenes send a false message to the nucleus that the cell is properly anchored

the defense cells, and some simply multiply so quickly that the defense cells cannot destroy them all. In either case, the tumor is able to grow and spread.

21a.3 Multiple Mutations

Healthy cells have interacting control systems that usually prevent cancer development. Normally, tumor-suppressor genes and proto-oncogenes regulate cell division so that it occurs only for growth and repair. If mutations occur, the cell will normally self-destruct before dividing and passing the genetic damage to its daughter cells. If those safeguards fail, the cell is usually prevented from dividing more than 50 to 60 times, because it lacks telomerase. If the telomerase gene is turned on, the cells will begin to starve due to lack of nutrients when the tumor consists of about a million cells. For a tumor to grow larger than this, the cells must attract a blood supply. To metastasize, cancer cells must break the molecules that anchor them in place.

With so many controls to prevent it, cancer development is a multistep process involving multiple mutations and changes in gene activity. The first mutation occurs and is passed on to all the descendant cells. Later—usually many years later—a second mutation occurs in one of the descendant cells containing the original mutation. Both mutations are passed to all the descendants of that cell. Many years later, a third mutation might occur in one of the daughter cells that already has two mutations. Each mutation brings the cell closer to becoming cancerous (Figure 21a.7).

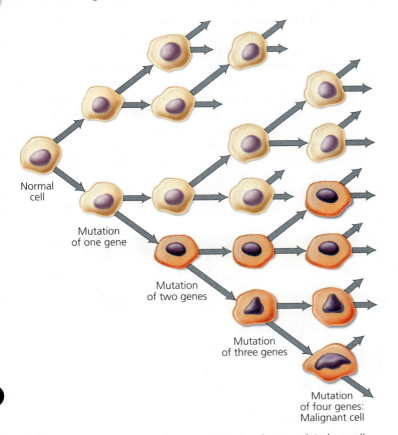

FIGURE 21a.7 *Multiple mutations must occur and accumulate in a cell before the cell becomes cancerous.*

Damage must occur in *at least* two genes (and most commonly more than six genes) before cancer occurs. For instance, colon cells must accumulate damage in at least one proto-oncogene and three tumor-suppressor genes before they become cancerous (Figure 21a.8). Furthermore, a tumor may contain different cell lines containing mutations in different sets of genes. In a recent study, researchers found 50,000 mutations in the lung tumor of a heavy smoker. The number of damaged genes needed to produce cancer explains why it is possible to inherit a predisposition to a certain form of cancer. A person who inherits only one mutant gene may be predisposed to cancer. But a second event, a mutation in at least one other gene, is required before uncontrolled cell division is unleashed.

21a.4 Cancer Stem Cell Hypothesis

All the cells of a cancerous tumor are genetically identical because they are descendants of a single cell that accumulated the mutations necessary for it to lose control over cell division. Do all the cells of the tumor have the ability to divide without restraint, metastasize, and form new tumors? Perhaps, but an increasing number of scientists are suggesting that only some cells within the tumor are capable of continually dividing. According to the cancer stem cell hypothesis, only a subpopulation of cells within the tumor—cancer stem cells—have the ability of unlimited self-renewal (replenishment) and give rise to the tumor.

Recall from Chapter 19a that stem cells are able to divide continually and to differentiate (specialize) into different types of cells. Embryonic stem cells taken from an early embryo can develop into all the cell types found in the body. In an adult, stem cells are sprinkled throughout tissues to give rise to new cells as needed for growth and repair. When an adult stem cell divides, one of the descendants remains a stem cell. The other daughter cell, sometimes called a *progenitor cell,* divides rapidly about five to eight times, giving rise to a population of cells that will then specialize to form a particular type of cell.

We do know that not all cells in a tumor have the same capacity for unlimited self-renewal. Those cells with the capacity for unlimited self-renewal—cancer stem cells—have protein markers on their surface that are different from those of other cells in the tumor. Populations of cancer stem cells have been isolated from bone marrow in the case of acute myeloid leukemia and from solid tumors in the breast or brain. Studies involving the injection of cancerous cells in mice have shown that far fewer cells are needed to initiate a tumor in a mouse when only cancer stem cells are used than if randomly selected cancerous cells are used. Some researchers suggest that the reason for the difference in the number of cells needed is that only cancer stem cells initiate tumor growth. Random samples of cells from a tumor may contain some cancer stem cells, but these represent only a fraction of the cells in a tumor. According to the cancer stem cell hypothesis, then, metastasis occurs when a cancer stem cell leaves the initial tumor and settles in a new location.

If the cancer stem cell hypothesis is correct, in which cells do the mutations occur? We don't know. Since stem cells are

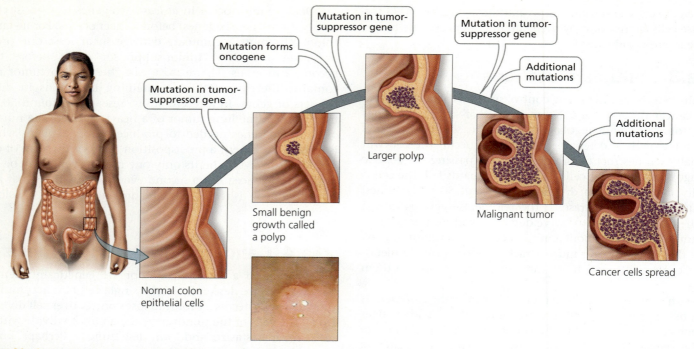

FIGURE 21a.8 *Multiple mutations must occur in a single cell before it becomes cancerous. In colon cells, at least one proto-oncogene and three tumor-suppressor genes must mutate before the cells become cancerous.*

already able to divide continually, the simplest explanation is that the mutations leading to cancer occur in adult stem cells. However, it is also possible that cancer mutations occur in the progenitor cells that multiply several times before specializing. In this case, the mutations would cause the progenitor cells to become the cancer stem cells, which would give rise to most of the cells of the tumor.

The cancer stem cell hypothesis might also explain why traditional methods of treating cancer—radiation and chemotherapy (discussed shortly)—may shrink tumors, although the tumors often return. Such treatments are aimed at killing rapidly dividing cells such as progenitor cells. Death of progenitor cells would shrink the tumor. However, lingering cancer stem cells could replenish the population. Researchers are looking for ways to cure cancer by killing the cancer stem cells.

21a.5 Known Causes of Cancer

We have seen that many of the tactics a cancer cell uses to evade normal cellular safeguards are consequences of changes in genes. Those genetic changes are often brought about by viruses or by mutations caused by exposure to certain chemicals or to radiation.

Viruses

It is estimated that viruses cause about 5% of the cancers in the United States (Table 21a.2). Some of the viruses that cause cancer have oncogenes among their genes. Once inside the host cell, the viral oncogene behaves as a host oncogene would, taking the cell one step closer to becoming cancerous. A viral oncogene is partly why the human papilloma virus that causes genital warts can cause cervical and penile cancer. If the viral genetic information is in the form of RNA, the enzyme reverse transcriptase uses viral RNA to synthesize viral DNA, which is then inserted into a host cell chromosome. Proteins produced by the viral DNA may then drive the cellular proto-oncogene to be expressed in abnormal levels or in the wrong place or time. RNA viruses can also pick up a proto-oncogene from one host cell and introduce it into a new host cell, thus promoting cell division. In other cases, a virus causes cancer because viral DNA becomes inserted into the host DNA in a location that disrupts the functioning of a gene that influences cell division. The viral DNA could, for example, be inserted into a regulatory gene that controls a proto-oncogene, breaking the switch that turns off the gene. Some viruses interfere with the function of the immune system, lessening its ability to find and destroy cancer cells as they arise.

TABLE 21a.2 Some Viruses Linked to Human Cancer

Virus	Types of Cancer
Human papillomaviruses (HPVs)	Cervical, penile, and other anogenital cancers in men and women
Hepatitis B and C viruses	Liver cancer
Epstein-Barr virus	B cell lymphomas, especially Burkitt's lymphoma; nasopharyngeal carcinoma
Human T cell leukemia virus (HTLV-1)	Adult T cell leukemia
Cytomegalovirus (CMV)	Lymphomas and leukemias
Kaposi sarcoma–herpesvirus	Kaposi's sarcoma

Chemicals

A *carcinogen* is an environmental agent that fosters the development of cancer. Some chemicals, especially certain organic chemicals, cause cancer by causing mutations. As we saw in Chapter 21, a mutation in as few as one nucleotide in a DNA sequence can alter a gene's message. Thus, even a small alteration in DNA can wreak havoc with a cell's regulatory mechanisms and lead to cancer.

Chemical carcinogens are around us most of the time—in air, food, water, and other substances in our surroundings. We can attempt to avoid contact with some of them. For instance, tobacco smoke contains a host of chemical carcinogens. Among the carcinogens in tobacco smoke is one that specifically mutates the tumor-suppressor gene *p53* and another that specifically mutates one of the *ras* proto-oncogenes. Tobacco smoke is responsible for 30% of all cancer deaths and may contribute to as many as 60% of all cancer deaths in the United States. Excessive alcohol consumption is another cancer risk factor that can be avoided. Other chemical carcinogens are more difficult to avoid. These include benzene, formaldehyde, hydrocarbons, certain pesticides, and chemicals in some dyes and preservatives.

Some chemicals contribute to the development of cancer by stimulating cell division, which increases the chance of additional mutations arising. If a cancerous cell has already formed, this stimulation will cause the cancer to progress. Certain hormones can promote cancer in this way. For example, the female hormone estrogen stimulates cell division in the tissues of the breast and in the lining of the uterus (the endometrium). Sustained high levels of estrogen are linked with breast cancer,[2] and the incidence of endometrial cancer is elevated in postmenopausal women whose hormone replacement therapy consists entirely of estrogen. Estrogen does not seem to promote endometrial cancer when taken in combination with progesterone. However, since 2002, several studies have shown that postmenopausal women who take estrogen in combination with progesterone have a slightly greater risk of developing breast cancer. Furthermore, breast tumors in women taking combined hormone pills were larger and more likely to have spread than were tumors in women not taking hormone pills.

Radiation

Radiation, too, can lead to cancer by causing mutations in DNA. It is impossible to avoid exposure to radiation from natural sources, such as the ultraviolet light from the sun, cosmic rays, radon, and uranium. However, we can take reasonable precautions to minimize our risks. For example, sunlight's ultraviolet rays cause skin cancer. Although we probably would not choose to spend our lives entirely indoors to reduce the risk of skin cancer, it is wise to avoid sunbathing, as well as tanning lamps and tanning parlors. It is also a good idea to use a sunscreen whenever exposure to the sun is unavoidable.

[2]The link between estrogen and breast cancer is discussed in Chapter 17.

stop and think

A cancer cluster is a greater number of cancer cases in an area than would be expected in a similar area over a given length of time. It usually involves the same type of cancer and is caused by common exposure. If you were asked to investigate a cancer cluster, which of the cancer risks would you look for?

21a.6 Reducing the Risk of Cancer

Although we tend to think of cancer as one disease, it is in fact a family of more than 200 diseases, usually named for the organ in which the tumor arises. Cancers of the epithelial tissues are *carcinomas*. *Leukemias* are cancers of the bone marrow. *Sarcomas* are cancers of the muscle, bone, cartilage, or connective tissues. *Lymphomas* are cancers of the lymphatic tissues. *Adenocarcinomas* are cancers of the glandular epithelia. Figure 21a.9 shows the estimated number of cases of cancer and cancer deaths for various types of cancer, and Table 21a.3 indicates where in this textbook certain types of cancer are discussed.

Cancer is the second leading cause of death in industrialized countries, but the good news is that some lifestyle changes can greatly decrease your risk of developing cancer (Table 21a.4). Tobacco use and unhealthy diet are responsible for two-thirds of all cancer deaths in the United States. Tobacco smoke is the leading carcinogen, and it is obvious how to modify that risk.

TABLE 21a.3 Some Discussions of Cancer in This Book

Cancer	Chapter
Skin	Chapter 4 (p. 77)
Leukemia	Chapter 11 (pp. 205–206)
Lung	Chapter 14 (pp. 282–283)
Colon, stomach, esophagus	Chapter 15 (pp. 291, 297, 298)
Testis, prostate	Chapter 17 (pp. 343–345)
Breast	Chapter 17 (pp. 354–355)
Cervical	Chapter 17a (p. 367)

TABLE 21a.4 Tips for Reducing Your Cancer Risk

1. Do not use tobacco. If you do, quit. Avoid exposure to secondhand smoke.
2. Reduce the amount of saturated fat in your diet, especially the fat from red meat.
3. Minimize your consumption of salt-cured, pickled, and smoked foods.
4. Eat at least five servings of fruit and vegetables every day.
5. Avoid excessive alcohol intake. If you consume alcohol, one or two drinks a day should be the maximum.
6. Watch your caloric intake, and maintain a healthy body weight.
7. Avoid excessive exposure to sunlight. Wear protective clothing. Use sunscreen.
8. Avoid unnecessary medical x-rays.
9. Have the appropriate screening exams on a regular basis. Women should have PAP tests and mammograms. Men should have prostate tests. All adults should have tests for colorectal cancer.

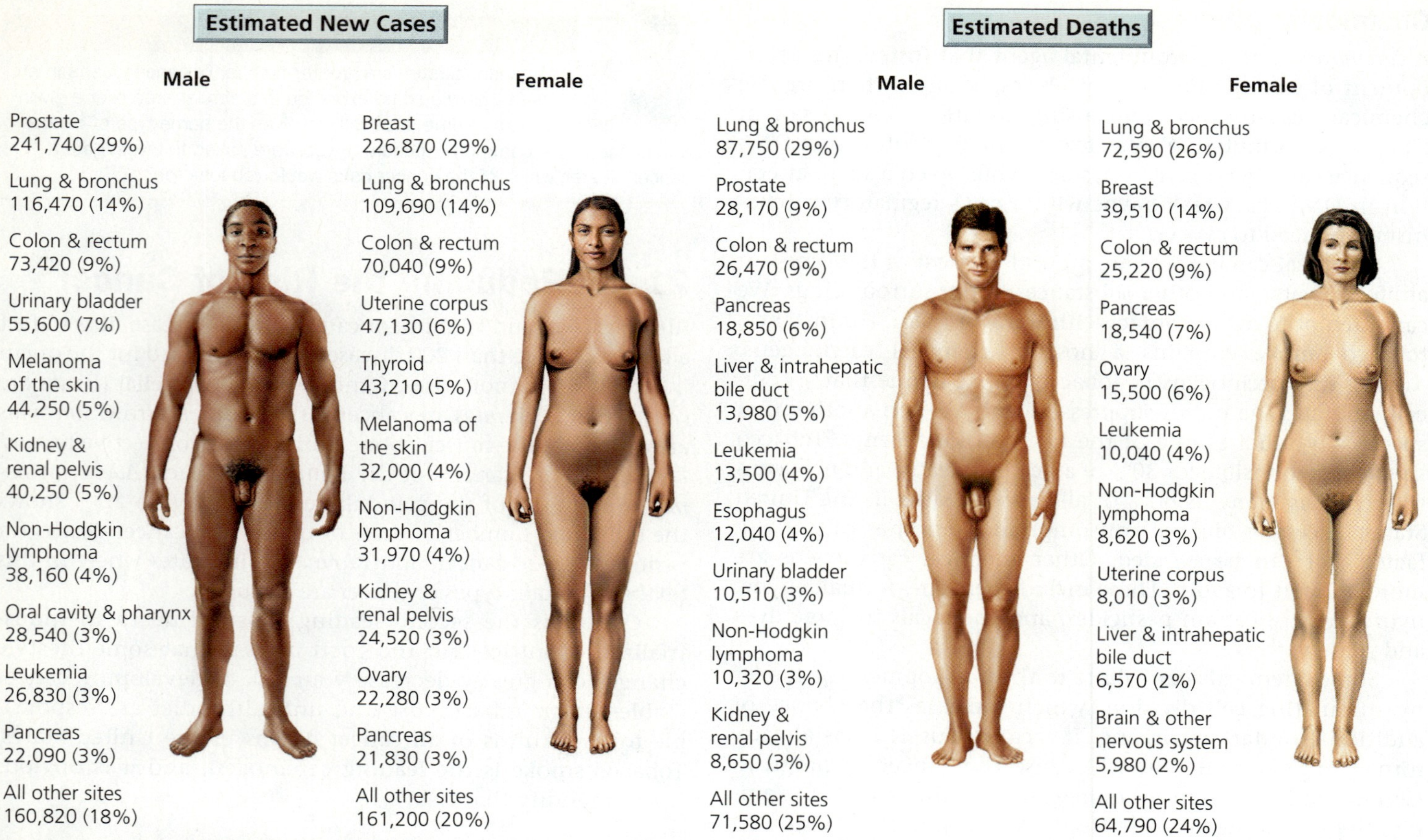

Estimated New Cases

Male
- Prostate 241,740 (29%)
- Lung & bronchus 116,470 (14%)
- Colon & rectum 73,420 (9%)
- Urinary bladder 55,600 (7%)
- Melanoma of the skin 44,250 (5%)
- Kidney & renal pelvis 40,250 (5%)
- Non-Hodgkin lymphoma 38,160 (4%)
- Oral cavity & pharynx 28,540 (3%)
- Leukemia 26,830 (3%)
- Pancreas 22,090 (3%)
- All other sites 160,820 (18%)

Female
- Breast 226,870 (29%)
- Lung & bronchus 109,690 (14%)
- Colon & rectum 70,040 (9%)
- Uterine corpus 47,130 (6%)
- Thyroid 43,210 (5%)
- Melanoma of the skin 32,000 (4%)
- Non-Hodgkin lymphoma 31,970 (4%)
- Kidney & renal pelvis 24,520 (3%)
- Ovary 22,280 (3%)
- Pancreas 21,830 (3%)
- All other sites 161,200 (20%)

Estimated Deaths

Male
- Lung & bronchus 87,750 (29%)
- Prostate 28,170 (9%)
- Colon & rectum 26,470 (9%)
- Pancreas 18,850 (6%)
- Liver & intrahepatic bile duct 13,980 (5%)
- Leukemia 13,500 (4%)
- Esophagus 12,040 (4%)
- Urinary bladder 10,510 (3%)
- Non-Hodgkin lymphoma 10,320 (3%)
- Kidney & renal pelvis 8,650 (3%)
- All other sites 71,580 (25%)

Female
- Lung & bronchus 72,590 (26%)
- Breast 39,510 (14%)
- Colon & rectum 25,220 (9%)
- Pancreas 18,540 (7%)
- Ovary 15,500 (6%)
- Leukemia 10,040 (4%)
- Non-Hodgkin lymphoma 8,620 (3%)
- Uterine corpus 8,010 (3%)
- Liver & intrahepatic bile duct 6,570 (2%)
- Brain & other nervous system 5,980 (2%)
- All other sites 64,790 (24%)

FIGURE 21a.9 *The American Cancer Society's 2012 estimates for the leading types of cancer in terms of new cases and deaths. The figures do not include basal and squamous cell skin cancer or in situ carcinomas other than those of the urinary bladder. The percentages may not total 100% because of rounding.*
Source: American Cancer Society, Cancer Facts and Figures 2012 (Atlanta: American Cancer Society, 2012). 2012 Cancer Statistics. Reprinted by permission of John Wiley & Sons, Inc.

A few simple diet changes may reduce your risk of developing cancer. The best rules are to eat a well-balanced diet and to eat all foods in moderation. For instance, a high-fat diet is linked to colon and breast cancers. Most people in the United States consume far too much fat. Thus it is wise to reduce fat intake, especially saturated fat, which comes from animal sources such as red meat. Consuming large quantities of smoked, salt-cured, and nitrite-cured foods, such as ham, bologna, and salami, increases the risk of cancers of the esophagus and stomach.

A diet rich in fruits and vegetables can reduce your cancer risk because these foods are high in fiber, which can dilute the contents of the intestines, bind to carcinogens, and reduce the amount of time the carcinogens spend in the intestine by speeding passage of intestinal contents. The so-called colorful vegetables—vegetables having colors other than green—are usually high in antioxidant vitamins, and these vitamins may play a role in protecting against cancer. As their name implies, antioxidants interfere with oxidation, a process that can result in the formation of molecules called *free radicals* that can damage DNA and thereby lead to cancer. The three major antioxidants are beta-carotene, vitamin E, and vitamin C. The first two are common in red, yellow, and orange fruits and vegetables, and the last abounds in citrus fruits, among other sources.

21a.7 Diagnosing Cancer

Early detection is critical to cancer survival because treatment is much more likely to be successful if the cancer has not yet spread. You know your own body better than anyone else does. The American Cancer Society suggests that you be alert for cancer's seven warning signs, the first letters of which spell the word CAUTION:

Change in bowel or bladder habit or function
A sore that does not heal
Unusual bleeding or bloody discharge
Thickening or lump in breast or elsewhere
Indigestion or difficulty swallowing
Obvious change in wart or mole
Nagging cough or hoarseness

But self-examination is not enough. There are additional ways to diagnose cancer—some more involved than others:

- **Routine screening.** Many routine tests can detect cancer in people who do not have symptoms. You can perform some of the tests on yourself; others require a visit to a medical professional (Table 21a.5).

TABLE 21a.5 Recommended Cancer Screening Tests

Guidelines suggested by the American Cancer Society for the early detection of cancer in people without symptoms, age 20 to 40

Cancer-related checkup every 3 years

Should include the procedures listed below plus health counselling (such as tips on quitting cigarette smoking) and examinations for cancers of the thyroid, testes, prostate, mouth, ovaries, skin, and lymph nodes. Some people are at higher than normal risk for certain cancers and may need to have tests more frequently.

Breast cancer	• Exam by doctor every 3 years • Self-exam every month • One baseline breast x-ray ages 35 to 40 Higher risk for breast cancer: Personal or family history of breast cancer; never had children; had first child after 30
Uterine cancer	• Women should report any vaginal bleeding or discharge to their doctors
Cervical cancer	• Yearly PAP test beginning at age 21 or 3 years after sexual activity begins Higher risk for cervical cancer: Early age at first intercourse; multiple sex partners

Guidelines suggested by the American Cancer Society for the early detection of cancer in people without symptoms, age 40 and over

Cancer-related checkup every year

Should include the procedures listed below plus health counseling (such as tips on quitting cigarette smoking) and examinations for cancers of the thyroid, testes, prostate, mouth, ovaries, skin, and lymph nodes. Some people are at higher than normal risk for certain cancers and may need to have tests more frequently.

Colon and rectal cancer	• Fecal occult blood test every year after age 50 • Flexible sigmoidoscopy beginning at age 50 and every 5 years thereafter, or colonoscopy every 10 years Higher risk for colorectal cancer: Personal or family history of colon or rectal cancer; personal or family history of polyps in the colon or rectum; ulcerative colitis
Breast cancer	• Exam by doctor every 3 years • Self-exam every month • Breast x-ray every year after 40 Higher risk for breast cancer: Personal or family history of breast cancer; never had children; had first child after 30
Uterine cancer	• Pelvic exam every year
Cervical cancer	• Yearly PAP test Higher risk for cervical cancer: Early age at first intercourse; multiple sex partners
Endometrial cancer	• Endometrial tissue sample at menopause if at risk Higher risk of endometrial cancer: Infertility, obesity, failure of ovulation, abnormal uterine bleeding, estrogen therapy
Prostate cancer	• Yearly prostate-specific antigen (PSA) blood test and digital rectal exam after age 50

- **Imaging.** Many imaging techniques allow physicians to look inside the body and identify tumors. These include x-rays, computerized tomography (CT) scans, magnetic resonance imaging (MRI), and ultrasound (Figure 21a.10).
- **Biopsy.** Biopsy is the removal and analysis of a small piece of tissue suspected to be cancerous. A biopsy is often done using a needle instead of surgery. In either case, cells are then examined under a microscope to see whether they have the characteristic appearance of cancer cells.
- **Tumor marker tests.** When cancer is suspected, certain blood tests can be used to look for tumor markers, which are chemicals produced either by the cells of the tumor or by body cells in response to a tumor. Prostate cells, for example, produce prostate-specific antigen (PSA). Men normally have low levels of PSA in their blood; however, abnormal proliferation of prostate cells, as would occur if a tumor were developing, can raise those levels. Thus, elevated blood PSA levels suggest the presence of prostate

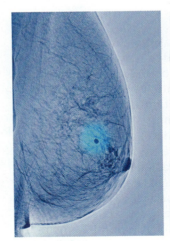
(a) Mammogram of a breast cancer tumor (white area)

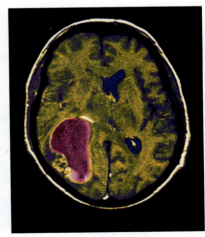
(b) MRI of a cancerous brain tumor (bright purple area)

FIGURE 21a.10 *Imaging techniques such as x-rays or MRIs can detect tumors.*

cancer. Currently, PSA is the only tumor marker that is useful in the original diagnosis of a cancer, but other tumor markers may reveal whether certain cancers have spread or returned. For example, blood levels of a marker called TA–90 can help determine whether melanoma (a type of skin cancer) has spread. The tumor marker CA 125 can identify ovarian cancer. CA–15–3 indicates a recurrence of breast cancer, and CEA indicates a recurrence of colon cancer.

- **Genetic tests.** DNA analysis of cells found in certain bodily fluids or excretions can identify gene mutations associated with certain cancers: sputum is examined for signs of lung cancer, urine for signs of bladder cancer, and feces for signs of colon cancer. Other signs of cancer can be detected by still other tests. For instance, as noted earlier, the enzyme telomerase is produced by cancer cells but rarely by normal ones. A test for telomerase appears to be helpful in diagnosing certain cancers, but this test is still in an experimental stage.

21a.8 Treating Cancer

The conventional cancer treatments—surgery, radiation therapy, and chemotherapy—are still the mainstays of cancer treatment. But many new treatments hold promise.

Surgery

When a cancerous tumor is accessible and can be removed without damaging vital surrounding tissue, surgery is usually performed to eradicate the cancer or remove as much as possible. If every cancer cell is removed, as can be done with early tumors (carcinoma in situ), a complete cure is possible. However, if cancer cells have begun to invade surrounding tissue or have spread to distant locations, surgery alone cannot "cure" the cancer. If the cancer has spread, surrounding tissue and perhaps even nearby lymph nodes may also be removed. Unfortunately, more than half of all tumors have already metastasized by the time of diagnosis, so further treatment is necessary.

Radiation

If cancer has spread from the initial site but is still localized, surgery is usually followed by radiation therapy. In some cases, such as cancer of the larynx (voice box), localized tumors that are difficult to remove surgically without damaging surrounding tissue may be treated with radiation alone.

As we have seen, radiation damages DNA, and extensive DNA damage triggers programmed cell death. The greatest damage caused by radiation is done to cells that are dividing rapidly. The intended targets of radiation, cancer cells, are dividing actively; but so are the cells of several types of tissues, called renewal tissues, whose cells normally continue dividing throughout life. These tissues include cells of the reproductive system, cells that replace layers of skin or the lining of the stomach, cells that give rise to blood cells, or cells that give rise to hair. Unfortunately, radiation cannot distinguish cancer cells from renewal tissue, so good cells are sacrificed in killing the harmful ones. The destruction of renewal tissues leads to the side effects of radiation, such as temporary sterility, nausea, anemia, and hair loss.

Advances in radiation treatment are reducing damage to healthy tissue near the tumor. Intensity-modulated radiation therapy, which is often used to treat cancers of the prostate, head, and neck, uses computers to deliver precise doses of radiation to three-dimensional shape of the tumor by controlling the intensity of multiple radiation beams. In radiation treatment for prostate cancer, initial studies show that damage to the rectum is reduced if a filler material that is also used in cosmetic surgery is injected into the tissue between the radiation source and the rectum to protect the healthy tissue.

Chemotherapy

When cancer is thought to have spread by the time of diagnosis, chemotherapy is often used. In general, the drugs used in chemotherapy reach all parts of the body and kill all rapidly dividing cells, just as radiation does. Some of the drugs block DNA synthesis, others damage DNA, and a few others prevent cell division by interfering with other cellular processes. The side effects are similar to those that accompany radiation therapy.

New chemotherapy techniques target the malignant tissue more precisely, thus sparing healthy cells. One technique uses magnetic fields to pull extremely small metallic beads coated with chemotherapy drugs into the tumor to kill the cancer cells. In another method of targeting only cancer cells, the doctor injects the patient with a light-activated chemotherapy drug. This drug is absorbed by all cells of the body but stays in cancer cells longer than in normal cells. After a few days, when the photosensitive drug remains only in cancer cells, light of a particular wavelength is directed precisely at the cancer cells. When the drug absorbs the light, it produces an active form of oxygen that kills the cells, damages blood vessels that deliver nutrients to the tumor, and causes the immune system to attack the cancer cells.

As we have seen, the idea underlying radiation and chemotherapy is that the damage they do to DNA in rapidly dividing cells will cause the cells to self-destruct. However, *p53*, the gene that detects DNA damage and initiates programmed cell death, is mutant in more than half of all cancers. As a result, even though the treatment succeeds in damaging the DNA in cancer cells, the cells do not always self-destruct, and treatment fails.

Immunotherapy

Cytotoxic T cells of the immune system continually search the body for abnormal cells, such as cancer cells, and kill any that they find (see Chapter 13). The goal of immunotherapy, then, is to boost the patient's immune system so that it becomes more effective in destroying cancer cells. One form of immunotherapy involves administering factors normally secreted by lymphocytes, including interleukin-2 (which stimulates lymphocytes that attack cancer cells), interferons (which stimulate the immune system and also directly affect the tumor cells), and tumor necrosis factor (which directly affects cancer cells, causing them to self-destruct).

Two types of vaccines can be used in the battle against cancer. One type is a vaccine against a virus that causes cancer.

For example, there is a vaccine against four of the human papillomaviruses that cause cervical cancer. Two of these viruses, HPV16 and HPV18, cause 70% of the cases of cervical cancer. Federal health officials now recommend that all girls 9 to 12 years of age and boys 11 to 12 years of age receive this vaccination.

The other type of vaccine is being tested in clinical trials with promising results. It is designed to work by stimulating T cells to attack and kill the cancer cells. Unlike most vaccines, these vaccines cannot *prevent* disease (cancer); they can only treat it. Cancer vaccines contain dead cancer cells, parts of cells, or proteins from cancer cells. Recall that vaccines cause the immune system to fight cells having the same characteristics as the cells in the vaccine. A vaccine strategy is now available to treat prostate cancer. Vaccines are being tested to treat melanoma (a deadly form of skin cancer) and leukemia, as well as cancers of the kidney, prostate, colon, and lung.

Inhibition of Blood Vessel Formation

In recent years, there has been a major change in the way doctors and scientists think about cancer. Gone are the days when the only aim was to kill the cancer cells. As researchers have learned more about the molecular biology of cancer, new ways of slowing its progression or dealing with it as a genetic disease have been developed.

The formation of blood vessels is a critical step in the life of a tumor, because blood vessels bring nourishment and provide a pathway for cell migration. Researchers are working on ways to cut off these lifelines and starve the tumor. Many drugs have been developed for this purpose, and some have been approved by the U.S. Food and Drug Administration (FDA). When a certain drug that blocks blood vessel formation is combined with chemotherapy, the survival rate of patients with colorectal cancer improves.

Gene Therapy

There are currently more than 400 clinical trials using gene therapy to treat cancer, but the FDA has not approved any means of gene therapy for cancer treatment. Nearly all these studies are in very early stages. One of the treatment strategies being tested is to insert normal tumor-suppressor genes into the cancerous cells. For instance, you may recall that gene *p53* normally triggers programmed cell death when DNA damage is detected but that this gene is often faulty in cancer cells. Researchers hope that inserting a healthy form of *p53* into cancer cells will lead to their death, causing the tumor to shrink.

Another strategy is to insert into a cancer cell a piece of DNA that will prevent an oncogene from exerting its effects. The inserted DNA is called *antisense DNA* because it is complementary to the mRNA produced by the oncogene. The antisense DNA would be expected to bind to the mRNA produced by the oncogene and prevent it from being translated into protein, thereby preventing the oncogene from exerting its effects. Currently, this method is being tried as a means of treating leukemia.

One of the most promising approaches using gene therapy is to insert a gene into tumor cells that makes the cells sensitive to a drug that will kill them. This method is now being evaluated as treatment for brain, ovarian, and prostate cancers. A viral gene for the enzyme thymidine kinase is inserted into the tumor cells, and when the gene is expressed, the resulting enzyme makes the cell sensitive to a drug called ganciclovir. The drug is inactive except inside the cells that produce the enzyme coded for by the inserted gene. Thus, only the tumor cells are killed.

Researchers have used gene therapy to cure two patients (out of 17 patients who were treated) of the aggressive skin cancer melanoma. The researchers used a virus to transfer a gene into body defense cells called T cells (discussed in Chapter 13). The inserted gene codes for a protein called *T-cell receptor,* or *TCR.* With this gene, the T cell can find tumor cells and destroy them.

Researchers delivered a gene-altered virus intravenously to treat colorectal, skin, ovarian, and lung cancers that had spread. The virus was administered in a single dose at five dosage levels. The virus reached and replicated within tumors throughout the body. Some patients at every level of dosage saw improvement; patients receiving the highest dosage showed the greatest improvement. Thus, gene therapy for cancer treatment holds promise for the future.

what would you do?
If a child has cancer, it is the parents' responsibility to decide whether to continue treatment. What factors do you think are ethical reasons to stop treatment?

looking ahead
In this and the previous few chapters, we considered genes and inheritance. We have seen that meiosis and mutations increase variability in a population. In the next chapter, we will see that this genetic variability in a population is important for evolution by natural selection.

22 Evolution and Our Heritage

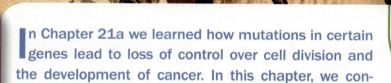

Did you Know?

- Humans lost their extensive body hair about 1.2 million years ago.
- Humans did not regularly wear clothing until sometime between 83,000 and 170,000 years ago.

The fossil record provides evidence of evolution by documenting that life on Earth has changed over time.

- 22.1 Evolution of Life on Earth
- 22.2 Scale of Evolutionary Change
- 22.3 Evidence of Evolution
- 22.4 Human Evolution

ETHICAL ISSUE
Conducting Research on Our Relatives

In Chapter 21a we learned how mutations in certain genes lead to loss of control over cell division and the development of cancer. In this chapter, we consider mutations in a different light. We focus on evolution and how mutations, along with processes such as genetic drift, gene flow, and natural selection, cause changes in allele frequencies within populations. We also address such questions as these: How did life arise and evolve on Earth? How has evolution shaped species, including our own? What were our ancestors like?

22.1 Evolution of Life on Earth

Evolution can broadly be defined as descent with modification from a common ancestor. It is the process by which Earth's life-forms have changed from their earliest beginnings to today. But how did life first arise on Earth?

Earth is estimated to be 4.5 billion years old. Evidence from physical and chemical changes in Earth's crust and atmosphere suggests that life has existed on Earth for about 3.8 billion years. The environment of the early Earth was very different from that of today and would have been an extremely hostile place for most organisms (Figure 22.1). Earth's crust was hot and volcanic. Intense lightning and ultraviolet radiation struck Earth's surface, and the atmosphere contained almost no gaseous oxygen (O_2). Scientists agree on the scarcity of oxygen but debate the other components of the early atmosphere. Most models suggest that carbon dioxide (CO_2), nitrogen (N_2), and water (H_2O) were present. Other gases that may have been present include carbon monoxide (CO), hydrogen (H_2), methane (CH_4), sulfur dioxide (SO_2), and hydrogen sulfide (H_2S). Once the crust cooled, water vapor condensed as rain, and runoff collected in depressions to form early seas.

FIGURE 22.1 *Representation of the early Earth*

How might life have evolved under these conditions? In the following sections, we present a plausible sequence of events that most scientists believe explains the origin of life on Earth. The sequence, known as **chemical evolution**, suggests that life evolved from chemicals slowly increasing in complexity over a period of perhaps 300 million years.

Small Organic Molecules

Scientists hypothesize that conditions of the early Earth favored the synthesis of small organic molecules from inorganic molecules. Specifically, the low-oxygen atmosphere of the primitive Earth encouraged the joining of simple molecules to form complex molecules. The low-oxygen atmosphere was important because oxygen attacks chemical bonds. Scientists further hypothesize that the energy required for the joining of simple molecules could have come from the lightning and intense ultraviolet (UV) radiation striking the primitive Earth. (UV radiation was likely more intense during those times than it is now, because young suns emit more UV radiation than do mature suns and because the early Earth lacked an ozone layer to shield it from UV radiation.)

In 1953, Stanley Miller and Harold Urey of the University of Chicago tested the hypothesis that organic molecules could be synthesized from inorganic ones. These scientists re-created in their laboratory conditions presumed similar to those of the early Earth (Figure 22.2). Miller and Urey discharged electric sparks (meant to simulate lightning) through an atmosphere that contained some of the gases thought to be present in the early atmosphere. They generated a variety of small organic molecules, supporting the hypothesis that organic molecules could be synthesized from inorganic ones.

The atmospheric composition used in the Miller–Urey experiment differed somewhat from the composition currently favored by scientists who study characteristics of the early Earth. Even so, many simulations conducted by other scientists using different gas mixtures (and energy sources) have produced organic compounds in varying amounts. Taken as a whole, these experiments demonstrate that organic molecules can be synthesized from inorganic molecules.

Macromolecules

Scientists hypothesize that these small organic molecules accumulated in the early oceans and over a long period formed a complex mixture. Somehow, perhaps by being washed onto clay or hot sand or lava, the small molecules joined to form macromolecules, such as proteins and nucleic acids. Some scientists suggest instead that deep-sea vents were important locations for the synthesis of small organic molecules and their joining to form larger molecules.

Which macromolecule led to the formation of the first cells? Some scientists suggest that RNA was the critical macromolecule, because it can act as an enzyme and because its

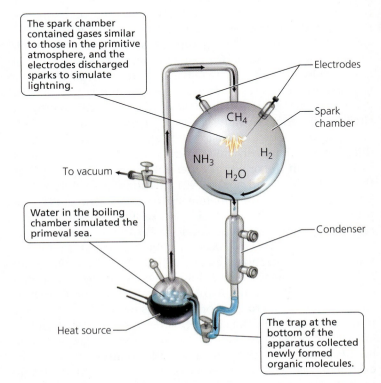

FIGURE 22.2 *Apparatus used by Miller and Urey to test the hypothesis that organic molecules could be synthesized from inorganic ones*

self-replicating abilities allowed information to be transferred from one generation to the next. Cells today store their genetic information as DNA. Thus, if RNA was the first genetic material, then DNA is likely to have evolved later from an RNA template. Other scientists suggest that proteins were the macromolecules that led to the first cells. These scientists point out that amino acids, peptides, and proteins are more chemically stable than nucleotides and nucleic acids when exposed to salt (such as would occur in primordial seas) and intense UV radiation (such as that presumed to have struck the early Earth).

Early Cells

Scientists hypothesize that the newly formed organic macromolecules aggregated into droplet-like structures that were the precursors of cells. The early droplets displayed some of the same properties as living cells, such as the ability to maintain an internal environment different from surrounding conditions.

Fossil evidence indicates that the earliest cells were prokaryotes. Recall from Chapter 3 that prokaryotic cells lack membrane-enclosed organelles, such as the nucleus, and are typically smaller than eukaryotic cells. These early prokaryotic cells relied on anaerobic metabolism (metabolism in the absence of oxygen). Eventually, some of these cells captured energy from sunlight and made their own complex organic molecules from CO_2 and H_2O in their environment. This process, known as *photosynthesis*, produced oxygen as a by-product. Oxygen began to accumulate in the environment. Next came cells that could use the now abundant oxygen to harness energy from stored organic molecules; these cells used cellular respiration (aerobic metabolism; see Chapter 3).

So the question arises: How did more complex cells arise from these early prokaryotic cells? It's possible that some of the organelles within eukaryotic cells appeared when other, smaller organisms became incorporated into the early, primitive cells. Mitochondria, for example, may be descendants of once free-living bacteria. These bacteria either invaded or were engulfed by the ancient cells and formed a mutually beneficial relationship with them. This idea is called the **endosymbiont theory**. Scientists generally accept this explanation for the origin of some features of eukaryotic cells, including mitochondria. Scientists currently are not certain whether endosymbiosis was a factor in the origin of other eukaryotic features, such as the membrane-bound nucleus. An additional possibility is that infolding of the plasma membrane of an ancient prokaryote produced some of the organelles (for example, the endoplasmic reticulum or Golgi complex) found in eukaryotic cells of today.

Fossils of prokaryotic cells have been dated at 3.5 billion years old, and scientists estimate that the first prokaryotic cells probably arose around 3.8 billion years ago. As we discuss later in the chapter, fossils are rarely formed and found, so the origin for a group of organisms is usually hypothesized to occur earlier than the oldest fossil evidence for that group. Eukaryotic cells evolved about 1.8 billion years ago. Multicellularity evolved in eukaryotes around 1.5 billion years ago and eventually led to organisms such as plants, fungi, and animals. Figure 22.3 summarizes the possible steps in the origin of life on Earth.

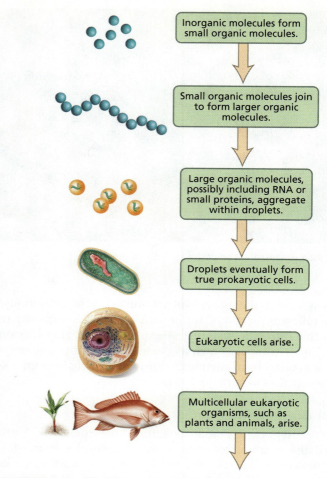

FIGURE 22.3 *Possible steps in the origin of life on Earth*

22.2 Scale of Evolutionary Change

Evolution occurs on two levels. One level, microevolution, is small; and the other, macroevolution, is large. **Microevolution** occurs through changes in allele frequencies within a population over a few generations. **Macroevolution**, conversely, consists of larger-scale evolutionary change over longer periods of time, such as the origin of groups of species (for example, mammals) and mass extinctions (the catastrophic disappearance of many species). We begin with microevolution, examining genetic variation within populations and describing the causes of microevolution. We then turn to the larger-scale phenomena of macroevolution.

Microevolution

Look around, and you will see variation in almost any group of individuals belonging to the same species. Consider your classmates. They do not all look alike, and, unless you are an identical twin, you probably do not precisely resemble your brothers or sisters. Before looking at what makes individuals in such a group different, let's define some basic terminology. A **population** is a group of individuals of the same species living in a particular area. A population of bluegill sunfish inhabits a pond, and a population of deer mice inhabits a small tract of forest. A **gene pool** consists of all the alleles of all the genes of all individuals in a population. (Recall from Chapter 20 that a *gene* is a segment of DNA on a chromosome that directs the synthesis of a specific protein, whereas *alleles* are different forms of a gene.) Now let's take a closer look at what makes individuals in a population different and examine some of the ways that variation can appear in populations.

Sexual reproduction and mutation produce variation in populations. Sexual reproduction shuffles alleles already present in the population. As discussed in Chapters 19 and 20, the gametes (eggs or sperm) of any one individual show substantial genetic variation that results from crossing over and independent assortment during meiosis. Also, the combination of gametes that unite at fertilization is a chance event. Of the millions of sperm produced by a male, only one fertilizes the egg. This union produces a new individual with a new combination of alleles.

New genes and *new* alleles originate by **mutation**, a change in the nucleotide sequence of DNA. Mutations occur at a low rate in any set of genes, so their contribution to genetic diversity in large populations is quite small. Mutations can appear spontaneously from mistakes in DNA replication, or they can be caused by outside sources such as radiation or chemical agents. Only mutations in cell lines that produce eggs or sperm can be passed to offspring.

Recall that microevolution involves changes in the frequency of certain alleles relative to others within the gene pool of a population. Some of the processes that produce those changes are genetic drift, gene flow, mutation, and natural selection.

Genetic drift **Genetic drift** occurs when allele frequencies within a population change randomly because of chance alone. This process is usually negligible in large populations. However, in populations with fewer than about 100 individuals, chance events can cause allele frequencies to drift randomly from one generation to the next. Two mechanisms that facilitate genetic drift in natural populations are the bottleneck effect and the founder effect.

Sometimes, dramatic reductions in population size occur because of natural disasters that kill many individuals at random. Consider a population that experiences a flood in which most members die. With fewer remaining individuals contributing to the gene pool, the genetic makeup of the survivors may not be representative of the original population. This change in the gene pool is the **bottleneck effect**, so named because the population experiences a dramatic decrease in size—much as the size of a bottle decreases at the neck. Certain alleles may be more or less common in the flood survivors than in the original population simply by chance. In fact, some alleles may be completely lost, thereby reducing overall genetic variability among survivors.

Genetic drift also occurs when a few individuals leave their population and establish themselves in a new, somewhat isolated place. By chance alone, the genetic makeup of the colonizing individuals is probably not representative of the full gene pool of the population they left. Genetic drift in new, small colonies is called the **founder effect**.

stop and think

Would you expect the founder effect to be associated with a relatively high or a relatively low frequency of inherited recessive disorders?

Gene flow Another cause of microevolution is **gene flow**, which occurs when individuals move into and out of populations. As individuals come and go, they carry with them their unique sets of genes. Gene flow occurs if these individuals successfully interbreed (mate and produce offspring) with the resident population, adding to the gene pool.

The cessation of gene flow can be important to the formation of new species. A **species** is a population or group of populations whose members are capable of successful interbreeding. Such interbreeding must occur under natural conditions and produce fertile offspring. But consider what happens when a population becomes geographically isolated from other populations of the same species. For example, over the course of geologic time, sea level fluctuates. During periods when sea level decreases, a previously offshore island may reconnect with a continent. When sea level increases, the island will redevelop offshore. Suppose that during the period of low sea level, a continuous population of frogs extends across the landscape. As sea level rises, however, and the island redevelops, the island population of frogs will be isolated from the mainland population. Frogs cannot cross saltwater, so the island population is effectively genetically isolated from the mainland population. The island population may take a separate evolutionary route as distinctly different sets of allele frequencies and mutations accumulate. Eventually, the isolated island population may become so different that it cannot successfully interbreed with the mainland population. At this point, there are two species of frogs instead of one. This process is called **speciation**.

Mutation Mutations, you will recall, are rare changes in the DNA of genes; they are the third way that the frequencies of certain alleles can change relative to others within gene pools. Mutations produce new alleles that, when transmitted in gametes, cause an immediate change in the gene pool. Essentially the new (mutant) allele is substituted for another allele. Like gene flow, mutation can introduce new alleles to a population that are then acted upon by natural selection. If the frequency of the mutant allele increases in a population, it is not because mutations are suddenly occurring more frequently. Instead, possession of the mutant allele might confer some advantage that enables individuals with the mutant allele to produce more offspring than can individuals who lack the allele. In other words, the increased frequency in a population of the mutant allele relative to others results from natural selection, our next focus.

Natural selection In his book *On the Origin of Species* (1859), Charles Darwin argued that species were not specially created, unchanging forms. He suggested that modern species are descendants of ancestral species. Put another way, present-day species evolved from past species. Darwin also proposed that evolution occurred by the process of natural selection. His ideas can be summarized briefly as follows:

1. Individual variation exists within a species. Some of this variation is inherited.
2. Some individuals have more surviving offspring than do others because their particular inherited characteristics make them better suited to their local environment; this is the process of **natural selection**.
3. Evolutionary change occurs as the traits of individuals that survive and reproduce become more common in the population. Traits of less successful individuals become less common.

According to Darwin's ideas, an individual's evolutionary success can be measured by fitness (sometimes called *Darwinian fitness*). **Fitness** compares the number of reproductively viable offspring among individuals. Individuals with greater fitness—that is, producing more successful offspring—have more of their genes represented in future generations. To succeed in terms of evolution, one must reproduce (Figure 22.4). Indeed, you could live more than 100 years, but your individual fitness would be zero if you did not reproduce. Some of the diseases or conditions discussed in earlier chapters are associated with zero fitness because they cause sterility (such as Turner's syndrome) or death before reproductive maturity (such as Tay-Sachs disease).

One result of natural selection is that populations become better suited to their particular environment. This transformation of the population toward better fitness in their environment is called **adaptation**. If the environment changes, however, the individuals who have become most finely attuned to the initial environment could lose their advantage, and other individuals with different alleles might be selected by nature to leave more offspring. If the environment once again stabilizes, the population reestablishes itself around a new set of allele frequencies that better meet the new environmental conditions.

The evolution of antibiotic resistance in bacteria provides an example of natural selection in action. Consider what can happen when you take an antibiotic to treat a bacterial infection. Many bacteria are killed by the treatment, but some may survive. Surviving bacteria have genes that confer resistance to the antibiotic. The survivors reproduce and pass on the trait for resistance to future generations of bacteria. Over time, antibiotic resistance becomes more common in bacterial populations than before. The evolution of drug-resistant strains can potentially become a serious threat to humankind (Chapter 13a).

Natural selection does not lead to perfect organisms. It can act on available variation only, and the available variation may not include the traits that would be ideal. Also, natural selection can only modify existing structures; it cannot produce completely new and different structures from scratch. Finally, organisms have to do many different things to survive and reproduce—such as escaping from predators and finding food, shelter, and mates—and it simply is not possible to be perfect at everything. Indeed, adaptations often are compromises between the many competing demands the organism faces.

Macroevolution

Large-scale evolutionary change is macroevolution. Whereas microevolution involves changes in the frequencies of alleles within populations, macroevolution produces changes in groups of species, such as might occur with major changes in climate. Our discussion of macroevolution begins with a description of how species are named. We then consider how their evolutionary histories can be analyzed and diagrammed.

Scientific names Systematic biology deals with the naming, classification, and evolutionary relationships of organisms. A universal system for naming and classifying organisms is essential for communicating information about them. Scientists use the naming system that Swedish naturalist Carl Linnaeus developed more than 200 years ago. Each organism is given a Latin binomial, or two-part name, consisting of the genus name followed by the specific epithet (the term *specific*, here, means "relating to species"). For example, humans are in the genus *Homo*, and our specific epithet is *sapiens*, so our binomial is *Homo sapiens*. By convention, the genus name and specific epithet are italicized, the first letter of the genus name is always capitalized, and the specific epithet is all lowercase. Sometimes, the genus name is abbreviated: *H. sapiens*.

Linnaeus also developed a system for classifying organisms using a series of increasingly broad categories: species, genus, family, order, class, phylum, and kingdom. Similar species were placed in the same genus; similar genera (the plural of *genus*) were placed in the same family; similar families in the same order, and so on. Above the kingdom level, scientists have added the

FIGURE 22.4 *Fitness is the number of offspring left by an individual. These parents clearly have high fitness.*

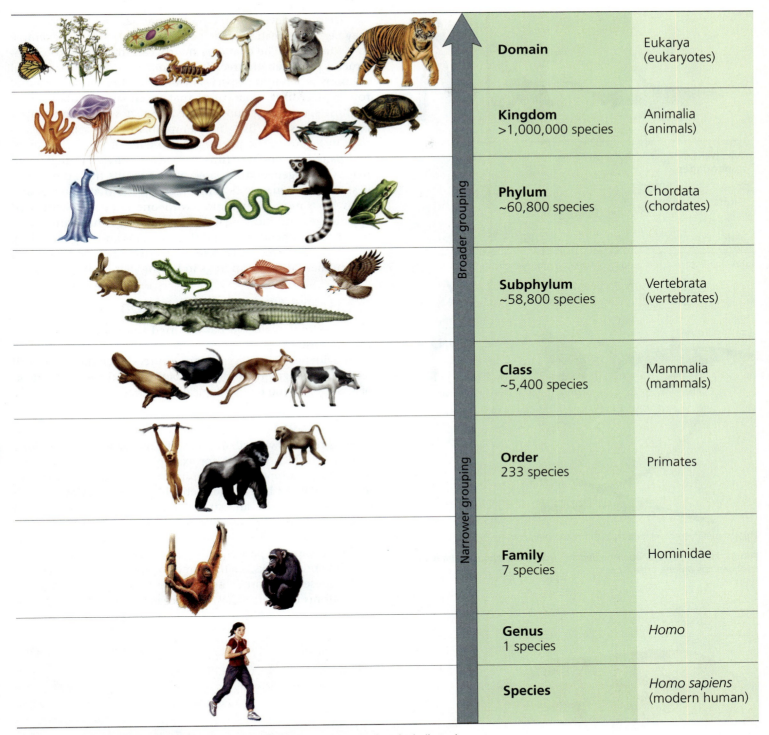

FIGURE 22.5 *Categories in the classification of living organisms. Any organism, including a human being, can be classified using a hierarchy of increasingly general categories. Modern humans are called Homo sapiens. We are placed in the family Hominidae, along with gorillas (two species), chimpanzees (two species, common chimpanzee and bonobo), and orangutans (two species). We are the only living species in our genus (Homo).*

category domain to Linnaeus's scheme (the three domains are Archaea, Bacteria, and Eukarya; see Chapter 1). Sometimes the Linnaean categories have subdivisions, for example, subphyla within phyla. Figure 22.5 presents the categories to which humans belong.

Phylogenetic trees Phylogenetic trees are branching diagrams used by scientists to depict hypotheses about evolutionary relationships among species or groups of species. Such trees can illustrate in simple graphic form concepts that are difficult to express in words.

	Shark	Frog	Human
Two paired appendages	Yes	Yes	Yes
Digits	No	Yes	Yes
Hair	No	No	Yes

(a) A character matrix is used to construct a phylogenetic tree.

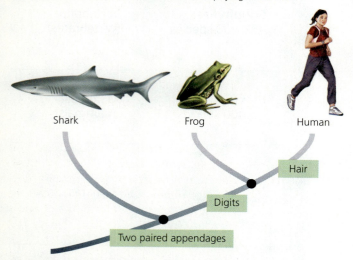

(b) A sample tree developed from the above character matrix

FIGURE 22.6 *A phylogenetic tree depicts hypotheses about evolutionary relationships among organisms.*

Scientists developing hypotheses about these relationships may begin by creating a character matrix, such as the example in Figure 22.6a. The character matrix can then be used to construct a phylogenetic tree. Typical matrices consist of vertical columns representing species or other groups, and horizontal rows representing "characters" that are either present or absent in those species or groups. Figure 22.6a shows a simple matrix comparing a shark (fish), frog (amphibian), and human (mammal) in regard to the presence or absence of three characters: two paired appendages (fins or limbs), individual digits, and hair. A phylogenetic tree based on the results from the matrix is shown in Figure 22.6b. This tree suggests that humans and frogs have more in common with one another (two paired appendages and individual digits) than either has with sharks (two paired appendages), and thus humans and frogs are likely to share a more recent common ancestor. The tree also shows that humans differ from frogs in having hair.

22.3 Evidence of Evolution

We know that evolution has occurred throughout Earth's history because the physical evidence of evolution surrounds us. Such evidence comes from many sources, including the fossil record, biogeography, molecular biology, and the comparison of anatomical and embryological structures.

Fossil Record

Earth is littered with silent relics of organisms that lived long ago (Figure 22.7). We find, for example, tiny spiders preserved in resin that dripped as sticky sap from some ancient tree. We find mineralized bones and teeth, hardened remains that tell us much about the ancestry of today's vertebrates (animals with a backbone). We also find impressions, such as footprints, of organisms that lived in the past. These preserved remnants and impressions of past organisms are **fossils**. Most fossils occur in sedimentary rocks, such as limestone, sandstone, shale, and chalk. These rocks form when sand and other particles settle to the bottoms of rivers, lakes, and oceans; accumulate in layers;

(a) A spider preserved in amber

(b) An ancient act of predation preserved in sedimentary rocks

FIGURE 22.7 *A sampling of past life in fossils*

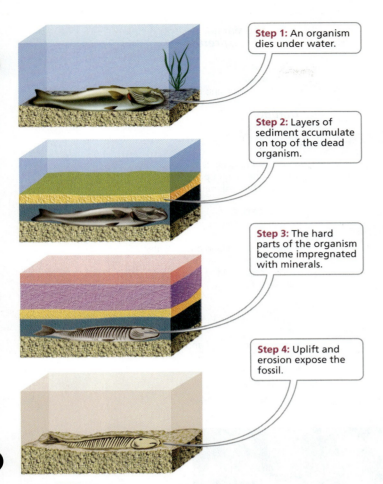

FIGURE 22.8 *A typical sequence for fossilization*

and harden. Fossils also occur in volcanic ash, tar pits, and a few other specialized conditions.

Fossilization is any process by which fossils form (Figure 22.8). In a typical case, an organism dies and settles to the bottom of a body of water. If not destroyed by scavengers, the organism is buried under accumulating layers of sediment. Soft parts usually decay. Hard parts such as bones, teeth, and shells may be preserved if they become impregnated with minerals from surrounding water and sediments. As new layers of sediment are added, the older (lower) layers solidify under the pressure generated by overlying sediments. Eventually, if geologic processes lift up the sediments and the water disappears, weather may erode the surface of the rock formation and expose the fossil.

Fossils provide strong evidence of evolution. Fossils of extinct organisms show both similarities to and differences from living species. Similarities to other fossil and modern species are used to assess degrees of evolutionary relationships. Often, fossils reveal combinations of features not seen in any living forms. Such combinations help us understand how major new adaptations arose. Sometimes we are lucky enough to find transitional forms that closely link ancient organisms to modern species. For example, several fossil whales have been discovered in the last 30 or so years; these fossils document a progression from terrestrial forms with forelimbs and hind limbs, to more aquatic forms with reduced hind limbs, to modern whales, which are fully aquatic and have no hind limbs (Figure 22.9a). Additionally, some fossil whale remains have been discovered that have an anklebone with a diagnostic shape; the bone is the astragalus, and the shape is that of a double pulley (Figure 22.9b). The double-pulley astragalus is characteristic of artiodactyls (an order of hoofed mammals that includes hippos, deer, cows, and pronghorn). Discovery of this anklebone with a double-pulley shape in fossil whales provides strong support for a close phylogenetic relationship between whales and artiodactyls.

Radiometric dating can be used to obtain estimates of the *absolute ages* of rocks and fossils. This technique relies on measuring the proportions of a radioactive isotope and its decay product. For example, radioactive potassium decays at a constant rate to form argon, so the ratio of radioactive potassium to argon in a fossil can be used to estimate the absolute age of the fossil. The *relative ages* of fossils can be determined because fossils found in deeper layers of rock are typically older than those found in layers closer to the surface. These kinds of observations enable scientists to piece together the chronological emergence of different kinds of organisms. For example, fishes are the first fossil vertebrates to appear in deep (old) layers of rock. Above those layers, fossils of amphibians appear, then reptiles, then mammals, and finally birds. This chronological sequence for the appearance of the major groups of vertebrates has been supported by other lines of evidence, some of which we consider later in this section.

Although the fossil record tells us much about past life, it has limitations. First, fossils are relatively rare. When most animals or plants die, their remains are eaten by predators or scavengers or are broken down by microorganisms, chemicals, or mechanical processes. Even if a fossil should form, the chances are small that it will be exposed by erosion or other forces and not be destroyed by those same forces before it is discovered. Second, the fossil record represents a biased sampling of past life. Aquatic plants and animals have a much higher probability of being buried in deep sediment than do terrestrial organisms. Thus, aquatic organisms are more likely to be preserved. Large animals with a hard skeleton are far more likely to be preserved than are small animals with soft parts. Organisms from large, enduring populations are more likely to be represented in the fossil record than are those from small, quickly disappearing populations. Despite these limitations, fossils document that life on Earth has not always been the same as it is today. The simple fact of these changes is potent evidence of evolution.

stop and think

Assuming equal population sizes and length of time in existence, which organism would most likely be represented in the fossil record: blue whale; elephant; earthworm? Explain your choice.

Geographic Distributions

Biogeography is the study of the geographic distribution of organisms. Geographic distributions often reflect evolutionary

(a) Several fossils reveal that whales evolved from terrestrial mammals that returned to the water.

Bowhead whale, *Balaena mysticetus*
- Extant
- 15 to 18 m long
- Fully marine
- Vestigial pelvis and hind limbs

Dorudon
- 40 million years ago
- 6 m long
- Fully marine
- Reduced hind limbs
- No connection of pelvis to vertebral column

Rodhocetus
- 46 million years ago
- 3 m long
- Spent time on land and in water
- Hind limbs
- Pelvis connected to vertebral column

Pakicetus
- 52 million years ago
- 1.8 m long
- Spent time on land and in water
- Hind limbs
- Pelvis connected to vertebral column

(b) Discovery in fossil whales of an ankle bone (astragalus) with a double-pulley shape links whales and artiodactyls (an order of hoofed mammals that includes hippos, cows, deer, and pronghorn). In the left photograph, the astragalus on the left is from the fossil whale *Rodhocetus,* whereas that on the right is from a modern-day pronghorn (shown in the right photograph).

FIGURE 22.9 The evolution of whales as revealed by transitional fossils

history and relationships because related species are more likely than are unrelated species to be found in the same geographic area. A careful comparison of the animals in a given place with those occurring elsewhere can yield clues about the relationship of the groups. If groups of animals have been separated, biogeography can tell us how long ago the separation occurred.

For example, today we find many species of marsupials—mammals such as opossums and kangaroos—in Australia but only a few in North and South America. The presence of so many species of marsupials in Australia suggests that they arose from distant ancestors whose descendants were not replaced by animals arriving from other regions. New distributions of organisms occur by two basic mechanisms. In one mechanism, the organisms disperse to new areas. In the other mechanism, the areas occupied by the organisms move or are subdivided. Australia is an island, remote from other major continental landmasses. The evolutionary history of the Australian marsupials involves both dispersal and the movement of continents. Fossil evidence suggests that marsupials evolved in China, and some later dispersed into North America. From North America, some marsupials dispersed southward into South America, then to Antarctica, and later to Australia, to which Antarctica was attached at the time (Figure 22.10). As Australia and other landmasses slowly shifted to form the modern continental arrangement, the ancestors of today's marsupials were carried away from their place of origin to evolve in isolation in Australia.

ETHICAL ISSUE

Conducting Research on Our Relatives

Chimpanzees look and behave somewhat differently from the way we do. Still, it is hard to look into their eyes and not see something of ourselves. Chimpanzees are our closest living relatives, sharing a remarkably high percentage of our DNA sequence. Nevertheless, we use them and other nonhuman primates in invasive scientific research that might benefit us. Is this ethical?

Using nonhuman primates in research is costly and controversial. Even so, they often are preferred as subjects because they are so similar to humans. For example, human and nonhuman primates possess brains with similar organization, develop comparable plaques in their arteries, and experience many of the same changes in anatomy, physiology, and behavior with age. In some cases, Nobel Prize–winning research has resulted from the contributions of research with nonhuman primates, including development of vaccines for yellow fever (1951) and polio (1954) and insight into how visual information is processed in the brain (1981). Research with nonhuman primates has also led to significant advances in our understanding of Alzheimer's disease, AIDS, and severe acute respiratory syndrome (SARS).

The care and use of nonhuman primates (and other vertebrate animals) in research is regulated by federal agencies such as the Public Health Service and the U.S. Department of Agriculture. Animal research also is regulated at the local level. Each college, university, or research center has an Institutional Animal Care and Use Committee whose members include veterinarians, researchers, and members of the public. In addition to federal and local oversight, scientists and animal care personnel are striving to improve housing for captive nonhuman primates and to consider their psychological well-being. Even with such efforts, controversy and questions remain.

Questions to Consider

- Should we ban the use of nonhuman primates in medical research? If we do, will such a ban slow the progress in fighting diseases such as AIDS and Alzheimer's? If you or a loved one had one of these illnesses, would you feel differently about research using nonhuman primates?
- Nonhuman primates represent a fraction of the animals used in research. More than 90% of research animals are rodents, such as rats, mice, and guinea pigs. Where would you draw the line when deciding which (if any) animals are acceptable for use in research that might benefit us?

Comparative Molecular Biology

Evidence of evolution can come from molecules that are the basic building blocks of life. For example, scientists can compare the sequences of amino acids in proteins or the nucleotide sequences in DNA. As described in Chapter 21, the Human Genome Project has provided information on the location of genes along our chromosomes and the order of the base pairs that make up our chromosomes. In 2005, the chimpanzee's genome was described and compared with that of humans. Such comparison revealed that the two genomes are strikingly similar; for example, the DNA sequence that can be directly compared between chimp and human genomes is about 99% identical. The genomes of the remaining great apes were published in 2011 (orangutan) and 2012 (gorilla). The gorilla data confirm the close relationship between chimps and humans, but they also reveal that about 15% of the human genome more closely resembles that of the gorilla than that of the chimpanzee.

Because of background radiation and errors in copying DNA, single-nucleotide changes in DNA, called *point mutations,* occur constantly over evolutionary time with clocklike regularity. The rates of these changes vary from gene to gene. Once calibrated against the fossil record, these **molecular clocks** allow scientists to compare DNA sequences in two species as a way to estimate the amount of time that has passed since the two species diverged from a single common ancestor. The more different the sequences, the more time that has elapsed since their common ancestor. For example, comparison of DNA sequences tells us that humans and chimpanzees diverged from a common ancestor about 6 million years ago, making chimpanzees our closest living relatives. (See the Ethical Issue essay, *Conducting Research on Our Relatives.*)

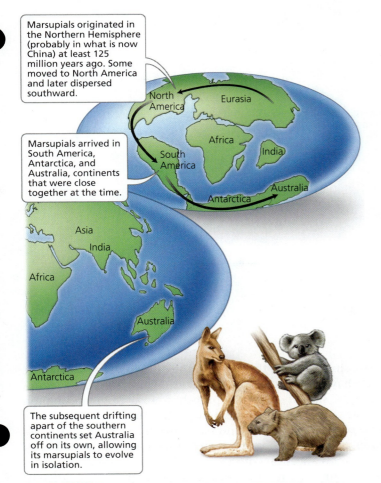

FIGURE 22.10 The story of marsupials and Australia involves both dispersal and movement of continents.

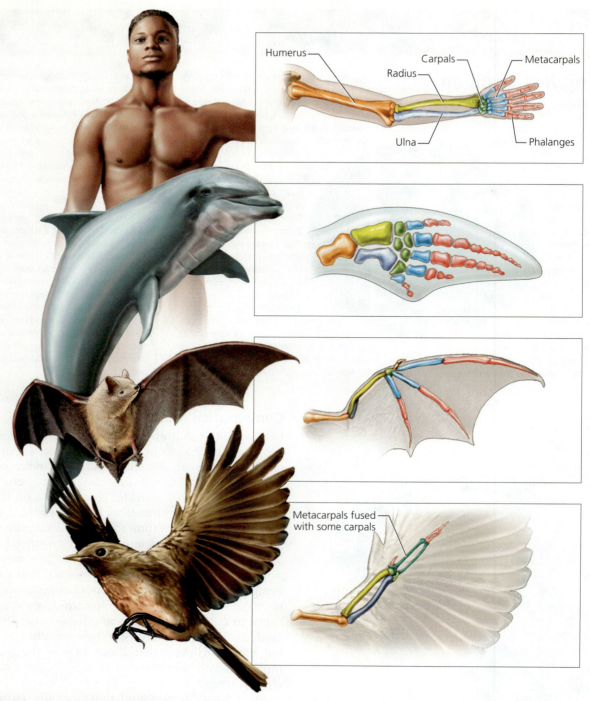

FIGURE 22.11 *Homologous structures. The similarity of the forelimb bones of humans, dolphins, bats, and birds suggests that these organisms share a common ancestry.*

Comparative Anatomy and Embryology

Comparative anatomy, as its name suggests, is the comparison of the anatomies (physical structures) of different species. Common, or shared, traits among different species have long been considered a measure of relatedness. Put simply, two species with more shared traits are considered more closely related than are two species without shared traits (this principle underlies the use of character matrices to construct phylogenetic trees). For example, many very different vertebrates share similarities in the bones of their forelimbs, showing that they have an ancestor in common (Figure 22.11). Structures that are similar and that probably arose from a common ancestry are called **homologous structures**. The forelimbs that support bird wings and bat wings are homologous structures. Sometimes, however, similarities are not inherited from a common ancestor. For example, bird wings and insect wings both permit flight, but they are made from entirely different structures. Whereas bird wings consist of forelimbs, insect

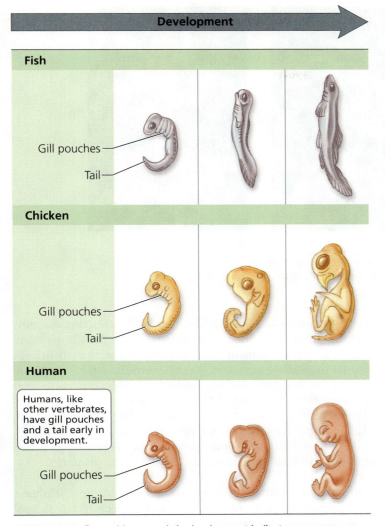

FIGURE 22.12 Resemblance early in development indicates common descent. Embryos in the different representative vertebrates are not to scale; they have been drawn to the same approximate size to permit comparison.

22.4 Human Evolution

We have learned about microevolution, macroevolution, and the evidence for evolution. Now let's look at our own past and see how humans evolved.

We begin our discussion of human evolution with the primates, an order of mammals that includes humans, apes, monkeys, and related forms (such as lemurs). Paleontologists (scientists who study fossils) believe that primates evolved about 65 million years ago. This estimate takes into account that the oldest primate fossils are about 55 million years old, and, as mentioned previously, organisms are assumed to evolve earlier than when they first appear in the fossil record. In contrast, molecular biologists have suggested that primates evolved about 90 million years ago. Their estimate is based on comparison of DNA sequences. Despite debate over when primates evolved, most scientists agree that the first primates probably arose from an arboreal (tree-living) mammal that ate insects. This ancestor to primates might have looked something like a modern tree shrew (Figure 22.13).

Primate Characteristics

Primates have several distinguishing characteristics. Many of these characteristics reflect an arboreal lifestyle specialized for the visual hunting and manual capture of insects. For example, primates have flexible, rotating shoulder joints and exceptionally mobile digits with sensitive pads on their ends. Flattened nails replace claws. In many primate species, the big toe is separated from the other toes, and thumbs are opposable to other fingers. Thus, primates have grasping feet *and* hands, features that help in the pursuit and capture of insects along branches. In addition, a complex visual system (forward-facing eyes with stereoscopic vision) and a large brain relative to body size provide the well-developed depth perception, hand–eye coordination, and neuromuscular control needed by arboreal insectivores. The relatively large brain is also associated with

wings are not true appendages; they are extensions of the insect's cuticle (exoskeleton). Thus, the wings of birds and insects do not reflect common ancestry. Instead, birds and insects independently evolved wings because of similar ecological roles and selection pressures, in a process known as **convergent evolution**. Structures that are similar because of convergent evolution are called **analogous structures**.

Homologous structures arise from the same kind of embryonic tissue. Hence, comparative embryology, the comparative study of early development, also can be a useful tool for studying evolution. Common embryological origins can be considered evidence of common descent. For example, 4-week-old human embryos closely resemble embryos of other vertebrates, including fish. Indeed, human embryos at 4 weeks' gestation come complete with a tail and gill pouches, as shown in Figure 22.12. As development proceeds, the gill pouches of fish develop into gills. The gill pouches of humans develop into other structures, such as the auditory tubes connecting the middle ear and throat. Nevertheless, the fact that human, fish, and all other vertebrate embryos look very similar at early stages of development indicates common descent from an ancient ancestor.

FIGURE 22.13 A tree shrew. These modern animals resemble the arboreal, insect-eating mammals from which primates probably evolved more than 65 million years ago.

(a) Ring-tailed lemur   **(b)** Slender loris   **(c)** Potto

FIGURE 22.14 *Examples of modern primates from the suborder whose members retain ancestral features. The female loris with young in part (b) illustrates many of the features characteristic of all primates, including grasping hands and feet, forward-facing eyes, and small litter size.*

complex social behavior (for example, members of social groups may form long-term alliances) and reliance on learned behavior (for example, tool use may be passed from one individual to the next through observation and imitation). Also, most primates give birth to only one infant at a time and provide extensive parental care—these characteristics may reflect the difficulty of carrying and rearing multiple infants in trees.

Modern primates are divided into two main groups (suborders). One suborder contains lemurs, lorises, and pottos, grouped together because they retain ancestral primate features such as small body size and nocturnal habits (Figure 22.14). The other suborder contains monkeys, apes, and humans (Figure 22.15).

In the past, the term *hominid* was used to describe members of the human lineage, such as species in the genera *Australopithecus* and *Homo* (discussed later in this chapter). At that time, human and prehuman species were the only members of the family Hominidae. Primate classification has changed, however, and apes now are included in the family Hominidae (refer again to Figure 22.5). Thus, the term *hominid* now includes apes and humans (members of the family Hominidae). The term **hominin** now is used for the human lineage and its immediate ancestors (members of the subfamily Homininae). As described earlier, molecular evidence suggests that the lines leading to modern humans and chimpanzees diverged about 6 million years ago. Molecular data further indicate that, after chimpanzees, gorillas are our next closest living relatives, followed by orangutans and then gibbons. Figure 22.16 shows these hypothesized relationships among living primates.

Comparison of human and chimp skeletal anatomy Humans are the most terrestrial of primates. Many aspects of our skeletal anatomy reflect this terrestrial lifestyle and our upright stance while walking. Walking on two feet is called **bipedalism**. Our S-shaped spine and relatively large patella (kneecap) reflect a bipedal gait. Although chimpanzees can walk on two feet, they typically use quadrupedal knuckle-walking when moving on the

(a) Black-headed spider monkey   **(b)** Chacma baboons   **(c)** Bornean orangutans   **(d)** Western gorilla

FIGURE 22.15 *Monkeys (such as spider monkeys and baboons) and apes (such as orangutans and gorillas) are placed with humans in another suborder.*

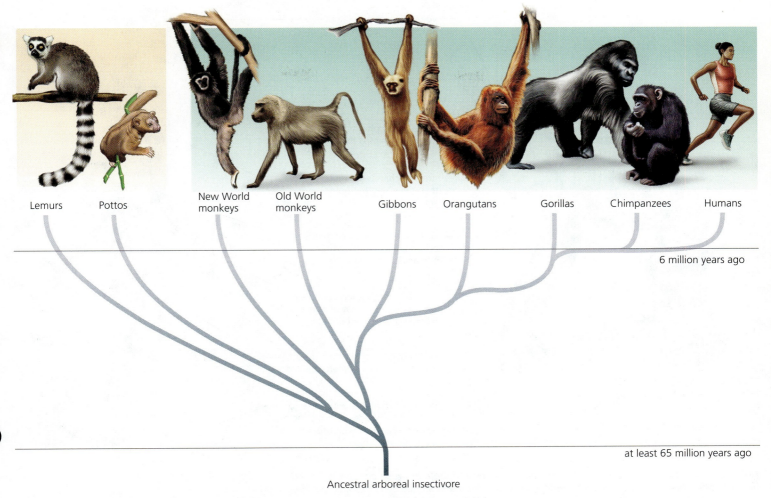

FIGURE 22.16 *Hypothesized relationships among living primates (Figures are not drawn to scale.)*

ground. Their hand bones are more robust than ours because they bear weight during terrestrial locomotion. Chimps also spend time in trees, where they use their arms in locomotion. Thus, the arms of chimps are long and exceptionally strong; the strength is reflected in the extensive areas on the scapula (shoulder blade) for the attachment of large arm muscles. Unlike humans, chimps have opposable big toes, another adaptation for climbing. Anatomical differences between chimpanzees and humans also are apparent in the skull and teeth. Chimps have a smaller braincase than humans, but more pronounced brow ridges and jutting jaws. The degree of *sexual dimorphism* (difference in appearance between the sexes) in canine teeth also is more pronounced in chimps than in humans; specifically, the canines are larger in males than in females in both species, but the sex difference is greater in chimps. Figure 22.17 shows some of the major differences between chimps and humans in their teeth and skeleton.

stop and think

In the following discussion of our ancestry, we describe the species that led to us. At some point the line *became* us. Before you start reading about our ancestry, what qualities would you now say are necessary for a species to qualify as human?

Misconceptions

Several popular misconceptions about human evolution exist among nonscientists. One misconception is the idea that we descended from chimpanzees or any of the other modern apes. Humans and chimpanzees represent separate phylogenetic branches that diverged about 6 million years ago. Thus the common ancestor of humans and chimpanzees was different from any modern species of ape.

Another misconception is that modern humans evolved in an orderly, stepwise fashion. We often see such a stepwise progression depicted in drawings, sometimes humorously, and its appeal lies in its simplicity. However, as is so often the case, the real story is far more complex. The path to modern humans has been fraught with unsuccessful phenotypes leading to dead end after dead end. In fact, the path looks more like a family "bush" than an orderly progression from primitive to modern.

A final misconception is that, over the course of human evolution, the various bones and organ systems evolved simultaneously and at the same rate. They did not. There is no reason to believe that the human brain evolved at the same rate as, say, the appendix or the foot. Instead, different traits evolved at different times and rates, by a phenomenon known as **mosaic evolution**.

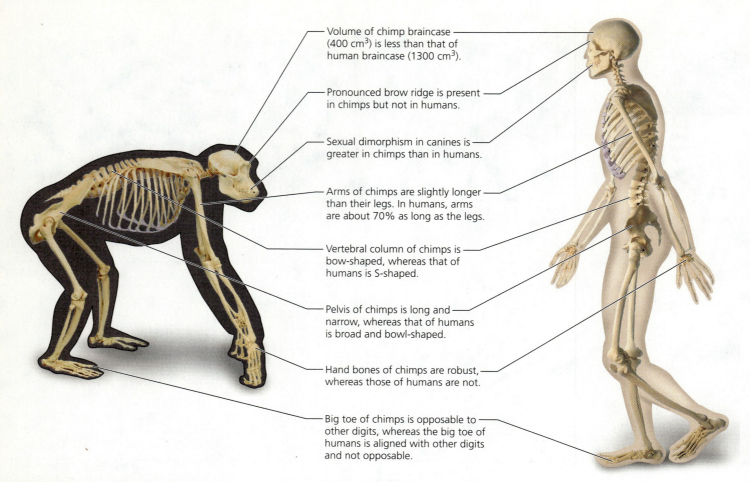

FIGURE 22.17 Some major differences in skeletal anatomy and teeth between chimpanzees and humans. Many of these traits reflect differences in locomotion and stance. Whereas chimpanzees are quadrupedal knuckle-walkers, humans are bipedal with an upright stance.

Trends in Hominin Evolution

Several evolutionary trends are apparent in the history of hominins. Bipedalism evolved early and probably set the stage for the evolution of other characteristics, such as increases in brain size. Cultural developments, such as tool use and language, are linked to increases in brain size. Once the hands were freed from the requirements of locomotion, they could be used for tasks such as making tools. Evidence that bipedalism preceded increases in brain size and cultural developments comes from fossilized hominin footprints found in Tanzania, Africa. These footprints, estimated to be about 3.6 million years old, were apparently made by two adults and a child (Figure 22.18). The footprints clearly predate the oldest stone tools from 2.6 million years ago. Other changes associated with upright posture include the S-shaped curvature of the vertebral column (the lumbar curve); modifications to the bones and muscles of the pelvis, legs, and feet; and positioning of the skull on top of the vertebral column (refer again to the human skeleton in Figure 22.17).

The faces of hominins also changed. For example, the forehead changed from sloping to vertical, and sites of muscle attachment, such as the brow ridges and crests on the skull, became smaller. The jaws became shorter, and the nose and chin more prominent. The overall size difference between males and females decreased. Males of our early ancestors appear to have been 1.5 times as heavy as females. Modern human males weigh about 1.2 times as much as females.

In the following discussion of our ancestors, we focus on the hominins that we know the most about—those in the genus *Australopithecus* and the genus *Homo*. We also mention some recent finds of apparently older hominins. Keep in mind that as new hominin fossils are found and genetic studies conducted, the dates for the origin of some species may be pushed back.

Australopithecines The first hominin remains discovered were given the genus name *Australopithecus*, meaning "southern ape." Species within *Australopithecus* are sometimes collectively termed australopithecines. *Australopithecus anamensis*, considered the earliest species of the australopithecine line, is known from a small number of fossils found in Kenya and Ethiopia and dated between 4.2 and 3.9 million years old. The most spectacular australopithecine fossil found to date is that of a young adult female of the species *Australopithecus afarensis* (because she was found in the Afar region of Ethiopia). The scientists who discovered her in 1974 named her Lucy. More than 60 pieces of Lucy's bones were found; when the bones were arranged, scientists estimated that, at death, she was about 1 m (3 ft) tall and weighed about 30 kg (66 lb; Figure 22.19). Her bones were determined to

FIGURE 22.18 These hominin footprints from Laetoli, Tanzania, predate the oldest known tools and thus provide evidence that bipedalism preceded increases in brain size and cultural trends such as toolmaking. The larger prints were made by two individuals, one following in the other's footsteps. The smaller prints may have been made by a child walking with the two individuals.

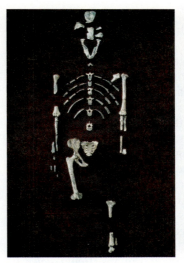

(a) The remarkably complete skeleton of Lucy

(b) Reconstruction of Lucy on display at the St. Louis Zoo

FIGURE 22.19 *Fossilized remains and reconstruction of Lucy, a young female of the hominin species Australopithecus afarensis. She was named Lucy because the Beatles' song "Lucy in the Sky with Diamonds" was playing the night Donald Johanson and his coworkers celebrated her discovery.*

be 3.2 million years old. As more remains were found, it became apparent that the males of Lucy's species were somewhat taller (about 1.5 m, or about 5 ft) and heavier (about 45 kg, or 99 lb). The brain of *A. afarensis* was similar in size to that of modern chimpanzees or gorillas—about 430 cm³ (26 in.³). Although many aspects of the anatomy of *A. afarensis* suggest adaptations for living in trees, the remains also indicate bipedalism. In *A. afarensis*, we see an example of mosaic evolution—bipedalism evolving before substantial increases in brain size.

In 1994, researchers found hominin remains in Ethiopia that were older than those of *A. anamensis*. Dated at 4.4 million years old, the fossils were assigned to the species *Ardipithecus ramidus*. The remains, which included parts from more than 30 individuals, took 15 years to fully excavate and analyze. The most striking find was a partial skeleton (125 pieces) of a female now called Ardi. Eleven papers published in 2009 detailed the anatomy of *A. ramidus* and its implications for human evolution. Because the shape of Ardi's pelvis indicated that it was good for both climbing and upright walking, the research team suggested that *A. ramidus* displayed facultative bipedalism. In other words, when moving along tree branches Ardi may have walked upright atop the branches as well as climbed with all four limbs. They further propose that she walked upright on the ground, but not as well as later hominins. It is possible that *Ardipithecus* gave rise to the genus *Australopithecus*, which most scientists think led to our own genus, the genus *Homo*. It also is possible that *Ardipithecus* was a side branch, not along the path that led to us.

About 3 million years ago, when *A. afarensis* had been in existence for nearly 1 million years, several new hominin species appeared in the fossil record. Scientists believe that *Australopithecus africanus*, one of these new species, was a hunting-and-gathering omnivore. *A. africanus*, like *A. afarensis*, was a gracile (or slender) hominin. Three more "robust" hominins (previously within *Australopithecus*, but now placed in a separate genus, *Paranthropus*) also appeared and are thought to have been savanna-dwelling vegetarians. The robust hominins had massive skulls, heavy facial bones, pronounced brows, and huge teeth. Whatever they ate, it required a lot of chewing. It is unclear whether Lucy's species, *A. afarensis*, gave rise to these other species or simply lived at the same time as they did. Whereas the robust hominins appear to have been evolutionary dead ends, descendants of *A. afarensis* may have led to genus *Homo*.

Homo habilis *Homo habilis* ("handy man"), the first member of the modern genus of humans, appeared in the fossil record about 2.5 million years ago. The remains classified as *H. habilis* are highly variable, causing some researchers to question their classification. Some researchers believe the remains are varied enough to represent more than one species. *H. habilis* differed from *A. afarensis* primarily in brain size. The cranial capacity of *H. habilis* has been estimated at between 500 and 800 cm³ (30 and 49 in.³). Some scientists hypothesize that *H. habilis* was the first hominin to use stone tools. Simple stone tools, dating from 2.5 million to 2.7 million years ago, have been found in Africa. Whether these tools were used by *H. habilis* or one of the species of robust hominins is unclear. *H. habilis* may have been capable of rudimentary speech. Casts of one brain made from reassembled skull fragments indicate a bulge in the area of the brain important in speech (see Chapter 8).

Homo ergaster and Homo erectus A new hominin, *Homo ergaster* ("working man"), appeared in the fossil record about 1.9 million years ago. The name reflects the many tools found with the remains. Traditionally, scientists had classified these remains as *Homo erectus* ("upright man"), but they now differentiate them from *H. erectus*. *H. ergaster* appears to have originated in East Africa and coexisted there for several thousand years with some of the robust hominins.

H. erectus is thought to have diverged from *H. ergaster* around 1.6 million years ago. *H. erectus* was a wanderer, believed by many to be the first hominin to migrate out of Africa, spreading to Asia. *H. erectus* was larger than earlier hominins (up to 1.85 m, or 6 ft tall, and weighing at least 65 kg, or 143 lb) and less sexually dimorphic. *H. erectus* had a brain volume of about 1000 cm^3 (61 in.^{3}). Evidence indicates that *H. erectus* used sophisticated tools and weapons and may have used fire. *H. erectus* disappeared from most locations about 400,000 years ago, but some remains from Java have been dated at only 50,000 years. The Java remains suggest that at least one population of *H. erectus* existed at the same time as modern humans (*Homo sapiens*).

Homo heidelbergensis, Homo sapiens, and Homo neanderthalensis

The origins of anatomically modern humans over the last 500,000 years are difficult to trace with certainty, and different interpretations exist. Traditionally, fossils that did not quite resemble modern humans were classified as Archaic *Homo sapiens*. Scientists now place these fossils in the species *Homo heidelbergensis;* the name refers to Heidelberg, Germany, where a fossil lower jaw intermediate to those of earlier forms and *H. sapiens* was found. Many scientists now believe that *H. heidelbergensis* evolved from *H. ergaster* and not *H. erectus*. *H. heidelbergensis* ranges from about 800,000 years ago to about 130,000 years ago, which is when the first anatomically modern human remains show up in the fossil record. Thus, scientists postulate that *H. sapiens* and *H. neanderthalensis* evolved from *H. heidelbergensis*.

The oldest fossil evidence for modern humans (*H. sapiens*, or "thinking man") comes from Africa and is about 130,000 years old. *H. sapiens* differs from earlier humans in having a larger brain (1300 cm^3, or 79 in.^{3}), flat forehead, absent or very small brow ridges, prominent chin, and a very slender body form.

Neanderthals, close evolutionary relatives of ours, are known to have been in Europe and Asia from about 200,000 years ago to 30,000 years ago. Neanderthals had distinct features apparently adapted for life in a cold climate. Some Neanderthals lived in caves (Figure 22.20a). Neanderthal burial sites have been discovered, making them the first hominins known to have buried their dead. Also, the discovery of 50,000-year-old remains of sick, injured, and elderly individuals suggests that Neanderthals cared for the less fortunate among them.

Interestingly, Neanderthals had a larger braincase than do *H. sapiens* and a slightly larger brain volume (about 1450 cm^3, or 88 in.^{3}). However, these features may not correlate with intelligence but rather with the Neanderthals' more massive body. They had larger bones, suggesting heavier musculature, and rather short legs. They also had a thick brow ridge, large nose, broad face, and well-developed incisors and canines. Some anthropologists consider Neanderthals to be a subspecies of *H. sapiens* and call them *H. sapiens neanderthalensis* (according to this scheme, modern humans are known as *H. sapiens sapiens*). Most anthropologists, however, assign Neanderthals species status and call them *Homo neanderthalensis*.

Neanderthals vanished from the fossil record some 30,000 years ago for still mysterious reasons. Some scientists suggest they were outcompeted or simply killed outright by a form of *H. sapiens* called Cro-Magnons (Figure 22.20b). Other scientists suggest that interbreeding between anatomically modern humans and Neanderthals might have resulted in the loss of the Neanderthal phenotype. Scientists now have a first version of the Neanderthal genome, developed from DNA recovered from the bones of three female Neanderthals whose remains were estimated at 38,000 years old. A comparison of the Neanderthal genome with those of present-day humans from different parts of the world suggests that following their migration

(a) The Neanderthals had an appearance adapted for life in a cold climate. Some anthropologists consider Neanderthals to be a subspecies of our own species; others consider them a separate species.

(b) The Cro-Magnons were quite similar to us in appearance, and some anthropologists believe they were responsible for the disappearance of the Neanderthals.

FIGURE 22.20 *Relatively recent representatives of the genus* Homo. *These photographs were taken at a museum display.*

TABLE 22.1 Review of Some Milestones in Human Evolution

Hominin	Years Ago	Milestone
Ardipithecus ramidus	6–4 million	Bipedalism (facultative)
Homo habilis	2.4–1.6 million	Tool use, speech
Homo erectus	1.6 million–50,000	Fire, migration
Homo neanderthalensis*	200,000–30,000	Buried their dead
Homo sapiens	130,000–present	Domestication of animals, agriculture

*Some scientists consider Neanderthals to be a subspecies of Homo sapiens and call them Homo sapiens neanderthalensis.

from Africa 50,000 to 80,000 years ago, modern humans bred with Neanderthals in the Middle East before extending their range into Eurasia. The scientists estimate that 1% to 4% of the genes of present-day non-Africans came from Neanderthals.

In 2003, hominin remains were discovered in a cave on the island of Flores in Indonesia. The remains display a peculiar mix of primitive features (for example, small brain size) and derived features (for example, small canines). Some researchers suggest the remains represent a new hominin species, *Homo floresiensis,* estimated to have lived between 95,000 and 17,000 years ago (note that this species would have overlapped in time with *H. sapiens*). Others suggest the remains are from individuals of *H. sapiens* who suffered from a pathological condition such as microcephaly (a neurodevelopmental disorder characterized by a small head) or hypothyroidism (reduced growth and development due to undersecretion of thyroid hormone). Debate continues, as does the study of stone artifacts from the cave and nearby sites.

About 12,000 years ago, *H. sapiens* changed from a nomadic lifestyle to a more sedentary one. Associated with this change were two major milestones in human history: the domestication of animals and the cultivation of crops. For example, mammals were domesticated for protection (dogs), food (cattle, pigs, and goats), transport (horses, camels, donkeys), wool (llamas and alpacas), and rodent control (cats and ferrets). Agriculture began with the cultivation of cereal grains about 9000 years ago.

Major milestones in human evolution are summarized in Table 22.1, and the major hominin species are shown in Figure 22.21.

Q Does this figure support or contradict the notion that humans evolved in an orderly, stepwise manner from primitive to modern? Explain your answer.

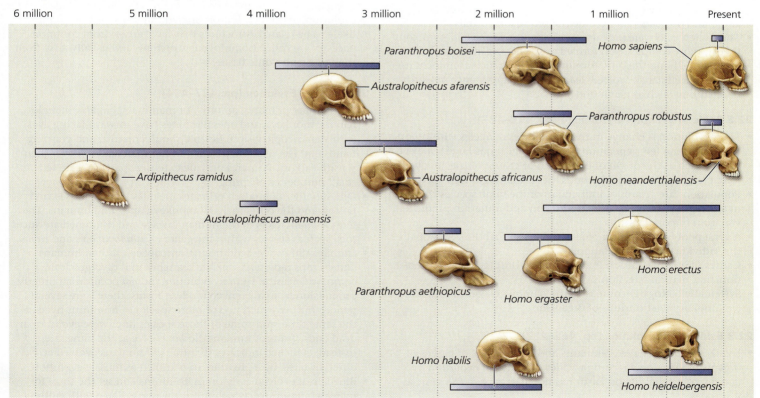

FIGURE 22.21 The major hominin species. Horizontal purple lines indicate the length of time each species existed.

A This figure contradicts the notion of an orderly progression from primitive to modern because it shows that more than one species of hominin existed simultaneously at some points in the past.

what would you do?

Human remains can tell scientists much about the diets, diseases, lifestyles, and genetic relationships of our ancestors. Such information can help piece together our evolutionary past. Sometimes, however, keeping human remains for scientific study conflicts with the wishes of modern-day descendants who wish to have their ancestors' remains returned to them for reburial. In the United States, many museums and universities are developing policies for the treatment and disposition of Native American and Native Hawaiian remains. How should human remains be treated, and who should get them? If you were a museum director developing a policy for the treatment of human remains, what would you do?

Did modern humans evolve several times in different regions? The idea that they did is known as the **multiregional hypothesis**. It suggests that modern humans evolved independently in locations such as Europe, Asia, and Africa from distinctive local populations of earlier humans. The alternative idea is known as the **Out of Africa hypothesis**. It suggests a single origin for all *H. sapiens*. According to the Out of Africa hypothesis, anatomically modern humans evolved from earlier humans in Africa and only later migrated to Europe, Asia, and other locations, where they replaced earlier human populations.

looking ahead

In Chapter 22, we learned about evolution. In Chapter 23, we focus on ecology, the study of the interactions between organisms and between organisms and their environments.

HIGHLIGHTING THE CONCEPTS

22.1 Evolution of Life on Earth (pp. 476–478)

- Earth is estimated to be 4.5 billion years old, and life is thought to have originated about 3.8 billion years ago.
- The first step in the origin of life may have been the synthesis of small organic molecules from inorganic molecules. Over time, the small organic molecules increased in complexity. A critical step was the origin of genetic material (possibly RNA and not DNA), which allowed information to be transferred from one generation to the next. Eventually, large organic molecules aggregated into droplets that became the precursors to cells.
- The earliest cells were prokaryotic. According to the endosymbiont theory, more complex cells formed as smaller organisms were incorporated into the cells, forming organelles such as mitochondria. Multicellularity evolved, leading to the appearance on Earth of organisms such as plants and animals.

22.2 Scale of Evolutionary Change (pp. 478–482)

- Microevolution is change in the frequencies of alleles within populations over a few generations. Macroevolution is large-scale evolution, such as the origin or extinction of groups of species over long periods of time. Speciation is the formation of new species.
- Populations are groups of individuals of the same species that live in a particular area. Sexual reproduction and mutation produce variation in populations.
- A gene pool is a collection of all the alleles of all the genes of all the individuals in a population.
- Causes of microevolution include genetic drift, gene flow, mutation, and natural selection.
- Systematic biology encompasses the naming, classification, and evolutionary relationships of organisms.

22.3 Evidence of Evolution (pp. 482–487)

- Fossils are the preserved remnants and impressions of past organisms. The fossil record provides evidence of evolution by documenting that life on Earth has not always been the same as it is today.
- Biogeography is the study of the geographic distribution of organisms. New distributions of organisms occur either when the organisms disperse to a new location or when the areas they occupy move or are subdivided. Related species are more likely than are unrelated species to be found in the same geographic area.
- Molecules that are the basic building blocks of life can be compared for evidence of evolution. For example, scientists compare the nucleotide sequences in DNA of different species to gauge relatedness and to estimate the time of divergence from a most recent common ancestor.
- Comparative anatomy and embryology also provide evidence of evolution. Species with more shared traits are considered more likely to be related. Structures that have arisen from common ancestry are called homologous structures and usually arise from the same embryonic tissue.

22.4 Human Evolution (pp. 487–494)

- Humans are primates, an order of mammals that also includes lemurs, monkeys, and apes. Primates have forward-facing eyes with stereoscopic vision, flexible shoulder joints, and grasping hands and feet. Flattened nails—rather than claws—cover their sensitive digits. Primates provide extensive parental care to a small number of offspring.
- One suborder of modern primates includes lemurs, lorises, and pottos, and the other includes monkeys, apes, and humans. The term *hominin* refers to the human lineage and its immediate ancestors, such as species within the genera *Australopithecus* and *Homo*.
- Humans did *not* descend from chimpanzees; rather, humans and chimps represent separate branches that diverged from a common ancestor. Human evolution did *not* occur in an orderly progression from ancient to modern forms; there were several periods during which two or more species of hominins lived at the same time, and some of those species lines were evolutionary dead ends. Traits of humans did *not* evolve at the same rate; instead, evidence indicates that traits of humans evolved at different rates, by a phenomenon known as mosaic evolution.
- Bipedalism evolved early on in hominins and set the stage for increased brain size, which in turn was associated with cultural trends such as tool use and language.
- The oldest hominin remains found to date are those of *Ardipithecus ramidus*, estimated to be at least 4.4 million years old. The remains

suggest facultative bipedalism. Other early human genera include *Australopithecus* and *Paranthropus*.
- *Homo habilis* exhibited speech and tool use. *H. erectus* used fire and was the first hominin to migrate out of Africa. *H. neanderthalensis* buried their dead. *H. sapiens* domesticated animals and cultivated crops.
- The multiregional hypothesis suggests that *H. sapiens* evolved independently in different regions from distinctive populations of early humans. The Out of Africa hypothesis suggests that modern humans evolved from early humans in Africa and then dispersed to other regions, where they replaced existing hominin species.

RECOGNIZING KEY TERMS

evolution *p. 476*
chemical evolution *p. 477*
endosymbiont theory *p. 478*
microevolution *p. 478*
macroevolution *p. 478*
population *p. 478*
gene pool *p. 478*

mutation *p. 479*
genetic drift *p. 479*
bottleneck effect *p. 479*
founder effect *p. 479*
gene flow *p. 479*
species *p. 479*
speciation *p. 479*
natural selection *p. 480*

fitness *p. 480*
adaptation *p. 480*
phylogenetic trees *p. 481*
fossil *p. 482*
biogeography *p. 483*
molecular clock *p. 485*
homologous structure *p. 486*
convergent evolution *p. 487*

analogous structure *p. 487*
hominin *p. 488*
bipedalism *p. 488*
mosaic evolution *p. 489*
multiregional hypothesis *p. 494*
Out of Africa hypothesis *p. 494*

REVIEWING THE CONCEPTS

1. How might life have evolved from inorganic molecules to complex cells? *pp. 477–478*
2. Distinguish microevolution from macroevolution. *p. 478*
3. What are four sources of variation within populations? *pp. 479–480*
4. How does genetic drift lead to microevolution? *p. 479*
5. Define *speciation*, and relate it to gene flow. *p. 479*
6. Define *natural selection*. How can variation within populations be maintained in the face of natural selection? *pp. 479–480*
7. Describe the binomial system by which organisms are named. *p. 480*
8. What is a phylogenetic tree? *pp. 481–482*
9. What is a fossil? Describe the process of fossilization, and relate it to limitations of the fossil record. *pp. 482–483*
10. How do new distributions of organisms arise? *pp. 483–485*
11. What is a molecular clock? *p. 485*
12. Distinguish homologous structures from analogous structures. *pp. 486–487*
13. Describe how comparative embryology provides evidence of evolution. *p. 487*
14. Which of the following is *not* characteristic of primates?
    a. high parental investment in small number of young
    b. forward-facing eyes
    c. claws
    d. grasping feet and hands
15. Choose the *incorrect* statement.
    a. Humans descended from chimpanzees.
    b. Whereas chimps are primarily quadrupedal knuckle-walkers, humans have a bipedal gait.
    c. Chimps are more arboreal than humans, and this is reflected in their longer and stronger arms.
    d. Chimps, but not humans, have opposable big toes.
16. The Out of Africa hypothesis suggests
    a. multiple origins for modern humans.
    b. that Cro-Magnons outcompeted Neanderthals and forced them out of Africa.
    c. that modern humans evolved in an orderly, stepwise fashion.
    d. a single origin for all *H. sapiens*.
17. Which of the following is *not* a trend in human evolution?
    a. increases in brain volume
    b. decreased sexual dimorphism in body size
    c. more prominent brow ridges and crests on skull
    d. change from arboreal to terrestrial existence
18. Which of the following does *not* produce variation in populations?
    a. bottleneck effect
    b. mutation
    c. crossing over
    d. independent assortment
19. Which of the following occurs when fertile individuals move into and out of populations?
    a. genetic drift
    b. speciation
    c. mutation
    d. gene flow
20. Which of the following was in very short supply in the environment of the early Earth?
    a. lightning
    b. volcanoes
    c. gaseous oxygen (O_2)
    d. UV radiation
21. The earliest cells
    a. were prokaryotic.
    b. were eukaryotic.
    c. evolved over 10 billion years ago.
    d. were part of multicellular organisms.
22. Which of the following primates is our closest living relative?
    a. gorilla
    b. orangutan
    c. chimpanzee
    d. gibbon
23. Which of the following features of an organism would promote fossilization?
    a. terrestrial existence
    b. member of a small population
    c. soft body parts
    d. aquatic existence

24. Which hominin was probably the first to migrate out of Africa?
    a. *Australopithecus afarensis*
    b. *Homo sapiens*
    c. *Homo erectus*
    d. *Australopithecus africanus*
25. Where did the major events of human evolution occur?
    a. Africa
    b. South America
    c. Europe
    d. North America
26. Which of the following hominin characteristics evolved the earliest?
    a. bipedalism
    b. large brain
    c. language
    d. use of fire

APPLYING THE CONCEPTS

1. A friend of yours believes that all organisms were specially created. You believe in evolution and want to present your case to your friend. What would you say?
2. Can life arise on Earth from inorganic material today? Why or why not?
3. You are a biologist exploring the dense tropical rain forests of Brazil. You notice an unfamiliar, medium-sized mammal moving above you in the trees. What characteristics should this animal display to be classified as a primate?
4. Tay-Sachs disease has zero fitness because it causes death before the individual reaches reproductive age. How does a trait with zero fitness persist in a population?

BECOMING INFORMATION LITERATE

Develop a PowerPoint presentation on tool use in living vertebrates. Include the following in your presentation: a definition of tool use, the species in which tool use has been reported, the type of tool use described and the conditions under which it was observed, and any anatomical or behavioral traits associated with species that make and use tools. Use at least three sources; list these sources in your last slide, and briefly explain why you found them to be reliable.

MasteringBiology®

Go to MasteringBiology for practice quizzes, activities, eText, videos, current events, and more.

Ecology, the Environment, and Us

23

Did you Know?

- Your body and about 10,000 species of microbes make up a human ecosystem.
- Every second, 1.5 acres of forest is cut down.

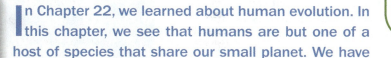

The best legacy we can give our children is a world in ecological balance.

- 23.1 Earth as an Ecosystem
- 23.2 Biosphere
- 23.3 Ecological Succession
- 23.4 Energy Flow
- 23.5 Chemical Cycles
- 23.6 Biodiversity

ETHICAL ISSUE
Maintaining Our Remaining Biodiversity

In Chapter 22, we learned about human evolution. In this chapter, we see that humans are but one of a host of species that share our small planet. We have many of the same needs as other species, and we face many of the same threats. If we are ever to understand the world around us and our place in it, we must have some knowledge of its physical characteristics and an understanding of our dependence on the other species we share it with.

23.1 Earth as an Ecosystem

Our focus in this chapter is ecology, a science whose name comes from the Greek words *oikos,* meaning "home," and *logos,* meaning "to study." Thus, ecology is the study of our home—Earth—encompassing both its living (biotic) components and nonliving (abiotic) components. More precisely, ecology is the study of the *interactions* between organisms and between organisms and the environment. Ecologists are the scientists who study these interactions.

If we could see Earth (Figure 23.1) from space, as an astronaut does, we would make two important observations about our home. First, we would see that Earth is isolated from other planets. So we could conclude that, aside from an occasional meteorite and other bits of debris from space, our planet has no source of new materials. In fact, many of the materials that came together to form it some 4.5 billion years ago cycle repeatedly from organism to organism and between the living and nonliving components of our world. A carbon atom in your body may once have been part of a dinosaur or part of Aristotle. Also, from our vantage point in space, we would be reminded by the light reflected from Earth's surface that this planet receives one very important contribution from without—energy, in the form of sunlight. As we will see, this energy is captured by green plants and transferred from organism to organism to sustain nearly all life on Earth.

FIGURE 23.1 *A view of Earth from space shows us that the planet is isolated. Because there is no regular input of materials, important elements must be cycled from organism to organism and between Earth's living and nonliving components. The only input to the system is energy from the sun, which sustains nearly all life on Earth.*

23.2 Biosphere

The part of the Earth where life exists is the **biosphere**. The biosphere includes many **ecosystems**, each made up of the organisms in a specific geographic area and their physical environment. All the living species in an ecosystem that can potentially interact form a **community**. A **population** is all the individuals of the *same species* that can potentially interact. Thus, populations of different species form a community.

When ecologists consider an individual species, they describe it according to its niche (sometimes called *ecological niche*). The **niche** is the organism's role in the ecosystem, defined by all the physical, chemical, and biological factors that keep the organism healthy and allow it to reproduce. Such factors include the nature of the organism's food and how it obtains this food; its predators; and its specific needs for shelter, as well as the temperature, light, water, and oxygen the organism requires to survive. Thus, the organism's **habitat**—the physical place where it lives—is also a part of its niche. For example, the habitat of an African lion is semi-open plains. Its niche includes the habitat, the times when it is active (any time of day), how it obtains food (predation on large animals), how it relates to other animals (lives in a social group called a pride; marks territory by roaring, urinating, and patrolling), and how it reproduces (gives birth to litters of one to four cubs after 110 days of pregnancy on average).

23.3 Ecological Succession

Ecosystems change over long periods. In fact, everything on Earth is constantly changing. The sequence of changes in the kinds of species making up a community is called ecological **succession**. There are two types of ecological succession: primary and secondary.

Primary succession occurs where no community previously existed (Figure 23.2). Such places may be found on

This thin mat of lichen is helping break down the bare rock, beginning soil formation.

After soil has begun to accumulate, plant species appear that often include shrubs and dwarf trees.

Trees often later become the dominant plant form, depending on elevation, annual rainfall, and average temperature.

FIGURE 23.2 *Primary succession is the sequence of changes in the species composition of a community that begins where no life previously existed.*

Temperate deciduous forest. These forests receive 75 to 125 cm (30 to 50 in.) of rainfall per year. Summers are hot, and winters are cold. Trees lose their leaves in the winter to avoid water loss when it is too cold to photosynthsize. Insects, mice, squirrels, and many species of birds are common in these forests.

Temperate grasslands. These grasslands receive 25 to 75 cm (10 to 30 in.) of rainfall per year. Long dry periods and fire are important factors in maintaining grasslands. Grazing animals, such as antelope, and burrowing animals, such as prairie dogs, are common.

Desert. Lack of water defines the desert community. Deserts receive less than 25 cm (10 in.) of rain each year. Most deserts are hot, but some are cold. Both plants and animals must be able to conserve water. Many desert plants are succulents with leaves that retain and store water. Animals may tend to avoid the sun by foraging at night.

Taiga. The taiga is composed of evergreen forests with variable rainfall of 50 to 100 cm (20 to 40 in.) per year. Winters are long and cold, and summers are short. The needles on evergreen trees help save water by providing little surface through which water can leave. Animals such as the grizzly bear, moose, wolf, and snowshoe hare are common.

Tropical rain forest. Tropical rain forests may receive 200 to 1000 cm (80 to 400 in.) of rain each year. It is hot throughout the year. Tropical rain forests have a tremendous diversity of life.

FIGURE 23.3 *Selected climax communities of Earth. Environmental conditions, particularly the temperature and the availability of water, affect the distribution of organisms. Each species is adapted to certain conditions.*

rocky outcroppings, or where lava has solidified into a new surface, or where land appears as a glacier recedes or ice caps melt. At the start of primary succession, no soil exists.

The first living things to invade such an area are called *pioneer species*. Among the most prominent of these are lichens, which are actually two species—a fungus and a photosynthetic organism, usually an alga—combined in a mutually beneficial relationship. The fungus provides the attachment to the barren surface, retains water, and releases minerals from the rock to be used by the alga; the alga provides the food. Lichens secrete acid, which helps break down the rock, beginning soil formation. In time, the lichens die and their remains mix with the rock particles, furthering soil formation in cracks and crevices.

Soil building is an extremely slow process, but once soil is in place, the rate of succession speeds up. Plants begin to appear. Their roots push into every crevice, and their leaves fall and decompose, accelerating the pace of soil building and new plant colonization. In some areas, trees eventually become the dominant plant form.

Each species that moves into the area changes the environmental conditions slightly, thus changing the available resources in ways that favor some species and hamper others. The community that eventually forms and remains, if no disturbances occur, is called the *climax community* (Figure 23.3). The nature of a climax community is determined by many factors, including temperature, rainfall, nutrient availability, and exposure to sun and wind.

When an existing ecological community is cleared away—either by natural means or by human activity—and is then left alone by humans, it undergoes a sequence of changes in species composition known as *secondary succession*. Secondary succession occurs in areas where soil is already in place. Such areas include old, deserted fields and farms (originally cleared and planted by humans but later abandoned) or areas damaged by catastrophic fire, flood, wind, or overgrazing (Figure 23.4). The initial invaders in secondary succession are likely to be grasses, weeds, and shrubs, but these are gradually outcompeted as other

Immediately after a fire there are no visible signs of life. Dead trees stand as ghostly reminders of the forest that had existed, and gray ash covers the forest floor.

The following spring young plants appear and begin the stages of secondary succession.

FIGURE 23.4 *Secondary succession following the 1988 fires in Yellowstone National Park.*

plants move in from the surrounding community. The area finally "heals," and the community that forms often resembles the one that existed before the disturbance.

23.4 Energy Flow

Virtually all the energy that propels life on this small planet comes from the sun. Only a small fraction of the sun's energy that reaches Earth's surface worldwide is captured and used by living organisms. Even so, the life that abounds on Earth owes its existence to that captured energy, which is shifted, shuffled, channeled, and scattered through various systems from one level to the next.

Food Chains and Food Webs

The flow of energy through the living world begins when light from the sun is absorbed by photosynthetic organisms, such as plants, algae, and cyanobacteria (a group of photosynthetic bacteria that usually live in water). *Photosynthesis* is a chemical process that essentially captures light energy and transforms it into the chemical energy of the sugar glucose, which is manufactured from carbon dioxide and water. The energy stored in the glucose molecules and oxygen released in the process of photosynthesis sustain nearly all life on Earth. The photosynthesizers themselves use some of the energy stored in these glucose molecules to fuel their own metabolic activities. Any remaining energy can be used for growth and reproduction. Once the photosynthesizer uses the energy in glucose to make new organic molecules, these molecules may become food for an animal.

The photosynthesizers, called the **producers**, form the lowest **trophic level** (*trophic* means "feeding"). All other organisms are **consumers** and use the energy that producers store (Figure 23.5).

Consumers belong to higher trophic levels and are grouped on the basis of their food source.

- **Herbivores**, or **primary consumers**, eat plants.
- **Carnivores**, or **secondary consumers**, feed on herbivores. Some carnivores eat other carnivores, forming still higher trophic levels—tertiary and quaternary consumers.
- **Omnivores** eat both plants and animals.
- **Decomposers**, such as bacteria, fungi, and worms, consume dead organic material for energy and release inorganic material that can then be reused by producers.

At one time, the pattern of feeding relationships responsible for the flow of energy through an ecological system was described as a **food chain**—a linear sequence in which A eats B, which then eats C, which then eats D, and so on. However, now we know that the chain analogy is too simplistic because many organisms eat at several trophic levels. To illustrate, consider the number of trophic levels on which humans feed. Whereas eating chickens—which are primary consumers—makes us secondary consumers, eating tuna (a predatory fish) makes us tertiary consumers (or quaternary consumers if the tuna has eaten another predatory fish). But we also can be considered primary consumers because we eat vegetables (Figure 23.6). More realistic patterns of feeding relationships, consisting of many interconnected food chains, are described as **food webs**. Figure 23.7 illustrates part of a food web in a community where land meets water.

FIGURE 23.5 *Trophic levels. The width of the arrow indicates the relative amount of energy transferred.*

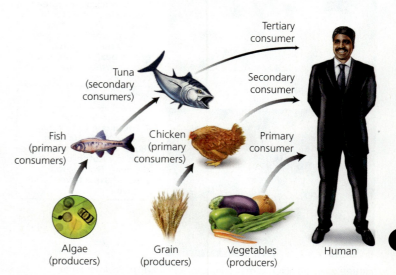

FIGURE 23.6 *Humans eat at several trophic levels. The width of the arrow indicates the relative amount of energy transferred.*

23.4 Energy Flow **501**

Q On which trophic level would the hawk be feeding if it ate the duck?

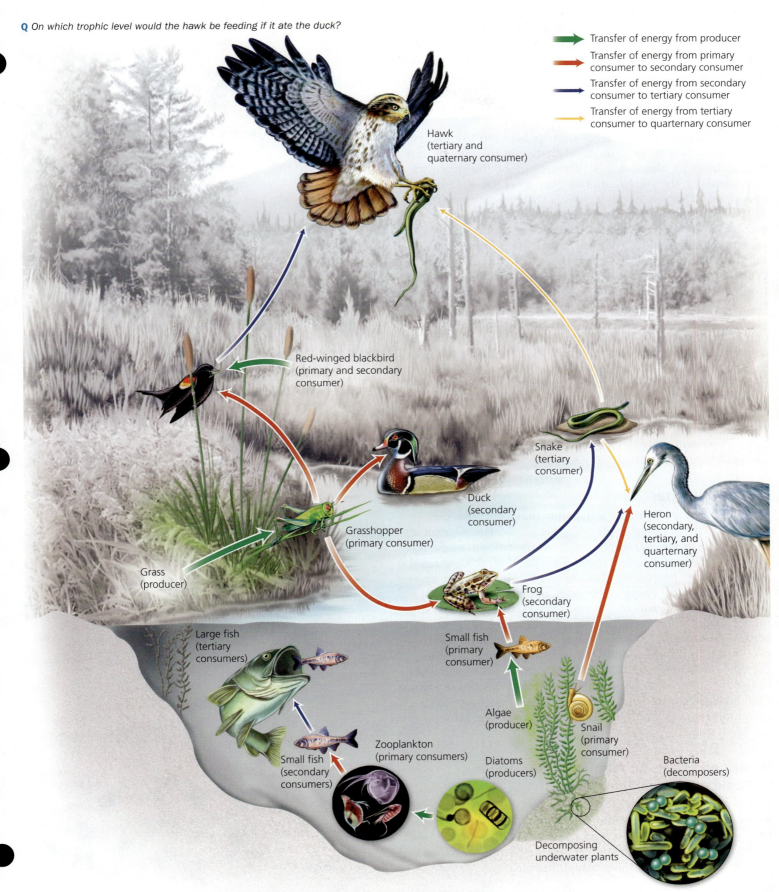

FIGURE 23.7 In this simplified food web, the arrows show the direction of flow of energy in the form of food. The width of the arrow indicates the relative amount of energy transferred.

A It would be a tertiary consumer.

Energy Transfer through Trophic Levels

Energy is lost as it is transferred from one trophic level to the next. Although the efficiency of transfer can vary greatly depending on the organisms in the system, on average only 10% of the energy available at one trophic level is transferred to the next higher level. As a result, ecosystems rarely have more than four or five trophic levels.

As we consider why energy transfer between trophic levels is so inefficient, keep in mind that only the energy that is converted to **biomass** (the dry weight of the organism) is available to the next higher trophic level. One reason for energy loss between trophic levels is that roughly two-thirds of the energy in the food that is digested is used by the animal for cellular respiration. In addition, an animal must first expend energy to obtain its food, usually by grazing or hunting (for example, the chicken in Figure 23.8 must do a lot of pecking). Furthermore, not all the food available at a given trophic level is captured and consumed. In addition, some of the food eaten cannot be digested and is lost as feces. The energy in the indigestible material is unavailable to the next higher trophic level. However, the remaining energy can be converted to biomass and will be available to the next higher trophic level. These losses become multiplied as the energy is transferred to successively higher trophic levels, until there simply is not enough food or energy left to feed another level. To put this another way, for every 100 calories of solar energy captured by producers, 10 calories are transferred to herbivores, and only 1 calorie is transferred to carnivores (secondary consumers).

Ecological Pyramids

An **ecological pyramid** is a diagram that compares certain properties in a series of related trophic levels. The *pyramid of energy*, for example, in Figure 23.9 shows the amount of energy available at each trophic level. The base of the pyramid reflects *primary productivity*—the amount of light energy converted to chemical energy in organic molecules, primarily by photosynthetic organisms. The tiers representing successively higher trophic levels grow smaller, forming a pyramid, because only about 10% of the energy contained in one trophic level is transferred to the next. *Pyramids of biomass* describe the number of individuals at each trophic level multiplied by their biomass. Because the energy available at a trophic level determines the biomass that can be supported, a pyramid of energy and a pyramid of biomass generally have the same shape.

Health and Environmental Consequences of Ecological Pyramids

Ecological pyramids convey important lessons. Let's take look at two of these lessons: (1) nondegradable substances accumulate to higher concentrations in organisms living at higher trophic levels; and (2) more humans could be nourished on a vegetarian diet than on a diet containing meat.

Biological magnification Chemicals that are essential to life—carbon, hydrogen, oxygen, nitrogen, and phosphorus—are

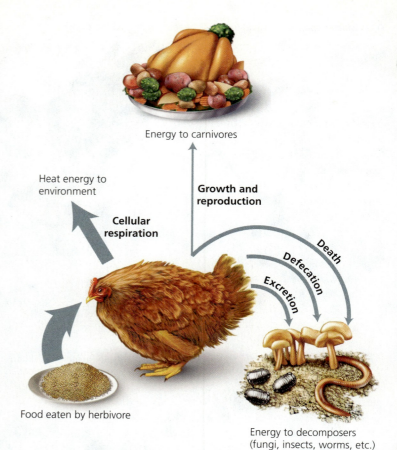

FIGURE 23.8 As energy flows through a food web, only a small amount of it is stored as body mass and becomes available to the next higher trophic level. Some of the food energy is simply not captured, some is undigested and lost in fecal waste, and some is used for cellular respiration. Only the remaining energy that has been converted to biomass becomes available to the next higher trophic level. The width of the arrow indicates the relative amount of energy transferred.

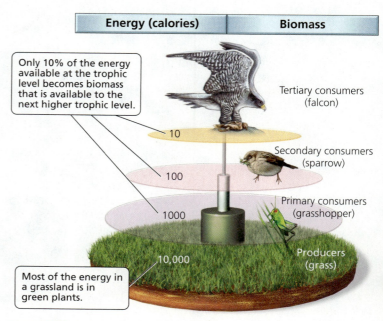

FIGURE 23.9 Ecological pyramids. Note that each trophic level contains less energy and less biomass than the level below it.

passed from one trophic level to the next, thus being continuously recycled from organism to organism. Organic molecules are broken down and then either metabolized for energy or put together to form the biomass of another individual.

However, certain substances—including chlorinated hydrocarbon pesticides such as DDT, heavy metals such as mercury, and radioactive isotopes—are broken down or excreted very slowly. Substances like these tend to stay in the body, often stored in fatty tissues such as liver, kidneys, and the fat around the intestines.

The concentrations of such substances become magnified at each higher trophic level. Remember that, on average, only 10% of the energy available at one trophic level transfers to the next. Consequently, consumers at one trophic level must eat many organisms from the previous trophic level to obtain enough energy to support life. If the organisms that are eaten are contaminated, a nondegradable pollutant will accumulate in the consumer. Consider a simplified example. Mercury, a nondegradable, potentially harmful substance, enters water through volcanic activity as well as through coal combustion and improper disposal of medical waste. Bacteria convert the mercury to a particularly toxic form, methyl mercury, which then works its way through the food web. If mercury pollutes the water in a certain aquatic ecosystem, phytoplankton (minute photosynthetic organisms that are abundant in aquatic environments) will absorb the mercury, and mercury levels in the phytoplankton will be in some dilute concentration that we will represent as 1. An herbivore such as a zooplankton (a minute non-photosynthetic organism) feeds on large numbers of phytoplankton. The mercury ingested from the phytoplankton thus becomes more concentrated in the zooplankton's tissues than it was in the phytoplankton—let's say about 10 times more. Large numbers of zooplankton may in turn be eaten by a small fish. Once in the fish's body, the mercury stays there and accumulates, let's say another 10 times, to reach a relative concentration of 100. When these mercury-containing fish are eaten by a tuna, the mercury passes to the tuna's body and accumulates. Because many small fish must be consumed to keep a tuna alive, the mercury concentration in the tuna's body might be 1000 times greater than that in water. This tendency of a nondegradable chemical to become more concentrated in organisms as it passes along the food chain is known as **biological magnification** (Figure 23.10).

Biological magnification is of more than theoretical concern for humans. We are often top carnivores, and we continue to pollute our environment with nondegradable, potentially harmful substances. If a human eats a long-lived predator—such as a tuna, swordfish, or shark—that has accumulated methyl mercury through biological magnification as described, that person could begin to accumulate dangerously high levels of mercury, too. High levels of mercury in humans affect the nervous system, causing muscle tremors, personality disorders, and birth defects. For this reason, the Food and Drug Administration now recommends that pregnant women, breast-feeding women, women of child-bearing age, and children stop eating shark, marlin, and swordfish altogether, and limit their intake of albacore (white) tuna to one 6 oz serving a week.

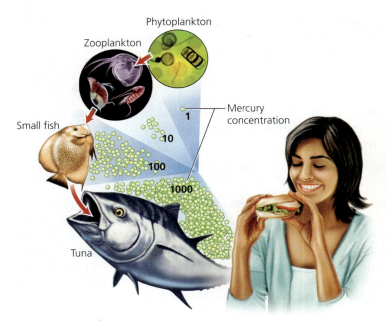

FIGURE 23.10 *Substances such as mercury that are broken down slowly or excreted slowly tend to accumulate in the body of an organism. The concentration becomes magnified at each successive trophic level, because a consumer must eat many individuals from a lower trophic level to stay alive. Because of this biological magnification, the mercury concentration in the tuna eaten by a human at the end of this food chain is about 1000 times greater than the mercury concentration in the phytoplankton at the beginning.*

Polychlorinated biphenyls (PCBs) are industrial chemicals suspected of causing cancer and nervous system damage. Like mercury, PCBs are nondegradable and accumulate in the food web. PCBs can be found in chicken, beef, and dairy products. Some farm-raised Atlantic salmon also have PCB levels high enough to trigger health warnings from the U.S. Environmental Protection Agency (EPA). (Wild salmon have lower levels of PCBs.) In addition, PCBs have been called environmental estrogens because they mimic the effects of the sex hormone estrogen or enhance estrogen's effects. Scientists, recognizing that PCBs have had feminizing effects on certain male animals, are investigating the possibility that PCBs may be reducing human fertility (for example, by reducing human sperm count) and affecting the rates of cancer in certain reproductive structures (breasts, ovaries, prostate gland, and testes).

Radioactivity can also pass along the food chain and accumulate in top predators. For example, bluefin tuna carried radioactivity from their spawning ground off the coast of Japan to the California coast. The contamination originated with the damaged nuclear reactor following the earthquake and tsunami in Japan in 2011. The bluefin tuna swam in the water and ate contaminated prey such as squid and krill. Their level of radioactivity is 10 times higher than in bluefin tuna before the disaster, but it is still not at levels that are harmful to humans.

World hunger The world's human population has grown at an alarming rate (a problem discussed in Chapter 24). The pyramid of energy suggests a way to feed the growing population more efficiently: persuade people to eat at lower trophic levels. Because only about 10% of the energy available on one trophic level is transferred to the next one, we see that

10,000 calories of corn energy ⟶ 1000 calories of beef energy ⟶ 100 calories of human energy

However, if humans began to eat one level lower on the food chain, about 10 times more energy would be available to them:

10,000 calories of grain (corn, wheat, rice and so on) ⟶ 1000 calories of human energy

In short, more people could be fed and less land would have to be cultivated if we adopted a largely, or exclusively, vegetarian diet (Figure 23.11). This is one reason that people in densely populated regions of the world, including China and India, are primarily vegetarians. As the human population continues to expand, meat is likely to become even more of a luxury throughout the world than it is today.

stop and think

The traditional diet of the Inuit, one of the native peoples of the North American coast, is part of a relatively long food chain:

Diatoms (producers) ⟶ Tiny marine animals ⟶ Fish ⟶ Seals ⟶ Inuits

How might the length of this food chain be one factor contributing to the small population size of Inuit groups?

23.5 Chemical Cycles

Earth's resources are limited. Life on Earth is demanding. Many of Earth's reserves would be depleted quickly if not for nature's cycling. Materials move through a series of transfers, from living to nonliving systems and back again (Figure 23.12). Let's look at some of the more important of these **biogeochemical cycles**, the recurring pathways through which certain materials travel between living and nonliving systems.

The Water Cycle

Each drop of rain reminds us that water recycles continuously, precipitating from the atmosphere and falling to Earth; collecting in ponds, lakes, or oceans; and then evaporating back into the atmosphere. Most of the rain or snow that falls over land returns to the sea at some point. This cycle provides us with a renewable source of drinking water. Because water is so critical to life, large amounts of it pause for a time in the bodies of living things. In living cells, water helps regulate temperature and acts as a solvent for biological reactions. The very

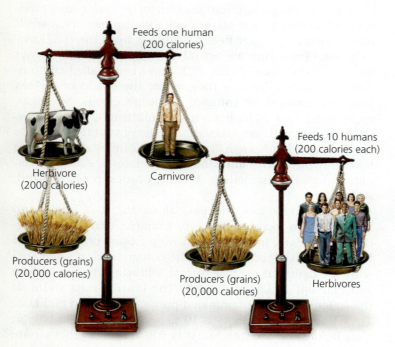

FIGURE 23.11 Energy pyramids may hold an important lesson for humans. Because only about 10% of the energy available at one trophic level is available at the next higher level, approximately 10 times more people could be fed if they ate a vegetarian diet rather than a diet containing only animal protein.

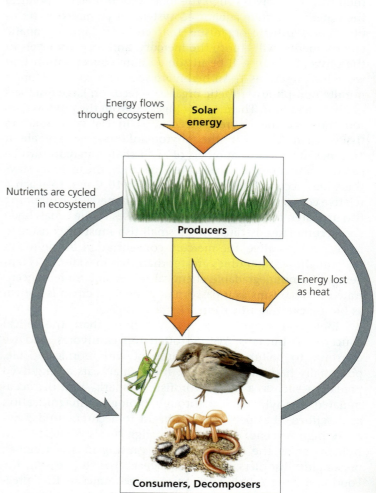

FIGURE 23.12 Through biogeochemical cycles, matter cycles between living organisms and the physical environment. This cycling of matter (gray arrows) is in contrast to the path of energy (yellow arrows), which flows through the ecosystem in one direction.

oxygen we breathe is produced from water through the reactions of photosynthesis. Water also cycles back to the environment from living things; plants return 99% of the water they absorb to the atmosphere in the process of transpiration (the evaporation of water from the leaves and stems of plants). All living things carry out cell respiration, which generates water that is exhaled as water vapor. The water cycle is shown in Figure 23.13.

Human activities and the water cycle Humans have disrupted the water cycle in several ways. We have cut down all the trees in vast areas of forest. This deforestation has reduced the amount of transpiration and, therefore, the amount of water vapor returned to the atmosphere. Our urban areas have altered runoff patterns. When rain falls on land, some water seeps down to replenish groundwater. In cities, most rainwater flows into sewers and becomes part of waste water.

The most significant way that humans disrupt the water cycle is by using more freshwater than is replenished (discussed in Chapter 24).

On a brighter side, we have developed some ways of recovering freshwater from the salty water of oceans or aquifers and from wastewater. Some coastal cities are building desalination plants to remove salts from ocean water and make it suitable for human consumption. However, desalination is an expensive process that uses a lot of energy. The Tampa Bay Seawater Desalination Plant in Florida produces about 95,000 cubic meters (25 million gallons) of freshwater a day, which accounts for 10% of the drinking water for the Tampa area. Texas recovers freshwater from salty water in aquifers. We have also developed technology to recover freshwater from wastewater. This technology, often called "toilet-to-tap," is the subject of an Environmental Issue essay in Chapter 2.

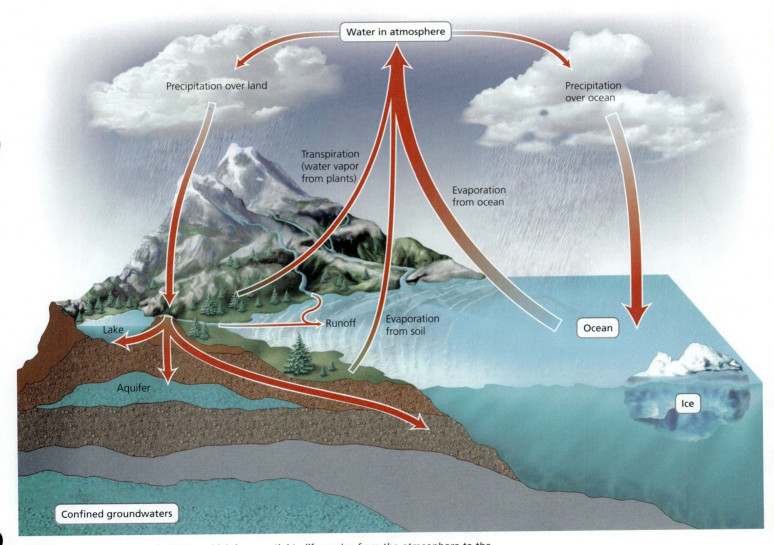

FIGURE 23.13 *The water cycle. Water, which is essential to life, cycles from the atmosphere to the land as precipitation, collects in oceans and other bodies of water, and evaporates back to the atmosphere. Water also returns to the atmosphere in the form of vapor lost from the leaves of plants (transpiration). This cycling gives us a renewable source of drinking water. The width of the arrow indicates the relative contribution of the process to the cycle.*

The Carbon Cycle

In the carbon cycle, carbon moves from the environment, into the bodies of living things, and back to the environment (Figure 23.14). Living organisms need carbon to build the molecules that give them life: proteins, carbohydrates, fats, and nucleic acids. Conversely, certain processes of living organisms—photosynthesis and respiration—are an integral part of the carbon cycle.

The primary movement of carbon from the environment into living organisms occurs during photosynthesis as plants, algae, and cyanobacteria use carbon dioxide (CO_2) to produce sugars and other organic molecules. When photosynthesizers are eaten by herbivores, these organic molecules serve as a carbon source for the herbivores. The herbivores use that carbon to produce their own organic molecules, which then serve as a carbon source for carnivores. When an organism dies, the organic molecules in its body will serve as a carbon source for decomposers. However, while alive, all organisms cycle carbon back to the environment through cellular respiration, which breaks down organic molecules to CO_2.

Some carbon is significantly delayed before being cycled back into the environment. For example, carbon may remain tied up in the wood of some trees for hundreds of years. Most of the carbon that is currently not cycling is thought to be stored in limestone, a type of sedimentary rock formed from the shells of marine organisms that sank to the bottom of the ocean floor and were covered and compressed by newer sediments. Other vast carbon stores are the fossil fuels (oil, coal, and natural gas), so named because they formed from the remains of organisms that lived millions of years ago.

Three processes release carbon from long-term storage and return it to the environment: decomposition, erosion, and combustion. When trees die, for example, the natural process of decomposition will make the carbon available for new organisms, which will respire and release CO_2 to the atmosphere. The carbon in limestone is recycled through erosion. Millions of years after it forms, sedimentary rock containing limestone can be lifted to Earth's surface by movement of tectonic plates, where it is eroded by chemical and physical weathering. These changes make the carbon available to cycle through the food web once again. Combustion, or burning, returns the carbon in fossil fuels to the environment. Today, fossil fuels such as coal, oil, and natural gas are being burned in huge amounts, and the carbon they contain is being returned to the atmosphere as CO_2.

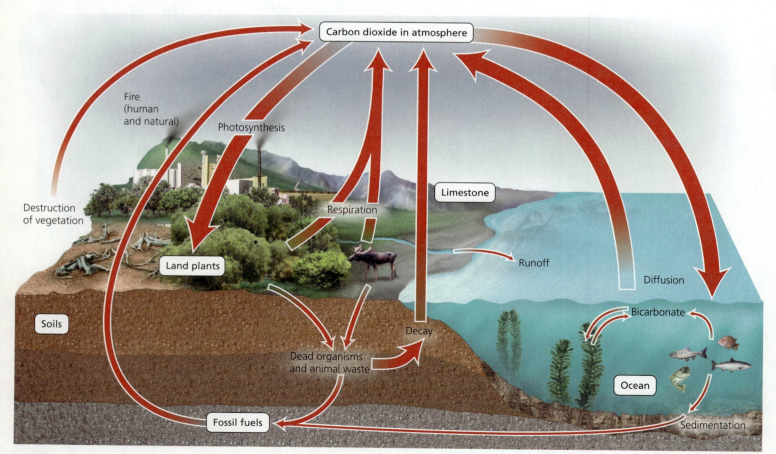

FIGURE 23.14 *The carbon cycle.* Carbon cycles between the environment and living organisms. Carbon dioxide (CO_2) is removed from the environment as producers use it to synthesize organic molecules by photosynthesis. The carbon in those organic molecules then moves through the food web, serving as a carbon source for herbivores, carnivores, and decomposers. Carbon is returned to the atmosphere as CO_2 when organisms use the organic compounds in cellular respiration. The width of the arrow indicates the relative contribution of the process to the cycle.

Increasing carbon dioxide levels Two human activities—burning fossil fuels and deforestation—are altering the carbon cycle and increasing the atmospheric carbon dioxide level. Recall that fossil fuels were formed from deposits of dead plants and animals that were buried in sediments and escaped decomposition between 345 million and 280 million years ago. High temperatures and pressure over millions of years converted the deposits to coal, oil, and natural gas. The burning of these fuels returns carbon to the environment in the form of CO_2. Deforestation—the removal of a forest without adequate replanting, as is occurring in areas of the Amazon rain forest and the U.S. Pacific Northwest—increases atmospheric CO_2 in two ways. First, it *reduces the removal* of CO_2 from the atmosphere by eliminating trees' photosynthesis. Second, the trees are often burned after cutting to clear the area, and burning *adds* CO_2 to the atmosphere.

The rise in atmospheric CO_2 raises concerns because CO_2 is one of the greenhouse gases that play a role in warming our atmosphere. Indeed, the scientific consensus is that the increase in atmospheric CO_2 is causing a rise in temperatures throughout the world—known popularly as *global warming*.

Global climate change and its many consequences are discussed in Chapter 24.

The rising level of atmospheric CO_2 is also making the oceans more acidic. When CO_2 dissolves in water, it forms carbonic acid. The acidity decreases the availability of calcium carbonate, which is needed by certain forms of marine life to form skeletons and shells. Some of these animals are at the bottom of the food chain, so their loss could have consequences for other animals that depend on them for food. Corals also build their skeletons from calcium carbonate, and coral reefs form the foundation for many fisheries. The arctic oceans are particularly vulnerable, because carbon dioxide is more soluble in cold water. Studies are under way to determine the possible effects of the acidification of both coral reefs and arctic waters.

The Nitrogen Cycle

Nitrogen is a principal constituent of several molecules needed for life, including proteins and nucleic acids. Nitrogen is often in short supply to living systems, so its cycling is particularly important (Figure 23.15).

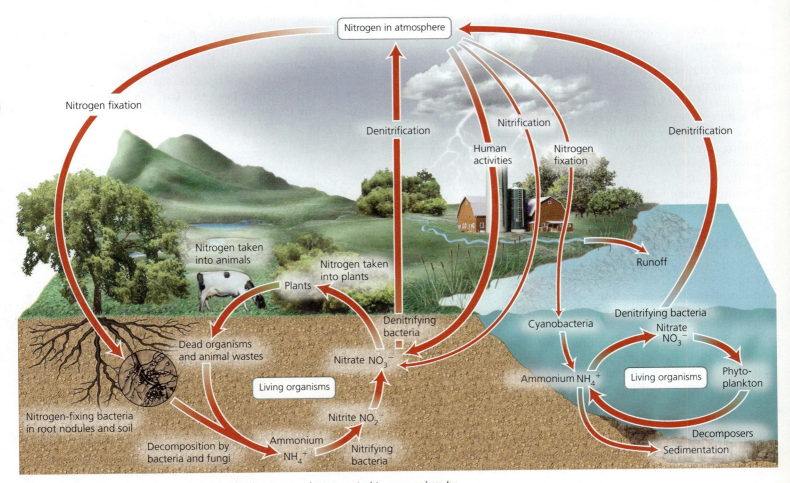

FIGURE 23.15 *The nitrogen cycle. Atmospheric nitrogen can be converted to ammonium by nitrogen-fixing bacteria, which then convert the ammonium to nitrate, the main form of nitrogen absorbed by plants. Plants use nitrate to produce proteins and nucleic acids. Animals eat the plants and use the plant's nitrogen-containing chemicals to produce their own proteins and nucleic acids. When plants and animals die, their nitrogen-containing molecules are converted to ammonium by bacteria. Denitrifying bacteria return nitrogen to the atmosphere. The width of the arrow indicates the relative contribution of the process to the cycle.*

The largest reservoir of nitrogen is the atmosphere, which is about 79% nitrogen gas (N_2). However, nitrogen gas cannot interact with life directly. (As you sit reading, you are bathed in nitrogen gas, but you do not interact with it.) Before it can be used, nitrogen gas must be converted to a form that living organisms can use—ammonium (NH_4^+). The process of converting nitrogen gas to ammonium is called **nitrogen fixation**. This is performed by nitrogen-fixing bacteria, many of which live in nodules on the roots of leguminous plants such as peas and alfalfa. Ammonium is then converted to nitrite (NO_2^-) and then to nitrate (NO_3^-) by nitrifying bacteria living in the soil, in a process called **nitrification**. The ammonium or nitrates are first absorbed by plants, which use the nitrogen to form plant proteins and nucleic acids. The nitrogen then passes through the food web and is incorporated into the nitrogen-containing compounds of animals.

When plants and animals die, decomposers such as bacteria break down the waste products and dead bodies of plants and animals, producing ammonium (NH_4^+). Much of the ammonium is converted to nitrate by nitrifying bacteria. Other bacteria balance the nitrogen cycle by performing a process called *denitrification*, in which the nitrates that are not assimilated into living organisms are converted to nitrogen gas. Denitrifying bacteria are found in wet soil, swamps, and estuaries.

Human activities and the nitrogen cycle Human burning of fossil fuels has affected the nitrogen cycle by adding nitrogen dioxide (NO_2) gas (as well as sulfur dioxide—SO_2—gas) to the atmosphere, where they react with water vapor to form acids. The acids eventually fall to Earth as acid rain, which is killing trees in northern forests and fish and amphibians in aquatic environments. (Acid rain is discussed in Chapter 2.) Sunlight can cause hydrocarbons and NO_2 to form photochemical smog, which contains ozone. Ozone is especially harmful to the respiratory system.

The Phosphorus Cycle

Phosphorus is an important component of many biological molecules, including the genetic material DNA, energy-transfer molecules such as adenosine triphosphate (ATP), and phospholipids found in membranes. Phosphorus is also essential in vertebrate bones and teeth.

Unlike the water, carbon, and nitrogen cycles, the phosphorus cycle does not have an atmospheric component (Figure 23.16). Instead, the reservoir for phosphorus is sedimentary rock, where it is found as phosphate ions. The cycle begins when erosion caused by rainfall or runoff from streams dissolves the phosphates in the rocks. The dissolved phosphate is readily absorbed by producers and incorporated into their biological molecules. When animals eat the plants, the phosphates are passed through food webs. Decomposers return the phosphates to the soil or water, where they become available to plants and animals once again. Much of the phosphate is lost to the sea. Although some of this phosphate may cycle through marine food webs, most of it becomes bound in sediment. The phosphates in sediment become unavailable to the biosphere unless geological forces bring the sediment to the surface.

Eutrophication Human activity is also disturbing the phosphorus and nitrogen cycles. Inadequate nitrogen and phosphorus in the soil can slow plant growth. For this reason, many crops are fertilized with synthetic preparations containing both nitrogen (as ammonium) and phosphates. Phosphates from fertilizer can wash into nearby streams, rivers, ponds, or lakes in runoff. Fertilizer runoff is one cause of *eutrophication*, the enrichment of water in a lake or pond by nutrients. The nutrient boost can lead to an explosive growth of photosynthetic organisms, such as algae in deeper water and weeds in shallow areas. When these organisms die, their remains accumulate at the bottom of the lake. Decomposers then thrive, depleting the water of dissolved oxygen. Gradually, fish that require higher oxygen concentrations in the water, such as pike, sturgeon, and whitefish, are replaced by species that can tolerate lower levels of dissolved oxygen, such as catfish and carp. Eventually, the oxygen depletion can kill all fish and bottom-dwelling life.

Eutrophication is not just a problem in lakes and ponds; there are now more than 400 coastal regions of oceanic dead zones, regions where marine life can't survive. The dead zones are caused by fertilizer runoff, which increases available nutrients for algae, and by global climate change, which increases the temperature of the top, well-oxygenated layer of the water and prevents mixing with the lower, less-oxygenated layers. One of the largest dead zones is in the Gulf of Mexico. This dead zone, which is sometimes the size of New Jersey, is caused by the fertilizer runoff from the farms in the Midwest that is carried to the Gulf by the Mississippi River.

what would you do?
The nitrates that cause the dead zone in the Gulf of Mexico come from farms along the upper Mississippi River, but the results are experienced by the people living along the coast of the Gulf. Who do you think should be financially responsible for addressing this problem? Why?

stop and think
A dead zone is caused by the expansion of a population of organisms due to an increase in nutrient supply and then the depletion of oxygen caused by the decomposition of those organisms. Do you predict that the *Deepwater Horizon* oil spill will increase or decrease the size of the dead zone in the Gulf of Mexico? Why?

23.6 Biodiversity

In this chapter we see that organisms are not randomly scattered over Earth. Instead, each is adapted to a particular niche, which allows it to remain healthy and to reproduce. Recall that an organism's niche includes physical, chemical, and biological factors. We also see in this chapter that the only input to nearly every ecosystem is solar energy, which producers use to manufacture organic molecules. Energy then flows through food webs. Resources, such as water, carbon,

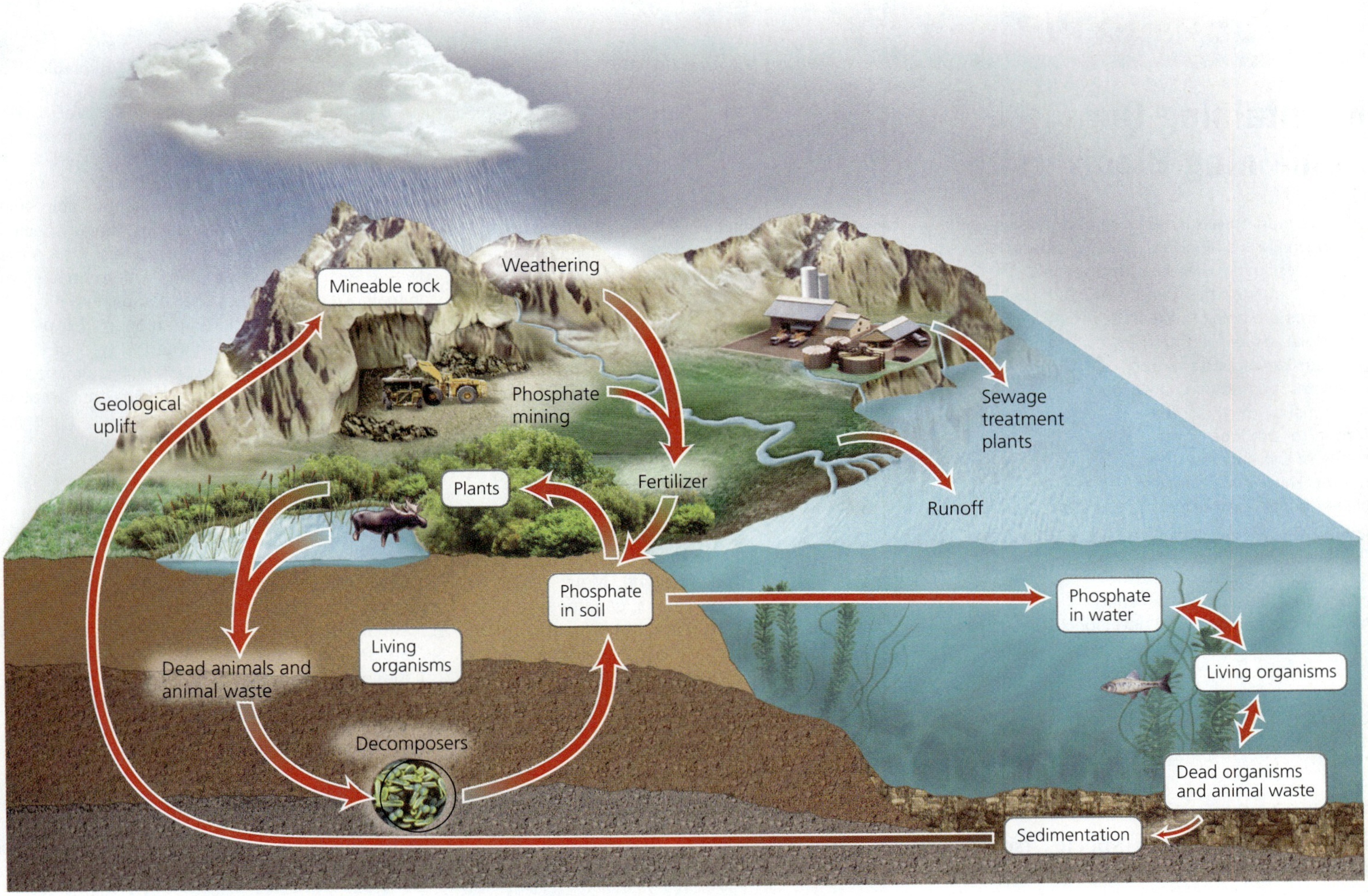

FIGURE 23.16 The phosphorus cycle. Phosphates dissolve from rocks in rainwater. They are then absorbed by producers and passed to other organisms in food webs. Decomposers return phosphates to the soil. Some phosphates are carried to the oceans and eventually deposited in marine sediments. The width of the arrow indicates the relative contribution of the process to the cycle.

nitrogen, and phosphorus, cycle through living and nonliving systems. It's no surprise, then, that even small changes to an organism's niche can have ripple effects that disturb the entire ecosystem.

A current example of this ripple effect is damage being done to coral reefs. The damage begins with humans upsetting the carbon cycle by burning fossil fuels and cutting down forests. These human activities have raised the level of carbon dioxide in the atmosphere. Elevated atmospheric carbon dioxide causes more carbon dioxide to diffuse into the oceans, where it dissolves, forming carbonic acid. As a result, the oceans are becoming more acidic, causing damage to coral reef ecosystems. This damage, in turn, damages many productive fisheries.

As the human population grows, it consumes and pollutes more resources (discussed in Chapter 24), which can have widespread effects on ecosystem Earth. Because living and nonliving systems in an ecosystem are profoundly interconnected, human disturbance may lead to changes in **biodiversity**—species richness. Biodiversity includes the number of species living in a given area and the abundance of each. Recently, scientists have raised an alarm worldwide in response to indications that globally, and especially in certain critical areas, biodiversity is decreasing. We are losing species, perhaps as many as 100 each day, largely because of human activity. *Mass extinctions,* the loss of many species from Earth, have occurred in previous eras, but never before because of humans. Today, many species are being forced to compete with us as we change the environment to suit ourselves, rendering it unsuitable for them.

How many species are there? No one knows. Scientific estimates range from 8 million to 30 million species now in existence, with about 90% of them living on land. However, only about 1.7 million of these species have been identified, and only 3% of those have been studied.

Most terrestrial species live in the tropics. Although tropical rain forests cover only 7% of Earth's surface, they are home to between 50% and 80%—estimated as 7 million to

ETHICAL ISSUE

Maintaining Our Remaining Biodiversity

Faced with powerful evidence of a marked reduction in Earth's biodiversity, some experts express guarded optimism that we may be able to slow or even stop this trend. Various measures are being suggested.

More developed (richer) and less developed (poorer) countries must take a careful census of their lifeforms to produce inventories of the species they harbor. This need is especially acute in tropical countries, which, unfortunately, are often least able to afford such programs. Advocates thus argue that the effort must be worldwide, with developed countries subsidizing the research in less affluent areas.

Nations must cooperate in linking the economic development of impoverished regions, particularly the tropics, to conservation. Lending agencies must stipulate that certain areas be set aside and left undeveloped. To minimize the effects on the world's biodiversity, nations should consult ecologists when deciding which locations are to remain undeveloped. (What effect do you think such rules would have had on the development of the United States if such restrictions had been in place during the industrial revolution?)

Educating the people whose decisions have the most immediate effect on biodiversity should be a priority. For example, recent studies have shown that native people can make more money through sustainable use of a forest (as in gathering nuts and fruits) than they would if the forest were cleared for agriculture or ranching. Many people in developing countries are also learning that scientists worldwide are interested in the medicines and healing knowledge they have, so, in a sense, the educational opportunities are reciprocal.

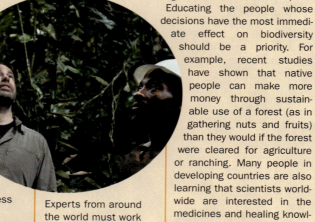

Experts from around the world must work together to make wise land-use policies.

Biologists worldwide must work more closely with zoning and land-use personnel to maximize the use of areas being cleared or cultivated. Such multiple land-use planning includes, for example, rotating crops to maintain healthy soil and increase biodiversity or replacing single-species forests planted by paper companies with more diverse forests that could harbor more species of animals. Of course, the paper companies may ask why they should be required to reduce their profits to increase biodiversity. This is the type of issue that will have to be negotiated in the process of bringing people to an understanding that what is good for the environment is ultimately good for everyone.

Questions to Consider

- Many of the issues regarding preserving biodiversity involve trade-offs between activities that are valuable to society and activities that reduce the biodiversity that allows ecosystems to provide a stable food supply. How can these trade-offs be evaluated? Who should be responsible for making these decisions?
- Should agencies that work to maintain biodiversity receive government funding or rely on private-sector funding?

---

8 million—of Earth's species (see the Ethical Issue essay, *Maintaining Our Remaining Biodiversity*).

Habitat destruction is largely responsible for the loss of biodiversity. The primary reasons habitats are destroyed are to create living space for humans and to provide room or resources for economic development. Calculations indicate that most of the original tropical rain forest will be destroyed by 2040. This means that in the next 30 years, we could lose a quarter of all the species on Earth. Evolutionary biologists tell us that, currently, we are losing species at the rate of 1000 to 10,000 times the average rate for the last 65 million years, since the extinction of the dinosaurs.

Rain forest is not the only type of habitat being destroyed. Lumbering and acid rain are destroying northern temperate forests. Also, marine ecosystems are being destroyed by pollution, overfishing, and coastal development. For example, shark populations have declined by more than 70% in the last 15 years. Commercial fishing is largely to blame for the reduction of shark populations. People eat shark meat and use the fins to make soup. Moreover, some sharks, hammerheads for example, are accidentally caught when they try to prey on the herring and squid being used as bait by people fishing for tuna or swordfish. The United States has federal regulations that restrict shark fishing; however, shark fishing is not regulated in Europe. If shark numbers continue to fall, we may not be able to save them from extinction. Sharks grow slowly, mature late in life, have few young at a time, and females do not reproduce every year.

Perhaps you are wondering why we should care about the loss of species, especially those we never knew of. Several materialistic reasons immediately come to mind. The first direct benefit to humans of maintaining biodiversity is that biodiversity preserves the genetic diversity encoded in the chromosomes of species. This genetic diversity is useful for crossbreeding. The crossbreeding of two strains of a species, each possessing different desirable traits, can combine those desirable traits in a single new strain. Interestingly, the usefulness of genetic diversity as a reservoir of genetic stores has increased dramatically with modern technology. Owing to recent advances in genetic engineering, specific genes can be isolated from their source organisms and moved into other kinds of organisms. Thus, maintaining genetic stores becomes even more important as we use genetic engineering to maximize crop yields or otherwise create new forms of organisms that might be useful to society (see Chapter 21).

The second direct benefit to humans of maintaining biodiversity lies in developing new kinds of medicines. About 25% of all known drugs come from plants. The rosy periwinkle from Madagascar, as discussed in Chapter 1, is a source of two anticancer drugs, vincristine and vinblastine; and the Mexican yam was once a source of oral contraceptives. Most of the plants with medicinal value have been found in the tropics, which, as we have noted, are being rapidly destroyed.

A third benefit is the many services that biodiversity provides in functioning ecosystems—cleansing water and air, enriching soil, cycling minerals, and pollinating crops, for example. Recall the important role that organisms play in the biogeochemical cycles discussed earlier in this chapter. For example, as pollinators, bees contribute $15 billion to the productivity of agriculture in the United States each year. This is one reason for great concern over the decline in bee populations in recent years. Most of the organisms that could become crop pests are kept under control by their own predators.

We see, then, that there are many practical reasons to preserve biodiversity. Earth and the organisms inhabiting it are interconnected through energy flow through food webs and biogeochemical cycles. The health of ecosystem Earth depends on biodiversity.

looking ahead

In this chapter, we learned about Earth, our home planet—the energy flow through ecosystems, chemical cycles, and biodiversity. In Chapter 24, we will consider the growth of the human population and how that growth affects Earth's resources.

HIGHLIGHTING THE CONCEPTS

23.1 Earth as an Ecosystem (p. 497)
- Ecology is the study of the interactions between organisms and between organisms and their environment.
- Materials on Earth are recycled among living organisms and the environment. The only outside input, or contribution, to Earth is energy from the sun.

23.2 Biosphere (p. 498)
- The biosphere consists of all the ecosystems on Earth. An ecosystem consists of a community of organisms and their physical environment. An organism's niche is the specific role it plays in the community; its habitat is the place where it lives.

23.3 Ecological Succession (pp. 498–500)
- Ecological succession is the sequence of changes in communities over long periods. Primary succession occurs where no community previously existed. The first settlers of a community are called pioneer species. A climax community eventually forms and remains as long as no disturbances occur. Secondary succession describes the sequences of changes that occur when an existing community is disturbed by human or natural means.

23.4 Energy Flow (pp. 500–504)
- Most of the energy in living systems comes from solar energy absorbed and stored in the molecules of photosynthetic organisms, called producers. The energy stored in producers' molecules enters the animal world through plant-eating herbivores (primary consumers), which may be eaten by carnivores (secondary consumers) or by animals that eat other carnivores (tertiary consumers). Decomposers consume dead organic material, releasing inorganic compounds that can be used by producers. The position of an organism in these feeding relationships is referred to as a trophic level.
- The feeding relationships that result in the one-way flow of energy through the ecosystem and in the cycling of materials among organisms are called food chains or food webs.
- Ecological pyramids depict the amount of energy or biomass at each trophic level. Pyramids of energy show the loss of energy from one trophic level to another. On average, only 10% of the energy available at one trophic level is available at the next higher level. Ecosystems generally have only four or five trophic levels. Pyramids of biomass have the same shape as pyramids of energy, because the available energy determines the biomass that can be formed.
- Biological magnification—the tendency for certain nondegradable substances to become more concentrated in organisms as it passes through a food web—is a consequence of the energy loss between trophic levels. Thus, top carnivores are most likely to be poisoned by nondegradable toxic substances in the environment.
- Because energy is lost with each transfer in the trophic scale, adopting a largely vegetarian diet would allow for more people to be fed.

23.5 Chemical Cycles (pp. 504–508)
- In biogeochemical cycles, materials cycle between organisms and the environment.
- The water cycle is the pathway of water as it falls as precipitation; collects in ponds, lakes, and seas; and returns to the atmosphere through evaporation.
- Humans disturb the water cycle by causing shortages through overuse of water supplies.
- The carbon cycle is the worldwide circulation of carbon from the carbon dioxide in air to the carbon in organic molecules of living organisms and back to the air. Carbon enters living systems when photosynthetic organisms incorporate carbon dioxide into organic materials. Carbon dioxide is formed again when the organic molecules are used by the living organism for cellular respiration.
- Humans have disturbed the carbon cycle through activities that increase carbon dioxide levels in the atmosphere: burning fossil fuels, which directly adds carbon dioxide, and deforestation, which decreases the removal of carbon dioxide from the atmosphere. Carbon dioxide is a greenhouse gas that traps heat in Earth's atmosphere and causes global warming.
- The nitrogen cycle is the worldwide circulation of nitrogen from nonliving to living systems and back again. Atmospheric nitrogen (N_2) cannot enter living systems. Nitrogen-fixing bacteria living in nodules on the roots of leguminous plants convert N_2 to ammonium (NH_4^+), which is converted to nitrites (NO_2^-) and then to nitrates (NO_3^-) by nitrifying bacteria. The ammonium and nitrates are then available to plants to use in their proteins and nucleic acids. Next, the nitrogen is transferred to organisms that consume the plants. Nitrates that are not assimilated into living organisms can be converted to nitrogen gas by denitrifying bacteria.
- Humans affect the nitrogen cycle by burning fossil fuels, which adds nitrogen dioxide to the atmosphere. Nitrogen dioxide reacts with water vapor in the atmosphere to form nitric oxide, which falls to Earth as acid rain. Sunlight converts nitrogen dioxide and hydrocarbons to photochemical smog, which damages the respiratory system.
- Phosphorus in the form of phosphates is washed from sedimentary rock by rainfall. The dissolved phosphates are used by

producers to produce important biological molecules, including DNA and ATP. When animals eat producers or other animals, phosphates are passed through the food webs. Decomposers release phosphates into the soil or water from dead organisms.
- Humans have disrupted the nitrogen and phosphorus cycles through the use of synthetic fertilizers containing nitrogen as ammonium and phosphates. Some of the phosphates wash into nearby waterways. Fertilizer runoff is one cause of eutrophication.

23.6 Biodiversity (pp. 508–511)
- Biodiversity, the number and variety of living things, is being reduced dramatically, largely because of human activity. Most of the loss is occurring in the tropics and is due to habitat destruction. Two practical reasons for concern over the loss of biodiversity are that the disappearing species could have genes that would someday prove useful or that they could be found to produce chemicals with medicinal qualities.

RECOGNIZING KEY TERMS

biosphere *p. 498*
ecosystem *p. 498*
community *p. 498*
population *p. 498*
niche *p. 498*
habitat *p. 498*
succession *p. 498*

producers *p. 500*
trophic level *p. 500*
consumer *p. 500*
herbivore *p. 500*
primary consumer *p. 500*
carnivore *p. 500*
secondary consumer *p. 500*

omnivore *p. 500*
decomposers *p. 500*
food chain *p. 500*
food web *p. 500*
biomass *p. 502*
ecological pyramid *p. 502*

biological magnification *p. 503*
biogeochemical cycle *p. 504*
nitrogen fixation *p. 508*
nitrification *p. 508*
biodiversity *p. 509*

REVIEWING THE CONCEPTS

1. What is ecological succession? How does primary succession differ from secondary succession? pp. 498–500
2. Explain how energy flows through an ecosystem. pp. 500–503
3. Define the following terms: *producer, primary consumer, secondary consumer,* and *decomposer.* Explain the role each plays in cycling nutrients through an ecosystem. p. 500
4. Explain why the feeding relationships in a community are more realistically portrayed as a food web than as a food chain. p. 500
5. Explain the reasons for the inefficiency of energy transfer from one trophic level to the next higher one. Why does this loss of energy limit the number of possible trophic levels? p. 502
6. Define *energy pyramid.* What causes the pyramidal shape? p. 502
7. What is meant by *biomass pyramid*? How does a biomass pyramid compare to an energy pyramid? p. 502
8. Define *biological magnification.* Explain how biological magnification is a consequence of the energy loss between trophic levels. Why should humans be concerned about biological magnification? pp. 502–503
9. Explain why more people could be fed on a vegetarian diet than on a diet that contains meat. p. 504
10. Describe the water cycle, the carbon cycle, the nitrogen cycle, and the phosphorus cycle. pp. 504–508
11. Explain some of the ways that humans are disturbing the water cycle. p. 505
12. Which human activities are primarily responsible for the rising level of atmospheric carbon dioxide? p. 507
13. Define *biodiversity.* Where is biodiversity greatest? Why should we be concerned about the loss of biodiversity? pp. 508–511
14. Which of the following is *not* constantly recycled in the biosphere?
    a. energy
    b. water
    c. carbon
    d. nitrogen
15. Secondary succession is most likely to be found
    a. on rocky outcroppings.
    b. in areas that were clear-cut for timber.
    c. where an island has appeared.
    d. in areas where estuaries have filled in to form land.
16. Eutrophication
    a. is caused by an overabundance in nutrients that leads to an increase in algae in aquatic environments.
    b. usually occurs in deserts.
    c. is a process in which nitrogen is fixed.
    d. is a necessary part of the nitrogen cycle.
17. What is the original source of almost all the energy in most ecosystems?
    a. carbon dioxide
    b. water
    c. carbohydrates
    d. sunlight
18. Animals that eat both producers and consumers are called
    a. autotrophs.
    b. herbivores.
    c. omnivores.
    d. primary consumers.
19. Biogeochemical cycles
    a. include only abiotic processes.
    b. include only processes conducted by living organisms.
    c. transfer energy from one trophic level to the next.
    d. none of the above
20. The nitrogen cycle is important and complex. Why is nitrogen important?
    a. It is essential to photosynthesis.
    b. It is poisonous.
    c. It is a component of proteins and nucleic acids.
    d. It is a greenhouse gas that causes climate warming.
21. In the nitrogen cycle, gaseous ammonia is converted to ammonia in the process of
    a. ammonification.
    b. nitrification.
    c. nitrogen fixation.
    d. eutrophication.
22. An organism's role in a community is its _____.
23. An animal that eats an herbivore is called a(n) _____.

APPLYING THE CONCEPTS

1. In a particular grassland ecosystem, energy flows in the following path:

    Grass ⟶ Crickets ⟶ Frogs ⟶ Herons

    Assuming that the efficiency of energy transfer is typical, how many calories will the herons receive if there were 100,000 calories of grass?

2. Explain why the mercury levels in a lake may be low enough for the water to be safe to drink, yet fish from the same lake may be poisonous to eat.

3. You are a crime scene investigator at a murder scene. Stuck to the bottom of the victim's shoe is a leaf unlike any others in the vicinity of the crime. What clues could you get from the leaf?

BECOMING INFORMATION LITERATE

Malaria is a disease transmitted by mosquitoes that infects 300 million people a year and kills about 200 million of them. DDT is a pesticide that is an affordable and effective way to kill mosquitoes. After being widely used in the 1950s, DDT was blamed for the alarming decline in certain species of birds during the 1960s and 1970s. The United States banned the use of DDT during the 1970s, and many other countries followed suit. Because of these bans, neither the United States nor the World Health Organization will fund the use of DDT to control malaria-causing mosquitoes. Some of the areas hit hardest by malaria—Africa, Asia, and Latin America—are too poor to afford other pesticides, which cost four to six times more than DDT.

Use at least three reliable sources (journals, newspapers, websites) to consider the social, health, economic, and political controversies regarding outside funding for the use of DDT to control mosquito populations in poor, malaria-stricken countries. Cite your sources, and explain why you chose the ones you used.

MasteringBiology®

Go to MasteringBiology for practice quizzes, activities, eText, videos, current events, and more.

24 Human Population, Limited Resources, and Pollution

Did you Know?

- If, on average, women decide to have children when they are young, the human population will grow dramatically faster than if they wait until their later reproductive years to have children.
- The world loses one acre of forest every second.

Hydroponic gardening can produce fruits and vegetables more quickly than conventional gardening and uses less space.

- 24.1 Population Changes
- 24.2 Patterns of Population Growth
- 24.3 Environmental Factors and Population Size
- 24.4 Earth's Carrying Capacity
- 24.5 Human Impacts on Earth's Carrying Capacity
- 24.6 Global Climate Change
- 24.7 Looking to the Future

ENVIRONMENTAL ISSUE

Air Pollution and Human Health

In Chapter 23, we learned about the functioning of ecosystems and how it is affected by human activities. In this chapter, we consider problems stemming from human population growth. We start by reviewing general principles governing population size. Then we consider how the rapid growth of the human population has led to pollution, depletion of Earth's resources, and global climate change. Last, we discuss how the choices we make today can affect the quality of life on Earth in the future.

24.1 Population Changes

There is no way to discuss human populations without arousing concern. The worries arise from two kinds of information. First, the increase in human numbers is startling, and we just do not know how long population growth can continue. Second, humans are using the world's natural resources much faster than they can be replaced (*if* they can be replaced). Although some resources are renewable, many are finite. We review the evidence for each of these trends and discuss possible consequences—and consider some solutions.

Population Growth Rate

Population dynamics describes how populations change in size. The human population, we can say, consists of all individuals alive at any time. This enormous population can be broken down into smaller groups. Thus we can speak of the population of the United States or Colombia or China. We can break it down further and consider, say, urban American populations versus rural ones.

Population size changes when individuals are added and removed at different rates. Usually, the rate is reported as the number of births or deaths per 1000 persons per year. Expressing a population's growth as a rate keeps the change in population size in perspective. For example, we would consider the addition of 1000 individuals to a population differently if it took place overnight than if it took place over a year. Furthermore, the addition of 1000 individuals would be more important if the initial population size were 100 than if it were 1000.

A population grows when more individuals are added than are subtracted. *Growth rate*, the difference between the birth rates and death rates of a population, is a measure of this event. Some individuals join a population or leave it by immigration or emigration, but most of a population's members are added by birth and are removed by death. Most likely, your own family has added to the population growth over the last few generations. How many living descendants do your grandparents have (Figure 24.1)?

In 2012, the estimated *birth rate* of the world's population was 19 births per 1000 individuals per year, and the *death rate* was 8 deaths per 1000 individuals per year. Thus, the growth rate of the world's population, an increase of 11 individuals per 1000, was 1.1%.

$$\frac{(\text{Births} - \text{Deaths})}{1000} = \frac{(19 - 8)}{1000} = \frac{11}{1000} = \frac{01.1}{100} = 1.1\%$$

However, knowing the growth rate is not enough to predict how quickly new individuals will be added to the population. We must also know the number of individuals in the starting population. Added individuals also reproduce, so population growth occurs much as compound interest accumulates in the bank—the more you start with, the more you end up with. At the same growth rate, then, the larger the size of the starting population, the more individuals are produced.

The age at which females have their first offspring dramatically affects the birth rate of a population. The younger women are when they begin to reproduce, the faster the population grows. In fact, the age when reproduction begins is the most important factor influencing a female's overall reproductive potential. For example, the increasing proportion of women who are postponing childbirth until their later reproductive years has been helping to slow the population growth rate of the United States. The birth rate in the United States is declining for women of all ages, but the largest drop in birth rate has occurred among women in their twenties. One reason may be that increasing numbers of women want to establish careers before beginning motherhood. Regardless of the reasons, this trend is contributing to slower U.S. population growth than in the past.

Besides birth rate, death rate is an important part of the equation that determines growth rate. Although the world birth rate has been steadily declining since 1970, the human population has continued to grow partly because the drop in death rate has been greater than the drop in birth rate. The discovery of antibiotics was certainly instrumental in the decline of death rates. However, even more important were improvements in sanitation, hygiene, and nutrition.

Although a growth rate of a few percent a year may not seem startling, its importance becomes clearer when it is expressed as **doubling time**—that is, the number of years it will take for a population to double at that rate of growth. To give you an idea of how important this measure is, if this page is 1/250 inch thick and were doubled (exponentially) only eight times, it would be an inch thick. At 12 more doublings, it would be as thick as a football field is long. At 42 times, it would reach to the moon; and at only 50 times, it would reach the 93 million miles to the sun. Obviously, knowing the doubling time of a population can yield some interesting predictions.

The simplest way to arrive at a doubling time is to divide the population's growth rate into 70 (a demographic constant value). Thus, if the world population continues to grow at the rate it did in 2012 (1.1%), the world population will double in just a little more than 65.5 years! This is how long it will be before we need two schools where there is now one and two roads, two dams, two telephone poles, two hospitals, two doctors, and two power plants where there is now one. With everything doubled, what happens to our quality of life in general? Doubling times vary widely in different countries, but in general, *less developed countries* (LDCs; poorer countries) have a much smaller doubling time than do *more developed countries* (MDCs; usually richer countries).

FIGURE 24.1 Gladys Davis is surrounded by her six great-grandchildren. She and her husband had four children, two of whom had children of their own, producing three grandchildren. The three grandchildren produced these six great-grandchildren. Two of the children shown here have started their own families, producing six great-great-grandchildren. Although the size of each individual family was small, there are currently 15 living descendants of Jack and Gladys Davis. How much has your family grown in the last few generations?

stop and think

A pair of cockroaches is placed on a small island. Assume that the doubling time for a population of cockroaches is 1 month. After 20 years, the island contains half of the total number of roaches it could possibly hold. How much longer would it take until the island was completely filled with roaches? (*Hint:* What is the doubling time for the population?)

Age Structure

When we want to predict future growth, a population's **age structure**—the relative number of individuals of each age—can be helpful. The age structure of the population is important because only individuals within a certain age range reproduce. Among humans, for instance, toddlers and octogenarians are counted as members of the population, but they do not reproduce. The ages are often grouped into prereproductive, reproductive, and postreproductive categories. Generally, only individuals of reproductive age add to the population size. However, the prereproductive group (those who are currently too young to reproduce) usually get older; and when they enter the reproductive class, they have children, thus adding more members to the population. Even if the birth rate remains the same, the overall population size will grow more rapidly if the size of the reproductive class increases.

stop and think

Health care has improved greatly in the last century and has increased population growth. Why would measures that decrease child mortality (death) contribute more to population growth than would measures that increase life expectancy?

We see, then, that the relative size of the base of a population's age structure (the prereproductive age category) determines how quickly numbers will be added to the population in the future. A wide base to any age structure diagram reflects a growing population, whereas a narrow base characterizes a population that is getting smaller (Figure 24.2). A large base to any such diagram, as exists in Afghanistan, spells trouble regarding future demands on the country's commodities. Moderately stable nations, such as the United States, show a more consistent age distribution—with some fluctuation, but a small prereproductive population. The prereproductive base of Canada's age structure is smaller than the reproductive base. Canada's population will probably decline in the future.

Immigration and Emigration

If we consider the human population of the world as a whole, the effects of immigration and emigration on population size are eliminated. However, immigration and emigration can dramatically affect the population size of countries, cities, and regions. People tend to migrate out of LDCs; the migration rate is −1/1000. In contrast, people tend to migrate into MDCs, where the average migration rate is 2/1000.

24.2 Patterns of Population Growth

Let's return to an idea we introduced earlier in this chapter—the importance of the initial population size to the number of new individuals added to a population. Recall that *when the*

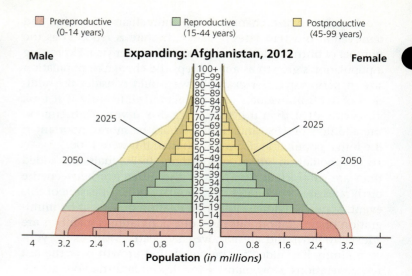

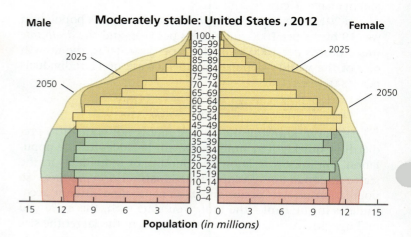

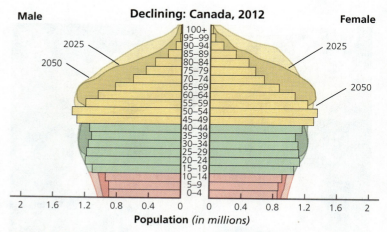

FIGURE 24.2 *Age structures of Afghanistan, the United States, and Canada. The age structures shown are estimates for 2012, 2025, and 2050. Source: U.S. Census Bureau, International Data Base.*

growth rate remains constant, the number of individuals added to the population increases more rapidly as the size of the population increases. This unrestricted growth at a constant rate is called *exponential growth.* When the size of a population growing in this fashion is shown graphically, we see a *J-shaped growth curve.*

Unrestrained growth occurs in environments that have plenty of resources and adequate waste removal. In the real world, however, these conditions rarely exist for long. What happens when the food supply begins to become depleted, when space runs out, or when wastes begin to accumulate? The answer to this question is more than theoretically interesting to humans.

The number of individuals of a given species that a particular environment can support for a prolonged period is called its **carrying capacity** (Figure 24.3). It is determined by such factors as availability of resources, including food, water, and space; the means of cleaning away wastes; disease; and predation pressure.

If environmental limits are placed on population growth, we may wonder what might happen if a population overshoots the carrying capacity of the environment. One possibility is that environmental resources may become critically depleted, and the population may decline precipitously. More commonly, however, population growth is rapid while resources are plentiful. But as the population size approaches the carrying capacity of its environment, growth becomes slower and slower because of environmental pressures such as limited resources or accumulating wastes.

Eventually, growth levels off. and the population size fluctuates slightly around the carrying capacity—a pattern that forms an *S-shaped growth curve.* This pattern is called *logistic growth.*

How do these growth patterns apply to human population growth? For most of the period following the Stone Age, the human population grew steadily but relatively slowly. Around the time of the industrial revolution, the human population size began to increase rapidly, its growth pattern resembling a J-shaped curve. Now, however, the rapidity of its growth is disturbing (Figure 24.4). Human population size did not reach 1 billion until 1804, but since then additional billions are being born at a much faster rate. In mid-2012, the world population was approximately 7 billion people. It is expected that, by 2025, our numbers are expected to reach nearly 8 billion people.

The growth rate for human populations in some parts of the world is greater than in other parts. The growth rate in LDCs—where most people live in poverty—greatly exceeds the growth rate in MDCs, where people are usually wealthier. During the next 40 years, 97% of the world's population growth is expected to be in LDCs. The growth rate of the world population has been slowing steadily over the past decade, suggesting that we may be reaching Earth's carrying capacity.

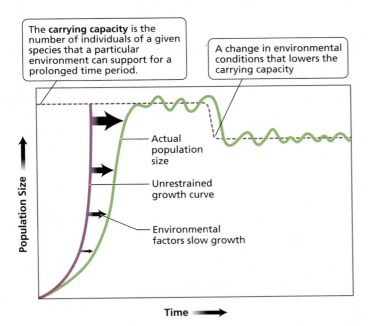

FIGURE 24.3 Population growth is often restrained by environmental factors, including the availability of food, water, and space and the accumulation of wastes. Environmental factors such as these determine the environment's carrying capacity. Eventually, growth levels off, and the population size fluctuates around the carrying capacity. If a change in the environment lowers the carrying capacity, the population will stabilize at a smaller size.

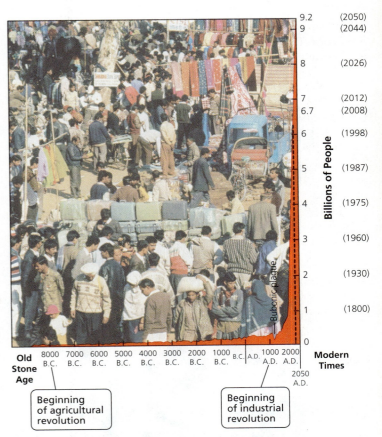

FIGURE 24.4 The human population has grown steadily throughout most of human history. It skyrocketed after the industrial revolution. The growth rate is now showing signs of slowing.

24.3 Environmental Factors and Population Size

The factors that reduce the size of populations can be described as *density-dependent* or *density-independent*. *Density-independent regulating factors* are causes of death that are not related to the density of individuals in a population (the number of individuals in a given area of habitat). These events include such natural disasters as floods, mudslides, earthquakes, hurricanes, and fires. For example, the earthquake that shook Haiti in 2010 killed more than 300,000 people. Such events are density-independent because adding or removing people from the population would not alter the mortality rate of the event itself.

Density-dependent regulating factors are events that have a greater impact on the population as conditions become more crowded. For example, famine can be a density-dependent factor. If only so much food is available, an increasing population would mean less food for each person until, finally, some people begin to starve (Figure 24.5). The greater the population density, the greater the effects of food scarcity on the population. Because disease-causing organisms and parasites spread more easily in crowded conditions than in uncrowded conditions, these too are density-dependent factors that regulate population sizes.

24.4 Earth's Carrying Capacity

Whatever your personal predictions for the future of humankind, one thing should be clear: the world is becoming an increasingly crowded place. You may recall from Chapter 23 that there is no input of new resources on Earth. Therefore, with our growing population, we run the risk of using up our resources and making our world unlivable because of our waste. In fact, human population size contributes to most of the problems we face today.

Earth's ability to support people depends on natural constraints, such as limited resource availability, as well as on human activities and choices. These, in turn, are influenced by economics, politics, technology, and values. Thus, Earth's carrying capacity for humans is uncertain and constantly changing. For instance, technological advances in agriculture and pollution control act to increase Earth's carrying capacity. At the same time, however, we have lowered the carrying capacity by consuming Earth's resources faster than they can be replaced—and, of course, some nonrenewable resources (such as fossil fuels) *cannot* be replaced on human time scales.

Not surprisingly, estimates of Earth's carrying capacity vary widely, ranging from 5 billion to 20 billion people. Those who accept the lower values point to resource depletion and pollution as evidence that, with a world population of more than 7 billion, we have already exceeded Earth's carrying capacity.

24.5 Human Impacts on Earth's Carrying Capacity

An **ecological footprint** is a measure of the amount of productive land and water required to support a person or population based on its consumption levels. Everything a person uses—food, energy, water, housing space—comes from somewhere. Calculating an ecological footprint includes everything that is consumed and the corresponding waste removal. Thus, ecological footprints describe the burden placed on Earth's carrying capacity.

Ecological footprints reflect lifestyle choices. Generally, more affluent people or populations have larger ecological footprints. Because wealth is not equally distributed around the globe, ecological footprint sizes vary among countries, with LDCs having smaller footprints than MDCs do (Figure 24.6).

Agricultural Advances

Agricultural production has increased in many countries, largely because of the so-called *green revolution* of the 1970s and 1980s—the development of high-yield varieties of crops and modern cultivation methods, including the use of farm machinery, fertilizers, pesticides, and irrigation. For example, Indonesia at one time imported more rice than any other country, but now it grows enough rice to feed its people, as well as some to export.

Nonetheless, the green revolution has a big price tag: high energy costs and environmental damage. Although crop yields may be four times greater than with more traditional methods, modern farming practices use up to 100 times more energy and mineral resources. It takes energy—usually obtained from fossil fuels—to produce fertilizers, power tractors and combines, and install and operate irrigation systems.

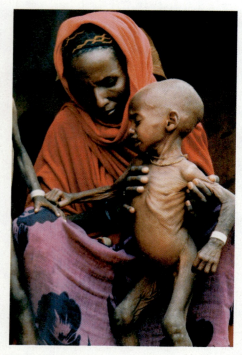

FIGURE 24.5 *Food shortages may reflect density-dependent population controls, because they have a greater impact as population density increases.*

24.5 Human Impacts on Earth's Carrying Capacity   **519**

Q What might account for the differences in the size of the ecological footprints of the nations shown here?

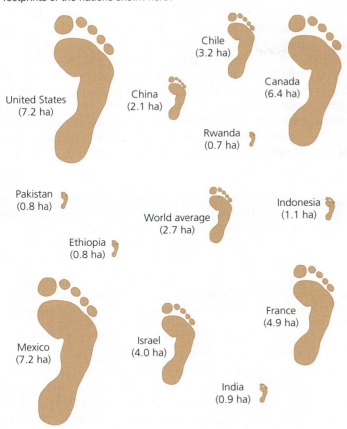

FIGURE 24.6 An ecological footprint is a measurement that includes all the direct and indirect impacts of the use of resources by a person or population. It is expressed as the amount of Earth's surface required for maintenance and is usually expressed in hectares (ha). The citizens of some nations have much larger footprints than others.
Source: Global Footprint Network, 2008.

A More-industrialized and affluent nations consume more of Earth's resources than poorer, less industrialized nations do.

Depletion of Resources

An intact natural ecosystem usually retains soil and remains fertile. However, overuse and misuse of resources can decrease Earth's carrying capacity.

Soil erosion As the human population grows, land is cultivated, grazed, and stripped of vegetation faster than it can recover. Wind and rain then carry away the topsoil, and the area's productivity declines. In this way, overfarming and overgrazing transform marginal farmland to desert, in a process called **desertification** (Figure 24.7). According to the United Nations, 25% of the land on Earth is currently threatened by desertification.

Deforestation People often do not value what they have until it is gone. This may be true for forests. Some of us live in areas where trees are so common that we take them for granted. However, trees, especially when they are grouped together to form forests, are essential to global ecosystems. As

FIGURE 24.7 Desertification is the conversion of farmland, or in this case rangeland, to desert through overuse or other harmful practices.

we learned in Chapter 23, forests play an important role in water, carbon, and nitrogen cycles. Tree roots also reduce erosion by holding soil in place. If trees are cut down, rainwater runs off the land, carrying away soil and causing floods. In addition, ecosystems containing trees support an incredible variety of life. Trees also influence local and global climate, including temperature and rainfall.

Why are forests disappearing? The answer is simple: the human population is expanding. People need space in which to live, and they need wood for homes and furniture. Livestock cannot graze in forests, and crops cannot grow in forests, so trees are cut for pasture and farmland. Most trees are not replanted.

Deforestation is the removal of trees from an area without replacing them (Figure 24.8). Deforestation is taking place in

FIGURE 24.8 Deforestation damages soil quality.

many regions of the world, including the United States. Before the colonization of North America, forests covered most of the eastern seacoast. Cities have now replaced many of those forests. In the U.S. Pacific Northwest, 80% of the forests are slated for logging.

Nonetheless, tropical forests—most of them located in less developed nations—are falling the fastest. Why? The first reason (based on impact) is that land is being cleared so that native families can feed themselves. The second reason is commercial logging. The third reason is cattle ranching. Cattle can graze for about 6 to 10 years after trees have been cleared from a tropical forest before shrubs take over and make the area unsuitable for rangeland. Foreign companies own most of these cattle ranches, and the beef is often exported to fast-food restaurant chains in other countries. Thus the decisions we make about what to eat for lunch may indirectly influence the rate of tropical deforestation. Other reasons include mining and the development of hydroelectric dams.

Whatever the reason for tropical deforestation, soil fertility declines rapidly when these forests are cut. As a result, native people who depend on the soil for their living become poorer.

Overfishing Overfishing is another practice that is lowering Earth's carrying capacity. Fishing has long provided humans with an abundant food supply, but now many fish populations are being depleted to such levels that some fish can no longer be caught in certain areas. Recent examples include Atlantic cod in New England and salmon in the northwestern United States. Worldwide, 70% of the fisheries are overexploited. The problem is that we have not given the fish enough time to replace their numbers at their reduced population size. As numbers of fish have dwindled, the human response has been to fish harder, using more boats and new techniques, including electronic searches. Some species, the Atlantic bluefin tuna for instance, may already be so overfished that they will become extinct. Overfishing is causing the loss of species and entire ecosystems, as well as a valuable food source.

Water shortage Earth has been called the Water Planet because water is so plentiful here. Still, relatively little is available for human use because 97% of Earth's water is in the oceans. You might think that at least all the freshwater would be available, but much of it is tied up in ice or clouds, hidden in underground rivers, or otherwise inaccessible for immediate human use. Indeed, freshwater scarcity is an emerging crisis.

In most regions, however, the underlying cause of water shortage is too many people drawing from a limited supply. Humans use water from two main sources: surface water, such as rivers and lakes, and groundwater, which is water found under Earth's surface in porous layers of rock, such as sand or gravel. When too much surface water is used in an area, ecosystems are affected. As much as 30% of a river's flow can usually be removed without affecting the natural ecosystem. However, in some parts of the southwestern United States, as much as 70% of the surface water has been removed.

Human use of groundwater is also depleting the aquifers (porous layers of underground rock where groundwater is found). Water is added to aquifers by rainfall or melting snow that filters through the topsoil. In some areas, however, water is being removed from aquifers faster than it can be replenished. The Ogallala Aquifer underlies eight states in the Great Plains region of the United States. Much of the water pumped up from the Ogallala Aquifer is used for and has made this region some of the most productive farmland in the country. However, the water level in the aquifer is dropping by nearly 3 feet a year. The water tables beneath the highly populated northeastern United States are also dwindling dangerously. In the Great Plains, the drought of 2011 to 2012 caused the largest annual decline in groundwater in 25 years.

what would you do?

A proposed crude oil pipeline from the tar sand hills of western Canada is expected to deliver 700,000 barrels of oil a day. The pipeline crosses 254 miles of Nebraska. The oil spill in the Gulf of Mexico in 2010 has some people worried that the pipeline might leak and endanger the Ogallala Aquifer. If you or someone you love lived in the region that gets its water from this aquifer, would you favor or oppose this pipeline? What factors would you consider in making your decision?

In some places drought, or lack of rain, has compounded the problem of water scarcity caused by the growing human population. A drought has affected the western United States for more than a decade. In some regions, farming and ranching have been severely affected. For example, some ranchers in the West have sold off parts of their herds because there is no grass to feed them. Drought also increases the risk of wildfires that kill plants, wildlife, and humans as well as destroy property. (Climate change, discussed later in the chapter, can also be blamed for the increased numbers of wildfires in recent years—and for the extended average fire season in the western United States, which is now 2 months longer than it was in 1970.) In Colorado, the summer of 2012 was the worst wildfire season in a decade.

A **water footprint** is an indicator of direct and indirect water use by a person, locality, or nation. *Direct use* refers to water use at home; *indirect use* refers to water used to produce goods and services. Because not all of the goods used in a country are produced there, a nation's water footprint includes the water needed to produce that product inside and outside that country's border. North Americans have an exceptionally large water footprint. The global average yearly use of water per person is 1240 cubic meters. In the United States, each person uses about 2500 cubic meters of water a year; in China, each person uses about 700 cubic meters of water a year.

These numbers are shocking, and you might be tempted to claim that *you* certainly do not use that much water. In one sense, you would be right. Personal use accounts for only a small percentage of total U.S. water use. Most of the water is used in agriculture. To feed its growing human population, our country increasingly relies on irrigation to make arid areas fruitful. The rest of the water is used primarily in industry for steam generation or the cooling of power plants.

Irrigation has pros and cons. It allows crops to grow in areas that would otherwise be barren. But it can also, in the long run, make the land unfit for agriculture. Irrigation water contains dissolved minerals. Whereas runoff from natural rainfall would carry these salts away, irrigation water soaks into the soil. Then, when the water evaporates from the soil, the salts are left behind. The resulting accumulation of salts in the soil is called *salinization*. Worldwide, salinization destroys the fertility of 5000 km² (1930 mi²) of irrigated land each year.

stop and think

Conservationists sometimes buy wetlands, hoping to preserve the habitats of endangered plants and wildlife. Despite this conservation strategy, the wetlands sometimes dry up. Why is it also important to buy the rights to the water that feeds the wetland?

What steps can be taken to reduce water shortages?

1. **Reduce water use.** Each of us could help by individual efforts. However, because most water is used in agriculture, the biggest benefit would come from improved irrigation methods. Between 1950 and 1980, for example, Israel reduced its water wastage from 83% to a mere 5%, primarily by changing from spray irrigation to drip irrigation. In the past, more than half the water diverted for irrigation in California's Imperial Valley was wasted because of poorly designed irrigation systems. However, each year additional efficient irrigation systems are installed.

2. **Raise the price of water.** The cost of water in the United States is substantially below its cost in European countries. Experience has shown that the consumer's interest in conserving water is directly related to its cost. Clearly, economic and legal incentives for conserving water are needed. Some experts suggest that water will be the most precious commodity in the twenty-first century, as oil was in the twentieth century.

Pollution

Another way we lower Earth's carrying capacity is by polluting resources, making the resource unfit for human use or damaging the environment in other ways.

Water pollution Factories, refineries, and waste treatment plants release fluids of varying quality directly into our water supplies. In Chapter 23, we discussed several contaminants responsible for surface water pollution. For example, mercury and PCBs in water become concentrated in organisms as these chemicals are passed up the food chain. We also considered how fertilizer runoff can lead to an overgrowth of algae and plants in bodies of water, which can deplete the oxygen content of the water so much that it no longer supports life. Although oil spills from oil rigs or tankers have become less common since the passage of the U.S. Oil Pollution Act of 1990, they can still occur with devastating consequences, as they did with the Deepwater Horizon oil spill in 2010.

The same materials that contaminate surface water can contaminate groundwater. Above- or belowground storage tanks can corrode and leak their contents, which may be oil, gasoline, or other hazardous liquids. Septic systems, which are designed to allow wastewater to slowly drain away from their sources, may leak bacteria, viruses, and household chemicals to groundwater. Hazardous waste sites and landfills can also leak harmful materials into groundwater.

Ozone pollution and ozone depletion Ozone, a gas whose molecules are composed of three atoms of oxygen (O_3), can be a good thing or a bad thing. At Earth's surface—say, anywhere in the first few miles of atmosphere—it is generally a bad thing. As an air pollutant, ozone is the primary component of photochemical smog, the noxious gas produced largely by sunlight interacting with air pollutants, such as hydrocarbons and nitrogen dioxide. Ozone irritates the eyes, skin, lungs, nose, and throat. (Effects of photochemical smog on the respiratory system are discussed in the Environmental Issue essay, *Air Pollution and Human Health*.) Ozone can also damage forests and crops and dissolve rubber.

In contrast, naturally produced ozone is an essential part of the stratosphere, a layer of the lower atmosphere that encircles the Earth about 10 to 45 km (6 to 28 mi) above its surface. The concentration of ozone in the stratosphere reaches levels of only about 1 molecule in 100,000. If all this ozone were compressed from top to bottom, its depth would equal only the diameter of a pencil lead. Still, the ozone is able to shield Earth from excessive ultraviolet (UV) radiation from the sun; without it, most terrestrial life-forms could not exist. UV radiation causes cataracts, aging of the skin, sunburn, and snow blindness; and it is the primary cause of skin cancer, which kills nearly 11,000 Americans each year. In addition, UV radiation inhibits the immune systems of animals and can interfere with plant growth.

In 1985, British scientists reported a sharp drop in the concentration of ozone over the Antarctic. This "hole" appears on a yearly cycle—at the beginning of the Antarctic spring (early September through mid-October; Figure 24.9). The primary cause was discovered to be the action of chlorofluorocarbons (CFCs)—chemicals once important in cooling systems, as aerosol propellants, in the manufacture of plastic foam such as Styrofoam, and as solvents in the electronic industry. As pollutants, the CFCs drift up to the stratosphere, where UV radiation breaks them down into chlorine, fluorine, and carbon. Then, under the conditions found in the stratosphere, chlorine can react with ozone, converting it to oxygen. Because chlorine is not altered in this reaction, a single chlorine molecule can destroy thousands of ozone molecules.

Several laws are now in place to prevent the production or release of CFCs. In 1990 and 1992, industrialized countries signed agreements to reduce CFC production and have now phased out CFC production. Substitutes for CFCs have been found for aerosol cans, refrigerators, and air conditioners. There is evidence that the reduction in CFC production is paying off. In 1996, two years after CFC use peaked, chlorine levels in the lower atmosphere were slightly lower than before. By 1999, chlorine levels in the stratosphere were reduced. Scientists predict that the ozone hole will recover by 2050–2070.

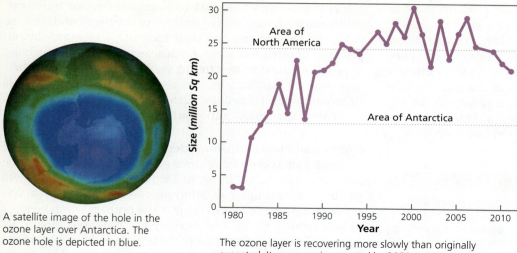

A satellite image of the hole in the ozone layer over Antarctica. The ozone hole is depicted in blue.

The ozone layer is recovering more slowly than originally expected. Its recovery is expected by 2050 to 2070.

The ozone layer absorbs ultraviolet (UV) radiation. A thinner ozone layer allows more UV radiation to reach Earth, increasing cancer risk.

FIGURE 24.9 *The thinning of the ozone layer.*

24.6 Global Climate Change

The word *climate* describes a region's average temperature, precipitation, humidity, barometric pressure, solar radiation, and other weather characteristics considered over a relatively long period of time. *Global climate change,* then, refers to changes in Earth's climate patterns—trends and variations. Although Earth's climate changes naturally over time, the rate of change today is extremely rapid.

The Intergovernmental Panel on Climate Change (IPCC), an international panel consisting of hundreds of scientists, was established by the United Nations Environment Programme and the World Meteorological Organization to assess technical and scientific information regarding climate change as well as its social and economical implications. These groups scrutinized thousands of scientific papers and measurements of physical conditions on Earth. Here are two key conclusions of the IPCC's *Fourth Assessment Report* (2007).

1. The global climate is unquestionably getting warmer. The global temperature has increased 0.74°C in the last 100 years. Seven of the eight warmest years on record have occurred since 2001.[1] During the last decades, the number of really hot days and heat waves increased, and the number of cold days decreased.

2. Most of the increase in global temperatures is due to human activity that releases greenhouse gases (carbon dioxide, methane, and nitrous oxide) into the atmosphere. The IPCC indicates that the "assessed likelihood, using expert judgment" that human activity is responsible for global warming is over 90%.

Nearly everyone agrees that the global climate is getting warmer, and most people agree that the elevated temperatures are due to human activities, primarily the burning of fossil fuels. We will look at the evidence supporting the link between greenhouse gases and climate change and then at the predicted consequences of global climate change.

[1]The hottest years on record were 2005 and 2010.

Global Warming and Greenhouse Gases

Recall from Chapter 23 that humans are disrupting the natural carbon cycle by burning fossil fuels, which increases atmospheric carbon dioxide (CO_2), and by clear-cutting forests, which reduces the removal of CO_2 from the atmosphere. Because of these activities, the level of atmospheric carbon dioxide is rising. Carbon dioxide, methane, nitrous oxides, water vapor, and CFCs are considered greenhouse gases. Like the glass on a greenhouse, these gases allow the sunlight to pass through to Earth's surface, where it is absorbed and radiated back to the atmosphere as heat—long-wave infrared radiation. Greenhouse gases then absorb the infrared radiation, thus trapping the heat (Figure 24.10). Because of this effect, a growing concern is that the increase in atmospheric CO_2 due to human activities is leading to a rise in temperatures throughout the world—a situation popularly known as **global warming**.

Scientists rely on direct and indirect evidence of the link between global warming and greenhouse gases. Direct evidence

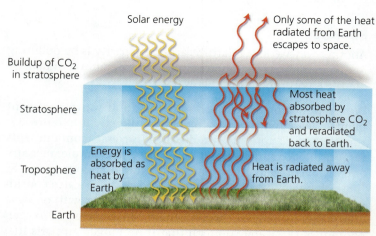

FIGURE 24.10 *Carbon dioxide and other greenhouse gases promote global warming by trapping heat (infrared radiation) in the atmosphere, much as greenhouse glass traps heat. Thus, the global warming that is believed to result from the accumulation of these gases is known as the greenhouse effect.*

comes from present-day measurement. Since 1958, atmospheric CO_2 has been measured at Mauna Loa Observatory in Hawaii. Indirect evidence comes from ice cores, long columns of ice taken from ice sheets, ice caps, or glaciers. The most recent ice is at the top of the core; the oldest is at the bottom. Scientists can examine these ice cores and determine greenhouse gas concentrations, environmental temperature, and snowfall to understand the region's climate history. Notice the correlation between rising atmospheric CO_2 and global temperature in Figure **24.11**.

Some people do not believe that rising atmospheric CO_2 from human impacts is the cause of increasing global temperature. Some point out that rising temperatures could be the result of a natural climate cycle. Others suggest that domesticated animals, such as cattle, sheep, and goats, are to blame. Bacteria in the guts of these animals produce methane, a greenhouse gas whose heat-trapping ability is 20 times greater than that of carbon dioxide. The collective belches and flatulence of all the grazing animals on Earth accounts for 20% of the methane gas released into the atmosphere.

In 2007, the IPCC predicted that Earth's surface temperatures will rise 1.8°C to 4.0°C (3.2°F to 7.2°F) during the twenty-first century. The range of the prediction for temperature increase is large because there are uncertainties concerning the effects of many of the interacting factors, such as cloud cover due to evaporation, cooling effects of atmospheric particulates such as sulfates, possible increased primary productivity due to the increase in CO_2, and ability of the ocean to absorb CO_2.

Why is there such commotion over global warming? Why should anyone care whether the atmosphere becomes a few degrees warmer? The answer is that global warming is closely related to global climate change, which has many social and economic consequences. Let's look at the current extent of global warming and possible consequences suggested in the 2007 IPCC report.

- **Melting ice caps, rising sea level, and warmer oceans.** Mountain glaciers and snow cover at both poles has declined. Land-based ice sheets in Greenland and Antarctica have been melting. When ice on land melts, sea level rises. In addition, the resulting reduction of white covering on Earth means that less of the sun's energy is reflected back to space. Less reflection further increases warming, which leads to increased glacier melting. During the decade from 1993 to 2003, sea level rose about 31 mm (1.22 in.), and the temperature of the oceans increased. Ocean water expands as it warms, contributing further to the rise in sea level.

  The temperature change, and therefore the melting of ice sheets, is greatest in the Arctic, where it has adversely affected the food web. Because of the melted ice sheets, polar bears must swim long distances (instead of walking on the ice) to capture the seals they rely on for food. Swimming requires more energy than walking does, so polar bears are starving. As a result, polar bears are becoming an endangered species. The melting ice sheets also affect human populations. The Inuit, a native people of the North American coast, also rely on seals for food, and sea ice is dangerous to hunt on.

- **Coastal cities could flood.** The rising sea level would have the greatest effect on low-lying coastal countries such as Bangladesh, Egypt, Vietnam, and Mozambique. Seawater could cover significant areas in these densely populated lands. In the United States, major cities such as New York, Miami, Jacksonville, Boston, San Francisco, and Los Angeles could someday be largely underwater. Do you live near a coastline?

- **Changing weather patterns.** Global warming is causing changes in both temperature and rainfall patterns. This climate change is responsible for the drought-causing water shortage in the U.S. Southwest. Most scientists agree that warmer oceans can fuel hurricanes. Hurricanes in the North Atlantic have intensified and will continue to intensify during the twenty-first century.

  Recall from Chapter 23 that the distribution of climax communities, each with its characteristic plant and animal life, is largely determined by temperature and rainfall. Because of global warming, some species will thrive, whereas others may become extinct. There will be a shift in the location of agricultural regions, meaning food production in the central plains of the United States and Canada could drop. Areas that now produce enough food for export may not be able to produce enough for the local population. Meanwhile, other regions—parts of India and Russia, for instance—might receive increased rainfall that would benefit agriculture. These changes will alter the distribution of "haves" and "have-nots." How will the world community deal with the shifts in economic and political power?

- **Human health.** The IPCC predicts that global warming will have various negative effects on human health. The change in climate is likely to increase the number of deaths due to heat waves, flooding, and droughts leading to starvation. Climate change could also increase air pollution. In addition, because mosquitoes and ticks, as well as other vectors of disease, would thrive in a warmer climate, we may see increases in malaria, West Nile virus encephalitis (inflammation of the brain), Lyme disease, and the like.

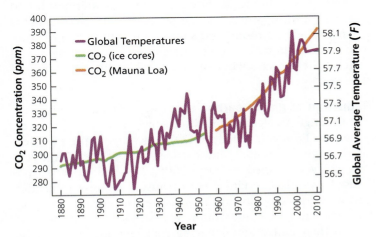

FIGURE 24.11 *The concentration of CO_2 in the atmosphere has been increasing for many years. Two major causes are the burning of fossil fuels, which releases CO_2, and deforestation, which leaves fewer trees to remove CO_2 from the atmosphere. Global temperatures have also risen during this period.*
Source: National Oceanic and Atmospheric Administration.

ENVIRONMENTAL ISSUE

Air Pollution and Human Health

Caution: The air you breathe may be hazardous to your health. It may even kill you—especially if you have heart or respiratory problems. Two major human sources of air pollution are motor vehicle exhaust and industrial emissions. The exhaust pipes on our cars and the smokestacks on factories spew oxides of sulfur, nitrogen, and carbon; a mixture of hydrocarbons; and many particulates (such as soot and smoke) into the air. In the atmosphere, sulfur dioxide and nitrogen dioxide dissolve in water vapor and form an aerosol of strong acids—sulfuric acid and nitric acid, respectively. On the one hand, some of this aerosol may drift upward, forming acidic clouds that can be blown hundreds of miles by the winds and then fall as acid rain. On the other hand, the acidic aerosol may remain close to its source as a component of the haze created by air pollution. In addition, sunlight can cause hydrocarbons and nitrogen dioxide to react with one another and form a mix of hundreds of substances called *photochemical smog*. The most harmful component of photochemical smog is ozone, which attacks cells and tissues, irritates the respiratory system, damages plants, and even erodes rubber. Paradoxically, the ozone that is so damaging when it is found at ground level in photochemical smog is the same compound that is beneficial in the upper layers of the atmosphere, where it prevents much of the sun's ultraviolet radiation from reaching Earth's surface. Unfortunately, the ozone found in smog does not make its way to the ozone layer of the upper atmosphere, because it is converted to oxygen within a few days.

When we breathe polluted air, the respiratory system is, not surprisingly, the first to be affected. As air containing toxic substances fills the lungs, cells lining the airways and those within the lungs are injured. Damaged cells release histamine, which causes nearby capillaries to widen and become more permeable to fluid. As a result, fluid leaks from the capillaries and accumulates within the tissues. Even a brief exposure to oxides of sulfur (5 parts per million, or ppm, for a few minutes), the oxides of nitrogen (2 ppm for 10 minutes), or ozone leads to fluid accumulation, increased mucus production, and muscle spasms (intense involuntary contractions) in the bronchioles. These effects obstruct air flow and reduce gas exchange. Long-term irritation of bronchi by pollutants is a cause of chronic bronchitis and emphysema. The process begins when irritation causes fluid accumulation that, in turn, stimulates mucus production and coughing. The cough and mucus, signs of chronic bronchitis, irritate the lungs even more. As the bronchitis continues, the air passageways become narrower, trapping air in the lungs. When the increased pressure accompanying a cough causes the overinflated alveoli to rupture, emphysema begins.

With so many documented health consequences of air pollution, you may wonder why the world's great minds haven't developed the technology to solve the problem. In fact, they have. But unfortunately, air pollution is not just a scientific problem; it is also a social, political, and economic problem. The technology to prevent air pollution is available. For example, scrubbers can remove 90% of the particulate matter and sulfur dioxide from industrial smokestacks. However, the technology is costly.

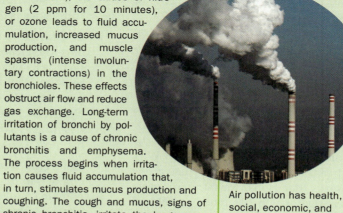

Air pollution has health, social, economic, and political consequences.

Questions to Consider

- Are you willing to pay more for goods and services, such as food, clothing, and transportation, to offset the cost of air pollution-reducing technology? How much more are you willing to pay for low-sulfur fuels, for instance?
- Some of the strategies for improving air quality are likely to impose a certain amount of inconvenience. Are you willing to carpool, take a bus, or—better yet—walk instead of drive to improve the air quality in your community?

---

stop and think

Scientists are working on a low-nitrogen diet for cattle and sheep to reduce the amount of methane these animals produce. How might such a diet affect global warming? On what factors might the outcome depend?

Carbon Footprint

A person's—or a nation's—**carbon footprint** is a measure of the amount of CO_2 entering the atmosphere as a result of that person's—or nation's—daily activities. It includes the carbon emissions directly from our activities as well as those from the whole life cycle—production through waste disposal—of the products we use.

Reduction of direct carbon footprint We have the most control over carbon emissions from our own activities. How can you reduce the size of your carbon footprint? One way is to walk, ride a bicycle, or take public transportation instead of driving. When driving is necessary, carpool with others whenever possible. When you purchase appliances, choose energy-efficient ones. Minimize the use of heating or cooling systems in your home by programming your thermostat and wearing clothing suited to the season. Turn off the lights when you leave a room, and shut down your computer after using it. In short, conserve energy. The environment will benefit; the exercise will improve your health; and you will save money.

Reduction of global carbon footprint To reduce future climate change, the leaders of 182 nations signed an agreement, the Kyoto Protocol, to reduce CO_2 emissions to pre-1990 levels through technology and reduced energy use. The Kyoto Protocol was ratified in early 2005. The United States has refused to ratify the agreement because it requires industrialized countries to reduce their emissions of greenhouse gases but does not require rapidly industrializing countries to reduce their emissions. Believing that technology can solve the problem of global warming, the United States and the European Union agreed to share information on technological advances. A United Nations conference on climate change was held in 2011 in Durban, South Africa. The talks had limited success.

Whether or not you believe energy usage and the burning of fossil fuels is affecting the global climate, one thing is certain: the supply of fossil fuels is limited. Oil, coal, and natural gas are *not* renewable resources. When they are gone, they are gone forever. Even if you are among the skeptics about the role of greenhouse gases in climate change, you should consider some of the possible social, political, economic, and environmental consequences of continued reliance on fossil fuels (Figure 24.12). We will have to think through our choices carefully. For example, some experts tout *biofuels*—fuels made from recently dead biological material—as the way to reduce our dependence on oil. Other experts argue that growing crops to make biofuel is driving up the cost of food. Sugarcane is sometimes used to make biofuel. Will increased use of biofuel speed deforestation of the Amazon rain forest to make room for growing sugarcane? How much stress will irrigating plants for biofuel place on our water supplies? Residents of several states in the Midwest are going to court to challenge the development of biofuel projects in their region.

Besides biofuel, what are our energy options to reduce reliance on fossil fuel? We have several choices, each best suited for different environments, and each with costs and benefits.

Nuclear energy is the energy that holds protons and neutrons together in the nucleus of an atom. In nuclear power plants, we split atomic nuclei and convert the energy released to thermal energy, which we then use to create electricity. The process, called *nuclear fission*, does not produce greenhouse gases, and in that respect it is good for environment. Mining uranium damages the environment less than mining for coal does. However, nuclear fission must be carefully controlled, and accidents have happened. In 2011, a nuclear disaster occurred at the Fukushima Daiichi plant in Japan when it was hit by a tsunami caused by an earthquake. The combined disasters caused the deaths of more than 10,000 people. It was more than a year before the radiation levels of fish living off the coast dropped enough to make them safe to eat. Accidents at nuclear power plants are extremely rare, but the results are disastrous. Nuclear power also involves other drawbacks: the creation of radioactive waste and the problem of its disposal. Even so, today more than 400 nuclear power plants are operating in more than 30 countries.

Hydroelectric power plants use the energy of falling water to produce electricity. Hydroelectric power does not produce greenhouse gases, it is efficient, and it is renewable (as long as it rains and snows). However, damming rivers damages affected habitats—both aquatic and terrestrial.

Other renewable energy technologies are not yet able to meet our electricity needs, but they show great promise. Solar power involves harnessing the sun's energy. In windy areas, giant windmills can convert wind power to electricity. Ocean currents can also be converted to electricity. Geothermal energy deep within Earth can be used to heat water that can, in turn, be used to generate electricity.

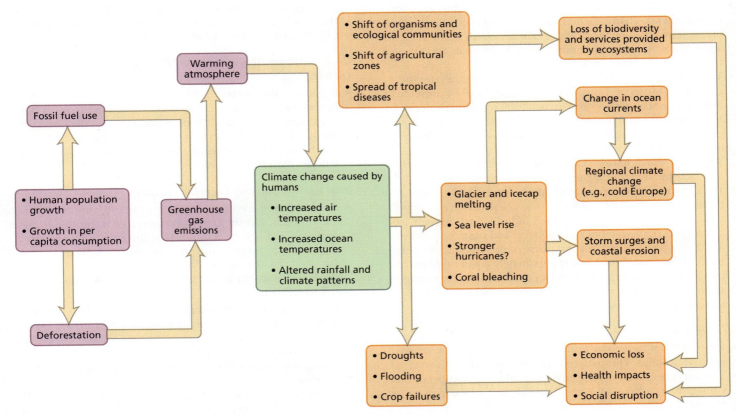

FIGURE 24.12 *Global climate change (shown in the green box) is related to human use of fossil fuels, as shown in the boxes on the left. The social and economic consequences of global climate change are shown in the boxes on the right.*

24.7 Looking to the Future

In many ways, our grandparents were right: older ideas *were* better—at least, some were. Today's economy is wasteful. Many of us behave as if the world exists for humans alone. We often measure progress as the *gross domestic product*, which measures the value of all the goods and services produced. However, a sustainable lifestyle requires that we use the *genuine progress indicator*, which includes the costs of production. Costs of production include depletion of resources and pollution—our ecological footprint—and negative social and economic costs.

In Chapter 23 and this chapter, we have seen that humans are a *small* part of the biosphere and that it must remain healthy if we are to remain healthy. We must realize that we are subject to the same principles that govern all other life—indeed, all of Earth. Instead of trying to conquer nature, we must work with natural laws. Above all, we must recognize that Earth's resources are limited and must be shared with all living organisms as well as preserved for future generations.

This dramatic turnabout in thinking and behavior will not be easy to accomplish. It will require changing the way governments and businesses operate as well as modifying the ingrained habits of individuals. However, humans are by nature problem solvers. We now have the opportunity to approach with intelligence one of the greatest problems ever to confront humankind.

Jane Goodall, famed animal behavior researcher and founder of the Jane Goodall Institute, a center for environmental studies, is optimistic about our future. In her words, "My reasons for hope are fourfold: (1) the human brain; (2) the resilience of nature; (3) the energy and enthusiasm that is found or can be kindled among young people worldwide; and (4) the indomitable human spirit."[2] We have to be optimistic that a solution is possible and then work together to achieve it.

looking ahead

In this book, we have explored many concepts and issues. You now have the background and learning skills to make wise decisions in the future about your health and societal issues.

[2]Jane Goodall (with Phillip Berman), *Reasons for Hope: A Spiritual Journey* (New York: Warner Books, 1999), 233.

HIGHLIGHTING THE CONCEPTS

24.1 Population Changes (pp. 514–516)

- Population dynamics describes how populations change in size. Population size changes when the number of individuals added differs from the number of individuals removed. Individuals are added through births and immigration. Individuals are removed through death and emigration.
- The change in population is usually expressed as a rate: the number of births or deaths for a certain number of individuals during a certain period (usually per 1000 individuals per year). If we rule out immigration and emigration, we can say a population's growth rate equals the birth rate minus the death rate.
- The age at which females have their first offspring is an important factor in determining a population's growth rate.
- The age structure of a population is an important determinant of future changes in population size. The age structure of a population consists of the relative numbers of individuals in each age category: prereproductive, reproductive, and postreproductive. A growing population has a large base of prereproductive individuals. A stable population has approximately equal numbers of individuals in each age category. A population that is getting smaller has a large proportion of individuals who are past reproductive age.

24.2 Patterns of Population Growth (pp. 516–517)

- When it is not limited by finite resources, a population grows without restraint. This pattern is called exponential growth.
- The carrying capacity of the environment is the number of individuals (of a particular species) it can support over a long period of time. Carrying capacity is determined by the amount of available resources.
- Generally, as a population grows and its size approaches the carrying capacity of the environment, the growth rate slows and the population size levels off, fluctuating around the carrying capacity. This pattern is called logistic growth.

24.3 Environmental Factors and Population Size (p. 518)

- Population size can be regulated by density-independent regulating factors (events, such as natural disasters, whose effects are not influenced by the size of the population) and by density-dependent regulating factors (factors that have a greater impact as the population becomes more crowded). Density-dependent factors include food availability and disease.

24.4 Earth's Carrying Capacity (p. 518)

- Humans have raised Earth's carrying capacity through technological advances, but they have also decreased it through resource depletion and pollution. The size of the human population contributes to many of the problems we face today.

24.5 Human Impacts on Earth's Carrying Capacity (pp. 518–521)

- An ecological footprint is a measure of the impact a person or population makes on the environment. It is expressed as the total amount of land and water needed to produce and dispose of the products that are consumed.
- Soil erosion due to overgrazing and overfarming is converting marginal farmlands to deserts, by a process called desertification.
- Deforestation—removing trees from an area without replacing them—is a growing problem throughout the world, especially in tropical rain forests.

- Overfishing is causing fish populations to fall below replacement levels.
- We are withdrawing more freshwater from available sources than can be replaced by precipitation.
- Pollution is making some freshwater sources unfit for human use.
- Air pollution in the form of chlorofluorocarbons (CFCs) has depleted the ozone layer in the lower atmosphere (the stratosphere). The ozone layer traps ultraviolet (UV) light, protecting life on Earth from the harmful effects of UV radiation.

24.6 Global Climate Change (pp. 522–525)
- Earth's surface temperature is getting warmer. It is more than 90% likely that this increase is due to an increase in greenhouse gases resulting from human activities, such as burning fossil fuels.
- Greenhouse gases (CO_2, methane, nitrous oxides, water vapor, and CFCs) trap heat in Earth's atmosphere.
- As Earth's atmosphere gets warmer, we can expect continued melting of ice caps and glaciers, rising oceans that may flood coastal cities, climate change, and human health problems related to heat and vector-borne diseases.
- A carbon footprint is the amount of CO_2 entering the atmosphere due to the daily activities of an individual or group of people. We can reduce our carbon footprint by reducing our reliance on fossil fuels. Alternative fuels each have their own costs and benefits.

24.7 Looking to the Future (p. 526)
- Humans are a small part of a large ecosystem. We must find ways to work within natural laws to preserve ecosystem Earth and its biodiversity.

RECOGNIZING KEY TERMS

population dynamics *p. 514*
doubling time *p. 515*
age structure *p. 516*
carrying capacity *p. 517*
ecological footprint *p. 518*
desertification *p. 519*
deforestation *p. 519*
water footprint *p. 520*
global warming *p. 522*
carbon footprint *p. 524*

REVIEWING THE CONCEPTS

1. Define the *growth rate* of a population. *pp. 514–515*
2. What is meant by the age structure of a population? How is a population's age structure related to predictions of its future growth? *p. 516*
3. Differentiate density-independent regulating factors from density-dependent regulating factors. Give examples of each. *p. 518*
4. What is an ecological footprint? *p. 518*
5. What is the carrying capacity of the environment? *p. 518*
6. What causes desertification? *p. 519*
7. Define *deforestation*. Where is it occurring most rapidly? What are the primary reasons for deforestation? *pp. 519–520*
8. What determines whether ozone is helpful or harmful? What is causing the destruction of the ozone layer? *p. 521*
9. What does the term *greenhouse gas* mean? Explain why there is concern that the rising level of atmospheric carbon dioxide could lead to global warming. *pp. 522–525*
10. Which of the following is an example of a density-dependent regulating factor?
    a. earthquake
    b. starvation
    c. flood
    d. drought
11. If a population reaches the carrying capacity of the environment,
    a. unrestrained growth will occur.
    b. the population will decline rapidly.
    c. food and other resources will increase.
    d. the population size will fluctuate around this level.
12. The ozone layer of the atmosphere absorbs
    a. carbon dioxide.
    b. chlorine.
    c. nitrogen dioxide.
    d. ultraviolet radiation.
13. An ecological footprint measures
    a. the number of people who can live in an area.
    b. the amount of natural resources used by each person in an area.
    c. the rate of population growth in an area.
    d. the degree of habitat destruction in an area.
14. Which pollutant is responsible for the hole in the ozone layer?
    a. carbon dioxide
    b. methane
    c. chlorofluorocarbon (CFC)
    d. sulfur dioxide
15. Which of the following is the primary reason for the decline in the population of bluefin tuna?
    a. overfishing
    b. radiation from nuclear disasters
    c. dams preventing the tuna from swimming upstream
    d. reduction in the tuna's food supply due to global warming
16. Choose the *incorrect* statement about water use.
    a. Indirect water use is greater than direct water use.
    b. Irrigation is responsible for a great deal of water use.
    c. Water use is greater is less developed countries.
    d. A water footprint is a measure of water use.
17. Which of the following is caused by heavy irrigation?
    a. salinization
    b. global climate change
    c. drought
    d. none of the above

APPLYING THE CONCEPTS

1. The populations of Kenya and Italy (shown in the figure) have approximately the same number of people, and the growth rate of the two populations is the same. Assuming that the growth rate stays the same, will the populations have the same number of people in 50 years? Explain.
2. Asian carp were introduced in 1963 to control underwater vegetation on fish farms, initially in Arkansas. A flood in 1993 allowed Asian carp to escape from the fish farms into rivers and eventually into the Mississippi River system. By 2010, Asian carp had expanded their range and were threatening to invade the Great Lakes. Asian carp are very large fish that eat plankton, which many other freshwater fish also consume.
   a. Explain why the Asian carp population was able to increase extremely rapidly in a new environment.
   b. Explain why scientists are concerned that Asian carp will reduce the population size of other freshwater fish, even though the carp do not eat these other fish.
3. When soldiers returned home after World War II, many of them began having children who came to be called the *baby boomers*, the largest generation in United States history. Born between 1946 and 1964, the baby boomers began to turn 60 in 2006. Many will soon reach retirement age. How would you expect this change in population structure to affect how tax money will be spent?
4. Explain why population growth is usually expressed as a *rate*.

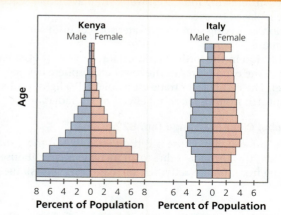

BECOMING INFORMATION LITERATE

Use at least three reliable sources (journals, newspapers, books, websites) to answer the following questions. List each source you considered, and explain why you chose the three sources used.
1. Choose a body of water (river, lake, pond, or ocean) near your home. Is it polluted? If so, what are the main pollutants and their sources? If not, how has it been protected?
2. Use an online calculator to determine the size of your water footprint. Keeping your lifestyle in mind, what three steps could you take to reduce the size of your water footprint?
3. Use an online calculator to determine the size of your carbon footprint. Keeping your lifestyle in mind, what three steps could you take to reduce the size of your carbon footprint?
4. Some government leaders in the United States are advocating drilling for oil in the Arctic National Wildlife Refuge in Alaska. Some of the native people are in favor of this project. List the costs and benefits of this endeavor from economic, environmental, and social points of view.

MasteringBiology®

Go to MasteringBiology for practice quizzes, activities, eText, videos, current events, and more.

Appendices

Appendix 1   *Answers to Reviewing the Concepts Questions*   A-2
Appendix 2   *Hints for Applying the Concepts Questions*   A-12

Appendix 1

Answers to Reviewing the Concepts Questions

Chapter 1

1. Seven traits that characterize life:
   a. Living things contain molecules of nucleic acids, proteins, carbohydrates, and lipids.
   b. Living things are composed of cells (smallest unit of life).
   c. Living things grow and reproduce.
   d. Living things use energy and raw materials for metabolism.
   e. Living things respond to their environment.
   f. Living things maintain homeostasis.
   g. Populations of living organisms evolve and have traits.

2. Mammals are unique in that they have hair and that they feed their young milk produced by mammary glands. In addition to mammalian traits, primates also have forward-looking eyes, a well-developed brain, and opposable thumbs.

3. A controlled experiment is an experiment that has two groups. One group is the control group, and the other group is the experimental group. Each group is treated in the same manner, except for the variable manipulated in the experiment. The control group is not treated/manipulated, and the experimental group is treated/manipulated.

4. Inductive reasoning uses accumulated facts to draw a logical conclusion. Deductive reasoning uses general statements to deduce other conclusions. Deductive reasoning is usually described as an "if-then" series of associations.

5. A placebo is generally used in human drug testing experiments to evaluate the effectiveness of a drug. The placebo is not the real drug being tested; it is usually a sugar pill that looks like the real drug and is given to the control group to determine whether the effects of the drug are real or imagined.

6. In a double-blind experiment, neither the participants nor the researcher evaluating the results knows who received the placebo until after the experiment is completed.

7. d
8. c
9. a
10. b
11. c
12. c
13. a
14. c

Chapter 2

1. An atom is the smallest unit of matter that cannot be broken down by simple chemical means. An element is a pure form of matter containing only one kind of atom. A compound is made of two or more elements. A molecule is a chemical structure held together by covalent bonds. Molecules may contain the same kinds of atoms or different kinds of atoms.

2. An isotope is an atom with a different number of neutrons than the most common form, resulting in a different mass. Carbon is usually found as carbon-12. Carbon isotopes, such as carbon-13 and carbon-14, have more mass with extra neutrons.

3. Medical x-ray imaging is used to diagnose illness. Radiation therapy is used directly at tumor sites to kill cancer cells.

4. Water is a polar molecule, so it makes a perfect solvent and medium for moving substances to and from cells in the body. Water has a high heat capacity, which allows for better temperature regulation of the body. Water has a high heat of vaporization, so evaporation of sweat removes heat from the body.

5. Polymers are chains of repeating molecules (monomers) formed by dehydration synthesis (loss of H_2O) and broken by hydrolysis (addition of H_2O). DNA is a polymer of nucleotides. Starch is a polymer of glucose. Proteins are polymers of amino acids.

6. Energy-storage polysaccharides are starch in plants and glycogen in humans and other animals. A structural polysaccharide is cellulose, found in plants.

7. A phospholipid is composed of a molecule of glycerol, two fatty acids, and a phosphate group. The end with the phosphate group is negatively charged and hydrophilic. The fatty acid tails are nonpolar and hydrophobic. Cell membranes consist of phospholipids in a bilayer: the hydrophobic fatty acid tails are on the inside of the membrane, and the hydrophilic heads are on the outside of the membrane.

8. ATP is the nucleotide adenosine triphosphate formed from ribose, the nitrogen-containing base adenine, and three phosphate groups. ATP is a high-energy molecule used in the human body to provide energy by the breaking of a phosphate bond.

9. Genes are made of DNA. RNA, in various forms, converts the genetic information in DNA into proteins. Proteins have a variety of functions within the cell, such as acting as enzymes, transporting molecules, and providing structure.

10. b
11. a
12. d
13. a
14. d
15. b
16. b
17. a
18. d
19. b
20. c
21. denaturation
22. Phospholipids

Chapter 3

1. Prokaryotic cells, which include bacteria and archaea, do not have membrane-bound organelles and are typically smaller than eukaryotic cells. Their DNA is circular and found in the cytoplasm; they do not have a nucleus. Eukaryotic cells, such as those found in plants and animals, are larger and more complex and have membrane-bound organelles; in addition, their DNA occurs in coiled linear strands in the nucleus. Prokaryotic cells and eukaryotic cells are similar in that they both contain ribosomes and cytoskeletal elements.

2. The plasma membrane is a selectively permeable membrane composed of a phospholipid bilayer embedded with proteins that can span the bilayer and carbohydrates that are found on the outer surface. The structure of the plasma membrane is often described as a fluid mosaic.

3. The plasma membrane maintains structural integrity of the cell, regulates movement of substances into and out of a cell, allows cell-to-cell recognition by glycoproteins, provides cell-to-cell communication, and sticks cells together to form tissues and organs.

4. Simple diffusion is the random movement of substances from a region of high concentration to a region of lower concentration. Oxygen crosses the cell membrane by simple diffusion. Molecules of glucose require facilitated diffusion, in which the molecules are moved with the help of carrier proteins, from a region of high concentration to a region of lower concentration.

5. The nucleus contains almost all the genetic information of a eukaryotic cell. Chromosomes consist of DNA and associated proteins.

6. Lysosomal storage diseases occur when lysosomal enzymes are absent and the cell cannot rid itself of certain molecules. The lysosomes fill up with these molecules, which then interfere with normal cellular functions.

7. The three types of fibers that make up the cytoskeleton are microtubules (hollow rods of tubulin that function in cell support and shape and movement of organelles and that form cilia and flagella); microfilaments (solid rods of actin that function in muscle contraction and pinching cells in two during cell division); and intermediate filaments (ropelike fibers composed of different proteins that function in cell shape and anchor organelles in place).

8. Cellular respiration provides energy to a cell by the breakdown of glucose through a series of chemical reactions that take place either in the cytoplasm or mitochondria of the cell.

9. b
10. d
11. d
12. c
13. a
14. d
15. c
16. b
17. a
18. a
19. c
20. Cytoplasm
21. Oxygen
22. fermentation

Chapter 4

1. The four types of human tissue are epithelial, connective, muscular, and nervous.

2. Epithelial tissue is formed from cells that are packed closely together, forming layers of cells of varying thickness. This type of organization makes epithelial tissue good for coverings such as the skin and for lining organs or body cavities. Connective tissue cells secrete a matrix that is found between the cells. The type of matrix dictates the function of the connective tissue. For example, the protein fibers in the gelatinous matrix of cartilage make it resilient and strong, which is ideal for strong, flexible support. The hard matrix of bone makes it a good support material. The liquid matrix of blood (plasma) allows for the tissue to flow.

3. Bone is well supplied by blood vessels whereas cartilage is not, so bone receives nutrients for growth and repair more rapidly than cartilage, which depends on diffusion for nutrients.

4. Blood is a connective tissue with a liquid matrix (plasma).

5. a. Skeletal muscle is composed of long, cylinder-shaped cells with many nuclei and mitochondria (for energy). It is called striated muscle because of the visible striations of actin and myosin filaments. Skeletal muscle tissue is usually attached to bone and has the ability to contract voluntarily.
   b. Cardiac muscle is composed of cells with only one nucleus that have striations and branch. They are found only in the heart and contract involuntarily continuously.
   c. Smooth muscle is composed of tapered cells with only one nucleus and lacks striations. Smooth muscle is found in blood vessels and airways or organs such as the stomach, and it contracts involuntarily when needed.

6. Nerve tissue is composed of neurons that conduct and transmit nerve impulses (via dendrites and axons) and of neuroglia cells that protect, support, and insulate the neurons.

7. Body temperature regulation is a negative feedback system. When body temperature rises above the set point, receptors in the skin send a message to the hypothalamus in the brain. The hypothalamus then sends a message to the sweat glands to produce sweat and cool the body. When body temperature drops below the set point, receptors in the skin send a message to the hypothalamus. The hypothalamus then sends a message to blood vessels to constrict, reducing blood flow to the arms and legs and thereby conserving heat. In addition, metabolism is increased to produce more heat.

8. b	15. c
9. a	16. d
10. a	17. c
11. b	18. c
12. a	19. a
13. d	20. a
14. a	21. a

Chapter 5

1. The osteon consists of a central canal surrounded by concentric rings of osteocytes (mature bone cells) in a rigid matrix. Osteocytes are located within a lacuna in the matrix. Canals connect the lacunae to each other and to the central canal. This allows for the transport of items between the cells and the blood vessels in the central canal.

2. During fetal development, most of the skeleton is first formed of cartilage. Cartilage cells actively divide, allowing the skeleton to grow as the fetus does. Beginning around the third month of development, osteoblasts (bone-forming cells) form a ring of bone around the cartilage and create the shaft of the bone; the cartilage cells degenerate, forming the center cavity of the bone. Osteoblasts then fill the cavity with spongy bone. After birth, bones grow longer as the cartilage cells in the growth plates at the ends of the bone divide. Bone replaces the newly formed cartilage. At puberty, hormones usually cause an increase in the rate of growth that lasts until the end of teenage years (usually about age 18), when cartilage cells slow their rate of division.

3. After a fracture, a blood clot is formed at the site of the fracture. Next, fibroblasts secrete collagen fibers, which form a fibrocartilaginous callus that is replaced by a bony callus. Over time, bone remodeling restores the bone to the original shape.

4. Bone remodeling is the continuous reshaping and replacing of bone during one's lifetime. Osteoblasts continuously form bone while osteoclasts continuously break down bone. Bone is built along the lines of stress on a bone, providing strength where it is needed the most.

5. The axial skeleton includes the bones of the skull, vertebral column, rib cage, and sternum. The appendicular skeleton includes the bones of the pectoral girdle and arms, and the bones of the pelvic girdle and legs.

6. A synovial joint is a freely movable joint, such as the knee joint. A thin layer of cartilage reduces friction on the surfaces of the bones that slide over one another. A synovial joint is surrounded by a two-layered joint capsule. The synovial membrane, which forms the inner layer of the capsule, secretes synovial fluid to lubricate the joint. Ligaments hold the joint together, support the joint, and determine the direction of movement at the joint.

7. c
8. d
9. d
10. c
11. a
12. c
13. d
14. b
15. c
16. b
17. c
18. c
19. calcium, phosphorus
20. osteoblasts
21. sprain

Chapter 6

1. Most muscles in the human body are arranged in antagonistic pairs so that the contraction of one muscle moves a bone in one direction and the contraction of another muscle moves the bone back. A good example of an antagonistic pair (Figure 6.1) is the biceps that pulls the forearm up to bend the arm at the elbow and the triceps that pulls the forearm down to extend the arm.

2. Skeletal muscles are attached to bone and are made up of fascicles, smaller bundles of muscle cells. Muscle cells (also called muscle fibers) contain specialized bundles of proteins called myofibrils. The orderly arrangement of myofibrils creates the striated (striped) appearance of muscle cells. The myofibrils in turn are made up of two types of myofilaments: actin and myosin filaments. Each myofibril is divided by bands of proteins, called Z lines, to form many sarcomeres, which are the contractile units of the muscle. The actin myofilaments are attached to the Z lines. The myosin filaments are positioned in the middle of the sarcomere, between the actin filaments.

3. The sliding filament model of muscle contraction proposes that muscles shorten when the thin actin filaments slide past the thick myosin filaments, increasing the degree of overlap between them.

4. The actin filaments are pulled across the myosin filaments by movements of the heads of myosin filaments. The head of a myosin filament, located at its end, has two important characteristics: it can bend, and it can bind and split ATP. The myosin head binds to the actin filament. The myosin head then bends pulling the actin filament to the center of the sarcomere. ATP provides energy to move the myosin head and the actin. ATP is also needed for the myosin head to release the actin.

5. Tropomyosin and troponin are proteins that cover the myosin-binding sites on actin during muscle relaxation. Calcium ions enter the sarcomere, bind to the tropomyosin–troponin complex, which changes shape and exposes the myosin-binding sites on the actin. The actin is then free to bind with the myosin again and cause contraction.

6. When a nerve impulse reaches the neuromuscular junction, acetylcholine is released, which creates an electrochemical message sent to the sarcoplasmic reticulum that releases calcium ions. Calcium ions enter the sarcomere and bind to the tropomyosin–troponin complex, which then changes shape and exposes the myosin-binding sites on the actin. The actin is then free to bind with the myosin again and cause contraction.

7. A motor unit is a motor neuron and all the muscle cells it stimulates. In general, the finer the movement, the fewer muscle cells per motor unit.

8. A muscle twitch (contraction) is a brief contraction of a muscle in response to a single stimulus. Wave summation of muscle contraction occurs when the muscle is stimulated again before the muscle can completely relax from the previous stimulus. Tetanus is a smooth, sustained contraction that occurs when stimuli arrive in such rapid succession that there is no time for muscle relaxation.

9. Sources of ATP for muscle contraction:
   a. Muscle cells—the first source of ATP, which is replenished during relaxation.
   b. Creatine phosphate—used after the first 6 seconds of exercise and supplies energy to convert ADP to ATP.
   c. Anaerobic respiration—occurs after about 10 minutes of exercise and does not need oxygen; glucose source is glycogen.
   d. Aerobic respiration—occurs when the heart begins to beat faster to deliver the oxygen needed.

10. Slow-twitch muscle cells contract slowly, but have more endurance than the fast-twitch muscle cells, which contract powerfully and quickly.

11. Resistance exercise builds the size of muscles when the muscle exerts more than 75% of its maximum force during the exercise.

12. c	19. b
13. a	20. d
14. c	21. a
15. a	22. d
16. c	23. actin, myosin
17. c	24. sarcomere
18. b	

Chapter 7

1. Neuroglial cells support, protect, insulate, and nurture neurons. There are several types of neuroglial cells.

2. a. Motor neurons conduct information away from the brain to muscles or glands.
   b. Sensory neurons conduct information toward the brain.
   c. Interneurons (association neurons) are found in the brain or spinal cord. They function in integrating information between motor and sensory neurons, and they are the most numerous type of neuron.

3. Refer to Figure 7.2, p. 115.

4. Schwann cells form the myelin sheath in the peripheral nervous system. A Schwann cell wraps around the axon many times, forming a spiral of membrane. The myelin sheath serves as an insulator for individual axons, allows the signals to move faster along the axon, and helps in the repair of a damaged axon.

5. In a resting neuron, sodium and potassium ions are unequally distributed across the plasma membrane. There are more sodium ions outside the membrane and more potassium ions on the inside of the membrane. There is a –70 mv difference between the inside and outside of the membrane during the resting potential. During an action potential, the sodium ions enter the cell, causing the loss of charge difference across the membrane and eventually creating a positive charge on the inside of the membrane.

6. When a neuron is resting (not conducting an impulse), energy is used to maintain the difference in the distribution of sodium ions and potassium ions across the membrane that causes the resting potential. When the neuron is stimulated, the potential difference across the membrane allows the neuron to respond quickly.

7. Sodium ions (with their positive charge) enter the cell when the sodium gates open. This movement causes the inside of the membrane to become less negative and eventually to become positive.

8. Potassium ions (with their positive charge) leave the cell, which returns the cell to its resting potential.

9. The sodium-potassium pump restores the initial ion distribution by using energy in ATP to pump sodium ions out of the cell and potassium ions back into the cell.

10. The refractory period is time following an action potential during which a neuron cannot be stimulated to generate another action potential.

11. Refer to Figure 7.6 on page 119.

12. When the neurotransmitter of an excitatory synapse binds to receptors, the sodium channels open, allowing sodium ions to enter the cell. This movement of sodium ions reduces the charge difference across the membrane, increasing the likelihood that an action potential will be generated. In an inhibitory synapse, the neurotransmitter binds to receptors that open other ion channels, often chloride channels. As a result of ion movements, the inside of the membrane becomes more negative, decreasing the likelihood that an action potential will be generated.

13. The neurotransmitter must be removed from the synapse to stop its action on the postsynaptic neuron. In some synapses, enzymes break down the neurotransmitter. In other synapses, the neurotransmitter is actively pumped back into the axon tip.

14. a
15. c
16. b
17. b
18. b
19. d
20. c
21. a
22. d
23. a
24. b
25. a
26. a
27. myelin sheath
28. sodium-potassium pump

Chapter 8

1. Parts of the nervous system:
   a. The central nervous system (CNS) consists of the brain and spinal cord.
   b. The peripheral nervous system (PNS) consists of the nerves and ganglia outside the CNS. The PNS is subdivided into the autonomic and somatic nervous systems.

2. Protection of the CNS:
   a. Meninges—protective outer coverings of the brain and spinal cord.
   b. Cerebrospinal fluid—fills internal cavities in the brain for protection and cushioning.
   c. Blood–brain barrier—selectively permeable barrier between the circulatory system and the cerebrospinal fluid.

3. The functions of the cerebrospinal fluid are shock absorption, support, and nourishment.

4. The gray matter is the thin cerebral cortex of each hemisphere of the cerebrum. It is composed of interneurons, which integrate information. The white matter consists of myelinated axons that lie below the cerebral cortex. Bands of white matter allow communication between different regions of the brain. The corpus callosum is a band of white matter that connects the two hemispheres.

5. Three types of functional areas of the cerebral cortex:
   a. Sensory—senses, including vision, hearing, olfaction
   b. Motor—movement
   c. Association—interpretation of sensations, language, thinking, decision making, memories, creativity

6. The general arrangement of the primary somatosensory area and the primary motor area is similar, but more of the sensory cortex is devoted to areas of greater sensitivity. The motor cortex has larger areas devoted to body parts with finer motor control.

7. The hypothalamus influences or regulates
   a. Blood pressure and heart rate
   b. Digestive activity
   c. Breathing rate
   d. Body temperature
   e. Coordination of endocrine system

8. The cerebellum is the part of the brain responsible for sensory-motor coordination.

9. The limbic system is the functional system of the brain responsible for emotions.

10. In a spinal reflex arc, a response to a stimulus can result from the activity of only three neurons. Sensory information enters the spinal cord over a sensory neuron. An interneuron receives the sensory information and communicates it to a motor neuron, which causes muscles to contract and remove the hand from the burner. The perception of pain occurs in the brain, which requires more neurons and synapses.

11. Two divisions of the peripheral nervous system:
   a. Somatic—controls the conscious, voluntary functions.
   b. Autonomic—controls the internal organs and maintains homeostasis.

12. a
13. d
14. d
15. c
16. d
17. b
18. b
19. d
20. a
21. a
22. c
23. d
24. a
25. a
26. d
27. sympathetic nervous system
28. cerebrum

Chapter 9

1. Sensory adaptation is the process by which sensory receptors stop responding when they are continuously stimulated. Olfactory receptors adapt quickly. When we enter a room with an odor, we generally smell it at first, then after a while the odor becomes less noticeable.

2. Five classes of receptors:
   a. Mechanoreceptors—general senses and special senses
   b. Chemoreceptors—special senses (taste and smell) and internal conditions
   c. Photoreceptors—special senses
   d. Thermoreceptors—general senses
   e. Pain receptors—general senses

3. Merkel disks are receptors for light touch on the skin.

4. The sclera protects and shapes the eye. The cornea provides a window for light to enter the eye and refracts light to the retina.

5. The cornea refracts (bends) light toward the retina. The lens changes shape to bend light in the appropriate direction to focus on close or far objects.

6. Light causes a photopigment within the receptor to split into its component parts, which causes a series of reactions that create a neural message.

7. Sound waves are pressure waves in air or water produced by vibrating objects. The amplitude (height) of the waves determines the loudness, and the frequency (cycles per second) determines the pitch of the sound.

8. The function of the tympanic membrane (eardrum) is to transfer vibrations caused by sound waves to the bones of the middle ear. The tympanic membrane forms the outer boundary of the middle ear and vibrates in response to sound waves. Vibrations of the tympanic membrane are transferred through the middle ear by three small bones (malleus, incus, and stapes; also known as the hammer, anvil, and stirrup) to the inner ear. There, hearing occurs when neural messages are generated in response to the pressure waves caused by the vibrations.

9. The amplification of the force of vibration in the middle ear is necessary to transfer the vibrations to the fluid of the inner ear. Amplification occurs because of the arrangement of the middle ear bones and the difference in size of the larger eardrum and the oval window.

10. The basilar membrane vibrates in different areas along its length according to pitch. The auditory nerve sends messages to the brain, which interprets the movement of the different areas of the membrane as differences in pitch.

11. Two types of hearing loss:
   a. Conductive hearing loss—sounds are not conducted through the auditory canal.
   b. Sensorineural hearing loss—is caused by damage to the hair cells of the inner ear.

12. Two structures for equilibrium:
   a. Semicircular canals are fluid-filled canals that are oriented at right angles to one another. At the base of each canal is an enlarged area called the ampulla, which contains a tuft of hair cells. The hair cells are embedded in a gelatinous material called the cupula. Movements of the body or head cause movements of the fluid in the semicircular canals. This movement of fluid causes the cupula to bend the hair cells, thereby stimulating them. Thus, the semicircular canals help us stay balanced as we move.
   b. The vestibule consists of two fluid-filled chambers—the utricle and the saccule. These chambers contain hair cells embedded in a gelatinous mass containing small granules of calcium carbonate called otoliths. When the position of the head changes, the otoliths cause the gelatinous mass to move and bend the hair cells. The brain interprets the pattern of stimulation from the hair cells to determine the position of the head relative to gravity.

13. Motion sickness is caused by mismatched sensory input from the eyes and the inner ear. Drugs designed for relieving motion sickness are effective, and they work by inhibiting messages from the vestibular apparatus. Some people say that staring at the horizon while moving can help.

14. Olfactory receptors are located at the roof of the nasal cavity. The receptors have long olfactory hairs that project into a coat of mucus that traps odor molecules. The odor molecules can then stimulate the olfactory receptors.

15. Taste buds are responsible for taste and are located on the tongue; a few are in the cheeks, the roof of the mouth, or the throat. Most taste buds are located in papillae on the tongue.

16. The five primary tastes are sweet, sour, bitter, salty, and umami (savory).

17. Taste buds are located on the papilla of the tongue and cheek. They are lemon-shaped. Each taste bud has taste hairs projecting out at the tip that have receptors for chemicals dissolved in water.

18. a
19. b
20. d
21. a
22. b
23. a
24. a
25. c
26. c
27. b
28. a
29. cones
30. lens
31. cochlea
32. maintain balance while we are moving

Chapter 10

1. Cells that are affected by a particular hormone are called target cells. Target cells have receptors for the hormone that is released. Other cells do not have the receptor and cannot be affected by the hormone.

2. Lipid-soluble (steroid) hormones can move through the lipid bilayer into a cell where the receptors reside. Steroid hormones combine with a receptor inside the cell, and then the hormone-receptor complex moves to the DNA, where it activates genes that direct the synthesis of specific proteins. Water-soluble hormones are composed of amino acids and cannot pass through the lipid bilayer. In this case, the hormone, considered the first messenger, attaches to a receptor on the surface of the cell, and this binding activates a molecule in the cytoplasm. This molecule inside the cell, considered the second messenger, sets off an enzyme cascade that affects cell activity.

3. In negative feedback systems, the outcome of a process feeds back on the system, shutting down the process. For example, the pancreas secretes the hormone insulin when blood glucose is high to stimulate glucose uptake and storage. When the blood glucose becomes low, the pancreas stops secreting insulin. In positive feedback systems, the outcome of a process feeds back to the system and stimulates the process to continue. For example, during childbirth the posterior pituitary secretes the hormone oxytocin. Oxytocin stimulates uterine contractions, which in turn stimulate the production of more oxytocin, and this increases the frequency and intensity of uterine contractions.

4. The anterior pituitary is larger than the posterior pituitary and is directly connected to the hypothalamus by a circulatory connection. The much smaller posterior pituitary is connected to the hypothalamus by neurosecretory cells.

5. In response to high levels of calcium in the blood, the thyroid gland secretes calcitonin, which promotes calcium storage by bones. When levels of calcium become too low, the parathyroid glands release parathyroid hormone, which increases calcium levels in the blood by stimulating bone-destroying cells (osteoclasts), reabsorption of calcium by the kidneys, and absorption into the bloodstream by the GI tract.

6. Glucocorticoids affect glucose homeostasis. Mineralocorticoids affect mineral homeostasis and water balance. The gonadocorticoids (sex hormones) probably have minimal effects, especially in normal adult males, because secretion by the testes far surpasses that by the adrenal cortex. In females, sex hormones from the adrenal cortex may somewhat alleviate menopausal symptoms brought on by the cessation of sex hormone production by the ovaries.

7. Hormones released by the adrenal medulla (epinephrine and norepinephrine) are responsible for the fight-or-flight response, which is prompted by stress or danger. The response includes increases in heart rate, respiratory rate, and blood glucose, as well as vasoconstriction (in areas that are not of immediate importance, such as the digestive tract) and dilation (in areas that are of immediate importance, such as skeletal and cardiac muscle).

8. Diabetes mellitus is a group of diseases characterized by problems with insulin production or insulin function. Diabetes insipidus results from a deficiency in antidiuretic hormone (ADH).

9. The thymus gland produces hormones that are involved in the maturation of the white blood cells known as T lymphocytes.

10. Local signaling molecules act at or near the site of their release within seconds or milliseconds. In contrast, true hormones travel in the blood to relatively distant sites in the body, so their effects take longer to materialize.

11. a
12. c
13. d
14. c
15. d
16. a
17. d
18. c
19. c
20. a
21. gigantism; acromegaly
22. adrenal cortex

Chapter 11

1. Plasma is the liquid matrix portion of blood. Plasma functions to transport dissolved substances, such as nutrients, ions, dissolved gasses, plasma proteins (for water balance) and hormones. Plasma functions in transporting cellular wastes from the cells to the kidneys.

2. The three categories of plasma proteins are albumins, globulins, and clotting proteins.

3. Three types of formed elements and their functions:
   a. Platelets are cell fragments that function in clotting.
   b. Red blood cells (erythrocytes) function in transporting oxygen to other cells in the body for cellular respiration and transporting a portion of the carbon dioxide waste from cells.
   c. White blood cells (leukocytes) function in defense against disease.

4. Erythrocytes compared with leukocytes:
   a. Erythrocytes are small, biconcave disk-shaped cells that lack a nucleus and are packed with hemoglobin. They account for about 45% of the blood volume.
   b. Leukocytes are larger than the erythrocytes, have a large nucleus, and account for less than 1% of the total blood volume.

5. Five types of white blood cells and the role each plays in body defense:
   a. Neutrophils phagocytize bacteria and other microbes.
   b. Eosinophils attack parasitic worms and phagocytize antibody–antigen complexes.
   c. Basophils release histamine, which attracts other white blood cells to the site of injury and causes blood vessels to dilate.
   d. Monocytes develop into macrophages.
   e. Lymphocytes—B lymphocytes give rise to plasma cells that produce antibodies, and T lymphocytes perform various functions of the immune response.

6. Each red blood cell is a biconcave disk, a shape that creates a large surface area for such a small cell. Red blood cells are packed with the protein hemoglobin, which has the ability to pick up oxygen in the lungs and release it in the tissues. Red blood cells lack a nucleus and mitochondria, so most of the interior is taken up by the hemoglobin molecules.

7. The function of hemoglobin is to bind to oxygen in the lungs and release it in the tissues. Hemoglobin is a globular-shaped protein consisting of four polypeptide chains. Each chain contains a heme group (with an iron atom) that can bind to an oxygen molecule.

8. Red blood cells are produced in the red bone marrow. When the oxygen-carrying capacity of the blood is low, the kidneys release the hormone erythropoietin. This stimulates the red bone marrow to produce more red blood cells. The kidneys reduce erythropoietin production when the oxygen-carrying capacity of the blood increases.

9. The liver and spleen destroy worn-out red blood cells. The red blood cells get stuck in the tiny circulatory channels of these organs, where macrophages destroy them. The hemoglobin is broken down into its component amino acids and bilirubin.

10. People with pernicious anemia cannot absorb vitamin B12 from the digestive system, so they need injections of this vitamin.

11. Antigens on the surface of red blood cells determine blood type. Type A blood has A antigens. Type B blood has B antigens. Type AB blood has both A and B antigens. Type O blood lacks these antigens on the red blood cell surfaces.

12. If a person is given an incompatible blood type, agglutination occurs because recipient's antibodies react to the antigens on the donor's red blood cells, causing the cells to clump and get stuck in blood vessels and block blood flow. The clumped cells may also break open and release hemoglobin, which can clog the filtering system of the kidneys and lead to death.

13. A person with type B blood can receive type B and type O blood. Type B blood is compatible because it has the same antigen as the recipient. Type O blood does not have any antigens, so it will not cause agglutination.

14. Hemolytic disease of the newborn occurs when anti-Rh antibodies from the mother cross the placenta and cause clumping of an Rh-positive fetus's blood cells. An Rh-negative mother may have formed anti-Rh antibodies if she previously gave birth to an Rh-positive baby and fetal blood entered the mother's body during the delivery. In subsequent pregnancies with an Rh-positive fetus, the Rh-negative antibodies enter the fetus's circulatory system and cause red blood cells to clump.

15. When a blood vessel is cut, it constricts and reduces blood flow. Next, platelets stick to the collagen fibers on the damaged blood vessel's wall to form a "plug." The platelets then produce thromboxane, which makes them stick together and attracts other platelets to the wound. Injured cells in the blood vessel release blood clotting factors and a clot forms.

16. The steps involved in blood clotting:
   a. A clot begins to form when clotting factors are released from injured cells of the damaged blood vessel.
   b. The clotting factors convert an inactive blood protein to prothrombin activator, which converts prothrombin to an active form, thrombin.
   c. Thrombin then causes a change in fibrinogen from the liver. The altered fibrinogen forms long strands of fibrin, which form a web that traps blood cells and forms the clot.

17. Blood clotting after an injury occurs when blood cells get trapped in a web of fibrin at the site of injury. Agglutination following a mismatched blood transfusion occurs because antibodies in the recipient's blood cause the donor's red blood cells to clump.

18. a
19. c
20. b
21. a
22. a
23. a
24. d
25. a
26. transport oxygen
27. Hemoglobin
28. fibrin

Chapter 12

1. A pulse is a pressure wave in the arteries created by the contraction of the ventricles of the heart.

2. Two important functions of arterioles:
   a. Control blood pressure.
   b. Regulate the amount of blood going into a capillary bed.

3. Blood flow into a capillary bed is controlled by a ring of smooth muscle called a precapillary sphincter that surrounds a capillary before it branches to form the capillary bed. When the sphincter contracts, blood is channeled through the capillary bed without filling it. The capillary bed opens when the sphincter relaxes. Input from hormones, the nervous system, and local conditions of the arterioles regulate the opening or closing of any particular capillary bed.

4. All blood vessels have a lumen, the hollow interior through which blood flows, and a smooth lining called the endothelium. In addition to these characteristics, an artery has a middle layer containing elastic fibers that allow the artery to expand as blood is pumped into it and recoil to its original size. The middle layer also contains circular smooth muscle, which allows the artery to contract. The outer, supporting layer of an artery is connective tissue containing elastic fibers and collagen. The smallest arteries—the arterioles—have the same three layers, but the middle layer is primarily smooth muscle. The smooth muscle allows arterioles to regulate blood pressure by constricting or dilating. Capillaries have only one cell layer; this facilitates exchange of materials between the blood and cells. Veins have the same three layers that arteries have, but the walls are thinner and have less smooth muscle. Thus, veins can expand and serve as blood reservoirs. Many veins have one-way valves that keep blood from "backing up."

5. Three mechanisms allow blood to return to the heart from the lower torso against the force of gravity. First, veins have valves that prevent the backflow of blood. Second, veins are surrounded by skeletal muscle. When the skeletal muscle contracts, it squeezes the vein pushing blood along. The valves ensure that blood flows one way—back to the heart. Third, pressure changes in the thoracic cavity that occur during breathing pull blood back toward the heart.

6. The human heart has four chambers—two atria and two muscular ventricles. Between each atrium and ventricle are the AV valves, which allow blood to move from the atrium to the ventricle. One-way valves exist between the ventricles and their arteries. The heart serves as two pumps. The right side of the heart pumps blood to the lungs through the pulmonary artery (pulmonary circuit). Blood is oxygenated in the lungs. The left atrium receives oxygenated blood from the pulmonary veins. The left side of the heart pumps oxygenated blood to the body cells (systemic circuit).

7. The path of blood from the left ventricle to the left atrium: left ventricle → aortic semilunar valve → aorta → body tissues → superior and inferior venae cavae → right atrium → right atrioventricular valve → right ventricle → pulmonary semilunar valve → pulmonary arteries → lungs → pulmonary veins → left atrium → left atrioventricular valve.

8. Each heartbeat (cardiac cycle) involves contraction (systole) and relaxation (diastole). First, all the chambers relax, and blood passes through the atria and enters the ventricles. Then, the atria contract and push the remaining blood into the ventricles. Next, the atria relax and the ventricles contract.

9. The sinoatrial (SA) node, located in the right atrium, consists of specialized cardiac muscle cells that initiate each heartbeat. The SA node generates an electrical impulse that travels through the wall of the right atrium. The signal reaches another cluster of specialized muscle cells called the atrioventricular (AV) node, located in the partition between the two atria. The AV node relays the stimulus by a bundle of specialized muscle fibers, called the atrioventricular bundle, located along the wall between the ventricles. The electrical signal is conducted to the ventricles through the atrioventricular bundle and quickly fans out through the ventricle walls through the Purkinje fibers. The rapid spread of the impulse through the ventricles ensures that they contract smoothly.

10. Three important functions of the lymphatic system:
    a. Return excess interstitial fluid to the bloodstream.
    b. Transport products of fat digestion from the small intestine to the bloodstream.
    c. Help defend against disease-causing organisms.

11. The lymph nodes are small nodular organs found along lymph vessels, which filter lymph. The lymph nodes contain macrophages and lymphocytes, cells that play an essential role in the body's defense system.

12. a
13. b
14. b
15. b
16. b
17. d
18. c
19. capillaries

Chapter 13

1. Nonspecific innate defenses are those you are born with. Nonspecific defenses target *any* foreign invaders; they are physical and chemical barriers. Specific defenses are our immune responses that target specific invaders of the body that manage to get by the nonspecific defenses. Specific defenses are adaptive, meaning that you acquire them to defend against specific invaders that you have experienced.

2. Seven innate, nonspecific defense mechanisms:
   a. Physical and chemical barriers prevent foreign cells from entering the body.
   b. Phagocytes engulf foreign cells.
   c. Inflammatory response—redness, warmth, swelling, and pain—destroys the invader and helps repair damaged tissue.
   d. Natural killer cells kill abnormal cells that are not recognized as belonging in the body.
   e. The complement system, a group of plasma proteins, assists other defense mechanisms.
   f. Interferons interfere with viral replication.
   g. Fever raises the body's temperature to become less hospitable to invaders.

3. Natural killer cells kill their target cells by releasing chemicals that form pores in the membranes of infected cells or tumor cells.

4. Interferons are proteins that are secreted by cells that are infected with a virus. They help protect healthy cells from becoming infected with a virus in two ways. First, they secrete chemicals that attract phagocytes and natural killer cells to kill infected cells. Second, interferons stimulate healthy cells to produce proteins that interfere with viral replication.

5. The complement system consists of plasma proteins that help the body's defense mechanisms. They are activated by an infection, which begins a series of different possible reactions. These include formation of protein complexes that create holes in bacterial cell walls, marking bacteria for destruction, and stimulating cells to release histamine or serve as chemical attractors for phagocytes.

6. The inflammatory response begins when injured tissues release chemicals that stimulate mast cells to release histamine. Histamine causes blood vessels to dilate, which brings more blood into the area, causing redness and heat. Increased blood flow brings in defensive cells and chemicals. The heat increases the metabolic rate of cells, which speeds healing. Histamine also causes capillaries to become more permeable. As a result, fluid seeps into the injured area, carrying defensive cells and chemicals and causing swelling. Pain hampers movement, which allows the injured area to heal.

7. An antigen-presenting cell engulfs invaders, digests them, and presents fragments of the antigen on its surface attached to a self-marker (MHC marker). Macrophages, activated B cells, and dendritic cells are antigen-presenting cells. They are recognized by an antigen-MHC complex on their surface that allows them to have self-markers along with the antigen fragments.

8. The antibody-mediated response is brought about by B lymphocytes, which transform to form plasma cells that secrete antibodies. The targets of the antibody-mediated response are antigens in the blood or lymph. These targets include viruses, bacteria, and foreign molecules.

9. An antibody is a Y-shaped protein with antigen-binding variable regions at the tips of the Y that are specific to a particular antigen. The antigen–antibody complex marks a cell for destruction by activating the complement system, attracting phagocytes, neutralizing the target, or agglutinating the target.

10. Cytotoxic T cells are responsible for cell-mediated responses. Their targets are cellular—bacteria, infected cells, and cancerous cells. Cytotoxic T cells kill their targets by releasing proteins that form holes in the target cell.

11. Natural killer cells are nonspecific defenders, whereas cytotoxic T cells are programmed to kill specific abnormal cells.

12. The primary immune response involves recognition of the antigen and production of B cells and T cells. The secondary response uses memory cells that remain in the system and can be activated as soon as a recognized antigen enters the body.

13. Active immunity involves vaccinating with an antigen-containing preparation. Passive immunity involves injecting prepared antibodies for a specific pathogen.

14. Monoclonal antibodies are antibodies produced in the laboratory from a single hybrid B cell. Some uses for these antibodies include home pregnancy tests; screening for prostate cancer; and diagnostic testing for hepatitis, AIDS, and influenza.

15. An autoimmune disorder is an immune response misdirected against the body's own tissues. In lupus erythematosus, connective tissue is attacked; in rheumatoid arthritis, synovial joints are attacked. There are no cures, but therapy to suppress the immune system can relieve some of the symptoms.

16. An allergy is an inappropriate immune response to an allergen that is not harmful. The symptoms are caused by release of histamine from activated mast cells and an inflammatory response.

17. a
18. d
19. b
20. a
21. c
22. d
23. b
24. a
25. d
26. c
27. natural killer cell
28. Histamine
29. plasma cells
30. Macrophages (or dendritic cells or activated B cells)

Chapter 14

1. We must breathe oxygen to obtain the maximal amount of ATP during cellular respiration.

2. The path of air from the nose to the cells that use the oxygen: nose → pharynx → larynx → trachea → bronchus → bronchioles → alveoli → capillary bed → bloodstream → capillary bed → cells that need oxygen.

3. Mucus in the respiratory tract traps debris and pathogens. Cilia in the respiratory tract constantly beat in an upward motion to force debris toward the pharynx. The debris can then be swallowed or spit out.

4. When we swallow, the larynx moves up and under the epiglottis, covering the opening to the larynx. This movement prevents food or drink from entering the trachea.

5. The sound of the voice is created when air passes through the larynx and causes the vocal cords to vibrate. The pitch can be altered by adjusting the tension on the vocal cords. The tone can be changed by changing the shape of the resonating chamber created by the mouth. Loudness is determined by the volume of air passing through the cords.

6. The cartilage rings in the trachea keep the trachea open at all times. As we breathe, the air passing through the trachea creates a negative pressure that pulls in on the wall of the trachea. The negative pressure would cause the trachea to collapse if it were not supported open.

7. The bronchial tree is the system of bronchi that splits at each lung to form treelike structures of bronchioles ending in alveoli sacs.

8. The pressure changes within the thoracic cavity that cause inhalation and exhalation are created by the diaphragm and muscles between the ribs. When the muscles of the rib cage contract, the ribs are lifted upward and outward, which increases the size of the thoracic cavity from front to back. The diaphragm flattens when it contracts, which increases the size of the thoracic cavity from top to bottom. The increase in size of the thoracic cavity causes the pressure within it to drop, which draws air into the lungs (inhalation). When these muscles relax, the size of the thoracic cavity decreases, and air moves outward (exhalation).

9. The vital capacity is larger than the tidal volume. Tidal volume is the volume we take in breathing normally (~500 ml). Vital capacity is the maximum amount of air we can exhale after a maximum inhalation (~4800 ml).

10. Most oxygen is transported in the bloodstream to the body cells by binding to hemoglobin in red blood cells.

11. Most carbon dioxide transported from the cells to the lungs is transported in the bloodstream as a bicarbonate ion.

12. The medulla oblongata of the brain contains the respiratory center that regulates breathing.

13. The medulla oblongata detects increased CO_2 levels in the blood by responding to increases in H^+ concentrations that result from carbonic acid that forms when CO_2 dissolves in plasma. Increases in H^+ result in an increased breathing rate, which lowers CO_2 levels in the blood.

14. Emphysema is the destruction of the alveolar walls. People with emphysema experience shortness of breath because they cannot get enough oxygen as a result of the loss of alveolar surface area.

15. a
16. b
17. b
18. d
19. b
20. b
21. b
22. a
23. d
24. a
25. a
26. epiglottis
27. carbonic anhydrase

Chapter 15

1. The structures of the gastrointestinal tract in the order that food passes through them: mouth → esophagus → stomach → small intestine → large intestine → colon → rectum.

2. The processing of food begins in the mouth, where food is mechanically broken down by the teeth. Saliva contains the enzyme salivary amylase, which begins the chemical breakdown of starch. The tongue helps keep the food in the mouth and then facilitates swallowing of the food.

3. The tooth is a hard structure covered by enamel. The inner portion is called dentin and surrounds the pulp. In the center there is a pulp cavity that houses blood vessels and nerves—structures that keep the tooth alive. Tooth decay is the destruction of the enamel coating of the tooth by acids produced by bacteria in the mouth. The result is a cavity.

4. The functions of the stomach are food storage, mechanical digestion, and chemical digestion. Gastric juice contains hydrochloric acid and pepsin, which begin the digestion of proteins.

5. Bile assists in the digestion of fats by coating small fat droplets created by mechanical digestion in the small intestine. This bile coating keeps the fat in small droplets (emulsifies it), which creates a larger surface area for lipase, a water-soluble enzyme, to chemically digest the fat into glycerol and fatty acids. Bile salts then combine with the glycerol and fatty acids to form micelles, which can be absorbed.

6. The primary site for digestion of carbohydrates, proteins, and fats is the small intestine. Carbohydrate digestion begins in the mouth, and protein digestion begins in the stomach.

7. The small intestine has many structures that increase the surface area for absorption. It has accordion-like folds (circular folds). The surface of the lining of the small intestine also has many fingerlike projections called villi. Each villus is covered with microvilli. The combined effect greatly increases the surface area for absorption.

8. The functions of the large intestine are to absorb water and store feces for removal from the body.

9. The sight, thought, or smell of food cause neural reflexes that stimulate the salivary glands to secrete salivary amylase. Chewing causes a neural influence that causes the stomach to produce gastric juice. The distention of the stomach by food and the partial digestion of protein cause the stomach to release the hormone gastrin, which circulates in the bloodstream and stimulates the stomach to produce gastric juices. Acidic chyme entering the small intestine triggers the release of the hormone secretin from the small intestine that causes the pancreas to release bicarbonate. Acidic chyme also causes the small intestine to release cholecystokinin, which causes the gallbladder to release bile and the pancreas to release digestive enzymes.

10. MyPlate shows the proportions of various types of food that you should eat as portions on a plate. It also describes nutrients and their functions.

11. Nutrients:

   a. Fats (lipids): Provide 9 calories of energy per gram; they are components of nerve sheaths and cell membranes; they insulate the body; they form protective cushions for body organs.
   b. Carbohydrates: Provide 4 calories per gram; primary fuel for all cell activities.
   c. Protein and amino acids: Provide 4 calories per gram; important component of all cells; structural proteins include muscle fibers; regulatory proteins include enzymes and certain hormones.
   d. Vitamins: Most function as regulatory molecules that allow the cellular reactions of the body to take place fast enough to support life.
   e. Minerals: Structural roles include providing hardness of bones and teeth; functional roles include oxygen transport in blood; electrolyte balance; proper nerve and muscle function.
   f. Water: Acts a solvent; transports materials; serves as a medium for all and participant in some chemical reactions; acts as a lubricant; acts as a protective cushion; helps regulate body temperature.
   g. Fiber (not absorbed from intestines): Soluble fiber is good for health of heart and blood vessels; insoluble fiber promotes intestinal health.

12. Someone who is overweight weighs more than is advised on a health and weight chart, but the weight may be due to muscle development. Someone who is obese is overweight because of body fat.

13. a
14. b
15. d
16. d
17. a
18. d
19. c
20. a
21. b
22. c
23. b
24. c
25. c
26. b
27. Chyme
28. liver
29. starch
30. Cholecystokinin

Chapter 16

1. Ammonia is formed in the liver when amino acids are broken down. It is converted to urea before leaving the liver. Urea is excreted by the kidneys and skin. Uric acid is formed from the recycling of nucleotides and is excreted by the kidneys and skin. Creatinine is produced in muscle cells as they use creatine phosphate as an alternate energy source; it is excreted by the kidneys.

2. The urinary system consists of the kidneys (filter blood and form urine), ureters (transport urine from kidneys to the bladder), urinary bladder (stores urine), and urethra (carries urine away from the body).

3. Nephrons contribute to the regulation of blood pH by secreting hydrogen ions into the filtrate (which will become urine) and by reabsorbing bicarbonate, which is critical to the carbonic acid buffer system in the blood.

4. Kidneys promote water conservation by producing concentrated urine; this is performed by the 20% of nephrons with long loops that dip into the renal medulla. Water conservation maintains blood volume and blood pressure.

5. Antidiuretic hormone (ADH) increases the permeability of the collecting duct to water so that more water is reabsorbed, increasing blood volume and pressure and resulting in the production of concentrated urine. Aldosterone increases reabsorption of sodium by the distal tubule and collecting duct, increasing blood volume and pressure (because water follows sodium) and resulting in the production of concentrated urine. Atrial natriuretic hormone (ANH) decreases the reabsorption of sodium, and this decreases blood volume and pressure and results in the production of dilute urine.

6. The solute concentration in the interstitial fluid of the kidney increases from the cortex to the medulla. The increasing concentration of salt allows for the reabsorption of more water in the loop of Henle and collecting duct, and production of concentrated urine.

7. The bladder is prevented from emptying by internal and external urethral sphincters. The

internal sphincter is made of smooth muscle and is involuntary, whereas the external sphincter is skeletal muscle and voluntary. When these sphincters relax, urine flows down the urethra to the external environment.

8. Males have a long urethra (~8 inches) and females have a short urethra (~1.5 inches). Females are more prone to urinary tract infections because their urethra is short.

9. c	15. a
10. b	16. c
11. a	17. d
12. c	18. Hemodialysis
13. a	19. internal; smooth; external; skeletal
14. c	

Chapter 17

1. The male gonads are testes, and the female gonads are the ovaries. They both function to produce gametes (eggs and sperm) and sex hormones. The ovaries produce eggs and the hormones estrogen and progesterone. The testes produce sperm and the hormone testosterone.

2. Temperature regulation in the testes depends on a muscle that contracts to bring the testes closer to the body for warmth when scrotal temperature is low or relaxes when the scrotal temperature is high, so the testes are farther away from the body. Sperm develop better in the testes at a few degrees lower than body temperature.

3. The path of sperm from their site of production to their release from the body: seminiferous tubules → epididymis → vas deferens → urethra.

4. The male accessory glands and their functions:

   a. Seminal vesicles—secrete most of the seminal fluids.
   b. Prostate gland—secretes watery alkaline fluid, which activates sperm and raises the vaginal pH.
   c. Bulbourethral glands—secrete lubricating mucus.

5. The function of the penis is to deliver sperm to the vagina. The penis becomes erect when neural activity dilates arterioles in the penis, allowing blood to enter the spongy tissue and causing the veins to drain less blood. The change in blood distribution allows blood to accumulate in the spongy tissue and cause an erection.

6. The three regions of a sperm cell:

   a. Head—contains the DNA and is covered by the acrosome, an enzyme-rich cap that helps egg penetration.
   b. Midpiece—contains mitochondria to power the tail for locomotion.
   c. Tail—locomotion to the egg.

7. Hormones from the hypothalamus, the anterior pituitary gland, and the testes that are important in the control of sperm production:

   a. Testosterone—produced in the testes. Testosterone is necessary for the maturation of sperm. When testosterone levels in the blood fall, the hypothalamus is stimulated to release GnRH.
   b. GnRH—produced in the hypothalamus; stimulates the anterior pituitary to release LH and FSH.
   c. LH—produced in the anterior pituitary; stimulates the testes to produce testosterone.
   d. FSH—produced in the anterior pituitary; stimulates sperm production (with testosterone).

8. The uterus has two layers. The inner layer is the endometrium, which is the lining that builds up each month in preparation for pregnancy. These preparations include cell division to thicken the endometrium, which is stimulated by estrogen, and an increase in glandular activity to nourish the embryo, which is stimulated by progesterone. The outer layer is the myometrium, which consists of smooth muscle. The myometrium allows the uterus to get larger as the fetus grows and provides the force to push the baby out during delivery.

9. The major structures of the female reproductive system:

   a. Ovaries—produce the eggs.
   b. Oviducts—path for egg/early embryo to travel to uterus.
   c. Uterus—site for implantation of the embryo and development of the fetus.
   d. Vagina—birth canal.

10. An ectopic pregnancy occurs if the embryo implants in a place other than the uterus. The most common type of ectopic pregnancy is a tubal pregnancy, in which the embryo implants in the oviduct. Because the oviduct cannot expand as the uterus does to accommodate embryonic growth, the oviduct may rupture, causing the mother to hemorrhage, which can be fatal.

11. The breasts contain milk glands that produce milk for offspring. The milk glands are connected to ducts that lead to the outside environment at the nipple. Most of the breast consists of fatty tissue.

12. The ovarian cycle starts with a primary follicle, which is a primary oocyte surrounded by a layer of follicle cells. As the primary follicle matures, the follicle cells divide and produce estrogen, which enters the bloodstream and accumulates in a fluid within the follicle. The accumulation of fluid causes the layer of follicle cells to split, and the fluid continues to accumulate in the newly formed cavity. The estrogen in the fluid causes the first meiotic division in the primary oocyte, forming a secondary oocyte. Estrogen-containing fluid continues to accumulate as the follicle grows. Eventually, the mature Graafian follicle forms. The Graafian follicle contains a secondary oocyte that is located at the edge and surrounded by follicle cells and a large fluid-filled cavity. At ovulation, the mature follicle pops, releasing the egg. The follicle cells remaining in the ovary are transformed into an endocrine structure called the corpus luteum, which secretes estrogen and progesterone. If pregnancy does not occur, the corpus luteum will degenerate within about 2 weeks.

13. Day 1 of the menstrual cycle is considered the first day of menstrual flow. At this point in the cycle, estrogen and progesterone levels are low, and a new follicle is beginning to develop. FSH from the anterior pituitary stimulates estrogen production by the developing follicle. As the follicle develops, estrogen levels rise and cause cell division in the endometrium, which thickens the endometrium for implantation of an embryo. As the follicle approaches maturity, the rapidly rising level of estrogen stimulates LH release from the anterior pituitary, which triggers ovulation. The follicle cells remaining in the ovary after ovulation are transformed into an endocrine structure, called the corpus luteum, that produces estrogen and progesterone. The estrogen continues to stimulate cell division in the endometrium, and progesterone causes the development of mucous glands that will nourish the embryo if one forms. Progesterone is also needed to maintain the endometrium. If pregnancy does not occur, the corpus luteum degenerates, and the level of estrogen and progesterone decrease. Without progesterone to maintain the endometrium, it decreases and the endometrium is no longer maintained. The endometrium breaks down and is lost as menstrual fluid.

14. Menopause is the cessation of the menstrual cycle and ovulation, because follicles are no longer developing. From the moment of birth, the number of primary follicles in a woman's ovaries begins to decrease. By the time she is about 45 to 55 years old, follicles remaining in the ovaries no longer respond to FSH and develop. Without egg development, ovulation cannot occur. Without the cycling of hormones, the menstrual cycle does not occur.

15. The stages of the human sexual response are excitement, plateau, orgasm, and resolution.

16. A vasectomy includes making incisions in the scrotum and sealing off each vas deferens so that sperm cannot enter the urethra. A tubal ligation includes cutting and sealing each oviduct so that the egg cannot reach the uterus and the sperm cannot reach the egg.

17. Birth control pills contain the hormones estrogen and progesterone, which inhibit the production of FSH and LH. This prevents the follicles from maturing, and, if a follicle should develop, it prevents ovulation.

18. Side effects of using the birth control pill may include acne and headaches. More serious side effects include high blood pressure and blood clots, which could lead to heart attack, stroke, or pulmonary embolism.

19. Use of progesterone-only means of contraception does not prepare the endometrium for embryo implantation and may prevent ovulation.

20. An IUD is an intrauterine device that can remain in the uterus for several years. It prevents pregnancy by interfering with fertilization and implantation.

21. The diaphragm, male condom, and female condom are barrier methods for birth control. They prevent pregnancy by preventing the sperm from reaching the egg.

22. c	28. b
23. b	29. b
24. a	30. c
25. a	31. epididymis
26. b	32. oviduct
27. a	

Chapter 18

1. The prenatal period is divided into the pre-embryonic period, embryonic period, and fetal period. The pre-embryonic period runs from fertilization through the second week and is the time when formation and implantation of the blastocyst take place. The extraembryonic membranes and placenta also begin to form at this time. The embryonic period extends from week 3 through week 8 and is the time when gastrulation takes place and the organs and organ systems form. The fetal period runs from the ninth week until birth and is a period of intense growth.

2. Secretions from the female reproductive tract alter the surface of the acrosome to destabilize the plasma membrane of the sperm.

3. During fertilization, fusion of the plasma membranes of the sperm and egg triggers granules near the plasma membrane of the oocyte to release enzymes. These enzymes cause the zona pellucida to quickly harden and thereby prevent passage of other sperm. (Polyspermy is the abnormal condition when more than one sperm fertilizes the egg.)

4. Implantation is the process by which the blastocyst becomes embedded in the endometrium. Implantation normally occurs high up on the back wall of the uterus. A blastocyst that implants outside the uterus may result in an ectopic pregnancy. Most implantations outside the uterus occur in the oviducts.

5. The four extraembryonic membranes are the amnion (which protects the embryo by enclosing it in a fluid-filled sac), yolk sac (which serves as a site of early blood cell formation and contains primordial germ cells that migrate to the gonads), allantois (which becomes part of the umbilical cord), and chorion (which becomes the embryo's major contribution to the placenta).

6. The placenta provides oxygen and nutrients to the fetus and removes wastes (such as carbon dioxide and urea). It is formed from the chorion of the embryo and the endometrium of the mother where implantation occurred. During implantation, cells of the outer layer of the trophoblast rapidly divide and invade the endometrium. Then, one layer of the trophoblast forms finger-like processes called chorionic villi that grow into the endometrium. Chorionic villi contain blood vessels connected to the developing embryo that provide exchange surfaces for diffusion of nutrients, oxygen, and wastes.

7. Gastrulation is the process early in prenatal development that forms the three germ layers—ectoderm, mesoderm, and endoderm. Ectoderm forms the nervous system and the outer layer of skin (including hair and nails). Mesoderm forms connective tissues and muscle, as well as organs such as the heart, kidneys, ovaries, and testes. Endoderm forms the linings of the urinary, respiratory, and digestive tracts. The pancreas, liver, thyroid, and parathyroid glands also form from endoderm.

8. Neurulation is formation of the CNS. During neurulation, the neural plate folds inward to form a groove extending the length of the embryo on its dorsal surface. The raised sides of the groove are called neural folds. These will meet and fuse to form the neural tube, which is a fluid-filled tube that becomes the central nervous system. The embryo during this period is called a neurula. Later, the anterior portion of the neural tube develops into the brain, and the posterior portion forms the spinal cord.

9. Sex is determined at fertilization by the sex chromosome carried by the sperm. If the sperm has an X chromosome, a female will be produced. If the sperm has a Y chromosome, a male will be produced.

10. Fetal circulation is designed to bypass the lungs and liver, which are not fully functional until after birth. There are two shunts that keep most blood away from the lungs. The foramen ovale is a shunt between the right and left atria, and the ductus arteriosis is a shunt between the pulmonary trunk and aorta. The ductus venosus shunts most blood past the liver. After birth these temporary shunts close so that more blood can go to the lungs and the liver.

11. The three stages of labor are the dilation stage (begins with regular contractions and ends when cervix is dilated to 10 cm), expulsion stage (begins with full dilation of the cervix and ends with delivery of the baby), and placental stage (begins with delivery of the baby and ends with expulsion of the placenta).

12. Environmental agents have the most drastic effects during critical periods, and most critical periods occur during the embryonic period when tissues and organs are forming (from week 3 to week 8). For example, limb development is most sensitive to environmental agents during weeks 4 through 6. Critical periods for the CNS are longer than those for most other organs and organ systems and span from week 3 to week 16.

13. Milk production begins after birth, when estrogen and progesterone levels decline to a level where prolactin can exert its effects.

14. Possible causes of aging include declines in the functioning of key organ systems, damage to cellular processes by free radicals, changes in proteins caused by glucose, and genetically programmed cessation of cell division. Besides medical advances, a healthy, moderate lifestyle can improve your chances of having a high-quality old age.

15. d
16. c
17. b
18. a
19. b
20. d
21. c
22. c
23. b
24. b
25. d
26. d

Chapter 19

1. A chromosome is composed of tightly coiled DNA and associated proteins, whereas a gene is a segment of the DNA in a chromosome that codes for a particular protein.

2. Mitosis is a type of division of the nucleus occurring in somatic cells in which two identical cells, called daughter cells, are generated from a single cell. The original cell first replicates its genetic material and then distributes a complete set of genetic information to each of its daughter cells. Mitosis is usually divided into prophase, metaphase, anaphase, and telophase. Cytokinesis is the division of the cytoplasm and organelles into two daughter cells during cell division. Cytokinesis usually occurs during telophase.

3. Meiosis is important because it is the type of nuclear division that produces haploid gametes (eggs or sperm) from diploid germ cells. The process is needed so that gametes have only one set of chromosomes. When fertilization occurs, the diploid number is restored. Meiosis is also important because it introduces genetic variability through independent assortment and crossing over.

4. In meiosis I, homologous chromosomes pair up at the equatorial plate so that each new cell will receive only one set of chromosomes during division. The orientation of the members of the pair (maternal or paternal) is random with respect to which member is closer to which pole. Each of the 23 pairs of chromosomes orients independently during metaphase I. The orientations of all 23 pairs will determine the assortments of maternal and paternal chromosomes in the daughter cells.

In meiosis II, the chromosomes line up on the equatorial plate in a manner similar to the way they do in mitosis. During anaphase II, the sister chromatids separate so that each gamete receives only one copy of each chromosome.

5. Independent assortment and crossing over create genetic variability in different ways. Crossing over is the breaking and rejoining of nonsister chromatids of homologous pairs of chromosomes during meiosis (specifically at prophase I, when homologous chromosomes pair up side by side). Crossing over results in the exchange of corresponding segments of chromatids and increases genetic variability in populations by changing the combination of alleles on a chromosome. Independent assortment is the process by which homologous chromosomes and the alleles they carry segregate randomly during meiosis, creating mixes of maternal and paternal chromosomes in gametes. During meiosis I, exchanges of genetic material between homologous chromosomes produce new combinations of genes. Independent assortment allows homologous chromosomes to assort independently of each other during meiosis I, thereby producing new combinations of genes for each gamete.

6. Nondisjunction is the failure of chromosomes to separate properly during anaphase of meiosis I or II. If a chromosome pair does not separate during meiosis I, one of the daughter cells will receive an extra chromosome, and one will lose a chromosome. This can happen in meiosis II with the same result: one gamete with an extra chromosome and one missing a chromosome. This can result in a trisomy ($n + 1$) or in monosomy ($n - 1$) after fertilization.

7. Down syndrome is caused by a trisomy of chromosome 21. The symptoms of Down syndrome include moderate to severe mental retardation, short stature or shortened body parts due to poor skeletal growth, characteristic facial features, and mild to severe heart conditions.

8. b
9. a
10. a
11. b
12. a
13. b
14. a
15. c
16. crossing over and independent assortment
17. Homologous
18. Synapsis
19. anaphase
20. anaphase II

Chapter 20

1. In a cross between a homozygous dominant individual and a homozygous recessive individual, all the offspring will be heterozygous for the trait. Their phenotypes will be the dominant phenotype.

2. A pedigree is a diagram showing all the known phenotypes for a particular trait of individuals in an extended family. It can be helpful in determining whether a person who has the dominant phenotype is homozygous or heterozygous. This information can be useful in determining whether a person is a carrier for a harmful recessive trait.

3. Codominance is the condition in which the effects of both alleles are separately expressed in a heterozygote. The human blood type AB is codominant because both the A and the B allele are expressed.

4. Multiple alleles occur when there are more than two different alleles for a trait. Human blood types (ABO) are an example of multiple alleles. Polygenic inheritance occurs when more than one gene controls the trait. The effects of polygenic inheritance are a range of phenotypes depending on the number of dominant or recessive alleles inherited. Human height is an example of polygenic inheritance.

5. Linked genes are genes located on the same chromosome. Independent assortment does not apply to linked genes. Crossing over can "unlink" the genes.

6. The pattern of inheritance for recessive X-linked genes in males is different from the pattern for recessive autosomal alleles in females because males inherit only one X chromosome. When a male inherits a recessive allele on the X chromosome, the allele is expressed without the other recessive allele present. In contrast, recessive autosomal alleles are not expressed unless they are homozygous.

7. Chorionic villi sampling and amniocentesis are two forms of prenatal genetic testing. In amniocentesis, a small amount of amniotic fluid is removed for genetic testing of living fetal cells in the fluid. Chorionic villi sampling involves removal of a small piece of a chorionic villus to be genetically tested. The villi have the same genetic makeup as the fetus.

8. b
9. a
10. a
11. c
12. c
13. a
14. c
15. c
16. polygenic
17. phenotype, genotype
18. homozygous, heterozygous
19. linked

Chapter 21

1. The structure of DNA is somewhat like a long ladder, twisted about itself like a spiral staircase. The DNA ladder is composed of two long strings of smaller molecules called nucleotides. Each nucleotide chain makes up one side of the ladder and half of each rung. A nucleotide consists of a phosphate, a nitrogen-containing base, and a sugar called deoxyribose. There are only four different nitrogen-containing bases used in DNA: adenine (A), thymine (T), cytosine (C), and guanine (G). Although DNA has only four different nucleotides, a DNA molecule is very long and has thousands of nucleotides. When forming a rung of the ladder, adenine must pair with thymine, and cytosine must pair with guanine.

2. During replication, the two DNA strands unwind, and new nucleotides are added to each side to make two new strands. New nucleotides are added following complementary base pairing rules: Adenine must pair with thymine, and cytosine must pair with guanine. The specificity of these base pairs is important, not only for the accurate production of new DNA molecules, but also for converting the information in the gene into a protein.

3. DNA replication is described as semiconservative because each strand of the original DNA molecule serves as a template for the formation of a new strand. This process is called semiconservative replication because in each of the new double-stranded DNA molecules, one original (parent) strand is saved (conserved), and the other (daughter) strand is new. Complementary base pairing creates two new DNA molecules that are identical to the parent molecule.

4. Transcription of DNA is the process by which a complementary single-stranded messenger RNA (mRNA) molecule is formed from a single-stranded DNA template. As a result, the information in DNA is transferred to RNA. Translation is the process of converting the nucleotide language of messenger RNA into the amino acid language of a protein. The mRNA template codes for the production of amino acid chains during translation.

5. RNA differs from DNA in three ways: (1) RNA is single-stranded, and DNA is double-stranded; (2) in RNA, the nucleotide sugar is ribose, and in DNA the nucleotide sugar is deoxyribose; and (3), in RNA, uracil replaces thymine found in DNA.

6. Messenger RNA, tRNA, and rRNA play different roles in protein synthesis. Messenger RNA (mRNA) is the type of RNA synthesized from and complementary to a region of DNA. It attaches to ribosomes in the cytoplasm and specifies the amino acid order in the protein. It carries the DNA's instructions for synthesizing a particular protein. Transfer (tRNA) is specialized to bring a specific amino acid to where it can be added to a polypeptide that is under construction. Ribosomal RNA (rRNA) combines with proteins to form ribosomes, which are the structures on which protein synthesis occurs.

7. A three-base sequence on messenger RNA (mRNA) specifies one of the 20 common amino acids or the beginning or end of the protein chain. Codons are read from the mRNA to produce amino acid chains. The order of the codons on mRNA determines the order of amino acids in the protein formed by translation. Codons also signal the start and end of a protein.

8. An anticodon is a three-base sequence on transfer RNA (tRNA) that binds to the complementary base pairs of a codon on the mRNA. Complementary base pairing ensures the correct amino acid is added to the amino acid chain.

9. The events in translation (protein synthesis):

   a. Initiation—small ribosomal subunit joins with the mRNA start codon. Complementary tRNA binds to the start codon on mRNA, and the large subunit joins to form a functional ribosome. tRNA continues to pair with the codons to produce a chain.

   b. Elongation—the amino acid chain continues to get longer.

   c. Termination—occurs when a stop codon is encountered. The ribosomes separate, and the process is terminated.

10. When a deletion occurs, it changes all the codons for the part of the mRNA molecule downstream.

11. Gene activity is regulated by chromosomal coiling and uncoiling, regulatory genes, and chemical signals such as hormones.

12. Genetic engineering is the manipulation of genetic material for human purposes. Restriction enzymes cut out pieces of DNA according to the nucleotide sequence they recognize. The "gene of interest" is identified in the mixture of pieces of DNA and added to a vector. Vectors are biological carriers of recombinant DNA to the new host cell.

13. In farming, genetically engineered crops are increasingly common. These crops are disease resistant, drought resistant, or pest resistant. In medicine, genetic engineering is being used to provide vaccines, therapeutic proteins, and some antibodies.

14. Gene therapy is the treatment of a genetic disease by inserting healthy functional genes into the body cells that are affected by the faulty gene. The usual method is to use a virus to deliver the DNA.

15. a
16. c
17. b
18. a
19. b
20. c
21. b
22. d
23. b
24. tRNA
25. uracil
26. restriction enzyme

Chapter 22

1. The following steps are thought to explain how life evolved from inorganic molecules to complex cells: Inorganic molecules formed organic molecules; small organic molecules joined to form complex organic molecules; genetic material originated; organic molecules and DNA aggregated into droplets, which eventually formed prokaryotic cells and then eukaryotic cells.

2. *Microevolution* refers to changes in allele frequencies in a population. *Macroevolution* describes large-scale changes in organisms over long periods of time.

3. Four sources of variation within populations are the chance union of a certain egg and sperm at fertilization in sexually reproducing species, mutation, crossing over, and independent assortment.

4. Genetic drift is a random change in allele frequencies so that the allele in question can become fixed or lost in the population.

5. Speciation is the formation of different species from a common ancestor. Disruption in gene flow between populations can cause speciation.

6. Natural selection is the process by which organisms differentially survive and reproduce. Changing environments can maintain variation in a population.

7. The binomial system of naming is a two-part unique scientific name that identifies a species. It is composed of the genus name and the specific epithet.

8. A phylogenetic tree is a hypothesis of organismal (evolutionary) relatedness; it is similar to a family tree.

9. A fossil is a preserved remnant or impression of past organisms. Dead organisms are covered by sediment, and minerals move into the parts that are not destroyed. Uplift or erosion may later expose the fossil. Organisms that have no "hard parts" generally do not fossilize well, so the fossil record is a biased sampling of past life.

10. New distributions of organisms arise when organisms disperse to new areas or when the areas occupied by organisms move or are subdivided.

11. The molecular clock is the constant rate of divergence of macromolecules from one another due to nucleotide changes in the genome.

12. Homologous structures have arisen from a common ancestry. Analogous structures have the same function but arose through convergent evolution, not shared ancestry.

13. Common embryological origins are evidence of common descent.

14. c
15. a
16. d
17. c
18. a
19. d
20. c
21. a
22. c
23. d
24. c
25. a
26. a

Chapter 23

1. Ecological succession is the sequence of change of species over time in a community. Primary succession occurs on bare rock. Secondary succession occurs on disturbed land (such as old abandoned fields) where soil already exists.

2. The source of energy in an ecosystem is the sun. Producers capture the sun energy by photosynthesis and use it to produce sugars for energy. Consumers eat producers for energy. Decomposers get energy from dead consumers and producers and then release inorganic materials that can be reused by producers.

3. Roles in an ecosystem:

   a. Producers—photosynthetic organisms capable of using light energy to produce sugars.
   b. Primary consumers—eat the producers.
   c. Secondary consumers—eat the primary consumers.
   d. Decomposers—eat dead producers and consumers and release inorganic materials that can be reused by producers.

4. A food chain is the successive series of organisms that energy (in the form of food) flows through in an ecosystem. Each organism in the series consumes the preceding one. It begins with the photosynthesizers and flows to herbivores and then to carnivores. The feeding relationships in a community are more realistically portrayed as a food web than as a food chain because an animal may feed at a number of different trophic levels. A food web describes all the interconnections in feeding and accounts for the varied diets of animals.

5. One reason why energy is lost between trophic levels is that an animal uses roughly two-thirds of the energy in the food it digests. In addition, an animal must first expend energy to obtain its food, usually by grazing or hunting. Furthermore, not all the food available at a given trophic level is captured and consumed. Finally, some of the food eaten cannot be digested and is lost as feces. The energy in the indigestible material is unavailable to the next higher trophic level. However, the remaining energy can be converted to biomass and will be available to the next higher trophic level.

6. An energy pyramid is a graphical representation in which blocks represent the decreasing amount of energy available at each trophic (feeding) level. It is a pyramid shape because energy is lost with each transfer, so less energy is available at each successive level.

7. A pyramid of biomass is a diagram in which blocks represent the amount of biomass (dry body mass of organisms) available at each trophic (feeding) level. The biomass pyramid is similar to the energy pyramid in that the producers contribute the most biomass and the tertiary consumers produce the least.

8. Biological magnification is the tendency of a nondegradable chemical to become increasingly concentrated in the bodies of organisms as it passes along the food chain. Humans are the top carnivore. If there is biological magnification of a harmful substance, the substance will be highly concentrated in the animals we eat on the next lower trophic level.

9. More people could be fed if humans ate at a lower trophic level because only 10% of the energy is transferred from one level to the next. Thus, if humans began to eat one level lower on the food chain, about 10 times more energy would be available to them.

10. Describe the water cycle, the carbon cycle, the nitrogen cycle, and the phosphorus cycle.

   a. The water cycle is the pathway of water as it falls as precipitation; collects in ponds, lakes, and seas; and returns to the atmosphere through evaporation.
   b. The carbon cycle is the worldwide circulation of carbon from the carbon dioxide in air to the carbon in organic molecules of living organisms and back to the air. Carbon enters living systems when photosynthetic organisms use carbon dioxide (and oxygen) to produce organic materials. Carbon dioxide is formed again when living organisms use the organic molecules for cellular respiration.
   c. The nitrogen cycle is the worldwide circulation of nitrogen from nonliving to living systems and back again. Atmospheric nitrogen (N_2) cannot enter living systems. Nitrogen-fixing bacteria living in nodules on the roots of leguminous plants convert N_2 to ammonium (NH_4^+), which is converted to nitrites (NO_2^-) and then to nitrates (NO_3^-) by nitrifying bacteria. The ammonium and nitrates are then available to plants to use in their proteins and nucleic acids. Next, the nitrogen is transferred to organisms that consume the plants. Nitrates that are not assimilated into living organisms can be converted to nitrogen gas by denitrifying bacteria.
   d. Phosphorus in the form of phosphates is washed from sedimentary rock by rainfall. Producers use the dissolved phosphates to produce important biological molecules, including DNA and ATP. When animals eat producers or other animals, phosphates are passed through the food webs. Decomposers release phosphates from dead organisms into the soil or water.

11. Humans are disturbing the water cycle by emptying aquifers, building dams, changing watersheds, and draining wetlands.

12. Burning fossil fuels and cutting down trees without replanting are the human activities that are primarily responsible for the rising level of atmospheric carbon dioxide.

13. Biodiversity, the number and variety of living things, is being reduced dramatically, largely because of human activity. Biodiversity is greatest in the tropics, and most biodiversity loss is occurring in the tropics as a result of habitat destruction. Two practical reasons for concern over the loss of biodiversity are that the disappearing species could have genes that would someday prove useful or that they could be found to produce chemicals with medicinal qualities.

14. a
15. b
16. a
17. d
18. c
19. d
20. c
21. c
22. niche
23. secondary consumer, carnivore

Chapter 24

1. A population's growth rate is its *birth rate* minus its *death rate* per 1000 individuals per year. Thus the growth rate of the world's population, an increase of 11 individuals per 1000, is 1.1%.

2. The age structure of a population is the number of males and females of each age in a population. The ages are often grouped into prereproductive, reproductive, and postreproductive categories. Generally, only individuals of reproductive age add to the size of the population. A population with a very large prereproductive age relative to the other groups will get larger in the future. The population with a large postreproductive group relative to the other groups will decline in size in the future. In a population that will remain stable in size, the categories are roughly equal in size.

3. Density-independent and density-dependent factors regulate population size.
   a. Density-dependent factors are dependent on population size. Competition for food is a density-dependent factor.
   b. Density-independent factors are not related to population size and include natural disasters such as fire and flood.

4. An ecological footprint is a measure of the impact a person or population makes on the environment. It is expressed as the total amount of biological productive land and water needed to produce and dispose of the products that are consumed. Calculation of an ecological footprint includes everything that is consumed and the corresponding waste removal. Thus, ecological footprints describe the burden placed on Earth's carrying capacity.

5. Overfarming and overgrazing can cause loss of topsoil, which results in desertification.

6. The carrying capacity of the environment is the number of individuals of a given species that a particular environment can support for a prolonged time period. The carrying capacity of the environment is determined by such factors as availability of resources, including food, water, and space; ability to clean away wastes; and predation pressure.

7. Deforestation is the removal of trees without replacement. Deforestation is occurring at an alarming rate in the tropical rain forests. The land is being used for growing crops to feed human populations and for logging.

8. Ozone in the upper atmosphere is helpful because it shields the earth from UV radiation. Ozone at low levels (in smog) can lead to breathing problems and irritated eyes, nose, and throat. CFCs released into the atmosphere can destroy beneficial ozone in the stratosphere in a chemical reaction.

9. The greenhouse effect is a process in which greenhouse gases trap heat in the atmosphere. Examples of greenhouse gases include carbon dioxide and methane. The greenhouse effect could lead to a rise in temperatures throughout the world.

10. b
11. d
12. d
13. b
14. c
15. a
16. c
17. a

Appendix 2 Hints for Applying the Concepts Questions

Chapter 1
1. Each part of the question asks about a different group of Swedish children. Choose the correct group from the key, and look at the line on the graph of the appropriate color.
2. Design an experiment with a control group and an experimental group.
3. Consider the questions on page 10.

Chapter 2
1. What functions do triglycerides perform in your body?
2. Cellulose is an important form of dietary fiber.
3. Consider the following qualities of water: polarity, heat of vaporization, and high heat capacity.
4. Consider the harmful effects of radiation on the body.

Chapter 3
1. Anesthetics need to cross the plasma membranes of nerve cells.
2. Mitochondria process energy for cells, and thus they occur in large numbers in cells with a high demand for energy.
3. Which element of the cytoskeleton is involved in cell division?

Chapter 4
1. Does cartilage have a blood supply?
2. What physiological mechanisms raise body temperature? What physiological mechanisms lower body temperature?
3. Where is connective tissue found?

Chapter 5
1. What happens to bone density if no stress is placed on the bone?
2. What kind of activity builds bone density?
3. What structure would indicate that the bone was still capable of growth?

Chapter 6
1. What role does acetylcholine play in muscle contraction?
2. What causes tendinitis? How is it treated?
3. What role do calcium ions play in muscle contraction?
4. Which fibers are darker in color: fast-twitch or slow-twitch fibers? How do the properties of fast-twitch fibers differ from those of slow-twitch fibers?

Chapter 7
1. What effect does an inhibitory neurotransmitter have on the postsynaptic neuron?
2. What would happen if a neurotransmitter were not removed from the synapse?
3. What factors cause ions to cross the membrane during an action potential? What role do potassium ions play in the action potential?
4. What role does the myelin sheath play in the conduction of an action potential?

Chapter 8
1. What are the functions of the spinal cord? Would the location of the injury affect which functions are lost?
2. What is the function of the sympathetic nervous system?
3. What could happen if the sensory nerves from the tongue were anesthetized?
4. What structure of the brain is important in transferring short-term memory to long-term memory?

Chapter 9
1. In what type of vision are distant objects seen more clearly than nearby objects?
2. What part of the ear is responsible for equilibrium?
3. What changes when you shift your focus from an object in the distance to a nearby object?

Chapter 10
1. Cortisone is a glucocorticoid.
2. Which internal system of communication is relatively slow? Which is relatively fast?
3. Theresa's symptoms and the timing of their onset suggest melatonin may be involved.
4. What hormones are produced by the adrenal cortex?

Chapter 11
1. Are the white blood cells that are produced in leukemia functional?
2. What regulates the circulating number of red blood cells?
3. What is the function of red blood cells? How is iron related to the ability of red blood cells to carry out their function?
4. Which antibodies against antigens on red blood cells does Raul have?
5. What could happen if either of Elizabeth's babies had Rh-positive blood?
6. What are the functions of platelets?

Chapter 12
1. What happens if the valves do not close properly?
2. The pressure that propels blood through the arteries is equal to the pressure against the arterial walls. What force generates this pressure? What is the relationship between high blood pressure and atherosclerosis?
3. What happens in the lymph nodes when you get sick?

Chapter 13
1. Is a vaccine effective immediately?
2. How specific is the immune response?
3. What role do helper T cells play in the immune response?
4. How long does it take antibodies to form after the first exposure to an antigen?
5. In an organ transplant, what would happen to the transplanted cells if they had very different self (MHC) markers from the recipient's?

Chapter 14
1. What factors regulate breathing rate and tidal volume?
2. What is the function of the cilia in the respiratory tubules?
3. What happens to Juan's blood level of carbon dioxide when he holds his breath? How would this affect his breathing?
4. Does inhalation or expiration involve muscle contraction? Which is a passive process?

Chapter 15
1. What role does the pancreas play in digestion?
2. What is the relationship between diarrhea and water reabsorption?
3. What would cause skin to develop a yellow tone? What might the cause of the yellow skin tone have to do with a tattoo?
4. What substance is released into the small intestine when fatty food enters the small intestine?

Chapter 16
1. Consider the role of the external urethral sphincter in urination.
2. Beer is a diuretic.
3. Consider the functions of the kidneys and the ways to replace these functions.
4. Which region of the nephron is involved in filtration?

Chapter 17
1. When making your recommendation for a means of contraception, consider its effectiveness in preventing pregnancy, this couple's need for protection against STDs, and the health effects of the means of contraception.
2. What happens to the endometrium during menstruation?
3. What does warm temperature have to do with sperm count?

Chapter 18
1. How does fetal circulation differ from circulation after birth?
2. Consider the concept of critical periods in development.
3. Consider the hormones involved in milk production and ejection.
4. Consider the potential role of free radicals in aging.

Chapter 19
1. What is nondisjunction?
2. What is the function of the spindle fibers?
3. Look at Figure 19.16.

Chapter 20
1. Color blindness is a sex-linked trait.
2. What are the genotypes of George and Sue? Use a Punnett square to determine the expected results from a cross with those genotypes.
3. A recessive trait is expressed only in the homozygous condition.
4. Recessive alleles are not expressed.
5. Use a Punnett square to determine the outcome of the cross.

Chapter 21
1. What is the start codon? What is the stop codon? What amino acids do the other codons code for?
2. Translate each of the mRNA strands. Remember that more than one codon can code for the same amino acid.
3. Which DNA fingerprint matches the one from the bloodstain at the scene?

Chapter 22
1. Describe the evidence in support of evolution.
2. Consider chemical evolution and the conditions of the early Earth.
3. What characteristics distinguish primates from other mammals?
4. Tay-Sachs disease is caused by a recessive allele.

Chapter 23
1. On average, what percentage of the energy available at one trophic level is available to the next level?
2. What happens to mercury as it moves up the food chain?
3. Average rainfall and temperature largely determine the organisms found in a given location.

Chapter 24
1. What will happen to the size of each population as the prereproductive individuals reach reproductive age?
2. a. What factors regulate population size?
   b. How might Asian carp affect factors regulating the population size of other species of fish?
3. How would spending differ between a young, growing population and an aging population?
4. The size of the population influences how quickly individuals are added to the population.

Glossary

A

Accommodation A change in the shape of the lens of the eye brought about by contraction of the smooth muscle of the ciliary body that changes the degree to which light rays are bent so that an image can be focused on the retina.

Acetylcholine A neurotransmitter found in both the central nervous system and the peripheral nervous system. It is the neurotransmitter released at neuromuscular junctions that causes muscle contraction.

Acetylcholinesterase An enzyme that breaks the neurotransmitter acetylcholine into its inactive components, acetate and choline. Acetylcholinesterase stops the action of acetylcholine at a synapse.

Acid Any substance that increases the concentration of hydrogen ions in solution.

Acinar cells Exocrine cells of the pancreas that secrete digestive enzymes into ducts that empty into the small intestine.

Acquired immune deficiency syndrome See *AIDS*.

Acromegaly A condition characterized by enlarged soft tissues and thickened bones of the extremities. It is caused by overproduction of growth hormone in adulthood.

Acrosome A membranous sac on the head of a sperm cell that contains enzymes that facilitate sperm penetration into the egg during fertilization.

ACTH See *adrenocorticotropic hormone*.

Actin The contractile protein that makes up the major portion of the thin filaments in muscle cells. An actin (thin) filament is composed of actin, troponin, and tropomyosin. In muscle cells, contraction occurs when actin interacts with another protein called myosin.

Actin filaments The thin filaments in muscle cells composed primarily of the protein actin and essential to muscle contraction. In addition to actin, thin filaments contain two other proteins important in the regulation of muscle contraction: tropomyosin and troponin.

Action potential A nerve impulse. An electrochemical signal conducted along an axon. A wave of depolarization caused by the inward flow of sodium ions followed by repolarization caused by the outward flow of potassium ions.

Active immunity Immune resistance in which the body actively participates by producing memory B cells and memory T cells after exposure to an antigen, either naturally or through vaccination.

Active site A specific location on an enzyme where the substrate binds.

Active transport The movement of molecules across the plasma membrane, usually against a concentration gradient (from a region of lower concentration to one of higher concentration) with the aid of a carrier protein and energy (usually in the form of adenosine triphosphate, or ATP) supplied by the cell.

Acute renal failure An abrupt, complete or nearly complete, cessation of kidney function.

Adaptation The process by which populations become better attuned to their particular environments as a result of natural selection.

Adaptive immune responses Body defense responses that are acquired by exposure to cells that do not belong in the body. It involves antibody-mediated responses and cell-mediated responses. Adaptive responses have memory for the pathogen that triggered them.

Adaptive trait A characteristic (structure, function, or behavior) of an organism that makes an individual better able to survive and reproduce in its natural environment. Adaptive traits arise through natural selection.

Addison's disease An autoimmune disorder characterized by fatigue, loss of appetite, low blood pressure, and increased skin pigmentation resulting from undersecretion of glucocorticoids and aldosterone.

Adenosine triphosphate (ATP) A nucleotide that consists of the sugar ribose, the base adenine, and three phosphate groups. ATP is the energy currency of all living cells.

ADH See *antidiuretic hormone*.

Adhesion junction A specialized junction between cells in which protein filaments hold together the plasma membranes of adjacent cells; a desmosome.

Adipose tissue A type of loose connective tissue that contains cells specialized for storing fat.

Adolescence The stage in postnatal development that begins with puberty. It is a period of rapid physical and sexual maturation during which the ability to reproduce is achieved. Adolescence ends with the cessation of growth in the late teens or early twenties.

Adrenal cortex The outer region of the adrenal gland that secretes glucocorticoids, mineralocorticoids, and gonadocorticoids.

Adrenal glands The body's two adrenal glands are located on top of the kidneys. The outer region of each adrenal gland secretes glucocorticoids, mineralocorticoids, and gonadocorticoids, and the inner region secretes epinephrine and norepinephrine.

Adrenal medulla The inner region of the adrenal gland that secretes epinephrine and norepinephrine.

Adrenaline See *epinephrine*.

Adrenocorticotropic hormone (ACTH) The anterior pituitary hormone that controls the synthesis and secretion of glucocorticoid hormones from the cortex of the adrenal glands.

Adulthood The stage in postnatal development that is generally reached somewhere between 18 and 21 years of age and during which bodily changes continue as part of the growth and aging process.

Advance directive A legal document that allows you to convey in advance your wishes for end-of-life care.

Afferent (sensory) neuron A nerve cell specialized to conduct nerve impulses from the sensory receptors *toward* the central nervous system.

Age structure The number of males and females of each age in a population. The ages are often grouped into prereproductive, reproductive, and postreproductive categories. Generally, only individuals of reproductive age add to the size of the population.

Agglutinate To clump together.

Aging The normal and progressive alteration in the structure and function of the body. Aging is possibly caused by declines in critical body systems, disruption of cell processes by free radicals, slowing or cessation of cell division, and decline in the ability to repair damaged DNA.

Agranulocytes The white blood cells without granules or with very small granules in their cytoplasm, including monocytes and lymphocytes.

AIDS Acquired immune deficiency syndrome. A diagnosis of AIDS is made when an HIV-positive person develops one of the following conditions: (1) a helper T cell count below 200 cells per mm^3 of blood; (2) one of 26 opportunistic infections, the most common of which are *Pneumocystis jiroveci* pneumonia and Kaposi's sarcoma, a cancer of connective tissue that affects primarily the skin; (3) a loss of more than 10% of body weight (wasting syndrome); or (4) dementia (mental incapacity, such as forgetfulness or inability to concentrate).

Albinism A genetic inability to produce the brown pigment melanin that normally gives color to the eyes, hair, and skin.

Aldosterone A hormone (the primary mineralocorticoid) released by the adrenal cortex that stimulates the reabsorption of sodium within kidney nephrons.

Allantois The extraembryonic membrane whose blood vessels become part of the umbilical cord, the ropelike connection between the embryo and the placenta.

Allele An alternative form of a gene. One of two or more slightly different versions of a gene that code for different forms of the same trait.

Allergen An antigen that stimulates an allergic response.

Allergy A strong immune response to an antigen (an allergen) that is not usually harmful to the body.

Allometric growth The change in the relative rates of growth of various parts of the body. Such growth helps shape developing humans and other organisms.

Alveolus (plural, alveoli) A thin-walled rounded chamber. In the lungs, the alveoli are the surfaces for gas exchange. They form clusters at the end of each bronchiole that are surrounded by a vast network of capillaries. The alveoli greatly increase the surface area for gas exchange.

Amino acid The building blocks of proteins consisting of a central carbon atom bound to a hydrogen atom, an amino group (NH_2), a carboxyl group (COOH), and a side chain designated by the letter R. There are 20 amino acids important to human life; our bodies can synthesize some amino acids (nonessential amino acids), whereas others cannot be synthesized and must be obtained from the foods we eat (essential amino acids).

Amniocentesis A method of prenatal testing for genetic problems in a fetus in which amniotic fluid is withdrawn through a needle so that the fluid can be tested biochemically and the cells can be cultured and examined for genetic abnormalities.

Amnion The extraembryonic membrane that encloses the embryo in a fluid-filled space called the amniotic cavity. Amniotic fluid forms a protective cushion around the embryo that later can be examined as part of prenatal testing in a procedure known as *amniocentesis*.

Ampulla A wider region in a canal or duct. In the inner ear, an ampulla is found at the base of each semicircular canal.

Anabolic steroids Synthetic hormones that mimic testosterone and stimulate the body to build muscle and increase strength. Steroid abuse can have many dangerous side effects.

Anabolism The building (synthetic) chemical reactions within living cells, as when cells build complicated molecules from simple ones. Compare with *catabolism*.

Anal canal The canal between the rectum and the anus. Feces pass through the anal canal.

Analgesic A substance, such as Demerol, that relieves pain.

Analogous structure A structure of one organism that is similar to that of another organism because of convergent evolution and not because the organisms share a common ancestry. Compare with *homologous structures*.

Anaphase In mitosis, the phase when the chromatids of each chromosome begin to separate, splitting at the centromere. Now separate entities, the chromatids are considered chromosomes, and they move toward opposite poles of the cell.

Anaphylactic shock An extreme allergic reaction that occurs within minutes after exposure to a substance to which a person is allergic. It can cause pooling of blood in capillaries, which causes dizziness, nausea, and sometimes unconsciousness and extreme difficulty in breathing. Anaphylactic shock can lead to death.

Androgen A steroid sex hormone secreted by the testes in males and produced in small quantities by the adrenal cortex in both sexes.

Anemia A condition in which the blood's ability to carry oxygen is reduced. It can result from too little hemoglobin, too few red blood cells, or both.

Anencephaly A neural tube defect that involves incomplete formation of the brain and results in stillbirth or death shortly after birth.

Anesthesia The drug-induced loss of the sensation of pain. It may be general, regional, or local.

Aneurysm A blood-filled sac in the wall of an artery caused by a weak area in the artery wall.

Angina pectoris Choking or strangling chest pain, usually experienced in the center of the chest or slightly to the left, that is caused by a temporary insufficiency of blood flow to the heart. It begins during physical exertion or emotional stress, when the demands on the heart are increased and the blood flow to the heart muscle can no longer meet the needs.

Angioplasty A procedure that widens the channel of an artery obstructed because of atherosclerosis. It involves inflating a tough, plastic balloon inside the artery.

Angiotensin I Renin converts angiotensinogen into this protein.

Angiotensin II A protein that stimulates the adrenal gland to release aldosterone.

ANH See *atrial natriuretic hormone*.

Anorexia nervosa An eating disorder characterized by deliberate self-starvation, a distorted body image, and low body weight.

Antagonistic pairs Muscles arranged in pairs so that the actions of the members of the pair are opposite to one another. This arrangement is characteristic of most skeletal muscles.

Antibody A Y-shaped protein produced by plasma cells during an adaptive immune response that recognizes and binds to a specific antigen because of the shape of the molecule. Antibodies defend against invaders in a variety of ways, including neutralization, agglutination and precipitation, or activation of the complement system.

Antibody-mediated immune responses Immune system responses conducted by B cells that produce antibodies and that defend primarily against enemies that are free in body fluids, including toxins or extracellular pathogens, such as bacteria or free viruses.

Anticodon A three-base sequence on transfer RNA (tRNA) that binds to the complementary base pairs of a codon on the mRNA.

Antidiuretic hormone (ADH) A hormone manufactured by the hypothalamus but stored in and released from the posterior pituitary. It regulates the amount of water reabsorbed by the distal convoluted tubules and collecting ducts of nephrons. ADH causes water retention at the kidneys and elevates blood pressure. It is also called *vasopressin*.

Antigen A substance that is recognized as foreign by the immune system. Antigens trigger an immune response.

Antigen-presenting cell (APC) A cell that presents an antigen to a helper T cell, initiating an immune response toward that antigen. An important type of antigen-presenting cell is a macrophage.

Aorta The body's main artery that conducts blood from the left ventricle toward the cells of the body. The aorta arches over the top of the heart and gives rise to the smaller arteries that feed the capillary beds of the body tissues.

Apoptosis A series of predictable physical changes in a cell that is undergoing programmed cell death. Apoptosis is sometimes used as a synonym for *programmed cell death*.

Appendicular skeleton The part of the skeleton that includes the pectoral girdle (shoulders), the pelvic girdle (pelvis), and the limbs (arms and legs).

Appendix A slender closed pouch that extends from the large intestine near the juncture with the small intestine.

Aqueous humor The fluid within the anterior chamber of the eye. It supplies nutrients and oxygen to the cornea and lens and carries away their metabolic wastes.

Arachnoid The middle layer of the meninges (the connective tissue layers that protect the central nervous system).

Areolar connective tissue A type of loose connective tissue composed of cells in a gelatinous matrix. It serves as a universal packing material between organs and anchors skin to underlying tissues and organs.

Arrector pili The tiny, smooth muscles attached to the hair follicles in the dermis.

Arteriole A small blood vessel located between an artery and a capillary. Arterioles serve to regulate blood flow through capillary beds to various regions of the body. They also regulate blood pressure. Arterioles are barely visible to the unaided eye.

Artery A large-diameter muscular tube (blood vessel) that transports blood away from the heart toward the cells of body tissues. Arteries conduct blood low in oxygen to the lungs and blood high in oxygen to the body tissues. Arteries typically have thick muscular and elastic walls that dampen the blood pressure pulsations caused by heart contractions.

Arthritis An inflammation of a joint.

Artificial insemination A treatment for infertility in which sperm are deposited in the woman's cervix or vagina at about the time of ovulation.

ASD See *autism spectrum disorder*.

Asperger's disorder See *autism spectrum disorder*.

Association neuron An interneuron. These neurons are located within the central nervous system between sensory and motor neurons and serve to integrate information.

Asthma A condition marked by spasms of the muscles of bronchioles, making airflow difficult. It is often triggered by allergy.

Astigmatism Irregularities in the curvature of the cornea or lens that cause distortions of a visual image because the irregularities cause light rays to converge unevenly.

Atherosclerosis A narrowing of the arteries caused by thickening of the arterial walls and a buildup of lipid (primarily cholesterol) deposits. Atherosclerosis reduces blood flow through the vessel, choking off the vital supply of oxygen and nutrients to the tissues served by that vessel.

Atom A unit of matter that cannot be further broken down by chemical means; it is composed of subatomic particles, which include protons (positively charged particles), neutrons (with no charge), and electrons (with negative charges).

Atomic number The number of protons in the nucleus of an atom.

ATP See *adenosine triphosphate*.

Atrial fibrillation Rapid, ineffective contractions of the atria of the heart.

Atrial natriuretic hormone (ANH) The hormone released by cells in the right atrium of the heart in response to stretching of the heart caused by increased blood volume and pressure. ANH decreases water and sodium reabsorption by the kidneys, resulting in the production of large amounts of urine.

Atrioventricular (AV) bundle A tract of specialized cardiac muscle cells that runs along the wall between the ventricles of the heart and conducts an electrical impulse that originated in the sinoatrial (SA) node and was conducted to the AV node to the ventricles. The bundle forks into right and left branches and then divides into many other specialized cardiac muscle cells, called Purkinje fibers, that penetrate the walls of the ventricles.

Atrioventricular (AV) node A region of specialized cardiac muscle cells located in the partition between the two atria. It receives an electrical signal that spreads through the atrial walls from the sinoatrial node and relays the stimulus to the ventricles by means of a bundle of specialized muscle fibers, called the atrioventricular bundle, that runs along the wall between the ventricles.

Atrioventricular (AV) valves Heart valves located between the atria and the ventricles that keep blood flowing in only one direction, from the atria to the ventricles. The right AV valve consists of three flaps of tissue and is also called the *tricuspid valve*. The left AV valve consists of two flaps of tissue and is also called the *bicuspid valve* or the *mitral valve*.

Atrium (plural, atria) An upper chamber of the heart that receives blood from veins and pumps it to a ventricle.

Auditory tubes Small tubes that join the upper region of the pharynx (throat) with the middle ear. They help to equalize the air pressure between the middle ear and the atmosphere. Also called *Eustachian tubes*.

Autism See *autism spectrum disorder*.

Autism spectrum disorder (ASD) A neurodevelopmental disorder characterized by deficits in social communication and interaction and by the performance of repetitive and restricted patterns of behavior. Hyper- or hypo-reactivity to sensory input is also characteristic of the disorder. Four previously distinct neurodevelopmental disorders—autistic disorder (also called *autism*), Asperger's disorder (sometimes called *Asperger's syndrome*), pervasive developmental disorder not otherwise specified (PDD-NOS), and childhood disintegrative disorder—now are subsumed under the single category ASD.

Autoimmune disorder An immune response misdirected against the body's own tissues.

Autonomic nervous system The part of the peripheral nervous system that governs the involuntary, unconscious activities that maintain a relatively stable internal environment. The autonomic nervous system has two branches: the sympathetic and the parasympathetic.

Autosomes The 22 pairs of chromosomes (excluding the pair of sex chromosomes) that determine the expression of most of the inherited characteristics of a person.

Autotroph An organism that makes its own food (organic compounds) from inorganic substances. The autotrophs include photoautotrophs, which use the energy of light, and chemoautotrophs, which use the energy in chemicals.

Axial skeleton The part of the skeleton that includes the skull, the vertebral column, the breastbone (sternum), and the rib cage.

Axon A long extension from the cell body of a neuron that carries an electrochemical message away from the cell body toward another neuron or effector (muscle or gland). The tips of the axon release a chemical called a neurotransmitter that can affect the activity of the receiving cell. Typically, there is one long axon on a neuron.

Axon terminal The tip of a branch of an axon that releases a chemical (neurotransmitter) that alters the activity of the target cell. A synaptic knob.

B

Bacterium A prokaryotic organism.

B cell See *B lymphocyte*.

B lymphocyte B cell. A type of white blood cell important in antibody-mediated immune responses that can transform into a plasma cell and produce antibodies.

Balanced polymorphism A phenomenon in which natural selection maintains two or more alleles for a trait in a population from one generation to the next. It occurs when the environment changes frequently or when the heterozygous condition is favored over either homozygous condition.

Ball-and-socket joint A joint, such as the shoulder and hip joints, that allows motion in all directions.

Barr body A structure formed by a condensed, inactivated X chromosome in the body cells of female mammals.

Basal body The structure that anchors the microtubules of a cilium or flagellum to a cell. It contains nine triplets of microtubules arranged in a ring.

Basal cell carcinoma The most common type of skin cancer, occurring in the rapidly dividing cells of the basal layer of the epidermis.

Basal metabolic rate (BMR) A measure of the minimum energy required to keep an awake, resting body alive. It generally represents between 60% and 75% of the body's energy needs.

Base Any substance that reduces the concentration of hydrogen ions in solution.

Basement membrane A noncellular layer beneath epithelial tissue that binds the epithelial cells to underlying connective tissue. It helps epithelial tissue resist stretching and forms a boundary.

Basilar membrane The floor of the central canal in the cochlea of the inner ear that supports the spiral organ (of Corti), which is the true site of hearing; when the basilar membrane vibrates in response to sound, hair cells on the spiral organ are bent, generating electrochemical messages that are interpreted as sound.

Basophil A white blood cell that releases histamine, a chemical that both attracts other white blood cells to the site and causes blood vessels to widen during an inflammatory response.

Benign tumor An abnormal mass of tissue that usually remains at the site where it forms.

Bicuspid valve A heart valve located between the left atrium and ventricle. It is also called the *mitral valve* or the left *atrioventricular (AV) valve*.

Bile A mixture of water, ions, cholesterol, bile pigments, and bile salts that emulsifies fat (keeps fat as small globules), facilitating digestion by lipase. Bile is produced by the liver, is stored in the gallbladder, and acts in the small intestine.

Bilirubin A yellow pigment produced from the breakdown of the heme portion of hemoglobin by liver cells. It is excreted by the liver in bile.

Binary fission A type of asexual reproduction in which the genetic information is replicated and then a cell, a bacterium for example, or organism divides into two equal parts.

Biodiversity The number and variety of all living things in a given area. It includes genetic diversity, species diversity, and ecological diversity.

Biofeedback The use of artificial signals to provide feedback about unconscious visceral and motor activity, particularly that associated with stress.

Biogeochemical cycle The recurring process by which materials (for example, carbon, water, nitrogen, and phosphorus) cycle between living and nonliving systems and back again.

Biogeography The study of the geographic distribution of organisms. New distributions of organisms occur when organisms move to new areas (dispersal) and when areas occupied by the organisms move or are subdivided.

Biological magnification The tendency of a nondegradable chemical to become more concentrated in organisms as it passes along a food chain.

Biomass In ecosystems, the dry weight of the body mass of a group of organisms in a particular habitat.

Biopsy The removal and examination, usually microscopic, of a piece of tissue to diagnose a disease, usually cancer.

Biosphere The part of Earth in which life is found. It encompasses all of Earth's living organisms.

Biotechnology The industrial or commercial use or alteration of living organisms, cells, or molecules to achieve specific useful goals.

Bioterrorism The use of biological agents, such as viruses, parasites, or bacteria, to intimidate or attack societies or governments. There is concern that biological agents could be intentionally introduced in to food or water supplies, for example.

Bipedalism Walking on two feet. This trait evolved early in hominin evolution and set the stage for the evolution of other characteristics, such as increases in brain size.

Bipolar neuron A neuron that has only two processes. The axon and the dendrite extend from opposite sides of the cell body. Bipolar neurons are receptor cells found only in some of the special sensory organs, such as in the retina of the eye and in the olfactory membrane of the nose.

Birth defects Developmental defects present at birth. Such defects involve structure, function, behavior, or metabolism and may or may not be hereditary.

Birth rate The number of births per a specified number of individuals in the population during a specific length of time.

Bladder A muscular saclike organ that receives urine from the two ureters and temporarily stores it until release into the urethra.

Blastocyst The stage of development consisting of a hollow ball of cells. It contains the inner cell mass, a group of cells that will become the embryo, and the trophoblast, a thin layer of cells that will give rise to part of the placenta.

Blind spot The region of the retina where the optic nerve leaves the eye and on which there are no photoreceptors. Objects focused on the blind spot cannot be seen.

Blood Connective tissue that consists of cells and platelets suspended in plasma, a liquid matrix.

Blood–brain barrier A mechanism that protects the central nervous system by selecting the substances permitted to enter the cerebrospinal fluid from the blood. The barrier results from the relative impermeability of the capillaries in the brain and spinal cord.

Blood pressure The force exerted by the blood against the walls of the blood vessels. It is caused by the contraction of the ventricles and is influenced by vasoconstriction.

Blood type A characteristic of a person's red blood cells determined on the basis of large molecules on the surface of the plasma membrane.

Blue baby Newborn whose foramen ovale, the fetal opening between the right and left atria of the heart, fails to close. As a result, much of the infant's blood still bypasses the lungs and is low in oxygen. The condition can be corrected with surgery.

BMR See *basal metabolic rate*.

Bolus A small, soft ball of food mixed with saliva.

Bone Strong connective tissue with specialized cells in a hard matrix composed of collagen fibers and mineral salts.

Bone marrow The soft material filling the cavities in bones. Yellow bone marrow serves as a fat-storage site. Red bone marrow is the site where blood cells are produced.

Bone remodeling The ongoing process of bone deposition and absorption in response to hormonal and mechanical factors.

Bottleneck effect The genetic drift associated with dramatic, unselective reductions in population size such that the genetic makeup of survivors is not representative of the original population.

Brain The organ composed of neurons and glial cells that receives sensory input and integrates, stores, and retrieves information and directs motor output.

Brain waves The patterns recorded in an EEG (electroencephalogram) that reflect the electrical activity of the brain and are correlated with the person's state of alertness.

Breast The front of the chest, especially either of the two protuberant glandular organs (mammary glands) that in human females and other female mammals produce milk to nourish newborns.

Breathing center A region in the medulla of the brain that controls the basic breathing rhythm.

Breech birth Delivery in which the baby is born buttocks first rather than head first. It is associated with difficult labors and umbilical cord accidents.

Bronchi (singular, bronchus) The respiratory passageways between the trachea and the bronchioles that conduct air into the lungs.

Bronchial tree The term given to the air tubules in the respiratory system because their repeated branching resembles the branches of a tree.

Bronchioles A series of small tubules branching from the smallest bronchi inside each lung.

Bronchitis Inflammation of the mucous membranes of the bronchi, causing excess mucus and a deep cough.

Brush border A fuzzy border of microvilli on the surface of absorptive epithelial cells of the small intestine.

Buffer A substance that prevents dramatic changes in pH by removing excess hydrogen ions from solution when concentrations increase and adding hydrogen ions when concentrations decrease.

Bulbourethral glands Male accessory reproductive glands that release a clear slippery liquid immediately before ejaculation. Also called *Cowper's glands*.

Bulimia nervosa An eating disorder characterized by binge eating followed by purging by means of enemas, laxatives, diuretics, or self-induced vomiting.

Bursa (plural, bursae) A flattened sac containing a thin film of synovial fluid that surrounds and cushions certain synovial joints. Bursae are common in locations where ligaments, muscles, skin, or tendons rub against bone.

Bursitis Inflammation of a bursa (a sac in a synovial joint that acts as a cushion). Bursitis causes fluid to build up within the bursa, resulting in intense pain that becomes worse when the joint is moved and cannot be relieved by resting in any position.

C

C-section See *cesarean section*.

Calcitonin (CT) A hormone secreted by the thyroid gland when blood calcium levels are high. It stimulates the removal of calcium from the blood and inhibits the breakdown of bone.

Callus A mass of repair tissue formed by collagen fibers secreted from fibroblasts or woven bone that forms around and links the ends of a broken bone.

Capillary A microscopic blood vessel between arterioles and venules with walls only one cell layer thick. It is the site where the exchange of materials between the blood and the tissues occurs.

Capillary bed A network of true capillaries servicing a particular area. Precapillary sphincters regulate blood flow through the capillary bed. When the sphincters relax, blood fills the capillary bed, and materials can be exchanged between the blood and the tissues. When the sphincters contract, blood flows directly from an arteriole to a venule, bypassing the capillaries.

Carbaminohemoglobin The compound formed when carbon dioxide binds to hemoglobin.

Carbohydrate An organic molecule that provides fuel for the human body. Carbohydrates, which we know as sugars and starches, can be classified by size into the monosaccharides, disaccharides, and polysaccharides.

Carbon footprint A measure of the amount of carbon dioxide that enters the atmosphere due to the activities of a person or population; it includes personal activities and the activities needed for production and waste removal of the products consumed.

Carbonic anhydrase An enzyme in the red blood cells that catalyzes the conversion of unbound carbon dioxide to carbonic acid.

Carcinogen A substance that causes cancer.

Carcinoma in situ A tumor that has not spread; "cancer in place."

Cardiac cycle The events associated with the flow of blood through the heart during a single heartbeat. It consists of systole (contraction) and diastole (relaxation) of the atria and then of the ventricles of the heart.

Cardiac muscle A contractile tissue that makes up the bulk of the walls of the heart. Cardiac muscle cells are cylindrical and have branching interconnections between them. Cardiac muscle cells are striped (striated) and have a single nucleus. Contraction of cardiac muscle is involuntary.

Cardiovascular Pertaining to the heart and blood vessels.

Cardiovascular system The organ system composed of the heart and blood vessels. The cardiovascular system distributes blood, delivers nutrients, and removes wastes.

Carnivore An animal that obtains energy by eating other animals. A secondary consumer.

Carpal tunnel syndrome A condition of the wrist and hand whose symptoms may include numbness or tingling in the affected hand, along with pain in the wrist, hand, and fingers that results from repeated motion in the hand or wrist, causing the tendons to become inflamed and press against the nerve.

Carrier An individual who displays the dominant phenotype but is heterozygous for a trait and can therefore pass the recessive allele to descendants.

Carrying capacity The number of individuals of a given species that a particular environment can support for a prolonged time period. The carrying capacity of the environment is determined by such factors as availability of resources, including food, water, and space; ability to clean away wastes; and predation pressure.

Cartilage A type of specialized connective tissue with a firm gelatinous matrix containing protein fibers for strength. The cartilage cells (chondrocytes) lie in small spaces (lacunae) within the matrix.

Catabolism Chemical reactions within living cells that break down complex molecules into simpler ones, releasing energy from chemical bonds. Compare with *anabolism*.

Cataract A cloudy or opaque lens in the eye. Cataracts reduce visual acuity and may be caused by glucose accumulation associated with type 1 and type 2 diabetes mellitus, excessive exposure to sunlight, and exposure to cigarette smoke.

CD4 cell See *helper T cell*.

Cecum A pouch that hangs below the junction of the small and large intestines; the appendix extends from the cecum.

Cell The smallest structure that shows all the characteristics of life.

Cell adhesion molecule (CAM) A molecule that pokes through the plasma membranes of most cells and helps hold cells together to form tissues and organs.

Cell body The part of a neuron that contains the organelles and nucleus needed to maintain the cells.

Cell cycle The entire sequence of events that a cell goes through from its origin in the division of a parent cell through its own division into two daughter cells. The cell cycle consists of two major phases: interphase and cell division.

Cell differentiation The process by which cells become specialized with respect to structure and function.

Cell-mediated immune responses Immune system responses conducted by T cells that protect against cellular threats, including body cells that have become infected with viruses or other pathogens and cancer cells.

Cell theory This fundamental organizing principle of biology states that (1) a cell is the smallest unit of life; (2) cells make up all living things, from unicellular to multicellular organisms; and (3) new cells arise from pre-existing cells.

Cellular respiration The oxygen-requiring pathway by which glucose is broken down by cells to yield carbon dioxide, water, and energy.

Cellulose A structural polysaccharide found in the cell walls of plants. Humans lack the enzymes necessary to digest cellulose, so it passes unchanged through our digestive tract. Although cellulose has no value as a nutrient, it is an important form of dietary fiber known to facilitate the passage of feces through the large intestines.

Cementum A calcified but sensitive part of a tooth that covers the root.

Central nervous system (CNS) The brain and the spinal cord.

Centriole A structure, found in pairs, within a centrosome. Each centriole is composed of nine sets of triplet microtubules arranged in a ring.

Centromere The region of a replicated chromosome at which sister chromatids are held together until they separate during cell division.

Centrosome The region near the nucleus that contains centrioles. It forms the mitotic spindle during prophase.

Cerebellum A region of the brain important in sensory–motor coordination. It is largely responsible for posture and smooth body movements.

Cerebral cortex The extensive area of gray matter covering the surfaces of the cerebrum. It is often referred to as the conscious part of the brain. The cerebral cortex has sensory, motor, and association areas.

Cerebral white matter A region of the cerebrum beneath the cortex consisting primarily of myelinated axons grouped into tracts that allow various regions of the brain to communicate with one another.

Cerebrospinal fluid The fluid bathing the internal and external surfaces of the central nervous system. It serves as a shock absorber, supports the brain, nourishes the brain, delivers chemical messengers, and removes waste products.

Cerebrovascular accident See *stroke*.

Cerebrum The largest and most prominent part of the brain, composed of the cerebral hemispheres. It is responsible for thinking, sensory perception, originating most conscious motor activity, personality, and memory.

Cervical cap A barrier means of contraception consisting of a small rubber dome that fits snugly over the cervix and is held in place partly by suction. It prevents sperm from reaching the egg.

Cervix The narrow neck of the uterus that projects into the vagina whose opening provides a passageway for materials moving between the vagina and the body of the uterus.

Cesarean section A procedure by which the baby and placenta are removed from the uterus through an incision in the abdominal wall and uterus. The term is often shortened to *C-section*.

Chancre A painless bump that forms during the first stage of syphilis at the site of contact, usually within 2 to 8 weeks of the initial contact.

Chemical digestion A part of the digestive process that involves breaking chemical bonds so that complex molecules are broken into their component subunits. Chemical digestion produces molecules that can be absorbed into the bloodstream and used by the cells.

Chemical evolution The sequence of events by which life evolved from chemicals slowly increasing in complexity over perhaps 300 million years.

Chemistry The branch of science concerned with the composition and properties of material substances, including their abilities to change into other substances.

Chemoreceptor A sensory receptor specialized to respond to chemicals. We describe the input from the chemoreceptors of the mouth as taste (gustation) and from those of the nose as smell (olfaction). Other chemoreceptors monitor levels of chemicals, such as carbon dioxide, oxygen, and glucose, in body fluids.

Childhood The stage in postnatal development that runs from about 13 months to 12 or 13 years of age. It is a time of continued growth during which gross and fine motor skills improve and coping skills develop. With the exception of the reproductive system, organ systems become fully functional.

Chitin A structural polysaccharide found in the exoskeletons (hard outer coverings) of animals such as insects, spiders, and crustaceans.

Chlamydia A genus of bacteria. In this text, it is an infection (usually sexually transmitted) caused by *Chlamydia trachomatis*, commonly causing urethritis and pelvic inflammatory disease.

Cholecystokinin A hormone secreted by the small intestine that stimulates the pancreas to release its digestive enzymes and the gallbladder to contract and release bile.

Chordae tendineae Strings of connective tissue that anchor the atrioventricular valves to the wall of the heart, preventing the backflow of blood.

Chorion The extraembryonic membrane that becomes the embryo's major contribution to the placenta.

Chorionic villi Fingerlike projections of the chorion of the embryo that grow into the uterine lining of the mother during formation of the placenta and become part of the placenta.

Chorionic villi sampling (CVS) A procedure for screening for genetic defects of a fetus by removing a piece of chorionic villi and examining the cells for genetic abnormalities.

Choroid The pigmented middle layer of the eyeball that contains blood vessels.

Chromatid One of the two identical replicates of a duplicated chromosome. The two chromatids that make up a chromosome are held together by a centromere and are referred to as sister chromatids. During cell division, the two strands separate and each becomes a chromosome in one of the two daughter cells.

Chromatin DNA and associated proteins in a dispersed, rather than condensed, state.

Chromosomal mutation A change in DNA in which a section of a chromosome becomes rearranged, duplicated, or deleted.

Chromosome DNA (which contains the genetic information of a cell) and specialized proteins, primarily histones.

Chronic renal failure A progressive and often irreversible decline in the rate of glomerular filtration.

Chylomicron A particle formed when proteins coat the surface of the products of lipid digestion, making lipids soluble in water and allowing them to be transported throughout the body.

Chyme The semifluid mass created during digestion once the food has been churned and mixed with the gastric juices of the stomach.

Cilia Extensions of the plasma membrane found on some cells, such as those lining the respiratory tract, that move in a back-and-forth motion. They are usually shorter and much more numerous than flagella but have the same 9 + 2 arrangement of microtubules at their core.

Ciliary body A portion of the middle coat of the eyeball near the lens that consists of smooth muscle and ligaments. Contractions of the smooth muscle of the ciliary body change the shape of the lens, which then focuses images on the retina.

Circulatory system An organ system composed of the cardiovascular system (heart and blood vessels) and the lymphatic system (lymphatic vessels and lymphoid tissues and organs).

Circumcision The surgical removal of the foreskin of the penis, usually performed when the male is an infant.

Cirrhosis A chronic disease of the liver in which the liver becomes fatty and the liver cells are gradually replaced with scar tissue.

Citric acid cycle The cyclic series of chemical reactions that follows the transition reaction and yields two molecules of adenosine triphosphate (one from each acetyl CoA that enters the cycle) and several molecules of nicotine adenine dinucleotide (NADH) and flavin adenine dinucleotide ($FADH_2$), carriers of high-energy electrons that enter the electron transport chain. This phase of cellular respiration occurs inside the mitochondrion and is sometimes called the *Krebs cycle*.

Cleavage A rapid series of mitotic cell divisions in which the zygote first divides into two cells, and then four cells, and then eight cells, and so on. Cleavage usually begins about one day after fertilization as the zygote moves along the oviduct toward the uterus.

Climax community The relatively stable community that eventually forms at the end of ecological succession and remains if no disturbances occur.

Clitoris A small, erectile body in the female that plays a role in sexual stimulation. It develops from the same embryological structure from which the glans penis develops in the male.

Clonal selection The hypothesis that, by binding to a receptor on a lymphocyte surface, an antigen selectively activates only those lymphocytes able to recognize that antigen and programs that lymphocyte to divide, forming an army of cells specialized to attack the stimulating antigen.

Clone A population of identical cells descended from a single ancestor.

CNS See *central nervous system*.

Cocaine A psychoactive drug extracted from the leaves of the coca plant that stimulates the central nervous system.

Cochlea The snail-shaped portion of the inner ear that contains the actual organ of hearing, the spiral organ (of Corti).

Codominance The condition in which the effects of both alleles are separately expressed in a heterozygote.

Codon A three-base sequence on messenger RNA (mRNA) that specifies one of the 20 common amino acids or the beginning or end of the protein chain.

Coenzyme An organic molecule such as a vitamin that functions as a cofactor and helps enzymes convert substrate to product.

Cofactor A nonprotein substance such as zinc, iron, and vitamins that helps enzymes convert substrate to product. It may permanently reside at the active site of the enzyme or may bind to the active site at the same time as the substrate.

Collagen fibers Strong insoluble protein fibers common in many connective tissues.

Collecting ducts Within the kidneys, the tubes that receive filtrate from the distal convoluted tubules of many nephrons and that eventually drain into the renal pelvis. Some tubular secretion occurs along collecting ducts.

Colon The division of the large intestine composed of the ascending colon, the transverse colon, and the descending colon.

Colostrum A cloudy yellowish fluid produced by the breasts in the interval after birth when milk is not yet available. Its composition is different from that of milk.

Columnar epithelium A type of epithelial tissue composed of tall, rectangular cells that are specialized for secretion and absorption.

Coma An unconscious state caused by trauma to neurons in regions of the brain responsible for stimulating the cerebrum, particularly those in the reticular activating system or thalamus. Coma can be caused by mechanical shock, such as might be caused by a blow to the head, tumors, infections, drug overdose (from barbiturates, alcohol, opiates, or aspirin), or failure of the liver or kidneys.

Combination birth control pill A means of hormonal contraception that consists of a series of pills with synthetic forms of estrogen and progesterone. The hormones in the pills mimic the effects of natural hormones ordinarily produced by the ovaries and inhibit FSH and LH secretion by the anterior pituitary gland and, therefore, prevent the development of an egg and ovulation.

Common cold An upper respiratory infection caused by one of the adenoviruses.

Community An assortment of organisms of various species interacting in a defined habitat.

Compact bone Very dense, hard bone, containing internal spaces of microscopic size and narrow channels that contain blood vessels and nerves. It makes up the shafts of long bones and the outer surfaces of all bones.

Complement system A group of about 20 proteins that enhances the body's defense mechanisms. The complement system destroys cellular pathogens by creating holes in the plasma membrane, making the cell leaky, enhancing phagocytosis, and stimulating inflammation.

Complementary base pairing The process by which specific bases are matched: adenine with thymine (in DNA) or with uracil (in RNA) and cytosine with guanine. Each base pair is held together by weak hydrogen bonds.

Complementary proteins A selection of foods, each containing incomplete proteins, that provide ample amounts of all essential amino acids when combined.

Complete dominance In genetic inheritance, the dominant allele in a heterozygote completely masks the effect of the recessive allele. Complete dominance often occurs because the dominant allele produces a functional protein and the recessive allele produces a less functional protein or none at all.

Complete protein A protein that contains ample amounts of all the essential amino acids. Animal sources of protein are generally complete proteins.

Compound A molecule that contains two or more different elements.

Computed tomography (CT scanning) A method of visualizing body structures, including the brain, using an x-ray source that moves in an arc around the body part to be imaged, thereby providing different views of the structure.

Concentration gradient A difference in the number of molecules or ions between two adjacent regions. Molecules or ions tend to move away from an area where they are more concentrated to an area where they are less concentrated. Each type of molecule or ion moves in response to its own concentration gradient.

Conclusion An interpretation of the data collected in an experiment.

Condom, female A barrier means of contraception used by a female that consists of a loose sac of polyurethane held in place by flexible rings (one at each end). It is used to prevent sperm from entering the female reproductive tract, and it also reduces the risk of spreading sexually transmitted infections.

Condom, male A barrier means of contraception consisting of a thin sheath of latex or animal intestines that is rolled onto an erect penis, where it prevents sperm from entering the vagina. A latex condom also helps prevent the spread of sexually transmitted diseases.

Cones Photoreceptors in the eye responsible for color vision. There are three types of cone cells: blue, green, and red.

Confounding variable In a controlled experiment, it is a second factor that differs between the control group and the experimental group.

Conjoined twins Individuals that develop from a single fertilized ovum that fails to completely split in two at an early stage of cleavage. Such twins have nearly identical genetic material and thus are always the same gender. They may be surgically separated after birth.

Connective tissue Tissue that binds together and supports other tissues of the body. All connective tissues contain cells in an extracellular matrix, which consists of protein fibers and a noncellular ground substance.

Constipation Infrequent, difficult bowel movements of hard feces. Constipation occurs when feces move through the large intestine too slowly and too much water is reabsorbed.

Consumer In ecosystems, an organism that obtains energy and raw materials by eating the tissues of other organisms.

Contact inhibition The phenomenon whereby cells placed in a dish in the presence of growth factors stop dividing once they have formed a monolayer. This may result from competition among the cells for growth factors and nutrients. Cancer cells do not display density-dependent contact inhibition and continue to divide, piling up on one another until nutrients run out.

Continuous ambulatory peritoneal dialysis (CAPD) A method of hemodialysis whereby the peritoneum, one of the body's own selectively permeable membranes, is used as the dialyzing membrane. It is an alternative to the artificial kidney machine during kidney failure.

Contraceptive sponge A barrier means of contraception consisting of a sponge containing spermicide.

Control group In a controlled scientific experiment, the control group is the one in which the variable is unaltered for comparison to the experimental group.

Controlled experiment An experiment in which the subjects are divided into two groups, usually called the *control group* and the *experimental group*. Ideally, the groups differ in only the factor(s) of interest.

Convergence In vision, the process by which the eyes are directed toward the midline of the body as an object moves closer. Convergence is necessary to keep the image focused on the fovea of the retina.

Convergent evolution The process by which two species become more alike because they have similar ecological roles and selection pressures. For example, birds and insects independently evolved wings.

Core temperature The temperature in body structures below the skin and subcutaneous layers.

Cornea A clear, transparent dome located in the front and center of the eye that both provides the window through which light enters the eye and helps bend light rays so that they focus on the retina.

Coronary arteries The arteries that deliver blood to cardiac muscle.

Coronary artery bypass A technique for bypassing a blocked coronary blood vessel to restore blood flow to the heart muscle. Typically, a segment of a leg vein is removed and grafted so that it provides a shunt between the aorta and a coronary artery past the point of obstruction.

Coronary artery disease (CAD) A condition in which fatty deposits associated with atherosclerosis form on the inside of coronary arteries, obstructing the flow of blood. It is the underlying cause of most heart attacks.

Coronary circulation The system of blood vessels that services the tissues of the heart itself.

Coronary sinus A vessel that returns deoxygenated blood collected from the heart muscle to the right atrium of the heart. The coronary sinus is formed from the merging of cardiac veins.

Corpus callosum A band of myelinated axons (white matter) that connects the two cerebral hemispheres so they can communicate with one another.

Corpus luteum A structure in the ovary that forms from the follicle cells remaining in the ovary after ovulation. The corpus luteum functions as an endocrine structure that secretes estrogen and progesterone.

Covalent bond A chemical bond formed when outer shell electrons are shared between atoms.

Cowper's glands See *bulbourethral glands*.

Cranial nerves Twelve pairs of nerves that arise from the brain and service the structures of the head and certain body parts such as the heart and diaphragm. Cranial nerves can be sensory, motor, or mixed.

Cranium The portion of the skull that forms the cranial (brain) case. It is formed from eight (sometimes more) flattened bones including the frontal bone, two parietal bones, the occipital bone, two temporal bones, the sphenoid bone, and the ethmoid bone.

Creatine phosphate A compound stored in muscle tissue that serves as an alternative energy source for muscle contraction.

Cretinism A condition characterized by dwarfism, mental retardation, and slowed sexual development. It is caused by undersecretion of thyroid hormone during fetal development or infancy.

Cristae Infoldings of the inner membrane of a mitochondrion.

Cross-bridges Myosin heads. Club-shaped ends of a myosin molecule that bind to actin filaments and can swivel, causing actin filaments to slide past the myosin filaments, which causes muscle contraction.

Cross-tolerance The development of tolerance for a drug that is not used, caused by the development of tolerance to another, usually similar, drug.

Crossing over The breaking and rejoining of nonsister chromatids of homologous pairs of chromosomes during meiosis (specifically at prophase I when homologous chromosomes pair up side by side). Crossing over results in the exchange of corresponding segments of chromatids and increases genetic variability in populations.

Crown The part of a tooth that is visible above the gum line. It is covered with enamel, a nonliving material that is hardened with calcium salts.

CT scanning See *computed tomography*.

Cuboidal epithelium A type of epithelial tissue composed of cube-shaped cells that are specialized for secretion and absorption.

Culture Social influences that produce an integrated pattern of knowledge, belief, and behavior.

Cupula A pliable gelatinous mass covering the hair cells within the ampulla of the semicircular canals of the inner ear and whose movement bends hair cells, triggering the generation of nerve impulses that are interpreted by the brain as movement of the head.

Cushing's syndrome A condition characterized by accumulation of fluid in the face and redistribution of body fat caused by prolonged exposure to cortisol.

Cutaneous membrane The skin. It is thick, relatively waterproof, and dry.

Cystitis Inflammation of the urinary bladder caused by bacteria.

Cytokinesis The division of the cytoplasm and organelles into two daughter cells during cell division. Cytokinesis usually occurs during telophase.

Cytoplasm The part of a cell that includes the aqueous fluid within the cell and all the organelles with the exception of the nucleus.

Cytoskeleton A complex network of protein filaments within the cytoplasm that gives the cell its shape, anchors organelles in place, and functions in the movement of entire cells or certain organelles or vesicles within cells. The cytoskeleton includes microtubules, microfilaments, and intermediate filaments.

Cytotoxic T cell A type of T lymphocyte that directly attacks infected body cells and tumor cells by releasing chemicals called perforins that cause the target cells to burst.

D

Darwinian fitness See *fitness*.

Decomposer An organism that obtains energy by consuming the remains or wastes of other organisms. Decomposers release inorganic materials that can then be used by producers. Bacteria and fungi are important decomposers.

Deductive reasoning A logical progression of thought proceeding from the general to the specific. It involves making specific deductions based on a larger generalization or premise. The statement is usually in the form of an "if ... then" premise.

Deforestation Removing trees from an area without replacing them.

Dehydration synthesis The process by which polymers are formed. Monomers are linked together through the removal of a water molecule.

Deletion Pertaining to chromosomes, the loss of a nucleotide or segment of a chromosome.

Denaturation The process by which changes in the environment of a protein, such as increased heat or changes in pH, cause it to unravel and lose its three-dimensional shape. Change in the shape of a protein results in loss of function.

Dendrite A process of a neuron specialized to pick up messages and transmit them toward the cell body. There are typically many short branching dendrites on a neuron.

Dense connective tissue Connective tissue that contains many tightly woven fibers and is found in ligaments, tendons, and the dermis.

Density-dependent regulating factor One of many factors that have a greater impact on the population size as conditions become more crowded. Such factors include disease and starvation.

Density-independent regulating factor One of many factors that regulate population size by causing deaths that are not related to the density of individuals in a population. Such factors include natural disasters.

Dentin A hard, bonelike substance that forms the main substance of teeth. It is covered by enamel on the crown and by cementum on the root.

Deoxyribonucleic acid (DNA) The molecular basis of genetic inheritance in all cells and some viruses. A category of nucleic acids that usually consists of a double helix of two nucleotide strands. The sequence of nucleotides carries the instructions for assembling proteins.

Depolarization A change in the difference in electrical charge across a membrane that moves it from a negative value toward 0 mV. During a nerve impulse (action potential), depolarization is caused by the inward flow of positively charged sodium ions.

Dermis The layer of the skin that lies just below the epidermis and is composed of connective tissue. The dermis contains blood vessels, oil glands, sensory structures, and nerve endings. The dermis does not wear away.

Desertification The process by which overfarming and overgrazing transform marginal farmlands and rangelands to deserts.

Desmosome A type of junction between cells that anchors adjacent cells together.

Detrital food web Energy flow begins with detritus (organic material from the remains of dead organisms) that is eaten by a primary consumer.

Detritivore An organism that obtains energy by consuming the remains or wastes of other organisms. Detritivores release inorganic materials that can then be used by producers.

Detrusor muscle A layer of smooth muscle within the walls of the urinary bladder. It plays a role in urination.

Diabetes insipidus A condition characterized by excessive urine production caused by inadequate antidiuretic hormone (ADH) production.

Diabetes mellitus A group of diseases characterized by excessive urine production, an abnormally high blood glucose level, and the presence of glucose in the urine. Caused by deficient production of insulin (type 1) or increased resistance to insulin (type 2 and gestational diabetes).

Diaphragm A broad sheet of muscle that separates the abdominal and thoracic cavities. When the diaphragm contracts, inhalation occurs.

Diaphragm, contraceptive A barrier means of contraception that consists of a dome-shaped soft rubber cup on a flexible ring. A diaphragm is used in conjunction with a spermicide to prevent sperm from reaching the egg.

Diarrhea Abnormally frequent, loose bowel movements. Diarrhea occurs when feces pass through the large intestine too quickly and too little water is reabsorbed. Diarrhea sometimes leads to dehydration.

Diastole Relaxation of the heart. Atrial diastole is the relaxation of the atria. Ventricular diastole is the relaxation of the ventricles.

Diastolic pressure The lowest blood pressure in an artery during the relaxation of the heart. In a typical, healthy adult, the diastolic pressure is about 80 mm Hg.

Dietary fiber The nondigestible carbohydrate part of plant foods that forms support structures of stems, seed, and leaves. Dietary fiber is important for heart and colon health.

Digestive system The organ system that breaks down and absorbs food. The digestive system includes the mouth, esophagus, stomach, small intestine, and large intestine. Associated structures include the teeth, tongue, salivary glands, liver, gallbladder, and pancreas.

Dilation stage The first stage of true labor. It begins with the onset of contractions and ends when the cervix has fully dilated to 10 cm (4 in.).

Diploid The condition of having two sets of chromosomes in each cell. Somatic (body) cells are diploid.

Disaccharide A molecule formed when two monosaccharides covalently bond to each other through dehydration synthesis. It is known as a double sugar.

Distal convoluted tubule The section of the renal tubule where reabsorption and secretion occur.

Diuretic A substance that promotes urine production. Alcohol is a diuretic.

DNA (deoxyribonucleic acid) The molecular basis of genetic inheritance in all cells and some viruses. A category of nucleic acids that usually consists of a double helix of two nucleotide strands. The sequence of nucleotides carries the instructions for assembling proteins.

DNA fingerprint The pattern of DNA fragments that have been cut by a restriction enzyme and sorted by size. Each person has a characteristic, individual DNA fingerprint.

DNA library A large collection of cloned recombinant DNA fragments containing the entire genome of an organism.

DNA ligase An enzyme that catalyzes the formation of bonds between the sugar and the phosphate molecules that form the sides of the DNA ladder during replication, repair, or the creation of recombinant DNA.

DNA polymerase Any one of the enzymes that catalyze the synthesis of DNA from free nucleotides using one strand of DNA as a template.

Dominant allele The allele that is fully expressed in the phenotype of an individual who is heterozygous for that gene. The dominant allele usually produces a functional protein, whereas the recessive allele does not.

Dopamine A neurotransmitter in the central nervous system thought to be involved in regulating emotions and in the brain pathways that control complex movements.

Dorsal nerve root The portion of a spinal nerve that arises from the back (posterior) side of the spinal cord and contains axons of sensory neurons. It joins with the ventral nerve root to form a single spinal nerve, which passes through the opening between the vertebrae.

Doubling time The number of years required for a population to double in size at a given, constant growth rate.

Down syndrome A collection of characteristics that tend to occur when an individual has three copies of chromosome 21. It is also known as trisomy 21.

Drug cocktail A combination of drugs used to treat people who are HIV positive. The combination usually includes a drug that blocks reverse transcription and a protease inhibitor.

Ductus arteriosus A small vessel in the fetus that connects the pulmonary artery to the aorta. It diverts blood away from the lungs.

Ductus venosus A small vessel in the fetus through which most blood from the placenta flows, bypassing the liver.

Duodenum The first region of the small intestine. The duodenum receives chyme from the stomach and digestive juices from the pancreas and liver.

Duplication Pertaining to chromosomes, the duplication of a region of a chromosome that often results from fusion of a fragment from a homologous chromosome.

Dura mater The tough, leathery outer layer of the meninges that protects the central nervous system. Around the brain, the dura mater has two layers that are separated by a fluid-filled space containing blood vessels.

Dysplasia The changes in shape, nuclei, and organization of adult cells. It is typical of precancerous cells.

E

Eardrum See *tympanic membrane*.

ECG See *electrocardiogram*.

Ecological footprint A measure of the amount of productive land and water used by a person or population to produce products consumed and to remove the waste of products consumed.

Ecological pyramid A diagram in which blocks represent each tropic (feeding) level.

Ecological succession The sequence of changes in the species making up a community over time.

Ecology The study of the interactions among organisms and between organisms and their environment.

Ecosystem All the organisms living in a certain area that can potentially interact, together with their physical environment.

Ectoderm The primary germ layer that forms the nervous system, epidermis, and epidermal derivatives such as hair, nails, and mammary glands.

Ectopic pregnancy A pregnancy in which the embryo (blastocyst) implants and begins development in a location other than the uterus, most commonly in an oviduct (a tubal pregnancy).

Edema Swelling caused by the accumulation of interstitial fluid.

EEG See *electroencephalogram*.

Effector A muscle or a gland that brings about a response to a stimulus.

Effector cells Lymphocytes that are responsible for the attack on cells or substances not recognized as belonging in the body.

Efferent (motor) neuron A neuron specialized to carry information *away from* the central nervous system to an effector, either a muscle or a gland.

Egg A mature female gamete. An ovum. The egg contains nutrients and the mother's genetic contribution to the next generation.

EKG See *electrocardiogram*.

Elastic cartilage The most flexible type of cartilage because of an abundance of wavy elastic fibers in its matrix.

Elastic fibers Coiled proteins found in connective tissues that allow the connective tissue to be stretched and recoil. Elastic fibers are common in tissues that require elasticity.

Electrocardiogram (ECG or EKG) A graphical record of the electrical activities of the heart.

Electroencephalogram (EEG) A graphical record of the electrical activity of the brain.

Electron transport chain A series of carrier proteins embedded in the inner membrane of the mitochondrion that receives electrons from the molecules of nicotine adenine dinucleotide (NADH) and flavin adenine dinucleotide ($FADH_2$) produced by glycolysis and the citric acid cycle. During the transfer of electrons from one molecule to the next, energy is released, and this energy is then used to make adenosine triphosphate (ATP). Oxygen is the final electron acceptor in the chain.

Element Any substance that cannot be broken down into simpler substances by ordinary chemical means.

Embolus A blood clot that drifts through the circulatory system and can lodge in a small blood vessel and block blood flow.

Embryo The developing human from week 3 through week 8 of gestation (the embryonic period).

Embryonic disk The flattened platelike structure that will become the embryo proper. It develops from the inner cell mass.

Embryonic period The period of prenatal development that extends from week 3 through week 8 of gestation. It is the period when tissues and organs form.

Emigration The departure of individuals from a population for some other area.

Emphysema A condition in which the alveolar walls break down, thicken, and form larger air spaces, making gas exchange difficult. This change results in less surface area for gas exchange and an increase in the volume of residual air in the lungs.

Enamel A nonliving material that is hardened with calcium salts and covers the crown of a tooth.

Encephalitis An inflammation of the meninges around the brain.

Endocardium A thin layer that lines the cavities of the heart.

Endocrine gland A gland that lacks ducts and releases its products (hormones) into the fluid just outside the cells.

Endocrine system The organ system that, along with the nervous system, functions in internal communication. It consists of endocrine glands, such as the pituitary gland and thyroid gland, and of organs, such as the kidneys and pancreas, that contain some endocrine tissue but have functions in addition to hormone secretion.

Endocytosis The process by which large molecules and single-celled organisms such as bacteria enter cells. It occurs when a region of the plasma membrane surrounds the substance to be ingested, then pinches off from the rest of the membrane, enclosing the substance in a saclike structure called a vesicle that is released into the cell. Two types of endocytosis are phagocytosis ("cell eating") and pinocytosis ("cell drinking").

Endoderm The primary germ layer that forms some organs and glands (for example, the pancreas, liver, thyroid gland, and parathyroid glands) and the epithelial lining of the urinary, respiratory, and gastrointestinal tracts.

Endometriosis A painful condition in which tissue from the lining of the uterus (the endometrium) is found outside the uterine cavity.

Endometrium The inner layer of the uterus consisting of connective tissue, glands, and blood vessels. The endometrium thickens and develops with each menstrual (uterine) cycle and is then lost as menstrual flow. It is the site of embryo implantation during pregnancy.

Endoplasmic reticulum The network of internal membranes within eukaryotic cells. Whereas rough endoplasmic reticulum (RER) has ribosomes attached to its surface, smooth endoplasmic reticulum (SER) lacks ribosomes and functions in the production of phospholipids for incorporation into cell membranes.

Endorphin A chemical released by nerve cells that binds to the so-called opiate receptors on the pain-transmitting neurons and quells the pain. The term is short for *endo*genous m*orphine*-like substance.

Endosymbiont hypothesis The hypothesis that organelles such as mitochondria were once free-living prokaryotic organisms that either invaded or were engulfed by primitive eukaryotic cells with which they established a symbiotic (mutually beneficial) relationship.

Endothelium The lining of the heart, blood vessels, and lymphatic vessels. It is composed of flattened, tight-fitting cells. The endothelium forms a smooth surface that minimizes friction and allows the blood or lymph to flow over it easily.

Enkephalin A chemical released by nerve cells that binds to the so-called opiate receptors on the pain-transmitting neurons and quells the pain.

Enzyme A substance (usually a protein, but sometimes an RNA molecule) that speeds up chemical reactions without being consumed in the process.

Enzyme–substrate complex Complex formed when a substrate binds to an enzyme at the active site.

Eosinophil The type of white blood cell important in the body's defense against parasitic worms. It releases chemicals that help counteract certain inflammatory chemicals released during an allergic response.

Epidermis The outermost layer of the skin, composed of epithelial cells.

Epidemiology The study of patterns of disease, including rate of occurrence, distribution, and control.

Epididymis A long tube coiled on the surface of each testis that serves as the site of sperm cell maturation and storage.

Epiglottis A part of the larynx that forms a movable lid of cartilage covering the opening into the trachea (the glottis).

Epinephrine Adrenaline. A hormone secreted by the adrenal medulla, along with norepinephrine, in response to stress. They initiate the physiological "fight-or-flight" reaction.

Epiphyseal plate A plate of cartilage that separates the head of the bone from the shaft, permitting the bone to grow. In late adolescence, the epiphyseal plate is replaced by bone, and growth stops. The epiphyseal plate is commonly called the *growth plate*.

Episiotomy An incision made to enlarge the vaginal opening, just before passage of the baby's head at the end of the second stage of labor.

Epithelial tissue One of the four primary tissue types. The tissue that covers body surfaces, lines body cavities and organs, and forms glands.

Equatorial plate A plane at the midline of a cell where chromosomes line up during mitosis or meiosis.

Erythrocyte A red blood cell. A nucleus-free biconcave cell in the blood that is specialized for transporting oxygen to cells and assists in transporting carbon dioxide away from cells.

Erythropoietin A hormone released by the kidneys when the oxygen content of the blood declines that stimulates red blood cell production.

Esophagus A muscular tube that conducts food from the pharynx to the stomach using peristalsis.

Essential amino acid Any of the eight amino acids that the body cannot synthesize and, therefore, must be supplied in the diet.

Estrogen A steroid sex hormone produced by the follicle cells and the corpus luteum in the ovary. Estrogen helps oocytes mature, stimulates cell division in the endometrium and the breast with each uterine cycle, and maintains secondary sex characteristics. The adrenal cortex also secretes estrogen.

Ethanol The alcohol in an alcoholic drink.

Eukaryotic cell A cell with a nucleus and extensive internal membranes that divide it into many compartments and enclose organelles. Eukaryotes include cells in plants, animals, and all other organisms except bacteria and archaea.

Eustachian tubes Auditory tubes. Small tubes that join the upper region of the pharynx (throat) with the middle ear. They help to equalize the air pressure between the middle ear and the atmosphere.

Eutrophication The enrichment of water in a lake or pond by nutrients. Eutrophication is often caused by nitrogen or phosphate that washes into bodies of water.

Evolution Descent with modification from a common ancestor. It is the process by which life forms on the earth have changed over time.

Excitatory synapse A synapse in which the response of the receptors for that neurotransmitter on the postsynaptic membrane increases the likelihood that an action potential will be generated in the postsynaptic neuron. The postsynaptic cell is excited because it becomes less negative than usual (slightly depolarized), usually because of the inflow of sodium ions.

Excretion The elimination of wastes and excess substances from the body.

Exhalation Breathing out (expiration) involves the movement of air out of the respiratory system into the atmosphere.

Exocrine glands Glands that secrete their product through ducts onto body surfaces, into the spaces within organs, or into a body cavity. Examples include the salivary glands of the mouth and the oil and sweat glands of the skin.

Exocytosis The process by which large molecules leave cells. It occurs when products packaged by cells in membrane-bound vesicles move toward the plasma membrane. Upon reaching the plasma membrane, the membrane of the vesicle fuses with it, spilling its contents outside the cell.

Exon The nucleotide sequences of a newly synthesized messenger RNA (mRNA) that are spliced together to form the mature mRNA that is ultimately translated into protein.

Exophthalmos A condition characterized by protruding eyes that is caused by the accumulation of interstitial fluid due to oversecretion of thyroid hormone.

Experimental group In a controlled scientific experiment, the experimental group is the one in which the variable is altered.

Expiration The process by which air is moved out of the respiratory system into the atmosphere. It is also called *exhalation*.

Expiratory reserve volume The additional volume of air that can be forcefully expelled from the lungs after normal exhalation.

Expulsion stage The second stage of true labor. It begins with full dilation of the cervix and ends with delivery of the baby.

External auditory canal The canal leading from the pinna of the ear to the eardrum (tympanic membrane).

External respiration The exchange of oxygen and carbon dioxide between the lungs and the blood.

External urethral sphincter A sphincter made of skeletal muscle that surrounds the urethra. This voluntary sphincter helps stop the flow of urine down the urethra when we wish to postpone urination.

Exteroceptor A sensory receptor that is located near the surface of the body and that responds to changes in the environment.

Extracellular fluid The watery solution outside cells. It is also called interstitial fluid.

Extraembryonic membranes Membranes that lie outside the embryo, where they protect and nourish the embryo and later the fetus. They include the amnion, yolk sac, chorion, and allantois.

F

Facial bones The bones that form the face. They include 14 bones that support several sensory structures and serve as attachments for most muscles of the face.

Facilitated diffusion The movement of a substance from a region of higher concentration to a region of lower concentration with the aid of a membrane protein that either transports the substance from one side of the membrane to the other or forms a channel through which it can move.

Farsightedness A condition in which distant objects are seen more clearly than near ones. Farsightedness occurs either because the eyeball is too short or the lens is too thin, causing the image to be focused behind the retina.

Fascicle A bundle of skeletal muscle fibers (cells) that forms a part of a muscle. Each fascicle is wrapped in its own connective tissue sheath.

Fast-twitch muscle cells Muscle fibers that contract rapidly and powerfully, with little endurance. They have few mitochondria and large glycogen reserves. They depend on anaerobic pathways to generate adenosine triphosphate (ATP) during muscle contraction.

Fat A triglyceride, which consists of glycerol and three fatty acids. Saturated fats come from animal sources and are solid at room temperature. Unsaturated fats come from plant sources and are liquid at room temperature.

Fatigue A state in which a muscle is physiologically unable to contract despite continued stimulation. Muscle fatigue results from a relative deficit of adenosine triphosphate (ATP).

Fat-soluble vitamin A vitamin that does not dissolve in water and is stored in fat. Examples are vitamins A, D, E, and K.

Fatty acid Chains of carbon atoms bonded to hydrogen atoms with an acidic group (COOH) at one end. Three fatty acids bond to a molecule of glycerol to form a triglyceride (fat).

Feces Waste material discharged from the large intestine during defecation. Feces consist primarily of undigested food, sloughed-off epithelial cells, water, and millions of bacteria.

Fermentation A pathway by which cells can harvest energy in the absence of oxygen. It nets only 2 molecules of ATP as compared with the approximately 36 molecules produced by cellular respiration.

Fertilization The union between an egg (technically a secondary oocyte) and a sperm. It takes about 24 hours from start to finish and usually occurs in a widened portion of the oviduct, not far from the ovary.

Fetal alcohol syndrome A group of characteristics in children born to mothers who consumed alcohol during pregnancy. It can include mental retardation, slow growth, and certain facial features such as an eye fold near the bridge of the nose.

Fetal period The period of prenatal development that extends from week 9 of gestation until birth. It is when rapid growth occurs.

Fetus The developing human from week 9 of gestation until birth (the fetal period).

Fever An abnormally elevated body temperature. Fever helps the body fight disease-causing invaders in a number of ways.

Fibrillation Rapid, ineffective contractions of the heart.

Fibrin A protein formed from fibrinogen by thrombin. It forms a web that traps blood cells, forming a blood clot.

Fibrinogen A plasma protein produced by the liver that is important in blood clotting. It is converted to fibrin by thrombin.

Fibroblasts Cells in connective tissue that secrete the protein fibers that are found in the matrix of the connective tissue. Fibroblasts also secrete collagen fibers for the repair of body tissues.

Fibrocartilage Cartilage with a matrix containing many collagen fibers. Fibrocartilage is found around the edges of joints and the intervertebral disks.

Fibrous joints Joints that are held together by connective tissue and lack a joint cavity. Most fibrous joints do not permit movement.

Fight-or-flight response The body's reaction to stress or threatening situations by the stimulation of the sympathetic division of the autonomic nervous system. Epinephrine and norepinephrine, hormones produced by the adrenal medulla, augment and prolong the response.

First messenger A water-soluble hormone that binds to a receptor on the plasma membrane of a target cell. This binding activates a molecule within the cell, called the second messenger, which influences enzyme activity there.

Fitness The average number of reproductively viable offspring left by an individual. It is sometimes called *Darwinian fitness*.

Flagellum A whiplike appendage of a cell that moves in an undulating manner. It is composed of an extension of the plasma membrane containing microtubules in a 9 + 2 array. In humans, it is found on sperm cells.

Floating ribs The last two ribs that do not attach directly to the sternum (breastbone).

Fluid mosaic A term used to describe the structure of the plasma membrane. Proteins interspersed throughout the lipid molecules give the membrane its mosaic quality. The ability of some proteins to move sideways gives the membrane its fluid quality.

Follicle A spherical structure in the ovary that contains an oocyte surrounded by one or more layers of follicle cells.

Follicle-stimulating hormone (FSH) A hormone secreted by the anterior pituitary gland that in females stimulates the development of the follicles in the ovaries, resulting in the development of ova (eggs) and the production of estrogen, and in males stimulates sperm production.

Fontanels The "soft spots" in the skull of a newborn. The membranous areas in the skull of a newborn that hold the skull bones together before and shortly after birth. The fontanels are gradually replaced by bone.

Food chain The successive series of organisms through which energy (in the form of food) flows in an ecosystem. Each organism in the series eats or decomposes the preceding one. It begins with the photosynthesizers and flows to herbivores and then to carnivores.

Food defense Precautions designed to prevent the *intentional* contamination of food.

Food safety Precautions designed to prevent the *unintentional* contamination of food.

Food web The interconnection of all the feeding relationships (food chains) in an ecosystem.

Foodborne illness An illness that results from ingesting contaminated food or water. Those caused by pathogens (bacteria, viruses, parasites) are considered infections while those caused by chemicals or toxins are considered poisonings. Symptoms first appear in the gastrointestinal tract and include diarrhea and vomiting. Treatment usually involves giving fluids to prevent dehydration; antibiotics may be prescribed if the illness is caused by bacteria. Many foodborne illnesses go unreported.

Foramen magnum The opening at the base of the skull (in the occipital bone) through which the spinal cord passes.

Foramen ovale In the fetus, a small hole in the wall between the right atrium and left atrium of the heart that allows most blood to bypass the lungs.

Formed elements Cells or cell fragments found in the blood. They include platelets, white blood cells, and red blood cells.

Fossilization The process by which fossils form. Typically an organism dies in water and is buried under accumulating layers of sediments. Hard parts become impregnated with minerals from surrounding water and sediments. Eventually the fossil may be exposed when sediments are uplifted and wind erodes the rock formation.

Fossils The preserved remnants or impressions of past organisms. Most fossils occur in sedimentary rocks.

Founder effect Genetic drift associated with the colonization of a new place by a few individuals so that by chance alone the genetic makeup of the colonizing individuals is not representative of the population they left.

Fovea A small region on the retina that contains a high density of cones but no rods. Objects are focused on the fovea for sharp vision.

Frameshift mutation A mutation that occurs when the number of nucleotides inserted or deleted is not a multiple of three, causing a change in the codon sequence on the mRNA molecule as well as a change in the resulting protein.

Fraternal twins Individuals that develop when two oocytes are released from the ovaries and fertilized by different sperm. Such twins may or may not be the same gender and are as similar genetically as any siblings. They are also called dizygotic twins.

Free radicals Molecular fragments that contain an unpaired electron.

FSH See *follicle-stimulating hormone*.

Full-term infant Baby born at least 38 weeks after fertilization.

G

Gallbladder A muscular pear-shaped sac that stores, modifies, and then concentrates bile. Bile is released from the gallbladder into the small intestine.

Gamete A reproductive cell (sperm or egg) that contains only one copy of each chromosome. A sperm and egg fuse at fertilization, producing a zygote.

Gamete intrafallopian transfer (GIFT) A procedure in which eggs and sperm are collected from a couple and then inserted into the woman's oviduct, where fertilization may occur. If fertilization occurs, resulting embryos drift naturally into the uterus.

Ganglion (plural, ganglia) A collection of nerve cell bodies outside the central nervous system.

Gap junction A type of junction between cells that links the cytoplasm of adjacent cells through small holes, allowing physical and electrical continuity between cells.

Gastric glands Any one of several glands in the stomach mucosa that contribute to the gastric juice (hydrochloric acid and pepsin).

Gastric juice The mixture of hydrochloric acid (HCl) and pepsin released into the stomach.

Gastrin A hormone released from the stomach lining in response to the presence of partially digested proteins. Gastrin triggers the production of gastric juice by the stomach.

Gastrointestinal (GI) tract A long tubular system specialized for the processing and absorption of food that begins at the mouth and continues to the esophagus, stomach, intestines, and anus. Several accessory glands empty their secretions into the GI tract to assist digestion.

Gastrula The embryo during gastrulation, when cells move to establish primary germ layers.

Gastrulation Cell movements that establish the primary germ layers of the embryo. The embryo during this period is called a gastrula.

Gated ion channel An ion channel that is opened to allow ions to pass through or closed to prevent passage by changes in the shape of a protein that functions as a gate.

Gene A segment of DNA on a chromosome that directs the synthesis of a specific polypeptide that will play a structural or functional role in the cell. Some genes have regulatory regions of DNA within their boundaries. Also, some genes code for RNA molecules that are needed for the production of the polypeptide but are not part of it.

Gene flow Movement of alleles between populations as a result of the movement of individuals. It is a cause of microevolution.

Gene pool All of the alleles of all of the genes of all individuals in a population.

Gene therapy Treating a genetic disease by inserting healthy functional genes into the body cells that are affected by the faulty gene.

General senses The sensations that arise from receptors in the skin, muscles, joints, bones, and internal organs and include touch, pressure, vibration, temperature, a sense of body and limb position, and pain.

Genetic code The base triplets in DNA that specify the amino acids that go into proteins or that function as start or stop signals in protein synthesis. It is used to convert the linear sequence of bases in DNA to the sequence of amino acids in proteins.

Genetic drift The random change in allele frequencies within a population due to chance alone. It is a cause of microevolution that is usually negligible in large populations.

Genetic engineering The manipulation of genetic material for human practical purposes.

Genital herpes A sexually transmitted disease caused by the herpes simplex virus (HSV) that is usually characterized by painful blisters on the genitals. It is usually caused by HSV-2 but can be caused by HSV-1.

Genital warts Warts that form in the genital area caused by the human papillomaviruses (HPV). These viruses also cause cervical cancer and penile cancer.

Genome The complete set of DNA of an organism, including all of its genes.

Genotype The genetic makeup of an individual. It refers to precise alleles that are present.

Gestational diabetes Diabetes mellitus that develops during pregnancy. It is characterized by insulin resistance and normally resolves after delivery of the baby and placenta.

Gigantism A condition characterized by rapid growth and unusual height caused by abnormally high levels of growth hormone in childhood; also called giantism.

GIFT See *gamete intrafallopian transfer*.

Gingivitis An inflammation of the gums.

Gland Epithelial tissue that secretes a product.

Glaucoma A condition in which the pressure within the anterior chamber of the eye increases as a result of the buildup of aqueous humor. It can cause blindness.

Glial cells Nonexcitable cells in the nervous system that are specialized to support, protect, and insulate neurons. Also called *neuroglial cells* or *neuroglia*.

Global warming A long-term increase in atmospheric temperatures caused by a buildup of carbon dioxide (CO_2) and other greenhouse gases in the atmosphere. Greenhouse gases trap heat in the atmosphere.

Glomerular (Bowman's) capsule A cuplike structure surrounding the glomerulus of a nephron.

Glomerular filtration The process by which water and dissolved substances move from the blood in the glomerulus to the inside of Bowman's capsule.

Glomerulus A tuft of capillaries within the renal corpuscle of a nephron.

Glottis The opening to the airways of the respiratory system from the pharynx into the larynx.

Glucagon The hormone secreted by the pancreas that elevates glucose levels in the blood.

Glucocorticoids Hormones secreted by the adrenal cortex that affect glucose homeostasis, thereby influencing metabolism and resistance to stress.

Gluconeogenesis The conversion of noncarbohydrate molecules to glucose.

Glycemic response A measure that describes how quickly a serving of food is converted to blood sugar and how much the level of blood sugar is affected.

Glycogen The storage polysaccharide of animals. This complex carbohydrate is stored in the liver and muscles where it serves as a short-term energy source that can be broken down to release energy-packed glucose molecules.

Glycolysis The splitting of glucose, a six-carbon sugar, into two three-carbon molecules called pyruvate. Glycolysis takes place in the cytoplasm of a cell and is the starting point for cellular respiration and fermentation.

GnRH See *gonadotropin-releasing hormone*.

Goiter, simple An enlarged thyroid gland caused by iodine deficiency.

Golgi complex An organelle consisting of flattened membranous disks that functions in protein processing and packaging.

Golgi tendon organs The highly branched nerve fibers located in the tendons that sense the degree of muscle tension.

Gonad An ovary in a female or a testis in a male. The gonads produce gametes (eggs or sperm) and sex hormones.

Gonadocorticoids The male and female sex hormones, androgens and estrogens, secreted by the adrenal cortex.

Gonadotropin-releasing hormone (GnRH) A hormone produced by the hypothalamus that causes the secretion of luteinizing hormone and follicle-stimulating hormone from the anterior pituitary gland.

Gonorrhea A sexually transmitted disease caused by the bacterium *Neisseria gonorrhoeae* that commonly causes urethritis and pelvic inflammatory disease. It may not cause symptoms, especially in women.

Graafian follicle A mature follicle in the ovary.

Graded potential A temporary local change in the membrane potential that varies directly with the strength of the stimulus.

Granulocytes White blood cells with granules in their cytoplasm. Examples are neutrophils, eosinophils, and basophils.

Graves' disease An autoimmune disorder caused by oversecretion of thyroid hormone. It is characterized by increased heart and metabolic rates, sweating, nervousness, and exophthalmos.

Gray matter Regions of the central nervous system that contain neuron cell bodies and unmyelinated axons. These regions are gray because they lack myelin. Gray matter is important in neural integration.

Greenhouse effect A process in which greenhouse gases trap heat in the atmosphere. Examples of greenhouse gases include carbon dioxide and methane. The greenhouse effect could lead to a rise in temperatures throughout the world.

Ground substance In connective tissue the ground substance forms the extracellular matrix that the cells are embedded in. It is composed of protein fibers and noncellular material.

Growth factor A type of signaling molecule that stimulates growth by stimulating cell division in target cells.

Growth hormone (GH) An anterior pituitary hormone with the primary function of stimulating growth through increases in protein synthesis, cell size, and rates of cell division. GH stimulates growth in general, especially bone growth.

Growth plate A plate of cartilage that separates each end of a long bone from its shaft that permits bone to grow. Also called an *epiphyseal plate*.

Gumma A large sore that forms during the third stage of syphilis.

H

Habitat The natural environment or place where an organism, population, or species lives.

Hair cells A type of mechanoreceptor that generates nerve impulses when bent or tilted. Hair cells in the inner ear are responsible for hearing and equilibrium.

Hair root plexus The nerve endings that surround the hair follicle and are sensitive to touch.

Hallucinogen A psychoactive drug that distorts sensory perception.

Haploid The condition of having one set of chromosomes, as in eggs and sperm.

HCG See *human chorionic gonadotropin*.

Health care agent A person selected in advance to make health care decisions for you should you be unable to make them for yourself. The selection is formalized by completing a health care proxy form.

Heart A muscular pump that keeps blood flowing through an animal's body. The human heart has four chambers: two atria and two ventricles.

Heart attack The death of heart muscle cells caused by an insufficient blood supply; a myocardial infarction.

Heart murmur Heart sounds other than "lub dup" that are created by turbulent blood flow. Heart murmurs can indicate a heart problem, such as a malfunctioning valve.

Heartburn A burning sensation behind the breastbone that occurs when acidic gastric juice backs up into the esophagus.

Heimlich maneuver A procedure intended to force a large burst of air out of the lungs to dislodge material lodged in the trachea.

Helper T cell The kind of T lymphocyte that serves as the main switch for the entire immune response by presenting the antigen to B cells and by secreting chemicals that stimulate other cells of the immune system. It is also known as a *T4 cell* or a *CD4 cell*, after the receptors on its surface.

Hematocrit The percentage of red blood cells in blood by volume. It is a measure of the oxygen-carrying ability of the blood.

Hemodialysis The use of artificial devices, such as the artificial kidney machine, to cleanse the blood during kidney failure.

Hemoglobin The oxygen-binding pigment in red blood cells. It consists of four subunits, each made up of an iron-containing heme group and a protein chain.

Hemolytic disease of the newborn A condition in which the red blood cells of an Rh-positive fetus or newborn are destroyed by anti-Rh antibodies previously produced in the bloodstream of an Rh-negative mother.

Hemophilia An inherited blood disorder in which there is insufficient production of blood-clotting factors. Hemophilia results in excessive bleeding in joints, deep tissues, and elsewhere. Hemophilia usually occurs in males.

Hepatitis An inflammation of the liver. Hepatitis is usually caused by one of six viruses.

Herbivore An animal that eats primary producers (green plants or algae). A primary consumer.

Herpes simplex virus See *HSV-1* and *HSV-2*.

Heterotroph An organism that cannot make its own food from inorganic substances and instead consumes other organisms or decaying material.

Heterozygote advantage The phenomenon in which the heterozygous condition is favored over either homozygous condition. It maintains genetic variation within a population in the face of natural selection.

Heterozygous The condition of having two different alleles for a particular gene.

High-density lipoprotein (HDL) A lipoprotein made in the liver and released into the blood that transports cholesterol away from the cells to the liver. HDLs are often called the "good" form of cholesterol.

Hinge joint A joint that permits motion in only one plane, such as the knee joint or the elbow.

Hippocampus The part of the limbic system of the brain that plays a role in converting short-term memory into long-term memory.

Histamine A substance released by basophils and mast cells during an inflammatory response that causes blood vessels to widen (dilate) and become more permeable.

Homeostasis The ability of living things to maintain a relatively constant internal environment in all levels of body organization.

Hominid A member of the family Hominidae, which now includes apes and humans. In the past, only humans and their immediate ancestors, such as species within the genus *Australopithecus*, were placed in Hominidae.

Hominin A member of the subfamily Homininae, which includes the human lineage and its immediate ancestors.

Homologous chromosomes A pair of chromosomes that bear genes for the same traits. One member of each pair came from each parent. Homologous chromosomes are the same size and shape and line up with one another during meiosis I.

Homologous structures Structures that have arisen from a common ancestry. Compare with *analogous structure*.

Homozygous The condition of having two identical alleles for a particular gene.

Hormonal implants A means of hormonal contraception consisting of silicon rods containing progesterone that are implanted under the skin in the upper arm and prevent pregnancy for up to 5 years.

Hormone A chemical messenger released by cells of the endocrine system that travels through the circulatory system to affect receptive target cells.

Host (vector) DNA The DNA that is recombined with pieces of DNA from another source (that might contain a desirable gene) in the formation of recombinant DNA.

HPV See *human papillomavirus*.

HSV-1 Herpes simplex virus 1. HSV-1 usually infects the upper half of the body and causes cold sores (fever blisters), but it can cause genital herpes if contact is made with the genital area.

HSV-2 Herpes simplex virus 2. HSV-2 causes genital herpes, but it can cause cold sores if contact is made with the mouth.

Human chorionic gonadotropin (HCG) A hormone produced by the cells of the early embryo (blastocyst) and the placenta that maintains the corpus luteum for approximately the first 3 months of pregnancy. HCG enters the bloodstream of the mother and is excreted in her urine. HCG forms the basis for many pregnancy tests.

Human papillomavirus (HPV) One of the group of viruses that commonly cause genital warts.

Hyaline cartilage A type of cartilage with a gel-like matrix that provides flexibility and support. It is found at the end of long bones as well as parts of the nose, ribs, larynx, and trachea.

Hydrocephalus A condition resulting from the excessive production or inadequate drainage of cerebrospinal fluid.

Hydrogen bond A weak chemical bond formed between a partially positively charged hydrogen atom in a molecule and a partially negatively charged atom in another molecule or in another region of the same molecule.

Hydrolysis The process by which polymers are broken apart by the addition of water.

Hydrophilic Water-loving. The heads of phospholipids (components of the plasma membrane) are hydrophilic.

Hydrophobic Water-fearing. The tails of phospholipids (components of the plasma membrane) are hydrophobic.

Hyperglycemia An elevated blood glucose level.

Hypertension High blood pressure. A high upper (systolic) value usually suggests that the person's arteries have become hardened and are no longer able to dampen the high pressure of each heartbeat. The lower (diastolic) value is generally considered more important because it indicates the pressure when the heart is relaxing.

Hyperthermia Abnormally elevated body temperature.

Hypertonic solution A solution with a higher concentration of solutes than plasma, for example.

Hypodermis The layer of loose connective tissue below the epidermis and dermis that anchors the skin to underlying tissues and organs.

Hypoglycemia Depressed levels of blood glucose often resulting from excess insulin.

Hypophysis See *pituitary gland*.

Hypothalamus A small brain region located below the thalamus that is essential to maintaining a stable environment within the body. The hypothalamus influences blood pressure, heart rate, digestive activity, breathing rate, and many other vital physiological processes. It acts as the body's "thermostat"; regulates food intake, hunger, and thirst; coordinates the activities of the nervous system and the endocrine system; and is part of the circuitry for emotions.

Hypothermia Abnormally low body temperature.

Hypothesis A testable explanation for a specific set of observations that serves as the basis for experimentation.

Hypotonic solution A solution with a lower concentration of solutes than plasma, for example.

I

ICSI See *intracytoplasmic sperm injection*.

Identical twins Individuals that develop from a single fertilized ovum that splits in two at an early stage of cleavage. Such twins have nearly identical genetic material and are always the same gender. They are also called monozygotic twins.

Immigration Movement of new individuals from other populations into an area.

Immune response The body's response to specific targets not recognized as belonging in the body.

Immune system The system of the body directly involved with body defenses against specific targets—pathogens, cells, or chemicals not recognized as belonging in the body.

Immunoglobulin (Ig) Any of the five classes of proteins that constitute the antibodies.

Implantation The process by which a blastocyst (pre-embryo) becomes embedded in the lining of the uterus. It normally occurs high up on the back wall of the uterus.

Impotence The inability to achieve or maintain an erection long enough for sexual intercourse.

In vitro fertilization (IVF) A procedure in which eggs and sperm are placed together in a dish in the laboratory. If fertilization occurs, early-stage embryos are then transferred to the woman's uterus, where it is hoped they will implant and complete development. It is a common treatment for infertility resulting from blocked oviducts.

Incomplete dominance In genetic inheritance, expression of the trait in a heterozygous individual is somewhere in between expression of the trait in a homozygous dominant individual and homozygous recessive individual.

Incomplete proteins Proteins that are deficient in one or more of the essential amino acids. Plant sources of protein are generally incomplete proteins.

Incontinence A condition characterized by the escape of small amounts of urine when sudden increases in abdominal pressure, perhaps caused by laughing, sneezing, or coughing, force urine past the external sphincter. This condition is common in women, particularly after childbirth, an event that may stretch or damage the external sphincter, making it less effective in controlling the flow of urine.

Incus The middle of three small bones in the middle ear that transmit information about sound from the eardrum to the inner ear. The incus is also known as the anvil.

Independent assortment Of chromosomes, the process by which homologous chromosomes and the alleles they carry segregate randomly during meiosis, creating mixes of maternal and paternal chromosomes in gametes. This is an important source of genetic variation in populations.

Independent assortment, law of In genetic inheritance, a principle that states that the alleles of unlinked genes (those that are located on different chromosomes) are randomly distributed to gametes.

Inductive reasoning A logical progression of thought proceeding from the specific to the general. It involves the accumulation of facts through observation until the sheer weight of the evidence forces some general statement about the phenomenon. A conclusion is reached on the basis of a number of observations.

Infancy The stage in postnatal development that roughly corresponds to the first year of life. It is a time of rapid growth when total body length usually increases by one-half and weight may triple.

Infertility The inability to conceive (become pregnant) or to cause conception (in the case of males).

Inflammatory response A nonspecific body response to injury or invasion by foreign organisms. It is characterized by redness, swelling, heat, and pain.

Influenza The flu. Influenza is a viral infection caused by any of the variants of influenza A or influenza B viruses.

Informed consent A document that a person must sign before participating in an experiment that lists all possible harmful consequences that might result from participation.

Inguinal canal A passage through the abdominal wall through which the testes pass in their descent to the scrotum.

Inhalation Breathing in. Inhalation (inspiration) is the movement of air into the respiratory system.

Inhibin A hormone produced in the testes that increases with sperm count and inhibits follicle-stimulating hormone secretion and, therefore, inhibits sperm production.

Inhibiting hormone A hormone that inhibits secretion of another hormone.

Inhibitory synapse A synapse in which the response of the receptors for that neurotransmitter on the postsynaptic membrane decreases the likelihood that an action potential will be generated in the postsynaptic neuron. The postsynaptic cell is inhibited because its resting potential becomes more negative than usual.

Innate defense responses Body defense responses that we are born with—barriers and chemical—to prevent entry of pathogens, and cells and chemicals that attack a pathogen if it breaks through outer barriers. These defense responses are nonspecific.

Inner cell mass A group of cells within the blastocyst that will become the embryo proper and some extraembryonic membranes.

Inner ear A series of passageways in the temporal bone that houses the organs for hearing (cochlea) and the sense of equilibrium (vestibular apparatus).

Inpatient A patient who is admitted to a hospital or clinic for treatment that requires a stay of at least one night.

Insertion The end of the muscle that is attached to the bone that moves when the muscle contracts.

Insoluble fiber A fiber that does not easily dissolve in water. These fibers include cellulose, hemicellulose, and lignin.

Inspiration Inhalation. The movement of air into the respiratory system.

Inspiratory reserve volume The volume of air that can be forcefully brought into the lungs after normal inhalation.

Insulin The hormone secreted by the pancreas that reduces glucose levels in the blood.

Insulin resistance The condition in which the body's cells fail to adequately respond to insulin. It characterizes type 2 diabetes mellitus and gestational diabetes.

Insulin shock A condition that results from severely depressed glucose levels in which brain cells fail to function properly, causing convulsions and unconsciousness. Often results when a diabetic injects too much insulin.

Integral proteins Proteins embedded in the plasma membrane, either completely or incompletely spanning the bilayer.

Integumentary system The skin.

Intercalated disks Thickenings of the plasma membranes of cardiac muscle cells that strengthen cardiac tissue and promote rapid conduction of impulses throughout the heart.

Intercostal muscles The layers of muscles between the ribs that raise and lower the rib cage during breathing.

Interferon A type of defensive protein produced by T lymphocytes that slows the spread of viruses already in the body by interfering with viral replication. Interferons also attract macrophages and natural killer cells, which kill the virus-infected cell.

Interleukin 1 A chemical secreted by a macrophage that activates helper T cells in an immune response.

Interleukin 2 A chemical released by a helper T cell that activates both B cells and T cells.

Intermediate filament A component of the cytoskeleton made from fibrous proteins. It maintains cell shape and anchors organelles such as the nucleus.

Internal respiration Movement of oxygen from the blood to the tissues, and movement of carbon dioxide from the tissues to the blood.

Internal urethral sphincter A thickening of smooth muscle at the junction of the bladder and the urethra. The action of this sphincter is involuntary and keeps urine from flowing into the urethra while the bladder is filling.

Interneuron An association neuron. Neurons located within the central nervous system between sensory and motor neurons that serve to integrate information.

Interoceptor A sensory receptor located inside the body, where it monitors conditions. Interoceptors play a vital role in maintaining homeostasis. They are an important part of the feedback loops that regulate blood pressure, blood chemistry, and breathing rate. Interoceptors may also cause us to feel pain, hunger, or thirst, thereby prompting us to take appropriate action.

Interphase The period between cell divisions when the DNA, cytoplasm, and organelles are duplicated and the cell grows in size. Interphase is also the time in the cell's life cycle when the cell carries out its functions in the body.

Interstitial cells Cells located between the seminiferous tubules in the testes that produce the steroid sex hormones, collectively called androgens.

Intervertebral disks Pads of cartilage that help cushion the bones of the vertebral column.

Intracytoplasmic sperm injection (ICSI) A procedure in which a tiny needle is used to inject a single sperm cell into an egg. It is an option for treating infertility when the man has few sperm or sperm that lack the strength or enzymes necessary to penetrate the egg.

Intrauterine device (IUD) A means of contraception consisting of a small plastic device that either is wrapped with copper wire or contains progesterone. It is inserted into the uterus to prevent pregnancy.

Intrinsic factor A protein secreted by the stomach necessary for the absorption of vitamin B_{12} from the small intestine.

Intron A portion of a newly formed mRNA that is cut out of the mature mRNA molecule and is not expressed (used to form a protein). An intervening sequence.

Ion An atom or group of atoms that carries an electric charge resulting from the loss or gain of electrons.

Ion channel A protein-lined pore or channel through a plasma membrane through which one type or a few types of ions can pass. Nerve cell ion channels are important in the generation and propagation of nerve impulses.

Ionic bond A chemical bond that results from the mutual attraction of oppositely charged ions.

Iris The colored portion of the eye. The iris regulates the amount of light that enters the eye.

Iron-deficiency anemia A reduction in the ability of the blood to carry oxygen due to an insufficient amount of iron in the diet, an inability to absorb iron from the digestive system, or blood loss.

Ischemia A temporary reduction in blood supply caused by obstructed blood flow. It causes reversible damage to heart muscle.

Islets of Langerhans See *pancreatic islets*.

Isotonic solution A solution with the same concentration of solutes as plasma, for example.

Isotopes Atoms that have the same number of protons but different numbers of neutrons.

IUD See *intrauterine device*.

IVF See *in vitro fertilization*.

J

Jaundice A condition in which the skin develops a yellow tone caused by the buildup of bilirubin in the blood and its deposition in certain tissues such as the skin. It is an indication that the liver is not handling bilirubin adequately.

Joint A point of contact between two bones; an articulation.

Junctional complexes Membrane specializations that attach adjacent cells to each other to form a contiguous sheet. There are three kinds of junctions between cells: tight junctions, desmosomes, and gap junctions.

Juxtaglomerular apparatus The region of the kidney nephron where the distal convoluted tubule contacts the afferent arteriole bringing blood into the glomerulus. Cells in this area secrete renin, an enzyme that triggers events eventually leading to increased reabsorption of sodium and water by the distal convoluted tubules and collecting ducts of nephrons.

K

Kaposi's sarcoma A cancer that forms tumors in connective tissue and manifests as pink or purple marks on the skin. It is common in people with a suppressed immune system, such as people living with HIV/AIDS, and is thought to be associated with a new virus in the herpes family, HHV-8.

Karyotype The arrangement of chromosomes based on physical characteristics such as length and location of the centromere. Karyotypes can be checked for defects in number or structure of chromosomes.

Keratinocytes Cells of the epidermis that undergo keratinization, the process in which keratin gradually replaces the contents of maturing cells.

Ketoacidosis A lowering of blood pH resulting from the accumulation of breakdown by-products of fats. This biochemical imbalance is characteristic of type 1 diabetes mellitus, where it is called diabetic ketoacidosis (DKA).

Kidney stones Small, hard crystals formed in the urinary tract when substances such as calcium (the most common constituent), uric acid, or magnesium ammonium phosphate precipitate from urine as a result of higher-than-normal concentrations. They are also called renal calculi.

Kidneys Reddish, kidney bean–shaped organs that filter wastes and excess materials from the blood, assist the respiratory system in regulating blood pH, and maintain fluid balance by regulating the volume and composition of blood and urine.

Klinefelter syndrome A genetic condition resulting from nondisjunction of the sex chromosomes in which a person inherits an extra X chromosome that results in an XXY genotype. The person has a male appearance.

Krebs cycle See *citric acid cycle*.

L

Labia majora Two elongated skin folds lateral to the labia minora. They are part of the female external genitalia.

Labia minora Two small skin folds on either side of the vagina and interior to the labia majora. They are part of the external genitalia of a female.

Labor The process by which the fetus is expelled from the uterus and moved through the vagina and into the outside world. During labor, uterine contractions occur at regular intervals, are often painful, and intensify with walking. Labor is usually divided into the dilation stage, expulsion stage, and placental stage.

Lactation The production and ejection of milk from the mammary glands. The hormone prolactin from the anterior pituitary gland promotes milk production, and the hormone oxytocin released from the posterior pituitary gland makes milk available to the suckling infant by stimulating milk ejection, or let-down.

Lacteal A lymphatic vessel in an intestinal villus that aids in the absorption of lipids.

Lactic acid fermentation The process by which glucose is broken down by muscle cells when oxygen is low during strenuous exercise.

Lacuna (plural, lacunae) A tiny cavity. It contains osteocytes (bone cells) in the matrix of bone and cartilage cells in the matrix of cartilage.

Lanugo Soft, fine hair that covers the fetus beginning about the third or fourth month after conception.

Large intestine The final segment of the gastrointestinal tract, consisting of the cecum, colon, rectum, and anal canal. The large intestine helps absorb water, forms feces, and plays a role in defecation.

Laryngitis An inflammation of the larynx in which the vocal cords become swollen and can no longer vibrate and produce sound.

Larynx The voice box or Adam's apple. A boxlike cartilaginous structure between the pharynx and the trachea held together by muscles and elastic tissue.

Latent period Pertaining to muscle contraction, the interval between the reception of the stimulus and the beginning of muscle contraction.

Latin binomial The two-part scientific name that consists of the genus name followed by the specific epithet.

Lens A transparent, semispherical body of tissue behind the iris and pupil that focuses light on the retina.

Leukemia A cancer of the blood-forming organs that causes white blood cell numbers to increase. The white blood cells are abnormal and do not effectively defend the body against infectious agents.

Leukocytes White blood cells. They are cells of the blood including neutrophils, eosinophils, basophils, monocytes, B lymphocytes, and T lymphocytes. Leukocytes are involved in body defense mechanisms and the removal of wastes, toxins, or damaged, abnormal cells.

LH See *luteinizing hormone*.

Life expectancy The average number of years a newborn is expected to live.

Ligament A strong band of connective tissue that holds the bones together, supports the joint, and directs the movement of the bones.

Limbic system A collective term for several structures in the brain involved in emotions and memory.

Linkage The tendency for a group of genes located on the same chromosome to be inherited together.

Lipid A compound, such as a triglyceride, phospholipid, or steroid, that does not dissolve in water. Dietary lipids include fats, oils, and cholesterol. They provide 9 calories per gram.

Lipid-soluble hormone A hormone that moves easily through the plasma membrane of cells and combines with receptors inside target cells to activate certain genes and stimulate protein synthesis. Steroid hormones are lipid-soluble hormones. Steroids are derived from cholesterol and secreted by the ovaries, testes, and adrenal glands.

Liver A large organ that functions mainly in the production of plasma proteins, the excretion of bile, the storage of energy reserves, the detoxification of poisons, and the interconversion of nutrients.

Local signaling molecules Chemical messengers, such as neurotransmitters, prostaglandins, growth factors, and nitric oxide, that act locally rather than traveling to distant sites within the body.

Locus The point on a chromosome where a particular gene is found.

Longitudinal fissure A deep groove that separates the cerebrum into two hemispheres.

Long-term memory Memory that stores a large amount of information for hours, days, or years.

Loop of the nephron (loop of Henle) A section of the renal tubule that resembles a hairpin turn. Reabsorption occurs here.

Loose connective tissue Connective tissue, such as areolar and adipose tissue, that contains many cells in which the fibers of the matrix are fewer in number and more loosely woven than those found in dense connective tissue.

Low-density lipoprotein (LDL) A protein carrier in the blood that transports cholesterol to the cells. LDLs are often called the "bad" form of cholesterol.

Lower esophageal sphincter A ring of muscle at the juncture of the esophagus and the stomach that controls the flow of materials between the esophagus and the stomach. It relaxes to allow food into the stomach and contracts to prevent too much food from moving back into the esophagus.

Lumen The hollow cavity or channel of a tubule through which transported material flows.

Luteinizing hormone (LH) A hormone secreted by the anterior pituitary gland that in females stimulates ovulation and the formation of the corpus luteum (which produces estrogen and progesterone) and prepares the mammary glands for milk production. In males, it stimulates testosterone production by the interstitial cells within the testes.

Lymph The fluid within the vessels of the lymphatic system. It is derived from the fluid that bathes the cells of the body (interstitial fluid).

Lymph nodes Small nodular organs found along lymph vessels that filter lymph. The lymph nodes contain macrophages and lymphocytes, cells that play an essential role in the body's defense system.

Lymphatic system A body system consisting of lymph, lymphatic vessels, and lymphatic tissue and organs. The lymphatic system helps return interstitial fluid to the blood, transports the products of fat digestion from the digestive system to the blood, and assists in body defenses.

Lymphatic vessels The vessels through which lymph flows. A network of vessels that drains interstitial fluid and returns it to the blood supply and transports the products of fat digestion from the digestive system to the blood supply.

Lymphocyte A type of white blood cell important in nonspecific and specific (immune) body defenses. The lymphocytes include B lymphocytes (B cells) that transform into plasma cells and produce antibodies and T lymphocytes (T cells) that are important in defense against foreign or infected cells.

Lymphoid organs Various organs that belong to the lymphatic system, including the tonsils, spleen, thymus, and Peyer's patches.

Lymphoid tissue The type of tissue that predominates in the lymphoid organs except the thymus. The organs in which lymphoid tissue is found are the lymph nodes, tonsils, and spleen. Lymphoid tissue is an important component of the immune system.

Lymphoma Cancer of the lymphoid tissues. In people with AIDS, it commonly affects the B cells.

Lysosomal storage diseases Disorders such as Tay-Sachs disease caused by the absence of lysosomal enzymes. In these disorders, molecules that would normally be degraded by the missing enzymes accumulate in the lysosomes and interfere with cell functioning.

Lysosome An organelle that serves as the principal site of digestion within the cell.

Lysozyme An enzyme present in tears, saliva, and certain other body fluids that kills bacteria by disrupting their cell walls.

M

Macroevolution Large-scale evolutionary changes such as those that might result from long-term changes in the climate or position of the continents. Examples include mass extinctions and the evolution of mammals.

Macromolecule A giant molecule of life such as a nucleic acid, protein, or polysaccharide. A macromolecule is formed by the joining together of smaller molecules.

Macrophage A large phagocytic cell derived from a monocyte that lives in loose connective tissue and engulfs anything detected as foreign.

Magnetic resonance imaging (MRI) A means of visualizing a region of the body, including the brain, in which the picture results from differences in the way the hydrogen nuclei in the water molecules within the tissues vibrate in response to a magnetic field created around the area to be pictured.

Malignant tumor A cancerous tumor. An abnormal mass of tissue with cells that can invade surrounding tissue and spread to multiple locations throughout the body.

Malleus The first of three small bones in the middle ear that transmit information about sound from the eardrum to the inner ear. The malleus is also known as the hammer.

Malnourishment A form of hunger that occurs when the diet is not balanced and the right foods are not eaten.

Mammary glands The milk-producing glands in the breasts.

Marijuana A psychoactive drug consisting of the leaves, stems, and flowers of the Indian hemp plant, *Cannabis sativa*.

Mass The measure of how much matter is in an object.

Mast cells Small, mobile connective tissue cells often found near blood vessels. In response to injury, mast cells release histamine, which dilates blood vessels and increases blood flow to an area, and heparin, which prevents blood clotting.

Matrix In connective tissue, the matrix is the material in which the cells are embedded. The matrix consists of ground substance, which is made up of protein fibers and noncellular material.

Matter Anything that takes up space and has mass. Matter is made up of atoms.

Mechanical digestion A part of the digestive process that involves physically breaking food into smaller pieces.

Mechanoreceptor A sensory receptor that is specialized to respond to distortions in the receptor itself or in nearby cells. Mechanoreceptors are responsible for the sensations we describe as touch, pressure, hearing, and equilibrium.

Medulla oblongata The part of the brain stem containing reflex centers for some of life's most vital physiological functions: the pace of the basic breathing rhythm, the force and rate of heart contraction, and blood pressure. The medulla oblongata connects the spinal cord to the rest of the brain.

Medullary cavity The cavity in the shaft of long bones that is filled with yellow marrow.

Medullary rhythmicity center The region of the brain stem controlling the basic rhythm of breathing.

Meiosis A type of cell division that occurs in the gonads and gives rise to gametes. As a result of two divisions (meiosis I and meiosis II), haploid gametes are produced from diploid germ cells.

Meissner's corpuscles Encapsulated nerve cell endings that are common on the hairless, sensitive areas of the skin, such as the lips, nipples, and fingertips, and tell us exactly where we have been touched.

Melanin A pigment produced by the melanocytes of the skin. There are two forms of melanin: black-to-brown and yellow-to-red.

Melanocytes Spider-shaped cells located at the base of the epidermis that manufacture and store melanin, a pigment involved in skin color and absorption of ultraviolet radiation.

Melanoma The least common and most dangerous form of skin cancer. It arises in the melanocytes, the pigment-producing cells of the skin.

Melatonin A hormone secreted by the pineal gland that reduces jet lag and promotes sleep.

Membranous epithelium A type of epithelial tissue that forms linings and coverings. Depending on its location, it may be specialized to protect, secrete, or absorb.

Memory cell A lymphocyte (B cell or T cell) of the immune system that forms in response to an antigen and that circulates for a long period of time; such cells are able to mount a quick immune response to a subsequent exposure to the same antigen.

Meninges Three protective connective tissue membranes that surround the central nervous system: the dura mater, the pia mater, and the arachnoid.

Meningitis An inflammation of the meninges (protective coverings of the brain and spinal cord).

Menopause The end of a female's reproductive potential when ovulation and menstruation cease.

Menstrual (uterine) cycle The sequence of events that occurs on an approximately 28-day cycle in the uterine lining (endometrium) that involves the thickening of and increased blood supply to the endometrium and the loss of the endometrium as menstrual flow.

Merkel cells Cells of the epidermis found in association with sensory neurons where the epidermis meets the dermis.

Merkel disk The Merkel cell–neuron combination that functions as a sensory receptor for light touch, providing information about objects contacting the skin. It is found on both the hairy and the hairless parts of the skin.

Mesoderm The primary germ layer that gives rise to muscle, bone, connective tissue, and organs such as the heart, kidneys, ovaries, and testes.

Messenger RNA (mRNA) A type of RNA synthesized from and complementary to a region of DNA that attaches to ribosomes in the cytoplasm and specifies the amino acid order in the protein. It carries the DNA instructions for synthesizing a particular protein.

Metabolism The sum of all chemical reactions within living cells.

Metaphase In mitosis, the phase when the chromosomes, guided by the fibers of the mitotic spindle, form a line at the center of the cell. As a result of this alignment, when the chromosomes split at the centromere, each daughter cell receives one chromatid from each chromosome and thus a complete set of the parent cell's chromosomes.

Metastasize To spread from one part of the body to another part not directly connected to the first part. Cancerous tumors metastasize and form new tumors in distant parts of the body.

MHC markers Molecules on the surface of body cells that label the cell as "self."

Microevolution Changes in populations at the genetic level. The causes include genetic drift, gene flow, natural selection, and mutation.

Microfilament A component of the cytoskeleton made from the globular protein actin. Microfilaments form contractile units in muscle cells and aid in pinching dividing cells in two.

Microtubule A component of the cytoskeleton made from the globular protein tubulin. Microtubules are responsible for the movement of cilia and flagella and serve as tracks for the movement of organelles and vesicles.

Microvilli Microscopic cytoplasm-filled extensions of the cell membrane that serve to increase the absorptive surface area of the cell.

Midbrain A region of the brain stem that coordinates reflex responses to auditory and visual stimuli.

Middle ear An air-filled space in the temporal bone that includes the tympanic membrane (eardrum) and three small bones (the malleus, incus, and stapes; sometimes called the hammer, anvil, and stirrup). It serves as an amplifier of sound pressure waves.

Minerals Inorganic substances that are not broken down during digestion and are important in regulating cellular processes.

Mineralocorticoids Hormones secreted by the adrenal cortex that affect mineral homeostasis and water balance.

Minipill A birth control pill that contains only progesterone.

Mitochondrion (plural, mitochondria) An organelle within which most of cellular respiration occurs in a eukaryotic cell. Cellular respiration is the process by which oxygen and an organic fuel such as glucose are consumed and energy is released and used to form ATP.

Mitosis A type of nuclear division occurring in somatic cells in which two identical cells, called daughter cells, are generated from a single cell. The original cell first replicates its genetic material and then distributes a complete set of genetic information to each of its daughter cells. Mitosis is usually divided into prophase, metaphase, anaphase, and telophase. Mitosis is essential to cell division.

Mitral valve A heart valve located between the left atrium and left ventricle that prevents the backflow of blood from the ventricle to the atrium. It is also called the *bicuspid valve* or the *left atrioventricular (AV) valve*.

Molecular clock The idea that there is a constant rate of divergence of macromolecules (such as proteins) from one another. This idea is based on the notion that single nucleotide changes in DNA (point mutations) and the amino acid changes in proteins that can be produced by some point mutations occur with steady, clocklike regularity. It permits comparison of molecular sequences to estimate the time of separation between species (the more differences in sequence, the more time that has elapsed since the common ancestor).

Molecule A chemical structure composed of atoms held together by covalent bonds.

Monoclonal antibodies Defensive proteins specific for a particular antigen secreted by a clone of genetically identical cells descended from a single cell.

Monocyte The largest white blood cell. Monocytes are active in fighting chronic infections, viruses, and intracellular bacterial infections. A monocyte can transform into a phagocytic macrophage.

Monohybrid cross A genetic cross that considers the inheritance of a single trait from individuals differing in the expression of that trait.

Monomer A small molecule that joins with identical molecules to form a polymer.

Mononucleosis A viral disease caused by the Epstein-Barr virus that results in an elevated level of monocytes in the blood. It causes fatigue and swollen glands, and there is no available treatment.

Monosaccharide The smallest molecular unit of a carbohydrate. Monosaccharides are known as simple sugars.

Monosomy A condition in which there is only one representative of a chromosome instead of two representatives.

Mons veneris A round fleshy prominence over the pubic bone in a female. Part of the female external genitalia.

Morning sickness The nausea and vomiting experienced by some women early in pregnancy. It is not restricted to the morning and may be caused, in part, by high levels of the hormone human chorionic gonadotropin (HCG).

Morphogenesis The development of body form that begins during the third week after fertilization.

Morula A solid ball of 12 or more cells produced by successive divisions of the zygote. The name reflects its resemblance to the fruit of the mulberry tree.

Mosaic evolution The phenomenon whereby various traits evolve at their own rates.

Motor (efferent) neuron A neuron specialized to carry information *away from* the central nervous system to an effector, either a muscle or a gland.

Motor unit A motor neuron and all the muscle fibers (cells) it stimulates.

MRI See *magnetic resonance imaging*.

mRNA See *messenger RNA*.

Mucosa The innermost layer of the gastrointestinal tract. It secretes mucus that helps lubricate the tube, allowing food to slide through easily.

Mucous membranes Sheets of epithelial tissue that line many passageways in the body that open to the exterior. Mucous membranes are specialized to secrete and absorb.

Mucus A sticky secretion that serves to lubricate body parts and trap particles of dirt and other secretions. It also helps protect the stomach from the action of gastric juice.

Multiple alleles Three or more alleles of a particular gene existing in a population. The alleles governing the ABO blood types provide an example.

Multiple sclerosis An autoimmune disease in which the body's own defense mechanisms attack myelin sheaths in the nervous system. As a patch of myelin is repaired, a hardened region called a sclerosis forms.

Multipolar neuron A neuron that has at least three processes, including an axon and a minimum of two dendrites. Most motor neurons and association neurons are multipolar.

Multipotent A term used to describe a cell that can differentiate into many cell types.

Multiregional hypothesis The idea that modern humans evolved independently in several different areas from distinctive local populations of *Homo erectus*. Compare with *Out of Africa hypothesis*.

Muscle fiber A muscle cell.

Muscle spindles Specialized muscle fibers with sensory nerve cell endings wrapped around them that report to the brain whenever a muscle is stretched.

Muscle tissue Tissue composed of muscle cells that contract when stimulated and passively lengthen to the resting state. There are three types of muscle tissue: skeletal, smooth, and cardiac.

Muscle twitch Contraction of a muscle in response to a single stimulus.

Muscularis The muscular layers of the gastrointestinal tract. These layers help move food through the gastrointestinal system and mix food with digestive secretions.

Mutation A change in the base sequence of the DNA of a gene. A mutation may occur spontaneously or be caused by outside sources, such as radiation or chemical agents.

Myelin sheath An insulating layer around axons that carry nerve impulses over relatively long distances that is composed of multiple wrappings of the plasma membrane of certain glial cells. Outside the brain and spinal cord, Schwann cells form the myelin sheath. The myelin sheath greatly increases the speed at which impulses travel and assists in the repair of damaged axons. The Schwann cells that form the myelin sheath are separated from one another by short regions of exposed axon called nodes of Ranvier.

Myocardial infarction A heart attack. A condition in which a part of the heart muscle dies because of an insufficient blood supply.

Myocardium Cardiac muscle tissue that makes up the bulk of the heart. The contractility of the myocardium is responsible for the heart's pumping action.

Myofibril A rodlike bundle of contractile proteins (myofilaments) found in skeletal and cardiac muscle cells essential to muscle contraction.

Myofilament A contractile protein within muscle cells. There are two types: myosin (thick) filaments and actin (thin) filaments.

Myoglobin An oxygen-binding pigment in muscle fibers.

Myometrium The smooth muscle layer in the wall of the uterus.

Myosin filaments The thick filaments in muscle cells composed of the protein myosin and essential to muscle contraction. A myosin molecule is shaped like a golf club with two heads.

Myosin heads Club-shaped ends of a myosin molecule that bind to actin filaments and can swivel, causing actin filaments to slide past the myosin filaments, which causes muscle contraction. They are also called *cross-bridges*.

Myxedema A condition characterized by swelling of the facial tissues because of the accumulation of interstitial fluids caused by undersecretion of thyroid hormone.

N

Nasal cavities Two chambers in the nose, separated by the septum.

Nasal conchae The three convoluted bones within each nasal cavity that increase surface area and direct airflow.

Nasal septum A thin partition of cartilage and bone that divides the inside of the nose into two nasal cavities.

Natural killer (NK) cells A type of cell in the immune system. These cells, probably lymphocytes, roam the body in search of abnormal cells and quickly kill them.

Natural selection The process by which some individuals live longer and produce more offspring than other individuals because their particular inherited characteristics make them better suited to their local environment.

Nearsightedness Myopia. Nearsightedness is a visual condition in which nearby objects can be seen more clearly than distant objects. Nearsightedness occurs because either the eyeball is elongated or the lens is too thick, causing the image to be focused in front of the retina.

Negative feedback mechanism The homeostatic mechanism in which the outcome of a process feeds back on the system, shutting down the process.

Nephrons Functional units of the kidneys responsible for the formation of urine. These microscopic tubules number 1 million to 2 million per kidney and perform filtration (only certain substances are allowed to pass out of the blood and into the nephron), reabsorption (some useful substances are returned from the nephron to the blood), and secretion (the nephron directly removes wastes and excess materials in the blood and adds them to the filtered fluid that becomes urine).

Nerve A bundle of parallel axons, dendrites, or both from many neurons. A nerve is usually covered with tough connective tissue.

Nervous tissue Tissue consisting of two types of cells, neurons and neuroglia, that make up the brain, spinal cord, and nerves.

Neural tube The embryonic structure that gives rise to the brain and spinal cord.

Neuroglia (neuroglial cells) Cells of the nervous system that support, insulate, and protect nerve cells; also called *glial cells*.

Neuromuscular junction The area of contact between the terminal end of a motor neuron and the cell membrane of a skeletal muscle fiber. When an action potential reaches the terminal end of the motor neuron, acetylcholine is released, triggering events that can lead to muscle contraction.

Neurons Nerve cells involved in intercellular communication. A neuron consists of a cell body, dendrites, and an axon. Neurons are excitable cells in the nervous system specialized to generate and transmit electrochemical signals called action potentials or nerve impulses.

Neurotransmitter A chemical released from the axon tip of a neuron that affects the activity of another cell (usually a nerve, muscle, or gland cell) by altering the electrical potential difference across the membrane of the receiving cell.

Neurula The embryo during neurulation (formation of the brain and spinal cord from ectoderm).

Neurulation A series of events during embryonic development when the central nervous system (brain and spinal cord) forms from the ectoderm. During this period, the embryo is called a neurula.

Neutrophil A phagocytic white blood cell important in defense against bacteria and removal of cellular debris. Most abundant of white blood cells.

Niche The role of a species in an ecosystem. The niche includes the habitat, food, nest sites, and so on. It describes how a member of a particular species uses materials in the environment and how it interacts with other organisms.

Nicotine The psychoactive component of tobacco products.

Nitric oxide A local signaling molecule that dilates blood vessels; also functions as a neurotransmitter.

Nitrification The conversion of ammonia to nitrate (NO_3^-) by nitrifying bacteria living in the soil.

Nitrogen fixation The process of converting nitrogen gas to ammonium (a nitrogen-containing molecule that can be used by living organisms). The process of nitrogen fixation is carried out by nitrogen-fixing bacteria living in nodules on the roots of leguminous plants such as peas and alfalfa.

Node of Ranvier A region of exposed axon between Schwann cells forming a myelin sheath. In myelinated nerves, the impulse jumps from one node of Ranvier to the next, greatly increasing the speed of conduction. This type of transmission is called saltatory conduction.

Nondisjunction Failure of the members of a pair of homologous chromosomes or the sister chromatids to separate during mitosis or meiosis. Nondisjunction results in cells with an abnormal number of chromosomes.

Norepinephrine Noradrenaline. A hormone secreted by the adrenal medulla, along with epinephrine, in response to stress. They initiate the physiological "fight-or-flight" reaction. Norepinephrine is also a neurotransmitter found in both the central and peripheral nervous systems. In the central nervous system, it is important in the regulation of mood, in the pleasure system of the brain, arousal, and dreaming sleep. Norepinephrine is thought to produce an energizing "good" feeling. It is also thought to be essential in hunger, thirst, and sex drive.

Notochord The flexible rod of tissue that develops during gastrulation and signals where the vertebral column will form. The notochord defines the axis of the embryo and gives the embryo some rigidity. During development, vertebrae form around the notochord. The notochord eventually degenerates, existing only as the pulpy, elastic material in the center of intervertebral disks.

Nuclear envelope The double membrane that surrounds the nucleus.

Nuclear pore An opening in the nuclear envelope that permits communication between the nucleus and the cytoplasm.

Nucleolus A specialized region within the nucleus that forms and disassembles during the course of the cell cycle. It plays a role in the generation of ribosomes, organelles involved in protein synthesis.

Nucleoplasm The chromatin and the aqueous environment within the nucleus.

Nucleotide A subunit of DNA composed of one five-carbon sugar (either ribose or deoxyribose), one phosphate group, and one of five nitrogen-containing bases. Nucleotides are the building blocks of nucleic acids (DNA and RNA).

Nucleus The command center of the cell containing almost all the genetic information.

Nutrient A chemical found in food that is essential for growth and function.

O

Obese Having excess body fat that negatively affects health.

Oil glands Glands associated with the hair follicles that produce sebum. Also called *sebaceous glands*.

Olfactory receptor Sensory receptors that respond to odorous molecules; sensory receptors for the sense of smell.

Oligosaccharide A chain of a few monosaccharides (simple sugars) joined together by dehydration synthesis. Disaccharides, formed by the joining of two monosaccharides, are an example.

Omnivore An organism that feeds on a variety of food types, such as plants and animals.

Oocyte A cell whose meiotic divisions will produce an ovum and up to three polar bodies.

Oogenesis The production of ova (eggs), including meiosis and maturation.

Oogonium (plural, oogonia) A germ cell in an ovary that divides, giving rise to oocytes.

Opiate A pain-relieving drug derived from the opium poppy.

Opsin One of several proteins that can be bound to retinal to form the visual pigments in rods and cones.

Optic nerve One of two nerves, one from each eye, responsible for bringing processed electrochemical messages from the retina to the brain for interpretation.

Organ A structure with a specific function composed of two or more different tissues.

Organ of Corti The spiral organ. The portion of the cochlea in the inner ear that contains receptor cells that sense vibrations caused by sound. It is most directly responsible for the sense of hearing.

Organ system A group of organs with a common function.

Organelle A component within a cell that carries out specific functions. Some organelles, such as the nucleus and mitochondria, have membranes, whereas others, such as ribosomes and microtubules, do not.

Origin In reference to a muscle, the end of the muscle that is attached to the bone that remains relatively stationary during a movement.

Osmosis A special case of diffusion in which water moves across the plasma membrane or any other selectively permeable membrane from a region of lower concentration of solute to a region of higher concentration of solute.

Osteoarthritis An inflammation in a joint that is caused by degeneration of the surfaces of the joint by wear and tear.

Osteoblast A bone-forming cell.

Osteoclast A large cell responsible for the breakdown and absorption of bone.

Osteocytes Mature bone cells found in lacunae that are arranged in concentric rings around the central canal. Cytoplasmic projections from osteocytes extend tiny channels that connect with other osteocytes.

Osteon The structural unit of compact bone that appears as a series of concentric circles of lacunae. The lacunae contain bone cells around a central canal containing blood vessels and nerves.

Osteoporosis A decrease in bone density that occurs when the destruction of bone outpaces the formation of new bone, causing bone to become thin, brittle, and susceptible to fracture.

OT See *oxytocin*.

Otoliths Granules of calcium carbonate embedded in gelatinous material in the utricle and saccule of the inner ear. Otoliths cause the gelatin to slide over and bend sensory hair cells when the head is moved. The bending generates nerve impulses that are sent to the brain for interpretation as the position of the head.

Out of Africa hypothesis The idea that modern humans evolved from *Homo erectus* in Africa and later migrated to Europe, Asia, and Australia. It suggests a single origin for *Homo sapiens*. Compare with *multiregional hypothesis*.

Outer ear The external appendage on the outside of the head (pinna) and the canal (the external auditory canal) that extends to the eardrum. It functions as the receiver for sound vibrations.

Outpatient A patient who receives health care without being admitted to a hospital or clinic for an overnight stay. Also referred to as ambulatory care.

Oval window A membrane-covered opening in the inner ear (cochlea) through which vibrations from the stirrup (stapes) are transmitted to fluid within the cochlea.

Ovarian cycle The sequence of events in the ovary that leads to ovulation. The cycle is approximately 28 days long and is closely coordinated with the menstrual cycle.

Ovary One of the female gonads. The female gonads produce the ova (eggs) and the hormones estrogen and progesterone.

Overweight Weighing more than is ideal on a standard height–weight chart. An athletic person may be overweight because of muscle development.

Oviduct One of two tubes that conduct the ova from the ovary to the uterus in the female reproductive system. It is also called a uterine tube or a fallopian tube.

Ovulation The release of the secondary oocyte from the ovary.

Ovum (plural, ova) A mature egg; a large haploid cell that is the female gamete.

Oxygen debt The amount of oxygen required after exercise to oxidize the lactic acid formed during exercise.

Oxyhemoglobin Hemoglobin bound to oxygen.

Oxytocin (OT) The hormone released at the posterior pituitary that stimulates uterine contractions and milk ejection.

P

Pacemaker See *sinoatrial (SA) node*.

Pacinian corpuscle A large encapsulated nerve cell ending that is located deep within the skin and near body organs that responds when pressure is first applied. It is important in sensing vibration.

Pain receptor A sensory receptor that is specialized to detect the physical or chemical damage to tissues that we sense as pain. Pain receptors are sometimes classified as chemoreceptors, because they often respond to chemicals liberated by damaged tissue, and occasionally as mechanoreceptors, because they are stimulated by physical changes, such as swelling, in the damaged tissue.

Palate The roof of the mouth. The front region of the palate, the hard palate, is reinforced with bone. The tongue pushes against the hard palate while mixing food with saliva. The soft palate is farther to the back of the mouth and consists of only muscle. The soft palate prevents food from entering the nose during swallowing.

Pancreas An accessory organ behind the stomach that secretes digestive enzymes, bicarbonate ions to neutralize the acid in chyme, and hormones that regulate blood sugar.

Pancreatic islets Small clusters of endocrine cells in the pancreas; also called *islets of Langerhans*.

Parasympathetic nervous system The branch of the autonomic nervous system that is active during restful conditions. Its effects generally oppose those of the sympathetic nervous system. The parasympathetic nervous system adjusts bodily functions so that energy is conserved during nonstressful times.

Parathormone See *parathyroid hormone*.

Parathyroid glands Four small, round masses at the back of the thyroid gland that secrete parathyroid hormone (parathormone).

Parathyroid hormone (PTH) A hormone released from the parathyroid glands that increases blood calcium levels by stimulating osteoclasts to break down bone. PTH, also called *parathormone*, is secreted when blood calcium levels are too low.

Parental generation The parents—the individuals in the earliest generation under consideration in a genetic cross.

Parkinson's disease A progressive disorder that results from the death of dopamine-producing neurons that lie in the heart of the brain's movement control center, the substantia nigra. Parkinson's disease is characterized by slowed movements, tremors, and muscle rigidity.

Parturition Birth, which usually occurs about 38 weeks after fertilization.

Passive immunity Temporary immune resistance that develops when a person receives antibodies that were produced by another person or animal.

Pathogen A disease-causing organism.

PCR See *polymerase chain reaction*.

PDD-NOS See *autism spectrum disorder*.

Pectoral girdle The bones that connect the arms to the rib cage. The pectoral girdle is composed of the shoulder blades (scapulae) and the collarbones (clavicles).

Pedigree A diagram showing the genetic connections among individuals in an extended family that is often used to trace the expression of a particular trait in that family.

Pelvic girdle The bones that connect the legs to the vertebral column. The pelvic girdle is composed of the paired hipbones.

Pelvic inflammatory disease (PID) A general term for any bacterial infection of a woman's pelvic organs. PID is usually caused by sexually transmitted bacteria, especially those that cause chlamydia and gonorrhea.

Penis The cylindrical external reproductive organ of a male through which most of the urethra extends and that serves to deliver sperm into the female tract during sexual intercourse.

Pepsin A protein-splitting enzyme initially secreted in the stomach in the inactive form of pepsinogen that is activated into pepsin by hydrochloric acid.

Peptic ulcer A local defect in the surface of the stomach or small intestine characterized by dead tissue and inflammation.

Peptide A chain containing only a few amino acids.

Perforin A type of protein released by a natural killer cell that creates numerous pores (holes) in the target cell, making it leaky. Fluid is then drawn into the leaky cell because of the high salt concentration within, and the cell bursts.

Pericardium The fibrous sac enclosing the heart that holds the heart in the center of the chest without hampering its movements.

Perichondrium The layer of dense connective tissue surrounding cartilage that contains blood vessels, which supply the cartilage with nutrients.

Periodontitis An inflammation of the gums and the tissues around the teeth.

Periosteum The membranous covering that nourishes bone.

Peripheral nervous system The part of the nervous system outside the brain and spinal cord. It keeps the central nervous system in continuous contact with almost every part of the body. It is composed of nerves and ganglia. The two branches are the somatic and the autonomic nervous systems.

Peripheral proteins Proteins attached to the inner or outer surface of the plasma membrane.

Peristalsis Rhythmic waves of muscular contraction and relaxation in the walls of hollow tubular organs, such as the digestive organs, that push contents through the tubes.

Pervasive developmental disorder not otherwise specified (PDD-NOS) See *autism spectrum disorder*.

PET scan See *positron emission tomography*.

pH A measure of hydrogen ion concentration of a solution; values range from 0 to 14 on the pH scale.

pH scale A scale for measuring the concentration of hydrogen ions. The scale ranges from 0 to 14. A pH of 7 is neutral, a pH of less than 7 is acidic, and a pH of greater than 7 is basic.

Phagocytes Scavenger cells specialized to engulf and destroy particulate matter, such as pathogens, damaged tissue, or dead cells.

Phagocytosis The process by which cells such as white blood cells ingest foreign cells or substances by surrounding the foreign material with cell membrane. It is a type of endocytosis.

Pharynx The space shared by the respiratory and digestive systems that is commonly called the throat. The pharynx is a passageway for air, food, and liquid.

Phenotype The observable physical and physiological traits of an individual. Phenotype results from the inherited alleles and their interactions with the environment.

Phospholipid An important component of cell membranes. It has a nonpolar "water-fearing" tail (made up of fatty acids) and a polar "water-loving" head (containing an R group, glycerol, and phosphate).

Photoreceptor A sensory receptor specialized to detect changes in light intensity. Photoreceptors are responsible for the sensation we describe as vision.

Phylogenetic trees Generalized descriptions of the history of life. They depict hypotheses about evolutionary relationships among species or higher taxa.

Physical dependence A condition in which continued use of a drug is needed to maintain normal cell functioning.

Pia mater The innermost layer of the meninges (the connective tissue layers that protect the central nervous system).

PID See *pelvic inflammatory disease*.

Piloerection Contraction of the arrector pili muscles causing hairs to stand on end and form a layer of insulation.

Pineal gland The gland that produces the hormone melatonin and is located at the center of the brain.

Pinna The visible part of the ear on each side of the head that gathers sound and channels it to the external auditory canal.

Pinocytosis A type of endocytosis in which cells engulf droplets of fluid and the dissolved substances therein.

Pituitary dwarfism A condition caused by insufficient growth hormone during childhood.

Pituitary gland The endocrine organ connected to the hypothalamus by a short stalk. It consists of the anterior and posterior lobes and is also called the *hypophysis*.

Pituitary portal system The system in which a capillary bed in the hypothalamus connects to veins that lead into a capillary bed in the anterior lobe of the pituitary gland. It allows hormones of the hypothalamus to control the secretion of hormones from the anterior pituitary gland.

Placebo In a controlled experiment to test the effectiveness of a drug, the placebo is a substance that appears to be identical to a drug but has no known effect on the condition for which it is taken.

Placenta The organ that delivers oxygen and nutrients to the embryo and later fetus and carries carbon dioxide and wastes away from each. The placenta is also called the afterbirth.

Placenta previa The condition in which the placenta forms in the lower half of the uterus, entirely or partially covering the cervix. It may cause premature birth or maternal hemorrhage and usually makes vaginal delivery impossible.

Placental stage The third (and final) stage of true labor. It begins with delivery of the baby and ends when the placenta detaches from the wall of the uterus and is expelled from the mother's body.

Plaque A bumpy layer consisting of smooth muscle cells filled with lipid material, especially cholesterol, that bulges into the channel of an artery and reduces blood flow. Another type of plaque is a buildup of food material and bacteria on teeth that leads to tooth decay.

Plasma A straw-colored liquid that makes up about 55% of blood. It serves as the medium for transporting materials within the blood. Plasma consists of water (91% to 93%) with substances dissolved in it (7% to 9%).

Plasma cell The effector cell, produced from a B lymphocyte, that secretes antibodies.

Plasma membrane The thin outer boundary of a cell that controls the movement of substances into and out of the cell.

Plasma proteins Proteins dissolved in plasma, including albumins, which are important in water balance between cells and the blood; globulins, which are important in transporting various substances in the blood; and antibodies, which are important in the immune response.

Plasmid A small, circular piece of self-replicating DNA that is separate from the chromosome and found in bacteria. Plasmids are often used as vectors in recombinant DNA research.

Plasmin An enzyme that breaks down fibrin and dissolves blood clots. Plasmin is formed from plasminogen.

Plasminogen A plasma protein. It is the inactive precursor of plasmin.

Platelet A cell fragment of a megakaryocyte that releases substances necessary for blood clotting. It is formed in the red bone marrow and is sometimes called a thrombocyte.

Platelet plug A mass of platelets clinging to the protein fiber collagen at a damaged blood vessel to prevent blood loss.

Pleiotropy One gene having many effects.

Pluripotent The ability of a cell to differentiate into nearly every cell type.

PMS See *premenstrual syndrome*.

Point mutation A mutation that involves changes in one or a few nucleotides in DNA.

Polar body Any of three small nonfunctional cells produced during the meiotic divisions of an oocyte. The divisions also produce a mature ovum (egg).

Polygenic inheritance Inheritance in which several independently assorting or loosely linked genes determine the expression of a trait.

Polymer A large molecule formed by the joining together of many smaller molecules of the same general type (monomers).

Polymerase chain reaction (PCR) A technique used to amplify (increase) the quantity of DNA in vitro using primers, DNA polymerase, and nucleotides.

Polypeptide A chain containing 10 or more amino acids.

Polysaccharide A complex carbohydrate formed when large numbers of monosaccharides (most commonly glucose) join together to form a long chain through dehydration synthesis. Most polysaccharides store energy or provide structure.

Polysome A cluster of ribosomes simultaneously translating the same messenger RNA (mRNA) strand.

Pons A part of the brain that connects upper and lower levels of the brain.

Population A group of potentially interacting individuals of the same species living in a distinct geographic area.

Population dynamics Changes in population size over time.

Portal system A system whereby a capillary bed drains to veins that drain to another capillary bed.

Positive feedback mechanism The mechanism by which the outcome of a process feeds back on the system, further stimulating the process.

Positron emission tomography (PET) A method that can be used to measure the activity of various brain regions. The person being scanned is injected with a radioactively labeled nutrient, usually glucose, that is tracked as it flows through the brain. The radioisotope emits positively charged particles, called positrons. When the positrons collide with electrons in the body, gamma rays are released. The gamma rays can be detected and recorded by PET receptors. Computers then use the information to construct a PET scan that shows where the radioisotope is being used in the brain.

Postnatal period The period of development after birth. It includes the stages of infancy, childhood, adolescence, and adulthood.

Postsynaptic neuron The neuron located after the synapse. The receiving neuron in a synapse. The membrane of the postsynaptic neuron has receptors specific for certain neurotransmitters.

Precapillary sphincter A ringlike muscle that acts as a valve that opens and closes a capillary bed. Contraction of the precapillary sphincter squeezes the capillary shut and directs blood through a thoroughfare channel to the venule. Relaxation of the precapillary sphincter allows blood to flow through the capillary bed.

Pre-embryo The developing human from fertilization through the second week of gestation (the pre-embryonic period).

Pre-embryonic period The period during prenatal development that extends from fertilization through the second week. Cleavage and implantation follow fertilization.

Prefrontal cortex An association area of the cerebral cortex that is important in decision making.

Premature infant A baby born before 37 weeks of gestation.

Premenstrual syndrome (PMS) A collection of uncomfortable symptoms, including irritability, stress, and bloating, that appears 7 to 10 days before a woman's menstrual period and is associated with hormonal cycling.

Prenatal period The period of development before birth. It is further subdivided into the pre-embryonic period (from fertilization through the second week), the embryonic period (from the third through the eighth weeks), and the fetal period (from the ninth week until birth).

Presynaptic neuron The neuron located before the synapse. The sending neuron in a synapse. The neuron in a synapse that releases neurotransmitter from its synaptic knobs into the synaptic cleft.

Primary consumer A herbivore. An animal that eats primary producers (green plants or algae).

Primary germ layers The layers produced by gastrulation from which all tissues and organs form. The primary germ layers are ectoderm, mesoderm, and endoderm.

Primary motor area A band of the frontal lobe of the cerebral cortex that initiates messages that direct voluntary movements.

Primary productivity In ecosystems, the gross primary productivity is the amount of light energy that is converted to chemical energy in the bonds of organic molecules during a given period. The net primary productivity is the amount of productivity left after the photosynthesizers have used some of the energy stored in organic molecules for their own metabolic activities.

Primary response The immune response that occurs during the body's first encounter with a particular antigen.

Primary somatosensory area A band of the parietal lobe of the cerebral cortex to which information is sent from receptors in the skin regarding touch, temperature, and pain and from receptors in the joints and skeletal muscles.

Primary spermatocyte The original large cell that develops from a spermatogonium during sperm development in the seminiferous tubules. It undergoes meiosis and gives rise to secondary spermatocytes.

Primary structure The precise sequence of amino acids of a protein. This sequence, determined by the genes, dictates a protein's structure and function.

Primary succession The sequence of changes in the species making up a community over time that begins in an area where no community previously existed and ends with a climax community.

Primordial germ cells Cells that migrate from the yolk sac of the developing human to the ovaries or testes, where they differentiate into immature cells that will eventually become oocytes or sperm.

Prion An infectious misfolded version of a host cell protein. It causes disease by causing a host protein to become misfolded and become a prion.

PRL See *prolactin*.

Prodrome The symptoms that precede recurring outbreaks of a disease such as genital herpes.

Producers In ecosystems, the producers are the organisms that convert energy from the physical environment into chemical energy in the bonds of organic molecules through photosynthesis or chemosynthesis. The producers form the first trophic level.

Product A material at the end of a chemical reaction.

Progesterone A sex hormone produced by the corpus luteum in the ovary. Progesterone helps prepare the endometrium (lining) of the uterus for pregnancy and maintains the endometrium.

Programmed cell death A genetically programmed series of events that causes a cell to self-destruct. Also called *apoptosis*.

Prokaryotic cell A cell that lacks a nucleus and other membrane-enclosed organelles. The prokaryotes include bacteria and archaea.

Prolactin (PRL) An anterior pituitary hormone that stimulates the mammary glands to produce milk.

Promoter A specific region on DNA next to the "start" sequence that controls the expression of the gene.

Prophase In mitosis, the phase when the chromosomes begin to thicken and shorten, the nucleolus disappears, the nuclear envelope begins to break down, and the mitotic spindle forms in the cytoplasm.

Prostaglandins The lipid molecules found in and released by the plasma membranes of most cells. They are often called local hormones (or local signaling molecules) because they exert their effects on the secreting cells themselves or on nearby cells.

Prostate gland An accessory reproductive gland in males that surrounds the urethra as it passes from the bladder. Its secretions contribute to semen and serve to activate the sperm and to counteract the acidity of the female reproductive tract.

Proteins The macromolecules composed of amino acids linked by peptide bonds. The functions of proteins include structural support, transport, movement, and regulation of chemical reactions.

Proto-oncogene A healthy gene that promotes cell division. An oncogene is a mutant proto-oncogene that can lead to the development of cancer.

Prothrombin A plasma protein synthesized by the liver that is important in blood clotting. It is converted to an active form (thrombin) by thromboplastin that is released from platelets.

Prothrombin activator A blood protein that converts prothrombin to thrombin as part of the blood-clotting process.

Protozoans A group of single-celled organisms with a well-defined eukaryotic nucleus. Protozoans can cause disease by producing toxins or by releasing enzymes that prevent host cells from functioning normally.

Proximal convoluted tubule The section of the renal tubule where reabsorption and secretion occur.

Psychoactive drug A drug that alters one's mood or emotional state.

PTH See *parathyroid hormone*.

Puberty The time when secondary sexual characteristics such as pubic and underarm hair develop. This period usually occurs slightly earlier in girls (from 12 to 15 years of age) than in boys (from 13 to 16 years of age).

Pulmonary arteries Blood vessels that carry blood low in oxygen from the right ventricle to the lungs, where it is oxygenated.

Pulmonary circuit (or circulation) The pathway that transports blood from the right ventricle of the heart to the lungs and back to the left atrium of the heart.

Pulmonary veins Blood vessels that carry oxygenated blood from the lungs to the left atrium of the heart.

Pulp The center of a tooth that contains the tooth's life-support systems.

Pulse The rhythmic expansion of an artery created by the surge of blood pushed along the artery by each contraction of the ventricles of the heart. With each beat of the heart, the wave of expansion begins, moving along the artery at the rate of 6 to 9 meters per second.

Punnett square A diagrammatic method used to determine the probable outcome of a genetic cross. The possible allele combinations in the gametes of one parent are used to label the columns, and the possible allele combinations of the other parent are used to label the rows. The alleles of each column and each row are then paired to determine the possible genotypes of the offspring.

Pupil The small hole through the center of the iris through which light passes to enter the eye. The size of the pupil is altered to regulate the amount of light entering the eye.

Purkinje fibers The specialized cardiac muscle cells that deliver an electrical signal from the atrioventricular bundle to the individual heart muscle cells in the ventricles.

Pyloric sphincter A ring of muscle between the stomach and small intestine that regulates the emptying of the stomach into the small intestine.

Pyramid of biomass A diagram in which blocks represent the amount of biomass (dry body mass of organisms) available at each trophic (feeding) level.

Pyramid of energy A diagram in which blocks represent the decreasing amount of energy available at each trophic (feeding) level.

Pyramid of numbers A diagram of the number of individuals at each trophic level.

Pyrogen A fever-producing substance.

Pyruvate The three-carbon compound produced by glycolysis, which is the first phase of cellular respiration.

Q

Quaternary structure The shape of an aggregate protein. It is determined by the mutually attractive forces between the protein's subunits.

R

Radioisotopes Isotopes that are unstable and spontaneously decay, emitting radiation in the form of gamma rays and alpha and beta particles.

RAS See *reticular activating system*.

RDS See *respiratory distress syndrome*.

Receptor A protein molecule located in the cytoplasm and on the plasma membrane of cells that is sensitive to chemical messengers.

Receptor potential An electrochemical message (a change in the degree of polarization of the membrane) generated in a sensory receptor in response to a stimulus. Receptor potentials vary in magnitude with the strength of the stimulus.

Recessive allele The allele whose effects are masked in the heterozygous condition. The recessive allele often produces a nonfunctional protein.

Recombinant DNA Segments of DNA from two sources that have been combined in vitro and transferred to cells in which their information can be expressed.

Recruitment A process of increasing the strength of muscle contraction by increasing the number of motor units being stimulated.

Rectum The final section of the gastrointestinal tract. The rectum receives and temporarily stores the feces.

Red blood cell See *erythrocyte*.

Red marrow Blood cell–forming connective tissue found in the marrow cavity of certain bones.

Reduction division The first meiotic division (meiosis I) that produces two cells, each of which contains one member of each homologous pair (23 chromosomes with replicates attached in humans).

Referred pain Pain felt at a site other than the area of origin.

Reflex A simple, stereotyped reaction to a stimulus.

Reflex arc A neural pathway consisting of a sensory receptor, a sensory neuron, usually at least one interneuron, a motor neuron, and an effector.

Refractory period The interval following an action potential during which a neuron cannot be stimulated to generate another action potential.

Relaxin The hormone released from the placenta and the ovaries. It initiates labor and facilitates delivery by dilating the cervix and relaxing the ligaments and cartilage of the pubic bones.

Releasing hormone A hormone that stimulates hormone secretion by another gland.

Renal corpuscle The portion of the nephron where water and small solutes are filtered from the blood. It consists of the glomerulus and Bowman's (glomerular) capsule.

Renal cortex The outer region of the kidney, containing renal columns.

Renal failure A decrease or complete cessation of glomerular filtration.

Renal medulla The inner region of the kidney. It contains cone-shaped structures called renal pyramids.

Renal pelvis The innermost region of the kidney; the chamber within the kidney.

Renal tubule The site of reabsorption and secretion by the nephron. It consists of the proximal convoluted tubule, the loop of the nephron (also called the loop of Henle), and the distal convoluted tubule.

Renin An enzyme released by cells of the juxtaglomerular apparatus of nephrons. Renin converts angiotensinogen, a protein produced by the liver and found in the plasma, into another protein, angiotensin I. These actions of renin initiate a series of hormonal events that leads to increased reabsorption of sodium and water by the distal convoluted tubules and collecting ducts of nephrons.

Rennin The gastric enzyme that breaks down milk proteins.

Replication Copying from a template, as occurs during the synthesis of new DNA from preexisting DNA.

Repolarization The return of the membrane potential to approximately its resting value. Repolarization of the nerve cell membrane during an action potential occurs because of the outflow of potassium ions.

RER See *rough endoplasmic reticulum*.

Residual volume The amount of air that remains in the lungs after a maximal exhalation.

Respiratory distress syndrome (RDS) A condition in newborns caused by an insufficient amount of surfactant, causing the alveoli of the lungs to collapse and thereby making breathing difficult.

Respiratory system The organ system that carries out gas exchange. The respiratory system includes the nose, pharynx, larynx, trachea, bronchi, and lungs.

Resting potential The separation of charge across the plasma membrane of a neuron when the neuron is not transmitting an action potential. It is caused primarily by the unequal distributions of sodium ions, potassium ions, and large negatively charged proteins on either side of the plasma membrane. The resting potential of a neuron is about −70 mV.

Restriction enzyme An enzyme that recognizes a specific sequence of bases in DNA and cuts the DNA at that sequence. Restriction enzymes are used to prepare DNA containing "sticky ends" during the creation of recombinant DNA. Their natural function in bacteria is to control the replication of viruses that infect the bacteria.

Reticular activating system (RAS) An extensive network of neurons that runs through the medulla and projects to the cerebral cortex. It filters out unimportant sensory information before it reaches the brain and controls changing levels of consciousness.

Reticular fibers Interconnected strands of collagen in certain connective tissues that branch extensively. Networks of reticular fibers support soft tissues, including the liver and spleen.

Retina The light-sensitive innermost layer of the eye containing numerous photoreceptors (rods and cones).

Retinal The light-absorbing portion of pigment molecules in the photoreceptors. Retinal combines with one of four opsins (proteins) to form the light-absorbing pigments in rods and cones.

Retrovirus Any one of the viruses that contain only RNA and carry out transcription from RNA to DNA (reverse transcription).

Rh factor A group of antigens found on the surface of the red blood cells of most people. A person who has these antigens is said to be Rh-positive. A person who lacks these antigens is Rh-negative.

Rheumatoid arthritis Inflammation of a joint caused by an autoimmune response. It is marked by inflammation of the synovial membrane and excess synovial fluid accumulation in the joints, causing swelling, pain, and stiffness.

Rhodopsin The light-absorbing pigment in the photoreceptors called rods. Rhodopsin is responsible for black and white vision.

Rhythm method of birth control A method of reducing the risk of pregnancy by avoiding intercourse on all days when sperm and egg might meet.

Ribonucleic acid (RNA) A single-stranded nucleic acid that contains ribose (a five-carbon sugar), phosphate, adenine, uracil, cytosine, or guanine. RNA plays a variety of roles in protein synthesis.

Ribosomal RNA (rRNA) A type of RNA that combines with proteins to form the ribosomes, structures on which protein synthesis occurs. The most abundant form of RNA.

Ribosome The site where protein synthesis occurs in a cell. It consists of two subunits, each containing ribosomal RNA and proteins.

RNA (ribonucleic acid) A single-stranded nucleic acid that contains ribose (a five-carbon sugar), phosphate, adenine, uracil, cytosine, or guanine. RNA plays a variety of roles in protein synthesis.

RNA polymerase One of the group of enzymes necessary for the synthesis of RNA from a DNA template. It binds with the promoter on DNA that aligns the appropriate RNA nucleotides and links them together.

Rods Photoreceptors containing rhodopsin and responsible for black and white vision. Rods are extremely sensitive to light.

Root The part of a tooth that is below the gum line. It is covered with a calcified, yet living and sensitive, connective tissue called cementum.

Root canal A channel through the root of a tooth that contains the blood vessels and nerves.

Rough endoplasmic reticulum (RER) Endoplasmic reticulum that is studded with ribosomes. It produces membrane.

Round window A membrane-covered opening in the cochlea that serves to relieve the pressure caused by the movements of the oval window.

rRNA See *ribosomal RNA*.

Rugae Folds in the mucosa of the lining of the empty stomach's walls that can unfold, allowing the stomach to expand as it fills.

S

SAD See *seasonal affective disorder*.

Salinization An accumulation of salts in soil caused by irrigation over a long period of time that makes the land unusable.

Saliva The secretion from the salivary glands that helps moisten and dissolve food particles in the mouth, facilitating taste and digestion. An enzyme in saliva (salivary amylase) begins the chemical digestion of starch.

Salivary amylase An enzyme in saliva that begins the chemical digestion of starches, breaking them into shorter chains of sugars.

Salivary glands Exocrine glands in the facial region that secrete saliva into the mouth to begin the digestion process.

Saltatory conduction The type of nerve transmission along a myelinated axon in which the nerve impulse jumps from one node of Ranvier to the next. Saltatory conduction greatly increases the speed of nerve conduction.

Sarcomere The smallest contractile unit of a striated or cardiac muscle cell.

Sarcoplasmic reticulum An elaborate form of smooth endoplasmic reticulum found in muscle fibers. The sarcoplasmic reticulum takes up, stores, and releases calcium ions as needed in muscle contraction.

Schizophrenia A mental illness characterized by hallucinations and disordered thoughts and emotions that is caused by excessive activity at dopamine synapses in the midbrain. As a result, dopamine is no longer in the proper balance with glutamate, a neurotransmitter released by neurons in the cerebral cortex.

Schwann cell A type of glial cell in the peripheral nervous system that forms the myelin sheath by wrapping around the axon many times. The myelin sheath insulates axons, increases the speed at which impulses are conducted, and assists in the repair of damaged neurons.

Science A systematic approach to acquiring knowledge through carefully documented investigation and experimentation.

Scientific method A procedure underlying most scientific investigations that involves observation, formulating a hypothesis, making predictions, experimenting to test the predictions, and drawing conclusions. Experimentation usually involves a control group and an experimental group that differ in one or very few factors (variables). New hypotheses may be generated from the results of experimentation.

Sclera The white part of the eye that protects and shapes the eyeball and serves as a site of attachment for muscles that move the eye.

Scrotum A loose fleshy sac containing the testes.

Seasonal affective disorder (SAD) A form of depression associated with winter months when overproduction of melatonin is triggered by short day length.

Sebaceous glands See *oil glands*.

Sebum An oily substance made of fats, cholesterol, proteins, and salts secreted by the oil glands.

Second messenger A molecule in the cytoplasm of a cell that is activated when a water-soluble hormone binds to a receptor on the surface of the cell. Second messengers influence the activity of enzymes and ultimately the activity of the cell to produce the effect of the hormone.

Secondary consumer A carnivore. An animal that obtains energy by eating other animals.

Secondary oocyte A haploid cell formed by meiotic division of a primary oocyte. It is released from an ovary at ovulation.

Secondary response The immune response during the body's second or subsequent exposures to a particular antigen. The secondary immune response is much quicker than is the primary response because memory cells specific for the antigen are present.

Secondary spermatocyte A haploid cell formed by meiotic division of a primary spermatocyte during sperm development in the seminiferous tubules.

Secondary structure The bending and folding of the chain of amino acids of a protein to produce shapes such as coils, spirals, and pleated sheets. These shapes form as a result of hydrogen bonding between different parts of the polypeptide chain.

Secondary succession The sequence of changes in the species making up a community over time that takes place after some disturbance destroys the existing life. Soil is already present.

Secretin A hormone released by the small intestine that inhibits the secretion of gastric juice and stimulates the release of bicarbonate ions from the pancreas and the production of bile in the liver.

Segregation, law of A genetic principle that states that the alleles for each gene separate (segregate) during meiosis and gamete formation, so half of the gametes bear one allele and the other half bear the other allele.

Selectively permeable A characteristic of the plasma membrane because it permits some substances to move across and denies access to others.

Semen The fluid expelled from the penis during male orgasm. Semen consists of sperm and the secretions of the accessory glands.

Semicircular canals Three canals in each ear that are oriented at right angles to one another and contain sensory receptors that precisely monitor any sudden movement of the head. They detect body position and movement.

Semiconservative replication Replication of DNA; the two strands of a DNA molecule become separated and each serves as a template to create a new double-stranded DNA. Each new double-stranded molecule consists of one new strand and one old strand.

Semilunar valves Heart valves located between each ventricle and its connecting artery that prevent the backflow of blood from the artery to the ventricle. Whereas the cusps of the atrioventricular (AV) valves are flaps of connective tissue, those of the semilunar valves are small pockets of tissue attached to the inner wall of their respective arteries.

Seminal vesicles A pair of male accessory reproductive glands located posterior to the urinary bladder. Their secretions contribute to semen and serve to nourish the sperm cells, reduce acidity in the vagina, and coagulate sperm.

Seminiferous tubules Coiled tubules within the testes where sperm are produced.

Sensory adaptation A gradual decline in the responsiveness of a sensory receptor that results in a decrease in awareness of the stimulus.

Sensory (afferent) neuron A nerve cell specialized to conduct nerve impulses from the sensory receptors *toward* the central nervous system.

Sensory receptors The structures specialized to respond to changes in their environment (stimuli) by generating electrochemical messages that are eventually converted to nerve impulses if the stimulus is strong enough. The nerve impulses are then conducted to the brain, where they are interpreted to build our perception of the world.

Septum (of heart) A wall that separates the right and left sides of the heart.

SER See *smooth endoplasmic reticulum*.

Serosa A thin layer of connective tissue that forms the outer layer of the gastrointestinal tract. It secretes a fluid that reduces friction between contacting surfaces.

Serotonin A neurotransmitter in the central nervous system thought to promote a generalized feeling of well-being.

Serous membranes Sheets of epithelial tissue that line the thoracic and abdominal cavities and the organs within them. Serous membranes secrete a fluid that lubricates the organs within these cavities.

Sex chromosomes The X and Y chromosomes. The pair of chromosomes involved in determining gender.

Sex-influenced inheritance An autosomal genetic trait that is expressed differently in males and females, usually because of the presence of sex hormones.

Sex-linked gene A gene located on the X chromosome.

Sexual dimorphism A difference in appearance between males and females within a species.

Short-term memory Memory of new information that lasts for a few seconds or minutes.

Sickle-cell anemia A type of anemia caused by a mutation that results in a change in one amino acid in a globin chain of hemoglobin (the iron-containing protein in red blood cells that transports oxygen). Such a change causes the red blood cell to assume a crescent (sickle) shape when oxygen levels are low. The sickle-shaped cells clog small blood vessels, leading to pain and tissue damage from insufficient oxygen.

Simple diffusion The spontaneous movement of a substance from a region of higher concentration to a region of lower concentration.

Simple epithelial tissue Epithelial tissue with only a single layer of cells.

Simple goiter An enlarged thyroid gland caused by iodine deficiency.

Sinoatrial (SA) node A region of specialized cardiac muscle cells located in the right atrium near the junction of the superior vena cava that sets the pace of the heart rate at about 70 to 80 beats a minute. Also known as the *pacemaker*. The SA node sends out an electrical signal that spreads through the muscle cells of the atria, causing them to contract.

Sinuses Large, air-filled spaces in the bones of the face.

Sinusitis Inflammation of the mucous membranes of the sinuses making it difficult for the sinuses to drain their mucous fluid.

Skeletal muscle A contractile tissue. One of three types of muscle in the body. Skeletal muscle cells are cylindrical, have many nuclei, and have stripes (striations). Skeletal muscle provides for conscious, voluntary control over contraction. It attaches to bones and forms the muscles of the body. Also called *striated muscle*.

Skeleton A framework of bones and cartilage that functions to support and protect internal organs and to permit body movement.

Sliding filament model A model of the mechanism of muscle contraction in which the myofilaments actin and myosin slide across one another, causing a sarcomere to shorten. When enough sarcomeres shorten, the muscle contracts.

Slow-twitch muscle cells Muscle fibers that are specialized to contract slowly but with incredible endurance when stimulated. They contain an abundant supply of myoglobin and mitochondria and are richly supplied with capillaries. They depend on aerobic pathways to generate adenosine triphosphate (ATP) during muscle contraction.

Small intestine The organ located between the stomach and large intestine responsible for the final digestion and absorption of nutrients.

Smooth endoplasmic reticulum (SER) Endoplasmic reticulum without ribosomes. It produces membrane and detoxifies drugs.

Smooth muscle A contractile tissue characterized by the lack of visible striations and by unconscious control over its contraction. It is found in the walls of blood vessels and airways and in organs such as the stomach, intestines, and bladder.

Sodium-potassium pump A molecular mechanism in a plasma membrane that uses cellular energy in the form of adenosine triphosphate (ATP) to pump ions against their concentration gradients. Typically, each pump ejects three sodium ions from the cell while bringing in two potassium ions.

Soluble fiber A type of dietary fiber that either dissolves or swells in water. This type of fiber includes the pectins, gums, mucilages, and some hemicelluloses. Soluble fiber has a gummy consistency.

Somatic cells All body cells except for gametes (egg and sperm). Somatic cells contain the diploid number of chromosomes, which in humans is 46.

Somatic nervous system The part of the peripheral nervous system that carries information to and from the central nervous system, resulting in voluntary movement and sensations.

Somites Blocks formed from mesoderm cells of the developing embryo that eventually form skeletal muscles of the neck and trunk, connective tissues, and vertebrae.

Source (donor) DNA DNA containing the "gene of interest" that will be combined with host DNA to form recombinant DNA.

Special senses The sensations of smell, taste, vision, hearing, and the sense of balance or equilibrium.

Speciation The formation of a new species.

Species A population or group of populations whose members are capable of successful interbreeding under natural conditions. Such interbreeding must produce fertile offspring.

Sperm A mature male gamete. A spermatozoon.

Spermatid A haploid cell that is formed by mitotic division of a haploid secondary spermatocyte and that develops into a spermatozoon.

Spermatocyte A cell developed from a spermatogonium during sperm development in the seminiferous tubules.

Spermatogenesis The series of events within the seminiferous tubules that gives rise to physically mature sperm from germ cells. It involves meiosis and maturation.

Spermatogonium (plural, spermatogonia) The undifferentiated male germ cells in the seminiferous tubules that give rise to spermatocytes.

Spermicides A means of contraception that consists of sperm-killing chemicals in some form of a carrier, such as foam, cream, jelly, film, or tablet.

Sphincter A ring of muscle between regions of a system of tubes that controls the flow of materials from one region to another past the sphincter.

Sphygmomanometer A device for measuring blood pressure. A sphygmomanometer consists of an inflatable cuff that wraps around the upper arm attached to a device that can measure the pressure within the cuff.

Spina bifida A birth defect in which the neural tube fails to develop and close properly. The mother's taking vitamins and folic acid before conception appears to reduce the chance of having a baby with spina bifida. Some cases can be improved with surgery.

Spinal cord A tube of neural tissue that is continuous with the medulla at the base of the brain and extends about 45 cm (17 in.) to just below the last rib. It conducts messages between the brain and the rest of the body and serves as a reflex center.

Spinal nerves Thirty-one pairs of nerves that arise from the spinal cord. Each spinal nerve services a specific region of the body. Spinal nerves carry both sensory and motor information.

Spiral organ (of Corti) The portion of the cochlea in the inner ear that contains receptor cells that sense vibrations caused by sound. It is most directly responsible for the sense of hearing.

Spleen The largest lymphoid organ; it contains a reservoir of lymphocytes and removes old or damaged red blood cells from the blood.

Spongy bone The bone formed from a latticework of thin struts of bone with marrow-filled areas between the struts. It is found in the ends of long bones and within the breastbone, pelvis, and bones of the skull. Spongy bone is less dense than compact bone and is made of an irregular network of collagen fibers surrounded by a calcium matrix.

Sprain Damage to a ligament (a strap of connective tissue that holds bones together).

Squamous cell carcinoma The second most common form of skin cancer that arises in the newly formed skin cells as they flatten and move toward the skin surface.

Squamous epithelium A type of epithelial tissue composed of flattened cells. It forms linings and coverings.

Stapes The last of three small bones in the middle ear that transmit information about sound from the eardrum to the inner ear. The stapes is also known as the stirrup.

Starch The storage polysaccharide in plants.

Statistical significance In a scientific experiment, statistical significance is a measure of the probability that the results are due to chance.

Stem cell A type of cell that divides continuously and can give rise to other types of cells.

Steroid A lipid, such as cholesterol, consisting of four carbon rings with functional groups attached.

Steroid hormones A group of closely related hormones chemically derived from cholesterol and secreted primarily by the ovaries, testes, and adrenal glands.

Stimulus Changes in the internal or external environment that a sensory receptor can detect and respond to by generating electrochemical messages.

Stomach A muscular sac that is well designed for the storage of food, the liquefaction of food, and the initial chemical digestion of proteins.

Stratified epithelial tissue Epithelial tissue with several layers of cells.

Strep throat A sore throat that is caused by *Streptococcus* bacteria.

Stress incontinence A mild form of incontinence characterized by the escape of small amounts of urine when sudden increases in abdominal pressure force urine past the external urethral sphincter.

Striated muscle See *skeletal muscle*.

Stroke A cerebrovascular accident. A condition in which nerve cells die because the blood supply to a region of the brain is shut off, usually because of hemorrhage or atherosclerosis. The extent and location of the mental or physical impairment caused by a stroke depend on the region of the brain involved.

Submucosa The layer of the digestive tract between the mucosa and the muscularis; the submucosa contains blood vessels, lymph vessels, and nerves.

Substrate The material on which an enzyme works.

Succession, ecological The sequence of changes in the species making up a community over time.

Summation (of muscle contraction) A phenomenon that results when a muscle is stimulated to contract before it has time to completely relax from a previous contraction. The response to each stimulation builds on the previous one.

Superovulation The ovulation of several oocytes. It is usually prompted by administration of hormones.

Suppressor T cell A type of T lymphocyte that turns off the immune response when the level of antigen falls by releasing chemicals that dampen the activity of both B cells and T cells.

Surface-to-volume ratio The physical relationship that dictates that increases in the volume of a cell occur faster than increases in its surface area. This relationship explains why most cells are small.

Surfactant Phospholipid molecules coating the alveolar surfaces that prevent the alveoli from collapsing.

Suture An immovable joint between the bones of the skull.

Sweat glands Exocrine glands found in the skin. One type of sweat gland is functional throughout life and releases sweat onto the surface of the skin. Another type releases its secretions into hair follicles and becomes functional at puberty.

Sympathetic nervous system The branch of the autonomic nervous system responsible for the "fight-or-flight" responses that occur during stressful or emergency situations. Its effects are generally opposite to those of the parasympathetic nervous system.

Synapse The site of communication between a neuron and another cell, such as another neuron or a muscle cell.

Synapsis The physical association of homologous pairs of chromosomes that occurs during prophase I of meiosis. The term literally means "bringing together."

Synaptic cleft The gap between two cells forming a synapse, for example, two communicating nerve cells.

Synaptic knob A small bulblike swelling of an axon terminal that releases neurotransmitter. An axon terminal.

Synaptic vesicle A tiny membranous sac containing molecules of a neurotransmitter. Synaptic vesicles are located in the synaptic knobs of axon endings, and they release their contents when an action potential reaches the synaptic knob.

Synergistic muscles Two or more muscles that work together to cause movement in the same direction.

Synovial cavity A fluid-filled space surrounding a synovial joint formed by a double-layered capsule. The fluid within the cavity is called synovial fluid.

Synovial fluid A viscous, clear fluid within a synovial cavity that acts as both a shock absorber and a lubricant between the bones.

Synovial joint A freely movable joint. A synovial joint is surrounded by a fluid-filled cavity. This is the most abundant type of joint in the body.

Synovial membranes Membranes that line the cavities of freely movable joints and secrete a fluid that lubricates the joint.

Syphilis A sexually transmitted disease caused by the bacterium *Treponema pallidum*. If untreated, it can progress through three stages and cause death. The first stage is characterized by a painless craterlike bump called a chancre that forms at the site where the bacterium entered the body. The second stage involves a rash covering the body, and the third stage is characterized by gummas.

Systematic biology The discipline that deals with the naming, classification, and evolutionary relationships of organisms. It is also called systematics.

Systemic circuit (of circulation) The pathway of blood from the left ventricle of the heart to the cells of the body and back to the right atrium.

Systole Contraction of the heart. Atrial systole is contraction of the atria. Ventricular systole is contraction of the ventricles.

Systolic pressure The highest pressure in an artery during each heartbeat. The higher of the two numbers in a blood pressure reading. In a typical, healthy adult, the systolic pressure is about 120 mm Hg.

T

T cell See *T lymphocyte*.

T lymphocyte T cell. A type of white blood cell. Some T lymphocytes attack and destroy cells that are not recognized as belonging in the body, such as an infected cell or a cancerous cell.

T4 cell A helper T cell. The kind of T lymphocyte that serves as the main switch for the entire immune response by presenting the antigen to B cells and by secreting chemicals that stimulate other cells of the immune system.

Tanning The buildup of melanin in the skin in response to ultraviolet (UV) exposure.

Target cell A cell with receptors that recognize and bind a specific hormone.

Taste bud A structure consisting of receptors responsible for the sense of taste surrounded by supporting cells. Taste buds are located primarily on the surface epithelium and certain papillae of the tongue.

Taste hairs Microvilli that project into a pore at the tip of the taste bud and bear the receptors for certain chemicals found in food.

Taxon (plural, taxa) A taxonomic group, such as a genus, family, or order.

TB See *tuberculosis*.

Tectorial membrane A membrane that forms the roof of the spiral organ of Corti (the actual organ of hearing). It projects over and is in contact with the sensory hair cells. Pressure waves caused by sound cause the sensory cells to push against the tectorial membrane and bend, resulting in nerve impulses that are carried to the brain by the auditory nerve.

Telomerase The enzyme that synthesizes telomeres.

Telomere Pieces of DNA at the tips of chromosomes that protect the ends of the chromosomes.

Telophase In mitosis, the phase when a nuclear envelope forms around the group of chromosomes at each pole, the mitotic spindle is disassembled, and nucleoli reappear. The chromosomes also become less condensed and more threadlike in appearance.

Temporomandibular joint (TMJ) syndrome A group of symptoms including headaches, toothaches, and earaches caused by physical stress on the mandibular joint.

Tendinitis Inflammation of a tendon caused by excessive stress on the tendon.

Tendon A band of connective tissue that connects muscle to bone.

Tertiary structure The three-dimensional shape of proteins formed by hydrogen, ionic, and covalent bonds between different side chains.

Testes The male gonads. The male reproductive organs that produce sperm and the hormone testosterone.

Testosterone A sex hormone needed for sperm production and the maintenance of male reproductive structures. Testosterone is produced primarily by the interstitial cells of the testes.

Tetanus A smooth, sustained contraction of muscle caused when stimuli are delivered in such rapid succession that there is no time for muscle relaxation.

TH See *thyroid hormone*.

Thalamus A brain structure located below the cerebral hemispheres that is important in sensory experience, motor activity, stimulation of the cerebral cortex, and memory.

Theory A broad-ranging explanation for some aspect of the universe that is consistent with many observations and experiments.

Thermoreceptor A sensory receptor specialized to detect changes in temperature.

Threshold The degree to which the voltage difference across the plasma membrane of a neuron or other excitable cell must change to trigger an action potential.

Thrombin A plasma protein important in blood clotting that is formed from prothrombin by thromboplastin. It converts fibrinogen to fibrin, which forms a web that traps blood cells and forms the clot.

Thrombocyte See *platelet*.

Thrombus A stationary blood clot that forms in the blood vessels. A thrombus can block blood flow.

Thymopoietin A hormone produced by the thymus gland that promotes the maturation of T lymphocytes.

Thymosin A hormone produced by the thymus gland that promotes the maturation of T lymphocytes.

Thymus gland A gland located on the top of the heart that secretes the hormones thymopoietin and thymosin. It decreases in size as we age.

Thyroid gland The shield-shaped structure at the front of the neck that synthesizes and secretes thyroid hormone and calcitonin.

Thyroid hormone (TH) A hormone released by the thyroid gland that regulates blood pressure and the body's metabolic rate and production of heat. It also promotes normal development of several organ systems.

Thyroid-stimulating hormone (TSH) The anterior pituitary hormone that acts on the thyroid gland to stimulate the synthesis and release of thyroid hormones.

Tidal volume The amount of air inhaled or exhaled during a normal breath.

Tight junction A type of junction between cells in which the membranes of neighboring cells are attached, forming a seal to prevent fluid from flowing across the epithelium through the minute spaces between adjacent cells.

Tissue A group of cells that work together to perform a common function.

Tolerance A progressive decrease in the effectiveness of a drug with continued use.

Tongue The large skeletal muscle studded with taste buds that aids in speech and eating.

Total lung capacity The total volume of air contained in the lungs after the deepest possible breath. It is calculated by adding the residual volume to the vital capacity.

Totipotent The ability of a cell to differentiate into any cell type in that organism. A fertilized egg is totipotent.

Trabecula (plural, trabeculae) A supporting bar or strand of spongy bone that forms an internal strut that braces the bone from within.

Trachea The tube that conducts air into the thoracic cavity toward the lungs. The trachea is reinforced with C-shaped rings of cartilage to prevent it from collapsing during inhalation and exhalation.

Trait A phenotypically expressed characteristic.

Transcription The process by which a complementary single-stranded messenger RNA (mRNA) molecule is formed from a single-stranded DNA template. As a result, the information in DNA is transferred to RNA.

Transfer RNA (tRNA) A type of RNA that binds to a specific amino acid and transports it to the appropriate region of messenger RNA (mRNA). Transfer RNA acts as an interpreter between the nucleic acid language of mRNA and the amino acid language of proteins.

Transgenic organism An organism that contains certain genes from another species that code for a desired trait. It can be created, for example, by injecting foreign DNA into an egg cell or an early embryo.

Transition reaction The phase of cellular respiration that follows glycolysis and involves pyruvate reacting with coenzyme A (CoA) in the matrix of the mitochondrion to form acetyl CoA. The acetyl CoA then enters the citric acid cycle.

Translation Protein synthesis. The process of converting the nucleotide language of messenger RNA (mRNA) into the amino acid language of a protein.

Transverse tubules T tubules. The tiny, cylindrical inpocketings of the muscle fiber's plasma membrane that carry nerve impulses to almost every sarcomere.

Tricuspid valve A heart valve located between the right atrium and right ventricle that prevents the backflow of blood from the ventricle to the atrium. It is also called the right *atrioventricular (AV) valve*.

Triglycerides The lipids composed of one molecule of glycerol and three fatty acids. They are known as fats when solid and oils when liquid.

Trisomy A condition in which there are three representatives of a chromosome instead of only two representatives.

tRNA See *transfer RNA*.

Trophic level The feeding level of one or more populations in a food web. The producers form the first trophic level. Herbivores, which eat the producers, form the second trophic level. Carnivores that eat herbivores form the third trophic level. Carnivores that eat other carnivores form the fourth trophic level.

Trophoblast A group of cells within the blastocyst that gives rise to the chorion, the extraembryonic membrane that will become part of the placenta.

Tropic hormone A hormone that influences another endocrine gland.

Tropomyosin A protein on the thin (actin) filaments in muscle cells that works with troponin to prevent actin and myosin from binding in the absence of calcium ions.

Troponin A protein on the thin (actin) filaments in muscle cells that works with tropomyosin to prevent actin and myosin from binding in the absence of calcium ions.

TSH See *thyroid-stimulating hormone*.

Tubal ligation A sterilization procedure in females in which each oviduct is cut and sealed to prevent sperm from reaching the eggs.

Tubal pregnancy An ectopic pregnancy in which the embryo implants in an oviduct. This is the most common type of ectopic pregnancy.

Tuberculosis (TB) A highly contagious disease caused by a rod-shaped bacterium, *Mycobacterium tuberculosis*. TB is spread by coughing.

Tubular reabsorption The process by which useful materials are removed from the filtrate within the renal tubule and returned to the blood.

Tubular secretion The process by which wastes and excess ions that escaped glomerular filtration are removed from the blood and added to the filtrate within the renal tubule.

Tumor A neoplasm. An abnormal growth of cells. A tumor forms from the new growth of tissue in which cell division is uncontrolled and progressive.

Tumor suppressor gene A gene that codes for a protein that suppresses cancer in its normal, healthy form.

Turner syndrome A genetic condition resulting from nondisjunction of the sex chromosomes in which a person has 22 pairs of autosomes and a single, unmatched X chromosome (XO). The person has a female appearance.

Tympanic membrane The eardrum. A membrane that forms the outer boundary of the middle ear and that vibrates in response to sound waves. Vibrations of the tympanic membrane are transferred through the middle ear by three small bones (malleus, incus, and stapes; also known as the hammer, anvil, and stirrup) to the inner ear, where hearing occurs when neural messages are generated in response to the pressure waves caused by the vibrations.

Type 1 diabetes mellitus An autoimmune disorder characterized by abnormally high glucose in the blood due to insufficient production of insulin by cells in the pancreas. It cannot be prevented, and there is no cure at this time.

Type 2 diabetes mellitus A condition characterized by increased resistance to insulin by body cells. It can be prevented or delayed through changes in lifestyle.

U

Umbilical cord The ropelike connection between the embryo (and later the fetus) and the placenta. It consists of blood vessels (two umbilical arteries and one umbilical vein) and supporting connective tissue.

Undernourishment Starvation. A form of hunger that occurs when inadequate amounts of food are eaten.

Ureters Tubular organs that carry urine from the kidneys to the urinary bladder.

Urethra The muscular tube that transports urine from the floor of the urinary bladder to the outside of the body. In males, it also conducts sperm from the vas deferens out of the body through the penis.

Urethritis Inflammation of the urethra caused by bacteria.

Urinalysis An analysis of the volume, microorganism content, and physical and chemical properties of urine.

Urinary bladder The muscular organ that temporarily stores urine until it is released from the body.

Urinary incontinence Lack of voluntary control over urination. Incontinence is the norm for infants and children younger than 2 or 3 years old, because nervous connections to the external urethral sphincter are incompletely developed. In adults, incontinence may result from damage to the external sphincter (often caused, in men, by surgery on the prostate gland), disease of the urinary bladder, and spinal cord injuries that disrupt the pathways along which travel impulses related to conscious control of urination. In any age group, urinary tract infection can result in incontinence.

Urinary retention The failure to completely or normally expel urine from the bladder. This condition may result from lack of the sensation to urinate, as might occur temporarily after general anesthesia, or from contraction or obstruction of the urethra, a condition caused, in men, by enlargement of the prostate gland. Immediate treatment for retention usually involves use of a urinary catheter to drain urine from the bladder.

Urinary system The system that consists of two kidneys, two ureters, one urinary bladder, and one urethra. Its main function is to regulate the volume, pressure, and composition of the blood.

Urinary tract infection (UTI) An infection caused by bacteria in the urinary system. Most bacteria enter the urinary system by moving up the urethra from outside the body.

Urination The process, involving both involuntary and voluntary actions, by which the urinary bladder is emptied. It is also called voiding or micturition.

Urine The yellowish fluid produced by the kidneys. It contains wastes and excess materials removed from the blood. Urine produced by the kidneys travels down the ureters to the urinary bladder, where it is stored until being released from the body through the urethra.

Uterine (menstrual) cycle The sequence of events that occurs on an approximately 28-day cycle in the uterine lining (endometrium) that involves the thickening of and increased blood supply to the endometrium and the loss of the endometrium as menstrual flow.

Uterus A hollow muscular organ in the female reproductive system in which the embryo implants and develops during pregnancy.

UTI See *urinary tract infection*.

V

Vaccination A procedure that introduces a harmless form of the disease-causing organism into the body to stimulate immune responses against that antigen.

Vagina A muscular tube in the female reproductive system that extends from the uterus to the vulva and serves to receive the penis during sexual intercourse and as the birth canal.

Vaginitis An inflammation of the vagina.

Variable In a controlled experiment, the factor that differs between the control group and experimental groups.

Varicose veins Veins that have become stretched and distended because blood is prevented from flowing freely and so accumulates, or "pools," in the vein. A common cause of varicose veins is weak valves within the veins.

Vas deferens A tubule that conducts sperm from the epididymis to the urethra.

Vasectomy A sterilization procedure in men in which the vas deferens on each side is cut and sealed to prevent sperm from leaving the man's body.

Vasoactive intestinal peptide (VIP) A hormone released by the small intestine into the bloodstream that triggers the small intestine to release intestinal juices.

Vasoconstriction A decrease in the diameter of blood vessels, commonly of the arterioles. Blood flow through the vessel is reduced, and blood pressure rises as a result of vasoconstriction.

Vasodilation An increase in the diameter of blood vessels, commonly of the arterioles. Blood flow through the vessels increases, and blood pressure decreases as a result of vasodilation.

Vasopressin See *antidiuretic hormone*.

Vector (disease) An organism that transports a pathogen between hosts.

Vector (DNA) A biological carrier, usually a plasmid or a virus, that ferries the recombinant DNA to the host cell.

Vein A blood vessel formed by the merger of venules that transports blood back toward the heart. Veins have walls that are easily stretched, so they serve as blood reservoirs, holding up to 65% of the body's total blood supply.

Vena cava One of two large veins that empty oxygen-depleted blood from the body to the right atrium of the heart. The superior vena cava delivers blood from regions above the heart. The inferior vena cava delivers blood from regions below the heart.

Ventilation In respiration, breathing.

Ventral nerve root The portion of a spinal nerve that arises from the front (anterior) side of the spinal cord and contains axons of motor neurons. It joins with the dorsal nerve root to form a single spinal nerve, which passes through the opening between the vertebrae.

Ventricle One of the two lower chambers of the heart that receive blood from the atria. The ventricles function as the main pumps of the heart. The right ventricle pumps blood to the lungs. The left ventricle pumps blood to body tissues.

Ventricular fibrillation Rapid, ineffective contractions of the ventricles of the heart, which render the ventricles useless as pumps and stop circulation.

Venule A small blood vessel that receives blood from the capillaries. Venules merge into larger vessels called veins. The exchange of materials between the blood and tissues across the walls of a venule is minimal.

Vertebra One of a series of joined bones that forms the vertebral column.

Vertebral column The "backbone." It is composed of 26 vertebrae (7 cervical, 12 thoracic, 5 lumbar, 1 sacrum, and 1 coccyx) and associated tissues. The spinal cord passes through a central canal within the vertebrae.

Vesicle A membrane-bound sac formed during endocytosis.

Vestibular apparatus A closed fluid-filled maze of chambers and canals within the inner ear that monitors the movement and position of the head and functions in the sense of balance.

Vestibule A space or cavity at the entrance to a canal. In the inner ear, the vestibule is a structure consisting of the utricle and saccule.

Villi (singular, villus) Small fingerlike projections on the small intestine wall that increase surface area for absorption.

VIP See *vasoactive intestinal peptide*.

Virus A minute infectious agent that consists of a nucleic acid encased in protein. A virus cannot replicate outside a living host cell.

Vital capacity The maximal amount of air that can be moved into and out of the lungs during forceful breathing.

Vitamin An organic (carbon-containing) compound that, although essential for health and growth, is needed only in minute quantities to regulate cellular processes.

Vitiligo A condition in which melanocytes disappear from areas of the skin, leaving white patches in their wake.

Vitreous humor The jellylike fluid filling the posterior cavity of the eye between the lens and the retina that helps to keep the eyeball from collapsing and holds the thin retina against the wall of the eye.

Vocal cords Folds of tissue in the larynx that vibrate when air passes through them, producing sound.

Vulva External genitalia of a female that surround the opening of the vagina and urethra.

W

Water footprint A measure of the amount of water used by a person or population for personal use and for the production of and waste removal of products consumed.

Water-soluble hormone A hormone that cannot pass through the plasma membrane on its own, so it influences target cells indirectly, through second messenger systems. Second messenger systems initiate enzyme cascades within the cell that ultimately activate certain enzymes. Water-soluble hormones include protein and peptide hormones, such as those secreted by the pancreas and pituitary gland.

Water-soluble vitamin A vitamin that dissolves in water. Water-soluble vitamins include vitamin C and the various B vitamins.

White blood cells See *leukocytes*.

White matter Regions of the central nervous system that are white owing to the presence of myelinated nerve fibers. White matter is important in neural communication over distances.

X

X-linked genes Genes located on the X chromosome. Most X-linked genes have no corresponding allele on the Y chromosome and will be expressed in a male, and in a female if she is homozygous.

Y

Yellow marrow A connective tissue found in the shaft of long bones that stores fat. It forms from red marrow, and, if the need arises, it can convert back to red marrow and form blood cells.

Yolk sac The extraembryonic membrane that is the primary source of nourishment for embryos in many species of vertebrates. In humans, however, the yolk sac does not provide nourishment (human embryos and fetuses receive nutrients from the placenta). In humans, the yolk sac is a site of blood cell formation and contains cells, called primordial germ cells, that migrate to the gonads, where they differentiate into immature cells that will eventually become sperm or oocytes.

Z

Zygote The diploid cell resulting from the joining of an egg nucleus and a sperm nucleus. The first cell of a new individual.

Zygote intrafallopian transfer (ZIFT) A procedure in which zygotes created by the union of egg and sperm in a dish in the laboratory are inserted into the woman's oviducts. Zygotes travel on their own from the oviducts to the uterus.

Credits

Chapter 1 C01 luoman/iStockphoto. **1.A** tulpahn/Shutterstock. **1.1(top left)** Lawrence Livermore National Lab. **1.1(top right)** A. Syred/Photo Researchers, Inc. **1.1(middle left)** ampwang/iStockphoto. **1.1(middle right)** ptaxa/iStockphoto. **1.1(middle left)** Cathy Keifer/Shutterstock. **1.1(middle right)** UpperCut Images/Alamy. **1.1(bottom)** Maria Bedacht/Fotolia. **1.7** "Ownership and Use of Wireless Telephones: A Population-Based Study of Swedish Children Aged 7-14 Years" by Fredrik Soderqvist et al. in BMC PUBLIC HEALTH 7 (2007): 105-13, fig. 1, p. 107. **1.8** Moodboard Premium/Glow Images.

Chapter 1A Special Topic C01A SOMOS/SuperStock. **1A.3** Alexander Raths/Fotolia.

Chapter 2 C02 Jim West/Alamy. **2.3** Norma Wilson/iStockphoto. **2.4a,b** Pasieka/Photo Researchers, Inc. **2.7a,b** Richard Megna /Fundamental Photographs. **2.7c** Charles D. Winters/Photo Researchers, Inc. **2.7d** John Holst/Shutterstock. **2.A** Clare Maxwell, Pearson Education. **2.13(battery acid)** Photosoup/iStockphoto. **2.13(beer)** Dr3amer/iStockphoto. **2.13(urine)** Clare Maxwell, Pearson Education. **2.13(human blood)** Andrew Syred/Photos Researchers, Inc. **2.13(baking soda)** Clare Maxwell, Pearson Education. **2.13(household ammonia)** Clare Maxwell, Pearson Education. **2.13(oven cleaner)** Clare Maxwell, Pearson Education. **2.14a** Simon Fraser/Photo Researchers, Inc. **2.14b** F. Harvey Pough. **2.18a** CNRI/Photo Researchers, Inc. **2.18b** DK Images. **2.18c** Biophoto Associates/Photo Researchers, Inc. **2.B** Renn Sminkey, CDV, Pearson Education. **2.21** iStockphoto.

Chapter 3 C03 Corbis Premium RF/Alamy. **3.1** Kari Lounatmaa/Photo Researchers, Inc. **3.2** Dr. Gopal Murti/Photo Researchers, Inc. **3.4a** Nina Zanetti, Pearson Education. **3.4b** Quest/Photo Researchers, Inc. **3.4c** Quest/Science Photo Library/Photo Researchers, Inc. **3.5a** David M. Phillps/Photo Researchers, Inc. **3.5b** David M. Phillips/Photo Researchers, Inc. **3.5c** Thomas Deerinck, NCMIR/Photo Researchers, Inc. **3.9** Sam Singer, Pearson Science. **3.13b** Secchi-Lecaque-Roussel-UCLAF/CNRI/Science Photo Library/Photo Researchers, Inc. **3.14a** Adrian T. Sumner/Stone/Getty Images. **3.14b** Brian Eyden/Science Photo Library/Photo Researchers, Inc. **3.15** Don Fawcett/Photo Researchers, Inc. **3.16b** Don W. Fawcett/Photo Researchers, Inc. **3.19b** Bill Longcore/Photo Researchers, Inc. **3.20b** Don W. Fawcett/Photo Researchers, Inc. **3.A** Charles D. Winters/Photo Researchers, Inc. **3.B** Eric V. Grave/Photo Researchers, Inc. **3.21a** SPL/Photo Researchers, Inc. **3.21b** Yorgos Nikas/Getty Images. **3.21c** Peter Satir/Photo Researchers, Inc.

Chapter 4 C04 YanLev/Shutterstock. **4.1(top left)** Steve Downing, Pearson Education. **4.1(top right)** Steve Downing, Pearson Education. **4.1(middle left)** Lisa Lee, Pearson Education. **4.1(middle right)** Steve Downing, Pearson Education. **4.1(bottom left)** Nina Zanetti, Pearson Science. **4.1(bottom right)** Gregory N. Fuller. M.D., Ph.D., Chief, Section of Neuropathology, M.D. Anderson Cancer Center, Houston, Texas. **4.2(top left and right)** Steve Downing, Pearson Education. **4.2(middle left)** Nina Zanetti, Pearson Education. **4.2(middle right)** Spike Walker (Microworld Science)/Dorling Kindersley Secondary Permissions. **4.2(bottom left)** Nina Zanetti, Pearson Education. **4.2(bottom right)** Nina Zanetti, Pearson Education. **4.3(top)** Nina Zanetti, Pearson Education. **4.3(middle and right)** Steve Downing, Pearson Education. **4.4** Prof. P. Motta, Department of Anatomy, University "La Sapienza," Rome/Science Photo Library/Photo Researchers, Inc. **4.5(top)** Philippa Claude. **4.5(middle)** From Douglas E. Kelly, The Journal Cell Biology 28 (1966): 51, fig. 7. Reproduced by copyright permission of The Rockefeller University Press. **4.5(bottom)** From C. Peracchia and A.F. Dulhunty, "The Journal Cell Biology" 70 (1976): 419, fig 6. Reproduced by copyright permission of The Rockefeller University Press. **4.9** Olivier Voision/Photo Researchers, Inc. **4.10** Dave Stamboulis/Alamy. **4.11** Maury Tannen/The Image Works. **4.A** Dr. P. Marazzi/Photo Researchers, Inc.

Chapter 5 C05 Superstock. **5.1c** Andrew Syred/Photo Researchers, Inc. **5.1e** SPL/Photo Researchers, Inc. **5.2** SPL/Photo Researchers, Inc. **5.Ab** Peter Sebastian/Photonica/Getty Images. **5.8** Martin Shields/Alamy. **5.15** Rakke, Terje/Getty Images, Inc. – Image Bank.

Chapter 6 C06 Patrick Green/Icon SMI/Newscom. **6.3** Nina Zanetti, Pearson Education. **6.7** Young-Jin Son, Drexel University College of Medicine. **6.10** Biophoto Associates. **6.11** Alamy. **(un. photo on p. 110)** Aurora Photos/Alamy.

Chapter 7 C07 Alamy. **7.2** David M. Phillips/Photo Researchers, Inc. **7.3** David M. Phillips/Photo Researchers, Inc. **7.6** Pro. S. Cinti/Photo Researchers, Inc. **7.8** Science Source/Photo Researchers, Inc. **7.A** Ron Sachs/CNP/Corbis.

Chapter 8 C08 Shutterstock. **8.3** Martin M. Rotker/Science Source/Photo Researchers, Inc. **(un. photo on p. 132)** Shutterstock. **8.7c** Ed Reschke.

Chapter 8A Special Topic C08A White Packert/Image Bank/Getty Images. **8A.4** Dr. Miles Herkenham, National Institute of Health. **8A.5** Martin M. Rotker/Science Source/Photo Researchers, Inc.

Chapter 9 C09 Purestock/AGE Fotostock. **9.5** Nataliva Hora/Shutterstock. **9.7** AP Photo/Gemunu Amarasinghe. **9.8a** Kevin Elsby/Alamy. **9.8b** WorldFoto/Alamy. **9.8c** Michael & Christine Denis-Huot/Photo Researchers, Inc. **9.10** Omikron/Photo Researchers, Inc. **9.11** Pearson Education. **9.14** Dr. Goran Bredberg/SPL/Photo Researchers, Inc. **9.A** Scanning electron micrographs by Dr. Robert S. Preston and Prof. Joseph E. Hawkins, Kresge Hearing Research Institute. Unversity of Michigan Medical School.

Chapter 10 C010 Randy Fairs/Corbis/Glow Images. **10.7** Bettmann/CORBIS. **10.8** Published in American Journal of Medicine, Vol. 20, p. 133, 1956, Dr. William H. Daughaday. Copyright Elsevier. With permission of Excerpta Media, Inc. **10.9** Huihe/Newscom. **10.12a** Alison Wright/CORBIS. **10.12b** Medical-on-Line/Alamy. **10.12c** Dr. P. Marazzi/Photo Researchers, Inc. **10.15** Bettmann/CORBIS. **10.16** Sharmyn McGraw. **10.17b** Parviz M. Pour/Photo Researchers, Inc.

Chapter 10A Special Topic C010A Juvenile Diabetes Research Foundation. **10A.2a** Kemter/iStockphoto. **10A.2b** iofoto/Shutterstock. **10A.2c** arekmalang/iStockphoto. **10A.2d** Superstock. **10A.3** Mark Hatfield/iStockphoto. **10A.4a** Svanblar/Shutterstock. **10A.4b** dlerick/iStockphoto. **10A.4c** GARO/PHANIE/Photo Researchers, Inc. **10A.5a** Pearson Education. **10A.5b** Dobs/Shutterstock. **10A.6** Wright Airforce Base.

Chapter 11 C011 Michael Tercha/MCT/Newscom. **11.4** Andres Syred/Photo Researchers, Inc. **(un. photo on p. 205)** Corbis. **11.9** CNRI/Photo Researchers, Inc.

Chapter 12 C012 Westend61 GmbH/Alamy. **12.2** Steve Gschmeissner/Photo Researchers, Inc. **12.3c** Photo Researchers, Inc. **12.4a** Biophoto Associates/Photo Researchers, Inc. **12.6a** Ed Reschke. **12.7a** Yiargo/Shutterstock. **12.8(left)** CNRI/Science Photo Library/Photo Researchers, Inc. **12.8(right)** Ralph T. Hutchings. **12.10b** Martin Dohrn/Royal College of Surgeons/Science Photo Library/Photo Researchers, Inc. **(un. photo on p. 226)** Andresr/Shutterstock. **12.15** R. Umesh Chandran/SPL/Photo Researchers, Inc.

Chapter 12A Special Topic C012A APHP-PSL-GARO/Photo Researchers, Inc. **12A.2** Living Art Enterprises/Photo Researchers, Inc. **12A.3a** Martin M. Rotker/Photo Researchers/Getty Images USA, Inc. **12A.3b** Ed Reschke. **12A.3c** Biophoto Associates/Photo Researchers, Inc. **12A.4** Science Photo Library/Glow Images. **12A.7** Lennart Nilsson/Scanpix.

Chapter 13 C013 arabianEye/Getty Images USA, Inc. **13.3** Lennart Nilsson/Scanpix. **13.4** Eye of Science/Photo Researchers, Inc. **13.5(left)** Lennart Nilsson/Scanpix. **13.5(middle)** Dr. Robert Dourmashkin/Queen Mary Unversity of London. **13.5(right)** Lennart Nilsson/Scanpix. **13.7** Greeland/Shutterstock. **13.16** Oliver Meckes/Photo Researchers, Inc.

Chapter 13A Special Topic C013A ScohAnam/iStockphoto. **13A.1a** Oliver Meckes/Photo Researchers, Inc. **13A.1b** Medical-on-Line/Alamy. **13A.1c** Science Source/Photo Researchers, Inc. **13A.3(left)** gaspr13/iStockphoto. **13A.3(right)** Eye of Science/Photo Researchers, Inc. **13A.4** Grapes/Michaud/Photo Researchers, Inc. **13A.5a(left)** Dariusz Majgier/Shutterstock. **13A.5b(right)** Larry Mulvehill/Photo Researchers, Inc. **13A.7** James Boardman/Alamy.

Chapter 14 C014 David Handschuh/NY Daily News Archive via Getty Images. **14.4a** CNRI/Science Photo Library/Photo Researchers, Inc. **14.4b** Dr. Andrew Evan, Indiana Unversity Medical Center, Department of Anatomy. **14.5** CNRI/Science Photo Library/Photo Researchers, Inc. **14.7** Ralph T. Hutchings. **14.8a** Oliver Meckes/Photo Researchers, Inc. **14.8b** Eye of Science/Photo Researchers, Inc. **14.9** Southern Illinois University/Photo Researchers, Inc. **14.14** Dr. Andrew Evan, Indiana University of Medical Center, Department of Anatomy. **14.15** James Stevenson/Photo Researchers, Inc.

Chapter 15 C015 Radius Images/Alamy. **15.7c** WG/Science Photo Library/Photo Researchers, Inc. **15.8a** David M. Martin/Photo Researchers, Inc. **15.8b,c** Eye of Science/Photo Researchers, Inc. **15.12** SPL/Photo Researchers, Inc. **15.A** Dr. E. Walker/Science Photo Library/Photo Researchers, Inc. **15.14** Gastrolab/Photo Researchers, Inc. **15.16(top left)** Ian O'Leary/Dorling Kindersley. **15.16(top right)** Clare Maxwell, Pearson Education. **15.16(bottom left and middle)** Clare Maxwell, Pearson Education. **15.16(bottom right)** Shebeko/Shutterstock. **15.17(left)** Peggy Greb/Agricultural Research Service/USDA. **15.17(right)** iStockphoto. **15.18(left)** Clare Maxwell, Pearson Education. **15.18(right)** EyeWire Collection/Photodisc/Getty Images. **15.19(left)** Ranald MacKechnie/Dorling Kindersley. **15.19(middle)** Iuri/Shutterstock. **15.19(right)** Mircea Bezergheanu/Shutterstock. **15.23** LYDIE/SIPA/Newscom.

Chapter 15A Special Topic C015A Jeff Holt/Bloomberg via Getty Images. **15A.1a** Arco Images/H. Reinhard/Alamy Images. **15A.1b** Jeremy Sutton-Hibbert/Alamy. **15A.2(top left)** Orientaly/Shutterstock. **15A.2(top right)** Klas Stolpe/AP Wide World Photos. **15A.2(middle left)** Stringer (China) China Out/Reuters Limited.

15A.2(bottom left) Getty Images, Inc. RF **15A.2(bottom right)** Alamy Images, Royalty Free. **15A.3** iStockphoto. **15A.4** Partnership for Food Safety Education.

Chapter 16 CO16 Erik Isakson/Getty Images. **16.13** Will & Deni McIntyre/Photo Researchers, Inc. **16.A** Olivier Matthysa/EPA/Newscom. **16.B** Michael Heron, Pearson Education. **16.15** Gallo Images/Adriaan Vorster/Getty Images.

Chapter 17 CO17 Judy Goodenough. **17.2** Professors P.M. Motta, K.R. Porter & P.M. Andrews/SPL/Photo Researchers, Inc. **17.3** Ed Reschke. **17.4** Juergen Berger/Photo Researchers, Inc. **17.8(top)** P. Bagavandoss/Photo Researchers, Inc. **17.8(bottom)** C. Edelmann/Photo Researchers, Inc. **17.10(top left)** RW Johnson Co./ZUMA Press/Newscom. **17.10(top right)** vario images GmbH & Co. KG/Alamy Images. **17.10(middle left)** Gary Parker/Photo Researchers, Inc. **17.10(middle right)** SIU/Photo Researchers, Inc. **17.10(bottom left)** Michael Newman/PhotoEdit Inc. **17.10(bottom right)** vario images GmbH & Co. KG/Alamy Images.

Chapter 17A Special Topic CO17A Denis Balabouse/CORBIS/All rights reserved/Reuters. **17A.2** Centers for Disease Control and Prevention (CDC). **17A.4(top)** Science Photo Library/Photo Researchers, Inc. **17A.4(middle)** Lester V. Bergman/CORBIS. **17A.4(bottom)** Centers for Disease Control and Prevention (CDC). **17A.5** Biophoto Associates/Photo Researchers, Inc. **17A.6** P. Marazzi/Photo Researchers, Inc. **17A.8** Chris Bjomberg/Photo Researchers, Inc.

Chapter 18 CO18 Andersen Ross/Getty Images. **18.1** William Bemis/Betty McGuire, Ph.D. **18.2a** Dr. Yorgos Nikas/Science Photo Library/Photo Researchers, Inc. **18.2b** Lennart Nilsson/Scanpix. **18.2c** Claude Edelmann/Photo Researchers, Inc. **18.2d** Lennart Nilsson/Scanpix. **18.5a** Barbara Penoya/Photodisc/Getty Images, Inc. **18.5b** Mein-Chun Jau/Dallas Morning News/CORBIS. **18.5c** Barbara Penoya/Photodisc/Getty Images, Inc. **18.15** Sally and Richard Greenhill/Alamy. **18.17** SuperStock, Inc. **18.18** AP Photo/George Gobet.

Chapter 18A Special Topic CO18A Rob Crandall/Alamy. **18A.1a** ZUMA Press, Inc./Alamy. **18A.1b** Karen Pulfer Focht/Landov Media. **18A.1c** Pete Jenkins/Alamy. **18A.2a** Mary Pato/Betty McGuire, Ph.D. **18A.2c** Betty McGuire, Ph.D. **18A.4** AP Photo/Knoxville News Sentinel, Michael Patrick. **18A.5** AP Photo/Dan Steinberg.

Chapter 19 CO19 Gregory Bull. **19.3** Biophoto Associates/Photo Researchers, Inc. **19.5(left)** Ed Reschke. **19.5(bottom right)** Centers for Disease Control and Prevention (CDC). **19.6** T.E. Schroeder/Biological Photo Service. **19.7** CNRI/SPL/Photo Researchers, Inc. **19.12** Dr. Judy Goodenough. **19.A** Kristy-Anne Glubish/Alamy Images.

Chapter 19A Special Topic CO19A Corfield/Alamy. **19A.3** Pascal Goetgheluck/Photo Researchers, Inc. **19A.4** AP Photo/Paul Clements. **19A.7a** The Lancet, HO/AP Wide World Photos. **19A.7b** The Lancet/Dr. Patrick H. Warnke. **19A.7c** Dr. Patrick Warnke. **19A.8** Eric Gay/AP Wide World Photos.

Chapter 20 CO20 Brian Milne/Stockbyte/Getty Images, Inc. **20.2(top left)** Beusbcus/iStockphoto. **20.2b(top right)** DennaBean/iStockphoto. **20.2(middle left)** Paco Romero/iStockphoto. **20.2(middle right)** George Doyle/Stockbyte/Getty Images. **20.2(middle left)** Radius Images/Alamy. **20.2(middle right)** Ariwasabui/Fotolia. **20.2(bottom)** Michael Newman/PhotoEdit Inc. **20.4(right)** DennaBean/iStockphoto. **20.4(left)** Beusbcus/iStockphoto. **20.9** Hattie Young/Science Photo Library/Photo Researchers, Inc. **20.10** Radu Sigheti/REUTERS/Landov Media. **20.12(left)** David M. Phillips/Photo Researchers, Inc. **20.12(right)** Eve of Science/Photo Researchers, Inc. **20.13b** University of Florida. **20.15(left)** Five P Minus Society. **20.15(right)** Douglas B. Chapman. **20.16b** Wellcome Trust Medical Photographic Library. **20.16c** Vince Bucci/Getty Images – GINS/Entertainment News & Sports.

Chapter 21 CO21 STR/EPA/Newscom. **21.6** Alfred Pasieka/Science Photo Library/Photo Researchers, Inc. **21.16** Will & Deni McIntyre/Photo Researchers, Inc. **21.19** National Cancer Institute. **(un. photo on p. 456)** Katie Stanton/Aqua Bounty Technologies, Inc. **21.A** Orchid Cellmark Inc. **(un. photo on p. 462)** Orchid Cellmark Inc.

Chapter 21A Special Topic CO21A Erik Hill/MCT/Newscom. **21A.1** Science Photo Library/Photo Researchers, Inc. **21A.2a** Department of Clinical Cytogenetics, Addenbrookes/Photo Researchers, Inc. **21A.2b** National Institute of Health Genetics. **21A.5** David McCarthy/SPL/Photo Researchers, Inc. **21A.6** Robert K. Moyzis. **21A.8** Judy Goodenough. **21A.10a** Deco/ Alamy Images. **21A.10b** BSIP/Photo Researchers, Inc.

Chapter 22 CO22 Kumar Sriskandan/Alamy. **22.4** Don Mason/Getty Images, Inc. **22.7a** Colin Keates/Dorling Kindersley Media Library. **22.7b** Alamy. **22.9b(right)** Robert Shantz/Alamy. **22.9b(left)** Dr. Philip D. Gingerich. **(un. photo on p. 485)** Shutterstock. **22.13** Frans Lanting/CORBIS. **22.14a** Gail Johnson/Shutterstock. **22.14b** Mr. Mayilvahnan/Hornbil Images Pvt. Ltd. **22.14c** Fulvio Eccardi Ambrosi/Bruce Coleman/Photoshot Holdings Ltd. **22.15a** dpenn/iStockphoto. **22.15b** Henk Bentlage/Shutterstock. **22.15c** CraigRJD/iStockphoto. **22.15d** Charles Taylor/Shutterstock. **22.18** John Reader/SPL/Photo Researchers, Inc. **22.19a** John Reader/Photo Researchers, Inc. **22.19b** AP Photo/Richard Drew. **22.20a** ©The Natural History Museum, London. **22.20b** Volker Steger/Photo Researchers, Inc.

Chapter 23 CO23 Alexey Stiop/Shutterstock. **23.1** M. Jentoft-Nilsen, F. Hasler, D. Chesters/GSFC & T. Neilsen/NASA Headquarters. **23.2(top)** anilakduygu/iStockphoto. **23.2(middle)** Chee-Onn Leong/Shutterstock. **23.2(bottom)** Jack Jelly/iStockphoto. **23.3(top)** Denis Tangney Jr./iStockphoto. **23.3 (second down from top)** mpiotii/iStockphoto. **23.3(third down from top)** jsnover/iStockphoto. **23.3(fourth down from top)** Clement Phillippe/Alamy Images. **23.3(bottom)** Erkki & Hanna/Shutterstock. **23.4(left)** Universal Images Group/DeAgostini/Alamy. **23.4(right)** Pat & Chuck Blackley/Alamy. **(un. photo on p. 510)** Jonathan Torgovnik/Getty Images.

Chapter 24 CO24 Getty Images Inc. – Stone Allstock. **24.1** Judy Goodenough. **24.4** Harald Sund/Image Bank/Getty Images Inc. **24.5** Associated Press (Wide World Photos). **24.7** Mark Boulton/Photo Researchers, Inc. **24.8** Paul Edmondson/Getty Images. **24.9(right)** Sergey Peterman/Shutterstock. **(un. photo on p. 524)** Tomas/Fotolia.

Index

Note: "f" indicates a figure and "t" indicates a table.

A

Abdominal cavity, 71, 72f
ABO blood types, 206–208, 432, 433, 434t
Abstinence, 356
Accessory glands
  digestive, 294f–296
  male reproductive system, 343–345
Accommodation, 156
Acetylcholine, 121, 122, 136
Acetylcholinesterase, 120
Acid, 28
Acid rain, 29, 508
Acid-base buffering system, 277–278
Acidification, of oceans, 509
ACL, see anterior cruciate ligament
Acne, 78
Acquired immune deficiency syndrome (AIDS), 367–371
Acromegaly, 177
Acrosome, 346f, 374
Actin, 102–103f, 108
Action potential, 116, 117f, 118f
Active immunity, 251–252
Active site, 36
Active transport, 48–49, 50t, 294f
Adam's apple, see larynx
Adaptation
  evolutionary, 4, 480
  sensory, 151
Adaptive immune response, 240, 244–251
Adaptive traits, 4
Addison's disease, 182, 183f
Adenocarcinoma, 471
Adenosine triphosphate (ATP), 38, 39, 104, 268
Adenovirus, 423, 455
Adhesion junctions, 70, 71f
Adipose connective tissue, 67t, 68f
Adrenal gland, 173f, 182–184, 436
  cortex, 176f, 182–183
  medulla, 184
Adrenaline, see epinephrine
Adrenocorticotropic hormone (ACTH), 173f, 176f, 178
Adult gene testing, 439
Adult stem cells, 419
Aerobic exercise, 109, 307, 355
Afferent neuron, see sensory neuron
Age structure (of population), 516f
Agglutination, 206–208
Aging, 177, 178, 186, 387–389
Agranulocytes, 200t, 201f, 202
AIDS (acquired immune deficiency syndrome), 367–371
Albinism, 432f, 434
Albumin, 199
Alcoholic beverage, 143–144, 292
  erectile dysfunction and, 345
Alcoholism, 146
Aldosterone, 173f, 175, 182, 331, 332, 333t
Allantois, 377
Alleles, 426, 427
Allergen, 254f
Allergy, 254f–255, 273
Allometric growth, 381
Alpha-fetoprotein, 437
ALS, see amyotrophic lateral sclerosis
Alveoli, 270f, 271t, 273, 274f
Alzheimer's disease, 121
Amino acids, 35–36, 294, 301, 303
Ammonium, 507f–508
Amniocentesis, 437–438f
Amnion, 377
Amphetamine, 148
Ampulla, 165, 166f
Amygdala, 133f, 134
Amyloid plaque, 121
Amyotrophic lateral sclerosis (ALS), 114
Anabolic pathway, 57
Anabolic steroid abuse, 110
Anal canal, 288f, 296f, 298
Analogous structure, 487
Anandamide, 146
Anaphase, 404, 405f, 412
Anaphylactic shock, 254–255
Androgen, 173f, 182
Anemia, 204
Anencephaly, 380, 386
Aneurysm, 215, 233–235f
Angina pectoris, 235
Angiotensin, 332
Anorexia nervosa, 309–310
Antagonistic pair of muscles, 101f
Anterior cruciate ligament (ACL), 95–96
Anterior pituitary gland, 88, 347f, 350–351
Antibiotics, 261, 363, 365, 515
  resistance to, 261, 364, 456, 480
Antibodies, 202, 206–208, 246, 247f, 249f–250ft
Antibody mediated immune responses, 246, 247f, 249f–250
Anticodon, 446–447f
Antidiuretic hormone (ADH), 173f, 176f, 179, 327, 330f, 331, 332, 333t
Antigen, 202, 245f, 247–251
Antigen-presenting cell (APC), 247f–248
Antigenic shift, 266
Antihistamine, 255
Antisense DNA, 475
Anus, 288f
Anvil, see incus
Aorta, 214f, 215, 220f, 221, 222f
Aortic bodies, respiration, 278f, 279
ApoE, 439
Apoptosis, see programmed cell death
Appendicitis, 296
Appendicular skeleton, 94–95
Appendix, 296f
Aqueous humor, 154f, 155t
Aquifers, 505, 520
Arachnoid, 127
Archaea, 4f, 42, 44t
Ardipithecus 491, 493t, 493f
Areolar connective tissue 67t, 68f
Arrector pili, 74f, 76
Arteries, 214f, 215–216f
Arterioles, 215–217f
Arthritis, 97
Artificial insemination, 384
Asbestos, 55
Ascending tracts, 134
Asperger syndrome, see autism spectrum disorder
Aspirin, 292
Assisted reproductive techniques (ARTs), 384
Association area, of cerebrum, 131
Association neuron, see interneuron
Asthma, 9, 254, 273
Astigmatism, 156t, 157f
Atherosclerosis, 234f–236, 237–238
  diet and, 301, 304
Athlete's foot, 264
Atom, 20–21
Atomic number, 21
Atomic weight, 21
ATP, see adenosine triphosphate
Atrial natriuretic hormone (ANH), 333
Atrioventricular bundle, see AV bundle
Atrioventricular node, see AV node
Atrioventricular valves, see AV valves
Atrium, 219–220f
Auditory nerve, 161f
Auditory tube, 160f, 161, 163t
Australopithecus, 490–491, 493f
Autism, see autism spectrum disorder
Autism spectrum disorder
  causes, 396–397, 451
  characterization, 393–394
  diagnosis, 395–396
  prevalence, 395
  prognosis, 398
  symptoms, 393–395
  treatment, 397–398
  vaccines not implicated, 398–399
Autoimmune disorders, 253
Autonomic nervous system, 126–127, 135–138

Autosome, 436, 401
AV bundle, 224f
AV node, 224f
AV valves, 219–220f, 221f
Axial skeleton, 91–94
Axon, 114, 115f

B

B cells, *see* B lymphocytes
B lymphocytes, 245–251
Bacillus, 260f
Bacteria, 42, 43f, 44t, 128, 209, 239, 240, 296–297, 451
   nitrifying, 507f–508
   nitrogen-fixing, 507f, 508
   STDs and, 363f–365t
Bacterial cloning, 451–452f
Balance, 165, 166f
Baldness, male pattern, 436
Ball-and-socket joint, 96
Basal metabolic rate (BMR), 307
Base, 28
Base pair substitution, 449f
Basement membrane, 64
Basilar membrane, 161f, 162f, 163t
Basophil, 200t, 201f, 202
Benign tumor, 463–464
Beta-thalassemia, 420
Bicarbonate ion, 277–278
Bicuspid valve, 220
Bile, 203, 293, 295–296
Bilirubin, 203, 295
Binary fission, 260
Biodiversity, 508–511
Biofuel, 525
Biogeochemical cycles, 504f–508
Biogeography, 483–484, 485f
Biological magnification 502–503f
Biomass, 502
Biopsy, 473
Biosphere, 5f, 498
Biotechnology, 450–457
Bioterrorism, 318–319
Bipedalism, 488, 489, 490, 491, 493t
Bipolar cell, 157, 158f
Birth, 383–384
Birth control, 356–359
Birth defects, 385–386
Birth rate, 515
Blastocyst, 373f, 374f, 376, 420f–421
Blind spot, 154f, 155t
Blood, 67t, 68f, 198–210
   clots, 215, 232
   clotting, 200, 208–210
Blood pressure, 215–216, 224–225f
   high, 233, 234
Blood type, 206–208
Blood vessels, 215–219
Blood-brain barrier, 128
Blue baby, 383
Body cavities, 70–72f
Bone, 67t, 68f
   fractures, 88–89
   remodeling, 89–91
Bone marrow, 86
   red, 86, 199, 201, 228
   stem cells, 423
   transplant, 206
   yellow, 86
Botox, 75, 76f
Bottleneck effect, 479
Botulism, 260–261
Bowman's capsule, *see* glomerular capsule
Brain, 128–134
Brain cancer, 9
Brain injury, 132
Brain stem, 129f, 133
BRCA genes, 354, 466
Breaking the water, 383
Breast, 348t, 347–349, 350f
Breast cancer, 354–355
Breast milk, 252
Breast-feeding, 386
Breathing, 218, 269f, 274–275f
   chemical regulation of, 279
   neural regulation of, 278–279
Breech birth, 383
Bronchial tree, 273f
Bronchiole, 270f, 271t
Bronchitis, 281
Bronchus, 270f, 271t, 273
Bruise, 203
Brush border, 292f
Budding, 262f, 263
Buffers, 28, 277–278
Bulbourethral glands, 343t, 344f, 345
Bulimia nervosa, 309–310
Bursa, 97
Bursitis, 96–97

C

Calcitonin (CT), 89, 91, 173f, 180–181, 182f
Calcium ions, 104, 105, 119–120
Calcium, 85, 87, 89
Callus, 89f
Calorie, 299–300
CAMs, *see* cellular adhesion molecules
Cancer stem cell hypothesis, 469–470
Cancer, 9, 163–475, 239, 241–242, *see also* specific type
   alcohol and, 145
   diet and, 301, 304
   viruses and, 263
Canines, 287, 289f
Cannabis sativa, 146–147
Capillaries, 215, 216f, 217f, 218f, 274f
   blood, 216–218
   lymphatic, 227f
Capsid, 261, 262f
Carbaminohemoglobin, 277
Carbohydrate, 31, 32f, 301, 302–303
Carbon cycle, 506f–507
Carbon dioxide, 277–278
   regulation of breathing rate, 279
   transport, 277–278
Carbon footprint, 524–525
Carbon monoxide, 203, 237
Carbonic anhydrase, 377–278
Carcinogen, 471
Carcinoma in situ, 464
Cardiac cycle, 222–223f
Cardiac muscle, 69ft–70
Cardiovascular disease, 232–238
Cardiovascular exercise, 226
Cardiovascular system, 73f, 213–225
   alcohol and, 145
   cocaine and, 147–148
Carnivores, 500f–501f, 506
Carotid arteries, 214f
Carotid bodies, respiration, 278f–279
Carrier, 430–431, 435–436
Carrying capacity, 517, 518–526
Carsickness, 165
Cartilage, 67t, 68f, 273
   model in bone development, 87f, 88
Cartilaginous joints, 95
Catabolic pathway, 57
Cataract, 156f, 521
CD4 receptor, 368, 371
Cecum, 288f, 296
Cell, 5f, 42–61
Cell cycle, 401–402f, 465
Cell differentiation, 379
Cell division, 400–415
Cell phone, 9
Cell theory, 42
Cell-mediated immune responses, 246, 247f, 250–251f
Cellular adhesion molecules (CAMs), 46, 468t
Cellular respiration, 56–59, 60f
Cellulose, 31, 32f
Central canal, 128
Central nervous system (CNS), 126, 127–135
   aging and, 387t
   development of, 379–380
Centriole, 43f, 55–56
Centromere, 402, 407
Cerebellum, 129f, 133
Cerebral cortex, 129f–132, 134
Cerebrospinal fluid, 128, 134
Cerebrovascular accident, 138
Cerebrum, 128–131, 133f
Cervical cancer, 367, 470
Cervical cap, 358
Cervix, 348
Cesarean section, 378, 383
CFCs (chlorofluorocarbons), 521
Chancre, 364f–365
Chemical evolution, 477
Chemistry, 20
Chemoreceptor, 151
Chemotherapy, 474

Chickenpox, 370
Childbirth
   hormones involved in, 173f, 174, 175f, 179
   stages of, 383
Chlamydia, 363–364, 365t
Chlorofluorocarbons (CFCs), 521
Cholecystokinin, 299t
Cholesterol, 6–8, 33–35, 301
Chondrocyte, 67
Chordae tendineae, 219–220f
Chorion, 377–378
Chorionic villi, 378
   sampling, 437, 438f
Choroid, 153, 154f, 155t, 158f
Chromatid, 402, 407, 410, 412
Chromatin, 50–51, 402, 403f
Chromosome, 50–51, 401–415, 426–429
Chronic wasting disease (CWD), 264–265
Chylomicron, 294
Chyme, 292, 296, 299
Cigarette smoking
   cardiovascular disease and, 237–238
   erectile dysfunction and, 345
   lung disease and, 271, 283
Ciliary body, 153, 154f, 155t
Cilium, 55–56, 271f
Circular folds, 293f
Circulation
   coronary, 222f
   fetal, 382–383
Circulatory system, 213–225
Circumcision, 345
Cirrhosis, 295
Citric acid cycle, 58, 59t
Cleavage, 374f, 375–376
Cleft palate, 93
Climax community, 499
Clinical trials, 8t–9
Clitoris, 348t, 349f
Clonal selection, 248f–249
Cloning,
   bacterial, 451–452f
   therapeutic, 421f–422
   reproductive, 422–423
Clostridium botulinum, 260–261, 315, 319
Clostridium difficile, 261
Clotting factor, 199, 209
Cocaine, 147–148
Coccus, 260f
Cochlea, 160f, 161f, 162f, 163t
Cochlear implant, 164
Codominance, 432, 433
Codons, 446f, 447–448
Coenzyme, 37
Cofactor, 37
Cold sore, 366
Cold, common, 279–280, 281
Collagen, 65, 75, 87, 424
Collateral circulation, 226
Collecting duct, 325, 327, 328f, 330f, 331–332
Colon, 288f, 296f–298

Colonoscopy, 298
Color blindness, 158–159, 435–436
Color vision, 158–159
Colorectal cancer, 298
Colostrum, 386
Columnar epithelium, 64, 65t, 66f
Coma, 138
Combustion, 506
Common cold, 279–280, 281
Community, 5t, 498
Compact bone, 86f
Complement system, 241t, 242, 243f
Complementary base pairing, 443, 444, 446
Complementary protein, 303, 304f
Complete dominance, 432
Complete protein, 303
Compound, 23, 24f
Concentration gradient, 47
Condom, 358
Conductive hearing loss, 163–164
Cones, 155t, 158f–159f
Confounding variable, 6
Connective tissue, 64, 65–67t, 68f, 430
Consciousness, 134
Constipation, 298, 303, 309
Consumers, 500f–501f
Contact inhibition, 466–467
Continuous ambulatory peritoneal dialysis (CAPD), 334
Contraception, *see* birth control
Contraceptive sponge, 358
Contraction, of muscle, 101–106
Control group, 6
Controlled experiment, 6
Convergent evolution, 487
Coral reef, 509
Cornea, 153, 154f, 155t
Coronary angiography, 235f
Coronary arteries, 222, 226
Coronary artery disease, 235f–236
Coronary artery spasm, 236
Coronary bypass, 236f, 237
Coronary circulation, 222f
Coronary thrombosis, 236
Corpus callosum, 129f
Corpus luteum, 350, 351f, 353
Corticosterone, 173f, 183
Cortisol, 173f, 183
Cortisone, 173f, 183
Covalent bond, 23–24, 25, 27t
Cramp, muscle, 15
Cranial cavity, 71, 72f
Cranial nerve, 135, 136f
Creatine phosphate, 107
Cretinism, 180–181
Creutzfeldt-Jakob disease, 264
Cri-du-chat syndrome, 436, 437f
Critical periods in development, 385f, 386
Critical thinking, 9–10
Cro-Magnons, 492
Cross bridges, 104f

Cross tolerance, 143
Crossing over, 410, 412f
Cuboidal epithelium, 64, 65t, 66f
Cupula, 165, 166f
Cushing's syndrome, 183–184
Cutaneous membrane, 72
Cystic fibrosis, 281, 426, 430, 432f, 437, 458
Cytokinesis, 402, 403f, 405f, 406f
Cytomegalovirus (CMV), 470t
Cytoplasm, 43f, 46
Cytoskeleton, 43f, 55–56
Cytotoxic T cells, 246t, 247f, 468–469

D

Darwin, Charles, 479–480
Daughter cells, 402
Dead zone, 508
Deafness, 162–164
Death rate, 515
Decomposers, 500–501f, 506, 507f–508
Deductive reasoning, 8
Deer tick, 265f
Defecation reflex, 298
Defense mechanisms, 239–255
Defibrillator, 224
Deforestation, 505, 507, 519f–520
Dehydration synthesis, 30
Delayed onset muscle soreness (DOMS), 110
Deletion, on chromosome, 436–437, 449–450
Denaturation, 36
Dendrite, 114–116
Dendritic cell, 246t, 247
Denitrification, 507f–508
Dense connective tissue, 67t, 68f
Density-dependent regulating factor, 518
Density-independent regulating factor, 518
Dentin, 288, 289f
Dependence, drug, 143
Dependent variable, 7, 8
Depolarization, 118
Depression, 121
Dermis, 74f, 75
Desalination (of water), 505
Descending tracts, 134
Desertification, 519f
Desmosomes, 70, 71f
Detrusor muscle, 337
Developmental milestones, 395–396
Diabetes insipidus, 179, 333
Diabetes mellitus, 179, 185, 303
   characterization, 190–191
   gestational, 196–197
   other specific types, 197
   prevalence, 190–191
   type 1, 191–195, 253
   type 2, 191–195

Diabetic ketoacidosis (DKA), 192–193
Diaphragm, 270f, 275f
　barrier contraception, 358
　in respiration, 270f, 275f
Diarrhea, 298
Diastole, 223
Diastolic pressure, 225f
Dietary fiber, 302–303
Diets, *see* weight loss programs
Digestive system, 73f, 286–299
Dihybrid cross, 428, 429, 430f
Dilation stage, 383
Diphtheria, 252
Diploid, 343, 401
Disaccharide, 31
Disparities in health and health care, 388
Displaced fracture, 89
Distal convoluted tubule, 324–325, 327, 328f, 331, 332
Diuretic, 333
Diverticulitis, 298
Diverticulosis, 298
Diverticulum, 298f
DNA (deoxyribonucleic acid), 37–38, 442–459
DNA chip, 458
DNA fingerprints, 459f
DNA ligase, 451
DNA polymerase, 444
DNA, replication of (synthesis), 402, 443–444f
Domain, 4
Dominant allele, 426, 429, 431–432
DOMS (delayed onset muscle soreness), 110
Dopamine, 121, 122
Dorsal cavity, 71,72f
Dorsal root, 135, 136f
Double-blind experiment, 9
Doubling time, 515
Down syndrome, *see* trisomy 21
Droplet infection, 265
Drought, 520
Drug labels, 17f
Duchenne muscular dystrophy, 435–436
Ductus arteriosus, 382–383
Ductus venosus, 382–383
Duodenum, 292
Duplication, on chromosome, 436–437
Dura mater, 127, 128f
Dwarfism, pituitary, 177
Dynamic equilibrium, 166f
Dysplasia, 464f

E

E. coli (*Escherichia coli*), 260, 297, 316, 319
Ear, 159–165
　infection, 164–165
Eardrum, 160f, 163t
Eastern equine encephalitis (EEE), 266
Eating disorders, 309–310
ECG, *see* electrocardiogram

Ecological footprint, 518–519f
Ecological niche, *see* niche
Ecological pyramid, 502f
Ecological succession, *see* succession
Ecology, 497–511
Ecosystem, 5f, 497, 498, 510–511
Ectoderm, 379
Ectopic pregnancy, 348, 364, 376
Effector cells, 245–251
Egg, 350, 400, 408
Ejaculation, 356–357
EKG, *see* electrocardiogram
Elastic cartilage, 67
Elastic fibers, 65, 75, 215
Electrocardiogram, 224, 225f
Electron transport chain, 57, 59
Electronic health records, 15
Electronic medical records, 15
Element, 21, 22f
Elephantiasis, 226f–227
Embolism, 210, 222–223
Embryo, 348, 373
Embryonic disk, 379
Embryonic period, 373, 378–381
Embryonic stem cell, 419, 420–423
Emergency contraception (morning after pill), 358–359
Emerging disease, 266
Emigration, 515, 516
Emotion, 133–134
Emphysema, 281–282f
Encephalitis, 128, 266
Endocardium, 219, 220f
Endocrine gland, 65, 171, 172f, *see* also specific glands.
Endocrine system, 73f, 171–189
Endocytosis, 49, 50t
Endoderm, 379
Endometriosis, 355
Endometrium, 347, 349, 352–353
Endoplasmic reticulum, 51–52, 54t
Endorphins, 355
Endoscopy, 291
Endosymbiont theory, 478
Endothelium, 216f
Energy balance, 307
Energy flow, 500–504
Enhancers, 450
Enkephalins, 355
Environment, 497–511
Environmental estrogen, 503
Enzyme, 36–37
Eosinophil, 200t, 201f, 202, 241t
Epidemic, 266, 267
Epidemiology, 267
Epidermis, 72, 74f
Epididymis, 343t, 344f, 345f
Epigenetics, 451
Epiglottis, 270f, 271t, 272f, 290f
Epinephrine, 173f, 175, 183f, 184
Epiphyseal plate, *see* growth plate
Epiphytes, 4
Episiotomy, 383

Epithelial tissue, 64, 65t, 66f
Epstein-Barr virus, 204, 470t
Equilibrium, 165, 166f
Erectile dysfunction (ED), 345
Erosion, 506, 519
Erythrocyte, *see* red blood cell
Erythropoietin, 204, 333
Escherichia coli, *see E. coli*
Esophageal cancer, 291
Esophagus, 288f, 290f
Essential amino acid, 301, 303
Essential fatty acid, 301, 302
Estrogen, 34f, 35, 88, 89, 90, 173f, 178, 182, 342, 348, 350, 351t, 352–354, 471
Ethanol, 143–144
Eukaryote, 4
Eukaryotic cell, 43, 44t
Eustachian tube, *see* auditory tube
Eutrophication, 508
Evolution, 4–5, 8, 476
　chemical, 477
　convergent, 487
　evidence of, 482–487
　human, 487–494
　mosaic, 489
　whale, 483, 484f
Excitatory synapse, 120
Excretion, 321, 323f
Exercise
　aerobic, 109, 307, 355
　cardiovascular, 226
　resistance, 109
　weight-bearing, 90
Exhalation, 275f
Exocrine gland, 65
Exocytosis, 49, 50t
Exons (expressed sequences), 445f
Exophthalmos, 181
Experimental group, 6
Expiration, *see* exhalation
Expiratory reserve volume, 276f
Exponential growth, 517f
Expressed sequences, 445f
Expulsion stage, 383
External auditory canal, 160f, 163t
External urethral sphincter, 337
Extracellular fluid, 46
Extraembryonic membranes, 377
Eye, 153–154

F

Facilitated diffusion, 47, 50t, 294f
Fallopian tubes, *see* oviducts
Farsightedness, 156t, 157f
Fascicles, 101, 103f
Fast-twitch muscle cells, 108–109f
Fat-soluble vitamins, 301, 304, 305t
Fat,
　absorption of, 293–294
　connective tissue, 67t, 68f
　dietary, 301–302

Fatty acid, 32–33, 294, 301
Fatty streak, 234
Feces, 296, 298
Female reproductive system, 348–354
Fermentation, 59–61
Fertility awareness, 358
Fertilization, 342, 372, 373–375
Fertilizer, 508, 518
Fetal period, 373, 381–383
Fetus, 348, 373
Fever, 241t, 243–244
Fever blister, 263
Fiber, 472
Fibrillin, 430
Fibrin, 209f
Fibrinogen, 199, 209f
Fibroblast, 65, 88–89
Fibrocartilage, 67
Fibrous joints, 95
Fight-or-flight response, 173f, 184
First messenger, 172, 174f
Fitness, 480
Flagellum, 55–56
Flu, 266, 280
Fluid mosaic, 46
Folic acid, 304
Follicle
    hair, 74f, 76–77
    ovarian, 349–350, 351f
Follicle stimulating hormone (FSH), 173f, 176f, 178, 350, 351t, 352, 347ft
Food
    contamination, 316–318
    defense, 318–319
    handling, 320–321
    oversight of, 318
    poisoning, 314–316
    safety, 314–320
    selection, 319–320
    storage, 320–321
Food chain, 500
Food labels, 304, 307f
Food web, 500, 501f
Foodborne illness, 314–316
Foramen ovale, 382, 383
Forensic science, 459
Foreskin, 345
Formed elements, 199f–204
Fossil, 482–483
Fossil fuels, 507, 525
Fossil record, 482–483
Fossilization, 482–483
Founder effect, 479
Fovea, 154f, 155t
Fragile X syndrome, 437f
Free radical, 388–389
Full-term infant, 383
Fungi, 4f, 264

G

Gallbladder, 288f, 294f, 295t–296
Gallstone, 296f
Gamete intrafallopian transfer (GIFT), 384
Gamma globulin, 252
Ganglia, 126, 136
Gap junction, 70, 71f
Gas transport, 269f, 276–278
Gastric glands, 291f, 292
Gastric juice, 292
Gastrin, 299t
Gastrointestinal tract, 286–294
Gastrula, 379
Gastrulation, 379
Gene, 426f, 444
Gene expression, 444–449f
Gene flow, 479
Gene mutation, 449–450
Gene pool, 478–479
Gene regulation, 450
Gene test, 458
Gene therapy, 454–455, 456f
General senses, 151–153
Genetic code, 446f
Genetic disorders, 458
Genetic drift, 479
Genetic engineering, 450–457
Genetic screening, 474
    adult, 439
    newborn, 438
    prenatal, 437–438f
Genetic variability, 406, 409–411
Genetically modified food, 454, 456–457
Genetics, 425–439
Genital herpes, 366f, 367t
Genital warts, 366–367t
Genomics, 457–459
Genotype, 426, 429
Genuine progress indicator, 526
Geothermal energy, 525
Gestation, 373
Giantism, see Gigantism
Giardiasis, 262–263f
Gigantism, 176–177
Gingivitis, 289
Gland, 65, 77–78
Glans penis, 344f, 345
Glaucoma, 154
Glial cells, see neuroglia
Global climate change, 508, 522–525
Global warming, 507, 522–525
Globulin, 199
Glomerular capsule, 324–326, 328f
Glomerular filtration, 326–327, 328f
Glottis, 272
Glucagon, 173f, 174, 175, 184–185
    diabetes mellitus and, 195
Glucocorticoids, 173f, 182–183
Glucose, 31, 107, 272f, 295
    diabetes mellitus and, 190–194
Glycemic response, 302–303
Glycerol, 294
Glycogen, 31, 32f, 107, 108, 295
Glycolysis, 57–58, 59t
GnRH, see gonadotropin-releasing hormone
Goiter, 180, 181f
Golden rice, 454
Golgi complex, 43f, 52, 54t
Gonadocorticoids, 173f, 182
Gonadotropin-releasing hormone (GnRH), 347ft
Gonads, 342
Gonorrhea, 363–364, 365t
Goose bump, 77
Graafian follicle, 350
Granulocytes, 200t, 201f, 202
Graph, 7
Grave's disease, 181
Gray matter, 129, 134, 135f
Green revolution, 518
Greenhouse gases, 507, 522–523
Gross domestic product, 526
Ground substance, 65
Growth factor, 138, 186, 466
Growth hormone (GH), 88, 176–177, 178, 387t, 388
Growth plate, 87, 88f
Growth rate, 514–515, 517
Gum disease, 288–289
Gumma, 365f, 365

H

H1N1 virus, 266
HAART (highly active antiretroviral therapy), 371
Habitat, 498
Hair, 76–77
Hair cell (in ear), 161f, 162f, 163t, 164f, 166f
Hair follicle, 74f, 78
Hallucinogens, 148–149
Haploid, 343, 406
Haplotype, 459
Hashimoto's thyroiditis, 253
Hay fever, 254–255
HCG, see human chorionic gonadotropin
HDL (high-density lipoprotein), 226, 237, 301, 303
Headache, 138, 271
Health care agent, 18
Health care disparities, 388
Hearing, 159–165
Hearing aid, 163–164
Hearing loss, 162–164
Heart, 213, 214f, 219–224, 226
Heart attack, 210, 222, 234, 236f–237
    diet and, 301
Heart failure, 237
Heart murmur, 221–222
Heart transplant, 237
Heartburn, 291
Heat stroke, 81

Heimlich maneuver, 272, 273f
Helicobacter pylori, 297
Helper T cells, 246t, 247f, 367–368, 369
Hematoma, 89f
Hemodialysis, 334
Hemoglobin, 202–203f, 204, 276–277f
Hemolytic disease of the newborn, 208
Hemophilia, 209, 435–436
Hemorrhoids, 303
Hepatic portal system, 295f
Hepatitis, 295, 470t
 A, 265
 B, 252
 C, 242, 295
Herbivores, 500f–501f, 506
Herpes simplex virus (HSV), 242, 263, 366
Heterozygous, 426, 427t
High-density lipoprotein, *see* HDL
Highly active antiretroviral therapy (HAART), 371
Hippocampus, 133f, 134
Histamine, 202, 242, 254
Histone, 401
Histoplasmosis, 264
HIV (human immunodeficiency virus), 367–371
HIV/AIDS, 267, 367–371
Homeostasis, 2, 4, 78–81
Hominid, 488
Hominin, 488
Homo, 480, 481f, 491–493
Homologous chromosomes, 401, 410
Homologous pair, *see* homologues
Homologous structure, 486
Homologues, 401, 407, 410, 426f
Homozygous, 426, 427t
Hormonal contraception, 356–357
Hormone replacement therapy, 354
Hormone, 172–175, 450, *see* also specific hormones.
 bone growth, 88
 inhibiting, 176, 176f
 interactions, 174–175
 lipid-soluble, 172, 174f
 regulation by feedback mechanisms, 172, 174, 175f
 releasing, 175–176
 steroid, 34f, 35, 172, 174f
 therapy, 178
 tropic, 178
 types of, 172, 174f
 water-soluble, 172, 174f
Hot flashes, 353, 354
HPV (human papilloma virus), 242, 366–367ft, 470t
Human chorionic gonadotropin (HCG), 353, 377
Human Genome Project, 457–458
Human growth hormone, 176–177, 178, 453
Human immunodeficiency virus, *see* HIV
Human inheritance, 425–439
Human papilloma virus, *see* HPV

Hunger, 503–504
Huntington's disease, 439
Hyaline cartilage, 67
Hydrochloric acid (HCl), 292
Hydroelectric power, 525
Hydrogen bond, 26, 27t
Hydrolysis, 30
Hydrophilic, 33, 34f
Hydrophobic, 33, 34f
Hyperglycemia, 192
Hypertension, 223
Hyperthermia, 81
Hypertonic solution, 48
Hypodermis 74f, 75–76
Hypoglycemia, 192
Hypothalamus, 129f, 132–133, 173f, 175–176, 243, 347f, 351
 body temperature control, 79–81
Hypothermia, 81
Hypothesis, 6, 8, 9
Hypotonic solution, 48

I

Ileum, 292
Immigration, 515, 516
Immunization, *see* vaccination
Immunoglobulins, 249–250
Immunotherapy, 474–475
Implantation, 374f, 376–377
Impotence, *see* erectile dysfunction
In vitro fertilization, 384
Incisors, 287, 289f
Incomplete dominance, 432–431, 433, 434f
Incomplete protein, 303
Incus, 160f–161, 162f, 163t
Independent assortment, 411, 412f, 426
Independent variable, 6, 7, 8
Induced pluripotent stem cells (iPSCs), 423
Inductive reasoning, 8
Infectious disease, 259–267
Infertility, 377, 384
Inflammation, 234, 241t, 242–243, 244f
Influenza, 280
Information literacy, 10
Information technology literacy, 10
Informed consent, 9
Inhalation, 275f
Inheritance of gender, 435
Inhibin, 347ft
Inhibitory synapse, 120
Innate body defenses, 240–244
Inner cell mass, 374f, 376, 377, 420f–421
Inner ear, 160f, 161f–163t
Insertion
 muscle, 101f
 mutation, 449–450
Inspiration, *see* inhalation
Inspiratory reserve volume, 276f
Insulin, 184–185, 303
 diabetes mellitus and, 190–197
 shock, 192

Integumentary system, 72, 73f, 74–78
Intercalated disk, 223
Interferons, 241t, 242, 243f, 474
Interkinesis, 408
Interleukin-2, 474
Intermediate filament, 55–56
Internal urethral sphincter, 337
Interphase, 402, 404f
Interstitial cell, 343
Interstitial cell stimulating hormone (ICSH), 347
Intervening sequences, *see* introns
Intervertebral disc, 93f
Intracytoplasmic sperm injection (ICSI), 384
Intrauterine device (IUD), 357–358
Intrinsic factor, 292
Introns (intervening sequences), 445f
Involuntary muscle, 70
Ion channels, 116, 117f–118, 119–120
Ionic bond, 24–25, 26f, 27t
Iris, 153, 154f, 155t
Iron, 204
Iron deficiency anemia, 204
Irradiated food, 23
Irrigation, 518, 520–521
Islets of Langerhans, *see* pancreatic islets
Isotonic solution, 48
Isotope, 21–22
IUD (intrauterine device), 357–358

J

J-shaped growth curve, 517f
Jaundice, 295
Jejunum, 292
Jet lag, 186
Joints, 95–97
Juxtaglomerular apparatus, 331–332

K

Kaposi's sarcoma, 242, 371
Karyotype, 405, 406f
Kidney
 donation, 334–336
 failure, 333
 functions, 322–333
 hormones that influence, 331–333
 stones, 329, 333
 structure, 324
 trafficking, 336
 transplantation, 334–336
Kilocalorie, 300
Klinefelter syndrome, 415

L

Labium majora, 349f
Labium minorum, 349f
Labor, 383
Lactation, 179–180, 386

Lacteal, 227, 293f, 294
Lactic acid fermentation, 60–61
Lactose intolerance, 37, 297
Language, 130f, 138
Laparoscopy, 356
Large intestine, 296f–298
Larynx, 270f, 271t, 272f
LASEK, 157
LASIK, 157
Latent infection, 263
Latin binomial, 480
Law of independent assortment, 426
Law of segregation, 426
LDCs, *see* less developed countries
LDL, *see* low-density lipoprotein
Lead poisoning, 205
Lens, 153, 154f, 155t, 156
Less developed countries, 515, 517, 518
Leukemia, 205–206, 242, 420, 455, 471
Leukocyte, *see* white blood cell
LH, *see* luteinizing hormone
Life, characteristics of, 1–4
Ligament, 95, 96–97
Limbic system, 133f–134
Limestone, 506
Linked genes, 435
Linnaeus, Carl, 480
Linoleic acid, 302
Lipase, 293
Lipid, 31–35, 301–302
Liposuction, 75–76
Liver, 203, 288f, 292, 294f, 295f, 301
Local signaling molecules, 186
Logistic growth, 517
Longitudinal fissure, 129
Loop of Henle, *see* loop of the nephron
Loop of the nephron, 324–325, 329–331
Loose connective tissue, 67t
Low-density lipoprotein (LDL), 7, 8, 234, 301, 303
Lumen
  blood vessel, 216f
  gastrointestinal tract, 287, 289f
Lumpectomy, 354
Lung, 270f, 271t, 274
Lung cancer, 282f
Lung volumes, 275, 276f
Lupus erythematosus, 253
Luteinizing hormone (LH), 173f, 176f, 178, 347ft, 350, 351t, 352
Lyme disease, 265, 267
Lymph, 225
Lymph node, 248, 227, 228f, 370
Lymphatic capillary, 227f
Lymphatic system, 73f, 225–228, 245
Lymphatic vessels, 225, 227f–228f
Lymphocyte, 200t, 201f, 202, 204, 227, 245–251, 253
Lymphoid organ, 227–228
Lymphoma, 471
Lysosome, 43f, 52, 53f, 54t, 55
Lysozyme, 240

M

Macroevolution, 478, 480–482
Macromolecule, 29
Macrophage, 72, 227, 241f, 242, 243, 246t, 247f–248f
Mad cow disease, 264–265
Major histocompatibility markers, *see* MHC markers
Male pattern baldness, 436
Malignant tumor, 464
Malleus, 160f–161, 163t
Mammal, 4
Mammary gland, *see* breast
Mammogram, 354
Marfan syndrome, 430
Marijuana, 146–147
  erectile dysfunction and, 345
Mast cell, 242
Mastectomy, 354
Matrix, 65, 87
Matter, 20
Mechanoreceptor, 151
Medical marijuana, 147
Medicinal plant, 2, 510
Medulla oblongata, 129f, 133
Megakaryocyte, 200, 201f
Meiosis, 343, 400, 405–415
Meissner's corpuscle, 152f
Melanin, 76, 432, 434
Melanocytes, 76
Melanoma, 77
Melatonin, 173f, 186
Memory, 134
Memory cells, 245–251
Meninges, 127–128f
Meningitis, 128, 266
Menopause, 182, 353–354
Menstrual cramps, 355
Menstrual cycle, 350–353
Menstruation, 352–353
Mercury, 503
Merkel disk, 152f
Mesoderm, 379
Messenger RNA, *see* mRNA
Metabolism, 2, 57
Metaphase, 404f
  Metaphase I, 408f
  Metaphase II, 409f
Metastasis, 464, 468
Methicillin-resistant *Staphylococcus aureus*, *see* MRSA
MHC markers (major histocompatibility markers), 245f, 419
Micelle, 294
Microarray analysis, 458f
Microevolution, 478–480
Microfilament, 43f, 55, 56
Microscopy, 44, 45f
Microtubule, 43f, 55–56
Midbrain, 129f, 133
Middle ear, 160f–161
Middle ear infection, 165
Milk ejection, 173f, 179–180

Mineralocorticoids, 173f, 182, 183f
Minerals, 294, 304, 306t
Mitochondrial diseases, 60–61
Mitochondrion, 43f, 54,
Mitosis, 343, 400–401f, 404f–405f, 409t, 410f
Mitral valve, 220
Molars, 288, 289f
Molecular clock, 485
Molecule, 5f, 24
Monoclonal antibodies, 252
Monocytes, 200t, 201f, 202
Monohybrid cross, 427–428, 429f
Monomer, 30
Mononucleosis, 204
Monosaccharide, 31, 294
Monosomy, 413f
More developed countries (MDCs), 515, 517, 518
Morning after pill, 358–359
Morning sickness, 377
Morphogenesis, 379
Morula, 374f, 375, 376
Mosaic evolution, 489
Motion sickness, 165
Motor neuron, 114f
Motor unit, 106f–107
Mouth, 287, 288f
mRNA (messenger RNA), 445t, 447–448
MRSA (methicillin-resistant *Staphylococcus aureus*), 261
Mucosa, 287, 289f, 291f, 293f
Mucous membrane, 71, 240
  STDs and, 363, 364
Mucus, 292
Multiple alleles, 434
Multiple sclerosis, 242
Multipotent stem cells, 419f–420
Multiregional hypothesis, 494
Muscle, 69ft–70
Muscle cell, *see* muscle fiber
Muscle contraction, 101–106
Muscle cramp, 105
Muscle fiber, 101, 103f
Muscle pull, 101
Muscle spasm, 15
Muscle spindle, 152, 153f
Muscle tissue, 64, 69–70
Muscle tone, 107
Muscle twitch, 107f
Muscular dystrophy, 105–106, 455, 458
Muscular system, 73f, 100–110
Muscularis, 287, 289f, 291f
Mutations, 449f–450, 479, 485
  cancer and, 469, 465–467, 471
Myelin sheath, 114, 115f, 116
Myocardium, 219, 220f
Myofibril, 102–103f
Myofilament, 102–103f
Myoglobin, 108
Myosin, 102–103f, 108
Myosin head, 103, 104f
MyPlate, 299, 300f
Myxedema, 181

N

Nails, 77
Nasal cavity, 269–270f, 271t
Natural family planning, 358
Natural killer cells, 241t–242f, 468, 469
Natural selection, 4, 8, 479–480
Neanderthals, 492–493
Nearsightedness, 156t, 157f
Negative feedback mechanisms, 79, 174, 204
    body temperature control, 79–81
    ovarian and uterine cycle and, 352–353
    testosterone and, 347f
Neoplasm, 463
Nephron, 324
    functions, 325–333
    structure, 324–325
Nerve, 135f
Nerve impulse, see action potential
Nervous system, 73f, 126–138
    effects of alcohol, 145
Nervous tissue, 64
Neural tube, 379–380
Neural tube defect, 379–380, 437
Neuroendocrine system, 132–133
Neurofibrillary tangle, 121
Neuroglial cell, 70, 113, 114–115f, 116
Neuromuscular junction, 104–105f, 106f
Neuron, 70, 113–122
Neurotransmitter, 118–122, 135–137
Neurula, 379
Neurulation, 379–380
Neutrophil, 200t, 201f, 202, 242
Newborn genetic testing, 438
Niche (ecological niche), 498, 509
Nicotine, 148, 215, 237
Nitrate, 507f–508
Nitric oxide (NO), 186, 345
Nitrification, 507f–508
Nitrifying bacteria, 507f–508
Nitrite, 507f–508
Nitrogen cycle, 507f–508
Nitrogen dioxide, 507f–508
Nitrogen fixation, 507f–508
Nitrogen-fixing bacteria, 507f, 508
Nociceptor, 151
Node of Ranvier, 115f–116
Noise pollution, 164
Nondisjunction, 412–415
Nondisplaced fracture, 89
Noradrenalin, see norepinephrine
Norepinephrine, 121, 122, 136, 173f, 182, 183f, 184
Nose, 269–270
Notochord, 379, 380f
Nuclear energy, 525
Nuclear envelope, 50
Nuclear pore, 50
Nucleic acid, 37–38
Nucleolus, 43f, 50f, 51
Nucleoplasm, 50f, 51
Nucleotide, 37–38, 442–443
Nucleus, 43f, 50–51, 54t

Nutrients, 299–306
Nutrition, 299–306

O

Obesity, 301, 307–308
    type 2 diabetes mellitus and, 191–192, 193t
Ogallala aquifer, 520
Oil gland, 77–78
Oils, 301
Olfaction, 165, 270
Olfactory bulb, 133f, 134, 167
Olfactory receptors, 167f
Oligosaccharide, 31
Omega-3 fatty acid, 302
Omega-6 fatty acid, 302
Omnivores, 500f–501f
Oncogenes, 466, 468t, 470, 475
Oogenesis, 348
Opiates, 149
Opportunistic infections, 368
Optic nerve, 154f, 155t, 158f
Oral cavity, 287
Organ donation, 335, 336
Organ of Corti, see spiral organ
Organ, 5f, 70
Organ system, 5f, 70
Organelle, 4, 43, 54t
Orgasm, 356
Origin (of muscle), 101f
Osmosis, 47–48, 50t, 199
Osteoarthritis, 97
Osteoblast, 87, 88–89, 91
Osteoclast, 89, 91
Osteocyte, 86f, 87
Osteon, 86f, 87
Osteoporosis, 90–91
Otoliths, 165, 166f
Out of Africa hypothesis, 494
Outer ear, 159–160f
Oval window, 160f, 161, 162f, 163t
Ovarian cycle, 349–350, 351f, 352f, 353t
Ovary, 342, 348t, 349f
Oviducts, 348, 349f
Ovulation, 178, 350, 351f, 353, 373, 374f
Ovum, see egg
Oxygen, 202, 268, 276–277f
    regulation of breathing rate, 279
Oxygen debt, 108
Oxygen transport, 276–277f
Oxyhemoglobin, 203, 276–277f
Oxytocin (OT)
    as treatment for autism spectrum disorder, 398
    childbirth and, 174, 175f, 179, 383
    gene and autism, 451
    lactation and, 179, 386
Ozone depletion, 521
Ozone hole, 521–522f
Ozone pollution, 521, 524

P

$p53$ gene, 465, 466, 467, 468, 474
Pacemaker, 223–224
Pacinian corpuscle, 152f
Pain, 152–153, 243
Pain receptor, 151
Palate, 287
Pancreas
    as endocrine organ, 173f, 184–185
    as exocrine organ, 288f, 292, 294f, 295t, 296f
Pancreatic islets, 184–185
Pandemic, 267
Pap test, 367
Parasitic worms, 264
Parasympathetic nervous system, 127, 135–138
Parathormone, see parathyroid hormone (PTH)
Parathyroid glands, 173f, 180f, 181–182
Parathyroid hormone (PTH), 89, 173f, 181–182
Parkinson's disease, 121
Partial pressure, 277
Parturition, 383–384
Passive immunity, 252
Pathogens, 239, 259–265
Pectoral girdle, 94f
Pedigree, 429–431f
Pelvic girdle, 94, 95f
Pelvic inflammatory disease (PID), 363
Penile cancer, 367
Penis, 343t, 344f, 345, 347
Pepsin, 292
Pepsinogen, 292
Peptic ulcer, 297f
Peptide, 35–36
Perception, 150–151
Perforins, 250
Pericardial cavity, 71, 72f
Pericardium, 219, 220f
Periodic table, 21, 22f
Periodontitis, 289
Periosteum, 86f
Peripheral nervous system, 126–127, 135–138
Peristalsis, 287, 290, 291f, 297–298
Peritonitis, 334
Periwinkle, 2, 510
Pervasive developmental disorder not otherwise specified (PDD-NOS), see autism spectrum disorder
Pesticides, 503, 518
Peyer's patch, 228
pH, 28–29
    regulation by kidneys, 329
    scale, 28–29
Phagocyte, 240–241ft, 249
Phagocytosis, 49, 202, 242
Pharming, 454
Pharynx, 270f, 271t, 272, 288f, 290
Phenotype, 426, 436, 444
Phenylketonuria (PKU), 438

Phosphates, 508–509f
Phospholipid, 32–34
Phosphorus, 85, 87
Phosphorous cycle, 508–509f
Photochemical smog, 508, 521, 524
Photopigment, 157–159
Photoreceptor, 151, 157–159f
Photosynthesis, 500, 505, 506, 507
Phylogenetic tree, 481–482
Pia mater, 127
PID, see pelvic inflammatory disease
Pili, 260
Piloerection, 80
Pimple, 78
Pineal gland, 173f, 186
Pinna, 160f, 163t
Pinocytosis, 49
Pioneer species, 498f–499
Pitch (of sound), 162
Pituitary gland, 132, 173f, 175–180
Placebo, 9
Placenta
   expulsion at birth, 383
   formation of, 377–378
Placenta previa, 378
Placental stage, 383
Placental stem cells, 419–420
Plantibodies, 454
Plaque
   coronary artery, 235
   dental, 288
Plasma, blood, 67, 199f, 277–278
Plasma cells, 246, 247f, 249f–250
Plasma membrane, 49–50
   functions, 46
   movement across, 47–49, 50t
   structure, 45–46
Plasma proteins, 199
Plasmids, 451
Plasmin, 210
Platelets, 67, 199f, 200t, 201f
Pleiotropy, 433, 434f
Pleural cavity, 71, 72f
Pleural membrane, 274
Pluripotent stem cells, 419f, 420–423
PMS, see premenstrual syndrome
Pneumocystis jiroveci, 371
Pneumonia, 280
Polar bodies, 350, 408, 411f
Pollution, 521
   air, 9
Polychlorinated biphenyls (PCBs), 503
Polygenic inheritance, 433–434, 435
Polymer, 30
Polymerase chain reaction, 451, 452–453f
Polyp, 298
Polysaccharide, 31, 32f
Polysome, 448f
Pons, 129f, 133
Population, 5f, 478
Population, human, 514–518
Portal system
   hepatic, 295f

Positive feedback mechanism, 174, 175f
Postnatal period, 372, 386–389
Pre-capillary sphincter, 217f–218
Pre-embryo, 373
Pre-embryonic period, 373–378
Prefrontal cortex, 131
Premature infant, 384
Premenstrual syndrome (PMS), 354–355
Premotor cortex, 130–131
Prenatal genetic testing, 437–438
Prenatal period, 372–383
Primary (immune) response, 250–251f
Primary care physician, 13–15
Primary germ layers, 379
Primary motor area, 130–131f
Primary somatosensory area, 130, 131f
Primary structure of protein, 36, 37f
Primary succession, 498f–499
Primate, 4–5
   characteristics, 487–488
   classification, 488, 489f
   use in research, 485
Primordial germ cells, 377
Prions, 264–265
Producers, 500f–501f
Product, 36–37, 57
Progesterone, 342, 348, 350, 351t, 352–354
Programmed cell death (apoptosis), 467f, 474
Prokaryote, 4, 259
Prokaryotic cell, 42, 43f, 44t
Prolactin (PRL), 176f, 177–178, 386
Promoter, 445
Prophase, 402, 404f
Prostaglandin, 186, 243, 297
Prostate cancer, 22–23, 344–345
Prostate gland, 343t, 344f
Prostate-specific antigen (PSA), 345
Protease, 371
Protease inhibitors, 371
Protein, 35–37, 303
Protein synthesis, see translation
Prothrombin, 209f
Protists, 4f
Proto-oncogenes, 466, 468t
Protozoans, 263–264
Provirus, 368, 371
Proximal convoluted tubule, 324, 325f, 327, 328f
PSA, see prostate-specific antigen
Psychoactive drugs, 142–149
PTH, see parathyroid hormone, 89
Puberty, 88, 347
Pubic symphysis, 95f
Pulmonary artery, 214f, 220f, 221, 222f
Pulmonary circuit, 221–222f
Pulmonary embolism, 222–223
Pulmonary vein, 214f, 220f, 221, 222f
Pulse, 215
Punnett square, 427, 428
Pupil, 154f, 155t
Purkinje fibers, 224

Pyramid of biomass, 502f
Pyramid of energy, 502f, 504f
Pyrogen, 243
Pyruvate, 57

Q

Quaternary structure of protein, 36, 37f

R

Radiation, 22–23, 474
Radioactivity, 503
Radioisotope, 23
Radiometric dating, 483
Radon, 27
Rain forest, 1, 2, 509–510
Ras oncogene, 466
Receptor
   for hormones, 46, 172, 174f
   lymphocytes and, 245–251
   sensory, 151
Recessive allele, 426, 429, 431–432
Recombinant DNA, 450–453
Recruitment, 106
Rectum, 288f, 296f, 298
Red blood cells, 67, 199f, 200t, 201f, 202f–204
Red bone marrow, 86, 199, 201, 228
Red-green colorblindness, 158–159
Reemerging disease, 266
Referred pain, 153f
Reflex
   defecation, 298
   spinal, 134–135, 136f
   swallowing, 290f
   withdrawal, 134–135, 136f
Reflex arc, 134–135, 136f
Refractory period, 118
Regenerative medicine, 423–424
Renal corpuscle, 324, 325f
Renal cortex, 324, 325, 330f
Renal failure, see kidney failure
Renal medulla, 324, 325, 329–330
Renal pelvis, 324, 325, 327
Renal tubule, 324
Renin, 331f, 332
Replication
   of DNA, 402, 443–444f
   of virus, 262f–263
Repolarization, 118
Reproductive cloning, 421f, 422–423
Reproductive system, 73f
   development of, 380–381
   female, 348–354
   male, 343–348
   marijuana and, 147
Residual volume, 276f
Resistance exercise, 109
Respiratory system, 73f, 268–283
   cocaine and, 147–148
   marijuana and, 146–147

Resting potential, 116, 117f
Restriction enzyme, 450–451, 452f
Reticular activating system, 133f, 134
Reticular fibers, 65
Retina, 154f, 155t, 158f
Retinoblastoma, 466
Retrovirus, 368, 423, 455
Reverse transcriptase, 368, 470
Reverse transcription inhibitor, 371
Reward center, of brain, 143
Rh factor, 206–208
Rheumatic fever, 253
Rheumatoid arthritis, 97
Rhinitis, allergic, 254
Rhodopsin, 157
RhoGAM, 208
Rhythm method, 358
Rib cage, 94f
Ribonucleic acid, see RNA
Ribosomal RNA, see rRNA
Ribosomes, 43f, 51, 447f–449
Rickets, 87
Rigor mortis, 104
Ringworm, 264
RNA (ribonucleic acid), 37–38, 444t, 445t
   synthesis of, 444–445f
RNA polymerase, 445
RNA synthesis, see transcription
RNA transcript, 445
Rods, 155t, 157–158f, 159f
Rough endoplasmic reticulum, 43f, 51–52, 54t
Round window, 160f, 161f, 163t
rRNA (ribosomal RNA), 445t, 447
Ruffini corpuscle, 152f
Runoff, 508, 521

S

S-shaped growth curve, 517
SA node, 223–224f
Saccule, 163t, 165, 166f
Salinization, 521
Saliva, 289, 299
Salivary amylase, 289
Salivary glands, 288f, 289f, 295t
Salmonella, 260, 316, 317, 319
Saltatory conduction, 115f, 116
Sarcoma, 471
Sarcomere, 103f
Sarcoplasmic reticulum, 104, 105
Saturated fat, 301, 472
Schwann cell, 114, 115f, 116
Sciatica, 93
SCID (severe combined immunodeficiency disease, 455
Scientific method, 5–9
Sclera, 153, 154f, 155t, 158f
SCNT (somatic cell nuclear transfer), 421–423
Scoliosis, 93
Scrapie, 264

Scrotum, 343
Seasickness, 165
Seasonal affective disorder (SAD), 186
Sebaceous gland, see oil gland
Second messenger, 172, 174f
Secondary (immune) response, 250–251f
Secondary sex characteristics, 347
Secondary structure of protein, 36, 37f
Secondary succession, 499f–500
Secretin, 299t
Segregation, law of, 426
Self markers (MHC markers), 419
Semen, 343–344
Semicircular canals, 163t, 165, 166f
Semiconservative replication, 444f
Semilunar valve, 220f, 221f
Seminal vesicles, 343t, 344f, 345
Seminiferous tubules, 343, 344f, 345f, 346f
Sensorineural hearing loss, 163–164
Sensory adaptation, 151
Sensory receptors, 151
Sensory systems, 150–168
Septum,
   heart, 219
   nasal, 269
Serosa, 287, 289f, 291f
Serotonin, 121, 122
Serous membrane, 71–72
Severe combined immunodeficiency disease (SCID), 455
Sex chromosomes, 380, 401, 436
Sex-influenced genes, 436
Sex-linked genes, 435
Sexual dimorphism, 489
Sexual response cycle, 355–356
Sexually transmitted disease (STD), 362–367
Sexually transmitted infection (STI), 362–367
Shedding, 262f, 263
Shingles, 370
Short tandem repeats (STRs), 459
Sickle-cell anemia, 204, 420
   genetics of, 432–433, 434f
Sickle-cell trait, 433
Sigmoidoscopy, 298
Simple diffusion, 47, 50t, 294f
Simple epithelium, 64, 65t, 66f
Single-nucleotide polymorphisms (SNPs or snips), 458
Sinoatrial node, see SA node
Sinus, respiratory, 270f, 271t
Sinusitis, 271
Sister chromatid, 402, 403f
Skeletal muscle, 69ft, 101–110, 218–219f
Skeletal system, 73f, 85–97
Skin, 72, 74–78, 240
   color, 76, 434
Skin cancer, 77
Skull, 92f
Sleep
   role of melatonin in, 186

Sliding filament model, 103f–104
Slipped disk, 93
Slow-twitch muscle cell, 108, 109f
Small intestine, 292–294
Smell, 165, 166
Smooth endoplasmic reticulum, 43f, 52, 54t
Smooth muscle, 69ft, 70, 215, 273, 287
SNPs (single-nucleotide polymorphisms or snips), 458–459
Sodium-potassium pump, 116, 117f
Solar power, 525
Somatic (body) cells, 401
Somatic cell nuclear transplant (SCNT), 421–423
Somatic nervous system, 126–127, 135
Somatosensory area, 130, 131f
Somite, 379–380
Sound, 159, 160f
Spasm
   coronary artery, 215
   muscle, 15
Special senses, 151
Specialized connective tissue, 67t, 68f
Speciation, 479
Species, 479
Sperm, 342, 346f, 400, 408
Spermatogenesis, 346f
Spermicidal preparations, 358
Sphincter
   digestive system, 291f
   precapillary, 217f–218
Sphygmomanometer, 225
Spina bifida, 304, 380, 386
Spinal cavity, 71, 72f
Spinal cord, 134–135f
Spinal cord injury, 138
Spinal nerve, 135f, 136f
Spiral organ (of Corti), 161f, 162f, 163t
Spirilla, 260f
Spleen, 203, 227
Spongy bone, 86f
Sprain, 96–97
Squamous epithelium, 64, 65t, 66f
Stapes, 160f–161, 162f, 163t
Staphylococcus, 260
Starch, 31, 303
Start signal (codon), 446f
Static equilibrium, 166f
Statistical significance, 7
STD (sexually transmitted disease), 362–367, 356
Stem cells, 138, 199–200, 201, 418–424
   cancer and, 469–470
Sterilization, 356
Steroid hormone, 301, 450
Steroid, 33–35
STI (sexually transmitted infection), 362–367
Sticky end, 451
Stimulants, 147–148
Stimuli, 150
Stomach, 288f, 290–292

Stop codon (signal), 445, 446f
Stratified epithelium, 64, 65t, 66f
Strep throat, 280
Stress, 184
Stress incontinence, 337
Striated muscle, 101–110
Stroke, 138, 210
   diet and, 301
Submucosa, 287, 289f, 291f, 293f
Substrate, 36, 57
Succession (ecological), 498f–500
Sulfur dioxide, 508
Summation, 107f, 120
Sunburn, 77
Suppressor T cells, 246t, 247f, 251
Surface-to-volume ratio, 44
Surfactant, 274
Sustainability, 526
Suture, 95
Swallowing, 290f
Sweat glands, 77, 78
Swimmer's ear, 164–165
Swine flu, 266
Sympathetic nervous system, 127, 135–138
Synapse, 118–122
Synapsis, 407
Synaptic cleft, 118, 122
Synaptic knob, 118–120,
Synaptic transmission, 118–120, 122
Synergistic muscles, 101
Synovial joint, 95–96f
Synovial membrane, 72, 95, 96f
Syphilis, 364f–365t
Systemic circuit, 221–222f
Systole, 222–223
Systolic pressure, 225f

T

T cells, *see* T lymphocytes
T lymphocytes, 245–251
T tubules (transverse tubules), 104, 105f
Tanning, 77
Tanning salon, 77
Target cell, 172
Taste, 165, 167–168f, 289–290
Taste bud, 167–168f, 289–290
Taste hairs, 168f
Tattoo, 75f
Tay-Sachs disease, 52, 437
Tectorial membrane, 161f, 162f, 163t
Teeth, 287, 289f
Telomerase, 467
Telomeres, 388, 467f–468t
Telophase, 404, 405f
Temperature regulation, 72, 79–81
Temporomandibular joint (TMJ) syndrome, 93
Tendinitis, 101
Tendon, 101
Tendon organ, 152, 253

Tertiary structure of protein, 36, 37f
Testes, 342, 343t, 344f, 345f, 347
Testicular cancer, 343
Testosterone, 34–35, 88, 342, 343, 347ft, 436, 450
Tetanus, 107f
Thalamus, 129f, 131–132, 133f, 157
THC (delta-9-tetrahydrocannabinol), 146
Theory, 8
Therapeutic cloning, 422f
Thermoreceptors, 151
Thimerosal, 398–399
Thoracic cavity, 71, 72f
Thoracic duct, 228f
Threshold, 118f
Thrombin, 209f
Thrombocyte, *see* platelet
Thrombus, 210
Thrush, 370
Thymopoietin, 173f, 185
Thymosin, 173f, 185
Thymus gland, 173f, 185, 227
Thyroid gland, 22–23, 88, 173f, 180–181
Thyroid hormone (TH), 88, 173f, 180–181
Thyroid-stimulating hormone (TSH), 173f, 176f, 178
Tick, deer, 265f
Tidal volume, 276f
Tight junctions, 70, 71f
Tissue, 64–70
Tissue plasminogen activator (tPA), 223
TMJ (temporomandibular joint) syndrome, 93
Tobacco smoke, 471
Toilet-to-tap program, 35, 505
Tongue, 289–290
Tonsils, 228
Tooth decay, 288
Total lung capacity, 276f
Totipotent stem cells, 419f
Touch, 152
Toxins, bacterial, 260
tPA (tissue plasminogen activator), 223
Trachea, 270f, 271t, 272f–273
Trans fat, 301–302
Transcription, 444–445f
Transcription factors, 450
Transfer RNA, *see* tRNA
Transgenic organisms, 452, 453–454, 455f
Transition reaction, 57–58, 59t
Translation, 446–449
Transmissible spongiform encephalopathies, *see* TSEs
Transverse tubules (T tubules), 104, 105f
Traumatic brain injury, 132
Tricuspid valve, 220f, 221f
Triglycerides, 32–33, 301
Trisomy, 412–413f
Trisomy 21, 413, 414
tRNA (transfer RNA), 445t, 446–448
Trophic levels, 500f–501f
Trophoblast, 374f, 376, 377

Tropomyosin, 104f
Troponin, 104f, 105
TSEs (transmissible spongiform encephalopathies), 264–265
Tubal ligation, 356
Tubal pregnancy, 348, 364
Tuberculosis, 266, 280–281
Tubular reabsorption, 325, 327, 328f
Tumor, 463–464
Tubular secretion, 325, 327, 328f
Tumor marker test, 473–474
Tumor suppressor genes, 465–469, 468t
Tuna, 503
Turner syndrome, 413, 414f
Twins, 375–376, 397
Tympanic canal, 161f
Tympanic membrane, *see* eardrum

U

Ulcer, peptic, 297f
Ultraviolet radiation, *see* UV radiation
Umami, 167, 168f
Umbilical cord, 377, 378, 382–383
Umbilical cord stem cells, 419–420, 423
Unipotent stem cells, 419f
Unsaturated fatty acid, 301
Ureter, 323
Urethra, 323, 337–339, 343t, 344f
Urinalysis, 329
Urinary bladder, 323, 336–338
Urinary incontinence, 337
Urinary retention, 337–338
Urinary system, 73f, 322–339
Urinary tract infection (UTI), 338–339
Urination, 336–338
Urine, 323, 327, 329
Uterine cycle, 350–353t
Uterus, 348t, 349f
Utricle, 163t, 165, 166f
UV radiation, 72, 76, 77

V

Vaccination, 251–252
Vaccines, 371, 454, 475
   not implicated in autism, 398–399
Vagina, 348t, 349f
Valve
   heart, 219–221f
   lymphatic system, 227
   vein, 216f, 218, 219f
Vancomycin, 261
Variable, 6
Vas deferens, 343t, 344f, 345f
Vasectomy, 356
Vasoactive intestinal peptide, 299t
Vasoconstriction, 215
Vasodilation, 215
Vector, of disease, 265–266
Vector DNA, 451
Veins, 214f, 215, 216f, 218

Vena cava, 214f, 220f, 221, 222f
Ventilating, see breathing
Ventral cavity, 71,72f
Ventral root, 135–136f
Ventricles
    brain, 128f
    heart, 219–220f
Ventricular fibrillation, 224
Venules, 215, 216f, 218
Vertebra, 93f
Vertebral column, 93–94, 134, 135f
Vertebrates, 4
Vesicle, 49
Vestibular apparatus, 163t
Vestibular canal, 161f, 162f
Vestibule, 163t, 165, 166f
Viagra, 345
Vibration, 152
Villus, 293f
Viral load, 370f
Viral STDs, 366–367t
Virulence, 259
Virus, 128, 204, 243, 261–263, 451, 455
    cancer and, 470
Vision, 153–159
Vital capacity, 276f

Vitamin D, 72, 76, 77, 333
Vitamin K, 209
Vitamins, 294, 296–297, 303–304, 305t, 472
Vitreous humor, 154f, 155t
Vocal cords, 272f
Voice, 347
Voice box, see larynx
Voluntary muscle, see skeletal muscle
Vulva, 348

W

Wasting syndrome, 370
Water
    conservation by the kidneys, 329–331
    cycle, 504–505f
    footprint, 520–521
    pollution, 521
    properties of, 27–28
    shortage of, 35
    recycling of, 35
Water-soluble vitamins, 304, 305t
Wax glands, 77, 78
Weight loss programs, 308–309
West Nile virus, 266

White blood cells, 67, 199f, 201f–202
White matter, 129, 134, 135f
Wind power, 525
Withdrawal reflex, 134–135, 136f
Wrinkles, 75, 76f

X

X chromosome, 435
X-axis, 7
X-linked genes, 435–436
X–SCID, 455

Y

Y chromosome, 435
Y-axis, 7, 8
Yellow bone marrow, 86
Yolk sac, 377, 379f, 380f

Z

Z line, 103f–104
Zygote, 373f, 374f, 375, 400, 408
Zygote intrafallopian transfer, 384

Metric-English System Conversions

Length
1 inch (in.) = 2.54 centimeters (cm)
1 centimeter (cm) = 0.3937 inch (in.)
1 foot (ft) = 0.3048 meter (m)
1 meter (m) = 3.2808 feet (ft) = 1.0936 yard (yd)
1 mile (mi) = 1.6904 kilometer (km)
1 kilometer (km) = 0.6214 mile (mi)

Area
1 square inch (in.^{2}) = 6.45 square centimeters (cm^2)
1 square centimeter (cm^2) = 0.155 square inch (in.^{2})
1 square foot (ft^2) = 0.0929 square meter (m^2)
1 square meter (m^2) = 10.7639 square feet (ft^2)
 = 1.1960 square yards (yd^2)
1 square mile (mi^2) = 2.5900 square kilometers (km^2)
1 acre (a) = 0.4047 hectare (ha)
1 hectare (ha) = 2.4710 acres (a) = 10,000 square meters (m^2)

Volume
1 cubic inch (in.^{3}) = 16.39 cubic centimeters (cm^3 or cc)
1 cubic centimeter (cm^3 or cc) = 0.06 cubic inch (in.^{3})
1 cubic foot (ft^3) = 0.028 cubic meter (m^3)
1 cubic meter (m^3) = 35.30 cubic feet (ft^3) = 1.3079 cubic yards (yd^3)
1 fluid ounce (oz) = 29.6 milliliters (mL) = 0.03 liter (L)
1 milliliter (mL) = 0.03 fluid ounce (oz) = $\frac{1}{4}$ teaspoon (approximate)
1 pint (pt) = 473 milliliters (mL) = 0.47 liter (L)
1 quart (qt) = 946 milliliters (mL) = 0.9463 liter (L)
1 gallon (gal) = 3.79 liters (L)
1 liter (L) = 1.0567 quarts (qt) = 0.26 gallon (gal)

Mass
1 ounce (oz) = 28.3496 grams (g)
1 gram (g) = 0.03527 ounce (oz)
1 pound (lb) = 0.4536 kilogram (kg)
1 kilogram (kg) = 2.2046 pounds (lb)
1 ton (tn), U.S. = 0.91 metric ton (t or tonne)
1 metric ton (t or tonne) = 1.10 tons (tn), U.S.

Metric Prefixes

Prefix	Abbreviation	Meaning
giga-	G	10^9 = 1,000,000,000
mega-	M	10^6 = 1,000,000
kilo-	k	10^3 = 1,000
hecto-	h	10^2 = 100
deka-	da	10^1 = 10
		10^0 = 1
deci-	d	10^{-1} = 0.1
centi-	c	10^{-2} = 0.01
milli-	m	10^{-3} = 0.001
micro-	μ	10^{-6} = 0.000001